The Nature of Mathematics

12th EDITION

Karl J. Smith

CENGAGE
Learning™

Australia • Brazil • Japan • Korea • Mexico • Singapore • Spain • United Kingdom • United States

CENGAGE
Learning™

The Nature of Mathematics: 12th Edition

Executive Editors:
Maureen Staudt
Michael Stranz

Senior Project Development Manager:
Linda deStefano

Marketing Specialist:
Courtney Sheldon

Senior Production/Manufacturing Manager:
Donna M. Brown

PreMedia Manager:
Joel Brennecke

Sr. Rights Acquisition Account Manager:
Todd Osborne

Cover Image:
Getty Images*

*Unless otherwise noted, all cover images used by Custom
Solutions, a part of Cengage Learning, have been supplied
courtesy of Getty Images with the exception of the Earthview
cover image, which has been supplied by the National
Aeronautics and Space Administration (NASA).

This book contains select works from existing Cengage Learning resources and
was produced by Cengage Learning Custom Solutions for collegiate use. As such,
those adopting and/or contributing to this work are responsible for editorial
content accuracy, continuity and completeness.

Compilation © 2011 Cengage Learning
ISBN-13: 978-1-133-35785-8

ISBN-10: 1-133-35785-7

Cengage Learning
5191 Natorp Boulevard
Mason, Ohio 45040
USA
Cengage Learning is a leading provider of customized learning solutions with
office locations around the globe, including Singapore, the United Kingdom,
Australia, Mexico, Brazil, and Japan. Locate your local office at:
international.cengage.com/region.

Cengage Learning products are represented in Canada by Nelson Education, Ltd.
For your lifelong learning solutions, visit **www.cengage.com/custom.**
Visit our corporate website at **www.cengage.com.**

Printed in the United States of America

Table of Contents

Preface

Like almost every subject of human interest, mathematics is as easy or as difficult as we choose to make it. Following this Preface, I have included a Fable, and have addressed it directly to you, the student. I hope you will take the time to read it, and then ponder why I call it a fable.

You will notice street sign symbols used throughout this book. I use this stop sign to mean that you should stop and pay attention to this idea, since it will be used as you travel through the rest of the book.

Caution means that you will need to proceed more slowly to understand this material.

A bump symbol means to watch out, because you are coming to some difficult material.

WWW means you should check this out on the Web.

I also use this special font to speak to you directly out of the context of the regular textual material. I call these author's notes; they are comments that I might say to you if we were chatting in my office about the content in this book.

I frequently encounter people who tell me about their unpleasant experiences with mathematics. I have a true sympathy for those people, and I recall one of my elementary school teachers who assigned additional arithmetic problems as punishment. This can only create negative attitudes toward mathematics, which is indeed unfortunate. If elementary school teachers and parents have positive attitudes toward mathematics, their children cannot help but see some of the beauty of the subject. I want students to come away from this course with the feeling that mathematics can be pleasant, useful, and practical—and enjoyed for its own sake.

Since the first edition, my goal has been, and continues to be, to create a positive attitude toward mathematics. But the world, the students, and the professors are very different today than they were when I began writing this book. This is a very different book from its first printing, and this edition is very different from the previous edition. The world of knowledge is more accessible today (via the Internet) than at any time in history. Supplementary help is available on the World Wide Web, and can be accessed at the following Web address: www.mathnature.com

All of the Web addresses mentioned in this book are linked to the above Web address. If you have access to a computer and the world wide Web, check out this Web address. You will find links to several search engines, history, and reference topics. You will find, for each section, homework hints, and a listing of essential ideas, projects, and links to related information on the Web.

This book was written for students who need a mathematics course to satisfy the general university competency requirement in mathematics. Because of the university requirement, many students enrolling in a course that uses my book have postponed taking this course as long as possible. They dread the experience, and come to class with a great deal of anxiety. Rather than simply presenting the technical details needed to proceed to the next course, I have attempted to give insight into what mathematics is, what it accomplishes, and how it is pursued as a human enterprise. However, at the same time, in this eleventh edition I have included a great deal of material to help students estimate, calculate, and solve problems *outside* the classroom or textbook setting.

This book was written to meet the needs of all students and schools. How did I accomplish that goal? First, the chapters are almost independent of one another, and can be covered in any order appropriate to a particular audience. Second, the problems are designed as the core of the course. There are problems that every student will find easy and this will provide the opportunity for success; there are also problems that are very challenging. Much interesting material appears in the problems, and students should get into the habit of reading (not necessarily working) all the problems whether or not they are assigned.

Level 1: Mechanical or drill problems

Level 2: Problems that require understanding of the concepts

Level 3: Problems that require problem solving skills or original thinking

What Are the Major Themes of This Book?

The major themes of this book are *problem solving* and estimation in the context of presenting the great ideas in the history of mathematics.

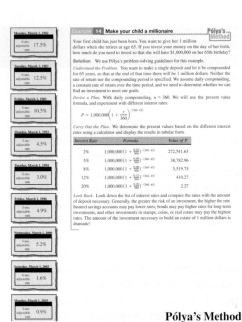

Pólya's Method

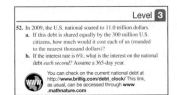

Level 3 Problem

In Your Own Words

**Historical
Note**

Historical Quest

I believe that *learning to solve problems is the principal reason for studying mathematics.* Problem solving is the process of applying previously acquired knowledge to new and unfamiliar situations. Solving word problems in most textbooks is one form of problem solving, but students also should be faced with non-text-type problems. In the first section of this edition I introduce students to Pólya's problem-solving techniques, and these techniques are used throughout the book to solve non-text-type problems. Problem-solving examples are found throughout the book (marked as **PÓLYA'S METHOD** examples).

You will find problems in *each* section that require Pólya's method for problem solving, and then you can practice your problem-solving skills with problems that are marked: **Level 3, Problem Solving.**

Students should learn the language and notation of mathematics. Most students who have trouble with mathematics do not realize that mathematics *does require hard work.* The usual pattern for most mathematics students is to open the book to the assigned page of problems, and begin working. Only after getting "stuck" is an attempt made to "find it in the book." The final resort is reading the text. In this book students are asked not only to "do math problems," but also to "experience mathematics." This means it is necessary to become involved with the **concepts** being presented, not "just get answers." In fact, the advertising slogan "Mathematics Is Not a Spectator Sport" is an invitation which suggests that the only way to succeed in mathematics is to become involved with it.

Students will learn to receive mathematical ideas through listening, reading, and visualizing. They are expected to present mathematical ideas by speaking, writing, drawing pictures and graphs, and demonstrating with concrete models. There is a category of problems in each section which is designated IN YOUR OWN WORDS, and which provides practice in communication skills.

Students should view mathematics in historical perspective. There is no argument that mathematics has been a driving force in the history of civilization. In order to bring students closer to this history, I've included not only Historical Notes, but a new category of problems called **Historical Quest** problems.

Students should learn to think critically. Many colleges have a broad educational goal of increasing critical thinking skills. Wikipedia defines **critical thinking** as "purposeful and reflective judgment about what to believe or do in response to observations, experience, verbal or written expressions or arguments." Critical thinking might involve determining the meaning and significance of what is observed or expressed, or, concerning a given inference or argument, determining whether there is adequate justification to accept the conclusion as true. Critical thinking begins in earnest in Section 1.1 when we introduce Pólya's problem solving method. These Pólya examples found throughout the book are not the usual "follow-the-leader"-type problems, but attempt, slowly, but surely, to teach critical thinking. The Problem Solving problems in almost every section continue this theme. The following sections are especially appropriate to teaching critical thinking skills: Problem Solving (1.1), Problem Solving with Logic (3.5), Cryptography (5.3), Modeling Uncategorized Problems (6.9), Summary of Financial Formulas (11.7), Probability Models (13.3), Voting Dilemmas (17.2), Apportionment Paradoxes (17.4), and What Is Calculus? (18.1).

A Note for Instructors

The prerequisites for this course vary considerably, as do the backgrounds of students. Some schools have no prerequisites, while other schools have an intermediate algebra prerequisite. The students, as well, have heterogeneous backgrounds. Some have little or no mathematics skills; others have had a great deal of mathematics. Even though the usual prerequisite for using this book is intermediate algebra, a careful selection of topics and chapters would allow a class with a beginning algebra prerequisite to study the material effectively.

Feel free to arrange the material in a different order from that presented in the text. I have written the chapters to be as independent of one another as possible. There is much more material than could be covered in a single course. This book can be used in classes designed for liberal arts, teacher training, finite mathematics, college algebra, or a combination of these.

Over the years, many instructors from all over the country have told me that they love the material, love to teach from this book, but complain that there is just too much material in this book to cover in one, or even two, semesters. In response to these requests, I have divided some of the material into two separate volumes:

The Nature of Problem Solving in Geometry and Probability
The Nature of Problem Solving in Algebra

The first volume, *The Nature of Geometry and Probability* includes chapters 1, 2, 3, 7, 8, 9, 11, 12, and 13 from this text.

The second volume, *The Nature of Algebra* includes chapters 1, 4, 5, 6, 9, 10, 14, 15, 16, and 17 from this text.

Since the first edition of this book, I have attempted to make the chapters as independent as possible to allow instructors to "pick and choose" the chapters to custom design the course. Because of advances in technology, it is now possible to design your own book for your class. The publisher offers a digital library, **TextChoice,** which helps you build your own custom version of *The Nature of Mathematics*. The details are included on the endpapers of this book.

One of the advantages of using a textbook that has traveled through many editions is that it is well seasoned. Errors are minimal, pedagogy is excellent, and it is easy to use; in other words, it works. For example, you will find that the sections and chapters are about the right length... each section will take about one classroom day. The problem sets are graded so that you can teach the course at different levels of difficulty, depending on the assigned problems. The problem sets are uniform in length (60 problems each), which facilitates the assigning of problems from day-to-day. The chapter reviews are complete and lead students to the type of review they will need to prepare for an examination.

Changes from the Previous Edition

As a result of extensive reviewer feedback, there are many new ideas and changes in this edition.

- The examples throughout the book have been redesigned. Each example now includes a title as well as a fresh easy-to read format.
- Each chapter now has a Chapter Challenge as an added problem-solving practice. These problems are out of context in order to give students additional challenge.
- Mathematical history has been an integral feature of this book since its inception, and we have long used Historical Notes to bring the human story into our venture through this text. In the last edition we experimented with a new type of problem called an Historical Quest and it has proved to be an overwhelming success, so we have greatly expanded its use in this edition. These problems are designed to *involve* the student in the historical development of the great ideas in mathematical history.

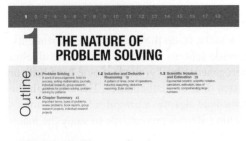

What in the World?

"Hey, Tom, what are you taking this semester?" asked Susan. "I'm taking English, history, and math. I can't believe my math teacher," responded Tom. "The first day we were there, she walked in, wrote her name on the board, and then she asked, 'How much space would you have if you, along with everyone else in the world, moved to California?' What a stupid question ... I would not have enough

Overview

There are many reasons for reading a book, but the best reason is because you want to read it. Although you are probably reading this first page because you were required to do so by your instructor, it is my hope that in a short while you will be reading this book because you want to read it. It was written for people who think they don't like mathematics, or people who think they can't work math problems, or

What in the World?

- The chapter openings have been redesigned, but continue to offer the popular "What in the World?" introduction. They use common conversations between two students to introduce chapter material, helping to connect the content with students' lives.

- The prologue and epilogue have been redesigned and offer unique "bookends" to the material in the book. The prologue asks, "Why Math?". This prologue not only puts mathematics into a historical perspective, but also is designed to get students thinking about problem solving. The problems accompanying this prologue could serve as a pre-test or diagnostic test, but I use these prologue problems to let the students know that this book will not be like other math books they may have used in the past. The epilogue, "Why Not Math?", is designed to tie together many parts of the book (which may or may not have been covered in the class) to show that there are many rooms in the mansion known as mathematics. The problems accompanying this epilogue could serve as a review to show that it would be difficult to choose a course of study in college without somehow being touched by mathematics. When have you seen a mathematics textbook that asks, "Why study mathematics?" and then actually produces an example to show it?*

*See Example 2, Section 11.5, 551.

Acknowledgments

I also appreciate the suggestions of the reviewers of this edition:

Vincent Edward Castellana, Eastern Kentucky University
Beth Greene Costner, Winthrop University
Charles Allen Matthews, Southeastern Oklahoma State University
James Waichiro Miller, Chaminade University of Honolulu
Tammy Potter, Gadsen State Community College
Jill S. Rafael, Sierra College
Leonora DePiola Smook, Suffolk Community College
Lynda Zenati, Robert Morris College

One of the nicest things about writing a successful book is all of the letters and suggestions I've received. I would like to thank the following people who gave suggestions for previous editions of this book: Jeffery Allbritten, Brenda Allen, Richard C. Andrews, Nancy Angle, Peter R. Atwood, John August, Charles Baker, V. Sagar Bakhshi, Jerald T. Ball, Carol Bauer, George Berzsenyi, Daniel C. Biles, Jan Boal, Elaine Bouldin, Kolman Brand, Chris C. Braunschweiger, Barry Brenin, T. A. Bronikowski, Charles M. Bundrick, T. W. Buquoi, Eugene Callahan, Michael W. Carroll, Joseph M. Cavanaugh, Rose Cavin, Peter Chen, James R. Choike, Mark Christie, Gerald Church, Robert Cicenia, Wil Clarke, Lynn Cleaveland, Penelope Ann Coe, Thomas C. Craven, Gladys C. Cummings, C.E. Davis, Steven W. Davis, Tony F. DeLia, Stephen DeLong, Ralph De Marr, Robbin Dengler, Carolyn Detmer, Maureen Dion, Charles Downey, Mickle Duggan, Samuel L. Dunn, Robert Dwarika, Beva Eastman, William J. Eccles, Gentil Estevez, Ernest Fandreyer, Loyal Farmer, Gregory N. Fiore, Robert Fliess, Richard Freitag, Gerald E. Gannon, Ralph Gellar, Sanford Geraci, Gary Gislason, Lourdes M. Gonzalez, Mark Greenhalgh, Martin Haines, Abdul Rahim Halabieh, Ward Heilman, John J. Hanevy, Michaael Helinger, Robert L. Hoburg, Caroline Hollingsworth, Scott Holm, Libby W. Holt, Peter Hovanec, M. Kay Hudspeth, Carol M. Hurwitz, James J. Jackson, Kind Jamison, Vernon H. Jantz, Josephine Johansen, Charles E. Johnson, Nancy J. Johnson, Judith M. Jones, Michael Jones, Martha C. Jordan, Ravindra N. Kalia, Judy D. Kennedy, Linda H. Kodama, Daniel Koral, Helen Kriegsman, Frances J. Lane, C. Deborah Laughton, William Leahey, John LeDuc, Richard Leedy, William A. Leonard, Beth Long, Adolf Mader, Winifred A. Mallam, John Martin, Maria M. Maspons, Cherry F. May, Paul McCombs, Cynthia L. McGinnis, George McNulty, Carol McVey, Max Melnikov, Valerie Melvin, Charles C. Miles, Allen D. Miller, Clifford D. Miller, Elaine I. Miller, Ronald H. Moore, John Mullen, Charles W. Nelson, Ann Ostberg, Barbara Ostrick, John Palumbo, Joanne V. Peeples, Gary Peterson, Michael Petricig, Mary Anne C. Petruska, Michael Pinter, Susan K. Puckett, Laurie Poe, Joan Raines, James V. Rauff, Richard Rempel, Pat Rhodes, Paul M. Riggs, Jane Rood, Peter Ross, O. Sassian, Mickey G. Settle, James R. Smart, Andrew Simoson, Glen T. Smith, Donald G. Spencer, Barb Tanzyus, Gustavo Valadez-Ortiz, John Vangor, Arnold Villone, Clifford H. Wagner, James Walters, Steve Warner, Steve Watnik, Pangyen Ben Weng, Barbara Williams, Carol E. Williams, Stephen S. Willoughby, Mary C. Woestman, Jean Woody, and Bruce Yoshiwara.

The creation of a textbook is really a team effort. My thanks to Beth Kluckhohn, Abigail Perrine, Carly Bergey, and Shaun Williams who led me through the process effortlessly. And I especially express my appreciation to Jack Morrell for carefully reading the entire manuscript while all the time offering me valuable suggestions.

I would especially like to thank Joe Salvati, from New School University in Manhattan, Robert J. Wisner of New Mexico State for his countless suggestions and ideas over the many editions of this book; Marc Bove, Shona Burke, John-Paul Ramin, Craig Barth, Jeremy Hayhurst, Paula Heighton, Gary Ostedt, and Bob Pirtle of Brooks/Cole; as well as Jack Thornton, for the sterling leadership and inspiration he has been to me from the inception of this book to the present.

Finally, my thanks go to my wife, Linda, who has always been there for me. Without her this book would exist only in my dreams, and I would never have embarked as an author.

Karl J. Smith
Sebastopol, CA
smithkjs@mathnature.com

Supplements

For the Student

Student Survival and Solutions Manual
AUTHOR: Karl Smith
ISBN: 0538495286
The Student Survival and Solutions Manual provides helpful study aids and fully worked-out solutions to all of the odd-numbered exercises in the text. It's a great way to check your answers and ensure that you took the correct steps to arrive at an answer.

Enhanced WebAssign
ISBN: 0538738103
Enhanced Webassign, used by over one million students at more than 1100 institutions, allows you to assign, collect, grade, and record homework assignments via the web. This proven and reliable homework system includes thousands of algorithmically generated homework problems, an eBook, links to relevant textbook sections, video examples, problem specific tutorials, and more.

For the Instructor

Annotated Instructor's Edition
AUTHOR: Karl Smith
ISBN: 0538738693
The Annotated Instructor's Edition features an appendix containing the answers to all problems in the book as well as icons denoting which problems can be found in Enhanced WebAssign.

Instructor's Manual
AUTHOR: Karl Smith
ISBN: 0538495278
Written by author Karl Smith, the Instructor's Manual provides worked-out solutions to all of the problems in the text. For instructors only.

Text-Specific DVDs
AUTHOR: Dana Mosely
ISBN: 1111571252
Hosted by Dana Mosley, these text-specific instructional videos provide students with visual reinforcement of concepts and explanations, presented in easy-to-understand terms with detailed examples and sample problems. A flexible format offers versatility for quickly accessing topics or catering lectures to self-paced, online, or hybrid courses. Closed captioning is provided for the hearing impaired.

Enhanced WebAssign
ISBN: 0538738103
Enhanced WebAssign, used by over one million students at more than 1,100 institutions, allows you to assign, collect, grade, and record homework assignments via the web. This proven and reliable homework system includes thousands of algorithmically generated homework problems, links to relevant textbook sections, video examples, problem-specific tutorials, and more.

New! Personal Study Plans and a Premium eBook
Diagnostic quizzing for each chapter identifies concepts that students still need to master, and directs them to the appropriate review material. Students will appreciate the interactive Premium eBook, which offers search, highlighting, and note-taking functionality, as well as links to multimedia resources, all available to students when you choose Enhanced WebAssign.

Note that the WebAssign problems for this text are highlighted by a ➤.

PowerLecture with ExamView®
ISBN: 0840053304
This CD-ROM provides the instructor with dynamic media tools for teaching. Create, deliver, and customize tests (both print and online) in minutes with ExamView® Computerized Testing Featuring Algorithmic Equations. Easily build solution sets for homework or exams using Solution Builder's online solutions manual. Microsoft® PowerPoint® lecture slides and figures from the book are also included on this CD-ROM.

Solution Builder
This online solutions manual allows instructors to create customizable solutions that they can print out to distribute or post as needed. This is a convenient and expedient way to deliver solutions to specific homework sets. Visit www.cengage.com/solutionbuilder.

Math Study Skills Workbook, 4th Edition
AUTHOR: Paul Nolting
ISBN-13: 978-0-840-05309-1
Paul Nolting's workbook will help you identify your strengths, weaknesses, and personal learning styles in math. Nolting offers proven study tips, test-taking strategies, a homework system, and recommendations for reducing anxiety and improving grades.

Book-companion Web site at **www.mathnature.com**
Author: Karl Smith
Created and updated by Karl Smith, the Web site offers supplementary help and practice for students. All of the Web addresses mentioned in the book are linked to the above Web address. You will find links to several search engines, history, and reference topics. You will find, for each section, homework hints, and a listing of essential ideas, projects, and links to related information on the Web.

To the Student

A FABLE

If people do not believe that mathematics is simple, it is only because they do not realize how complicated life is.

JOHN VAN NEUMANN

Once upon a time, two young ladies, Shelley and Cindy, came to a town called Mathematics. People had warned them that this was a particularly confusing town. Many people who arrived in Mathematics were very enthusiastic, but could not find their way around, became frustrated, gave up, and left town.

Shelley was strongly determined to succeed. She was going to learn her way through the town. For example, in order to learn how to go from her dorm to class, she concentrated on memorizing this clearly essential information: she had to walk 325 steps south, then 253 steps west, then 129 steps in a diagonal (south-west), and finally 86 steps north. It was not easy to remember all of that, but fortunately she had a very good instructor who helped her to walk this same path 50 times. In order to stick to the strictly necessary information, she ignored much of the beauty along the route, such as the color of the adjacent buildings or the existence of trees, bushes, and nearby flowers. She always walked blindfolded. After repeated exercising, she succeeded in learning her way to class and also to the cafeteria. But she could not learn the way to the grocery store, the bus station, or a nice restaurant; there were just too many routes to memorize. It was so overwhelming! Finally, she gave up and left town; Mathematics was too complicated for her.

Cindy, on the other hand, was of a much less serious nature. To the dismay of her instructor, she did not even intend to memorize the number of steps of her walks. Neither did she use the standard blindfold which students need for learning. She was always curious, looking at the different buildings, trees, bushes, and nearby flowers or anything else not necessarily related to her walk. Sometimes she walked down dead-end alleys in order to find out where they were leading, even if this was obviously superfluous. Curiously, Cindy succeeded in learning how to walk from one place to another. She even found it easy and enjoyed the scenery. She eventually built a building on a vacant lot in the city of Mathematics.*

*My thanks to Emilio Roxin of the University of Rhode Island for the idea for this fable.

A HISTORICAL OVERVIEW

Whether you love or loathe mathematics, it is hard to deny its importance in the development of the main ideas of this world! Read the BON VOYAGE invitation on the inside front cover. The goal of this text is to help you to discover an answer to the question, "Why study math?"

The study of mathematics can be organized as a history or story of the development of mathematical ideas, or it can be organized by topic. The intended audience of this book dictates that the development should be by topic, but mathematics involves real people with real stories, so you will find this text to be very historical in its presentation. This overview rearranges the material you will encounter in the text into a historical timeline. It is not intended to be read as a history of mathematics, but rather as an overview to make you want to do further investigation. Sit back, relax, and use this overview as a *starting place* to expand your knowledge about the beginnings of some of the greatest ideas in the history of the world!

We have divided this history of mathematics into seven chronological periods:

Babylonian, Egyptian, and Native American Period	3000 B.C. to 601 B.C.
Greek, Chinese, and Roman Periods	600 B.C. to A.D. 499
Hindu and Persian Period	500 to 1199
Transition Period	1200 to 1599
Age of Reason	1600 to 1699
Early Modern Period	1700 to 1799
Modern Period	1800 to present

Babylonian, Egyptian, and Native American Period: 3000 B.C. to 601 B.C.

Mesopotamia is an ancient region located in southwest Asia between the lower Tigris and Euphrates Rivers and is historically known as the birthplace of civilization. It is part of modern Iraq. Mesopotamian mathematics refers to the mathematics of the ancient Babylonians, and this mathematics is sometimes referred to as Sumerian mathematics. Over 50,000 tablets from Mesopotamia have been found and are exhibited at major museums around the world.

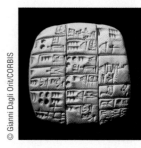
© Gianni Dagli Orti/CORBIS

Sumerian clay tablet

Babylonian, Egyptian, and Native American Period: 3000 BC to 601 BC

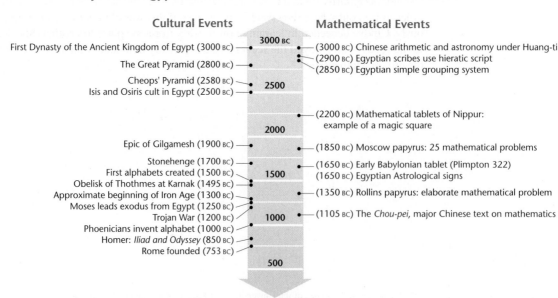

Cultural Events

First Dynasty of the Ancient Kingdom of Egypt (3000 BC)
The Great Pyramid (2800 BC)
Cheops' Pyramid (2580 BC)
Isis and Osiris cult in Egypt (2500 BC)
Epic of Gilgamesh (1900 BC)
Stonehenge (1700 BC)
First alphabets created (1500 BC)
Obelisk of Thothmes at Karnak (1495 BC)
Approximate beginning of Iron Age (1300 BC)
Moses leads exodus from Egypt (1250 BC)
Trojan War (1200 BC)
Phoenicians invent alphabet (1000 BC)
Homer: *Iliad and Odyssey* (850 BC)
Rome founded (753 BC)

3000 BC
2500
2000
1500
1000
500

Mathematical Events

(3000 BC) Chinese arithmetic and astronomy under Huang-ti
(2900 BC) Egyptian scribes use hieratic script
(2850 BC) Egyptian simple grouping system
(2200 BC) Mathematical tablets of Nippur: example of a magic square
(1850 BC) Moscow papyrus: 25 mathematical problems
(1650 BC) Early Babylonian tablet (Plimpton 322)
(1650 BC) Egyptian Astrological signs
(1350 BC) Rollins papyrus: elaborate mathematical problem
(1105 BC) The *Chou-pei*, major Chinese text on mathematics

Interesting readings about Babylon can be found in a book on the history of mathematics, such as *An Introduction to the History of Mathematics,* 6th edition, by Howard Eves (New York: Saunders, 1990), or by looking at the many sources on the World Wide Web. You can find links to such Web sites, as well as all Web sites in this book, by looking at the Web page for this text:

 **www.mathnature.com**

This Web page allows you to access a world of information by using the links provided.

The mathematics of this period was very practical and it was used in construction, surveying, recordkeeping, and in the creation of calendars. The culture of the Babylonians reached its height about 2500 B.C., and about 1700 B.C. King Hammurabi formulated a famous code of law. In 330 B.C., Alexander the Great conquered Asia Minor, ending the great, Persian (Achaemenid) Empire. Even though there was a great deal of political and social upheaval during this period, there was a continuity in the development of mathematics from ancient time to the time of Alexander.

The main information we have about the civilization and mathematics of the Babylonians is their numeration system, which we introduce in the text in Section 4.1 (p. 140). The Babylonian numeration system was positional with base 60. It did not have a 0 symbol, but it did represent fractions, squares, square roots, cubes, and cube roots. We have evidence that the Babylonians knew the quadratic formula and they had stated algebraic problems verbally. The base 60 system of the Babylonians led to the division of a circle into 360 equal parts that today we call degrees, and each degree was in turn divided into 360 parts that today we call seconds. The Greek astronomer Ptolemy (A.D. 85–165) used this Babylonian system, which no doubt is why we have minutes, seconds, and degree measurement today.

The Egyptian civilization existed from about 4000 B.C., and was less influenced by foreign powers than was the Babylonian civilization. Egypt was divided into two kingdoms until about 3000 B.C., when the ruler Menes unified Egypt and consequently became known as the founder of the first dynasty in 2500 B.C. This was the egyptians' pyramid-building period, and the Great Pyramid of Cheops was built around 2600 B.C. (Chapter 7, p. 369; see The Riddle of the Pyramids).

The Egyptians developed their own pictorial way of writing, called *hieroglyphics,* and their numeration system was consequently very pictorial (Chapter 4).

The Egyptian numeration system is an example of a simple grouping system. Although the Egyptians were able to write fractions, they used only unit fractions. Like the Babylonians, they had not developed a symbol for zero. Since the writing of the Egyptians was on papyrus, and not on tablets as with the Babylonians, most of the written history has been lost. Our information comes from the Rhind papyrus, discovered in 1858 and dated to about 1700 B.C., and the Moscow papyrus, which has been dated to about the same time period.

The mathematics of the Egyptians remained remarkably unchanged from the time of the first dynasty to the time of Alexander the Great who conquered Egypt in 332 B.C. The Egyptians did surveying using a unique method of stretching rope, so they referred to their surveyors as "rope stretchers." The basic unit used by the Egyptians for measuring length was the *cubit,* which was the distance from a person's elbow to the end of the middle finger. A *khet* was defined to equal 100 cubits; khets were used

@Bettmann/CORBIS

Egyptian hieroglyphics: Inscription and relief from the grave of Prince Rahdep (ca. 2800 B.C.)

by the Egyptians when land was surveyed. The Egyptians did not have the concept of a variable, and all of their problems were verbal or arithmetic. Even though they solved many equations, they used the word *AHA* or *heap* in place of the variable. For an example of an Egyptian problem, see Ahmes' dilemma in Chapter 1 and the statement of the problem in terms of Thoth, an ancient Egyptian god of wisdom and learning.

The Egyptians had formulas for the area of a circle and the volume of a cube, box, cylinder, and other figures. Particularly remarkable is their formula for the volume of a truncated pyramid of a square base, which in modern notation is

$$V = \frac{h}{3}\left(a^2 + ab + b^2\right)$$

where h is the height and a and b are the sides of the top and bottom. Even though we are not certain the Egyptians knew of the Pythagorean theorem, we believe they did because the rope stretchers had knots on their ropes that would form right triangles. They had a very good reckoning of the calendar, and knew that a solar year was approximately $365\frac{1}{4}$ days long. They chose as the first day of their year the day on which the Nile would flood.

Contemporaneous with the great civilizations in Mesopotamia was the great Mayan civilization in what is now Mexico. A Mayan timeline is shown in Figure 1.

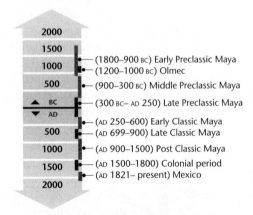

FIGURE 1 Mayan timeline

Just as with the Mesopotamian civilizations, the Olmeca and Mayan civilizations lie between two great rivers, in this case the Grijalva and Papaloapan Rivers. Sometimes the Olmecas are referred to as the Tenocelome. The Olmeca culture is considered the mother culture of the Americas. What we know about the Olmecas centers around their art. We do know they were a farming community. The Mayan civilization began around 2600 B.C. and gave rise to the Olmecas. The Olmecas had developed a written hieroglyphic language by 700 B.C., and they had a very accurate solar calendar. The Mayan culture had developed a positional numeration system.

You will find the influences from this period discussed throughout the book.

Greek, Chinese, and Roman Periods: 600 B.C. to A.D. 499

Greek mathematics began in 585 B.C. when Thales, the first of the Seven Sages of Greece (625–547 B.C.) traveled to Egypt.[*]

The Greek civilization was most influential in our history of mathematics. So striking was its influence that the historian Morris Kline declares, "One of the great problems of the history of civilization is how to account for the brilliance and creativity of the ancient Greeks."[†] The Greeks settled in Asia Minor, modern Greece, southern Italy, Sicily, Crete, and North Africa. They replaced the various hieroglyphic systems with the Phoenician alphabet, and with that they were able to become more literate and more capable of recording history and ideas. The Greeks had their own numeration system. They had fractions and some irrational numbers, including π.

The great mathematical contributions of the Greeks are Euclid's *Elements* and Apollonius' *Conic Sections* (p. 732, Figure 15.26). Greek knowledge developed in several centers or schools. (See Figure 3 on page P5 for depiction of one of these centers of learning.) The first was founded by Thales (ca. 640–546 B.C.) and known as the Ionian in Miletus. It is reported that while he was traveling and studying in Egypt, Thales calculated the heights of the pyramids by using similar triangles (see Section 7.4). You can read about these great Greek mathematicians in *Mathematics Thought from Ancient to Modern Times,* by Morris Kline.[‡] You can also refer to the World Wide Web at **www.mathnature.com.**

Between 585 B.C. and 352 B.C., schools flourished and established the foundations for the way knowledge is organized today. Figure 2 shows each of the seven major schools, along with each school's most notable contribution. Links to textual discussion are shown within each school of thought, along with the principal person for each of these schools. Books have been written about the importance of each of these Greek schools, and several links can be found at **www.mathnature.com.**

One of the three greatest mathematicians in the entire history of mathematics was Archimedes (287–212 B.C.). His accomplishments are truly remarkable, and you should seek out other sources about the magnitude of his accomplishments. He invented a pump (the Archimedean screw), military engines and weapons, and catapults; in addition, he used a parabolic mirror as a weapon by concentrating the sun's rays on the invading Roman ships. "The most famous of the stories about Archimedes is his discovery of the method of testing the debasement of a crown of gold. The king of Syracuse had ordered the crown. When it was delivered, he suspected that it was filled with baser metals and sent it to Archimedes to

---

[*]The Seven Sages in Greek history refer to Thales of Miletus, Bias of Priene, Chilo of Sparta, Cleobulus of Rhodes, Periander of Corinth, Pittacus of Mitylene, and Solon of Athens; they were famous because of their practical knowledge about the world and how things work.

[†]p. 24, *Mathematical Thought from Ancient to Modern Times* by Morris Kline (New York: Oxford University Press, 1972).

[‡] New York: Oxford University Press, 1972.

Greek, Chinese, and Roman Period: 600 BC to AD 499

Cultural Events

Mathematical Events

600	
500	(585 BC) Thales, founder of Greek geometry
Persians capture Babylon (538 BC)	(540 BC) The teachings of Pythagoras
Pindar's *Odes* (500 BC)	(500 BC) Sulvasutras: Pythagorean numbers
400	(450 BC) Zeno: paradoxes of motion
Siddhartha, the Buddha, delivers his sermons in Deer Park (480 BC)	(425 BC) Theodorus of Cyrene: irrational numbers
	(384 BC) Aristotle: logic
300	(380 BC) Plato's Academy: logic
Alexander the Great completes his conquest of the known world (323 BC)	(323 BC) Euclid: geometry, perfect numbers
	(300 BC) First use of Hindu numeration system
200	(230 BC) Sieve of Eratosthenes
Hannibal crosses the Alps (218 BC)	(225 BC) Archimedes: circle, pi, curves, series
Rosetta Stone engraved (200 BC)	(180 BC) Hypsicles: number theory
100	
Birth of Julius Caesar (100 BC)	
Virgil: *Aeneid* (20 BC)	(60 BC) Geminus: parallel postulate
Birth of Christ (4 BC)	BC / AD
	(AD 50) Negative numbers used in China
100	(AD 75) Heron: measurements, roots, surveying
	(AD 100) Nichomachus: number theory
	(AD 150) Ptolemy: trigonometry
Goths invade Asia Minor (AD 200)	(AD 200) Mayan calendar
200	(AD 250) Diophantus: number theory, algebra
Founding of Constantinople (AD 324)	(AD 300) Pappus: *Mathematical Collection*
300	
Augustine, *Confessions* (AD 400)	(AD 410) Hypatia of Alexandria: first woman mentioned in the history of mathematics
400	
Fall of Rome (AD 476)	(AD 480) Tsu Ch'ung-chi approximates π as 355/113
500	

devise some method of testing the contents without, of course, destroying the workmanship. Archimedes pondered the problem; one day while bathing he observed that his body was partly buoyed up by the water and suddenly grasped the principle that enabled him to solve the problem. He was so excited by this discovery that he ran out into the street naked shouting, `Eureka!' ('I have found it!') He had discovered that a body immersed in water is buoyed up by a force equal to the weight of the water displaced, and by means of this principle was able to determine the contents of the crown."[*]

The Romans conquered the world, but their mathematical contributions were minor. We introduce the Roman numerals in Section 4.1, their fractions were based on a duodecimal (base 12) system and are still used today in certain circumstances. The unit of weight was the *as* and one-twelfth of this was the *uncia,* from which we get our measurements of *ounce* and *inch,* respectively. The Romans improved on our calendar, and set up the notion of leap year every four years. The Julian calendar was adopted in 45 B.C. The Romans conquered Greece and Mesopotamia, and in 47 B.C., they set fire to the Egyptian fleet in the harbor of Alexandria. The fire spread to the city and burned the library, destroying two and a half centuries of book-collecting, including all the important knowledge of the time.

Another great world civilization existed in China and also developed a decimal numeration system and used a decimal system with symbols 1, 2, 3, . . . , 9, 10, 100, 1000, and 10000. Calculations were performed using small bamboo counting rods, which eventually evolved into the abacus. Our first historical reference to the Chinese culture is the yin-yang symbol, which has its roots in ancient cosmology. The original meaning is representative of the mountains, both the bright side and the dark side. The "yin" represents the female, or shaded, aspect, the earth, the darkness, the moon, and passivity. The "yang" represents the male, light, sun, heaven, and the active principle in nature. These words can be traced back to the Shang and Chou Dynasty (1550–1050 B.C.), but most scholars credit them to the Han Dynasty (220–206 B.C.). One of the first examples of a magic square comes from Lo River around 200 B.C., where legend tells us that the emperor Yu of the Shang dynasty received a magic square on the back of a tortoise's shell.

From 100 B.C. to A.D. 100 the Chinese described the motion of the planets, as well as what is the earliest known proof of the Pythagorean theorem. The longest surviving and most influential Chinese math book is dated from the beginning of the Han Dynasty around A.D. 50. It includes measurement and area problems, proportions, volumes, and some approximations for π. Sun Zi (ca. A.D. 250) wrote his mathematical manual, which included the "Chinese remainder problem": Find *n* so that upon division by 3 you obtain a remainder of 2; upon division by 5 a remainder of 3; and upon division by 7 you get a remainder of 2. His solution: Add 140, 63, 30 to obtain 233,

[*]pp. 105–106, *Mathematical Thought from Ancient to Modern Times* by Morris Kline (New York: Oxford University Press, 1972).

Greek Schools (585–352 BC)

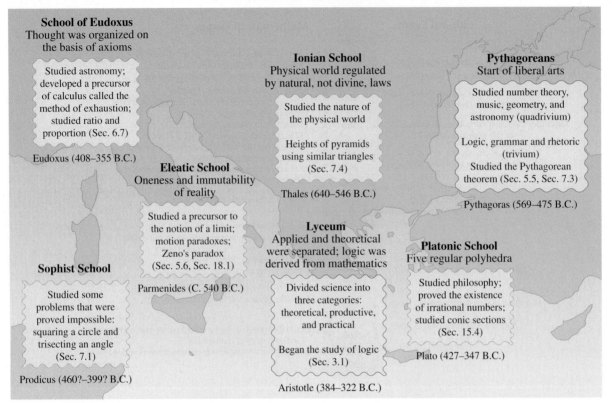

School of Eudoxus
Thought was organized on the basis of axioms

Studied astronomy; developed a precursor of calculus called the method of exhaustion; studied ratio and proportion (Sec. 6.7)

Eudoxus (408–355 B.C.)

Ionian School
Physical world regulated by natural, not divine, laws

Studied the nature of the physical world

Heights of pyramids using similar triangles (Sec. 7.4)

Thales (640–546 B.C.)

Pythagoreans
Start of liberal arts

Studied number theory, music, geometry, and astronomy (quadrivium)

Logic, grammar and rhetoric (trivium)
Studied the Pythagorean theorem (Sec. 5.5, Sec. 7.3)

Pythagoras (569–475 B.C.)

Eleatic School
Oneness and immutability of reality

Studied a precursor to the notion of a limit; motion paradoxes; Zeno's paradox (Sec. 5.6, Sec. 18.1)

Parmenides (C. 540 B.C.)

Lyceum
Applied and theoretical were separated; logic was derived from mathematics

Divided science into three categories: theoretical, productive, and practical

Began the study of logic (Sec. 3.1)

Aristotle (384–322 B.C.)

Platonic School
Five regular polyhedra

Studied philosophy; proved the existence of irrational numbers; studied conic sections (Sec. 15.4)

Plato (427–347 B.C.)

Sophist School

Studied some problems that were proved impossible: squaring a circle and trisecting an angle (Sec. 7.1)

Prodicus (460?–399? B.C.)

FIGURE 2 Greek schools from 585 B.C. to 352 B.C.

FIGURE 3 The School at Athens by Raphael, 1509. This fresco includes portraits of Raphael's contemporaries and demonstrates the use of perspective. Note the figures in the lower right, who are, no doubt, discussing mathematics.

and subtract 210 to obtain 23. Zhang Qiujian (ca. A.D. 450) wrote a mathematics manual that included a formula for summing the terms of an arithmetic sequence, along with the solution to a system of two linear equations in three unknowns. The problem is the "One Hundred Fowl Problem," and is included in Problem Set 5.7 (p. 240). At the end of this historic period, the mathematician and astronomer Wang Xiaotong (ca. A.D. 626) solved cubic equations by generalization of an algorithm for finding the cube root.

Check **www.mathnature.com** for links to many excellent sites on Greek mathematics.

Hindu and Persian Period: 500 to 1199

Much of the mathematics that we read in contemporary mathematics textbooks ignores the rich history of this period. Included on the World Wide Web are some very good sources for this period. Check our Web site **www.mathnature.com** for some links. The Hindu civilization dates back to 2000 B.C., but the first recorded mathematics was during the Śulvasūtra period from 800 B.C. to 200 A.D. In the third century, Brahmi symbols were used for 1, 2, 3, . . . , 9 and are significant because there was a single symbol for each number. There was no zero or positional notation at this time, but by A.D. 600 the

Hindus used the Brahmi symbols with positional notation. In Chapter 4, we will discuss a numeration system that eventually evolved from these Brahmi symbols. For fractions, the Hindus used sexagesimal positional notation in astronomy, but in other applications they used a ratio of integers and wrote $\frac{3}{4}$ (without the fractional bar we use today). The first mathematically important period was the second period, A.D. 200–1200. The important mathematicians of this period are Āryabhata (A.D. 476–550), Brahmagupta (A.D. 598–670), Mahāvīra (9th century), and Bhāskara (1114–1185). In Chapter 6, we include some historical questions from Bhāskara and Brahmagupta.

The Hindus developed arithmetic independently of geometry and had a fairly good knowledge of rudimentary algebra. They knew that quadratic equations had two solutions, and they had a fairly good approximation for π. Astronomy motivated their study of trigonometry. Around 1200, scientific activity in India declined, and mathematical progress ceased and did not revive until the British conquered India in the 18th century.

The Persians invited Hindu scientists to settle in Baghdad, and when Plato's Academy closed in A.D. 529, many scholars traveled to Persia and became part of the Iranian tradition of science and mathematics. Omar Khayyám (1048–1122) and Nasîr-Eddin (1201–1274), both renowned Persian scholars, worked freely with irrationals, which contrasts with the Greek idea of number. What we call Pascal's triangle dates

Hindu and Arabian Period: AD 500 to 1199

Cultural Events

Mathematical Events

First plans of the Vatican Palace in Rome (500) — AD 500

600

Mohammed's vision (610) —

(630) Brahmagupta: algebra, astronomy

Northern Irish submit to Catholicism (697) — 700
Charlemagne crowned emperor of Holy Roman Empire (800) —
Utrecht Psalter (832) —
Beginning of Carolinian dynasties (832) —
First printed book (870) — 800
Alfred the Great (871) —
Schism of the Church (871) —
Vikings discover Greenland (900) —
Emperor Otto I (The Great Emperor) (912–973) —
Beginning of the Dark Ages (950) — 900
Emperor Otto II (973–983) —
Development of systematic musical notation (990) —
First canonization of saints (993) —
Leif Erickson crosses Atlantic to Vinland (1003) —
World's first novel, *Tale of Genji* (1008) — 1000
School of Chartres (1028) —
Normans penetrate England (1050) —
Macbeth defeated at Dunsinane (1054) —
Consecration of Westminster Abbey (1065) —
Chinese use movable type to print books (1086) — 1100
First modern university (1088) —
Start of first Crusade (1096) —
Chinese invent playing card (1110) —
Commencement of troubadour music (1125) — 1200
Beginning of Plantagenet reign (1154) —
Maimonides: *Mishneh Torah* (1165) —
Domesday Book; tax census ordered
by William the Conqueror (1186)

(710) Bede: calendar, finger arithmetic

(750) First use of zero symbol

(810) Mohammed ibn Mûsâ al-Khwârizmî coins term *algebra*
(810) Hindu numerals
(850) Mahavira: arithmetic, algebra
(870) Iâbit ibn Qorra: algebra, magic squares, amicable numbers
(900) Abû Kâmil: Algebra, Bakhshali manuscript

(976) Oldest example of written numerals in Europe
(980) Abu'wefa: constructions, trig tables
(999) Pope Sylvester II (Gerbert): arithmetic,
pi approximated as $\sqrt{8}\approx2.83$
(1000) Sridhara recognizes the importance of zero
(1020) Al-Karkhî: algebra
(1075) Game of rithmomachia
(1110) Persian scholar Omar Khayyám: cubic equations, Pascal's Triangle
(1120) Bhāskara
(1125) Earliest account of mariner's compass
(1150) Bhāskara: algebra
(1175) Averroës: trigonometry, astronomy

back to this period. The word *algebra* comes from the Persians in a book by the Persian astronomer Mohammed ibn Musa al-Khâwarizmî (780–850) entitled *Hisâb al-jabr w'al muqâbala*. Due to the Arab conquest of Persia, Persian scholars (notably Nasir-Eddin and al-Khwarizmi) were obliged to publish their works in the Arabic language and not Persian, causing many historians to falsely label the texts as products of Arab scholars. Al-Khwarizmi solved quadratic equations and knew there are two roots, and even though the Persians gave algebraic solutions of quadratic equations, they explained their work geometrically. They solved some cubics, but could solve only simple trigonometric problems.

Check **www.mathnature.com** for links to many excellent sites on Hindu and Arabian mathematics.

Transition Period: 1200 to 1599

Mathematics during the Middle Ages was transitional between the great early civilizations and the Renaissance.

Transition Period: 1200 to 1599

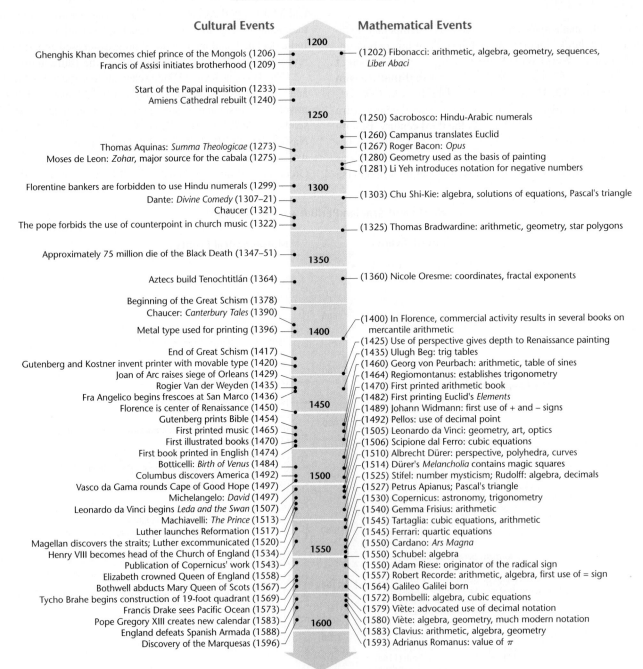

Cultural Events

Ghenghis Khan becomes chief prince of the Mongols (1206)
Francis of Assisi initiates brotherhood (1209)

Start of the Papal inquisition (1233)
Amiens Cathedral rebuilt (1240)

Thomas Aquinas: *Summa Theologicae* (1273)
Moses de Leon: *Zohar*, major source for the cabala (1275)

Florentine bankers are forbidden to use Hindu numerals (1299)
Dante: *Divine Comedy* (1307–21)
Chaucer (1321)
The pope forbids the use of counterpoint in church music (1322)

Approximately 75 million die of the Black Death (1347–51)

Aztecs build Tenochtitlán (1364)

Beginning of the Great Schism (1378)
Chaucer: *Canterbury Tales* (1390)
Metal type used for printing (1396)

End of Great Schism (1417)
Gutenberg and Kostner invent printer with movable type (1420)
Joan of Arc raises siege of Orleans (1429)
Rogier Van der Weyden (1435)
Fra Angelico begins frescoes at San Marco (1436)
Florence is center of Renaissance (1450)
Gutenberg prints Bible (1454)
First printed music (1465)
First illustrated books (1470)
First book printed in English (1474)
Botticelli: *Birth of Venus* (1484)
Columbus discovers America (1492)
Vasco da Gama rounds Cape of Good Hope (1497)
Michelangelo: *David* (1497)
Leonardo da Vinci begins *Leda and the Swan* (1507)
Machiavelli: *The Prince* (1513)
Luther launches Reformation (1517)
Magellan discovers the straits; Luther excommunicated (1520)
Henry VIII becomes head of the Church of England (1534)
Publication of Copernicus' work (1543)
Elizabeth crowned Queen of England (1558)
Bothwell abducts Mary Queen of Scots (1567)
Tycho Brahe begins construction of 19-foot quadrant (1569)
Francis Drake sees Pacific Ocean (1573)
Pope Gregory XIII creates new calendar (1583)
England defeats Spanish Armada (1588)
Discovery of the Marquesas (1596)

1200
1250
1300
1350
1400
1450
1500
1550
1600

Mathematical Events

(1202) Fibonacci: arithmetic, algebra, geometry, sequences, *Liber Abaci*

(1250) Sacrobosco: Hindu-Arabic numerals
(1260) Campanus translates Euclid
(1267) Roger Bacon: *Opus*
(1280) Geometry used as the basis of painting
(1281) Li Yeh introduces notation for negative numbers
(1303) Chu Shi-Kie: algebra, solutions of equations, Pascal's triangle
(1325) Thomas Bradwardine: arithmetic, geometry, star polygons

(1360) Nicole Oresme: coordinates, fractal exponents

(1400) In Florence, commercial activity results in several books on mercantile arithmetic
(1425) Use of perspective gives depth to Renaissance painting
(1435) Ulugh Beg: trig tables
(1460) Georg von Peurbach: arithmetic, table of sines
(1464) Regiomontanus: establishes trigonometry
(1470) First printed arithmetic book
(1482) First printing Euclid's *Elements*
(1489) Johann Widmann: first use of + and − signs
(1492) Pellos: use of decimal point
(1505) Leonardo da Vinci: geometry, art, optics
(1506) Scipione dal Ferro: cubic equations
(1510) Albrecht Dürer: perspective, polyhedra, curves
(1514) Dürer's *Melancholia* contains magic squares
(1525) Stifel: number mysticism; Rudolff: algebra, decimals
(1527) Petrus Apianus; Pascal's triangle
(1530) Copernicus: astronomy, trigonometry
(1540) Gemma Frisius: arithmetic
(1545) Tartaglia: cubic equations, arithmetic
(1545) Ferrari: quartic equations
(1550) Cardano: *Ars Magna*
(1550) Schubel: algebra
(1550) Adam Riese: originator of the radical sign
(1557) Robert Recorde: arithmetic, algebra, first use of = sign
(1564) Galileo Galilei born
(1572) Bombelli: algebra, cubic equations
(1579) Viète: advocated use of decimal notation
(1580) Viète: algebra, geometry, much modern notation
(1583) Clavius: arithmetic, algebra, geometry
(1593) Adrianus Romanus: value of π

In the 1400s the Black Death killed over 70% of the European population. The Turks conquered Constantinople, and many Eastern scholars traveled to Europe, spreading Greek knowledge as they traveled. The period from 1400 to 1600, known as the Renaissance, forever changed the intellectual outlook in Europe and raised up mathematical thinking to new levels. Johann Gutenberg's invention of printing with movable type in 1450 changed the complexion of the world. Linen and cotton paper, which the Europeans learned about from the Chinese through the Arabians, came at precisely the right historical moment. The first printed edition of Euclid's *Elements* in a Latin translation appeared in 1482. Other early printed books were Apollonius' *Conic Sections*, Pappus' works, and Diophantus' *Arithmetica*.

The first breakthrough in mathematics was by artists who discovered mathematical perspective. The theoretical genius in mathematical perspective was Leone Alberti (1404–1472). He was a secretary in the Papal Chancery writing biographies of the saints, but his work *Della Pictura* on the laws of perspective (1435) was a masterpiece. He said, "Nothing pleases me so much as mathematical investigations and demonstrations, especially when I can turn them into some useful practice drawing from mathematics and the principles of painting perspective and some amazing propositions on the moving of weights." He collaborated with Toscanelli, who supplied Columbus with maps for his first voyage. The best mathematician among the Renaissance artists was Albrecht Dürer (1471–1528). The most significant development of the Renaissance was the breakthrough in astronomical theory by Nicolaus Copernicus (1473–1543) and Johannes Kepler (1571–1630). There were no really significant new results in mathematics during this period of history.

It is interesting to tie together some of the previous timelines to trace the history of algebra. It began around 2000 B.C. in Egypt and Babylon. This knowledge was incorporated into the mathematics of Greece between 500 B.C. and A.D. 320, as well as into the Persian civilization and Indian mathematics around A.D. 1000. By the Transition Period, the great ideas of algebra had made their way to Europe, as shown in Figure 4. Additional information can be found on the World Wide Web; check our Web page at **www.mathnature.com.**

Age of Reason 1600 to 1699

From Shakespeare and Galileo to Peter the Great and the great Bernoulli family, this period, also called the Age of Genius, marks the growth of intellectual endeavors; both technology and knowledge grew as never seen before in history. A great deal of the content of this book focuses on discoveries from this period of time, so instead of providing a commentary in this overview, we will simply list the references to this period in world history. Other sources and links are found on our Web page **www.mathnature.com.**

Age of Reason: 1600 to 1699

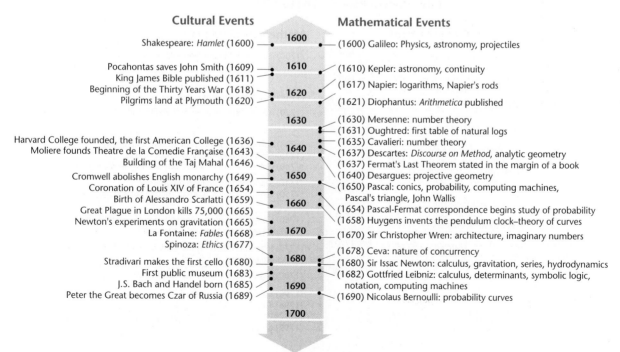

Cultural Events

Shakespeare: *Hamlet* (1600)
Pocahontas saves John Smith (1609)
King James Bible published (1611)
Beginning of the Thirty Years War (1618)
Pilgrims land at Plymouth (1620)

Harvard College founded, the first American College (1636)
Moliere founds Theatre de la Comedie Française (1643)
Building of the Taj Mahal (1646)
Cromwell abolishes English monarchy (1649)
Coronation of Louis XIV of France (1654)
Birth of Alessandro Scarlatti (1659)
Great Plague in London kills 75,000 (1665)
Newton's experiments on gravitation (1665)
La Fontaine: *Fables* (1668)
Spinoza: *Ethics* (1677)

Stradivari makes the first cello (1680)
First public museum (1683)
J.S. Bach and Handel born (1685)
Peter the Great becomes Czar of Russia (1689)

Mathematical Events

(1600) Galileo: Physics, astronomy, projectiles
(1610) Kepler: astronomy, continuity
(1617) Napier: logarithms, Napier's rods
(1621) Diophantus: *Arithmetica* published
(1630) Mersenne: number theory
(1631) Oughtred: first table of natural logs
(1635) Cavalieri: number theory
(1637) Descartes: *Discourse on Method*, analytic geometry
(1637) Fermat's Last Theorem stated in the margin of a book
(1640) Desargues: projective geometry
(1650) Pascal: conics, probability, computing machines, Pascal's triangle, John Wallis
(1654) Pascal-Fermat correspondence begins study of probability
(1658) Huygens invents the pendulum clock–theory of curves
(1670) Sir Christopher Wren: architecture, imaginary numbers
(1678) Ceva: nature of concurrency
(1680) Sir Issac Newton: calculus, gravitation, series, hydrodynamics
(1682) Gottfried Leibniz: calculus, determinants, symbolic logic, notation, computing machines
(1690) Nicolaus Bernoulli: probability curves

1600
1610
1620
1630
1640
1650
1660
1670
1680
1690
1700

Early Modern Period: 1700 to 1799

This period marks the dawn of modern mathematics. The Early Modern Period was characterized by experimentation and formalization of the ideas germinated in the previous century. There is so much that we could say about the period from 1700 to 1799. The mathematics that you studied in high school represents, for the most part, the ideas formulated during this period. Take a look at the mathematical events in the following timeline, and you will see an abundance of discoveries, often embodied in the contents of entire books. One of the best sources of information about this period is found at these Web sites:

www.mathnature.com and www-history.mcs.st-and.ac.uk/ ~history/Indexes/Full_Chron.html

There are a multitude of historical references to this period documented throughout the book.

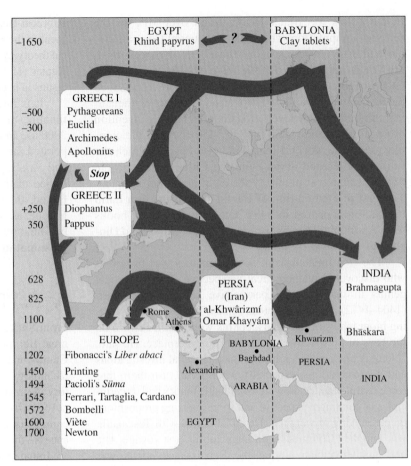

FIGURE 4 Mainstreams in the flow of algebra

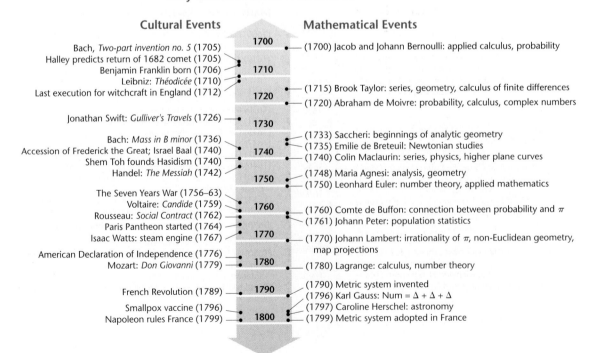

Early Modern Period: 1700 to 1799

Cultural Events		Mathematical Events
Bach, *Two-part invention no. 5* (1705)	**1700**	(1700) Jacob and Johann Bernoulli: applied calculus, probability
Halley predicts return of 1682 comet (1705)		
Benjamin Franklin born (1706)	**1710**	
Leibniz: *Théodicée* (1710)		
Last execution for witchcraft in England (1712)	**1720**	(1715) Brook Taylor: series, geometry, calculus of finite differences
		(1720) Abraham de Moivre: probability, calculus, complex numbers
Jonathan Swift: *Gulliver's Travels* (1726)	**1730**	
Bach: *Mass in B minor* (1736)		(1733) Saccheri: beginnings of analytic geometry
Accession of Frederick the Great; Israel Baal (1740)	**1740**	(1735) Emilie de Breteuil: Newtonian studies
Shem Toh founds Hasidism (1740)		(1740) Colin Maclaurin: series, physics, higher plane curves
Handel: *The Messiah* (1742)	**1750**	(1748) Maria Agnesi: analysis, geometry
		(1750) Leonhard Euler: number theory, applied mathematics
The Seven Years War (1756–63)		
Voltaire: *Candide* (1759)	**1760**	(1760) Comte de Buffon: connection between probability and π
Rousseau: *Social Contract* (1762)		(1761) Johann Peter: population statistics
Paris Pantheon started (1764)	**1770**	
Isaac Watts: steam engine (1767)		(1770) Johann Lambert: irrationality of π, non-Euclidean geometry, map projections
American Declaration of Independence (1776)	**1780**	
Mozart: *Don Giovanni* (1779)		(1780) Lagrange: calculus, number theory
		(1790) Metric system invented
French Revolution (1789)	**1790**	(1796) Karl Gauss: Num $= \Delta + \Delta + \Delta$
Smallpox vaccine (1796)		(1797) Caroline Herschel: astronomy
Napoleon rules France (1799)	**1800**	(1799) Metric system adopted in France

Modern Period: 1800 to Present

What we call the Modern Period includes all of the discoveries of the last two centuries. Students often think that all the important mathematics has been done, and there is nothing new to be discovered, but this is not true. Mathematics is alive and constantly changing. There is no way a short

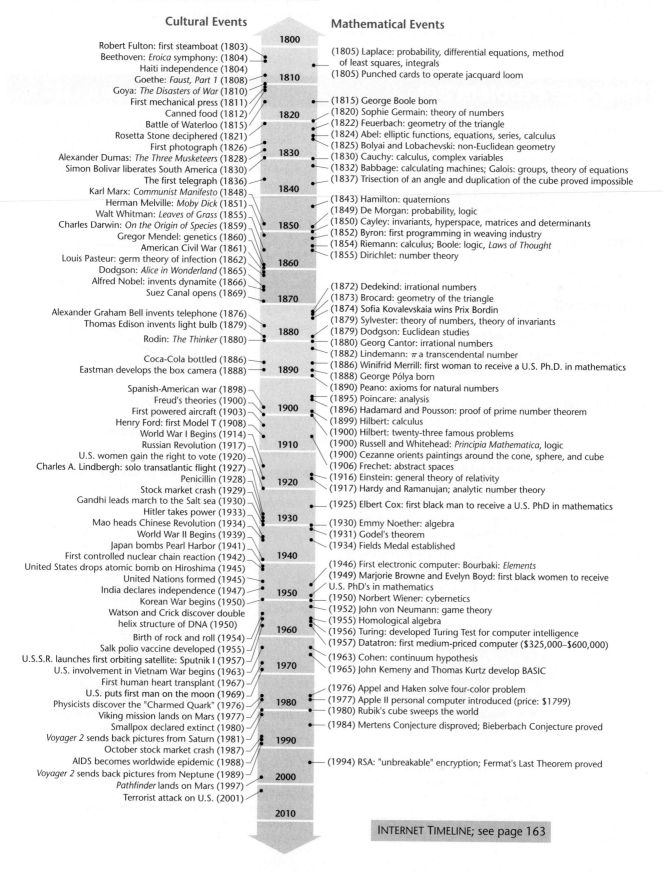

Cultural Events

Robert Fulton: first steamboat (1803)
Beethoven: *Eroica* symphony: (1804)
Haiti independence (1804)
Goethe: *Faust, Part 1* (1808)
Goya: *The Disasters of War* (1810)
First mechanical press (1811)
Canned food (1812)
Battle of Waterloo (1815)
Rosetta Stone deciphered (1821)
First photograph (1826)
Alexander Dumas: *The Three Musketeers* (1828)
Simon Bolivar liberates South America (1830)
The first telegraph (1836)
Karl Marx: *Communist Manifesto* (1848)
Herman Melville: *Moby Dick* (1851)
Walt Whitman: *Leaves of Grass* (1855)
Charles Darwin: *On the Origin of Species* (1859)
Gregor Mendel: genetics (1860)
American Civil War (1861)
Louis Pasteur: germ theory of infection (1862)
Dodgson: *Alice in Wonderland* (1865)
Alfred Nobel: invents dynamite (1866)
Suez Canal opens (1869)
Alexander Graham Bell invents telephone (1876)
Thomas Edison invents light bulb (1879)
Rodin: *The Thinker* (1880)
Coca-Cola bottled (1886)
Eastman develops the box camera (1888)
Spanish-American war (1898)
Freud's theories (1900)
First powered aircraft (1903)
Henry Ford: first Model T (1908)
World War I Begins (1914)
Russian Revolution (1917)
U.S. women gain the right to vote (1920)
Charles A. Lindbergh: solo transatlantic flight (1927)
Penicillin (1928)
Stock market crash (1929)
Gandhi leads march to the Salt sea (1930)
Hitler takes power (1933)
Mao heads Chinese Revolution (1934)
World War II Begins (1939)
Japan bombs Pearl Harbor (1941)
First controlled nuclear chain reaction (1942)
United States drops atomic bomb on Hiroshima (1945)
United Nations formed (1945)
India declares independence (1947)
Korean War begins (1950)
Watson and Crick discover double helix structure of DNA (1950)
Birth of rock and roll (1954)
Salk polio vaccine developed (1955)
U.S.S.R. launches first orbiting satellite: Sputnik I (1957)
U.S. involvement in Vietnam War begins (1963)
First human heart transplant (1967)
U.S. puts first man on the moon (1969)
Physicists discover the "Charmed Quark" (1976)
Viking mission lands on Mars (1977)
Smallpox declared extinct (1980)
Voyager 2 sends back pictures from Saturn (1981)
October stock market crash (1987)
AIDS becomes worldwide epidemic (1988)
Voyager 2 sends back pictures from Neptune (1989)
Pathfinder lands on Mars (1997)
Terrorist attack on U.S. (2001)

1800
1810
1820
1830
1840
1850
1860
1870
1880
1890
1900
1910
1920
1930
1940
1950
1960
1970
1980
1990
2000
2010

Mathematical Events

(1805) Laplace: probability, differential equations, method of least squares, integrals
(1805) Punched cards to operate jacquard loom
(1815) George Boole born
(1820) Sophie Germain: theory of numbers
(1822) Feuerbach: geometry of the triangle
(1824) Abel: elliptic functions, equations, series, calculus
(1825) Bolyai and Lobachevski: non-Euclidean geometry
(1830) Cauchy: calculus, complex variables
(1832) Babbage: calculating machines; Galois: groups, theory of equations
(1837) Trisection of an angle and duplication of the cube proved impossible
(1843) Hamilton: quaternions
(1849) De Morgan: probability, logic
(1850) Cayley: invariants, hyperspace, matrices and determinants
(1852) Byron: first programming in weaving industry
(1854) Riemann: calculus; Boole: logic, *Laws of Thought*
(1855) Dirichlet: number theory
(1872) Dedekind: irrational numbers
(1873) Brocard: geometry of the triangle
(1874) Sofia Kovalevskaia wins Prix Bordin
(1879) Sylvester: theory of numbers, theory of invariants
(1879) Dodgson: Euclidean studies
(1880) Georg Cantor: irrational numbers
(1882) Lindemann: π a transcendental number
(1886) Winifrid Merrill: first woman to receive a U.S. Ph.D. in mathematics
(1888) George Pólya born
(1890) Peano: axioms for natural numbers
(1895) Poincare: analysis
(1896) Hadamard and Pousson: proof of prime number theorem
(1899) Hilbert: calculus
(1900) Hilbert: twenty-three famous problems
(1900) Russell and Whitehead: *Principia Mathematica*, logic
(1900) Cezanne orients paintings around the cone, sphere, and cube
(1906) Frechet: abstract spaces
(1916) Einstein: general theory of relativity
(1917) Hardy and Ramanujan; analytic number theory
(1925) Elbert Cox: first black man to receive a U.S. PhD in mathematics
(1930) Emmy Noether: algebra
(1931) Godel's theorem
(1934) Fields Medal established
(1946) First electronic computer: Bourbaki: *Elements*
(1949) Marjorie Browne and Evelyn Boyd: first black women to receive U.S. PhD's in mathematics
(1950) Norbert Wiener: cybernetics
(1952) John von Neumann: game theory
(1955) Homological algebra
(1956) Turing: developed Turing Test for computer intelligence
(1957) Datatron: first medium-priced computer ($325,000–$600,000)
(1963) Cohen: continuum hypothesis
(1965) John Kemeny and Thomas Kurtz develop BASIC
(1976) Appel and Haken solve four-color problem
(1977) Apple II personal computer introduced (price: $1799)
(1980) Rubik's cube sweeps the world
(1984) Mertens Conjecture disproved; Bieberbach Conjecture proved
(1994) RSA: "unbreakable" encryption; Fermat's Last Theorem proved

INTERNET TIMELINE; see page 163

commentary or overview can convey the richness or implications of the mathematical discoveries of this period. As we enter the new millennium, we can only imagine and dream about what is to come!

One of the major themes of this text is problem solving. The following problem set is a potpourri of problems that should give you a foretaste of the variety of ideas and concepts that we will consider in this book. Although none of these problems is to be considered routine, you might wish to attempt to work some of them before you begin, and then return to these problems at the end of your study in this book.

Prologue Problem Set

1. **HISTORICAL QUEST**   What are the seven chronological periods into which the prologue divided history? Which period seems the most interesting to you, and why?

2. **HISTORICAL QUEST**   Select what you believe to be the most interesting cultural event and the most interesting mathematical event of the Babylonian, Egyptian, and Native American Period.

3. **HISTORICAL QUEST**   Select what you believe to be the most interesting cultural event and the most interesting mathematical event of the Greek, Chinese, and Roman Period.

4. **HISTORICAL QUEST**   Select what you believe to be the most interesting cultural event and the most interesting mathematical event of the Hindu and Persian Period.

5. **HISTORICAL QUEST**   Select what you believe to be the most interesting cultural event and the most interesting mathematical event of the Transition Period.

6. **HISTORICAL QUEST**   Select what you believe to be the most interesting cultural event and the most interesting mathematical event of the Age of Reason.

7. **HISTORICAL QUEST**   Select what you believe to be the most interesting cultural event and the most interesting mathematical event of the Early Modern Period.

8. **HISTORICAL QUEST**   Select what you believe to be the most interesting cultural event and the most interesting mathematical event of the Modern Period.

9. A long, straight fence having a pole every 8 ft is 1,440 ft long. How many fence poles are needed for the fence?

10. How many cards must you draw from a deck of 52 playing cards to be sure that at least two are from the same suit?

11. How many people must be in a room to be sure that at least four of them have the same birthday (not necessarily the same year)?

12. Find the units digit of $3^{2007} - 2^{2007}$.

13. If a year had two consecutive months with a Friday the thirteenth, which months would they have to be?

14. On Saturday evenings, a favorite pastime of the high school students in Santa Rosa, California, is to cruise certain streets. The selected routes are shown in the following illustration. Is it possible to choose a route so that all of the permitted streets are traveled exactly once?

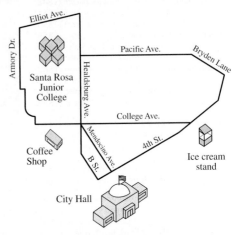

Santa Rosa street problem

15. What is the largest number that is a divisor of both 210 and 330?

16. The News Clip shows a letter printed in the "Ask Marilyn" column of *Parade* magazine (Sept. 27, 1992). How would you answer it?

> **Dear Marilyn,**
>
> I recently purchased a tube of caulking and it says a 1/4-inch bead will yield about 30 feet. But it says a 1/8-inch bead will yield about 96 feet — more than three times as much. I'm not a math genius, but it seems that because 1/8 inch is half of 1/4 inch, the smaller bead should yield only twice as much. Can you explain it?
>
> Norm Bean, St. Louis, Mo.

Hint: We won't give you the answer, but we will quote one line from Marilyn's answer: "So the question should be not why the smaller one yields that much, but why it yields that little."

17. If the population of the world on October 12, 2002 was 6.248 billion, when do you think the world population will reach 7 billion? Calculate the date (to the nearest month) using the information that the world population reached 6 billion on October 12, 1999.

18. The Pacific 12 football conference consists of the following schools:

Arizona
Arizona State
Cal Berkeley
Colorado
Oregon
Oregon State
Stanford (CA)
UCLA
USC
Utah
Washington
Washington State

a. Is it possible to visit each of these schools by crossing each common state border exactly once? If so, show the path.
b. Is it possible to start the trip in any given state, cross each common state border exactly once, and end the trip in the state in which you started?

19. If $(a, b) = a \times b + a + b$, what is the value of $((1, 2), (3, 4))$?

20. If it is known that all Angelenos are Venusians and all Venusians are Los Angeles residents, then what must necessarily be the conclusion?

21. If 1 is the first odd number, what is the 473rd odd number?

22. If $1 + 2 + 3 + \cdots + n = \frac{n(n + 1)}{2}$, what is the sum of the first 100,000 counting numbers beginning with 1?

23. A four-inch cube is painted red on all sides. It is then cut into one-inch cubes. What fraction of all the one-inch cubes are painted on one side only?

24. If slot machines had two arms and people had one arm, then it is probable that our number system would be based on the digits 0, 1, 2, 3, and 4 only. How would the number we know as 18 be written in such a number system?

25. If $M(a, b)$ stands for the larger number in the parentheses, and $m(a, b)$ stands for the lesser number in the parentheses, what is the value of $M(m(1, 2), m(2, 3))$?

26. If a group of 50 persons consists of 20 males, 12 children, and 25 women, how many men are in the group?

27. There are only five regular polyhedra, and Figure 5 shows the patterns that give those polyhedra. Name the polyhedron obtained from each of the patterns shown.

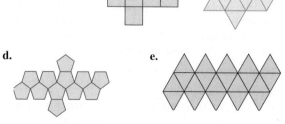

a.   b.   c.

d.   e.

FIGURE 5 Five regular polyhedra patterns

28. Jack and Jill decided to exercise together. Jack walks around their favorite lake in 16 minutes and Jill jogs around the lake

in 10 minutes. If Jack and Jill start at the same time and at the same place, and continue to exercise around the lake until they return to the starting point at the same time, how long will they be exercising?

29. What is the 1,000th positive integer that is not divisible by 3?

30. A frugal man allows himself a glass of wine before dinner on every third day, an after-dinner chocolate every fifth day, and a steak dinner once a week. If it happens that he enjoys all three luxuries on March 31, what will be the date of the next steak dinner that is preceded by wine and followed by an after-dinner chocolate?

31. How many trees must be cut to make a trillion one-dollar bills? To answer this question you need to make some assumptions. Assume that a pound of paper is equal to a pound of wood, and also assume that a dollar bill weighs about one gram. This implies that a pound of wood yields about 450 dollar bills. Furthermore, estimate that an average tree has a height of 50 ft and a diameter of 12 inches. Finally, assume that wood yields about 50 lb/ft³.

32. Estimate the volume of beer in the six-pack shown in the photograph.

33. You are given a square with sides equal to 8 inches, with two inscribed semicircles of radius 4. What is the area of the shaded region?

34. Critique the statement given in the News Clip.

> **Smoking ban**
> Judy Green, owner of the White Restaurant and an adamant opponent of a smoking ban, went so far as to survey numerous restaurants. She cited one restaurant that suffered a 75% decline in business after the smoking ban was activated.

35. The two small circles have radii of 2 and 3. Find the ratio of the area of the smallest circle to the area of the shaded region.

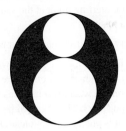

36. A large container filled with water is to be drained, and you would like to drain it as quickly as possible. You can drain the container with either one 1-in. diameter hose or two $\frac{1}{2}$-in.-diameter hoses. Which do you think would be faster (one 1-in. drain or two $\frac{1}{2}$-in. drains), and why?

37. A gambler went to the horse races two days in a row. On the first day, she doubled her money and spent $30. On the second day, she tripled her money and spent $20, after which she had what she started with the first day. How much did she start with?

38. The map shows the percent of children age 19–35 months who are immunized by the state. What conclusions can you draw from this map?

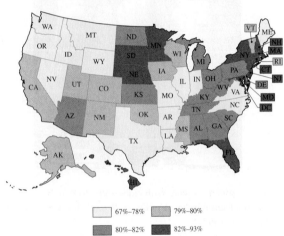

	67%–78%		79%–80%
	80%–82%		82%–93%

Source: Centers for Disease Control
www.cdc.gov/Vaccines/states-surv/his/tables/07/tab03_atigen_state.xls.

39. A charter flight has signed up 100 travelers. The travelers are told that if they can sign up an additional 25 persons, they can save $78 each. What is the cost per person if 100 persons make the trip?

40. Find $\lim_{n\to\infty}\left(1 + \frac{1}{n}\right)^n$.

41. Suppose that it costs $450 to enroll your child in a 10-week summer recreational program. If this cost is prorated (that is, reduced linearly over the 10-week period), express the cost as a function of the number of weeks that have elapsed since the start of the 10-week session. Draw a graph to show the cost at any time for the duration of the session.

42. Candidates Rameriz (R), Smith (S), and Tillem (T) are running for office. According to public opinion polls, the preferences are (percentages rounded to the nearest percent):

Ranking	38%	29%	24%	10%
1st choice	R	S	T	R
2nd choice	S	R	S	T
3rd choice	T	T	R	S

a. Who will win the plurality vote?
b. Who will win Borda count?
c. Does a strategy exist that the voters in the 24% column could use to vote insincerely to keep Rameriz from winning?

43. Suppose the percentage of alcohol in the blood t hours after consumption is given by

$$C(t) = 0.3e^{-t/2}$$

What is the rate at which the percentage of alcohol is changing with respect to time?

44. If a megamile is one million miles and a kilomile is one thousand miles, how many kilomiles are there in 2.376 megamiles?

45. A map of a small village is shown in Figure 6. To walk from A to B, Sarah obviously must walk at least 7 blocks (all the blocks are the same length). What is the number of shortest paths from A to B?

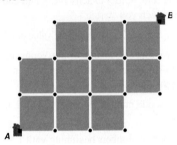

FIGURE 6 A village map

46. A hospital wishes to provide for its patients a diet that has a minimum of 100 g of carbohydrates, 60 g of protein, and 40 g of fats per day. These requirements can be met with two foods:

Food	Carbohydrates	Protein	Fats
A	6 g/oz	3 g/oz	1 g/oz
B	2 g/oz	2 g/oz	2 g/oz

It is also important to minimize costs; food A costs $0.14 per ounce and food B costs $0.06 per ounce. How many ounces of each food should be bought for each patient per day to meet the minimum daily requirements at the lowest cost?

47. On July 24, 2010, the U.S. national debt was $13 trillion and on that date there were 308.1 million people. How long would it take to pay off this debt if *every* person pays $1 per day?

48. Find the smallest number of operations needed to build up to the number 100 if you start at 0 and use only two operations: doubling or increasing by 1. *Challenge*: Answer the same question for the positive integer n.

49. If $\log_2 x + \log_4 x = \log_b x$, what is b?

50. Supply the missing number in the following sequence: 10, 11, 12, 13, 14, 15, 16, 17, 20, 22, 24, _____, 100, 121, 10,000.

51. How many different configurations can you see in Figure 7?

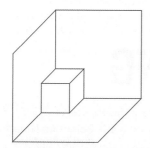

FIGURE 7 Count the cubes

52. Answer the question asked in the News Clip from the "Ask Marilyn" column of *Parade* magazine (July 16, 1995).

Dear Marilyn,

Three safari hunters are captured by a sadistic tribe of natives and forced to participate in a duel to the death. Each is given a pistol and tied to a post the same distance from the other two. They must take turns shooting at each other, one shot per turn. The worst shot of the three hunters (1 in 3 accuracy) must shoot first. The second turn goes to the hunter with 50–50 (1 in 2) accuracy. And (if he's still alive!) the third turn goes to the crack shot (100% accuracy). The rotation continues until only one hunter remains, who is then rewarded with his freedom. Which hunter has the best chance of surviving, and why?

From "Ask Marilyn," by Marilyn vos Savant, *Parade Magazine*, July 16, 1992. Reprinted with permission from Parade, © 1995.

53. Five cards are drawn at random from a pack of cards that have been numbered consecutively from 1 to 104 and have been thoroughly shuffled. What is the probability that the numbers on the cards as drawn are in increasing order of magnitude?

54. What is the sum of the counting numbers from 1 to 104?

55. The Kabbalah is a body of mystical teachings from the Torah. One medieval inscription is shown at the left:

ד	ט	ב
ג	ה	ז
ח	א	ו

4	9	2
3	5	7
8	1	6

The inscription on the left shows Hebrew characters that can be translated into numbers, as shown at the right. What can you say about this pattern of numbers?

56. What is the maximum number of points of intersection of n distinct lines?

57. The equation $P = 153,000e^{0.05t}$ represents the population of a city t years after 2000. What is the population in the year 2000? Show a graph of the city's population for the next 20 years.

58. The Egyptians had an interesting, pictorial numeration system. Here is how you would count using Egyptian numerals:

$$|,\ ||,\ |||,||||,\ \ ||||,\ \ |||||,\ \ ||||||,\ \ |||||||,\ \ ||||||||,$$
$$\cap,\ \cap|,\ \cap||,\ \cap|||,\ \ldots$$

Write down your age using Egyptian numerals. The symbol "|" is called a stroke, and the "$\cap$" is called a heel bone. The Egyptians used a scroll for 100, a lotus flower for 1,000, a pointing finger for 10,000, a polliwog for 100,000, and an astonished man for the number 1,000,000. *Without* doing any research, write what you think today's date would look like using Egyptian numerals.

59. If you start with $1 and double your money each day, how much money would you have in 30 days?

60. Consider two experiments and events defined as follows:

Experiment A: Roll one die 4 times and keep a record of how many times you obtain at least one 6. Event $E = \{$obtain at least one 6 in 4 rolls of a single die$\}$

Experiment B: Roll a pair of dice 24 times and keep a record of how many times you obtain at least one 12. Event $F = \{$obtain at least one 12 in 24 rolls of a pair of dice$\}$

Do you think event E or event F is more likely? You might wish to experiment by rolling dice.

1

THE NATURE OF PROBLEM SOLVING

What in the World?

"Hey, Tom, what are you taking this semester?" asked Susan. "I'm taking English, history, and math. I can't believe my math teacher," responded Tom. "The first day we were there, she walked in, wrote her name on the board, and then she asked, 'How much space would you have if you, along with everyone else in the world, moved to California?' What a stupid question ... I would not have enough room to turn around!"

"Oh, I had that math class last semester," said Susan. "It isn't so bad. The whole idea is to give you the ability to solve problems *outside* the class. I want to get a good job when I graduate, and I've read that because of the economy, employers are looking for people with problem-solving skills. I hear that working smarter is more important than working harder."

Overview

There are many reasons for reading a book, but the best reason is because you want to read it. Although you are probably reading this first page because you were required to do so by your instructor, it is my hope that in a short while you will be reading this book because you *want* to read it. It was written for people who think they don't like mathematics, or people who think they can't work math problems, or people who think they are never going to use math. The common thread in this book is *problem solving*—that is, strengthening your ability to solve problems—not in the classroom, but outside the classroom. This first chapter is designed to introduce you to the nature of problem solving. Notice the first thing you see on this page is the question, "What in the world?" Each chapter begins with such a real world question that appears later in the chapter. This first one is considered in Problem 59, page 43.

As you begin your trip through this book, I wish you a BON VOYAGE!

1.1 | Problem Solving

The idea that aptitude for mathematics is rarer than aptitude for other subjects is merely an illusion which is caused by belated or neglected beginners.

J.F. HERBART

CHAPTER **CHALLENGE**	
At the beginning of each chapter we present a puzzle which represents some pattern. See if you can fill in the question mark.	$A + B = C$ $A + C = D$ $B + C = E$ $F + H = N$ $G + J = ?$

A Word of Encouragement

Do you think of mathematics as a difficult, foreboding subject that was invented hundreds of years ago? Do you think that you will never be able (or even want) to use mathematics? If you answered "yes" to either of these questions, then I want you to know that I have written this book for you. I have tried to give you some insight into how mathematics is developed and to introduce you to some of the people behind the mathematics. In this book, I will present some of the great ideas of mathematics, and then we will look at how these ideas can be used in an everyday setting to build your problem-solving abilities. *The most important prerequisite for this course is an openness to try out new ideas—a willingness to experience the suggested activities rather than to sit on the sideline as a spectator.* I have attempted to make this material interesting by putting it together differently from the way you might have had mathematics presented in the past. You will find this book difficult if you wait for the book or the teacher to give you answers—instead, *be willing to guess, experiment, estimate, and manipulate,* and try out problems *without fear of being wrong!*

There is a common belief that mathematics is to be pursued only in a clear-cut logical fashion. This belief is perpetuated by the way mathematics is presented in most textbooks. Often it is reduced to a series of definitions, methods to solve various types of problems, and theorems. These theorems are justified by means of proofs and deductive reasoning. I do not mean to minimize the importance of proof in mathematics, for it is the very thing that gives mathematics its strength. But the power of the imagination is every bit as important as the power of deductive reasoning. As the mathematician Augustus De Morgan once said, "The power of mathematical invention is not reasoning but imagination."

Hints for Success

Mathematics is different from other subjects. One topic builds upon another, and you need to make sure that you understand *each* topic before progressing to the next one.

You must make a commitment to attend each class. Obviously, unforeseen circumstances can come up, but you must plan to attend class regularly. Pay attention to what

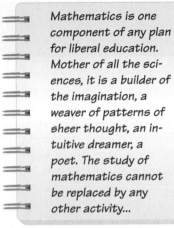

Mathematics is one component of any plan for liberal education. Mother of all the sciences, it is a builder of the imagination, a weaver of patterns of sheer thought, an intuitive dreamer, a poet. The study of mathematics cannot be replaced by any other activity...

your teacher says and does, and take notes. If you must miss class, write an outline of the text corresponding to the missed material, including working out each text example on your notebook paper.

You must make a commitment to daily work. Do not expect to save up and do your mathematics work once or twice a week. It will take a daily commitment on your part, and you will find mathematics difficult if you try to "get it done" in spurts. You could not expect to become proficient in tennis, soccer, or playing the piano by practicing once a week, and the same is true of mathematics. Try to schedule a regular time to study mathematics each day.

Read the text carefully. Many students expect to get through a mathematics course by beginning with the homework problems, then reading some examples, and reading the text only as a desperate attempt to find an answer. This procedure is backward; do your homework only *after* reading the text.

Writing Mathematics

The fundamental objective of education has always been to prepare students for life. A measure of your success with this book is a measure of its usefulness to you in your life. What are the basics for your knowledge "in life"? In this information age with access to a world of knowledge on the Internet, we still would respond by saying that the basics remain "reading, 'riting, and 'rithmetic." As you progress through the material in this book, we will give you opportunities to read mathematics and to consider some of the great ideas in the history of civilization, to develop your problem-solving skills ('rithmetic), and to communicate mathematical ideas to others ('riting). Perhaps you think of mathematics as "working problems" and "getting answers," but it is so much more. Mathematics is a way of thought that includes all three Rs, and to strengthen your skills you will be asked to communicate your knowledge in written form.

Journals

To begin building your skills in writing mathematics, you might keep a journal summarizing each day's work. Keep a record of your feelings and perceptions about what happened in class. Ask yourself, "How long did the homework take?" "What time of the day or night did I spend working and studying mathematics?" "What is the most important idea that I should remember from the day's lesson?" To help you with your journals or writing of mathematics, you will find problems in this text designated **"IN YOUR OWN WORDS."** (For example, look at Problems 1–4 of the problem set at the end of this section.) There are no right answers or wrong answers to this type of problem, but you are encouraged to look at these for ideas of what you might write in your journal.

Journal Ideas

Write in your journal every day.
Include important ideas.
Include new words, ideas, formulas, or concepts.
Include questions that you want to ask later.
If possible, carry your journal with you so you can write in it anytime you get an idea.

Reasons for Keeping a Journal

It will record ideas you might otherwise forget.
It will keep a record of your progress.
If you have trouble later, it may help you diagnose areas for change or improvement.
It will build your writing skills.

Individual Research

At the end of each chapter you will find problems requiring some library research. I hope that as you progress through the course you will find one or more topics that interest you so much that you will want to do additional reading on that topic, even if it is not assigned.

Your instructor may assign one or more of these as term papers. One of the best ways for you to become aware of all the books and periodicals that are available is to log onto the Internet, or visit the library to research specific topics.

Preparing a mathematics paper or project can give you interesting and worthwhile experiences. In preparing a paper or project, you will get experience in using resources to find information, in doing independent work, in organizing your presentation, and in communicating ideas orally, in writing, and in visual demonstrations. You will broaden your background in mathematics and encounter new mathematical topics that you never before knew existed. In setting up an exhibit you will experience the satisfaction of demonstrating what you have accomplished. It may be a way of satisfying your curiosity and your desire to be creative. It is an opportunity for developing originality, craftsmanship, and new mathematical understandings. If you are requested to do some individual research problems, here are some suggestions.

1. *Select a topic that has interest potential.* Do not do a project on a topic that does not interest you. Suggestions are given on the Web at **www.mathnature.com.**

2. *Find as much information about the topic as possible.* Many of the Individual Research problems have one or two references to get you started. In addition, check the following sources:

   Periodicals: *The Mathematics Teacher, Teaching Children Mathematics* (formerly *Arithmetic Teacher),* and *Scientific American;* each of these has its own cumulative index; also check the *Reader's Guide.*

   Source books: *The World of Mathematics* by James R. Newman is a gold mine of ideas. *Mathematics,* a Time-Life book by David Bergamini, may provide you with many ideas. Encyclopedias can be consulted after you have some project ideas; however, I do not have in mind that the term project necessarily be a term paper.

   Internet: Use one or more search engines on the Internet for information on a particular topic. The more specific you can be in describing what you are looking for, the better the engine will be able to find material on your topic. The most widely used search engine is Google, but there are others that you might use. You may also check the Web address for this book to find specific computer links:

 www.mathnature.com

   If you do not have a computer or a modem, then you may need to visit your college or local library for access to this research information.

3. *Prepare and organize your material into a concise, interesting report.* Include drawings in color, pictures, applications, and examples to get the reader's attention and add meaning to your report.

4. *Build an exhibit that will tell the story of your topic.* Remember the science projects in high school? That type of presentation might be appropriate. Use models, applications, and charts that lend variety. Give your paper or exhibit a catchy, descriptive title.

5. *A **term** project cannot be done in one or two evenings.*

Group Research

Working in small groups is typical of most work environments, and being able to work with others to communicate specific ideas is an important skill to learn. At the end of each chapter is a list of suggested group projects, and you are encouraged to work with three or four others to submit a single report.

Guidelines for Problem Solving

We begin this study of **problem solving** by looking at the *process* of problem solving. As a mathematics teacher, I often hear the comment, "I can do mathematics, but I can't solve word problems." There *is* a great fear and avoidance of "real-life" problems because they do not fit into the same mold as the "examples in the book." Few practical problems from everyday life come in the same form as those you study in school.

To compound the difficulty, learning to solve problems takes time. All too often, the mathematics curriculum is so packed with content that the real process of problem solving is slighted and, because of time limitations, becomes an exercise in mimicking the instructor's steps instead of developing into an approach that can be used long after the final examination is over.

Before we build problem-solving skills, it is necessary to build certain prerequisite skills necessary for problem solving. It is my goal to develop your skills in the mechanics of mathematics, in understanding the important concepts, and finally in applying those skills to solve a new type of problem. I have segregated the problems in this book to help you build these different skills:

IN YOUR OWN WORDS	This type of problem asks you to discuss or rephrase main ideas or procedures using your own words.
Level 1 Problems	These are mechanical and drill problems, and are directly related to an example in the book.
Level 2 Problems	These problems require an understanding of the concepts and are loosely related to an example in the book.
Level 3 Problems	These problems are extensions of the examples, but generally do not have corresponding examples.
Problem Solving	These require problem-solving skills or original thinking and generally do not have direct examples in the book. These should be considered Level 3 problems.
Research Problems	These problems require Internet research or library work. Most are intended for individual research but a few are group research projects. You will find these problems for research in the chapter summary and at the Web address for this book:

www.mathnature.com

The model for problem solving that we will use was first published in 1945 by the great, charismatic mathematician George Pólya. His book *How to Solve It* (Princeton University Press, 1973) has become a classic. In Pólya's book you will find this problem-solving model as well as a treasure trove of strategy, know-how, rules of thumb, good advice, anecdotes, history, and problems at all levels of mathematics. His problem-solving model is as follows.

Guidelines for Problem Solving

Step 1   *Understand the problem.* Ask questions, experiment, or otherwise rephrase the question in your own words.

Step 2   *Devise a plan.* Find the connection between the data and the unknown. Look for patterns, relate to a previously solved problem or a known formula, or simplify the given information to give you an easier problem.

Step 3   *Carry out the plan.* Check the steps as you go.

Step 4   *Look back.* Examine the solution obtained. In other words, check your answer.

Pólya's original statement of this procedure is reprinted in the following box.*

UNDERSTANDING THE PROBLEM

First

You have to understand the problem.

What is the unknown? What are the data? What is the condition? Is it possible to satisfy the condition? Is the condition sufficient to determine the unknown? Or is it insufficient? Or redundant? Or contradictory?

Draw a figure. Introduce a suitable notation.

Separate the various parts of the condition. Can you write them down?

DEVISING A PLAN

Second

Find the connection between the data and the unknown. You may be obliged to consider auxiliary problems if an immediate connection cannot be found.

Have you seen it before? Or have you seen the same problem in a slightly different form?

Do you know a related problem? Do you know a theorem that could be useful?

Look at the unknown! And try to think of a familiar problem having the same or a similar unknown.

Is the problem related to one you have solved before? Could you use it?

Could you use its result? Could you use its method? Should you introduce some auxiliary element in order to make its use possible?

Could you restate the problem? Could you restate it still differently? Go back to definitions.

If you cannot solve the proposed problem try to solve first some related problem. Could you imagine a more accessible related problem? A more general problem? A more special problem? An analogous problem? Could you solve a part of the problem? Keep only a part of the condition, drop the other part; how far is the unknown then determined, how can it vary? Could you derive something useful from the data? Could you think of other data appropriate to determine the unknown? Could you change the unknown or the data, or both if necessary, so that the new unknown and the new data are nearer to each other? Did you use all the data? Did you see the whole condition? Have you taken into account all essential notions involved in the problem?

CARRYING OUT THE PLAN

Third

Carry out your plan.

Carrying out your plan of the solution, *check each step*. Can you see clearly that the step is correct? Can you prove that it is correct?

LOOKING BACK

Fourth

Examine the solution.

Can you *check the result*? Can you check the argument?

Can you derive the result differently? Can you see it at a glance?

Let's apply this procedure for problem solving to the map shown in Figure 1.1; we refer to this problem as the **street problem.** Melissa lives at the YWCA (point *A*) and works at Macy's (point *B*). She usually walks to work. How many different routes can Melissa take?

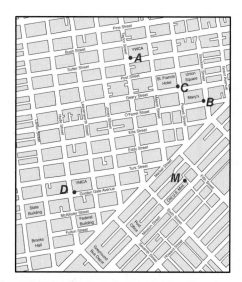

FIGURE 1.1 Portion of a map of San Francisco

*This is taken word for word as it was written by Pólya in 1941. It was printed in *How to Solve It* (Princeton, NJ: Princeton University Press, 1973).

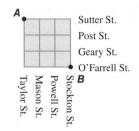

FIGURE 1.2 Simplified portion of Figure 1.1

Where would you begin with this problem?

Step 1   *Understand the Problem.* Can you restate it in your own words? Can you trace out one or two possible paths? What assumptions are reasonable? We assume that Melissa will not do any backtracking—that is, she always travels toward her destination. We also assume that she travels along the city streets—she cannot cut diagonally across a lot or a block.

Step 2   *Devise a Plan.* Simplify the question asked. Consider the simplified drawing shown in Figure 1.2.

Step 3   *Carry Out the Plan.* Count the number of ways it is possible to arrive at each point, or, as it is sometimes called, a *vertex.*

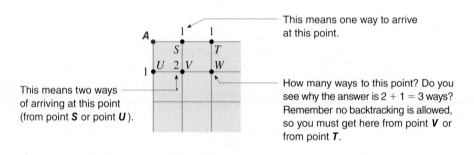

Now fill in all the possibilities on Figure 1.3, as shown by the above procedure.

Step 4   *Look Back.* Does the answer "20 different routes" make sense? Do you think you could fill in all of them?

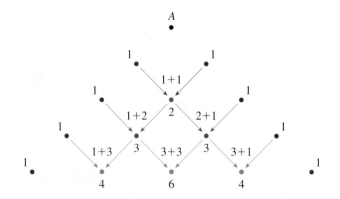

FIGURE 1.3 Map with solution

Example **1** **Problem solving—from here to there**

In how many different ways could Melissa get from the YWCA (point *A*) to the St. Francis Hotel (point *C* in Figure 1.1), using the method of Figure 1.3?

Solution   Draw a simplified version of Figure 1.3, as shown. There are 6 different paths.

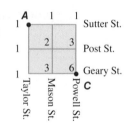

Problem Solving by Patterns

Let's formulate a general solution. Consider a map with a starting point *A*:

In the Sherlock Holmes mystery *The Final Solution*, Moriarty is a mathematician who wrote a treatise on Pascal's triangle.

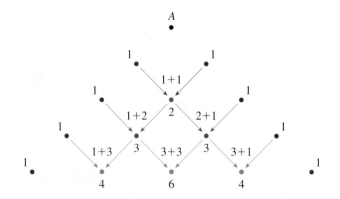

Do you see the pattern for building this figure? Each new row is found by adding the two previous numbers, as shown by the arrows. This pattern is known as **Pascal's triangle.** In Figure 1.4 the rows and diagonals are numbered for easy reference.

www.mathnature.com
There is an online interactive version of Pascal's triangle.

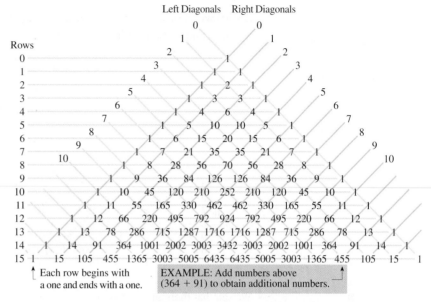

FIGURE 1.4 Pascal's triangle

How does this pattern apply to Melissa's trip from the YWCA to Macy's? It is 3 blocks down and 3 blocks over. Look at Figure 1.4 and count out these blocks, as shown in Figure 1.5.

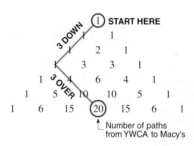

FIGURE 1.5 Using Pascal's triangle to solve the street problem

Historical NOTE

**Blaise Pascal
(1623–1662)**

Described as "the greatest 'might-have-been' in the history of mathematics," Pascal was a person of frail health, and because he needed to conserve his energy, he was forbidden to study mathematics. This aroused his curiosity and forced him to acquire most of his knowledge of the subject by himself. At 18, he had invented one of the first calculating machines. However, at 27, because of his health, he promised God that he would abandon mathematics and spend his time in religious study. Three years later he broke this promise and wrote *Traite du triangle arithmétique*, in which he investigated what we today call Pascal's triangle. The very next year he was almost killed when his runaway horse jumped an embankment. He took this to be a sign of God's displeasure with him and again gave up mathematics—this time permanently.

Example **2** **Pascal's triangle to track paths**

In how many different ways could Melissa get from the YWCA (point *A* in Figure 1.1) to the YMCA (point *D*)?

Solution   Look at Figure 1.1; from point *A* to point *D* is 7 blocks down and 3 blocks left. Use Figure 1.4 as follows:

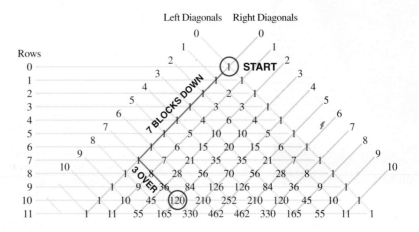

We see that there are 120 paths.

Pascal's triangle applies to the street problem only if the streets are rectangular. If the map shows irregularities (for example, diagonal streets or obstructions), then you must revert back to numbering the vertices.

Example **3** **Travel with irregular paths**

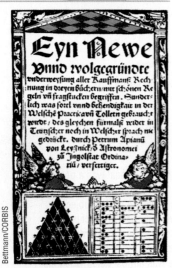
In how many different ways could Melissa get from the YWCA (point *A*) to the Old U.S. Mint (point *M*)?

Solution   If the streets are irregular or if there are obstructions, you cannot use Pascal's triangle, but you can still count the blocks in the same fashion, as shown in the figure.

There are 52 paths from point *A* to point *M* (if, as usual, we do not allow backtracking).

Problem solving is a difficult task to master, and you are not expected to master it after one section of this book (or even after several chapters of this book). However, you must make building your problem-solving skills an ongoing process. One of the most important aspects of problem solving is to relate new problems to old problems. The problem-solving techniques outlined here should be applied when you are faced with a new problem. When you are faced with a problem similar to one you have already worked, you can apply previously developed techniques (as we did in Examples 1–3). Now, because Example 4 seems to be a new type of problem, we again apply the guidelines for problem solving.

Example 4 | Cows and chickens

Pólya's Method

A jokester tells you that he has a group of cows and chickens and that he counted 13 heads and 36 feet. How many cows and chickens does he have?

Solution   Let's use Pólya's problem-solving guidelines.

Understand the Problem. A good way to make sure you understand a problem is to attempt to phrase it in a simpler setting:

one chicken and one cow:	2 heads and 6 feet (chickens have two feet; cows have four)
two chickens and one cow:	3 heads and 8 feet
one chicken and two cows:	3 heads and 10 feet

Devise a Plan. How you organize the material is often important in problem solving. Let's organize the information into a table:

No. of chickens	No. of cows	No. of heads	No. of feet
0	13	13	52

Do you see why we started here? The problem says we must have 13 heads. There are other possible starting places (13 chickens and 0 cows, for example), but an important aspect of problem solving is to start with *some* plan.

No. of chickens	No. of cows	No. of heads	No. of feet
1	12	13	50
2	11	13	48
3	10	13	46
4	9	13	44

Carry Out the Plan. Now, look for patterns. Do you see that as the number of cows decreases by one and the number of chickens increases by one, the number of feet must decrease by two? Does this make sense to you? Remember, step 1 requires that you not just push numbers around, but that you understand what you are doing. Since we need 36 feet for the solution to this problem, we see

$$44 - 36 = 8$$

so the number of chickens must increase by an additional four. The answer is 8 chickens and 5 cows.

Look Back.

No. of chickens	No. of cows	No. of heads	No. of feet
8	5	13	36

Check: 8 chickens have 16 feet and 5 cows have 20 feet, so the total number of heads is $8 + 5 = 13$, and the number of feet is 36.

Example **5** **Number of birth orders**

 Pólya's Method

If a family has 5 children, in how many different birth orders could the parents have a 3-boy, 2-girl family?

Solution

Understand the Problem. Part of understanding the problem might involve estimation. For example, if a family has 1 child, there are 2 possible orders (B or G). If a family has 2 children, there are 4 orders (BB, BG, GB, GG); for 3 children, 8 orders; for 4 children, 16 orders; and for 5 children, a total of 32 orders. This means, for example, that an answer of 140 possible orders is an unreasonable answer.

Devise a Plan. You might begin by enumeration:

BBBGG, BBGBG, BBGGB, . . .

This would seem to be too tedious. Instead, rewrite this as a simpler problem and look for a pattern.

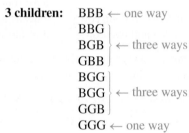

1 child:	B ← one way	**2 children:**	BB ← one way
	G ← one way		BG ⎫
			GB ⎬ ← two ways
			GG ← one way

3 children: BBB ← one way
BBG ⎫
BGB ⎬ ← three ways
GBB ⎭
BGG ⎫
BGG ⎬ ← three ways
GGB ⎭
GGG ← one way

Look at the possibilities:

| 1 child | | 1B | | 1G | |
| 2 children | 1BB | | 2 | 1GG | |

ways for 1 boy and 1 girl

| 3 children | 1BBB | | 3 | 3 | 1GGG | Look familiar? |

ways for 2 boys and 1 girl

ways for 1 boy and 2 girls

Look at Pascal's triangle in Figure 1.4; for 5 children, look at row 5.

Carry Out the Plan.

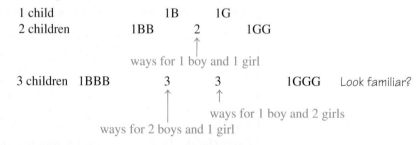

1	5	10	10	5	1
↑	↑	↑	↑	↑	↑
5 boys	4 boys and 1 girl	3 boys and 2 girls	2 boys and 3 girls	1 boy and 4 girls	5 girls

The family could have 3 boys and 2 girls in a total of 10 ways.

Look Back. We predicted that there are a total of 32 ways a family could have 5 children; let's sum the number of possibilities we found in carrying out the plan to see if it totals 32:

$$1 + 5 + 10 + 10 + 5 + 1 = 32$$

In this book, we are concerned about the thought process and not just typical "manipulative" mathematics. Using common sense is part of that thought process, and the following example illustrates how you can use common sense to analyze a given situation.

Example 6   Birthday dilemma

"I'm nine years old," says Adam.
"I'm ten years old," says Eve.
"My tenth birthday is tomorrow," says Adam.
"My tenth birthday was yesterday," says Eve.
"But I'm older than Eve," says Adam.
How is this possible if both children are telling the truth?

Solution   We use Pólya's problem-solving guidelines for this example.

Understand the Problem. What do we mean by the words of the problem? A birthday is the celebration of the day of one's birth. Is a person's age always the same as the number of birthdays?

Devise a Plan. The only time the number of birthdays is different from the person's age is when we are dealing with a leap year. Let's suppose that the day of this conversation is in a leap year on February 29.

Carry Out the Plan. Eve was born ten years ago (a nonleap year) on February 28 and Adam was born ten years ago on March 1. But if someone is born on March 1 then that person is younger than someone born on February 28, right? Not necessarily! Suppose Adam was born in New York City just after midnight on March 1 and that Eve was born before 9:00 P.M in Los Angeles on February 28.

Look Back. Since Adam was born before 9:00 P.M on February 28, he is older than Eve, even though his birthday is on March 1.

The following example illustrates the necessity of carefully reading the question.

Example 7   Meet for dinner

Nick and Marsha are driving from Santa Rosa, CA, to Los Angeles, a distance of 460 miles. They leave at 11:00 A.M. and average 50 mph. On the other hand, Mary and Dan leave at 1:00 P.M in Dan's sports car. Who is closer to Los Angeles when they meet for dinner in San Luis Obispo at 5:00 P.M.?

Solution

Understand the Problem. If they are sitting in the same restaurant, then they are all the same distance from Los Angeles.

The last example of this section illustrates that problem solving may require that you change the conceptual mode.

Example 8   Pascal's triangle—first time in print

If you have been reading the historical notes in the margins, you may have noticed that Blaise Pascal was born in 1623 and died in 1662. You may also have noticed that the first time Pascal's triangle appeared in print was in 1527. How can this be?

Solution   It was a reviewer of this book who brought this apparent discrepancy to my attention. However, the facts are all correct. How could Pascal's triangle have been in print almost 100 years before he was born? The fact is, the number pattern we call Pascal's triangle is *named after* Pascal, but was not *discovered* by Pascal. This number pattern seems to have been discovered several times, by Johann Scheubel in the 16th century and by the Chinese mathematician Nakone Genjun; and recent research has traced the triangle pattern as far back as Omar Khayyám (1048–1122).

"Wait!" you exclaim. "How was I to answer the question in Example 8—I don't know all those facts about the triangle." You are not expected to know these facts, but you are expected to begin to think critically about the information you are given and the assumptions you are making. It was never stated that Blaise Pascal was the first to think of or publish Pascal's triangle!

Problem Set 1.1

Level 1

1. **IN YOUR OWN WORDS**   In the text it was stated that "the most important prerequisite for this course is an openness to try out new ideas—a willingness to experience the suggested activities rather than to sit on the sideline as a spectator." Do you agree or disagree that this is the *most* important prerequisite? Discuss.

2. **IN YOUR OWN WORDS**   What do you thin the primary goal of mathematics education should be? What do you think it is in the United States? Discuss the differences between what it is and what you think it should be.

3. **IN YOUR OWN WORDS**   In the chapter overview (did you read it?), it was pointed out that this book was written for people who think they don't like mathematics, or people who think they can't work math problems, or people who think they are never going to use math. Do any of those descriptions apply to you or to someone you know? Discuss.

4. **IN YOUR OWN WORDS**   At the beginning of this section, three hints for success were listed. Discuss each of these from your perspective. Are there any other hints that you might add to this list?

5. Describe the location of the numbers 1, 2, 3, 4, 5, . . . in Pascal's triangle.

6. Describe the location of the numbers 1, 4, 10, 20, 35, . . . in Pascal's triangle.

7. **IN YOUR OWN WORDS**   In Example 2, the solution was found by going 7 blocks down and 3 blocks over. Could the solution also have been obtained by going 3 blocks over and 7 blocks down? Would this Pascal's triangle solution end up in a different location? Describe a property of Pascal's triangle that is relevant to an answer for this question.

8.

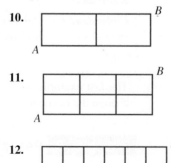

© Tony Freeman/PhotoEdit

  a. If a family has 5 children, in how many ways could the parents have 2 boys and 3 girls?
  b. If a family has 6 children, in how many ways could the parents have 3 boys and 3 girls?

9. **a.** If a family has 7 children, in how many ways could the parents have 4 boys and 3 girls?
  b. If a family has 8 children, in how many ways could the parents have 3 boys and 5 girls?

In Problems 10–13, what is the number of direct routes from point A to point B?

10.

11.

12.

13.
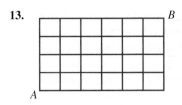

Use the map in Figure 1.6 to determine the number of different paths from point A to the point indicated in Problems 14–17. Remember, no backtracking is allowed.

FIGURE 1.6 Map of a portion of San Francisco

14. *E*   **15.** *F*   **16.** *G*   **17.** *H*

Level 2

18. If an island's only residents are penguins and bears, and if there are 16 heads and 34 feet on the island, how many penguins and how many bears are on the island?

19. Below are listed three problems. Do not solve these problems; simply tell which one you think is most like Problem 18.
  a. A penguin in a tub weighs 8 lb, and a bear in a tub weighs 800 lb. If the penguin and the bear together weigh 802 lb, how much does the tub weigh?
  b. A bottle and a cork cost $1.10, and the bottle is a dollar more than the cork. How much does the cork cost?
  c. Bob has 15 roses and 22 carnations. Carol has twice as many roses and half as many carnations. How many flowers does Carol have?

20. Ten full crates of walnuts weigh 410 pounds, whereas an empty crate weighs 10 pounds. How much do the walnuts alone weigh?

21. There are three separate, equal-size boxes, and inside each box there are two separate small boxes, and inside each of the small boxes there are three even smaller boxes. How many boxes are there all together?

22. Jerry's mother has three children. The oldest is a boy named Harry, who has brown eyes. Everyone says he is a math whiz. The next younger is a girl named Henrietta. Everyone calls her Mary because she hates her name. The youngest child has green eyes and can wiggle his ears. What is his first name?

23. A deaf-mute walks into a stationery store and wants to purchase a pencil sharpener. To communicate this need, the customer pantomimes by sticking a finger in one ear and rotating the other hand around the other ear. The very next customer is a blind person who needs a pair of scissors. How should this customer communicate this idea to the clerk?

24. If you expect to get 50,000 miles on each tire from a set of five tires (four and one spare), how should you rotate the tires so that each tire gets the same amount of wear, and how far can you drive before buying a new set of tires?

25. **a.** What is the sum of the numbers in row 1 of Pascal's triangle?
  b. What is the sum of the numbers in row 2 of Pascal's triangle?
  c. What is the sum of the numbers in row 3 of Pascal's triangle?
  d. What is the sum of the numbers in row 4 of Pascal's triangle?

26. What is the sum of the numbers in row n of Pascal's triangle?

Use the map in Figure 1.6 to determine the number of different paths from point A to the point indicated in Problems 27–30. Remember, no backtracking is allowed.

27. *J*   **28.** *I*   **29.** *L*   **30.** *K*

Problems 31–44 are not typical math problems but are problems that require only common sense (and sometimes creative thinking).

31. How many 3-cent stamps are there in a dozen?

32. Which weighs more—a ton of coal or a ton of feathers?

33. If you take 7 cards from a deck of 52 cards, how many cards do you have?

34. Oak Park cemetery in Oak Park, New Jersey, will not bury anyone living west of the Mississippi. Why?

35. If posts are spaced 10 feet apart, how many posts are needed for 100 feet of straight-line fence?

36. At six o'clock the grandfather clock struck 6 times. If it was 30 seconds between the first and last strokes, how long will it take the same clock to strike noon?

37. A person arrives at home just in time to hear one chime from the grandfather clock. A half-hour later it strikes once. Another half-hour later it chimes once. Still another half-hour later it chimes once. What time did the person arrive home?

38. Two girls were born on the same day of the same month of the same year to the same parents, but they are not twins. Explain how this is possible.

39. How many outs are there in a baseball game that lasts the full 9 innings?

40. Two U.S. coins total $0.30, yet one of these coins is not a nickel. What are the coins?

41. Two volumes of Newman's *The World of Mathematics* stand side by side, in order, on a shelf. A bookworm starts at page i of Volume I and bores its way in a straight line to the last page of Volume II. Each cover is 2 mm thick, and the first volume is $\frac{17}{19}$ as thick as the second volume. The first volume is 38 mm thick without its cover. How far does the bookworm travel?

42. A farmer has to get a fox, a goose, and a bag of corn across a river in a boat that is large enough only for him and one of these three items. If he leaves the fox alone with the goose, the fox will eat the goose. If he leaves the goose alone with the corn, the goose will eat the corn. How does he get all the items across the river?

43. Can you place ten lumps of sugar in three empty cups so that there is an odd number of lumps in each cup?

44. Six glasses are standing in a row. The first three are empty, and the last three are full of water. By handling and moving only one glass, it is possible to change this arrangement so that no empty glass is next to another empty one and no full glass is next to another full one. How can this be done?

Level 3

45. **IN YOUR OWN WORDS** Suppose you have a long list of numbers to add, and you have misplaced your calculator. Discuss the different approaches that could be used for adding this column of numbers.

46. **IN YOUR OWN WORDS** You are faced with a long division problem, and you have misplaced your calculator. You do not

remember how to do long division. Discuss your alternatives to come up with the answer to your problem.

47. **IN YOUR OWN WORDS** You have 10 items in your grocery cart. Six people are waiting in the express lane (10 items or less); one person is waiting in the first checkout stand and two people are waiting in another checkout stand. The other checkout stands are closed. What additional information do you need in order to decide which lane to enter?

48. **IN YOUR OWN WORDS** You drive up to your bank and see five cars in front of you waiting for two lanes of the drive-through banking services. What additional information do you need in order to decide whether to drive through or park your car and enter the bank to do your banking?

49. A boy cyclist and a girl cyclist are 10 miles apart and pedaling toward each other. The boy's rate is 6 miles per hour, and the girl's rate is 4 miles per hour. There is also a friendly fly zooming continuously back and forth from one bike to the other. If the fly's rate is 20 miles per hour, by the time the cyclists reach each other, how far does the fly fly?

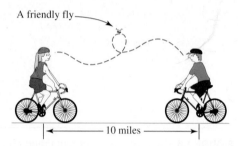

A friendly fly

|← 10 miles →|

50. Two race cars face each other on a single 30-mile track, and each is moving at 60 mph. A fly on the front of one car flies back and forth on a zigzagging path between the cars until they meet. If the fly travels at 25 mph, how far will it have traveled when the cars collide?

51. Alex, Beverly, and Cal live on the same straight road. Alex lives 10 miles from Beverly and Cal lives 2 miles from Beverly. How far does Alex live from Cal?

52. In a different language, *liro cas* means "red tomato." The meaning of *dum cas dan* is "big red barn" and *xer dan* means "big horse." What are the words for "red barn" in this language?

53. Assume that the first "gh" sound in *ghghgh* is pronounced as in *hiccough,* the second "gh" as in *Edinburgh,* and the third "gh" as in *laugh.* How should the word *ghghgh* be pronounced?

54. Write down a three-digit number. Write the number in reverse order. Subtract the smaller of the two numbers from the larger to obtain a new number. Write down the new number. Reverse the digits again, but add the numbers this time. Complete this process for another three-digit number. Do you notice a pattern, and does your pattern work for all three-digit numbers?

55. Start with a common fraction between 0 and 1. Form a new fraction, using the following rules: *New denominator*: Add the numerator and denominator of the original fraction. *New numerator*: Add the new denominator to the original numerator. Write the new fraction and use a calculator to find a decimal equivalent to four decimal places. Repeat these steps again, this time calling the new fraction the original. Continue the process until a pattern appears about the decimal equivalent. What is the decimal equivalent (correct to two decimal places)?

56. The number 6 has four divisors—namely, 1, 2, 3, and 6. List all numbers less than 20 that have exactly four divisors.

Problem Solving 3

Each section of the book has one or more problems designated by "Problem Solving." These problems may require additional insight, information, or effort to solve. True problem-solving ability comes from solving problems that "are not like the examples" but rather require independent thinking. I hope you will make it a habit to read these problems and attempt to work those that interest you, even though they may not be part of your regular class assignment.

57. Consider the routes from *A* to *B* and notice that there is now a barricade blocking the path. Work out a general solution for the number of paths with a blockade, and then illustrate your general solution by giving the number of paths for each of the following street patterns.

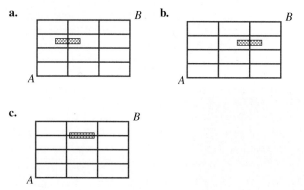

a.   **b.**

c.

58. HISTORICAL QUEST   Thoth, an ancient Egyptian god of wisdom and learning, has abducted Ahmes, a famous Egyptian scribe, in order to assess his intellectual prowess. Thoth places Ahmes before a large funnel set in the ground (see Figure 1.7). It has a circular opening 1,000 ft in diameter, and its walls are quite slippery. If Ahmes attempts to enter the funnel, he will slip down the wall. At the bottom of the funnel is a sleep-inducing liquid that will instantly put Ahmes to sleep for eight hours if he touches it.* Thoth hands Ahmes two objects: a rope 1,006.28 ft in length and the skull of a chicken. Thoth says to Ahmes, "If you are able to get to the central tower and touch it, we will live in harmony for the next millennium. If not, I will detain you for further testing. Please note that with each passing hour, I will decrease the rope's length by a foot." How can Ahmes reach the central ankh tower and touch it?

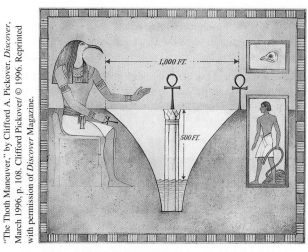

FIGURE 1.7 Ahmes's dilemma. Note that there are two ankh-shaped towers. One stands on a cylindrical platform in the center of the funnel. The platform's surface is at ground level. The distance from the surface to the liquid is 500 ft. The other ankh tower is on land, at the edge of the funnel.

"The Thoth Maneuver," by Clifford A. Pickover, *Discover*, March 1996, p. 108. Clifford Pickover © 1996. Reprinted with permission of *Discover* Magazine.

59. A magician divides a deck of cards into two equal piles, counts down from the top of the first pile to the seventh card, and shows it to the audience without looking at it herself. These seven cards are replaced faced down in the same order on top of the first pile. She then picks up the other pile and deals the top three cards up in a row in front of her. If the first card is a six, then she starts counting with "six" and counts to ten, thus placing four more cards on this pile as shown. In turn, the magician does the same for the next two cards. If the card is a ten or a face card, then no additional cards are added. The remainder of this pile is placed on top of the first pile.

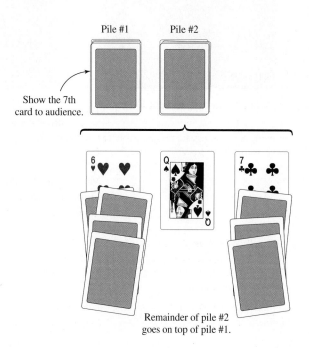

Pile #1   Pile #2

Show the 7th
card to audience.

Remainder of pile #2
goes on top of pile #1.

*From "The Thoth Maneuver," by Clifford A. Pickover, *Discover*, March 1996, p. 108. Clifford Pickover © 1996. Reprinted with permission of Discover Magazine. Nenad Jakesevic and Sonja Lamut © 1996. Reprinted with permission of Discover Magazine.

Next, the magician adds the values of the three face-up cards (6 + 10 + 7 for this illustration) and counts down in the first deck this number of cards. That card is the card that was originally shown to the audience. Explain why this trick works.

60. A very magical teacher had a student select a two-digit number between 50 and 100 and write it on the board out of view of the instructor. Next, the student was asked to add 76 to the number, producing a three-digit sum. If the digit in the hundreds place is added to the remaining two-digit number and this result is subtracted from the original number, the answer is 23, which was predicted by the instructor. How did the instructor know the answer would be 23? *Note:* This problem is dedicated to my friend Bill Leonard of Cal State, Fullerton. His favorite number is 23.

1.2 | Inductive and Deductive Reasoning

Studying numerical patterns is one frequently used technique of problem solving.

Magic Squares

A magic square is an arrangement of numbers in the shape of a square with the sums of each vertical column, each horizontal row, and each diagonal all equal. One of the most famous ones appeared in a 1514 engraving *Melancholia* by Dürer, as shown in Figure 1.8. (Notice that the date appears in the magic square).

Historical NOTE

Burstein Collection/CORBIS

Melancholia by Albrecht Dürer

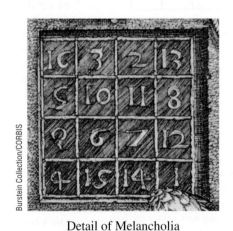

Burstein Collection/CORBIS

Detail of Melancholia

IXOHOXI

FIGURE 1.8 Early magic squares

Magic squares can be constructed using the first 9, 16, 25, 36, 49, 64, and 81 consecutive numbers. You may want to try some of them. One with the first 25 numbers is shown here.

23	12	1	20	9
4	18	7	21	15
10	24	13	2	16
11	5	19	8	22
17	6	25	14	3

There are formal methods for finding magic squares, but we will not describe them here. Figure 1.8 shows a rather interesting magic square called IXOHOXI because it is a magic square when it is turned upside down and also when it is reflected in a mirror. You are asked to find another magic square that can be turned upside down and is still a magic square.

Let's consider some other simple patterns.

A Pattern of Nines

A very familiar pattern is found in the ordinary "times tables." By pointing out patterns, teachers can make it easier for children to learn some of their multiplication tables. For example, consider the multiplication table for 9s:

$$1 \times 9 = 9$$
$$2 \times 9 = 18$$
$$3 \times 9 = 27$$
$$4 \times 9 = 36$$
$$5 \times 9 = 45$$
$$6 \times 9 = 54$$
$$7 \times 9 = 63$$
$$8 \times 9 = 72$$
$$9 \times 9 = 81$$
$$10 \times 9 = 90$$

What patterns do you notice? You should be able to see many number relationships by looking at the totals. For example, notice that the sum of the digits to the right of the equality is 9 in all the examples ($1 + 8 = 9, 2 + 7 = 9, 3 + 6 = 9$, and so on). Will this always be the case for multiplication by 9? (Consider $11 \times 9 = 99$. The sum of the digits is 18. However, notice the result if you add the digits of 18.) Do you see any other patterns? Can you explain why they "work"? This pattern of adding after multiplying by 9 generates a sequence of numbers: 9, 9, 9, ... We call this the *nine pattern*. The two number tricks described in the news clip in the margin use this nine pattern.

> Nine is one of the most fascinating of all numbers. Here are two interesting tricks that involve the number nine. You need a calculator for these.
>
> Mix up the serial number on a dollar bill. You now have two numbers, the original serial number and the mixed-up one. Subtract the smaller from the larger. If you add the digits of the answer, you will obtain a 9 or a number larger than 9; if it is larger than 9, add the digits of this answer again. Repeat the process until you obtain a single digit as with the *nine pattern*. That digit will *always* be 9.
>
> Here is another trick. Using a calculator keyboard or push-button phone, choose any three-digit column, row, or diagonal, and arrange these digits in any order. Multiply this number by another row, column, or diagonal. If you repeatedly add the digits the answer will *always* be nine.

| Example | **1** | **Eight pattern** | **Pólya's Method** |

Find the *eight pattern*.

Solution   We use Pólya's problem-solving guidelines for this example.

Understand the Problem.   What do we mean by the *eight pattern?* Do you understand the example for the *nine pattern?*

Devise a Plan.   We will carry out the multiplications of successive counting numbers by 8 and if there is more than a single-digit answer, we add the digits.* What we are looking for is a pattern for these single-digit numerals (shown in blue).

Carry Out the Plan.

$$1 \times 8 = 8, \quad 2 \times 8 = \underbrace{16}, \quad 3 \times 8 = \underbrace{24}, \quad 4 \times 8 = \underbrace{32}, \quad 5 \times 8 = \underbrace{40}, \ldots$$
$$\qquad\quad 8 \qquad\quad 1 + 6 = 7 \qquad 2 + 4 = 6 \qquad 3 + 2 = 5 \qquad 4 + 0 = 4$$

Continue with some additional terms:

$$6 \times 8 = 48 \quad \text{and} \quad 4 + 8 = 12 \quad \text{and} \quad 1 + 2 = 3$$
$$7 \times 8 = 56 \quad \text{and} \quad 5 + 6 = 11 \quad \text{and} \quad 1 + 1 = 2$$

*****Counting numbers** are the numbers we use for counting—namely, 1, 2, 3, 4, ... Sometimes they are also called **natural numbers**. The integers are the counting numbers, their opposites, and 0, namely, . . . , $-3, -2, -1, 0, 1, 2, 3, \ldots$. We assume a knowledge of these numbers.

$$8 \times 8 = 64 \quad \text{and} \quad 6 + 4 = 10 \quad \text{and} \quad 1 + 0 = 1$$
$$9 \times 8 = 72 \quad \text{and} \quad 7 + 2 = 9$$
$$10 \times 8 = 80 \quad \text{and} \quad 8 + 0 = 8$$

We now see the eight pattern: 8, 7, 6, 5, 4, 3, 2, 1, 9, 8, 7, 6, ...

Look Back. Let's do more than check the arithmetic, since this pattern seems clear. The problem seems to be asking whether we understand the concept of a *nine pattern* or an *eight pattern.* Verify that the *seven pattern* is 7, 5, 3, 1, 8, 6, 4, 2, 9, 7, 5, 3, 1, ...

Order of Operations

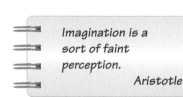

Imagination is a sort of faint perception.

Aristotle

Complicated arithmetic problems can sometimes be solved by using patterns. Given a difficult problem, a problem solver will often try to *solve a simpler, but similar, problem.*

The second suggestion for solving using Pólya's problem-solving procedure stated, "If you cannot solve the proposed problem, look around for an appropriate related problem (a simpler one, if possible)." For example, suppose we wish to compute the following number:

$$10 + 123{,}456{,}789 \times 9$$

Instead of doing a lot of arithmetic, let's study the following pattern:

$$2 + 1 \times 9$$
$$3 + 12 \times 9$$
$$4 + 123 \times 9$$

Do you see the next entry in this pattern? Do you see that if we continue the pattern we will eventually reach the desired expression of $10 + 123{,}456{,}789 \times 9$? Using Pólya's strategy, we begin by working these easier problems. Thus, we begin with $2 + 1 \times 9$. There is a possibility of ambiguity in calculating this number:

Left to right	Multiplication first
$2 + 1 \times 9 = 3 \times 9 = 27$	$2 + 1 \times 9 = 2 + 9 = 11$

Although either of these might be acceptable in certain situations, it is not acceptable to get two different answers to the same problem. We therefore agree to do a problem like this by multiplying first. If we wish to change this order, we use parentheses, as in $(2 + 1) \times 9 = 27$. We summarize with a procedure known as the **order-of-operations agreement.**

Order of Operations

Step 1 Perform any operations enclosed in parentheses.

Step 2 Perform multiplications and divisions as they occur by working from left to right.

Step 3 Perform additions and subtractions as they occur by working from left to right.

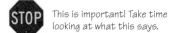

This is important! Take time looking at what this says.

Thus, the correct result for $2 + 1 \times 9$ is 11. Also,

$$3 + 12 \times 9 = 3 + 108 = 111$$
$$4 + 123 \times 9 = 4 + 1{,}107 = 1{,}111$$
$$5 + 1{,}234 \times 9 = 5 + 11{,}106 = 11{,}111$$

Do you see a pattern? If so, then make a prediction about the desired result.

If you do not see a pattern, continue with this pattern to see more terms, or go back and try another pattern. For this example, we predict

$$10 + 123,456,789 \times 9 = 1,111,111,111$$

The most difficult part of this type of problem solving is coming up with a correct pattern. For this example, you might guess that

$$2 + 1 \times 1$$
$$3 + 12 \times 2$$
$$4 + 123 \times 3$$
$$5 + 1,234 \times 4$$
$$\vdots$$

leads to $10 + (123,456,789 \times 9)$. Calculating, we find

$$2 + 1 \times 1 = 2 + 1 = 3$$
$$3 + 12 \times 2 = 3 + 24 = 27$$
$$4 + 123 \times 3 = 4 + 369 = 373$$
$$5 + 1,234 \times 4 = 5 + 4,936 = 4,941$$

If you begin a pattern and it does not lead to a pattern of answers, then you need to remember that part of Pólya's problem-solving procedure is to work both backward and forward. Be willing to give up one pattern and begin another.

We also point out that the patterns you find are not necessarily unique. One last time, we try a pattern for $10 + 123,456,789 \times 9$:

$$10 + 1 \times 9 = 10 + 9 = 19$$
$$10 + 12 \times 9 = 10 + 108 = 118$$
$$10 + 123 \times 9 = 10 + 1,107 = 1,117$$
$$10 + 1,234 \times 9 = 10 + 11,106 = 11,116$$
$$\vdots$$

We do see a pattern here (although not quite as easily as the one we found with the first pattern for this example):

$$10 + 123,456,789 \times 9 = 1,111,111,111$$

Inductive Reasoning

The type of reasoning used here and in the first sections of this book—first observing patterns and then predicting answers for more complicated problems—is called **inductive reasoning.** It is a very important method of thought and is sometimes called the *scientific method.* It involves reasoning from particular facts or individual cases to a general **conjecture**—a statement you think may be true. That is, a generalization is made on the basis of some observed occurrences. The more individual occurrences we observe, the better able we are to make a correct generalization. Peter in the *B.C.* cartoon makes the mistake of generalizing on the basis of a single observation.

AMAZING HOW THEY CAN TAKE SUCH DEVASTATING FALLS, THEN JUST WALK AWAY.

I HOPE I CAN MAKE IT TO THE CEMETERY!

| Example **2** | **Sum of 100 odd numbers** |

Pólya's Method

What is the sum of the first 100 consecutive odd numbers?

Solution   We use Pólya's problem-solving guidelines for this example.

Understand the Problem. Do you know what the terms mean? Odd numbers are $1, 3, 5, \ldots,$ and *sum* indicates addition:

$$1 + 3 + 5 + \cdots + ?$$

The first thing you need to understand is what the last term will be, so you will know when you have reached 100 consecutive odd numbers.

1 + 3 is two terms.
1 + 3 + 5 is three terms.
1 + 3 + 5 +7 is four terms.

It seems as if the last term is always one less than twice the number of terms. Thus, the sum of the first 100 consecutive odd numbers is

$$1 + 3 + 5 + \cdots + 195 + 197 + 199$$

This is one less than 2(100).

Devise a Plan. The plan we will use is to look for a pattern:

$$1 = 1 \quad \text{One term}$$
$$1 + 3 = 4 \quad \text{Sum of two terms}$$
$$1 + 3 + 5 = 9 \quad \text{Sum of three terms}$$

Do you see a pattern yet? If not, continue:

$$1 + 3 + 5 + 7 = 16$$
$$1 + 3 + 5 + 7 + 9 = 25$$

Carry Out the Plan. It appears that the sum of 2 terms is $2 \cdot 2$; of 3 terms, $3 \cdot 3$; of 4 terms, $4 \cdot 4$; and so on. The sum of the first 100 consecutive odd numbers is therefore $100 \cdot 100$.

Looking Back. Does $100 \cdot 100 = 10,000$ seem correct?

The numbers $2 \cdot 2, 3 \cdot 3, 4 \cdot 4,$ and $100 \cdot 100$ from Example 2 are usually written as $2^2, 3^2, 4^2,$ and 100^2. The number b^2 means $b \cdot b$ and is pronounced ***b*** **squared,** and the number b^3 means $b \cdot b \cdot b$ and is pronounced ***b*** **cubed.** The process of repeated multiplication is called **exponentiation.** Numbers that are multiplied are called **factors,** so we note that b^2 means we have two factors and one multiplication, where as b^3 indicates three factors (two multiplications).

Deductive Reasoning

Another method of reasoning used in mathematics is called **deductive reasoning.** This method of reasoning produces results that are *certain* within the logical system being developed. That is, deductive reasoning involves reaching a conclusion by using a formal structure based on a set of **undefined terms** and a set of accepted unproved **axioms** or **premises.** For example, consider the following argument:

1. If you read the *Times,* then you are well informed.

2. You read the *Times.*

3. Therefore, you are well informed.

Statements 1 and 2 are the *premises* of the argument; statement 3 is called the **conclusion.** If you accept statements 1 and 2 as true, then you *must* accept statement 3 as true. Such reasoning is called *deductive reasoning;* and if the conclusion follows from the premises, the reasoning is said to be **valid.**

Deductive Reasoning

Deductive reasoning consists of reaching a conclusion by using a formal structure based on a set of *undefined terms* and on a set of accepted unproved *axioms* or *premises*. The conclusions are said to *be proved* and are called **theorems.**

Reasoning that is not valid is called **invalid** reasoning. Logic accepts no conclusions except those that are inescapable. This is possible because of the strict way in which concepts are defined. Difficulty in simplifying arguments may arise because of their length, the vagueness of the words used, the literary style, or the possible emotional impact of the words used.

Consider the following two arguments:

1. If George Washington was assassinated, then he is dead. Therefore, if he is dead, he was assassinated.

2. If you use heroin, then you first used marijuana. Therefore, if you use marijuana, then you will use heroin.

Logically, these two arguments are exactly the same, and both are *invalid* forms of reasoning. Nearly everyone would agree that the first is invalid, but many people see the second as valid. The reason lies in the emotional appeal of the words used.

To avoid these difficulties, we look at the *form* of the arguments and not at the independent truth or falsity of the statements. One type of logic problem is called a **syllogism.** A syllogism has three parts: two *premises,* or hypotheses, and a *conclusion.* The premises give us information from which we form a conclusion. With the syllogism, we are interested in knowing whether the conclusion *necessarily follows* from the premises. If it does, it is called a *valid syllogism*; if not, it is called *invalid.* Consider the following examples:

Valid Forms of Reasoning		*Invalid Forms of Reasoning*
All Chevrolets are automobiles.	*Premise*	Some people are nice.
All automobiles have four wheels.	*Premise*	Some people are broke.
All Chevrolets have four wheels.	*Conclusion*	There are some nice broke people.
All teachers are crazy.	*Premise*	All dodos are extinct.
Karl Smith is a teacher.	*Premise*	No dinosaurs are dodos.
Karl Smith is crazy.	*Conclusion*	Therefore, all dinosaurs are extinct.

To analyze such arguments, we need to have a systematic method of approach. We will use **Euler circles,** named after one of the most famous mathematicians in all of mathematics, Leonhard Euler.

For two sets p and q, we make the interpretations shown in Figure 1.9.

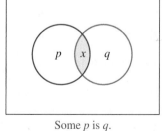

Some p is q.

(x means that intersection is not empty)

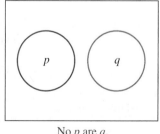

No p are q.

Disjoint sets

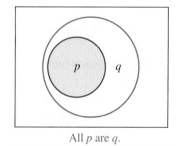

All p are q.

Subsets

FIGURE 1.9 Euler circles for syllogisms

Historical NOTE

**Leonhard Euler
(1707–1783)**

Euler's name is attached to every branch of mathematics, and we will visit his work many times in this book. His name is pronounced "Oiler" and it is sometimes joked that if you want to give a student a one-word mathematics test, just ask the student to pronounce Leonhard's last name. He was the most prolific writer on the subject of mathematics, and his mathematical textbooks were masterfully written. His writing was not at all slowed down by his total blindness for the last 17 years of his life. He possessed a phenomenal memory, had almost total recall, and could mentally calculate long and complicated problems.

Example 3 Testing for valid arguments

Test the validity of the following arguments.

a. All dictionaries are books.
  This is a dictionary.
  Therefore, this is a book.
b. If you like potato chips, then you will like Krinkles.
  You do not like potato chips.
  Therefore, you do not like Krinkles.

Solution

a. Begin by drawing Euler circles showing the first premise:

  All dictionaries are books.

  Let   *p*: dictionaries
     q: books

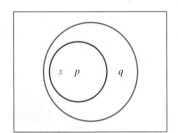

For the second premise, we place *x* (this object) inside the circle of dictionaries (labeled *p*). The conclusion, "This object is a book," cannot be avoided (since *x must* be in *q*), so it is valid.

b. Again, begin by using Euler circles:

  If you like potato chips, then you will like Krinkles.

  The first premise is the same as

  All people who like potato chips like Krinkles.

  Let   *p*: people who like potato chips
     q: people who like krinkles

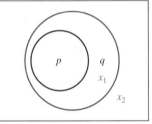

For the second premise, you will place the *x* (you) outside the circle labeled *p*. Notice that you are not forced to place *x* into a single region; it could be placed in either of two places—those labeled x_1 and x_2. Since the stated conclusion is not forced, the argument is not valid.

Problem Set 1.2

Level 1

1. **IN YOUR OWN WORDS** Discuss the nature of *inductive* and *deductive reasoning*.

2. **IN YOUR OWN WORDS** Explain what is meant by the *seven pattern*.

3. **IN YOUR OWN WORDS** What do we mean by *order of operations*?

4. **IN YOUR OWN WORDS** What is the scientific method?

5. **IN YOUR OWN WORDS** Explain inductive reasoning. Give an original example of an occasion when you have used inductive reasoning or heard it being used.

6. **IN YOUR OWN WORDS** Explain deductive reasoning. Give an original example of an occasion when you have used deductive reasoning or heard it being used.

Perform the operations in Problems 7–18.

7. **a.** $5 + 2 \times 6$ **b.** $7 + 3 \times 2$

8. **a.** $14 + 6 \times 3$ **b.** $30 \div 5 \times 2$

9. **a.** $3 \times 8 + 3 \times 7$ **b.** $3(8 + 7)$

10. **a.** $(8 + 6) \div 2$ **b.** $8 + 6 \div 2$

11. **a.** $12 + 6/3$ **b.** $(12 + 6)/3$

12. **a.** $450 + 550/10$ **b.** $\dfrac{450 + 550}{10}$

13. **a.** $20/2 \cdot 5$ **b.** $20/(2 \cdot 5)$

14. **a.** $1 + 3 \times 2 + 4 + 3 \times 6$
  b. $3 + 6 \times 2 + 8 + 4 \times 3$

15. **a.** $10 + 5 \times 2 + 6 \times 3$
  b. $4 + 3 \times 8 + 6 + 4 \times 5$

16. a. $8 + 2(3 + 12) - 5 \times 3$
  b. $25 - 4(12 - 2 \times 6) + 3$

17. a. $3 + 9 \div 3 \times 2 + 2 \times 6 \div 3$
  b. $[(3 + 9) \div 3] \times 2 + [(2 \times 6) \div 3]$

18. a. $3 + [(9 \div 3) \times 2] + [(2 \times 6) \div 3]$
  b. $[(3 + 9) \div (3 \times 2)] + [(2 \times 6) \div 3]$

19. Does the B.C. cartoon illustrate inductive or deductive reasoning? Explain your answer.

B.C. reprinted by permission of Johnny Hart and Creators Syndicate.

20. Does the news clip below illustrate inductive or deductive reasoning? Explain your answer.

> The old fellow in charge of the checkroom in a large hotel was noted for his memory. He never used checks or marks of any sort to help him return coats to their rightful owners.
>
> Thinking to test him, a frequent hotel guest asked him as he received his coat, "Sam, how did you know this is my coat?"
>
> "I don't, sir," was the calm response.
>
> "Then why did you give it to me?" asked the guest.
>
> "Because," said Sam, "it's the one you gave me, sir."

Lucille J. Goodyear

Problems 21–24 are modeled after Example 1.
Find the requested pattern.

21. three pattern   **22.** four pattern

23. five pattern   **24.** six pattern

25. a. What is the sum of the first 25 consecutive odd numbers?
  b. What is the sum of the first 250 consecutive odd numbers?

26. a. What is the sum of the first 50 consecutive odd numbers?
  b. What is the sum of the first 1,000 consecutive odd numbers?

27. HISTORICAL QUEST   The first known example of a magic square comes from China. Legend tells us that around the year 200 B.C. the emperor Yu of the Shang dynasty received the following magic square etched on the back of a tortoise's shell:

The incident supposedly took place along the Lo River, so this magic square has come to be known as the Lo-shu magic square. The even numbers are black (female numbers) and the odd numbers are white (male numbers). Translate this magic square into modern symbols.

This same magic square (called *wafq* in Arabic) appears in Islamic literature in the 10th century A.D. and is attributed to Jabir ibn Hayyan.

28. Consider the square shown in Figure 1.10.

10	7	8	11
14	11	12	15
13	10	11	14
15	12	13	16

FIGURE 1.10 Magic square?

a. Is this a magic square?
b. Circle any number; cross out all the numbers in the same row and column. Then circle any remaining number and cross out all the numbers in the same row and column. Circle the remaining number. The sum of the circled numbers is 48. Why?

Level 2

Use Euler circles to check the validity of the arguments in Problems 29–40.

29. All mathematicians are eccentrics.
  All eccentrics are rich.
  Therefore, all mathematicians are rich.

30. All snarks are fribbles.
  All fribbles are ugly.
  Therefore, all snarks are ugly.

31. All cats are animals.
  This is not an animal.
  Therefore, this is not a cat.

32. All bachelors are handsome.
Some bachelors do not drink lemonade.
Therefore, some handsome men do not drink lemonade.

33. No students are enthusiastic.
You are enthusiastic.
Therefore, you are not a student.

34. No politicians are honest.
Some dishonest people are found out.
Therefore, some politicians are found out.

35. All candy is fattening.
All candy is delicious.
Therefore, all fattening food is delicious.

36. All parallelograms are rectangles.
All rectangles are polygons.
Therefore, all parallelograms are polygons.

37. No professors are ignorant.
All ignorant people are vain.
Therefore, no professors are vain.

38. No monkeys are soldiers.
All monkeys are mischievous.
Therefore, some mischievous creatures are not soldiers.

39. All lions are fierce.
Some lions do not drink coffee.
Therefore, some creatures that drink coffee are not fierce.

40. All red hair is pretty.
No pretty things are valuable.
Therefore, no red hair is valuable.

Level 3

Problems 41–44 refer to the lyrics of "By the Time I Get to Phoenix." Tell whether each answer you give is arrived at inductively or deductively.

By the Time I Get to Phoenix

By the time I get to Phoenix she'll be risin'.
She'll find the note I left hangin' on her door.
She'll laugh when she reads the part that says I'm leavin',
'Cause I've left that girl so many times before.

By the time I make Albuquerque she'll be workin'.
She'll probably stop at lunch and give me a call.
But she'll just hear that phone keep on ringin'
Off the wall, that's all.

By the time I make Oklahoma she'll be sleepin'.
She'll turn softly and call my name out low.
And she'll cry just to think I'd really leave her,
'tho' time and time I've tried to tell her so,
She just didn't know
I would really go.

41. In what basic direction (north, south, east, or west) is the person traveling?

42. What method of transportation or travel is the person using?

43. What is the probable starting point of this journey?

44. List five facts you know about each person involved.

Problems 45–48 refer to the lyrics of "Ode to Billy Joe." Tell whether each answer you give is arrived at inductively or deductively.

Ode to Billy Joe

It was the third of June, another sleepy, dusty, delta day.
I was choppin' cotton and my brother was balin' hay.
And at dinnertime we stopped and walked back to the house to eat,
And Mama hollered at the back door, "Y'all remember to wipe your feet."
Then she said, "I got some news this mornin' from Choctaw Ridge,
Today Billy Joe McAllister jumped off the Tallahatchee Bridge."

Papa said to Mama, as he passed around the black-eyed peas,
"Well, Billy Joe never had a lick o' sense, pass the biscuits please,
There's five more acres in the lower forty I've got to plow,"
And Mama said it was a shame about Billy Joe anyhow.
Seems like nothin' ever comes to no good up on Choctaw Ridge,
And now Billy Joe McAllister's jumped off the Tallahatchee Bridge.

Brother said he recollected when he and Tom and Billy Joe,
Put a frog down my back at the Carroll County picture show,
And wasn't I talkin' to him after church last Sunday night,
"I'll have another piece of apple pie, you know, it don't seem right,
I saw him at the sawmill yesterday on Choctaw Ridge,
And now you tell me Billy Joe's jumped off the Tallahatchee Bridge."

Mama said to me, "Child, what's happened to your appetite?
I been cookin' all mornin' and you haven't touched a single bite,
That nice young preacher Brother Taylor dropped by today,
Said he'd be pleased to have dinner on Sunday, Oh, by the way,
He said he saw a girl that looked a lot like you up on Choctaw Ridge
And she an' Billy Joe was throwin' somethin' off the Tallahatchee Bridge."

A year has come and gone since we heard the news 'bout Billy Joe,
Brother married Becky Thompson, they bought a store in Tupelo,
There was a virus goin' round, Papa caught it and he died last spring.
And now Mama doesn't seem to want to do much of anything.
And me I spend a lot of time pickin' flowers up on Choctaw Ridge,
And drop them into the muddy water off the Tallahatchee Bridge.

45. How many people are involved in this story? List them by name and/or description.

46. Who "saw him at the sawmill yesterday"?

47. In which state is the Tallahatchee Bridge located?

48. On what day or days of the week could the death not have taken place?

On what day of the week was the death most probable?

49. Which direction is the bus traveling?*

Did you arrive at your answer using inductive or deductive reasoning?

50. Which is larger—the number of all seven-letter English words ending in *ing*, or the number of seven-letter words with "*i*" as the fifth letter?[†] Did you arrive at your answer using inductive or deductive reasoning?

51. Consider the following pattern:

$$9 \times 1 - 1 = 8$$
$$9 \times 21 - 1 = 188$$
$$9 \times 321 - 1 = 2{,}888$$
$$9 \times 4{,}321 - 1 = 38{,}888$$

a. Use this pattern and inductive reasoning to find the next problem and the next answer in the sequence.

b. Use this pattern to find

$$9 \times 987{,}654{,}321 - 1$$

c. Use this pattern to find

$$9 \times 10{,}987{,}654{,}321 - 1$$

52. Consider the following pattern:

$$123{,}456{,}789 \times 9 = 1{,}111{,}111{,}101$$
$$123{,}456{,}789 \times 18 = 2{,}222{,}222{,}202$$
$$123{,}456{,}789 \times 27 = 3{,}333{,}333{,}303$$

a. Use this pattern and inductive reasoning to find the next problem and the next answer in the sequence.

b. Use this pattern to find

$$123{,}456{,}789 \times 9{,}000$$

c. Use this pattern to find

$$123{,}456{,}789 \times 81{,}000$$

53. What is the sum of the digits in

$$333333334^2$$

Did you arrive at your answer using inductive or deductive reasoning?

54. Enter 999999 into your calculator, then divide it by seven. Now toss a die (or randomly pick a number from 1 through 6) and multiply this number by the displayed calculator number. Arrange the digits of the product from lowest to highest (from left to right). What is this six-digit number?

Explain how you arrived at your answer, and discuss whether you arrived at your result empirically, by induction, or by deduction.

Problem Solving 3

55. How many squares are there in Figure 1.11?

FIGURE 1.11 How many squares?

Hint:

□	1 1-by-1 square TOTAL: 1

4   1-by-1 squares
1   2-by-2 square
TOTAL: 5 (by addition)

9   1-by-1 squares
4   2-by-2 squares
1   3-by-3 square
TOTAL: 14 (by addition)

56. How many triangles are there in Figure 1.12?

FIGURE 1.12 How many triangles?

57. You have 9 coins, but you are told that one of the coins is counterfeit and weighs just a little more than an authentic coin. How can you determine the counterfeit with 2 weighings on a two-pan balance scale? (This problem is discussed in Chapter 2).

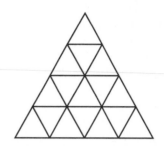

58.

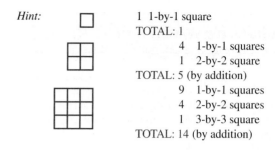

In 1987 Martin Gardner (long-time math buff and past editor of the "Mathematical Games" department of *Scientific American*) offered $100 to anyone who could find a 3×3 magic square made with consecutive primes. The prize was won by Harry Nelson of Lawrence Livermore Laboratories. He produced the following simplest such square:

1,480,028,201	1,480,028,129	1,480,028,183
1,480,028,153	1,480,028,171	1,480,028,189
1,480,028,159	1,480,028,213	1,480,028,141

Prove this is a magic square.

59.

$100 REWARD (FOR REAL)

Now, a real $100 offer: *Find a 3 × 3 magic square with nine distinct square numbers*. If you find such a magic square, write to me and I will include it in the next edition and pay you a $100 reward . Show that the following magic squares do not win the award.

a.

127^2	46^2	58^2
2^2	113^2	94^2
74^2	82^2	97^2

b.

35^2	3495^2	2958^2
3642^2	2125^2	1785^2
2775^2	2058^2	3005^2

60. Find a 5 × 5 magic square whose magic number is 44. That is, the sum of the rows, columns, and diagonals is 44. Furthermore, the square can be turned upside down without changing this property. *Hint:* The magic number is XLIV, and the entries of the square use Roman numerals.

1.3   Scientific Notation and Estimation

"How Big Is the Cosmos?" There is a dynamic Web site demonstrating the answer to this question (go to "Powers of Ten" at **www.mathnature.com**). Figure 1.13 illustrates the size of the known cosmos.

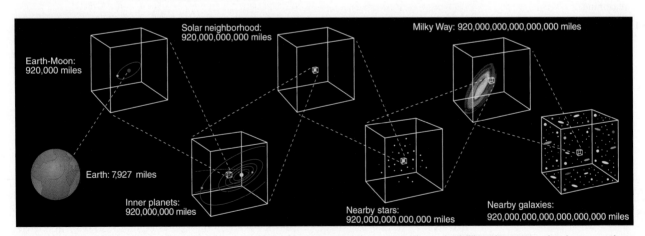

Earth-Moon: 920,000 miles

Solar neighborhood: 920,000,000,000 miles

Milky Way: 920,000,000,000,000,000 miles

Earth: 7,927 miles

Inner planets: 920,000,000 miles

Nearby stars: 920,000,000,000,000 miles

Nearby galaxies: 920,000,000,000,000,000,000 miles

FIGURE 1.13 Size of the universe.* Each successive cube is a thousand times as wide and a billion times as voluminous as the one before it.

How can the human mind comprehend such numbers? Scientists often work with very large numbers. Distances such as those in Figure 1.13 are measured in terms of the distance that light, moving at 186,000 miles per second, travels in a year. In this section, we turn to patterns to see if there is an easy way to deal with very large and very small numbers. We will also discuss estimation as a problem-solving technique.

*Illustration is adapted from *The Universe*, Life Nature Library (1962).

Exponential Notation

We often encounter expressions that comprise multiplication of the same numbers. For example,

$$10 \cdot 10 \cdot 10 \qquad \text{or} \qquad 6 \cdot 6 \cdot 6 \cdot 6 \cdot 6 \cdot 6 \cdot 6 \qquad \text{or}$$
$$15 \cdot 15 \cdot 15 \cdot 15 \cdot 15 \cdot 15 \cdot 15 \cdot 15 \cdot 15 \cdot 15 \cdot 15 \cdot 15 \cdot 15 \cdot 15$$

These numbers can be written more concisely using what is called **exponential notation:**

$$\underbrace{10 \cdot 10 \cdot 10}_{\textit{3 factors}} = 10^3 \qquad \underbrace{6 \cdot 6 \cdot 6 \cdot 6 \cdot 6 \cdot 6 \cdot 6}_{\textit{7 factors}} = 6^7$$

How would you use exponential notation for the product of the 15s?   *Answer:* 15^{14}.

Exponential Notation

For any nonzero number b and any counting number n,

$$b^n = \underbrace{b \cdot b \cdot b \cdots \cdot b,}_{n \text{ factors}} \qquad b^0 = 1, \qquad b^{-n} = \frac{1}{b^n}$$

The number b is called the **base**, the number n in b^n is called the **exponent**, and the number b^n is called a **power** or **exponential.**

Example 1 Write numbers without exponents

Write without exponents.   **a.** 10^5 **b.** 6^2 **c.** 7^5 **d.** 2^{63} **e.** 3^{-2} **f.** 8.9^0

Solution
a. $10^5 = 10 \cdot 10 \cdot 10 \cdot 10 \cdot 10 = 100,000$
b. $6^2 = 6 \cdot 6 = 36$
c. $7^5 = 7 \cdot 7 \cdot 7 \cdot 7 \cdot 7$ or $16,807$
d. $2^{63} = \underbrace{2 \cdot 2 \cdot 2 \cdots \cdot 2 \cdot 2}_{\text{63 factors}}$ or $9,223,372,036,854,775,808$

 CAUTION *Note:* 10^5 *is not five multiplications, but rather five factors of 10.*

Note: You are not expected to find the form at the right; the factored form is acceptable.
e. $3^{-2} = \dfrac{1}{3^2} = \dfrac{1}{9}$
f. $8.9^0 = 1$ By definition, any nonzero number to the zero power is 1.

Since an exponent indicates a multiplication, the proper procedure is first to simplify the exponent, and then to carry out the multiplication. This leads to an **extended order-of-operations agreement.**

Extended Order of Operations

Step 1 Perform any operations enclosed in parentheses.

Step 2 Perform any operations that involve raising to a power.

Step 3 Perform multiplications and divisions as they occur by working from left to right.

Step 4 Perform additions and subtractions as they occur by working from left to right.

Scientific Notation

There is a similar pattern for multiplications of any number by a power of 10. Consider the following examples, and notice what happens to the decimal point.

$9.42 \times 10^1 = 94.2$ *We find these answers by direct multiplication.*

$9.42 \times 10^2 = 942.$

$9.42 \times 10^3 = 9,420.$

$9.42 \times 10^4 = 94,200.$

Do you see the pattern? Look at the decimal point (which is included for emphasis). If we multiply 9.42×10^5, how many places to the right will the decimal point be moved?

$9.42 \times 10^5 = 9\ 42,000.$
$$\downarrow \qquad \uparrow$$
5 places to the right

Using this pattern, can you multiply the following *without direct calculation*?

$9.42 \times 10^{12} = 9,420,000,000,000$

This answer is found by observing the pattern, not by direct multiplication.

The pattern also extends to smaller numbers:

$9.42 \times 10^{-1} = 0.942$ *These numbers are found by direct multiplication.*

$9.42 \times 10^{-2} = 0.0942$ *For example,*

$9.42 \times 10^{-3} = 0.00942$ $9.42 \times 10^{-2} = 9.42 \times \frac{1}{100} = 9.42 \times 0.01 = 0.0942$

Do you see that the same pattern also holds for multiplying by 10 with a negative exponent? Can you multiply the following *without direct calculation*?

$9.42 \times 10^{-6} = 0.\ 000009\ 42$
$$\uparrow \qquad\qquad \downarrow$$
Moved six places to the left

These patterns lead to a useful way for writing large and small numbers, called *scientific notation*.

Scientific Notation

The **scientific notation** for a nonzero number is that number written as a power of 10 times another number x, such that x is between 1 and 10, including 1; that is, $1 \leq x < 10$.

Example 2 Write in scientific notation

Write the given numbers in scientific notation.
a. 123,400   **b.** 0.000035   **c.** 1,000,000,000,000   **d.** 7.35

Solution
a. $123,400 = 1.234 \times 10^5$
b. $0.000035 = 3.5 \times 10^{-5}$
c. $1,000,000,000,000 = 10^{12}$; technically, this is 1×10^{12} with the 1 understood.
d. 7.35 (or 7.35×10^0) is in scientific notation.

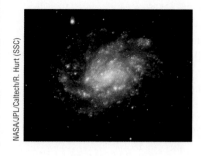

NASA/JPL/Caltech/R. Hurt (SSC)

| Example | **3** | **Miles in a light-year** |

Assuming that light travels at 186,000 miles per second, what is the distance (in miles) that light travels in 1 year? This is the unit of length known as a *light-year*. Give your answer in scientific notation.

Solution   One year is 365.25 days $= 365.25 \times 24$ hours
$$= 365.25 \times 24 \times 60 \text{ minutes}$$
$$= 365.25 \times 24 \times 60 \times 60 \text{ seconds}$$
$$= 31{,}557{,}600 \text{ seconds}$$

Since light travels 186,000 miles each second and there are 31,557,600 seconds in 1 year, we have

$$186{,}000 \times 31{,}557{,}600 = 5{,}869{,}713{,}600{,}000 \approx 5.87 \times 10^{12}$$

Thus, light travels about 5.87×10^{12} miles in 1 year.

Calculators

Throughout the book we will include calculator comments for those of you who have (or expect to have) a calculator. Calculators are classified according to their ability to perform different types of calculations, as well as by the type of logic they use to do the calculations. The problem of selecting a calculator is further complicated by the multiplicity of brands from which to choose. Therefore, choosing a calculator and learning to use it require some sort of instruction.

For most nonscientific purposes, a four-function calculator with memory is sufficient for everyday use. If you anticipate taking several mathematics and/or science courses, you will find that a scientific calculator is a worthwhile investment. These calculators use essentially three types of logic: *arithmetic, algebraic,* and *RPN.* In the previous section, we discussed the correct order of operations, according to which the correct value for

$$2 + 3 \times 4$$

is 14 (multiply first). An algebraic calculator will "know" this and will give the correct answer, whereas an arithmetic calculator will simply work from left to right and obtain the incorrect answer, 20. Therefore, if you have an arithmetic-logic calculator, you will need to be careful about the order of operations. Some arithmetic-logic calculators provide parentheses, $\boxed{(}\,\boxed{)}$, so that operations can be grouped, as in

$$\boxed{2}\,\boxed{+}\,\boxed{(}\,\boxed{3}\,\boxed{\times}\,\boxed{4}\,\boxed{)}\,\boxed{=}$$

but then you must remember to insert the parentheses.

The last type of logic is RPN. A calculator using this logic is characterized by $\boxed{\text{ENTER}}$ or $\boxed{\text{SAVE}}$ keys and does not have an equal key $\boxed{=}$. With an RPN calculator, the operation symbol is entered after the numbers have been entered. These three types of logic can be illustrated by the problem $2 + 3 \times 4$:

Arithmetic logic: $\boxed{3}\,\boxed{\times}\,\boxed{4}\,\boxed{=}\,\boxed{+}\,\boxed{2}\,\boxed{=}$ *Input to match order of operations.*

Algebraic logic: $\boxed{2}\,\boxed{+}\,\boxed{3}\,\boxed{\times}\,\boxed{4}\,\boxed{=}$ *input is the same as the problem.*

RPN logic: $\boxed{2}\,\boxed{\text{ENTER}}\,\boxed{3}\,\boxed{\text{ENTER}}\,\boxed{4}\,\boxed{\times}\,\boxed{+}$ *Operations input last.*

In this book, we will illustrate the examples using algebraic logic. If you have a calculator with RPN logic, you can use your owner's manual to change the examples to RPN. We do not recommend using an arithmetic logic calculator. We will also indicate the keys to be pushed by drawing boxes around the numerals and operational signs as shown.

Example 4 Calculator addition

Show the calculator steps for $14 + 38$.

Solution Be sure to turn your calculator on, or clear the machine if it is already on. A clear button is designated by $\boxed{C}$, and the display will show 0 after the clear button is pushed. You will need to check these steps every time you use your calculator, but after a while it becomes automatic. We will not remind you of this in each example.

Press	Display	
$\boxed{1}$	1	Here we show each numeral in a single box, which means you key
$\boxed{4}$	14	in one numeral at a time, as shown. From now on, this will be shown as $\boxed{14}$.
$\boxed{+}$	14	
$\boxed{38}$	38	Some calculators display all of keystrokes: 14 + 38
$\boxed{=}$	52	

After completing Example 4, you can either continue with the same problem or start a new problem. If the next button pressed is an operation button, the result 52 will be carried over to the new problem. If the next button pressed is a numeral, the 52 will be lost and a new problem started. For this reason, it is not necessary to press $\boxed{C}$ to clear between problems. The button $\boxed{CE}$ is called the *clear entry* key and is used if you make a mistake keying in a number and do not want to start over with the problem. For example, if you want $2 + 3$ and accidentally push $\boxed{2}\ \boxed{+}\ \boxed{4}$ you can then push $\boxed{CE}\ \boxed{3}\ \boxed{=}$ to obtain the correct answer. This is especially helpful if you are in the middle of a long calculation. Some models have a $\boxed{\leftarrow}$ key instead of a $\boxed{CE}$ key.

Example 5 Mixed operations using a calculator

Show the calculator steps and display for $4 + 3 \times 5 - 7$.

Solution

Press:	$\boxed{4}$	$\boxed{+}$	$\boxed{3}$	$\boxed{\times}$	$\boxed{5}$	$\boxed{-}$	$\boxed{7}$	$\boxed{=}$ or $\boxed{ENTER}$
Display:	4	4+	4+3	4+3·	4+3·5	4+3·5−	4+3·5−7	12

If you have an algebraic-logic calculator, your machine will perform the correct order of operations. If it is an arithmetic-logic calculator, it will give the incorrect answer 28 unless you input the numbers using the order-of-operations agreement.

Example 6 Calculator multiplication

Repeat Example 3 using a calculator; that is, find

$$365.25 \times 24 \times 60 \times 60 \times 186{,}000$$

Solution When you press these calculator keys and then press $\boxed{=}$ the display will probably show something that looks like:

5.86971 12   or   5.86971 + 12   or   5.86971E12

This display is a form of scientific notation. The 12 or +12 at the right (separated by one or two blank spaces) is the exponent on the 10 when the number in the display is written in scientific notation. That is,

5.86971 12   means 5.86971×10^{12}

Suppose you have a particularly large number that you wish to input into a calculator—say, 920,000,000,000,000,000,000 miles divided by 7,927 miles (from Figure 1.13). You can input 7,927, but if you attempt to input the larger number you will be stuck when you fill up the display (9 or 12 digits). Instead, you will need to write

$$920{,}000{,}000{,}000{,}000{,}000{,}000 = 9.2 \times 10^{20}$$

This may be entered by pressing an $\boxed{\text{EE}}$, $\boxed{\text{EEx}}$, or $\boxed{\text{EXP}}$ key:

$\boxed{9.2}$ $\boxed{\text{EE}}$ $\boxed{20}$ $\boxed{\div}$ $\boxed{7927}$ $\boxed{=}$ *Display:* 1.160590387E 17

This means that the last cube in Figure 1.13 is about 1.2×10^{17} times larger than the earth.

Scientific notation is represented in a slightly different form on many calculators and computers, and this new form is sometimes called **floating-point form.** When representing very large or very small answers, most calculators will automatically output the answers in floating-point notation. The following example compares the different forms for large and small numbers with which you should be familiar.

Do not confuse the scientific notation keys on your calculator with the exponent key. Exponent keys are labeled $\boxed{y^x}$, $\boxed{\wedge}$ and key or $\boxed{10^x}$.

Example 7 | Find scientific and calculator notation

Write each given number in scientific notation and in calculator notation.
a. 745   **b.** 1,230,000,000   **c.** 0.00573   **d.** 0.00000 06239

Solution   The form given in this example is sometimes called **fixed-point form** or **decimal notation** to distinguish it from the other forms.

	Fixed-Point	*Scientific Notation*	*Floating-Point*
a.	745	7.45×10^2	7.45 02
b.	1,230,000,000	1.23×10^9	1.23 09
c.	0.00573	5.73×10^{-3}	5.73 −03
d.	0.00000 06239	6.239×10^{-7}	6.239 −07

Estimation

The scientist in Figure 1.14 is making an estimate of the velocity of the stream.

Part of problem solving is using common sense about the answers you obtain. This is even more important when using a calculator, because there is a misconception that if a calculator or computer displays an answer, "it must be correct." Reading and understanding the problem are parts of the process of problem solving.

When problem solving, you must ask whether the answer you have found is reasonable. If I ask for the amount of rent you must pay for an apartment, and you do a calculation and arrive at an answer of $16.25, you know that you have made a mistake. Likewise, an answer of $135,000 would not be reasonable. As we progress through this course you will be using a calculator for many of your calculations, and with a calculator you can easily press the wrong button and come up with an outrageous answer. One aspect of *looking back* is using common sense to make sure the answer is reasonable. The ability to recognize the difference between reasonable answers and unreasonable ones is important not only in mathematics, but whenever you are problem solving. This ability is even more important when you use a calculator, because pressing the incorrect key can often cause outrageously unreasonable answers.

Whenever you try to find an answer, you should ask yourself whether the answer is reasonable. How do you decide whether an answer is reasonable? One way is to **estimate** an answer. Webster's *New World Dictionary* tells us that as a verb, to *estimate* means "to form an opinion or a judgment about" or to calculate "approximately."

FIGURE 1.14 Estimating the velocity of a stream

The National Council of Teachers of Mathematics emphases the importance of estimation:

> The broad *mathematical context* for an estimate is usually one of the following types:
>
> A. An exact value is known but for some reason an estimate is used.
> B. An exact value is possible but is not known and an estimate is used.
> C. An exact value is impossible.

We will work on building your estimation skills throughout this book.

Example 8 **Estimate annual salary**

If your salary is $14.75 per hour, your annual salary is approximately

A. $5,000   B. $10,000   C. $15,000   D. $30,000   E. $45,000

Solution Problem solving often requires some assumptions about the problem. For this problem, we are not told how many hours per week you work, or how many weeks per year you are paid. We assume a 40-hour work-week, and we also assume that you are paid for 52 weeks per year. *Estimate:* Your hourly salary is about $15 per hour. A 40-hour week gives us $40 \times \$15 = \$600$ per week. For the estimate, we calculate the wages for 50 weeks instead of 52: 50 weeks yields $50 \times \$600 = \$30,000$. The answer is D.

Example 9 **Estimate map distance**

Use the map in Figure 1.15 to estimate the distance from Orlando International Airport to Disney World.

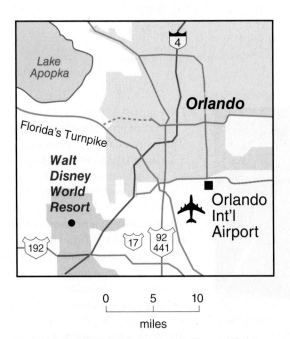

FIGURE 1.15 Map around Walt Disney World

Solution Note that the scale is 10 miles to 1 in. Looking at the map, you will note that it is approximately 1.5 in. from the airport to Disney World. This means that we estimate the distance to be 15 miles.

There are two important reasons for estimation: (1) to form a reasonable opinion or (2) to check the reasonableness of an answer. We will consider estimation for measurements in Chapter 9; if reason (1) is our motive, we should not think it necessary to follow an estimation by direct calculation. To do so would defeat the purpose of the estimation. On the other hand, if we are using the estimate for reason (2)—to see whether an answer is reasonable—we might perform the estimate as a check on the calculated answer for Example 3 (the problem about the speed of light):

$$186,000 \text{ miles per second} \approx 2 \times 10^5 \text{ miles per second}$$

and

$$
\begin{aligned}
\text{one year} \;&\approx\; 4 \times 10^2 \text{ days} \\
&\approx\; \underbrace{4 \times 10^2}_{days} \times \underbrace{2 \times 10}_{hr\ per\ day} \times \underbrace{6 \times 10}_{min\ per\ hr} \times \underbrace{6 \times 10}_{sec\ per\ hr} \\
&\approx\; \underbrace{(4 \times 2 \times 6 \times 6) \times 10^5}_{seconds\ per\ year} \\
&\approx\; (10 \times 36) \times 10^5 \\
&\approx\; 3.6 \times 10^7
\end{aligned}
$$

Thus, one light year is about

$$\left(2 \times 10^5 \, \frac{\text{miles}}{\text{second}}\right)\left(3.6 \times 10^7 \, \frac{\text{seconds}}{\text{year}}\right) \approx 7.2 \times 10^{12} \, \frac{\text{miles}}{\text{year}}$$

This estimate seems to confirm the reasonableness of the answer 5.87×10^{12} we obtained in Example 3.

Laws of Exponents

In working out the previous estimation for Example 3, we used some properties of exponents that we can derive by, once again, turning to some patterns. Consider

$$10 \cdot 10 \cdot 10 \cdot 10 \cdot 10 = 10^5$$

and

$$10^2 \cdot 10^3 = (10 \cdot 10) \cdot (10 \cdot 10 \cdot 10) = 10^5$$

When we *multiply powers* of the same base, we *add* exponents. This is called the **addition law of exponents.**

$$
\begin{aligned}
2^3 \cdot 2^4 &= (2 \cdot 2 \cdot 2) \cdot (2 \cdot 2 \cdot 2 \cdot 2) \\
&= 2^{3+4} \\
&= 2^7
\end{aligned}
$$

Suppose we wish to raise a power to a power. We can apply the addition law of exponents. Consider

$$(2^3)^2 = 2^3 \cdot 2^3 = 2^{3+3} = 2^{2 \cdot 3} = 2^6$$
$$(10^2)^3 = 10^2 \cdot 10^2 \cdot 10^2 = 10^{2+2+2} = 10^{3 \cdot 2} = 10^6$$

When we *raise a power to a power*, we *multiply* the exponents. This is called the **multiplication law of exponents.**

A third law is needed to raise products to powers. Consider

$$
\begin{aligned}
(2 \cdot 3)^2 &= (2 \cdot 3) \cdot (2 \cdot 3) \\
&= (2 \cdot 2) \cdot (3 \cdot 3) \\
&= 2^2 \cdot 3^2
\end{aligned}
$$

Thus, $(2 \cdot 3)^2 = 2^2 \cdot 3^2$.

Another result, called the **distributive law of exponents,** says that to *raise a product to a power*, raise each factor to that power and then multiply. For example,

$$(3 \cdot 10^4)^2 = 3^2 \cdot (10^4)^2 = 3^2 \cdot 10^8 = 9 \cdot 10^8$$

Similar patterns can be observed for quotients.

We now summarize the five laws of exponents.*

STOP

Theorems and laws of mathematics are highlighted in a box that looks like this.

Laws of Exponents

Addition law:	$b^m \cdot b^n = b^{m+n}$
Multiplication law:	$(b^n)^m = b^{mn}$
Subtraction law:	$\dfrac{b^m}{b^n} = b^{m-n}$
Distributive laws:	$(ab)^m = a^m b^m \qquad \left(\dfrac{a}{b}\right)^m = \dfrac{a^m}{b^m}$

Example 10 Estimate a speed

Pólya's Method

Under $\frac{3}{4}$ impulse power, the starship *Enterprise* will travel 1 million kilometers (km) in 3 minutes.† Compare full impulse power with the speed of light, which is approximately $1.08 \cdot 10^9$ kilometers per hour (km/hr).

Solution We use Pólya's problem-solving guidelines for this example.

Understand the Problem. You might say, "I don't know anything about *Star Trek*," but with most problem solving in the real world, the problems you are asked to solve are often about situations with which you are unfamiliar. Finding the necessary information to understand the question is part of the process. We assume that full impulse is the same as 1 impulse power, so that if we multiply $\frac{3}{4}$ impulse power by $\frac{4}{3}$ we will obtain $\left(\frac{3}{4} \cdot \frac{4}{3} = 1\right)$ full impulse power.

Devise a Plan. We will calculate the distance traveled (in kilometers) in one hour under $\frac{3}{4}$ power, and then will multiply that result by $\frac{4}{3}$ to obtain the distance in kilometers per hour under full impulse power.

Carry Out the Plan.

$$\frac{3}{4} \text{ impulse power} = \frac{1{,}000{,}000 \text{ km}}{3 \text{ min}} \qquad \textit{Given}$$

$$= \frac{10^6 \text{ km}}{3 \text{ min}} \cdot \frac{20}{20} \qquad \textit{Multiply by } 1 = \frac{20}{20} \textit{ to change 3 minutes to 60 minutes.}$$

$$= \frac{10^6 \cdot 2 \cdot 10 \text{ km}}{60 \text{ min}}$$

$$= \frac{2 \cdot 10^7 \text{ km}}{1 \text{ hr}}$$

$$= 2 \cdot 10^7 \text{ km/hr}$$

We now multiply both sides by $\frac{4}{3}$ to find the distance under full impulse.

$$\frac{4}{3}\left(\frac{3}{4} \text{ impulse power}\right) = \frac{4}{3} \cdot 2 \cdot 10^7 \text{ km/hr}$$

$$\text{full impulse power} = \frac{8}{3} \cdot 10^7 \text{ km/hr}$$

$$\approx 2.666666667 \cdot 10^7 \text{ km/hr}$$

*You may be familiar with these laws of exponents from algebra. They hold with certain restrictions; for example, division by zero is excluded. We will discuss different sets of numbers in Chapter 5.

†*Star Trek, The Next Generation* (episode that first aired the week of May 15, 1993).

Comparing this to the speed of light, we see

$$\frac{\text{IMPULSE SPEED}}{\text{SPEED OF LIGHT}} = \frac{2.666666667 \cdot 10^7}{1.08 \cdot 10^9} = \frac{2.666666667}{1.08} \cdot 10^{7-9} \approx 0.025$$

Look Back. We see that full impulse power is about 2.5% of the speed of light.

Comprehending Large Numbers

We began this section by looking at the size of the cosmos. But just how large is large? Most of us are accustomed to hearing about millions, billions (budgets or costs of disasters), or even trillions (the national debt is about $15 trillion), but how do we really understand the magnitude of these numbers?

**You may have seen the worldwide show,
*Who Wants to Be a Millionaire?***

A **million** is a fairly modest number, 10^6. Yet if we were to count one number per second, nonstop, it would take us about 278 hours or approximately $11\frac{1}{2}$ days to count to a million. Not a million days have elapsed since the birth of Christ (a million days is about 2,700 years). A large book of about 700 pages contains about a million letters. How large a room would it take to hold 1,000,000 inflated balloons?

The next big number is a **billion,** which is defined to be 1,000 millions.

However, with the U.S. government bailout in early 2009 we have entered the age of trillions. How large is a **trillion**? How long would it take you to count to a trillion?

**Congressional leaders said that as much as $1 trillion will
be needed to avoid an imminent meltdown in the U.S.
financial system.**

Go ahead—make a guess. To get some idea about how large a trillion is, let's compare it to some familiar units:

- If you gave away $1,000,000 *per day*, it would take you more than 2,700 *years* to give away a trillion dollars.
- A stack of a trillion $1 bills would be more than 59,000 miles high.
- At 5% interest, a trillion dollars would earn you $219,178,080 interest *per day*!
- A trillion seconds ago, Neanderthals walked the earth (31,710 years ago).

But a trillion is only 10^{12}, a mere nothing when compared with the real giants. Keep these magnitudes (sizes) in mind. Earlier in this section we noticed that a cube containing our solar neighborhood is 9.2×10^{11} miles on a side. (See Figure 1.13.) This is less than a billion times the size of the earth. (Actually, it is $9.2 \cdot 10^{11} \div 7,927 \approx 1.2 \times 10^{8}$.)

There is an old story of a king who, being under obligation to one of his subjects, offered to reward him in any way the subject desired. Being of mathematical mind and modest tastes, the subject simply asked for a chessboard with one grain of wheat on the first square, two on the second, four on the third, and so forth. The old king was delighted with this modest request! Alas, the king was soon sorry he granted the request.

Example **11** **Estimate a large number**

Pólya's Method

Estimate the magnitude of the grains of wheat on the last square of a chessboard.

Solution We use Pólya's problem-solving guide-lines for this example.*

Understand the Problem. Each square on the chessboard has grains of wheat placed on it. To answer this question you need to know that a chessboard has 64 squares. The first square has $1 = 2^0$ grains, the next has $2 = 2^1$ grains, the next $4 = 2^2$, and so on. Thus, he needed 2^{63} grains of wheat for the last square alone. We showed this number in Example 1d.

Devise a Plan. We know (from Example 1) that $2^{63} \approx 9.22337 \times 10^{18}$. We need to find the size of a grain of wheat, and then convert 2^{63} grains into bushels. Finally, we need to state this answer in terms we can understand.

Carry Out the Plan. I went to a health food store, purchased some raw wheat, and found that there are about 250 grains per cubic inch (in.³). I also went to a dictionary and found that a bushel is 2,150 in.³. Thus, the number of grains of wheat in a bushel is

$$2,150 \times 250 = 537,500 = 5.375 \times 10^5$$

To find the number of bushels in 2^{63} grains, we need to divide:

$$\left(9.922337 \times 10^{18}\right) \div \left(5.375 \times 10^5\right) = \frac{9.22337}{5.375} \times 10^{18-5}$$
$$\approx 1.72 \times 10^{13}$$

Look Back. This answer does not mean a thing without looking back and putting it in terms we can understand. I googled "U.S. wheat production" and found that in 2001 the U.S. wheat production was 2,281,763,000 bushels. To find the number of years it would take the United States to produce the necessary wheat for the last square of the chessboard, we need to divide the production into the amount needed:

$$\frac{1.72 \times 10^{13}}{2.28 \times 10^9} = \frac{1.72}{2.28} \times 10^{13-9} \approx 0.75 \times 10^4 \text{ or } 7.5 \times 10^3$$

This is 7,500 years!

---

*A chessboard has 64 alternating black and red squares arranged into an 8-by-8 pattern.

What is the name of the largest number you know? Go ahead—answer this question.
Recently we have heard about the national debt, which exceeds $15 **trillion.** Table 1.1
shows some large numbers.

TABLE 1.1			Some Large Numbers
Number	Name		Meaning
1	one	1	← Basic counting unit
2	two	2	← Number of computer states (on/off)
2^3	byte	8	← A basic unit on a computer; a string of eight binary digits
10	ten	10	← Number of fingers on two normal hands
10^2	hundred	100	← Number of pennies in a dollar
10^3	thousand	1,000	← About 5,000 of these dots ● would fit 50 per row on this page
2^{10}	kilobyte	1,024	← Computer term for 1,024 bytes, abbreviated K
10^6	million	1,000,000	← Number of letters in a large book.
2^{20}	megabyte	1,048,576	← A unit of computer storage capacity; MB
10^9	billion	1,000,000,000	← Discussed in text
2^{30}	gigabyte	1,073,741,824	← Approximately 1,000 MB; abbreviated GB
10^{12}	trillion	1,000,000,000,000	← National debt is about $15 trillion
10^{15}	quadrillion	1,000,000,000,000,000	← Number of words ever printed
10^{18}	quintillion	1,000,000,000,000,000,000	← Estimated number of insects in the world
10^{21}	sextillion	1,000,000,000,000,000,000	← Cups of water in all the oceans
10^{63}	vigintillion	1 followed by 63 zeros	← Chessboard problem; cubic inches in Milky Way
10^{100}	googol	1 followed by 100 zeros	← 10^{128} is the number of neutrons in the universe.
10^{googol}	googolplex	10 to the power of a googol	← This is really too large to comprehend.

Do Things Really Change?

"Students today can't prepare bark to calculate their problems. They depend upon their slates
which are more expensive. What will they do when their slate is dropped and it breaks? They
will be unable to write!"

Teacher's Conference, 1703

"Students depend upon paper too much. They don't know how to write on slate without chalk
dust all over themselves. They can't clean a slate properly. What will they do when they run
out of paper?"

Principal's Association, 1815

"Students today depend too much upon ink. They don't know how to use a pen knife to
sharpen a pencil. Pen and ink will never replace the pencil."

National Association of Teachers, 1907

"Students today depend upon store bought ink. They don't know how to make their own.
When they run out of ink they will be unable to write word or ciphers until their next trip to
the settlement. This is a sad commentary on modern education."

The Rural American Teacher, 1929

"Students today depend upon these expensive fountain pens. They can no longer write with a
straight pen and nib (not to mention sharpening their own quills). We parents must not allow
them to wallow in such luxury to the detriment of learning how to cope in the real business
world, which is not so extravagant."

PTA Gazette, 1941

"Ball point pens will be the ruin of education in our country. Students use these devices and
then throw them away. The American virtues of thrift and frugality are being discarded.
Business and banks will never allow such expensive luxuries."

Federal Teacher, 1950

"Students today depend too much on hand-held calculators."

Anonymous, 1995

Problem Set 1.3

Level 1

1. **IN YOUR OWN WORDS** What do we mean by *exponent*?

2. **IN YOUR OWN WORDS** Define *scientific notation* and discuss why it is useful.

3. **IN YOUR OWN WORDS** Do you plan to use a calculator for working the problems in this book? If so, what type of logic does it use?

4. **IN YOUR OWN WORDS** Describe the differences in evaluating exponents and using scientific notation on a calculator.

5. **IN YOUR OWN WORDS** What is the largest number whose name you know? Describe the size of this number.

6. **IN YOUR OWN WORDS** What is a *trillion*? Do not simply define this number, but discuss its magnitude (size) in terms that are easy to understand.

Write each of the numbers in Problems 7–10 in scientific notation and in floating-point notation (as on a calculator).

7. **a.** 3,200 **b.** 0.0004 **c.** 64,000,000,000

8. **a.** 23.79 **b.** 0.000001 **c.** 35,000,000,000

9. **a.** 5,629 **b.** 630,000 **c.** 0.00000 0034

10. **a.** googol **b.** 1,200,300 **c.** 0.00000 123

Write each of the numbers in Problems 11–14 in fixed-point notation.

11. **a.** 7^2 **b.** 7.2×10^{10} **c.** $4.56 + 3$

12. **a.** 2^6 **b.** 2.1×10^{-3} **c.** $4.07 + 4$

13. **a.** 6^3 **b.** 4.1×10^{-7} **c.** $4.8 \ -7$

14. **a.** 6^{-2} **b.** 3.217×10^7 **c.** $8.89 \ -11$

Write each of the numbers in Problems 15–18 in scientific notation

15.

16. The velocity of light in a vacuum is about 30,000,000,000 cm/sec.

17. The distance between Earth and Mars (220,000,000 miles) when drawn to scale is 0.00000 25 in.

18. In 2010, the national debt was approximately $12 trillion. It has been proposed that this number be used to define a new monetary unit, a *light buck*. That is, one light buck is the amount necessary to generate domestic goods and services at the rate of $186,000 per second. What is the national debt in terms of light bucks?

Write each of the numbers in Problems 19–22 in fixed-point notation.

19. A kilowatt-hour is about 3.6×10^6 joules.

20. A ton is about 9.06×10^2 kilograms.

21. The volume of a typical neuron is about 3×10^{-8} cm^3.

22. If the sun were a light bulb, it would be rated at 3.8×10^{25} watts.

23. Estimate the distance from Los Angeles International Airport to Disneyland.

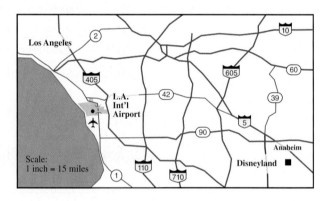

24. Estimate the distance from Fish Camp to Yosemite Village in Yosemite National Park.

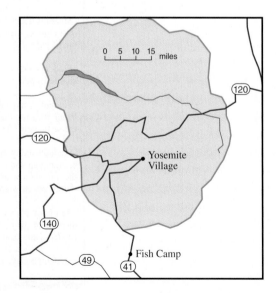

In Problems 25–30, first estimate your answer and then calculate the exact answer.

25. How many pages are necessary to make 1,850 copies of a manuscript that is 487 pages long? (Print on one side only.)

26. If you are paid $16.25 per hour, what is your annual salary?

27. If your car gets 23 miles per gallon, how far can you go on 15 gallons of gas?

28. If your car travels 280 miles and uses 10.2 gallons, how many miles per gallon did you get?

29. In the musical *Rent* there is a song called "Seasons of Love" that uses the number 525,600 minutes. How long is this?

30. It has been estimated that there are 107 billion pieces of mail per year. If the postage rates are raised 2¢, how much extra revenue does that generate?

39.

Alan Schein Photography/Corbis Edge/Corbis

40.

Robert F. Sisson/National Geographic/Getty Images

Level 2

Compute the results in Problems 31–36. Leave your answers in scientific notation.

31. a. $(6 \times 10^5)(2 \times 10^3)$ **b.** $\dfrac{6 \times 10^5}{2 \times 10^3}$

32. a. $\dfrac{(5 \times 10^4)(8 \times 10^5)}{4 \times 10^6}$ **b.** $\dfrac{(6 \times 10^{-3})(7 \times 10^8)}{3 \times 10^7}$

33. a. $\dfrac{(6 \times 10^7)(4.8 \times 10^{-6})}{2.4 \times 10^5}$ **b.** $\dfrac{(2.5 \times 10^3)(6.6 \times 10^8)}{8.25 \times 10^4}$

34. a. $\dfrac{(2xy^{-2})(2^{-1}x^{-1}y^4)}{x^{-2}y^2}$ **b.** $\dfrac{x^2y(2x^3y^{-5})}{2^{-2}x^4y^{-8}}$

35. a. $\dfrac{0.00016 \times 500}{2,000,000}$ **b.** $\dfrac{15,000 \times 0.0000004}{0.005}$

36. a. $\dfrac{4,500,000,000,000 \times 0.00001}{50 \times 0.0003}$

b. $\dfrac{0.0348 \times 0.00000\ 00000\ 00002}{0.000058 \times 0.03}$

Estimate the number of items in each photograph in Problems 37–40.

37.

klenger/iStockphoto.com

38.

Lori Sparkia, 2010/Used under license from Shutterstock.com

In Problems 41–48, you need to make some assumptions before you estimate your answer. State your assumptions and then calculate the exact answer.

41. How many classrooms would be necessary to hold 1,000,000 inflated balloons?

42. Carrie Dashow, the "say hello" woman, is trying to personally greet 1,000,000 people. In her first year, which ended on January 3, 2000, she had greeted 13,688 people. At this rate, how long will it take her to greet one million people?

43. Approximately how high would a stack of 1 million $1 bills be? (Assume there are 233 new $1 bills per inch.)

44. Estimate how many pennies it would take to make a stack 1 in. high. Approximately how high would a stack of 1 million pennies be?

45. If the U.S. annual production of sugar is 30,000,000 tons, estimate the number of grains of sugar produced in a year in the United States. Use scientific notation. (*Note:* There are 2,000 lb per ton; assume there are 2,260,000 grains in a pound of sugar.)

46. The *San Francisco Examiner* (Feb. 6, 2000, Travel Section) reported that David Phillips, a civil engineer at University of California, Davis, was pushing his shopping cart when he noticed a promotion of Healthy Choice®.

Eric Francis/Bloomberg via Getty Images.

He could earn 1,000 airline miles for every 10 bar codes from Healthy Choice products he sent to the company by the end of the month. Frozen entries are about $2 apiece, but with a little work he found individual servings of chocolate pudding for 25 cents each. He was able to accumulate 1,215,000 airline miles. How much did it cost him?

47. A school in Oakland, California, spent $100,000 in changing its mascot sign. If the school had used this amount of money for chalk, estimate the length of the chalk laid end-to-end.

48. "Each year Delta serves 9 million cans of Coke®. Laid end-to-end, they would stretch from Atlanta to Chicago."* Is this a reasonable estimate? What is the actual length of 9 million Coke cans laid end-to-end?

Level 3

49. a. What is the largest number you can represent on your calculator?
b. What is the largest number you can think of using only three digits?
c. Use scientific notation to estimate this number. Your calculator may help, but will probably not give you the answer directly.

50. HISTORICAL QUEST Zerah Colburn (1804–1840) toured America when he was 6 years old to display his calculating ability. He could instantaneously give the square and cube roots of large numbers. It is reported that it took him only a few seconds to find 8^{16}. Use your calculator to *help you* find this number exactly (not in scientific notation).

51. HISTORICAL QUEST Jedidiah Buxion (1707–1772) never learned to write, but given any distance he could tell you the number of inches, and given any length of time he could tell you the number of seconds. If he listened to a speech or a sermon, he could tell the number of words or syllables in it. It reportedly took him only a few moments to mentally calculate the number of cubic inches in a right-angle block of stone

23,451,789 yards long, 5,642,732 yards wide, and 54,465 yards thick. Estimate this answer on your calculator. You will use scientific notation because of the limitations of your calculator, but remember that Jedidiah gave the *exact* answer by working the problem in his head.

52. HISTORICAL QUEST George Bidder (1806–1878) not only possessed exceptional power at calculations but also went on to obtain a good education. He could give immediate answers to problems of compound interest and annuities. One question he was asked was, If the moon is 238,000 miles from the earth and sound travels at the rate of 4 miles per minute, how long would it be before the inhabitants of the moon could hear the Battle of Waterloo? By calculating *mentally,* he gave the answer in less than one minute! First make an estimate, and then use your calculator to give the answer in days, hours, and minutes, to the nearest minute.

53. Estimate the number of bricks required to build a solid wall 100 ft by 10 ft by 1 ft.

54. A sheet of notebook paper is approximately 0.003 in. thick. Tear the sheet in half so that there are 2 sheets. Repeat so that there are 4 sheets. If you repeat again, there will be a pile of 8 sheets. Continue in this fashion until the paper has been halved 50 times. If it were possible to complete the process, how high would you guess the final pile would be? After you have guessed, *compute* the height.

Problem Solving 3

55. If it takes one second to write down each digit, how long will it take to write down all the numbers from 1 to 1,000?

56. If it takes one second to write down each digit, how long will it take to write down all the numbers from 1 to 1,000,000?

57. Imagine that you have written down the numbers from 1 to 1,000. What is the total number of zeros you have recorded?

58. Imagine that you have written down the numbers from 1 to 1,000,000. What is the total number of zeros you have recorded?

59. a. If the entire population of the world moved to California and each person were given an equal amount of area, how much space would you *guess* that each person would have (multiple choice)?
A. 7 in.^{2}
B. 7 ft^2
C. 70 ft^2
D. 700 ft^2
E. 1 mi^2
b. If California is $158,600$ mi^2 and the world population is 6.3 billion, calculate the answer to part **a**.

60. It is known that a person's body has about one gallon of blood in it, and that a cubic foot will hold about 7.5 gallons of liquid. It is also known that Central Park in New York has an area of 840 acres. If walls were built around the park, how tall would those walls need to be to contain the blood of all 6,300,000,000 people in the world?

© Rafael Macia/Photo Researchers

1.4 | CHAPTER SUMMARY

Numeracy is the ability to cope confidently with the mathematical demands of adult life.
MATHEMATICS COUNTS

Important Ideas

Guidelines for problem solving [1.1]
Order of operations [1.2]
Extended order of operations [1.3]
Laws of exponents [1.3]
Inductive vs deductive reasoning [1.3]
Euler circles [1.3]

Take some time getting ready to work the review problems in this section. First review the listed important ideas. Look back at the definition and property boxes in this chapter. If you look online, you will find a list of important terms introduced in this chapter, as well as the types of problems that were introduced in this chapter. You will maximize your understanding of this chapter by working the problems in this section only after you have studied the material.

You will find some review help online at **www.mathnature.com.** There are links giving general test help in studying for a mathematics examination, as well as specific help for reviewing this chapter.

Chapter 1 Review Questions

1. In your own words, describe Pólya's problem-solving model.

2. In how many ways can a person walk 5 blocks north and 4 blocks west, if the streets are arranged in a standard rectangular arrangement?

3. A chessboard consists of 64 squares, as shown in Figure 1.17. The rook can move one or more squares horizontally or vertically. Suppose a rook is in the upper-left-hand corner of a chessboard. Tell how many ways the rook can reach the point marked "X". Assume that the rook always moves toward its destination.

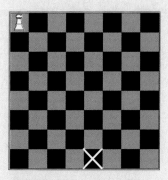

FIGURE 1.17 Chessboard

4. Compute 111,111,111 × 111,111,111. Do not use direct multiplication; show all your work.

5. What is meant by "order of operations"?

6. What is scientific notation?

7. Does the story in the news clip illustrate inductive or deductive reasoning?

> Q: What has 18 legs and catches flies?
> A: I don't know, what?
> Q: A baseball team. What has 36 legs and catches flies?
> A: I don't know that, either.
> Q: Two baseball teams. If the United States has 100 senators, and each state has 2 senators, what does...
> A: I know this one!
> Q: Good. What does each state have?
> A: Three baseball teams!

8. Show the calculator keys you would press as well as the calculator display for the result. Also state the brand and model of the calculator you are using.

   a. 2^{63}

   b. $\dfrac{9.22 \times 10^{18}}{6.34 \times 10^{6}}$

9. Assume that there is a $281.9 billion budget "windfall." Which of the following choices would come closest to liquidating this windfall?
   A. Buy the entire U.S. population a steak dinner.
   B. Burn one dollar per second for the next 1,000 years.
   C. Give $80,000 to every resident of San Francisco and use the remainder of the money to buy a $200 iPod for every resident of China.

10. What is wrong, if anything, with the following "Great Tapes" advertisement?

> Instead of reading the 100 greatest books of all time, buy these beautifully transcribed books on tape. If you listen to only one 45-minute tape per day, you will complete the greatest books of all time in only one year.

11. In 1995, it was reported that an iceberg separated from Antarctica. The size of this iceberg was reported equal to to 7×10^{16} ice cubes. Convert this size to meaningful units.

Paul & Lindamarie Ambrose/Taxi/Getty Images

12. The national debt in 2010 soared to $13,800,000,000,000. Write this number in scientific notation. Suppose that in 2010 there were 310,000,000 people in the United States. If the debt is divided equally among these people, how much is each person's share?

13. Rearrange the cards in the formulation shown here so that each horizontal, vertical, and diagonal line of three adds up to 15.

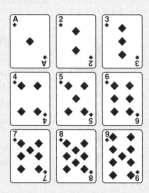

14. Assume that your classroom is 20 ft × 30 ft × 10 ft. If you fill this room with dollar bills (assume that a stack of 233 dollar bills is 1 in. tall), how many classrooms would it take to contain the 2010 national debt of $13,800,000,000,000?

Use Euler circles to check the validity of each of the arguments given in Problems 15–18.

15. All birds have wings.
All flies have wings.
Therefore, some flies are birds.

16. No apples are bananas.
All apples are fruit.
Therefore, no bananas are fruit.

17. All artists are creative.
Some musicians are artists.
Therefore, some musicians are creative.

18. All rectangles are polygons.
All squares are rectangles.
Therefore, all squares are polygons.

19. Consider the following pattern:

1 is happy.
10 is happy because $1^2 + 0^2 = 1$, which is happy.
13 is happy because $1^2 + 3^2 = 10$, which is happy.
19 is happy because $1^1 + 9^2 = 82$ and $8^2 + 2^2 = 68$
and $6^2 + 8^2 = 100$ and $1^2 + 0^2 + 0^2 = 1$, which is happy.

On the other hand,

2, 3, 4, 5, 6, 7, 8, and 9 are unhappy.
11 is unhappy because $1^2 + 1^2 = 2$, which is unhappy.
12 is unhappy because $1^2 + 2^2 = 5$, which is unhappy.

Find one unhappy number as well as one happy number.

20. Suppose you could write out 7^{1000}. What is the last digit?

BOOK REPORTS

Write a 500-word report on one of these books:

Mathematical Magic Show, Martin Gardner (New York: Alfred A. Knopf, 1977).

How to Solve It: A New Aspect of Mathematical Method, George Pólya (New Jersey: Princeton University Press, 1945, 1973).

Group RESEARCH PROJECTS

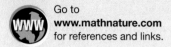

Go to
www.mathnature.com
for references and links.

Working in small groups is typical of most work environments, and learning to work with others to communicate specific ideas is an important skill. Work with three or four other students to submit a single report based on each of the following questions.

G1. It is stated in the Prologue that "Mathematics is alive and constantly changing." As we complete the second decade of this century, we stand on the threshold of major changes in the mathematics curriculum in the United States.

Report on some of these recent changes.

REFERENCES Lynn Steen, *Everybody Counts: A Report to the Nation on the Future of Mathematics Education* (Washington, DC: National Academy Press, 1989). See also *Curriculum and Evaluation Standards for School Mathematics* from the National Council of Teachers of Mathematics (Reston, VA: NCTM, 1989). A reassessment of these standards was done by Kenneth A. Ross, *The MAA and the New NCTM Standards* (© 2000 The Mathematical Association of America.)

G2. Do some research on Pascal's triangle, and see how many properties you can discover. You might begin by answering these questions:
   a. What are the successive powers of 11?
   b. Where are the natural numbers found in Pascal's triangle?
   c. What are triangular numbers and how are they found in Pascal's triangle?
   d. What are the tetrahedral numbers and how are they found in Pascal's triangle?
   e. What relationships do the patterns in Figure 1.18 have to Pascal's triangle?

Multiples of 2

Multiples of 3

FIGURE 1.18 Patterns in Pascal's triangle

REFERENCES James N. Boyd, "Pascal's Triangle," *Mathematics Teacher,* November 1983, pp. 559–560.

Dale Seymour, *Visual Patterns in Pascal's Triangle,* (Palo Alto, CA: Dale Seymour Publications, 1986).

Karl J. Smith, "Pascal's Triangle," *Two-Year College Mathematics Journal*, Volume 4, pp. 1–13 (Winter 1973).

Individual RESEARCH PROJECTS

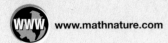
www.mathnature.com

Learning to use sources outside your classroom and textbook is an important skill, and here are some ideas for extending some of the ideas in this chapter. You can find references to these projects in a library or at www.mathnature.com.

PROJECT 1.1 Find some puzzles, tricks, or magic stunts that are based on mathematics.

PROJECT 1.2 Write a short paper about the construction of magic squares. Figure 1.19 shows a magic square created by Benjamin Franklin.

52	61	4	13	20	29	36	45
14	3	62	51	46	35	30	19
53	60	5	12	21	28	37	44
11	6	59	54	43	38	27	22
55	58	7	10	23	26	39	42
9	8	57	56	41	40	25	24
50	63	2	15	18	31	34	47
16	1	64	49	48	33	32	17

FIGURE 1.19 Benjamin Franklin magic square*

PROJECT 1.3 Design a piece of art based on a magic square.

PROJECT 1.4 An *alphamagic square*, invented by Lee Sallows, is a magic square so that not only do the numbers spelled out in words form a magic square, but the numbers of letters of the words also form a magic square. For example,

five	twenty-two	eighteen
twenty-eight	fifteen	two
twelve	eight	twenty-five

gives rise to two magic squares:

5	22	18
28	15	2
12	8	25

and

4	9	8
11	7	3
6	5	10

The first magic square comes from the numbers represented by the words in the alphamagic square, and the second magic square comes from the numbers of letters in the words of the alphamagic square.

a. Verify that this is an alphamagic square.   **b.** Find another alphamagic square.

PROJECT 1.5 Answer the question posed in Problem 59, Section 1.3 for your own state. If you live in California, then use Florida.

PROJECT 1.6 Read the article "Mathematics at the Turn of the Millennium," by Philip A. Griffiths, *The American Mathematical Monthly*, January 2000, pp. 1–14. Briefly describe each of these famous problems:

a. Fermat's last theorem

b. Kepler's sphere packing conjecture

c. The four-color problem

Which of these problems are discussed later in this text, and where?

The objective of this article was to communicate something about mathematics to a general audience. How well did it succeed with you? Discuss.

*"How Many Squares Are There, Mr. Franklin?" by Maya Mohsin Ahmed, *The Mathematical Monthly*, May 2004, p. 394.

2 THE NATURE OF SETS

What in the World?

"Hey, James!" said Tony. "Have you made up your mind yet where you are going next year?"

"Nah, my folks are on my case, but I'm in no hurry," James responded. "There are plenty of spots for me in college. I know I will get in someplace."

Many states, including California, Florida, and Kentucky, have state-mandated assessment programs. The following question is found on the California Assessment Program Test as an open-ended problem. We consider it in Problem Set 2.2, Problem 2.

James knows that half the students from his school are accepted at the public university nearby. Also, half are accepted at the local private college. James thinks that this adds up to 100%, so he will surely be accepted at one or the other institution. Explain why James may be wrong. If possible, use a diagram in your explanation.

A Question of Thinking: A First Look at Students Performance on Open-ended Questions in Mathematics

Overview

Sets are considered to be one of the most fundamental building blocks of mathematics. In fact, most mathematics books from basic arithmetic to calculus must introduce the concept early in the book. Small children learn to categorize sets when they learn numbers, colors, shapes, and sizes. The PBS show *Sesame Street* teaches the concept of set building with the song, "One of These Things Is Not Like the Others." A quick search of the Internet will show you that the ideas of set theory can be as elementary as counting and as complex as logic, calculus, and abstract algebra. In this chapter we will consider the basic ideas of set theory and of counting.

© Norbert von der Groeben/The Image Works

Shelter Gate, UC Berkeley Campus

2.1   Sets, Subsets, and Venn Diagrams

The first time I met eminent proof theorist Gaisi Takeuti I asked
him what set theory was really about. "We are trying to get [an]
exact description of thoughts of infinite mind," he said. And then
he laughed, as if filled with happiness by this impossible task.

RUDY RUCKER

CHAPTER **CHALLENGE**

See if you can fill in the question mark.

2	7	9
5	4	0
7	11	?

A fundamental concept in mathematics—and in life, for that matter—is the sorting of objects into similar groupings. Every language has an abundance of words that mean "a collection" or "a grouping." For example, we speak of a *herd* of cattle, a *flock* of birds, a *school* of fish, a track *team*, a stamp *collection*, and a *set* of dishes. All these grouping words serve the same purpose, and in mathematics we use the word *set* to refer to any collection of objects.

A herd of buffalo at Yellowstone National Park is an example of a set.

The study of sets is sometimes called **set theory,** and the first person to formally study sets was Georg Cantor (1845–1918). In this chapter, we introduce many of Cantor's ideas.

Denoting Sets

In Section 1.2 we introduced the idea of *undefined terms*. The word **set** is a perfect example of why there must be some undefined terms. Every definition requires other terms, so some *undefined terms* are necessary to get us started. To illustrate this idea, let's try to define the word *set* by using dictionary definitions:

"*Set:* a *collection* of objects." What is a collection?

"*Collection:* an *accumulation.*" What is an accumulation?

"*Accumulation:* a *collection*, a *pile*, or a *heap.*" We see that the word *collection* gives us a **circular definition.** What is a *pile*?

"*Pile:* a *heap.*" What is a heap?

"*Heap:* a *pile.*"

Do you see that a dictionary leads us in circles? In mathematics, we do not allow circular definitions, and this forces us to accept some words without definition. The term *set* is undefined. Remember, the fact that we do not define *set* does not prevent us from having an intuitive grasp of how to use the word.

Sets are usually specified in one of two ways. The first is by *description*, and the other is by the *roster* method. In the **description method,** we specify the set by describing it in such a way that we know exactly which elements belong to it. An example is the set of 50 states in the United States of America. We say that this set is **well defined,** since there is no doubt that the state of California belongs to it and that the state of Germany does not; neither does the state of confusion. Lack of confusion, in fact, is necessary in using sets. The distinctive property that determines the inclusion or exclusion of a particular element is called the *defining property* of the set.

Consider the example of *the set of good students in this class.* This set is not well defined, since it is a matter of opinion whether a student is a "good" student. If we agree, however, on the meaning of the words *good students,* then the set is said to be *well defined.* A better (and more precise) formulation is usually required—for example, *the set of all students in this class who received a C or better on the first examination.* This is well defined, since it can be clearly determined exactly which students received a C or better on the first test.

In the **roster method,** the set is defined by listing the members. The objects in a set are called **members** or **elements** of the set and are said to **belong to** or **be contained in** the set. For example, instead of defining a set as *the set of all students in this class who received a C or better on the first examination,* we might simply define the set by listing its members: {Howie, Mary, Larry}.

Sets are usually denoted by capital letters, and the notation used for sets is braces. Thus, the expression

$$A = \{4, 5, 6\}$$

means that A is the name for the set whose members are the numbers 4, 5, and 6.

Sometimes we use braces with a defining property, as in the following examples:

{states in the United States of America}

{all students in this class who received an A on the first test}

If S is a set, we write $a \in S$ if a is a member of the set S, and we write $b \notin S$ if b is not a member of the set S. Thus, "$a \in \mathbb{Z}$" means that the variable a is an integer, and the statement "$b \in \mathbb{Z}, b \neq 0$" means that the variable b is a nonzero integer.

Example 1 Set member notation

Let C = cities in California, a = city of Anaheim, and b = city of Berlin. Use set membership notation to describe relations among a, b, and C.

Solution $a \in C; b \notin C$

A common use of set terminology is to refer to certain sets of numbers. In Chapter 5, we will discuss certain sets of numbers, which we list here:

$\mathbb{N} = \{1, 2, 3, 4, \ldots\}$ Set of **natural,** or **counting, numbers**

$\mathbb{W} = \{0, 1, 2, 3, 4, \ldots\}$ Set of **whole numbers**

$\mathbb{Z} = \{\ldots, -2, -1, 0, 1, 2, \ldots\}$ Set of **integers**

$\mathbb{Q} = \left\{\dfrac{a}{b} \,\middle|\, a \in \mathbb{Z}, b \in \mathbb{Z}, b \neq 0\right\}$ Set of **rational numbers**

The notation we use for the set of rational numbers may be new to you. If we try to list the set of rational numbers by roster, we will find that this is a difficult task (see Problem 56).

A new notation called **set-builder notation** was invented to allow us to combine both the roster and the description methods. Consider:

The set of all x

$\{\overbrace{x}\,|\,x$ is an even counting number$\}$

$\uparrow$

such that

We now use set-builder notation for the set of rational numbers:

$\{\frac{a}{b}\,|\,a$ is an integer and b is a nonzero integer$\}$

Read this as: "The set of all $\frac{a}{b}$ such that a is an integer and b is a nonzero integer."

Example 2 Writing sets by roster

Specify the given sets by roster. If the set is not well defined, say so.
a. {counting numbers between 10 and 20}
b. $\{x\,|\,x$ is an integer between -20 and $20\}$
c. {distinct letters in the word *happy*}
d. {U.S. presidents arranged in chronological order}
e. {good U.S. presidents}

Solution
a. {11, 12, 13, 14, 15, 16, 17, 18, 19}
   Notice that *between* does not include the first and last numbers.
b. $\{-19, -18, -17, \ldots, 17, 18, 19\}$
   Notice that *ellipses* (three dots) are used to denote some missing numbers. When using ellipses, you must be careful to list enough elements so that someone looking at the set can see the intended pattern.
c. $\{h, a, p, y\}$
   Notice that *distinct* means "different", so we do not list two *ps*.
d. {Washington, Adams, Jefferson, . . . , Clinton, Bush, Obama}
e. Not well defined

Example 3 Writing sets by description

Specify the given sets by description.
a. $\{1, 2, 3, 4, 5, \ldots\}$
b. $\{0, 1, 2, 3, 4, 5, \ldots\}$
c. $\{x\,|\,x \in \mathbb{Z}\}$
d. $\{12, 14, 16, \ldots, 98\}$
e. $\{4, 44, 444, 4444, \ldots\}$
f. $\{m, a, t, h, e, i, c, s\}$

Solution   Answers may vary.
a. Counting (or natural) numbers
b. Whole numbers
c. We would read this as "The set of all x such that x belongs to the set of integers." More simply, the answer is integers.
d. {Even numbers between 10 and 100}
e. {Counting numbers whose digits consist of fours only}
f. {Distinct letters in the word *mathematics*}

Equal and Equivalent Sets

We say that two sets A and B are **equal,** written $A = B$, if the sets contain exactly the same elements. Thus, if $E = \{2, 4, 6, 8, \ldots\}$, then

$$\{x \mid x \text{ is an even counting number}\} = \{x \mid x \in E\}$$

The order in which you represent elements in a set has no effect on set membership. Thus,

$$\{1, 2, 3\} = \{3, 1, 2\} = \{2, 1, 3\} = \ldots$$

Also, if an element appears in a set more than once, it is not generally listed more than a single time. For example,

$$\{1, 2, 3, 3\} = \{1, 2, 3\}$$

Another possible relationship between sets is that of *equivalence*. Two sets A and B are **equivalent,** written $A \leftrightarrow B$, if they have the same *number* of elements. Don't confuse this concept with equality. Equivalent sets do not need to be equal sets, but equal sets are always equivalent.

Example **4** **Equal and equivalent sets**

Which of the following sets are equivalent? Are any equal?
$\{\bigcirc, \triangle, \square, \odot\}$, $\{5, 8, 11, 14\}$, $\{\heartsuit, \clubsuit, \blacklozenge, \spadesuit\}$, $\{\bullet, \odot, \star, \boxplus\}$, $\{1, 2, 3, 4\}$

Solution   All of the given sets are equivalent. Notice that no two of them are equal, but they all share the property of "fourness."

The number of elements in a set is often called its **cardinality.** The cardinality of the sets in Example 4 is 4; that is, the common property of the sets is the **cardinal number** of the set. The cardinality of a set S is denoted by $|S|$. Equivalent sets with four elements each have in common the property of "fourness," and thus we would say that their cardinality is 4.

Example **5** **Finding cardinality**

Find the cardinality of each of the following sets.
a. $R = \{5, \triangle, Y, \pi\}$ **b.** $S = \{\ \ \}$ **c.** $T = \{\text{states of the United States}\}$

Solution
a. The cardinality of R is 4, so we write $|R| = 4$.
b. The cardinality of S (the empty set) is 0, so we write $|S| = 0$.
c. The cardinality of T is 50, or $|T| = 50$.

Universal and Empty Sets

We now consider two important sets in set theory. The first is the set that contains every element under consideration, and the second is the set that contains no elements. A **universal set,** denoted by U, contains all the elements under consideration in a given discussion; and the **empty set** contains no elements, and thus has cardinality 0. The empty set is denoted by $\{\ \ \}$ or $\varnothing$. Do not confuse the notations $\varnothing$, 0, and $\{\varnothing\}$. The symbol $\varnothing$

denotes a *set* with no elements; the symbol 0 denotes a *number*; and the symbol $\{\varnothing\}$ is a set with one element (namely, the set containing $\varnothing$).

For example, if $U = \{1, 2, 3, 4, 5, 6, 7, 8, 9\}$, then all sets we would be considering would have elements only among the elements of U. No set could contain the number 10, since 10 is not in that agreed-upon universe.

For every problem, a universal set must be specified or implied, and it must remain fixed for that problem. However, when a new problem is begun, a new universal set can be specified.

Notice that we defined *a* universal set and *the* empty set; that is, a universal set may vary from problem to problem, but there is only one empty set. After all, it doesn't matter whether the empty set contains no numbers or no people—it is still empty. The following are examples of descriptions of the empty set:

{living saber-toothed tigers}   {counting numbers less than 1}

Venn Diagrams

A *set* is a collection, and a useful way to depict a set is to draw a circle or an oval as a representation for the set. The elements are depicted inside the circle, and objects not in the set are shown outside the circle. The *universal set* contains all the elements under consideration in a given discussion and is depicted as a rectangle. This representation of a set is called a **Venn diagram,** after John Venn (1834–1923). As we saw in Chapter 1, the Swiss mathematician Leonhard Euler (1707–1783) also used circles to illustrate principles of logic, so sometimes these diagrams are called *Euler circles*, which we discussed in Chapter 1. However, Venn was the first to use them in a general way.

Example   **6**   **Venn diagram for a given set**

Let the universal set be all of the cards in a deck of cards.[*] Draw a Venn diagram for the set of hearts.

Solution   It is customary to represent the universal set as a rectangle (labeled U) and the set of hearts (labeled H) as a circle, as shown in Figure 2.1. Note that

$$H = \{\heartsuit A, \heartsuit 2, \heartsuit 3, \heartsuit 4, \heartsuit 5, \heartsuit 6, \heartsuit 7, \heartsuit 8, \heartsuit 9, \heartsuit 10, \heartsuit J, \heartsuit Q, \heartsuit K\}$$

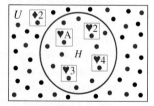

FIGURE 2.1 Venn diagram for a deck of cards

In the Venn diagram, the sets involved are too large to list all of the elements individually in either H or U, but we can say that the two of hearts (labeled $\heartsuit 2$) is a member of H, whereas the two of diamonds (labeled $\diamondsuit 2$) is not a member of H. We write

$$\heartsuit 2 \in H, \quad \text{whereas} \quad \diamondsuit 2 \notin H$$

[*]See Figure 12.2, p. 584 if you are not familiar with a deck of cards.

The set of elements that are not in H is referred to as the **complement** of H, and this is written using an overbar. In Example 6, $\overline{H} = \{spades, diamonds, clubs\}$. A Venn diagram for complement is shown in Figure 2.2.

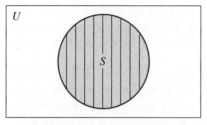

a. To represent a set S, shade the interior of S; the answer is everything shaded (shown as a color screen).

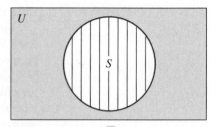

b. To represent a set $\overline{S}$, shade S; the answer is everything not shaded (shown as a color screen).

FIGURE 2.2 Venn diagrams for a set and for its complement

Notice that any set S divides the universe into two regions as shown in Figure 2.3.

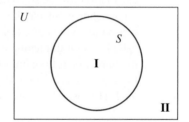

FIGURE 2.3 General representation of a set S

Notice that the cardinality of the deck of cards in Example 6 is 52, and the cardinality of H is 13. Since the deck of cards is the universal set for Example 6, we can symbolize the cardinality of these sets as follows:

$$|H| = 13 \quad \text{and} \quad |U| = 52$$

Note that $H \neq |H|$. In words, a set is **not** the same as its cardinality.

Subsets and Proper Subsets

Most applications will involve more than one set, so we begin by considering the relationships between two sets A and B. The various possible relationships are shown in Figure 2.4. We say that A is a **subset** of B, which in set theory is written $A \subseteq B$, if every element of A is also an element of B (see Figure 2.4a). Similarly, $B \subseteq A$ if every element of B is also an element of A (Figure 2.4b). Figure 2.4c shows two equal sets. Finally, A and B are **disjoint** if they have no elements in common (Figure 2.4d).

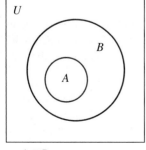

a. $A \subseteq B$

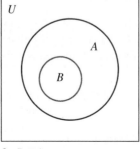

b. $B \subseteq A$

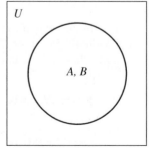

c. $A = B$

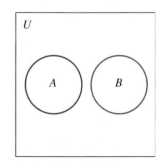

d. A and B are disjoint

FIGURE 2.4 Relationships between two sets A and B

Example 7 Set notation

Answer each of the given true/false questions.

A = {American manufactured automobiles}
C = {Chevrolet, Cobolt, Corvette}
R = {Colors in the rainbow}
S = {red, orange, yellow, green, blue, indigo, violet}

a. $A \subseteq C$ **b.** $C \subseteq A$ **c.** Chevrolet $\subseteq A$ **d.** red $\in R$ **e.** {red} $\in R$
f. $R \subseteq S$ **g.** $R \subset S$ **h.** $\varnothing \subseteq A$ **i.** { } $\subset C$ **j.** {$\varnothing$} $\subset R$

Solution

a. "A is a subset of C" is *false* because there is at least one American manufactured automobile—say the Cadillac—that is not listed in the set C.
b. "C is a subset of A" is *true* because each element of C is also an element of A.
c. "Chevrolet is a subset of A" is *false* since "Chevrolet" is an element, not a set.
d. "Red is an element of the set R" is *true*. Contrast the notation in parts **c** and **d**. If part **c** were {Chevrolet} $\subseteq A$, it would have been true.
e. "The set consisting of the color red is an element of the set R" is *false* since, even though the color red *is* an element of R (see part **d**), the *set* containing red—namely, "{red}"—is not an element of R.
f. "R is a subset of S" is *true* since every color of the rainbow is listed in the set S.
g. "R is a proper subset of S" is *false* because $R = S$. To be a proper subset, there must be some element of S that is not in the set R.
h. "The empty set is a subset of A" is *true*, since the empty set is a subset of every set.
i. "The empty set is a proper subset of C" is *true*. The empty set is a proper subset of every *nonempty* set. $\varnothing \subset \varnothing$ is false, but $\varnothing \subseteq \varnothing$ is true.
j. "The set containing the empty set is a subset of R" is *false* since "$\varnothing$" is not listed inside the set R.

This might seem like a simple example, but it shows some technical differences in the symbolism of set theory. Do not skip over this example.

Example 8 Subsets of a given set

Find all possible subsets of C = {5, 7}.

Solution
{5}, {7} are obvious subsets.

{5, 7} is also a subset, since both 5 and 7 are elements of C.

{ } is also a subset of C. It is a subset because all of its elements belong to C. Stated a different way, if it were not a subset of C, we would have to be able to find an element of { } that is not in C. Since we cannot find such an element, we say that the empty set is a subset of C (and, in fact, *the empty set is a subset of every set*).

We see that there are four subsets of C from Example 8, even though C has only two elements. We must therefore be careful to distinguish between a subset and an element. Remember, 5 and {5} mean different things.

The subsets of C can be classified into two categories: *proper* and *improper*. Since every set is a subset of itself, we immediately know one subset for *any* given set: the set itself. A *proper subset* is a subset that is not equal to the original set; that is, A is a **proper subset** of a subset B, written $A \subset B$, if A is a subset of B and $A \neq B$. An **improper subset** of a set A is the set A.

We see there are three proper subsets of C = {5, 7}: $\varnothing$, {5}, and {7}. There is one improper subset of C: {5, 7}.

Example 9 Classifying proper and improper subsets

Find the proper and improper subsets of $A = \{2, 4, 6, 8\}$. What is the cardinality of A?

Solution   The cardinality of A is 4 (because there are 4 elements in A). There is one improper subset: $\{\mathbf{2, 4, 6, 8}\}$. The proper subsets are as follows:

$\{ \ \}$,
$\{2\}, \{4\}, \{6\}, \{8\}$,
$\{2, 4\} \{2, 6\}, \{2, 8\}, \{4, 6\}, \{4, 8\}, \{6, 8\}$,
$\{2, 4, 6\}, \{2, 4, 8\}, \{2, 6, 8\}, \{4, 6, 8\}$

From Example 8, we see that a set of 2 elements has 4 subsets, and from Example 9, we note that a set of cardinality four has 16 subsets. In Problems 31–32, the following property is discovered.

CAUTION

Number of Subsets

The number of subsets of a set of size n is

$$2^n$$

Sometimes we are given two sets X and Y, and we know nothing about the way they are related. In this situation, we draw a general figure, such as the one shown in Figure 2.5.

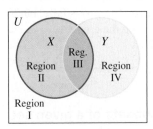

X is regions II and III.
Y is regions III and IV.
$\overline{X}$ is regions I and IV.
$\overline{Y}$ is regions I and II.
If $X \subseteq Y$, then region II is empty.
If $Y \subseteq X$, then region IV is empty.
If $X = Y$, then regions II and IV are empty.
If X and Y are disjoint, then region III is empty.

FIGURE 2.5 General Venn diagram for two sets

We can generalize for more sets. The general Venn diagram for three sets divides the universe into eight regions, as shown in Figure 2.6.

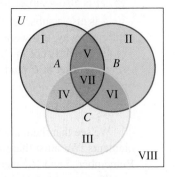

FIGURE 2.6 General Venn diagram for three sets

Example 10 Regions in a Venn diagram

Name the regions in Figure 2.6 described by each of the following.
a. A **b.** C **c.** $\overline{A}$ **d.** $\overline{B}$ **e.** $A \subseteq B$ **f.** A and C are disjoint

Solution
a. A is regions I, IV, V, and VII.
b. C is regions III, IV, VI, and VII.
c. $\overline{A}$ is regions II, III, VI, and VIII.
d. $\overline{B}$ is regions I, III, IV, and VIII.
e. $A \subseteq B$ means that regions I and IV are empty.
f. A and C are disjoint means that regions IV and VII are empty.

Problem Set 2.1

Level 1

1. **IN YOUR OWN WORDS**   Why do you think mathematics accepts the word *set* as an undefined term?

2. **IN YOUR OWN WORDS**   Distinguish between equal and equivalent sets.

3. **IN YOUR OWN WORDS**   What is a universal set?

4. **IN YOUR OWN WORDS**   What is the empty set?

5. **IN YOUR OWN WORDS**   Give three descriptions of the empty set.

6. **IN YOUR OWN WORDS**   Give an example of a set with cardinality 0.

7. **IN YOUR OWN WORDS**   Give an example of a set with cardinality greater than 1 thousand, but less than 1 million.

8. **IN YOUR OWN WORDS**   Give an example of a set with cardinality greater than 1 million.

Tell whether each set in Problems 9–12 is well defined. If it is not well defined, change it so that it is well defined.

9. a. The set of students attending the University of California
   b. {Grains of sand on earth}

10. a. The set of counting numbers between 3 and 4
    b. The set of happy people in your country

11. a. {Good bets on the next race at Hialeah}
    b. {Years that will be bumper years for growing corn in Iowa}

12. a. The set of people with pointed ears

Zachary Quinto as Spock

b. {Counting numbers less than 0}

Specify the sets in Problems 13–18 by roster.

13. a. {Distinct letters in the word *Mathematics*}
    b. {Current U.S. president}

14. a. {Odd counting numbers less than 11}
    b. {Positive multiples of 3}

15. a. $\{A \,|\, A$ is a counting number greater than 6$\}$
    b. $\{B \,|\, B$ is a counting number less than 6$\}$

16. a. $\{C \,|\, C$ is an integer greater than 6$\}$
    b. $\{B \,|\, B$ is an integer less than 6$\}$

17. a. {Distinct letters in the word *pipe*}
    b. {Counting numbers greater than 150}

18. a. {Counting numbers containing only 1s}
    b. {Even counting numbers between 5 and 15}

Specify the sets in Problems 19–24 by description.

19. $\{1, 2, 3, 4, 5, 6, 7, 8, 9\}$

20. $\{1, 11, 121, 1331, 14641, \ldots\}$

21. $\{10, 20, 30, \ldots, 100\}$

22. $\{50, 500, 5000, \ldots\}$

23. $\{101, 103, 105, \ldots, 169\}$

24. $\{m, i, s, p\}$

Write out in words the description of the sets given in Problems 25–30, and then list each set in roster form.

25. $\{x \,|\, x$ is an odd counting number$\}$

26. $\{x \,|\, x$ is a natural number between 1 and 10$\}$

27. $\{x \,|\, x \in \mathbb{N}, x \neq 8\}$

28. $\{x \,|\, x \in \mathbb{W}, x \leq 8\}$

29. $\{x \,|\, x \in \mathbb{W}, x < 8\}$

30. $\{x \,|\, x \in \mathbb{W}, x \notin E\}$ where $E = \{2, 4, 6, \ldots\}$

31. List all possible subsets of the given set.
    a. $A = \varnothing$
    b. $B = \{1\}$
    c. $C = \{1, 2\}$.
    d. $D = \{1, 2, 3\}$.
    e. $E = \{1, 2, 3, 4\}$.

32. List all possible subsets of the set given set.
   a. $G = \varnothing$
   b. $H = \{6\}$.
   c. $I = \{6, 7\}$,
   d. $J = \{6, 7, 8\}$
   e. $K = \{6, 7, 8, 9\}$

33. Look for a pattern in Problems 31. Can you guess how many subsets the set $F = \{1, 2, 3, 4, 5\}$ has? Does this guess match the formula?

34. Look for a pattern in Problem 32. Can you guess how many subsets the set $L = \{6, 7, 8, 9, 10\}$ has? Does this guess match the formula?

Level 2

35. Draw a Venn diagram showing people who are over 30, people who are 30 or under, and people who drive a car.

36. Draw a Venn diagram showing males, females, and those people who ride bicycles.

37. Draw a Venn diagram showing that all Chevrolets are automobiles.

38. Draw a Venn diagram showing that all cell phones are communication devices.

39. Consider the sets

$A = \{\text{distinct letters in the word } pipe\}$
$B = \{4\}$
$C = \{p, i, e\}$
$D = \{2 + 1\}$
$E = \{three\}$
$F = \{3\}$

   a. What is the cardinality of each set?
   b. Which of the sets are equivalent?
   c. Which of the given sets are equal?

40. Consider the sets

$A = \{16\}$
$B = \{10 + 6\}$
$C = \{10, 6\}$
$D = \{2^5\}$
$E = \{2, 5\}$

   a. What is the cardinality of each set
   b. Which of the sets are equivalent?
   c. Which of the given sets are equal?

Decide whether each statement in Problems 41–54 is true or false. Give reasons for your answers.

41. a. $\{m, a, t, h\} \subseteq \{m, a, t, h, e, i, c, s\}$
   b. $\{math\} \in \{m, a, t, h\}$

42. a. $m \in \{m, a, t, h\}$
   b. $\{m\} \in \{m, a, t, h\}$

43. a. $\{m, a, t, h\} \subseteq \{h, t, a, m\}$
   b. $\{m, a, t, h\} \subset \{h, t, a, m\}$

44. a. $\{white\} \in \{colors\ of\ the\ rainbow\}$
   b. $\{white\} \in \{colors\ of\ the\ U.S.\ flag\}$

45. $\{math, history\} \subset \{high\ school\ subjects\}$

46. $\{\ \ \} \subseteq \{Jeff, Maureen, Terry\}$

47. $1 \in \{1, 2, 3, 4, 5\}$

48. $\{1\} \in \{1, 2, 3, 4, 5\}$

49. $1 \in \{\{1\}, \{2\}, \{3\}, \{4\}\}$

50. $\{1\} \in \{\{1\}, \{2\}, \{3\}, \{4\}\}$

51. $\{1\} \subset \{\{1\}, \{2\}, \{3\}, \{4\}\}$

52. $0 = \{\ \ \}$

53. $\varnothing = \{\ \ \}$

54. $\{\varnothing\} = \{\ \ \}$

Level 3

55. IN YOUR OWN WORDS Give an example of a set that cannot be written using the roster method.

56. IN YOUR OWN WORDS Is it possible to list the set of rational numbers between 0 and 1 by roster? If you think so, then list them, and if you do not think so, explain why.

57. Five people plan to meet after school, and if they all show up, there will be one group of five people. However, if only four of them show up, in how many ways is this possible?

58. Five people plan to meet after school, and if they all show up, there will be one group of five people. However, if only three of them show up, in how many ways is this possible?

Problem Solving 3

59. In Section 1.2 we used Euler circles to represent expressions such as "All Chevrolets are automobiles." Rephrase this using set terminology.

60. In Section 1.2 we used Euler circles to represent expressions such as "Some people are nice." Represent this relationship using Venn diagrams. Be sure to label the circles and the universe.

2.2 | Operations with Sets

Fundamental Operations

Suppose we consider two general sets, B and Y, as shown in Figure 2.7. If we show the set Y using a yellow highlighter and the set B using a blue highlighter, it is easy to visualize two operations. The *intersection* of the sets is the region shown in green (the parts that are *both* yellow and blue). We see this is region III, and we describe this using the word "**and.**" The *union* of the sets is the part shown in *any* color (the parts that are yellow or blue or green). We see this is regions II, III, and IV, and we describe this using the word "**or.**"

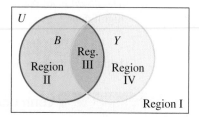

FIGURE 2.7 Venn diagram showing intersection and union

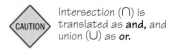

Intersection (∩) is translated as **and,** and union (∪) as **or.**

Operations on Sets: Intersection and Union

The **intersection** of sets A and B, denoted by $A \cap B$, is the set consisting of all elements common to A and B. The **union** of sets A and B, denoted by $A \cup B$, is the set consisting of all elements of A or B or both.

Example 1 | Venn diagram for union and intersection

Draw Venn diagrams for union and intersection.
a. $B \cap Y$ **b.** $B \cup Y$

Solution   Highlighter pens work well when drawing Venn diagrams, but instead of using two colors as shown in Figure 2.7, we use one highlighter (any color) to indicate the final result as shown in this example.

a. "$B \cap Y$" is the intersection of the sets B and Y. Draw two circles as shown in Figure 2.8. First shade B using horizontal lines, and then shade the second set, Y, using vertical lines.

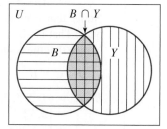

FIGURE 2.8 Intersection of two sets

The *intersection* is all parts that are shaded twice (both horizontal and vertical), as shown with the pink highlighter.

b. "$B \cup Y$" is the union of the sets B and Y. In Figure 2.9, first shade B using horizontal lines and then shade the second set, Y, using vertical lines.

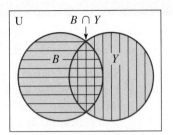

FIGURE 2.9 Union of two sets

The *union* is all parts that are shaded (either once or twice), as shown with the pink highlighter.

In the next example we are given more than two sets.

Example 2 | Venn diagram with three sets

Let $U = \{1, 2, 3, 4, 5, 6, 7, 8, 9\}$, $A = \{2, 4, 6, 8\}$, $B = \{1, 3, 5, 7\}$, and $C = \{5, 7\}$. A Venn diagram showing these sets is shown in Figure 2.10.

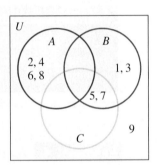

FIGURE 2.10 Venn diagram showing three sets

Find: **a.** $A \cup C$ **b.** $B \cup C$ **c.** $B \cap C$ **d.** $A \cap C$

Solution
a. $A \cup C = \{2, 4, 6, 8\} \cup \{5, 7\}$
$\quad\quad\quad = \{2, 4, 5, 6, 7, 8\}$
Notice that the union consists of all elements in A or in C or in both. Also note that the order in which the elements are listed is not important.
b. $B \cup C = \{1, 3, 5, 7\} \cup \{5, 7\}$
$\quad\quad\quad = \{1, 3, 5, 7\}$
Notice that, even though the elements 5 and 7 appear in both sets, they are listed only once. That is, the sets $\{1, 3, 5, 7\}$ and $\{1, 3, 5, 5, 7, 7\}$ are equal (exactly the same). Notice that the resulting set has a name (it is called B), and we write

$\quad\quad B \cup C = B$

c. $B \cap C = \{1, 3, 5, 7\} \cap \{5, 7\}$
$\quad\quad\quad = \{5, 7\}$ *The intersection contains the elements common to both sets.*
$\quad\quad\quad = C$
d. $A \cap C = \{2, 4, 6, 8\} \cap \{5, 7\} = \{\quad\}$
These sets have no elements in common, so we write $\{\quad\}$ or $\varnothing$.

Suppose we consider the cardinality of the various sets in Example 2:

$$|U| = 9, \quad |A| = 4, \quad |B| = 4, \quad \text{and} \quad |C| = 2$$

$$\begin{aligned}
|A \cup C| &= 6 \quad (\text{part } \mathbf{a}) \\
|B \cup C| &= 4 \quad (\text{part } \mathbf{b}) \\
|B \cap C| &= 2 \quad (\text{part } \mathbf{c}) \\
|A \cap C| &= 0 \quad (\text{part } \mathbf{d}) \\
|\varnothing| &= 0 \quad (\text{part } \mathbf{d})
\end{aligned}$$

Consider the set S formed from the sets in Example 2:

$$S = \{U, A, B, C, \varnothing\}$$

This is a set of sets; there are five sets in S, so $|S| = 5$. Furthermore, if we remove sets from S, one-by-one, we find

$$\begin{aligned}
T &= \{A, B, C, \varnothing\}, \quad \text{so} \quad |T| = 4 \\
U &= \{B, C, \varnothing\}, \quad \text{so} \quad |U| = 3 \\
V &= \{C, \varnothing\}, \quad \text{so} \quad |V| = 2
\end{aligned}$$

Finally,

$$W = \{\varnothing\}, \quad \text{so} \quad |W| = 1$$

Thus $|\{\varnothing\}| = 1$, but $|\varnothing| = 0$, so $\{\varnothing\} \neq \varnothing$.

Did you get that?
$\{\varnothing\} \neq \varnothing$

Cardinality of Intersections and Unions

The *cardinality of an intersection* is easy; it is found by looking at the number of elements in the intersection.

The *cardinality of a union* is a bit more difficult. For sets with small cardinalities, we can find the cardinality of the unions by direct counting, but if the sets have large cardinalities, it might not be easy to find the union and then the cardinality by direct counting. Some students might want to find $|B \cup C|$ by adding $|B|$ and $|C|$, but you can see from Example 2 that $|B \cup C| \neq |B| + |C|$. However, if you look at the Venn diagram for the number of elements in the union of two sets, the situation becomes quite clear, as shown in Figure 2.11.

$|X| + |Y|$ adds this region twice

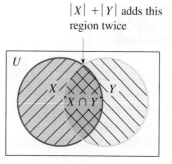

FIGURE 2.11 Venn diagram for the number of elements in the union of two sets

Formula for the Cardinality of the Union of Two Sets

For any two sets X and Y,

$$|X \cup Y| \;=\; |X| + |Y| \;-\; |X \cap Y|$$

The elements in the intersection are counted twice.

This corrects for the "error" introduced by counting those elements in the intersection twice.

STOP This formula will be used later in the book.

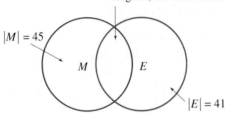

Example 3 | **Cardinality of a union**

Suppose a survey indicates that 45 students are taking mathematics and 41 are taking English. How many students are taking math or English?

Solution At first, it might seem that all you do is add 41 and 45, but such is not the case. Let $M = \{$persons taking math$\}$ and $E = \{$persons taking English$\}$.

To find out how many students are taking math and English, we need to know the number in this intersection.

$|M| = 45$

M E

$|E| = 41$

As you can see, you need further information. Problem solving requires that you not only recognize what known information is needed when answering a question, but also recognize when additional information is needed. Suppose 12 students are taking both math and English. In this case we see that

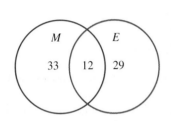

By diagram:
First, fill in 12 in $M \cap E$. Then,
$|M| = 45$, so fill in 33 since $45 - 12 = 33$
$|E| = 41$, so fill in 29 since $41 - 12 = 29$
The total number is $33 + 12 + 29 = 74$.

By formula:
$|M \cup E| = |M| + |E| - |M \cap E|$
$= 45 + 41 - 12$
$= 74$

Example 3 looks very much like the open-ended examination question we posed at the beginning of this chapter. You will find that open-ended question in the problem set. In the next section, we will consider survey questions that involve more than two sets.

Problem Set | 2.2

Level 1

1. **IN YOUR OWN WORDS** What do we mean by the operations of *union, intersection,* and *complementation*?

2. **IN YOUR OWN WORDS** This section began with an open-ended question from the 1987 Examination of the California Assessment Program:

   James knows that half the students from his school are accepted at the public university nearby. Also, half are accepted at the local private college. James thinks that this adds up to 100%, so he will surely be accepted at one or the other institution. Explain why James may be wrong. If possible, use a diagram in your explanation.

3. **a.** What English word is used to describe union?
   b. What English word is used to describe intersection?
   c. What English word is used to describe complement?

4. State formulas for:
   a. cardinality of an intersection
   b. cardinality of a union

Perform the given set operations in Problems 5–18.
Let $U = \{1, 2, 3, 4, 5, 6, 7, 8, 9, 10\}$.

5. $\{2, 6, 8\} \cup \{6, 8 \ 10\}$

6. $\{2, 6, 8\} \cap \{6, 8, 10\}$

7. $\{2, 5, 8\} \cup \{3, 6, 9\}$

8. $\{2, 5, 8\} \cap \{3, 6, 9\}$

9. $\{1, 2, 3, 4, 5\} \cap \{3, 4, 5, 6, 7\}$

10. $\{1, 2, 3, 4, 5\} \cup \{3, 4, 5, 6, 7\}$

11. $\overline{\{2, 8, 9\}}$

12. $\overline{\{1, 2, 5, 7, 9\}}$

13. $\overline{\{1,3,5,7,9\}}$

14. $\overline{\{2,4,6,8,10\}}$

15. $\{x \mid x \text{ is a multiple of } 2\} \cup \{x \text{ is a multiple of } 3\}$

16. $\{x \mid x \text{ is a multiple of } 2\} \cap \{x \mid \text{ is a multiple of } 3\}$

17. $\overline{\{x \mid x \text{ is even}\}}$

18. $\overline{\{y \mid y \text{ is odd}\}}$

Perform the given set operations in Problems 19–28.

19. $\{x \mid x \text{ is a positive integer}\} \cup \{x \mid x \text{ is a negative integer}\}$

20. $\{x \mid x \text{ is a positive integer}\} \cap \{x \mid x \text{ is a negative integer}\}$

21. $\mathbb{N} \cup \mathbb{W}$

22. $\mathbb{N} \cap \mathbb{W}$

23. $\overline{U}$

24. $\overline{\varnothing}$

25. $X \cap \varnothing$ for any set X

26. $X \cup \varnothing$ for any set X

27. $U \cup \varnothing$

28. $U \cap \varnothing$

Let $U = \{1, 2, 3, 4, 5, 6, 7\}$, $A = \{1, 2, 3, 4\}$, $B = \{1, 2, 5, 6\}$, and $C = \{3, 5, 7\}$. List all the members of each of the sets in Problems 29–44.

29. $A \cup B$ **30.** $A \cap B$

31. $A \cup C$ **32.** $A \cap C$

33. $B \cap C$ **34.** $B \cup C$

35. $\overline{A}$ **36.** $\overline{B}$

37. $\overline{C}$ **38.** $\overline{U}$

39. $\overline{\{x \mid x \text{ is greater than } 4\}}$

40. $\overline{\{y \mid y \text{ is between 4 and 10}\}}$

41. $\{x \mid x \text{ is less than } 5\} \cup \{x \mid x \text{ is greater than } 5\}$

42. $\{x \mid x \text{ is less than } 5\} \cap \{x \mid x \text{ is greater than } 5\}$

43. $\varnothing \cup A$ **44.** $\varnothing \cap B$

In Problems 45–48, use set notation to identify the shaded region.

45.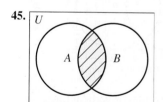

46.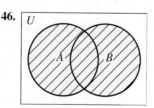

47.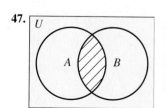

48.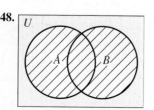

Draw Venn diagrams for each of the relationships in Problems 49–52.

49. $X \cup Y$ **50.** $X \cap Z$

51. $\overline{Y}$ **52.** $\overline{Z}$

Level 2

53. Montgomery College has a 50-piece band and a 36-piece orchestra. If 14 people are members of both the band and the orchestra, can the band and orchestra travel in two 40-passenger buses?

Cengage Learning

54. From a survey of 100 college students, a marketing research company found that 75 students owned Ipods, 45 owned cars, and 35 owned both cars and Ipods.
  a. How many students owned either a car or an Ipod (but not both)?
  b. How many students do not own either a car or an Ipod?

55. In a survey of a TriDelt chapter with 50 members, 18 were taking mathematics, 35 were taking English, and 6 were taking both. How many were not taking either of these subjects?

56. In a senior class at Rancho Cotati High School, there were 25 football players and 16 basketball players. If 7 persons played both sports, how many different people played in these sports?

57. The fire department wants to send booklets on fire hazards to all teachers and homeowners in town. How many booklets does it need, using these statistics?

     50,000 homeowners

     4,000 teachers

     3,000 teachers who own their own homes

58. Santa Rosa Junior College enrolled 29,000 students in the fall of 1999. It was reported that of that number, 58% were female and 42% were male. In addition, 62% were over the age of 25. How many students were there in each category if 40% of those over the age of 25 were male? Draw a Venn diagram showing these relationships.

Problem Solving 3

59. The general Venn diagram for two sets has four regions (Figure 2.7), and the one for three sets has eight regions (Figure 2.10). Use patterns to develop a formula for the number of regions in a Venn diagram with n sets.

60. Each of the circles in Figure 2.12 is identified by a letter, each having a number value from 1 to 9. Where the circles overlap, the number is the sum of the values of the letters in the overlapping circles. What is the number value for each letter?*

* From "Perception Puzzles," by Jean Moyer, Sky, January 1995, p. 120. *Math Puzzles and Logic Problems* © 1995 Dell Magazines, a division of Penny Marketing Limited Partnership, reprinted by permission of Dell Magazines.

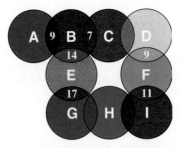

FIGURE 2.12 Circle intersection puzzle

2.3 | Applications of Sets

In the previous section, we introduced three operations: *intersection, union,* and *complement.* These are known as the *fundamental set operations.* In this section, we consider mixed operations with more than two sets, as well as some additional applications with sets.

Combined Operations with Sets

For sets, we perform operations from left to right; however, if there are parentheses, we perform operations within them first.

Example 1 Order of operations

Verbalize the correct order of operations and then illustrate the combined set operations using Venn diagrams:
a. $\overline{A} \cup \overline{B}$ **b.** $\overline{A \cup B}$

Solution
a. This is a combined operation that should be read from left to right. First find the complements of A and B and *then* find the union. This is called a *union of complements.*

Step 1 Shade $\overline{A}$ (vertical lines), then shade $\overline{B}$ (horizontal lines).

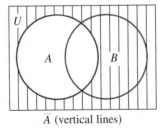

$\overline{A}$ (vertical lines)

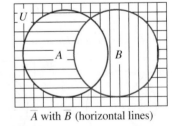

$\overline{A}$ with $\overline{B}$ (horizontal lines)

Step 2 $\overline{A} \cup \overline{B}$ is every portion that is shaded with horizontal or vertical lines. We show that here using a color highlighter.

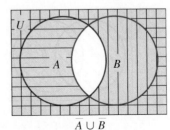

$\overline{A} \cup \overline{B}$

b. This is a combined operation that should be interpreted to mean $(\overline{A \cup B})$, which is the complement of the union. *First* find $A \cup B$ (vertical lines), and *then* find the complement (color highlighter). This is called the *complement of a union*. We show only the final result.

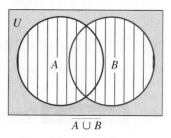

$$\overline{A \cup B}$$

De Morgan's Laws

Notice from Example 1 that $\overline{A \cup B} \neq \overline{A} \cup \overline{B}$. If they were equal, the final highlighted color portions of the Venn diagrams would be the same. The next example takes us a step further by showing what $\overline{A \cup B}$ does equal.

Example 2 **DE MORGAN'S LAW**
　　　　　　　　　　　　　　　　　　　　　　　　　　Pólya's
　　　　　　　　　　　　　　　　　　　　　　　　　　Method

Prove $\overline{A \cup B} = \overline{A} \cap \overline{B}$.

Solution　We use Pólya's problem-solving guidelines for this example.

Understand the Problem.　We wish to prove the given statement is true *for all* sets A and B, so we cannot work with a *particular* example.

Devise a Plan.　The procedure is to draw separate Venn diagrams for the left and the right sides, and then to compare them to see if they are identical.

Carry Out the Plan.

Step 1　Draw a diagram for the expression on the left side of the equal sign, namely $\overline{A \cup B}$. The final result is shown with color highlighter. (See Example 1 for details.)

Step 2　Draw a diagram for the expression on the right side of the equal sign. First draw $\overline{A}$ (vertical lines) and then $\overline{B}$ (horizontal lines). The final result, $\overline{A} \cap \overline{B}$, is the part with both vertical and horizontal lines (as shown with the color highlighter) in the right portion of Figure 2.13.

Step 3　Compare the portions shaded by the color highlighter in the two Venn diagrams.

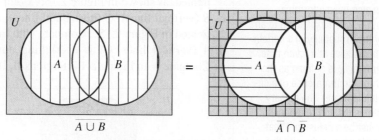

FIGURE 2.13 De Morgan's Law

Look Back.　They are the same, so we have proved $\overline{A \cup B} = \overline{A} \cap \overline{B}$.

The result proved in Example 2 is called **De Morgan's law** for sets. In the problem set you are asked to prove the second part of De Morgan's law.

De Morgan's Laws for Sets

For any sets X and Y

$$\overline{X \cup Y} = \overline{X} \cap \overline{Y} \qquad \overline{X \cap Y} = \overline{X} \cup \overline{Y}$$

Venn Diagrams with Three Sets

Next, we find the regions described by some combined operations for three sets.

Example **3** | **Operations with three sets**

Using the eight regions labeled in Figure 2.14, describe each of the following sets.

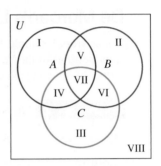

FIGURE 2.14 Venn diagram for three general sets showing eight regions

a. $A \cup B$
b. $A \cap C$
c. $B \cap C$
d. $\overline{A}$
e. $\overline{A \cup B}$
f. $A \cap B \cap C$
g. $A \cup (B \cap C)$
h. $\overline{A \cup B} \cap C$

Solution

a. A (vertical lines); B (horizontal lines); $A \cup B$ is everything shaded and is highlighted, as shown in Figure 2.15a. Compare with Figure 2.14 to see the answer is Regions I, II, IV, V, VI, and VII.

b. A (vertical lines); C (horizontal lines); $A \cap C$ is everything shaded twice and is highlighted, as shown in Figure 2.15b. Compare with Figure 2.14; Regions IV and VII.

c. B (vertical lines); C (horizontal lines); $B \cap C$ is everything shaded twice, as highlighted in Figure 2.15c. Compare with Figure 2.14; Regions VI and VII.

d. A (vertical lines); $\overline{A}$ is everything not shaded, as highlighted in Figure 2.15d. Compare with Figure 2.14; Regions II, III, VI, and VIII.

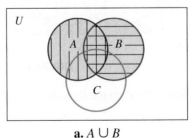

a. $A \cup B$ **b.** $A \cap C$

FIGURE 2.15

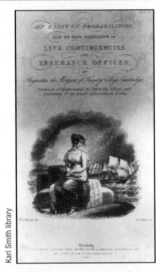

Historical NOTE

The photo shows the frontispiece of a book written by Augustus De Morgan (1806–1871) in 1838.

Not only was Augustus De Morgan a distinguished mathematician and logician, but in 1845 he suggested the slanted line we sometimes use when writing fractions, such as $\frac{1}{2}$ or $\frac{2}{3}$. De Morgan also used a formula involving π when explaining to an actuary the probability of a certain group being alive at a given time. (The number π is defined as the ratio of the circumference of a circle to its diameter.) The actuary was astonished and responded: "That must be a delusion. What can a circle have to do with the number of people alive at a given time?" Of course, actuaries today are more sophisticated. The number π is used in a wide variety of applications.

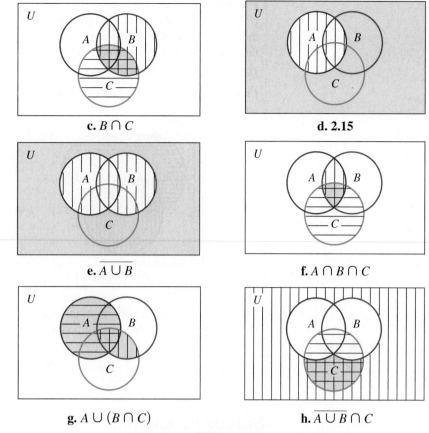

FIGURE 2.15 Working with Venn diagrams

e. $A \cup B$ (vertical lines); $\overline{A \cup B}$ is everything not shaded, as highlighted in Figure 15e. Compare with Figure 2.14; Regions III and VIII.

f. $A \cap B \cap C$ is $(A \cap B) \cap C$, so we show $A \cap B$ (vertical lines) and C (horizontal lines); highlight regions shaded twice, as shown in Figure 2.15f. Compare with Figure 2.14; Region VII.

g. Parentheses first, $B \cap C$ (vertical lines); A (horizontal lines); $A \cup (B \cap C)$ is everything shaded, as highlighted in Figure 2.15g. Compare with Figure 2.14; Regions I, IV, V, VI, and VII.

h. $\overline{A \cup B}$ (vertical lines); and C (horizontal lines); everything shaded twice is highlighted in Figure 2.15h. Compare with Figure 2.14; Region III.

Example 4 Union is associative   **Pólya's Method**

If $(P \cup Q) \cup R = P \cup (Q \cup R)$, we say that the operation of union is **associative.** Is the operation of union for sets an associative operation?

Solution   We use Pólya's problem-solving guidelines for this example.

Understand the Problem. Even though we cannot answer this question by using a particular example, we can use one to help us understand the question. Let $U = \{1, 2, 3, 4, 5, 6, 7, 8, 9, 10\}$, $P = \{1, 4, 7\}$, $Q = \{2, 4, 9, 10\}$, and $R = \{6, 7, 8, 9\}$. Is the following true?

$$(P \cup Q) \cup R = P \cup (Q \cup R)?$$

$$(P \cup Q) \cup R = \{1, 2, 4, 7, 9, 10\} \cup \{6, 7, 8, 9\}$$
$$= \{1, 2, 4, 6, 7, 8, 9, 10\}$$

$$P \cup (Q \cup R) = \{1, 4, 7\} \cup \{2, 4, 6, 7, 8, 9, 10\}$$
$$= \{1, 2, 4, 6, 7, 8, 9, 10\}$$

For this example, the operation of union for sets is associative. If we had observed $(P \cup Q) \cup R \neq P \cup (Q \cup R)$, then we would have had a counterexample. Although they are equal in this example, we cannot say that the property is true for all possibilities. However, all is not lost because it did help us to understand the question.

Devise a Plan. Use Venn diagrams.

Carry Out the Plan. Recall that the union is the entire shaded area.

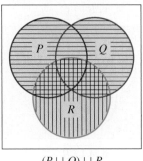
$(P \cup Q) \cup R$

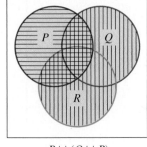
$P \cup (Q \cup R)$

Look Back. The operation of union for sets is an associative operation since the parts shaded in yellow are the same for both diagrams.

Survey Problems

In the previous section, we used the property of the cardinality of a union to find the number of elements in various regions formed by two sets. For three sets, the situation is a little more involved. There is a formula for the number of elements, but it is easier to use Venn diagrams, as illustrated by Example 5. Remember, the usual procedure is to fill in the number in the innermost region first and work your way outward through the Venn diagram using subtraction.

Example **5** **Survey problem**

Pólya's
Method

A survey of 100 randomly selected students gave the following information.

45 students are taking mathematics.
41 students are taking English.
40 students are taking history.
15 students are taking math and English.
18 students are taking math and history.
17 students are taking English and history.
 7 students are taking all three.

a. How many are taking only mathematics?
b. How many are taking only English?
c. How many are taking only history?
d. How many are not taking any of these courses?

Solution We use Pólya's problem-solving guidelines for this example.

Understand the Problem. We are considering students who are members of one or more of three sets. If U represents the universe, then $|U| = 100$. We also define the three sets: $M = \{\text{students taking mathematics}\}$, $E = \{\text{students taking English}\}$, $H = \{\text{students taking history}\}$.

Devise a Plan. The plan is to draw a Venn diagram, and then to fill in the various regions. We fill in the innermost region first, and then work our way outward (using subtraction) until the numbers of elements of the eight regions formed by the three sets are known.

Carry Out the Plan.

Step 1 We note $|M \cap E \cap H| = 7$ in Figure 2.16a.

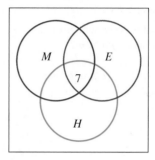

FIGURE 2.16a First step is inner region

Step 2 Fill in the other inner portions.
$|E \cap H| = 17$, but 7 have previously been accounted for, so an additional 10 members $(17 - 7 = 10)$ are added to the Venn diagram (see Figure 2.16b). $|M \cap H| = 18$; fill in $18 - 7 = 11$; $|M \cap E| = 15$; fill in $15 - 7 = 8$.

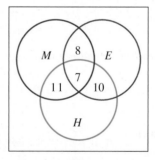

FIGURE 2.16b Second step is two-region intersections

Step 3 Fill in the other regions (see Figure 2.16c).
$|H| = 40$, but 28 have previously been accounted for in the set H, so there are an additional 12 members $(40 - 11 - 7 - 10 = 12)$. $|E| = 41$; fill in $41 - 8 - 7 - 10 = 16$; $|M| = 45$; fill in $45 - 11 - 7 - 8 = 19$.

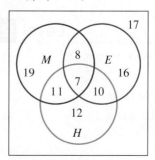

FIGURE 2.16c Third step is the one-region parts

Step 4   Add all the numbers listed in the sets of the Venn diagram to see that 83 students have been accounted for. Since 100 students were surveyed, we see that 17 are not taking any of the three courses. (We also filled this number in Figure 16c.) We now have the answers to the questions directly from the Venn diagram: **a.** 19   **b.** 16   **c.** 12   **d.** 17

Look Back.   Does our answer make sense? Add all the numbers in the Venn diagram as a check to see that we have accounted for the 100 students.

Problem Set   2.3

Level 1

1. What do we mean by *De Morgan's laws?*

2. IN YOUR OWN WORDS   What is the general procedure for drawing a Venn diagram for a survey problem?

Let $U = \{1, 2, 3, 4, 5, 6, 7\}$, $A = \{1, 2, 3, 4\}$, $B = \{1, 2, 5, 6\}$, and $C = \{3, 5, 7\}$. List all the members of each of the sets in Problems 3–10.

3. $(A \cup B) \cap C$

4. $A \cup (B \cap C)$

5. $\overline{A \cup B} \cap C$

6. $A \cup \overline{B \cap C}$

7. $\overline{A} \cup (B \cap C)$

8. $(A \cup B) \cap \overline{C}$

9. $\overline{(A \cup B) \cap C}$

10. $\overline{A} \cup (\overline{B} \cap \overline{C})$

Consider the sets X and Y. Write each of the statements in Problems 11–18 in symbols.

11. A union of complements

12. A complement of a union

13. A complement of an intersection

14. An intersection of complements

15. The complement of the union of X and Y

16. The union of the complements of X and Y

17. The intersection of the complements of X and Y

18. The complement of the intersection of X and Y

Level 2

Draw Venn diagrams for each of the relationships in Problems 19–34.

19. $\overline{A} \cup B$

20. $A \cap \overline{B}$

21. $\overline{A \cap B}$

22. $\overline{A} \cup B$

23. $A \cap (B \cup C)$

24. $A \cup (B \cup C)$

25. $A \cap \overline{B} \cup C$

26. $\overline{A} \cup B \cup C$

27. $\overline{A} \cup B \cap C$

28. $A \cup \overline{B \cap C}$

29. $\overline{A} \cup (B \cap C)$

30. $(A \cup B) \cap \overline{C}$

31. $\overline{(A \cup B) \cap C}$

32. $\overline{A} \cup (\overline{B} \cap \overline{C})$

33. $\overline{(A \cup B)} \cup C$

34. $(A \cap B) \cup (A \cap C)$

35. Draw a Venn diagram showing the relationship among cats, dogs, and animals.

36. Draw a Venn diagram showing the relationship among trucks, buses, and cars.

37. Draw a Venn diagram showing people in a classroom wearing some black, people wearing some blue, and people wearing some brown.

38. Draw a Venn diagram showing birds, bees, and living creatures.

39. In 1995 the United States population was approximately 263 million. It was reported that of that number, 79% were white, 12% were black, and 9% were Hispanic. If $\frac{1}{2}$% have one black and one white parent, 2% have one black and one Hispanic parent, and 1% have one white and one Hispanic parent, how many people are there in each category? Draw a Venn diagram showing these relationships.

40. Here are the results of the voting by eight senators on voting by three bills in the U.S. Senate during the 107th Congress, second session (2002):

Senator	Bill A	Bill B	Bill C
	S. 2414	H.R. 2356	Amen. 3017
Clinton (D)	yea	yea	yea
Kennedy (D)	yea	yea	yea
Bayh (D)	nay	yea	nay
Inouye (D)	yea	yea	nay
Campbell (R)	yea	nay	nay
Helms (R)	yea	nay	nay
Thurmond (R)	yea	nay	nay
Lott (R)	yea	nay	nay

Place each senator in the appropriate region of Figure 2.17. A "yea" vote is recorded as supporting the bill and consequently places the senator inside the circle representing that particular bill.

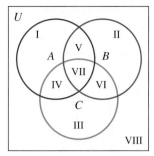

FIGURE 2.17 Senatorial votes

41. Listed below are five female and five male Wimbledon tennis champions, along with their country of citizenship and handedness.

Female

Stephi Graf, Germany, right
Martina Navratilova, U.S., left
Chris Evert Lloyd, U.S., right
Evonne Goolagong, Australia, right
Virginia Wade, Britain, right

Male

Michael Stich, Germany, right
Stefan Edberg, Sweden, right
Boris Becker, Germany, right
Pat Cash, Australia, right
John McEnroe, U.S., left

Using Figure 2.18, indicate in which region each of the above individuals would be placed.

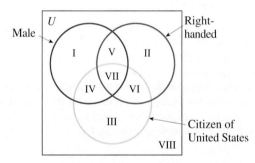

FIGURE 2.18 Wimbledon tennis champions

In Problems 42–45, use set notation to identify the shaded region.

42.

43.

44.

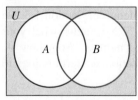

45.

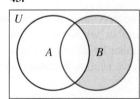

In Problems 46–51, use Venn diagrams to prove or disprove each statement. Remember to draw a diagram for the left side of the equation and another for the right side. If the final shaded portions are the same, then you have proved the result. If the final shaded portions are not identical, then you have disproved the result.

46. $\overline{X \cup Y} = \overline{X} \cup \overline{Y}$

47. $\overline{A} \cap \overline{B} = \overline{A \cup B}$

48. $(A \cap B) \cap C = A \cap (B \cap C)$

49. $A \cup (B \cup C) = (A \cup B) \cup (A \cup C)$

50. $A \cap (B \cup C) = (A \cap B) \cap (A \cap C)$

51. $\overline{X \cap Y} = \overline{X} \cup \overline{Y}$

52. In a recent survey of 100 persons, the following information was gathered:

59 use shampoo A.
51 use shampoo B.
35 use shampoo C.
24 use shampoos A and B.
19 use shampoos A and C.
13 use shampoos B and C.
11 use all three.

Let

$A = \{\text{persons who use shampoo A}\}$
$B = \{\text{persons who use shampoo B}\}$
$C = \{\text{persons who use shampoo C}\}$

Use a Venn diagram to show how many persons are in each of the eight possible categories.

53. Matt E. Matic was applying for a job. To determine whether he could handle the job, the personnel manager sent him out to poll 100 people about their favorite types of TV shows. His data were as follows:

59 preferred comedies.
38 preferred variety shows.
42 preferred serious drama.
18 preferred comedies and variety programs.
12 preferred variety and serious drama.
16 preferred comedies and serious drama.
7 preferred all types.
2 did not like any of these TV show types.

If you were the personnel manager, would you hire Matt on the basis of this survey? Why or why not?

54. A poll was taken of 100 students at a commuter campus to find out how they got to campus. The results were as follows:

42 said they drove alone.
28 rode in a carpool.
31 rode public transportation.
9 used both carpools and public transportation.
10 used both a carpool and sometimes their own cars.
6 used buses as well as their own cars.
4 used all three methods.

How many used none of the above-mentioned means of transportation?

55. In an interview of 50 students,

> 12 liked Proposition 8 and Proposition 13.
> 18 liked Proposition 8, but not Proposition 5.
> 4 liked Proposition 8, Proposition 13, and Proposition 5.
> 25 liked Proposition 8.
> 15 liked Proposition 13.
> 10 liked Proposition 5, but not Proposition 8 nor Proposition 13.
> 1 liked Proposition 13 and Proposition 5, but not Proposition 8.

a. Show the completed Venn diagram.
b. Of those surveyed, how many did not like any of the three propositions?
c. How many liked Proposition 8 and Proposition 5?

Problem Solving 3

56. HISTORICAL QUEST On the *NBC Nightly News* on Thursday, May 25, 1995, Tom Brokaw read a brief report on computer use in the United States. The story compared computer users by ethnic background, and Brokaw reported that 14% of blacks and 13% of Hispanics use computers, but 27% of whites use computers. Brokaw then commented that computer use by whites was equal to that of blacks and Hispanics combined.

a. Draw a Venn diagram showing the white, black, Hispanic, Asian, and Native American populations of the United States, along with U.S. computer users.
b. Use the Venn diagram from part **a** to show that percentages cannot be added as was done by Brokaw.

57. Draw a Venn diagram for four sets, and label the 16 regions.

58. The Venn diagram in Figure 2.19 shows five sets. It was drawn by Allen J. Schwenk of the U.S. Naval Academy. As you can

see, there are 32 separate regions. Describe the following regions:

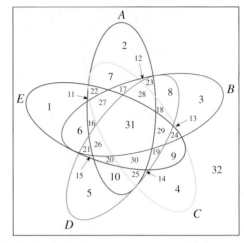

FIGURE 2.19 General Venn diagram for five sets

a. 1   **b.** 11   **c.** 21   **d.** 31   **e.** 16

59. Using the Venn diagram in Figure 2.19, specify which region is described by
a. $\overline{A \cup B \cup C \cup D \cup E}$
b. $A \cap \overline{B \cup C \cup D \cup E}$

60. Human blood is typed Rh^+ (positive blood) or Rh^- (negative blood). This Rh factor is called an *antigen*. There are two other antigens known as A and B types. Blood lacking both A and B types is called type O. Sketch a Venn diagram showing the three types of antigens A, B, and Rh, and label each of the eight regions. For example, O^+ is inside the Rh set (anything in Rh is positive) but outside both A and B. On the other hand, O^- is that region outside all three circles.

2.4 Finite and Infinite Sets

In Chapter 1, we discussed some very large numbers. In Problem 5 of Section 1.3, we asked, "What is the largest number whose name you know?" No matter what your answer, we can find a larger number by adding 1 to that "largest number." Somehow we know that "there is no end," but we need to introduce some notation and terminology to tackle the "really big" numbers!

Photo with infinitely many reflections

Infinite Sets

Certain sets, such as $\mathbb{N}$, $\mathbb{Z}$, $\mathbb{W}$, and $A = \{1000, 2000, 3000, \ldots\}$, have a common property. We call these *infinite sets*. If the cardinality of a set is 0 or a counting number, we say the set is **finite.** Otherwise, we say it is **infinite.** We can also say that a set is finite if it has a cardinality less than some counting number, even though we may not know its precise cardinality. For example, we can safely assert that the set of students attending the University of Hawaii is finite even though we may not know its cardinality, because the cardinality is certainly less than a million.

You use the ideas of cardinality and of equivalent sets every time you count something, even though you don't use the term *cardinality.* For example, a set of sheep is counted by finding a set of consecutive counting numbers starting with 1 that has the same cardinality as the set of sheep. These equivalent sets are found by using an idea called **one-to-one correspondence**. If the set of sheep is placed into a one-to-one correspondence with a set of pebbles, the set of pebbles can then be used to represent the set of sheep.

One-to-One Correspondence

Two sets A and B are said to be in a **one-to-one correspondence** if we can find a pairing so that

1. Each element of A is paired with precisely one element of B; and

2. Each element of B is paired with precisely one element of A.

Example 1 | Finding a one-to-one correspondence

Which of the following sets can be placed into a one-to-one correspondence?

$$K = \{3, 4\}, L = \{4\}, M = \{3\}, N = \{\text{three}\}, P = \{\heartsuit, \clubsuit, \diamondsuit, \spadesuit\}, Q = \{t, h, r, e\}$$

Solution   Notice that

$$L = \{4\} \qquad\qquad P = \{\heartsuit, \clubsuit, \diamondsuit, \spadesuit\}$$
$$\updownarrow \qquad\qquad\qquad \updownarrow \; \updownarrow \; \updownarrow \; \updownarrow$$
$$M = \{3\} \qquad\qquad Q = \{t, \; h, \; r, \; e\}$$
$$\updownarrow$$
$$N = \{\text{three}\}$$

We see that L, M, and N all have the same cardinality, and using that knowledge we see that any pair of them can be placed into a one-to-one correspondence; we write $L \leftrightarrow M$, $L \leftrightarrow N$, and $M \leftrightarrow N$. We also see that P and Q can be placed into a one-to-one correspondence.

Since infinity is not a number, we cannot correctly say that the cardinality of the counting numbers is infinity. In the late 18th century, Georg Cantor assigned a cardinal number $\aleph_0$ (pronounced "aleph-naught") to the set of counting numbers. That is, $\aleph_0$ is the cardinality of the set of counting numbers

$$\mathbb{N} = \{1, 2, 3, 4, \ldots\}$$

The set

$$E = \{2, 4, 6, 8, \ldots\}$$

also has cardinality $\aleph_0$, since it can be put into a one-to-one correspondence with set $\mathbb{N}$:

$$\mathbb{N} = \{1, 2, 3, 4, \ldots, n, \ldots\}$$
$$\updownarrow \; \updownarrow \; \updownarrow \; \updownarrow \qquad \updownarrow$$
$$E = \{2, 4, 6, 8, \ldots, 2n, \ldots\}$$

Notice that all elements of E are elements of the set $\mathbb{N}$, and that $E \neq \mathbb{N}$. Recall that in such a case, we say that E is a *proper subset* of $\mathbb{N}$. Cantor used this property as a defining property for infinite sets.

> ## Infinite Set
>
> An **infinite set** is a set that can be placed into a one-to-one correspondence with a proper subset of itself.

Example 2 Infinite set

Show that the set of integers $\mathbb{Z} = \{\ldots, -3, -2, -1, 0, 1, 2, 3, \ldots\}$ is infinite.

Solution We can show that the set of integers can be placed into a one-to-one correspondence with the set of counting numbers:

$$\mathbb{N} = \{1, \quad 2, \quad 3, \quad 4, \quad 5, \quad 6, \quad 7, \ldots, \quad 2n, \; 2n + 1, \ldots\}$$
$$\updownarrow \; \updownarrow \; \updownarrow \; \updownarrow \; \updownarrow \; \updownarrow \; \updownarrow \qquad \updownarrow \qquad \updownarrow$$
$$\mathbb{Z} = \{0, +1, -1, +2, -2, +3, -3, \ldots, \quad +n, \quad -n, \ldots\}$$

Since $\mathbb{N}$ is a proper subset of $\mathbb{Z}$, we see that the set of integers is infinite.

Example 2 shows not only that the set of integers is infinite, but also that the cardinality of $\mathbb{Z}$ is $\aleph_0$. You might think that $\aleph_0 + \aleph_0 = 2\aleph_0$, but Example 2 shows that the cardinality of the integers is the same as the cardinality of the set of counting numbers; similarly, the set of counting numbers has the same cardinality as the set of negative counting numbers, so

$$\aleph_0 + \aleph_0 = \aleph_0 \quad \text{or} \quad 2\aleph_0 = \aleph_0$$

If a set is finite or if it has cardinality $\aleph_0$, we say that the set is **countable.** Thus, both the set of counting numbers and the set of integers are countable. A set that is not countable is said to be **uncountable.** We have shown that $\mathbb{N}$ and $\mathbb{Z}$ are countable, and in the problem set you are asked to show that $\mathbb{W}$ is countable. We can now show that $\mathbb{Q}$ is countable.

$$\frac{0}{1} \quad \frac{1}{1} \quad \frac{2}{1} \quad \frac{3}{1} \quad \frac{4}{1} \quad \frac{5}{1} \quad \frac{6}{1} \quad \frac{7}{1} \quad \ldots$$

$$\frac{1}{2} \quad \frac{2}{2} \quad \frac{3}{2} \quad \frac{4}{2} \quad \frac{5}{2} \quad \frac{6}{2} \quad \frac{7}{2} \quad \ldots$$

$$\frac{1}{3} \quad \frac{2}{3} \quad \frac{3}{3} \quad \frac{4}{3} \quad \frac{5}{3} \quad \frac{6}{3} \quad \frac{7}{3} \quad \ldots$$

$$\frac{1}{4} \quad \frac{2}{4} \quad \frac{3}{4} \quad \frac{4}{4} \quad \frac{5}{4} \quad \frac{6}{4} \quad \frac{7}{4} \quad \ldots$$

$$\frac{1}{5} \quad \frac{2}{5} \quad \frac{3}{5} \quad \frac{4}{5} \quad \frac{5}{5} \quad \frac{6}{5} \quad \frac{7}{5} \quad \ldots$$

$$\frac{1}{6} \quad \frac{2}{6} \quad \frac{3}{6} \quad \frac{4}{6} \quad \frac{5}{6} \quad \frac{6}{6} \quad \frac{7}{6} \quad \ldots$$

$$\frac{1}{7} \quad \frac{2}{7} \quad \frac{3}{7} \quad \frac{4}{7} \quad \frac{5}{7} \quad \frac{6}{7} \quad \frac{7}{7} \quad \ldots$$

$$\vdots \quad \vdots \quad \vdots \quad \vdots \quad \vdots \quad \vdots \quad \vdots$$

Example 3 Show that rationals are countable

Show that the set of rational numbers is countable.

Solution What we need to is to put the set of rationals into a one-to-one correspondence with the set of counting numbers. This is not an easy task, it was first done by Georg Cantor (see Historical Note on page 50).

First, consider the positive rational numbers and arrange them in rows so that all the numbers with denominators of 1 are written in row 1, those with denominator 2 are written in row 2, and so on. Notice that some numbers are highlighted because those numbers, when reduced, are found somewhere else in the list.

We see that every nonnegative rational number will appear on this list. For example, 103/879 will appear in row 103 and column 879. To set up a one-to-one correspondence between the set of nonnegative rationals and the set of counting numbers, we follow the path shown below:

$$\frac{0}{1} \rightarrow \frac{1}{1} \rightarrow \frac{2}{1} \quad \frac{3}{1} \rightarrow \frac{4}{1} \quad \frac{5}{1} \rightarrow \frac{6}{1} \quad \frac{7}{1} \quad \cdots$$

$$\frac{1}{2} \rightarrow \frac{2}{2} \quad \frac{3}{2} \quad \frac{4}{2} \quad \frac{5}{2} \quad \frac{6}{2} \quad \frac{7}{2} \quad \cdots$$

$$\frac{1}{3} \quad \frac{2}{3} \quad \frac{3}{3} \quad \frac{4}{3} \quad \frac{5}{3} \quad \frac{6}{3} \quad \frac{7}{3} \quad \cdots$$

$$\frac{1}{4} \quad \frac{2}{4} \quad \frac{3}{4} \quad \frac{4}{4} \quad \frac{5}{4} \quad \frac{6}{4} \quad \frac{7}{4} \quad \cdots$$

$$\frac{1}{5} \quad \frac{2}{5} \quad \frac{3}{5} \quad \frac{4}{5} \quad \frac{5}{5} \quad \frac{6}{5} \quad \frac{7}{5} \quad \cdots$$

$$\frac{1}{6} \quad \frac{2}{6} \quad \frac{3}{6} \quad \frac{4}{6} \quad \frac{5}{6} \quad \frac{6}{6} \quad \frac{7}{6} \quad \cdots$$

$$\frac{1}{7} \rightarrow \frac{2}{7} \quad \frac{3}{7} \quad \frac{4}{7} \quad \frac{5}{7} \quad \frac{6}{7} \quad \frac{7}{7} \quad \cdots$$

$$\vdots \qquad \vdots \qquad \vdots \qquad \vdots \qquad \vdots \qquad \vdots \qquad \vdots$$

Now we set up the one-to-one correspondence: $\frac{0}{1} \leftrightarrow 1, \frac{1}{1} \leftrightarrow 2, \frac{1}{2} \leftrightarrow 3, \frac{1}{3} \leftrightarrow 4$ (we skip $\frac{2}{2}$ because it is equal to $\frac{1}{1}$, and is already in our list), $\frac{1}{4} \leftrightarrow 5, \frac{2}{3} \leftrightarrow 6, \ldots$ Do you see the pattern? What we have shown is that the set of positive rational numbers has cardinality $\aleph_0$.

Finally, we use the method shown in Example 2; that is, we let each negative number follow its positive counterpart, so we can extend this correspondence to include all negative rational numbers. Thus, the set of all rational numbers has cardinality $\aleph_0$ and is therefore countable.

You may be saying to yourself, "All infinite sets are countable" (that is, have cardinality $\aleph_0$). This is not the case, as we see in the next example, where we show that the set of real numbers is uncountable.

Example 4 Show that reals are uncountable

Show that the set of real numbers is uncountable.

Solution In Chapter 5, we will see that the set of real numbers, $\mathbb{R}$, can be characterized as the set of all possible repeating and terminating decimals. In order to prove that the set of real numbers is uncountable, we note that every set is either countable or uncountable. We will assume that the set is countable, and then we will arrive at a contradiction. This contradiction, in turn, leads us to the conclusion that our assumption is incorrect, thus establishing that the set is uncountable. This process is called **proof by contradiction.**

Let's suppose that $\mathbb{R}$ is countable. Then, there is some one-to-one correspondence between $\mathbb{R}$ and $\mathbb{N}$, say:

$$1 \leftrightarrow \quad 5.285938 \ldots$$
$$2 \leftrightarrow -1,200.444444 \ldots$$
$$3 \leftrightarrow \quad 630.500000 \ldots$$
$$4 \leftrightarrow \quad 0.123456 \ldots$$
$$\vdots$$

Now, if we assume that there is a one-to-one correspondence between $\mathbb{N}$ and $\mathbb{R}$, then *every* decimal number is in the above list. To show this is not possible, we construct a new decimal as follows. The first digit of this new decimal is *any* digit different from the first digit of the entry corresponding to the first correspondence. (That is, anything other than 2 using the above-listed correspondence). The second digit is any digit different from the second digit of the entry corresponding to the second correspondence (4 in this example). Do the same for *all* the numbers in the one-to-one correspondence. Because of the way we have constructed this new number, it is not on the list. But we began by assuming that all numbers are on the list (i.e., that all decimal numbers are part of the one-to-one correspondence). Since both these statements cannot be true, the original assumption has led to a contradiction. This forces us to accept the only possible alternative to the original assumption. That is, it is not possible to set up a one-to-one correspondence between $\mathbb{R}$ and $\mathbb{N}$, which means that $\mathbb{R}$ is uncountable.

It can also be shown that the irrational numbers (real numbers that are not rational) are uncountable. Furthermore, the set $\{x \mid a \le x \le b\}$, where a and D are any real numbers, is also uncountable. Individual Project 2.4 considers this set for a particular choice of a and b. What we have done here is to show that there are cardinalities that are greater than $\aleph_0$. Cantor called the cardinality of the real numbers c, which is greater than $\aleph_0$ and is sometimes called the *power of the continuum*. The cardinality c is sometimes called $\aleph_1$, because there is speculation (yet unproved) that there is no set with cardinality between that of $\mathbb{N}$ and $\mathbb{R}$. Cantor was able to show, however, that there are even larger orders of infinity, and, in fact, infinitely many larger alephs.

Cartesian Product of Sets

There is an operation of sets called the *Cartesian product* that provides a way of generating new sets when given the elements of two sets. Suppose we have the sets

$$A = \{a, b, c\} \qquad \text{and} \qquad B = \{1, 2\}$$

The **Cartesian product** of sets A and B, denoted by $A \times B$, is the set of all *ordered pairs* (x, y) where $x \in A$ and $y \in B$. For this example,

$$A \times B = \{(a, 1), (a, 2), (b, 1), (b, 2), (c, 1), (c, 2)\}$$

Notice that $A \times B \ne B \times A$.

Example 5   Cardinality of a Cartesian product

How many elements are in the Cartesian product of the given sets?
a. $A = \{$Frank, George, Hazel$\}$ and $B = \{$Alfie, Bogie, Calvin, Doug, Ernie$\}$
b. $C = \{$U.S. Senators$\}$ and $D = \{$U.S. President, U.S. Vice President, Secretary of State$\}$

Solution
a. One of the ways to find a Cartesian product is to represent the sets as an array.

$A \times B$	Alfie, a	Bogie, b	Calvin, c	Doug, d	Ernie, e
Frank, f	(f, a)	(f, b)	(f, c)	(f, d)	(f, e)
George, g	(g, a)	(g, b)	(g, c)	(g, d)	(g, e)
Hazel, h	(h, a)	(h, b)	(h, c)	(h, d)	(h, e)

Since a Cartesian product can be arranged as a rectangular array, we see that the number of elements is the product of the number of elements in the sets. That is, since $|A| = 3$ and $|B| = 5$, we see $|A \times B| = 3 \times 5 = 15$.

b. We are looking for $|C \times D|$, but we see it is not practical to form the rectangular array because $|C| = 100$ and $|D| = 3$. We still can visualize the size of the array even without writing it out, so we conclude

$$|C \times D| = |C| \times |D| = 100 \times 3 = 300$$

We will frequently find it convenient to use the multiplication principle illustrated in Example 5, so we restate it as a general property, called the **fundamental counting principle.**

Fundamental Counting Principle

The fundamental counting principle gives the number of ways of performing two or more tasks. If task A can be performed in m ways, and if after task A is performed, a second task B can be performed in n ways, then task A followed by task B can be performed in $m \times n$ ways.

Example 6   Classify as finite or infinite

Classify each of the sets as finite or infinite.
a. Set of people on Earth
b. Set of license plates that can be issued using three letters followed by three numerals
c. Set of drops of water in all the oceans of the world

Solution
a. We do not know the size of this set, but we do know that there are fewer than 10 billion people on Earth. If the cardinality of a set is less than a finite number (and 10 billion is finite), then it must be a finite set.
b. We can use the fundamental counting principle to calculate the number of possible license plates:

$$26 \times 26 \times 26 \times 10 \times 10 \times 10 = 17{,}576{,}000$$

Note that each "26" counts the number of letters of the alphabet, and each "10" counts the number of numerals in each position. Since this is a particular number, we see that this set of license plates is finite.
c. Certainly this must be an infinite set . . . who could count the number of drops? Let's consider this a bit more carefully. What is a drop? Is that well defined? If you are told by your doctor to use two drops to dilate the pupils of your eyes, can you do that? You have a device called an "eyedropper" and you squeeze out two drops. Now, there is a finite number of drops that make up an ounce, and there are a certain number of ounces in a gallon, and a certain number of gallons in a cubic foot. The earth will fit entirely inside a cube of a certain size, and if this cube were filled with water, it would contain a finite number of gallons. The number of drops of water in all the oceans of the world is certainly less than this number, and hence the set is finite.

Problem Set

Level 1

1. IN YOUR OWN WORDS   Why do you think the fundamental counting principle is so "fundamental"?

2. IN YOUR OWN WORDS   Describe your preconceived notions about "infinity." Next, describe your understanding of "infinity" in light of the material in this section.

3. **IN YOUR OWN WORDS** What is the Cartesian product of two sets?

4. **IN YOUR OWN WORDS** What does the cardinality of the Cartesian product have to do with the fundamental counting principle?

Find the Cartesian product of the sets given in Problems 5–10.

5. $A = \{c, d, f\}$ and $B = \{w, x\}$; find $A \times B$.

6. $A = \{c, d, f\}$ and $B = \{w, x\}$; find $B \times A$.

7. $F = \{1, 2, 3, 4, 5\}$ and $C = \{a, b, c\}$; find $F \times C$.

8. $F = \{1, 2, 3, 4, 5\}$ and $C = \{a, b, c\}$; find $C \times F$.

9. $X = \{x | x \in \mathbb{N}, x \text{ is between 1 and 5}\}$ and $V = \{v | v \text{ is a vowel}\}$; find $X \times V$.

10. $X = \{x | x \in \mathbb{N}, x \text{ is between 1 and 5}\}$ and $V = \{v | v \text{ is a vowel}\}$; find $V \times X$.

Give the cardinality of each set in Problems 11–24.

11. $\{48, 49, \ldots, 189, 190\}$

12. $\{16, 20, 24, \ldots, 400, 404\}$

13. $\{-42, -41, \ldots, 41, 42\}$

14. $\{-998, -997, \ldots, 85, 86\}$

15. $\{c, d, e, \ldots, v, w\}$

16. $\{0\}$

17. $\varnothing$

18. $\{\varnothing\}$

19. $\{100, 200, 300, \ldots\}$

20. $\left\{\frac{1}{2}, \frac{1}{3}, \frac{1}{4}, \frac{1}{5}, \cdots\right\}$

21. $T \times W$ where $T = \{1, 2, \ldots, 19, 20\}$, $W = \{a, b, \ldots, v, w\}$

22. $S \times V$ where $S = \{\text{U.S. states}\}$, $V = \{a, b, c\}$

23. $\{x | x \text{ is an odd integer}\}$

24. $\{y | y \text{ is an even integer}\}$

Level 2

25. If $A = \{\text{letters of the alphabet}\}$ and $B = \{0, 1, 2, \ldots, 9\}$, what is $|A \times B|$?

26. If $A = \{\text{letters of the alphabet}\}$, what is $|A \times A|$?

27. If $C = \{500, 501, 502, \ldots, 599\}$ and $D = \{\text{U.S. state capitals}\}$, what is $|C \times D|$?

28. If $M = \{\text{distinct letters in the word } mathematics\}$ and $N = \{\text{distinct letters in the word } nevertheless\}$, what is $|M \times N|$?

29. Show that there is more than one one-to-one correspondence between the sets $\{m, a, t\}$ and $\{1, 2, 3\}$.

30. Show that the following sets have the same cardinality, by placing the elements into a one-to-one correspondence.
$\{1, 2, 3, 4, \ldots, 999, 1000\}$
$\{7964, 7965, 7966, 7967, \ldots, 8962, 8963\}$

31. Show that the following sets do not have the same cardinality.
$\{1, 2, 3, 4, \ldots, 586, 587\}$
$\{550, 551, 552, 553, \ldots, 902, 903\}$

32. Do the following sets have the same cardinality?
$\{48, 49, 50, \ldots, 783, 784\}$
$\{485, 487, 489, \ldots, 2053, 2055\}$

33. Classify the following sets as finite or infinite.
    a. The set of stars in the Milky Way

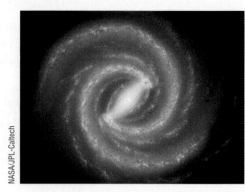

Milky Way Galaxy

    b. The set of counting numbers greater than one million

34. Classify the following sets as finite or infinite.
    a. The set of people who are living or who have ever lived.
    b. The set of grains of sand on all the beaches on Earth

Level 3

Show that each set in Problems 35–44 has cardinality $\aleph_0$.

35. $\{-1, -2, -3, \ldots\}$

36. $\{1000, 3000, 5000, \ldots\}$

37. $\left\{\frac{1}{x} | x \in \mathbb{N}\right\}$

38. $\{y | y \in \mathbb{Z} \text{ and is divisible by 2}\}$

39. $\mathbb{W}$

40. $\mathbb{Z}$

41. Set of odd counting numbers

42. Set of positive multiples of 5

43. $\{1, 3, 9, 27, 81, \ldots\}$

44. $\{-85, -80, -75, \ldots\}$

Show that each set in Problems 45–48 is infinite by placing it into a one-to-one correspondence with a proper subset of itself.

45. $\mathbb{W}$

46. $\mathbb{N}$

47. $\{12, 14, 16, \ldots\}$

48. $\{4, 44, 444, 4444, \ldots\}$

*An infinite set is said to be **countably infinite** if it can be placed into a one-to-one correspondence with the set of counting numbers, $\mathbb{N}$. If the set cannot be placed into such a correspondence, it is said to be **uncountably infinite**. Decide whether the sets in Problems 49–54 are countably infinite.*

49. Set of whole numbers

50. Set of integers

51. Set of rationals

52. Set of reals

53. $\{2, 4, 8, 16, 32, \ldots\}$

54. $\{1, 4, 9, 16, 25, \ldots\}$

Determine whether each statement in Problems 55–58 is true or false. Give reasons for your answers.

55. All uncountable sets are infinite.

56. All infinite sets are uncountable.

57. If two sets are infinite, then they are equivalent.

58. If two sets are equivalent, then they are infinite.

Problem Solving 3

59. Make up an example to show that
$$\aleph_0 + \aleph_0 = \aleph_0$$

60. Insert "=" or "≠" in each box.
   a. $\aleph_0 \;\square\; \aleph_0 + \aleph_0$
   b. $2\aleph_0 \;\square\; \aleph_0 + \aleph_0 + \aleph_0$
   c. $\aleph_0 + 2 \;\square\; \aleph_0 - 2$
   d. $|\mathbb{N}| \;\square\; \aleph_0$

2.5 CHAPTER SUMMARY

Numeracy is the ability to cope confidently with the mathematical demands of adult life.
MATHEMATICS COUNTS

Important Ideas

Denoting sets [2.1]
Sets of numbers [2.1]
Universal and empty sets [2.1]
Equal and equivalent sets [2.1]
Venn diagrams [2.1]
Operations with sets [2.1, 2.2, 2.3]
Cardinality of unions and intersections [2.2]
De Morgan's laws [2.3]
Survey problems [2.3]
One-to-one correspondence [2.4]
Fundamental counting principle [2.4]

Take some time getting ready to work the review problems in this section. First review the listed important ideas. Look back at the definition and property boxes. If you look online, you will find a list of important terms introduced in this chapter, as well as the types of problems that were introduced in this chapter. You will maximize your understanding of this chapter by working the problems in this section only after you have studied the material.

You will find some review help online at **www.mathnature.com.** There are links giving general test help in studying for a mathematics examination, as well as specific help for reviewing this chapter.

Chapter **2** Review Questions

In Problems 1–8, let U = {1, 2, 3, 4, 5, 6, 7, 8, 9, 10},
A = {1, 3, 5, 7, 9}, and B = {2, 4, 6, 9, 10}. Find:

1. a. $A \cup B$ **b.** $A \cap B$ **2. a.** $\overline{B}$ **b.** $|\varnothing|$

3. a. $|U|$ **b.** $|A|$ **4. a.** $|A \times B|$ **b.** $|U \times B|$

5. a. $\overline{A \cap B}$ **b.** $\overline{A} \cup \overline{B}$

   c. Does $\overline{A \cap B} = \overline{A} \cup \overline{B}$? Why or why not?

6. $\overline{(A \cup B)} \cap A$

7. $\overline{A} \cap (B \cup A)$

8. Replace the question mark with one of the symbols
$=, \subset, \subseteq,$ or $\in$ to make the statement true.
   a. $\{2\} \, ? \, B$ **b.** $2 \, ? \, B$ **c.** $\varnothing \, ? \, \{ \ \}$
   d. $\{5, 3, 7, 9, 1\} \, ? \, A$ **e.** $B \, ? \, U$

9. If $U = \{1, 2, 3, \ldots, 49, 50\}$, $N = \{$odd numbers$\}$, and
$P = \{$two-digit numbers$\}$, find:
   a. $|N|$ **b.** $|P|$ **c.** $|N \cap P|$ **d.** $|N \cup P|$

Draw Venn diagrams for the sets in Problems 10–13.

10. a. $X \cap Y$ **b.** $X \cup Y$

11. a. $\overline{X}$ **b.** $\overline{\overline{X}}$

12. a. $\overline{X} \cap Y$ **b.** $\overline{X \cup Y}$

13. a. $(X \cap Y) \cap Z$
   b. $(X \cup Y) \cap Z$

Prove or disprove the statements given in
Problems 14–15.

14. $A \cup (B \cap C) = (A \cup B) \cap C$

15. $(A \cup B) \cap C = (A \cap C) \cup (B \cap C)$

16. The set of rational numbers is $\mathbb{Q} = \{\frac{a}{b} | a \in \mathbb{Z}, b \in \mathbb{N}\}$.
   a. Write this statement in words.
   b. Give an example for the number q where $q \in \mathbb{Q}$.

17. Show that the set $F = \{5, 10, 15, 20, 25, \ldots\}$ is infinite.

18. a. Write the set $\{-1, -2, -3, \ldots\}$ using set-builder
     notation.
   b. Show that this set has cardinality u_0.

19. A survey of 70 college students showed the following data:

    42 had a car; 50 had a TV; 30 had a bicycle; 17 had a car
    and a bicycle; 35 had a car and a TV; 25 had a TV and a
    bicycle; 15 had all three

   How many students had none of the three items?

20. Human blood is typed Rh$^+$ (positive blood) or Rh$^-$ (negative
blood). This Rh factor is called an *antigen*. There are two
other antigens known as A and B types. Blood lacking both A
and B types is called type O. In Problem 60 of Section 2.3,
you were asked to sketch a Venn diagram showing the three
types of antigens A, B, and Rh. One possible diagram is
shown in Figure 2.20.

Recipient Blood Types	Donor Blood Types
A$^+$	A$^+$, A$^-$, O$^+$, O$^-$
B$^+$	B$^+$, B$^-$, O$^+$, O$^-$
AB$^+$	A$^+$, A$^-$, B$^+$, B$^-$,
	AB$^+$, AB$^-$, O$^+$, O$^-$
O$^+$	O$^+$, O$^-$
A$^-$	A$^-$, O$^-$
B$^-$	B$^-$, O$^-$
AB$^-$	A$^-$, B$^-$, AB$^-$, O$^-$
O$^-$	O$^-$

FIGURE 2.20 The eight blood types and compatibility table

a. List the eight possible blood types.
b. A *universal donor* is a person who can give blood to any of the
blood types, and this is a person with type O$^-$ blood. The reason for this is that the region labeled O$^-$ includes all parts of
the Venn diagram outside the circle labeled O$^+$. Name these
regions.
c. A *universal recipient* is a person who can receive blood from
any of the blood types, and this is a person with type AB$^+$
blood. The reason for this is that the region labeled AB$^+$ is inside all parts of the Venn diagram except the region labeled O$^+$.
Name the regions in which AB$^+$ is located.

BOOK REPORTS

Write a 500-word report on this book:

**Innumeracy: Mathematical Illiteracy and Its
Consequences,** John Allen Paulos (New York: Hill
and Wang, 1988).

Group RESEARCH PROJECTS

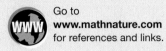

Go to
www.mathnature.com
for references and links.

Working in small groups is typical of most work environments, and learning to work with others to communicate specific ideas is an important skill. Work with three or four other students to submit a single report based on each of the following questions.

G3. A teacher assigned five problems, A, B, C, D, and E. Not all students turned in answers to all of the problems. Here is a tally of the percentage of students turning in various combinations of problems:

A; 46%	A, B; 25%	A, B, C; 13%	A, B, C, D; 7%	A, B, C, D, E; 4%
B; 40%	B, C; 26%	A, B, E; 19%	A, B, C, E; 8%	
C; 43%	C, D; 26%	A, D, E; 16%	A, B, D, E; 9%	
D; 38%	D, E; 22%	B, C, D; 12%	A, C, D, E; 11%	
E; 41%	A, E; 30%	C, D, E; 14%	B, C, D, E; 6%	

What percent of the students did not turn in any problems? Assume that no students turned in combinations not listed.

G4. Symbolically name the 32 regions formed by a Venn diagram with five sets.

G5. **HISTORICAL QUEST**   A famous mathematician, Bertrand Russell, created a whole series of paradoxes by considering situations such as the following *barber's rule*: "Suppose in the small California town of Ferndale it is the practice of many of the men to be shaved by the barber. Now, the barber has a rule that has come to be known as the barber's rule: *He shaves those men and only those men who do not shave themselves.* The question is: Does the barber shave himself?" If he does shave himself, then, according to the barber's rule, he does not shave himself. On the other hand, if he does not shave himself, then, according to the barber's rule, he shaves himself. We can only conclude that there can be no such barber's rule. But why not? Write a paper explaining what is meant by a *paradox*. Use the Historical Note for some suggestions about mathematicians who have done work in this area. You might begin with the links for this problem at **www.mathnature.com.**

Historical NOTE

About the time that Cantor's work began to gain acceptance, certain inconsistencies began to appear. One of these inconsistencies, called Russell's paradox, is what Project **G5** is about. Other famous paradoxes in set theory have been studied by many famous mathematicians, including Zermelo, Frankel, von Neumann, Bernays, and Poincaré. These studies have given rise to three main schools of thought concerning the foundation of mathematics.

Individual RESEARCH PROJECTS

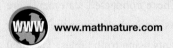

www.mathnature.com

Learning to use sources outside your classroom and textbook is an important skill, and here are some ideas for extending some of the ideas discussed in this chapter. You can find references to these projects in a library or at **www.mathnature.com.**

PROJECT 2.1 What is the millionth positive integer that is not a square or a cube?

PROJECT 2.2 What is the millionth positive integer that is not a square, cube, or fifth power?

PROJECT 2.3 Write a report discussing the creation of colors using additive color mixing and subtractive color mixing.

PROJECT 2.4 Prove that $\mathbb{S} = \{x \mid x \in \mathbb{R} \text{ and } 0 < x < 1\}$ is not a countable set.

3 THE NATURE OF LOGIC

What in the World?

"What's up?" asked Shannon. "I'm going nuts filling out these applications!" exclaimed Missy. "They must make these forms obscure just to test us. Look at this question on the residency questionnaire for the California State University system."

Complete and submit this residency questionnaire only if your response to item 34a of Part A was other than "California" or your response to item 37 was "no."

"Maybe I can help," added Shannon.

"What are the questions in items 34a and 37?" Missy responded. "Well, for what it's worth, just take a look."

Question 34a: Were you born in California? Answer yes or no. Question 37: . . .

"I quit!" shouted Missy.

"Wait a minute," urged Shannon. "We know that Question 34a can be answered by a yes or no response, so we can figure this out. Let's look at the possibilities."

Overview

In the first chapter we studied patterns that allow us to form conjectures using inductive reasoning. Inductive reasoning is the process of forming conjectures that are based on a number of observations of specific instances.

There is, however, a type of reasoning that produces *certain* results. It is based on accepting certain *premises* and then *deducing* certain inescapable conclusions. Not only much of mathematics, but also much of the reasoning we do in the world, is based on the principles of logic introduced in this chapter.

We begin with *simple statements*, which are either true or false, and then we combine those statements using certain connectives, such as *and, or, not, because, either . . . or, neither . . . nor*, and *unless*. These connected statements are called compound statements, and we discover the truth or falsity of compound statements using truth tables. The final step in our journey through this chapter is to begin with an argument, translate it into symbols, derive logical conclusions, and then translate back into English to complete a proof of a logical argument.

3.1 | Deductive Reasoning

Symbolic Logic is Mathematics, Mathematics is Symbolic Logic,
the twain are one.

J. F. KEYSER

CHAPTER **CHALLENGE**

See if you can fill in the question mark.

● ‡ — — ‡ — — ‡ ■ ■ ?

Historical NOTE

Karl Smith library

Aristotle
(384–322 B.C.)

Logic began to flourish during the classical Greek period. Aristotle was the first person to study the subject systematically, and he and many other Greeks searched for universal truths that were irrefutable. The logic of this period, referred to as Aristotelian logic, is still used today and is based on the syllogism, which we will study later in this chapter.

In Eastern culture, the yin-yang symbol shows the duality of nature: male-female, light-dark, good-evil

Terminology

In Chapter 1, we introduced a type of reasoning that produces results based on certain laws of logic. For example, consider the following argument.

1. All humans are mortal.
2. Socrates is a human.
3. Therefore, Socrates is mortal.

Statements 1, 2, and 3 are called an **argument.** If you accept statements 1 and 2 of the argument as true, then you *must* accept statement 3 as true. Statements 1 and 2 are called the **hypotheses** or **premises** of the argument, and statement 3 is called the **conclusion.** Such reasoning is called **deductive reasoning** and, if the conclusion follows from the hypotheses, the reasoning is said to be **valid.**

The purpose of this chapter is to build a logical foundation to aid you, not only in your study of mathematics and other subjects, but also in your day-to-day contact with others.

Logic is a method of reasoning that accepts only those conclusions that are inescapable. This is possible because of the strict way in which every concept is defined. That is, everything must be defined in a way that leaves no doubt or vagueness in meaning. Nothing can be taken for granted, and dictionary definitions are not usually sufficient. For example, in English one often defines a sentence as "a word or group of words stating, asking, commanding, requesting, or exclaiming something; a conventional unit of connected speech or writing, usually containing a subject and predicate, beginning with a capital letter, and ending with an end mark." **Symbolic logic** uses symbols to represent various logical ideas, and it bridges the gap between philosophy and logic. Philosophy needs logic in order to establish that a philosophical doctrine is true. To demonstrate that everyone would accept a proof that a particular position is true, the philosopher uses logic and symbols.

We begin with tiny steps. In symbolic logic, the word *statement* refers to a declarative sentence.

Statement

A **statement** is a declarative sentence that is either true or false, but not both true and false. A **simple statement** is one without any connective.

All of the following are statements, since they are either true or false.

1. School starts on Friday the 13th.
2. $5 + 6 = 16$
3. Fish swim.
4. Mickey Mouse is president.

If the sentence is a question or a command, or if it is vague or nonsensical, then it cannot be classified as true or false; thus, we would not call it a statement. For example:

5. Go away!
6. What are you doing?
7. This sentence is false.

These are not statements by our definition, since they cannot possibly be either true or false. Difficulty in simplifying arguments may arise because of their length, the vagueness of the words used, the literary style, or the possible emotional impact of the words used. Consider the following two arguments first stated in Chapter 1.

First Argument	*Second Argument*
If you use heroin, then you first used marijuana.	If George Washington was assassinated, then he is dead.
Therefore, if you use marijuana, then you will use heroin.	Therefore, if he is dead, he was assassinated.

Logically, these two arguments are exactly the same, and both are **invalid** forms of reasoning. Nearly everyone would agree that the second is invalid, but many people see the first as valid, because of the emotional appeal of the words used.

One way of avoiding these difficulties is to use Euler circles as we did in Chapter 1. Another is to set up an *artificial symbolic language.* This procedure was first suggested by Leibniz in his search to unify all of mathematics. What we will do is invent a notational shorthand. We denote simple statements with letters such as *p, q, r, s, . . . ,* and then define certain connectives. The problem, then, is to *translate the English statements into symbolic form, simplify the symbolic form,* and then *translate the symbolic form back into English statements.*

The key to simplifying complicated logical arguments is to consider only simple statements connected by certain **operators** (**connectives**), each of which is defined precisely. Some of these operators are *not, and, or, neither . . . nor, if . . . then, unless,* and *because.*

A **compound statement** is formed by combining simple statements with one or more of these operators. Because of our basic definition of a statement, we see that the *truth value* of any statement is either true (T) or false (F), but we should point out that the newest chapter of logic is presently being written. This new logic, called **fuzzy logic,** defines gradations of "T" and "F" to describe real-world concepts that are, by nature, often vague. The difference between the logic we study in this chapter and fuzzy logic is something Aristotle called the **law of the excluded middle,** which says that every statement is either true or false (never both).

As we saw in Chapter 2, an object either does or does not belong to a set. There is no middle ground: The number 5 belongs fully to the set of counting numbers or it does not. This is true because of the law of the excluded middle. Sets that are fuzzy break the law of the excluded middle, to some degree. The study of fuzzy logic requires concepts of probability (Chapter 13), as well as an understanding of the logic we introduce in this chapter. For purposes of this course, we accept the law of the excluded middle as an axiom, and work within a framework that forces every statement to be either true or false.

Truth Value

The **truth value** of a *simple statement* is either true (T) or false (F). The **truth value** of a *compound statement* is true or false and depends only on the truth values of its simple component parts. It is determined by using the rules for connecting those parts with well-defined operators.

It is not sufficient to assume that we know the meanings of the operators *and, or, not,* and so on, even though they may seem obvious and simple. The strength of logic is that it does not leave any meanings to chance or to individual interpretation. In defining the truth values of these words, we will, however, try to conform to common usage.

Conjunction

If p and q represent two simple statements, then "p and q" is the compound statement using the operator called **conjunction.** The word **and** is symbolized by $\wedge$. For example,

I have a penny and a quarter in my pocket.

When is this compound statement true? Let p and q represent two simple statements as follows:

p: I have a penny q: I have a quarter.

There are four logical possibilities (only one of which is true in any specific instance). The four possibilities can be shown in table form:

		p	q
1. I have a penny. I have a quarter.	1.	T	T
2. I have a penny. I do not have a quarter.	2.	T	F
3. I do not have a penny. I have a quarter.	3.	F	T
4. I do not have a penny. I do not have a quarter.	4.	F	F

If p and q are both true, we would certainly say that the compound statement is true; otherwise, we say it is false. Thus, we define "$p \wedge q$" according to Table 3.1. The common usage of the word *and* conforms to the technical definition of conjunction.

TABLE 3.1

Definition of Conjunction

p	q	$p \wedge q$
T	T	T
T	F	F
F	T	F
F	F	F

 The definitions of conjunction, disjunction, and negation are fundamental to understanding the material of this chapter.

It is worth noting that statements p and q need not be related. For example,

Fish swim and there are 100 U.S. senators.

is a true compound statement.

Disjunction

The operator **or,** denoted by $\vee$, is called **disjunction.** The meaning of this simple word is ambiguous, as we can see by considering the following examples.

I. I have a penny or a quarter in my pocket.

II. The president is speaking in New York or in California at 7:00 P.M. tonight.

Let's represent four statements as follows:

p: I have a penny in my pocket.

q: I have a quarter in my pocket.

n: The president is speaking in New York at 7:00 P.M. tonight.

c: The president is speaking in California at 7:00 P.M. tonight.

In everyday terms, what do we mean by each of these examples? Statement I may mean:

I have a penny in my pocket. I have a quarter in my pocket.

I have both a penny and a quarter in my pocket.

Statement II may mean:

The president is speaking in New York at 7:00 P.M. tonight.

The president is speaking in California at 7:00 P.M. tonight.

It does *not* mean that he will do both.

These statements illustrate different usages of the word *or*. Now we are forced to select a single meaning for the operator, so we choose a definition that conforms to statement I. That is, "$p \vee q$" means "p or q, perhaps both." Thus, we define "$p \vee q$" according to Table 3.2.

TABLE 3.2		
Definition of Disjunction		
p	*q*	*p* $\vee$ *q*
T	T	T
T	F	T
F	T	T
F	F	F

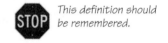 *This definition should be remembered.*

In logic, the statement defined in Table 3.2 is called the **inclusive or.** The second meaning of the word *or* ("*p* or *q*, but not both") is called the **exclusive or.** In this book, we will translate the exclusive or by using the operator **either . . . or.** Thus, in this book, we would translate the two examples as:

I. *I have a penny or a quarter in my pocket.*

II. *The president is speaking either in New York or in California at 7:00 P.M. tonight.*

Negation

The operator **not,** denoted by $\sim$, is called **negation.** In sign language, this operator is easy to express, as shown in Figure 3.1. The sign can use either a polite double motion or a firmer or definite single motion.

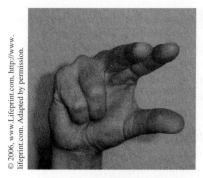

FIGURE 3.1 Sign language for "no"

Table 3.3 serves as a straightforward definition of negation. The negation of a true statement is false, and the negation of a false statement is true.

TABLE 3.3

Definition of Negation

p	$\sim p$
T	F
F	T

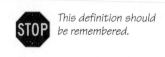

This definition should be remembered.

Example 1 Finding the negation

For the statement t: Otto is telling the truth, translate $\sim t$.

Solution   We can translate $\sim t$ as "It is not the case that Otto is telling the truth."

In everyday language we use the words *all*, *none*, and *some* (at least one). You must be careful when negating statements containing the words *all*, *none*, or *some*. For example, write the negation of

All students have pencils.

Go ahead; write it down before reading on. Did you write "No students have pencils" or "All students do not have pencils"? Remember that if a statement is false, then its negation must be true. The correct negations are:

It is not the case that all students have pencils.
Not all students have pencils.
At least one student does not have a pencil.
Some students do not have pencils.

In mathematics, the word *some* is used to mean "at least one." Table 3.4 gives some of the common negations.

TABLE 3.4

Negation of All, Some, and No

Statement	Negation
All	Some . . . not
Some	No
Some . . . not	All
No	Some

Example **2** **Finding the negation of *all, some*, and *no***

Write the negation of each statement.
a. All people have compassion.
b. Some animals are dirty.
c. Some students do not take Math 10.
d. No students are enthusiastic.

Solution
a. Some people do not have compassion.
b. No animal is dirty.
c. All students take Math 10.
d. Some students are enthusiastic.

We consider the negation of compound statements in the next section.

Order of Operations

Parentheses are used to indicate the order of operations. Thus, $\sim(n \wedge c)$ means the negation of the statement "*n* and *c*," and is read as "it is not the case that *n* and *c*." On the other hand, $\sim n \wedge c$ means "the negation of *n* and the statement *c*," which is read "not *n* and *c*."

Example **3** **Translate into words**

Translate the given statements into words.
a. $p \wedge q$ **b.** $\sim p$ **c.** $\sim(p \wedge q)$ **d.** $\sim p \wedge q$ **e.** $\sim(\sim q)$
Let   *p*: I eat spinach;   *q*: I am strong.

Solution
a. A translation of $p \wedge q$ is "I eat spinach and I am strong."
b. A translation of $\sim p$ is "I do not eat spinach."
c. A translation for $\sim(p \wedge q)$ is "It is not the case that I eat spinach and am strong."
d. A translation for $\sim p \wedge q$ is "I do not eat spinach and I am strong."
e. A translation of $\sim(\sim q)$ is "I am not not strong" or, if we assume that "not strong" is the same as "weak," then we can translate this as "I am not weak."

We can now consider the truth value of a compound statement using parentheses.

Example **4** **Finding the truth value**

Suppose *n* is a true statement, and *c* is a false statement. What is the truth value of the compound statement $(n \vee c) \wedge \sim(n \wedge c)$?

Solution   Begin with the symbolic statement, fill in the truth values given in the problem, and then simplify using the correct order of operations and the definitions in Tables 3.1–3.3.

$$(n \vee c) \wedge \sim(n \wedge c) \quad \text{Start with given statement.}$$
$$(T \vee F) \wedge \sim(T \wedge F) \quad \text{Fill in the truth values.}$$
$$T \wedge \sim F \quad \text{Simplify.}$$
$$T \wedge T$$
$$T$$

Thus, the compound statement is true. This result does not depend on the particular statements *n* and *c*. As long as *n* is true and *c* is false, the result of the compound statement is the same—namely, true.

We conclude this section by finding the truth values for the statements in Example 3.

| Example | 5 | **Truth values of compound statements** |

Suppose that p is T and q is F. Find the truth values for the statements in Example 3.
a. $p \wedge q$ **b.** $\sim p$ **c.** $\sim(p \wedge q)$ **d.** $\sim p \wedge q$ **e.** $\sim(\sim q)$
Let p: I eat spinach; q: I am strong.

Solution

a. $p \wedge q$ *Given*
    $\quad$ T $\wedge$ F   *Substitute truth values.*
    $\qquad$ F   *Definition of conjunction*
Thus, the statement "I eat spinach, and I am strong" is false when p (I eat spinach) is true and q (I am strong) is false.

b. $\sim p$ *Given*
    $\quad \sim$T   *Substitute truth value.*
    $\qquad$ F   *Definition of negation*
The statement "I do not eat spinach" is false when p (I eat spinach) is true.

c. $\sim(p \wedge q)$ *Given*
    $\sim$(T $\wedge$ F)   *Substitute truth values.*
    $\quad \sim$F   *Definition of conjunction*
    $\qquad$ T   *Definition of negation*
The statement "It is not the case that I eat spinach and am strong" is true when p (I eat spinach) is true and q (I am strong) is false.

d. $\sim p \wedge q$ *Given*
    $\sim$T $\wedge$ F   *Substitute truth values.*
    $\quad$ F $\wedge$ F   *Definition of negation*
    $\qquad$ F   *Definition of conjunction*
The statement "I do not eat spinach, and I am strong" is false when p (I eat spinach) is true and q (I am strong) is false.

e. $\sim(\sim q)$ *Given*
    $\sim(\sim$F)   *Substitute truth values.*
    $\quad \sim$(T)   *Definition of negation*
    $\qquad$ F   *Definition of negation again*
The statement "I am not weak" is false when q (I am strong) is false.

Problem Set   3.1

Level 1

1. **IN YOUR OWN WORDS**   What is an operator?

2. **IN YOUR OWN WORDS**   What do we mean by *conjunction*? Include as part of your answer the definition.

3. **IN YOUR OWN WORDS**   What do we mean by *disjunction*? Include as part of your answer the definition.

4. **IN YOUR OWN WORDS**   What do we mean by *negation*? Include as part of your answer the definition.

According to the definition, which of the examples in Problems 5–8 are statements?

5. **a.** Hickory, Dickory, Dock, the mouse ran up the clock.
   b. $3 + 5 = 9$

  c. Is John ugly?
  d. John has a wart on the end of his nose.

6. **a.** March 18, 2011, is a Friday.
   b. Division by zero is impossible.
   c. Logic is difficult.
   d. $4 - 6 = 2$

7. **a** $6 + 9 \neq 7 + 8$
   b. Thomas Jefferson was the 23rd president.
   c. Sit down and be quiet!
   d. If wages continue to rise, then prices will also rise.

8. **a.** Dan and Mary were married on August 3, 1979.
   b. $6 + 12 \neq 10 + 8$
   b. Do not read this sentence.
   c. Do you have a cold?

In the children's game called "Guess Who?"® the players are asked to guess a card by asking questions about the faces. Answer the questions in Problems 9–12 about the faces shown in Figure 3.2.

FIGURE 3.2 Guess Who?

 9. Who has glasses and a hat?

10. Who is male and wears a hat?

11. Who is bald or is blond?

12. Who is female or does not wear glasses?

Figure 3.3 shows some additional faces from the "Guess Who?"® game. Decide whether each of the statements in Problems 13–19 is true or false.

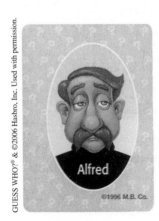

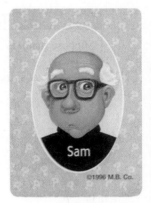

FIGURE 3.3 Guess Who?, part II

13. All are males or have blue eyes.

14. There are two faces that have white hair and blue eyes.

15. Not all faces have eyebrows.

16. At least one is a female.

17. No person has glasses and a mustache.

18. Not everyone has white hair.

19. No person does not have white hair.

Level 2

Write the negation of each statement in Problems 20–27.

20. All mathematicians are ogres.

21. All dogs have fleas.

22. No even integers are divisible by 5.

23. No triangles are squares.

24. All squares are rectangles.

25. All counting numbers are divisible by 1.

26. Some apples are rotten.

27. Some integers are not odd.

28. Let p: Prices will rise; q: Taxes will rise. Translate each of the following statements into symbols.
   a. Prices will rise, or taxes will not rise.
   b. Prices will rise, and taxes will not rise.
   c. Prices will rise, and taxes will rise.
   d. Prices will not rise, and taxes will rise.

29. Assume that prices rise and taxes also rise. Under these assumptions, which of the statements in Problem 28 are true?

30. Assume that prices rise and taxes do not rise. Under these assumptions, which of the statements in Problem 28 are true?

31. Let p: Prices will rise; q: Taxes will rise. Translate each of the following statements into words.
   a. $p \vee q$
   b. $\sim p \wedge q$
   c. $p \vee \sim q$
   d. $\sim p \vee \sim q$

32. Let p: Today is Friday; q: There is homework tonight. Translate each of the following statements into words.
   a. $p \wedge q$
   b. $\sim p \wedge q$
   c. $p \vee \sim q$
   d. $\sim p \vee \sim q$

33. Assume p is T and q is T. Under these assumptions, which of the statements in Problem 31 are true?

34. Assume p is F and q is T. Under these assumptions, which of the statements in Problem 32 are true?

Find the truth value for each of the compound statements in Problems 35–40.

35. Assume r is F, s is T, and t is T.
   a. $(r \vee s) \vee t$ **b.** $(r \wedge s) \wedge \sim t$

36. Assume r is T, s is T, and t is T.
   a. $r \wedge (s \vee t)$ **b.** $(r \wedge s) \vee (r \wedge t)$

37. Assume p is T, q is T, and r is F.
   a. $(p \vee q) \wedge r$ **b.** $(p \wedge q) \wedge \sim r$

38. Assume p is T, q is T, and r is T.
   a. $p \wedge (q \vee r)$ **b.** $(p \wedge q) \vee (p \wedge r)$

39. Assume p is T, q is T, and r is F.
   a. $(p \vee q) \vee (r \wedge \sim q)$ **b.** $\sim(\sim p) \vee (p \wedge q)$

40. Assume p is T, q is F, and r is T.
   a. $(p \wedge q) \vee (p \wedge \sim r)$ **b.** $(\sim p \vee q) \wedge \sim p$

Level 3

Translate the statements in Problems 41–49 into symbols. For each simple statement, be sure to indicate the meanings of the symbols you use. Answers are not unique.

41. W. C. Fields is eating, drinking, and having a good time.

42. Sam will not seek and will not accept the nomination.

43. Jack will not go tonight, and Rosamond will not go tomorrow.

44. Fat Albert lives to eat and does not eat to live.

45. The decision will depend on judgment or intuition, and not on who paid the most.

46. The successful applicant for the job will have a B.A. degree in liberal arts or psychology.

47. The winner must have an A.A. degree in drafting or three years of professional experience.

48. a. Dinner includes soup and salad, or the vegetable of the day.
   b. Dinner includes soup, and salad or the vegetable of the day.

49. a. Marsha finished the sign and table, or a pair of chairs.
   b. Marsha finished the sign, and the table or a pair of chairs.

In Problems 50–57, find the truth value when p is T, q is F, and r is F.

50. $(p \wedge q) \wedge r$ **51.** $p \vee (q \wedge r)$

52. $(p \vee q) \wedge \sim (p \vee \sim q)$ **53.** $(p \wedge \sim q) \vee (r \wedge \sim q)$

54. $\sim(\sim p) \wedge (q \vee p)$ **55.** $\sim(r \wedge q) \wedge (q \vee \sim q)$

56. $(p \wedge q) \vee [(p \vee \sim q) \vee (\sim r \wedge p)]$

57. $(q \vee \sim q) \wedge [(p \wedge \sim q) \vee (\sim r \vee r)]$

Problem Solving 3

58. Prove the law of double negation. That is, prove that for any statement $\sim(\sim p)$ *has the same truth value as p.*

59. Smith received the following note from Melissa: "Dr. Smith, I wish to explain that I was really joking when I told you that I didn't mean what I said about reconsidering my decision not to change my mind." Did Melissa change her mind or didn't she?

60. Their are three errers in this item. See if you can find all three.

3.2 Truth Tables and the Conditional

In the last section, we defined the connectives *and, or,* and *not* by listing all possibilities in the form of a table. We focus on and expand these ideas in this section.

Constructing Truth Tables

The connectives *and, or,* and *not* are called the **fundamental operators.** To go beyond these and consider additional operators, we need a device known in logic as a **truth table.** A truth table shows how the truth values of compound statements depend on the fundamental operators. Tables 3.1, 3.2, and 3.3 from the previous section should be memorized. They are summarized here in Table 3.5.

TABLE 3.5

Truth Table of the Fundamental Operators

p	q	Conjunction $p \wedge q$	Disjunction $p \vee q$	Negation $\sim p$	$\sim q$
T	T	T	T	F	F
T	F	F	T	F	T
F	T	F	T	T	F
F	F	F	F	T	T

Example **1** **Find a truth table**

Construct a truth table for the compound statement:

　　Alfie did not come last night and did not pick up his money.

Solution　First identify the operators: *not . . . and . . . not.* Next, choose variables to represent the simple statements. Let *p*: Alfie came last night. *q*: Alfie picked up his money. Translate the English statement into symbols: $\sim p \wedge \sim q.$

　　Finally, construct a truth table. List all possible combinations of truth values for the simple statements (see columns A and B).

A	B
p	q
T	T
T	F
F	T
F	F

Insert the truth values for and (see $\sim p \ \sim q$ columns C and D)

A	B	C	D
p	q	$\sim p$	$\sim q$
T	T	F	F
T	F	F	T
F	T	T	T
F	F	T	T

Finally, insert the truth values for $\sim p \wedge \sim q$ (see column E).

A	B	C	D	E
p	q	$\sim p$	$\sim q$	$\sim p \wedge \sim q$
T	T	F	F	F
T	F	F	T	F
F	T	T	F	F
F	F	T	T	T

The only time the compound statement is true is when *both p* and *q* are false.

The following example provides a truth table solution for Problem 58 of the previous problem set.

Example 2 Truth table for a double negative

Construct a truth table for $\sim(\sim p)$.

Solution

p	$\sim p$	$\sim(\sim p)$
T	F	T
F	T	F

> There is a story about a logic professor who was telling her class that a double negative is known to mean a negative in some languages and a positive in others (as in English). She continued by saying that there is no spoken language in which a double positive means a negative. Just then she heard from the back of the room a sarcastic "Yeah, yeah."

Notice from Example 2 that $\sim(\sim p)$ and p have the same truth values. (See also Problem 58 of Section 3.1.) If two statements have the same truth values, one can replace the other in any logical expression; this is called the **law of double negation.** This means that the double negation of a statement is the same as the original statement.

Law of Double Negation

$\sim(\sim p)$ may be replaced by p in any logical expression.

Example 3 Translate a news statement

Use the law of double negation to rewrite the following statement made by an Iraq official and reported on a national news report (December 3, 2002): "All that the U.S. has said about Iraq is a false lie."

Solution   If we assume that a lie is not the truth, then a "false lie" is the negation of "not the truth." This means a correct equivalent translation is "All that the U.S. has said about Iraq is the truth."

Example 4 Determine the truth of a statement

Construct a truth table to determine when the following statement is true.

$$\sim(p \wedge q) \wedge [(p \vee q) \wedge q]$$

Solution   We begin as before and move from left to right with parentheses taking precedence, focusing our attention on no more than two columns at a time (refer to Table 3.5 to find the correct entries).

A	B	C	D	E	F	G
p	q	$p \wedge q$	$\sim(p \wedge q)$	$p \vee q$	$(p \vee q) \wedge q$	$\sim(p \wedge q) \wedge [(p \vee q) \wedge q]$
T	T	T	F	T	T	F
T	F	F	T	T	F	F
F	T	F	T	T	T	T
F	F	F	T	F	F	F

The compound statement is true only when p is false and q is true.

Conditional

We can now use truth tables to prove certain useful results and to introduce some additional operators. The first one we'll consider is called the **conditional.** The statement "if p, then q" is called a *conditional statement.* It is symbolized by $p \rightarrow q$, where p is called the **antecedent,** and q is called the **consequent.** There are several ways of using a conditional, as illustrated by the following examples.

Uses of the IF-THEN format

1. We use "if-then" to indicate a *logical* relationship—one in which the consequent follows logically from the antecedent:

    If $\sim(\sim p)$ has the same truth value as p, then p can replace $\sim(\sim p)$.

2. We can use "if-then" to indicate a causal relationship:

    If John drops that rock, then it will land on my foot.

3. We can use "if-then" to report a decision on the part of the speaker:

    If John drops that rock, then I will hit him.

4. We can use "if-then" when the consequent follows from the antecedent by the very definition of the words used:

    If John drives a Geo, then John drives a car.

5. Finally, we can use "if-then" to make a *material implication.* There is no logical, causal, or definitional relationship between the antecedent and consequent; we use the expression simply to convey humor or emphasis:

    If John gets an A on that test, then I'm a monkey's uncle.

The consequent is obviously false, and the speaker wishes to emphasize that the antecedent is also false.

Our task is to state a definition of the conditional that applies to all of these *if-then* statements. We will approach the problem by asking under what circumstances a given conditional would be false. Let's consider another example. Suppose I make you a promise: "*If I receive my check tomorrow, then I will pay you the $10 that I owe you.*" If I keep my promise, let's agree to say the statement is true; if I don't, then it is false. Let

p: I receive my check tomorrow.
q: I will pay you the $10 that I owe you.

We symbolize the promise *by* $p \rightarrow q$. There are four possibilities:

$$p \quad q$$

Case 1: T T *I receive my check tomorrow, and I pay you the $10.*
In this case, the promise, or conditional, is true: $T \rightarrow T$ is T.

Case 2: T F *I receive my check tomorrow, and I do not pay you the $10.*
In this case, the conditional is false, since I did not fulfill my promise: $T \rightarrow F$ is F.

Case 3: F T *I do not receive my check tomorrow, but I pay you the $10.*
In this case, I certainly didn't break my promise, so the conditional is true: $F \rightarrow T$ is T.

Case 4: F F *I do not receive my check tomorrow, and I do not pay you the $10.*
Here, again, I did not break my promise, so the conditional is true: $F \rightarrow F$ is T.

Actually, the promise was not tested for cases 3 and 4, since I didn't receive my check. Assume the principle of "innocent until proven guilty." The only time that I will

Historical NOTE

**Albert Einstein
(1879–1955)**

In regard to the real nature of scientific truth, Einstein said, "As far as the laws of mathematics refer to reality, they are not certain, and as far as they are certain, they do not refer to reality." Einstein was one of the intellectual giants of the 20th century. He was a shy, unassuming man who was told as a child that he would never make a success of anything. However, he was able to use the tools of logic and mathematical reasoning to change our understanding of the universe.

have broken my promise is in case 2. The test for a conditional is to determine when it is false. In symbols,

$p \to q$ is false whenever $p \land \sim q$ is true (case 2)

or

$p \to q$ is true whenever $\sim(p \land \sim q)$ is true.

Construct a truth table for $\sim(p \land \sim q)$.

A	B	C	D	E
p	q	$\sim q$	$p \land \sim q$	$\sim(p \land \sim q)$
T	T	F	F	T
T	F	T	T	F
F	T	F	F	T
F	F	T	F	T

Notice that the entries in column E are the same as those we found using the promise viewpoint. Thus, we use this truth table to define the conditional $p \to q$ as shown in Table 3.6.

The following examples illustrate the definition of the conditional.

TABLE 3.6

Definition of Conditional

p	q	$p \to q$
T	T	T
T	F	F
F	T	T
F	F	T

STOP *Add this to your list of important definitions.*

Example **5**   **Variations of the conditional**

Make up an example illustrating each of the four cases for the conditional.

Solution   Examples can vary.

a. **Case 1:** $T \to T$ *If $7 < 14$, then $7 + 2 < 14 + 2$.*
     This is true, since both component parts are true.
b. **Case 2:** $T \to F$ *If $7 + 5 = 12$, then $7 + 10 = 15$.*
     This is false, since the antecedent is true, but the consequent is false.
c. **Case 3:** $F \to T$ *If $7 + 5 = 15$, then $8 + 2 = 10$.*
     This is true, since the antecedent is false.
d. **Case 4:** $F \to F$ *If $7 + 5 = 25$, then $7 = 20$.*
     This is true, since the antecedent is false.

Example 5 shows that the conditional, in mathematics, does not mean that there is any cause-and-effect relationship. *Any* two statements can be joined with the if-then connective, and the result must be true or false.

| Example | **6** | **1040 tax form** |

The following sentence is found on a tax form:

> If you do not itemize deductions on Schedule A and you have charitable contributions, then complete the worksheet on page 14 and enter the allowable part on line 36b.

Use symbolic form to analyze this sentence.

Solution   We use Pólya's problem-solving guidelines for this example.

Understand the Problem.   We need to rephrase, or analyze, this statement so that we can determine an action to take under all possible circumstances.

Devise a Plan.   We will translate the statement into symbolic form. It could then be analyzed using a truth table.

Carry Out the Plan.   First identify the operators:

> **If** you do **not** itemize deductions on Schedule A **and** you have charitable contributions, **then** complete the worksheet on page 14 **and** enter the allowable part on line 36b.

Next, assign variables to simple statements:

> d: You itemize deductions on Schedule A.
>
> c: You have charitable contributions.
>
> w: You complete the worksheet on page 14.
>
> b: You enter the allowable part on line 36b.

Rewrite the sentence, making substitutions for the variables:

> **If not** d **and** c, **then** (w **and** b).

Complete the translation into symbols:

> $(\sim d \wedge c) \rightarrow (w \wedge b)$

Look Back. We check, for example, to see whether the statement is true if d, c, w, and b are all true. To check all the possible circumstances, we would construct a truth table with 16 rows.

$$(\sim d \land c) \to (w \land b)$$
$$(\sim T \land T) \to (T \land T)$$
$$(F \land T) \to T$$
$$F \to T$$
$$T$$

The given statement is true in this case.

Translations for the Conditional

The *if* part of a conditional need not be stated first. All of the following statements have the same meaning:

Conditional Translation	Example
If p, then q.	If you are 18, then you can vote.
q, if p.	You can vote, if you are 18.
p, only if q	You are 18 only if you can vote.
All p are q.	All 18-year-olds can vote.

In addition to these translations for the conditional, there are related statements.

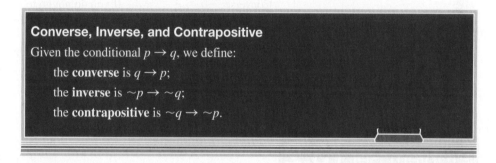

Converse, Inverse, and Contrapositive

Given the conditional $p \to q$, we define:

   the **converse** is $q \to p$;

   the **inverse** is $\sim p \to \sim q$;

   the **contrapositive** is $\sim q \to \sim p$.

Example **7** **Find the converse, inverse, and contrapositive**

Write the converse, inverse, and contrapositive of the statement:

   If it is a 380Z, then it is a car.

Solution   Let p: It is a 380Z; q: It is a car. The given statement is symbolized as $p \to q$.

Converse: $q \to p$	If it is a car, then it is a 380Z.
Inverse: $\sim p \to \sim q$	If it is not a 380Z, then it is not a car.
Contrapositive: $\sim q \to \sim p$	If it is not a car, then it is not a 380Z.

As you can see from Example 7, not all these statements are equivalent in meaning. The converse and inverse always have the same truth values, as do the contrapositive and the original statement. The latter result is called the **law of contraposition,** which is summarized in the following box.

Law of Contraposition

A conditional may always be replaced by its contrapositive without having its truth value affected.

Table 3.7 shows truth tables for the converse, inverse, and contrapositive.

TABLE 3.7

Truth Table for Variations of the Conditional

p	q	$\sim p$	$\sim q$	Statement $p \to q$	Converse $q \to p$	Inverse $\sim p \to \sim q$	Contrapositive $\sim q \to \sim p$
T	T	F	F	T	T	T	T
T	F	F	T	F	T	T	F
F	T	T	F	T	F	F	T
F	F	T	T	T	T	T	T

Example **8** Converse, inverse, and contrapositive

Assume that the following statement is true:

$p \to \sim q$ If you obey the law, then you will not go to jail.

Write the converse, inverse, and contrapositive.

Solution   We note that p: You obey the law; q: You go to jail.

Converse: $\sim q \to p$ If you do not go to jail, then you obey the law.
Inverse: $\sim p \to q$ If you do not obey the law, then you will go to jail.
Note: $\sim(\sim q)$ is replaced by q (double negation).
Contrapositive: $q \to \sim p$ If you go to jail, then you did not obey the law.
This has the same truth value as the given statement.

Problem Set   **3.2**

Level **1**

1. **IN YOUR OWN WORDS** What is a truth table?
2. **IN YOUR OWN WORDS** What is a conditional? Discuss.
3. **IN YOUR OWN WORDS** What is the law of double negation?
4. **IN YOUR OWN WORDS** What is the law of contraposition?

Construct a truth table for the statements given in Problems 5–22.

5. $\sim p \vee q$
6. $\sim p \wedge \sim q$
7. $\sim(p \wedge q)$
8. $\sim r \vee \sim s$
9. $\sim(\sim r)$
10. $(r \wedge s) \vee \sim s$
11. $p \wedge \sim q$
12. $\sim p \vee \sim q$
13. $(\sim p \wedge q) \vee \sim q$
14. $(p \wedge \sim q) \wedge p$

15. $p \vee (p \to q)$
16. $(p \wedge q) \to p$
17. $[p \wedge (p \vee q)] \to p$
18. $[p \vee (p \wedge q)] \to p$
19. $(p \vee q) \vee r$
20. $(p \wedge q) \wedge \sim r$
21. $[(p \vee q) \wedge \sim r] \wedge r$
22. $[p \wedge (q \vee \sim p)] \vee r$
23. If the sign on the left means no parking, what do you think the sign on the right means?

24. Repeat Example 6, except this time assume that you do not complete the worksheet, but all other statements are true.

Level 2

Write the converse, inverse, and contrapositive of the statements in Problems 25–30.

25. $\sim p \to \sim q$

26. $\sim r \to t$

27. $\sim t \to \sim s$

28. If you break the law, then you will go to jail.

29. I will go Saturday if I get paid.

30. If you brush your teeth with Smiles toothpaste, then you will have fewer cavities.

Translate the sentences in Problems 31–38 into if-then form.

31. All triangles are polygons.

32. All prime numbers greater than 2 are odd numbers.

33. All good people go to heaven.

34. Everything happens to everybody sooner or later if there is time enough. (G. B. Shaw)

35. We are not weak if we make a proper use of those means which the God of Nature has placed in our power. (Patrick Henry)

36. All useless life is an early death. (Goethe)

37. All work is noble. (Thomas Carlyle)

38. Everything's got a moral if only you can find it. (Lewis Carroll)

First decide whether each simple statement in Problems 39–42 is true or false. Then state whether the given compound statement is true or false.

39. If $5 + 10 = 16$, then $15 - 10 = 3$.

40. The moon is made of green cheese only if Mickey Mouse is president.

41. If $1 + 1 = 10$, then the moon is made of green cheese.

42. $3 \cdot 2 = 6$ only if water runs uphill.

Tell which of the statements in Problems 43–46 are true.

43. Let p: $2 + 3 = 5$; q: $12 - 7 = 5$.
    a. $\sim p \lor q$
    b. $\sim p \land \sim q$
    c. $\sim(p \land q)$

44. Let p: 2 is prime; q: 1 is prime.
    a. $\sim p \lor \sim q$
    b. $\sim(\sim p)$
    c. $(p \land q) \lor \sim q$

45. Let p: $5 + 8 = 10$; q: $4 + 4 = 8$.
    a. $(p \land \sim q) \land p$
    b. $p \lor (p \to q)$
    c. $(p \land q) \to p$

46. Let p: $1 + 1 = 2$; q: $9 - 3 = 5$.
    a. $(p \to \sim q) \to (q \to \sim p)$
    b. $(p \to q) \to (\sim q \to \sim p)$

Translate the statements in Problems 47–52 into symbolic form.

47. If the qualifying person is a child and not your dependent, enter this child's name.

48. If the amount on line 31 is less than \$26,673 and a child lives with you, turn to page 27.

49. If line 32 is \$86,025 or less, multiply \$2,500 by the total number of exemptions claimed on line 6e.

50. If married and filing a joint return, enter your spouse's earned income.

51. If you are a student or disabled, see line 6 of instructions.

52. If the income on line 1 was reported to you on Form W-2 and the "Statutory employee" box on that form was checked, see instructions for line 1 (Schedule C) and check here.

Level 3

In Problems 53–58, fill in the blanks with a symbolic statement that follows from the given statement.

53. The applicant for the position must have a two-year college degree in drafting or five years of experience in the field. Let

    a: You are an applicant for the position.
    q: You are qualified for the position.
    e: You have a two-year college degree in drafting.
    f: You have five years of experience in the field.

    a. $a \to$ _____
    b. $(a \land e) \to$ _____
    c. $(a \land f) \to$ _____

54. To qualify for a loan, the applicant must have a gross income of at least \$35,000 if single or combined income of \$50,000 if married. Let

    q: You qualify for a loan.
    m: You are married.
    i: You have an income of at least \$35,000.
    b: Your spouse has an income of at least \$35,000.

    a. $(\sim m \land \sim i) \to$ _____
    b. $(\sim m \land i) \to$ _____
    c. $[m \land (i \land b)] \to$ _____

55. To qualify for the special fare, you must fly on Monday, Tuesday, Wednesday, or Thursday, and you must stay over a Saturday evening. Let

    q: You qualify for the special fare.
    m: You fly on Monday.
    t: You fly on Tuesday.
    w: You fly on Wednesday.
    h: You fly on Thursday.
    s: You stay over a Saturday evening.

    _____ $\to q$

56. This contract is noncancellable by tenant for 60 days. Let

  t: You are a tenant.

  d: It is within 60 days.

  c: This contract can be canceled.

  $(t \wedge d) \rightarrow$ _____

57. The tenant agrees to lease the premises for 12 months beginning on September 1 and at a monthly rental charge of $800. Let

  t: You are a tenant.

  m: You lease the premises for 12 months.

  s: You will begin on September 1.

  p: You will pay $800 per month.

  $t \rightarrow$ _____

58. Utilities, except for water and garbage, are paid by the tenant. Let

  t: You are a tenant.

  w: You pay water.

  g: You pay garbage.

  u: You pay the other utilities.

  $t \rightarrow$ _____

59. If Apollo can do anything, could he make an object that he could not lift?

60. Decide about the truth or falsity of the following statement:

  If wishes were horses, then beggars could ride.

3.3 | Operators and Laws of Logic

Jack Sullivan/Alamy

In the previous sections of this chapter, we considered simple statements *p* and *q* and the three fundamental operators *and, or*, and *not*. We also looked at an operator called the *conditional*; in this section, we continue our investigation of the conditional.

Biconditional, Implication, and Logical Equivalence

In the previous section we took great care to point out that a general statement $p \rightarrow q$ and its converse $q \rightarrow p$ do not have the same truth values. However, it may be the case that $p \rightarrow q$ *and also* $q \rightarrow p$ have the same truth values for specific instances of *p* and *q*. In this case, we write

  $p \leftrightarrow q$

and call this operator the **biconditional.** To determine the truth values of the biconditional, we construct a truth table for

  $(p \rightarrow q) \wedge (q \rightarrow p)$

p	*q*	$p \rightarrow q$	$q \rightarrow p$	$(p \rightarrow q) \wedge (q \rightarrow p)$
T	T	T	T	T
T	F	F	T	F
F	T	T	F	F
F	F	T	T	T

This leads us to define the biconditional so that it is true only when both p and q are true or when both p and q are false (that is, whenever they have the same truth values). This definition is shown in Table 3.8.

TABLE 3.8		
Definition of Biconditional		
p	q	$p \leftrightarrow q$
T	T	T
T	F	F
F	T	F
F	F	T

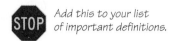 *Add this to your list of important definitions.*

The biconditional can be a somewhat perplexing connective (see Figure 3.4), but in mathematics and logic, $p \leftrightarrow q$ is precisely defined and can be translated in several ways, all of which have the same meaning:

Biconditional Translation

1. p if and only if q
2. q if and only if p
3. If p then q, and conversely.
4. If q then p, and conversely.

FIGURE 3.4 What do these signs mean?

Example 1 Write an "if and only if" statement

Rewrite the following as one statement:

    1. If a polygon has three sides, then it is a triangle.

    2. If a polygon is a triangle, then it has three sides.

Solution   A polygon is a triangle if and only if it has three sides.

The set of logical possibilities for which a given statement is true is called its **truth set.** A logical statement which is necessarily true is called a **tautology.** This means that a compound statement is a tautology if you obtain only Ts on a truth table.

Example 2 Verify a tautology

Is $(p \lor q) \to (\sim q \to p)$ a tautology?

Solution

p	q	$p \lor q$	$\sim q$	$\sim q \to p$	$(p \lor q) \to (\sim q \to p)$
T	T	T	F	T	T
T	F	T	T	T	T
F	T	T	F	T	T
F	F	F	T	F	T

Thus, the given statement is a tautology.

If a conditional is a tautology, as in Example 2, then it is called an **implication** and is symbolized by $\Rightarrow$. That is, Example 2 can be written

$$(p \vee q) \Rightarrow (\sim q \to p)$$

The implication symbol $p \Rightarrow q$ is pronounced "p implies q."

A biconditional statement $p \leftrightarrow q$ that is also a tautology (that is, always true) is a **logical equivalence,** written $p \Leftrightarrow q$ and read "p is logically equivalent to q."

Example **3** **Verify an equivalence**

Show that $(p \to q) \Leftrightarrow \sim(p \wedge \sim q)$.

Solution We begin by constructing a truth table for

$(p \to q) \leftrightarrow \sim(p \wedge \sim q)$

p	q	$p \to q$	$\sim q$	$p \wedge \sim q$	$\sim(p \wedge \sim q)$	$(p \to q) \leftrightarrow \sim(p \wedge \sim q)$
T	T	T	F	F	T	T
T	F	F	T	T	F	T
F	T	T	F	F	T	T
F	F	T	T	F	T	T

Since all possibilities are true (it is a tautology), we see it is a logical equivalence and we write

$$(p \to q) \Leftrightarrow \sim(p \wedge \sim q)$$

Notice the difference between the symbols $\to$ and $\Rightarrow$ and between $\leftrightarrow$ and $\Leftrightarrow$. The conditional ($\to$) and biconditional ($\leftrightarrow$) are used as logical connectives and the truth values for these may be true or false. If, because of the statements being connected, the truth table gives *only true values*, then we use the symbols $\Rightarrow$ and $\Leftrightarrow$ in place of $\to$ and $\leftrightarrow$ and call them implication ($\Rightarrow$) and logical equivalence ($\Leftrightarrow$).

Laws of Logic

We can use the idea of logical equivalence to write two previously stated laws:

Law of double negation: $\sim(\sim p) \Leftrightarrow p$
Law of contraposition: $(\sim q \to \sim p) \Leftrightarrow (p \to q)$

You will be asked to prove these laws in the problem set. Other important laws are developed in the next section. Two additional laws, called **De Morgan's laws,** will be considered in this section.

De Morgan's laws

$\sim(p \vee q) \Leftrightarrow \sim p \wedge \sim q$
$\sim(p \wedge q) \Leftrightarrow \sim p \vee \sim q$

There is a strong tie between set theory (Chapter 2) and symbolic logic, so it is no coincidence that there are De Morgan's laws both in set theory and in symbolic logic. We prove the second one here and leave the first for the problem set. The proof, by truth table, is shown:

p	q	$p \wedge q$	$\sim(p \wedge q)$	$\sim p$	$\sim q$	$\sim p \vee \sim q$	$\sim(p \wedge q) \leftrightarrow (\sim p \vee \sim q)$
T	T	T	F	F	F	F	T
T	F	F	T	F	T	T	T
F	T	F	T	T	F	T	T
F	F	F	T	T	T	T	T

Since the last column is all Ts, we can write $\sim(p \wedge q) \Leftrightarrow \sim p \vee \sim q$.

Negation of a Compound Statement

De Morgan's laws can be used to write the negation of a compound statement using conjunction and disjunction.

Example 4   Negation of a compound

Write the negation of the compound statements.
a. John went to work or he went to bed.
b. Alfie didn't come last night and didn't pick up his money.

Solution   Begin by writing the statement in symbolic form, then find the negation, and finally use the rules of logic to simplify before translating back into English.

a. Let w: John went to work; b: John went to bed.

The symbolic form is:	$w \vee b$
Negation:	$\sim(w \vee b)$
Simplify (De Morgan's law):	$\sim w \wedge \sim b$

Translate back into English:

John did not go to work and he did not go to bed.

b. Let p: Alfie came last night; q: Alfie picked up his money.

The symbolic form is:	$\sim p \wedge \sim q$
Negation:	$\sim(\sim p \wedge \sim q)$
Simplify (De Morgan's law):	$\sim(\sim p) \vee \sim(\sim q)$
Simplify (law of double negation):	$p \vee q$

Translate: Alfie came last night or he picked up his money.

Sometimes it is necessary to find the **negation of a conditional,** $p \rightarrow q$. It can be shown (see Problem 41) that

$$(p \rightarrow q) \Leftrightarrow (\sim p \vee q)$$

Therefore, the negation of $p \rightarrow q$ is equivalent to the negation of $\sim p \vee q$:

$\sim(p \rightarrow q)$	$\Leftrightarrow \sim(\sim p \vee q)$	*Given*
	$\Leftrightarrow \sim(\sim p) \wedge \sim q$	*De Morgan's law*
	$\Leftrightarrow p \wedge \sim q$	*Law of double negation*

Negation of a Conditional

$$\sim(p \rightarrow q) \Leftrightarrow (p \wedge \sim q)$$

Example 5 Language analysis

Pólya's Method

The officer said to the detective, "If John was at the scene of the crime, then he knows that Jean could not have done it." "No, Colombo," answered the detective, "that is not correct. John was at the scene of the crime and Jean did it." Was the detective's statement a correct negation of Colombo's statement?

Solution We use Pólya's problem-solving guidelines for this example.

Understand the Problem. We are asking for the negation of a given statement, which we then wish to compare with the proposed negation.

Devise a Plan. Translate the given statement into symbolic form, find the negation, simplify, and then translate back into English.

Carry Out the Plan. Let p: John was at the scene of the crime; q: Jean committed the crime. Then $p \rightarrow \sim q$ is a symbolic statement of the given statement, "If John was at the scene of the crime, then he knows that Jean could not have done it." We now form the negation and then simplify:

$$\sim(p \rightarrow \sim q) \Leftrightarrow p \wedge \sim(\sim q) \quad \text{Negation of a conditional}$$
$$\Leftrightarrow p \wedge q \quad \text{Law of double negation}$$

Translate back into English: John was at the scene of the crime and Jean committed the crime.

Look Back. The detective's negation was correct.

Miscellaneous Operators

Occasionally, we encounter other operators, and it is necessary to formulate precise definitions of these additional operators. As Table 3.9 shows, they are all defined in terms of our previous operators.

TABLE 3.9 **Additional Operators**

p	q	Either p or q $(p \vee q) \wedge \sim(p \wedge q)$	Neither p nor q $\sim(p \vee q)$	p unless q $\sim q \rightarrow p$	p because q $(p \wedge q) \wedge (q \rightarrow p)$	No p is q $p \rightarrow \sim q$
T	T	F	F	T	T	F
T	F	T	F	T	F	T
F	T	T	F	T	F	T
F	T	F	T	F	F	T

 CAUTION Use this table for reference.

Example 6 *Because* as a conjunction

Show that the statement "p because q" as defined in Table 3.9 is equivalent to conjunction.

Solution According to Table 3.9, "p because q" means $(p \wedge q) \wedge (q \rightarrow p)$ and conjunction means $p \wedge q$ (Table 3.1).

		From Table 3.9 $\downarrow$		From Table 3.1 $\downarrow$
p	q	$(p \wedge q) \wedge (q \to p)$	$p \wedge q$	$(p \wedge q) \wedge (q \to p) \leftrightarrow (p \wedge q)$
T	T	T	T	T
T	F	F	F	T
F	T	F	F	T
F	F	F	F	T

The last column is all Ts, so we can write $(p \wedge q) \wedge (q \to p) \Leftrightarrow p \wedge q$.

Problem Set | 3.3

Level 1

1. **IN YOUR OWN WORDS**   Discuss the difference between the conditional and the biconditional.

2. **IN YOUR OWN WORDS**   Discuss the procedure for finding the negation of compound statements.

3. **IN YOUR OWN WORDS**   Discuss when you use the symbols $\leftrightarrow$ and $\Leftrightarrow$.

4. **IN YOUR OWN WORDS**   Discuss when you use the symbols $\to$ and $\Rightarrow$.

5. **IN YOUR OWN WORDS**   Make up five good English statements, each using one of the additional operators shown in Table 3.9.

Use the parking signs in this photograph to answer the questions in Problems 6–7. Assume it is not Sunday.

© Austin MacRae

6. **a.** How long can you legally park if you arrive at midnight?
   b. How long can you legally park if you arrive at 2:00 P.M.?

7. **a.** How long can you legally park if you arrive at 5:00 P.M.?
   b. How long can you legally park if you arrive at 4:00 P.M.?

Level 2

Use truth tables in Problems 8–15 to determine whether the given compound statement is a tautology.

8. $(p \wedge q) \vee (p \to \sim q)$

9. $(p \vee q) \wedge (q \to \sim p)$

10. $(\sim p \to q) \to p$

11. $(\sim q \to p) \to q$

12. $(p \wedge q) \leftrightarrow (p \vee q)$

13. $(p \to q) \leftrightarrow (\sim p \vee q)$

14. $(p \vee q) \leftrightarrow (q \vee p)$

15. $(p \vee \sim q) \leftrightarrow (\sim p \vee q)$

Verify the indicated definition in Problems 16–19 from Table 3.9 using a truth table.

16. either p or q

17. neither p nor q

18. p unless q

19. no p is q

Translate the statements in Problems 20–29 into symbols. For each simple statement, indicate the meanings of the symbols you use.

20. Neither smoking nor drinking is good for your health.

21. I will not buy a new house unless all provisions of the sale are clearly understood.

22. To obtain the loan, I must have an income of $85,000 per year.

23. I am obligated to pay the rent because I signed the contract.

24. I cannot go with you because I have a previous engagement.

25. No man is an island.

26. Either I will invest my money in stocks or I will put it in a savings account.

27. Be nice to people on your way up 'cause you'll meet 'em on your way down. (Jimmy Durante)

28. No person who has once heartily and wholly laughed can be altogether irreclaimably bad. (Thomas Carlyle)

29. If by the mere force of numbers a majority should deprive a minority of any clearly written constitutional right, it might, in a moral point of view, justify revolution. (Abraham Lincoln)

Use the tautology $(p \rightarrow q) \Leftrightarrow (\sim p \lor q)$ to write each statement in Problems 30–35 in an equivalent form.

30. If I go, then I paid $100.

31. If the cherries have turned red, then they are ready to be picked.

32. We will not visit New York or we will visit the Statue of Liberty.

33. Hannah will not watch Jon Stewart or she will watch the NBC late-night orchestra.

34. The sun is shining or I will not go to the park.

35. The money is available or I will not take my vacation.

Level 3

36. Show that the definition for *neither p nor q* could also be $\sim p \land \sim q$.

37. Prove the law of double negation by using a truth table.

38. Prove the law of contraposition by using a truth table.

39. Prove De Morgan's law:

$$\sim(p \lor q) \Leftrightarrow \sim p \land \sim q$$

40. In the text we used laws of logic to prove that

$$\sim(p \rightarrow q) \Leftrightarrow (p \land \sim q)$$

Use a truth table to prove this result.

41. Prove $(p \rightarrow q) \Leftrightarrow (\sim p \lor q)$.

Write the negation of each compound statement in Problems 42–55.

42. $p \rightarrow q$

43. $p \rightarrow \sim q$

44. $\sim p \rightarrow q$

45. $\sim p \rightarrow \sim q$

46. Cole went to Macy's or Sears.

47. Jane went to the basketball game or to the soccer game.

48. Theron is not here and he is not at home.

49. Missy is not on time and she missed the boat.

50. If I can't go with you, then I'll go with Bill.

51. If you're out of Schlitz, you're out of beer.

52. If $x + 2 = 5$, then $x = 3$.

53. If $x - 5 = 4$, then $x = 1$.

54. If $x = -5$, then $x^2 = 25$.

55. $2x + 3y = 8$ if $x = 1$ and $y = 2$.

Problem Solving 3

56. To qualify for a loan of $200,000 an applicant must have a gross income of $72,000 if single, or $100,000 combined income if married. It is also necessary to have assets of at least $50,000. Write these statements symbolically, and decide whether Liz, who is single and has assets of $125,000, satisfies the conditions for obtaining a loan. She has an income of $58,000.

57. An airline advertisement states, "OBTAIN 40% OFF REGULAR FARE."

Read the fine print: "You must purchase your tickets between January 5 and February 15 and fly round trip between February 20 and May 3. You must also depart on a Monday, Tuesday, or Wednesday, and return on a Tuesday, Wednesday, or Thursday. You must also stay over a Saturday night." Write out these conditions symbolically.

58. The contract states, "No alterations, redecorating, tacks, or nails may be made in the building, unless written permission is obtained." Write this statement symbolically.

59. The contract states, "The tenant shall not let or sublet the whole or any portion of the premises to anyone for any purpose whatsoever, unless written permission from the landlord is obtained." Write this statement symbolically.

60. Translate into symbolic form: *Either Alfie is not afraid to go, or Bogie and Clyde will have lied.*

3.4 | The Nature of Proof

One of the greatest strengths of mathematics is its concern with the logical proof of its propositions. Any logical system must start with some undefined terms, definitions, and postulates or axioms. We have seen several examples of each of these. From here, other assertions can be made. These assertions are called theorems, and they must be proved using the rules of logic. In this section, we are concerned not so much with "proving mathematics" as with investigating the nature of proof in mathematics since the idea of proof does occupy a great portion of a mathematician's time.

Cartoon by Sidney Harris - American Scientist Magazine. 1977. Reprinted by permission of ScienceCartoonsPlus.com.

"I think you should be more explicit here in step two."

In Chapter 1 we discussed inductive reasoning. Experimentation, guessing, and looking for patterns are all part of inductive reasoning. After a conjecture or generalization has been made, it needs to be proved using the rules of logic. After the conjecture is proved, it is called a **theorem.** Often several years will pass from the conjectural stage to the final proved form. Certain definitions must be made, and certain axioms or **postulates** must be accepted. Perhaps the proof of the conjecture will require the results of some previous theorems.

A **syllogism** is a form of reasoning in which two statements or premises are made and a *logical conclusion* is drawn from them. In this book, we will consider three types of syllogisms: direct reasoning, indirect reasoning, and transitive reasoning.

Direct Reasoning

The simplest type of syllogism is **direct reasoning.** This type of argument consists of two *premises*, or *hypotheses*, and a *conclusion*. For example,

$p \to q$	If you receive an A on the final, then you will pass the course.
p	You receive an A on the final.
$\therefore q$	Therefore, you pass the course.

The three dot symbol $\therefore$ is used to symbolize the word *therefore*, which is used to separate the conclusion from the premises. We can use a truth table to prove direct reasoning. We begin by noting that the argument form can be rewritten as

$$[(p \to q) \land p] \to q$$

p	q	$p \to q$	$(p \to q) \land p$	$[(p \to q) \land p] \to q$
T	T	T	T	T
T	F	F	F	T
F	T	T	F	T
F	F	T	F	T

Since the argument is always true, we can write $[(p \rightarrow q) \wedge p] \Rightarrow q$, which proves the reasoning form called direct reasoning. It is also called **modus ponens, law of detachment,** or **assuming the antecedent.**

Direct Reasoning

Major premise: $p \rightarrow q$
Minor premise: p
Conclusion: $\therefore q$

CAUTION   *This is one of the important rules of reasoning.*

Example   **1**   **Formulate a conclusion**

Use direct reasoning to formulate a conclusion for each of the given arguments.
a. If you play chess, then you are intelligent.
   You play chess.
b. If $x + 2 = 3$, then $x = 1$.
   $x + 2 = 3$.
c. If you are a logical person, then you will understand this example.
   You are a logical person.

Solution
a. You are intelligent.
b. $x = 1$.
c. You understand this example.

Indirect Reasoning

The following syllogism illustrates **indirect reasoning.**

$p \rightarrow q$ If you receive an A on the final, then you will pass the course.
$\sim q$ You did not pass the course.
$\therefore \sim p$ Therefore, you did not receive an A on the final.

We can prove this is valid by using a truth table (see Problem 34) or by using direct reasoning as follows:

$[(\sim q \rightarrow \sim p) \wedge \sim q] \Rightarrow \sim p$ Direct reasoning
$[(p \rightarrow q) \wedge \sim q] \Rightarrow \sim p$ Law of contraposition

This type of reasoning is also called **modus tollens** or **denying the consequent.**

Indirect Reasoning

Major premise: $p \rightarrow q$
Minor premise: $\sim q$
Conclusion: $\therefore \sim p$

CAUTION   *This is another of the important rules of reasoning.*

Example 2 | Use indirect reasoning

Formulate a conclusion for each statement by using indirect reasoning.
a. If the cat takes the rat, then the rat will take the cheese.
   The rat does not take the cheese.
b. If x is an even number, then $3x$ is an even number.
   $3x$ is not an even number.
c. If you received an A on the test, then I am Napoleon.
   I am not Napoleon.

Solution
a. The cat does not take the rat.
b. x is not an even number.
c. You did not receive an A on the test.

Transitive Reasoning

Sometimes we must consider some extended arguments. Transitivity allows us to reason through several premises to some conclusion. The argument form is given in the box.

Transitive Reasoning

Major premise:	$p \rightarrow q$
Minor premise:	$q \rightarrow r$
Conclusion:	$\therefore p \rightarrow r$

CAUTION — This is the last of the three important rules of reasoning.

We can prove transitivity by using a truth table. Notice that for three statements we need a truth table with eight possibilities.

p	q	r	$p \rightarrow q$	$q \rightarrow r$	$p \rightarrow r$	$(p \rightarrow q) \wedge (q \rightarrow r)$	$[(p \rightarrow q) \wedge (q \rightarrow r)] \rightarrow (p \rightarrow r)$
T	T	T	T	T	T	T	T
T	T	F	T	F	F	F	T
T	F	T	F	T	T	F	T
T	F	F	F	T	F	F	T
F	T	T	T	T	T	T	T
F	T	F	T	F	T	F	T
F	F	T	T	T	T	T	T
F	F	F	T	T	T	T	T

↑
All Ts

Since transitivity is always true, we may write

$$[(p \rightarrow q) \wedge (q \rightarrow r)] \Rightarrow (p \rightarrow r)$$

Example 3 Formulate a conclusion

Formulate a conclusion for each argument.
a. If you attend class, then you will pass the course.
    If you pass the course, then you will graduate.
b. If you graduate, then you will get a good job.
    If you get a good job, then you will meet the right people.
    If you meet the right people, then you will become well known.
c. If $x + 2x + 3 = 9$, then $3x + 3 = 9$.
    If $3x + 3 = 9$, then $3x = 6$.
    If $3x = 6$, then $x = 2$.

Solution
a. If you attend class, then you will graduate.
b. The transitive law can be extended to several premises:
    If you graduate, then you will become well known.

Notice that we can apply the transitive law to both parts **a** and **b**: If you attend class, then you will become well known.

c. If $x + 2x + 3 = 9$, then $x = 2$.

Logical Proof

The point of studying these various argument forms is to acquire the ability to apply them to longer and more involved arguments.

Example 4 Given premises, form a valid conclusion

Form a valid conclusion using all these statements. We number the premises for easy reference.
1. If I receive a check for $500, then we will go on vacation.
2. If the car breaks down, then we will not go on vacation.
3. The car breaks down.

Solution   Begin by changing the argument into symbolic form.
1. $c \to v$ where c: I receive a $500 check.
                      v: We will go on vacation.
2. $b \to \sim v$ b: The car breaks down.
3. b
Next, simplify the argument. For this example, we rearrange the premises. Notice the numbers to help you keep track of the premises.

2.	$b \to \sim v$	*Given*
1′.	$\underline{\sim v \to \sim c}$	*Contrapositive of the first premise*
	$\therefore b \to \sim c$	*Transitive*
3.	$\underline{b}$	*Given*
	$\therefore \sim c$	*Direct reasoning*

Finally, we translate the conclusion back into words: I did not receive a check for $500.

Example 5 Conclusion from a puzzle

Pólya's Method

Form a valid conclusion using all the statements.[*]
1. All unripe fruit is unwholesome.
2. All these apples are wholesome.
3. No fruit grown in the shade is ripe.

[*]From Lewis Carroll, *Symbolic Logic and The Game of Logic* (New York: Dover Publications, 1958).

Solution We use Pólya's problem-solving guidelines for this example.

Understand the Problem. The forms in this argument are not exactly like those we are used to seeing. We recall that the statement *all p is q* can be translated as *if p then q.*

Devise a Plan. The procedure we will use is to (step 1) translate into symbols, (step 2) simplify using logical arguments, and finally (step 3), translate the symbolic form back into English.

Carry Out the Plan.

Step 1

1. $\sim r \rightarrow (\sim w)$ where r: This fruit is ripe.
2. $a \rightarrow w$ w: This fruit is wholesome.
3. $s \rightarrow (\sim r)$ a: This fruit is an apple.
                                      s: This fruit is grown in the shade.
Note: The sentence "No *p* is *q*" is translated as $p \rightarrow (\sim q)$.

Step 2

Let $(1')$, $w \rightarrow r$, replace (1) by the law of contraposition.
Let $(3')$, $r \rightarrow \sim s$, replace (3) by the law of contraposition.
Rearranging the premises, we have:

2. $a \rightarrow w$
1'. $w \rightarrow r$
$\overline{}$
     $\therefore a \rightarrow r$ Transitive
3'. $r \rightarrow (\sim s)$
$\overline{}$
     $a \rightarrow (\sim s)$ Transitive

Step 3

Conclusion: All these apples were not grown in the shade.
If we assume that $\sim s$ represents grown in the sun, then the conclusion can be stated more simply: All these apples were grown in the sun.

Look Back. Does this conclusion seem reasonable?

What is this?
(circular reasoning)

Fallacies

Sometimes *invalid arguments* are given. The remainder of this section is devoted to some of the more common **logical fallacies.**

Fallacy of the Converse

Example **6** **Invalid argument**

Show that the following argument is not valid.

> If a person reads the *Times*, then she is well informed.
> This person is well informed.
> Therefore, this person reads the *Times*.

Solution This argument has the following form:

$$p \rightarrow q$$
$$q$$
$$\overline{\therefore p}$$

By considering the associated truth table, we can test the validity of our argument.

p	q	$p \rightarrow q$	$(p \rightarrow q) \wedge q$	$[(p \rightarrow q) \wedge q] \rightarrow p$
T	T	T	T	T
T	F	F	F	T
F	T	T	T	F
F	F	T	F	T

We see that the result is not always true; thus the argument is invalid.

If $p \rightarrow q$ were replaced by $q \rightarrow p$, the argument in the preceding example would become valid. That is, the argument would be valid if the direct statement and the converse had the same truth values, which in general they do not. For this reason the argument is sometimes called the **fallacy of the converse** or the fallacy of **assuming the consequent.**

We can often show that a given argument is invalid by finding a **counterexample.** In the preceding example we found a counterexample by looking at the truth table. The entry in the third row is false, so the argument can be shown to be false in the case in which p is false and q is true. In terms of this example, a person could never see the *Times* (*p* false) and still be well informed (*q* true), which shows the argument form is invalid.

Fallacy of the Inverse

Consider the following argument:

> If a person reads the *Times*, then he is well informed.
> This person does not read the *Times*.
> Therefore, this person is not well informed.

As we have seen in Example 6, a person who never sees the *Times* might also be well informed. This line of reasoning is called the **fallacy of the inverse** (sometimes also called the fallacy of **denying the antecedent**). A truth table for

$$[(p \rightarrow q) \wedge (\sim p)] \rightarrow (\sim q)$$

shows that the fallacy of the inverse is not valid. (The truth table is left as a problem.)

Example 7 Test validity

Test the validity of the following argument.

> If a person goes to college, he will make a lot of money.
> You do not go to college.
> Therefore, you will not make a lot of money.

Solution This argument has the following form:

$$p \rightarrow q$$
$$\underline{\sim p}$$
$$\therefore \sim q$$

We recognize this reasoning as the fallacy of the inverse (the fallacy of denying the antecedent).

False Chain Pattern

In certain areas of this country, it is thought that thunderstorms cause milk to sour. This belief is an example of the fallacy we will call the **false chain pattern.** It can be shown as follows:

Hot, humid weather favors thunderstorms.
Hot, humid weather favors bacterial growth, which causes milk to sour.
Therefore, thunderstorms cause milk to sour.

The false chain pattern is illustrated by

$$\begin{array}{l} p \to q \\ p \to r \\ \hline \therefore q \to r \end{array}$$

and can be proved invalid by construction of an appropriate truth table.

The three types of common fallacies that we have discussed can be summarized as follows:

Assuming the Consequent Fallacy of the Converse	Denying the Antecedent Fallacy of the Inverse	False Chain Pattern
$p \to q$	$p \to q$	$p \to q$
q	$\sim p$	$p \to r$
$\therefore p$	$\therefore \sim q$	$\therefore q \to r$

Study these fallacies, and notice how they differ from direct reasoning, indirect reasoning, and transitivity.

Problem Set 3.4

Level 1

1. Explain what we mean by direct reasoning.

2. Explain what we mean by indirect reasoning.

3. Explain what we mean by transitive reasoning.

4. **IN YOUR OWN WORDS**   What do we mean by logical fallacies?

5. **IN YOUR OWN WORDS**   What is a syllogism?

6. **IN YOUR OWN WORDS**   There are similarities between Pólya's problem-solving method and the steps a detective will go through in solving a case. Rewrite each of these detective problem-solving steps using the language of Pólya's problem-solving method. (See Problems 55–56 below for examples of detective-type problems.)

Understand the case.
   What are you looking for?

Investigate the case.
   Have you solved a similar case?
   What are the facts?

Analyze the facts/data.
   What information is important?
   What information is not important?

What pieces of information do not seem to fit together logically?
Which data are inconsistent with the given information?

Reexamine the facts.
   Do the facts support the solution?
   Can we obtain a conviction?

Determine whether each argument in Problems 7–10 is valid or invalid. Give reasons for your answer.

7. **a.** $p \to q$
   $\dfrac{\sim q}{\therefore \sim p}$

   b. $p \to q$
   $\dfrac{q}{\therefore p}$

8. **a.** $p \lor q$
   $\dfrac{\sim p}{\therefore q}$

   b. $p \lor q$
   $\dfrac{\sim q}{\therefore p}$

9. **a.** $p \to q$
   $\dfrac{\sim p}{\therefore \sim q}$

   b. $p \to q$
   $\dfrac{p}{\therefore q}$

10. **a.** $p \to \sim q$
   $\dfrac{q}{\therefore \sim p}$

   b. $\sim p \to q$
   $\dfrac{\sim p}{\therefore q}$

Determine whether each argument in Problems 11–33 is valid or invalid. If valid, name the type of reasoning, and if invalid, determine the error in reasoning.

11. If I inherit $1,000, I will buy you a cookie.
I inherit $1,000.
Therefore, I will buy you a cookie.

12. If a^2 is even, then a must be even.
a is odd.
Therefore, a^2 is odd.
Note: Assume that if a number is odd, then it is not even.

13. All snarks are fribbles.
All fribbles are ugly.
Therefore, all snarks are ugly.

14. All cats are animals.
This is not an animal.
Therefore, this is not a cat.

15. If I don't get a raise in pay, I will quit.
I don't get a raise in pay.
Therefore, I quit.

16. If Fermat's Last Theorem is ever proved,
then my life is complete.
Fermat's Last Theorem was proved in 1994.
Therefore, my life is complete.

17. If Congress appropriates the money, the project can be completed.
Congress appropriates the money.
Therefore, the project can be completed.

18. If Alice drinks from the bottle marked "poison,"
she will become sick.
Alice does not drink from a bottle that is marked "poison."
Therefore, she does not become sick.

19. Blue-chip stocks are safe investments.
Stocks that pay a high rate of interest are safe investments.
Therefore, blue-chip stocks pay a high rate of interest.

20. If Al understands logic, then he enjoys this sort of problem.
Al does not understand logic.
Therefore, Al does not enjoy this sort of problem.

21. If Al understands a problem, it is easy.
This problem is not easy.
Therefore, Al does not understand this problem.

22. If Mary does not have a little lamb, then she has a big bear.
Mary does not have a big bear.
Therefore, Mary has a little lamb.

23. If $2x - 4 = 0$, then $x = 2$.
$x \neq 2$
Therefore, $2x - 4 \neq 0$.

24. If Todd eats Krinkles cereal, then he has "extra energy."
Todd has "extra energy."
Therefore, Todd eats Krinkles cereal.

25. If Ron uses Slippery oil, then his car is in good running condition.
Ron's car is in good running condition.
Therefore, Ron uses Slippery oil.

26. If you get a fill-up of gas, you will get a free car wash.
You get a fill-up of gas.
Therefore, you will get a free car wash.

27. If the San Francisco 49ers lose, then the Dallas Cowboys win.
If the Dallas Cowboys win, then they will go to the Super Bowl.
Therefore, if the San Francisco 49ers lose, then the Dallas Cowboys will go to the Super Bowl.

28. If Missy uses Smiles toothpaste, then she has fewer cavities.
Therefore, if Missy has fewer cavities, then she uses Smiles toothpaste.

29. All mathematicians are eccentrics.
All eccentrics are rich.
Therefore, all mathematicians are rich.

30. (Let x be an integer and y a nonzero integer for this argument.)
If Q is a rational number, then $Q = \frac{x}{y}$, where $\frac{x}{y}$ is a reduced fraction.
$Q \neq \frac{x}{y}$, where $\frac{x}{y}$ is a reduced fraction.

Therefore, Q is not a rational number.

31. If you like beer, you'll like Bud.
You don't like Bud.
Therefore, you don't like beer.

32. No students are enthusiastic.
You are enthusiastic.
Therefore, you are not a student.

33. If the crime occurred after 4:00 A.M., then Smith could not have done it.
If the crime occurred at or before 4:00 A.M., then Jones could not have done it.
The crime involved two persons, if Jones did not commit the crime.
Therefore, if Smith committed the crime, it involved two persons.

Level

34. Prove $[(p \rightarrow q) \wedge {\sim}q] \rightarrow {\sim}p$ by constructing a truth table. What is the name we give to this type of reasoning?

35. Show that $[(p \rightarrow q) \wedge (p \rightarrow r)] \rightarrow (q \rightarrow r)$ is an invalid argument by constructing a truth table. What is the name of this fallacy?

36. Prove that $[(p \rightarrow q) \wedge {\sim}p] \rightarrow {\sim}q$ is an invalid argument. What is the name of this fallacy?

In Problems 37–54, form a valid conclusion, using all the premises given for each argument. Give reasons.

37. If you learn mathematics, then you are intelligent.
If you are intelligent, then you understand human nature.

38. If I am idle, then I become lazy.
I am idle.

39. If we go to the concert, then we are enlightened.
We are not enlightened.

40. If you climb the highest mountain, then you feel great.
If you feel great, then you are happy.

41. $a = 0$ or $b = 0$
$a \neq 0$

42. If $a \cdot b = 0$, then $a = 0$ or $b = 0$.
$a \cdot b = 0$

43. If we interfere with the publication of false information, we are guilty of suppressing the freedom of others.
We are not guilty of suppressing the freedom of others.

44. If a nail is lost, then a shoe is lost.
If a shoe is lost, then a horse is lost.
If a horse is lost, then a rider is lost.
If a rider is lost, then a battle is lost.
If a battle is lost, then a kingdom is lost.

Reprinted with special permission. © King Features Syndicate.

45. If I eat that piece of pie, I will get fat.
I will not get fat.

46. If 2 divides a positive integer N and if N is greater than 2, then N is not a prime number.
N is a prime number.

47. If we win first prize, we will go to Europe.
If we are ingenious, we will win first prize.
We are ingenious.

48. If I am tired, then I cannot finish my homework.
If I understand the homework, then I can finish my homework.

Level 3

HISTORICAL QUEST *Problems 49–52 are problems written by Charles Dodgson, better known as Lewis Carroll. You might wish to read the Historical Note and Example 5 in this section.*

49. Babies are illogical.
Nobody is despised who can manage a crocodile.
Illogical persons are despised.

50. All hummingbirds are richly colored.
No large birds live on honey.
Birds that do not live on honey are dull in color.

51. No ducks waltz.
No officers ever decline to waltz.
All my poultry are ducks.

52. Everyone who is sane can do logic.
No lunatics are fit to serve on a jury.
None of your sons can do logic.

53. If the government awards the contract to Airfirst Aircraft Company, then Senator Firstair stands to earn a great deal of money.
If Airsecond Aircraft Company does not suffer financial setbacks, then Senator Firstair does not stand to earn a great deal of money.
The government awards the contract to Airfirst Aircraft Company.

54. If you go to college, then you get a good job.
If you get a good job, then you make a lot of money.
If you do not obey the law, then you do not make a lot of money.
You go to college.

Problem Solving 3

55. THE CASE OF THE DEAD PROFESSOR* The detective, Columbo, had just arrived at the scene of the crime and found that the professor had been at the lab working for hours. He seemed to have electrocuted himself and ended up blowing the fuses for the whole building. Later in the night, the janitor came to clean up the lab and found the professor's body. Columbo suspected the janitor, but when he was questioned, he vehemently denied killing the professor. He said, "I came to work late. When I got off the elevator, I found the professor dead with his head on the lab table. I wish I could tell you more, but that is all I know."

Then Columbo said, "I think you can tell us more at headquarters." Downtown under grueling interrogation, the janitor confessed. What made Columbo suspect the janitor?

56. THE CASE OF THE TUMBLED TOWER[†] Dwayne got up at 6:00 A.M. and was watching the sunrise from his bedroom window. After the sun came up, he started working on his toothpick tower in his room. The tower was very fragile. While he was working on his tower, his little brother came into the room bugging him. Dwayne's brother wanted to be more like Dwayne and he wanted to build something out of toothpicks, too. He was very jealous of Dwayne.

In the afternoon, Dwayne went out to buy candy at the candy store and he left his little brother home (even though he was supposed to be babysitting). On his way out of the house, he heard on the radio that there was going to be a slight westerly wind coming. He then left the house, forgetting that he had left his bedroom window open. When he returned and saw that he had left the window open, he thought that the wind had blown the tower over. But then he remembered something and said that his little brother must have knocked over the tower. What did he remember?

*My thanks to Don Gernes of Ponderosa High School for this idea.
[†]Problems 55 and 56 were adapted from "Solving the Mystery," by Frances R. Curicio and J. Lewis McNeece, *The Mathematics Teacher*, November 1993, pp. 682–685.

HISTORICAL QUEST   *Form a valid conclusion in Problems 57–60 using all of the given statements. These problems were written by Charles Dodgson, better known as Lewis Carroll. You might wish to read the Historical Note and Example 5 in this section.*

57. Nobody who really appreciates Beethoven fails to keep silent while the *Moonlight Sonata* is being played. Guinea pigs are hopelessly ignorant of music. No one who is hopelessly ignorant of music ever keeps silent while the *Moonlight Sonata* is being played.

58. No kitten that loves fish is unteachable. No kitten with a tail will play with a gorilla. Kittens with whiskers always love fish. No teachable kitten has green eyes. Kittens have tails unless they have whiskers.

59. When I work a logic problem without grumbling, you may be sure it is one that I can understand. These problems are not arranged in regular order, like the problems I am used to. No easy problem ever makes my head ache. I can't understand problems that are not arranged in regular order, like those I am used to. I never grumble at a problem unless it gives me a headache.

60. Every idea of mine that cannot be expressed as a syllogism is really ridiculous. None of my ideas about rock stars is worth writing down. No idea of mine that fails to come true can be expressed as a syllogism. I never have any really ridiculous idea that I do not at once refer to my lawyer. All my dreams are about rock stars. I never refer any idea of mine to my lawyer unless it is worth writing down.

3.5   Problem Solving Using Logic

In the previous chapter we introduced the distinction between inductive and deductive reasoning. In mathematics, we often build a deductive system using the principles of logic we have developed. Let us begin with a simple example.

Example   1   Undefined terms, defined terms, axioms, and theorems

You are probably familiar with the game of Monopoly®. Just as in mathematics, a game often has some undefined terms or materials used in the game. In addition, there are definitions used in the game (just as in mathematics), and the game usually has rules, which correspond to mathematical axioms or postulates. Finally, the game has outcomes, just as in mathematics there are theorems that result from the given definitions and rules.* Classify each of the following components from the game of Monopoly as an undefined term, defined term, axiom, or theorem.

a. Park Place
b. If you land on an unowned property, then you may buy it from the bank.
c. bank
d. *Community Chest*
e. jail

Solution
a. Park Place is a property location in the game, so we would called this a *defined* term.
b. This is a rule of the game, so we call this an *axiom*.
c. The bank is a *defined term*.
d. *Community Chest* is a *definition* in the game.
e. In the game of Monopoly, the rules talk of "landing in jail" or "getting out of jail" but the term itself is undefined.

*My thanks to Don Gernes of Ponderosa High School for this idea.

In Chapter 1, we laid the foundation for problem solving. The key to building problem-solving skills is to encounter problem solving in a variety of contexts. We will now apply another method of proof to solving some logic puzzles that have been around for a long time but that nevertheless continue to challenge the lay reader. Example 2 gives an analysis of Problem 57 of Problem Set 1.2.

Example **2** **Two-pan problem**

Pólya's Method

You are given nine steel balls of the same size and color. One of the nine balls is slightly heavier in weight; the others all weigh the same. Using a two-pan balance, what is the minimum number of weighings necessary to find the ball of different weight?

Solution   We use Pólya's problem-solving guidelines for this example.

Understand the Problem.   Do you understand the terminology? Are you familiar with a two-pan balance?

Devise a Plan.   One of the techniques for Pólya's method is to guess and test. That is the method we will use here. We will find a solution by trial and error (that is, guess), and then test it. Next we will see whether we can find a solution with fewer weighings. We will accomplish this by first solving some simpler problems.

Carry Out the Plan.   Suppose we had two steel balls. Then one weighing would suffice. What about three steel balls? One weighing would still suffice. How? Put one ball on each pan. If it doesn't balance, then you have found the heavier one. If it balances, then the one not weighed is the heavier one. Does this give you an idea for the nine balls? Let's try two weighings.

1. Divide the 9 steel balls into 3 groups of 3. First weighing: Balance 3 balls against 3 balls.
   a. The weighing either balances or doesn't balance (law of excluded middle).
   b. If it balances, then the heavier one is in the group not weighed. If it doesn't balance, then take the group with the heavier ball.
2. Divide the 3 steel balls from the heavier group. Second weighing: Balance 1 ball against 1 ball.
   a. The weighing either balances or doesn't balance. (Why?)
   b. If it doesn't balance, then the heavier ball is the one that tips the scale. If it does balance, then the heavier ball is the one not weighed.

Look Back.   Two weighings is the solution.

Example **3** **The letter game**

Consider the "letter game" patterned after an article by Don Gernes from Ponderosa High School.[*] We have three undefined terms, the letters *M, I,* and *U,* and one definition, "*x* means any string of *I*'s and *U*'s." Finally, there are four postulates for this letter game.

Postulates
1. If a string of letters ends in *I*, you may add a *U* at the end. This is the *addition rule.*
2. If you have *Mx*, then you may add *x* to get *Mxx*, by what we call the *doubling rule.*
3. If three *I*'s occur—that is, *III*—then you may replace that string by *U.* We call this the *substitution rule.*
4. If *UU* occurs, you drop it. This is the *deletion rule.*

Given: MI,
Prove: MUIU

*[*]"The Rules of the Game," from the Sharing Teaching Ideas section of *The Mathematics Teacher*, May 1999, pp. 424–426.

Solution We begin with the given and justify each step.

MI	Given
MII	Doubling rule
MIII	Doubling rule
MIIII	Doubling rule
MIIIIU	Addition rule
MUIU	Substitution rule

We can shorten this sequence of steps (without reasons) by

$$MI \rightarrow MII \rightarrow MIII \rightarrow MIIII \rightarrow MIIIIU \rightarrow MUIU$$

The remainder of this section requires some additional effort. Some people like and enjoy logic puzzles, and others avoid them. If you spend the time necessary to understand the rest of the examples in this section, your logical abilities in everyday life should be enhanced. Sit back, relax, and enjoy!

Example **4** **Liar problem**

Pólya's Method

On the South Side, a member of the mob had just knocked off a store.
Since the boss had told them all to lay low, he was a bit mad. He decided to have a talk with the boys.

From the boys' comments, can you help the boss figure out who committed the crime? Assume that the Boss is telling the truth.

Solution We use Pólya's problem-solving guidelines for this example.

Understand the Problem.

> Let a: Alfie did it.
>
> b: Bogie did it.
>
> c: Clyde did it.
>
> d: Dirty Dave did it.
>
> f: Fingers did it.

Translate the sentences into symbolic form:

Alfie said:	$b \lor c$
Bogie said:	$\sim(f \lor b)$
Clyde said:	$[\sim(b \lor c)] \land \sim[\sim(f \lor b)]$
Dirty Dave said:	$[(b \lor c) \land (f \lor b)] \lor [\sim(b \lor c) \land \sim(f \lor b)]$
Fingers said:	$\sim\{[(b \lor c) \land (f \lor b)] \lor [\sim(b \lor c) \land \sim(f \lor b)]\}$

Assume that a is true, and check all these statements. Next, assume that b is true, and check all the statements. Do the same for c, d, and f. Since we assume the boss is telling the truth, we look for those cases (if any) in which three are truthful and two are lying.

Devise a Plan. We can summarize our analysis by using a matrix (a rectangular array) with vacant cells for all possible pairings of the elements in each set.

	a	b	c	d	f
a					
b					
c					
d					
f					

Let us assume that the *a* at the left means that we are making the assumption that Alfie did it. Then we place a 1 or a 0 in each position to indicate the truth or falsity, respectively, of each witness's statement with the assumption that person *a* (as listed at the left) is guilty.

Carry Out the Plan.

		Witness Statements				
Conclusion		a	b	c	d	f
Alfie did it.	*a*	0	1	0	1	0
Bogie did it.	*b*	1	0	0	1	0
Clyde did it.	*c*	1	1	0	0	1
Dave did it.	*d*	0	1	0	1	0
Fingers did it.	*f*	0	0	1	0	1

Next we assume that Bogie did it, and we fill in the second row of the matrix. We continue until the matrix is complete, as shown. Now if we take the boss's statement as the major premise, we see that in only one case are there three truthful mobsters and two liars. Thus we can say that Clyde knocked off the store.

Look Back. Assume that Clyde knocked off the store, and check each mobster's comments to see that in this case there are three truthful mobsters and two liars.

Example **5** **Prisoner problem**   Pólya's Method

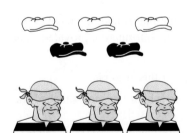

During an ancient war three prisoners were brought into a room. In the room was a large box containing three white hats and two black hats. Each man was blindfolded, and one of the hats was placed on his head.

The men were lined up, one behind the other, facing the wall. The blindfold of the man farthest from the wall (man C) was removed, and he was permitted to look at the hats of the two men in front of him. If he knew (not guessed) the color of the hat on his head, he would be freed. However, he was unable to tell. The blindfold was then taken from the head of the next man (man B), who could see only the hat of the one man in front of him. This man had the same chance for freedom, but he, too, was unable to tell the color of his hat. The remaining man (man A) then told the guards the color of the hat he was wearing and was released. What color hat was he wearing, and how did he know?

Solution We use Pólya's problem-solving guidelines for this example.

Understand the Problem. The situation looks like this:

Devise a Plan. We consider what the last man in line could see, and then decide which of these possibilities fit the conditions of what the middle man in the line saw.

Carry Out the Plan.

1. What did man C see? There are four possibilities:

	I	II	III	IV
A	white	white	black	black
B	white	black	white	black

If he had seen possibility IV, he would have known that he had on a white hat. (Why?) He did not know, so it must be possibility I, II, or III.

2. What did man B see? If he had seen a black hat, then he would have known that his hat was white. (Why?) This rules out possibility III.

3. Since both of the remaining possibilities, I and II, have a white hat on the front man, man A *knew* he had a white hat.

Look Back. Reread the question to see whether all the conditions of the problem are satisfied.

Example 6 Reporter's error problem

Pólya's Method

Emor D. Nilap, a news reporter who was a little backward, was sent to cover a billiards tournament. Since he wanted to do a good job, he rounded up some human-interest facts to make his story a little more interesting. He gathered the following information:

1. The men competing were Pat, Milt, Dick, Joe, and Karl.
2. Milt had once beaten the winner at tennis.
3. Pat and Karl frequently played cards together.
4. Joe's finish ahead of Karl was totally unexpected.
5. The man who finished fourth left the tournament after his match and did not see the last two games.
6. The winner had not met the man who came in fifth prior to the day of the tournament.
7. The winner and the runner-up had never met until Joe introduced them just before the final game.

When Emor returned to his office, he found that he had forgotten the order in which the men had finished, but he was able to figure it out from the information provided. Can you figure it out?

Solution   We use Pólya's problem-solving guidelines for this example.

Understand the Problem. We need to find the order in which the men finished the tournament.

Devise a Plan. The procedure we will use is to form a matrix of all possibilities and then use the rules of logic to eliminate the parts that are impossible; what remains is the solution.

Carry Out the Plan. From (1) we are able to make the following matrix of the possible solutions:

The steps are listed with the "work" shown after the last step.

Pat	Milt	Dick	Joe	Karl
1	1	1	1	1
2	2	2	2	2
3	3	3	3	3
4	4	4	4	4
5	5	5	5	5

a. We are given seven premises. We begin by the process of elimination; that is, we cross out those positions that it would be impossible for the men to occupy. The order in which we work the problem is not unique.

b. From (7), Joe is not the winner or the runner-up. Thus we cross out 1 and 2 under Joe's name and label those places *b*, as shown below.

c. By (5), Joe did not finish fourth; cross out and label *c*.

d. Joe finished ahead of Karl, by (4); therefore Karl was not first, second, or third. By the same premise, Joe could not have been last. (Why?)

e. Therefore Joe finished third, which means that Pat, Milt, and Dick are not in third place.

f. By (2), Milt was not the winner.

g. By (7), Milt was not the runner-up.

h. By (6), Milt was not fifth.

i. Therefore Milt finished fourth, which means that Pat, Dick, and Karl are not in fourth place.

j. Therefore Karl finished fifth, which means that the others did not finish fifth.

k. By (6) and the fact that Karl finished last, along with (3), we see that Pat could not have finished first.

l. Therefore Pat is second, which means that Dick was not second.

m. Therefore Dick finished first, and the problem is finished.

Pat	Milt	Dick	Joe	Karl
$\cancel{1}^k$	$\cancel{1}^f$	1^m	$\cancel{1}^b$	$\cancel{1}^d$
2^ℓ	$\cancel{2}^g$	$\cancel{2}^\ell$	$\cancel{2}^b$	$\cancel{2}^d$
$\cancel{3}^e$	$\cancel{3}^e$	$\cancel{3}^e$	3^e	$\cancel{3}^d$
$\cancel{4}^i$	4^i	$\cancel{4}^i$	$\cancel{4}^c$	$\cancel{4}^i$
$\cancel{5}^j$	$\cancel{5}^h$	$\cancel{5}^j$	$\cancel{5}^d$	5^j

Look Back. We have found that Dick finished first, Pat second, Joe third, Milt fourth, and Karl last. Reread each of the statements to make sure this solution does not cause any inconsistencies.

Example **7** **Turkey problem**

A man I know once owned a number of turkeys. One day, one of his gobblers flew over the man's fence and laid an egg on a neighbor's property. To whom did the egg belong—to the man who owned the gobbler, to the gobbler, or to the neighbor?

Solution   To understand the question, you must understand the terminology. Since a turkey gobbler is a male turkey, it could not lay an egg. You must be careful to use common sense along with the rules of logic when answering puzzle problems.

Problem Set 3.5

Level 1

For each of the situations in Problems 1–8, classify each item as an undefined term, a defined term, a axiom, or a theorem.

1. The game of Monopoly®
  a. playing board
  b. If you pass go, then collect $200.
  c. passing "GO"
  d. "chance" card
  e. race car playing piece

2. The game of Monopoly®
  a. dice
  b. luxury tax
  c. doubles

d. If you roll three doubles in a row, then you go directly to jail.
  e. jail

3. The game of Sorry®
  a. pawn
  b. You take turns drawing cards from a special deck and moving your pawn around the board.
  c. home
  d. Pawns may move forward or backward.
  e. backward

4. The game of Clue®
  a. playing board
  b. Professor Plum
  c. detective notebook

d. You may not enter or land on a square that's already occupied by another suspect.

e. weapon

5. The game of Clue®
   a. The pack of cards consists of three groups: Suspects, Rooms, and Weapons.
   b. Roll the die and move your token the number of squares you rolled.
   c. a die
   d. a door
   e. Miss Scarlet always plays first.

6. The game of baseball
   a. base
   b. ball
   c. home run
   d. If the player accumulates three strikes, then the player is called out.
   e. If an opposing player catches a foul ball, then the batter is called out.

7. The game of soccer
   a. players
   b. If a player is fouled, then that player's team gets a free kick.
   c. free kick
   d. goal
   e. The referee objectively applies the rules of the game to each play.

8. The game of basketball
   a. If a player travels, then the other team gets possession of the ball.
   b. ball
   c. traveling
   d. basket
   e. free throw

Level 2

Use the definitions and postulates given in Example 3 to prove the theorems in Problems 9–14. Give both statements and reasons.

9. Given: *MIII*;   Prove: *M*
10. Given: *MI*;   Prove: *MUI*
11. Given: *MI*;   Prove: *MIUIU*
12. Given: *MI*;   Prove: *MIIUIIU*
13. Given: *MIIIUUIIIII*;   Prove: *MIIU*
14. Given: *MIIIUII*;   Prove: *MIUIU*

Here is another variation of the letter game.
Undefined terms: A, E, and I.
Definition: x means any string of E's and I's.
Postulates:
1. *If a string of letters begins with A, then you may add x to get Ax.*
2. *If a string of letters ends with A, then you may add an E to get EA.*
3. *EA = AE*
4. *If a string 3 I's occurs, that is III, then you may substitute E in its place.*
5. *If AA or EE occur, you can drop it.*

Use these rules to prove the theorems in Problems 15–20. Give both statements and reasons.

15. Given: *EA*;   Prove: A
16. Given: *AIII*;   Prove: *EA*
17. Given: *AI*;   Prove: A
18. Given *IAEA*;   Prove: *I*
19. Given: *AE*;   Prove: A
20. Given: *AA*;   Prove: E

In a certain kingdom there were knights and knaves. The knights always tell the truth and the knaves always lie. Everyone in the kingdom is either a knight or a knave. Use this information to answer the questions in Problems 21–24.

21. There are two people, and the first one says, "Either I am a knave or the other person is a knight." In which category does each belong?

22. There are two people, Ed and Ted. Ed says, "Ted and I are different." Ted claims, "Only a knave would say that Ed is a knave." In which category does each belong?

23. Three people—call them Al, Bob, and Cary—were standing together at a bus stop. A stranger asked Al, "Are you a knight or a knave?" Al answered, but the stranger could not make out what he said. The stranger then asked Bob, "What did Al say?" Bob replied, "Al said that he is a knave." At this point, Cary said, "Don't believe Bob; he is lying!" What are Bob and Cary?

24. Three people, Al, Bob, and Cary, are standing together at the mall. Al says, "All of us are knaves," and Bob says, "Exactly one of us is a knight." What are all three?

25. **BEAR PROBLEM** A fox, hunting for a morsel of food, spotted a huge bear about 100 yards due east of him. Before the hunter could become the hunted, the crafty fox ran due north for 100 yards but then realized the bear had not moved. Thus he stopped and remained hidden. At this point the bear was due south of the fox. What was the color of the bear?

26. **PARENT PROBLEM** Two doctors—an old doctor and a young doctor—were discussing an interesting case. The young doctor is the son of the older doctor, but the older doctor is not the father of the younger doctor. How can you explain this?

27. **SOCK PROBLEM** I have a habit of getting up before the sun rises. My socks are all mixed up in the drawer, which contains 10 black and 20 blue socks. I reach into the drawer and grab some socks in the dark. How many socks do I need to take from the drawer to be sure that I have a matched pair?

28. **TEACHER CONVENTION** A group of 50 teachers and school administrators attended a convention. If any two persons are picked at random, at least one of the two would be a teacher. From this information, is it possible to determine what percentage of the people at the convention were teachers?

29. **FLOWER GARDEN** I visited a beautiful flower garden yesterday and counted exactly 50 flowers. Each flower was either red or yellow, and the flowers were not all the same color. My friend made the following observation: No matter which two flowers you might have picked, at least one was bound to be

red. From this can you determine how many were red and how many were yellow?

30. SISTERS' GARDEN   Three sisters visited a garden of red, white, blue, and yellow flowers. One sister observed that if any four flowers were picked, one of them would be red. Another observed that if any four were picked, at least one of them would be blue. The third sister noted that if any four were picked, at least one would be yellow. Does this necessarily mean that if any four were picked, one would be white?

Level 3

Consider a silly game called Bluffhead *in which each of three players takes a card from a shuffled deck and holds it, face out, to his or her forehead, as shown in the figure.*[*]

Undefined term: forehead
Defined terms:
    deck of cards
    ace, the highest card
Postulates:

1. *The players' names are Alice, Ben, and Cole, who always speak in that order, and then it goes back to Alice again.*
2. *Each player makes one of the following statements:*
    "I win." (I have a higher card than everyone else.)
    "I lose." (Someone has a higher card than I have.)
    "I tie." (I and at least one other player tie with the highest card.)
    "I don't win." (I either tie as a winner or lose.)
    "I don't lose." (I either win or tie as a winner.)
    "I don't know."
3. *The players play with perfect logic and reveal information through acceptable statements.*
4. *Each player says the strongest thing he or she can—that is, the player chooses the statement that is true and highest on the list.*

In Problems 31–36, state what you can infer about the cards from the given statements.

31. Alice: "I don't know."
    Ben: "I lose."
    Cole: "I lose."

32. Alice: "I don't know."
    Ben: "I don't win."
    Cole: "I win."

33. Alice: "I don't know."
    Ben: "I don't win."
    Cole: "I tie as a winner."

34. Alice: "I don't know."
    Ben: "I don't win."
    Cole: "I don't win."

35. Alice: "I don't know."
    Ben: "I don't know."
    Cole: "I don't know."
    Alice: "I lose."
    What will Ben and Cole say next?

36. Alice: "I don't know."
    Ben: "I don't know."
    Cole: "I don't know."
    Alice: "I don't know."
    Ben: "I don't know."
    Cole: "I don't know."
    Alice: "I don't know."
    Ben: "I don't know."
    Cole: "I win."

In Problems 37–54, fill in each blank with a digit so that every statement is true.[*] *That is,* ***after*** *you have replaced each problem number with the correct answer for that problem number, then the statements will be true.*

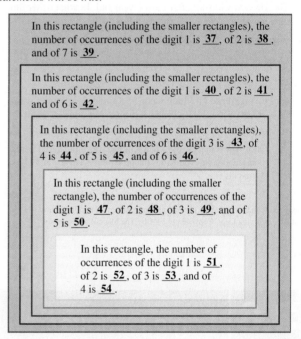

In this rectangle (including the smaller rectangles), the number of occurrences of the digit 1 is **37**, of 2 is **38**, and of 7 is **39**.

In this rectangle (including the smaller rectangles), the number of occurrences of the digit 1 is **40**, of 2 is **41**, and of 6 is **42**.

In this rectangle (including the smaller rectangles), the number of occurrences of the digit 3 is **43**, of 4 is **44**, of 5 is **45**, and of 6 is **46**.

In this rectangle (including the smaller rectangle), the number of occurrences of the digit 1 is **47**, of 2 is **48**, of 3 is **49**, and of 5 is **50**.

In this rectangle, the number of occurrences of the digit 1 is **51**, of 2 is **52**, of 3 is **53**, and of 4 is **54**.

Problem Solving 3

55. THE HAT GAME   Harry, Larry, and Moe are sitting in a circle so that they see each other. The game is played by a judge placing either a black or a white hat on each person's head so that each person can see the others' hats, but not the hat on his or her own head. Now when the judge says "GO," all players who see a black hat must raise their hands. The first player to deduce (not guess) the color of his own hat wins the game. Now, the judge put a black hat on each player's head, and then says, "GO." All players raised their hands and after a few moments Moe said he knew his hat was black. How did he deduce this?

---

[*]From "Bluffhead" by Dennis E. Shasha, *Scientific American*, April 2004, p. 108.

[*]By Guney Mentes in *Games*, August 1996, p. 43.

56. BROKEN WINDOW PROBLEM Three children were playing baseball and their names were Alice, Ben, and Cole. One of them hit a home run and broke your expensive plate glass window, and you went out to question the children. Each child made two statements.

> Alice: "Cole didn't do it. Ben did it."
> Ben: "I didn't do it. Cole did it."
> Cole: "I didn't do it. Alice did it."

Now if you know that one of the three always tells the truth, one always lies, and the other tells the truth half the time, can you point to the guilty party?

57. THE MARBLE PLAYERS Four boys were playing marbles; their names were Gary, Harry, Iggy, and Jack. One of the boys had 9 marbles, another had 15, and each of the other two had 12. Their ages were 3, 10, 17, and 18, but not respectively. Gary shot before Harry and Jack. Jack was older than the boy with 15 marbles. Harry had fewer than 15 marbles. Jack shot before Harry. Iggy shot after Harry. If Iggy was 10 years old, he did not have 15 marbles. Gary and Jack together had an even number of marbles. The youngest boy was not the one with 15 marbles. If Harry had 12 marbles, he wasn't the youngest. The 10-year-old shot after the 17-year-old. In what order did they shoot, and how old was each boy?

58. WHODUNIT? Daniel Kilraine was killed on a lonely road, two miles from Pontiac, at 3:30 A.M. on March 3, 1997. Otto, Curly, Slim, Mickey, and The Kid were arrested a week later in Detroit and questioned. Each of the five made four statements, three of which were true and one of which was false. One of these men killed Kilraine. Whodunit? Their statements were:[*]

> Otto: "I was in Chicago when Kilraine was murdered. I never killed anyone. The Kid is the guilty man. Mickey and I are pals."

Curly: "I did not kill Kilraine. I never owned a revolver in my life. The Kid knows me. I was in Detroit the night of March 3rd."

Slim: "Curly lied when he said he never owned a revolver. The murder was committed on March 3rd. One of us is guilty. Otto was in Chicago at the time."

Mickey: "I did not kill Kilraine. The Kid has never been in Pontiac. I never saw Otto before. Curly was in Detroit with me on the night of March 3rd."

The Kid: "I did not kill Kilraine. I have never been in Pontiac. I never saw Curly before. Otto lied when he said I am guilty."

59. COIN PROBLEM Suppose you are given 12 coins, one of which is counterfeit and weighs a little more or less than the real coins. Using a two-pan balance scale, what is the minimum number of weighings necessary to find the counterfeit coin? (If you are interested in a general solution to this type of problem, see T. H. O'Beirne's book, *Puzzles and Paradoxes*, Oxford University Press, New York, 1965, Chapters 2 and 3.)

60. MIXED BAG PROBLEM Suppose I have three bags, one with two peaches, another with two plums, and a third mixed bag with one peach and one plum. Now I give the bags (in mixed-up order) to Alice, Betty, and Connie. I tell the three to look into their bags and that I want each to make a false statement about the contents of her bag. Here is what they say:

> Alice: I have two peaches.
> Betty: I have two plums.
> Connie: I have one peach and one plum.

Now here is the game. I want you to develop a strategy of asking one of the three to reach into her bag, pull out one fruit, and show it to you. The fruit is then returned to the bag and you ask another to do the same thing. Continue until you can deduce which bag is the mixed bag. What is the minimum possible number of necessary moves? Explain.

---

[*]Reprinted with permission of The Macmillan Company from *Introduction to Logic* (2nd ed.), by Irving Copi. Copyright © 1961 by The Macmillan Company.

3.6 Logic Circuits

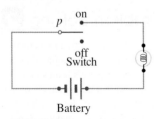

FIGURE 3.5 Schematic diagram showing how a light might be connected to a switch

One of the applications of symbolic logic that is easiest to understand is that of an electrical **circuit** (see Figure 3.5).

Consider the following symbols that we use for the parts of a circuit.

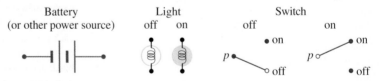

Basic Circuits

There are two types of circuits for connecting two switches together; the first is called a **series circuit,** and the other is called a **parallel circuit.**

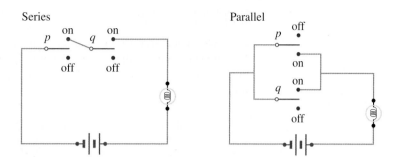

Example   1   **Circuit design**

Pólya's Method

How can symbolic logic be used to describe circuits?

Solution   We use Pólya's problem-solving guidelines for this example.

Understand the Problem. When two switches are connected in *series*, current will flow only when both switches are closed. When they are connected in *parallel*, current will flow if either switch is closed.

Devise a Plan. Let switch *p* be considered as a logical proposition that is either true or false. If *p* is true, then the switch will be considered closed (current flows), and if *p* is false, then the switch will be considered open (current does not flow).

Carry Out the Plan. A series circuit of two switches *p* and *q* can be specified as a logical statement $p \wedge q$. A parallel circuit can be specified as a logical statement $p \vee q$.

Look Back. Each circuit has four possible states, as shown in Figure 3.6. Note that some circuits show the light on, and others show the light off.

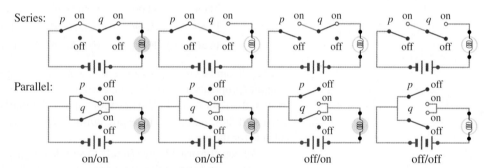

FIGURE 3.6 Series and parallel circuits showing four possible states for each

The three fundamental operators are conjunction, disjunction, and negation. In Example 1 we saw that a series circuit can be represented as a conjunction, and a parallel circuit as a disjunction. The circuit for negation is relatively easy, as is shown in Figure 3.7. Note that the light is on when the switch is off, and the light is off when the switch is on.

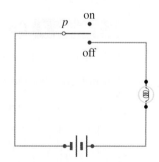

FIGURE 3.7 Negation circuit

Logical Gates

It is often necessary to combine circuits. We symbolically represent these as **gates.** The series circuit is called an **AND-gate,** the parallel circuit is called an **OR-gate,** and the negation circuit is called a **NOT-gate.** The notation for each of these is shown in Figure 3.8.

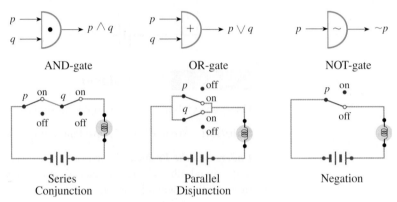

FIGURE 3.8 Circuits and gates

The main concept when dealing with circuits is that the light is ON if the compound statement is true, and the light is OFF if the compound statement is false. Study the following table until you are certain you understand why the light is on or why it is off.

p	q	$p \wedge q$	$p \vee q$	$\sim p$
T; on	T; on	T; light on	T; light on	F; light off
T; on	F; off	F; light off	T; light on	F; light off
F; off	T; on	F; light off	T; light on	T; light on
F; off	F; off	F; light off	F; light off	T; light on

Notice (from Figure 3.8) that the NOT-gate has a single input (proposition p) and a single output (proposition $\sim p$). That is, current will flow out of the NOT-gate in those cases in which the light is on for the negation circuit. Similarly, the AND-gate and OR-gate have two switches, p and q. These are symbolized as two inputs, p and q. The single output stands for the light.

We can construct logic circuits to simulate logical truth tables. For example, suppose we wish to find the truth values for $\sim(p \vee q)$. We should design the following circuit (remember that parentheses indicate the operation to be performed first):

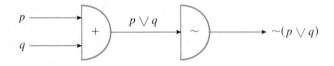

Example **2** **Compare circuit design and gate diagram**

Design a circuit for $\sim p \wedge (\sim q)$. Show both the circuit and the simplified gate diagram.

Solution For the circuit we begin with a truth table:

p	q	$\sim p$	$\sim q$	$\sim p \wedge \sim q$
T	T	F	F	F
T	F	F	T	F
F	T	T	F	F
F	**F**	T	T	**T**

The light must be on only when both switches are off (see boldface in table), as shown in Figure 3.9a.:

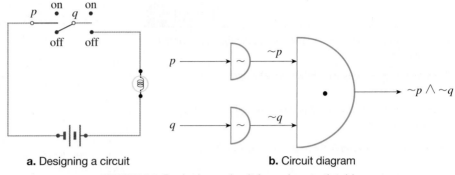

a. Designing a circuit   **b.** Circuit diagram

FIGURE 3.9   Designing a circuit for a given truth table

We can simulate this circuit by drawing the gate diagram, as shown in Figure 3.9b.

Example   **3**   **Design a circuit to test truth values**

Design a circuit that will find the truth values for $(p \vee q) \wedge q$.

Solution   We use gates to symbolize $(p \vee q) \wedge q$:

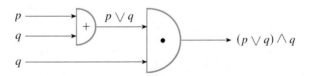

Next we must have some way of determining when the output values of $(p \vee q) \wedge q$ in Example 3 are true and when they are false. We can do this by again connecting a light to the circuit. When $(p \vee q) \wedge q$ is true, the light should be on; when it is false, the light should be off. We represent this connection as follows:

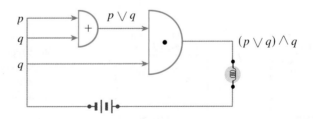

Problem Set **3.6**

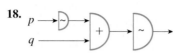

18.

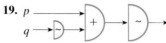

19.

Level **1**

1. **IN YOUR OWN WORDS** Explain when the light on a circuit is on and when it is off.

2. **IN YOUR OWN WORDS** When will the light be on for a series circuit?

3. **IN YOUR OWN WORDS** When will the light be on for a parallel circuit?

What do each of the circuit symbols in Problems 4–9 mean?

4.

5.

6.

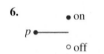

7.

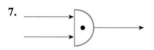

8.

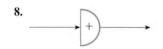

9.

Write a symbolic statement for each circuit shown in Problems 10–15.

10.

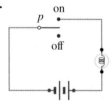

11.

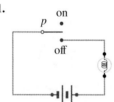

12.

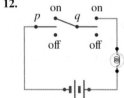

13.

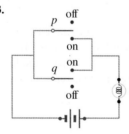

14.

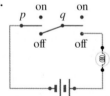

15.

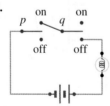

Build a truth table for each of the gate circuits in Problems 16–19.

16.

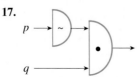

17.

Level **2**

Using both switches and simplified gates, design a circuit that would find the truth values for the statements in Problems 20–25 (answers are not unique).

20. $p \wedge q$ 21. $p \vee q$
22. $\sim p \wedge q$ 23. $p \wedge \sim q$
24. $\sim(p \vee q)$ 25. $\sim(p \wedge q)$

First build a truth table and then, using switches, design a circuit that would find the truth values for the statements in Problems 26–31 (answers are not unique).

26. $p \rightarrow q$ 27. $q \rightarrow p$
28. $p \rightarrow \sim q$ 29. $q \rightarrow \sim p$
30. $\sim(p \rightarrow \sim q)$ 31. $\sim(q \rightarrow p)$

Using AND-gates, OR-gates, and NOT-gates, design circuits that would find the truth values for the statements in Problems 32–37.

32. $p \wedge \sim q$ 33. $(p \wedge q) \vee r$
34. $(p \wedge q) \vee (p \wedge r)$ 35. $\sim(\sim p \vee q)$
36. $\sim[(p \wedge q) \vee r]$ 37. $(\sim p \wedge q) \vee (p \vee \sim q)$

First build a truth table and then design a circuit using switches that would find the truth values for the statements in Problems 38–45 (answers are not unique).

38. $(p \wedge q) \wedge r$ 39. $(p \vee q) \vee r$
40. $(p \vee q) \wedge r$ 41. $p \vee (q \wedge r)$
42. $p \wedge (q \vee r)$ 43. $(p \wedge q) \vee r$
44. $p \vee (q \vee r)$ 45. $p \wedge (q \wedge r)$

Level **3**

46. Suppose you wish to design a machine that dispenses 10¢ gum balls. The machine will not give change, but will accept only nickels and dimes.
    a. Write a compound logical statement that represents the condition for dispensing a gum ball.
    b. Construct a truth table for the statement.
    c. Draw a circuit diagram.

47. Suppose you wish to design a notification system for three air-plane rest rooms. A light should go on at the attendant's post when at least one of the rooms is in use.
    a. Write a compound logic statement that represents the condition for notifying the attendant.
    b. Construct a truth table for the statement.
    c. Draw a circuit diagram.

48. Suppose you wish to design a security system that requires that three people in different locations turn the keys simultaneously.
    a. Write a compound logic statement that represents the condition for activating the security system.
    b. Construct a truth table for the statement.
    c. Draw a circuit diagram.

49. Suppose you wish to design a thermostat that will heat a room. The system will activate the furnace under either of two conditions:

    I. The system is set on automatic mode and all the doors are closed.
    II. The system is set on manual mode and an authorization code has been entered.

    Write a compound logic statement that represents the condition for heating the room.

50. Write a truth table for the thermostat described in Problem 49.

51. Draw a circuit diagram representing the thermostat described in Problem 49.

52. The conditional $p \to q$ was defined to be equivalent to $\sim(p \wedge \sim q)$. Use gates to represent this conditional.

53. The cost of the circuit is often a factor in work with switching circuits. The circuit you constructed in Problem 52, by using the definition of the conditional, required three gates. Construct a truth table for $p \to q$ and $\sim p \vee q$. What do you notice about the truth values of the result? Design circuits for $p \to q$ by using only one OR-gate and one NOT-gate.

Using the results of Problems 52 and 53, design circuits for the conditional statements given in Problems 54–56.

54. $\sim p \to q$ 55. $p \to \sim q$ 56. $\sim q \to \sim p$

Problem Solving 3

57. a. Suppose an engineer designs the following circuit:

    $$\sim\{\sim[(p \vee q) \wedge (\sim p)] \wedge (\sim p)\}$$

    What will the circuit look like?
    b. If each gate costs 2¢ and a company is going to manufacture 1 million items using this circuit, how much will the circuits cost?

58. Use truth tables to find the following.
    a. $\sim\{\sim[(p \vee q) \wedge \sim p] \wedge (\sim p)\}$
    b. $p \vee q$
    c. What do you notice about the final column of each of these truth tables?
    d. Draw a simpler circuit that will output the same values as those in Problem 57.
    e. If the company in Problem 57b is going to manufacture 1 million items using this simpler circuit, how much will the circuits cost? How much will the company save by using this simpler circuit rather than the more complicated one?

59. Alfie, Bogie, and Clyde are the members of a Senate committee. Design a circuit that will output *yea* or *nay* depending on the way the majority of the committee members voted.

60. Suppose that the Senate committee of Problem 59 has five members. Design a circuit that will output the result of the majority of the committee's vote.

3.7 CHAPTER SUMMARY

Students should be able to construct cogent arguments in support of their claims.

NCTM Standards

Important Ideas

Truth table of fundamental operators [3.1, 3.2]
Law of double negation [3.2]
Law of contraposition [3.2]
De Morgan's laws [3.3]
Negation of a conditional [3.3]
Direct reasoning [3.4]
Indirect reasoning [3.4]
Transitive reasoning [3.4]
Relationship between logic and circuits [3.6]

Take some time getting ready to work the review problems in this section. First review these important ideas. Look back at the definition and property boxes. If you look online, you will find a list of important terms introduced in this chapter, as well as the types of problems that were introduced. You will maximize your understanding of this chapter by working the problems in this section only after you have studied the material.

You will find some review help online at **www.mathnature.com**. There are links giving general test help in studying for a mathematics examination, as well as specific help for reviewing this chapter.

Chapter 3 Review Questions

1. a. What is a logical statement?
   b. What is a tautology?
   c. What is the law of contraposition?

2. Complete the following truth table.

p	q	$\sim p$	$p \wedge q$	$p \vee q$	$p \to q$	$p \leftrightarrow q$

Construct truth tables for the statements in Problems 3–6.

3. $\sim(p \wedge q)$

4. $[(p \vee \sim q) \wedge \sim p] \to \sim q$

5. $[(p \wedge q) \wedge r] \to p$

6. $\sim(p \wedge q) \leftrightarrow (\sim p \vee \sim q)$

7. State and prove the principle of direct reasoning.

8. Give an example of indirect reasoning.

9. Give an example of a common fallacy of logic, and show why it is a fallacy.

10. Is the following valid?

$$p \to q$$
$$\underline{\sim q}$$
$$\therefore \sim p$$

Can you support your answer?

11. Write the negation of each of the following statements.

   a. All birds have feathers.
   b. Some apples are rotten.
   c. No car has two wheels.
   d. Not all smart people attend college.
   e. If you go on Tuesday, then you cannot win the lottery.

12. Which of the following statements are true?

   a. If $111 + 1 = 1,000$, then I'm a monkey's uncle.
   b. If $6 + 2 = 10$ and $7 + 2 = 9$, then $5 + 2 = 7$.
   c. If $6 + 2 = 10$ or $5 + 2 = 7$, then $7 + 2 = 9$.
   d. If $5 + 2 = 7$ and $7 + 2 \neq 9$, then $6 + 2 = 10$.
   e. If the moon is made of green cheese, then $5 + 7 = 57$.

13. Let p: P is a prime number; q: $P + 2$ is a prime number. Translate the following statements into verbal form.
   a. $p \to q$ **b.** $(p \vee q) \wedge [\sim(p \wedge q)]$

14. Consider this statement: "All computers are incapable of self-direction."
   a. Translate this statement into symbolic form.
   b. Write the contrapositive of the statement.

15. Translate into symbols and identify the type of argument.

   If there are a finite number of primes, then there is some natural number, greater than 1, that is not divisible by any prime.
   Every natural number greater than 1 is divisible by a prime number.
   Therefore, there are infinitely many primes.

16. a. Using switches, design a circuit that would find the truth values for $\sim p \vee \sim q$.
   b. Using AND-gates, OR-gates, or NOT-gates, design a circuit that would find the truth values for $[(p \wedge q) \wedge p]$.

In Problems 17–19, form a valid conclusion using all the premises given in each problem.

17. If I attend to my duties, I am rewarded.
   If I am lazy, I am not rewarded.
   I am lazy.

18. All organic food is healthy.
   All artificial sweeteners are unhealthy.
   No prune is nonorganic.

19. All squares are rectangles.
   All rectangles are quadrilaterals.
   All quadrilaterals are polygons.

20. TABLE PUZZLE The mathematics department of a very famous two-year college consists of four people: Josie, who is department chairperson, Maureen, Terry, and Warren. For department meetings they always sit in the same seats around a square table. Their hobbies are (in alphabetical order) baking, gardening, hiking, and surfing. Josie, whose hobby is gardening, sits on Terry's left. Maureen sits at the hiker's right. Warren, who faces Terry, is not the baker. Make a drawing showing who sits where, and state the hobby of each person.

BOOK REPORTS

Write a 500-word report on one of these books:

Overcoming Math Anxiety, Sheila Tobias (Boston: Houghton Mifflin Co., 1978).

Mathematical Puzzles for Beginners & Enthusiasts, Geoffrey Mott-Smith (New York: Dover, 1954).

Group | RESEARCH PROJECTS

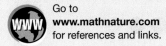

Go to
www.mathnature.com
for references and links.

Working in small groups is typical of most work environments, and learning to work with others to communicate specific ideas is an important skill. Work with three or four other students to submit a single report based on each of the following questions.

G6. Write a symbolic statement for each of the following verbal statements. (1) Either Donna does not like Elmer because Elmer is bald, or Donna likes Frank and George because they are handsome twins. (2) Either neither you nor I am honest, or Hank is a liar because Iggy did not have the money.

G7. Consider the following question: "Of all possible collections of states that yield 270 or more electoral votes—enough to win a presidential election—which collection has the smallest geographical area?" *Hint*: Let O be the set consisting of the optimal collection of states. Your group should choose a state and *prove* mathematically that the state is *not* in O or prove that the state is in O.

References This problem is found in the article "Proof by Contradiction and the Electoral College," by Charles Redmond, Michael P. Federici, and Donald M. Platte in *The Mathematics Teacher*, November 1998. The U.S. Electoral College Calculator
http://www.archives.gov/federal_register/electoral_college/calculator.html
National Archives and Records Administration Federal Register
http://www.archives.gov/federal_register/electoral_college.html

G8. Five cabbies have been called to pick up five fares at the Hilton Towers. On arrival, they find that their passengers are slightly intoxicated. Each man has a different first and last name, a different profession, and a different destination; in addition, each man's wife has a different first name. Unable to determine who's who and who's going where, the cabbies want to know: **Who is the accountant? What is Winston's last name? Who is going to Elm Street?** Use only the following facts to answer these questions:

1. Sam is married to Donna.
2. Barbara's husband gets into the third cab.
3. Ulysses is a banker.
4. The last cab goes to Camp St.
5. Alice lives on Denver Street.
6. The teacher gets into the fourth cab.
7. Tom gets into the second cab.
8. Eve is married to the stock broker.
9. Mr. Brown lives on Denver St.
10. Mr. Camp gets into the cab in front of Connie's husband.
11. Mr. Adams gets into the first cab.
12. Mr. Duncan lives on Bourbon St.
13. The lawyer lives on Anchor St.
14. Mr. Duncan gets into the cab in front of Mr. Evans.
15. The lawyer is three cabs in front of Victor.
16. Mr. Camp is in the cab in front of the teacher.

G9. This problem is similar to the previous taxicab problem, only instead of five taxi drivers we consider nine persons who play positions on a baseball team. You will need to get the details from this books's website (**www.mathnature.com**) for a description of this problem.

G10. Consider the apparatus shown in Figure 3.10.

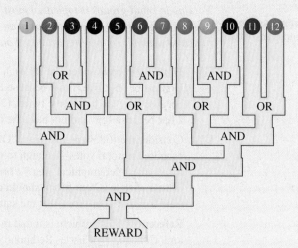

FIGURE 3.10 Reward Game

Note that there are 12 chutes (numbered 1 to 12), and a ball dropped into a chute will slide down the tube until it reaches an AND-gate or an OR-gate. If two balls reach an AND-GATE, then one ball will pass through, but if only one reaches an AND-gate, it will not pass through. If one or two balls reach an OR-gate, then one ball will pass through. The object is to obtain a reward by having a ball reach the location called REWARD. What is the fewest number of balls that can be released in order to gain the reward?

Individual RESEARCH PROJECTS

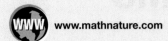
www.mathnature.com

Learning to use sources outside your classroom and textbook is an important skill, and here are some ideas for extending some of the ideas in this chapter. You can find references to these projects in a library or at **www.mathnature.com.**

PROJECT 3.1 Deck of Cards Puzzle from *Games* magazine.
See **www.mathnature.com**, Section 3.1, for a statement of this puzzle.

PROJECT 3.2 Do some research to explain the differences between the words *necessary* and *sufficient.*

PROJECT 3.3 Sometimes statements *p* and *q* are described as *contradictory, contrary,* or *inconsistent.* Consult a logic text, and then define these terms using truth tables.

PROJECT 3.4 Between now and the end of the course, look for logical arguments in newspapers, periodicals, and books. Translate these arguments into symbolic form. Turn in as many of them as you find. Be sure to indicate where you found each argument.

PROJECT 3.5 Suppose a prisoner must make a choice between two doors. One door leads to freedom and the other door is booby-trapped so that it leads to death. The doors are labeled as follows:

Door 1:   This door leads to freedom and the other door leads to death.
Door 2:   One of these doors leads to freedom and the other of these doors leads to death.

If exactly one of the signs is true, which door should the prisoner choose? Give reasons for your answer.

PROJECT 3.6 Convention Problem   A mathematician attended a convention of fifty male and female scientists. The mathematician observed that if any two of them were picked at random, at least one of the two would be male. From this information, is it possible to deduce what percentage of the attendees were women?

PROJECT 3.7 A man is about to be electrocuted but is given a chance to save his life. In the execution chamber are two chairs, labeled 1 and 2, and a jailer. One chair is electrified; the other is not. The prisoner must sit on one of the chairs, but before doing so, he may ask the jailer one question, to which the jailer must answer "yes" or "no." The jailer is a consistent liar or else a consistent truth teller, but the prisoner does not know which. Knowing that the jailer either deliberately lies or faithfully tells the truth, what question should the prisoner ask?

PROJECT 3.8 Build a simple device that will add single-digit numbers.

4

THE NATURE OF NUMERATION SYSTEMS

What in the World?

"It's just not fair!" exclaimed José. "Half my class has computers at home. I just went to the student store and the cheapest computer and printer I can find is over $500. Forget that!"

"You're right, I gave up on getting my own computer," said Sheri. "But I use the library computer all the time. Using Google™, I can look up almost anything! Last week, I was able to access some information at the Stanford library."

"Do you think that library computer will help me with my paper? I need to *invent* my own numeration system," said José. "How am I supposed to do what took centuries for others to do? As I've said before, it's just not fair!"

Overview

It is nearly impossible to graduate from college today and not have some computer skills. We are all affected by computers, and even though we do not need to understand all about what makes them operate any more than we need to know about an internal combustion engine to drive a car, it is worthwhile to learn just a little about how they do what they do. We study numeration systems in this chapter so that we can better understand our own numeration system, as well as gain some insight into the internal workings of a computer. In addition to looking at computers today, we take a peek back into the history of computers.

Some knowledge about computers and how they work may help you to understand how better to deal with computers in your day-to-day life.

© Dennis MacDonald/Photo Edit

4.1 Early Numeration Systems

To err is human, but to really foul things up requires a computer.

ANONYMOUS

CHAPTER **CHALLENGE**

See if you can fill in the question mark.

2		5		7		8
	3		8		11	
		7		15		
			?			

"Do you know your numbers?" is a question you might hear one child asking another. At an early age, we learn our numbers as we learn to count, but if we are asked to define what we mean by *number*, we are generally at a loss. There are many different kinds of numbers, and one type of number is usually defined in terms of more primitive types of numbers. The word *number* is taken as one of our primitive (or undefined) words, but the following definition will help us distinguish between a number and the symbol that we use to represent a number.

> ### Number/Numeral
> A **number** is used to answer the question "How many?" and usually refers to numbers used to count objects:
>
> $$\{1, 2, 3, 4, 5, 6, 7, 8, 9, 10, \ldots\}$$
>
> This set of numbers is called the set of **counting numbers.**
>   A **numeral** is a symbol used to represent a number, and a **numeration system** consists of a set of basic symbols and some rules for making other symbols from them, the purpose being the identification of all numbers.

The invention of a precise and "workable" numeration system is one of the greatest inventions of humanity. It is certainly equal to the invention of the alphabet, which enabled humans to carry the knowledge of one generation to the next. It is simple for us to use the symbol 17 to represent this many objects:

● ● ● ● ● ● ● ● ● ● ● ● ● ● ● ● ●

However, this is not the first and probably will not be the last numeration system to be developed; it took centuries to arrive at this stage of symbolic representation. Here are some of the ways that 17 has been written:

Tally: 卌 卌 卌 || Egyptian: ∩||||||| Roman: XVII

Mayan: (symbol)   Linguistic: *seventeen, seibzehn, dix-sept*

The concept represented by each of these symbols is the same, but the symbols differ. The concept or idea of "seventeenness" is called a *number*; the symbol used to represent the concept is called a *numeral.*

Three of the earliest civilizations known to use numerals were the Egyptian, Babylonian, and Roman. We shall examine these systems for two reasons. First, it will help us to more fully understand our own numeration (decimal) system; second, it will help us to see how the ideas of these other systems have been incorporated into our system.

There is a story of a woman who decided she wanted to write to a prisoner in a nearby state penitentiary. However, she was puzzled over how to address him, since she knew him only by a string of numerals. She solved her dilemma by beginning her letter: "Dear 5944930, may I call you 594?"

Egyptian Numeration System

Perhaps the earliest type of written numeration system developed was a **simple grouping system.** The Egyptians used such a system by the time of the first dynasty, around 2850 B.C. The symbols of the Egyptian system were part of their *hieroglyphics* and are shown in Table 4.1.

TABLE 4.1	Egyptian Hieroglyphic Numerals	
Usual Numeral (decimal)	Egyptian Numeral	Descriptive Name
1	I	Stroke
10	∩	Heel bone
100	9	Scroll
1,000	⚘	Lotus flower
10,000	⌒	Pointing finger
100,000	⌒	Polliwog
1,000,000	⍦	Astonished man

Any number is expressed by using these symbols *additively*; each symbol is repeated the required number of times, but with no more than nine repetitions. This additivity is called the **addition principle.**

Historical NOTE

There are two important sources for our information about the Egyptian numeration system. The first is the Moscow papyrus, which was written in about 1850 B.C. and contains 25 mathematical problems. The second is the Rhind papyrus (1650 B.C.). The chronology of this papyrus is interesting. Around 1575 B.C. the scribe Ahmose copied (and most likely edited) material from an older manuscript. In 1858, A. Henry Rhind acquired the papyrus in Luxor (he found it in a room of a small building near the Ramesseum). After this, the manuscript became known as the Rhind papyrus. It contains 85 problems and was a type of mathematical handbook for the Egyptians. From these two sources, we conclude that most of the problems were of a practical nature (although *some* were of a theoretical nature).

Example **1** Translate hieroglyphics

Write ⌒ ⚘ ⚘ 99 9 ∩∩ ∩∩ ||||| using decimal numerals.

Solution ⌒ ⚘ ⚘ 99 9 ∩∩ ∩∩ ||||| $= 10,000 + 2 \times 1,000 + 3 \times 100 + 4 \times 10 + 5 \times 1$

$$= 12,345$$

The position of the individual symbols is not important. For example,

⌒ ⚘ ⚘ 99 9 ∩∩ ∩∩ ||||| $=$ ⚘ ⚘ 99 9 ⌒ ∩∩ ∩∩ |||||

$=$ ||||| ⌒ ⚘ 99 ∩∩∩∩ ⚘ 9

"*No, no!* ⚘ *as in* ⍦︎⌒⌒⚘⌒︎⍦."

The Egyptians had a simple **repetitive-type** numeration system. Addition and subtraction were performed by repeating the symbols and regrouping.

Example **2** **Egyptian arithmetic**

Use Egyptian arithmetic to find

a. $245 + 457$ **b.** $142 - 67$

Solution

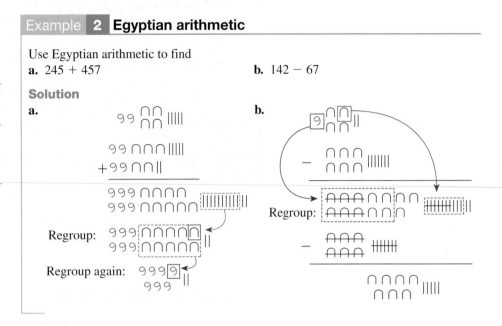

In Egyptian arithmetic, multiplication and division were performed by successions of additions that did not require the memorization of a multiplication table and could easily be done on an abacus-type device (see links on this topic at **www.mathnature.com**).

The Egyptians also used unit fractions (that is, fractions of the form $1/n$, for some counting number n). These were indicated by placing an oval over the numeral for the denominator. Thus,

$$\frac{1}{3} \qquad \frac{1}{4} \qquad \frac{1}{10} \qquad \frac{1}{33}$$

The fractions $\frac{1}{2}$ and $\frac{2}{3}$ were exceptions:

$$\text{or} \quad = \frac{1}{2} \qquad \text{or} \quad = \frac{2}{3}$$

Fractions that were not unit fractions (with the exception of $\frac{2}{3}$ just mentioned) were represented as sums of unit fractions. For example,

$$\frac{2}{7} \text{ could be expressed as } \frac{1}{4} + \frac{1}{28}$$

Repetitions were not allowed, so $\frac{2}{7}$ would not be written as $\frac{1}{7} + \frac{1}{7}$. Why so difficult? We don't know, but the Rhind papyrus (see Figure 4.1) includes a table for all such decompositions for odd denominators from 5 to 101.

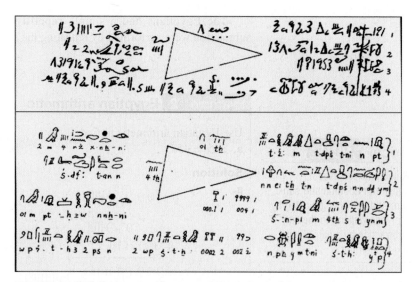

FIGURE 4.1 Rhind papyrus. The top portion is part of the actual papyrus and the bottom portion is a translation. Look for the Egyptian symbols; can you see the polliwog and the astonished man?

Roman Numeration System

The Roman numeration system was used in Europe until the 18th century, and is still used today in outlining, in numbering certain pages in books, and for certain copyright dates. The Roman numerals were selected letters from the Roman alphabet, as shown in Table 4.2.

TABLE 4.2							Roman Numerals
Decimal Numeral	1	5	10	50	100	500	1,000
Roman Numeral	I	V	X	L	C	D	M

The Roman numeration system is *repetitive*, like the Egyptian system, but has two additional properties, called the *subtraction principle* and the *multiplication principle*.

Subtraction Principle

One property of the Roman numeration system is based on the **subtraction principle.** Reading from the left, we add the value of each numeral unless its value is smaller than the value of the numeral to its right, in which case we subtract it from that number. Only the numbers 1, 10, 100, . . . , can be subtracted, and only from the next two higher numbers. For example,

   I, II, III, IIII or IV, V, VI, VII, VIII, VIIII or IX, X, XI, . . .

are the successive counting numbers. Instead of IIII we write IV, which (using the subtraction principle) means $5 - 1 = 4$. The I can be subtracted from V or X, and the X can be subtracted from L or C; thus

   XL   means $50 - 10 = 40$

The C can be subtracted only from D or M. For example,

   DC means $500 + 100 = 600$ but   CD means $500 - 100 = 400$

Multiplication Principle

A bar is placed over a Roman numeral or a group of numerals to mean 1,000 times the value of the Roman numerals themselves. When one symbol (such as a bar) is used to multiply the value of an associated symbol, it is called the **multiplication principle.** For example,

$$\overline{V} \text{ means } 5 \times 1,000 = 5,000 \qquad \text{and} \qquad \overline{XL} \text{ means } 40 \times 1,000 = 40,000$$

The bar is not used if there is an existing symbol. For example, M and not $\overline{I}$ is used for 1,000.

Example 3 Translate Roman numerals

Write each Roman numeral as a decimal numeral.
a. CCLXIII   **b.** MXXIV   **c.** MCMXCIX   **d.** $\overline{CDCLIX}$

Solution
a. CCLXIII + 100 + 100 + 50 + 10 + 1 + 1 + 1 = 263
b. MXXIV = 1,000 + 10 + 10 + (5 − 1) = 1,024
c. MCMXCIX = 1,000 + (1,000 − 100) + (100 − 10) + (10 − 1) = 1,999
d. $\overline{CDCLIX}$ = (500 − 100) × 1,000 + 100 + 50 + (10 − 1) = 400,159

Example 4 Write Roman numerals

Write each decimal numeral as a Roman numeral.
a. 49   **b.** 379   **c.** 345,123   **d.** 2,003

Solution
a. 49 = (50 − 10) + (10 − 1) = XLIX
b. 379 = 100 + 100 + 100 + 50 + 20 + (10 − 1) = CCCLXXIX
c. 345,123 = 345 × 1,000 + 123
$$= [100 + 100 + 100 + (50 − 10) + 5] \times 1,000 + (100 + 20 + 3)$$
$$= \overline{CCCXLV}CXXIII$$
d. 2,003 = 2,000 + 3
$$= MMIII$$

Fields Medal

The Fields Medal is often referred to as the "Nobel Prize of Mathematics." The existence of the award is related to the fact that Alfred Nobel chose not to include mathematics in his areas of recognition. There is no documentary evidence to explain this exclusion, but the general gossip attributes it to a personal conflict between Nobel and Mittag-Leffler. John Charles Fields (1863–1932) was disturbed by the lack of such a mathematical award. so he worked toward the establishment of these awards from 1922 until his death in 1932. It was Fields' last will and testament that provided the necessary funds for the establishment of the award, which was finalized on January 4, 1934. The first Fields Medals were awarded in 1936 to L. V. Ahlfors (Harvard) and Jesse Douglas (MIT). Because of World War II the next award was not until 1950; it has been awarded every 4 years at the International Congress of Mathematics.
My thanks to Henry S. Tropp, Humboldt State University, for this information.
Professor Tropp was my first mathematics professor and was
instrumental in my interest in mathematics. I thank him not only for his
research on the Fields Medal, but also for his influence in my life.

Stefan Zachow/The International Mathematics Union

Babylonian Numeration System

The Babylonian numeration system differed from the Egyptian in several respects. Whereas the Egyptian system was a simple grouping system, the Babylonians employed a much more useful **positional system.** Since they lacked papyrus, they used mostly clay as a writing medium, and thus the Babylonian cuneiform was much less pictorial than the Egyptian system. They employed only two wedge-shaped characters, which date from 2000 B.C. and are shown in Table 4.3.

TABLE 4.3	
	Babylonian Cuneiform Numerals
Decimal Numeral	**Babylonian Numeral**
1	▼
2	▼▼
9	▼▼▼▼▼▼▼▼▼
10	◁
59	◁◁◁▼▼▼▼▼ ◁◁▼▼▼▼

Notice from the table that, for numbers 1 through 59, the system is *repetitive.* However, unlike in the Egyptian system, the position of the symbols is important. The ◁ symbol *must* appear to the left of any ▼s to represent numbers smaller than 60. For numbers larger than 60, the symbols ◁ and ▼ are written to the left of ◁ (as in a positional system), and now take on a new value that is 60 times larger than the original value. That is,

▼ ◁◁◁ ▼▼▼▼▼   means $(1 \times 60) + 35$

Study the following example.

Example 5 | Translate Babylonian numerals

Write each of the Babylonian numerals in decimal form.

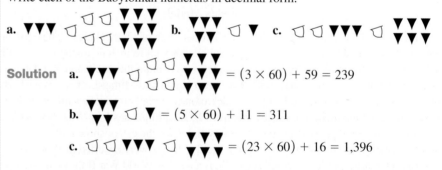

Solution a. ▼▼▼ ◁◁◁ ▼▼▼▼▼ = $(3 \times 60) + 59 = 239$

b. ▼▼▼▼▼ ◁ ▼ = $(5 \times 60) + 11 = 311$

c. ◁◁ ▼▼▼ ◁ ▼▼▼▼▼▼ = $(23 \times 60) + 16 = 1{,}396$

The Babylonian numeration system is called a *sexagesimal system* and uses the principle of position. However, the system is not fully positional, because only numbers larger than 60 use the position principle; numbers within each basic 60-group are written by a simple grouping system. A true positional sexagesimal system would require 60 different symbols.

Historical NOTE

Cuneiform tablet

More than 50,000 clay tablets are available to us, and the work of interpreting them is continuing. Most of our knowledge of the tablets is less than 130 years old and has been provided by Professors Otto Neugebauer and F. Thureau-Dangin.

The Babylonians carried their positional system a step further. If any numerals were to the left of the second 60-group, they had the value of 60×60 or 60^2. Thus,

 $= (2 \times 60^2) + (45 \times 60) + 24$
$= 7,200 + 2,700 + 24$
$= 9,924$

The Babylonians also made use of a *subtractive symbol,* . That is, 38 could be written

or

Example 6 Complex Babylonian numerals

Change each Babylonian numeral to decimal form.

a.

b.

c.

Solution   **a.** $(3 \times 60) + 20 - 1 = 199$
b. $(2 \times 60^2) + (23 \times 60) + 16 = 7,200 + 1,380 + 16 = 8,596$
c. Note the ambiguity; this might be 54 or it might be $(20 \times 60) + 34 = 1,234$

Example 7 Babylonian arithmetic

Write 4,571 as a Babylonian numeral.

Solution $1 \times 60^2 = 3,600$
$16 \times 60 = 960$
$\underline{11 \times 1 = 11}$
$4,571$

Thus, is the Babylonian numeral.

Although this positional numeration system was in many ways superior to the Egyptian system, it suffered from the lack of a zero or placeholder symbol. For example, how is the number 60 represented? Does ▼ ▼ mean 2 or 61? In Example 6c the value of the number had to be found from the context. (Scholars tell us that such ambiguity can be resolved only by a careful study of the context.) However, in later Babylon, around 300 B.C., records show that there is a zero symbol, an idea that was later used by the Hindus.

Arithmetic with the Babylonian numerals is quite simple, since there are only two symbols. A study of the Babylonian arithmetic will be left for the reader (see Problem 6).

Properties of Numeration Systems

A *numeration system* consists of a set of basic symbols. Because the set of counting numbers is infinite and the set of symbols in any numeration system is finite, some rules for using the set of symbols are necessary. Different historical systems use one or more of these properties:

(continued)

A *simple grouping* system invents a symbol to represent the number of objects in a predetermined group of objects. For example, the symbol

3 is used to represent   ● ● ●
V is used to represent   ● ● ● ● ●
or   ∩ is used to represent   ● ● ● ● ● ● ● ● ● ●

A *positional system* reuses a symbol to represent different numbers of objects by changing the position in its symbolic (numerical) representation. For example,

in 35 the "3" is used to represent

● ● ● ● ● ● ● ● ● ●
● ● ● ● ● ● ● ● ● ●
● ● ● ● ● ● ● ● ● ●

whereas in 53 the "3" is used to represent

● ● ●

The *addition principle* means that the various values of individual symbols (numerals) are added, as in ∩∩∩ |||| meaning

$$10 + 10 + 10 + 1 + 1 + 1 + 1$$

The *subtraction principle* means that certain values are subtracted, as in IX meaning $10 - 1 = 9$.

The *multiplication principle* means that certain symbolic values are multiplied, as in $\overline{V}$ meaning $1{,}000 \times 5$.

A *repetitive system* reuses the same number over and over in an additive manner, as in XXX meaning $10 + 10 + 10$.

Other Historical Systems

One of the richest areas in the history of mathematics is the study of historical numeration systems. In this book, we discuss the Egyptian numeration system because it illustrates a simple grouping system, the Babylonian numeration system because it illustrates a positional numeration system, and the Roman system because it is still in limited use today.

Other numeration systems of interest from a historical point of view are the Mayan, Greek, and Chinese numeration systems. These are summarized in Table 4.4.

Karl Smith library

Historical NOTE

The Aztec nation used chocolate not only as a food, but also as a counting unit. Rather than counting by tens, they used a system based on 20. For quantities up to 28, the system was repetitive and used dots; the next unit was a flag that was repeated for numbers up to 400; next, a fir tree was repeated for numbers up to 800. The largest symbol used by the Aztecs was a sack that represented 8,000 because each sack of cocoa contains about 8,000 beans.

TABLE 4.4

		Historical Numeration Systems						
Numeration System	**Classification**	**Selected Numerals**						
		1	2	5	10	50	100	500	1,000
Decimal system	positional (10s)	1	2	5	10	50	100	500	1,000
Egyptian system	grouping	|	||	|||||	∩	∩∩∩ ∩∩	9	999 99	𝄐
Roman system	positional	I	II	V	X	L	C	D	M
Babylonian (Sumerian)	positional (60s)	▼	▼▼	▼▼▼ ▼▼	◁	◁◁◁ ◁◁	▼◁◁ ◁◁	▼▼▼▼◁ ▼▼▼▼◁	◁▼▼▼◁◁ ▼▼▼◁
Greek	grouping	α	β	ε	ι	ν	ρ	φ	α′
Mayan	positional (vertical)	●	● ●	—	=	≝	⊖	⊖̇	⊖̈
Chinese	positional (vertical)	〳	〴	五	十	五十	百	五百	千

Problem Set 4.1

Level 1

1. **IN YOUR OWN WORDS** Explain the difference between *number* and *numeral*. Give examples of each.

2. **IN YOUR OWN WORDS** Discuss the similarities and differences between a simple grouping system and a positional system. Give examples of each.

3. **IN YOUR OWN WORDS** What do you regard as the shortcomings and contributions of the Egyptian numeration system?

4. **IN YOUR OWN WORDS** What do you regard as the shortcomings and contributions of the Roman numeration system?

5. **IN YOUR OWN WORDS** What do you regard as the shortcomings and contributions of the Babylonian numeration system?

6. **IN YOUR OWN WORDS** Discuss addition and subtraction for Babylonian numerals. Show examples.

7. Tell which of the named properties apply to the Egyptian, Roman, and Babylonian numeration systems.
   a. grouping system
   b. positional system
   c. repetitive system
   d. additive system
   e. subtractive system
   f. multiplicative system

Write each numeral in Problems 8–21 as a decimal numeral. By decimal numeral, we mean the system we use in everyday arithmetic.

8. a. ![Egyptian numeral] b. ![Egyptian numeral]

9. a. ![Egyptian numeral] b. ![Egyptian numeral]

10. a. ![Egyptian numeral] b. ![Egyptian numeral]

11. a. ![Egyptian numeral] b. ![Egyptian numeral]

12. a. ![Egyptian numeral] b. ![Egyptian numeral]

13. a. ![Egyptian numeral] b. ![Egyptian numeral]

14. a. XLVIII b. DCCIX

15. a. MCMXCIX b. MMI

16. a. $\overline{\text{DL}}$ b. $\overline{\text{CD}}$

17. a. $\overline{\text{V}}$MMDC b. $\overline{\text{IX}}$DCCXII

18. a. ![Babylonian numeral] b. ![Babylonian numeral]

19. a. ![Babylonian numeral] b. ![Babylonian numeral]

20. a. ![Babylonian numeral] b. ![Babylonian numeral]

21. a. ![Babylonian numeral] b. ![Babylonian numeral]

22. Kathy had a dream in which she was selling roses in an international market. She started out with 143 roses. Then her first customer, an Egyptian, bought ![Egyptian numeral] of them. Soon afterward, a Babylonian asked to buy ![Babylonian numeral] roses; later XIV roses were bought by (you guessed it) a Roman. How many roses did she have left to sell at the end of her dream?

23. In a strange mix-up, several people (no two of whom spoke the same language) needed to pool their resources to survive the scorching desert heat. Eric had 45 bottles of water; Dimetrius had ![Egyptian numeral] bottles; Sparticus had ![Babylonian numeral] bottles. How many bottles of water did they have all together?

The Rhind papyrus contains many problems. Two are symbolized on the following simulated papyri. The one in Problem 25 is an 18th-century Mother Goose rhyme. Answer the question posed in each problem.

24.
> In each of 7 houses are 7 cats.
>
> Each cat kills 7 mice.
>
> Each mouse would have eaten 7 spelt (wheat).
>
> Each ear of spelt would have produced 7 kehat of grain.
>
> Query: How much grain is saved by the 7 houses' cats?

25.
> As I was going to St. Ives I met a man with seven wives.
>
> Every wife had seven sacks,
>
> Every sack had seven cats.
>
> Every cat had seven kits.
>
> Kits, cats, sacks, and wives,
>
> How many were there going to St. Ives?

Level 2

Write each of the numerals in Problems 26–31 in the Egyptian numeration system.

26. 47   27. 75

28. 258   29. 521

30. 852   31. 2,007

Write each of the numerals in Problems 32–37 in the Roman numeration system.

32. 47   33. 75

34. 258   35. 521

36. 852   37. 2,007

Write each of the numerals in Problems 38–43 in the Babylonian numeration system.

38. 47   39. 75

40. 258   41. 521

42. 852   43. 2,007

Perform the indicated operations in Problems 44–49.

44.

$$\text{𒐕}\cap\cap\text{II} \atop +99\cap\cap\cap\cap \, \text{IIIII} \, \cap\cap\cap$$

45.

$$\text{𒐕}\text{𒐕}\cap\text{I} \atop -\cap \, \cap\text{IIII}$$

46.

$$99\cap\cap\text{II} \atop -9\cap\cap\cap\cap \, \text{IIIII} \, \cap\cap\cap\cap$$

47.

$$\triangleleft\triangleleft\triangleleft\blacktriangledown\blacktriangledown\blacktriangledown\blacktriangledown \atop +\triangleleft\blacktriangledown\blacktriangledown\blacktriangledown \, \blacktriangledown\blacktriangledown\blacktriangledown$$

48.

$$\triangleleft\triangleleft\blacktriangledown\blacktriangledown\blacktriangledown\blacktriangledown \atop -\triangleleft\blacktriangledown\blacktriangledown\blacktriangledown \, \blacktriangledown\blacktriangledown\blacktriangledown$$

49.

$$\blacktriangledown\triangleleft\blacktriangledown\blacktriangledown \atop -\triangleleft\triangleleft\triangleleft\blacktriangledown\blacktriangledown\blacktriangledown\blacktriangledown \, \triangleleft \quad \blacktriangledown\blacktriangledown\blacktriangledown$$

What is the largest number that begins with the Roman numeral symbol in Problems 50–55?

50. I

51. V

52. X

53. L

54. C

55. M

Problem Solving 3

56. In Example 3c, we wrote 1,999 as MCMXCIX, which according to the National Institute of Standards and Technology, is the preferred representation. Which of the following does not also represent 1,999?
A. MCMXCXI
B. MCMXCVIIII
C. MDCCCCLXXXXVIIII

57. a. What is the largest number that uses each of the seven Roman numerals exactly once?
b. What is the smallest number that uses each of the seven Roman numerals exactly once?

58. HISTORICAL QUEST The Fields Medal (discussed in this section) has a Roman numeral misprint. The date is written in Roman numerals: MCNXXXIII. Translate this number into decimal numerals. *Hint:* There is no typo in this problem. . . . any conclusion? Write a history of the Fields Medal.

59. HISTORICAL QUEST The Yale tablet from the Babylonian collection, written between 1900 B.C. and 1600 B.C., contains the following algebra problem:

> The length of a rectangle exceeds the width by 7, and the area is 60. What are the dimensions of the rectangle?

Answer the question in the Babylonian numeration system.

60. HISTORICAL QUEST Answer the following question from the Yale tablet (see Problem 59):

> The length of a rectangle exceeds the width by 10 and the area is 600. What are the dimensions of the rectangle?

Give your answer using Babylonian notation.

4.2 Hindu-Arabic Numeration System

The numeration system in common use today (the one we have been calling the decimal system) has ten symbols—namely, 0, 1, 2, 3, 4, 5, 6, 7, 8, and 9. The selection of ten digits was no doubt a result of our having ten fingers (digits).

The symbols originated in India in about 300 B.C. However, because the early specimens do not contain a zero or use a positional system, this numeration system offered no advantage over other systems then in use in India.

The date of the invention of the zero symbol is not known. The symbol did not originate in India but probably came from the late Babylonian period via the Greek world.

By the year A.D. 750 the zero symbol and the idea of a positional system had been brought to Baghdad and translated into Arabic. We are not certain how these numerals were introduced into Europe, but it is likely that they came via Spain in the 8th century. Gerbert, who later became Pope Sylvester II in 999, studied in Spain and was the first European scholar known to have taught these numerals. Because of their origins, these numerals are called the **Hindu-Arabic numerals.** Since ten basic symbols are used, the Hindu-Arabic numeration system is also called the *decimal numeration system*, from the Latin word *decem*, meaning "ten."

Although we now know that the decimal system is very efficient, its introduction met with considerable controversy. Two opposing factions, the "algorists" and the "abacists," arose. Those favoring the Hindu-Arabic system were called algorists, since the symbols were introduced into Europe in a book called (in Latin) *Liber Algorismi de Numero Indorum*, by the Arab mathematician al-Khowârizmî. The word *algorismi* is the origin of our word *algorithm*. The abacists favored the status quo—using Roman numerals and doing arithmetic on an abacus. The battle between the abacists and the algorists lasted for 400 years. The Roman Catholic Church exerted great influence in commerce, science, and theology. The church criticized those using the "heathen" Hindu-Arabic numerals and consequently kept the world using Roman numerals until 1500. Roman numerals were easy to write and learn, and addition and subtraction with them were easier than with the "new" Hindu-Arabic numerals. It seems incredible that our decimal system has been in general use only since about the year 1500.

Decimal System—Grouping by Tens

Let's examine the Hindu-Arabic or **decimal numeration system** a little more closely:

1. It uses ten symbols, called digits.

2. Larger numbers are expressed in terms of powers of 10.

3. It is positional.

Consider how we count objects:

■　■　■　■　■　■　■　■　■
1　2　3　4　5　6　7　8　9　?

At this point we could invent another symbol as the Egyptians did (you might suggest 0, but remember that 0 represents no objects), or we could reuse the digit symbols by repeating them or by altering their positions. We agree to use 10 to mean 1 group of

We call this group a **ten**. The symbol 0 was invented as a placeholder to show that the 1 here is in a different position from the 1 representing ■. We continue to count:

▢■■■■■■■■■▢ ■	**11**	This is 1 group and 1 extra.
▢■■■■■■■■■▢ ■■	**12**	This is 1 group and 2 extra.
⋮		
▢■■■■■■■■■▢ ▢■■■■■■■■■▢	**20**	This is 2 groups.
▢■■■■■■■■■▢ ▢■■■■■■■■■▢ ■	**21**	This is 2 groups and 1 extra.
⋮		

We continue in the same fashion until we have 9 groups and 9 extra. What's next? It is 10 groups or a group of groups:

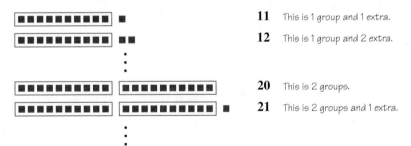

We call this group of groups a $10 \cdot 10$ or 10^2 or a **hundred.** We again use position and repeat the symbol 1 with yet a different meaning: 100.

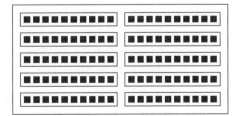

Example 1 Find the meaning of a number

What does 134 mean?

Solution

134 means that we have **1** group of 100, **3** groups of 10, and **4** extra. In symbols,

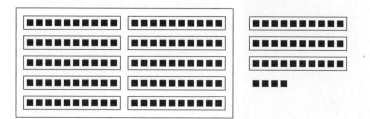

We denote this more simply by writing:

These represent the number in each group.
$$\downarrow \qquad\qquad \downarrow$$
$$(\mathbf{1} \times \underbrace{10^2}) + (\mathbf{3} \times \underbrace{10}) + \mathbf{4}$$
$$\uparrow \qquad\qquad \uparrow$$
These are the names of the groups.

This leads us to the meaning *one hundred, three tens, four ones.*

Expanded Notation

The representation, or the meaning, of the number 134 in Example 1 is called **expanded notation.**

Example 2 Write in expanded form

Write 52,613 in expanded form.

Solution $52{,}613 = 50{,}000 + 2{,}000 + 600 + 10 + 3$
$$= 5 \times 10^4 + 2 \times 10^3 + 6 \times 10^2 + 1 \times 10 + 3$$

Example 3 Write in decimal form

Write $4 \times 10^8 + 9 \times 10^7 + 6 \times 10^4 + 3 \times 10 + 7$ in decimal form.

Solution You can use the order of operations and multiply out the digits, but you should be able to go directly to decimal form if you remember what place value means:

49 00600 37 = 490,060,037
⇈ ⇈
Notice that there were no powers of 10^6, 10^5, 10^3, or 10^2.

A period, called a **decimal point** in the decimal system, is used to separate the fractional parts from the whole parts. The positions to the right of the decimal point are fractions:

$$\frac{1}{10} = 10^{-1}, \quad \frac{1}{100} = 10^{-2}, \quad \frac{1}{1{,}000} = 10^{-3}$$

To complete the pattern, we also sometimes write $10 = 10^1$ and $1 = 10^0$.

Example **4** **Expanded form with a decimal**

Write 479.352 using expanded notation.

Solution $479.352 = 400 + 70 + 9 + 0.3 + 0.05 + 0.002$

$$= 400 + 70 + 9 + \frac{3}{10} + \frac{5}{100} + \frac{2}{1,000}$$

$$= 4 \times 10^2 + 7 \times 10^1 + 9 \times 10^0 + 3 \times 10^{-1} + 5 \times 10^{-2} + 2 \times 10^{-3}$$

Problem Set 4.2

Level 1

1. **IN YOUR OWN WORDS** Discuss the difference between "number" and "numeral."

2. **IN YOUR OWN WORDS** What is expanded notation?

3. **IN YOUR OWN WORDS** What is an abacus, and what does it have to do with expanded notation?

4. **IN YOUR OWN WORDS** Illustrate the meaning of 123 by showing the appropriate groupings.

5. **IN YOUR OWN WORDS** Illustrate the meaning of 145 by showing the appropriate groupings.

6. **IN YOUR OWN WORDS** Illustrate the meaning of 1,134 by showing the appropriate groupings.

7. **IN YOUR OWN WORDS** Define b^n for b any nonzero number and n any counting number.

8. **IN YOUR OWN WORDS** Define b^{-n} for b any nonzero number and n any counting number.

Give the meaning of the numeral 5 in each of the numbers in Problems 9–14.

9. 805

10. 508

11. 0.00765

12. 5×10^4

13. 0.00567

14. 58,000,000

Write the numbers in Problems 15–31 in decimal notation.

15. **a.** 10^5 **b.** 10^3

16. **a.** 10^6 **b.** 10^4

17. **a.** 10^{-4} **b.** 10^{-3}

18. **a.** 10^{-2} **b.** 10^{-6}

19. **a.** 5×10^3 **b.** 5×10^2

20. **a.** 8×10^{-4} **b.** 7×10^{-3}

21. **a.** 6×10^{-2} **b.** 9×10^{-5}

22. **a.** 5×10^{-6} **b.** 2×10^{-9}

23. $1 \times 10^4 + 0 \times 10^3 + 2 \times 10^2 + 3 \times 10^1 + 4 \times 10^0$

24. $6 \times 10^1 + 5 \times 10^0 + 0 \times 10^{-1} + 8 \times 10^{-2} + 9 \times 10^{-3}$

25. $5 \times 10^5 + 2 \times 10^4 + 1 \times 10^3 + 6 \times 10^2 + 5 \times 10^1 + 8 \times 10^0$

26. $6 \times 10^7 + 4 \times 10^3 + 1 \times 10^0$

27. $7 \times 10^6 + 3 \times 10^{-2}$

28. $6 \times 10^9 + 2 \times 10^{-3}$

29. $5 \times 10^5 + 4 \times 10^2 + 5 \times 10^1 + 7 \times 10^0 + 3 \times 10^{-1} + 4 \times 10^{-2}$

30. $3 \times 10^3 + 2 \times 10^1 + 8 \times 10^0 + 5 \times 10^{-1} + 4 \times 10^{-2} + 2 \times 10^{-4}$

31. $2 \times 10^4 + 6 \times 10^2 + 4 \times 10^{-1} + 7 \times 10^{-3} + 6 \times 10^{-4} + 9 \times 10^{-5}$

Write each of the numbers in Problems 32–43 in expanded notation.

32. **a.** 741   **b.** 728,407

33. **a.** 0.096421   **b.** 27.572

34. **a.** 47.00215   **b.** 521

35. **a.** 6,245   **b.** 2,305,681

36. **a.** 428.31   **b.** 5,245.5

37. 0.00000527

38. 100,000.001

39. 893.0001

40. 8.00005

41. 678,000.01

42. 143,912.1743

43. 57,285.9361

Level 2

HISTORICAL QUEST *One of the oldest devices used for calculation is the abacus, as shown in Figure 4.2. Each rod names one of the positions we use in counting. Each bead on the bottom of the central bar represents one unit in that column and each bead on the top represents 5 units in that column. The number illustrated in Figure 4.2 is 1,734. What number is illustrated by the drawings in Problems 44–49?*

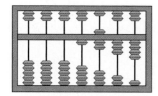

FIGURE 4.2 Abacus

Go to **www.mathnature.com** for a couple of helpful links on using an abacus.

44.

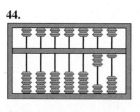

45.

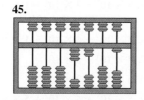

46.

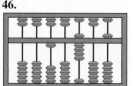

47.

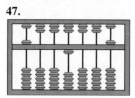

48.

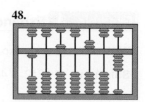

49.

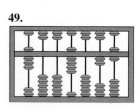

Sketch an abacus to show the numbers given in Problems 50–57.

50. 132   **51.** 849

52. 3,214   **53.** 9,387

54. 1,998   **55.** 2,001

56. 3,000,400 **57.** 8,007,009

Problem Solving 3

58. Can you find a pattern?

$$0, 1, 2, 10, 11, 12, 20, 21, 22, 100, \ldots$$

59. Can you find a pattern?

$$0, 1, 2, 3, 4, 10, 11, 12, 13, 14, 20, 21, \ldots$$

60. Notice the following pattern for multiplication by 11:

$$14 \times 11 = 1\_4 \qquad 51 \times 11 = 5\_1$$

First, copy the first and last digits of the number to be multiplied by 11. Leave a space between these digits. Then, insert the sum of the original two digits between those original digits:

$$14 \times 11 = 1\mathbf{5}4 \qquad 51 \times 11 = 5\mathbf{6}1$$

$$\uparrow \qquad\qquad\qquad \uparrow$$

$$1 + 4 = 5 \qquad\qquad 5 + 1 = 6$$

Use expanded notation to show why this pattern "works."

4.3   Different Numeration Systems

In the previous section, we discussed the Hindu-Arabic numeration system and grouping by tens. However, we could group by twos, fives, twelves, or any other counting number. In this section, we summarize numeration systems with bases other than ten. This not only will help you understand our own numeration system, but will give you insight into the numeration systems used with computers, namely, base 2 (**binary**), base 8 (**octal**), and base 16 (**hexadecimal**).

Number of Symbols

The number of symbols used in a particular base depends on the method of grouping for that base. For example, in base ten the grouping is by tens, and in base five the grouping is by fives. Suppose we wish to count ■■■■■■■■■■■ in various bases. Let's look for patterns in Table 4.5. Note the use of the subscript following the numeral to keep track of the base in which we are working.

TABLE 4.5				
				Grouping in Various Bases
Base	**Symbols**		**Method of Grouping**	**Notation**
two	0, 1			1011_{two}
three	0, 1, 2			102_{three}
four	0, 1, 2, 3			23_{four}
five	0, 1, 2, 3, 4			21_{five}
six	0, 1, 2, 3, 4, 5			15_{six}
seven	0, 1, 2, 3, 4, 5, 6			14_{seven}
eight	0, 1, 2, 3, 4, 5, 6, 7			13_{eight}
nine	0, 1, 2, 3, 4, 5, 6, 7, 8			12_{nine}
ten	0, 1, 2, 3, 4, 5, 6, 7, 8, 9			11_{ten}

Historical NOTE

Karl Smith library

Karok from the Yukis nation

A study of the Native Americans of California uncovered the use of a wide variety of number bases. Several of these bases are discussed by Barnabus Hughes in an article entitled "California Indian Arithmetic" in *The Bulletin of the California Mathematics Council* (Winter 1971/1972). According to Professor Hughes, the Yukis, who lived north of Willits, used both the quaternary (base four) and the octonary (base eight) systems. Instead of counting on their fingers, these Native Americans enumerated the spaces between the fingers.

The sketch above shows a Karok woman because the Yukis were nearly eliminated at the beginning of the 20th century and consequently there are no known photographs of the Yukis. The Native Americans in the western corner of California use the decimal system, while those in the rest of the state use the quinary (base five) system for small numbers, but switch to a vigesimal (base twenty) system for large numbers.

Do you see any patterns? Suppose we wish to continue this pattern. Can we group by elevens or twelves? We can, provided new symbols are "invented." For base eleven (or higher bases), we use the symbol T to represent ■■■■■■■■■■. For base twelve (or higher bases), we use E to stand for ■■■■■■■■■■■. For bases larger than twelve, other symbols can be invented.

For example, $2T_{twelve}$ means that there are two groupings of twelve and T (ten) extra:

We continue with the pattern from Table 4.5 by continuing beyond base ten in Table 4.6.

TABLE 4.6			
			Grouping in Various Bases
Base	**Symbols**	**Method of Grouping**	**Notation**
eleven	0, 1, 2, 3, 4, 5, 6, 7, 8, 9, T		10_{eleven}
twelve	0, 1, 2, 3, 4, 5, 6, 7, 8, 9, T, E		E_{twelve}
thirteen	0, 1, 2, 3, 4, 5, 6, 7, 8, 9, T, E, U		$E_{thirteen}$
fourteen	0, 1, 2, 3, 4, 5, 6, 7, 8, 9, T, E, U, V		$E_{fourteen}$

Do you see more patterns? Can you determine the number of symbols in the **base b system**? Remember that b stands for some counting number greater than 1. Base two has two symbols, base three has three symbols, and so on.

Change from Base *b* to Base 10

To change from base *b* to base ten, we write the numerals in expanded notation. The resulting number is in base ten.

Example **1** **Write in base 10**

Change each number to base ten.
a. 1011.01_{two}
b. 1011.01_{four}
c. 1011.01_{five}

Solution

a. $1011.01_{two} = 1 \times 2^3 + 0 \times 2^2 + 1 \times 2^1 + 1 \times 2^0 + 0 \times 2^{-1} + 1 \times 2^{-2}$
$= 8 + 0 + 2 + 1 + 0 + 0.25$
$= 11.25$

b. $1011.01_{four} = 1 \times 4^3 + 0 \times 4^2 + 1 \times 4^1 + 1 \times 4^0 + 0 \times 4^{-1} + 1 \times 4^{-2}$
$= 64 + 0 + 4 + 1 + 0 + 0.065$
$= 69.0625$

c. $1011.01_{five} = 1 \times 5^3 + 0 \times 5^2 + 1 \times 5^1 + 1 \times 5^0 + 0 \times 5^{-1} + 1 \times 5^{-2}$
$= 125 + 0 + 5 + 1 + 0 + 0.04$
$= 131.04$

Change from Base 10 to Base *b*

To see how to change from base ten to any other valid base, let's again look for a pattern:

To change from base ten to base two, group by twos.
To change from base ten to base three, group by threes.
To change from base ten to base four, group by fours.
To change from base ten to base five, group by fives.

The groupings from this pattern are summarized in Table 4.7.

TABLE 4.7

	Place-Value Chart				
Base	**Place Value**				
2	$2^5 = 32$	$2^4 = 16$	$2^3 = 8$	$2^2 = 4$	$2^1 = 2$	$2^0 = 1$
3	$3^5 = 243$	$3^4 = 81$	$3^3 = 27$	$3^2 = 9$	$3^1 = 3$	$3^0 = 1$
4	$4^5 = 1{,}024$	$4^4 = 256$	$4^3 = 64$	$4^2 = 16$	$4^1 = 4$	$4^0 = 1$
5	$5^5 = 3{,}125$	$5^4 = 625$	$5^3 = 125$	$5^2 = 25$	$5^1 = 5$	$5^0 = 1$
8	$8^5 = 32{,}768$	$8^4 = 4{,}096$	$8^3 = 512$	$8^2 = 64$	$8^1 = 8$	$8^0 = 1$
10	$10^5 = 100{,}000$	$10^4 = 10{,}000$	$10^3 = 1{,}000$	$10^2 = 100$	$10^1 = 10$	$10^0 = 1$
12	$12^5 = 248{,}832$	$12^4 = 20{,}736$	$12^3 = 1{,}728$	$12^2 = 144$	$12^1 = 12$	$12^0 = 1$

The next example shows how we can interpret this grouping process in terms of a simple division.

Example **2** **Write in base two**   **Pólya's Method**

Convert 42 to base two.

Solution   We use Pólya's problem-solving guidelines for this example.

Understand the Problem. Using Table 4.7, we see that the largest power of two smaller than 42 is 2^5 so we begin with $2^5 = 32$:

$$42 = 1 \times 2^5 + 10$$
$$10 = 0 \times 2^4 + \mathbf{10}$$
$$\mathbf{10} = 1 \times 2^3 + 2$$
$$2 = 0 \times 2^2 + 2$$
$$2 = 1 \times 2^1 + \mathbf{0}$$
$$\mathbf{0} = 0 \times 2^0$$

We could now write out 42 in expanded notation.

Devise a Plan. Instead of carrying out the steps by using Table 4.7, we will begin with 42 and carry out repeated division, saving each remainder as we go.

Carry Out the Plan. We are changing to base 2, so we do repeated division by 2:

$$\begin{array}{r} 21 \\ \hline 2)42 \end{array} \quad \text{r. } 0 \leftarrow \textbf{Save remainder.}$$

Next we need to divide 21 by 2, but instead of rewriting our work we work our way up:

$$\begin{array}{r} 10 \\ \hline 2)21 \\ \hline 2)42 \end{array} \quad \begin{array}{l} \text{r. } \mathbf{1} \leftarrow \textbf{Save all remainders.} \\ \text{r. } 0 \leftarrow \text{Save remainder.} \end{array}$$

Continue by doing repeated division.

Stop when you get a zero here.
↓

$$\begin{array}{r} 0 \\ \hline 2)\ 1 \\ \hline 2)\ 2 \\ \hline 2)\ 5 \\ \hline 2)10 \\ \hline 2)21 \\ \hline 2)42 \end{array} \quad \begin{array}{l} \text{r. } 1 \\ \text{r. } 0 \\ \text{r. } 1 \\ \text{r. } 0 \\ \text{r. } 1 \\ \text{r. } 0 \ \downarrow \end{array} \quad \textbf{Answer is found by reading down.}$$

Thus, $42 = 101010_{two}$.

Look Back. You can check by using expanded notation:

$$101010_{two} = 1 \times 2^5 + 1 \times 2^3 + 1 \times 2 = 32 + 8 + 2 = 42$$

Example **3** **Write in base b**

Write 42 in   **a.** base three   **b.** base four

Solution
a. Begin with $3^3 = 27$ (from Table 4.7):

$$\begin{array}{ll} 42 = 1 \times 3^3 + 15 & \text{or} \\ 15 = 1 \times 3^2 + 6 & \\ 6 = 2 \times 3^1 + 0 & \\ 0 = 0 \times 3^0 & \end{array} \qquad \begin{array}{r} 0 \\ \hline 3)\ 1 \\ \hline 3)\ 4 \\ \hline 3)14 \\ \hline 3)42 \end{array} \quad \begin{array}{l} \text{r. } 1 \\ \text{r. } 1 \\ \text{r. } 2 \\ \text{r. } 0 \end{array}$$

Thus, $42 = 1120_{three}$.

b. Begin with $4^2 = 16$ (from Table 4.7):

$$42 = 2 \times 4^2 + 10 \qquad \text{or}$$
$$10 = 2 \times 4^1 + 2$$
$$2 = 2 \times 4^0$$

$$\begin{array}{r} 0 \quad \text{r. } 2 \\ 4\overline{)2} \quad \text{r. } 2 \\ 4\overline{)10} \quad \text{r. } 2 \\ 4\overline{)42} \end{array}$$

Thus, $42 = 222_{four}$.

How would you change from base ten to base seven? To base eight? To base b? We see from the above pattern that we group by b's or perform repeated division by b.

Example **4** **Minimize cost**   Pólya's Method

Suppose you need to purchase 1,000 name tags and can buy them by the gross (144), the dozen (12), or individually. The name tags cost \$0.50 each, \$4.80 per dozen, and \$56.40 per gross. How should you order to minimize the cost?

Solution   We use Pólya's problem-solving guidelines for this example.

Understand the Problem. If you purchase 1,000 tags individually, the cost is $\$0.50 \times 1,000 = \$500$. This is not the least cost possible, because of the bulk discounts. We must find the maximum number of gross, the number of dozens, and then purchase the remainder individually.

Devise a Plan. We will proceed by repeated division by 12, which we recognize as equivalent to changing the number to base twelve.

Carry Out the Plan. Change 1,000 to base 12:

$$\begin{array}{r} 0 \quad \text{r. } 6 \\ 12\overline{)6} \quad \text{r. } 11, \text{ or } E \text{ in base twelve} \\ 12\overline{)83} \quad \text{r. } 4 \\ 12\overline{)1,000} \end{array}$$

Thus, $1,000 = 6E4_{twelve}$ so you must purchase 6 gross, 11 dozen, and 4 individual tags.

Look Back. The cost is $6 \times \$56.40 + 11 \times \$4.80 + 4 \times \$0.50 = \$393.20$. As you can see, this is considerably less expensive than purchasing the individual name tags.

Problem Set | **4.3**

Level **1**

1. **IN YOUR OWN WORDS**   Explain the process of changing from base eight to base ten.

2. **IN YOUR OWN WORDS**   Explain the process of changing from base sixteen to base ten.

3. **IN YOUR OWN WORDS**   Explain the process of changing from base ten to base eight.

4. **IN YOUR OWN WORDS**   Explain the process of changing from base ten to base sixteen.

5. Count the number of people in the indicated base.

a. base ten   **b.** base five
c. base three   **d.** base eight
e. base two   **f.** base nine

6. Count the number of people in the indicated base.

a. base ten **b.** base five **c.** base thirteen
d. base eight **e.** base two **f.** base twelve

In Problems 7–16, write the numbers in expanded notation.

7. 643_{eight} **8.** 5387.9_{twelve}

9. 110111.1001_{two} **10.** 5411.102_{six}

11. 64200051_{eight} **12.** 1021.221_{three}

13. 323000.2_{four} **14.** 234000_{five}

15. 3.40231_{five} **16.** 2033.1_{four}

Change the numbers in Problems 17–30 to base ten.

17. 527_{eight} **18.** 527_{twelve}

19. $25TE_{teelve}$ **20.** 1101.11_{two}

21. 431_{five} **22.** 65_{eight}

23. 1011.101_{two} **24.** 11101000110_{two}

25. 573_{twelve} **26.** 4312_{eight}

27. 2110_{three} **28.** 4312_{five}

29. 537.1_{eight} **30.** 3721_{eight}

31. a. Change 724 to base five.
　　b. Change 628 to base four.

32. a. Change 256 to base two.
　　b. Change 427 to base twelve.

33. a. Change 412 to base five.
　　b. Change 615 to base eight.

34. a. Change 5,133 to base twelve.
　　b. Change 615 to base two.

35. a. Change 512 to base two.
　　b. Change 795 to base three.

36. a. Change 52 to base three.
　　b. Change 4,731 to base twelve.

37. a. Change 602 to base eight.
　　b. Change 76 to base four.

Level 2

Use number bases to answer the questions in Problems 38–51.

38. Change 158 hours to days and hours.

39. Change 52 days to weeks and days.

40. Change 39 ounces to pounds and ounces.

41. Change 55 inches to feet and inches.

42. Change $4.59 to quarters, nickels, and pennies.

43. Change 500 to gross, dozens, and units.

44. Suppose you have two quarters, four nickels, and two pennies. Use base five to write a numeral to indicate your financial status.

45. Using only quarters, nickels, and pennies, what is the minimum number of coins needed to make $0.84?

46. Change $8.34 to the smallest number of coins consisting of quarters, nickels, and pennies.

47. A bookstore ordered 9 gross, 5 dozen, and 4 pencils. Write this number in base twelve and in base ten.

48. Change 54 months to years and months.

49. Change 44 days to weeks and days.

50. Change 49 hours to days and hours.

51. Change 29 hours to days and hours.

52. Add 5 years, 7 months to 6 years, 8 months.

53. Add 3 years, 10 months to 2 years, 5 months.

54. Add 10 ft, 7 in. to 7 ft, 10 in.

55. Add 6 ft, 8 in. to 9 ft, 5 in.

56. Add 2 gross, 3 dozen, 4 units to 5 gross, 9 dozen, 10 units.

57. Add 1 gross, 9 dozen, 7 units to 2 gross, 8 dozen, 8 units.

58. If you keep one investment for 1 year, 7 months, 11 days, and then roll it over into another investment for 1 year, 6 months, 26 days, how long was the total investment (if you assume that all months are 30 days long)?

59. If you keep one investment for 3 years, 4 months, 21 days, and then roll it over into another investment for 1 year, 6 months, 15 days, how long was the total investment (if you assume that all months are 30 days long)?

Problem Solving 3

60. HISTORICAL QUEST The *duodecimal numeration system* refers to the base twelve system, which uses the symbols 0, 1, 2, 3, 4, 5, 6, 7, 8, 9, *T, E*. Historically, a numeration system based on 12 is not new. There were 12 tribes in Israel and 12 Apostles of Christ. In Babylon, 12 was used as a base for the numeration system before it was replaced by 60. In the 18th and 19th centuries, Charles XII of Sweden and Georg Buffon (1707–1788) advocated the adoption of the base twelve system. Even today there is a Duodecimal Society of America that advocates the adoption of this system. According to the Society's literature, no one "who thought long enough—three to 17 minutes—to grasp the central idea of the duodecimal system ever failed to concede its superiority." Study the duodecimal system from 3 to 17 minutes, and comment on whether you agree with the Society's statement.

4.4 | Binary Numeration System

One of the biggest surprises of our time has been the impact of the personal computer on everyone's life. We have just completed a section on different numeration systems, and it is this study that enables us understand how computers work (see Section 4.5). In this section, we focus on one of those systems—namely, the one that uses only two symbols (*binary numerals*), called the **binary numeration system.**

Base Two

For a computer to be of use in calculating, the human mind had to invent a way of representing numbers and other symbols in terms of electronic devices. In fact, it is the simplest of electronic devices, the switch, that is at the heart of communication between machines and human beings. A switch can be only "on" or "off," as illustrated in Figure 4.3.

FIGURE 4.3 Two-state device. Light bulbs serve as a good example of two-state devices. The 1 is symbolized by "on," and the 0 is symbolized by "off."

TABLE 4.8		
Binary Numerals		
Decimal		Binary
0	↔	0
1	↔	1
2	↔	10
3	↔	11
4	↔	100
5	↔	101
⋮		⋮
8	↔	1000
9	↔	1001
⋮		⋮
15	↔	1111
16	↔	10000
⋮		⋮
31	↔	11111
32	↔	100000
⋮		⋮

We can see that it would be easy to represent the numbers "one" and "zero" by "on" and "off," but that won't get us very far. Binary numerals represent the numbers 1, 2, 3, . . . using only the symbols 1 and 0, as shown in Table 4.8. Can you see the pattern?

Recall that expanded notation provides the link between the binary numeration and decimal numeration systems, and that for a binary numeration system the grouping is by twos.

Example 1 | Base 2 to base 10

What does 1110_{two} mean? Change it to base 10.

Solution

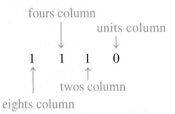

We see this means: $1 \times 8 + 1 \times 4 + 1 \times 2 + 0 = 8 + 4 + 2 = 14$

As we see in Example 1, to change a number from binary to decimal we write it in expanded notation and then carry out the arithmetic. To change a number from decimal to binary, it is necessary to find how many units (1 if odd, 0 if even), how many twos, how many fours, and so on are contained in that given number. This can be accomplished by repeated division by 2, as shown in the following example.

Example 2 | Base 10 to base 2

Change 47 to a binary numeral.

Solution

$$\begin{array}{rl} 0 & \text{r. 1} \quad \textit{There is one thirty-two.}\\ 2\overline{)\ 1} & \text{r. 0} \quad \textit{There are zero sixteens.}\\ 2\overline{)\ 2} & \text{r. 1} \quad \textit{There is one eight.}\\ 2\overline{)\ 5} & \text{r. 1} \quad \textit{There is one four.}\\ 2\overline{)11} & \text{r. 1} \quad \textit{There is one two.}\\ 2\overline{)23} & \text{r. 1} \quad \textit{There is one unit.}\\ 2\overline{)47} & \leftarrow \text{Start here} \end{array}$$

If you read down the remainders, you obtain the binary numeral 101111.

Check: 101111_{two} means

$$1 \times 32 + 0 \times 16 + 1 \times 8 + 1 \times 4 + 1 \times 2 + 1 \times 1 = 47$$

ASCII Code

In a computer, each switch, called a **bit** (from <u>bi</u>nary dig<u>it</u>), represents a 1 or 0, depending on whether it is on or off. A group of eight of these switches is lined up to form a **byte,** which can represent any number from 0 to $11111111_{two} = 255_{ten}$.

Of course, when we use computers we need to represent more than numerals. There must be a representation for every letter and symbol that we wish to use. For example, we will need to represent statements such as

$$x = 3.45 - y$$

The code that has been most commonly used is the American Standard Code for Information Interchange, developed in 1964 and called **ASCII** (pronounced "ask-key") **code.** A partial list of this code is shown in Table 4.9.

TABLE 4.9

Partial Listing of ASCII Code. The entire ASCII code has 127 symbols							
ASCII Code	**Symbol**	**ASCII Code**	**Symbol**	**ASCII Code**	**Symbol**	**ASCII Code**	**Symbol**
32	Space	53	5	67	C	95	-
33	!	54	6	68	D	96	`
34	"	55	7	69	E	97	a
35	#	56	8	70	F	98	b
⋮	⋮	57	9	⋮	⋮	⋮	⋮
48	0	58	:	90	Z	119	w
49	1	59	;	91	[	120	x
50	2	⋮	⋮	92	\	121	y
51	3	65	A	93	]	122	z
52	4	66	B	94	^	⋮	⋮

Historical NOTE

You are no doubt aware of the special numerals at the bottom of your bank checks:

0 1 2 3 4 5 6 7 8 9

These are called MICR (Magnetic Ink Character Recognition) numerals and are printed in magnetic ink so they can be read mechanically. Although they may look futuristic, modern-day computers do not use MICR, which was designed in the late 1950s, but instead use an alphabet called OCR (Optical Character Recognition):

ABCDEFGHIJKLM
NOPQRSTUVWXYZ
0123456789
.,:;=+/?'"
¥♪₦*%|&*{}

Another optical scanner is used at your supermarket to read bar codes. All packages (not only those at the supermarket) have or will have a bar code.

From this table you can find that the word *cat* would be represented by the ASCII code

99 97 116

In the machine, this would be represented by binary numerals, each one in a different byte. Thus, *cat* is represented as shown:

Symbol	ASCII Code	Binary	Byte
c	99	01100011	off on on off off off on on
a	97	01100001	off on on off off off off on
t	116	01110100	off on on on off on off off

Each of these bytes is stored at some location in the computer called its address. Each address (billions of them) stores one byte of information. To store the information 3 *cats* would require six addresses:

Address	Information	
1000	00110011	← **Code for 3**
1001	00100000	← **Code for space**
1010	01100011	← **Code for *c***
1011	01100001	← **Code for *a***
1100	01110100	← **Code for *t***
1101	01110011	← **Code for *s***

During the process of solving a problem, the storage unit will contain not only the set of instructions that specify the sequence of operations required to complete the problem, but also the input data. Notice that an address number has nothing to do with the contents of that address. The early computers stored these individual bits with switches, vacuum tubes, or cathode ray tubes, but computers today use much more sophisticated means for storing bits of information. These advances in storing information have enabled computers to use codes that use many more than eight bits. For example, a location might look like the following:

Address 00101   00011111000110110110000000001101

Binary Arithmetic

The binary numeration system has only two symbols, 0 and 1, which makes counting in the system simple. Arithmetic is particularly easy in the binary system, since the only arithmetic "facts" one needs are the following:

Addition			**Multiplication**		
+	0	1	×	0	1
0	0	1	0	0	0
1	1	10	1	0	1

You simply must remember when adding $1 + 1$ in binary that you put down 0 and "carry" the 1, as shown in the following example.

Example 3 Binary arithmetic

Carry out the following operations: **a.** $11_{two} + 101_{two}$ **b.** $11_{two} \times 101_{two}$

Solution a.
$$\begin{array}{r} 11_{two} \\ +101_{two} \\ \hline 1000_{two} \end{array}$$

Note $1 + 1 = 10$, carry 1;
$1 + 1 = 10$, carry 1 again;
$1 + 1 = 10$ for the last time.

b.
$$\begin{array}{r} 11_{two} \\ \times 101_{two} \\ \hline 11_{two} \\ 00_{two} \\ 11_{two} \\ \hline 1111_{two} \end{array}$$

Problem Set 4.4

Level 1

1. IN YOUR OWN WORDS Describe Figure 4.3 on page 154.

2. IN YOUR OWN WORDS Describe a procedure for converting a decimal number into binary representation.

What decimal number is represented by the light bulbs shown in Problems 3–6?

3.

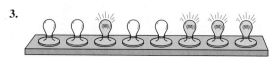

4.

5.

6.

Write each number given in Problems 7–18 as a decimal numeral.

7. 1101_{two} **8.** 1001_{two} **9.** 1011_{two}

10. 1111_{two} **11.** 11101_{two} **12.** 10111_{two}

13. 11011_{two} **14.** 11111_{two} **15.** 1100011_{two}

16. 1110111_{two} **17.** 10111000_{two} **18.** 11111111_{two}

Write each number given in Problems 19–30 as a binary numeral.

19. 13 **20.** 15 **21.** 35 **22.** 46

23. 51 **24.** 63 **25.** 64 **26.** 256

27. 128 **28.** 615 **29.** 795 **30.** 803

Level 2

Write the ASCII codes for the capital letters in Problems 31–34.

31. DO **32.** PRINT **33.** END **34.** SAVE

Write the words for the ASCII codes given in Problems 35–38.

35. 72 65 86 69 **36.** 70 85 78

37. 83 84 85 68 89 **38.** 72 65 82 68

Perform the indicated operations in Problems 39–46.

39.
$$\begin{array}{r} 11_{two} \\ + \ 10_{two} \\ \hline \end{array}$$
40.
$$\begin{array}{r} 1011_{two} \\ + \ 111_{two} \\ \hline \end{array}$$

41.
$$\begin{array}{r} 110_{two} \\ + \ 111_{two} \\ \hline \end{array}$$
42.
$$\begin{array}{r} 1101_{two} \\ + \ 1100_{two} \\ \hline \end{array}$$

43.
$$\begin{array}{r} 101_{two} \\ - \ 11_{two} \\ \hline \end{array}$$
44.
$$\begin{array}{r} 11011_{two} \\ - \ 10110_{two} \\ \hline \end{array}$$

45.
$$\begin{array}{r} 111_{two} \\ \times \ 101_{two} \\ \hline \end{array}$$
46.
$$\begin{array}{r} 10110_{two} \\ \times \ 1001_{two} \\ \hline \end{array}$$

Level 3

47. IN YOUR OWN WORDS There is a slogan among computer operators: **GIGO** (*garbage in, garbage out*). This phrase reflects the fact that a computer will do exactly what it is told to do. Consider the following humorous story.

A very large computer system has been created for the military. It was built and staffed by the best computer people in the country. "The system is now ready to answer questions," said the spokesperson for the project. A four-star general bit off the end of a cigar, looked whimsically at his comrades, and said, "Ask the machine if there will be war or peace." The machine replied: YES. "Yes *what*?" bellowed the general. The operator typed in this question, and the machine answered: YES, SIR!

In terms of what you know about logic, why did the computer correctly answer the first question YES?

48. IN YOUR OWN WORDS The octal numeration system refers to the base eight system, which uses the symbols 0, 1, 2, 3, 4, 5, 6, 7. Describe a process for converting octal numbers to decimal and decimal to octal.

Computer programmers often use octal numerals (see Problem 48) instead of binary numerals because it is easy to convert from binary to octal directly. Table 4.10 shows this conversion.

TABLE 4.10

Octal Binary Equivalence	
Octal	Binary
0	000
1	001
2	010
3	011
4	100
5	101
6	110
7	111

A binary number 1101110111 is separated into three-digit groupings by starting at the right end of the number and supplying leading zeros at the left if necessary: 001 101 110 111. The binary groups are then replaced by their octal equivalents:

$$001_{two} = 1_{eight} \qquad 101_{two} = 5_{eight}$$
$$110_{two} = 6_{eight} \qquad 111_{two} = 7_{eight}$$

and the binary number is converted to its octal equivalent: 1567. Conversely, an octal number can be expanded to a binary number using the same table of equivalents:

$$5307_{eight} = 101\ 011\ 000\ 111_{two}$$

Use this information for Problems 49–58.

Convert the numbers in Problems 49–52 to the binary system.

49. a. 5_{eight}
  b. 6_{eight}
50. a. 14_{eight}
  b. 56_{eight}
51. a. 167_{eight}
  b. 624_{eight}
52. a. 5700_{eight}
  b. 04320_{eight}

Convert the numbers in Problems 53–58 to the octal system.

53. a. 101_{two} **b.** 100_{two}
54. a. 011_{two} **b.** 001_{two}
55. $000\ 000\ 111\ 111\ 101\ 000_{two}$
56. $100\ 000\ 000\ 101\ 110\ 111_{two}$
57. $111\ 111\ 011\ 101\ 010\ 001_{two}$
58. $100\ 101\ 011\ 001\ 010\ 111_{two}$

Problem Solving 3

59. Age Finder

Table 4.11 will enable you to determine anyone's age. Just hand Table 4.11 to a person and ask that person to tell you in which column or columns their age appears. Using this information only, how can you determine the person's age?

TABLE 4.11

					Age Chart
FARMER'S MANUAL					
1	2	4	8	16	32
3	3	5	9	17	33
5	6	6	10	18	34
7	7	7	11	19	35
9	10	12	12	20	36
11	11	14	14	21	37
13	14	14	14	22	38
17	18	20	24	24	40
19	19	21	25	25	41
21	22	22	26	26	42
23	23	23	27	27	43
27	27	29	29	29	45
29	30	30	30	30	46
31	31	31	31	31	47
33	34	36	40	48	48
35	35	37	41	49	49
37	38	38	42	50	50
39	39	39	43	51	51
41	42	44	44	52	52
43	43	45	45	53	53
45	46	46	46	54	54
47	47	47	47	55	55
49	50	52	56	56	56
51	51	53	57	57	57
53	54	54	58	58	58
55	55	55	59	59	59
57	58	60	60	60	60
59	59	61	61	61	61
63	63	63	63	63	63

60. Magic Number Cards   You will need to use a pair of scissors for this problem. Construct the five cards shown here.
  Cut out those parts of the cards that are white. Ask someone to select a number between 1 and 31. Hold the cards face up on top of one another, and ask the person if his or her number is on the card. If the answer is yes, place the card on the table face up with the word *yes* at the upper left corner. If the answer is no, place the card face up with the word *no* at the upper left corner. Repeat this procedure with each card, placing the cards on top of one another on the table. After you have gone through the five cards, turn the entire stack over. The chosen number will be seen. Explain why this trick works.

Card 1
YES		
1	3	
5	7	
9	11	
13	15	
	17	19
	21	23
	25	27
	29	31
		ON

Card 2
YES		
2	3	
6	7	
10	11	
14	15	
	18	19
	22	23
	26	27
	30	31
		ON

Card 3
YES		
4	5	
6	7	
12	13	
14	15	
	20	21
	22	23
	28	29
	30	31
		ON

Card 4
YES		
8	9	
10	11	
12	13	
14	15	
	24	25
	26	27
	28	29
	30	31
		ON

"YES" is on back of this corner
Back of Card 5

Card 5
NO YES		
16	17	
18	19	
20	21	
22	23	
	24	25
	26	27
	28	29
	30	31

4.5 | History of Calculating Devices

Few people foresaw that computers would jump the boundaries of scientific and engineering communities and create a revolution in our way of life in the last half of the 20th century. No one expected to see a computer sitting on a desk, much less on desks in every type of business, large and small, and even in our homes. Today no one is ready to face the world without some knowledge of computers. What has created this change in our lives is not only the advances in technology that have made computers small and affordable, but the tremendous imagination shown in developing ways to use them. In this section we will consider the historical achievements leading to the easy availability of calculators and computers.

First Calculating Tool

The first "device" for arithmetic computations is finger counting. It has the advantages of low cost and instant availability. You are familiar with addition and subtraction on your fingers, but here is a method for multiplication by 9. Place both hands as shown at the left. To multiply 9 × 3, simply bend the fourth finger from the left as shown here.

"It says, 'don't fold, spindle, or mutilate.'"

Historical NOTE

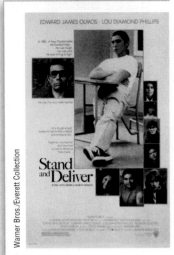

In the movie *Stand and Deliver* (Warner Brothers, 1988), the teacher Jaime Escalante (played by actor Edward James Olmos) confronts a gang member, Chuco:

Escalante: "Oh. You know the times tables?"
Chuco: "I know the ones . . . twos . . . three.
[On "three" Chuco flips the bird to Escalante.]
Escalante: "Finger man. I heard about you. Are you The Finger Man? I'm the Finger Man, too. Do you know what I can do? I know how to multiply by nine! Nine times three. What you got? Twenty-seven. Six times nine. One, two, three, four, five, six, seven eight. What you got? Seventy two."*

* *Stand and Deliver,* Warner Brothers, 1988, p. 9.

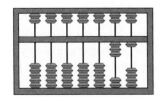

FIGURE 4.5 Abacus: The number shown is 31.

The answer is read as 27, the bent finger serving to distinguish between the first and second digits in the answer. (See the Historical Note about "the finger man" in the margin.)

What about 9×36? Separate the third and fourth fingers from the left (as shown at the left), since 36 has 3 tens. Next, bend the sixth finger from the left, since 36 has 6 units. Now the answer can be read directly from the fingers:

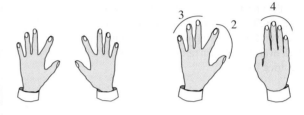

The answer is 324.

Aristophanes devised a complicated finger-calculating system in about 500 B.C., but it was very difficult to learn. Figure 4.4 shows an illustration from a manual published about two thousand years later, in 1520.

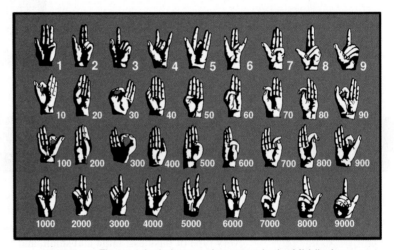

FIGURE 4.4 Finger calculation was important in the Middle Ages.

Early Calculating Devices

Numbers can be represented by stones, slip knots, or beads on a string. These devices evolved into the abacus, as shown in Figure 4.5 (see also Problems 44–49 of Section 4.2).

Abaci (plural of abacus) were used thousands of years ago, and are still used today. In the hands of an expert, they even rival some calculators for speed in performing certain calculations. An abacus consists of rods that contain sliding beads, four or five in the lower section, and two in the upper that equal one in the lower section of the next higher denomination. The abacus is useful today for teaching mathematics to youngsters. One can actually see the "carry" in addition and the "borrowing" in subtraction.

In the early 1600s, John Napier invented a device similar to a multiplication table with movable parts, as shown in Figure 4.6. These rods are known as Napier's rods or Napier's bones.

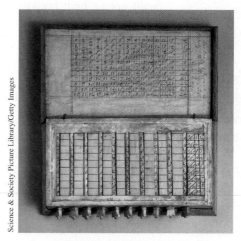

FIGURE 4.6　Napier's rods (1617)

A device used for many years was a slide rule (see Figure 4.7), which was also invented by Napier. The answers given by a slide rule are only visual approximations and do not have the precision that is often required.

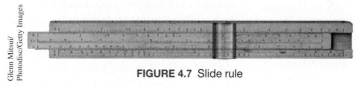

FIGURE 4.7　Slide rule

Mechanical Calculators

The 17th century saw the beginnings of calculating machines. When Blaise Pascal (1623–1662) was 19, he began to develop a machine to add long columns of figures (see Figure 4.8a). He built several versions, and since all proved to be unreliable, he considered this project a failure; but the machine introduced basic principles that are used in modern calculators.

　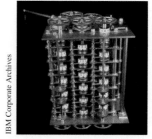

a. Pascal's calculator (1642)　　　**b.** Babbage's calculating machine

FIGURE 4.8　Early mechanical calculators

The next advance in mechanical calculators came from Germany in about 1672, when the great mathematician Gottfried Leibniz studied Pascal's calculators, improved them, and drew up plans for a mechanical calculator. In 1695 a machine was finally built, but this calculator also proved to be unreliable.

In the 19th century, an eccentric Englishman, Charles Babbage, developed plans for a grandiose calculating machine, called a "difference engine," with thousands of gears, ratchets, and counters (see Figure 4.8b).

Historical NOTE

Karl Smith library

**Charles Babbage
(1792–1871)**

Babbage continued for 40 years trying to build his "engines." He was constantly seeking financial help to subsidize his projects. However, his eccentricities often got him into trouble. For example, he hated organ-grinders and other street musicians, and for this reason people ridiculed him. He was a no-nonsense man and would not tolerate inaccuracy. Had he been willing to compromise his ideas of perfection with the technical skill available, he might have developed a workable machine.

Four years later, in 1826, even though Babbage had still not built his difference engine, he began an even more elaborate project—the building of what he called an "analytic engine." This machine was capable of an accuracy to 20 decimal places, but it, too, could not be built because the technical knowledge to build it was not far enough advanced. Much later, International Business Machines (IBM) built both the difference and analytic engines based on Babbage's design, but using modern technology, and they work perfectly.

Hand-Held Calculators

In the past few years pocket calculators have been one of the fastest-growing items in the United States. There are probably two reasons for this increase in popularity. Most people (including mathematicians!) don't like to do arithmetic, and a good calculator is very inexpensive. It is assumed that you have access to a calculator to use with this book, and the basics of their use were introduced in Chapter 1.

First Computers

The devices discussed thus far in this section would all be classified as calculators. With some of the new programmable calculators, the distinction between a calculator and a computer is less well defined than it was in the past. Stimulated by the need to make ballistics calculations and to break secret codes during World War II, researchers made great advances in calculating machines. During the late 1930s and early 1940s, John Vincent Atanasoff, a physics professor at Iowa State University, and his graduate student Clifford E. Berry built an electronic digital computer, but could not get it to work properly. The first working electronic computers were invented by J. Presper Eckert and John W. Mauchly at the University of Pennsylvania. All computers now in use derive from the original work they did between 1942 and 1946. They built the first fully electronic digital computer, called ENIAC (Electronic Numerical Integrator and Calculator), as shown in Figure 4.9.

Jerry Cooke/Historical Premium/Corbis

FIGURE 4.9 ENIAC computer (1946)

The ENIAC filled 30 feet × 50 feet of floor space, weighed 30 tons, had 18,000 vacuum tubes, cost $487,000 to build, and used enough electricity to operate three 150-kilowatt radio stations. Vacuum tubes generated a lot of heat, and the room where computers were installed had to be kept at a carefully monitored temperature and humidity level; still, the tubes had a substantial failure rate, and in order to "program" the computer, many switches and wires had to be adjusted.

There is some controversy over who actually invented the first computer. In the early 1970s there was a lengthy court case over a patent dispute between Sperry Rand (who had acquired the ENIAC patent) and Honeywell Corporation, who represented those who originally worked on the ENIAC project. The judge in that case ruled that "between 1937 and 1942, Atanasoff . . . developed and built an automatic electronic digital computer for solving large systems of linear equations . . . . Eckert and Mauchly did not themselves invent the automatic electronic digital computer, but instead derived that subject matter from one Dr. John Vincent Atanasoff." On the other hand, others believe that a court of law is not the place for deciding questions about the history of science and give Eckert and Mauchly the honor of inventing the first computer. In 1980, the Association for Computing Machinery honored them as founders of the computer industry.

The UNIVAC I, built in 1951 by the builders of the ENIAC, became the first commercially available computer. Unlike the ENIAC, it could handle alphabetic data as well as numeric data. The invention of the transistor in 1947 and solid-state devices in the 1950s provided the technology for smaller, faster, more reliable machines.

Present-Day Computers

In 1958, Seymour Cray developed the first **supercomputer,** sometimes called the Cray computer, which could handle at least 10 million instructions per second. It is used in major scientific research and military defense installations. The newest version of this computer, the X-1, operates a million times faster than the first supercomputer. It is difficult to keep track of the world's fastest computer, so if you want up-to-date information, you should check the *Top500 Supercomputer Sites* (**www.top500.org/lists**). One of the fastest, "The Roadrunner," is an IBM supercomputer at Los Alamos National Laboratory (Figure 4.10). In June 2008 it was declared the world's fastest computer. It is the first computer to break the pentaflop barrier. One pentaflop is one quadrillion operations per second.

> Keep in mind some of the deception that goes on in the name of technology. People sometimes think that "if a calculator or a computer did it, then it must be correct." This is false, and the fact that a "machine" is involved has no effect on the validity of the results. Have you ever been told that "The computer requires...," or "There is nothing we can do; it is done by computer," or "The computer won't permit it." What they really mean is they don't want to instruct the *programmer* to do it.

Courtesy of Los Alamos National Laboratory

FIGURE 4.10 "Roadrunner" supercomputer

Throughout the 1960s and early 1970s, computers continued to become faster and more powerful; they became a part of the business world. The "user" often never saw the machines. A "job" would be submitted to be "batch processed" by someone trained to run the computer. The notion of "time sharing" developed, so that the computer could handle more than one job, apparently simultaneously, by switching quickly from one job to another. Using a computer at this time was often frustrating; an incomplete job would be returned because of an error and the user would have to resubmit the job. It often took days to complete the project.

The large computers, known as mainframes, were followed by the **minicomputers,** which took up less than 3 cubic feet of space and no longer required a controlled atmosphere. These were still used by many people at the same time, though often directly through the use of terminals.

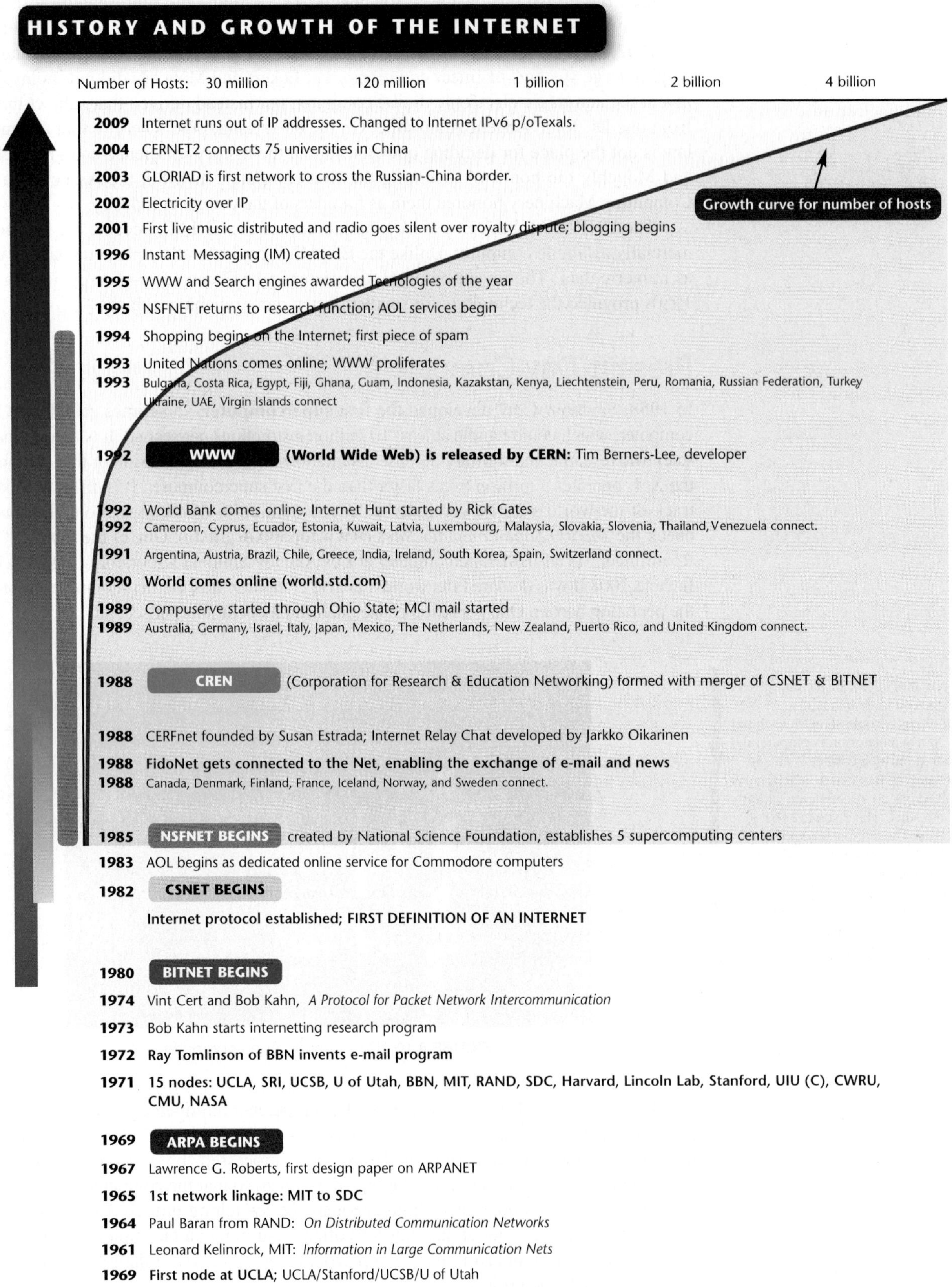

HISTORY AND GROWTH OF THE INTERNET

Number of Hosts: 30 million   120 million   1 billion   2 billion   4 billion

Growth curve for number of hosts

2009 Internet runs out of IP addresses. Changed to Internet IPv6 p/oTexals.

2004 CERNET2 connects 75 universities in China

2003 GLORIAD is first network to cross the Russian-China border.

2002 Electricity over IP

2001 First live music distributed and radio goes silent over royalty dispute; blogging begins

1996 Instant Messaging (IM) created

1995 WWW and Search engines awarded Tecnologies of the year

1995 NSFNET returns to research function; AOL services begin

1994 Shopping begins on the Internet; first piece of spam

1993 United Nations comes online; WWW proliferates

1993 Bulgaria, Costa Rica, Egypt, Fiji, Ghana, Guam, Indonesia, Kazakstan, Kenya, Liechtenstein, Peru, Romania, Russian Federation, Turkey Ukraine, UAE, Virgin Islands connect

1992 **WWW** **(World Wide Web) is released by CERN:** Tim Berners-Lee, developer

1992 World Bank comes online; Internet Hunt started by Rick Gates

1992 Cameroon, Cyprus, Ecuador, Estonia, Kuwait, Latvia, Luxembourg, Malaysia, Slovakia, Slovenia, Thailand, Venezuela connect.

1991 Argentina, Austria, Brazil, Chile, Greece, India, Ireland, South Korea, Spain, Switzerland connect.

1990 **World comes online (world.std.com)**

1989 Compuserve started through Ohio State; MCI mail started

1989 Australia, Germany, Israel, Italy, Japan, Mexico, The Netherlands, New Zealand, Puerto Rico, and United Kingdom connect.

1988 **CREN** (Corporation for Research & Education Networking) formed with merger of CSNET & BITNET

1988 CERFnet founded by Susan Estrada; Internet Relay Chat developed by Jarkko Oikarinen

1988 **FidoNet gets connected to the Net, enabling the exchange of e-mail and news**

1988 Canada, Denmark, Finland, France, Iceland, Norway, and Sweden connect.

1985 **NSFNET BEGINS** created by National Science Foundation, establishes 5 supercomputing centers

1983 AOL begins as dedicated online service for Commodore computers

1982 **CSNET BEGINS**

Internet protocol established; FIRST DEFINITION OF AN INTERNET

1980 **BITNET BEGINS**

1974 Vint Cert and Bob Kahn, *A Protocol for Packet Network Intercommunication*

1973 Bob Kahn starts internetting research program

1972 **Ray Tomlinson of BBN invents e-mail program**

1971 **15 nodes: UCLA, SRI, UCSB, U of Utah, BBN, MIT, RAND, SDC, Harvard, Lincoln Lab, Stanford, UIU (C), CWRU, CMU, NASA**

1969 **ARPA BEGINS**

1967 Lawrence G. Roberts, first design paper on ARPANET

1965 **1st network linkage: MIT to SDC**

1964 Paul Baran from RAND: *On Distributed Communication Networks*

1961 Leonard Kelinrock, MIT: *Information in Large Communication Nets*

1969 **First node at UCLA;** UCLA/Stanford/UCSB/U of Utah

FIGURE 4.11 The growth and history of the Internet

Historical NOTE

Karl Smith library

**Steven Jobs
(1955–)**

Steven Jobs cofounded Apple Computer, Inc. and codesigned the Apple II computer. He also led the development of the Macintosh computer and is credited with helping to define key technology trends in the personal computer industry. In October 1985 he (along with five others) founded NeXT, Inc. to develop innovative, personal, and affordable computer solutions for the 1990s. In 1997 he returned to Apple Computers.

The term **personal computer** was first used by Steward Brand in *The Whole Earth Catalog* (1968), which, oddly enough, was before any such computers existed. In 1976, Steven Jobs and Stephen Wozniak designed and built the Apple II, the first personal computer to be commercially successful. This computer proved extremely popular. Small businesses could afford to purchase these machines that, with a printer attached, took up only about twice the space of a typewriter. Technology "buffs" could purchase their own computers. This computer was designed so that innovations that would improve or enlarge the scope of performance of the machine could be added with relatively little difficulty. Owners could open up their machines and install new devices. One such device, the **mouse,** was introduced to the world by Douglas Engelbart in 1968. This increased contact between the user and the machine produced an atmosphere of tremendous creativity and innovation. In many cases, individual owners brought their own machines to work to introduce their superiors to the potential usefulness of the personal computer.

In the fast-moving technology of the computer world, the invention of language to describe it changes as fast as teenage slang. Today there are battery-powered **laptops,** weighing about 2 pounds, that are more powerful than the huge ENIAC. Today, laptop, iPods, and smart phones are changing the nature of computers every year. The next trend linked together several personal computers so that they could share software and different users can easily access the same documents. Such arrangements are called LANs, or **local area networks**. The most widely known **networks** are the **Internet,** which was first described by Paul Baran of the RAND Corp. in 1962, and the **World Wide Web** (www), developed by Tim Berners-Lee based on Baran's ideas and released in 1992. The growth of the Internet and the World Wide Web could not have been predicted. Figure 4.11 shows some of the significant events in the growth of the Internet from 1962 to the present. In 1962 there were 4 hosts, and in 1992 there were 727,000. By 2010 there were almost 2 billion.

Computer Hardware

The machine, or computer itself, is referred to as **hardware.** A personal computer might sell for under $500, but many extras, called **peripherals,** can increase the price by several thousand dollars. In this section we will familiarize you with the various components of a typical personal computer system that you might find in a home, classroom, or office. We will also discuss the link between the electronic signals of the computer and the human mind that enables us to harness the power and speed of the computer. The basic parts of a computer are shown in Figure 4.12.

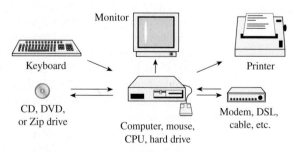

FIGURE 4.12 Input and output with a computer

We will focus our attention in this section on the terminology you will need to know to purchase a computer. The central unit of a computer contains a small microchip, called the central processing unit, where all processing takes place, and the computer memory, where the information is stored while the computer is completing a given task. The microchip and the unit that contains this chip are collectively referred to as the CPU (Central Processing Unit). Almost daily new and more popular chips are announced.

Although you do not need to understand electronic circuitry, a simple conceptual model of the inside of a computer can be very useful. The memory of a computer can be thought of as rows and rows of little storage boxes, each one with an address—yes, an address, just like your address. These boxes are of two types: **ROM** (read-only memory) and **RAM** (random-access memory). ROM is very special: You can never change what is in it and it is not erased when the computer is turned off. It is programmed by the manufacturer to contain routines for certain processes that the computer must always use. The CPU can fetch information from it, but cannot send new results back. It is well protected because it is at the heart of the machine's capabilities. On the other hand, programs and data are stored in RAM while the computer is completing a task. If the power to the computer is interrupted for any time, however brief, everything in RAM is lost forever.

Your computer will also have memory. When referring to memory, we used to talk in terms of a thousand bytes, called a *kilobyte* (KB or K), or in terms of a thousand kilobytes, called a *megabyte* (MB or MEG). However, today, we refer to *gigabytes* (GB; one gigabyte is about one billion bytes) or *terabytes* (TB; one trillion bytes). Your computer will come with 512 MEG or 1 GB (or even more) of RAM.

You will also need a device that reads and stores data, called a **hard drive,** which is used as additional memory to store the programs you purchase. Hard drives in sizes of 40, 80, 160, or more gigabytes are commonly installed in **microcomputers.**

To use a computer, you will need **input** and **output** devices. A **keyboard** and a mouse are the most commonly used input devices. The most common output devices are a **monitor** and a **printer.** One of the monitor's main functions is to enable you to keep track of what is going on in the computer. The letters or pictures seen on a monitor are displayed by little dots called **pixels,** just as they are on a television. As the number of dots is increased, we say that the **resolution** of the monitor increases. The better the resolution, the easier on your eyes.

Computer Software

Communication with computers has become more sophisticated, and when a computer is purchased it often comes with several *programs* that allow the user to communicate with the computer. A **computer program** is a set of step-by-step directions that instruct a computer how to carry out a certain task. Programs that allow the user to carry out particular tasks on a computer are referred to as **software.**

The business world has welcomed the advances made possible by the use of personal computers and powerful programs, known as **software packages,** that enable the computer to do many diverse tasks. The most widely used computer applications are **word processing, database management, gaming,** and **spreadsheets.**

There are several sources for obtaining software. You can buy it in a store or you can download *shareware,* which is try-before-you-buy software. Some software is available on the Internet free of charge (*freeware* or *public domain software*). The most widely used software is from Microsoft, which owns *Windows*®, *Word*®, *Internet Explorer*®, and *Excel*®. Bill Gates (1956-) is the cofounder of Microsoft and is known as the world's wealthiest person. In 1975, Ed Roberts built the first personal computer (the Altair 8800) in Albuquerque, New Mexico; it was featured on the cover of *Popular Electronics,* and was named because the word *Altair* was used in an episode of *Star Trek.* Bill Gates and his partner, Paul Allen, were students at Harvard at the time, and traveled to New Mexico to look at this new machine. The story about the beginning of the huge Microsoft empire is told by Stephen Segaller in *Nerds 2.0.1*: "He [Bill Gates] kept postponing and postponing actually writing the code [for the *Altair*]. He said, 'I know how to write it, I have a design in my head, I'll get it done, don't worry about it, Paul.' Four days before he was due to go back to Harvard he checked into a hotel and he was incommunicado for three days. Bill came back three or four days later with this huge sheet of paper. He'd written 4K of code in three days, and typed the whole thing in, got it working, and went back to school, just barely. It was really one of the most amazing displays of programming I've ever seen."

Uses of Computers

Today, computers are used for a variety of purposes (see Figure 4.13).

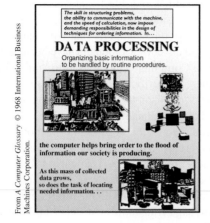

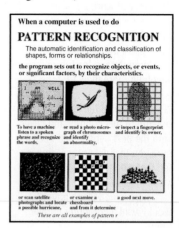

FIGURE 4.13 Common computer uses

Data processing involves large collections of data on which relatively few calculations need to be made. For example, a credit card company could use the system to keep track of the balances of its customers' accounts. **Information retrieval** involves locating and displaying material from a description of its content. For example, a police department may want information about a suspect, or a realtor may want listings that meet the criteria of a client. **Pattern recognition** is the identification and classification of shapes, forms, or relationships. For example, a chess-playing program will examine the chessboard carefully to determine a good next move. The computer carries out these tasks by accepting information, performing mathematical and logical operations with the information, and then supplying the results of the operations as new information. Finally, a computer can be used for **simulation** of a real situation. For example, a prototype of a new airplane can be tested under various circumstances to determine its limitations.

By using a **modem,** installed internally or attached as a peripheral, and a program called a **communications package,** a computer can transfer information to and from other computers directly, via satellite, a fiber optic line, or even over a phone line. If you send software or data from your computer to another, this is called **uploading**; if you get software or data from another computer, this is called **downloading.** Networks have been designed specifically to enable people to send messages to each other via computer. Such networks, known as *electronic mail* or **e-mail,** have become commonplace. Individuals or user groups set up networks to share information about a common interest. The more common ones are **Facebook** and **Twitter,** but many people use **bulletin boards** or **chat rooms** to communicate with others **online** (connected to a computer network).

Misuses of Computers

Most companies and bulletin boards require a **password** or special procedures to be able to go online with their computers. A major problem for certain computer installations has been to make sure that it is impossible for the wrong person to get into their computer. There have been many instances where people have "broken into" computers just as a safecracker breaks into a safe. They have been clever enough to discover how to access the system. A few of these people have been caught and prosecuted. A security break of this nature can be very serious. By transferring funds in a bank computer, someone can steal money. By breaking into a military installation, a spy could steal secrets or commit sabotage. By breaking into a university's computers, one could change grades or records. **Computer abuse** includes *illegal use* (as described above, but also includes illegal copying of programs, called *pirating*) and abuse caused by *ignorance*. Ignorance of computers leads to the assumption that the output data are always correct or that the use of a

I have a spelling checker.
It came with my PC.
It clearly marks four my
revue Mistakes
I cannot sea.

I've run this poem threw it.
I'm sure your pleased
* two no*
Its letter perfect in
* its weigh,*
My checker tolled me sew.

My thanks to John Martin of Santa Rosa Junior College for this quotation.

Historical **NOTE**

Famous last words. . .
"I think there is a world market for maybe five computers."
Thomas Watson, chairman of IBM, 1943.

"Computers in the future may weigh no more than 1.5 tons."
Popular Mechanics, 1949.

"I have traveled the length and breadth of this country and talked with the best people, and I can assure you that data processing is a fad that won't last out the year."
A Prentice Hall editor, 1957.

"There is no reason anyone would want a computer in their home."
Ken Olson, president, chairman, and founder of Digital Equipment Corp., 1977

computer is beyond one's comprehension. Breaking into computers, called *hacking,* has become a subculture in what is known as the *computer underground.*

With the tremendous advances in computer technology, financial investors and analysts can simulate various possible outcomes and select strategies that best suit their goals. Designers and engineers use computer-aided design, known as CAD. Using computer graphics, sculptors can model their work on a computer that will rotate the image so that the proposed work can be viewed from all sides. Computers have been used to help people with disabilities learn certain skills. Computers can be used to "talk" for those without speech. For those without sight, there are artificial vision systems that translate a visual image to a tactile "image" that can be "seen" on their back. A person who has never seen a candle burn can learn the shape and dynamic quality of a flame.

Along with the advances in computer technology came new courses, and even new departments, in colleges and universities. New specialties developed involving the design, use, and impact of computers on our society. The feats of the computer captured the imagination of some who thought that a computer could be designed to be as intelligent as a human mind, and the field known as **artificial intelligence** was born. Many researchers worked on natural language translation with very high hopes initially, but the progress has been slow. Progress continues in "expert systems," programs that involve the sophisticated strategy of pushing the computer ever closer to humanlike capabilities. The quest has produced powerful uses for computers.

What does the future hold? Earlier editions of this book attempted to make predictions of the future, and in every case those predictions became outdated during the life of the edition. Instead of making a prediction, we invite you to visit a computer store.

Problem Set 4.5

Level 1

1. **IN YOUR OWN WORDS** Describe some of the computing devices that were used before the invention of the electronic computer.

2. **IN YOUR OWN WORDS** Distinguish between hardware and software.

3. **IN YOUR OWN WORDS** What does the CPU in a computer do?

4. **IN YOUR OWN WORDS** What are input and output devices?

5. **IN YOUR OWN WORDS** What is the difference between ROM and RAM?

6. **IN YOUR OWN WORDS** Give at least one example of each of the indicated computer functions: data processing, information retrieval, pattern recognition, and simulation.

7. **IN YOUR OWN WORDS** What are five common uses for a computer at home? Describe the purpose of each.

8. **IN YOUR OWN WORDS** What are five common uses for a computer in the office? Describe the purpose of each.

9. **IN YOUR OWN WORDS** It has been said that "computers influence our lives increasingly every year, and the trend will continue." Do you see this as a benefit or a detriment to humanity? Explain your reasons.

10. **IN YOUR OWN WORDS** A heated controversy rages about the possibility of a computer actually thinking. Do you believe that is possible? Do you think a computer can eventually be taught to be truly creative?

Use finger multiplication to do the calculations shown in Problems 11–14.

11. **a.** 3×9	**b.** 7×9	**c.** 6×9
12. **a.** 5×9	**b.** 8×9	**c.** 9×9
13. **a.** 27×9	**b.** 48×9	**c.** 56×9
14. **a.** 35×9	**b.** 47×9	**c.** 68×9

15. Arrange the development of the following computers in chronological order of their first appearance.

Altair, Cray, ENIAC, Apple, and UNIVAC

HISTORICAL QUEST *Many names were mentioned in this section. Give a brief description for each person named in Problems 16–33, telling each person contributed to the development of computers.*

16. Paul Allen	17. Aristophanes
18. John Atanasoff	19. Charles Babbage
20. Paul Baran	21. Clifford Berry
22. Steward Brand	23. Seymour Cray
24. J. Presper Eckert	25. Douglas Engelbart
26. Bill Gates	27. Steven Jobs
28. Tim Berners-Lee	29. Gottfried Leibniz
30. John Mauchly	31. John Napier
32. Blaise Pascal	33. Stephen Wozniak

Level 2

Problems 34–45 list a specific task. Decide whether a computer should be used, and, if so, is it because of the computer's speed, complicated computations, repetition, or some other reason?

34. Guiding a missile to its target

35. Controlling the docking of a spacecraft

36. Tabulating averages for a teacher's classes

37. Identifying the marital status indicated on tax returns reporting income more than $100,000

38. Sorting the mail according to zip code

39. Assembling automobiles on an assembly line

40. Turning on and off the lights in a public office building

41. Typing and printing a term paper

42. Teaching a person to solve quadratic equations

43. Teaching a person to play the piano

44. Teaching a person how to repair a flat tire

45. Compiling a weather report and forecast

Problems 46–50 are multiple-choice questions designed to get you to begin thinking about computer use and abuse.

46. As part of a patient monitoring system in a local hospital, the computer (a) constantly monitors the patient's pulse and respiration rates; (b) sounds an alarm if the rates fall outside a preset range; and (c) notifies the nearest nursing station if either or both of its systems fail to function properly. The computer's capabilities are suited to these tasks because the tasks call for:
A. Speed
B. Repetition
C. Program modification
D. Simulation
E. Data processing

47. Computers are capable of organizing large amounts of data. Which of the following situations would call for a computer having that capability?
A Printing the names and addresses of all employees
B. Turning lights on and off at specified times
C. Running a statistical analysis of family incomes as taken from the 2010 census
D. Monitoring the temperature and humidity in a large office building
E. All of the above

48. Karlin Publishing purchased a program for its computer that would enable it to streamline the printing process and decrease the time it takes to print each page. The program was too fast for Karlin's printer, so the company hired Shannon Foley to eliminate the problem. The human function required by Shannon for this job was:
A. Performing calculations
B. Modifying the hardware
C. Interpreting output
D. Modifying the software
E. None of the above

49. Melissa knew that if she flunked mathematics she would be ineligible to play in her team's last five basketball games. When she found out that one of her classmates had been able to change someone else's grade on the school computer, she had this same person change her math grade from F to C. This example of computer abuse was:
A. Invasion of privacy
B. Stealing funds
C. Pirating software
D. Falsifying information
E. Stealing information

50. One example of computer abuse is categorized as assuming that computers are largely incomprehensible. Which of the following situations would indicate this abuse?
A. A secretary refuses promotion because he will have to interact with the company's computer.
B. A clerk inputs the wrong information into the computer to defraud the company.
C. A student changes the grade of a classmate.
D. A hacker (computer hobbyist) copies a copyrighted program for three of her friends.
E. A credit card customer wants to receive his bill on the 25th of the month instead of on the 10th, but when he calls to have it changed he is told that there is nothing that can be done since that is the way the computer was programmed.

Level 3

51. IN YOUR OWN WORDS   There is a great deal of concern today about **invasion of privacy** by computer. With more and more information about all of us being kept in computerized databases, there is the increasing possibility that a computer may be used to invade our privacy. Discuss this issue.

52. IN YOUR OWN WORDS   Have you ever been told that something cannot be changed because "that is the way the computer does it"? Discuss this issue.

53. IN YOUR OWN WORDS   Have you ever been told that something is right because the computer did it? For example, suppose I have my computer print, THE VALUE OF π IS 3.141592. Is this statement necessarily correct? What confidence do you place in computer results? What confidence do you think any of us should place in computer results? Discuss.

54. IN YOUR OWN WORDS   Definitions for "thinking" are given. *According to each of the given definitions*, answer the question, "Can computers think?"
a. To remember
b. To subject to the process of logical thought
c. To form a mental picture of
d. To perceive or recognize
e. To have feeling or consideration for
f. To create or devise
g. To have the ability to learn

55. IN YOUR OWN WORDS   What do you mean by "thinking" and "reasoning"? Try to formulate these ideas as clearly as possible, and then discuss the following question: "Can computers think or reason?"

56. IN YOUR OWN WORDS   In his book *Future Shock*, Alvin Toffler divided humanity's time on earth into 800 lifetimes. The 800th lifetime, in which we now live, has produced more knowledge than the previous 799 combined, and this has been made possible by computers. How have computers made this possible? What impact has this had on our lives? Do people actually know more today because they have ready access to vast amounts of knowledge?

57. **IN YOUR OWN WORDS**   In his article "Toward an Intelligence Beyond Man's" (*Time*, February 20, 1978), Robert Jastrow claimed that by the 1990s computer intelligence would match that of the human brain. He quoted Dartmouth President Emeritus John Kemeny as saying we would "see the ultimate relation between man and computer as a symbiotic union of two living species, each completely dependent on the other for survival." He called the computer a new form of life. Now, with the advantage of a historical perspective, do you think his predictions came to pass? Now that we have completed the decade of the 1990s, and twenty years of perspecitve, do you agree or disagree with the hypothesis that a computer could be a "new form of life"?

Problem Solving 3

58. **IN YOUR OWN WORDS**   One of the things that can go wrong when we use a computer is due to *machine error*. Such errors can be caused by an electrical surge or a power failure. Talk to some users of computers and gather some anecdotal information about computer problems due to machine errors.

59. **IN YOUR OWN WORDS**   It is possible that a computer can make a mistake due to a *programming error*. There are three types of programming errors: *syntax errors, run-time errors,* and *logic errors*. Do some research and write a paragraph about each of these types of programming error.

60. **IN YOUR OWN WORDS**   One category of possible computer errors includes *data errors* or *input errors*. What is meant by this type of error?

4.6   CHAPTER SUMMARY

The availability of low-cost calculators, computers, and related new technology has already dramatically changed the nature of business, industry, government, sciences, and social sciences.

NCTM STANDARDS

Important Ideas

Properties of numeration systems [4.1]
Decimal numeration system [4.2]
Converting between numeration systems [4.3]
Use a binary numeration system to store "data" in a computer [4.4]
The history of calculating devices [4.5]
Distinguish between computer hardware and software [4.5]

Take some time getting ready to work the review problems in this section. First review these important ideas. Look back at the definition and property boxes in this chapter. If you look online, you will find a list of important terms introduced, as well as the types of problems that were introduced in this chapter. You will maximize your understanding of this chapter by working the problems in this section only after you have studied the material.

You will find some review help online at **www.mathnature.com**. There are links giving general test help in studying for a mathematics examination, as well as specific help for reviewing this chapter.

Chapter **4** Review Questions

1. **IN YOUR OWN WORDS** What do we mean when we say that a numeration system is positional? Give examples.

2. **IN YOUR OWN WORDS** Is addition easier in a positional system or in a grouping system? Discuss and show examples.

3. **IN YOUR OWN WORDS** What are some of the characteristics of the Hindu-Arabic numeration system?

4. **IN YOUR OWN WORDS** Briefly discuss some of the events leading up to the invention of the computer.

5. **IN YOUR OWN WORDS** Discuss some computer abuses.

6. **IN YOUR OWN WORDS** Briefly describe each of the given computer terms.
   - **a.** hardware
   - **b.** software
   - **c.** word processing
   - **d.** network
   - **e.** e-mail
   - **f.** RAM
   - **g.** computer bulletin board
   - **h.** hard drive

Write the numbers given in Problems 7–10 in expanded notation.

7. one billion

8. 436.20001

9. 523_{eight}

10. 1001110_{two}

11. Write $4 \times 10^6 + 2 \times 10^4 + 5 \times 10^0 + 6 \times 10^{-1} + 2 \times 10^{-2}$ in decimal notation.

Write the numbers in Problems 12–15 in base ten.

12. 11101_{two}

13. 1111011_{two}

14. 122_{three}

15. 821_{twelve}

Write the numbers in Problems 16–19 in base two.

16. 12

17. 52

18. 2007

19. one million

20. **a.** Write 1,331 in base twelve.
    b. Write 100 in base five.

BOOK REPORTS

Write a 500-word report on one of these books:

How Computers Work, Ron White (Emeryville, CA: Ziff-Davis Press, 1993).

Nerds 2.0.1, Stephen Segaller (New York: TV Books, 1998).

Group RESEARCH PROJECTS

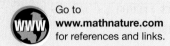

Go to
www.mathnature.com
for references and links.

Working in small groups is typical of most work environments, and learning to work with others to communicate specific ideas is an important skill. Work with three or four other students to submit a single report based on each of the following questions.

G11. Invent an original numeration system.

G12. Organize a debate. One side represents the algorists and the other side the abacists. The year is 1400. Debate the merits of the Roman numeration system and the Hindu-Arabic numeration system.

REFERENCE: Barbara E. Reynolds, "The Algorists vs The Abacists: An Ancient Controversy on the Use of Calculators," *The College Mathematics Journal,* Vol. 24, No. 3, May 1993, pp. 218–223. Includes additional references.

G13. Organize a debate. The issue: "Resolved: Computers can think."

G14. In a now famous paper, Alan Turing asked, "What would we ask a computer to do before we would say that it could think?" In the 1950s Turing devised a test for "thinking" that is now known as the **Turing test**. Dr. Hugh Loebner, a New York philanthropist, has offered $100,000 for the first machine that fools a judge into thinking it is a person. In 1991, the Computer Museum in Boston held a contest in which 10 judges at the museum held conversations on terminals with eight respondents around the world, including six computers and two humans. The conversations of about 15 minutes each were limited to particular subjects, such as wine, fishing, clothing, and Shakespeare, but in a true Turing test, the questions could involve any topic. Work as a group to decide the questions you would ask. Do you think a computer will ever be able to pass the test?

REFERENCES: Betsy Carpenter, "Will Machines Ever Think?" *U.S. News & World Report,* October 17, 1988, pp. 64–65.

Stanley Wellborn, "Machines That Think," *U.S. News & World Report,* December 5, 1983, pp. 59–62.

G15. Construct an exhibit on ancient computing methods. Some suggestions for your exhibit are charts of sample computations by ancient methods, pebbles, tally sticks, tally marks in sand, Roman number computations, abaci, Napier's bones, and old computing devices. You should consider answering the following questions as part of your exhibit: How do you multiply with Roman numerals? What is the scratch system? What is the lattice method of computation? What changes in our methods of long multiplication and long division have taken place over the years? How did the old computing machines work? Who invented the slide rule?

Individual　RESEARCH PROJECTS

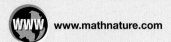

Learning to use sources outside your classroom and textbook is an important skill, and here are some ideas for extending some of the ideas in this chapter. You can find references to these projects in a library or at **www.mathnature.com.**

PROJECT 4.1 Write a paper discussing the Egyptian method of multiplication.

PROJECT 4.2 What are some of the significant events in the development of mathematics? Who are some of the famous people who have contributed to mathematical knowledge?

PROJECT 4.3 The people from the Long Lost Land had only the following four symbols:

□, △, ★, ○

Write out the first 20 numbers. Find

a. □ + ★　　　**b.** ★ × ★

(Use your imagination to invent a system to answer these questions.)

PROJECT 4.4 Is it possible to have a numeration system with a base that is negative?

PROJECT 4.5 "I became operational at the HAL Plant in Urbana, Ill., on January 12, 1997," the computer HAL declares in Arthur C. Clarke's 1968 novel, *2001: A Space Odyssey*. Now that time has passed and many advances have been made in computer technology since 1968, write a paper showing the similarities and differences between HAL and the computers of today.

PROJECT 4.6 Software bugs (for example, the Y2KMillennium Bug) can have devastating effects. Write a paper on some famous software bugs and some of the problems that they have caused.

PROJECT 4.7 Build a working model of Napier's rods.

PROJECT 4.8 Write a paper and prepare a classroom demonstration on the use of an abacus. Build your own device as a project.

PROJECT 4.9 Write a paper on the invention of the first electronic computer.

PROJECT 4.10 In Chapter 1, we introduced Pascal's triangle. The reproduction here is from a 14th century Chinese manuscript, and in this form is sometimes called *Yang Hui's triangle*. Even though we have not discussed these ancient Chinese numerals, see if you can reconstruct the basics of their numeration system. (Use your imagination.)

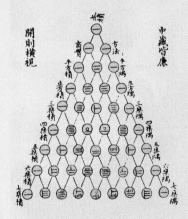

5 THE NATURE OF NUMBERS

Outline

What in the World?

"I'm never going to have children!" exclaimed Shelly. "Sometimes my little brother drives me nuts!"

"I know what you mean," added Mary. "Just yesterday, my brother asked me to explain what 5 means. The best I could do was to show him my fist."

"Stop! That's beginning to sound too much like my math class—just yesterday, Ms. Jones asked us whether $\sqrt{2}$ is rational or irrational, and we need to justify our answer," Shelly said with a bit of anger. "I don't have the faintest; the only radical I know is my brother. I don't know why we should know about weird numbers like $\sqrt{2}$."

Overview

Before we can do mathematical work, we need to have some building blocks for our journey. Those building blocks are sets of numbers. We assume that you know about some of these sets of numbers, but to form a common basis for the rest of the textbook, we will discuss some sets of numbers and their properties in this chapter.

The first set of numbers we encounter as children is the set of counting numbers. As we build more and more complex sets of numbers in this chapter, we move from the counting numbers to the integers (which include the counting numbers, zero, and their opposites), to the fractions, which we characterize as numbers whose decimal representations terminate or repeat. Even though this is a very useful set of numbers, there are certain applications (finding the length of a diagonal of a square, for example) that require numbers that cannot be written as terminating or repeating decimals. This set is called the set of irrational numbers.

Jeff Maloney/Getty Images

5.1 | Natural Numbers

In most sciences one generation tears down what another has built, and what one has established another undoes. In mathematics alone each generation builds a new story to the old structure.

HERMAN HANKEL

CHAPTER **CHALLENGE**

See if you can fill in the question mark.   3   12   27   48   ?

The most basic set of numbers used by any society is the set of numbers used for counting:

$$\mathbb{N} = \{1, 2, 3, 4, 5, 6, 7, 8, 9, 10, 11, \ldots\}$$

This set of numbers is called the set of **counting numbers** or **natural numbers.** Let's assume that you understand what the numbers in this set represent, and you understand the operation of **addition,** $+$. That is, we assume, without definition, knowledge of the operation of addition of natural numbers.

There are a few self-evident properties of addition for this set of natural numbers. They are called "self-evident" because they almost seem too obvious to be stated explicitly. For example, if you jump into the air, you expect to come back down. That assumption is well founded in experience and is also based on an assumption that jumping has certain undeniable properties. But astronauts have found that some very basic assumptions are valid on earth and false in space. Recognizing these assumptions (properties, axioms, laws, or postulates) is important.

Historical NOTE

The word **add** comes from the Latin word **adhere,** which means "to put to." Johannes Widman (1462–1498) first used "+" and "−" signs in 1489 when he stated, "What is −, that is minus, what is +, that is more." The symbol "+" is believed to be a derivation of the Latin **et** ("and ").*

Closure Property

When we add or multiply any two natural numbers, we know that we obtain a natural number. This "knowing" is an assumption based on experiences (inductive reasoning), but we have actually experienced only a small number of cases for all the possible sums and products of numbers. The scientist—and the mathematician in particular—is very skeptical about making assumptions too quickly. The assumption that the sum or product of two natural numbers is a natural number is given the name *closure* and is referred to as the **closure property.** The property is phrased in terms of sets and operations. Think of a set as a "box"; there is a label on the box—say, addition. If *all* additions of numbers in the box have answers that are already *in* the box, then we say the set is **closed** for addition. If there is at least one answer that is not contained in the box, then the set is said to be **not closed** for that operation.

Slavoljub Pantelic/iStockphoto.com

A closed box   A box that is not closed

*Behende und hupsche Rechnung auf allen Kauffmanschaff, 1489, which became widely popular and was reprinted in 1508, 1519, and 1526.

> **Closure for + in $\mathbb{N}$**
>
> Let $\mathbb{N}$ be the set of natural (or counting) numbers. Let a and b be any natural numbers. Then
>
> $a + b$ is a natural number
>
> We say $\mathbb{N}$ is **closed for addition.**

Example   1   Test closure for addition

Is the set $A = \{1, 2, 3, 4, 5, 6, 7, 8, 9, 10\}$ closed for addition?

Solution   The set A is *not closed* for addition because

$5 + 7 = 12$ and $12 \notin A$

Note: The fact that $5 + 3 = 8$ is in the set does not change the fact that A is not closed for addition.

You need find only one *counterexample* to show that a property does not hold.

Commutative and Associative Properties

The word commute can mean to travel back and forth from home to work; this back-and-forth idea can help you remember that the commutative property applies if you read from left to right or from right to left.

Another self-evident property of the natural numbers concerns the order in which they are added. It is called the **commutative property for addition** and states that the *order* in which two numbers are added makes no difference; that is (if we read from left to right),

$a + b = b + a$

for any two natural numbers a and b. The commutative property allows us to rearrange numbers; it is called a *property of order.* Together with another property, called the *associative property,* it is used in calculation and simplification.

The **associative property for addition** allows us to group numbers for addition. Suppose you wish to add three numbers—say, 2, 3, and 8:

$2 + 3 + 8$

The word associate can mean "connect", "join", or "unite"; with this property you associate two of the added numbers.

To add these numbers you must first add two of them and then add this sum to the third. The associative property tells us that, no matter which two numbers are added first, the final result is the same. If parentheses are used to indicate the numbers to be added first, then this property can be symbolized by

$(2 + 3) + 8 = 2 + (3 + 8)$

The parentheses indicate the numbers to be added first. This associative property for addition holds for *any* three or more natural numbers.

Add the column of numbers in the margin. How long does it take? Five seconds is long enough if you use the associative and commutative properties for addition:

$(9 + 1) + (8 + 2) + (7 + 3) + (6 + 4) + 5 = 10 + 10 + 10 + 10 + 5 = 45$

However, it takes much longer if you don't rearrange (commute) and regroup (associate) the numbers:

9
8
7
6
5
4
3
2
+1

$$
\begin{aligned}
(9 + 8) + (7 + 6 + 5 + 4 + 3 + 2 + 1) &= (17 + 7) + (6 + 5 + 4 + 3 + 2 + 1) \\
&= (24 + 6) + (5 + 4 + 3 + 2 + 1) \\
&= (30 + 5) + (4 + 3 + 2 + 1) \\
&= (35 + 4) + (3 + 2 + 1) \\
&= (39 + 3) + (2 + 1) \\
&= (42 + 2) + 1 \\
&= 44 + 1 \\
&= 45
\end{aligned}
$$

The properties of associativity and commutativity are not restricted to the operation of addition. For example, these properties also hold for $\mathbb{N}$ and multiplication, as we will now discuss.

Multiplication is defined as repeated addition.

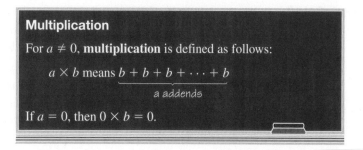

Multiplication

For $a \neq 0$, **multiplication** is defined as follows:

$$a \times b \text{ means } \underbrace{b + b + b + \cdots + b}_{a \text{ addends}}$$

If $a = 0$, then $0 \times b = 0$.

We now consider the property of **closure for multiplication.**

Closure for · in $\mathbb{N}$

Let $\mathbb{N}$ be the set of natural (or counting) numbers. Let a and b be any natural numbers. Then

  ab is a natural number.

We say $\mathbb{N}$ is **closed for multiplication.**

Example 2 | Test closure for multiplication

Is the set $B = \{0, 1\}$ closed for multiplication?

Solution   The set B is *closed* for the operation of multiplication, because all possible products are in B:

  $0 \times 0 = 0$ $0 \times 1 = 0$ $1 \times 0 = 0$ $1 \times 1 = 1$

We can now consider commutativity and associativity for multiplication in the set $\mathbb{N}$ of natural numbers:

  Commutativity: $2 \times 3 \stackrel{?}{=} 3 \times 2$
  Associativity: $(2 \times 3) \times 4 \stackrel{?}{=} 2 \times (3 \times 4)$

The question mark above the equal sign signifies that we should not assume the conclusion (namely, that the expressions on both sides are equal) until we check the arithmetic. Even though we can check these properties for particular natural numbers, it is impossible to check them for *all* natural numbers, so we accept the following properties as axioms.

Commutative and Associative Properties

For any natural numbers a, b, and c:

Commutative properties	*Associative properties*
Addition: $a + b = b + a$	Addition: $(a + b) + c = a + (b + c)$
Multiplication: $ab = ba$	Multiplication: $(ab)c = a(bc)$

To distinguish between the commutative and associative properties, remember the following:

1. When the *commutative property* is used, the *order* in which the elements appear from left to right is changed, but the grouping is not changed.

2. When the *associative property* is used, the elements are *grouped* differently, but the order in which they appear is not changed.

3. If *both* the order and the grouping have been changed, then both the commutative and associative properties have been used.

We are not confined to $\mathbb{N}$ when discussing the associative and commutative properties (or any of the properties, for that matter). Nor are we restricted to addition and multiplication for our operations. Indeed, it is often fun to form your own "group" of numbers and see whether the properties hold for your group under your designated operation.

Distributive Property

Are there properties in $\mathbb{N}$ that involve both operations? Consider an example.

Example 3 | **Distributive property discovery**

Suppose you are selling tickets for a raffle, and the tickets cost $2 each. You sell 3 tickets on Monday and 4 tickets on Tuesday. How much money did you collect?

Solution I   You sold a total of $3 + 4 = 7$ tickets, which cost $2 each, so you collected $2 \times 7 = 14$ dollars. That is,

$$2 \times (3 + 4) = 14$$

Solution II   You collected $2 \times 3 = 6$ dollars on Monday and $2 \times 4 = 8$ dollars on Tuesday for a total of $6 + 8 = 14$ dollars. That is,

$$(2 \times 3) + (2 \times 4) = 14$$

Since these solutions are equal, we see

$$2 \times (3 + 4) = (2 \times 3) + (2 \times 4)$$

Do you suppose this would be true if the tickets cost a dollars and you sold b tickets on Monday and c tickets on Tuesday? Then the equation would be

$$a \times (b + c) = (a \times b) + (a \times c)$$

or simply

$$a(b + c) = ab + ac$$

This example illustrates the **distributive property for multiplication over addition.**

STOP If there is one property to remember, it is this one!

Distributive Property

$$a(b + c) = ab + ac$$

In the set $\mathbb{N}$ of natural numbers, is addition distributive over multiplication? We wish to check

$$3 + (4 \times 5) \stackrel{?}{=} (3 + 4) \times (3 + 5)$$

Checking:

$$3 + (4 \times 5) = 3 + 20 = 23 \quad \text{and} \quad (3 + 4) \times (3 + 5) = 7 \times 8 = 56$$

Thus, addition is not distributive over multiplication in the set of natural numbers.

The distributive property can also help to simplify arithmetic. Suppose you wish to multiply 9 by 71. You can use the distributive property to do the following mental multiplication:

$$9 \times 71 = 9 \times (70 + 1) = (9 \times 70) + (9 \times 1) = 630 + 9 = 639$$

This allows you to do the problem quickly and simply in your head.

Definition of Subtraction

Since these properties hold for the operations of addition and multiplication, we might reasonably ask whether they hold for other operations. *Subtraction* is defined as the opposite of addition.

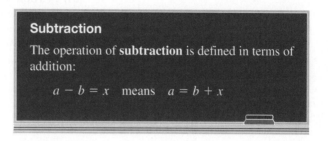

Subtraction

The operation of **subtraction** is defined in terms of addition:

$$a - b = x \quad \text{means} \quad a = b + x$$

To test the commutative property for subtraction, we check a particular example:

$$3 - 2 \overset{?}{=} 2 - 3$$

Now, $3 - 2 = 1$, but $2 - 3$ doesn't even exist in the set of natural numbers. Therefore, the commutative property does not hold for subtraction in $\mathbb{N}$.

Furthermore, to provide the result of the operation of subtraction for $2 - 3$, we must find a number that when added to 3 gives the result 2. But there is *no such natural number.* Thus, the set of natural numbers is *closed* for addition and multiplication, but is *not closed* for subtraction. In Section 5.3, we will add elements to the set of natural numbers to create the set of *integers*, which will be closed for subtraction as well as for addition and multiplication.

To make sure you understand the properties discussed in this section, this problem set focuses on the properties of closure, commutativity, associativity, and distributivity rather than on the set of natural numbers and the operations of addition, multiplication, and subtraction. Since you are so familiar with the set of natural numbers and with these operations, you could probably answer questions about them without much reflection on the concepts involved. Therefore, in the problem set we work with some operations other than addition, multiplication, and subtraction. When we refer to a table, rows are horizontal and columns are vertical.

Problem Set 5.1

Level 1

IN YOUR OWN WORDS   *Explain each of the words or concepts in Problems 1–9.*

1. Natural number   **2.** Multiplication

3. Subtraction   **4.** Closure for addition

5. Commutativity   **6.** Associativity

7. Distributivity   **8.** Closure for multiplication

9. Contrast commutativity and associativity.

Use the definition of multiplication to show what each expression in Problems 10–15 means.

10. a. $2 \cdot 3$ **b.** $3 \cdot 2$

11. a. $3 \cdot 4$ **b.** $4 \cdot 3$

12. a. $5 \cdot 2$ **b.** $2 \cdot 5$

13. a. $2 \cdot 184$ **b.** $184 \cdot 2$

14. a. $3 \cdot 145$ **b.** $145 \cdot 3$

15. a. xy **b.** yx

In Problems 16–26, classify each as an example of the commutative property, the associative property, or both.

16. $3 + 5 = 5 + 3$

17. $2 + 3 + 5 = 2 + 5 + 3$

18. $2 + (3 + 5) = (2 + 3) + 5$

19. $6 + (2 + 3) = (6 + 2) + 3$

20. $6 + (2 + 3) = (6 + 3) + 2$

21. $6 + (2 + 3) = 6 + (3 + 2)$

22. $6 + (2 + 3) = (2 + 3) + 6$

23. $(4 + 5)(6 + 9) = (4 + 5)(9 + 6)$

24. $(4 + 5)(6 + 9) = (6 + 9)(4 + 5)$

25. $(3 + 5) + (2 + 4) = (3 + 5) + (4 + 2)$

26. $(3 + 5) + (2 + 4) = (3 + 4) + (5 + 2)$

27. "Isn't this one just too sweet, dear?" asked the wife as she tried on a beautiful diamond ring. "No," the husband replied. "It's just too dear, sweet." Does this story remind you of the associative or the commutative property?

28. Is the operation of putting on your shoes and socks commutative?

29. In the English language, the meanings of certain phrases can be very different depending on the association of the words. For example,

(MAN EATING) TIGER

is not the same as

MAN (EATING TIGER)

Decide whether each of the following groups of words is associative.

a. HIGH SCHOOL STUDENT

b. SLOW CURVE SIGN

c. BARE FACTS PERSON

d. RED FIRE ENGINE

e. TRAVELING SALESMAN JOKE

f. BROWN SMOKING JACKET

30. Think of three nonassociative word triples as shown in Problem 29.

Level 2

31. IN YOUR OWN WORDS   Why do you think *addition* of natural numbers was left undefined? Try to write a definition. Look in one or more dictionaries. What problems do you find with these definitions?

32. Consider the set $A = \{1, 4, 7, 9\}$ with an operation $\otimes$ defined by the table.

$\otimes$	1	4	7	9
1	9	7	1	4
4	7	9	4	1
7	1	4	7	9
9	4	1	9	7

$a \otimes b$ means find the entry in row a and column b; for example,

$$7 \otimes 9 = 9$$

(the entry in row 7 and column 9). Find each of the following.

a. $7 \otimes 4$ **b.** $9 \otimes 1$

c. $1 \otimes 7$ **d.** $9 \otimes 9$

e. $4 \otimes 7$ **f.** $1 \otimes 9$

g. $7 \otimes 1$ **h.** $7 \otimes 7$

33. Consider the set

$$F = \{1, -1, i, -i\}$$

with an operation $\times$ defined by the table.

$\times$	1	-1	i	$-i$
1	1	-1	i	$-i$
-1	-1	1	$-i$	i
i	i	$-i$	-1	1
$-i$	$-i$	i	1	-1

$a \times b$ means find the entry in row a and column b; for example, $-1 \times (-i) = i$ (the entry in row -1 and column $-i$). Find each of the following.

a. $-1 \times i$ **b.** $i \times i$ **c.** $-i \times i$

d. $-i \times 1$ **e.** $1 \times i$ **f.** $-i \times -i$

34. Consider the set A and the operation $\otimes$ from Problem 32. Is the set A closed for the operation of $\otimes$? Give reasons for your answer.

35. Consider the set F and the operation of $\times$ from Problem 33. Is the set F closed for the operation of $\times$? Give reasons for your answer.

36. Consider the set A and the operation $\otimes$ from Problem 32. Does the set A satisfy the given property for the operation of $\otimes$? Give reasons for your answer.
 a. Associative
 b. Commutative

37. Consider the set F and the operation of $\times$ from Problem 33. Does the set F satisfy the given property for the operation of $\times$? Give reasons.
 a. Associative
 b. Commutative

38. Let a be the process of putting on a shirt; let b be the process of putting on a pair of socks; and let c be the process of putting on a pair of shoes. Let $\star$ be the operation of "followed by."
 a. Is $\star$ commutative for $\{a, b, c\}$?
 b. Is $\star$ associative for $\{a, b, c\}$?

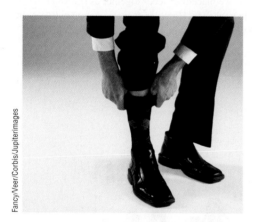

39. Consider the set $\mathbb{N}$ of natural numbers and an operation $\triangleleft$ which means "select the first of the two." That is,

$$4 \triangleleft 3 = 4; \quad 3 \triangleleft 4 = 3;$$
$$5 \triangleleft 7 = 5; \quad 6 \triangleleft 6 = 6$$

Is the set $\mathbb{N}$ closed for the operation of $\triangleleft$? Give reasons.

40. Consider the set $\mathbb{N}$ of natural numbers and an operation $\triangleleft$ defined in Problem 39.
 a. Is $\triangleleft$ associative for $\mathbb{N}$? Give reasons.
 b. Is $\triangleleft$ commutative for $\mathbb{N}$? Give reasons.

41. Consider the operation $\bullet$ defined by the following table.

$\bullet$	$\square$	$\triangle$	$\circ$
$\square$	$\circ$	$\square$	$\triangle$
$\triangle$	$\square$	$\triangle$	$\circ$
$\circ$	$\triangle$	$\square$	$\circ$

 a. Find $\square \bullet \triangle$.
 b. Find $\triangle \bullet \circ$.
 c. Does $\circ \bullet \square = \square \bullet \circ$?
     Is the set commutative for $\bullet$?
 d. Does $(\circ \bullet \triangle) \bullet \triangle = \circ \bullet (\triangle \bullet \triangle)$?

42. Let $\downarrow$ mean "select the smaller number" and $\rightarrow$ mean "select the second of the two." Is $\rightarrow$ distributive over $\downarrow$ in the set of natural numbers, $\mathbb{N}$?

43. Let $\downarrow$ mean "select the smaller number" and $\rightarrow$ mean "select the second of the two." Is $\downarrow$ distributive over $\rightarrow$ in the set of natural numbers, $\mathbb{N}$?

44. Do the following problems mentally using the distributive property.
 a. 6×82 **b.** 8×41 **c.** 7×49

45. Do the following problems mentally using the distributive property.
 a. 5×99 **b.** 4×88 **c.** 8×52

46. Is the set $\{0, 1\}$ commutative for the operation of multiplication? Give reasons.

47. Is the set $\{0, 1\}$ associative for the operation of multiplication? Give reasons.

48. Is the set $\{-1, 0, 1\}$ commutative for the operation of multiplication? Give reasons.

49. Is the set of even natural numbers closed for the operation of addition? Give reasons.

50. Is the set of even natural numbers closed for the operation of multiplication? Give reasons.

51. Is the set of odd natural numbers closed for the operation of multiplication? Give reasons.

52. Is the set of odd natural numbers closed for the operation of addition? Give reasons.

Level 3

53. Let $S = \{1, 2, 3, \ldots, 9, 10\}$. Define an operation $\odot$ as

$$a \odot b = 0 \cdot a + 1 \cdot b$$

Is S closed for the operation of $\odot$?

54. Let $S = \{1, 2, 3, \ldots, 99, 100\}$. Define an operation $\oplus$ as

$$a \oplus b = 2a + b$$

Check the commutative and associative properties.

55. Let $S = \{1, 2, 3, \ldots, 999, 1000\}$. Define an operation $\oslash$ as

$$a \oslash b = 2(a + b)$$

Check the commutative and associative properties.

Problem Solving 3

56. Consider a soldier facing in a given direction (say north). Let us denote

"left face" by ℓ,

"right face" by r,

"about face" by a, and

"stand fast" by f.

(The element f means "don't move from your present position." It does not mean "return to your original position.") Then we define

$$H = \{\ell, r, a, f\}$$

and an operation $\star$ meaning "followed by." Thus, $\ell \star \ell = a$ means "left face" followed by "left face" and is the same as the single command "about face." Complete the following table.

$\star$	ℓ	r	a	f
ℓ	a			
r				
a				
f				

57. Is the set H from Problem 56 closed for the operation of $\star$?

58. Does the set H and operation $\star$ from Problem 56 satisfy the following properties? Give reasons.
a. Associative
b. Commutative

59. Mensa is an association for people with high IQs. An advertisement for the organization offers the following challenge: "Take this instant test to see if you're a genius."

> *Put the appropriate plus or minus signs between the numbers, in the correct places, so that the sum total will equal 1.*
> $$0\ 1\ 2\ 3\ 4\ 5\ 6\ 7\ 8\ 9 = 1$$

The advertisement also states, "This problem stumps 45% of the Mensa members who try it. And they *all* have IQs in the top 2% nationwide. See if you can pass this test. If you can do it, you might have what it takes to join us. To find out, . . . write to American Mensa, 2626 East 14th St., Brooklyn, NY 11235."

60. The Vanishing Leprechaun Puzzle The puzzle shown in Figure 5.1 consists of three pieces and was originally published by W. A. Elliott Company, 212 Adelaide St. W., Toronto, Canada M5H 1W7.

If we place the pieces together as shown at the top, we see 15 leprechauns. However, if we *commute* pieces A and B as shown at the bottom, we count 14 leprechauns!

Clearly, from Figure 5.1 we see

$$AB \neq BA$$

Can you explain where the vanishing leprechaun went?

FIGURE 5.1 Leprechaun Puzzle: How many leprechauns?

5.2 | Prime Numbers

A set of numbers that is important, not only in algebra but in all of mathematics, is the set of prime numbers. To understand prime numbers, you must first understand the idea of divisibility, along with some new terminology and notation.

Divisibility

The natural number 10 is divisible by 2, since there is a natural number 5 so that $10 = 2 \cdot 5$; it is not divisible by 3, since there is no natural number k such that $10 = 3 \cdot k$. This leads us to the following definition of **divisibility.**

 CAUTION Do not confuse this notation with the notation sometimes used for fractions: "5/30" means 5 divided by 30, which is a fraction; "5|30" means "5 divides 30," which is a statement.

> **Divisibility**
>
> If m and d are natural numbers, and if there is a natural number k so that $m = d \cdot k$, we say that **d is a divisor of m, d is a factor of m, d divides m,** and **m is a multiple of d.** We denote this relationship by $d \mid m$.

That is, $5 \mid 30$ is read "5 divides 30" and means that there exists some natural number k—namely, 6—such that $30 = 5 \cdot k$.

Example 1 | Test divisibility

Tell whether each of the following is true or false, and give the meaning of each.
a. $7 \mid 63$ **b.** $8 \mid 104$ **c.** $14 \mid 2$ **d.** $6 \mid 15$

Solution
a. $7 \mid 63$ is read "7 divides 63" and is true since we can find a natural number k—namely, 9—such that $63 = 7 \cdot k$.
b. $8 \mid 104$ is true, since $104 = 8 \cdot 13$.
c. $14 \mid 2$ is false because we can find no natural number k so that $2 = 14 \cdot k$. We write $14 \nmid 2$ to say that 14 does not divide 2.
d. $6 \mid 15$ is false, because we can find no natural number k so that $15 = 6 \cdot k$.

It is easy to see that 1 divides every natural number m, since $m = 1 \cdot m$. Also, by the commutative property of multiplication, $m = m \cdot 1$; thus every natural number m divides itself. We have proved the following theorem.

STOP This property is basic to understanding what follows.

> **Number of Divisors**
>
> Every natural (counting) number greater than 1 has at least two distinct divisors, itself and 1.

Example 2 | Divisibility rule

If $x \mid (a + b)$ and $x \mid a$, then $x \mid b$.

Solution   Since $x \mid (a + b)$, there must exist a natural number K so that $xK = a + b$. Also, since $x \mid a$, there must exist a natural number k so that $xk = a$. Now, since a and b are natural numbers, we see $a + b > a$ so that $K > k$, or in other words, $K - k$ is a natural number. By substitution, $xK = xk + b$ so that $b = xK - xk = x(K - k)$. Since $K - k$ is a natural number, we see that $x \mid b$.

Consider the number 341,592. Is this number divisible by 2? By 3? By 4? By 5? You may know some ways of answering these questions without actually doing the division.

Example 3 | Divisibility rule for 2

Find a rule for the divisibility of any number M by 2.

Solution   We use Pólya's problem-solving guidelines for this example.

Understand the Problem.   You might already know the rule for divisibility by 2. It says that *if the last digit of the number is even, then the number is divisible by 2.* That is, if the number M ends in 0, 2, 4, 6, or 8, it is divisible by 2. This example asks us to show why this rule "works."

Devise a Plan.   We will begin with a simpler example, say 341,592. We write the number in expanded notation.

$$341{,}592 = 3 \times 10^5 + 4 \times 10^4 + 1 \times 10^3 + 5 \times 10^2 + 9 \times 10^1 + 2$$

The question to answer is "When will this number be divisible by 2?" Associate all the digits except the last one:

$$341{,}592 = \underbrace{3 \times 10^5 + 4 \times 10^4 + 1 \times 10^3 + 5 \times 10^2 + 9 \times 10^1}_{\text{This is divisible by 2.}} + 2$$

The associated part is *always* divisible by 2, since $2 \mid 10$, $2 \mid 10^2$, $2 \mid 10^3$, $2 \mid 10^4$, and $2 \mid 10^5$.

Carry Out the Plan.   For the number M we see that since $2 \mid 10$, $2 \mid 10^2, \ldots, 2 \mid 10^n$, the divisibility of M by 2 depends solely on whether 2 divides the last digit.

Look Back.   We see that $2 \mid 341{,}592$ since $2 \mid 2$. Also $2 \mid 838$ since $2 \mid 8$, and $2 \nmid 839$ since $2 \nmid 9$.

Similar rules apply for divisibility by 4 or 8. You might even expect to try the same type of rule for divisibility by 3, but this situation is not quite so easy, as shown by the following example.

Example 4 | Divisibility rule for 3

Find a rule for divisibility by 3.

Solution   We use Pólya's problem-solving guidelines for this example.

Understand the Problem.   Try a simple example. We see that $3 \mid 84$ since we can find a natural number—namely, 28—so that $84 = 3 \cdot k$. We also note that $3 \nmid 4$, so the same type of rule that worked for divisibility by 2 will not work for divisibility by 3.

Devise a Plan.   We will once again look at the expanded notation. Consider a simpler problem—say, the divisibility of 341,592 by 3:

$$341{,}592 = 3 \times 10^5 + 4 \times 10^4 + 1 \times 10^3 + 5 \times 10^2 + 9 \times 10^1 + 2$$

The plan is to make each product divisible by 3. We do this by adding and subtracting 1 from each term containing 10^b, where b is a natural number.

Carry Out the Plan.

$$341{,}592 = 3 \times (10^5 - \mathbf{1 + 1}) + 4 \times (10^4 - \mathbf{1 + 1}) + 1 \times (10^3 - \mathbf{1 + 1})$$
$$+ 5 \times (10^2 - \mathbf{1 + 1}) + 9 \times (10^1 - \mathbf{1 + 1}) + 2$$

We now use the distributive and associative properties to rewrite this expression:

$$341{,}592 = [3 \times (10^5 - 1) + 3] + [4 \times (10^4 - 1) + 4] + [1 \times (10^3 - 1) + 1]$$
$$+ [5 \times (10^2 - 1) + 5] + [9 \times (10^1 - 1) + 9] + 2$$
$$= \underbrace{[3(10^5 - 1) + 4(10^4 - 1) + 1(10^3 - 1) + 5(10^2 - 1) + 9(10^1 - 1)]}_{\text{This is divisible by 3.}}$$
$$+ \underbrace{[3 + 4 + 1 + 5 + 9 + 2]}_{\text{This is the sum of the digits.}}$$

Notice what we have done:

$$10^1 - 1 = 9$$
$$10^2 - 1 = 99$$
$$10^3 - 1 = 999$$
$$10^4 - 1 = 9{,}999$$
$$10^5 - 1 = 99{,}999$$

These are all divisible by 3, and hence we see that

if $3 \mid (3 + 4 + 1 + 5 + 9 + 2)$, then $3 \mid 341{,}592$

Checking, we see that $3 \mid 24$ (sum of digits), so $3 \mid 341{,}592$. Furthermore, since $3 \mid (10^n - 1)$ for any natural number n, we see that the divisibility of a number N by 3 depends on whether 3 divides the sum of the digits.

Look Back.   If the sum of the digits of a number N is divisible by 3, then N is divisible by 3.

Some of the more common rules of divisibility are shown in Table 5.1.

TABLE 5.1	
Rules of Divisibility for a Natural Number N	
N is divisible by	**Test**
1	all N
2	if the last digit is divisible by 2.
3	if the sum of the digits is divisible by 3.
4	if the number formed by the last two digits is divisible by 4.
5	if the last digit is 0 or 5.
6	if the number is divisible by 2 and by 3.
8	if the number formed by the last three digits is divisible by 8.
9	if the sum of the digits is divisible by 9.
10	if the last digit is 0.
12	if the number is divisible by 3 and by 4.

Finding Primes

Since every natural number greater than 1 has at least two divisors, can any number have more than two?

Checking: 2 has exactly two divisors: 1, 2
3 has exactly two divisors: 1, 3
4 has more than two divisors: 1, 2, and 4

Thus, some numbers (such as 2 and 3) have exactly two divisors, and some (such as 4 and 6) have more than two divisors. Do any natural numbers have fewer than two divisors?

We now state a definition that classifies each natural number according to the number of divisors it has.

> **Prime Number**
>
> A **prime number** is a natural number that has exactly two divisors. A natural number that has more than two divisors is called a **composite number.**

We see that 2 is prime, 3 is prime, 4 is composite (since it is divisible by three natural numbers), 5 is prime, 6 is composite (since it is divisible by 1, 2, 3, and 6). Note that every natural number greater than 1 is either prime or composite. The number 1 is neither prime nor composite.

One method for finding primes smaller than some given number was first used by a Greek mathematician, Eratosthenes, more than 2,000 years ago. The technique is known as the **sieve of Eratosthenes.** Suppose we wish to find the primes less than 100. We prepare a table of natural numbers 1–100 using the following procedure.

Sieve of Eratosthenes

Step 1   Write down a list of numbers from 1 to 100 (see Table 5.2 page 187).

Step 2   Cross out 1, since it is not classified as a prime number.

Step 3   Draw a circle around 2, the smallest prime number. Then cross out every following multiple of 2, since each is divisible by 2 and thus is not prime.

Step 4   Draw a circle around 3, the next prime number. Then cross out each succeeding multiple of 3. Some of these numbers, such as 6 and 12, will already have been crossed out because they are also multiples of 2.

Step 5   Circle the next open prime, 5, and cross out all subsequent multiples of 5.

Step 6   The next prime number is 7; circle 7 and cross out multiples of 7.

Step 7   Since 7 is the largest prime less than $\sqrt{100} = 10$, we now know that all the remaining numbers are prime.

Historical NOTE

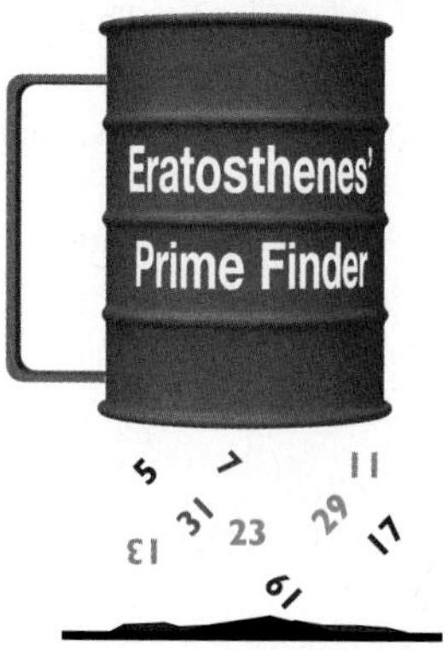

Eratosthenes was a highly talented and versatile person: athlete, astronomer, geographer, historian, poet, and mathematician. In addition to his sieve for prime numbers, he invented a mechanical "mean finder" for duplicating the cube, and he made some remarkably accurate measurements of the size of the earth. Unfortunately, most of his mathematical contributions are lost. It is said that he committed suicide by starving himself to death upon discovering that he had ophthalmia.

TABLE 5.2

Age Chart

Finding Primes Using the Sieve of Eratosthenes

1	2	3	4	5	6	7	8	9	10
11	12	13	14	15	16	17	18	19	20
21	22	23	24	25	26	27	28	29	30
31	32	33	34	35	36	37	38	39	40
41	42	43	44	45	46	47	48	49	50
51	52	53	54	55	56	57	58	59	60
61	62	63	64	65	66	67	68	69	70
71	72	73	74	75	76	77	78	79	80
81	82	83	84	85	86	87	88	89	90
91	92	93	94	95	96	97	98	99	100

The process is a simple one, since you do not have to cross out the multiples of 3 (for example) by checking for divisibility by 3 but can simply cross out every third number. Thus, anyone who can count can find primes by this method. Also, notice that in finding the primes under 100, we had crossed out all the composite numbers by the time we crossed out the multiples of 7. That is, to find all primes less than 100: (1) Find the largest prime smaller than or equal to $\sqrt{100} = 10$ (7 in this case); (2) cross out multiples of primes up to and including 7; and (3) all the remaining numbers in the chart are primes.

This result generalizes. If you wish to find all primes smaller than n:

1. Find the *largest* prime less than or equal to $\sqrt{n}$.
2. Cross out the multiples of primes less than or equal to $\sqrt{n}$.
3. All the remaining numbers in the chart are primes.

Phrasing this another way, if n is composite, then one of its factors must be less than or equal to $\sqrt{n}$. That is, if $n = ab$, then it can't be true that *both a and b* are greater than $\sqrt{n}$. (otherwise $ab > \sqrt{n}\sqrt{n} = n = ab$, so $ab > ab$ is a contradiction). Thus, one of the factors must be less than or equal to $\sqrt{n}$.

Prime Factorization

Prime numbers are fundamental to many mathematical processes. In particular, we use prime numbers in working with rational numbers later in this chapter. You will need to understand *prime factorization, greatest common factor,* and *least common multiple.*

The operation of **factoring** is the reverse of the operation of multiplying. For example, multiplying 3 by 6 yields $3 \cdot 6 = 18$, and this answer is unique (only one answer is possible). In the reverse process, called factoring, you are given the number 18 and asked for numbers that can be multiplied together to give 18. This process is *not* unique; we list several different factorizations of 18:

$$18 = 1 \cdot 18 = 18 \cdot 1 = 2 \cdot 9 = 9 \cdot 2 = 1 \cdot 1 \cdot 2 \cdot 9 = 3 \cdot 6 = 2 \cdot 3 \cdot 3 = \cdots$$

There are, in fact, infinitely many possibilities. We make some agreements, so the process gives a unique answer:

1. We will not consider the order in which the factors are listed as important. That is, $2 \cdot 9$ and $9 \cdot 2$ are considered the same factorization.
2. We will not consider 1 as a factor when writing out any factorizations. That is, prime numbers do not have factorizations.
3. Recall that we are working in the set of natural numbers; thus, $18 = 36 \cdot \frac{1}{2}$ and $18 = (-2)(-9)$ are *not* considered factorizations of 18.

   With these agreements, we have greatly reduced the possibilities:

   $$18 = 2 \cdot 9 = 3 \cdot 6 = 2 \cdot 3^2$$

Finding prime factorizations is a process that is used in a multitude of mathematical applications.

These are the only three possible factorizations. Notice that the last factorization contains only prime factors; thus it is called the **prime factorization** of 18.

It should be clear that, if a number is composite, it can be factored as the product of two natural numbers greater than 1. Each of these two numbers will be prime or composite. If both are prime, then we have a prime factorization. If one or more is composite, we repeat the process, and continue until we have written the original number as a product of primes. It is also true that this representation is unique. This is one of the most important results in arithmetic, and it carries the impressive title **fundamental theorem of arithmetic.**

As you might guess from the name, this result is important for the following material.

Fundamental Theorem of Arithmetic

Every natural number greater than 1 is either a prime or a product of primes, and its prime factorization is unique (except for the order in which the factors appear).

Historical   NOTE

A statement equivalent to the fundamental theorem of arithmetic is found in Book IX of Euclid's Elements. This work is not only the earliest known major Greek mathematical book, but it is also the most influential textbook of all time. It was composed around 300 B.C. and first printed in 1482. Except for the Bible, no other book has been through so many printings. Euclid, the first professor of mathematics at the Museum of Alexandria, was the author of at least ten other books.

Example 5 Find prime factorizations

Find the prime factorizations:
a. 385   **b.** 1,400

Solution

a. One of the easiest ways to find the prime factors of a number is to try division by each of the prime numbers in order: 2, 3, 5, 7, ... The rules of divisibility in Table 5.1 may help. For this example, we need to check primes up to $\sqrt{385} \approx 19$. If none of the primes up to 19 is a factor of 385, then 385 is prime. We see by inspection that 385 is not divisible by 2 or 3. It is divisible by 5, so

$$385 = 5 \cdot 77$$

Since 77 is composite ($77 = 7 \cdot 11$), we write

$$385 = 5 \cdot 7 \cdot 11$$

We are now finished since all the factors are prime numbers.

   Many people prefer to find the prime factorization using a **factor tree.**

b. Using a factor tree, we may begin with *any* factors of 1,400:

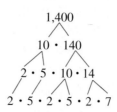

We write the prime factorization using exponents:

$$1,400 = 2^3 \cdot 5^2 \cdot 7$$

The answers to Example 5 lead us to what is called **canonical form.** The *canonical representation* of a number is the representation of that number as a product of consecutive primes using exponential notation with the factors arranged in order of increasing magnitude. For example, the canonical forms of the numbers in Example 5 are $385 = 2^0 \cdot 3^0 \cdot 5^1 \cdot 7^1 \cdot 11^1$ and $1,400 = 2^3 \cdot 3^0 \cdot 5^2 \cdot 7^1$. Remember that any nonzero base to the zero power is 1.

Example 6 Find canonical representation

Find the canonical representation of 3,465.

Solution We use a factor tree.

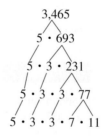

The prime factorization in canonical form is $2^0 \cdot 3^2 \cdot 5^1 \cdot 7^1 \cdot 11^1$.

Greatest Common Factor

Suppose we look at the set of factors common to a given set of numbers:

Factors of 18:   {1, 2, 3,   6, 9, 18}

Factors of 12:   {1, 2, 3, 4, 6, 12}

Common factors: {1, 2, 3, 6}

The *greatest common factor* is the largest number in the set of common factors.

> ## Greatest Common Factor
>
> The **greatest common factor** (**g.c.f.**) of a set of numbers is the largest number that divides (evenly) into each of the numbers in the given set.

The procedure for finding the greatest common factor involves the canonical form of the given numbers. For example, suppose we want to find the greatest common factor of 24 and 30. We first find all factors, then the intersection (common factors), and finally the greatest one in that set:

Factors of 24:	$\{1, 2, 3, 4, 6, 8, 12, 24\}$
Factors of 30:	$\{1, 2, 3, 5, 6, 10, 15, 30\}$
Common facors:	$\{1, 2, 3, 6\}$
g.c.f. $= 6$	

This process can be rather tedious, so we outline a more refined procedure:

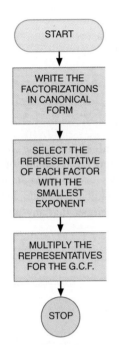

START

WRITE THE FACTORIZATIONS IN CANONICAL FORM

SELECT THE REPRESENTATIVE OF EACH FACTOR WITH THE SMALLEST EXPONENT

MULTIPLY THE REPRESENTATIVES FOR THE G.C.F.

STOP

FIGURE 5.2 Flowchart for finding the greatest common factor (g.c.f.)

Procedure for Finding the g.c.f

Step 1 Write the factorizations in canonical form.

Step 2 Select the representative of each factor with the SMALLEST exponent.

Step 3 Multiply the representatives to find the greatest common factor (g.c.f.).

We illustrate this procedure for the numbers 24 and 30. First, find the canonical representation of each:

$$24 = 2^3 \cdot 3 \quad = 2^3 \cdot 3^1 \cdot 5^0$$
$$30 = 2 \cdot 3 \cdot 5 = 2^1 \cdot 3^1 \cdot 5^1$$

Select one representative from each of the columns in the factorizations. The representative we select when finding the g.c.f. is the one with the smallest exponent. The g.c.f. is the product of these representatives.

$$\text{g.c.f.} = 2^1 \cdot 3^1 \cdot 5^0 = 6$$

The procedure for finding the g.c.f. is summarized in Figure 5.2.

Example 7 Find the greatest common factor

Find the greatest common factor of the given sets of numbers.
a. 300, 144   **b.** 15, 28   **c.** 3150, 588, 280

Solution
a. $300 = \mathbf{2^2 \cdot 3^1} \cdot 5^2$
$\quad 144 = 2^4 \cdot 3^2 \cdot \mathbf{5^0}$
$\quad \textbf{g.c.f.} = 2^2 \cdot 3^1 \cdot 5^0$
$\qquad\quad = 4 \cdot 3 \cdot 1$
$\qquad\quad = 12$

b. $15 = \mathbf{2^0} \cdot 3^1 \cdot 5^1 \cdot \mathbf{7^0}$
$\quad 28 = 2^2 \cdot \mathbf{3^0} \cdot \mathbf{5^0} \cdot 7^1$
$\quad \textbf{g.c.f.} = 2^0 \cdot 3^0 \cdot 5^0 \cdot 7^0$
$\qquad\quad = 1 \cdot 1 \cdot 1 \cdot 1$
$\qquad\quad = 1$

c. $3{,}150 = \mathbf{2^1} \cdot 3^2 \cdot 5^2 \cdot \mathbf{7^1}$
$\quad\; 588 = 2^2 \cdot 3^1 \cdot \mathbf{5^0} \cdot 7^2$
$\quad\; 280 = 2^3 \cdot \mathbf{3^0} \cdot 5^1 \cdot 7^1$
$\quad\; \textbf{g.c.f.} = 2^1 \cdot 3^0 \cdot 5^0 \cdot 7^1$
$\qquad\qquad = 2 \cdot 1 \cdot 1 \cdot 7$
$\qquad\qquad = 14$

If the greatest common factor of two numbers is 1, we say that the numbers are **relatively prime.** Notice that 15 and 28 are relatively prime, but they themselves are not prime. It is possible for relatively prime numbers to be composite numbers.

Least Common Multiple

The greatest common factor is the largest number in the intersection of the factors of a set of given numbers. On the other hand, the *least common multiple* is the smallest number in the intersection of the multiples of a set of given numbers.

> ### Least Common Multiple
> The **least common multiple (l.c.m.)** of a set of numbers is the smallest number that each of the numbers in the set divides into evenly.

For example, suppose we want to find the least common multiple of 24 and 30.

Multiples of 24: $\{24, 48, 72, 96, 120, \ldots\}$

Multiples of 30: $\{30, 60, 90, 120, \ldots\}$

Common factors: $\{120, 240, \ldots\}$

l.c.m. $= 120$

An algorithm for finding the least common multiple is very much like the one for finding the g.c.f.

Procedure for Finding the l.c.m.

Step 1   Write the factorizations in canonical form.

Step 2   Select the representative of each factor with the **LARGEST** exponent.

Step 3   Multiply the representatives to find the least common multiple (l.c.m.).

The process, as before, begins by finding the canonical representations of the numbers involved. For example, to obtain the l.c.m. of the numbers 24 and 30, write each in canonical form:

$$24 = 2^3 \cdot 3^1 \cdot 5^0 \qquad 30 = 2^1 \cdot 3^1 \cdot 5^1$$

For the l.c.m., we choose the representative of each factor with the largest exponent. The l.c.m. is the product of these representatives.

$$\text{l.c.m.} = \mathbf{2^3 \cdot 3^1 \cdot 5^1 = 120}$$

The procedure for finding the least common multiple is shown in Figure 5.3.

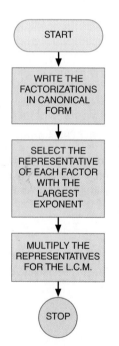

FIGURE 5.3 Procedure for finding the least common multiple (l.c.m.)

START

↓

WRITE THE FACTORIZATIONS IN CANONICAL FORM

↓

SELECT THE REPRESENTATIVE OF EACH FACTOR WITH THE LARGEST EXPONENT

↓

MULTIPLY THE REPRESENTATIVES FOR THE L.C.M.

↓

STOP

Example 8 Find the least common multiple

Find the least common multiple of 300, 144, and 108.

Solution

$$
\begin{aligned}
300 &= 2^2 \cdot 3^1 \cdot \mathbf{5^2} \\
144 &= \mathbf{2^4} \cdot 3^2 \cdot 5^0 \\
108 &= 2^2 \cdot \mathbf{3^3} \cdot 5^0 \\
\textbf{l.c.m.} &= 2^4 \cdot 3^3 \cdot 5^2 \\
&= 16 \cdot 27 \cdot 25 = 10{,}800
\end{aligned}
$$

In Pursuit of Primes

The method of Eratosthenes gives a finite list of primes, but it is not very satisfactory to use if we wish to determine whether a given number n is a prime. For centuries, mathematicians have tried to find a formula that would yield *every* prime. Let's try to find a formula that results in giving only primes. A possible candidate is

$$n^2 - n + 41$$

If we try this formula for $n = 1$, we obtain $1^2 - 1 + 41 = 41$.

For $n = 2$: $2^2 - 2 + 41 = 43$, a prime
For $n = 3$: $3^2 - 3 + 41 = 47$, a prime

So far, so good; that is, we are obtaining only primes. Continuing, we keep finding only primes for n up to 40:

For $n = 40$: $40^2 - 40 + 41 = 1{,}601$, a prime

Inductively, we might conclude that the formula yields only primes, but the next value provides a counterexample:

For $n = 41$: $41^2 - 41 + 41 = 41^2 = 1{,}681$, not a prime!

A more serious attempt to find a prime number formula was made by Pierre de Fermat, who tried the formula

$$2^{2^n} + 1$$

For $n = 1$: $2^{2^1} + 1 = 5$, a prime
For $n = 2$: $2^{2^2} + 1 = 2^4 + 1 = 17$, a prime
For $n = 3$: $2^{2^3} + 1 = 2^8 + 1 = 257$, a prime
For $n = 4$: $2^{2^4} + 1 = 2^{16} + 1 = 65{,}537$, a prime
For $n = 5$: $2^{2^5} + 1 = 2^{32} + 1 = 4{,}294{,}967{,}297$ Is 4,294,967,297 a prime?

The answer is not easy. See the Historical Note in the margin. It turns out this number is not prime! It is divisible by 641. Whether this formula generates any other primes is unknown.

In 1644, the French priest and number theorist Marin Mersenne (1588–1648) stated without proof that the number

$$2^{251} - 1$$

is composite. In the 19th century, mathematicians finally proved Mersenne correct when they discovered that this number was divisible by both 503 and 54,217. Mersenne did discover, however, that

$$2^{257} - 1$$

is a prime number.

In 1970, a young Russian named Yuri Matyasevich discovered several explicit polynomials (such as $n^2 - n + 41$) of this sort that generate only prime numbers, but all of those he discovered are too complicated to reproduce here. The largest known prime number at that time was

$$2^{11{,}213} - 1$$

which was discovered at the University of Illinois through the use of number theory and computers. The mathematicians were so proud of this discovery that the following postmark was used on the university's postage meter:

Courtesy of Donald B. Gillies, University of Illinois

Some other large prime numbers and the dates of their discovery are shown in Table 5.3. The Great Internet Mersenne Prime Search, or GIMPS, is a computing project that uses volunteers' computers to hunt for primes. The last source shown in Table 5.3 claimed a $100,000 prize from the Electronic Frontier Foundation for being the first to find a prime number that has more than ten million digits.

TABLE 5.3

Some Large Prime

Prime Number	Date of Discovery	Source
$2^{257} - 1$	1644	Marin Mersenne
$2^{11,213} - 1$	1970	University of Illinois
$2^{19,937} - 1$	1971	Bryant Tuckerman
$2^{21,701} - 1$	1978	Laura Nickel and Curt Knoll
$2^{86,243} - 1$	1983	David Slowinski
$2^{216,091} - 1$	1985	Amdal Benchmark Center
$2^{858,433} - 1$	1994	David Slowinski and Paul Gage
$2^{1,398,269} - 1$	1996	Joel Armengaud
$2^{3,021,377} - 1$	1998	Clarkson and Kurowski
$2^{6,972,593} - 1$	1999	Hajratwala and Kurowski
$2^{13,466,917} - 1$	2001	Cameron and Kurowski
$2^{30,402,457} - 1$	2005	Cohen and Boone, GIMPS
$2^{32,582,657} - 1$	2007	Cooper and Boone, GIMPS
$2^{37,156,667} - 1$	2008	Hans-Michael, George Woltman, et. al., GIMPS
$2^{43,112,609} - 1$	2008	Edison Smith, George Woltman, et. al., GIMPS

If you are interested in finding out more about the search for large primes, you might wish to check out GIMPS. GIMPS is a worldwide project coordinated by George Woltman, who wrote a program for the PC to find primes. The hunt for record prime numbers used to be the exclusive domain of supercomputers, but today by using thousands of individual machines, it is possible to collectively surpass even the most powerful computers.

Infinitude of Primes

Table 5.3 shows some very large primes. Is there a largest prime? If there is no largest prime, then there must be infinitely many primes.

http://www.mersenne.org/
prime.htm or find this link
at: **www.mathnature.com**

Example **9**   **No largest prime**

Pólya's
Method

Show that there is no largest prime.

Solution   We use Pólya's problem-solving guidelines for this example.

Understand the Problem.   Is there a prime larger than the largest known prime shown in Table 5.3? If we find one, then we are finished. If we can't find one, is there a way we can still show it is not the largest prime?

Devise a Plan. We will consider a simpler problem. Suppose we believe that 19 is the largest prime. The task is to show that there exists a prime larger than 19. There are two ways to proceed. We could simply find a larger prime—say, 23. But what if we can't find a larger one? Without actually finding the largest prime, we will proceed by showing that 19 can't be the largest prime. Consider the number

$$M = (2 \cdot 3 \cdot 5 \cdot 7 \cdot 11 \cdot 13 \cdot 17 \cdot 19) + 1$$

This number is larger than 19. Is it a prime?

Carry Out the Plan. According to our assumption, M must be composite, since it is larger than 19. But if it is composite, it has a prime divisor. Check all the primes:

2 does not divide M since $2 | (2 \cdot 3 \cdot 5 \cdot 7 \cdot 11 \cdot 13 \cdot 17 \cdot 19)$, and thus 2 does not divide 1 more than this number.

3 does not divide M for the same reason.

Repeat for every known prime.

Thus, if M is not divisible by any known prime, then either it must be prime or there is a prime divisor larger than 19. In either case, we have found a prime larger than 19.

Look Back. If *anyone* claims to be in possession of the largest prime, we need only carry out an argument like the above to find a larger prime. Thus, we are saying that there are infinitely many primes, since it is impossible to have a largest prime.

Historical **NOTE**

The only standing ovation ever given at a meeting of the American Mathematical Association was given to a mathematician named Cole in 1903. It was commonly believed that $2^{67} - 1$ was a prime. Cole first multiplied out 2^{67} and then subtracted 1. Moving to another board, he wrote

761,838,257,287
$\times$ 193,707,721

He multiplied it out and came up with the same result as on the first blackboard. He had factored a number thought to be a prime! Here is what it looks like (using a computer):

$2^{67} - 1 = 147573952589676412927$ $761838257287 \times 193707721 = 147573952589676412927$

Problem Set **5.2**

Level **1**

1. **IN YOUR OWN WORDS**   What is a prime number?

2. **IN YOUR OWN WORDS**   Describe a process for finding a prime factorization.

3. **IN YOUR OWN WORDS**   What is the canonical representation of a number?

4. **IN YOUR OWN WORDS**   What does g.c.f. mean, and what is the procedure for finding the g.c.f. of a set of numbers?

5. **IN YOUR OWN WORDS**   What does l.c.m. mean, and what is the procedure for finding the l.c.m. of a set of numbers?

6. **IN YOUR OWN WORDS**   Compare and contrast finding the g.c.f. and l.c.m. of a set of numbers.

Which of the numbers in Problems 7–10 are prime?

7. **a.** 59   **b.** 57
   c. 1   **d.** 1,997

8. **a.** 63   **b.** 73
   c. 79   **d.** 1,999

9. **a.** 43   **b.** 97
   c. 171   **d.** 2,007

10. **a.** 91   **b.** 87
    c. 111   **d.** 2,008

Are the statements in Problems 11–14 true or false?

11. a. 6|48   **b.** 7|48
    c. 8|48   **d.** 9|48

12. a. 6∤39   **b.** 5∤30
    c. 16∤576   **d.** 3|7,823

13. a. 15|5   **b.** 5|83,410
    c. 2|628,174   **d.** 10|148,729,320

14. a. 15|4,814   **b.** 17|255
    c. 9|7,823,572   **d.** 10∤148,729,320

15. Find all prime numbers less than or equal to 300.

16. Determine the largest prime you need to consider to be sure that you have excluded, in the sieve of Eratosthenes, all primes less than or equal to:
    a. 200
    b. 500
    c. 1,000
    d. 1,000,000

Write the prime factorization for each of the numbers in Problems 17–20. If the number is prime, so state.

17. a. 24   **b.** 30
    c. 300   **d.** 144

18. a. 108   **b.** 740
    c. 699   **d.** 123

19. a. 120   **b.** 90
    c. 75   **d.** 975

20. a. 490   **b.** 4,752
    c. 143   **d.** 51

Find the canonical representation for each of the numbers in Problems 21–36.

21. 83   **22.** 97

23. 127   **24.** 113

25. 377   **26.** 151

27. 105   **28.** 187

29. 67   **30.** 229

31. 315   **32.** 111

33. 567   **34.** 568

35. 2,869   **36.** 793

Find the g.c.f. and l.c.m. of the sets of numbers in Problems 37–44.

37. {60, 72}

38. {95, 1425}

39. {12, 54, 171}

40. {11, 13, 23}

41. {9, 12, 14}

42. {3, 6, 15, 54}

43. {75, 90, 120}

44. {85, 100, 240}

Level 2

45. Bill and Sue both work at night. Bill has every sixth night off and Sue has every eighth night off. If they are both off tonight, how many nights will it be before they are both off again at the same time?

46. Two movie theaters, UAI and UAII, start their movies at 7:00 P.M. The movie at UAI takes 75 minutes and the movie at UAII takes 90 minutes. If the shows run continuously, when will they again start at the same time?

47. IN YOUR OWN WORDS We used a sieve of Eratosthenes in Table 5.2 by arranging the first 100 numbers into 10 rows and 10 columns. Repeat the sieve process for the first 100 numbers by arranging the numbers in the following patterns.

    a. By 6:   1   2   3   4   5   6
                    7   8   9   10   11   12
                    13   14   15   ...

    b. By 7:   1   2   3   4   5   6   7
                    8   9   10   11   12   13   14
                    15   16   17   ...

    c. By 21:   1   2   3   4   5   6   ...

As you are using these sieves, look for patterns. Describe some of the patterns you notice. Do you think that any of these sieves are better than the one shown in Table 5.2? Why or why not?

48. Use the sieve in Problem 47a to make a conjecture about primes and multiples of 6.

49. Lucky Numbers Set up a sieve similar to the one illustrated here:

1	2̸	3	4̸	5̸	6̸
7	8̸	9	1̸0	1̸1	1̸2
13	1̸4	15	1̸6	1̸7	1̸8
1̸9	2̸0	21	2̸2	2̸3	2̸4
25	2̸6	27	2̸8	2̸9	3̸0
31	3̸2	33	3̸4	3̸5	3̸6
37	3̸8	39	4̸0	4̸1	4̸2
43	4̸4	45	4̸6	4̸7	4̸8

Follow these directions:

Step 1 Start counting with 1 each time.

Step 2 Cross out every second number (shown as).

Step 3 The next uncrossed number is 3, so cross out every 3rd number that remains (shown as ×).

Step 4 The next uncrossed number is 7, so cross out every 7th number that remains (shown as —).

Continue in the same fashion. The numbers that are not crossed out are called *lucky numbers*. What are the lucky numbers less than 100?

Level 3

50. Pairs of consecutive odd numbers that are primes are called *prime twins*. For example, 3 and 5, 11 and 13, and 41 and 43 are prime twins. Can you find any others?

51. **IN YOUR OWN WORDS** Three consecutive odd numbers that are primes are called *prime triplets*. It is easy to show that 3, 5, and 7 are the only prime triplets. Can you explain why this is true?

52. **HISTORICAL QUEST** In 1742 the mathematician Christian Goldbach observed that every even number (except 2) seemed representable as the sum of two primes. Goldbach could not prove this result, known today as *Goldbach's conjecture*, so he wrote to his friend, the world-famous mathematician Leonhard Euler (see historical note on page 762). Euler was unable to prove or disprove this conjecture, and it remains unsolved to this day. Write the following numbers as the sum of two primes (the first three are worked for you):

$$4 = 2 + 2 \qquad 6 = 3 + 3 \qquad 8 = 5 + 3$$
$$10 = \qquad 12 = \qquad 14 =$$
$$16 = \qquad 18 = \qquad 20 =$$
$$40 = \qquad 80 = \qquad 100 =$$

To date, this conjecture has not been proved, but the Russian mathematician L. Schnirelmann (1905–1938) proved that every positive integer can be represented as the sum of not more than 300,000 primes. That may seem like a long way off from Goldbach's conjecture, but at least 300,000 is a finite number! Later, another mathematician, I. M. Vinogradoff, proved that there exists a number N such that all numbers larger than N can be written as the sum of, at most, four primes.

53. Let $S = \{1, 2, 3, 5, 6, 10, 15, 30\}$, and define an operation $\mathcal{M}$ meaning *least common multiple*. For example,

$$5 \mathcal{M} 10 = 10, \qquad 5 \mathcal{M} 6 = 30, \qquad 10 \mathcal{M} 30 = 30$$

   a. Is S closed for the operation of $\mathcal{M}$?
   b. Is S associative for $\mathcal{M}$?
   c. Is S commutative for $\mathcal{M}$?

54. Use an argument similar to the one in the text to show that 23 is not the largest prime.

Problem Solving 3

55. In the text, we showed that 19 is not the largest prime by considering

$$M = 2 \cdot 3 \cdot 5 \cdot 7 \cdot 11 \cdot 13 \cdot 17 \cdot 19 + 1$$

Now, M is either prime or composite. If it is prime, then since it is larger than 19, we have a prime larger than 19. If it is composite, it has a prime divisor larger than 19. In either case, we find a prime larger than 19. Show that this number M does not always generate primes. That is,

$$2 + 1 = 3, \text{ a prime}$$
$$2 \cdot 3 + 1 = 7, \text{ a prime}$$
$$2 \cdot 3 \cdot 5 + 1 = 31, \text{ a prime}$$

Find an example in which the product of consecutive primes plus 1 does not yield a prime.

56. What is the smallest natural number that is divisible by the first 20 counting numbers?

57. Some primes are 1 more than a square. For example, $5 = 2^2 + 1$. Can you find any other primes p so that $p = n^2 + 1$?

58. Some primes are 1 less than a square. For example, $3 = 2^2 - 1$. Can you find any other primes p so that $p = n^2 - 1$?

59. **HISTORICAL QUEST** The Pythagoreans studied numbers to find certain mystical properties in them. Certain numbers they studied were called *perfect numbers*. A *perfect number* is a natural number that is equal to the sum of all its divisors that are less than the number itself. A divisor that is less than the number itself is called a *proper divisor*. The proper divisors of 6 are $\{1, 2, 3\}$ and

$$1 + 2 + 3 = 6$$

so 6 is a perfect number. It is not hard to show that 6 is the smallest perfect number. On the other hand, 24 is not perfect, since its proper divisors are $\{1, 2, 3, 4, 6, 8, 12\}$, which have the sum

$$1 + 2 + 3 + 4 + 6 + 8 + 12 = 36$$

The Pythagoreans discovered the first four perfect numbers. Fourteen centuries later the fifth perfect number was discovered. The 43rd perfect number is

$$2^{N-1}(2^N - 1)$$

where N is the largest prime listed in Table 5.3. It is known that all even perfect numbers are of the form shown for the 43rd perfect number. Show that if $N = 5$, the resulting number is perfect.

60. **HISTORICAL QUEST** The Pythagoreans studied numbers that they called *amicable* or *friendly*. A pair of numbers is *friendly* if each number is the sum of the proper divisors of the other (a proper divisor includes the number 1, but not the number itself). The Pythagoreans discovered that 220 and 284 are friendly. The proper divisors of 220 are $\{1, 2, 4, 5, 10, 11, 20, 22, 44, 55, 110\}$, and

$$1 + 2 + 4 + 5 + 10 + 11 + 20 + 22 + 44 + 55 + 110 = 284$$

Also, the proper divisors of 284 are $\{1, 2, 4, 71, 142\}$, and

$$1 + 2 + 4 + 71 + 142 = 220$$

The next pair of friendly numbers was found by Pierre de Fermat (1601–1665): 17,296 and 18,416. In 1638 the French mathematician René Descartes (1596–1650) found a third pair, and the Swiss mathematician Leonhard Euler (1707–1783) found more than 60 pairs. In 1866, a 16-year-old Italian schoolboy, Nicolo Pagonini, found another relatively small pair of friendly numbers that had been overlooked by the great mathematicians. He found the pair of numbers 1,184 and 1,210. Show that 1,184 and 1,210 are friendly.

5.3 | Integers

B.C. cartoon reprinted by permission of Johnny Hart and Creators Syndicate.

Historically, an agricultural-type society would need only natural numbers, but what about a subtraction such as

$$5 - 5 = ?$$

Certainly society would have a need for a number representing $5 - 5$, so a new number, called **zero,** was invented, so that $5 = 5 + 0$ (remember the definition of subtraction). If this new number is annexed to the set of natural numbers, a set called the set of **whole numbers** is formed:

$$\mathbb{W} = \{0, 1, 2, 3, 4, \ldots\}$$

This one annexation to the existing numbers satisfied society's needs for several thousand years.

However, as society evolved, the need for bookkeeping advanced, and eventually the need to answer this: Can we annex new numbers to the set $\mathbb{W}$ so that it is possible to carry out *all* subtractions? The numbers that need to be annexed are the **opposites** of the natural numbers. The opposite of 3, which is denoted by -3, is the number that when added to 3 gives 0. If we add these opposites to the set $\mathbb{W}$ we have the following set:

$$\mathbb{Z} = \{\ldots, -3, -2, -1, 0, 1, 2, 3, \ldots\}$$

This set is known as the set of **integers.** It is customary to refer to certain subsets of $\mathbb{Z}$ as follows:

1. Positive integers: $\{1, 2, 3, 4, \ldots\}$

2. Zero: $\{0\}$

3. Negative integers: $\{-1, -2, -3, \ldots\}$

Now with this new enlarged set of numbers, are we able to carry out all possible additions, subtractions, and multiplications? Before we answer this question, let's review the process by which we operate within the set of integers. It is assumed that you have had an algebra course, so the following summary is intended only as a review.

You might recall that the process for describing the operations with integers requires the notion of *absolute value*, which represents the distance of a number from the origin when plotted on a number line. We give an algebraic definition.

STOP This definition may be difficult for you to understand; stop for a few moments to make sure you understand what it says.

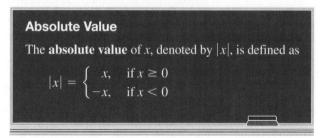

Absolute Value

The **absolute value** of x, denoted by $|x|$, is defined as

$$|x| = \begin{cases} x, & \text{if } x \geq 0 \\ -x, & \text{if } x < 0 \end{cases}$$

Example   **1**   **Find absolute value**

Find the absolute value of each number: **a.** $|5|$ **b.** $|-5|$ **c.** $|-(-3)|$

Solution
a. $|5| = 5$, since $5 \geq 0$
b. $|-5| = 5$, since $-5 < 0$ and $-(-5) = 5$
c. $|-(-3)| = |3| = 3$, since $3 \geq 0$

Addition of Integers

⬦ CAUTION

You may feel you already know how to add integers, but this is one of the fundamental ideas of mathematics

If one (or both) of the integers is 0, then we use the identity property to write $x + 0 = 0 + x = x$, for all x. We could introduce the addition of nonzero integers in terms of number lines, and we note that if the numbers we're adding have the same sign, the result is the same as the sum of the absolute values, except for a plus or a minus sign (since their directions on a number line are the same). If we're adding numbers with different signs, their directions are opposite, so the net result is the difference of the absolute values with a sign to indicate final position. This is summarized by the following procedure for adding integers.

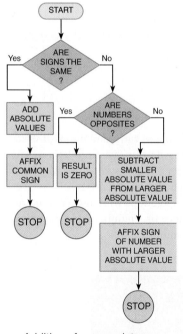

Addition of nonzero integers

Addition of Integers

To add integers x and y, look at the signs of x and y.

Step 1   Check to see if at least one of the integers is 0; if so, then
$x + 0 = 0 + x = x$.

Step 2   Check to see if the signs are the same. If so, then

POSITIVE + POSITIVE   = POSITIVE
NEGATIVE + NEGATIVE = NEGATIVE $\Big\{ |x| + |y|$

Sign part   Whole-number part

Step 3   Check to see if the signs are different. If so, then

POSITIVE + NEGATIVE = Sign of the larger $\Big\{$ Subtract smaller absolute
NEGATIVE + POSITIVE = absolute value   value from larger one

Sign part   Whole-number part

Notice that all integers consist of two parts: a sign part and a whole-number part.

Example   **2**   **Add integers**

Add the integers:
a. $41 + 13$
b. $-41 + (-13)$
c. $41 + (-13)$
d. $-41 + 13$

Solution
a. POSITIVE + POSITIVE: $41 + 13 = 54$ Add absolute values.
b. NEGATIVE + NEGATIVE: $-41 + (-13) = -54$ Add absolute values.
c. POSITIVE + NEGATIVE: $41 + (-13) = 28$ Subtract absolute values.
d. NEGATIVE + POSITIVE: $-41 + 13 = -28$ Subtract absolute values.

COMPUTATIONAL WINDOW

When using a calculator, you must distinguish between a negative sign, as in -5, and a subtraction sign, as in $8 - 5$. For entering negative numbers into a calculator, you'll find a key marked

$\boxed{+/-}$ or $\boxed{\text{CHS}}$ or $\boxed{(-)}$

These keys change the sign of a number. For example, $5 + (-6)$ is entered as

$\boxed{5}\ \boxed{+}\ \boxed{6}\ \boxed{+/-}\ \boxed{=}$ or $\boxed{5}\ \boxed{+}\ \boxed{(-)}\ \boxed{6}\ \boxed{\text{ENTER}}$

Example **3** **Calculator addition**

Indicate the sequence of keys to enter $(-8) + (-5)$ into a calculator.

Solution

$\boxed{8}\ \boxed{+/-}\ \boxed{+}\ \boxed{5}\ \boxed{+/-}\ \boxed{=}$ or $\boxed{(-)}\ \boxed{8}\ \boxed{+}\ \boxed{(-)}\ \boxed{5}\ \boxed{\text{ENTER}}$
  ↑   ↑
Change sign key   Change sign key

Notice that a calculator has separate keys for subtraction $\boxed{-}$ and opposite $\boxed{+/-}$ or $\boxed{(-)}$. The latter keys change the sign of a number to the opposite of its present sign. For example, $5 - (-2)$ would be entered as

$\boxed{5}\ \boxed{-}\ \boxed{2}\ \boxed{+/-}\ \boxed{=}$ or $\boxed{5}\ \boxed{-}\ \boxed{(-)}\ \boxed{2}\ \boxed{\text{ENTER}}$

Because the calculator assumes that all numbers entered are positive, a negative number is obtained by taking the opposite of a positive.

Multiplication of Integers

For whole numbers, multiplication is defined as repeated addition, since we say that $5 \cdot 4$ means

$$\underbrace{4 + 4 + 4 + 4 + 4}_{5\ addends}$$

However, we cannot do this for the integers, since $(-5) \cdot 4$ or

$$\underbrace{4 + 4 + 4 + \cdots + 4}_{-5\ addends\ does\ not\ make\ sense.}$$

Even though you may remember how to multiply integers, we consider four patterns.

POSITIVE · POSITIVE We know how to multiply positive numbers since these are natural numbers. *The product of two positive numbers is a positive number.*

POSITIVE · NEGATIVE Consider, for example, $3 \cdot (-4)$. We look at the pattern:

$3 \cdot 4 = 12$
$3 \cdot 3 = 9$
$3 \cdot 2 = 6$
$3 \cdot 1 = 3$
$3 \cdot 0 = 0$

What comes next? **Answer this question before reading further.**

$$3 \cdot (-1) = -3$$
$$3 \cdot (-2) = -6$$
$$3 \cdot (-3) = -9$$
$$3 \cdot (-4) = -12$$

Do you know how to continue? Try building a few more such patterns using different numbers. What did you discover about the product of a positive number and a negative number? *The product of a positive number and a negative number is a negative number.*

NEGATIVE · POSITIVE Since we now know how to multiply a positive by a negative, and if we assume the commutative property holds, the result here must be the same for a negative times a positive. *The product of a negative number and a positive number is a negative number.*

NEGATIVE · NEGATIVE Consider the example $-3 \cdot (-4)$. Let's build another pattern.

$$-3 \cdot 4 = -12$$
$$-3 \cdot 3 = -9$$
$$-3 \cdot 2 = -6$$
$$-3 \cdot 1 = -3$$
$$-3 \cdot 0 = 0$$

What comes next? **Answer this question before reading further.**

$$-3 \cdot (-1) = 3$$
$$-3 \cdot (-2) = 6$$
$$-3 \cdot (-3) = 9$$
$$-3 \cdot (-4) = 12$$

Take a few moments with this idea and make sure you know how to multiply integers. Can you explain this idea to someone else?

Thus, as the pattern indicates: *The product of two negative numbers is a positive number.* We summarize our discussion in the following box.

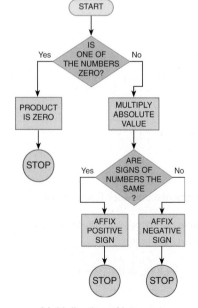

Multiplication of integers

Multiplication of Integers
To multiply integers x and y, look at the signs of x and y.

Step 1 Check to see if at least one of the integers is 0; if so, then $x \cdot 0 = 0 \cdot x = 0$.

Step 2

POSITIVE × POSITIVE	= POSITIVE
POSITIVE × NEGATIVE	= NEGATIVE
NEGATIVE × POSITIVE	= NEGATIVE
NEGATIVE × NEGATIVE	= POSITIVE

$|x| \times |y|$

Sign part   Whole-number part

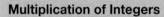

Example 4 **Multiply integers**

Multiply the given integers.

a. $(41)(13)$ **b.** $(-41)(-13)$ **c.** $(41)(-13)$ **d.** $(-41)(13)$

Solution

a. POSITIVE · POSITIVE: $(41)(13) = 533$ Positive
b. NEGATIVE · NEGATIVE: $(-41)(-13) = 533$ Positive
c. POSITIVE · NEGATIVE: $(41)(-13) = -533$ Negative
d. NEGATIVE · POSITIVE: $(-41)(13) = -533$ Negative

Subtraction of Integers

What about subtracting negative numbers? Negative already indicates "going back." Does subtraction of a negative indicate "going ahead?" Consider the following pattern:

$$4 - 4 = 0$$
$$4 - 3 = 1$$
$$4 - 2 = 2$$
$$4 - 1 = 3$$
$$4 - 0 = 4$$

Stop and look for patterns:

$$4 - (-1) = 5$$
$$4 - (-2) = 6$$
$$4 - (-3) = 7$$

Guided by these results, we make the following procedure for subtraction of integers.

This is saying that we do not subtract integers, but rather we change subtraction problems to addition by adding the opposite.

Subtraction of Integers

To subtract, add the opposite (of the number being subtracted). In symbols,

$$x - y = x + (-y)$$

Example 5 Subtract integers

Subtract the given integers.
a. $41 - 13$ **b.** $41 - (-13)$ **c.** $-41 - 13$ **d.** $-41 - (-13)$

Solution
a. POSITIVE − POSITIVE: $41 - 13 = 41 + (-13) = 28$

Subtraction ↑   Addition ↑
                 Opposite
Complete the addition.

b. POSITIVE − NEGATIVE: $41 - (-13) = 41 + 13 = 54$
c. NEGATIVE − POSITIVE: $-41 - 13 = -41 + (-13) = -54$
d. NEGATIVE − NEGATIVE: $-41 - (-13) = -41 + 13 = -28$

Division of Integers

Let's take an overview of what has been done in this chapter. We began with the *natural numbers*, which are closed for addition and multiplication. Next, we defined subtraction and created a situation where it was impossible to subtract some numbers from others. After looking at the prime numbers and factorization, we then "created" another set (called the *integers*) that includes not only the natural numbers, but also zero and the opposite of each of its members.

Since the subtraction of integers is defined in terms of addition, we can easily show that the integers are closed for subtraction. You are asked to do this in Problem 57. We now will define division and then ask, "Is the set of integers closed for division?"

Division is defined as the opposite operation of multiplication.

This completes our definitions of the fundamental operations. Can you define multiplication, subtraction, and division?

> **Division**
>
> If a and b are integers, where $b \neq 0$, then **division** $a \div b$ is written as $\frac{a}{b}$ and is defined in terms of multiplication.
>
> $$\frac{a}{b} = z \qquad \text{means} \qquad a = bz$$

Since division is defined in terms of multiplication, the rules for dividing integers are identical to those for multiplication. We summarize the procedure for $x \div y$, but first we must make sure $y \neq 0$, because division by zero is not defined.

NEVER DIVIDE BY 0.

> **Division of Integers**
>
> To divide integers x and y, look at the signs of x and y.
>
> **Step 1** $0 \div x = 0$ for all numbers x
>
> $x \div 0$ is not permitted.
>
> **Step 2** If $y \neq 0$,
>
> $$\left. \begin{array}{l} \text{POSITIVE} \div \text{POSITIVE} = \text{POSITIVE} \\ \text{POSITIVE} \div \text{NEGATIVE} = \text{NEGATIVE} \\ \text{NEGATIVE} \div \text{POSITIVE} = \text{NEGATIVE} \\ \text{NEGATIVE} \div \text{NEGATIVE} = \text{POSITIVE} \end{array} \right\} \quad \dfrac{|x|}{|y|}$$
>
> Sign part   Whole-number part

Example 6 Divide integers

Divide the given integers.
  a. $12 \div 6$ **b.** $-18 \div 2$ **c.** $10 \div (-2)$ **d.** $-65 \div (-13)$

Solution

a. POSITIVE ÷ POSITIVE: $\dfrac{12}{6} = 2$ *Positive*

b. NEGATIVE ÷ POSITIVE: $\dfrac{-18}{2} = -9$ *Negative*

c. POSITIVE ÷ NEGATIVE: $\dfrac{10}{-2} = -5$ *Negative*

d. NEGATIVE ÷ NEGATIVE: $\dfrac{-65}{-13} = 5$ *Positive*

Notice that for $\frac{a}{b}$, we require $b \neq 0$. Why do we not allow **division by zero?** We consider two possibilities.

1. Division of a nonzero number by zero:

$$a \div 0 \quad \text{or} \quad \frac{a}{0} = x$$

What does this mean? Is there such a number x so that this makes sense? We see that any number x would have to be such that $a = 0 \cdot x$. But $0 \cdot x = 0$ for all x, and since $a \neq 0$, we see that such a situation is impossible. That is, $a \div 0$ does not exist.

2. Division of zero by zero:

$$0 \div 0 \quad \text{or} \quad \frac{0}{0} = x$$

What does this mean? Is there such a number x? We see that *any x* makes this true, since $0 \cdot x = 0$ for all x. But this leads to certain absurdities, for example:

If $\frac{0}{0} = 2$, then this checks since $0 \cdot 2 = 0$; also

if $\frac{0}{0} = 5$, then this checks since $0 \cdot 5 = 0$.

But since both 2 and 5 are equal to the *same* number, we would conclude that $2 = 5$. This is absurd, so we say that division of zero by zero is excluded (or that it is *indeterminate*).

> Abraham Lincoln used a biblical reference (Mark 3:25) to initiate his campaign in 1858. He said, "A house divided against itself cannot stand." I offer a corollary to Lincoln's statement: "A house divided by itself is one (provided, of course, that the house is not zero)."

Is the set of integers closed for division? Certainly we can find many examples in which an integer divided by an integer is an integer. Does this mean that the set of integers is closed? What about $1 \div 2$ or $4 \div 5$? These numbers do not exist in the set of integers; thus, the set is *not* closed for division. Now, as long as society has no need for such division problems, the question of inventing new numbers will not arise. However, as the need to divide one into two or more parts arises, some new numbers will have to be invented so that the set will be closed for division. We'll do this in the next section. The problem with inventing such new numbers is that it must be done in such a way that the properties of the existing numbers are left unchanged. That is, closure for addition, subtraction, and multiplication must be retained.

Historical NOTE

Karl Smith library

Srinivasa Ramanujan (1887–1920)

Ramanujan, who lived only 33 years, was a mathematical prodigy of great originality. He was largely self-taught, but was "discovered" in 1913 by the eminent British mathematician G. H. Hardy. Hardy brought Ramanujan to Cambridge, and in 1918 Ramanujan became the first Indian to become a Fellow of the Royal Society. An often-told story about Hardy and Ramanujan is that when Hardy visited Ramanujan in the hospital, he came in a taxi bearing the number 1729. He asked Ramanujan if there was anything interesting about this number. Without hesitation, Ramanujan said there was: It is the smallest positive integer that can be represented in two different ways as a sum of two cubes:

$$1{,}729 = 1^3 + 12^3 = 9^3 + 10^3$$

Problem Set 5.3

Level 1

1. IN YOUR OWN WORDS Explain how to add integers.

2. IN YOUR OWN WORDS Explain how to subtract integers.

3. IN YOUR OWN WORDS Explain how to multiply integers.

4. IN YOUR OWN WORDS Explain how to divide integers.

5. IN YOUR OWN WORDS Explain the difference between $0 \div 5$ and $5 \div 0$.

6. IN YOUR OWN WORDS Why is division by 0 not defined?

Evaluate each absolute value expression in Problems 7–8.

7. a. $|30|$ **b.** $|-30|$
    c. $-|30|$ **d.** $|30| - |-30|$

8. a. $|18|$ **b.** $|-18|$
    c. $|-(-18)|$ **d.** $|-18| + |18|$

Simplify the expressions in Problems 9–37.

9. a. $5 + 3$ **b.** $-5 + 3$

10. a. $4 + (-7)$ **b.** $-2 + (-4)$

11. a. $7 + 3$ **b.** $-9 + 5$

12. a. $-10 + 4$ **b.** $-8 + (-10)$

13. a. $-15 + 8$ **b.** $|-14 + 2|$

14. a. $|8 + (-10)|$ **b.** $-38 + (-14)$

15. a. $10 - 7$ **b.** $7 - 10$

16. a. $6 - (-4)$ **b.** $0 - (-15)$

17. a. $3(-6)$ **b.** $-5(4)$

18. a. $\frac{-4}{-2}$ **b.** $\frac{-6}{-3}$

19. a. $14(-5)$ **b.** $-14(-5)$

20. a. $-5(8 - 12)$ **b.** $[-5(8)] - 12$

21. **a.** $\frac{-12}{4}$ **b.** $\frac{-63}{-9}$

22. **a.** $\frac{12}{-4}$ **b.** $(-6)\frac{14}{-2}$

23. **a.** $\frac{-528}{-4}$ **b.** $(-1)^3$

24. **a.** $(-1)^4$ **b.** $10\left(\frac{-8}{-2}\right)$

25. **a.** $7(-8)$ **b.** $-5(15)$

26. **a.** $-5(-6)$ **b.** $\frac{-42}{3}$

27. **a.** -2^2 **b.** $(-2)^2$

28. **a.** $(-3)^2$ **b.** -3^2

29. **a.** $-4 - 8$ **b.** $31 + (-16)$

30. **a.** $-14 - 21$ **b.** $-9 + 16 + (-11)$

31. **a.** $162 + (-12)$ **b.** $-12 + [(-4) + (-3)]$

32. **a.** $-46 - (-46)$ **b.** $|7 - (-3)|$

33. **a.** $|-5 - (-10)|$ **b.** $|5 - (-5)|$

34. **a.** $-7 - (-18)$ **b.** $62 - (-112)$

35. **a.** $|-8| + (-8)$ **b.** $-9 - (4 - 5)$

36. **a.** $-(23 + 14)$ **b.** $-7 - (6 - 4)$

37. **a.** $-6 - (-6)$ **b.** $-18 - 5$

Level 2

Simplify the expressions in Problems 38–50.

38. **a.** $-8 + 7 + 16$ **b.** $14 + (-10) - 8 - 11$

39. **a.** $6 + (-8) - 5$ **b.** $-2(-3) + (-1)(6)$

40. **a.** $-5(7) - (-9)$ **b.** $-2(3) - (-8)$

41. **a.** $6 - (-2)$ **b.** $-5 - (-3)$

42. **a.** $-15 - (-6)$ **b.** $-4 + |6 - 8|$

43. **a.** $\frac{-32}{-8} - 5 - (-7)$ **b.** $\frac{-15}{-5} - 4 - (-8)$

44. **a.** $-3 - [(-6) - 4]$ **b.** $5 + (-19) + |15|$

45. **a.** $|-3| - [-(-2)]$ **b.** $15 - (-7)$

46. **a.** $[-54 \div (-9)] \div 3$ **b.** $-54 \div [(-9) \div 3]$

47. **a.** $[48 \div (-6)] \div (-2)$ **b.** $48 \div [(-6) \div (-2)]$

48. **a.** $15 - (-3) - |4 - 11|$
     b. $-12 + (-7) - 10 - 14$

49. **a.** $-5(2) + (-3)(-4) - 6(-7)$
     b. $-8(3) - 6(-4) - 2(-8)$

50. **a.** $(-2)(3)(-4)(5)(-6)(7)(-8)(9)(-10)$
     b. $1 + (-2) + 3 + (-4) + 5 + (-6) + 7 + (-8) + 9 + (-10)$

Level 3

51. Perform the indicated operations. Let k be a natural number.
   a. 1^6 **b.** 1^{67} **c.** 1^{2007}
   d. 1^{2k} **e.** 1^{2k+1} **f.** 1^{2k-1}

52. Perform the indicated operations. Let k be a natural number.
   a. 2^2 **b.** 2^3 **c.** 2^4 **d.** 2^5
   e. Is 2^{2k+1} positive or negative?

53. Perform the indicated operations. Let k be a natural number.
   a. $(-1)^6$ **b.** $(-1)^{67}$ **c.** $(-1)^{2007}$
   d. $(-1)^{2k}$ **e.** $(-1)^{2k+1}$ **f.** $(-1)^{2k-1}$

54. Perform the indicated operations. Let k be a natural number.
   a. $(-2)^2$ **b.** $(-2)^3$ **c.** $(-2)^4$ **d.** $(-2)^5$
   e. Is $(-2)^{2k+1}$ positive or negative?

55. **a.** State the commutative property.
   b. Is $\mathbb{Z}$ commutative for addition? Give reasons.
   c. Is $\mathbb{Z}$ commutative for subtraction? Give reasons.
   d. Is $\mathbb{Z}$ commutative for multiplication? Give reasons.
   e. Is $\mathbb{Z}$ commutative for division? Give reasons.

56. **a.** State the associative property.
   b. Is $\mathbb{Z}$ associative for addition? Give reasons.
   c. Is $\mathbb{Z}$ associative for subtraction? Give reasons.
   d. Is $\mathbb{Z}$ associative for multiplication? Give reasons.
   e. Is $\mathbb{Z}$ associative for division? Give reasons.

57. **IN YOUR OWN WORDS**   Show that the set $\mathbb{Z}$ is closed for subtraction.

Problem Solving 3

58. Find a finite subset of $\mathbb{Z}$ that is closed for multiplication.

59. Multiply $1{,}234{,}567 \times 9{,}999{,}999$
   a. using a calculator
   b. using patterns

60. **Four Fours**   *B.C.* apparently has a mental block against fours, as we can see from the cartoon.
     See if you can handle fours by writing the numbers from 1 to 10 using four 4s, operation symbols, or possibly grouping symbols for each. Here are the first three completed for you:

$$\frac{4}{4} + 4 - 4 = 1$$

$$\frac{4}{4} + \frac{4}{4} = 2$$

$$\frac{4 + 4 + 4}{4} = 3$$

5.4 | Rational Numbers

Historical NOTE

The symbol "÷" was adopted by John Wallis (1616–1703) and was used in Great Britain and in the United States [but not on the European continent, where the colon (:) was used]. In 1923, the National Committee on Mathematical Requirements stated: "Since neither ÷ nor : as signs of division play any part in business life, it seems proper to consider only the needs of algebra, and to make more use of the fractional form and (where the meaning is clear) of the symbol ' / ' and to drop the symbol ÷ in writing algebraic expressions."

—From Report of the *National Committee on Mathematical Requirements* under the auspices of the Mathematical Association of America, Inc. (1923), p. 81.

Historically, the need for a closed set for division came before the need for closure for subtraction. We need to find some number k so that

$$1 \div 2 = k$$

As we saw in Section 4.1, the ancient Egyptians limited their fractions by requiring the numerators to be 1. The Romans avoided fractions by the use of subunits; feet were divided into inches and pounds into ounces, and a twelfth part of the Roman unit was called an *uncia*.

However, people soon felt the practical need to obtain greater accuracy in measurement and the theoretical need to close the number system with respect to the operation of division. In the set $\mathbb{Z}$ of integers, some divisions are possible:

$$\frac{10}{-2}, \quad \frac{-4}{2}, \quad \frac{-16}{-8}, \quad \ldots$$

However, certain others are not:

$$\frac{1}{2}, \quad \frac{-16}{5}, \quad \frac{5}{12}, \quad \ldots$$

Just as we extended the set of natural numbers by creating the concept of opposites, we can extend the set of integers. That is, the number $\frac{5}{12}$ is defined to be that number obtained when 5 is divided by 12. This new set, consisting of the integers as well as the quotients of integers, is called the set of *rational numbers*.

> **Rational Number**
> The set of **rational numbers,** denoted by $\mathbb{Q}$, is the set of all numbers of the form
> $$\frac{a}{b}$$
> where a and b are integers, and $b \neq 0$.

Notice that a rational number has fractional form. In arithmetic you learned that if a number is written in the form $\frac{a}{b}$ it means $a \div b$ and that a is called the **numerator** and b the **denominator.** Also, if a and b are both positive, $\frac{a}{b}$ is called

a **proper fraction** if $a < b$;

an **improper fraction** if $a > b$; and

a **whole number** if b divides evenly into a.

It is assumed that you know how to perform the basic operations with fractions, but we will spend the next few pages reviewing those operations.

Fundamental Property

If the greatest common factor of the numerator and denominator of a given fraction is 1, then we say the fraction is in lowest terms or **reduced.** If the greatest common factor is not 1, then divide both the numerator and denominator by this greatest common factor using the **fundamental property of fractions.**

3 OUT OF 2
PEOPLE
— HAVE —
TROUBLE
— WITH —
FRACTIONS

Historical NOTE

The Arabic word for fraction is *al-kasr* and is derived from the Latin word *fractus* meaning "to break." The English word *fraction* was first used by Chaucer in 1321 and the fraction bar was used in 1556 by Tartaglia, who said, "We write the numerator above a little bar and the denominator below it."

Fundamental Property of Fractions

If $\frac{a}{b}$ is any rational number and x is any nonzero integer, then

$$\frac{a \cdot x}{b \cdot x} = \frac{x \cdot a}{x \cdot b} = \frac{a}{b}$$

 The fundamental property works only for **products** and NOT for sums.

That is, given some fraction that you wish to simplify:

1. Find the g.c.f. of the numerator and denominator (this is x in the fundamental property).
2. Use the fundamental property to simplify the fraction.

Example 1 Reduce fractions

Reduce the given fractions: **a.** $\frac{24}{30}$ **b.** $\frac{300}{144}$

Solution

a. First, find the greatest common factor: $24 = 2^3 \cdot 3^1 \cdot 5^0$
$$30 = 2^1 \cdot 3^1 \cdot 5^1$$
$$\text{g.c.f.} = 2^1 \cdot 3^1 \cdot 5^0 = 6$$

Next, use the fundamental property to simplify the fraction. Thus,

$$\frac{24}{30} = \frac{6 \cdot 2^2}{6 \cdot 5} = \frac{2^2}{5} = \frac{4}{5}$$
$$\uparrow$$
$$\text{g.c.f.}$$

In this book, we agree to leave all fractional answers in reduced form.

CAUTION

b. $300 = 2^2 \cdot 3^1 \cdot 5^2$
$144 = 2^4 \cdot 3^2 \cdot 5^0$
$\text{g.c.f.} = 2^2 \cdot 3^1 \cdot 5^0 = 12$

$$\frac{300}{144} = \frac{12 \cdot 5^2}{12 \cdot 2^2 \cdot 3} = \frac{5^2}{2^2 \cdot 3} = \frac{25}{12}$$

Note that $\frac{25}{12}$ is reduced because the g.c.f. of the numerator and denominator is 1. Notice that a reduced fraction may be an improper fraction.

Operations with Rationals

Now that we have defined rational numbers, we need to review how to add, subtract, multiply, and divide them. The procedure for each of these operations is given in algebraic form.

Operations with Rational Numbers

If $\frac{a}{b}$ and $\frac{c}{d}$ are rational numbers, then

ADDITION $\qquad \dfrac{a}{b} + \dfrac{c}{d} = \dfrac{ad}{bd} + \dfrac{bc}{bd} = \dfrac{ad + bc}{bd}$

SUBTRACTION $\qquad \dfrac{a}{b} - \dfrac{c}{d} = \dfrac{ad}{bd} - \dfrac{bc}{bd} = \dfrac{ad - bc}{bd}$

MULTIPLICATION $\qquad \dfrac{a}{b} \times \dfrac{c}{d} = \dfrac{ac}{bd}$

DIVISION $\qquad \dfrac{a}{b} \div \dfrac{c}{d} = \dfrac{ad}{bc} \quad (c \neq 0)$

You will note that both addition and subtraction require that we first obtain *common denominators*. That process requires a multiplication of fractions, so we begin with an example reviewing multiplication of rational numbers.

Example 2 | Multiply rationals

Multiply the given rational numbers.

a. $\frac{1}{3} \times \frac{2}{5}$ **b.** $\frac{2}{3} \times \frac{-4}{7}$ **c.** $-5 \times \frac{-2}{3}$ **d.** $3\frac{1}{2} \times 2\frac{3}{5}$ **e.** $\frac{3}{4} \times \frac{2}{3}$

Solution

a. $\dfrac{1}{3} \times \dfrac{2}{5} = \boxed{\dfrac{1 \times 2}{3 \times 5}} = \dfrac{2}{15}$

             ↑

    This step is often done in your head

b. $\dfrac{2}{3} \times \dfrac{-4}{7} = \dfrac{-8}{21}$

c. When multiplying a whole number and a fraction, write the whole number as a fraction, and then multiply:

$$-5 \times \frac{-2}{3} = \frac{-5}{1} \times \frac{-2}{3} = \frac{10}{3}$$

d. When multiplying mixed numbers, write the mixed numbers as improper fractions, and *then* multiply:

$$3\frac{1}{2} \times 2\frac{3}{5} = \frac{7}{2} \times \frac{13}{5} = \frac{91}{10}$$

e. For most complicated fractions, the best procedure is to reduce *before* multiplying rather than after, as illustrated here.

$$\overset{1}{\underset{2}{\cancel{3}}} \times \overset{1}{\underset{1}{\cancel{2}}} = \frac{1 \times 1}{2 \times 1} = \frac{1}{2}$$

Example 3 | Division of rational numbers

Justify the rule for division of rational numbers.

Solution   We use Pólya's problem-solving guidelines for this example.

Understand the Problem.   Given $\frac{a}{b} \div \frac{c}{d}$ where $c \neq 0$. Note that we don't say $b \neq 0$ and $d \neq 0$, because the definition of rational number excludes these possibilities, but does not exclude $c = 0$, so this is the condition that must be stated. The example asks us to show where the rule for division as stated in the previous box comes from. Let

$$\frac{a}{b} \div \frac{c}{d} = \boxed{}$$

We are looking for the value of $\boxed{}$. To understand what we are doing here, look at a more familiar problem:

$$\frac{2}{3} \div \frac{4}{5} = \boxed{} \quad \text{means} \quad \frac{4}{5} \times \boxed{} = \frac{2}{3}$$

What do we put into the box to get the answer? Do it in two steps. First multiply $\frac{4}{5}$ by $\frac{5}{4}$ to obtain 1, and *then* multiply by $\frac{2}{3}$ to obtain the result that makes the equation true:

$$\frac{4}{5} \times \boxed{\frac{5}{4} \times \frac{2}{3}} = \frac{2}{3}$$

Thus, $\frac{2}{3} \div \frac{4}{5} = \boxed{\frac{5}{4} \times \frac{2}{3}}$. This seems to suggest that we "invert" the fraction we are dividing by, and then multiply.

Devise a Plan. We will use the definition of division, and then the fundamental property of fractions to multiply the numerator and denominator by the same number.

Carry Out the Plan.

$$\frac{a}{b} \div \frac{c}{d} = \frac{\frac{a}{b}}{\frac{c}{d}}$$

Write the division using fractional notation.

$$= \frac{\frac{a}{b}}{\frac{c}{d}} \times \frac{\frac{d}{c}}{\frac{d}{c}}$$

Multiply numerator and denominator by $\frac{d}{c}$; fundamental property of fractions

$$= \frac{\frac{ad}{bc}}{\frac{cd}{dc}}$$

Multiplication of the "big" fractions

$$= \frac{\frac{ad}{bc}}{1}$$

Reduce the fraction $\frac{cd}{dc}$.

$$= \frac{ad}{bc}$$

Look Back. The result here, $\dfrac{a}{b} \div \dfrac{c}{d} = \dfrac{ad}{bc}$, checks with the entry in the box.

Example 4 | Divide rational numbers

Divide the given rational numbers.

a. $\frac{-3}{4} \div \frac{-4}{7}$ **b.** $\frac{4}{3} \div \frac{8}{9}$

Solution

a. $\dfrac{-3}{4} \div \dfrac{-4}{7} = \dfrac{-3}{4} \times \dfrac{7}{-4} = \dfrac{-21}{-16} = \dfrac{21}{16}$

b. $\dfrac{4}{3} \div \dfrac{8}{9} = \dfrac{\overset{1}{\cancel{4}}}{\underset{1}{\cancel{3}}} \times \dfrac{\overset{3}{\cancel{9}}}{\underset{2}{\cancel{8}}} = \dfrac{1 \times 3}{1 \times 2} = \dfrac{3}{2}$

To carry out addition or subtraction of fractions, you must find the **least common denominator.** The least common denominator is the same as the least common multiple.

Example 5 | Simplify rational expressions

This example is intentionally long. Stick with it because it was designed to illustrate many of the procedures of this section.

Simplify the given expressions.

a. $\frac{5}{24} + \frac{7}{30}$ **b.** $\frac{19}{300} + \frac{55}{144} + \frac{25}{108}$ **c.** $\frac{7}{18} - \frac{-5}{24}$ **d.** $\frac{-1}{15} - \frac{27}{50}$

Solution

a. The denominators are 24 and 30, so we find the l.c.m. of these numbers.

$$24 = 2^3 \cdot 3^1 \cdot 5^0$$
$$30 = 2^1 \cdot 3^1 \cdot 5^1$$
$$\text{l.c.m.} = 2^3 \cdot 3^1 \cdot 5^1 = 120$$

$$\frac{5}{24} = \frac{5}{24} \cdot \frac{5}{5} = \frac{25}{120}$$

We are simply multiplying each fraction by the identity 1, since $\frac{5}{5}$ and $\frac{4}{4}$ are both equal to 1.

$$+\frac{7}{30} = \frac{7}{30} \cdot \frac{4}{4} = \frac{28}{120}$$

$$\frac{53}{120}$$

The answer is in reduced form, since 53 and 120 are relatively prime.

b. If the numbers are complicated, it is easier to work in factored form. In Section 5.2 we found the l.c.m. of 300, 144, and 108 to be $2^4 \cdot 3^3 \cdot 5^2$. Thus,

$$\frac{19}{300} = \frac{19}{2^2 \cdot 3 \cdot 5^2} \cdot \frac{2^2 \cdot 3^2}{2^2 \cdot 3^2} = \frac{19 \cdot 2^2 \cdot 3^2}{2^4 \cdot 3^3 \cdot 5^2}$$

This is the step during which we multiply each fraction by 1.

$$\frac{55}{144} = \frac{5 \cdot 11}{2^4 \cdot 3^2} \cdot \frac{3 \cdot 5^2}{3 \cdot 5^2} = \frac{3 \cdot 5^3 \cdot 11}{2^4 \cdot 3^3 \cdot 5^2}$$

$$+ \frac{25}{108} = \frac{5^2}{2^2 \cdot 3^3} \cdot \frac{2^2 \cdot 5^2}{2^2 \cdot 5^2} = \frac{2^2 \cdot 5^4}{2^4 \cdot 3^3 \cdot 5^2}$$

Adding: $\dfrac{19 \cdot 2^2 \cdot 3^2}{2^4 \cdot 3^3 \cdot 5^2} + \dfrac{3 \cdot 5^3 \cdot 11}{2^4 \cdot 3^3 \cdot 5^2} + \dfrac{2^2 \cdot 5^4}{2^4 \cdot 3^3 \cdot 5^2}$

$$= \frac{684}{2^4 \cdot 3^3 \cdot 5^2} + \frac{4,125}{2^4 \cdot 3^3 \cdot 5^2} + \frac{2,500}{2^4 \cdot 3^3 \cdot 5^2} = \frac{7,309}{2^4 \cdot 3^3 \cdot 5^2}$$

Now 7,309 is not divisible by 2, 3, or 5; thus, 7,309 and $2^4 \cdot 3^3 \cdot 5^2$ are relatively prime, and the solution is complete:

$$\frac{7,309}{2^4 \cdot 3^3 \cdot 5^2} \quad \text{or} \quad \frac{7,309}{10,800}$$

c.
$$18 = 2^1 \cdot 3^2$$
$$24 = 2^3 \cdot 3^1$$
$$\text{l.c.m.} = 2^3 \cdot 3^2 = 72$$

$$\frac{7}{18} = \frac{7}{18} \cdot \frac{4}{4} = \frac{28}{72}$$

$$-\frac{5}{24} = +\frac{5}{24} \cdot \frac{3}{3} = \frac{15}{72}$$

$$\frac{43}{72}$$

d.
$$15 = 3^1 \cdot 5^1$$
$$24 = 2^1 \cdot 5^2$$
$$\text{l.c.m.} = 2^1 \cdot 3^1 \cdot 5^2 = 150$$

$$\frac{-1}{15} = \frac{-1}{15} \cdot \frac{10}{10} = \frac{-10}{150}$$

$$-\frac{27}{50} = +\frac{-27}{50} \cdot \frac{3}{3} = \frac{-81}{150}$$

$$\frac{-91}{150}$$

Notice with subtraction it is usually easier to add the opposite.

The set $\mathbb{Q}$ is closed for the operations of addition, subtraction, multiplication, and nonzero division. As an example, we will show that the rationals are closed for addition. We need to show that, given any two elements of $\mathbb{Q}$, their sum is also an element of $\mathbb{Q}$. Suppose

$$\frac{x}{y} \quad \text{and} \quad \frac{w}{z}$$

are any two rational numbers. By definition of addition,

$$\frac{x}{y} + \frac{w}{z} = \frac{xz + wy}{yz}$$

We now need to show that $\frac{xz + wy}{yz}$ is a rational number. Since x and w are integers and y and z are nonzero integers, we know from closure of the integers for multiplication that xz and wy are also integers. Since the set of integers is closed for addition, we know that $xz + wy$ is also an integer. This means that, since yz is a nonzero integer,

$$\frac{xz + wy}{yz}$$

is a rational number. Thus, the rational numbers are closed for addition.

Problem Set 5.4

Level 1

1. **IN YOUR OWN WORDS**   What does it mean for a fraction to be reduced? Describe a process for reducing a fraction.

2. **IN YOUR OWN WORDS**   Describe the process for multiplying fractions.

3. **IN YOUR OWN WORDS**   Describe the process for dividing fractions.

4. **IN YOUR OWN WORDS**   Describe the process for adding fractions.

5. **IN YOUR OWN WORDS**   Use algebra to show where the formula for subtracting fractions comes from.

Completely reduce the fractions in Problems 6–17.

6. **a.** $\frac{3}{9}$ **b.** $\frac{6}{9}$ 7. **a.** $\frac{2}{10}$ **b.** $\frac{3}{12}$

8. **a.** $\frac{4}{12}$ **b.** $\frac{6}{12}$ 9. **a.** $\frac{14}{7}$ **b.** $\frac{38}{19}$

10. **a.** $\frac{92}{20}$ **b.** $\frac{72}{15}$ 11. **a.** $\frac{42}{14}$ **b.** $\frac{16}{24}$

12. **a.** $\frac{18}{30}$ **b.** $\frac{70}{105}$ 13. **a.** $\frac{50}{400}$ **b.** $\frac{140}{420}$

14. **a.** $\frac{150}{1,000}$ **b.** $\frac{2,500}{10,000}$ 15. **a.** $\frac{78}{455}$ **b.** $\frac{75}{500}$

16. **a.** $\frac{240}{672}$ **b.** $\frac{5,670}{12,150}$ 17. **a.** $\frac{2,431}{3,003}$ **b.** $\frac{47,957}{54,808}$

Perform the indicated operations in Problems 18–37. (Recall that negative exponents are sometimes used to denote fractions. For example, $\frac{1}{7} = 7^{-1}$.)

18. **a.** $\frac{2}{3} + \frac{7}{9}$ **b.** $\frac{-5}{7} + \frac{4}{3}$

19. **a.** $\frac{-12}{35} - \frac{8}{15}$ **b.** $2^{-1} + 3^{-1}$

20. **a.** $3 + 3^{-1}$ **b.** $2 + 2^{-1}$

21. **a.** $\frac{7}{9} - \frac{2}{3}$ **b.** $\frac{4}{7} - \frac{-5}{9}$

22. **a.** $\frac{-3}{5} - \frac{-6}{9}$ **b.** $\frac{6}{7} \div \frac{-3}{7}$

23. **a.** $\frac{5}{3} \div \frac{7}{12}$ **b.** $\frac{-2}{9} \div \frac{6}{7}$

24. **a.** $\frac{2}{3} \times 9$ **b.** $6 \times \frac{5}{9}$

25. **a.** $3^{-1} + 5^{-1}$ **b.** $2^{-1} + 5^{-1}$

26. **a.** $\frac{1}{2} + \frac{1}{3} + \frac{1}{5}$ **b.** $2^{-1} + 3^{-1} + 5^{-1}$

27. **a.** $\frac{2}{3} \times \frac{5}{7}$ **b.** $\frac{-1}{8} \times \frac{2}{-5}$

28. **a.** $\frac{4}{9} \div \frac{2}{3}$ **b.** $\frac{105}{-11} \div \frac{-15}{33}$

29. **a.** 7×7^{-1} **b.** -14×14^{-1}

30. **a.** 8×8^{-1} **b.** -12×12^{-1}

31. **a.** $6 \div 6^{-1}$ **b.** $-5 \div 5^{-1}$

32. **a.** $\frac{1}{10} \cdot \frac{-2}{5}$ **b.** $\frac{-5}{18} \cdot \frac{9}{25}$

33. **a.** $\left(\frac{3}{7} \cdot \frac{3}{5}\right) \div \frac{1}{2}$ **b.** $\frac{2}{-3} \cdot \frac{-2}{15}$

34. **a.** $\frac{4}{5}\left(\frac{17}{95}\right) + \frac{4}{5}\left(\frac{78}{95}\right)$ **b.** $\frac{4}{5}\left(\frac{17}{95} + \frac{78}{95}\right)$

35. **a.** $\frac{28}{9} - \frac{4}{27}$ **b.** $\frac{1}{-8} + \frac{1}{-7}$

36. **a.** $\frac{2}{3} - \frac{7}{12}$ **b.** $\frac{-7}{24} + \frac{-13}{16}$

37. **a.** $-5 + (-5)^2$ **b.** $-3 + (-3)^2$

Level 2

38. **a.** $\dfrac{\frac{1}{2} + \frac{-2}{3}}{\frac{5}{6} - \frac{-3}{5}}$ **b.** $\dfrac{\frac{1}{3} - \frac{-1}{4}}{\frac{7}{8} - \frac{3}{16}}$

39. **a.** $\dfrac{2^{-1} + 3^{-2}}{2^{-1} + 3^{-1}}$ **b.** $\dfrac{6 + 2^{-1}}{\frac{1}{2} + \frac{1}{3}}$

40. **a.** $\frac{-2}{15} + \frac{3}{5} + \frac{7}{12}$ **b.** $-2\frac{3}{5} + 4\frac{1}{8} - 7\frac{1}{10}$

41. **a.** $\frac{-3}{4} \cdot \frac{119}{200} + \frac{-3}{4} \cdot \frac{81}{200}$ **b.** $\frac{-3}{4}\left(\frac{119}{200} + \frac{81}{200}\right)$

42. **a.** $3 + 3^{-1} + 3^{-2}$ **b.** $5^{-1} + 5^{-2} + 5$

43. $\frac{11}{144} + \frac{17}{300} + \frac{7}{108}$ 44. $\frac{11}{108} - \frac{7}{144} + \frac{23}{300}$

45. $\frac{7}{60} - \frac{19}{90} + \frac{21}{51}$ 46. $\frac{14}{90} + \frac{7}{60} - \frac{11}{50}$

47. $\frac{143}{210} + \frac{15}{124} + \frac{11}{1,085}$ 48. $\frac{15}{484} - \frac{5}{234} + \frac{27}{200}$

Level 3

49. **IN YOUR OWN WORDS**   Show that the set $\mathbb{Q}$ of rationals is closed for subtraction.

50. **IN YOUR OWN WORDS**   Show that the set $\mathbb{Q}$ of rationals is closed for nonzero division.

51. **IN YOUR OWN WORDS**   Is the set $\mathbb{Z}$ of integers closed for addition, subtraction, multiplication, and nonzero division? Explain.

52. IN YOUR OWN WORDS   Is $\mathbb{Q}$, the set of rationals, associative for addition? Explain.

53. IN YOUR OWN WORDS   Is $\mathbb{Q}$, the set of rationals, commutative for addition? Explain.

54. IN YOUR OWN WORDS   Is $\mathbb{Q}$, the set of rationals, associative and/or commutative for multiplication? Explain.

Problem Solving 3

A unit fraction (a fraction with a numerator of 1) is sometimes called an Egyptian fraction. In Section 4.1, we said that the Egyptians expressed their fractions as a sum of distinct (different) unit fractions. How might the Egyptians have written the fractions in Problems 55–58?

55. $\frac{3}{4}$

56. $\frac{47}{60}$

57. $\frac{67}{120}$

58. $\frac{7}{17}$

59. HISTORICAL QUEST   Recall that the Egyptians used only unit fractions to represent numbers. Is the answer given on the papyrus correct?

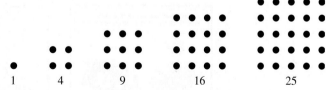

A quantity and its two-thirds and its half and its one-seventh together make 33. Find the quantity.

Answer: 14 $+ \frac{1}{4} + \frac{1}{56} + \frac{1}{97} + \frac{1}{194} + \frac{1}{388} + \frac{1}{679} + \frac{1}{776}$

60. The sum and difference of the same two squares may be primes, as in this example:

$$9 - 4 = 5 \quad \text{and} \quad 9 + 4 = 13$$

Can the sum and difference of the same two primes be squares? Can you find more than one example?

5.5 | Irrational Numbers

We have been considering numbers as they relate to practical problems. However, numbers can be appreciated for their beauty and interrelationships. The Pythagoreans were a Hellenic group of astronomers, musicians, mathematicians, and philosophers who believed that all things are essentially numeric. To our knowledge, they were among the first to investigate numbers for their own sake.

Pythagorean Theorem

Much of the Pythagoreans' lifestyle was embodied in their beliefs about numbers. They considered the number 1 the essence of reason; the number 2 was identified with opinion; and 4 was associated with justice because it is the first number that is the product of equals (the first perfect squared number, other than 1). Of the numbers greater than 1, odd numbers were masculine and even numbers were feminine; thus, 5 represented marriage, since it was the union of the first masculine and feminine numbers $(2 + 3 = 5)$.

The Pythagoreans were also interested in special types of numbers that had mystical meanings: perfect numbers, friendly numbers, deficient numbers, abundant numbers, prime numbers, triangular numbers, square numbers, and pentagonal numbers. Other than the prime numbers we have already considered, the perfect square numbers are probably the most interesting. They are called **perfect squares** because they can be arranged into squares (see Figure 5.4). They are found by squaring the natural numbers.

$$\begin{array}{ccccc} \bullet & \begin{matrix}\bullet\bullet\\\bullet\bullet\end{matrix} & \begin{matrix}\bullet\bullet\bullet\\\bullet\bullet\bullet\\\bullet\bullet\bullet\end{matrix} & \begin{matrix}\bullet\bullet\bullet\bullet\\\bullet\bullet\bullet\bullet\\\bullet\bullet\bullet\bullet\\\bullet\bullet\bullet\bullet\end{matrix} & \begin{matrix}\bullet\bullet\bullet\bullet\bullet\\\bullet\bullet\bullet\bullet\bullet\\\bullet\bullet\bullet\bullet\bullet\\\bullet\bullet\bullet\bullet\bullet\\\bullet\bullet\bullet\bullet\bullet\end{matrix}\\ 1 & 4 & 9 & 16 & 25 \end{array}$$

FIGURE 5.4 Square numbers 1, 4, 9, 16, and 25.
Other square numbers (not pictured) are 36, 49, 64, 81, 100, 121, 144, 169, . . .

**Pythagoras of Samos
(569 B.C.–475 B.C.)**

Very little is known about Pythagoras, but he is an extremely important person in the history of mathematics. He founded a school (see the Prologue) in the 6th century B.C. and his followers were called Pythagoreans. They had their own philosophy, religion, and way of life. This secret society investigated music, astronomy, geometry, and number properties. Because of their strict secrecy, much of what we know about the Pythagoreans is legend, and it is difficult to tell just what work can be attributed to Pythagoras himself. We also know that it was considered impious for a member of the Pythagorean Society to claim any discovery for himself. Instead, each new idea was attributed to their founder, Pythagoras. Every evening each member of the Pythagorean Society had to reflect on three questions:

1. What good have I done today?
2. What have I failed at today?
3. What have I not done today that I should have done?

The Pythagoreans discovered the famous property of **square numbers** that today bears Pythagoras' name. They found that if they constructed any right triangle and then constructed squares on each of the legs of the triangle, the area of the larger square was equal to the sum of the areas of the smaller squares (see Figure 5.5).

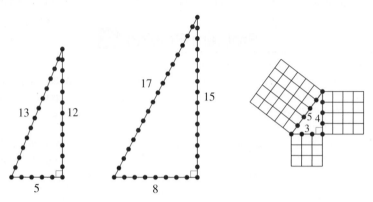

FIGURE 5.5 Relationships of sides of right triangles

Today we state the **Pythagorean theorem** algebraically by saying that, if a and b are the lengths of the **legs** (or sides) of a right triangle and c is the length of the **hypotenuse** (the longest side), then the square of the length of the hypotenuse is equal to the sum of the squares of the lengths of the other two sides.

 STOP The following result, called the **Pythagorean theorem,** is one of the most famous (and important) results in all of mathematics.

Pythagorean Theorem

For a right triangle ABC, with sides of length a, b, and hypotenuse c,

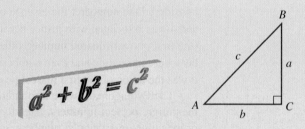

$$a^2 + b^2 = c^2$$

Also, if $a^2 + b^2 = c^2$ for a triangle with sides a, b, and c, then $\triangle ABC$ is a right triangle.

Square Roots

The Pythagoreans were overjoyed with this discovery, but it led to a revolutionary idea in mathematics—one that caused the Pythagoreans many problems.

Legend tells us that one day while the Pythagoreans were at sea, one of their group came up with the following argument. Suppose each leg of a right triangle is 1, then

$$a^2 + b^2 = c^2$$
$$1^2 + 1^2 = c^2$$
$$2 = c^2$$

If we denote the number whose square is 2 by $\sqrt{2}$, we have $\sqrt{2} = c$. The symbol $\sqrt{2}$ is read "square root of two." This means that $\sqrt{2}$ is that number such that, if multiplied by itself, is exactly equal to 2; that is,

$$\sqrt{2} \times \sqrt{2} = 2$$

Square Root

The square root of a nonnegative number n is a number so that its square is equal to n. In symbols, the **positive square root of n**, denoted by $\sqrt{n}$, is defined as that number for which

$$\sqrt{n}\sqrt{n} = n$$

Example 1   Use the definition of square root

Use the definition of square root to find the indicated products.

a. $\sqrt{3} \times \sqrt{3}$

b. $\sqrt{4} \times \sqrt{4}$

c. $\sqrt{5} \times \sqrt{5}$

d. $\sqrt{15} \times \sqrt{15}$

e. $\sqrt{16} \times \sqrt{16}$

f. $\sqrt{144} \times \sqrt{144}$

g. $\sqrt{200} \times \sqrt{200}$

> CAUTION   What is $\sqrt{3}$? The definition says that it is the number with the property that $\sqrt{3}\sqrt{3} = 3$.

Solution   **a.** 3   **b.** 4   **c.** 5   **d.** 15   **e.** 16   **f.** 144   **g.** 200

Some square roots are rational. For example, $\sqrt{4} \times \sqrt{4} = 4$ and we also know $2 \times 2 = 4$, so it seems that $\sqrt{4} = 2$. But wait! We also know $(2) \times (-2) = 4$, so isn't it just as reasonable to say $\sqrt{4} = -2$? Mathematicians have agreed that the square root symbol may be used only to denote positive numbers, so that $\sqrt{4} = 2$ and NOT -2.

What about square roots of numbers that are not perfect squares? Is $\sqrt{2}$, for example, a rational number? Remember, if $\sqrt{2} = a/b$, where a/b is some fraction so that

$$\frac{a}{b} \cdot \frac{a}{b} = 2$$

then it is *rational*. The Pythagoreans were among the first to investigate this question. Now remember that, for the Pythagoreans, mathematics and religion were one; they asserted that all natural phenomena could be expressed by whole numbers or ratios of whole numbers. Thus, they believed that $\sqrt{2}$ must be some whole number or fraction (ratio of two whole numbers). Suppose we try to find such a rational number:

$$\frac{7}{5} \times \frac{7}{5} = \frac{49}{25} = 1.96 \qquad \text{or, try again:} \qquad \frac{707}{500} \times \frac{707}{500} = \frac{499,849}{250,000} = 1.999396$$

We are "getting closer" to 2, but we are still not quite there, so we really get down to business and use a calculator:

$$\boxed{2}\,\boxed{\sqrt{}} \qquad \textit{Display:} \quad 1.414213562$$

If you square this number, do you obtain 2? Notice that the last digit of this multiplication will be 4; what should it be if it were the square root of 2? Even if we use a computer to find the following possibility for $\sqrt{2}$, we see that its square is still not 2:

1.41421356237309504880168872420969807856967187537694807317667973799907324784

Can you give a brief argument showing why that can't be $\sqrt{2}$? Project 5.9 asks for *a proof* that $\sqrt{2}$ is not rational. Such a number is called an *irrational number*. The set of **irrational numbers** is the set of numbers whose decimal representations do not terminate, nor do they repeat.

It can be shown that not only $\sqrt{2}$ is irrational, but also $\sqrt{3}$, $\sqrt{5}$, $\sqrt{6}$, $\sqrt{7}$, $\sqrt{8}$, $\sqrt{10}$; in fact, the square root of any whole number that is not a perfect square is irrational. Also the cube root of any whole number that is not a perfect cube, and so on, is an irrational number. The **number π**, which is the ratio of the circumference of any circle to its diameter, is also not rational. The number π cannot be written in exact decimal form (which is why we use the symbol "π"), but if you press the $\boxed{\pi}$ key on your calculator, you will see the approximation 3.141592654. In everyday work, you will use irrational numbers when finding the circumference and area of a circle.

Another commonly used irrational number is the **number e**, which is defined in Chapter 11, and is related to natural growth or decay. Like π, the number e cannot be written in exact decimal form, but if you press $\boxed{e^x}\,\boxed{1}$ keys on your calculator, you will see the approximation 2.718281828.

Example **2** **Exponentials by calculator**

Approximate e^2 and e^{-3} using your calculator.

Solution   This example is a check to make sure you know how to use your calculator. Check the outputs of your calculator with those shown here. Look for the $\boxed{e^x}$ key on your calculator. Some calculators require input of the value of x first, whereas others require input of the exponent after the e^x key is pressed.

$$e^2 \approx 7.389056099 \qquad e^{-3} \approx 0.0497870684$$

We also use irrational numbers when applying the Pythagorean theorem. Since the Pythagorean theorem asserts $a^2 + b^2 = c^2$, then

$$c = \sqrt{a^2 + b^2}$$

Also, if you wish to find the length of one of the legs of a right triangle, say a, when you know both b and c, you can use the formula

$$a = \sqrt{c^2 - b^2}$$

Example   3   **Find the reach of a ladder**

If a 13-ft ladder is placed against a building so that the base of the ladder is 5 ft away from the building, how high up does the ladder reach?

Solution: Consider Figure 5.6. Let h be the height of the ladder on the building. Since h is one of the legs of a right triangle, use the formula

$$a = \sqrt{c^2 - b^2}$$

$$h = \sqrt{13^2 - 5^2} \qquad \text{Substitute.}$$

↑
unknown

$$= \sqrt{169 - 25}$$

$$= \sqrt{144}$$

$$= 12$$

Thus, the ladder reaches 12 ft up the side of the building.

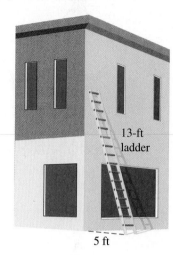

13-ft ladder

5 ft

FIGURE 5.6 Ladder problem

© Myrleen F. Cate/PhotoEdit

The Pythagorean theorem is of value only when dealing with a right triangle. Carpenters often use this property when they want to construct a right angle. That is, if $a^2 + b^2 = c^2$, then an angle of the triangle must be a right angle.

Example   4   **Squaring the corner of a room**

A carpenter wants to make sure that the corner of a room is square (is a right angle). If she measures out sides (legs) of 3 ft and 4 ft, how long should she make the diagonal (hypotenuse) in order to make sure the corner is square?

Solution   The triangle (corner of the room) is shown in Figure 5.7.
The hypotenuse is the unknown, so use the formula

$$c = \sqrt{a^2 + b^2}$$

Thus,

$$c = \sqrt{3^2 + 4^2} = \sqrt{9 + 16} = \sqrt{25} = 5$$

If she makes the diagonal 5 ft long, then by the Pythagorean theorem, the angle is a right angle, which forces the corner of the room to be square.

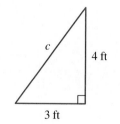

c

4 ft

3 ft

FIGURE 5.7 Building a right angle

Your answers to Examples 3 and 4 are rational. But suppose the result were irrational. You can either leave your result in **radical form** or estimate your result, as shown in Example 5.

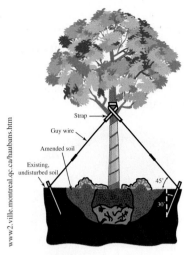

FIGURE 5.8 Guy wires

| Example | 5 | **Length of guy wire** |

Suppose that you need to attach three guy wires to support a large tree, as shown in Figure 5.8. If one guy wire is attached 10 ft away from the center of tree and it is attached 6 ft up on the tree, what is the exact length of one wire? How much wire would you need to purchase for three guy wires?

Solution   The length of the guy wire is the length of the hypotenuse of a right triangle:

$$c = \sqrt{a^2 + b^2}$$
$$= \sqrt{10^2 + 6^2}$$
$$= \sqrt{136}$$

The exact length of the guy wire is $\sqrt{136}$; it is irrational, since 136 is not a perfect square. We use a calculator to find $3\sqrt{136}$: *Display:*   34.985711   You need to purchase 35 ft.

Operations with Square Roots

There are times when a square root is irrational and yet we do not want a rational approximation. We begin by defining when a square root is *simplified.*

> ### Simplified Square Root
>
> A square root is **simplified** if:
> - The radicand (the number under the radical sign) has no factor with an exponent larger than 1 when it is written in factored form.
> - The radicand is not written as a fraction or by using negative exponents.
> - There are no square root symbols used in the denominators of fractions.

In order to make sure that a square root is simplified, we will need to know certain **laws of square roots.**

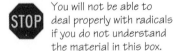

You will not be able to deal properly with radicals if you do not understand the material in this box.

> ### Laws of Square Roots
>
> Let a and b be positive numbers. Then:
>
> **1.** $\sqrt{0} = 0$ **2.** $\sqrt{a^2} = a$ **3.** $\sqrt{ab} = \sqrt{a}\sqrt{b}$ **4.** $\sqrt{\dfrac{a}{b}} = \dfrac{\sqrt{a}}{\sqrt{b}}$

| Example | 6 | **Simplify a square root** |

Simplify $\sqrt{8}$.

Solution

Step 1   Factor the radicand: $\sqrt{8} = \sqrt{2^3}$

Step 2   Write the radicand as a product of as many factors with exponents of 2 as possible; if there is a remaining factor, it will have an exponent of 1: $\sqrt{2^3} = \sqrt{2^2 \cdot 2^1}$.

Step 3   Use Law 3 for square roots: $\sqrt{2^2 \cdot 2^1} = \sqrt{2^2} \cdot \sqrt{2^1}$.

Step 4   Use Law 2 for square roots: $\sqrt{2^2} \cdot \sqrt{2^1} = 2\sqrt{2}$.

Notice that the simplified form in Example 6 still contains a radical, so $\sqrt{8}$ is an irrational number. We call $2\sqrt{2}$ the *exact* simplified representation for $\sqrt{8}$, and the calculator representation 2.828427125 is an *approximation*. The whole process of simplifying radicals depends on factoring the radicand and separating out the square factors and is usually condensed, as shown by the following example.

Example 7 Simplify radical expressions

Simplify the radical expressions and assume that the variables are positive. If the expression is simplified, so state.

a. $\sqrt{441}$ **b.** $3\sqrt{2,100}$ **c.** $\sqrt{(x+y)^2}$ **d.** $\sqrt{x^2+y^2}$

e. $\sqrt{\dfrac{2}{5}}$ **f.** $\dfrac{7\sqrt{2}}{\sqrt{6}}$ **g.** $\sqrt{0.1}$ **h.** $\dfrac{4}{\sqrt{20}}$

i. $\sqrt{\dfrac{5x^2}{27y}}$ **j.** $\sqrt{2^2-4(1)(-5)}$ **k.** $3+\sqrt{5}$ **l.** $\dfrac{3+\sqrt{5}}{3}$

m. $\dfrac{6+3\sqrt{5}}{3}$ **n.** $\dfrac{-(-2)+\sqrt{(-2)^2-4(2)(-1)}}{2(2)}$

> **CAUTION** There are many parts to this example, and you should review them carefully because they lay the groundwork for future sections in this book.

Solution

a. If you do not see any factors that are square numbers, you can use a factor tree:

$$\sqrt{441}=\sqrt{3^2\cdot 7^2}=3\cdot 7=21$$

(factor tree for 441: 441 = 3 · 147; 147 = 3 · 49; 49 = 7 · 7)

b. $3\sqrt{2,100}=3\sqrt{10^2\cdot 3\cdot 7}=3(10\sqrt{3\cdot 7})=30\sqrt{21}$

c. $\sqrt{(x+y)^2}=x+y$

d. $\sqrt{x^2+y^2}$ is simplified; remember that we are looking for square factors, not square terms.

e. $\sqrt{\dfrac{2}{5}}=\dfrac{\sqrt{2}}{\sqrt{5}}\cdot\dfrac{\sqrt{5}}{\sqrt{5}}=\dfrac{\sqrt{10}}{5}$

f. This example came from a *Peanuts* cartoon, and the solution shown in the cartoon is correct.

g. $\sqrt{0.1}=\sqrt{\dfrac{1}{10}}=\dfrac{\sqrt{1}}{\sqrt{10}}\cdot\dfrac{\sqrt{10}}{\sqrt{10}}=\dfrac{\sqrt{10}}{10}$ This is also written as $\dfrac{1}{10}\sqrt{10}$ or $0.1\sqrt{10}$.

h. $\dfrac{4}{\sqrt{20}}=\dfrac{4}{2\sqrt{5}}\cdot\dfrac{\sqrt{5}}{\sqrt{5}}=\dfrac{2\sqrt{5}}{5}$

i. $\sqrt{\dfrac{5x^2}{27y}}=\sqrt{\dfrac{5x^2}{3^2\cdot 3y}}=\dfrac{x\sqrt{5}}{3\sqrt{3y}}\cdot\dfrac{\sqrt{3y}}{\sqrt{3y}}=\dfrac{x\sqrt{5\cdot 3y}}{3(3y)}=\dfrac{x\sqrt{15y}}{9y}$

j. $\sqrt{2^2-4(1)(-5)}=\sqrt{24}=2\sqrt{6}$

k. $3+\sqrt{5}$ is simplified.

l. $\dfrac{3 + \sqrt{5}}{3}$ is simplified.

m. $\dfrac{6 + 3\sqrt{5}}{3} = \dfrac{3(2 + \sqrt{5})}{3} = 2 + \sqrt{5}$

CAUTION

Do not cancel terms:
$\dfrac{\cancel{6} + 3\sqrt{5}}{\cancel{3}} = 2 + 3\sqrt{5}$

is NOT correct.

n. $\dfrac{-(-2) + \sqrt{(-2)^2 - 4(2)(-1)}}{2(2)} = \dfrac{2 + \sqrt{4 + 4(2)}}{4}$

$= \dfrac{2 + \sqrt{12}}{4}$

$= \dfrac{2 + 2\sqrt{3}}{4}$

$= \dfrac{2(1 + \sqrt{3})}{4}$

$= \dfrac{1 + \sqrt{3}}{2}$

Problem Set 5.5

Level 1

1. **IN YOUR OWN WORDS**   What is the Pythagorean theorem?

2. **IN YOUR OWN WORDS**   Explain the two meanings of the square root symbol—as an operation and as a number.

3. **IN YOUR OWN WORDS**   Discuss the exact and decimal approximations for an irrational number.

4. **IN YOUR OWN WORDS**   What does it mean for a square root to be simplified?

5. **IN YOUR OWN WORDS**   A computer approximation for $\sqrt{2}$ is

   1.4142135623730950488016887242096980785696718753769 4

   Give a brief argument showing why that can't be $\sqrt{2}$.

6. **IN YOUR OWN WORDS**   What do you think is meant by a triangular number?

7. Write the following limerick in symbols, and then tell whether it is true or false:[*]

   > *A dozen, a gross, and a score*
   > *Plus three times the square root of four*
   > *Divided by seven*
   > *Plus five times eleven*
   > *Is nine squared and not a bit more.*

Use the definition of square root to find the indicated products in Problems 8–11. Assume the variables are positive.

8. **a.** $\sqrt{6} \times \sqrt{6}$ **b.** $\sqrt{7} \times \sqrt{7}$
   c. $\sqrt{9} \times \sqrt{9}$ **d.** $\sqrt{14} \times \sqrt{14}$

9. **a.** $\sqrt{30} \times \sqrt{30}$ **b.** $\sqrt{36} \times \sqrt{36}$
   c. $\sqrt{807} \times \sqrt{807}$ **d.** $\sqrt{169} \times \sqrt{169}$

10. **a.** $\sqrt{400} \times \sqrt{400}$ **b.** $\sqrt{2.5} \times \sqrt{2.5}$
    c. $\sqrt{2.4} \times \sqrt{2.4}$ **d.** $\sqrt{0.25} \times \sqrt{0.25}$

11. **a.** $\sqrt{a} \times \sqrt{a}$ **b.** $\sqrt{xy} \times \sqrt{xy}$
    c. $2\sqrt{b} \times 2\sqrt{b}$ **d.** $5\sqrt{w} \times 8\sqrt{w}$

Classify each number in Problems 12–17 as rational or irrational. If it is rational, write it without a square root symbol. If it is irrational, approximate it with a rational number correct to the nearest thousandth.

12. **a.** $\sqrt{9}$ **b.** $\sqrt{25}$ **c.** e **d.** π

13. **a.** $\sqrt{10}$ **b.** $\sqrt{30}$ **c.** π^2 **d.** e^2

14. **a.** $\sqrt{36}$ **b.** $\sqrt{50}$ **c.** $\sqrt{e}$ **d.** $\frac{\pi}{2}$

15. **a.** $\sqrt{169}$ **b.** $\sqrt{400}$ **c.** e^π **d.** π^e

16. **a.** $\sqrt{500}$ **b.** $\sqrt{1,000}$ **c.** $\frac{\pi}{6}$ **d.** $e\pi$

17. **a.** $\sqrt{1,024}$ **b.** $\sqrt{1,936}$ **c.** $\sqrt{\pi}$ **d.** $\sqrt{\frac{\pi}{2}}$

Simplify the expressions in Problems 18–37. Assume that the variables are positive.

18. **a.** $-\sqrt{16}$ **b.** $-\sqrt{144}$
    c. $\sqrt{125}$ **d.** $\sqrt{96}$

19. **a.** $\sqrt{1,000}$ **b.** $\sqrt{2,800}$
    c. $\sqrt{2,240}$ **d.** $\sqrt{4,410}$

20. **a.** $3\sqrt{75}$ **b.** $2\sqrt{90}$
    c. $5\sqrt{48}$ **d.** $3\sqrt{96}$

21. **a.** $\sqrt{\frac{1}{2}}$ **b.** $\sqrt{\frac{1}{3}}$
    c. $\sqrt{\frac{3}{5}}$ **d.** $\sqrt{\frac{3}{7}}$

22. **a.** $-\sqrt{0.1}$ **b.** $-\sqrt{0.4}$
    c. $\sqrt{0.75}$ **d.** $\sqrt{0.05}$

23. a. $\dfrac{1}{\sqrt{2}}$ **b.** $\dfrac{-1}{\sqrt{3}}$

   c. $\dfrac{2}{\sqrt{5}}$ **d.** $\dfrac{5}{\sqrt{10}}$

24. a. $\sqrt{(a+b)^2}$ **b.** $\sqrt{a^2 + b^2}$

25. a. $\sqrt{x^2 + 4}$ **b.** $\sqrt{(x+2)^2}$

26. a. $\sqrt{5^2 - 4(3)(2)}$ **b.** $\sqrt{7^2 - 4(5)(2)}$

27. a. $\sqrt{10^2 - 4(5)(-5)}$ **b.** $\sqrt{12^2 - 4(3)(12)}$

28. a. $\sqrt{6^2 - 4(3)(-2)}$ **b.** $\sqrt{2^2 - 4(1)(-1)}$

29. a. $\dfrac{6 + 2\sqrt{5}}{2}$ **b.** $\dfrac{8 - 4\sqrt{3}}{4}$

30. a. $\dfrac{12 - 3\sqrt{2}}{6}$ **b.** $\dfrac{6 - 2\sqrt{5}}{4}$

31. a. $\dfrac{3 - 9\sqrt{x}}{3}$ **b.** $\dfrac{9 + 3\sqrt{x}}{-3}$

32. a. $\dfrac{3}{\sqrt{x}}$ **b.** $\dfrac{-7}{\sqrt{y}}$

33. a. $\sqrt{\dfrac{4x^2}{25y}}$ **b.** $\sqrt{\dfrac{5y}{16x}}$

34. $\dfrac{-7 + \sqrt{7^2 - 4(2)(3)}}{2(2)}$

35. $\dfrac{-(-2) - \sqrt{(-2)^2 - 4(6)(-3)}}{2(6)}$

36. $\dfrac{-10 - \sqrt{10^2 - 4(3)(6)}}{2(3)}$

37. $\dfrac{-(-12) + \sqrt{(-12)^2 - 4(1)(-1)}}{2(1)}$

Level 2

38. How far from the base of a building must a 26-ft ladder be placed so that it reaches 10 ft up the wall?

39. How high up on a wall does a 26-ft ladder reach if the bottom of the ladder is placed 10 ft from the base of the building?

40. If a carpenter wants to make sure that the corner of a room is square and measures out 5 ft and 12 ft along the walls, how long should he make the diagonal?

41. If a carpenter wants to be sure that the corner of a building is square and measures out 6 ft and 8 ft along the sides, how long should she make the diagonal?

42. An antenna is to be erected and held by guy wires. If the guy wires are 15 ft from the base of the antenna and the antenna is 10 ft high, what is the exact length of each guy wire? What is the length of each guy wire to the nearest foot? If three guy wires are to be attached, how many feet of wire should be purchased if it can't be bought by a fraction of a foot?

43. What is the exact length of the hypotenuse if the legs of a right triangle are 2 in. each?

44. What is the exact length of the hypotenuse if the legs of a right triangle are 3 ft each?

45. An empty lot is 400 ft by 300 ft. How many feet would you save by walking diagonally across the lot instead of walking the length and width?

46. A diagonal brace is to be placed in the wall of a room. The height of the wall is 8 ft and the wall is 20 ft long. What is the exact length of the brace? What is the length of the brace to the nearest foot?

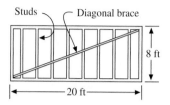

47. A balloon rises at a rate of 4 ft per second when the wind is blowing horizontally at a rate of 3 ft per second. After three seconds, how far away from the starting point, in a direct line, is the balloon?

Level 3

48. Consider a square inch as shown in Figure 5.9.

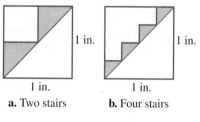

a. Two stairs   **b.** Four stairs

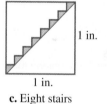

c. Eight stairs

FIGURE 5.9 Find the length of steps

a. Find the total length of the segments making up the stairs in each part of Figure 5.9.

b. Find the length of the diagonal of each square.

c. It seems that if we continue the progression started with the three parts of Figure 5.9, at some point the stairs will become indistinguishable from the diagonal. This seems to say that your answers for parts **a** and **b** should be the same (if there are enough stairs). Discuss.

49. Find an irrational number between 1 and 3.

50. Find an irrational number between 0.53 and 0.54.

51. Find an irrational number between $\frac{1}{11}$ and $\frac{1}{10}$.

52. Without using a radical symbol, write an irrational number using only 2s and 3s.

53. Suppose three squares of uniform thickness are made of gold plate and you are offered either the large square or the two smaller ones. Which choices would you make for each of the squares having sides whose lengths are given below?

   a. 1 in. and 1 in. or 2 in.   **b.** 3 in. and 4 in. or 5 in.
   c. 4 in. and 5 in. or 7 in.   **d.** 5 in. and 7 in. or 9 in.
   e. 10 in. and 11 in. or 15 in.

Problem Solving 3

54. The Pythagorean theorem tells us that the sum of the squares of the lengths of the legs of a right triangle is equal to the square of the length of the hypotenuse. Verify that the theorem is true by tracing the squares and fitting them onto the pattern of the upper two squares shown in Figure 5.10.

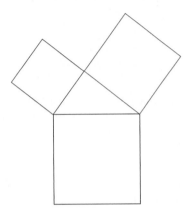

Next, cut the squares along the lines and rearrange the pieces so that the pieces all fit into the large square in Figure 5.10.

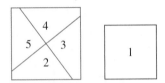

FIGURE 5.10 Pythagorean theorem squares

55. Repeat Problem 54 for the following squares.

56. A man wanted to board a plane with a 5-ft-long steel rod, but airline regulations say that the maximum length of any object or parcel checked on board is 4 ft. Without bending or cutting the rod, or altering it in any way, how did the man check it through without violating the rule?

57. HISTORICAL QUEST The Historical Note on page 222 introduces the great mathematician Karl Gauss. Gauss kept a scientific diary containing 146 entries, some of which were independently discovered and published by others. On July 10, 1796, he wrote

What do you think this meant? Illustrate with some numerical examples.

58. What mathematical property is illustrated in Figure 5.11?

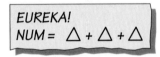

FIGURE 5.11 Can you discover this mathematical property?

59. How are the square numbers embedded in Pascal's triangle?

60. How are the triangular numbers embedded in Pascal's triangle?

5.6 Groups, Fields, and Real Numbers

You are familiar with the rational numbers (fractions, for example) and the irrational numbers (π or square roots of certain numbers, for example), and now we wish to consider the most general set of numbers to be used in elementary mathematics. This set consists of the annexation of the irrational numbers to the set of rational numbers and is called the set of *real numbers*.

Definition of Real Numbers

The set of real numbers is the most common set used in elementary math.

Real Numbers

The set of **real numbers,** denoted by $\mathbb{R}$, is defined as the union of the set of rationals and the set of irrationals.

Decimal Representation

Let's consider the decimal representation of a real number. If a number is *rational*, then its decimal representation is either **terminating** or **repeating.**

Example 1   Convert fractional form to decimal form

Find the decimal representation of each of the given rational numbers.

a. $\frac{1}{4}$ **b.** $\frac{5}{8}$ **c.** $\frac{58}{10}$ **d.** $\frac{2}{3}$ **e.** $\frac{1}{6}$ **f.** $\frac{5}{11}$ **g.** $\frac{1}{7}$

Solution

a. $\frac{1}{4} = 0.25$ $\boxed{1}\ \boxed{\div}\ \boxed{4}$ *This is a terminating decimal.*

b. $\frac{5}{8} = 0.625$ $\boxed{5}\ \boxed{\div}\ \boxed{8}$ *This is a terminating decimal.*

c. $\frac{58}{10} = 5.8$ $\boxed{58}\ \boxed{\div}\ \boxed{10}$ *This is a terminating decimal.*

d. $\frac{2}{3} = 0.666\ldots$ $\boxed{2}\ \boxed{\div}\ \boxed{3}$ *Display:* .6666666667
This is a repeating decimal.

 CAUTION Notice that calculators always represent decimals as terminating decimals, so you need to *interpret* the calculator display as a repeating decimal. If you look at the long division, you can see that the division never terminates:

$$\begin{array}{r} .666\ldots \\ 3\overline{)2.000\ldots} \\ \underline{1.8} \\ 20 \\ \underline{18} \\ 2\ldots \end{array}$$

e. $\frac{1}{6} = 0.166\ldots$ $\boxed{1}\ \boxed{\div}\ \boxed{6}$ *This is a repeating decimal.*

f. $\frac{5}{11} = 0.4545\ldots$ $\boxed{5}\ \boxed{\div}\ \boxed{11}$ *This is a repeating decimal.*

g. $\frac{1}{7} \approx 0.143$ $\boxed{1}\ \boxed{\div}\ \boxed{7}$ *Display:* .1428571429

 CAUTION It may happen that you do not recognize a pattern by looking at the calculator display. For most of our work, an approximation of the result will suffice. Can you use long division to show that the decimal representation must terminate (have a 0 remainder) or *must* repeat? For $\frac{1}{7}$ it repeats after six digits.

When a decimal repeats, we sometimes use an overbar to indicate the numerals that repeat. For Example 1,

One digit repeats: $\frac{2}{3} = 0.\overline{6}$, $\frac{1}{6} = 0.1\overline{6}$

Two digits repeat: $\frac{5}{11} = 0.\overline{45}$

Six digits repeat: $\frac{1}{7} = 0.\overline{142857}$

Real numbers that are *irrational* have decimal representations that are *nonterminating* and *nonrepeating*:

$$\sqrt{2} = 1.414213\ldots \qquad \pi = 3.141592\ldots \qquad e = 2.71828\ldots$$

In each of these examples, the numbers exhibit no repeating pattern and are irrational. Other decimals that do not terminate or repeat are also irrational:

0.12345678910111213 . . .   0.101101110111110111110 . . .

We now have some different ways to classify real numbers:

1. Positive, negative, or zero

2. A rational number or an irrational number
   a. If the decimal representation terminates, it is rational.
   b. If the decimal representation repeats, it is rational.
   c. If it has a nonterminating and nonrepeating decimal, it is irrational.

We have illustrated the procedure for changing from a fraction to a decimal: Divide the numerator by the denominator. To reverse the procedure and to change from a terminating decimal representation of a rational number to a fractional representation, use expanded notation. Recall that $10^{-1} = \frac{1}{10}$, $10^{-2} = \frac{1}{100}$, $10^{-3} = \frac{1}{1,000}, \ldots,$ $10^{-n} = \frac{1}{10^n}$. Thus 0.5 means $5 \times 10^{-1} = 5 \cdot \frac{1}{10} = \frac{5}{10} = \frac{1}{2}$.

Example 2 Convert terminating form to fractional form

Change the terminating decimals to fractional form.
a. 0.123   **b.** 56.28   **c.** 0.3479

Solution Write each in expanded notation.
a. $0.123 = 1 \times 10^{-1} + 2 \times 10^{-2} + 3 \times 10^{-3}$

$$= \frac{1}{10} + \frac{2}{100} + \frac{3}{1,000} = \frac{123}{1,000}$$

b. $56.28 = \frac{5,628}{100} = \frac{1,407}{25}$ *The steps shown in part a can often be done mentally.*

c. $0.3479 = \frac{3,479}{10,000}$ *Make sure the fraction is reduced.*

It is assumed that you can carry out the basic operations with real numbers written in decimal form. You are asked to carry out the basic operations of addition, subtraction, multiplication, and division of decimal fractions.

Real Number Line

If we consider a line and associate the numbers 0 and 1 with two points situated so that the 1 is to the right of 0, we call the distance between these points a **unit distance.** Next, if we mark off equal distances to the right and associate the successive points with the natural numbers, and mark equal distances to the left and associate those points successively with the opposites of the natural numbers, we have drawn a **number line,** as shown in Figure 5.12.

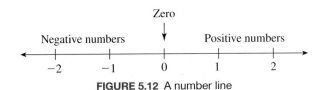

FIGURE 5.12 A number line

If we now associate points on the number line with rational numbers, it appears that the number line is just about "filled up" by the rationals. The reason for this feeling of "fullness" is that the rationals form what is termed a **dense set.** That is, between every two rationals we can find another rational (see Figure 5.13).

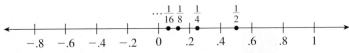

FIGURE 5.13 A number line with rationals

If we plot *all* the points of this dense set called the rationals, are there still any "holes"? In other words, is there any room left for any of the irrationals? We have shown that $\sqrt{2}$ is irrational. We can show that there is a place on the number line representing this length by using the Pythagorean theorem, as shown in Figure 5.14.

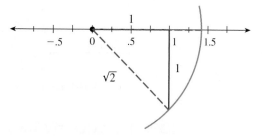

FIGURE 5.14 Finding an irrational "hole" on a dense number line with rationals plotted

We could show that other irrationals have their places on the number line (see Project G16 of the Group Research at the end of the chapter as well as Figure 5.14). These points (corresponding to both the rational and irrational numbers), when plotted on a line, form what is known as the **real number line,** as shown in Figure 5.15.

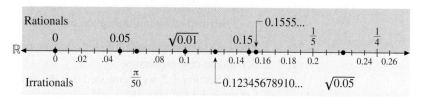

FIGURE 5.15 Real number line showing some rationals and some irrationals

Historical NOTE

The Most Famous Irrational Number, π

It is not at all obvious that π is an irrational number. In the Bible (I Kings 7:23), π is given as 3. In 1892, *The New York Times* printed an article showing π as 3.2. In 1897, House Bill No. 246 of the Indiana State Legislature defined π as 4; this definition was "offered as a contribution to education to be used only by the State of Indiana free of cost. . . ." Actually, the bill's author had 3.2 in mind but worded the bill so that it came out to be 4. The bill was never passed. In a book written in 1934, π was proposed to equal $3\frac{13}{81}$. (The symbol itself was first published in 1706, but was not used widely until Euler adopted it in 1737.) By 1873 William Shanks had calculated π to 700 decimal places. It took him 15 years to do so; however, computer techniques have shown that his last 100 or so places are incorrect.

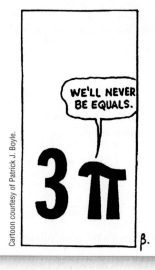

A mnemonic for remembering π:

"Yes, I have a number."

Notice that the number of letters in the words gives an approximation for π (the comma is the decimal point).

June 11, 1989–Two researchers, David Chudnovsky and his brother Gregory, both of Columbia University, have cut themselves a record slice of pi: 480 billion decimal places. This exceeded the previous calculation by more than 200 places. This newly calculated number would stretch for 600 miles if printed, a Columbia University official stated.

Why would someone want 480 billion decimal places of π? It is not the actual calculation that is important, but rather the algorithm, which also has self-check features to check for mistakes.

The relationships among the various sets of numbers we have been discussing are shown in Figure 5.16.

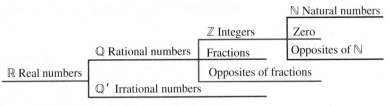

STOP *Spend some time studying these relationships.*

FIGURE 5.16 Classifications within the set of real numbers, $\mathbb{R}$

Identity and Inverse Properties

At the beginning of this chapter, we stated some properties of natural numbers that apply to sets of real numbers as well. We repeat them here for easy reference. Let a, b, and c be real numbers; we denote this by $a, b, c \in \mathbb{R}$. Then,

	Addition	*Multiplication*
Closure:	$(a + b) \in \mathbb{R}$	$ab \in \mathbb{R}$
Associative:	$(a + b) + c = a + (b + c)$	$(ab)c = a(bc)$
Commutative:	$a + b = b + a$	$ab = ba$

Distributive for multiplication over addition: $a(b + c) = ab + ac$

There are two additional properties that are important in the set of real numbers: the identity and inverse properties.

In Section 4.2, we mentioned that the development of the concept of zero and a symbol for it did not take place at the same time as the development of the natural numbers. The Greeks were using the letter "oh" for zero as early as A.D. 150, but the predominant system in Europe was the Roman numeration system, which did not include zero. It was not until the 15th century, when the Hindu-Arabic numeration system finally replaced the Roman system, that the zero symbol came into common usage.

The number 0 (zero) has a special property for addition that allows it to be added to any real number without changing the value of that number. This property is called the **identity property for addition** of real numbers.

Historical NOTE

The Egyptians did not have a zero symbol, but their numeration system did not require such a symbol. On the other hand, the Babylonians, with their positional system, had a need for a zero symbol but did not really use one until around A.D. 150. The Mayan Indians' numeration system was one of the first to use a zero symbol, not only as a placeholder but as a number zero. The exact time of its use is not known, but the symbol was noted by the early 16th century Spanish expeditions into Yucatan. Evidently, the Mayans were using the zero long before Columbus arrived in America.

Identity for Addition

There exists in $\mathbb{R}$ a number 0, called **zero,** so that

$$0 + a = a + 0 = a$$

for any $a \in \mathbb{R}$. The number zero is called the **identity for addition** or the **additive identity.**

Remember that when you are studying algebra, you are studying ideas and not just rules about specific numbers. A mathematician would attempt to isolate the *concept* of an identity. First, does an identity property apply for other operations?

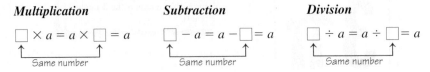

Multiplication
$\square \times a = a \times \square = a$
Same number

Subtraction
$\square - a = a - \square = a$
Same number

Division
$\square \div a = a \div \square = a$
Same number

Is there a real number that will satisfy any of the blanks for multiplication, subtraction, or division?

The second property involves another special, important number in $\mathbb{R}$, namely the number 1 (one). This number has the property that it can multiply any real number without changing the value of that number. This property is called the **identity property for multiplication** of real numbers.

Identity for Multiplication

There exists in $\mathbb{R}$ a number 1, called **one,** so that

$$1 \times a = a \times 1 = a$$

for any $a \in \mathbb{R}$. The number one is called the **identity for multiplication** or the **multiplicative identity.**

Notice that there is no real number that satisfies the identity property for subtraction or division. There may be identities for other operations or for sets other than the set of real numbers.

In the Section 5.3 we also spoke of opposites when we were adding and subtracting integers. Recall the property of opposites:

$$5 + (-5) = 0 \quad -128 + 128 = 0 \quad a + (-a) = 0$$

When opposites are added, the result is zero, the identity for addition. This idea, which can be generalized, is called the **inverse property for addition.**

Inverse Property for Addition

For each $a \in \mathbb{R}$, there is a unique number $(-a) \in \mathbb{R}$, called the **opposite** (or **additive inverse**) of a, so that

$$a + (-a) = -a + a = 0$$

Recall that the product of a number and its reciprocal is 1, the identity for multiplication. The reciprocal of a number, then, is the multiplicative inverse of the number, as we will now show.

Inverse for multiplication
$$\downarrow$$
$$5 \times \boxed{} = \boxed{} \times 5 = 1$$

$$5 \times \boxed{\frac{1}{5}} = \frac{1}{5} \times 5 = 1$$
$$\uparrow$$
Since $\frac{1}{5} \in \mathbb{R}$, $\frac{1}{5}$ is an inverse of 5 for multiplication.

Inverse for multiplication
$$\downarrow$$
$$-128 \times \boxed{} = \boxed{} \times (-128) = 1$$

$$-128 \times \boxed{\frac{1}{-128}} = \frac{1}{-128} \times (-128) = 1$$
$$\uparrow$$
Since $\frac{1}{-128} \in \mathbb{R}$, $\frac{1}{-128}$ is an inverse of -128 for multiplication.

To show this inverse property for multiplication in a general way, we seek to find a replacement for the box for each and every real number a:

$$a \times \boxed{} = \boxed{} \times a = 1$$

Does the inverse property for multiplication hold for every real number a? No, because if $a = 0$, then

$$0 \times \boxed{} = \boxed{} \times 0 = 1$$

does not have a replacement for the box in $\mathbb{R}$. However, the inverse property for multiplication holds for all *nonzero* replacements of a, and we adopt this condition as part of the inverse property for multiplication of real numbers.

Inverse Property for Multiplication

For *each* number $a \in \mathbb{R}$, $a \neq 0$, there exists a number $a^{-1} \in \mathbb{R}$, called the **reciprocal** (or **multiplicative inverse**) of a, so that

$$a \times a^{-1} = a^{-1} \times a = 1$$

Example 3   **Inverse property for multiplication**

Given the set $A = \{-1, 0, 1\}$. Does this set A have an element that satisfies the inverse property for multiplication?

Solution   Before we can talk about the inverse property, we need to find the identity for the operation, in this case, multiplication. Without an identity for multiplication, we cannot have an inverse for multiplication. The operation of multiplication has the identity 1. Check to see that *each* element of A has an inverse:

Member of set	*Operation*	*Inverse*	*Identity*	
↓	↓	↓		
-1	$\times$	-1	$= 1$	
0	$\times$	?	$= 1$	No such number for "?"; does not matter because it is 0.
1	$\times$	1	$= 1$	

Since every nonzero element in the set has an inverse, we say the inverse property for multiplication is satisfied.

Algebraic Structure

The idea of using a variable to represent an operation will take some extra effort to understand.

We have now introduced several properties that we can apply to a given set with a given operation. If every element of a set satisfies the closure, associative, identity, and inverse properties for a particular operation, then we call that set a *group*. We use the symbol $\circ$ to stand for any operation. This operation might be $+$, $\times$, or any other *given* operation.

Group

Let $\mathbb{S}$ be any set, let $\circ$ be any operation, and let a, b, and c be any elements of $\mathbb{S}$. We say that $\mathbb{S}$ is a **group** for the operation of $\circ$ if the following properties are satisfied:

1. *Closure:* $(a \circ b) \in \mathbb{S}$

2. *Associative:* $(a \circ b) \circ c = a \circ (b \circ c)$

3. *Identity:*   There exists a number $\iota \in \mathbb{S}$ so that

$$x \circ \iota = \iota \circ x = x \text{ for } every\ x \in \mathbb{S}$$

4. *Inverse:*   For *each* $x \in \mathbb{S}$, there exists a corresponding $x^{-1} \in \overline{\mathbb{S}\ \text{so that}}$

$$x \circ x^{-1} = x^{-1} \circ x = \iota, \text{ where } \iota \text{ is the identity element in } \mathbb{S}.$$

Furthermore, $\mathbb{S}$ is called a **commutative group** (or **Abelian group**) if the following property is satisfied:

5. *Commutative:* $a \circ b = b \circ a$

The idea of a group is an important unifying idea in more advanced mathematics.

A commutative group is called Abelian to honor the mathematician Niels Abel (see Historical Note on page 255).

Example **4** **Natural numbers is a group**

Is the set $\mathbb{N}$ of natural numbers a group for multiplication?

Solution   We have already studied the properties of $\mathbb{N}$, so we can form hasty conclusions:

1. **Closure:**   The product of any two natural numbers is a natural number.

2. **Associative:**   Yes

3. **Identity:**   Yes, namely, 1

4. **Inverse:**   No, since there is no number $\square$ in $\mathbb{N}$ so that $3 \times \square = \square \times 3 = 1$
   All we need to do to show that a property doesn't hold is to come up with one counterexample. We might also note that an inverse exists, namely $\frac{1}{3}$, but $\frac{1}{3} \notin \mathbb{N}$.
   Therefore, $\mathbb{N}$ does not form a group for multiplication.

Example **5** **Test group properties**

Pólya's Method

This is a lengthy example, but if you stick with it, you will have a better understanding of the ideas of this section.

Let us partition the set of natural numbers into two sets, E (even) and O (odd). Consider the operations of addition $(+)$ and multiplication $(\times)$ in the set $\{E, O\}$. Does the set form a group for either or both of these operations?

Solution   We use Pólya's problem-solving guidelines for this example.

Understand the Problem.   In this example, we are considering a set consisting of two elements E and O (never mind that each of these elements is also a set). How could we possibly define addition? $E + O$ means that we take *any* even number and add to it *any* odd number, and then ask whether the result is even, odd, or something else; in this case we conclude that the answer must be an odd number, so we write $E + O = O$. Consider the operations of addition and multiplication:

Addition	*Multiplication*
even + even = even	even $\times$ even = even
even + odd = odd	even $\times$ odd = even
odd + even = odd	odd $\times$ even = even
odd + odd = even	odd $\times$ odd = odd

These operations can be summarized in table format:

+	E	O
E	E	O
O	O	E

×	E	O
E	E	E
O	E	O

Devise a Plan.   We will check the properties for each of these operations one at a time. If and when we find a counterexample for one of the group properties, we will have the conclusion that it is not a group for that operation.

Carry Out the Plan.

Property	Addition	Multiplication
1. Closure	Yes	Yes

Every entry in the tables is either an E or O, both of which are in the set.

2. Associative	Yes	Yes

We prove this by checking all possibilities (there are several). We show a few here.

$(E + E) + E = E + (E + E)$	$(E \times E) \times E = E \times (E \times E)$
$(E + E) + O = E + (E + O)$	$(E \times E) \times O = E \times (E \times O)$
$(E + O) + E = E + (O + E)$	$(E \times O) \times E = E \times (O \times E)$
$\vdots$	$\vdots$

Property	Addition	Multiplication
3. Identity	Yes, it is E.	Yes, it is O.
4. Inverse	Yes	No
	Inverse of E is E $(E + E = E)$.	E does not have an inverse
	Inverse of O is O $(O + O = E)$.	because $E \times ? = O$ has no value for "?".

Conclusion: The set $\{E, O\}$ is a *group* for $+$, but not for $\times$.

5. Commutative	Yes	Yes

The tables are symmetric with respect to the principal diagonal. That is, $E + O = O + E$ and $E \times O = O \times E$.

Look Back. The set $\{E, O\}$ is a commutative group for addition.

Example 5 showed that a given set was a commutative group for one operation but not for another. If a set is a commutative group for two operations and *also* satisfies the distributive property, it is called *a field*. We define a field in terms of the set $\mathbb{R}$ and the operations of $+$ and x.

The field properties summarize the main properties used with the set of real numbers. We call this the structure of the real numbers.

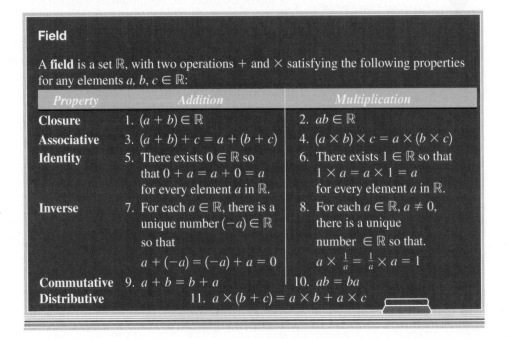

Field

A **field** is a set $\mathbb{R}$, with two operations $+$ and $\times$ satisfying the following properties for any elements $a, b, c \in \mathbb{R}$:

Property	Addition	Multiplication
Closure	1. $(a + b) \in \mathbb{R}$	2. $ab \in \mathbb{R}$
Associative	3. $(a + b) + c = a + (b + c)$	4. $(a \times b) \times c = a \times (b \times c)$
Identity	5. There exists $0 \in \mathbb{R}$ so that $0 + a = a + 0 = a$ for every element a in $\mathbb{R}$.	6. There exists $1 \in \mathbb{R}$ so that $1 \times a = a \times 1 = a$ for every element a in $\mathbb{R}$.
Inverse	7. For each $a \in \mathbb{R}$, there is a unique number $(-a) \in \mathbb{R}$ so that $a + (-a) = (-a) + a = 0$	8. For each $a \in \mathbb{R}$, $a \neq 0$, there is a unique number $\in \mathbb{R}$ so that. $a \times \frac{1}{a} = \frac{1}{a} \times a = 1$
Commutative	9. $a + b = b + a$	10. $ab = ba$
Distributive	11. $a \times (b + c) = a \times b + a \times c$	

The set of real numbers is a field, but there are other fields. In our definition of a field we used $\mathbb{R}$ and the operations of addition and multiplication for the sake of understanding, but for the mathematician, a field is defined as a set with *any two* operations satisfying the 11 stated properties.

Problem Set 5.6

Level 1

1. **IN YOUR OWN WORDS**   What is the distinguishing characteristic between the rational and irrational numbers?

2. **IN YOUR OWN WORDS**   A segment of length $\sqrt{2}$ is shown in Figure 5.14. Describe a process you might use to draw a segment of length $\sqrt{3}$.

3. **IN YOUR OWN WORDS**   Explain the identity property.

4. **IN YOUR OWN WORDS**   Explain the inverse property.

5. What is a group?

6. What is a field?

7. Tell whether each number is an element of $\mathbb{N}$ (a natural number), $\mathbb{Z}$ (an integer), $\mathbb{Q}$ (a rational number), $\mathbb{Q}'$ (an irrational number), or $\mathbb{R}$ (a real number). You may need to list more than one set for each answer.

a. 7   **b.** 4.93
c. 0.656656665 . . .   **d.** $\sqrt{2}$
e. 3.14159   **f.** $\frac{17}{43}$
g. $0.00\overline{27}$ **h.** $\sqrt{9}$
i. 0

8. Tell whether each number is an element of $\mathbb{N}$ (a natural number), $\mathbb{Z}$ (an integer), $\mathbb{Q}$ (a rational number), $\mathbb{Q}'$ (an irrational number), or $\mathbb{R}$ (a real number). Since these sets are not all disjoint, you may need to list more than one set for each answer.

a. 19   **b.** 6.48
c. 1.868686868 . . .   **d.** $\sqrt{8}$
e. π **f.** $0.001\overline{2}$
g. $\sqrt{16}$ **h.** $\sqrt{1,000}$
i. $\frac{1}{7}$

Express each of the numbers in Problems 9–12 as a decimal.

9. a. $\frac{3}{2}$ **b.** $\frac{7}{10}$ **c.** $\frac{3}{5}$ **d.** $\frac{27}{15}$

10. a. $\frac{5}{6}$ **b.** $\frac{2}{7}$ **c.** $\frac{3}{25}$ **d.** $2\frac{1}{6}$

11. a. $\frac{2}{3}$ **b.** $2\frac{2}{13}$ **c.** $\frac{15}{3}$ **d.** $\frac{12}{11}$

12. a. $\frac{-4}{5}$ **b.** $-\frac{2}{3}$ **c.** $-\frac{17}{6}$ **d.** $\frac{-14}{5}$

Change the terminating decimals in Problems 13–20 to fractional form.

13. a. 0.5   **b.** 0.8   **14. a.** 0.25   **b.** 0.75

15. a. 0.45   **b.** 0.234   **16. a.** 0.111   **b.** 0.52

17. a. 98.7   **b.** 0.63   **18. a.** 0.24   **b.** 16.45

19. a. 15.3   **b.** 6.95   **20. a.** 0.64   **b.** 6.98

Carry out the operations with decimal forms in Problems 21–25.

21. a. $6.28 - 3.101$ **b.** $-6.824 + 1.32$
     c. $1.36 + 0.541$ **d.** $6.31 - 12.62$

22. a. $-4.2 - 0.921$ **b.** $8.23 + (-0.005)$
     c. $-6.03 \times (-4.6)$ **d.** 5.002×9.009

23. a. -0.44×0.298 **b.** $-10.5(6.23)$
     c. $3.72 \div 0.3$ **d.** $(-5.95) \div (-7.00)$

24. a. $13.06 \div 0.02$ **b.** $8 \div 4.002$
     c. $0.5(6.2 + 3.4)$ **d.** $0.25(5.03 - 4.005)$

25. a. $5.2 \times 2.3 - 4.5$ **b.** $5.2 - 2.3 \times 4.5$
     c. $8.2 + 2.8 \times 23$ **d.** $8.2 \times 2.8 + 23$

Identify each of the properties illustrated in Problems 26–31.

26. a. $5 + 7 = 7 + 5$ **b.** $5 \cdot \frac{1}{5} = 1$

27. a. $5 \cdot 1 = 1 \cdot 5$
     b. $3(4 + 8) = 3(4) + 3(8)$

28. a. $a + (10 + b) = (a + 10) + b$
     b. $a + (10 + b) = (10 + b) + a$

29. a. mustard + catsup = catsup + mustard
     b. (red + blue) + yellow = red + (blue + yellow)

30. $15 + [a + (-a)] = 15 + 0$

31. $\dfrac{x^2 + x - 1}{x^2 - 1} = \dfrac{x^2 + x - 1}{x^2 - 1} \cdot \dfrac{\frac{1}{x^2}}{\frac{1}{x^2}}$

Figure 5.17 shows the cover of a leading mathematics journal. It depicts the symbols for the twelve animals in the Chinese zodiac. (2010, for example, is the year of the tiger and 2011 is the year of the rabbit.) Use this cover illustration to answer the questions in Problems 32–35.

32. Find 卯  申 = 丑

33. If x is any element in the Chinese zodiac, then x ⊚ $x = ?$

34. Is the set of Chinese zodiac elements with the operation of ⊚ commutative?

35. Is there an identity element?

Level 2

Check whether each of the sets and operations in Problems 36–44 forms a group.

36. $\mathbb{N}$ for $+$ **37.** $\mathbb{N}$ for $-$ **38.** $\mathbb{W}$ for $+$

39. $\mathbb{W}$ for $\times$ **40.** $\mathbb{Z}$ for $+$ **41.** $\mathbb{Z}$ for $\times$

42. $\mathbb{Q}$ for $+$ **43.** $\mathbb{Q}$ for $\times$ **44.** $\mathbb{Q}$ for $\div$

45. a. Given the set $\{1, 2, 3, 4\}$ and the operation $\times$, construct a multiplication table showing all possible answers for the numbers in the set.
     b. Given the set $\{1, 2, 3, 4\}$ and the operation $*$ defined by $a * b = 2a$, construct a table for $*$ showing all possible answers for numbers in the set.
     c. Verify as many of the field properties as possible for the operations of $\times$ and $*$.

Level **3**

Find a rational number and an irrational number between each of the given pairs of numbers in Problems 46–50.

46. 1 and 10   **47.** 2 and 3

48. 3 and 4   **49.** 4.5 and 4.6

50. 8.00 and 8.01

51. List the rationals between 0 and 1 by roster.

Let ○ be an arbitrary operation in Problems 52–59. Describe the operation ○ for each problem.

52. 5 ○ 3 = 8; 7 ○ 2 = 9; 9 ○ 1 = 10; 8 ○ 2 = 10; . . .

53. 5 ○ 3 = 15; 7 ○ 2 = 14; 9 ○ 1 = 9; 8 ○ 2 = 16; . . .

54. 5 ○ 3 = 2; 7 ○ 2 = 5; 9 ○ 1 = 8; 8 ○ 2 = 6; . . .

55. 1 ○ 9 = 11; 2 ○ 7 = 10; 9 ○ 0 = 10; 9 ○ 8 = 18; . . .

56. 8 ○ 0 = 1; 5 ○ 4 = 21; 1 ○ 0 = 1; 5 ○ 6 = 31; . . .

57. 4 ○ 6 = 20; 8 ○ 2 = 20; 7 ○ 9 = 32; 6 ○ 8 = 28; . . .

58. 4 ○ 7 = 1; 4 ○ 5 = 3; 7 ○ 3 = 11; 12 ○ 9 = 15; . . .

59. 4 ○ 7 = 17; 5 ○ 6 = 26; 6 ○ 4 = 37; 2 ○ 8 = 5; . . .

Problem Solving **3**

60. Symmetries of a Square   Cut out a small square and label it as shown in Figure 5.18.

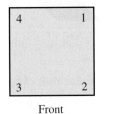

FIGURE 5.18 Square construction

Be sure that 1 is in front of 1′, 2 is in front of 2′, 3 is in front of 3′, and 4 is in front of 4′. We will study certain *symmetries* of this square—that is, the results that are obtained when a square is moved around according to certain rules that we will establish.

Hold the square with the front facing you and the 1 in the top right-hand corner as shown. This is called the *basic position*.

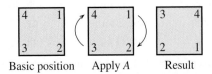

Basic position   Apply *A*   Result

Now rotate the square 90° clockwise so that 1 moves into the position formerly held by 2 and so that 4 and 3 end up on top. We use the letter *A* to denote this rotation of the square. That is, *A* indicates a clockwise rotation of the square through 90°. Other symmetries can be obtained similarly according to Table 5.4. You should be able to tell how each of the results in the table was found. Do this before continuing with the problem.

We now have a set of elements: {*A, B, C, D, E, F, G, H*}. We must define an operation that combines a pair of these

symmetries. Define an operation ★ which means "followed by." Consider, for example, *A* ★ *B*: Start with the basic position, apply *A*, *followed by B* (without returning to basic position) to obtain the result shown:

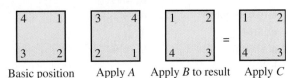

Basic position   Apply *A*   Apply *B* to result   Apply *C*

Thus, we say *A* ★ *B* = *C*, meaning "*A* followed by *B* is the same as applying the single element *C*."

a. Complete a table for the operation ★ and the set {*A, B, C, D, E, F, G, H*}.

b. Is the set closed for ★?

c. Is the set associative for ★?

d. Is the set commutative for ★?

e. Does the set have an identity for ★?

f. Does the inverse property hold for the set and the operation ★?

g. Is the set of symmetries of a square with the operation ★ a group?

TABLE 5.4		
Symmetries of a Square		
Element	**Description**	**Result**
A	90° clockwise rotation	3 4 / 2 1
B	180° clockwise rotation	2 3 / 1 4
C	270° clockwise rotation	1 2 / 4 3
D	360° clockwise rotation	4 1 / 3 2
E	Flip about a **horizontal** line through the middle of the square	3′ 2′ / 4′ 1′
F	Flip about a **vertical** line through the middle of the square	1′ 4′ / 2′ 3′
G	Flip along a line drawn from **upper** left to lower right	4′ 3′ / 1′ 2′
H	Flip along a line drawn from **lower** left to upper right	2′ 1′ / 3′ 4′

5.7 | Discrete Mathematics

FIGURE 5.19 A digitized image of the *Mona Lisa*

Discrete mathematics is that part of mathematics that deals with sets of objects that can be counted or processes that consist of a sequence of individual steps. That is, a *discrete* set is one that is *finite* (the cardinality of the set is a natural number). Computers have enhanced the importance of discrete mathematics because many natural phenomena are idealized in terms of discrete (or countable) steps or numbers of objects. For example, television sets today are termed *digital* TVs, which means that the screen is divided into a large number of cells, and each cell is associated with a number designating its color, brightness, and other characteristics of that picture in that cell. CDs and digital music utilize the same ideas from discrete mathematics. Figure 5.19 shows a famous image that has been *digitized* by computer and used on a cover of a *Finite Mathematics* textbook.

When you hear the word *algebra,* you probably think of the algebra you encountered in high school. The dictionary says that *algebra* is a generalization of arithmetic. Most people speak of algebra in the singular, but to a mathematician **algebra** refers to a structure, a set of symbols that operate according to certain agreed-upon properties, as we saw in the last section. A mathematical dictionary reveals that we can speak of an *algebra with a unit element*, a *simple algebra*, an *algebra of* subsets, a *commutative algebra*, or an *algebra over a field*. In the next chapter we consider the algebra studied in high school, but in this section we will look at a different, discrete algebra that focuses on the definition of operations and on the properties we have studied in this chapter.

Clock Arithmetic

FIGURE 5.20 A 12-hour clock

We now consider a mathematical system based on a 12-hour clock (see Figure 5.20).

We'll need to define some operations for this set of numbers, and we'll use the way we tell time as a guide to our definitions. For example, if you have an appointment at 4:00 P.M. and you are two hours late, you arrive at 6:00 P.M. On the other hand, if your appointment is at 11:00 A.M. and you're two hours late, you arrive at 1:00 P.M. That is, on a 12-hour clock,

$$4 + 2 = 6 \quad \text{and} \quad 11 + 2 = 1$$

Using the clock as a guide, we define an operation called "clock addition," which is different from ordinary addition because it consists of a *finite* set, $\{1, 2, 3, 4, 5, 6, 7, 8, 9, 10, 11, 12\}$, and is *closed* for addition, as you can see by looking at Table 5.5.

TABLE 5.5

Addition on a 12-Hour Clock

+	1	2	3	4	5	6	7	8	9	10	11	12
1	2	3	4	5	6	7	8	9	10	11	12	1
2	3	4	5	6	7	8	9	10	11	12	1	2
3	4	5	6	7	8	9	10	11	12	1	2	3
4	5	6	7	8	9	10	11	12	1	2	3	4
5	6	7	8	9	10	11	12	1	2	3	4	5
6	7	8	9	10	11	12	1	2	3	4	5	6
7	8	9	10	11	12	1	2	3	4	5	6	7
8	9	10	11	12	1	2	3	4	5	6	7	8
9	10	11	12	1	2	3	4	5	6	7	8	9
10	11	12	1	2	3	4	5	6	7	8	9	10
11	12	1	2	3	4	5	6	7	8	9	10	11

Example 1 Find clock sums

Use a 12-hour clock to find the following sums.
a. $7 + 5$ **b.** $9 + 5$ **c.** $4 + 11$ **d.** $12 + 3$

Solution   Look at Table 5.5. Find the first number in the column at the left and then move across that row until you find the second number in the row at the top. The answer is at the intersection of the indicated row and column.
a. $7 + 5 = 12$ **b.** $9 + 5 = 2$ **c.** $4 + 11 = 3$ **d.** $12 + 3 = 3$

Subtraction, multiplication, and division may be defined as they are in ordinary arithmetic.

Note that these definitions of the basic operations are consistent with the usual definitions.

Elementary Operations

Subtraction: $a - b = x$ means $a = b + x$
Multiplication: $a \times b = ab$ means $\underbrace{b + b + \cdots + b}_{a \text{ addends}}$

Zero multiplication: If $a = 0$, then $a \times b = 0 \times b = 0$
Division: $a \div b = \frac{a}{b} = x$ means $a = bx$ provided b has an inverse for multiplication.

Example 2 Clock operations

Carry out the given operations on a 12-hour clock by using the definitions for the elementary operations.
a. $4 - 9$ **b.** 4×9 **c.** $4 \div 7$ **d.** $\frac{4}{9}$

Solution
a. $4 - 9 = x$ means $4 = 9 + x$. That is, what number when added to 9 produces 4? From Table 5.5, we see $x = 7$.

b. 4×9 means $9 + 9 + 9 + 9 = 12$ (from Table 5.5).

c. $4 \div 7 = t$ means $4 = 7t$. That is, what number can be multiplied by 7 to obtain the result 4? We can proceed by trial and error:

$$7 \times 1 = 7; \quad 7 \times 2 = 2; \quad 7 \times 3 = 9; \quad 7 \times 4 = 4; \ldots$$

We see $t = 4$.

d. $\frac{4}{9}$ means $4 \div 9 = s$ or $4 = 9s$. We need to find the number that, when multiplied by 9, produces 4. Once again, proceed by checking each possibility:

$$\begin{array}{llll}
9 \times 1 = 9; & 9 \times 2 = 6; & 9 \times 3 = 3; & 9 \times 4 = 12; \\
9 \times 5 = 9; & 9 \times 6 = 6; & 9 \times 7 = 3; & 9 \times 8 = 12; \\
9 \times 9 = 9; & 9 \times 10 = 6; & 9 \times 11 = 3; & 9 \times 12 = 12
\end{array}$$

Even though this process is tedious, it is complete because we are dealing with a finite set, and since we have checked all possibilities we see there is no such number. We say $\frac{4}{9}$ does not exist.

We see from Example 2 that it might be worthwhile to construct a multiplication table. However, since the addition and multiplication tables for a 12-hour clock are rather large, we shorten our arithmetic system by considering a clock with fewer than 12 hours.

Modulo Five Arithmetic

Consider a mathematical system based on a 5-hour clock, numbered 0, 1, 2, 3, and 4, as shown in Figure 5.21. Clock addition on this 5-hour clock is the same as on an ordinary clock, except that the only numbers in this set are $\{0, 1, 2, 3, 4\}$.

FIGURE 5.21 A five-hour clock

We define the operations of addition, subtraction, multiplication, and division just as we did for a 12-hour clock. Since subtraction and division are defined in terms of addition and multiplication, respectively, we need only the two operation tables that are shown in Table 5.6.

Instead of speaking about "arithmetic on the 5-hour clock" mathematicians usually speak of "modulo 5 arithmetic." The set $\{0, 1, 2, 3, 4\}$, together with the operations defined in Table 5.6, is called a **modulo 5** or **mod 5** system. Suppose it is 4 o'clock on a 5-hour clock. What time will the clock show 9 hours later? We could write

$$4 + 9 = 3 \qquad \text{and} \qquad 2 + 1 = 3$$

thus $4 + 9 = 2 + 1$. Since we do not wish to confuse this with ordinary arithmetic in which

$$4 + 9 \neq 2 + 1$$

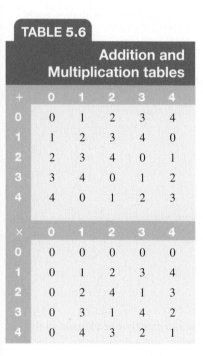

TABLE 5.6

Addition and Multiplication tables

+	0	1	2	3	4
0	0	1	2	3	4
1	1	2	3	4	0
2	2	3	4	0	1
3	3	4	0	1	2
4	4	0	1	2	3

×	0	1	2	3	4
0	0	0	0	0	0
1	0	1	2	3	4
2	0	2	4	1	3
3	0	3	1	4	2
4	0	4	3	2	1

we use the following notation:

$$4 + 9 \equiv 2 + 1, \text{(mod 5)}$$

which is read "4 + 9 is congruent to 2 + 1, mod 5." We define *congruence* as follows.

Congruence Mod *m*

The real numbers a and b are **congruent modulo *m*,** written $a \equiv b, \text{(mod } m)$, if a and b differ by a multiple of m.

Example 3 Checking congruence

Decide whether each statement is true or false.
a. $3 \equiv 8, \text{(mod 5)}$ **b.** $3 \equiv 53, \text{(mod 5)}$ **c.** $3 \equiv 19, \text{(mod 5)}$

Solution
a. $3 \equiv 8, \text{(mod 5)}$ because $8 - 3 = 5$, and 5 is a multiple of 5.
b. $3 \equiv 53, \text{(mod 5)}$ because $53 - 3 = 50$, and 50 is a multiple of 5.
c. $3 \not\equiv 19, \text{(mod 5)}$ because $19 - 3 = 16$, and 16 is not a multiple of 5.

Another way to determine whether two numbers are congruent mod m is to divide each by m and check the remainders. If the remainders are the same, then the numbers are congruent mod m. For example, $3 \div 5$ gives a remainder 3, and $53 \div 5$ gives a remainder of 3, so $3 \equiv 53, \text{(mod 5)}$.

Example 4 Solve modular equations

Solve each equation for x. In other words, find a replacement for x that makes each equation true.
a. $4 + 9 \equiv x, \text{(mod 5)}$
b. $15 + 92 \equiv x, \text{(mod 5)}$
c. $2 + 4 \equiv x, \text{(mod 5)}$
d. $2 - 4 \equiv x, \text{(mod 5)}$
e. $7 \times 5 \equiv x, \text{(mod 7)}$
f. $3 - 5 \equiv x, \text{(mod 12)}$

Solution
a. $4 + 9 = 13 \equiv 3, \text{(mod 5)}$
b. $15 + 92 = 107 \equiv 2, \text{(mod 5)}$
c. $2 + 4 = 6 \equiv 1, \text{(mod 5)}$
d. $2 - 4 \equiv 7 - 4, \text{(mod 5)}$ Since $2 \equiv 7, \text{(mod 5)}$, we can replace 2 by 7.
   $\equiv 3, \text{(mod 5)}$
e. $7 \times 5 = 35 \equiv 0, \text{(mod 7)}$
f. $3 - 5 \equiv 15 - 5, \text{(mod 12)}$ Since $3 \equiv 15, \text{(mod 12)}$
   $\equiv 10, \text{(mod 12)}$

Notice that every whole number is congruent modulo 5 to exactly one element in the set $I = \{0, 1, 2, 3, 4\}$, which contains all possible remainders when dividing by 5. Since the number of elements in I is rather small, we can easily solve equations by trying the numbers 0, 1, 2, 3, and 4. In modulo 7 try numbers in the set $\{0, 1, 2, \ldots, 6\}$ and in modulo 10 try numbers in the set $\{0, 1, 2, \ldots, 9\}$. Consider the following example.

Example 5 | Modular arithmetic application

The manager of a TV station hired a college student, U. R. Stuck, as an election-eve runner. The request for reimbursement turned in is shown in the margin. The station manager refused to pay the bill, since the mileage was not honestly recorded. How did he know? *Note:* The answer has nothing to do with the illegible digits in the mileage report, and you should also assume that the mileage is rounded to the nearest mile.

Solution   Let x = the number of miles in one round trip. For the reimbursement request, we see there are 8 trips, so the total mileage must be $8x$. Since the last digits are legible, we see the difference is 5, (mod 10). Thus,

$$8x \equiv 5, (\text{mod } 10)$$

We solve this by checking each possible value:

$x = 0$: $8(0) \equiv 0, (\text{mod } 10)$
$x = 1$: $8(1) \equiv 8, (\text{mod } 10)$
$x = 2$: $8(2) = 16 \equiv 6, (\text{mod } 10)$
$x = 3$: $8(3) = 24 \equiv 4, (\text{mod } 10)$
$x = 4$: $8(4) = 32 \equiv 2, (\text{mod } 10)$
$x = 5$: $8(5) = 40 \equiv 0, (\text{mod } 10)$
$x = 6$: $8(6) = 48 \equiv 8, (\text{mod } 10)$
$x = 7$: $8(7) = 56 \equiv 6, (\text{mod } 10)$
$x = 8$: $8(8) = 64 \equiv 4, (\text{mod } 10)$
$x = 9$: $8(9) = 72 \equiv 2, (\text{mod } 10)$

These are the only possible values for x in modulo 10, and none gives a mileage reading for 5, (mod 10) in the units digit. Therefore, Stuck did not report the mileage honestly.

Example 6 | Modular arithmetic application   Pólya's Method

Suppose you are planning to buy paper to cover some shelves. You need to cover 1 shelf 100 inches long, and in order to minimize the amount of waste you need to cover at least 30 11-inch shelves, but not more than 50 11-inch shelves. The paper can be purchased in multiples of 36 inches. How much paper should you buy to minimize the waste?

Solution   We use Pólya's problem-solving guidelines for this example.

Understand the Problem. Suppose I need to cover 30 shelves; then I will need $30 \times 11 + 100$ inches of paper. Since $30 \times 11 + 100 = 430$ inches and the paper comes in 36-inch lengths, I can find $430 \div 36 = 11.9\overline{4}$. This means the *minimum* purchase is $12 \times 36 = 432$ inches of paper. On the other hand, if I need to cover 50 shelves, then I will need $50 \times 11 + 100 = 650$ inches; $650 \div 36 = 18.0\overline{5}$. This requires 19 units of length 36 inches: $36 \times 19 = 684$ inches.

Devise a Plan. We know the minimum purchase and we also know the maximum purchase. We need to find a solution that specifies a general solution (for any number of shelves between 30 and 50). Since the paper comes in multiples of 36 inches, we must buy $36x$ inches of paper, where x represents the number of multiples we buy. Suppose we need to cover k shelves at 11 inches and an extra shelf at 100 inches for a total of $11k + 100$ inches. We will write this as a congruence modulo 11 and then solve for x.

Carry Out the Plan. The amount of shelf paper required is $11k + 100$, which means

$$36x = 11k + 100$$
$$36x = 100, \text{(mod 11)}$$
$$3x = 1, \text{(mod 11)} \quad \text{Note: } 36 \equiv 3, \text{(mod 11) and } 100 \equiv 1, \text{(mod 11)}$$

Since we are working in modulo 11, and if we assume no waste, we can check all possibilities (since the set of possibilities is *finite*—namely, 0, 1, 2, 3, 4, 5, 6, 7, 8, 9, and 10; all these congruences are modulo 11):

$x = 0$: $3(0) = 0 \not\equiv 1$ $x = 1$: $3(1) = 3 \not\equiv 1$ $x = 2$: $3(2) = 6 \not\equiv 1$
$x = 3$: $3(3) = 9 \not\equiv 1$ **$x = 4$: $3(4) = 12 \equiv 1$** $x = 5$: $3(5) = 15 \not\equiv 1$
$x = 6$: $3(6) = 18 \not\equiv 1$ $x = 7$: $3(7) = 21 \not\equiv 1$ $x = 8$: $3(8) = 24 \not\equiv 1$
$x = 9$: $3(9) = 27 \not\equiv 1$ $x = 10$: $3(10) = 30 \not\equiv 1$

The solution is $x \equiv 4, \text{(mod 11)}$.

Look Back. We know that the minimum is 12 multiples of the 36-inch paper, and the maximum is 19 multiples. To *minimize the waste* we see that $x \equiv 4, \text{(mod 11)}$ must be $x = 4, 15, 26, \ldots$. This means that to minimize the waste we should buy 15 multiples of the 36-inch paper:

$$36(15) = 540$$

inches of paper. This amount allows us to cover the 100-inch board and 40 11-inch boards.

Group Properties for a Modulo System

> There was a young fellow named Ben
> Who could only count modulo 10
> He said when I go
> Past my last little toe
> I shall have to start over again.

It appears that some modulo systems "behave" like ordinary algebra and others do not. The distinction is found by determining which systems form a field. We assume that modular arithmetic follows the usual order of operations.

Let's first explore the group properties for the set $I = \{0, 1, 2, 3, 4\}$ and the operation of addition, modulo 5.

Closure for $+$:   The set I of elements modulo 5 is closed with respect to addition. That is, for any pair of elements, there is a unique element that represents their sum, and that is also a member of the original set.

Associative for $+$:   Addition of elements modulo 5 satisfies the associative property. That is,

$$(a + b) + c \equiv a + (b + c)$$

for all elements a, b, and $c \in I$. As a specific example, we evaluate $2 + 3 + 4$ in two ways:

$$(2 + 3) + 4 = 9 \equiv 4, \text{(mod 5)}$$
$$2 + (3 + 4) = 9 \equiv 4, \text{(mod 5)}$$

Identity for $+$:   The set I of elements modulo 5 includes an identity element for addition. That is, the set contains an element 0 such that the sum of any given element and zero is the given element. In modulo 5,

$$0 + 0 \equiv 0, 1 + 0 \equiv 1, 2 + 0 \equiv 2, 3 + 0 \equiv 3, 4 + 0 \equiv 4$$

Inverse for $+$:   Each element in arithmetic modulo 5 has an inverse with respect to addition. That is, for each element $a \in I$, there exists a unique element $a' \in I$ such that $a + a' \equiv a' + a \equiv 0$. The element a' is said to be the *inverse* of a. Specifically (in mod 5):

Elements in the set   Identity
↓   ↓

The inverse of 0 is 0: $0 + 0 \equiv 0$

The inverse of 1 is 4: $1 + 4 \equiv 0$

The inverse of 2 is 3: $2 + 3 \equiv 0$

The inverse of 3 is 2: $3 + 2 \equiv 0$

The inverse of 4 is 1: $4 + 1 \equiv 0$
↑
Inverses

Since the closure, associative, identity, and inverse properties are satisfied for addition modulo 5, we conclude that it is a *group*.

Example 7 Testing group properties, modulo 6

Is the set $\{1, 2, 3, 4, 5\}$ a group for multiplication modulo 6?

Solution   This set has an identity element 1 for multiplication. However, the inverse property is not satisfied (in mod 6):

Elements in the set   Identity
↓   ↓

The inverse of 1 is 1: 1×1 $\equiv 1$

The inverse of 2 does not exist: $2 \times ?$ $\equiv 1$

$2 \times 1 \equiv 2, 2 \times 2 \equiv 4,$ ↑

$2 \times 3 \equiv 0, 2 \times 4 \equiv 2, 2 \times 5 \equiv 4$ Inverses

None of the possible products gives the answer 1.

However, the best way to proceed is always to check the closure property first. Note that $2 \times 3 = 6 \equiv 0 \pmod{6}$, which is not in the set, so the set is not closed.

The set is *not* a group for multiplication modulo 6.

Example 8 Testing field properties, modulo 5

Is the set $I = \{0, 1, 2, 3, 4\}$ a field for $+$ and $\times$ modulo 5?

Solution   We have shown (just prior to Example 7) that four of the 11 field properties are satisfied.

Commutative for $+$:   Addition in arithmetic modulo 5 satisfies the commutative property. That is,

$$a + b \equiv b + a$$

where a and b are any elements in I. Specifically, we see that the entries in Table 5.6 on page 233 are symmetric with respect to the principal diagonal.

We can now say that I is a commutative group for addition. We continue by verifying other properties for multiplication.

Closure for $\times$:   The set I is closed for multiplication, as we can see from Table 5.6.

Associative for $\times$:   This property is satisfied, and the details are left for you to verify.

Identity for ×:   The identity element for multiplication is 1, since (in mod 5):

$$0 \times 1 \equiv 0, 1 \times 1 \equiv 1, 2 \times 1 \equiv 2, 3 \times 1 \equiv 3, 4 \times 1 \equiv 4$$

Inverse for ×:   The inverse for multiplication can be checked by finding the inverse of each element (mod 5).

Elements in the set   Identity

↓   ↓

There is no inverse of 0: $0 \times ? \equiv 1$
The inverse of 1 is 1: $1 \times 1 \equiv 1$
The inverse of 2 is 3: $2 \times 3 \equiv 1$
The inverse of 3 is 2: $3 \times 2 \equiv 1$
The inverse of 4 is 4: $4 \times 4 \equiv 1$

↑

Inverses

Since 0 does not have an inverse, we say the inverse property for multiplication is not satisfied, so *I* is not a group. We note, however, that for a field we are checking to see whether all *nonzero* elements have inverses.

Commutative for ×:   The set is commutative for multiplication, as we can see by looking at Table 5.6.

Distributive for × over +:   We need to check $a(b + c) \equiv ab + ac$.

Check some particular examples:

$$2(3 + 4) = 2(7) = 14 \equiv 4, \text{(mod 5)}$$
$$2(3) + 2(4) = 6 + 8 = 14 \equiv 4, \text{(mod 5)}$$

Thus, $2(3 + 4) \equiv 2(3) + 2(4)$. Also,

$$4(1 + 2) \equiv 4(1) + 4(2) \text{ and}$$
$$3(2 + 4) \equiv 3(2) + 3(4)$$

These examples seem to imply that the distributive property holds.

The set *I* is a field for the operations of addition and multiplication, modulo 5.

Problem Set 5.7

Level 1

1. IN YOUR OWN WORDS   What do we mean by clock arithmetic?

2. IN YOUR OWN WORDS   How can the operations of addition and multiplication be defined for a 24-hour clock?

3. IN YOUR OWN WORDS   Discuss the meaning of the definition of congruence modulo *m*.

4. Define precisely the concept of congruence modulo *m*.

Perform the indicated operations in Problems 5–10 using arithmetic for a 12-hour clock.

5. a. $9 + 6$ **b.** $5 - 7$
    c. 5×3 **d.** 2×7

6. a. $7 + 10$ **b.** $7 - 9$
    c. 6×7 **d.** $1 \div 5$

7. a. $5 + 7$ **b.** $4 - 8$
    c. $2 - 6$ **b.** 9×3

8. a. 4×8 **b.** 2×3
    c. $1 \div 12$ **b.** $10 + 6$

9. a. $3 \times 5 - 7$ **b.** $7 + 3 \times 2$

10. a. $5 \times 2 - 11$ **b.** $5 \times 8 + 5 \times 4$

Which of the statements in Problems 11–16 are true?

11. a. $5 + 8 \equiv 1, \text{(mod 6)}$ **b.** $4 + 5 \equiv 1, \text{(mod 7)}$

12. a. $5 \equiv 53, \text{(mod 8)}$ **b.** $102 \equiv 1, \text{(mod 2)}$

13. a. $47 \equiv 2, \text{(mod 5)}$ **b.** $108 \equiv 12, \text{(mod 8)}$

14. **a.** $5,670 \equiv 270$, (mod 365)
    b. $2,001 \equiv 39$, (mod 73)

15. **a.** $2,007 \equiv 0$, (mod 2,007)
    b. $246 \equiv 150$, (mod 6)

16. **a.** $126 \equiv 1$, (mod 7)
    b. $144 \equiv 12$, (mod 144)

Perform the indicated operations in Problems 17–22.

17. **a.** $9 + 6$, (mod 5)
    b. $7 - 11$, (mod 12)
    c. 4×3, (mod 5)
    d. $1 \div 2$, (mod 5)

18. **a.** $5 + 2$, (mod 4)
    b. $2 - 4$, (mod 5)
    a. 6×6, (mod 8)
    b. $5 \div 7$, (mod 9)

19. **a.** $4 + 3$, (mod 5)
    b. $6 - 12$, (mod 8)

20. **a.** $2 \div 3$, (mod 7)
    b. 121×47, (mod 121)

21. **a.** $7 + 41$, (mod 5)
    b. $4 - 5$, (mod 11)

22. **a.** 62×4, (mod 2)
    b. $7 \div 12$, (mod 13)

Solve each equation for x in Problems 23–31. Assume that k is any natural number.

23. **a.** $x + 3 \equiv 0$, (mod 7)   **b.** $4x \equiv 1$, (mod 5)

24. **a.** $x + 5 \equiv 2$, (mod 9)   **b.** $4x \equiv 1$, (mod 6)

25. **a.** $x - 2 \equiv 3$, (mod 6)   **b.** $3x \equiv 2$, (mod 7)

26. **a.** $5x \equiv 2$, (mod 7)   **b.** $7x + 1 \equiv 3$, (mod 11)

27. **a.** $x^2 \equiv 1$, (mod 4)   **b.** $x \div 4 \equiv 5$, (mod 9)

28. **a.** $x^2 \equiv 1$, (mod 5)   **b.** $4 \div 6 \equiv x$, (mod 13)

29. **a.** $4k \equiv x$, (mod 4)   **b.** $4k + 2 \equiv x$, (mod 4)

30. $2x^2 - 1 \equiv 3$, (mod 7)

31. $5x^3 - 3x^2 + 70 \equiv 0$, (mod 2)

Level 2

32. Assume that today is Monday (day 2). Determine the day of the week it will be at the end of each of the following periods. (Assume no leap years.)
    a. 24 days   **b.** 155 days
    c. 365 days   **d.** 2 years

33. Assume that today is Friday (day 6). Determine the day of the week it will be at the end of each of the following periods. (Assume no leap years.)
    a. 30 days   **b.** 195 days
    c. 390 days   **d.** 3 years

34. Your doctor tells you to take a certain medication every 8 hours. If you begin at 8:00 A.M., show that you will not have to take the medication between midnight and 7:00 A.M.

35. Suppose you make six round trips to visit a sick aunt and wish to record your mileage to the nearest mile. You forget the original odometer reading, but you do remember that the units digit has increased by 8 miles. What are the possible distances between your house and your aunt's house?

36. Suppose you are planning to purchase some rope. You need between 15 and 20 pieces that are 7 inches long and one piece that is 80 inches long. The rope can be purchased in multiples of 12 inches. How much rope should you buy to minimize waste?

37. If you know that your aunt in Problem 35 lives somewhere between 10 and 15 miles from your house, how far exactly is her house, given the information in Problem 41?

38. Is the set {0, 1, 2, 3, 4, 5} a group for addition modulo 6?

39. Is the set {0, 1, 2, 3, 4, 5, 6} a group for addition modulo 7?

40. Is the set {0, 1, 2, 3} a field for addition and multiplication modulo 4?

Problems 41–46 involve the set {0, 1, 2, 3, 4, 5, 6, 7, 8, 9, 10} and addition and multiplication modulo 11.

41. Make a table for addition and multiplication.

42. Is the set a group for addition?

43. Is the set a group for multiplication?

44. Is the set a commutative group for addition?

45. Is the set a commutative group for multiplication?

46. Is the set a field for the operations of addition and multiplication?

When constructing addition and multiplication tables for various modular systems, you have probably noticed some patterns. For example, if you consider multiplication mod 19, we find (mod 19):

$$9 \times 1 = 9, 9 \times 2 = 18, 9 \times 3 = 27 \equiv 8, 9 \times 4 = 36 \equiv 17, \ldots$$

Now, draw a circle and divide it into n − 1 parts. In this example, 19 − 1 = 18, so the circumference of the circle in Figure 5.22 is divided into 18 equal parts. Label those parts 1, 2, 3, ... Connect the points formed by the products; that is connect 9 and 1, 2 and 18, 3 and 8, and so on. We call this the (19, 9) modular design.

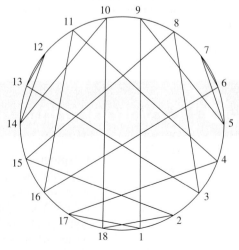

FIGURE 5.22 The (19, 9) design

By shading in alternate regions, interesting patterns can be found, as shown in Figure 5.23.

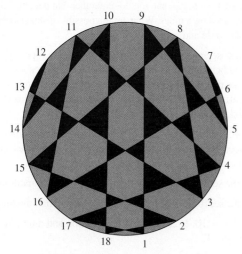

FIGURE 5.23 Shaded (19, 9) design

Create the modular designs in Problems 47–52.

47. (19, 2)   **48.** (19, 18)

49. (21, 5)   **50.** (21, 10)

51. (65, 2)   **52.** (65, 3)

Problem Solving 3

An International Standard Book Number (ISBN) is used to identify books. The ISBN for the 9th edition of The Nature of Mathematics was 0-534-36890-5. The first digit, 0, indicates the book is published in an English-speaking country. The next three digits, 534, identify the publisher (Brooks/Cole), and the next five digits identify the particular book. The last digit is a check digit, which is used as follows. Multiply the first digit by 10, the next digit by 9, and so on:

$$10(0) + 9(5) + 8(3) + 7(4) + 6(3) + 5(6) + 4(8) + 3(9) + 2(0) + 1(5) = 209$$

This number is congruent to 0, (mod 11). This is not by chance. The check digit, 5, is chosen so that this sum is 0, (mod 11). In other words, the check digit x for the book with ISBN 0-534-34015-x is found by considering

$$10(0) + 9(5) + 8(3) + 7(4) + 6(3) + 5(4) + 4(0) + 3(1) + 2(5) = 148 \equiv 5$$

This means that for check digit x, $5 + x \equiv 0$, (mod 11). The one-digit solution to this equation is x = 6, so the check digit is 6. What is the check digit for a book with the given ISBN in Problems 53–54?

53. **a.** 0-534-13728-*x*   **b.** 0-691-02356-*x*

54. **a.** 0-8028-1430-*x*   **b.** 9-68-7270-82-*x*

55. What are the possible check digits for ISBNs? What is the check digit for the 10th edition of *The Nature of Mathematics* with ISBN 0-534-40023-X? What do you think the X stands for in this ISBN?

56. If it is now 2 P.M., what time will it be 99,999,999,999 hours from now?

57. Write a schedule for 12 teams so that each team will play every other team once and no team will be idle.

58. **One Hundred Fowl Problem** (an old Chinese puzzle) A man buys 100 birds for $100. A rooster is worth $10, a hen is worth $3, and chicks are worth $1 a pair. How many roosters, hens, and chicks did he buy if he bought at least one of each type?

59. **Chinese Remainder Problem**   A band of 17 pirates decided to divide their doubloons into equal portions. When they found that they had 3 coins remaining, they agreed to give them to their Chinese cook, Wun Tu. But 6 of the pirates were killed in a fight. Now when the treasure was divided equally among them, there were 4 coins left that they considered giving to Wun Tu. Before they could divide the coins, there was a shipwreck and only 6 pirates, the coins, and the cook were saved. This time equal division left a remainder of 5 coins for the cook. Now Wun Tu took advantage of his culinary position to concoct a poison mushroom stew so that the entire fortune in doubloons became his own. What is the smallest number of coins that the cook would have finally received?*

60. What is the next smallest number of coins that would satisfy the conditions of Problem 59?

---

*This problem is from Sun Zi (ca. AD 250), who wrote a mathematical manual during the Three Kingdoms period in China.

5.8 | Cryptography

Cryptography is the art of writing or deciphering messages in code. Work in cryptography combines theoretical foundations with practical applications. It involves, at some level, ways to generate random-looking sequences and to detect and evaluate nonrandom effects. You might think of governments, covert operations, and spies when you think of cryptography, but there are many applications outside government. Cable television companies, whose signals are easily intercepted from satellite relays, use encryption schemes to prevent unauthorized use of their transmissions. Banks encrypt financial transactions and records in their computers for purposes of authenticity and integrity—that is, to make sure that the sender and contents are really as they appear—as well as for privacy. Cryptography figured prominently in a recent movie on the war against the drug cartel in Colombia. A cryptologist needs skills in communications, engineering, speech research, signals processing, and the design of specialized computers.

Simple Codes

Games magazine, December 1985

Simple codes can be formed by replacing one letter by another. You may have seen this type of code in a children's magazine, or as a puzzle problem in a newspaper. The television show *Wheel of Fortune* uses a variation of this simple code-breaking skill. The cartoon at the left is from a popular game magazine. Codes such as this can easily be broken by using some logic and a knowledge of our language. Here is an analysis of how this code could be broken.

RAE WA XLHH XDL WAYLTSAT XDNX VM DL KALZS'X

PAIIEXL IR LGLPEXVAS, V FVHH SLYLT YAXL MAT DVI

NWNVS!

We have color-coded the solution by showing the new material at each step in color.

1. The hint for this puzzle was "A four-letter word that starts and ends with the same letter is often THAT." From this, we conclude that the coded word XDNX is that, so that X is T, D is H, N is A. Fill in these letters above the letters of the puzzle:

      T   **TH**   **THAT** **H**   **T**
   RAE WA XLHH XDL WAYLTSAT XDNX VM DL KALZS'X
     T   **T**   **T**   **H**
   PAIIEXL IR LGLPEXVAS, V FVHH SLYLT YAXL MAT DVI
   A A
   NWNVS!

             TH

2. The three-letter word XDL is probably the word *the*, so fill in E for L:

      TE **THE** **E**   **THAT** **HE** **E** **T**
   RAE WA XLHH XDL WAYLTSAT XDNX VM DL KALZS'X
     TE **E E** **T**   **E E** **TE**   **H**
   PAIIEXL IR LGLPEXVAS, V FVHH SLYLT YAXL MAT DVI
   A A
   NWNVS!

3. The single letter V is not A since N is A, so we guess that it is I:

      TE   THE   E   THAT **I** HE E T
   RAE WA XLHH XDL WAYLTSAT XDNX VM DL KALZS'X
     TE E E T**I** **I** **I** E E TE H**I**
   PAIIEXL IR LGLPEXVAS, V FVHH SLYLT YAXL MAT DVI
   A A**I**
   NWNVS!

4. The coded word XLHH begins with TE_ _; since the last two letters are the same, we assume that H is the letter L:

      TE**LL** THE   E   THAT I HE E T
   RAE WA XLHH XDL WAYLTSAT XDNX VM DL KALZS'X
      TE E E TI I **I**LL E E TE H I
   PAIIEXL IR LGLPEXVAS, V FVHH SLYLT YAXL MAT DVI
   A A I
   NWNVS!

European mathematicians Pomerance, Rumely, and Adleman, building on recent work by theorists in the United Sates, have solved a 3,000-year-old problem in mathematics, an achievement that raises questions about the security of recently developed secret codes.

The problem, one of the oldest in mathematics, is how to determine whether a number is prime—that is, whether it can be divided by any number other than itself and 1. The smallest prime numbers are 2, 3, 5, 7, and 11.

Using the new method, a test on a 97-digit number, which previously could have taken as long as 100 years with the fastest computer, was done in 77 seconds.

It's possible that the method may help determine the divisors of a large number that is not prime. One way to tell whether a number is prime is to try to divide it by other numbers. If the number is very large, trying to divide it by other numbers is hopeless. In 1640, Pierre de Fermat invented a test for determining primes without doing division. In the last decade, mathematicians decided that they could run Fermat's test on a suspected prime number many times, and if it consistently passed, there was a high probability that the number was prime.

5. The coded word FVHH ends with _ILL; assume F is W. Also we see the coded word VM which is I_, so we also assume that the coded M is decoded as F:

TELL THE   E   THAT IF HE   E   T
RAE WA XLHH XDL WAYLTSAT XDNX VM DL KALZS'X

  TE   E E T I   I WILL E E   TE F   H I
PAIIEXL IR LGLPEXVAS, V FVHH SLYLT YAXL MAT DVI

A A I
NWNVS!

6. Now, observing the context of the cartoon and making some assumptions about what a prisoner might be saying, we make some trial-and-error guesses: He wants a pardon from the governor, so we count letters and fill in the blanks for the word *governor*:

O GO TELL THE GOVERNOR THAT IF HE OE N T
RAE WA XLHH XDL WAYLTSAT XDNX VM DL KALZS'X

O   TE   E E   TION I WILL NEVER VOTE F OR HI
PAIIEXL IR LGLPEXV AS, V F VHH S LYLT YAXL MA T DVI

AGAIN
NWNVS!

7. We can now fill in the completed deciphered message. Note that spaces translate as spaces:

YOU GO TELL THE GOVERNOR THAT IF HE DOESN'T COMMUTE
MY EXECUTION, I WILL NEVER VOTE FOR HIM AGAIN!

Modular Codes*

One common code is based on modular arithmetic, and it is known as a **modular code.** Rather than simply substituting one letter for another (which is an easily broken code), more sophisticated codes can be developed. To **encrypt** a message means to scramble it by something called an **encoding key.** The result is a secret or coded message, called **ciphertext.** The coded message is then unscrambled using a secret **decoding key.** This coding procedure requires that both the sender and the receiver know, and conceal, the encoding and decoding keys. Suppose, for example, that we wish to send the following secret message:

THE FBI HAS BED BUGS.

Suppose we select some encoding key—say, multiply by 2. Then we code the message according to some given modular number, say 29, as shown in Figure 5.24 on page 243. Notice that we use 29 instead of 0; this is still considered a modulo 29 system since $29 \equiv 0$, (mod 29).

We look up the numerical value for each letter, space, and punctuation mark according to

20	8	5	29	6	2	9	29	8	1	19	29	2	5	4	29	2	21	7	19	28

Message: T H E   F B I   H A S   B E D   B U G S .

Note that the "code" 20-8-5-29-6-2-9-29-8-1-19-29-2-5-4-29-2-21-7-19-28 would be "easy" to break. Why?

Now we encode by multiplying each of these numbers by the encoding key; then we *modulate*, or write each of these answers modulo 29:

*This code requires Section 5.7.

On October 11, 1988, Mark S. Manasse of Digital Corporation's Systems and Arjen K. Lenstra of the University of Chicago linked over a dozen users of some 400 computers on three continents to find the factors of a 100-digit number.

Several of the most secure cipher systems invented in the past decade are based on the fact that large numbers are extremely difficult to factor, even using the most powerful computers for a long period of time. The accomplishment of factoring a 100-digit number "is likely to prompt cryptographers to reconsider their assumptions about cipher security," Lenstra said in a telephone interview. ... Using larger numbers makes the work of cryptographers more cumbersome.

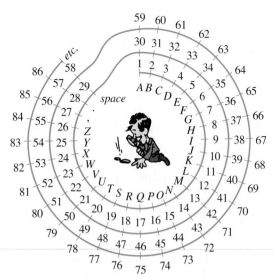

FIGURE 5.24. Modular 29 code

	20	8	5	29	6	2	9	29	8	1	19	29	2	5	4	29	2	21	7	19	28
Message:	T	H	E		F	B	I		H	A	S		B	E	D		B	U	G	S	.

Numerical value: 20- 8- 5- 29- 6- 2- 9- 29- 8- 1- 19- 29- 2- 5- 4- 29- 2- 21- 7- 19- 28

Encode (encoding key 2):   40- 16- 10- 58- 12- 4- 18- 58- 16- 2- 38- 58- 4- 10- 8- 58- 4- 42- 14- 38- 56

Modulate: 11- 16- 10- 29- 12- 4- 18- 29- 16- 2- 9- 29- 4- 10- 8- 29- 4- 13- 14- 9- 27

Coded message: K P J   L D R   P B I   D J H   D M N I ,

The coded message is

KPJ LDR PBI DJH DMNI,

The decoding key is the inverse of the encoding key. Since the encoding key is to multiply by 2, the decoding key is to divide by 2.

Example 1 Decode message

Decode the following message for which the encoding key is to multiply by 2:
NBXXMC RI B CAXX FBK,

Solution   First use Figure 5.24 to find the numerical value:

14-2-24-24-13-3-29-18-9-29-2-29-3-1-24-24-29-6-2-11-27

The decoding key is to divide by 2, so we look at the code and write the odd numbers so that they are evenly divisible by 2, (mod 29):

14 - 2 - 24 - 24 - **42** - **32** - **58** - 18 - **38** - 58 - 2 - **58** - **32** - 30 - 24 - 24 - **58** - 6 - 2 - **40** - 56

Decoding key (divide by 2):   7 - 1 - 12 - 12 - 21 - 16 - 29 - 9 - 19 - 29 - 1 - 29 - 16 - 15 - 12 - 12 - 29 - 3 - 1 - 20 - 28

Decoded message (Figure 5.24):   G - A - L - L - U - P - - - I - S - - - A - - - P - O - L - L - - - C - A - T - .

The decoded message is: GALLUP IS A POLL CAT.

Unbreakable Codes

In 1970, mathematicians Whitfield Diffie and Martin Hellman showed a way to make the keys public. Suppose a code has two keys, an encoding key and a decoding key, and also suppose it is impossible to compute one key from the other in the sense that no person or

It is easy to multiply two large prime numbers to obtain a large number as the answer. But the reverse process—factoring a large number to determine its components—presents a formidable challenge. The problem appears so hard that the difficulty of factoring underlies the so-called RSA method of encrypting digital information. An international team of computer scientists recently spent 8 months finding the factors of a 129-digit number that was suggested 17 years ago as a test of the security of the RSA cryptographic scheme. This effort demonstrates the strength of the RSA crypto-system. However, this effort also demonstrates that significantly larger numbers may be necessary in the future to ensure security.

Science News, May 7, 1994

computer would be able to do it; then this would constitute an unbreakable code. Here is the way such a code works. Everyone owns a unique pair of keys, one of which remains private, but the other is public in the sense that it is listed in a readily available directory. To send a message, you look up the public key for the person to whom the message is to be sent. You use the public key to scramble the message. The receiver then uses his or her private key to decode the message with total secrecy and privacy.

Three mathematicians, Ron Rivest, Adi Shamir, and Leonard Adleman, created a public key algorithm known as RSA, from their initials. This method depends on ideas of prime numbers and factoring studied in this chapter. Consider two prime numbers, say, 11 and 7. Now if I hand you their product, 77, this product can be made public, while the factors 11 and 7 remain secret. To *encode* you need the number 77, but to *decode* you need the factors. Now, *if* the factors are large enough (say, 200 digits each), then the product is so large that it can never be factored, and the result is an unbreakable code.[*]

Why would we need an unbreakable public-key code? Some applications include protecting money transfers from tampering, shielding sensitive business data from competitors, and protecting computer software from viruses. The government has published a *digital signature algorithm* (DSA), which depends on a single very large prime number.

[*]RSA Laboratories, Redwood City, CA, tests its ability to create difficult ciphers by establishing a series of cryptographic contests. Its first challenge required 140 days to solve. The coded message was "Strong cryptography makes the world a better place." If you are interested in this type of challenge, check out **www.rsa.com** for additional information.

Problem Set 5.8

Level 1

Number the letters of the alphabet from 1 to 26; code a comma as 27, period as 28, and space as 29. Encode the messages in Problems 1–4.

1. NEVER SAY NEVER.

2. THE EAGLE HAS LANDED.

3. YOU BET YOUR LIFE.

4. MY BANK BALANCE IS NEGATIVE.

Number the letters of the alphabet from 1 to 26 and code a blank as 29. Decode the messages in Problems 5–8.

5. 1-18-5-29-23-5-29-8-1-22-9-14-7-29-6-21-14-29-25-5-20

6. 9-29-12-15-22-5-29-13-1-20-8-5-13-1-20-9-3-19

7. 6-1-9-12-21-18-5-29-20-5-1-3-8-5-19-29-19-21-3-3-5-19-19

8. 19-17-21-1-18-5-29-13-5-1-12-19-29-13-1-11-5-29-18-15-21-14-4-29-16-5-15-16-12-5

Give the decoding key for the encoding keys in Problems 9–14.

9. Multiply by 8.

10. Divide by 6.

11. Times 20, add 2.

12. Divide by 4, minus 3.

13. Multiply by 4 and add 2, then double the result.

14. Multiply by 3 and subtract 3, then divide the result by 2.

Level 2

Use Figure 5.24 to encode or decode the messages in Problems 15–23.

15. Encoding key: Multiply by 3: NEVER SAY NEVER.

16. Encoding key: Multiply by 5. THE EAGLE HAS LANDED.

17. Encoding key: Multiply by 2 and add 5. YOU BET YOUR LIFE.

18. Encoding key: Multiply by 4 and subtract 10. MY BANK BALANCE IS NEGATIVE.

19. Encoding key: Multiply by 3.
    XEJSBQ LEJSBQ,. C RCGG UEQZ

20. Encoding key: Multiply by 2 and add 10.
    LJMQKVTSSKQJ.SJKITJ,ZKJULECSJ.
    IJSKGTKITJTESTSJSETTM

21. Encoding key: Multiply by 2 and subtract 7.
    SKAXBVIXAVXVGKQDV.WSYQCLT

22. Encoding key: multiply by 2 and subtract 11.
    ALL PERSONS BY NATURE DESIRE KNOWLEDGE.

23. Encoding key: Multiply by 3 and add 5. SHOW ME A DROPOUT FROM A DATA PROCESSING SCHOOL AND I WILL SHOW YOU A NINCOMPUTER.

Problem Solving 3

The ciphers in Problems 24–27 are taken from Games December 1985.

24. L FPHWDJ QJPQFJ TDP MJJQ BPKR. WDJU HVJ IPTHVBR TDP DHXJG'W KPW WDJ KSWR WP ALWJ QJPQFJ WDJNRJFXJR. (*Hint*: Ciphertext pattern QJPQFJ often represents PEOPLE.)

25. GHJWHY NDW LOGKL TGRLBK, UBLRGXV, GHV XYOZLD WH DZL DWR VWS ZL RXBOJ G UGH MWX GOO LYGLWHZHSL. (*Hint*: A three letter word after a series of words set off by commas is often AND.)

26. GRAPE MA QLMVFGCB, BLLCRQU XRUX GPMRQUB, TFQBMPQMEK BTXLYNEL DLBM BXFVB PUPRQBM LPTX FMXLG. (*Hint:* Ciphertext B represents S. Note its high frequency as a first and last letter. *Bonus hint*: The fifth word is *not* THAT.)

27. MZDGBWV-MVCWZ WTZ HR ZHMAD XHKEBWTEG NZHLTUCG TUCWV CELTZHEXCEB RHZ PCWBSCZ GBWBTHE. (*Hint*: The five vowels, A to U, are represented by C, H, K, T, and W, but not necessarily in that order.)

Cryptic arithmetic is a type of mathematics puzzle in which letters have been replaced by digits. Replace each letter by a digit (the same digit for the same letter throughout; different digits for different letters), and the arithmetic will be performed correctly. Break the codes in Problems 28–30.

28.
$$
\begin{array}{r}
\text{SEND} \\
+ \text{MORE} \\
\hline
\text{MONEY}
\end{array}
$$

29.
$$
\begin{array}{r}
\text{THIS} \\
\text{IS} \\
+ \text{VERY} \\
\hline
\text{EASY}
\end{array}
$$

30.
$$
\begin{array}{r}
\text{DAD} \\
\text{SEND} \\
+ \text{MORE} \\
\hline
\text{MONEY}
\end{array}
$$

5.9   CHAPTER SUMMARY

More new jobs will require more postsecondary mathematics education.

A CHALLENGE OF NUMBERS,
MSEB, 1990

Important Ideas

Properties of numbers: closure property [5.1]; commutative property [5.1]; associative property [5.1]; distributive property [5.1]; identity [5.6]; inverse [5.6]; groups [5.6]; fields [5.6]

Relationships among the following sets of numbers:
natural numbers [5.1]; whole numbers [5.3]; integers [5.3]; rational numbers [5.4]; irrational numbers [5.5]; real numbers [5.6]

Meanings of fundamental operations:
addition, multiplication [5.1]; subtraction [5.1]; division [5.2]; the difference between a square root and an irrational number [5.5]

Rules of divisibility [5.2]

Operations among the following sets of numbers: integers [5.3]; rationals [5.4]; irrationals [5.5]; reals [5.6]

Least number of divisors [5.3]

Fundamental theorem of arithmetic [5.3]

Fundamental property of fractions [5.4]

Pythagorean theorem [5.5]

Elementary operations [5.7]

Take some time getting ready to work the review problems in this section. First review these important ideas. Look back at the definition and property boxes in this chapter, as well as the types of problems that were introduced. You will maximize your understanding of this chapter by working the problems in this section only after you have studied the material.

You will find some review help online at **www.mathnature.com**. There are links giving general test help in studying for a mathematics examination, as well as specific help for reviewing this chapter.

Chapter **5** Review Questions

Perform the indicated operations in Problems 1–8.

1. $-4 + 5(-3)$

2. $\frac{4}{7} + \frac{5}{9}$

3. $\frac{7}{30} + \frac{5}{42} + \frac{5}{99}$

4. $-\sqrt{10} \cdot \sqrt{10}$

5. $\left(\frac{11}{12} + 2\right) + \frac{-11}{12}$

6. $\frac{3^{-1} + 4^{-1}}{6}$

7. $\frac{-7}{9} \cdot \frac{99}{174} + \frac{-7}{9} \cdot \frac{75}{174}$

8. $\frac{-3 + \sqrt{3^2 + 4(2)(3)}}{2(2)}$

Reduce each fraction in Problems 9–13. If it is reduced, so state.

9. $\frac{8}{3}$

10. $\frac{16}{18}$

11. $\frac{100}{825}$

12. $\frac{184}{207}$

13. $\frac{1,209}{2,821}$

Find the decimal representation for each number in Problems 14–18, and classify it as rational or irrational.

14. $\frac{3}{8}$

15. $\frac{7}{3}$

16. $\frac{3}{7}$

17. $\frac{125}{20}$

18. $\frac{3}{13}$

Find the prime factorization of each number given in Problems 19–23. If it is prime, so state.

19. 89

20. 101

21. 349

22. 1,001

23. 6,825

Solve the equations in Problems 24–26 for x.

24. $\frac{x}{5} \equiv 2$, (mod 8)

25. $2x \equiv 3$, (mod 7)

26. $2x^2 + 7x + 1 \equiv 0$, (mod 2)

27. Find the greatest common factor for the set of numbers {49, 1001, 2401}.

28. Find the least common multiple for the set of numbers {49, 1001, 2401}.

29. If $a \odot b = a \times b + a + b$, what is the value of $(1 \odot 2) \odot (3 \odot 4)$?

30. If $a \uparrow b$ stands for the larger number of the pair and $a \downarrow b$ stands for the lesser number of the pair, what is the value of $(1 \downarrow 2) \uparrow (2 \downarrow 3)$

Define each of the operations given in Problems 31–33.

31. multiplication

32. subtraction

33. division

34. **IN YOUR OWN WORDS**   Explain why we cannot divide by zero.

35. Find an irrational number between 34 and 35.

36. What is a field? Describe each property.

37. Consider the set $\mathbb{N}$ and let $a, b \in \mathbb{N}$. Let D be an operation defined by $a \, D \, b =$ g.c.f. of a and b. Is $\mathbb{N}$ a commutative group for the operation of D?

38. **Do You Weigh Too Much?** In the United States, a fat person is defined as anyone with a "body mass index" (or BMI) of 27.6 or higher. The World Health Organization defines a fat person as one with a BMI of 25 or higher. To determine BMI, multiply your weight in pounds by 703, and then divide that result by your height (in inches) squared. Write this in formula form using w for weight in pounds and h for height in inches. Calculate the BMI (rounded to the nearest tenth) for each of these people:

   a. 5 ft 5 in., 165 lb   **b.** 6 ft, 185 lb

   c. What is the weight (rounded to the nearest pound) for a 5-ft, 6-in. tall person who wants a BMI of 25?

39. The news clip is from the column "Ask Marilyn," *Parade Magazine*, April 15, 1990. Answer the question asked in the article.

> There are 1,000 tenants and 1,000 apartments. The first tenant opens every door. The second closes every other door. The third tenant goes to every third door, opening it if it is closed and closing it if it is open. The fourth tenant goes to every fourth door, closing it if it is open and opening it if it is closed. This continues with each tenant until the 1,000th tenant closes the 1,000th door. Which doors are open?
>
> Anita Mueller Arnold, MD

40. Suppose you are building a stairway (such as shown in Figure 5.25). Assume that the vertical rise is 8 ft, the horizontal run is 12 ft, and the maximum rise for each step is 8 inches. How many steps are necessary? What is the total length of the segments making up the stairs, and what is the length of the diagonal?

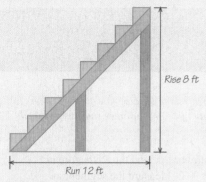

FIGURE 5.25 Building a stairway

BOOK REPORTS

Write a 500-word report on one of these books:

Estimation and Mental Computation, Harold L. Schoen and Marilyn J. Zweng (Reston, VA: Yearbook of the National Council of Teachers of Mathematics, 1986).

The Man Who Knew Infinity, S. Kanigel (New York: Charles Scribners, 1991).

Group RESEARCH PROJECTS

Go to
www.mathnature.com
for references and links.

Working in small groups is typical of most work environments, and learning to work with others to communicate specific ideas is an important skill. Work with three or four other students to submit a single report based on each of the following questions.

G16. With only a straightedge and compass, use a number line and the Pythagorean theorem to construct a segment whose length is $\sqrt{2}$. Measure the segment as accurately as possible, and write your answer in decimal form. *Do not use a calculator or any tables.* Now, continue your work to construct segments whose lengths are $\sqrt{3}, \sqrt{4}, \sqrt{5}, \ldots$.

G17. Four Fours Write the numbers from 1 to 100 (inclusive) using exactly four fours. See Problem 60, Problem Set 5.3, to help you get started.

G18. Pythagorean Theorem Write out three different proofs of the Pythagorean theorem.

G19. Modular Art Many interesting designs such as those shown in Figure 5.26 can be created using patterns based on modular arithmetic. (You might also wish to look at Problem 46, Section 5.7, p. 93. Prepare a report for class presentation based on the article "Using Mathematical Structures to Generate Artistic Designs" by Sonia Forseth and Andrea Price Troutman, *The Mathematics Teacher*, May 1974, pp. 393–398. Another source is "Mod Art: The Art of Mathematics" by Susan Morris, *Technology Review,* March/April 1979.

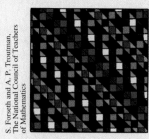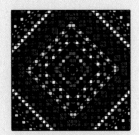

S. Forseth and A. P. Troutman,
The National Council of Teachers
of Mathematics

FIGURE 5.26 Modular art

G20. The Babylonians estimated square roots using the following formula:

$$\text{If } n = a^2 + b, \text{then } \sqrt{n} \approx a + \frac{b}{2a}$$

For example, if $n = 11$, then $n = 9 + 2$, so that $n = 11$, $a = 3$, and $b = 2$.
$$\uparrow$$
perfect square

This Babylonian approximation for $\sqrt{11}$ is found by

$$\sqrt{11} \approx 3 + \frac{2}{2(3)} \approx 3.3333 \ldots$$

With a calculator, we obtain $\sqrt{11} \approx 3.31662479$. Write a paper about this approximation method. Here are some questions you might consider:

a. Can b be negative?

b. Consider the following possibilities:

$$|b| < a^2 \qquad |b| = a^2 \qquad |b| > a^2$$

Can you formulate any conclusions about the appropriate hypotheses for the Babylonian approximation?

c. Put this formula into a historical context.

Individual | RESEARCH PROJECTS

www.mathnature.com

Learning to use sources outside your classroom and textbook is an important skill, and here are some ideas for extending some of the ideas in this chapter. You can find references to these projects in a library or at **www.mathnature.com.**

PROJECT 5.1 In the text we tried some formulas that might have generated only primes, but, alas, they failed. Below are some other formulas. Show that these, too, do not generate only primes.

a. $n^2 + n + 41$ **b.** $n^2 - 79n + 1{,}601$ **c.** $2n^2 + 29$ **d.** $9n^2 - 498n + 6{,}683$
e. $n^2 + 1$, n an even integer

PROJECT 5.2 For what values of n is $11 \cdot 14^n + 1$ a prime?

PROJECT 5.3 A formula that generates all prime numbers is given by David Dunlop and Thomas Sigmund in their book *Problem Solving with the Programmable Calculator* (Englewood Cliffs, N.J.: Prentice-Hall, 1983). The authors claim that the formula $\sqrt{1 + 24n}$ produces every prime number except 2 and 3, but give no proof or reference to a proof. Create a table, and give an argument to support or find a counterexample to disprove their claim.

PROJECT 5.4 A large prime, $2^{30{,}402{,}457} - 1$, is a number that has 9,152,052 digits. A number this large is hard to comprehend. Write a paper making the size of this number meaningful to a nonmathematical reader.

PROJECT 5.5 Investigate some of the properties of primes not discussed in the text. Why are primes important to mathematicians? Why are primes important in mathematics? What are some of the important theorems concerning primes?

PROJECT 5.6 Historical Quest   We mentioned that the Egyptians wrote their fractions as sums of unit fractions. Show that every positive fraction less than 1 can be written as a sum of unit fractions.

PROJECT 5.7 Historical Quest   The Egyptians had a very elaborate and well-developed system for working with fractions. Write a paper on Egyptian fractions.

PROJECT 5.8 Form a group using a geoboard. Go to **www.mathnature.com** for some ideas about writing this paper.

PROJECT 5.9 Prove that $y\sqrt{2}$ is irrational.

PROJECT 5.10 Historical Quest   Write a paper or prepare an exhibit illustrating the Pythagorean theorem.

PROJECT 5.11 Write a paper on the symmetries of a cube. Go to **www.mathnature.com** for some ideas about writing this paper.

PROJECT 5.12 What is a Diophantine equation?

PROJECT 5.13 Prepare an exhibit on cryptography. Include devices or charts for writing and deciphering codes, coded messages, and illustrations of famous codes from history. For example, codes are found in literature in *Before the Curtain Falls* by J. Rives Childs, *The Gold Bug* by Edgar Allan Poe, and *Voyage to the Center of Earth* by Jules Verne.

PROJECT 5.14 Write a paper on the importance of cryptography for the Internet.

7

THE NATURE OF GEOMETRY

What in the World?

"Carol, where have you been? I haven't seen you in over a year!" said Tom.

"I've been on an archeological dig," Carol responded. "Donald and I have been looking at the Garden Houses of Ostia. They were built in the second century as part of the Roman Empire, and they were excavated in the first part of this century."

"I've never heard of Ostia," Tom said with a swing in his arm. "Where or what is Ostia?"

"Ostia was a boom town in the second century," Carol said with enthusiasm. "Take a look at the map. The Tiber river was the lifeblood of the city

Map of Ostia, from 'Civitates Orbis Terrarum' by Georg Braun (1541–1622) and Frans Hogenberg (1535–90) c.1572–1617 (coloured engraving), Hoefnagel, Joris (1542–1600) (after) / Private Collection / The Stapleton Collection / The Bridgeman Art Library International

Ostia, whose population reached 50,000 at its peak. They constructed an artificial harbor on the Tyrrhenian Sea, and the city became a major port of Rome, which is about 25 kilometers away. Many apartment buildings were built, many four stories high, and they included not only living space, but also shops and gardens. We are trying to reconstruct life in the city. You can see plans and a three-dimensional reconstruction of the city at the following website, www.ostia-antica.org/indexes.htm.

Overview

Geometry, or "earth measure," was one of the first branches of mathematics. Both the Egyptians and the Babylonians needed geometry for construction, land measurement, and commerce. They both discovered the Pythagorean theorem, although it was not proved until the Greeks developed geometry formally. This formal development utilizes deductive logic (which was briefly discussed in Chapter 3), beginning with certain assumptions, called **axioms** or **postulates.** Historically, the first axioms that were accepted seemed to conform to the physical world. In this chapter, we look at that body of mathematics known as *geometry*.

7.1 | Geometry

Geometry is the science created to give understanding and mastery of the external relations of things; to make easy the explanation and description of such relations and the transmission of this mastery.

G. B. HALSTED

CHAPTER **CHALLENGE**

See if you can fill in the question mark.

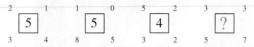

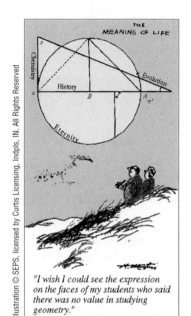

"I wish I could see the expression on the faces of my students who said there was no value in studying geometry."

When we refer to Euclidean geometry, we are talking about the geometry known to the Greeks as summarized in a work known as Euclid's *Elements*. This 13-volume set of books collected all the material known about geometry and organized it into a logical deductive system. It is the most widely used and studied book in history, with the exception of the Bible. An overview of the history of mathematics, and of geometry in particular, is presented in the prologue of this book, and it includes the geometry of both the Egyptians and the Greeks. We do not know very much about the life of Euclid except that he was the first professor of mathematics at the University of Alexandria (which opened in 300 B.C.).

Greek (Euclidean) Geometry

Geometry involves **points** and sets of points called **lines, planes,** and **surfaces.** Certain concepts in geometry are called **undefined terms.** For example, what is a line? You might say, "I know what a line is!" But try to define a line. Is it a set of points? Any set of points? What is a point?

1. A point is something that has no length, width, or thickness.

2. A point is a location in space.

Certainly these are not satisfactory definitions because they involve other terms that are not defined. We will therefore take the terms *point, line*, and *plane* as undefined.

Author's **NOTE**

Early civilizations observed from nature certain simple shapes such as triangles, rectangles, and circles. The study of geometry began with the need to measure and understand the properties of these simple shapes.

Historical **NOTE**

Euclid
(ca. 300 B.C.)

We know very little about the actual life of Euclid, but we do know that he was the first professor of mathematics at the University of Alexandria. The *Elements* was a 13-volume masterpiece of mathematical thinking. It was described by Augustus De Morgan (1806–1871) as follows: "The thirteen books of Euclid must have been a tremendous advance, probably even greater than that contained in the *Principia* of Newton."

We often draw physical models or pictures to represent these concepts; however, we must be careful not to try to prove assertions by looking at pictures, since a picture may contain hidden assumptions or ambiguities. For example, consider Figure 7.1. Do you see an old woman or a young woman? Is the fly in Figure 7.2 on the cube or in the cube? The point we are making is that although we may use a figure in geometry to help us understand a problem, we cannot use what we see in a figure as a basis for our reasoning.

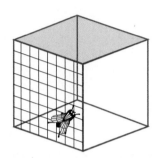

FIGURE 7.2 Is the fly on the cube or in it? On which face?

FIGURE 7.1 An old woman or a young woman?

Geometry can be separated into two categories:

1. Traditional (which is the geometry of Euclid)
2. Transformational (which is more algebraic than the traditional approach)

When Euclid was formalizing traditional geometry, he based it on five postulates, which have come to be known as **Euclid's postulates.** A **postulate** or **axiom** is a statement accepted without proof. In mathematics, a result that is proved on the basis of some agreed-upon postulates is called a **theorem.**

Euclid's Postulates

1. A straight line can be drawn from any point to any other point.
2. A straight line extends infinitely far in either direction.
3. A circle can be described with any point as center and with a radius equal to any finite straight line drawn from the center.
4. All right angles are equal to each other.
5. Given a straight line and any point not on this line, there is one and only one line through that point that is parallel to the given line.*

The first four of these postulates were obvious and noncontroversial, but the fifth one was different. This fifth postulate looked more like a theorem than a postulate. It was much more difficult to understand than the other four postulates, and for more than

---

*The fifth postulate stated here is the one usually found in high school geometry books. It is sometimes called Playfair's axiom and is equivalent to Euclid's original statement as translated from the original Greek by T. L. Heath: "If a straight line falling on two straight lines makes the interior angle on the same side less than two right angles, the two straight lines, if produced infinitely, meet on that side on which the angles are less than the two right angles."

20 centuries mathematicians tried to derive it from the other postulates or to replace it by a more acceptable equivalent. Two straight lines in the same plane are said to be **parallel** if they do not intersect.

Today we can either accept the fifth postulate as a postulate (without proof) or deny it. If it is denied, it turns out that no contradiction results; in fact, if it is not accepted, other geometries called **non-Euclidean geometries** result. If it is accepted, then the geometry that results is consistent with our everyday experiences and is called **Euclidean geometry.**

Let's look at each of Euclid's postulates. The first one says that a straight line can be drawn from any point to any other point. To connect two points, you need a device called a **straightedge** (a device that we assume has no markings on it; you will use a ruler, but not to measure, when you are treating it as a straightedge). The portion of the line that connects points A and B in Figure 7.3 is called a **line segment.** We write $\overline{AB}$ (or $\overline{BA}$). We contrast this notation with $\overleftrightarrow{AB}$, which is used to name the line passing through the points A and B. We use the symbol $|\overline{AB}|$ for the length of segment $\overline{AB}$.

The second postulate says that we can draw a straight line. This seems straightforward and obvious, but we should point out that we will indicate a line by putting arrows on each end. If we consider a point on a line, that point separates the line into parts: two **half-lines** and the point itself. If the arrow points in only one direction, the figure is called a **ray.** We write $\overrightarrow{AB}$ (or $\overleftarrow{BA}$) for the ray with endpoint A passing through B. These definitions are illustrated in Figure 7.3.

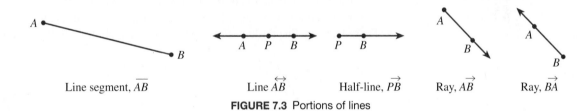

Line segment, $\overline{AB}$ Line $\overleftrightarrow{AB}$ Half-line, $\overrightarrow{PB}$ Ray, $\overrightarrow{AB}$ Ray, $\overrightarrow{BA}$

FIGURE 7.3 Portions of lines

To construct a line segment of length equal to the length of a given line segment, we need a device called a **compass.** Figure 7.4 shows a compass, which is used to mark off and duplicate lengths, but not to measure them.

If objects have exactly the same size and shape, they are called **congruent.** We can use a straightedge and compass to **construct** a figure so that it meets certain requirements. To *construct a line segment congruent to a given line segment*, copy a segment $\overline{AB}$ on any line ℓ. First fix the compass so that the pointer is on point A and the pencil is on B, as shown in Figure 7.5a. Then, on line ℓ, choose a point C. Next, without changing the compass setting, place the pointer on C and strike an arc at D, as shown in Figure 7.5b.

Pointer   Pencil

FIGURE 7.4 A compass

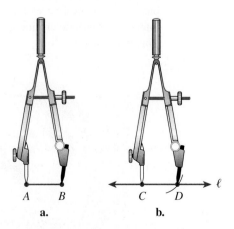

A B C D ℓ

a.   **b.**

FIGURE 7.5 Constructing a line segment

Euclid's third postulate leads us to a second construction. The task is to construct a circle, given its center and radius. These steps are summarized in Figure 7.6.

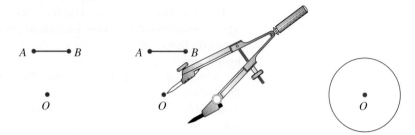

a. Given, a point and a radius of length $|\overline{AB}|$.

b. Set the legs of the compass on the ends of radius $\overline{AB}$; move the pointer to point O without changing the setting.

c. Hold the pointer at point O and move the pencil end to draw the circle.

FIGURE 7.6 Construction of a circle

We will demonstrate the fourth postulate in the next section when we consider angles.

The final construction of this section will demonstrate the fifth postulate. The task is to construct a line through a point P parallel to a given line ℓ, as shown in Figure 7.7a. First, draw any line through P that intersects ℓ at a point A, as shown in Figure 7.7b.

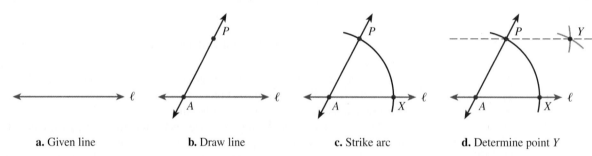

a. Given line

b. Draw line

c. Strike arc

d. Determine point Y

FIGURE 7.7 Construction of a line parallel to a given line through a given point

Now draw an arc with the pointer at A and radius $\overline{AP}$, and label the point of intersection of the arc and the line X, as shown in Figure 7.7c. With the same opening of the compass, draw an arc first with the pointer at P and then with the pointer at X. Their point of intersection will determine a point Y (Figure 7.7d). Draw the line through both P and Y. This line is parallel to ℓ.

Escher's *Drawing Hands*

See www.mathnature.com for some Escher links.

Transformational Geometry

FIGURE 7.8 A reflection

We now turn our attention to the second category of geometry. **Transformational geometry** is quite different from traditional geometry in that it deals with the study of *transformations*.

A **transformation** is the passage from one geometric figure to another by means of reflections, translations, rotations, contractions, or dilations. For example, given a line *L* and a point *P*, as shown in Figure 7.8, we call the point *P′* the **reflection** of *P* about the line *L* if $\overline{PP'}$ is perpendicular to *L* and is also bisected by *L*.

Each point in the plane has exactly one reflection point corresponding to a given line *L*. A reflection is called a *reflection transformation*, and the line of reflection is called the **line of symmetry.** The easiest way to describe a line symmetry is to say that if you fold a paper along its line of symmetry, then the figure will fold onto itself to form a perfect match, as shown in Figure 7.9.

Snowflakes

FIGURE 7.9 Line symmetry on the maple leaf of Canada

Many everyday objects exhibit a line of symmetry. From snowflakes in nature, to the Taj Mahal in architectural design, to many flags and logos, we see examples of lines of symmetry (see Figure 7.10).

Other transformations include *translations, rotations, dilations*, and *contractions*, which are illustrated in Figure 7.11.

Taj Mahal

Chrysler logo

FIGURE 7.10 Examples of line symmetry

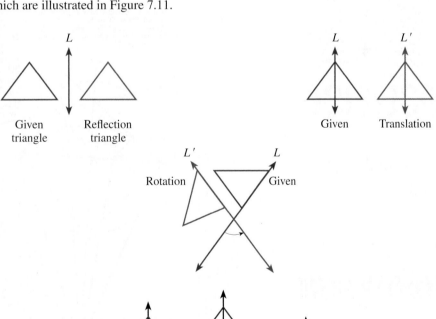

FIGURE 7.11 Transformations of a fixed geometric figure

Similarity

Geometry is also concerned with the study of the relationships between geometric figures. A primary relationship is that of *congruence*. A second relationship is called **similarity.** Two figures are said to be **similar** if they have the same shape, although not necessarily the same size. These ideas are considered in Sections 7.3 and 7.4.

If we were to develop this text formally, we would need to be very careful about the statement of our postulates. The notion of mathematical proof requires a very precise formulation of all postulates and theorems. However, since we are not formally developing geometry, we simply accept the general drift of the statement. Remember, though, that you cannot base a mathematical proof on general drift. On the other hand, there are some facts that we *know to be true* that tell us that certain properties are impossible. For example, if someone claims to be able to trisect an angle with a straightedge and compass, we know *without looking at the construction* that the construction is wrong. This can be frustrating to someone who believes that he or she has accomplished the impossible. In fact, it was so frustrating to Daniel Wade Arthur that he was motivated to take out a paid advertisement in the *Los Angeles Times*, which is reproduced in the News Clip.

Calligraphers sometimes experiment with some inversion transformations.

John Longdon has been creating such word designs for about 15 years, and the two shown here are from the book Wordplay. *The following inversion was found in the April 1992 issue of* OMNI *magazine.*

ΖΤΊΟΗΜΊΞΗ ΝΟΛ ΟΝΛΝΙΟ
DINAND LUDWIG FER-
HERMANN LUDWIG FER-
RAL AND MORAL FORCES."
OF THE ACTION OF NATU-
FECT UNDERSTANDING
KNOWLEDGE AND A PER-
ACHIEVE IS A PERFECT
"ALL THAT SCIENCE CAN

science

"THE WHOLE OF SCIENCE IS NOTHING MORE THAN A RE-FINEMENT OF EVERYDAY THINKING." *ALBERT EINSTEIN*

John Langdon, OMNI, April 1992, p. 4.

Here is a favorite of mine from Longdon's book:

minimum

See **www.mathnature.com** for one student's attempt at trisecting an angle.

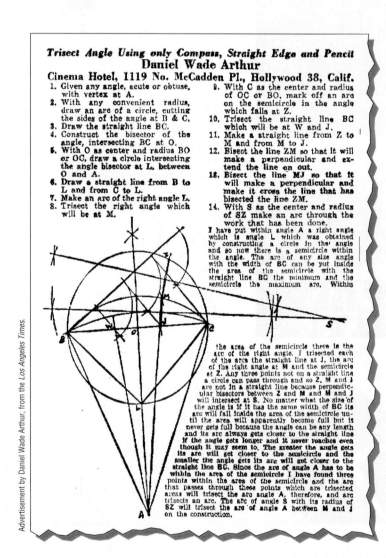

Advertisement by Daniel Wade Arthur, from the *Los Angeles Times.*

Problem Set **7.1**

Level **1**

1. Is the woman in the figure a young woman or an old woman?

2. IN YOUR OWN WORDS   Describe what you see in the following illustration.

3. IN YOUR OWN WORDS   Why do you think Problem 1 is included in this problem set? How does this question relate to working problems in geometry?

4. IN YOUR OWN WORDS   Describe a procedure for constructing a line segment congruent to a given segment.

5. IN YOUR OWN WORDS   Describe a procedure for constructing a circle with a radius congruent to a given segment.

6. IN YOUR OWN WORDS   Discuss what it means to be an undefined term. What is the difference between an axiom and a theorem?

7. IN YOUR OWN WORDS   Describe line symmetry.

8. IN YOUR OWN WORDS   What are the two categories into which geometry is usually separated?

In Problems 9–10, label each cartoon as illustrating a translation, reflection, rotation, dilation, or a contraction.

9.

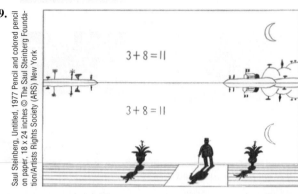

10.

Use the illustration in Figure 7.12 to draw the figures requested in Problems 11–19.

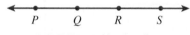

FIGURE 7.12 Number line

11. $\overline{PQ}$	**12.** $\overline{RS}$	**13.** $\overleftrightarrow{PQ}$
14. $\overleftrightarrow{RS}$	**15.** $\overrightarrow{PQ}$	**16.** $\overrightarrow{SR}$
17. $\overleftrightarrow{PQ}$	**18.** $\overrightarrow{RS}$	**19.** $\overline{PS}$

Find at least one line of symmetry for each of the illustrations in Problems 20–27, if possible. If there is no line of symmetry, so state.

20.

21.

22.

23.

24.

25.

26.

27.

Level 2

Carry out the constructions requested in Problems 28–39.

28. A line segment congruent to $\overline{AB}$

A B

29. A line segment congruent to $\overline{CD}$

C

D

30. A line segment congruent to $\overline{EF}$

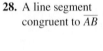

E F

31. A line segment congruent to $\overline{GH}$

G

H

32. Circle with radius congruent to $\overline{WX}$.

W X

33. Circle with radius congruent to $\overline{UV}$.

U V

34. Circle with radius congruent to $\overline{ST}$.

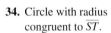

S T

35. Circle with radius congruent to $\overline{YZ}$.

Y Z

36. Line through P parallel to ℓ

ℓ

P

37. Line through Q parallel to m

m

Q

38. Line through R parallel to n.

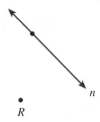

n

R

39. Line through S parallel to k.

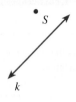

S

k

Which of the pictures in Problems 40–47 illustrate a line symmetry?

40. Chambered nautilus

FlamingPumpkin/iStockphoto.com

41. Butterfly

© Tim Zurowski/CORBIS

42. Human brain

DNY59/iStockphoto.com

43. Human face

Amanda Mack/iStockphoto.com

44. Wallpaper by: Walter Crane

'Swan, Rush and Iris' wallpaper design, Crane, Walter (1845–1915) / Victoria & Albert Museum, London, UK / The Stapleton Collection / The Bridgeman Art Library International

45. Sculpture by: Indian School

Avolokitesvara in the form of Padmapani, 11th century (copper), Indian School (11th century) / Detroit Institute of Arts, USA / Gift of H. Kevorkian / The Bridgeman Art Library International

46. Human circulatory system

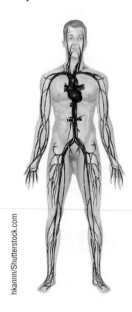

47. Empire State Building

52.

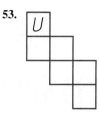

53.

Start with the given cube and then rotate this cube 90° in the directions indicated by the arrows, as shown in Figure 7.14. In Problems 54–58, select the cube marked A, B, C, or D to indicate which most closely matches the result.

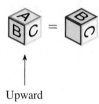

To the right   To the left

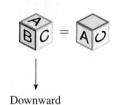

Upward   Downward

Downward, then to the right

FIGURE 7.14 Rotating a cube

Level 3

Study the patterns shown in Figure 7.13. When folded they will form cubes spelling CUBE. Letter each pattern in Problems 48–53.

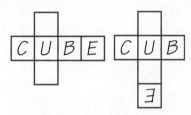

FIGURE 7.13 CUBE pattern

48.

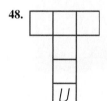

49.

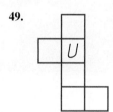

50.

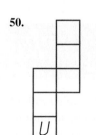

51.

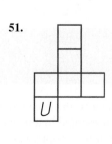

54. Rotation

A   B   C   D

55. Rotation

A   B   C   D

56. Rotation

A   B   C   D

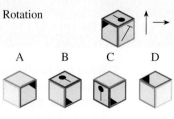

57. Rotation

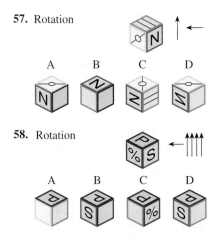

A   B   C   D

58. Rotation

A   B   C   D

59. The letters of the alphabet can be sorted into the following categories: (1) FGJLNPQRSZ; (2) BCDEK; (3) AMTUVWY; and (4) HIOX. What defines these categories?

60. The "Mirror Image" illustration shows a problem condensed from *Games* magazine. Answer the questions asked in that problem.

Mirror Image
by Diane Dawson

In the Fabulous Kingdom, anything can happen—and usually does. When we held a looking glass up to this page, we noticed a few incongruities between it and its "reflection" on the facing page. Can you spot thirty *substantial* differences here, with or without the help of a mirror? (Since both illustrations were hand-drawn there are bound to be hairline differences—these should be ignored.)

7.2 Polygons and Angles

When the Euclidean geometry introduced in the preceding section is studied in high school as an entire course, it is usually presented in a *formal* manner using definitions, axioms, and theorems. The development of this chapter is *informal*, which means that we base our results on observations and intuition. We begin by assuming that you are familiar with the ideas of *point, line,* and *plane.*

Angles

A connecting point of two sides is called a **vertex** (plural **vertices**) and is usually designated by a capital letter. An **angle** is composed of two rays or segments with a common endpoint. The angles between the sides of a polygon are sometimes also denoted by a capital letter, but other ways of denoting angles are shown in Figure 7.15.

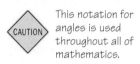

This notation for angles is used throughout all of mathematics.

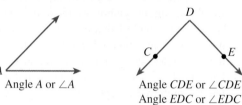

Angle A or $\angle A$

Angle CDE or $\angle CDE$
Angle EDC or $\angle EDC$

Angle 2 or $\angle 2$

FIGURE 7.15 Ways of denoting angles

Example 1 Find angles

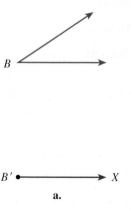

FIGURE 7.16

Locate each of the following angles in Figure 7.16.
a. $\angle AOB$ **b.** $\angle COB$ **c.** $\angle BOA$ **d.** $\angle DOC$ **e.** $\angle 3$ **f.** $\angle 4$

Solution

a. $\angle AOB$

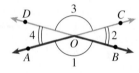

b. $\angle COB$

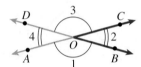

c. $\angle BOA$

d. $\angle DOC$

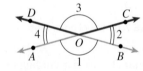

e. $\angle 3$

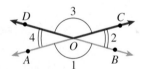

f. $\angle 4$

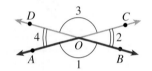

We return to the constructions first introduced in the last section. To construct an angle with the same size as a given angle B, first draw a ray from B', as shown in Figure 7.17a.

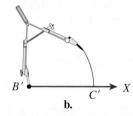

a.

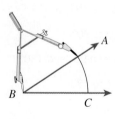

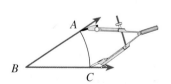

b.

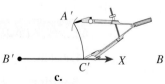

c.

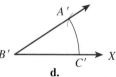

d.

FIGURE 7.17 Construction of an angle congruent to a given angle

Next, mark off an arc with the pointer at the vertex of the given angle and label the points A and C. Without changing the compass, mark off a similar arc with the pointer at B', as shown in Figure 7.17b. Label the point C' where this arc crosses the ray from B'. Place the pointer at C and set the compass to the distance from C to A. Without changing the compass, put the pointer at C' and strike an arc to make a point of intersection A' with the arc from C', as shown in Figure 7.17c. Finally, draw a ray from B' through A'.

Two angles are said to be **equal** if they describe the same angle. If we write m in front of an angle symbol, we mean the measure of the angle rather than the angle itself. Notice in Example 1 that parts **a** and **c** name the *same angle,* so $\angle AOB = \angle BOA$. Also notice the single and double arcs used to mark the angles in Example 1; these are used to denote angles with equal measure, so $m\angle COB = m\angle AOD$ and $m\angle COD = m\angle AOB$, but $\angle COD \neq \angle AOB$ (since they are not the same angle). Denoting an angle by a single letter is preferred except in the case (as shown by Example 1) where several angles share the same vertex.

Angles are often measured using a unit called a **degree,** which is defined to be $\frac{1}{360}$ of a full revolution. The symbol $°$ is used to designate degrees. To measure an angle, you can use a **protractor,** but in this book the angle whose measures we need will be labeled as in Figure 7.18.

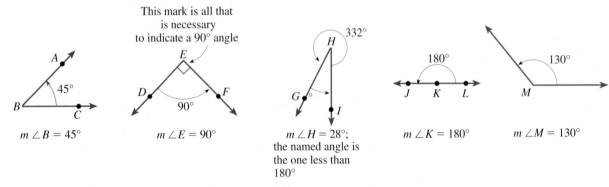

FIGURE 7.18 Labeling angles

Angles are sometimes classified according to their measures, as shown in Table 7.1.

TABLE 7.1	
	Types of Angles
Angle Measure	**Classification**
Less than 90°	**Acute**
Equal to 90°	**Right**
Between 90° and 180°	**Obtuse**
Equal to 180°	**Straight**

STOP Remember these names for types of angles.

Experience leads us to see the plausibility of Euclid's fourth postulate that all right angles are congruent to one another.

Example 2 **Classify angles**

Label the angles $B, E, H, K,$ and M in Figure 7.18 by classification.

Solution $\angle B$ is acute; $\angle E$ is right; $\angle H$ is acute; $\angle K$ is straight; $\angle M$ is obtuse.

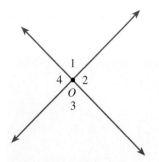

FIGURE 7.19 Angles formed by intersecting lines

Two angles with the same measure are said to be **congruent.** If the sum of the measures of two angles is 90°, they are called **complementary angles,** and if the sum is 180°, they are called **supplementary angles.**

Consider any two distinct (different) intersecting lines in a plane, and let O be the point of intersection as shown in Figure 7.19.

These lines must form four angles. Angles with a common ray, common vertex, and on opposite sides of their common sides are called **adjacent angles.** We say that $\angle 1$ and $\angle 2$, $\angle 2$ and $\angle 3$, $\angle 3$ and $\angle 4$, as well as $\angle 4$ and $\angle 1$ are pairs of adjacent angles. We also say that $\angle 1$ and $\angle 3$ as well as $\angle 2$ and $\angle 4$ are pairs of **vertical angles**—that is, two angles for which each side of one angle is a prolongation through the vertex of a side of the other.

Example   3   Classify angle pairs

Classify the named angles or pairs of angles shown in Figure 7.20.

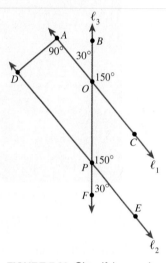

FIGURE 7.20 Classifying angles

a. $\angle AOB$ **b.** $\angle BOC$ **c.** $\angle DAO$ **d.** $\angle AOB$ and $\angle BOC$ **e.** $\angle BOC$ and $\angle OPE$
f. $\angle AOB$ and $\angle COP$ **g.** $\angle FPE$ and $\angle EPO$

Solution   **a.** Acute   **b.** Obtuse   **c.** Right   **d.** Supplementary and adjacent
e. Congruent   **f.** Vertical   **g.** Supplementary and adjacent

Polygons

A **polygon** is a geometric figure that has three or more straight sides, all of which lie on a flat surface or plane so that the starting point and the ending point are the same. Polygons can be classified according to their number of sides, as shown in Figure 7.21. We say any polygon is **regular** if its sides are the same length.

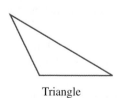
Triangle

Quadrilateral

Pentagon

Hexagon

Polygons (not pictured)
Heptagon: 7 sides
Octagon: 8 sides
Nonagon: 9 sides
Decagon: 10 sides
Dodecagon: 12 sides
n-gon: n sides

FIGURE 7.21 Polygons classified according to number of sides

A polygon with four sides is a **quadrilateral.** Some other classifications are given in the following box.

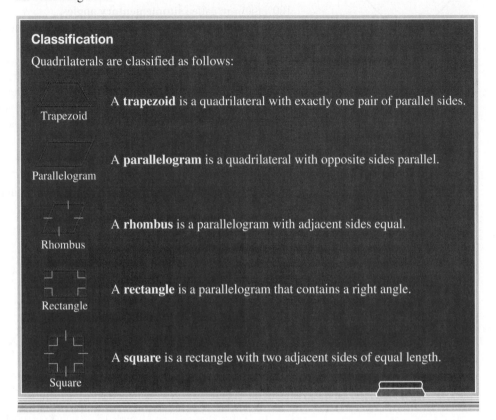

Classification

Quadrilaterals are classified as follows:

A **trapezoid** is a quadrilateral with exactly one pair of parallel sides.

Trapezoid

A **parallelogram** is a quadrilateral with opposite sides parallel.

Parallelogram

A **rhombus** is a parallelogram with adjacent sides equal.

Rhombus

A **rectangle** is a parallelogram that contains a right angle.

Rectangle

A **square** is a rectangle with two adjacent sides of equal length.

Square

Angles with Parallel Lines

Consider three lines arranged similarly to those shown in Example 3. Suppose that two of the lines, say, ℓ, and ℓ_2, are *parallel* (that is, they lie in the same plane and never intersect), and also that a third line ℓ_3 intersects the parallel lines at points P and Q, as shown in Figure 7.22. The line ℓ_3 is called a **transversal.** The notation we use for parallel lines is $\ell_1 \| \ell_2$.

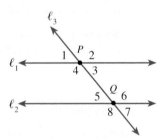

FIGURE 7.22 Parallel lines cut by a transversal

We make some observations about angles:

Vertical angles are congruent.

Alternate interior angles are pairs of angles whose interiors lie between the parallel lines, but on opposite sides of the transversal, each having one of the lines for one of its sides. Alternate interior angles are congruent.

Alternate exterior angles are pairs of angles that lie outside the parallel lines, but on opposite sides of the transversal, each with one side adjacent to each parallel. Alternate exterior angles are congruent.

Corresponding angles are two nonadjacent angles whose interiors lie on the same side of the transversal such that one angle lies between the parallel lines and the other does not. Corresponding angles are congruent.

Example **4**   **Angle pair terminology**

Consider Figure 7.23.

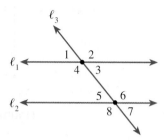

FIGURE 7.23 Angles formed by parallel lines and a transversal

a. Name the vertical angles.
b. Name the alternate interior angles.
c. Name the corresponding angles.
d. Name the alternate exterior angles.

Solution

a. The vertical angles are: $\angle 1$ and $\angle 3$; $\angle 2$ and $\angle 4$; $\angle 5$ and $\angle 7$; $\angle 6$ and $\angle 8$
b. The alternate interior angles are: $\angle 4$ and $\angle 6$; $\angle 3$ and $\angle 5$
c. The corresponding angles are: $\angle 1$ and $\angle 5$; $\angle 2$ and $\angle 6$; $\angle 3$ and $\angle 7$; $\angle 4$ and $\angle 8$
d. The alternate exterior angles are: $\angle 1$ and $\angle 7$; $\angle 2$ and $\angle 8$

To summarize the results from Example 4, notice that the following angles are congruent, written $\simeq$:

$$\angle 1 \simeq \angle 3 \simeq \angle 5 \simeq \angle 7 \quad \text{and} \quad \angle 2 \simeq \angle 4 \simeq \angle 6 \simeq \angle 8$$

Also, all pairs of adjacent angles are supplementary.

Example **5**   **Find angles with parallel lines and transversal**

Find the measures of the eight numbered angles in Figure 7.23, where ℓ_1 and ℓ_2 are parallel. Assume that $m\angle 5 = 50°$.

Solution
$m\angle 1 = 50°$ *Corresponding angles*
$m\angle 2 = 180° - 50° = 130°$ $\angle 1$ and $\angle 2$ are supplementary.
$m\angle 3 = 50°$ $\angle 3$ and $\angle 1$ are vertical angles.
$m\angle 4 = 130°$ *Vertical angles*
$m\angle 5 = 50°$ *Given*
$m\angle 6 = 130°$ *Supplementary angles*
$m\angle 7 = 50°$ *Vertical angles*
$m\angle 8 = 130°$ *Supplementary angles*

Perpendicular Lines

If two lines intersect so that the adjacent angles are equal, then the lines are **perpendicular.** Simply, lines that intersect to form angles of 90° (right angles) are called perpendicular lines. In Figure 7.24, lines ℓ_3 and ℓ_4 intersect to form a right angle, and therefore they are perpendicular lines.

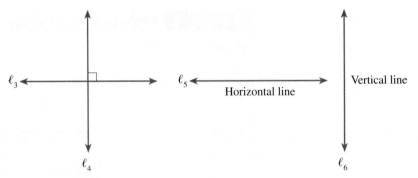

FIGURE 7.24 Horizontal, vertical, and perpendicular lines

In a diagram on a printed page, any line that is parallel to the top and bottom edge of the page is considered **horizontal.** Lines that are perpendicular to a horizontal line are considered to be **vertical.** In Figure 7.24, line ℓ_5 is a horizontal line and line ℓ_6 is a vertical line.

Example 6 Horizontal and vertical angles

The diagram in Figure 7.25 shows lines in the same plane. Which of the given statements are true?

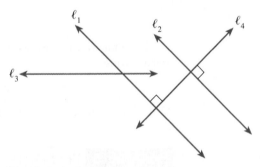

FIGURE 7.25 Lines in the same plane

a. Lines ℓ_1 and ℓ_2 are parallel and horizontal.
b. Lines ℓ_4 and ℓ_2 are intersecting and not perpendicular.
c. Lines ℓ_2 and ℓ_3 are intersecting lines.

Solution
a. Lines ℓ_1 and ℓ_2 are parallel, but they are not horizontal. Thus, the statement (which is a conjunction) is false.
b. Lines ℓ_4 and ℓ_2 are both intersecting and perpendicular. The statement is false.
c. Lines ℓ_2 and ℓ_3 are intersecting. Recall that the arrows on ℓ_3 indicate that it goes on without end in both directions, so it will intersect ℓ_2. This statement is true.

Problem Set | 7.2

Level 1

1. IN YOUR OWN WORDS What is an angle?

2. IN YOUR OWN WORDS Distinguish between equal angles and congruent angles.

3. IN YOUR OWN WORDS Distinguish between a half-line and a ray.

4. IN YOUR OWN WORDS What is a quadrilateral? Describe five different classifications of quadrilaterals.

5. IN YOUR OWN WORDS   Distinguish between horizontal and vertical lines.

6. IN YOUR OWN WORDS   Describe right angles, acute angles, and obtuse angles.

7. IN YOUR OWN WORDS   Describe adjacent, vertical, and corresponding angles.

8. IN YOUR OWN WORDS   Describe parallel lines.

Name the polygons in Problems 9–14 according to the number of sides.

9. a. 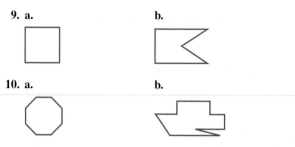   **b.**

10. a. 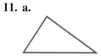   **b.**

11. a. 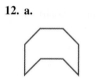   **b.**

12. a.   **b.**

13. a. 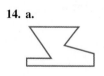   **b.**

14. a.   **b.**

Determine whether each sentence in Problems 15–19 is true or false.

15. a. Every square is a rectangle.
   b. Every square is a parallelogram.

16. a. Every square is a rhombus.
   b. Every rhombus is a square.

17. a. Every square is a quadrilateral.
   b. Every parallelogram is a rectangle.

18. a. A rectangle is a parallelogram.
   b. A trapezoid is a quadrilateral.

19. a. A quadrilateral is a trapezoid.
   b. A parallelogram is a trapezoid.

For the quadrilaterals named in Problems 20–24, answer "yes" or "no" to indicate whether each of the following properties is satisfied:

   a. Opposite sides are parallel.
   b. Opposite sides have equal length.
   c. Opposite angles have equal measure.
   d. Interior angles are right angles.
   e. Diagonals have equal length.

20. rectangle   **21.** square

22. parallelogram   **23.** trapezoid

24. rhombus

Level 2

Using only a straightedge and a compass, reproduce the angles given in Problems 25–30.

25.   **26.**

27.   **28.**

29.   **30.**

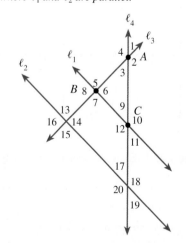

In Problems 31–35 classify the requested angles shown in Figure 7.26, where ℓ_1 and ℓ_2 are parallel.

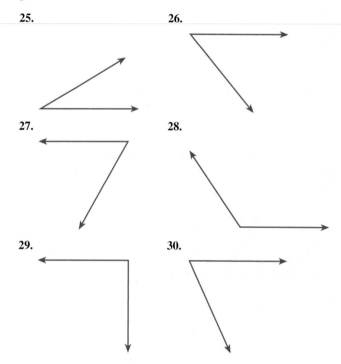

FIGURE 7.26 Parallel lines cut by transversals

31. a. $\angle BAC$ if $m\angle 1$ is $30°$
   b. $\angle ABC$ if $m\angle 5$ is $90°$

32. a. $\angle 18$ if $m\angle 17$ is $105°$
   b. $\angle 19$ if $m\angle 11$ is $70°$

33. a. $\angle 10$ if $m\angle 11$ is $90°$
   b. $\angle 16$ if $m\angle 15$ is $30°$

34. a. $\angle CBA$ if $m\angle 16$ is $120°$
   b. $\angle BCA$ if $m\angle 19$ is $110°$

35. a. $\angle 1$ if $m\angle 2$ is $130°$
   b. $\angle 5$ if $m\angle 15$ is $88°$

In Problems 36–43 classify the pairs of angles shown in Figure 7.26.

36. $\angle 2$ and $\angle 4$

37. $\angle 13$ and $\angle 14$

38. $\angle 9$ and $\angle 12$

39. $\angle 9$ and $\angle 10$

40. $\angle 9$ and $\angle 17$

41. $\angle 9$ and $\angle 11$

42. $\angle 12$ and $\angle 18$

43. $\angle 7$ and $\angle 13$

Each of the four diagrams in Figure 7.27 shows lines in the same plane. Classify each of the statements in Problems 44–49 as true or false.

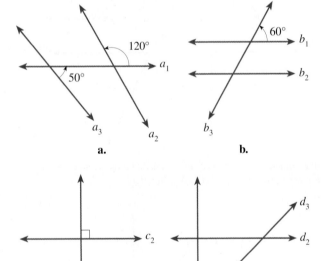

FIGURE 7.27 Intersecting lines in the same plane

44. The lines a_1 and a_3 are both intersecting and parallel.

45. The lines b_1 and b_2 are vertical.

46. The lines c_1 and c_2 are not intersecting.

47. The lines d_1 and d_3 are intersecting.

48. The lines c_2 and d_2 are horizontal.

49. The lines a_2 and a_3 are both vertical and parallel.

In Problems 50–55, find the measures of all the angles in Figure 7.28.

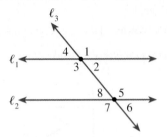

FIGURE 7.28 ℓ_1 is parallel to ℓ_2

50. Given $m\angle 7 = 110°$

51. Given $m\angle 2 = 65°$

52. Given $m\angle 6 = 19°$

53. Given $m\angle 1 = 153°$

54. Given $m\angle 5 = 120°$

55. Given $m\angle 3 = 163°$

56. Consider the angles in the accompanying figure. Classify the named angles as acute, right, straight, or obtuse.

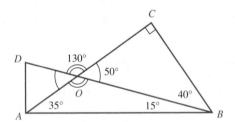

a. $\angle DOC$ **b.** $\angle AOB$
c. $\angle DBC$ **d.** $\angle CAB$
e. $\angle DOB$ **f.** $\angle C$
g. $\angle COB$ **h.** $\angle AOC$
i. $\angle DOA$
j. Name an angle congruent to $\angle DOB$.

57. Consider the angles in the accompanying figure. Classify the named angles as acute, right, straight, or obtuse.

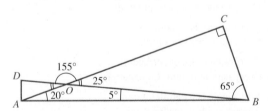

a. $\angle DOC$ **b.** $\angle AOB$
c. $\angle DBC$ **d.** $\angle CAB$
e. $\angle DOB$ **f.** $\angle C$
g. $\angle COB$ **h.** $\angle AOC$
i. $\angle DOA$
j. Name an angle congruent to $\angle DOB$.

58. This problem anticipates a major result derived in the next section. Show that *the sum of the measures of the interior angles of any triangle is 180°.*

    a. Draw three triangles, one with all acute angles, one with a right angle, and a third with an obtuse angle. A triangle with all acute angles is shown here:

    b. Tear apart the angles of each triangle you've drawn, and place them together to form a straight angle at the bottom, as shown here:

Problem Solving 3

59. The first illustration in the accompanying figure shows a cube with the top cut off. Use solid lines and shading to depict seven other different views of a cube with one side cut off.

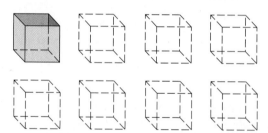

60. HISTORICAL QUEST Allow me to start you on a journey in Golygon City. You can take a similar trip in Manhattan, Tokyo, or almost any large city whose streets form a grid of squares. Here are your directions. Stroll down a city block, and at the end turn left or right. Walk two more blocks, turn left or right, then walk another three blocks, turn once more, and so on. Each time you turn, you must walk straight one block farther than before. If after a number of turns you arrive at your starting point, you have traced a golygon, as shown in Figure 7.29. A *golygon* consists of straight-line segments that have lengths (measured in miles, meters, or whatever unit you prefer) of one, two, three, and so on units. Draw some golygons. Can you make a conjecture about golygons?[*]

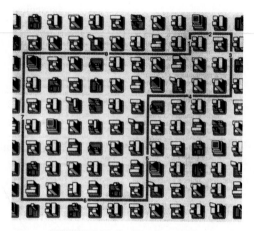

FIGURE 7.29 A sample golygon

[*]From "Mathematical Recreations," by A. K. Dewdney, *Scientific American*, July 1990, p. 118. Copyright © 1990 by Scientific American, Inc. All rights reserved. Illustration by Slim Films.

7.3 | Triangles

One of the most frequently encountered polygons is the **triangle.** In this section we take a closer look at them.

Terminology

Every triangle has six parts: three sides and three angles. We name the sides by naming the endpoints of the line segments, and we name the angles by identifying the vertex (see Figure 7.30).

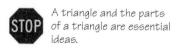 A triangle and the parts of a triangle are essential ideas.

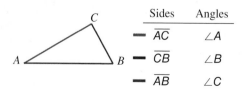

	Sides	Angles
—	$\overline{AC}$	$\angle A$
—	$\overline{CB}$	$\angle B$
—	$\overline{AB}$	$\angle C$

FIGURE 7.30 A standard triangle showing the six parts

Triangles are classified both by sides and by angles (single, double, and triple marks are used to indicate segments of equal length):

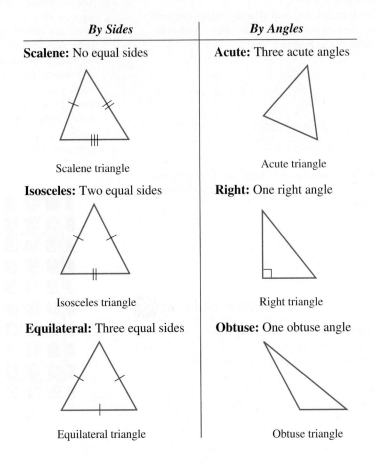

By Sides	*By Angles*
Scalene: No equal sides	**Acute:** Three acute angles
Scalene triangle	Acute triangle
Isosceles: Two equal sides	**Right:** One right angle
Isosceles triangle	Right triangle
Equilateral: Three equal sides	**Obtuse:** One obtuse angle
Equilateral triangle	Obtuse triangle

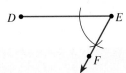

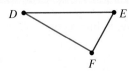

FIGURE 7.31 Constructing congruent triangles

We say that two triangles are **congruent** if they have the same size and shape. Suppose that we wish to construct a triangle with vertices *D*, *E*, and *F*, congruent to △*ABC* as shown in Figure 7.30. We would proceed as follows (as shown in Figure 7.31):

1. Mark off segment $\overline{DE}$ so that it is congruent to $\overline{AB}$. We write this as $\overline{DE} \simeq \overline{AB}$.
2. Construct angle *E* so that it is congruent to angle *B*. We write this as $\angle E \simeq \angle B$.
3. Mark off segment $\overline{EF} \simeq \overline{BC}$.

You can now see that, if you connect points *D* and *F* with a straightedge, the resulting △*DEF* has the same size and shape as △*ABC*. The procedure we used here is called SAS, meaning we constructed two sides and an *included angle* (an angle between two sides) congruent to two sides and an included angle of another triangle. We call these **corresponding parts.** There are other procedures for constructing congruent triangles; some of these are discussed in the problem set. For this example, we say △*ABC* ≃ △*DEF*. From this we conclude that all six corresponding parts are congruent.

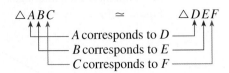

Example 1 | Corresponding angles with congruent triangles

Name the corresponding parts of the given triangles.
a. $\triangle ABC \simeq \triangle A'B'C'$ **b.** $\triangle RST \simeq \triangle UST$

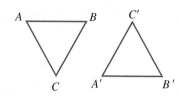 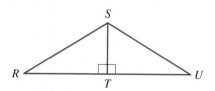

Solution

a. $\overline{AB}$ corresponds to $\overline{A'B'}$
$\overline{AC}$ corresponds to $\overline{A'C'}$
$\overline{BC}$ corresponds to $\overline{B'C'}$
$\angle A$ corresponds to $\angle A'$
$\angle B$ corresponds to $\angle B'$
$\angle C$ corresponds to $\angle C'$

b. $\overline{RS}$ corresponds to $\overline{US}$
$\overline{RT}$ corresponds to $\overline{UT}$
$\overline{ST}$ corresponds to $\overline{ST}$
$\angle R$ corresponds to $\angle U$
$\angle RTS$ corresponds to $\angle UTS$
$\angle RST$ corresponds to $\angle UST$

Angles of a Triangle

One of the most basic properties of a triangle involves the sum of the measures of its angles. To discover this property for yourself, place a pencil with an eraser along one side of any triangle as shown in Figure 7.32**a.**

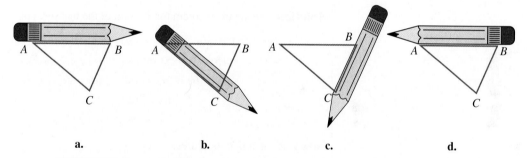

FIGURE 7.32 Demonstration that the sum of the measures of the angles in a triangle is 180°

Now rotate the pencil to correspond to the size of $\angle A$ as shown in Figure 7.32**b.** You see your pencil is along $\overline{AC}$. Next, rotate the pencil through $\angle C$, as shown in Figure 7.32**c.** Finally, rotate the pencil through $\angle B$. Notice that the pencil has been rotated the same amount as the sum of the angles of the triangle. Also notice that the orientation of the pencil is exactly reversed from the starting position. This leads us to the following important theorem.

You will frequently need to use this property. **STOP**

Angles in a Triangle

The sum of the measures of the angles in any triangle is 180°.

Example 2 Find angles in a triangle

Find the missing angle measure in the triangle in Figure 7.33.

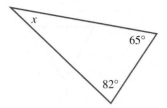

FIGURE 7.33 What is *x*?

Solution   Let *x* represent the missing angle measure.

$$65 + 82 + x = 180$$
$$147 + x = 180$$
$$x = 33$$

The missing angle's measure is 33°.

Example 3 Use algebra to find angles in a triangle

Find the measures of the angles of a triangle if it is known that the measures are x, $2x - 15$, and $3(x + 7)$ degrees.

Solution   Using the theorem for the sum of the measures of angles in a triangle, we have

$$x + (2x - 15) + 3(x + 17) = 180 \quad \text{Sum of the measures of the angles is 180°.}$$
$$x + 2x - 15 + 3x + 51 = 180 \quad \text{Eliminate parentheses.}$$
$$6x + 36 = 180 \quad \text{Combine similar terms.}$$
$$6x = 144 \quad \text{Subtract 36 from both sides.}$$
$$x = 24 \quad \text{Divide both sides by 6.}$$

Now find the angle measures:

$$x = 24$$
$$2x - 15 = 2(24) - 15 = 33$$
$$3(x + 17) = 3(24 + 17) = 123$$

The angles have measures of 24°, 33°, and 123°.

An **exterior angle** of a triangle is the angle on the other side of an extension of one side of the triangle. An example is the angle whose measure is marked as *x* in Figure 7.34.

Notice that the following relationships are true for any $\triangle ABC$ with exterior angle *x*:

$$m\angle A + m\angle B + m\angle C = 180° \quad \text{and} \quad m\angle C + x = 180°$$

Thus,

$$m\angle A + m\angle B + m\angle C = m\angle C + x$$
$$m\angle A + m\angle B = x \quad \text{Subtract } m\angle C \text{ from both sides.}$$

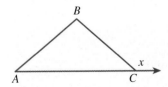

FIGURE 7.34 Exterior angle *x*

Exterior Angle Property

The measure of the exterior angle of a triangle equals the sum of the measures of the two interior angles.

Example 4 Find the exterior angle

Find the value of x in Figure 7.35.

FIGURE 7.35 What is x?

Solution

$$\underbrace{63 + 42}_{\text{Sum of interior angles}} = \underset{\downarrow}{x} \quad \text{Exterior angle}$$

$$105 = x$$

The measure of the exterior angle is $105°$.

Isosceles Triangle Property

In an isosceles triangle, there are two sides of equal length and the third side is called its **base.** The angle included by its legs is called the **vertex angle,** and the angles that include the base are called **base angles.**

There is an important theorem in geometry that is known as the **isosceles triangle property.**

Isosceles Triangle Property

If two sides of a triangle have the same length, then angles opposite them are congruent.

In other words, if a triangle is isosceles, then the base angles have equal measures. The converse is also true; namely, if two angles of a triangle are congruent, the sides opposite them have equal length.

Example 5 Equiangular implies equilateral

Give a reasonable argument to prove that if a triangle is equiangular, it is also equilateral.

Solution If $\triangle ABC$ is equiangular, then $m\angle A = m\angle B = m\angle C$. Since $m\angle A = m\angle B$, from the converse of the isosceles triangle property, we have $|\overline{BC}| = |\overline{AC}|$. Again, since $m\angle B = m\angle C$, we have $|\overline{AC}| = |\overline{AB}|$. Thus, we see that all three sides have the same length, and consequently $\triangle ABC$ is equilateral.

Problem Set 7.3

Level **1**

1. **IN YOUR OWN WORDS** What is a triangle?

2. **IN YOUR OWN WORDS** What is the sum of the measures of the angles of a triangle?

3. **IN YOUR OWN WORDS** Explain the notation $\triangle ABC \simeq \triangle DEF$.

4. **IN YOUR OWN WORDS** Explain why the musical instrument called a "triangle" is not a good example of a geometric triangle.

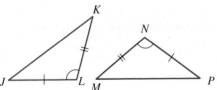

Name the corresponding parts of the triangles in Problems 5–10.

5.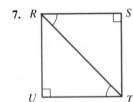

6.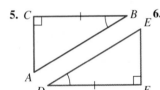

7.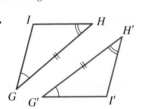

8.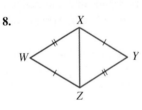

9.

10.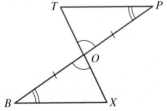

In Problems 11–16, find the measure of the third angle in each triangle.

11.

12.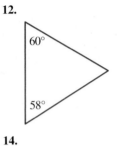

13.

14.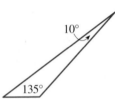

15.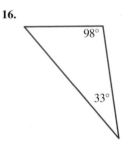

16.

Find the measure of the indicated exterior angle in each of the triangles in Problems 17–22.

17.

18.

19.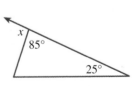

20.

21.

22.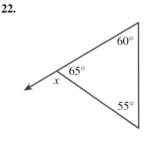

Using only a straightedge and a compass, reproduce the triangles given in Problems 23–28.

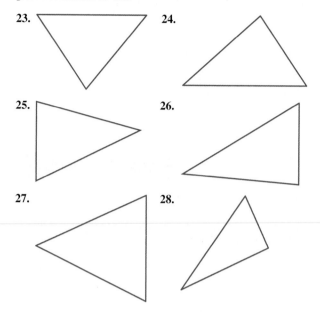

23.

24.

25.

26.

27.

28.

Use algebra to find the value of x in each of the triangles in Problems 29–34. Notice that the measurement of the angle is not necessarily the same as the value of x.

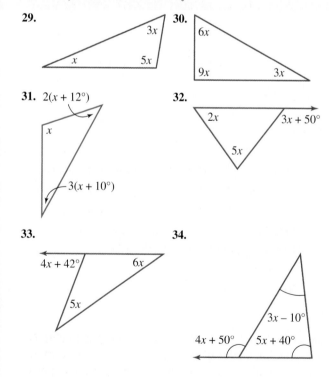

29.

$3x$

x $5x$

30.

$6x$

$9x$ $3x$

31. $2(x + 12°)$

x

$3(x + 10°)$

32.

$2x$ $3x + 50°$

$5x$

33.

$4x + 42°$ $6x$

$5x$

34.

$3x - 10°$

$4x + 50°$ $5x + 40°$

Find the numerical measures of the angles of the triangle whose angle measures are given in Problems 35–40.

35. x, x, x

36. $x, x, 2x$

37. $x, x + 10°, x + 20°$

38. $x, x - 10°, x + 10°$

39. $x, 14° + 3x, 3(x + 25°)$

40. $x, 3x - 10°, 3(55° - x)$

In Problems 41–44, assume that the given angles have been measured, and calculate the angles (marked as x) that could not be found by direct measurement.

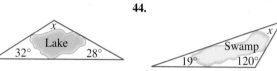

41.

x

$40°$

42.

x

$38°$

43.

x

Lake

$32°$ $28°$

44.

x

Swamp

$19°$ $120°$

Draw an example of each of the triangles described in Problems 45–52. If the figure cannot exist, write "impossible."

45. acute scalene

46. acute isosceles

47. acute equilateral

48. right scalene

49. right isosceles

50. right equilateral

51. obtuse scalene

52. obtuse isosceles

53. In the text we constructed congruent triangles by using SAS. Reproduce the triangle shown in Problem 25 by using SSS. This means to construct the triangle by using the lengths of the three sides.

54. In the text we constructed congruent triangles by using SAS. Reproduce the triangle shown in Problem 26 by using SSS. This means to construct the triangle by using the lengths of the three sides.

55. In the text we constructed congruent triangles by using SAS. Reproduce the triangle shown in Problem 27 by using ASA. This means to construct the triangle by using a side included between two angles.

56. In the text we constructed congruent triangles by using SAS. Reproduce the triangle shown in Problem 28 by using ASA. This means to construct the triangle by using a side included between two angles.

57. The legs of a picnic table form a triangle where $\overline{AC}$ and $\overline{BC}$ have the same length, as shown in Figure 7.36.

Find the measures of the angles x and y so that the top of the table will be parallel to the ground, given the following measurements.

a. $m \angle ACB = 90°$

b. $m\angle ACB = 85°$

c. $m\angle ACB = 92°$

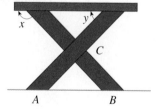

FIGURE 7.36 Picnic table

58. IN YOUR OWN WORDS Show that *the sum of the measures of the interior angles of any quadrilateral is 360°* by carrying out the following steps:

a. Draw any quadrilateral, as illustrated in Figure 7.37 (but draw your quadrilateral so it has a different shape from the one shown here).

b. Divide the quadrilateral into two triangles by drawing a diagonal (a line segment connecting two nonadjacent vertices). Label the angles of your triangles as shown in Figure 7.37.

c. What is the sum $m\angle 1 + m\angle 2 + m\angle 3$?

d. What is the sum $m\angle 4 + m\angle 5 + m\angle 6$?

e. The sum of the measures of the angles of the quadrilateral is

$$(m\angle 1 + m\angle 2 + m\angle 3) + (m\angle 4 + m\angle 5 + m\angle 6)$$

What is this sum?

f. Do you think this argument will apply for *any* quadrilateral?

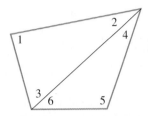

FIGURE 7.37 Quadrilateral

Problem Solving **3**

59. Look at Problem 58. What is the sum of the measures of the interior angles of any pentagon?

60. Look at Problem 58. What is the sum of the measures of the interior angles of any octagon?

7.4 | Similar Triangles

Congruent figures have exactly the same size and shape. However, it is possible for figures to have exactly the same shape without necessarily having the same size. Such figures are called *similar figures.* In this section we will focus on **similar triangles.** If $\triangle ABC$ is similar to $\triangle DEF$, we write

$$\triangle ABC \sim \triangle DEF$$

Similar triangles are shown in Figure 7.38.

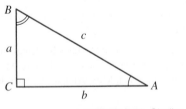

FIGURE 7.38 Similar triangles

$m\angle A = m\angle D$, so these are corresponding angles.

$m\angle B = m\angle E$, so these are corresponding angles.

$m\angle C = m\angle F$, so these are corresponding angles.

Side $\overline{BC}$ is opposite $\angle A$ and side $\overline{EF}$ is opposite $\angle D$, so we say that $\overline{BC}$ corresponds to $\overline{EF}$.

$\overline{AC}$ corresponds to $\overline{DF}$.

$\overline{AB}$ corresponds to $\overline{DE}$.

Since these figures have the same shape, we talk about **corresponding angles** and **corresponding sides.** The corresponding angles of similar triangles are those angles that have equal measure. The corresponding sides are those sides that are opposite equal angles.

Even though corresponding angles are equal, corresponding sides do not need to have the same length. If they do have the same length, the triangles are congruent.

However, when they are not the same length, we can say they are proportional. From Figure 7.38 we see that the lengths of the sides are labeled *a, b, c* and *d, e, f.* When we say the sides are proportional, we mean

Primary ratios:

$$\frac{a}{b} = \frac{d}{e} \qquad \frac{a}{c} = \frac{d}{f} \qquad \frac{b}{c} = \frac{e}{f}$$

Reciprocals:

$$\frac{b}{a} = \frac{e}{d} \qquad \frac{c}{a} = \frac{f}{d} \qquad \frac{c}{b} = \frac{f}{e}$$

We summarize with an important property of similar triangles called the **similar triangle theorem.**

CAUTION

This result is used in many applications.

Similar Triangle Theorem

Two triangles are similar if two angles of one triangle are congruent to two angles of the other triangle. If the triangles are similar, then their corresponding sides are proportional.

Example 1   Identify similar triangles

Identify pairs of triangles that are similar in Figure 7.39.

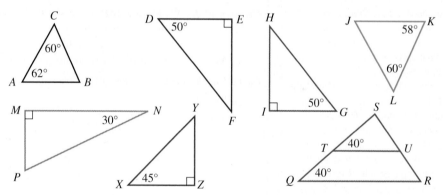

FIGURE 7.39 Which of these triangles are similar?

Solution $\triangle ABC \sim \triangle JKL$; $\triangle DEF \sim \triangle GIH$; $\triangle SQR \sim \triangle STU$

Example 2   Find lengths given similar triangles

Given the similar triangles in Figure 7.40, find the unknown lengths marked b' and c'.

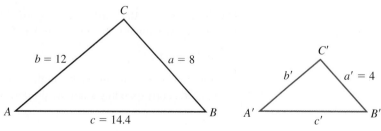

FIGURE 7.40 Given $\triangle ABC \sim \triangle A'B'C'$

Solution   Since corresponding sides are proportional (other proportions are possible), we have

$$\frac{a}{b} = \frac{a'}{b'} \qquad\qquad \frac{a}{c} = \frac{a'}{c'}$$

$$\frac{8}{12} = \frac{4}{b'} \qquad\qquad \frac{8}{14.4} = \frac{4}{c'}$$

$$b' = \frac{4(12)}{8} \qquad\qquad c' = \frac{14.4(4)}{8}$$

$$b' = 6 \qquad\qquad c' = 7.2$$

Identifying similar triangles is simplified even further if we know that the triangles are right triangles, because then the triangles are similar if one of the acute angles of one triangle has the same measure as an acute angle of the other.

Example **3** **Find tree height**

Find the height of a tree that is difficult to measure directly.

Solution   We use Pólya's problem-solving guidelines for this example.

Understand the Problem. We need to find the height of some tree without measuring it directly.

Devise a Plan. We assume that it is a sunny day, and we will measure the height of the tree by measuring its shadow on the ground. For reference, we also measure the length of the shadow of an object of known height (say our own height, a meterstick, or a yardstick). We will then use similar triangles and proportions to find the height of the tree.

Carry Out the Plan. Suppose that a tree and a yardstick are casting shadows as shown in Figure 7.41.

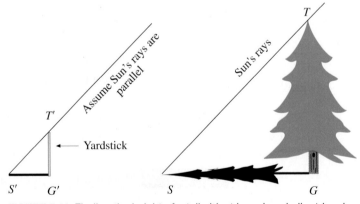

FIGURE 7.41 Finding the height of a tall object by using similar triangles

If the shadow of the yardstick is 3 yards long and the shadow of the tree is 12 yards long, we use similar triangles to estimate h, the height of the tree, if we know that

$$m\angle S = m\angle S'$$

Since $\angle G$ and $\angle G'$ are right angles, and since $m\angle S = m\angle S'$, $\triangle SGT \sim \triangle S'G'T'$. Therefore, corresponding sides are proportional.

$$\frac{1}{3} = \frac{h}{12}$$ You solved proportions like this in the previous chapter.

$$h = \frac{1(12)}{3}$$

$$h = 4$$

Look Back. The tree is 4 yards, or 12 ft, tall.

There is a relationship between the sizes of the angles of a right triangle and the ratios of the lengths of the sides. In a right triangle, the side opposite the right angle is called the **hypotenuse.** Each of the acute angles of a right triangle has one side that is the hypotenuse; the other side of that angle is called the **adjacent side.** (See Figure 7.42.)

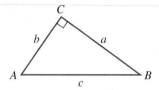

FIGURE 7.42 A right triangle

In $\triangle ABC$ with right angle at C:

The hypotenuse is c;

The side adjacent to $\angle A$ is b;

The side adjacent to $\angle B$ is a.

We also talk about an **opposite side.** The side opposite $\angle A$ is a, and the side opposite $\angle B$ is b.

Problem Set　7.4

Level 1

1. IN YOUR OWN WORDS   Contrast congruent and similar triangles.

2. IN YOUR OWN WORDS   What does it mean when we say the corresponding sides of two congruent triangles are proportional?

In Problems 3–8, tell whether it is possible to conclude that the triangles are similar.

3.

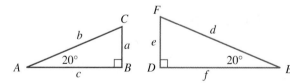

4.

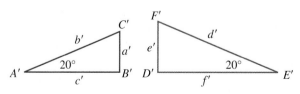

5.

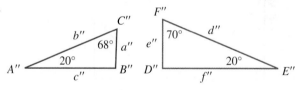

6.

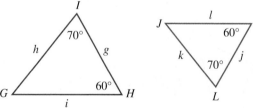

7.

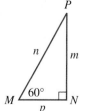
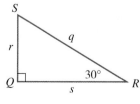

8.

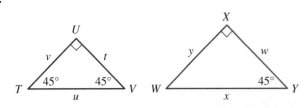

List the measures for all six angles for the figures given in Problems 9–14.

9.

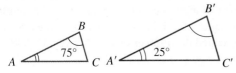

10.

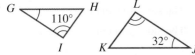

11.

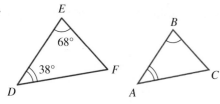

12.

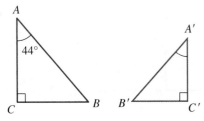

13.

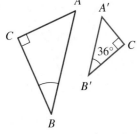

14.

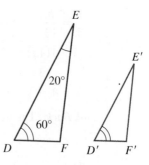

List the lengths of all six sides for the figures given in Problems 15–20.

15.

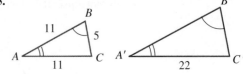

16.

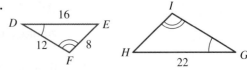

17.

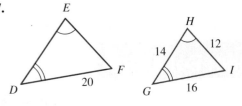

18.

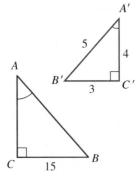

19.

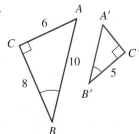

20.

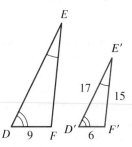

Given two similar triangles, as shown in Figure 7.43, find the unknown lengths in Problems 21–28.

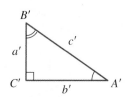

FIGURE 7.43 Similar triangles

21. $a = 4$, $b = 8$; find c.

22. $a' = 7$, $b' = 3$; find c'.

23. $a = 4$, $b = 8$, $a' = 2$; find b'.

24. $b = 5$, $c = 15$, $b' = 3$; find c'.

25. $c = 6$, $a = 4$, $c' = 8$; find a'.

26. $a' = 7$, $b' = 3$, $a = 5$; find b.

27. $b' = 8$, $c' = 12$, $c = 4$; find b.

28. $c' = 9$, $a' = 2$, $c = 5$; find a.

Each figure in Problems 29–34 contains two similar triangles. Find the unknown measure indicated by a variable. Answer to the nearest tenth.

29.

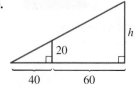

30.

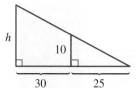

31.

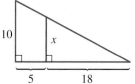

32.

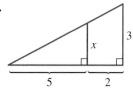

33.

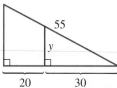

34.

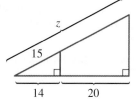

Level **2**

35. Given $\overline{AC}$ is perpendicular to $\overline{MB}$, and $\triangle ABC$ is an equilateral triangle. Show that $\triangle ABM \sim \triangle CBM$.

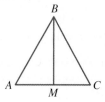

36. Given $m\angle D = m\angle E$. Show that $\triangle ABD \sim \triangle CBE$.

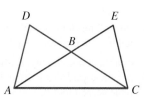

37. Given that $\overline{CT}$ bisects both $\angle ACO$ and $\angle ATO$. Show that $\triangle CAT \sim \triangle COT$.

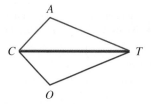

38. Given that $\overline{AW}$ bisects $\angle R_1AR_2$ and $\angle R_1OR_2$. Show that $\triangle AOR_1 \sim \triangle AOR_2$.

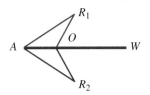

39. Use similar triangles and a proportion to find the length of the lake shown in Figure 7.44.

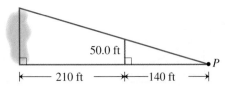

FIGURE 7.44 Determining the length of a lake

40. Suppose the distances in Problem 39 are changed as follows: 150 ft instead of 210 ft and 90 ft instead of 140 ft. What is the length of this lake (to the nearest foot)?

41. Use similar triangles and a proportion to find the height of the house siding shown in Figure 7.45.

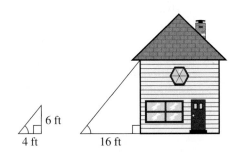

FIGURE 7.45 Determining the height of a building

42. Suppose the 6 ft distance in Problem 41 is 5 ft 8 in. Use this information to find the height of the house (to the nearest inch) shown in Figure 7.45.

43. A building casts a shadow 75 ft long. At the same time, the shadow cast by a vertical yardstick is 5 ft long. How tall is the building?

44. Suppose the shadow of the building in Problem 43 is 75 ft 3 in. How tall is the building (to the nearest inch)?

45. A bell tower casts a shadow 45 ft long. At the same time, the shadow cast by a vertical yardstick is 23 in. long. How tall is the bell tower (to the nearest foot)?

46. If a tree casts a shadow of 12 ft at the same time that a 6-ft person casts a shadow of $2\frac{1}{2}$ ft, find the height of the tree (to the nearest foot).

47. If a tree casts a shadow of 10 ft at the same time that a 5-ft person casts a shadow of 3 ft, find the height of the tree (to the nearest foot).

48. If a tree casts a shadow of 8 ft 3 in. at the same time that a 5-ft 10-in. person casts a shadow of 2 ft 7 in., find the height of the tree (to the nearest inch).

49. If a tree casts a shadow of 4 ft 5 in. at the same time that a 5-ft 9-in. person casts a shadow of 3 ft 10 in., find the height of the tree (to the nearest inch).

50. If lines are drawn on a map, a triangle can be formed by the cities of New York City, Washington, D.C., and Buffalo, New York. On the map, the distance between New York City and Washington, D.C., is 2.1 cm, New York to Buffalo is 2.5 cm, and Buffalo to Washington, D.C., is approximately 2.85 cm. If the actual straight-line distance from Buffalo to Washington is 285 miles, how far is it between the other pairs of cities?

51. If lines are drawn on a map, a triangle can be formed by the cities of New Orleans, Louisiana; Denver, Colorado; and Chicago, Illinois. On the map, the distance between New Orleans and Denver is 10.8 cm, Chicago to New Orleans is 9.5 cm, and Chicago to Denver is 10.2 cm. If the actual straight-line distance from Chicago to New Orleans is 950 miles, how far is it between the other pairs of cities?

52. Suppose a 6-ft person wishes to determine the height of a bridge above the bottom of a canyon, as shown in Figure 7.46.

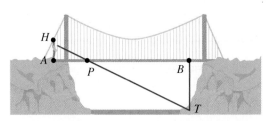

FIGURE 7.46 Height of a bridge

To do this, this person stands at one end of the bridge (point A) and looks down to a point directly below the other end (point B). With the help of a companion, point P is determined to form two triangles.
a. Identify the two triangles.
b. Are the triangles similar? Why or why not?
c. Find the distance to the bottom of the canyon, if $|\overline{AP}| = 10$ ft and $|\overline{PB}| = 40$ft

53. Suppose a 6-ft person wishes to determine the height of a foot-bridge connecting two buildings, as shown in Figure 7.47.

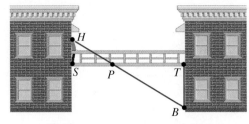

FIGURE 7.47 Height of a footbridge

To do this, this person stands at one end of the bridge (point S) and looks down to a point directly below the other end (point T). With the help of a companion, point P is determined to form two triangles.

a. Identify the two triangles.
b. Are the triangles similar? Why or why not?
c. Find the height of the footbridge, if $|\overline{SP}| = 10$ ft and $|\overline{PT}| = 35$ ft

Level 3

54. Given $|\overline{AB}| = |\overline{BC}|$ and M midpoint of $\overline{AC}$. Show that $\triangle ABM \sim \triangle CBM$.

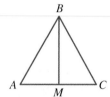

55. Given $m\angle D = m\angle E$ and $|\overline{AB}| = |\overline{BC}|$. Show that $\triangle ADC \sim \triangle CEA$.

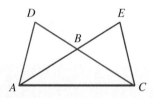

56. A useful theorem that uses proportionality, but not triangles, is the following:

If two lines connect the endpoints of parallel segments of different lengths, then a line through the intersection of the two lines connecting those endpoints divides the parallel segments proportionally. (See Figure 7.48.)

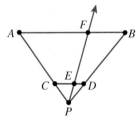

FIGURE 7.48 Suppose $\overline{AB} \parallel \overline{CD}$; then $\dfrac{|\overline{AF}|}{|\overline{BF}|} = \dfrac{|\overline{CE}|}{|\overline{DE}|}$

Using this result, divide the following line segment into a 2-to-3 ratio:

57. Divide the segment given in Problem 56 into two parts in a 3-to-7 ratio.

58. Present an argument showing that if two triangles are equilateral, then they are similar triangles.

59. For any right triangle ABC (right angle at C), drop a perpendicular from point C to base $\overline{AB}$ at the point D. Show that the two triangles thus formed are similar.

60. Present an argument showing that two triangles similar to a third triangle are similar to each other.

7.5 | Right-Triangle Trigonometry

An important theorem from geometry, the **Pythagorean theorem,** has an important algebraic representation, and is important in our study of triangles.

You should be familiar with this theorem.

Pythagorean Theorem

For any right triangle with sides of lengths a and b and hypotenuse of length c,

$$a^2 + b^2 = c^2$$

Also, if a, b, and c are the lengths of the sides of a triangle so that $a^2 + b^2 = c^2$, then the triangle is a right triangle.

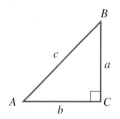

FIGURE 7.49 Right triangle

A correctly labeled right triangle is shown in Figure 7.49. In a right triangle, the sides that are not the hypotenuse are sometimes called **legs.**

Example **1** **Build a right angle**

A carpenter wants to make sure that the corner of a closet is square (a right angle). If she measures out sides of 3 feet and 4 feet, how long should she make the diagonal (hypotenuse)?

Solution The length of the hypotenuse is the unknown, so use the Pythagorean theorem:

$$c = \sqrt{a^2 + b^2}$$
$$= \sqrt{3^2 + 4^2} \qquad \text{The sides are 3 and 4.}$$
$$= \sqrt{25}$$
$$= 5$$

She should make the diagonal 5 feet long.

Trigonometric Ratios

Of interest to us in this section is the value found by forming the ratios of the lengths of the sides of a right triangle. There are six possible ratios for the triangle shown in Figure 7.49.

Primary Ratios	Reciprocal Ratios
$\dfrac{a}{b}, \dfrac{a}{c}, \dfrac{b}{c}$	$\dfrac{b}{a}, \dfrac{c}{a}, \dfrac{c}{b}$

In this section, we study the relationship of the primary ratios with the angles A and B. These ratios, called the **trigonometric ratios,** and are defined in the box.

CAUTION

This is the essential definition of this section. It also forms the basis for a trigonometry course.

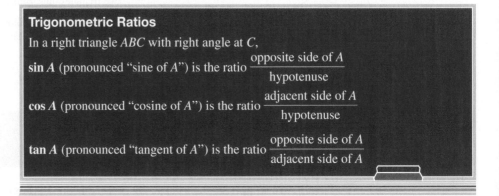

Trigonometric Ratios

In a right triangle ABC with right angle at C,

$\sin A$ (pronounced "sine of A") is the ratio $\dfrac{\text{opposite side of } A}{\text{hypotenuse}}$

$\cos A$ (pronounced "cosine of A") is the ratio $\dfrac{\text{adjacent side of } A}{\text{hypotenuse}}$

$\tan A$ (pronounced "tangent of A") is the ratio $\dfrac{\text{opposite side of } A}{\text{adjacent side of } A}$

Example **2** **Find angles in a triangle using trigonometry**

Given a right triangle with sides of length 5 and 12, find the trigonometric ratios for the angles A and B. Show your answers in both common fraction and decimal fraction form, with decimals rounded to four places.

Solution First use the Pythagorean theorem to find the length of the hypotenuse.

$$c = \sqrt{5^2 + 12^2}$$
$$= \sqrt{25 + 144}$$
$$= \sqrt{169}$$
$$= 13$$

$\sin A = \frac{5}{13} \approx 0.3846; \quad \cos A = \frac{12}{13} \approx 0.9231; \quad \tan A = \frac{5}{12} \approx 0.4167$

$\sin B = \frac{12}{13} \approx 0.9231; \quad \cos B = \frac{5}{13} \approx 0.3846; \quad \tan B = \frac{12}{5} \approx 2.4$

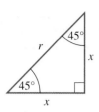

Example 3 | Exact values for 45° angle

Find the cosine, sine, and tangent of 45°.

Solution   If one of the angles of a right triangle is 45°, then the other acute angle must also be 45° (because the sum of the angles of a triangle is 180°). Furthermore, since the base angles have the same measure, the triangle is isosceles. By the Pythagorean theorem,

$$x^2 + x^2 = r^2$$
$$2x^2 = r^2$$
$$\sqrt{2}x = r \qquad \text{Note that } x \text{ is positive.}$$

Now use the definition of the trigonometric ratios.

$$\cos 45° = \frac{x}{r} = \frac{x}{\sqrt{2}\,x} = \frac{1}{\sqrt{2}}$$

$$\sin 45° = \frac{x}{r} = \frac{1}{\sqrt{2}}$$

$$\tan 45° = \frac{x}{x} = 1$$

Thus, $\cos 45° = \sin 45° = \dfrac{1}{\sqrt{2}}$ $\left(\text{or } \frac{1}{\sqrt{2}} = \frac{1}{\sqrt{2}} \cdot \frac{\sqrt{2}}{\sqrt{2}} = \frac{\sqrt{2}}{2}\right)$, and $\tan 45° = 1$.

Tables of ratios for different angles are available and most calculators have keys for the sine, cosine, and tangent ratios. Check with your calculator owner's manual to see how to use your calculator for these problems.

Example 4 | Trigonometric ratios by calculator

Find the trigonometric ratios by using a calculator. Round your answers to four decimal places.

a. sin 45°   **b.** cos 32°   **c.** tan 19°

Solution
a. $\sin 45° \approx 0.7071$; press $\boxed{\sin}$ $\boxed{45}$.*

Compare with Example 3: $\sin 45° = \frac{\sqrt{2}}{2} \approx 0.7071$.

b. $\cos 32° \approx 0.8480$; press $\boxed{\cos}$ $\boxed{32}$.
c. $\tan 19° \approx 0.3443$; press $\boxed{\tan}$ $\boxed{19}$.

Trigonometric ratios are useful in a variety of situations, as illustrated in the next example.

Example 5 | Find the height of a pyramid

The angle from the ground to the top of the Great Pyramid of Cheops is 52° if a point on the ground directly below the top is 351 ft away. (See Figure 7.50.) What is the height of the pyramid?

Solution   From Figure 7.50, we see that for height h

$$\tan 52° = \frac{h}{351}$$

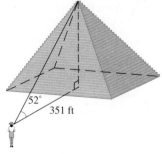

FIGURE 7.50 Calculating the height of the Great Pyramid of Cheops

*Or $\boxed{45}$ $\boxed{\sin}$, depending on brand and model of calculator. Also, be sure your calculator is set to DEGREE mode.

Solving for h by multiplying both sides by 351, we obtain

$$h = 351 \tan 52°$$

Now we need to know the ratio for tan 52°. By calculator, press $\boxed{351}$ $\boxed{\times}$ $\boxed{\tan}$ $\boxed{52}$ $\boxed{=}$, which displays 449.2595129. The height of the Great Pyramid of Cheops is about 449 ft.

Inverse Trigonometric Ratios

We can also use right-triangle trigonometry to find one of the acute angles if we know the trigonometric ratio. For example, suppose we know (as we do from Example 3) that

$$\tan \theta = 1$$

Also suppose that we do not know the angle θ. In other words, we ask, "What is the angle θ?" We answer by saying, "θ is the angle whose tangent is 1." In mathematics, we call this the **inverse tangent** and we write

$$\theta = \tan^{-1} 1$$

To find the angle θ, we turn to a calculator. Find the button labeled $\boxed{\tan^{-1}}$, and press

$$\boxed{\tan^{-1}} \quad \boxed{1} \quad \boxed{=}$$

The display is 45, which means $\theta = 45°$.

We now define the **inverse trigonometric ratios** for a right triangle.

This definition is used to find angles in a right triangle when the sides are known.

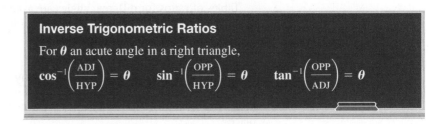

Inverse Trigonometric Ratios

For θ an acute angle in a right triangle,

$$\cos^{-1}\left(\frac{\text{ADJ}}{\text{HYP}}\right) = \theta \qquad \sin^{-1}\left(\frac{\text{OPP}}{\text{HYP}}\right) = \theta \qquad \tan^{-1}\left(\frac{\text{OPP}}{\text{ADJ}}\right) = \theta$$

We return to Example 2.

Example 6 | Find angles of a triangle using inverse trigonometric ratios

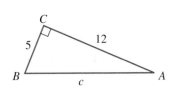

Given a right triangle with sides of length 5 and 12, find the measures of the angles of this triangle.

Solution   From Example 2, we see the hypotenuse is 13. Also from Example 2,

$$\sin A = \frac{5}{13} \approx 0.3846 \quad \text{and} \quad \sin B = \frac{12}{13} \approx 0.9231$$

Our task here is to find the measures of angles A and B. What is A? We might say, "A is the measure of the angle whose sine is $\frac{5}{13}$." This is the inverse sine, and we write[†]

$$A = \sin^{-1}\frac{5}{13} \approx 22.6° \quad \text{and} \quad B = \sin^{-1}\frac{12}{13} \approx 67.4°$$

---

[*]On some calculators, you press $\boxed{1}$ $\boxed{\tan^{-1}}$ $\boxed{=}$ or $\boxed{\tan^{-1}}$ $\boxed{1}$ $\boxed{\text{ENTER}}$. Also make sure your calculator is set DEGREE mode.

[†]There is usually more than one ratio that can be used to find a particular angle. For this example, we could also have used $A = \cos^{-1}\frac{12}{13} \approx 22.6°$ or $A = \tan^{-1}\frac{5}{12} \approx 22.6°$ and $B = \cos^{-1}\frac{5}{13} \approx 67.4°$ or $B = \tan^{-1}\frac{12}{5} \approx 67.4°$.

Right-triangle trigonometry can be used in a variety of applications. One of the most common has to do with an observer looking at an object.

The **angle of elevation** is the acute angle measured up from a horizontal line to the line of sight, whereas if we take the climber's viewpoint, and measure from a horizontal down to the line of sight, we call this angle the **angle of depression.**

Remember to measure from the horizontal.

Line of sight
Angle of elevation
Horizontal line

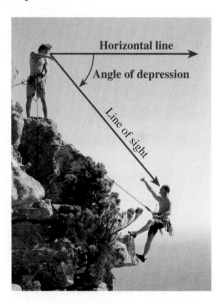
Horizontal line
Angle of depression
Line of sight

Example 7 | Find the height of a tree from angle of elevation

The angle of elevation to the top of a tree from a point on the ground 42 ft from its base is 33°. Find the height of the tree (to the nearest foot).

Solution   Let θ = angle of elevation and h = height of tree. Then

$$\tan \theta = \frac{h}{42}$$
$$h = 42 \tan 33°$$
$$\approx 27.28$$

The tree is 27 ft tall.

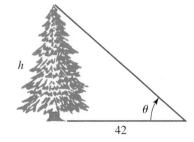
h
42
θ

Problem Set 7.5

Level 1

1. What is the Pythagorean theorem?

2. What is a sine?

3. What is a cosine?

4. What is a tangent?

5. **HISTORICAL QUEST**   The mathematics of the early Egyptians was practical and centered around surveying, construction, and recordkeeping. They used a simple device to aid surveying—a rope with 12 equal divisions marked by knots, as shown in Figure 7.51.

FIGURE 7.51 Egyptian measuring rope

When the rope was stretched and staked so that a triangle was formed with sides 3, 4, and 5, the angle formed by the shorter sides was a right angle. This method was extremely useful in Egypt, where the Nile flooded the rich lands close to the river each year. The lands needed to be resurveyed when the waters subsided.

Which of the following ropes would form right triangles?

Rope *A*:   30 knots (sides 5, 12, and 13)

Rope *B*:   9 knots (sides 2, 3, and 4)

Use the right triangle in Figure 7.52 to answer the questions in Problems 6–16.

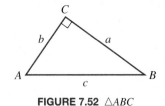

FIGURE 7.52 △*ABC*

6. What is the side opposite ∠*A*?

7. What is the side opposite ∠*B*?

8. What is the side adjacent ∠*A*?

9. What is the side adjacent ∠*B*?

10. What is the hypotenuse?

11. What is sin *A*?

12. What is sin *B*?

13. What is cos *A*?

14. What is cos *B*?

15. What is tan *A*?

16. What is tan *B*?

Find the trigonometric ratios in Problems 17–28 by using a calculator. Round your answers to four decimal places.

17. $\sin 56°$ 18. $\sin 15°$

19. $\sin 61°$ 20. $\sin 18°$

21. $\cos 54°$ 22. $\cos 8°$

23. $\cos 90°$ 24. $\cos 34°$

25. $\tan 24°$ 26. $\tan 52°$

27. $\tan 75°$ 28. $\tan 89°$

Find the angles (to the nearest degree) for the information given in Problems 29–37.

29. $\sin^{-1}\frac{1}{2}$ 30. $\cos^{-1}1$

31. $\tan^{-1}1$ 32. $\cos^{-1}\frac{\sqrt{2}}{2}$

33. $\tan^{-1}\sqrt{3}$ 34. $\sin^{-1}0$

35. $\tan^{-1}1.5$ 36. $\sin^{-1}0.35$

37. $\cos^{-1}0.8$

Level 2

In Problems 38–47, decide if the given triangle is a right triangle. If it is, find the sine, cosine, and tangent for the angle A.

38.

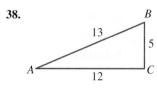

39.

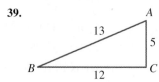

40.

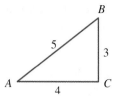

41.

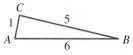

42.

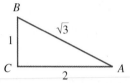

43.

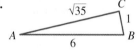

44.

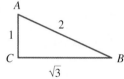

45.

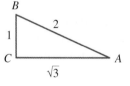

46.

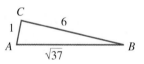

47.

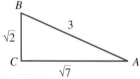

48. If the angle from the horizontal to the top of a building is 38° and the horizontal distance from its base is 90 ft, what is the height of the building (to the nearest foot)?

49. If the angle from the horizontal to the top of a tower is 52° and the horizontal distance from its base is 85 ft, what is the height of the tower (to the nearest foot)?

50. A building's angle of elevation from a point on the ground 30 m from its base is 35°. Find the height of the building (to the nearest meter).

51. From a cliff 150 m above the shoreline, a ship's angle of depression is 37°. Find the distance of the ship (to the nearest meter) from a point directly below the observer.

52. From a police helicopter flying at 1,000 ft, a stolen car is sighted at an angle of depression of 71°. Find the distance of the car (to the nearest ft) from a point directly below the helicopter.

53. A 16-ft ladder on level ground is leaning against a house. If the ladder's angle of elevation is 52°, how far above the ground (to the nearest inch) is the top of the ladder?

54. Find the height of the Barrington Space Needle (to the nearest foot) if its angle of elevation at 1,000 ft from a point on the ground directly below the top is 58.15°.

55. The world's tallest chimney is the stack at the International Nickel Company. Find its height (to the nearest foot) if its angle of elevation at 1,000 ft from a point on the ground directly below the top stack is 51.36°.

56. HISTORICAL QUEST   The angle of elevation of the top of the Great Pyramid of Khufu (or Cheops) from a point on the ground 351 ft from a point directly below the top is 52.0°. Find the height of the pyramid (to the nearest foot).

Historical **NOTE**

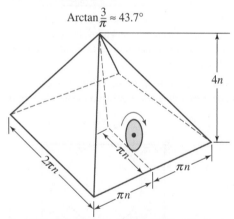

Adapted from "Angles of Elevation of the Pyramids of Egypt," in *The Mathematics Teacher*, February 1982, pp. 124–127.

The angle of elevation for a pyramid is the angle between the edge of the base and the slant height, the line from the apex of the pyramid to the midpoint of any side of the base. It is the maximum possible ascent for anyone trying to climb the pyramid to the top. In an article, "Angles of Elevation to the Pyramids of Egypt," in *The Mathematics Teacher* (February 1982, pp. 124–127), author Arthur F. Smith notes that the angle of elevation of these pyramids is either about 44° or 52°. Why did the Egyptians build pyramids using these angles of elevation? (See Project 7.5, p. 387).

Smith states that (according to Kurt Mendelssohn, *The Riddle of the Pyramids*, New York: Praeger Publications, 1974) Egyptians might have measured long horizontal distances by means of a circular drum with some convenient diameter such as one cubit. The circumference would then have been π cubits. In order to design a pyramid of convenient and attractive proportions, the Egyptians used a 4:1 ratio

(continued)

for the rise relative to revolutions of the drum. Smith then shows that the angle of elevation of the slant height is

$$\tan^{-1} \frac{4}{\pi} \approx 51.9°$$

If a small angle of elevation was desired (as in the case of the Red Pyramid), a 1:3 ratio might have been used. In that case, the angle of elevation of the slant height is

$$\tan^{-1} \frac{3}{\pi} \approx 43.7°$$

© Richard T. Nowitz/CORBIS

Problem Solving 3

57. In a 30°–60°–90° triangle, the length of the side adjacent to the 30° angle is $\sqrt{3}$ times the length of the side opposite the 30° angle. Use this information to find the exact values for cos 30°, sin 30°, and tan 30°.

58. What is the radius of the largest circle you can cut from a rectangular poster board with measurements 11 in. by 17 in.? (See Figure 7.53.)

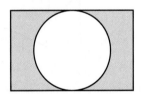

FIGURE 7.53 Circle in a rectangle

59. What is the width of the largest rectangle with length 16 in. you can cut from a circular piece of cardboard having a radius of 10 in.? (See Figure 7.54.)

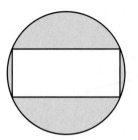

FIGURE 7.54 Rectangle in a circle

60. a. If the distance from the earth to the sun is 92.9 million miles, and the angle formed between Venus, the earth, and the sun is 47° (as shown in Figure 7.55), find the distance from the sun to Venus (to the nearest hundred thousand miles).

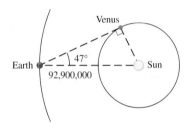

FIGURE 7.55 Earth, Venus, and the Sun

b. Find the distance from the earth to Venus (to the nearest hundred thousand miles).

7.6 | Mathematics, Art, and Non-Euclidean Geometries

The topic of this section is certainly ambitious, and if treated thoroughly would take volumes. In this section, we will simply introduce some of the connections between art and mathematics. Others are found elsewhere in this text, while even more are linked on the website.

Golden Rectangles

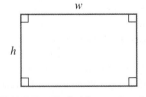

FIGURE 7.56 Golden rectangle

Certain rectangles hold some special interest because of the relationship between their height and width. Consider a rectangle with height h and width w, as shown in Figure 7.56.

Consider the proportion

$$\frac{h}{w} = \frac{w}{h + w}$$

This relationship is called the **divine proportion.** If $h = 1$, we can solve the resulting equation for w:

$$\frac{1}{w} = \frac{w}{1 + w}$$

$$1 + w = w^2$$

$$w^2 - w - 1 = 0$$

$$w = \frac{1 \pm \sqrt{(-1)^2 - 4(1)(-1)}}{2(1)} \qquad \textit{Quadratic formula}$$

$$= \frac{1 \pm \sqrt{5}}{2}$$

Since w is a length, we disregard the negative value to find

$$w = \frac{1 + \sqrt{5}}{2} \approx 1.618033989$$

This number is called the **golden ratio** and is denoted by ϕ (pronounced *phi*) or τ (pronounced *tau*). We will use τ in this book.

A rectangle that satisfies this proportion for finding the golden ratio is called a **golden rectangle** and can easily be constructed using a straightedge and a compass. Consider the proportion

$$\frac{h}{w} = \frac{w}{h + w}$$

which means the ratio of the height (h) to the width (w) is the same as the ratio of the width (w) to the sum of its height and width. To draw such a rectangle, we can begin with *any* square $CDHG$, as shown in Figure 7.57.

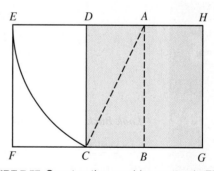

FIGURE 7.57 Constructing a golden rectangle *EFGH*

The square in Figure 7.57 is shown with its interior shaded. Now divide the square into two equal parts, as shown by the dashed segment $\overline{AB}$. Set your compass so that it measures the length of $\overline{AC}$. Draw an arc, with center at A and radius equal to the length of $\overline{AC}$, so that it intersects the extension of side $\overline{HD}$; label this point E. Now draw side $\overline{EF}$. The resulting rectangle $EFGH$ is a golden rectangle; $CDEF$ is also a golden rectangle.

There are many interesting properties associated with the golden ratio τ. Consider the pattern of numbers

$$1, 1, 2, 3, 5, 8, 13, 21, 34, 55, 89, 144, \ldots$$

which is formed by adding the first two terms for the third term, and then continues by adding successive pairs of numbers. Suppose we consider the ratios of the successive terms in this list:

$$\frac{1}{1} = 1.0000; \quad \frac{2}{1} = 2.0000; \quad \frac{3}{2} = 1.5000; \quad \frac{5}{3} \approx 1.667; \quad \frac{8}{5} = 1.6000;$$

$$\frac{13}{8} = 1.625; \quad \frac{21}{13} \approx 1.615; \quad \frac{34}{21} \approx 1.619; \quad \frac{55}{34} \approx 1.618; \quad \frac{89}{55} \approx 1.618$$

If you continue to find these ratios, you will notice that the sequence oscillates about a number approximately equal to 1.618, which is τ.

Suppose we repeat this same procedure, except we start with *any* two nonzero numbers, say 4 and 7:

4, 7, 11, 18, 29, 47, 76, 123, 199, 322, . . .

Next, form the ratios of the successive terms:

$$\frac{7}{4} = 1.750; \quad \frac{11}{7} \approx 1.571; \quad \frac{18}{11} \approx 1.636; \quad \frac{29}{18} \approx 1.611; \quad \frac{47}{29} \approx 1.621; \ldots$$

These ratios are oscillating about the same number, τ.

Example 1 **Number puzzle**

Pólya's Method

Find some numbers with the property that the number and its reciprocal differ by 1.

Solution We use Pólya's problem-solving guidelines for this example.

Understand the Problem. We know, for example, that if x is such a number, then $\frac{1}{x} + 1$ must be equal to the original number.

Devise a Plan. We let x be any nonzero number. Then we need to set $\frac{1}{x} + 1$ equal to x, and solve the resulting equation.

Carry Out the Plan.
$$x = \frac{1}{x} + 1$$
$$x^2 = 1 + x \quad \text{Multiply both sides by x.}$$
$$x^2 - x - 1 = 0$$
$$x = \frac{1 \pm \sqrt{(-1)^2 - 4(1)(-1)}}{2(1)}$$
$$= \frac{1 \pm \sqrt{5}}{2}$$

Look Back. We recognize the positive value as τ:

$$\tau = \frac{1 + \sqrt{5}}{2} \approx 1.618033989$$

Use your calculator to find the reciprocal $\frac{1}{\tau} = \frac{\sqrt{5} - 1}{2} \approx 0.618033989$.

It has been said that many everyday rectangular objects have a length-to-width ratio of about 1.6:1, as illustrated in Figure 7.58.

Mathematics and Art

Psychologists have tested individuals to determine the rectangles they find most pleasing; the results are those rectangles whose length-to-width ratios are near the golden ratio. George Markowsky, on the other hand, in the article "Misconceptions About the Golden Ratio,"[*]

FIGURE 7.58 An 18-oz box of Kellogg's Rice Krispies is 32 cm by 20 cm, for a ratio of 1:6; a 1-lb box of C&H sugar is 17 cm by 10 cm, for a ratio of 1.7.

David Lee / Alamy

suggests that among rectangles with greatly different length-to-width ratios, the golden rectangle is the most pleasing; but when confronted with rectangles with ratios "close" to the golden ratio, subjects are unable to select the "best" rectangle.

We can see evidence of golden rectangles in many works of art. Whether the artist had such rectangles in mind is open to speculation, but we can see golden rectangles in the work of Albrecht Dürer, Leonardo da Vinci, George Bellows, Pieter Mondriaan, and Georges Seurat (see Figure 7.59).

The Metropolitan Museum of Art / Art Resource, NY

FIGURE 7.59 *La Parade* by the French impressionist Georges Seurat

The Parthenon in Athens has been used as an example of a building with a height-to-width ratio that is almost equal to the golden ratio.

| Example | 2 | Golden ratio in the Parthenon |

Pólya's Method

If the Parthenon is 101 feet wide, what is its height (to the nearest foot) if we assume the dimensions are in a golden ratio?

Solution We use Pólya's problem-solving guidelines for this example.

James M. House/Shutterstock.com

The Parthenon in Nashville's
Centennial Park is a replica of the
original Parthenon in Athens

Understand the Problem. First, understand the problem. Since the Parthenon is built to satisfy the golden ratio, the height h and the width w satisfy the following proportion:

$$\frac{h}{w} = \frac{w}{h + w}$$

Devise a Plan. The width is 101 feet, so

$$\frac{h}{101} = \frac{101}{h + 101}$$

We will solve this equation for h.

Carry Out the Plan. There is only one unknown, which is written in variable form, so we now solve the equation for *h*:

$$h(h + 101) = 101^2$$
$$h^2 + 101h - 101^2 = 0$$
$$h = \frac{-101 \pm \sqrt{101^2 - 4(1)(-101^2)}}{2(1)}$$
$$= \frac{-101 \pm 101\sqrt{1 + 4}}{2}$$
$$\approx 62.4 \quad \text{\small Disregard the negative solution, since distances are nonnegative.}$$

Look Back. The Parthenon is about 62 feet high.

Many studies of the human body itself involve the golden ratio (remember $\tau \approx 1.62$). Figure 7.60 shows a drawing of an idealized athlete.

David (1501–1504) by Michelangelo illustrates many golden ratios.

Nat Brandt/Photo Researchers

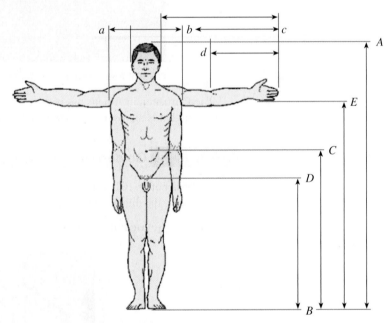

FIGURE 7.60 Proportions of the human body

Let

$|AC|$ = distance from the top of the head to the navel;

$|CB|$ = distance from the navel to the floor; and

$|AB|$ = height.

Then

$$\frac{|CB|}{|AC|} \approx \tau \quad \text{and} \quad \frac{|AB|}{|CB|} \approx \tau$$

Also, let $|ab|$ = shoulder width and $|bc|$ = arm length. Then

$$\frac{|bc|}{|ab|} \approx \tau$$

See if you can find other ratios on the human body that approximate τ. Figure 7.61 shows a study of the human face by da Vinci, in which the rectangles approximate golden rectangles.

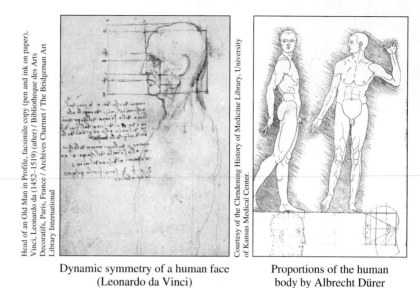

Dynamic symmetry of a human face
(Leonardo da Vinci)

Proportions of the human
body by Albrecht Dürer

FIGURE 7.61 Golden rectangles used in art

The last application of the golden rectangle we will consider in this section is related to the manner in which a chambered nautilus grows. The spiral of the chambered nautilus can be seen in the photograph below, and can be constructed by following the steps in Figure 7.62.

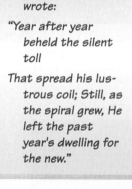

In The Chambered
  Nautilus, Oliver
  Wendell Holmes
  wrote:

"Year after year
  beheld the silent
  toll

That spread his lus-
  trous coil; Still, as
  the spiral grew, He
  left the past
  year's dwelling for
  the new."

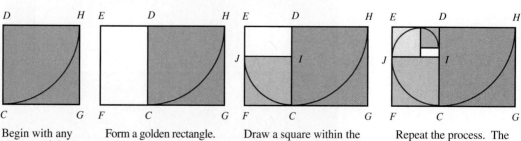

Begin with any
square and draw a
quarter circle as
shown.

Form a golden rectangle.

Draw a square within the
new rectangle *CDEF*; draw
a quarter circle as shown.

Repeat the process. The
resulting curve is called a
logarithmic spiral.

FIGURE 7.62 A spiral is constructed using a golden rectangle

Projective Geometry

As Europe passed out of the Middle Ages and into the Renaissance, artists were at the forefront of the intellectual revolution. No longer satisfied with flat-looking scenes, they wanted to portray people and objects as they looked in real life. The artists' problem was one of dimension, and dimension is related to mathematics, so many of the Renaissance artists had to solve some original mathematics problems. How could a flat surface be made to look three dimensional?

One of the first (but rather unsuccessful) attempts at portraying depth in a painting is shown in Figure 7.63, Duccio's *Last Supper*.

Notice that the figures are in a boxed-in room. This technique is characteristic of the period, and was an attempt to make perspective easier to define.

Most errors in perspective were attributed to carelessness of the artists. In 1754, the artist William Hogarth forced artists to confront their inadequacies with his engraving *Perspective Absurdities* (Figure 7.64), in which he challenged the viewer to find all the perspective mistakes.

FIGURE 7.63 Duccio's *Last Supper* illustrates perspective that is incorrect.

FIGURE 7.64 Hogarth's *Perspective Absurdities*

Contemporary examples of false perspective are shown in Figure 7.65.

FIGURE 7.65 The cartoon (above) pokes fun at the idea of perspective, and the photo shows three people of the same height

The first great painter of the Italian Renaissance was Masaccio, whose painting *The Holy Trinity* (Figure 7.66) shows mathematical perspective. The pane at the right is another drawing by Masaccio (1401–1428) showing the structure of perspective.* The ideas of shapes, rectangles, triangles, and circles, and how these figures relate to each other, were important not only to mathematicians, but also to artists and architects.

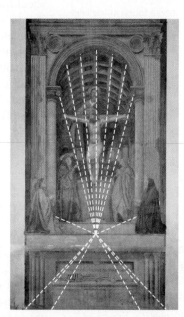

FIGURE 7.66 Masaccio's *The Holy Trinity* is a study in perspective

Artists finally solved the problem of perspective by considering the surface of the picture to be a window through which the object was painted. This technique was pioneered by Paolo Uccello (1397–1475), Piero della Francesca (1416–1492), Leonardo da Vinci (1452–1519), and Albrecht Dürer (1471–1528). As the lines of vision from the object converge at the eye, the picture captures a cross section of them, as shown in Figure 7.67.

The mathematical study of vanishing points and perspective is part of a branch of mathematics called **projective geometry,** the study of those properties of geometric configurations that do not change (are invariant) under projections.

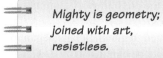

Mighty is geometry; joined with art, resistless.

FIGURE 7.67 Albrecht Dürer's *Draughtsman Drawing a Recumbent Woman* shows how the problem of perspective can be overcome. The point in front of the artist's eye fixes the point of viewing the painting. The grid on the window corresponds to the grid on the artist's canvas.

*Perspective in Masaccio's *Trinity* was first analyzed in *The Invention of Infinity*, J. V. Field, Oxford University Press, 1997. A summary of the argument about how to find the ideal viewing position is found in this book.

Non-Euclidean Geometry

An interesting interactive site exploring some of the ideas of non-Euclidean geometry is found in the links for this section at **www.mathnature.com**.

Euclid's so-called fifth postulate (see Section 7.1) caused problems from the time it was stated. Several different formulations of this postulate are given in the Historical Note shown in the margin.

It somehow doesn't seem like the other postulates but rather like a theorem that should be proved. In fact, this postulate even bothered Euclid himself, since he didn't use it until he had proved his 29th theorem. Many mathematicians tried to find a proof for this postulate.

One of the first serious attempts to prove Euclid's fifth postulate was made by Girolamo Saccheri (1667–1733), an Italian Jesuit. Saccheri's plan was simple. He constructed a quadrilateral, later known as a **Saccheri quadrilateral,** with base angles A and B right angles, and with sides $\overline{AC}$ and $\overline{BD}$ the same length, as shown in Figure 7.68.

FIGURE 7.68 A Saccheri quadrilateral

Historical NOTE

There are many ways of stating Euclid's fifth postulate, and several historical ones are repeated here.

1. **Poseidonius** (ca. 135–51 B.C.): Two parallel lines are equidistant.
2. **Proclus** (A.D. 410–485): If a line intersects one of two parallel lines, then it also intersects the other.
3. **Legendre** (1752–1833): A line through a point in the interior of an angle other than a straight angle intersects at least one of the arms of the angle.
4. **Bolyai** (1802–1860): There is a circle through every set of three noncollinear points.
5. **Playfair** (1748–1819): Given a straight line and any point not on this line, there is one and only one line through that point that is parallel to the given line.

Playfair's statement is the one most often used in high school geometry textbooks. All of these statements are *equivalent*, and if you accept any one of them, you must accept them all.

As you may know from high school geometry, the summit angles C and D are also right angles. However, this result uses Euclid's fifth postulate. Now it is also true that *if* the summit angles are right angles, *then* Euclid's fifth postulate holds. The problem, then, was to establish the fact that angles C and D are right angles. Here is the plan:

1. Assume that the angles are obtuse and deduce a contradiction.
2. Assume that the angles are acute and deduce a contradiction.
3. Therefore, by the first two steps, the angles must be right angles.
4. From step 3, Euclid's fifth postulate can be deduced.

It turned out not to be as easy as Saccheri thought, because he was not able to establish a contradiction. He gave up the search because, he said, it "led to results that were repugnant to the nature of a straight line."

Saccheri never realized the significance of what he had started, and his work was forgotten until 1889. However, in the meantime, Johann Lambert (1728–1777) and Adrien-Marie Legendre (1752–1833) similarly investigated the possibility of eliminating Euclid's fifth postulate by proving it from the other postulates.

By the early years of the 19th century, three accomplished mathematicians began to suspect that the parallel postulate was independent and could not be eliminated by deducing it from the others. The great mathematician Karl Gauss, whom we've mentioned before, was the first to reach this conclusion, but since he didn't publish this finding of his, the credit goes to two others. In 1811, an 18-year-old Russian named Nikolai Lobachevsky pondered the possibility of a "non-Euclidean" geometry—that is, a geometry that did not assume Euclid's fifth postulate. In 1840, he published his ideas in an article entitled "Geometrical Researches on the Theory of Parallels." The postulate he used was subsequently named after him.

The Lobachevskian Postulate

The summit angles of a Saccheri quadrilateral are acute.

This axiom, in place of Euclid's fifth postulate, leads to a geometry that we call **hyperbolic geometry.** If we use the plane as a model for Euclidean geometry, what model could serve for hyperbolic geometry? A rough model for this geometry can be seen by placing two trumpet bells together as shown in Figure 7.69a. It is called a **pseudosphere** and is generated by a curve called a *tractrix* (as shown in Figure 7.69b). The tractrix is rotated about the line $\overleftrightarrow{AB}$. The pseudosphere has the property that, through a point not on a line, there are many lines parallel to a given line.

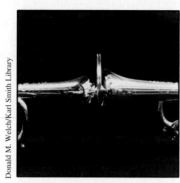

Donald M. Welch/Karl Smith Library

Two trumpets placed end to end model of a pseudosphere

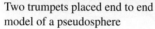

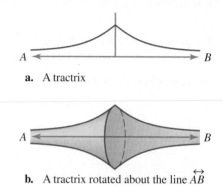

a. A tractrix

b. A tractrix rotated about the line $\overleftrightarrow{AB}$

FIGURE 7.69 A tractrix can be used to generate a pseudosphere.

Georg Riemann (1826–1866), who also worked in this area, pointed out that, although a straight line may be extended indefinitely, it need not have infinite length. It could instead be similar to the arc of a circle, which eventually begins to retrace itself. Such a line is called *reentrant.* An example of a re-entrant line is found by considering a great circle on a sphere. A **great circle** is a circle on a sphere with a diameter equal to the diameter of the sphere. With this model, a Saccheri quadrilateral is constructed on a sphere with the summit angles obtuse. The resulting geometry is called **elliptic geometry.** The shortest path between any two points on a sphere is an arc of the great circle through those points; these arcs correspond to line segments in Euclidean geometry. In 1854, Riemann showed that, with some other slight adjustments in the remaining postulates, another consistent non-Euclidean geometry can be developed. Notice that the fifth, or parallel, postulate fails to hold because any two great circles on a sphere must intersect at two points (see Figure 7.70).

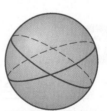

FIGURE 7.70 A sphere showing the intersection of two great circles

We have not, by any means, discussed all possible geometries. We have merely shown that the Euclidean geometry that is taught in high school is not the only possible model. A comparison of some of the properties of these geometries is shown in Table 7.2.

TABLE 7.2

Comparison of Major Two-Dimensional Geometries

Euclidean Geometry	Hyperbolic Geometry	Elliptic Geometry
Euclid (about 300 B.C.)	Gauss, Bolyai, Lobachevsky (ca. 1830)	Riemann (ca. 1850)
Given a point not on a line, there is one and only one line through the point parallel to the given line.	Given a point not on a line, there are an inifinite number of lines through the point that do not intersect the given line.	There are no parallels.

A representative line in each geometry is shown in color for each model, and the shaded portion showing a Saccheri quadrilateral is shown directly below the representative models.

Geometry is on a plane:

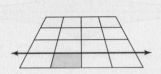

Lines are infinitely long.

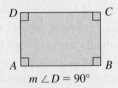

$m \angle D = 90°$

The sum of the angles of a triangle is 180°.

Geometry is on a pseudosphere:

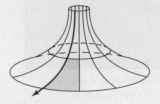

Lines are infinitely long.

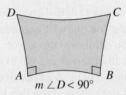

$m \angle D < 90°$

The sum of the angles of a triangle is less than 180°.

Geometry is on a sphere:

Lines are finite length.

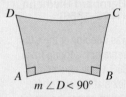

$m \angle D < 90°$

The sum of the angles of a triangle is more than 180°.

Problem Set | 7.6

Level 1

1. **IN YOUR OWN WORDS** What is a non-Euclidean geometry?
2. **IN YOUR OWN WORDS** Compare the major two-dimensional geometries.
3. **IN YOUR OWN WORDS** Why do you think Euclidean geometry remains so prevalent today even though we know there are other valid geometries?
4. What is a golden ratio?
5. What is the divine proportion?

6. What is a golden rectangle?
7. What is τ?
8. **a.** Pick any two nonzero numbers. Construct a set of numbers, in order, by adding these two numbers to find the next number. Continue by adding the previous two numbers to form your list of terms.
   b. Form the ratios of successive terms, and show that after a while they oscillate around the golden ratio.
9. Repeat Problem 8 for two other nonzero numbers.

10. Start with 1 and 3, and add these numbers to get the next number. Continue by adding two successive numbers to get the next number until you have ten terms, in order, on your list. Next, find the ratios of the successive terms. How do these compare with τ?

11. HISTORICAL QUEST If the Parthenon in Greece is 60 ft tall (at the apex) and 97 ft wide, find the ratio of width to height and compare it with τ.

12. HISTORICAL QUEST The Great Pyramid of Giza has dimensions as follows: height, $h = 481$ ft; base, $b = 756$ ft; and slant height, $s = 612$ ft. Is the ratio of any of these dimensions related to τ?

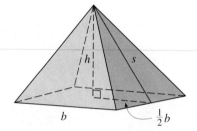

13. HISTORICAL QUEST The American Pyramid of the Sun at Teotihuacán, Mexico, has a height of 216 ft, a base of 700 ft, and a slant height of 411 ft. Does the ratio of any of these measurements approximate τ?

14. How closely does the ratio of the dimensions of a polo field (160 yd by 300 yd) approximate τ?

15. How closely does the ratio of the dimensions of a basketball court (36 ft by 78 ft) approximate τ?

16. Associate each of the given names with one of the following geometries: Euclidean, hyperbolic, elliptic.
  a. Karl Gauss
  b. Georg Riemann
  c. Euclid
  d. Janos Bolyai
  e. Nikolai Lobachevsky

Name the geometry (Euclidean, elliptic, or hyperbolic) in which each of the statements in Problems 17–24 is possible.

17. The sum of the measures of the angles of a triangle is 180°.

18. The sum of the measures of the angles of a triangle is greater than 180°.

19. The summit angles of a Saccheri quadrilateral are right angles.

20. No line can be drawn through a given point parallel to a given line.

21. Lines have finite length.

22. Used to measure distances in the construction of the pyramids in ancient Egypt.

23. The summit angles of a Saccheri quadrilateral are acute.

24. The measures of the summit base angles of a Saccheri quadrilateral are greater than 90°.

Which of the figures in Problems 25–28 are Saccheri quadrilaterals?

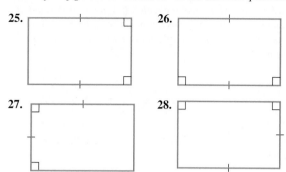

25.   **26.**

27.   **28.**

29. a. What are the length and width of a standard index card?
  b. Write the ratio of length to width as a decimal and compare it with τ.

30. a. What are the length and width of a standard brick?
  b. Write the ratio of length to width as a decimal and compare it with τ.

31. a. What are the length and width of this textbook?
  b. Write the ratio of length to width as a decimal and compare it with τ.

Use Figure 7.60 to find the requested measurements in Problems 32–34 using your own body as the model.

32. a. $|CB| \div |AC|$ **b.** $|bc| \div |ab|$
33. a. $|AB| \div |CB|$ **b.** $|ac| \div |bc|$
34. a. $|AD| \div |AE|$ **b.** $|dc| \div |bd|$

35. A plant grows for two months and then adds a new branch. Each new branch grows for two months, and then adds another branch. After the second month, each branch adds a new branch every month. Assume that the growth begins in January.

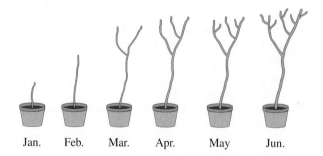

Jan.   Feb.   Mar.   Apr.   May   Jun.

  a. How many branches will there be in March?
  b. How many branches will there be in June?
  c. How many branches will there be after 12 months?
  d. Form a sequence of ratios of successive terms by considering the number of branches for the first 12 months. How do these numbers compare with τ?

Leo Moser studied the effect that two face-to-face panes of glass have on light reflected through the panes. If a ray is unreflected, it has just one path through the glass. If it has one reflection, it can be reflected two ways. For two reflections, it can be reflected three ways. Use this information in Problems 36–37.

0 reflections,   1 reflection,   2 reflections,
1 path   2 paths   3 paths

36. a. Show the possible paths for three reflections.
   b. Show the possible paths for four reflections.
   c. Make a conjecture about the number of paths for *n* reflections.

37. Form a sequence of ratios of the number of successive paths. How do these numbers compare with τ?

38. If a rectangle is to have sides with lengths in the golden ratio, what is the width if the height (the shorter side) is two units?

39. If a window is to be 5 feet wide, how high should it be, to the nearest tenth of a foot, to be a golden rectangle?

40. If a canvas for a painting is 18 inches wide, how high should it be, to the nearest inch, to be in the divine proportion?

41. A photograph is to be printed on a rectangle in the divine proportion. If it is 9 cm high, how wide is it, to the nearest centimeter?

42. If the Parthenon is 60 ft high, what is its width, to the nearest foot, if we assume the building conforms to the golden ratio?

In Problems 43–46, choose one of the following statements to complete the sentence, and discuss the reasoning for your selection.

43. IN YOUR OWN WORDS Artists of the Renaissance discovered how to draw a three-dimensional-looking world on a flat canvas by . . .
   A. . . . reading about projective geometry from a classic book on the subject by Leonardo da Vinci.
   B. . . . discovering the use of a vanishing point for parallel lines.
   C. . . . using non-Euclidean geometry to accurately represent spherical (or three-dimensional) points on a flat surface (the canvas).

44. IN YOUR OWN WORDS Non-Euclidean geometry . . .
   A. . . . has no practical applications.
   B. . . . has helped us understand the difference between provable facts and assumptions.
   C. . . . was a false step in helping us understand the nature of the world in which we live.

45. IN YOUR OWN WORDS The knowledge that the sum of the measures of the angles of a triangle is 180° is . . .
   A. . . . known because of Figure 7.32, p. 351.
   B. . . . certain, so the foundations of geometry would be destroyed if it were proved false.
   C. . . . sometimes used to decide whether the geometry is on a plane, a pseudosphere, or a sphere.

46. IN YOUR OWN WORDS Euclid's treatment of the fifth axiom . . .
   A. . . . shows that he was really smart.
   B. . . . is now known to be wrong.
   C. . . . shows that it is important to question the nature of the assumptions that are made.

47. On a globe, locate San Francisco, Miami, and Detroit. Connect these cities with the shortest paths to form a triangle. Next, use a protractor to measure the angles. What is their sum?

48. On a globe, locate Tokyo, Seattle, and Honolulu. Connect these cities with the shortest paths to form a triangle. Next, use a protractor to measure the angles. What is their sum?

Level **3**

49. Consider the spheres shown in Figure 7.71.

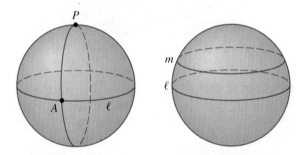

FIGURE 7.71 A "line" on a sphere is a great circle

   a. Explain what is meant by a *great circle* on a sphere.
   b. Do one or both of the lettered curves in the figure at the right seem to be lines?
   c. In Euclidean geometry, we say that a line has no endpoints. Does this property apply to lines on a sphere?

50. IN YOUR OWN WORDS In Figure 7.71, the sphere at the left shows two lines. In Euclidean geometry we know that two lines are either parallel or cross at exactly one point. Discuss this property in relation to what you observe in Figure 7.71.

51. IN YOUR OWN WORDS In Figure 7.70, the lines on the sphere at the right appear to be parallel. In Table 7.2, we said that there are no parallels on a sphere. What's wrong with our reasoning?

52. Walking north or south on the earth is defined as walking along the meridians, and walking east or west is defined as walking along parallels. There are points on the surface of the earth from which it is possible to walk 300 ft south, then a mile east, then 300 ft north, and be right back where you started. Find one such point.

53. In Lobachevskian geometry, the sum of the measures of the angles of a triangle is less than 180°. Discuss the following statement: In Lobachevskian geometry, the sum of the measures of the angles of a quadrilateral is less than 360°.

54. In Euclidean geometry, if two triangles are similar, then they may or may not also be congruent. In Lobachevskian geometry, if two triangles are similar, do you think that they must also be congruent?

Problem Solving 3

55. "When is an open book a closed book?"* If the length-to-width ratio of a book remains unchanged when that book is opened, then that book is said to be in the librarian's ratio.

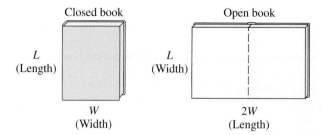

Closed book

L
(Length)

W
(Width)

Open book

L
(Width)

2W
(Length)

(Note that the length, L, of the closed book becomes the width of the opened book, and twice the width of the closed book becomes the length of the opened book.) Find an approximate and an exact representation for the constant L/W.

56. Does this book satisfy the librarian's ratio (see Problem 55)?

57. a. Show that the solutions to the equation

$$x^2 - x - 1 = 0$$

are $\tau = \frac{1 + \sqrt{5}}{2}$ and $\bar{\tau} = \frac{1 - \sqrt{5}}{2}$.

b. Is there a relationship between the decimal representations of these numbers? If so, explain.

58. If you form a list of numbers where the first two numbers are ones, and then find successive terms (in order) by adding these two numbers to obtain the next, you will find

1, 1, 2, 3, 5, 8, 13, . . .

Now, we see that the first number on the list is 1, and the seventh number is 13. It is known that the nth number on this list is

$$\frac{\tau^n - (-\bar{\tau})^n}{\sqrt{5}}$$

for τ defined in Problem 57. Use this formula to find the requested terms.
a. $n = 5$ **b.** $n = 10$ **c.** $n = 20$

59. Find two additional points satisfying the conditions stated in Problem 52.

*My thanks to Monte J. Zerger of Friends University in Wichita, Kansas, for this problem. I found it on pp. 17–27 in Vol. 18 (1985–1986) of *The Journal of Recreational Mathematics*.

60. IN YOUR OWN WORDS   Escher's work Circle Limit IV is a tessellation (see Section 8.3) based on Lobachevskian geometry.

Circle Limit IV

Consider the following model. Draw a circle.

Now draw a "perpendicular circle," which is a circle whose tangents to the original circle at the points of intersection are perpendicular. The two circles are called **orthogonal circles.**

Next, set up another model of Lobachevskian geometry. Points of the plane are points inside the circle. Lines are both diameters of the circle and arcs of orthogonal circles as well as being inside the original circle. On the circle you have drawn, draw several lines for this model.

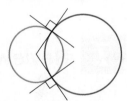

Show how these orthogonal circles are used in Escher's work, *Circle Limit IV.*

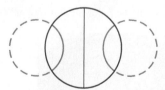

7.7   CHAPTER SUMMARY

Today's society expects schools to ensure that all students have an opportunity to become mathematically literate . . .

NCTM STANDARDS

Important Ideas

Euclidean postulates [7.1]
Terminology associated with polygons, angles, and parallel/perpendicular lines [7.2]
Sum of the measures of angles in a triangle [7.3]
Exterior angles of a triangle [7.3]
Isosceles triangle property [7.3]
Similar triangle theorem [7.4]
Pythagorean theorem [7.5]
Golden rectangle and golden ratio [7.6]
Lobachevskian postulate [7.6]
Comparison of major two-dimensional geometries [7.6]

STOP Take some time getting ready to work the review problems in this section. First review these important ideas. Look back at the definition and property boxes. If you look online, you will find a list of important terms introduced in this chapter, as well as the types of problems that were introduced. You will maximize your understanding of this chapter by working the problems in this section only after you have studied the material.

WWW You will find some review help online at **www.mathnature.com**. There are links giving general test help in studying for a mathematics examination, as well as specific help for reviewing this chapter.

Chapter 7 Review Questions

1. If a 4-cm cube is painted green and then cut into 64 1-cm cubes, how many of those cubes will be painted on
   a. 4 sides?
   b. 3 sides?
   c. 2 sides?
   d. 1 side?
   e. 0 sides?

2. What do you see in the following illustration?

From The Playful Eye, edited by Julian Rothenstein and Mel Gooding, Chronicle Books, 2000, courtesy Hordern/Dalgety collection

3. Which of the following ideas (if any) does the illustration for Problem 2 show: inversion, reflection, contraction, dilation, or rotation.

4. Which of the following pictures illustrates a reflection?

Saul Steinberg, Untitled, 1977 Pencil and colored pencil on paper, 18 x 24 inches © The Saul Steinberg Foundation/Artists Rights Society (ARS) New York

Photo by Olga Leontieva/Courtesy of Games Magazine, April 2005, p. 80, www.gamesmagazine-online.com

5. Which of the pictures in Problem 4 is upside down?

6. Which of the following illustrate at least one line of symmetry? If a figure illustrates symmetry, show a line of symmetry. If it does not, tell why.

a. 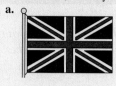   **b.**

c.    **d.**

7. Bourbon Palace (Figure 7.72) is home to the Assemblée Nationale, the French parliament's lower house.

© Steven Le Vourc'h, www. stevenlevourch.com

FIGURE 7.72 Bourbon Palace

a. What is a golden rectangle?
b. Does it look to you like the front of this building forms a golden rectangle?

8. If the width of Burbon's Palace building is 280 ft, what is the height, assuming that the front forms a golden rectangle?

9. One method for estimating the capacity of a boat is to divide the product of the length and width of the boat by 15. Write a formula to represent this idea.

10. Suppose ℓ_1 and ℓ_2 are parallel lines. Classify the angle or pairs of angles shown Figure 7.73.

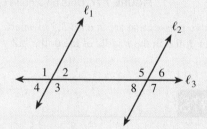

FIGURE 7.73 Two parallel lines cut by a transversal

a. $\angle 1$
b. $\angle 5$ and $\angle 6$
c. $\angle 2$ and $\angle 4$
d. $\angle 2$
e. $\angle 7$
f. $\angle 1$ and $\angle 3$
g. Can you classify any of the lines ℓ_1, ℓ_2, or ℓ_3 as perpendicular, horizontal, or vertical?

11. If an angle is 49°, then ...
a. ... what is its complement?   **b.** ... what is its supplement?
c. ... if it is an angle in a right triangle, what is the size of the other acute angle?

12. If $3x + 20°$, $2x - 40°$, and $x - 16°$ are angles of a triangle, then what is x?

13. If a box is 10 in. wide, what is its height (to the nearest inch) if we assume the dimensions are in a golden ratio?

14. In a right triangle with one leg of length 12 in. and a hypotenuse of 13 in., what is the length of the other leg?

15. In an isosceles right triangle, if the hypotenuse is 10 in., how long are the other legs?

16. Find the value of the given trigonometric ratios. Round your answers to four decimal places.
a. $\sin 59°$ **b.** $\tan 0°$ **c.** $\cos 18°$ **d.** $\tan 82°$

17. The world's most powerful lighthouse is on the coast of Brittany, France, and is about 160 ft tall. Suppose you are in a boat just off the coast, as shown in Figure 7.74. Determine your distance (to the nearest foot) from the base of the lighthouse if $\angle B = 12°$.

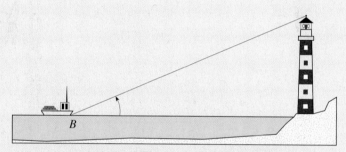

FIGURE 7.74 Distance from ship to shore

18. For the die shown in Figure 7.75, fill in the blanks in each of the given dice.

FIGURE 7.75 A die problem

a.   **b.**   **c.**   **d.**

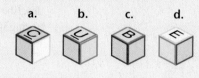

19. What is a Saccheri quadrilateral?

20. Discuss why this quadrilateral leads to different kinds of geometries.

Group RESEARCH PROJECTS

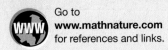

Go to
www.mathnature.com
for references and links.

Working in small groups is typical of most work environments, and learning to work with others to communicate specific ideas is an important skill. Work with three or four other students to submit a single report based on each of the following questions.

G26. In Figure 7.76 there are eight square rooms making up a maze. Each square room has two walls that are mirrors and two walls that are open spaces. Identify the mirrored walls, and then solve the maze by showing how you can pass through all eight rooms consecutively without going through the same room twice. If that is not possible, tell why.

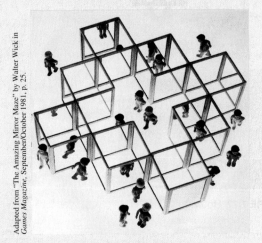

Adapted from "The Amazing Mirror Maze" by Walter Wick in *Games Magazine*, September/October 1981, p. 25.

FIGURE 7.76 Mirror Maze

G27. At the beginning of this chapter, we included a picture of a fly on or in a cube, and we saw that figures can sometimes be ambiguous. Figure 7.77a shows a cube with a dot in the middle of each face. Draw a cube around the dots in Figure 7.77b so that each dot is in the middle of a face.

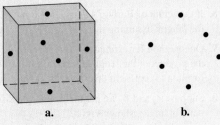

FIGURE 7.77 Dots on a cube puzzle

G28. Place a dollar bill across the top of two glasses that are at least 3.5 in. apart. Now, describe how you can place a quarter dollar in the middle of the dollar bill without having it fall.

BOOK REPORTS

Write a 500-word report on one of these books:

Sphereland, a Fantasy About Curved Spaces and an Expanding Universe, Dionys Burger (New York: Thomas Crowell Company, 1965).

Mathematical Recreations and Essays, W. W. R. Ball and H. S. M. Coxeter (New York: Macmillan, 1962).

Individual RESEARCH PROJECTS

www.mathnature.com

Learning to use sources outside your classroom and textbook is an important skill, and here are some ideas for extending some of the ideas in this chapter. You can find references to these projects in a library or at **www.mathnature.com.**

PROJECT 7.1 Write a report on optical illusions.

PROJECT 7.2 Create an aestheometry design beginning with an angle.

PROJECT 7.3 Create an aestheometry design beginning with a circle.

PROJECT 7.4 Write a paper discussing the nature of an unsolved problem as compared with an impossible problem. As part of your paper discuss the problems of trisecting an angle and squaring a circle.

PROJECT 7.5 Historical Quest　The historical note on page 369 asks, "Why did the Egyptians build the pyramids using a slant height angle of about 44° or 52°?" Write a paper answering this question.

PROJECT 7.6 Do some research on the length-to-width ratios of the packaging of common household items. Form some conclusions. Find some examples of the golden ratio in art. Do some research on dynamic symmetry.

PROJECT 7.7 In Example 2 of Section 7.6, we assumed the width of the Parthenon to be 101 ft and found the height to be 62.4 ft (assuming the golden ratio). If you worked Problem 11 of Section 7.6, you assumed the height to be 60 ft and the width to be 97 ft. Are the numbers from Example 2 and from Problem 11 consistent? Can you draw any conclusions?

PROJECT 7.8 Write a paper on perspective. How are three-dimensional objects represented in two dimensions?

PROJECT 7.9 The discovery and acceptance of non-Euclidean geometries had an impact on all of our thinking about the nature of scientific truth. Can we ever know truth in general? Write a paper on the nature of scientific laws, the nature of an axiomatic system, and the implications of non-Euclidean geometries.

PROJECT 7.10 Find the one composite number in the following set:

31
331
3331
33331
333331
3333331
33333331
333333331

8 THE NATURE OF NETWORKS AND GRAPH THEORY

What in the World?

"Where do you live, Fritz?" asked Lisel.

"I live at 45 Heimelstraße. I'm looking forward to seeing you on Sunday!" proclaimed Fritz. "I will introduce you to my family and some of my friends on the island. We can take a stroll and I'll buy you a sundae at my favorite shop."

"Take a stroll?!!" screamed Lisel. "You KNOW I don't get it! Everybody thinks they can cross those bridges without doubling back, but I've never been able to do it."

"That's my secret! I can show you if you come on Sunday," said Fritz with a grin.

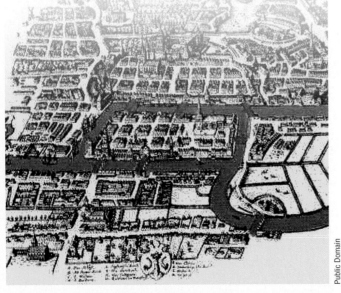

Public Domain

Overview

When we use the word "network" today, we probably think of a computer network, but we use networks every time we make a phone call, send an e-mail or "snail mail," drive on an interstate, or fly in an airplane.

In this chapter we look at mathematical ideas called *circuits, cycles,* and *trees,* which are part of geometry known as *graph theory.* When used in the context of graph theory, the word *graph* means something different from the way we use the term in referring to the coordinate plane or the way we use it to represent statistical data.

Even though these topics sound abstract, they have many interesting and useful applications such as finding the minimum cost of traveling to a number of locations (known as the *traveling salesperson problem*), installing an irrigation system, using a search engine on the Internet, or even coloring maps or making realistic-looking original landscapes in movies.

Many of the applications of this chapter are part of a new branch of mathematics (new in the sense that it began in the late 1930s) and grew out of mathematical problems associated with World War II. This branch of mathematics, called **operations research** (or *operational research*), deals with the application of scientific methods to management decision making, especially for the allocation of resources. Some examples include forecasting water pollution or predicting the scope of the AIDS epidemic.

8.1 | Euler Circuits and Hamiltonian Cycles

There is nothing in the world except empty space. Geometry bent one way here describes gravitation. Rippled another way somewhere else it manifests all the qualities of an electromagnetic wave. Excited at still another plane, the magic material that is space shows itself as a particle. There is nothing that is foreign or "physical" immersed in space. Everything that is, is constructed out of geometry.

JOHN A. WHEELER

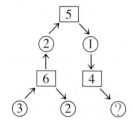

CHAPTER **CHALLENGE**

See if you can fill in the question mark.

We now turn to one of the newer branches of geometry known as *graph theory*, which includes *circuits, cycles*, and *trees*.

Euler Circuits

In the 18th century, in the German town of Königsberg (now the Russian city of Kaliningrad), a popular pastime was to walk along the bank of the Pregel River and cross over some of the seven bridges that connected two islands, as shown in Figure 8.1.

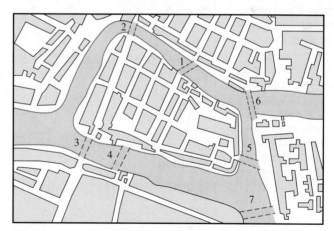

FIGURE 8.1 Königsberg bridges

One day a native asked a neighbor this question, "How can you take a walk so that you cross each of our seven bridges once and only once and end up where you started?" The problem intrigued the neighbor and soon caught the interest of many other people of Königsberg as well. Whenever people tried it, they ended up either not crossing a bridge at all or else crossing one bridge twice. This problem was brought to the attention of the Swiss mathematician Leonhard Euler, who was serving at the court of the Russian empress Catherine the Great in St. Petersburg. The method of solution we discuss here was first developed by Euler, and it led to the development of two major topics in geometry. The first is *networks*, which we discuss in this section, and the second is *topology*, which we discuss in Section 8.3.

We will use Pólya's problem-solving method for the Königsberg bridge problem.

Understand the Problem. To understand the problem, Euler began by drawing a diagram for the problem, as shown in Figure 8.2a.

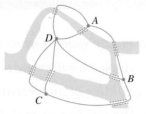 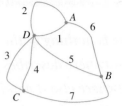

a. Crossing the bridges   **b.** Labeling network

FIGURE 8.2 Königsberg bridge problem

Next, Euler used one of the great problem-solving procedures—namely, to change the conceptual mode. That is, he **let the land area be represented as points (sometimes called *vertices* or *nodes*), and let the bridges be represented by arcs or line segments (sometimes called *edges*) connecting the given points.** As part of understanding the problem, we can do what Euler did—we can begin by tracing a diagram like the one shown in Figure 8.2b.

Devise a Plan. To solve the bridge problem, we need to draw the figure without lifting the pencil from the paper. Figures similar to the one in Figure 8.2b are called **networks** or **graphs.** In a network, the points where the line segments meet (or cross) are called **vertices,** and the lines representing bridges are called **edges** or **arcs.** Each separated part of the plane formed by a network is called a **region.** We say that a graph is **connected** if there is at least one path between each pair of vertices.

Example	1	Count edges, vertices, and regions of a network

Complete the table for each of the given networks.

a.   **b.**   **c.**

d.   **e.**   **f.**

Solution

Graph	Edges (E)	Vertices (V)	Regions (R)	V + R − 2
a.	3	3	2	3
b.	4	3	3	4
c.	5	3	4	5
d.	4	4	2	4
e.	5	4	3	5
f.	6	4	4	6

Note that $V + R - 2 = E$. Do you think this will always be true?

A network is said to be **traversable** if it can be traced in one sweep without lifting the pencil from the paper and without tracing the same edge more than once. Vertices may be passed through more than once. The **degree** of a vertex is the number of edges that meet at that vertex.

Example **2** | **Test traversablity**

List the number of edges and the degree of each vertex shown in Example 1. Find the sum of the degrees of the vertices, and tell whether each network is traversable.

Solution

Graph	Number of Edges	Degree of Each Vertex	Sum	Traversable
a.	3	2; 2; 2	6	yes
b.	4	3; 2; 3	8	yes
c.	5	4; 3; 3	10	yes
d.	4	2; 2; 2; 2	8	yes
e.	5	2; 3; 2; 3	10	yes
f.	6	3; 3; 3; 3	12	no

**William Rowan Hamilton
(1805–1865)**

William Rowan Hamilton has been called the most renowned Irish mathematician. He was a child prodigy who read Greek, Hebrew, and Latin by the time he was five, and by the age of ten he knew over a dozen languages. He was appointed Professor of Astronomy and Royal Astronomer of Ireland at the age of 22. The problem discussed in this section that bears his name was discussed by Leonhard Euler and C. A. Vandermonde in 1771. The problem is named after him because of his invention of a puzzle called "Traveler's Dodecahedron," or "A Voyage 'Round the World." We discuss this problem in the problem set.

First, note that *the sum of the degrees of the vertices in Example 2 equals twice the number of edges.* Do you see why this must always be true? Consider any graph. Each edge must be connected at both ends, so the sum of all of those ends must be twice the number of vertices.

Now, consider a second observation regarding traversability. It is assumed that you worked Example 2 by actually tracing out the networks. However, a more complicated network, such as the Königsberg bridge problem, will require some analysis. The goal is to begin at some vertex, travel on each edge exactly once, and then return to the starting vertex. Such a path is called an **Euler circuit.** We can now rephrase the Königsberg bridge problem: "Does the network in Figure 8.2 have an Euler circuit?"

To answer this question, we will follow Euler's lead and classify vertices. Vertex *A* in Figure 8.2 is degree 3, so the vertex *A* is called an **odd vertex.** In the same way, *D* is an odd vertex, because it is degree 5. A vertex with even degree is called an **even vertex.** Euler discovered that only a certain number of odd vertices can exist in any network if you are to travel it in one journey without retracing any edge. You may start at any vertex and end at any other vertex, as long as you travel the entire network. Also the network must connect each point (this is called a **connected network**).

Let's examine networks more carefully and look for a pattern, as shown in Table 8.1.

TABLE 8.1

	Arrivals and Departures for a Vertex in a Network	
Number of Arcs	**Description**	**Possibilities**
1	1 departure (starting point) 1 arrival (ending point)	
2	1 arrival (arrive then depart) and 1 departure (depart then arrive)	
3	1 arrival, 2 departures 2 arrivals, 1 departure	
4	2 arrivals, 2 departures	
5	2 arrivals, 3 departures (starting point) 3 arrivals, 2 departures (ending point)	

We see that, if the vertex is odd, then it must be a starting point or an ending point. What is the largest number of starting and ending points in any network? [*Answer*: Two—one starting point and one ending point.] This discussion allows us to now formulate the step we have called devise a plan, which we now state without proof.

Count the number of odd vertices:

If there are no odd vertices, the network is traversable and any point may be a starting point. The point selected will also be the ending point.

If there is one odd vertex, the network is not traversable. A network cannot have only one starting or ending point without the other.

If there are two odd vertices, the network is traversable; one odd vertex must be a starting point and the other odd vertex must be the ending point.

If there are more than two odd vertices, the network is not traversable. A network cannot have more than one starting point and one ending point.

Carry Out the Plan. Classify the vertices; there are four odd vertices, so the network is not traversable.

Look Back. We have solved the Königsberg bridge problem, but you should note that saying it cannot be done is not the same thing as saying "I can't do the problem." We can do the problem, and the solution is certain.

We summarize this investigation.

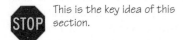

This is the key idea of this section.

Euler's Circuit Theorem

Every vertex on a graph with an Euler circuit has an even degree, and, conversely, if in a connected graph every vertex has an even degree, then the graph has an Euler circuit.

Example **3** Find an Euler circuit

Which of the following networks have an Euler circuit? Do not answer by trial and error, but by analyzing the number of odd vertices.

a.

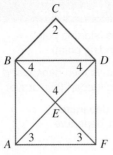

b.

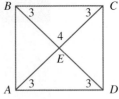

Solution

a. I first saw this network as a child's puzzle in elementary school. It has four even vertices (B, C, D, and E) and two odd vertices (A and F), and it is therefore traversable. To traverse it, you must start at A or F (that is, at an odd vertex). The path is shown.

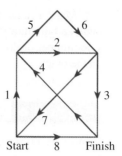

Note that we do not end at the beginning point, so this network does not have an Euler circuit (even though it is traversable).

b. This network has one even vertex and four odd vertices, so it is not traversable and does not have an Euler circuit.

Applications of Euler Circuits

Euler circuits have a wide variety of applications. We will mention a few.

- *Supermarket problem*   Set up the shelves in a market or convenience store so that it is possible to enter the store at the door and travel in each aisle exactly once (once and only once) and leave by the same door.
- *Police patrol problem*   Suppose a police car needs to patrol a gated subdivision and would like to enter the gate, cruise all the streets exactly once, and then leave by the same gate.
- *Floor-plan problem*   Suppose you have a floor plan of a building with a security guard who needs to go through the building and lock each door at the end of the day.
- *Water-pipe problem*   Suppose you have a network of water pipes, and you wish to inspect the pipeline. Can you pass your hand over each pipe exactly once without lifting your hand from a pipe, and without going over some pipe a second time?

We will examine one of these applications and leave the others for the problem set. Let's look at the *floor-plan problem*. This problem, which is related to the Königsberg bridge problem, involves taking a trip through all the rooms and passing through each door only once. There is, however, one important difference between these two problems. The Königsberg bridge problem requires an Euler circuit, but the floor-plan problem does not. In other words, with the bridges we must end up where we started, but the floor plan problem seeks only traversability. Let's draw a floor-plan problem as shown in Figure 8.3a.

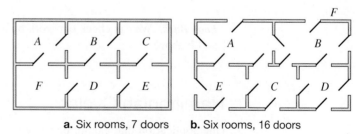

a. Six rooms, 7 doors   **b.** Six rooms, 16 doors

FIGURE 8.3 Floor-plan problem

Label the rooms as *A, B, C, D, E,* and *F*. In Figure 8.3a, rooms *A, C, E,* and *F* have two doors, and rooms *B* and *D* have three doors; in Figure 8.3b, it looks as if there are five rooms, but since there are doors that lead to the "outside," we must count the outside as a room. So this figure also has six rooms labeled *A, B, C, D, E,* and *F*. Rooms *A, B,* and *C* each have 5 doors, rooms *D* and *E* each have 4 doors, and room *F* has 9 doors.

Make a conjecture about the solution to the floor-plan problem. If there are no rooms with an odd number of doors, then it will be traversable. If there are two rooms with an odd number of doors, then it will be traversable: Start in one of those rooms, and end up in the other.

Example **4** **Floor-plan puzzle**

Solve the floor-plan problems:

a. 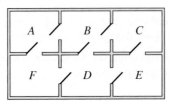   **b.**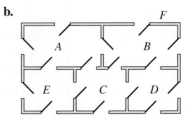

Solution   We use Pólya's problem-solving guidelines for this example.

Understand the Problem. The floor-plan problem asks, "Can we travel into each room and pass through every door once?"

Devise a Plan. Classify each room as even or odd, according to the number of doors in that room. A solution will be possible if there are no rooms with an odd number of doors, or if there are exactly two rooms with an odd number of doors.

Carry Out the Plan.

a. There are six rooms, and rooms *B* and *D* are odd, so this floor plan can be traversed. The solution requires that we begin in either room *B* or room *D*, and finish in the other.

b. There are six rooms, and rooms *A*, *B*, *C*, and *F* are odd (with *D* and *E* even). Since there are more than two odd rooms, this floor plan cannot be traversed. If one of the doors connecting two of the odd rooms is blocked, then the floor plan could be traversed.

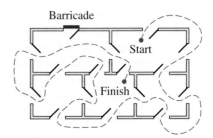

Look Back. We can check our work by actually drawing possible routes, as shown for part **b** above.

Hamiltonian Cycles

One application that cannot be solved using Euler circuits is the so-called **traveling salesperson problem:** A salesperson starts at home and wants to visit several cities without going through any city more than once, and then return to the starting city. This problem is so famous with so many people working on its solution that it is often referred to in the literature as **TSP.** The salesperson would like to do this in the most efficient way (that is, least distance, least time, smallest cost, . . .). To answer this question, we reverse the roles of the vertices and edges of an Euler circuit. Now, we ask whether we can visit each vertex exactly once and end at the original vertex. Such a path is called a **Hamiltonian cycle.**

Example 5 | Find a Hamiltonian cycle

Find a Hamiltonian cycle for the network in Figure 8.4.

FIGURE 8.4 Network

Solution   Note that this network is the one given in Example 1f. We found in Example 2f, that there was not an Euler circuit for this network. On the other hand, it is easy to find a Hamiltonian cycle:

$$A \rightarrow C \rightarrow B \rightarrow D \rightarrow A$$

Example 6 | Identifying cycles

Use Figure 8.5 to decide if the given cycle is Hamiltonian. If it is not, tell why.

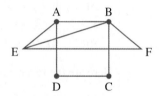

FIGURE 8.5 Network with 6 vertices

a. $A \rightarrow B \rightarrow C \rightarrow D \rightarrow A$ **b.** $C \rightarrow D \rightarrow A \rightarrow B \rightarrow F \rightarrow E \rightarrow A$
c. $D \rightarrow C \rightarrow B \rightarrow A \rightarrow E \rightarrow F$ **d.** $B \rightarrow C \rightarrow D \rightarrow A \rightarrow E \rightarrow F \rightarrow B$

Solution
a. This is not a Hamiltonian cycle because it does not visit each vertex, but repeats the same vertices, which is sometimes called a **loop.**
b. This is not a Hamiltonian cycle because it does not return to the starting point.
c. This is not a Hamiltonian cycle because it does not return to the starting point.
d. This is a Hamiltonian cycle.

See **www.mathnature. com** for links for solving the TSP.

It seems as if the problem of deciding whether a network has a Hamiltonian cycle should have a solution similar to that of the Euler circuit problem, but such is not the case. In fact, no solution is known at this time, and it is one of the great unsolved problems of mathematics. In this book, the best we will be able to do is a trial-and-error solution. If you are interested in seeing some of the different attempts at finding a solution to this problem, you can check the web address shown in the margin.

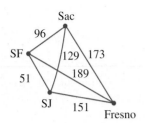

FIGURE 8.6 TSP for four cities

Example **7**   **Find the shortest trip**

Pólya's Method

A salesman wants to visit four California cities, San Francisco, Sacramento, San Jose, and Fresno. Driving distances are shown in Figure 8.6. What is the shortest trip starting and ending in San Francisco that visits each of these cities?

Solution   Since the best we can do is to offer some possible methods of attack, we will use this example to help build your problem-solving skills. We use Pólya's problem-solving guidelines for this example.

Understand the Problem. Part of understanding the problem is to decide what we mean by the "best" solution. For this problem, let us assume it is the least miles traveled. We also note that, in terms of miles traveled, each route and its reverse are equivalent. That is,

$$SF \rightarrow \text{San Jose} \rightarrow \text{Fresno} \rightarrow \text{Sacramento} \rightarrow SF$$

is the same as

$$SF \rightarrow \text{Sacramento} \rightarrow \text{Fresno} \rightarrow \text{San Jose} \rightarrow SF$$

Devise a Plan. There are several possible methods of attack for this problem: for example *brute force* (listing all possible routes) and *nearest neighbor* (at each city, go to the nearest neighbor that has not been previously visited). Sometimes the nearest-neighbor plan will form a loop without going to some city, so we repair this problem using a method called the *sorted-edge method*. In the sorted-edge method, we sort the choices by selecting the nearest neighbor that does not form a loop.

Carry Out the Plan.
BRUTE FORCE:

$$SF \xrightarrow{96} S \xrightarrow{173} F \xrightarrow{151} SJ \xrightarrow{51} SF \qquad \text{Total:}\quad 471 \text{ miles}$$
$$SF \xrightarrow{96} S \xrightarrow{129} SJ \xrightarrow{151} F \xrightarrow{189} SF \qquad \text{Total:}\quad 565 \text{ miles}$$
$$SF \xrightarrow{51} SJ \xrightarrow{129} S \xrightarrow{173} F \xrightarrow{189} SF \qquad \text{Total:}\quad 542 \text{ miles}$$

Here are the reverse trips (so we don't need to calculate these).

$$SF \rightarrow SJ \rightarrow F \rightarrow S \rightarrow SF$$
$$SF \rightarrow F \rightarrow SJ \rightarrow S \rightarrow SF$$
$$SF \rightarrow F \rightarrow S \rightarrow SJ \rightarrow SF$$

We see that 471 is the minimum number of miles.

NEAREST NEIGHBOR:

$$SF \xrightarrow{51} SJ \xrightarrow{129} S \xrightarrow{96} SF$$ A loop is formed; Fresno is not included because it is never the nearest neighbor if we start in San Francisco.

SORTED EDGE:

For this method, we sort the distances (edges of the graph) from smallest to largest: 51, 96, 129, 151, 173, and 189. This gives the following trip (skipping 96 and 151 because these choices would form a loop):

$$SF \xrightarrow{51} SJ \xrightarrow{129} S \xrightarrow{173} F \xrightarrow{189} SF \qquad \text{Total:}\quad 542 \text{ miles}$$

Look Back. With this simple problem, it is easy to see that the best overall solution is a trip with 471 miles, but as you can imagine, for a larger number of cities the solution may not be at all obvious.

We summarize the **sorted-edge method** for finding an approximate solution to a traveling salesperson problem.

Sorted-Edge Method

Draw a graph showing the cities and the distances; identify the starting vertex.

Step 1   Choose the edge attached to the starting vertex that has the shortest distance or the lowest cost. Travel along this edge to the next vertex.

Step 2   At the second vertex, travel along the edge with the shortest distance or lowest cost. Do not choose a vertex that would lead to a vertex already visited.

Step 3   Continue until all vertices are visited and arriving back at the original vertex.

The sorted-edge method may not produce the optimal solution, so you should also check other methods. Since the brute-force method requires that we check all the routes, it is worthwhile to find a formula that tells us the number of routes we need to check. Note in Example 7 we found three possible routes (along with three reversals). Consider the next example which generalizes the number of routes we found by brute force in Example 7.

Example   8   Find the number of routes

a. How many routes are there for four cities, say, San Francisco, Sacramento, San Jose, and Fresno?

b. How many routes are there for n cities?

Solution

a. If we start in San Francisco, there are 3 cities to which we can travel. Then, by the fundamental counting principle, we have

$$3 \cdot 2 \cdot 1 = 6 \text{ routes}$$

Since half the routes are reversals of the others, we have

$$\frac{3 \cdot 2 \cdot 1}{2} = 3 \text{ routes}$$

b. Following the steps in part **a,** we note that from the first city there are $n - 1$ cities to visit, so (from the fundamental counting principle) there are

$$(n - 1)(n - 2)(n - 3) \cdot \cdots \cdot 3 \cdot 2 \cdot 1 \text{ routes}$$

and if we disregard reversals there are

$$\frac{(n - 1)(n - 2)(n - 3) \cdot \cdots \cdot 3 \cdot 2 \cdot 1}{2} \text{ routes}$$

 This formula will be used later in the text. CAUTION

Problem Set   8.1

Level 1

1. IN YOUR OWN WORDS   Describe the Königsberg bridge problem.

2. IN YOUR OWN WORDS   Describe the floor-plan problem.

3. IN YOUR OWN WORDS   Describe the solution to the Königsberg bridge problem.

4. IN YOUR OWN WORDS   Describe the traveling salesperson problem.

5. IN YOUR OWN WORDS   Contrast Euler circuits and Hamiltonian cycles.

Which of the networks in Problems 6–11 are Euler circuits?
If a network can be traversed, show how.

6.
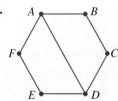

7.
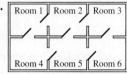

8.
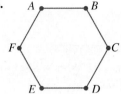

9.
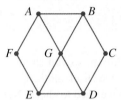

10.
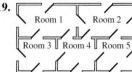

11.
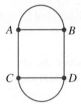

Which of the networks in Problems 12–17 have Hamiltonian cycles? If a network has one, describe it.

12.

13.

14.
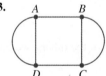

15.
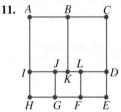

16.

17.

For which of the floor plans in Problems 18–23 can you pass through all the rooms while going through each door exactly once? If it is possible, show how it might be done.

18.
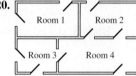

19.
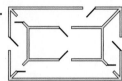

20.
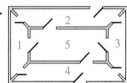

21.
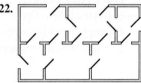

22.

23.

Level 2

Which of the networks in Problems 24–27 are Euler circuits? If a network can be traversed, show how. Note these are the same networks as those given in Problems 28–31.

24.
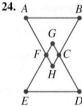

25.
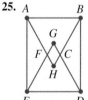

26.
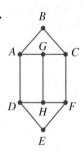

27.
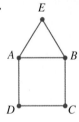

Which of the networks in Problems 28–31 have Hamiltonian cycles? If a network has one, describe it. Notice these are the same networks as those given in Problems 24–27.

28.
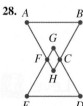

29.
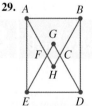

30.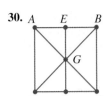

31.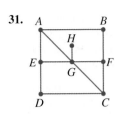

32. After Euler solved the Königsberg bridge problem, an eighth bridge was built as shown in Figure 8.7. Is this network traversable? If so, show how.

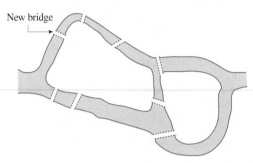

FIGURE 8.7 Königsberg with eight bridges

33. HISTORICAL QUEST   Traveler's Dodecahedron   This problem was sold in the last half of the 19th century as a puzzle known as the "Traveler's Dodecahedron" or "A Voyage 'Round the World." It consisted of 20 pegs (called *cities*), and the point of the puzzle was to use string to connect each peg only once, arriving back at the same peg you started from. Find a route (starting at Brussels—labeled 1) that visits each of the 20 cities on the dodecahedron shown in Figure 8.8.

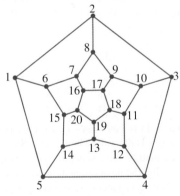

FIGURE 8.8 Hamilton's "traveler's dodecahedron"

34. HISTORICAL QUEST   Is there an Euler circuit for the Traveler's Dodecahedron shown in Figure 8.8? If so, show it.

35. Is there an Euler circuit for the graph shown in Figure 8.9?

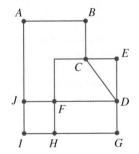

FIGURE 8.9 Network problem

36. Suppose you want to get from point *A* to point *B* in New York City (see Figure 8.10), and also suppose you wish to cross over each of the six bridges exactly once. Is it possible? If so, show one such path.

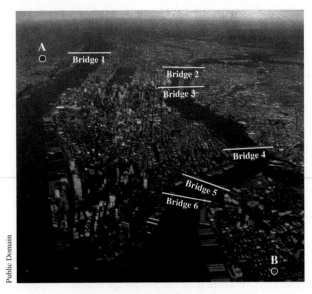

Bridge 1: George Washington Bridge
Bridge 2: Triborough Bridge
Bridge 3: Queensboro (59th Street) Bridge
Bridge 4: Williamsburg Bridge
Bridge 5: Manhattan Bridge
Bridge 6: Brooklyn Bridge

FIGURE 8.10 New York City

37. A simplified map of New York City, showing the subway connections between Manhattan and The Bronx, Queens, and Brooklyn, is shown in Figure 8.11. Is it possible to travel on the New York subway system and use each subway exactly once? You can visit each borough (The Bronx, Queens, Brooklyn, or Manhattan) as many times as you wish.

FIGURE 8.11 New York City subways

38. A portion of London's Underground transit system is shown in Figure 8.12. Is it possible to travel the entire system and visit each station while taking each route exactly once?

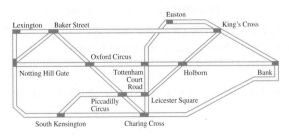

FIGURE 8.12 London Underground

39. In Massachusetts there is a re-creation of an 1830s New England village called Old Sturbridge Village. A map is shown in Figure 8.13. Is it possible to stroll the streets marked in color? Give reasons for your answer.

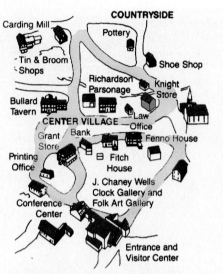

FIGURE 8.13 Old Sturbridge Village

40. Reconsider the question in Problem 39 if the church across from the Knight Store is opened.

Level **3**

41. The edges of a cube form a three-dimensional network. Are the edges of a cube traversable?

42. A saleswoman wants to visit each of the cities New York City, Boston, Cleveland, and Washington, D.C. Driving distances are as shown in Figure 8.14. What is the shortest trip starting in New York that visits each of these cities?

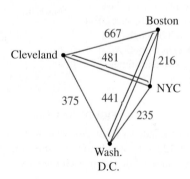

FIGURE 8.14 TSP for four cities

Find a solution using the brute-force method.

43. Repeat Problem 42 using the indicated method.
  a. Find a solution if possible using the nearest-neighbor method.
  b. Find a solution if possible using the sorted-edge method.

44. A salesperson wants to visit each of the cities Denver, St. Louis, Los Angeles, and New Orleans. Driving distances are as shown in Figure 8.15. What is the shortest trip starting in Denver that visits each of these cities?

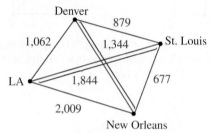

FIGURE 8.15 TSP for four cities

  a. Find a solution if possible using the nearest-neighbor method.
  b. Find a solution if possible using the sorted-edge method.

45. Repeat Problem 44 using the brute-force method.

46. Count the number of vertices, edges (arcs), and regions for each of Problems 6–17. Let V = number of vertices, E = number of edges, and R = number of regions. Compare $V + R$ with E. Make a conjecture relating V, R, and E. This relationship is called *Euler's formula for networks*.

Problem Solving **3**

47. The saleswoman in Problem 42 needs to add Atlanta to her itinerary. Driving distances are shown. What is the shortest trip starting in New York that visits each of these cities?

	A	B	C	NYC	D.C.
A	—	1,115	780	887	634
B	1,115	—	667	216	441
C	780	667	—	481	375
NYC	887	216	481	—	235
D.C.	634	441	375	233	—

48. A quality control inspector must visit franchises in Atlanta, Boston, Chicago, Dallas, and Minneapolis. Since this inspection must be monthly, the inspector, who lives in Chicago, would like to find the most efficient route (in terms of distances). Driving distances are shown. What is the most efficient route?

	A	B	C	D	M
A	—	1,115	717	691	1,131
B	1,115	—	1,013	1,845	1,619
C	717	1,013	—	937	420
D	691	1,845	937	—	963
M	1,131	1,619	420	963	—

49. What is the sum of the measures of the angles of a tetrahedron? *Hint:* Consider the sum of the measures of the face angles of a cube. A cube has six square faces, and since each face has four right angles, the sum of the measures of the angles on each face is $360°$; hence, the sum of the measures of the face angles of a cube is $6(360°) = 2,160°$.

50. What is the sum of the measures of the angles of a pentagonal prism? (See Figure 8.16.)

FIGURE 8.16 Pentagonal prism

51. On a planet far, far away, Luke finds himself in a strange building with hexagon-shaped rooms as shown in Figure 8.17.

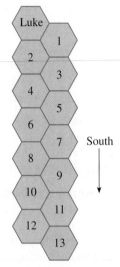

FIGURE 8.17 Strange room arrangement

In his search for the princess, Luke always moves to an adjacent room and always in a southerly direction.
a. How many paths are there to room 1?
to room 2?
to room 3?
to room 4?
b. How many paths are there to room 10?
c. How many paths are there to room 13?

52. How many paths are there to room *n* in Problem 51?

53. Emil Torday told the story of seeing some African children playing with a pattern in the sand as shown in Figure 8.18.

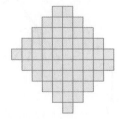

FIGURE 8.18 African sand game

*The children were drawing, and I was at once asked to perform certain impossible tasks; great was their joy when the white man failed to accomplish them.**

One task was to trace the figure in the sand with one continuous sweep of the finger.

*Quoted by Claudia Zaslavsky in *Africa Counts* (Boston: Prindle, Weber, & Schmidt, 1973) from *On the Trail of the Bushongo* by Emil Torday.

a. What is the children's secret for successfully drawing this pattern?
b. Draw this figure; why is it difficult to do this without knowing something about networks?

54. HISTORICAL QUEST
About a century ago, August Möbius made the discovery that, if you take a strip of paper (see Figure 8.19), give it a single half-twist, and paste the ends together, you will have a piece of paper

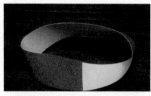

FIGURE 8.19 Möbius strip

with only one side! Construct a Möbius strip, and verify that it has only one side. How many edges does it have?

55. Construct a Möbius strip (see Problem 54). Cut the strip in half down the center. Describe the result.

56. Construct a Möbius strip (see Problem 54). Cut the strip in half down the center. Cut it in half again. Describe the result.

57. Construct a Möbius strip (see Problem 54). Cut the strip along a path that is one-third the distance from the edge. Describe the result.

58. Construct a Möbius strip (see Problem 54). Mark a point *A* on the strip. Draw an arc from *A* around the strip until you return to the point *A*. Do you think you could connect *any* two points on the sheet of paper without lifting your pencil?

59. Take a strip of paper 11 in. by 1 in., and give it three half-twists; join the ends together. How many edges and sides does this band have? What happens if you cut down the center of this piece?

60. What is a Klein bottle? Examine the bottle shown in Figure 8.20.

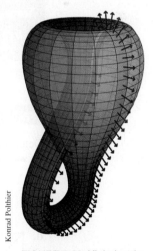

FIGURE 8.20 Klein bottle

Can you build or construct a physical model? You can use the limerick as a hint.

A mathematician named Klein
Thought the Möbius strip was divine.
Said he, "If you glue the edges of two,
You'll get a weird bottle like mine."

8.2 | Trees and Minimum Spanning Trees

YIELD

A circuit could be defined as a path/route that begins and ends at the same vertex.

In the last section, we considered graphs with circuits: an Euler circuit (which is a round trip path traveling all the edges) and a Hamiltonian cycle (a path that visits each vertex exactly once). In this section, we consider another kind of graph, called a *tree*, which does not have a circuit.

Trees

Let us begin with an example.

Example 1 | Draw a family tree

Suppose you wish to draw a family tree showing yourself, your parents, and your maternal and paternal grandparents.

Solution   One possibility for showing this family tree is shown in Figure 8.21.

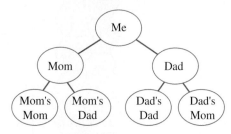

FIGURE 8.21 Personal family tree

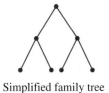

Simplified family tree

The family tree shown in Example 1 has two obvious properties. It is a connected graph because there is at least one path between each pair of vertices, and there are no circuits in this family tree. A simplified tree diagram for Example 1 is shown in the margin.

> **Tree**
>
> A *tree* is a graph that is connected and has no circuits.

Example 2 | Determine if a network is a tree

Determine which of the given graphs are trees.

a.   **b.**   **c.**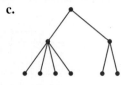

d.    **e.**

CAUTION

In a tree, there is always exactly one path from each vertex in the graph to any other vertex in the graph. We illustrate this property of trees with the following example.

Solution
a. This is not a tree because it is not connected.
b. This is a tree.
c. This is a tree.
d. This is not a tree because there is at least one circuit.
e. There is a circuit, so it is not a tree.

Example	3	**Sprinkler system design**

Ben wishes to install a sprinkler system to water the areas shown in Figure 8.22. Show how this might be done.

FIGURE 8.22 Locations of a faucet and sprinkler heads

Solution   We know there is at least one way to build a tree from each vertex (in this case, the faucet, labeled *F*). We show one such way in Figure 8.23.

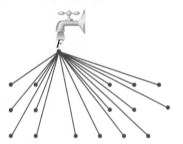

FIGURE 8.23 Sprinkler system

The solution shown for Example 3 may or may not be an efficient solution to the sprinkler system problem. Suppose we connect the vertices in Example 3 without regard to whether the graph is a tree, as shown in Figure 8.24a. Next, we remove edges until the resulting graph is a tree. A tree that is created from another graph by removing edges but keeping a path to each vertex is called a **spanning tree.** Can you form a spanning tree for the graph in Figure 8.24a? If you think about it for a moment, you will see that any connected graph will have a spanning tree, and that if the original graph has at least one circuit, then it will have several different spanning trees. Figure 8.24b shows a spanning tree for the sprinkler problem of Example 3.

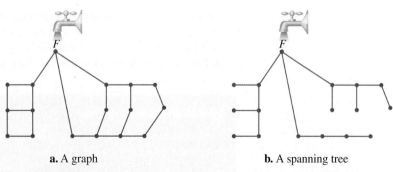

a. A graph   **b.** A spanning tree

FIGURE 8.24 Comparison of a graph and spanning tree for Example 3

Example 4 | Find spanning trees

Find two different spanning trees for each of the given graphs.

a.

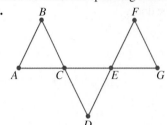

b.

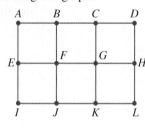

Solution   Since a spanning tree must have a path connecting all vertices, but cannot have any circuits, we remove edges (one at a time) without moving any of the vertices and without creating a disconnected graph. We show two different possibilities for each of the given graphs, while noting that others are possible.

a. This graph has three circuits: $A \to B \to C \to A$, $C \to D \to E \to C$, and $E \to F \to G \to E$. To obtain a spanning tree, we must break up each of these circuits, but at the same time not disconnect the graph. There are many ways we could do this. In the first tree we remove edges $\overline{BC}$, $\overline{CE}$, and $\overline{EF}$. In the second tree, we remove edges $\overline{AB}$, $\overline{CD}$, and $\overline{FG}$, as shown in Figure 8.25.

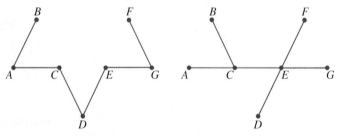

FIGURE 8.25 Spanning tree for graph a

b. There are many circuits and two possible spanning trees, as shown in Figure 8.26.

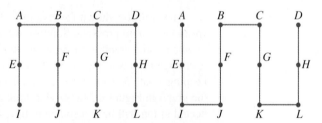

FIGURE 8.26 Spanning tree for graph b

Minimum Spanning Trees

As we can see from Example 4, there may be several spanning trees. Sometimes the length of each edge is associated with a cost or a length, called the edge's **weight.** In such cases we are often interested in minimizing the cost or the distance. If the edges of a graph have weight, then we refer to the graph as a **weighted graph.**

> **Minimum Spanning Tree**
>
> A **minimum spanning tree** is a spanning tree for which the sum of the numbers associated with the edges is a minimum.

| Example | 5 | **Find minimum cost** |

Pólya's Method

A portion of the Santa Rosa Junior College campus, along with some walkways (lengths shown in feet) connecting the buildings, is shown in Figure 8.27.

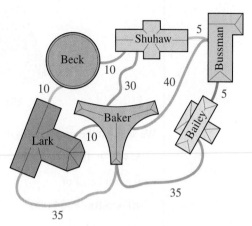

FIGURE 8.27 Portion of campus

Suppose the decision is made to connect each building with a brick walkway, and the only requirement is that there be one brick walkway connecting each of the buildings. We assume that the cost of installing a brick walkway is $100/ft. What is the minimum cost for this project?

Solution   We use Polya's problem-solving guidelines for this example.

Understand the Problem. To make sure we understand the problem, we consider a simpler problem. Consider the simple graph shown in the margin with costs shown in color. We consider the vertices to be buildings since the only requirement is that there be one brick walkway connecting each of the buildings. To find the best way to construct the walkways, we consider minimum spanning trees. Since this is a circuit, we can break this circuit in one of three ways: eliminate one of the sides, *AB*, *BC*, or *AC*.

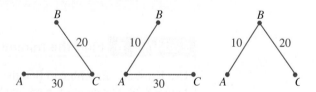

We see that the cost associated with each of these trees is:

$$20 + 30 = 50 \qquad 10 + 30 = 40 \qquad 10 + 20 = 30$$

The minimal cost for this simplified problem is 30.

Devise a Plan. We will carry out the steps for the Santa Rosa campus as follows: Look at Figure 8.27 and find the side with the smallest weight (because we wish to keep the smaller weights). We see there are two sides labeled 5; select either of these. Next, select a side with the smallest remaining weight (it is also 5). Continue by each time selecting the smallest remaining weight until every vertex is connected, but *do not select any edge that creates a circuit.*

Carry Out the Plan. Following this procedure, we select both of the edges labeled 5, as well as the three labeled 10. The resulting pathways are shown in Figure 8.28.

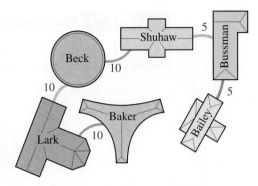

FIGURE 8.28 Minimum spanning tree

We see that the graph in Figure 8.28 is a minimum spanning tree, so the total distance is

$$5 + 5 + 10 + 10 + 10 = 40$$

with a total cost of

$$40 \times \$100 = \$4,000$$

Look Back. We can try other possible routes connecting all of the buildings, but in each case, the cost is more than $4,000.

The process used in Example 5 illustrates a procedure called **Kruskal's algorithm.**

Pay attention to this procedure (algorithm).

There are many websites that illustrate Kruskal's algorithm. You might wish to explore this idea by checking the links at **www.mathnature.com.**

Kruskal's Algorithm

To construct a minimum spanning tree from a weighted graph, use the following procedure:

Step 1   Select any edge with minimum weight.

Step 2   Select the next edge with minimum weight among those not yet selected.

Step 3   Continue to choose edges of minimum weight from those not yet selected, but make sure not to select any edge that forms a circuit.

Step 4   Repeat the process until the tree connects all the vertices of the original graph.

Example   6   Find the minimum spanning tree

Use Kruskal's algorithm to find the minimum spanning tree for the weighted graph in Figure 8.29. The numbers represent hundreds of dollars.

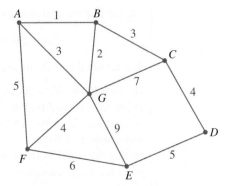

FIGURE 8.29 Find the minimum spanning tree

Solution

Step 1 Choose the side with weight 1 (see Figure 8.30a).

Step 2 Choose the side with weight 2 (Figure 8.30b).

Step 3 Choose the side with weight 3 (Figure 8.30c). Do not connect *AG* because that would form a circuit. Continue with this process to select the two sides with weights of 4 (Figure 8.30c).

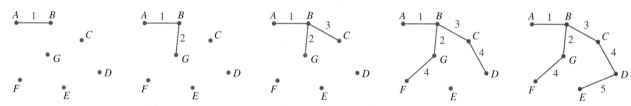

FIGURE 8.30 Steps in finding a minimum spanning tree

Step 4 When we connect the side with the next lowest weight, namely, *ED* with weight 5, we know we are finished because we now have a tree with all of the vertices connected.

We can now calculate the weight of this tree:

$$1 + 2 + 3 + 4 + 4 + 5 = 19$$

Since these weights are in hundreds of dollars, the weight of the minimum spanning tree is $1,900.

The next example is adapted from a standardized test given in the United Kingdom in 1995.

Example **7** **Cost of building a pipeline**

A company is considering building a gas pipeline network to connect seven wells (*A, B, C, D, E, F, G*) to a processing plant *H*. The possible pipelines that it can construct and their costs (in hundreds of thousands of dollars) are listed in the following table.

Pipeline	AB	AD	AE	BC	BE	BF	CG	DE	DF	EH	FG	FH
Cost	23	19	17	15	30	27	10	14	20	28	11	35

What pipelines do you suggest be built and what is the total cost of your suggested pipeline network?

Solution   Begin by drawing a graph to represent the data. This graph is shown in Figure 8.31.

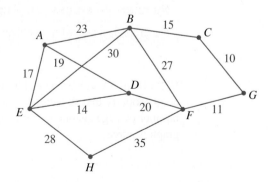

FIGURE 8.31 Building a gas pipeline

We now apply Kruskal's algorithm (in table form).

	Link	*Cost*	*Decision*
Step 1	*CG*	10	Smallest value: add to tree
Step 2	*FG*	11	Next smallest value; add to tree
Step 3	*DE*	14	Add to tree; note that the graph does not need to be connected at this step.
	BC	15	Add to tree
	AE	17	Add to tree
	AD	19	Reject; it forms a circuit *ADE*
	DF	20	Add to tree
	AB	23	Reject; it forms a circuit *ABCGFDE*
	BF	27	Reject; it forms a circuit *BFGC*
Step 4	*EH*	28	Add to tree; stop because all vertices are now included.

The completed minimal spanning tree is shown in Figure 8.32.

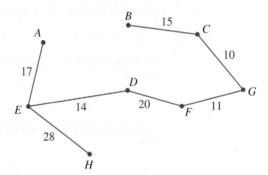

FIGURE 8.32 Minimal spanning tree for pipeline problem

The minimal cost is

$$10 + 11 + 14 + 15 + 17 + 20 + 28 = 115$$

so, the cost of the pipeline is $115,000.

Note in Example 7 that there were eight given vertices, and that there were seven links added to form the minimal spanning tree. This is a general result.

Number-of-vertices-and-edges-in-a-tree theorem

If a graph is a tree with n vertices, then the number of edges is $n - 1$.

In Problem 59, you are asked to explain why this seems plausible. There is another related property that says the converse of this property holds for connected graphs. If the number of edges is one less than the number of vertices in a connected graph, then the graph is a tree.

Problem Set | 8.2

1. **IN YOUR OWN WORDS**   What do we mean by a tree?

2. **IN YOUR OWN WORDS**   What do we mean by a spanning tree?

3. **IN YOUR OWN WORDS**   State Kruskal's algorithm. When would you use this algorithm?

Determine whether each of the graphs in Problems 4–11 is a tree. If it is not, explain why.

4.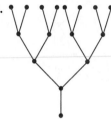

5.

6.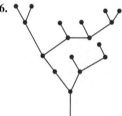

7.

8.

9.

10.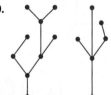

11.

Find two different spanning trees for each graph in Problems 12–19.

12.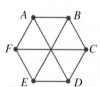

13.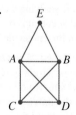

14.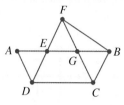

15.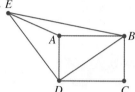

16.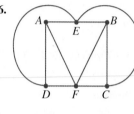

17.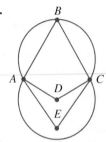

18.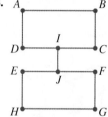

19.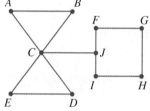

Find all the spanning trees for the graphs in Problems 20–25.

20.

21.

22.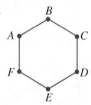

23.

24.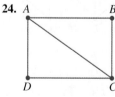

25.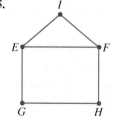

In chemistry, molecules are represented as ball-and-stick models or by bond-line diagrams. For example, the ball-and-stick model for ethane is shown in Figure 8.33a and the bond-line diagram is shown in Figure 8.33b. A corresponding tree diagram is shown in Figure 8.33c.

   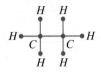

a. Ball-and-stick model   **b.** Bond-line drawing   **c.** Tree diagram

FIGURE 8.33 Ethane molecule

Draw graphs for each of the molecules in Problems 26–31. If the graph does not form a tree, tell why.

26. Methane

27. Propane

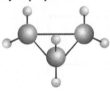

28. Butane

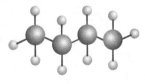

29. Isobutane

30. Cyclopropane

31. Cyclohexane

Find the minimum spanning tree for each of the graphs in Problems 32–41.

32.

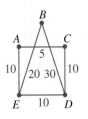

33.

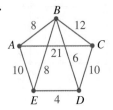

34.

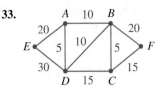

35.

36.

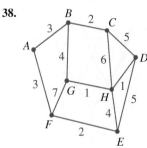

▶ **37.**

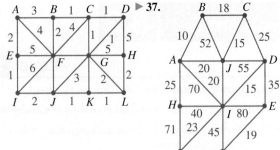

38.

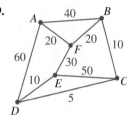

39.

40.

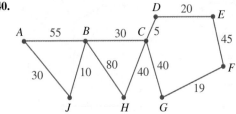

41.

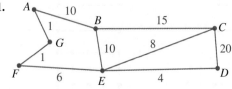

Level 2

42. A chemist is studying a chemical compound with a treelike structure that contains 39 atoms. How many chemical bonds are there in this molecule?

43. A chemist is studying a chemical compound with a treelike structure that contains 65 atoms. How many chemical bonds are there in this molecule?

44. A website called "The Oracle of Bacon" is hosted by the University of Virginia. (Go to **www.mathnature.com** for a link to this site, if you wish.) It links other actors to Kevin Bacon by the movies in which the actors appeared. For example, I entered the movie star from *Gone with the Wind*, Clark Gable, and the Oracle told me that Clark Gable had a Bacon number of two because Clark Gable was in the 1953 movie *Mogambo* with Donald Siden, and Donald Siden was in the 1995 movie *Balto* with Kevin Bacon. I tried it again with Will Smith, and the Oracle said that he also had a Bacon number of two since he was in the 2000 movie *The Legend of Bagger Vance* with Charlize Theron, who in turn was in the 2002 movie *Trapped* with Kevin Bacon. Robert De Niro has a Bacon number of one because he was in the 1996 movie *Sleepers* with Kevin Bacon. This trivia game was started by Albright College students Craig Fass, Brian Turtle, and Mike Ginelli, who hypothesized that all actors—living or dead— have a Bacon number of 6 or less. If you could draw a diagram of the relationships between Kevin Bacon and other actors, would the result be a tree?

45. Suppose you use the Yahoo search engine to do research for a term paper. You start on the Yahoo page and follow the links. You decide to keep a record of the sites you visit. If you keep the record as a graph, will it form a tree?

46. Use a tree to show the following family tree. You have two children, Shannon and Melissa. Shannon has three children, Søren, Thoren, and Floren. Melissa has two children, Hannah and Banana.

47. Use a tree to show the following management relationships for a college. The positions are a college president; an academic vice president, who reports directly to the president and who supervises six academic deans, each of whom supervises three departments; a vice president for business services, who reports directly to the president and who supervises the personnel office, scholarships and grants, as well as the bookstore and food services; and finally, a vice president of operations, who is in charge of supervising facilities, grounds, and certified staff, and who also reports directly to the president.

48. How many edges are there in a tree with 15 vertices?

49. How many vertices are there in a tree with 48 edges?

50. Suppose you wish to install a drip sprinkler system and need to run a drip water line to five areas, as shown in the given graph. The numbers show the distances in feet. What is the smallest number of feet of drip hose necessary to install?

51. Suppose your college is planning to install some covered walkways connecting six buildings as shown in the following map. The plan is to allow a person to walk to any building under cover, and the numbers shown on the map represent distances measured in feet. If the covered walkway costs $350/ft, what is the minimum cost for this project?

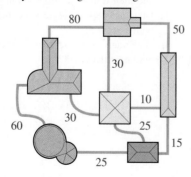

52. A mutual water system obtained estimates for installing water pipes among its respective properties (labeled *A*, *B*, *C*, *D*, and *E*). These amounts (in dollars) are shown in color in Figure 8.34. Which lines should the mutual water system install to minimize the cost?

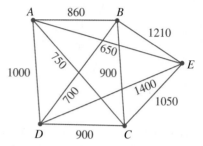

FIGURE 8.34 Water line cost estimates

53. The map in Figure 8.35 shows driving distances and times between California and Nevada cities. Use Kruskal's algorithm to find the minimum spanning tree for the following cities: Santa Rosa, San Francisco, Oakland, Manteca, Yosemite Village, Merced, Fresno, and San Jose.

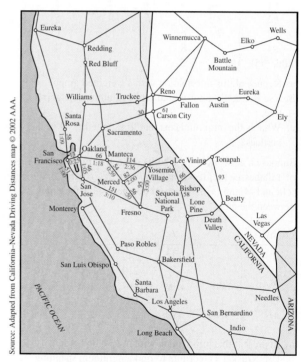

Source: Adapted from California–Nevada Driving Distances map © 2002 AAA.

FIGURE 8.35 California-Nevada driving distances

54. Use the map in Figure 8.35 and Kruskal's algorithm to construct the minimum spanning tree for the cities of Reno, Carson City, Lee Vining, Fallon, Austin, Tonopah, Bishop, Beatty, Death Valley, and Lone Pine.

Level 3

55. Suppose *XYZ* drilling has four oil wells that must be connected via pipelines to a storage tank. The cost of each pipeline (in millions of dollars) is shown in the following table:

From\To	#1	#2	#3	#4	Tank
#1	—	1	4	2	3
#2	1	—	3	1	2
#3	4	3	—	1	4
#4	2	1	1	—	1
Tank	3	2	4	1	—

a. Represent this information with a weighted graph.
b. Use Kruskal's algorithm to find a minimum spanning tree.
c. What is the minimum cost that links together all of the wells and the tank?

56. Consider the following table showing costs between vertices.

From\To	#1	#2	#3	#4	Tank (#5)
#1	—	3	7	3	2
#2	3	—	6	8	5
#3	7	6	—	10	9
#4	3	8	10	—	4
Tank (#5)	2	5	9	4	—

a. Represent this information with a weighted graph.
b. Use Kruskal's algorithm to find a minimum spanning tree.
c. What is the minimum cost that links together all of the vertices?

57. Suppose a network is to be built connecting the Florida cities of Tallahassee (*T*), Jacksonville (*J*), St. Petersburg (*P*), Orlando (*O*), and Miami (*M*). The given numbers show the miles between the cities.

From\To	*T*	*J*	*P*	*O*	*M*
T	—	172	249	253	476
J	172	—	219	160	410
P	249	219	—	105	237
O	253	160	105	—	252
M	476	410	237	252	—

a. Represent this information with a weighted graph.
b. Use Kruskal's algorithm to find a minimum spanning tree.
c. What is the minimum cost that links together all of the cities if the cost is $85/mi?

58. Suppose a network is to be built connecting the cities of Norfolk (*N*), Raleigh (*R*), Charlotte (*C*), Atlanta (*A*), and Savannah (*S*). The given numbers show the miles between cities.

From\To	*N*	*R*	*C*	*A*	*S*
N	—	170	333	597	519
R	170	—	175	427	358
C	333	175	—	250	249
A	597	427	250	—	256
S	519	358	249	256	—

a. Represent this information with a weighted graph.
b. Use Kruskal's algorithm to find a minimum spanning tree.
c. What is the minimum cost that links together all of the cities if the cost is $205/mi?

Problem Solving 3

59. *Number-of-Vertices-and-Edges-in-a-Tree Property*
a. State the number-of-vertices-and-edges-in-a-tree theorem.
b. Consider a tree with one vertex. What is the number of edges? Does the property hold in this case?
c. Consider a tree with two vertices. How many edges can you have and still have a tree? Explain why you cannot have two or more edges.
d. Consider a tree with three vertices. How many edges can you have and still have a tree? Explain why you cannot have three or more edges.

60. **HISTORICAL QUEST** In 1889, Arthur Cayley proved that a complete graph with *n* vertices has n^{n-2} spanning trees.
a. How many spanning trees are there for a complete graph with 3 vertices, according to Cayley's theorem? Verify this number by drawing a complete graph with 3 vertices and then finding all the spanning trees.
b. How many spanning trees are there for a complete graph with 4 vertices, according to Cayley's theorem? Verify this number by drawing a complete graph with 4 vertices and then finding all the spanning trees.
c. How many spanning trees are there for a complete graph with 5 vertices?
d. How many spanning trees are there for a complete graph with 6 vertices?

8.3 | Topology and Fractals

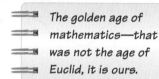

The golden age of mathematics—that was not the age of Euclid, it is ours.

A brief look at the history of geometry illustrates, in a very graphical way, the historical evolution of many mathematical ideas and the nature of changes in mathematical thought. The geometry of the Greeks included very concrete notions of space and geometry. They considered space to be a locus in which objects could move freely about, and their geometry, known as Euclidean geometry, was a geometry of congruence. In this section we investigate two very different branches of geometry that question, or alter, the way we think of space and dimension.

Topology

In the 17th century, space came to be conceptualized as a set of points, and, with the non-Euclidean geometries of the 19th century, mathematicians gave up the notion that geometry had to describe the physical universe. The existence of multiple geometries was accepted, but space was still thought of as a geometry of congruence. The emphasis shifted to sets, and geometry was studied as a mathematical system. Space could be conceived as a set of points together with an abstract set of relations in which these points are involved. The time was right for geometry to be considered as the theory of such a space, and in 1895 Jules-Henri Poincaré published a book using this notion of space and geometry in a systematic development. This book was called *Vorstudien zur Topologie (Introductory Studies in Topology).* However, topology was not the invention of any one person, and the names of Cantor, Euler, Fréchet, Hausdorff, Möbius, and Riemann are associated with the origins of **topology.** Today it is a broad and fundamental branch of mathematics.

To obtain an idea about the nature of topology, consider a device called a *geoboard,* which you may have used in elementary school. Suppose we stretch one rubber band over the pegs to form a square and another to form a triangle, as shown in Figure 8.36.

Historical NOTE

The set theory of Cantor (see the Historical Note on page 50) provided a basis for topology, which was presented for the first time by Jules-Henri Poincaré (1854–1912) in *Analysis Situs.* A second branch of topology was added in 1914 by Felix Hausdorff (1868–1942) in *Basic Features of Set Theory.* Earlier mathematicians, including Euler, Möbius, and Klein, had touched on some of the ideas we study in topology, but the field was given its major impetus by L. E. J. Brouwer (1882–1966). Today much research is being done in topology, which has practical applications in astronomy, chemistry, economics, and electrical circuitry.

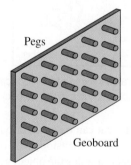
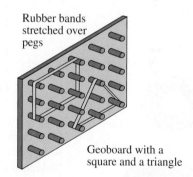

FIGURE 8.36 Creating geometric figures with a geoboard

In high school geometry, the square and the triangle in Figure 8.36 would be seen as different. However, in topology, these figures are viewed as the same object. Topology is concerned with discovering and analyzing the essential similarities and differences between sets and figures. One important idea is called *elastic motion,* which includes bending, stretching, shrinking, or distorting the figure in any way that allows the points to remain distinct. It does not include cutting a figure unless we "sew up" the cut *exactly* as it was before.

> ## Topologically Equivalent Figures
>
> Two geometric figures are said to be **topologically equivalent** if one figure can be elastically twisted, stretched, bent, shrunk, or straightened into the same shape as the other. One can cut the figure, provided at some point the cut edges are "glued" back together again to be exactly the same as before.

Rubber bands can be stretched into a wide variety of shapes. All forms in Figure 8.37 are topologically equivalent. We say that a curve is **planar** if it lies flat in a plane.

FIGURE 8.37 Topologically equivalent curves

Karl J. Smith

The children and their distorted images are topologically equivalent.

All of the curves in Figure 8.37 are *planar simple closed curves*. A curve is **closed** if it divides the plane into three disjoint subsets: the set of points on the curve itself, the set of points *interior* to the curve, and the set of points *exterior* to the curve. It is said to be **simple** if it has only one interior. Sometimes a simple closed curve is called a **Jordan curve.** Notice that, to pass from a point in the interior to a point in the exterior, it is necessary to cross over the given curve an odd number of times. This property remains the same for any distortion and is therefore called an *invariant* property.

Two-dimensional surfaces in a three-dimensional space are classified according to the number of cuts possible without slicing the object into two pieces. The number of cuts that can be made without cutting the figure into two pieces is called its **genus.** The genus of an object is the number of holes in the object. (See Figure 8.38.)

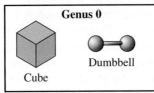
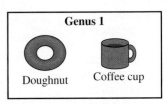
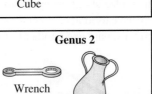
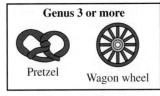

FIGURE 8.38 Genus of the surfaces of some everyday objects. Look at the number of holes in the objects.

FIGURE 8.39 A doughnut is topologically equivalent to a coffee cup

For example, no cut can be made through a sphere without cutting it into two pieces, so its genus is 0. In three dimensions, you can generally classify the genus of an object by looking at the number of holes the object has. A doughnut, for example, has genus 1 since it has 1 hole. In mathematical terms, we say it has genus 1 since only one closed cut can be made without dividing it into two pieces. All figures with the same genus are topologically equivalent. Figure 8.39 shows that a doughnut and a coffee cup are topologically equivalent, and Figure 8.40 shows objects of genus 0, genus 1, and genus 2.

A sphere has genus 0.

A doughnut has genus 1.

Two holes allow two cuts, so this form has genus 2.

FIGURE 8.40 Genus of a sphere, a doughnut, and a two-holed doughnut

Four-Color Problem

One of the earliest and most famous problems in topology is the **four-color problem.** It was first stated in 1850 by the English mathematician Francis Guthrie. It states that any map on a plane or a sphere can be colored with at most four colors so that any two countries that share a common boundary are colored differently. (See Figure 8.41, for example.)

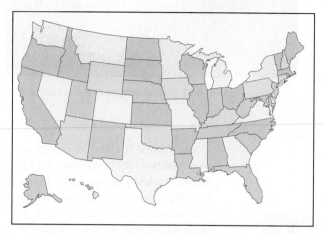

FIGURE 8.41 Every map can be colored with four colors.

All attempts to prove this conjecture had failed until Kenneth Appel and Wolfgang Haken of the University of Illinois announced their proof in 1976. The university honored their discovery using the illustrated postmark.

Since the theorem was first stated, many unsuccessful attempts have been made to prove it. The first published "incorrect proof" is due to Kempe, who enumerated four cases and disposed of each. However, in 1990, an error was found in one of those cases, which it turned out was subcategorized as 1,930 different cases. Appel and Haken reduced the map to a graph as Euler did with the Königsberg bridge problem. They reduced each country to a point and used computers to check every possible arrangement of four colors for each case, requiring more than 1,200 hours of computer time to verify the proof.

Fractal Geometry

One of the newest and most exciting branches of mathematics is called **fractal geometry.** Fractals have been used recently to produce realistic computer images in the movies, and the new supercrisp high-definition television (HDTV) uses fractals to squeeze the HDTV signal into existing broadcast channels. In February 1989, Iterated System, Inc., began marketing a $32,500 software package for creating models of biological systems from fractals. Today you can find hundreds of fractal generators online, most of them free.

Fractals were created to Benoit B. Mandelbrot over 30 years ago, but have become important only in the last few years because of computers. Mandelbrot's first book on fractals appeared in 1975; in it he used computer graphics to illustrate the fractals. The

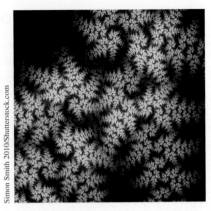

FIGURE 8.42 Fractal images of a mountainscape (generated by Richard Voss) and fern

book inspired Richard Voss, a physicist at IBM, to create stunning landscapes, earthly and otherworldly (see Figure 8.42). "Without computer graphics, this work could have been completely disregarded," Mandelbrot acknowledges.

What exactly is a fractal? We are used to describing the dimension of an object without having a precise definition: A point has 0 dimension; a line, 1 dimension; a plane, 2 dimensions; and the world around us, 3 dimensions. We can even stretch our imagination to believe that Einstein used a four-dimensional model. However, what about a dimension of 1.5? Fractals allow us to define objects with noninteger dimension. For example, a jagged line is given a fractional dimension between 1 and 2, and the exact value is determined by the line's "jaggedness."

We will illustrate this concept by constructing the most famous fractal curve, the so-called "snowflake curve."* Start with a line segment $\overline{AB}$:

$$\overline{\hspace{3cm}}$$
$A \hspace{3cm} B$

Here we say the number of segments is $N = 1$; $r = 1$ is the length of this segment.

Divide this segment into thirds, by marking locations C and D:

$A \qquad C \qquad D \qquad B$ — $N = 3$; $r = \frac{1}{3}$ is the length of each segment.

Now construct an equilateral triangle $\triangle CED$ on the middle segment and then remove the middle segment $\overline{CD}$.

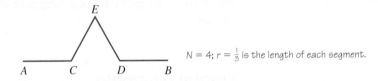

$N = 4$; $r = \frac{1}{3}$ is the length of each segment.

Now, repeat the above steps for each segment:

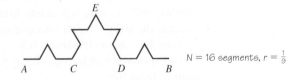

$N = 16$ segments, $r = \frac{1}{9}$

---

* If you want to investigate the snowflake curve and fractal further, see Group Project G29, p. 423.

Again, repeat the process:

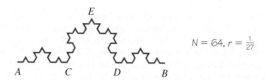

$$N = 64, r = \frac{1}{27}$$

If you repeat this process (until you reach any desired level of complexity), you have a fractal curve with dimension between 1 and 2.

The actual description of the dimension is more difficult to understand.

Mandelbrot defined the dimension as follows:

$$D = \frac{\log N}{\log \frac{1}{r}}$$

where N is any integer and r is the length of each segment.[*] For the illustrations above, we can calculate the dimension:

$$N = 3, \ r = \frac{1}{3}; \ D = 1$$

$$N = 4, \ r = \frac{1}{3}; \ D \approx 1.26$$

$$N = 16, \ r = \frac{1}{9}; \ D \approx 1.26$$

$$N = 64, \ r = \frac{1}{27}; \ D \approx 1.26$$

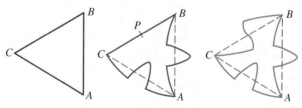

FIGURE 8.43 Escher print: *Circle Limit III*

Tessellations

The construction of the snowflake curve reminds us of another interesting mathematical construction, called a **tessellation.** By skillfully altering a basic polygon (such as a triangle, rectangle, or hexagon), the artist Escher was able to produce artistic tessellations such as that shown in Figure 8.43.

We can describe a procedure for reproducing a simple tessellation based on the Escher print in Figure 8.43.

Step 1   Start with an equilateral triangle $\triangle ABC$. Mark off the same curve on sides $\overline{AB}$ and $\overline{AC}$, as shown in Figure 8.44. Mark off another curve on side $\overline{BC}$ that is symmetric about the midpoint P. If you choose the curves carefully, as Escher did, an interesting figure suitable for tessellating will be formed.

FIGURE 8.44 Tessellation pattern

Step 2   Six of these figures accurately fit together around a point, forming a hexagonal array. If you trace and cut one of these basic figures, you can continue the tessellation over as large an area as you wish, as shown in Figure 8.45.

FIGURE 8.45 Tessellation pattern

[*]Mandelbrot defined r as the ratio L/N, where L is the sum of the lengths of the N line segments. We will discuss logarithms in Chapter 10, but at this point, it is not necessary that you understand this formula.

From "Geometrical Forms Known as Fractals Find Sense in Chaos," by Jeanne McDermott, *Smithsonian*, December 1983, p. 116.

Guest Essay: WHAT GOOD ARE FRACTALS?

Okay, fractals can make sense out of chaos (see the Guest Essay in Section 6.4), but what can you do with them? It is a question currently being asked by physicists and other scientists at many professional meetings, says Alan Norton, a former associate of Mandelbrot who is now working on computer architectures at IBM. For a young idea still being translated into the dialects of each scientific discipline, the answer, Norton says, is: Quite a bit. The fractal dimension may give scientists a way to describe a complex phenomenon with a single number.

Harold Hastings, professor of mathematics at Hofstra University on Long Island, is enthusiastic about modeling the Okefenokee Swamp in Georgia with fractals. From aerial photographs, he has studied vegetation patterns and found that some key tree groups, such as cypress, are patchier and show a larger fractal dimension than others.

Shaun Lovejoy, a meteorologist who works at Météorologie Nationale, the French national weather service in Paris, confirmed that clouds follow fractal patterns. Again, by analyzing satellite photographs, he found similarities in the shapes of many cloud types that formed over the Indian Ocean. From tiny puff-like clouds to an enormous mass that extended from Central Africa to Southern India, all exhibited the same fractal dimension. Prior to Mandelbrot's discovery of fractals, cloud shapes had not been candidates for mathematical analysis and meteorologists who theorize about the origin of weather ignored them. Lovejoy's work suggests that the atmosphere on a small-scale weather pattern near the Earth's surface resembles that on a large-scale weather pattern extending many miles away, an idea that runs counter to current theories.

The occurrence of earthquakes. The surfaces of metal fractures. The path a computer program takes when it scurries through its memory. The way our own neurons fire when we go searching through *our* memories.   The wish list for fractal description grows. Time will tell whether the fractal dimension becomes invaluable to scientists interested in building mathematical models of the world's workings.

Problem Set 8.3

Level 1

1. **IN YOUR OWN WORDS**   What do we mean by topology?

2. **IN YOUR OWN WORDS**   What is the four-color problem?

3. Group the figures into classes so that all the elements within each class are topologically equivalent, and no elements from different classes are topologically equivalent.

A. 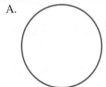   B.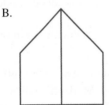

C.   D.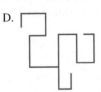

E.   F.

G.   H.

4. Group the figures into classes so that all the elements within each class are topologically equivalent, and no elements from different classes are topologically equivalent.

A. 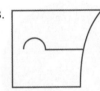   B.

C.   D.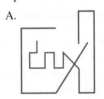

E. 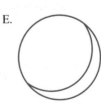   F.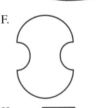

G. 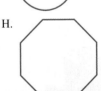   H.

5. Which of the figures in Problem 3 are simple closed curves?

6. Which of the figures in Problem 4 are simple closed curves?

7. Group the letters of the alphabet into classes so that all the elements within each class are topologically equivalent and no elements from different classes are topologically equivalent.

A   B   C   D   E   F   G   H   I   J   K   L   M

N   O   P   Q   R   S   T   U   V   W   X   Y   Z

8. Group the objects into classes so that all the elements within each class are topologically equivalent, and no elements from different classes are topologically equivalent.
  A. a glass
  B. a bowling ball
  C. a sheet of typing paper
  D. a sphere
  E. a ruler
  F. a banana
  G. a sheet of two-ring-binder paper

9. Group the objects into classes so that all the elements within each class are topologically equivalent, and no elements from different classes are topologically equivalent.
  A. a bolt
  B. a straw
  C. a horseshoe
  D. a sewing needle
  E. a brick
  F. a pencil
  G. a funnel with a handle

Level 2

In Problems 10–13, determine whether each of the points A, B, and C is inside or outside of the simple closed curve.

10.

11.

12.

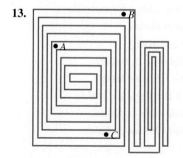

13.

14. a. Let X be a point obviously outside the figure given in Problem 10. Draw $\overline{AX}$. How many times does it cross the curve? Repeat for $\overline{BX}$ and $\overline{CX}$.
  b. Repeat part **a** for the figure given in Problem 11.
  c. Repeat part **a** for the figure given in Problem 12.
  d. Repeat part **a** for the figure given in Problem 13.
  e. Make a conjecture based on parts **a–d**. This conjecture involves a theorem called the *Jordan curve theorem*.

15. One of the simplest map-coloring rules of topology involves a map of "countries" with straight lines as boundaries. How many colors would be necessary for the four-corner area on a U.S. map? Note that a common point is not considered a common boundary.

How many colors are needed for the 4-corners area?

16. IN YOUR OWN WORDS If a map (on a plane or on the surface of a sphere) is partitioned into two or more regions with each vertex of even degree, then the resulting map can be colored with exactly two colors. Draw a map illustrating this fact.

17. IN YOUR OWN WORDS If a map (on a plane or on the surface of a sphere) is partitioned into regions, each with an even number of edges, and if each vertex is of degree 3, the resulting map can be colored with exactly three colors. Draw a map illustrating this fact.

18. IN YOUR OWN WORDS If a map (on a plane or on the surface of a sphere) is partitioned into at least five regions, each sharing its borders with exactly three neighboring regions, the resulting map can be colored in three colors. Draw a map illustrating this fact.

19. Color the eight vertices of a cube in two colors (say red and blue) so that any plane containing three points of one color contains one point of the other color.

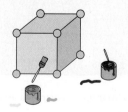

20. Construct a fractal curve by forming squares (rather than triangles as shown in the text). The first step is shown here.

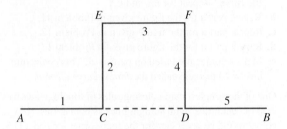

21. Two identical squares joined so that they have a common edge is called a *domino*, as shown in Figure 8.46. Three identical squares can be joined together to form a *tromino*, and they come in two different shapes. *Pentominos*, composed of five squares, come in 12 different shapes.

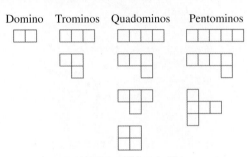

Domino   Trominos   Quadominos   Pentominos

FIGURE 8.46 Mosaic tiles

Complete Figure 8.46 by showing the other 9 pentomino shapes.

Level **3**

22. IN YOUR OWN WORDS Construct a tessellation using triangles.

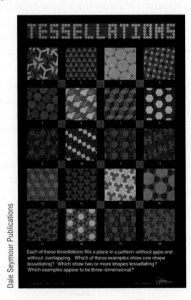

23. IN YOUR OWN WORDS Construct a tessellation using rectangles.

Forming a mosaic pattern using polygons, like those shown in Figure 8.46, is called tiling. A tiling and a tessellation are really the same idea, although the word tessellation is usually used when the pattern is more complicated than a polygon. For example, most classroom floors are tiled with simple squares, a very simple pattern with dominoes. An early example of a tromino pattern is found in an 18th century painting. This pattern is shown in Figure 8.47.

FIGURE 8.47 A tiling pattern for a 6 × 6 square using trominos

Use pentominos to tile a rectangle of the size requested in Problems 24–25.

24. 3 × 20

25. 4 × 15

26. IN YOUR OWN WORDS Design a mosaic using a pentagon.

27. IN YOUR OWN WORDS Design a mosaic using a hexagon.

28.

© Kelly M. Houle 1999

This is an example showing a young girl playing on the grass (look at the cylinder). The painting, entitled "Golden Afternoon" (look at the canvas), shows a distorted image that becomes complete when viewed as a reflection; this is called *anamorphosis*. What topological relation does the picture on canvas have with the image in the reflecting glass?

Problem Solving **3**

29. Answer the questions after reading the poem in the News Clip.

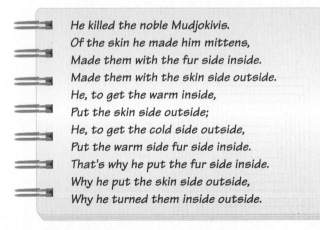

He killed the noble Mudjokivis.
Of the skin he made him mittens,
Made them with the fur side inside.
Made them with the skin side outside.
He, to get the warm inside,
Put the skin side outside;
He, to get the cold side outside,
Put the warm side fur side inside.
That's why he put the fur side inside.
Why he put the skin side outside,
Why he turned them inside outside.

a. If a right-handed mitten is turned inside out, as is suggested in the poem, will it still fit a right hand?

b. Is a right-handed mitten topologically equivalent to a left-handed mitten?

30. Some mathematicians were reluctant to accept the proof of the four-color problem because of the necessity of computer verification. The proof was not "elegant" in the sense that it required the computer analysis of a large number of cases. Study the map in Figure 8.48 and determine for yourself whether it is the *first five-color map,* providing a counterexample for the computerized "proof."

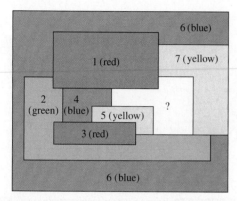

FIGURE 8.48 Is this the world's first five-color map?

What is the color of the white region? Consider the following table:

Region	Blue	Yellow	Green	Red
1	X	X	X	
2	X	X		X
3	X	X	X	
4		X	X	X
5	X		X	X
6		X	X	X
7	X		X	X
?	X	X	X	X

The x's indicate the colors that share a boundary with the given region. As you can see, the white region is bounded by all four colors, so therefore requires a fifth color.

8.4 CHAPTER SUMMARY

Mathematics is an aspect of culture as well as a collection of algorithms.

CARL BOYER

Important Ideas

Königsberg bridge problem [8.1]
Euler's circuit theorem [8.1]
Hamiltonian cycles and the traveling salesperson problem (TSP) [8.1]
Kruskal's algorithm [8.2]
Topologically equivalent figures, four-color problem, and fractals [8.3]

Take some time getting ready to work the review problems in this section. First review these important ideas. Look back at the definition and property boxes. If you look online, you will find a list of important terms introduced in this chapter, as well as the types of problems that were introduced. You will maximize your understanding of this chapter by working the problems in this section only after you have studied the material.

You will find some review help online at **www.mathnature.com**. There are links giving general test help in studying for a mathematics examination, as well as specific help for reviewing this chapter.

Chapter 8 Review Questions

In Problems 1–4, tell whether the network is traversable. If the network is traversable, show how.

1.
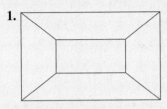

2.
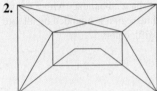

3.

4.
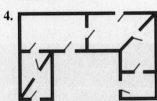

5. Consider the network shown in Figure 8.49.

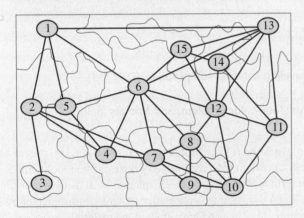

FIGURE 8.49 City network map

a. Is there a path that is an Euler circuit?
b. Is there a path that is a Hamiltonian cycle?

6. Draw a map with seven regions such that the indicated number of colors is required so that no two bordering regions have the same color.
a. Two colors   **b.** Three colors
c. Four colors   **d.** Five colors

The San Francisco Chronicle *reported that two Stanford graduates, Dave Kaval and Brad Null, set a goal to see a game in every major league baseball stadium. (Note: There are 16 National League and 14 American League teams.) They began in San Francisco and selected the route shown in Figure 8.50. Use this information in Problems 7–10.*

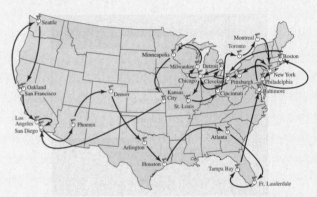

FIGURE 8.50 Tour of major league stadiums (TSP)

7. In how many ways could Kaval and Null start in San Francisco and visit the cities of Minneapolis, Milwaukee, Chicago, St. Louis, Detroit, Cleveland, Cincinnati, and Pittsburgh?

8. If we use the brute-force method for solving the TSP of Kaval and Null, the number of possible routes for the 30 major league cities is the astronomical number 4.4×10^{30}. Use the brute-force method for the simplified problem of finding the best way to begin in San Francisco and visit Los Angeles, San Diego, and Phoenix. The mileage chart is shown below. What is the mileage using the brute-force method?

From\To	SF	LA	SD	P
SF	—	369	502	755
LA	369	—	133	372
SD	502	133	—	353
P	755	372	353	—

9. Use the nearest-neighbor method to approximate the optimal route for the mileage chart shown in Problem 8. What is the mileage when using this route?

10. Show a complete, weighted graph for these cities. Is there a minimum spanning tree? If so, what is the mileage using this method?

11. The Big 10 football conference consists of the following schools:

Ohio State, Penn State, Michigan, Michigan State, Wisconsin, Iowa, Illinois, Northwestern (Ill.), Indiana, Purdue (Indiana), Minnesota

Is it possible to visit each of these schools by crossing each common state border exactly once? If so, find an Euler path.

12. Is it possible to find an Euler circuit for the Big 10 football conference described in Problem 11? That is, is it possible to start the trip in any given state and end the trip in the state in which you started?

In Problems 13–16, consider the graph in Figure 8.51.

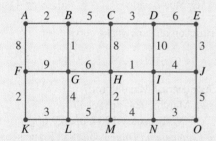

FIGURE 8.51 Weighted graph

13. Find the cost of the nearest-neighbor tour, starting at *A*.

14. Find the cost of the nearest-neighbor tour, starting at *M*.

15. How many tours would be necessary to find the most efficient solution by using the brute-force method starting at *K*?

16. Use Kruskal's algorithm to find the cost of the minimum spanning tree.

In Problems 17–19, consider the graph in Figure 8.52.

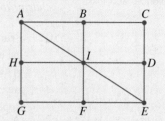

FIGURE 8.52 Graph with 9 vertices

17. Is there an Euler circuit for the graph in Figure 8.52?

18. Find a Hamiltonian circuit for the graph in Figure 8.52.

19. Find a spanning tree for the graph in Figure 8.52.

20. Take a strip of paper 11 in. by 1 in. and give it four half-twists; join the edges together. How many edges and sides does this band have? Cut the band down the center. What is the result?

BOOK REPORTS

Write a 500-word report on one of these books:

The Traveling Salesman Problem: A Guided Tour of Combinatorial Optimization, E. L. Lawler, J. K. Lenstra, A. H. G. Rinnooy Kan, D. B. Shmoys (New York: John Wiley and Sons, 1987).

Graph Theory and Its Applications, Jonathan Gross and Jay Yellen (Boca Raton: CRC Press, LLC, 1998).

Group RESEARCH PROJECTS

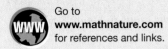 Go to
www.mathnature.com
for references and links.

Working in small groups is typical of most work environments, and learning to work with others to communicate specific ideas is an important skill. Work with three or four other students to submit a single report based on each of the following questions.

G29. Fractals To get you started on your paper, we ask the following question that relates the ideas of series and fractals using the *snowflake curve*. Draw an equilateral triangle with side length a (Figure 8.53a). Next, three equilateral triangles, each of side $\frac{a}{3}$, are cut out and placed in the middle of each side of the first triangle (Figure 8.53b). Repeat this process. As part of the work on this paper, find the perimeter and the area of the snowflake curve formed if you continue this process indefinitely.

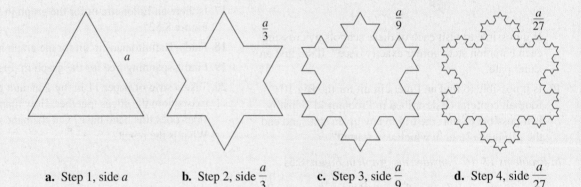

a. Step 1, side a **b.** Step 2, side $\frac{a}{3}$ **c.** Step 3, side $\frac{a}{9}$ **d.** Step 4, side $\frac{a}{27}$

FIGURE 8.53 Construction of a snowflake curve

REFERENCES

Anthony Barcellos, "The Fractal Geometry of Mandelbrot,"
   The College Mathematics Journal, March 1984, pp. 98–114.
"Interview, Benoit B. Mandelbrot," *OMNI*, February 1984, pp. 65–66.
Benoit Mandelbrot, *Fractals: Form, Chance, and Dimension* (San Francisco:
   W. H. Freeman, 1977).
Benoit Mandelbrot, *The Fractal Geometry of Nature* (San Francisco:
   W. H. Freeman, 1982).

G30. Anamorphic Art In Problem 28, Section 8.3 we showed an example of a young girl playing on the grass which was difficult to see on the plane at the bottom, but was easy to see in the cylinder. For example, can you guess what you will see in a reflective cylinder placed in the marked spot in Figure 8.54?

Write a paper about *anamorphic art,* which refers to artwork that is indistinct when viewed from a normal viewpoint, but becomes recognizable when the image is viewed as a reflection. Discuss the two main techniques for creating anamorphic art.

REFERENCES

Linda Bolton, *Hidden Pictures* (New York: Dial Books, 1993).
"The Secret of Anamorphic Art," Art Johnson and Joan D. Martin, *The Mathematics Teacher, January 1998.*
Ivan Moscovich, *The Magical Cylinder* (Norfork, England: Tarquin Publications, 1988).
Marion Walter, *The Mirror Puzzle* Book, (Norfolk, England: Tarquin Publication, 1985).

Dover Publications.

FIGURE 8.54 What do you see here?

Individual RESEARCH PROJECTS

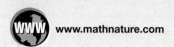

www.mathnature.com

Learning to use sources outside your classroom and textbook is an important skill, and here are some ideas for extending some of the ideas in this chapter. You can find references to these projects in a library or at **www.mathnature.com.**

PROJECT 8.1 Write a report on Ramsey theory.

PROJECT 8.2 Historical Quest   Write a report on the geometry of the Garden Houses of the second-century city of Ostia. See the "What in the World" commentary on page 330.

PROJECT 8.3 Historical Quest   The German artist Albrecht Dürer (1471–1528) was not only a Renaissance artist, but also somewhat of a mathematician. Do some research on the mathematics of Dürer.

PROJECT 8.4 Prepare a classroom demonstration of topology by drawing geometric figures on a piece of rubber inner tube, and illustrate the ways these figures can be distorted.

PROJECT 8.5 The problem shown in the News Clip was first published by John Jackson in 1821. Without the poetry, the puzzle can be stated as follows: **Arrange nine trees so they occur in ten rows of three trees each.** Find a solution.

PROJECT 8.6 Celebrate the Millennium
   a. Consider the product $1 \cdot 2 \cdot 3 \cdot \cdots \cdot 1{,}998 \cdot 1{,}999 \cdot 2{,}000$. What is the last digit?
   b. Consider the product $1 \cdot 2 \cdot 3 \cdot \cdots \cdot 1{,}998 \cdot 1{,}999 \cdot 2{,}000$. From the product, cross out all even factors, as well as all multiples of 5. Now, what is the last digit of the resulting product?

Your aid I want,
nine trees to plant
In rows just half a
score;
And let there be in
each row three.
Solve this: I ask no
more.

9

THE NATURE OF MEASUREMENT

What in the World?

"I can't wait until I move into my new home," said Susan. "I've been living out of boxes for longer than I want to remember."

"Have you picked out your carpet and tile, and made the other choices about color schemes yet?" asked Laurie.

"Heavens, no! They are just doing the framing and I have to build the decks, but I will get around to those selections soon," said Susan. "I hope I don't run out of money before I've finished. My contractor gave me allowances for carpet, tile, light fixtures, and the like, but after looking around a bit I think those amounts are totally unrealistic. I think the goal was to come up with the lowest bid, and the dimensions of the house plan do not seem to match the amounts allowed. Can you help me make some estimates this Saturday?"

Overview

Numbers are used to count and to measure. In counting, the numbers are considered exact unless the result has been rounded. When used for measurement, the numbers are never exact. In this chapter, we study measurement in more detail.

Dimension refers to those properties called length, area, and volume. A figure having length only is said to be *one dimensional*. A figure having area is said to be *two dimensional*, and an object having volume is said to be *three dimensional*.

In this chapter, we use both the United States and the metric measurement systems, not as they relate to each other, but as independent systems used to measure the size of objects in our world. The goal of this chapter is to give you the ability both to measure and to estimate the size, weight, capacity, and temperature of objects.

David Papazian/Corbis

9.1 | Perimeter

Neglect of mathematics works injury to all knowledge, since he who is ignorant of it cannot know the other sciences or the things of this world.

ROGER BACON

CHAPTER **CHALLENGE**

See if you can fill in the question mark.   B,   E,   I,   N,   ?

In this chapter we are concerned with the notion of measurement. To **measure** an object is to assign a number to its size. Measurement is never exact, and you therefore need to decide how **precise** the measure should be. For example, the measurement might be to the nearest inch, nearest foot, or nearest mile. The precision of a measurement depends not only on the instrument used but also on the purpose of your measurement. For example, if you are measuring the size of a room to lay carpet, the precision of your measurement might be different than if you are measuring the size of an airport hangar.

The **accuracy** refers to your answer. Suppose that you measure with an instrument that measures to the nearest tenth of a unit. You find one measurement to be 4.6 and another measurement to be 2.1. If, in the process of your work, you need to multiply these numbers, the result you obtain is

$$4.6 \times 2.1 = 9.66$$

This product is calculated to two decimal places, but it does not seem quite right that you obtain an answer that is more accurate (two decimal places) than the instrument you are using to make your measurements (one decimal place). In this book, we will require that the accuracy of your answers not exceed the precision of the measurement. This means that after the calculations are completed, the final answer should be rounded. The principle we will use is stated in the following box.

> CAUTION — Spend a few moments thinking about the ideas of precision and accuracy.

Accuracy Procedure Used in This Book

All measurements are as precise as given in the text. **If you are asked to make a measurement, the precision will be specified.**

Step 1   Carry out all calculations without rounding.

Step 2   After you obtain a final answer, round this answer to be as accurate as *the least precise* measurement.

Step 3   If you are working with money, round your final answer to the nearest cent.

 The answers we give will conform to this procedure.

This accuracy procedure specifies that, to avoid round-off error, you should round only once (at the end). This is particularly important if you are using a calculator, which will display 8, 10, 12, or even more decimal places.

You will also be asked to *estimate* the size of many objects in this chapter. As we introduce different units of measurement, you should remember some reference points so that you can make intelligent estimates. Many comparisons will be mentioned in the text, but you need to remember only those that are meaningful for you to estimate other sizes or distances. You will also need to choose appropriate units of measurement. For example, you would not measure your height in yards or miles, or the distance to New York City in inches.

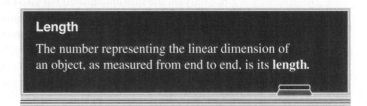

FIGURE 9.1 The metric world in 2010

Measuring Length

What is your height? To answer that question, you must take a measurement.

> **Length**
>
> The number representing the linear dimension of
> an object, as measured from end to end, is its **length.**

The most common system of measurement used in the world is what we call the **metric system.** There have been numerous attempts to make the metric system mandatory in the United States. Today, big business is supporting the drive toward metric conversion, and it appears inevitable that the metric system will eventually come into use in the United States. In the meantime, it is important that we understand how to use both the United States and metric systems.

The most difficult problem in changing from the **United States system** (the customary system of measurement in the United States) to the metric system is not mathematical, but psychological. Many people fear that changing to the metric system will require complex multiplying and dividing and the use of confusing decimal points. For example, in a recent popular article, James Collier states:

For instance, if someone tells me it's 250 miles up to Lake George, or 400 out to Cleveland, I can pretty well figure out how long it's going to take and plan accordingly. Translating all of this into kilometers is going to be an awful headache. A kilometer is about 0.62 miles, so to convert miles into kilometers you divide by six and multiply by ten, and even that isn't accurate. Who can do that kind of thing when somebody is asking me are we almost there, the dog is beginning to drool and somebody else is telling you you're driving too fast?

Of course, that won't matter, because you won't know how fast you're going anyway. I remember once driving in a rented car on a superhighway in France, and every time I looked down at the speedometer we were going 120. That kind of thing can give you the creeps. What's it going to be like when your wife keeps shouting, "Slow down, you're going almost 130"? But if you think kilometers will be hard to calculate. . . .

The author of this article has missed the whole point. Why are kilometers hard to calculate? How does he know that it's 400 miles to Cleveland? He knows because the odometer on his car or a road sign told him. Won't it be just as easy to read an odometer calibrated to kilometers or a metric road sign telling him how far it is to Cleveland?

The real advantage of using the metric system is the ease of conversion from one unit of measurement to another. How many of you remember the difficulty you had in learning to change tablespoons to cups? Or pints to gallons?

In this book we will work with both the U.S. and the metric measurement systems. You should be familiar with both and be able to make estimates in both systems. The following box gives the standard units of length.

Standard Length Units

U.S. System	Metric System
inch (in.)	**meter (m)**
foot (ft, 12 in)	centimeter $\left(\text{cm}, \dfrac{1}{100}\,\text{m}\right)$
yard (yd, 36 in.)	kilometer (km; 1,000 m)
mile (mi, 63,360 in.)	

To understand the size of any measurement, you need to see it, have experience with it, and take measurements using it as a standard unit.

The basic unit of measurement for the U.S. system is the inch; it is shown in Figure 9.2. You can remember that an inch is about the distance from the joint of your thumb to the tip of your thumb.

You should commit the approximate size of both the inch and the centimeter to memory.

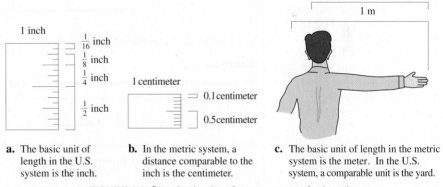

a. The basic unit of length in the U.S. system is the inch.

b. In the metric system, a distance comparable to the inch is the centimeter.

c. The basic unit of length in the metric system is the meter. In the U.S. system, a comparable unit is the yard.

FIGURE 9.2 Standard units of measurement for length

The basic unit of measurement for the metric system is the meter; it is also shown in Figure 9.2. You can remember that a meter is about the distance from your left ear to the tip of the fingers on the end of your outstretched right arm.

For the larger distances of a mile and a kilometer, you will need to look at maps, or the odometer of your car. However, you should have some idea of these distances.*

It might help to have a visual image of certain prefixes as you progress through this chapter. Greek prefixes **kilo-, hecto-,** and **deka-** are used for measurements larger than

*We could tell you that a mile is 5,280 ft or that a kilometer is 1,000 m, but to do so does not give you any feeling for what these distances really are. You need to get into a car and watch the odometer to see how far you travel in going 1 mile. Most cars in the United States do not have odometers set to kilometers, and until they do it is difficult to measure in kilometers. You might, however, be familiar with a 10-kilometer race. It takes a good runner about 30 minutes to run 10 kilometers and an average runner about 45 minutes. You can walk a kilometer in about 6 minutes.

the basic metric unit, and Latin prefixes **deci-, centi-,** and **milli-** are used for smaller quantities (see Figure 9.3). As you can see from Figure 9.3, a centimeter is $\frac{1}{100}$ of a meter; this means that 1 meter is equal to 100 centimeters.

Historical NOTE

Definition of one yard

Early measurements were made in terms of the human body (digit, palm, cubit, span, and foot). Eventually, measurements were standardized in terms of the physical measurements of certain monarchs. King Henry I, for example, decreed that one yard was the distance from the tip of his nose to the end of his thumb. In 1790 the French Academy of Science was asked by the government to develop a logical system of measurement, and the original **metric system** came into being. By 1900 it had been adopted by more than 35 major countries (see Figure 9.1 for the metric countries in the world today.). In 1906 there was a major effort to convert to the metric system in the United States, but it was opposed by big business and the attempt failed. In 1960 the metric system was revised and simplified to what is now known as the **SI system** (an abbreviation of *Système International d'Unités*). The term metric in this book will refer to the SI system.

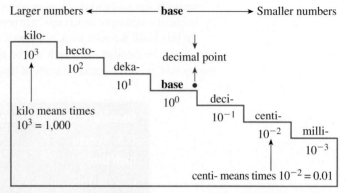

FIGURE 9.3 Metric prefixes

Now we will measure given line segments with different levels of precision. We will consider two different rulers, one marked to the nearest centimeter and another marked to the nearest $\frac{1}{10}$ centimeter.

Example 1 Use a ruler to find a length of a segment

Measure the given segment

$$B \; \text{———————————}$$

a. to the nearest centimeter. **b.** to the nearest $\frac{1}{10}$ centimeter.

Solution

a. B ———————————— End of B is nearer to 5 than to 6.

b. B ———————————— B is 5.3 cm long.

Perimeter

One application of both measurement and geometry involves finding the distance around a polygon. This distance is called the *perimeter* of the polygon.

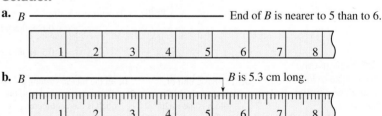

Remember this term.

Perimeter
The *perimeter* of a polygon is the sum of the lengths of the sides of that polygon.

Following are some formulas for finding the perimeters of the most common polygons. An **equilateral triangle** is a triangle with all sides equal in length (see Figure 9.4a.) A **rectangle** is a quadrilateral with opposite sides parallel and equal in length (Figure 9.4b.) A **square** is a quadrilateral with all sides equal in length (Figure 9.4c).

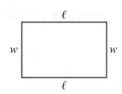

Equilateral triangle

PERIMETER = 3(SIDE)

$P = 3s$

Rectangle

PERIMETER = 2(LENGTH) + 2(WIDTH)

$P = 2\ell + 2w$

Square

PERIMETER = 4(SIDE)

$P = 4s$

FIGURE 9.4 Figures and formulas for an equilateral triangle, rectangle, and square

Example 2 Find the perimeter

Find the perimeter of each polygon.

a.

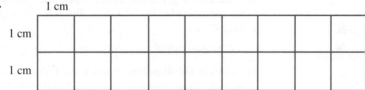

b.  **c.**

d. 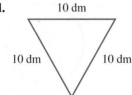 **e.**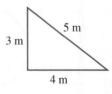

Solution

a. Rectangle is 2 cm by 9 cm, so $P = 2\ell + 2w$

$= 2(9) + 2(2)$

$= 18 + 4$

$= 22$ cm

b. Rectangle is 2 ft by 4 ft, so $P = 2\ell + 2w$

$= 2(4) + 2(2)$

$= 8 + 4$

$= 12$ ft

c. Square, so $P = 4s$

$= 4(5)$

$= 20$ mi

d. Equilateral triangle, so $P = 3s$

$= 3(10)$

$= 30$ dm

e. Triangle (add lengths of sides), so $P = 3 + 4 + 5$

$= 12$ m

| Example | 3 | Find the length of a pen |

Suppose you have enough material for 70 ft of fence and want to build a rectangular pen 14 ft wide. What is the length of this pen?

Solution   PERIMETER = 2(LENGTH) + 2(WIDTH)   This is the formula for perimeter.
                                                 Fill in the given information.

$$70 = 2(\text{LENGTH}) + 2(14)$$
$$70 = 2\ell + 28 \qquad \text{Let } \ell = \text{LENGTH OF PEN}$$
$$42 = 2\ell$$
$$21 = \ell$$

The pen will be 21 ft long.

Circumference

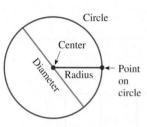

FIGURE 9.5 Circle

A **circle** is the set of all points in a plane a given distance, called the **radius,** from a given point, called the **center.** Although a circle is not a polygon, sometimes we need to find the distance around a circle. This distance is called the **circumference.** For *any circle,* if you divide the circumference by the diameter, you will get the *same number* (see Figure 9.5). This number is given the name π (**pi**). The number π is an irrational number and is sometimes approximated by 3.14 or $\frac{22}{7}$. We need this number π to state a formula for the circumference C:

$$C = d\pi \quad \text{where } d = \text{DIAMETER} \qquad \text{or} \qquad C = 2\pi r \quad \text{where } r = \text{RADIUS}$$

For a circle, the **diameter** is twice the radius. This means that, if you know the radius and want to find the diameter, you simply multiply by 2. If you know the diameter and want to find the radius, divide by 2.

| Example | 4 | Find distance around a figure |

Find the distance around each figure.

a. 4 ft

b. 28 cm

c. 5 m

d. $\frac{1}{2}$ of a circle

2 dm

e. 30 cm

$\frac{1}{8}$ of a circle

Solution

a. $C = 4\pi$; a decimal approximation is $C \approx 4(3.14) = 12.56$. The circumference is 4π ft, which is about 13 ft.

b. $C = 28\pi$; this is about 88 cm.

c. $C = 2\pi(5) = 10\pi$; this is about 31 m.

d. This is half of a circle (called a **semicircle**); thus, the curved part is half of the circumference ($C = 2\pi \approx 6.28$), or about 3.14, which is added to the diameter:

$$3.14 + 2 = 5.14$$

The distance around the figure is about 5 dm.

My wife and I stopped for lunch in a Nebraska town on our way to California, and I asked the waitress how much snow the area usually got. "About as deep as a meter," she replied. Impressed by her use of the metric system, I asked where she had learned it. She was momentarily baffled, then said, "That's the one I mean," and pointed out the window to the parking meter in front of the restaurant.

N. A. Norris

e. This is one-eighth of a circle. The curved part is one-eighth of the circumference $[C = 2\pi(30) \approx 188.5]$, or about 23.6, which is added here to the radius (on both sides):

$$23.6 + 30 + 30 \approx 84 \text{ cm}$$

Many calculators have a single key marked π. If you press it, the display shows an approximation correct to several decimal places (the number of places depends on your calculator):

Press: $\boxed{\pi}$ *Display:* 3.141592654

If your calculator doesn't have a key marked π, you may want to use this approximation to obtain the accuracy you want. In this book, the answers are found by using the π key on a calculator and then rounding the answer.

Problem Set 9.1

Level 1

1. **IN YOUR OWN WORDS** Contrast precision and accuracy.

2. What is the agreement about the accuracy of answers in this book?

3. State the perimeter formulas for a square, rectangle, equilateral triangle, and regular pentagon.

4. What is the formula for the circumference of a circle?

5. Define π.

From memory, and without using any measuring devices, draw a line segment with approximate length as indicated in Problems 6–8.

6. **a.** 1 in.   **b.** $\frac{1}{2}$ in.   **c.** 5 cm

7. **a.** 2 in.   **b.** $\frac{1}{4}$ in.   **c.** 1 cm

8. **a.** 3 in.   **b.** 10 cm   **c.** 3 cm

Pick the best choices in Problems 9–33 by estimating. Do not measure. For metric measurements, do not attempt to convert to the U.S. system. The hardest part of the transition to the metric system is the transition to thinking in metrics.

9. The length of this math textbook is about
   A. 10 in.   B. 10 cm   C. 1 ft

10. The length of a car is about
   A. 1 m   B. 4 m   C. 10 m

11. The length of a dollar bill is about
   A. 3 in.   B. 6 in.   C. 9 in.

12. The width of a dollar bill is about
   A. 1.9 cm   B. 6.5 cm   C. 0.65 m

13. The perimeter of a dollar bill is
   A. 18 in.   B. 6 in.   C. 46 in.

14. The perimeter of a five-dollar bill is
   A. 18 cm   B. 6 cm   C. 46 cm

15. The length of a new pencil is
   A. 4 in.   B. 18 in.   C. 7 in.

16. The diameter of a new pencil is
   A. 0.25 cm   B. 0.25 in.   C. 0.25 ft

17. The circumference of an automobile tire is
   A. 60 cm   B. 60 in.   C. 1 m

18. The perimeter of this textbook is
   A. 36 in.   B. 36 cm   C. 11 in.

19. The perimeter of a VISA credit card is
   A. 30 in.   B. 30 cm   C. 1 m

20. The perimeter of a sheet of notebook paper is
   A. 1 cm   B. 10 cm   C. 1 m

21. The perimeter of the screen of a wall-mount TV set is
   A. 30 in.   B. 100 cm   C. 100 in.

22. The perimeter of a classroom is
   A. 100 ft   B. 100 m   C. 100 yd

23. The distance from your home to the nearest grocery store is most likely to be
   A. 1 cm   B. 1 m   C. 1 km

24. Your height is closest to
   A. 5 ft   B. 10 ft   C. 25 in.

25. An adult's height is most likely to be about
   A. 6 m   B. 50 cm   C. 170 cm

26. The distance around your waist is closest to
   A. 10 in.   B. 36 in.   C. 30 cm

27. The distance from floor to ceiling in a typical home is about
   A. 2.5 m   B. 0.5 m   C. 4.5 m

28. The length of a diagonal on a typical computer monitor is about
   A. 14 cm   B. 14 in.   C. 31 in.

29. The length of a 100-yard football field is
   A. 100 m
   B. more than 100 m
   C. less than 100 m

30. The distance from San Francisco to New York is about 3,000 miles. This distance is
   A. less than 3,000 km
   B. more than 3,000 km
   C. about 3,000 km

31. Suppose someone could run the 100-meter dash in 10 seconds flat. At the same rate, this person should be able to run the 100-yard dash in
   A. less than 10 sec
   B. more than 10 sec
   C. 10 sec

32. The prefix *centi-* means
   A. one thousand
   B. one-thousandth
   C. one-hundredth

33. The prefix *kilo-* means
   A. one thousand
   B. one-thousandth
   C. one-hundredth

Measure the segments given in Problems 34–36 with the indicated precision.

34. _____
   a. to the nearest centimeter
   b. to the nearest $\frac{1}{10}$ centimeter
   c. to the nearest inch
   d. to the nearest eighth of an inch

35. _____
   a. to the nearest centimeter
   b. to the nearest $\frac{1}{10}$ centimeter
   c. to the nearest inch
   d. to the nearest eighth of an inch

36. _____
   a. to the nearest centimeter
   b. to the nearest $\frac{1}{10}$ centimeter
   c. to the nearest inch
   d. to the nearest eighth of an inch

Find the perimeter, circumference, or distance around the figures given in Problems 37–51 by using the appropriate formula. Round approximate answers to the least accurate of the given measurements.

37.
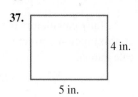
4 in.
5 in.

38.
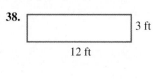
3 ft
12 ft

39.
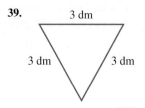
3 dm
3 dm   3 dm

40.
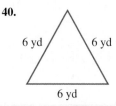
6 yd   6 yd
6 yd

41.
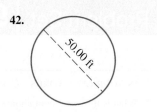
2.40 m

42.
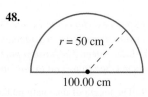
50.00 ft

43.
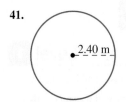
r r
50.00 ft
120.00 ft

44.
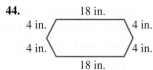
18 in.
4 in.   4 in.
4 in.   4 in.
18 in.

45.
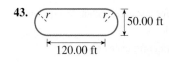
14 ft
9 ft   9 ft
14 ft

46.
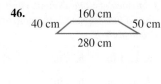
160 cm
40 cm   50 cm
280 cm

47.
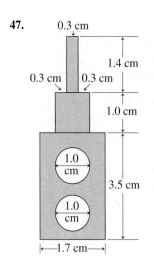
0.3 cm
1.4 cm
0.3 cm   0.3 cm
1.0 cm
1.0 cm
3.5 cm
1.0 cm
1.7 cm

48.
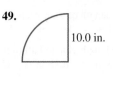
$r = 50$ cm
100.00 cm

49.
10.0 in.

50.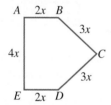

2.00 cm

|← 3.00 cm →|

51.

2.00 cm

|←3.00 cm →|

52. What is the width of a rectangular lot that has a perimeter of 410 ft and a length of 140 ft?

53. What is the length of a rectangular lot that has a perimeter of 750 m and a width of 75 m?

54. Find the dimensions of a rectangle with a perimeter of 54 cm if the length is 5 less than three times the width.

If we define a *cubit* as the distance from your elbow to your fingertips, then we may find that the cubit defined for your body is a little different from the Egyptian cubit on display at the Louvre. How does your cubit compare with this Egyptian cubit?

58. Find your own metric measurements.

Women	Men
a. height	**a.** height
b. bust	**b.** chest
c. waist	**c.** waist
d. hips	**d.** seat
e. waist to hemline	**e.** neck

Level **3**

55. The perimeter of $\triangle ABC$ is 117 in. Find the lengths of the sides.

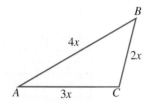

56. The perimeter of this pentagon is 280 cm. Find the lengths of the sides.

57. HISTORICAL QUEST   One of the first recorded units of measurement is the **cubit,** shown in Figure 9.6.

Réunion des Musées Nationaux / Art Resource, NY

FIGURE 9.6 An ancient cubit, from the Louvre, Paris. The original measures 52.5 cm

---

Problem Solving **3**

59. HISTORICAL QUEST   In the sixth chapter of Genesis, the dimensions of Noah's ark are given as 300 cubits long, 50 cubits wide, and 30 cubits high. Use the Egyptian cubit in Figure 9.6 to convert these measurements to the nearest meter.

60. Suppose that we fit a band tightly around the earth at the equator. We wish to raise the band so that it is uniformly supported 6 ft above the earth at the equator.
   a. Guess how much extra length would have to be added to the band (not the supports) to do this.
   b. Calculate the amount of extra material that would be needed.

Equator

9.2 | Area

The measurement of length, discussed in the previous section, is sometimes considered to be a *one-dimensional measurement* (back and forth, i.e., end to end). In this section, we consider a *two-dimensional measurement* (end to end and top to bottom, i.e., back/forth and up/down).

Rectangles

Suppose that you want to carpet your living room. The price of carpet is quoted as a price per square yard. A square yard is a measure of **area.** To measure the area of a plane figure, you fill it with **square units.** (See Figure 9.7.)

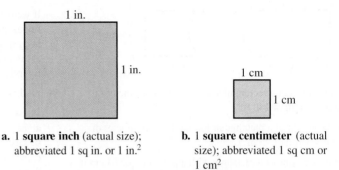

a. 1 **square inch** (actual size); abbreviated 1 sq in. or 1 in.^{2}

b. 1 **square centimeter** (actual size); abbreviated 1 sq cm or 1 cm^2

FIGURE 9.7 Common units of measurement for area

Example 1 | Find an area

What is the area of the shaded region?

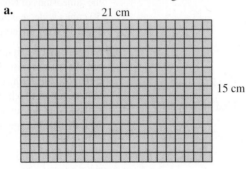

a. 21 cm / 15 cm

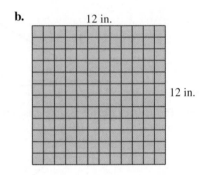

b. 12 in. / 12 in.

Solution To find the area, count the number of square units in each region.

a. You can count the number of square centimeters in the shaded region; there are 315 squares. Also notice:

Across   Down
21 cm × 15 cm = 21 × 15 cm × cm = 315 cm^2

b. The shaded region is a **square foot.** You can count 144 square inches inside the region. Also notice:

Across   Down
12 in. × 12 in. = 144 in.^{2}

As you can see from Example 1, the area of a rectangular or square region is the product of the distance across (length) and the distance down (width).

You will need these formulas not only for this text, but also for real-world problems.

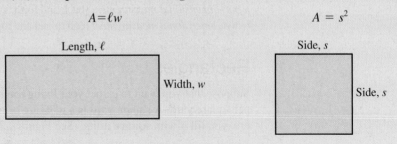

Area of Rectangles and Squares

The formulas for the area, A, of a rectangle of length ℓ and width w and the area, A, of a square of side with length s are:

$$A = \ell w \qquad\qquad A = s^2$$

Length, ℓ

Width, w

Side, s

Side, s

| Example | 2 | **Area of a square yard in square feet** |

How many square feet (see Figure 9.8) are there in a square yard?

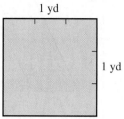

1 yd

1 yd

FIGURE 9.8 $1 \text{ yd}^2 = 9 \text{ ft}^2$

Solution   Since 1 yd = 3 ft, we see from Figure 9.8 that

$$1 \text{ yd}^2 = (3 \text{ ft})^2 \quad \textit{Substitute.}$$
$$= 9 \text{ ft}^2$$

Parallelograms

A **parallelogram** is a quadrilateral with two pairs of parallel sides, as shown in Figure 9.9.

FIGURE 9.9 Parallelograms

To find the area of a parallelogram, we can estimate the area by counting the number of square units inside the parallelogram (which may require estimation of partial square units), or we can show that the formula for the area of a parallelogram is the same as the formula for the area of a rectangle.

Area of a Parallelogram

The area, A, of a parallelogram with base b and height h is

$$A = bh$$

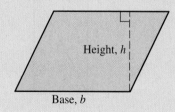

Height, h

Base, b

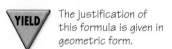

The justification of this formula is given in geometric form.

Move this triangular piece to form a rectangle.

cut here

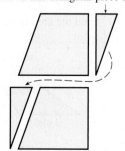

Cut this off and move it to the other side.

This is the rectangle that has been formed from the parallelogram. This means that the formula for the area of a parallelogram is the same as that for a rectangle.

Example 3 | Find the area of a parallelogram

Find the area of each shaded region.

a.

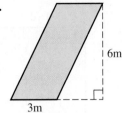

6m

3m

b.

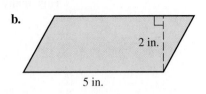

2 in.

5 in.

Solution

a. $b = 3$ m and $h = 6$ m

$A = bh$

$= 3$ m $\times 6$ m

$= 18$ m^2

b. $b = 5$ in. and $h = 2$ in.

$A = bh$

$= 2$ in. $\times 5$ in.

$= 10$ in.^{2}

Triangles

You can find the area of a triangle by filling in and approximating the number of square units, by rearranging the parts, or by noticing that *every* triangle has an area that is exactly half that of a corresponding parallelogram.

Area of a Triangle

The area, A, of a triangle with base b and height h is[*]

$A = \frac{1}{2}bh$

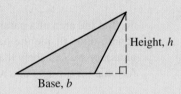

Height, h

Base, b

The justification of this formula is given in geometric form.

These triangles have the same area.

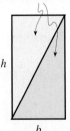

h

b

These triangles have the same area.

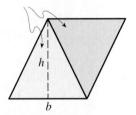

h

b

These triangles have the same area.

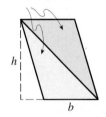

h

b

Example 4 | Find the area of a triangle

Find the area of each shaded region.

a.

4 mm

3 mm

b.

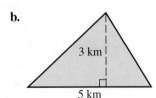

3 km

5 km

[*]The height of a triangle is the perpendicular distance from the vertex to the base.

Solution

a. $b = 3$ mm, $h = 4$ mm

$$A = \frac{1}{2}bh$$

$$= \frac{1}{2} \times 3 \text{ mm} \times 4 \text{ mm}$$

$$= 6 \text{ mm}^2$$

b. $b = 5$ km, $h = 3$ km

$$A = \frac{1}{2}bh$$

$$= \frac{1}{2} \times 5 \text{ km} \times 3 \text{ km}$$

$$= \frac{15}{2} \text{ km}^2 \text{ or } 7\frac{1}{2} \text{ km}^2$$

Trapezoids

A **trapezoid** is a quadrilateral with two sides parallel. These sides are called the *bases,* and the perpendicular distance between the bases is the *height.* We can find the area of a trapezoid by finding the sum of the areas of two triangles, as shown in Figure 9.10.

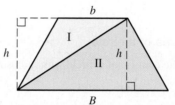

Area of triangle I: $\frac{1}{2}bh$

Area of triangle II: $\frac{1}{2}Bh$

Total area: $\frac{1}{2}bh + \frac{1}{2}Bh$

FIGURE 9.10 Trapezoid

If we use the distributive property, we obtain the area formula for a trapezoid, as shown in the following box.

Area of a Trapezoid

The area, A. of a trapezoid with bases b and B and height h is

$$A = \tfrac{1}{2}h(b + B)$$

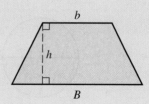

Example 5 Find the area of a trapezoid

Find the area of each shaded region.

a.

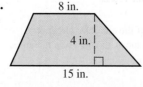

b.

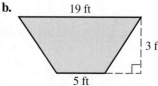

Solution

a. $h = 4; b = 8; B = 15$

$$A = \frac{1}{2}h(b + B)$$

$$= \frac{1}{2}(4)(8 + 15)$$

$$= 2(23)$$

$$= 46$$

The area is 46 in.^{2}.

b. $h = 3; b = 19; B = 5$

$$A = \frac{1}{2}h(b + B)$$

$$= \frac{1}{2}(3)(19 + 5)$$

$$= \frac{3}{2}(24)$$

$$= 36$$

The area is 36 ft^2.

Circles

The last of our area formulas is for the area of a circle. Historically, we know from the Rhind papyrus that the Egyptians knew of the formula for the area of a circle. We state this formula using modern notation.

Area of a circle

The area, A, of a circle with radius r is

$$A = \pi r^2$$

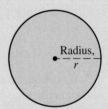

Radius, r

Even though it is beyond the scope of this course to derive a formula for the area of a circle, we can give a geometric justification that may appeal to your intuition.

Consider a circle with radius r

$C = 2\pi r$

Cut the circle in half:

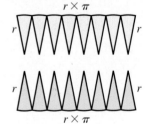

Half of the circumference is πr.

Cut along the dashed lines so that each half lies flat when it is opened up, as shown below.

$r \times \pi$

$r \times \pi$

Fit these two pieces together:

$r \times \pi$

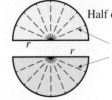

$r \times \pi$

Therefore, it looks as if the area of a circle of radius r is about the same as the area of a rectangle of length πr and width r—that is, πr^2.

Example **6** **Area of a circle or semicircle**

Find the area (to the nearest tenth unit) of each shaded region.

a.   **b.**

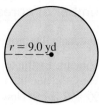

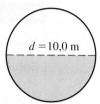

Solution

a. Using a calculator: $9^2\pi \approx 254.4690049$. To the nearest tenth, the area is 254.5 yd^2.

b. Notice that the shaded portion is only half the area of the circle. Using a calculator: $5^2\pi/2 \approx 39.26990817$. To the nearest tenth, the area is 39.3 m^2.

Applications

Sometimes we measure area using one unit of measurement and then we want to convert the result to another.

Example **7** **Carpet purchase**

Suppose your living room is 12 ft by 15 ft and you want to know how many square yards of carpet you need to cover this area.

Solution

Method I. $A = 12 \text{ ft} \times 15 \text{ ft}$

$= 180 \text{ ft}^2$

$= 180 \times 1 \text{ ft}^2$

$= 180 \times \left(\dfrac{1}{9} \text{ yd}^2\right)$

$= 20 \text{ yd}^2$

Since 1 yd = 3 ft
1 yd² = (1 yd) × (1 yd)
1 yd² = (3 ft) × (3 ft)
1 yd² = 9 ft²
$\frac{1}{9}$ yd² = 1 ft²   Divide both sides by 9.

Method II. Change feet to yards to begin the problem:

$12 \text{ ft} = 4 \text{ yd}$ and $15 \text{ ft} = 5 \text{ yd}$

$A = 4 \text{ yd} \times 5 \text{ yd}$

$= 20 \text{ yd}^2$

If the area is large, as with property, a larger unit is needed. This unit is called an *acre*.

> **Acre**
>
> An **acre** is 43,560 ft^2.

This definition leads us to a procedure for changing square feet to acres.

To Change to Acres

To convert square feet to acres, divide by 43,560.

When working with acres, you usually need a calculator to convert square feet into acres, as shown in Example 8.

Example 8 Find the number of acres

How many acres are there in a rectangular property measuring 363 ft by 180 ft?

Solution   Begin with an estimate: Estimate: $363 \approx 400$ and $180 \approx 200$, so area is about $400 \times 200 = 80,000$. Thus, the number is acres is under (since our estimate numbers are over) $80,000 \div 40,000 = 2$ acres. Now, we carry out the actual computation:

$$A = 363 \times 180 \text{ ft}$$
$$= 65,340 \text{ ft}^2$$
$$= 65,340 \div 43,560 \text{ acres}$$
$$= 1.5 \text{ acres} \quad \text{By calculator}$$

Problem Set   9.2

Level 1

1. **IN YOUR OWN WORDS**   What do we mean by area?

2. **IN YOUR OWN WORDS**   How do you find the area of a circle? Contrast with the procedure for finding the circumference of a circle.

Estimate the area of each figure in Problems 3–8 to the nearest square centimeter.

3. **a.** 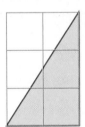   **b.**

4. **a.** 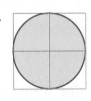   **b.**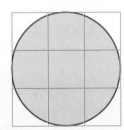

5.

6.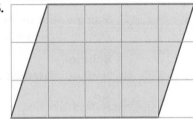

7.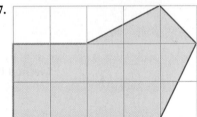

8.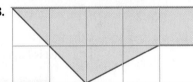

Pick the best choices in Problems 9–18 by estimating. Do not measure. For metric measurements, do not attempt to convert to the U.S. system.

9. The area of a dollar bill is
   A. 18 in.   B. 6 in.^{2} C. 18 in.^{2}

10. The area of a five-dollar bill is
    A. 100 cm   B. 10 cm^2 C. 100 cm^2

11. The area of the front cover of this textbook is
    A. 80 in.^{2} B. 80 cm^2 C. 70 in.

12. The area of a VISA credit card is
    A. 8 in.   B. 8 in.^{2} C. 8 cm^2

13. The area of a sheet of notebook paper is
    A. 90 cm^2 B. 10 in.^{2} C. 600 cm^2

14. The area of a sheet of notebook paper is
    A. 90 in.^{2} B. 10 cm^2 C. 600 in.^{2}

15. The area of the screen of a plasma TV set is
    A. 10 ft^2 B. 48 in.^{2} C. 100 in.^{2}

16. The area of a classroom is
    A. 100 ft^2 B. 1,000 ft^2 C. 0.5 acre

17. The area of the bottom of your feet is
    A. 1 m^2 B. 400 in.^{2} C. 400 cm^2

18. It is known that the area of the skin covering your entire body is about 100 times the area that you will find if you trace your hand on a sheet of paper. Using this estimate, the area of your skin is
    A. 300 in.^{2} B. 3,000 in.^{2} C. 3,000 cm^2

Level 2

Find the area of each shaded region in Problems 19–40. (Assume that given measurements are exact, and round approximate answers to the nearest tenth of a square unit.)

19.
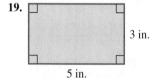
3 in.
5 in.

20.
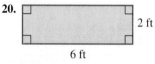
2 ft
6 ft

21.
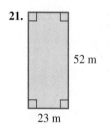
52 m
23 m

22.
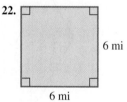
6 mi
6 mi

23.
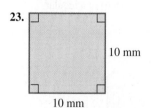
10 mm
10 mm

24.
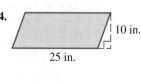
10 in.
25 in.

25.
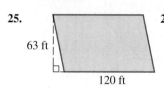
63 ft
120 ft

26.
3 m
9 m

27.
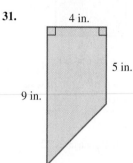
13 dm
21 dm

28.
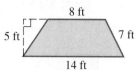
8 ft   10 ft

29.
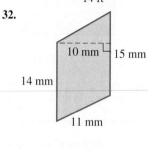
160 cm
30 cm   50 cm
210 cm

30.
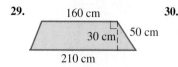
8 ft
5 ft   7 ft
14 ft

31.
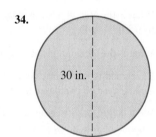
4 in.
5 in.
9 in.

32.
10 mm   15 mm
14 mm
11 mm

33.
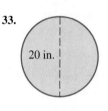
20 in.

34.
30 in.

35.
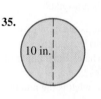
10 in.

36.
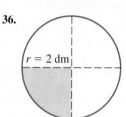
r = 2 dm

37.
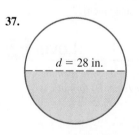
d = 28 in.

38.
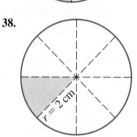
r = 2 cm

39.
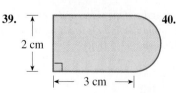
2 cm
3 cm

40.
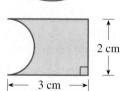
2 cm
3 cm

41. Which property costs less per square foot?

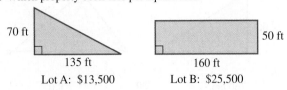

Lot A: $13,500   Lot B: $25,500

42. Which property costs less per square foot?

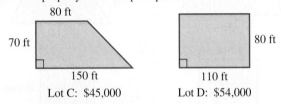

Lot C: $45,000   Lot D: $54,000

43. If a rectangular piece of property is 750 ft by 1,290 ft, what is the acreage (to the nearest tenth of an acre)?

44. How many square feet are there in $4\frac{1}{2}$ acres?

45. What is the area of a television screen that measures 12 in. by 18 in.?

46. What is the area of a rectangular lot that measures 185 ft by 75 ft?

47. What is the area of a piece of $8\frac{1}{2}$-in. by 11-in. typing paper?

48. If a certain type of fabric comes in a bolt 3 feet wide, how long a piece must be purchased to have 24 square feet?

49. What is the cost of seeding a rectangular lawn 100 ft by 30 ft if 1 pound of seed costs $5.85 and covers 150 square feet? Use estimation to decide whether your answer is reasonable.

50. a. If a mini pizza has a 6-in. diameter, what is the number of square inches (to the nearest square inch)?
  b. If a small pizza has a 10-in. diameter, what is the number of square inches (to the nearest square inch)?
  c. If a medium pizza has a 12-in. diameter, what is the number of square inches (to the nearest square inch)?
  d. If a large pizza has a 14-in. diameter, what is the number of square inches (to the nearest square inch)?

Level 3

If the lengths of three sides of a triangle are known, then the following formula, known as Hero's (or Heron's) formula, is sometimes used:

$$A = \sqrt{s(s-a)(s-b)(s-c)}$$

where a, b, and c are the given lengths of the sides, and $s = 0.5(a + b + c)$. Use this formula to find the areas (correct to the nearest square foot) in Problems 51–53.

51. $a = 5$ ft, $b = 8$ ft, $c = 10$ ft

52. $a = 180$ ft, $b = 200$ ft, $c = 350$ ft

53.

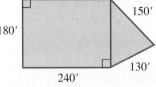

54. What is the area of a square whose diagonal is equal to 7 in.?

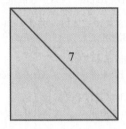

55. What is the area of the regular pentagon with a side equal to 10 in.?

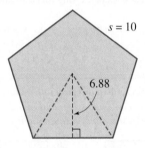

s = 10

6.88

56. What is the area (to the nearest square inch) of the regular hexagon with a side equal to 10 in.?

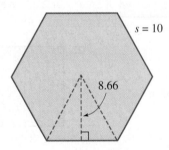

s = 10

8.66

57. What is the area (to the nearest square inch) of a regular octagon with a side equal to 10 in.?

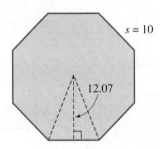

s = 10

12.07

Problem Solving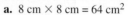

Find the area (to the nearest square inch) of the shaded region contained in the 10-in. squares in Problems 58–59. Assume that the arcs intersect the midpoints of the sides.

58.

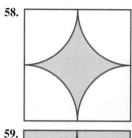

59.

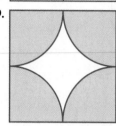

60. Figure 9.11 illustrates a strange and interesting relationship. The square in part **a** has an area of 64 cm² (8 cm by 8 cm). When *this same figure* is cut and rearranged as shown in part **b**, it appears to have an area of 65 cm². Where did this "extra" square centimeter come from?

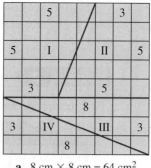

a. 8 cm × 8 cm = 64 cm²

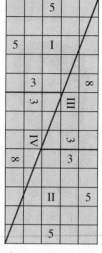

b. 13 cm × 5 cm = 65 cm²

FIGURE 9.11 Extra square centimeter?

[*Hint:* Construct your own square 8 cm on a side, and then cut it into the four pieces as shown. Place the four pieces together as illustrated. Be sure to do your measuring and cutting very carefully. Satisfy yourself that this "extra" square centimeter has appeared. Can you explain this relationship?]

9.3 | Surface Area, Volume, and Capacity

In this section we complete our trilogy of dimensions by looking at a *three-dimensional measurement.* We have considered length (a *one-dimensional measurement*, say back and forth), area (a *two-dimensional measurement*, say back/forth and up/down). Now, the measurement of volume has a third direction of measurement: back/forth; up/down; in/out.

Before we consider volume, we begin by measuring the surface area of three dimensional objects, and after considering volume, we measure the contents of a three-dimensional object.

Surface Area

Suppose you wish to paint a box whose edges are each 3 ft, and you need to know how much paint to buy. To determine this you need to find the sum of the areas of all the faces—this is called the **surface area.**

Example 1 | Find the surface area of a box

Find the amount of paint needed for a box with edges 3 ft.

Solution   A box (cube) has 6 faces of equal area.

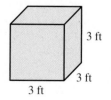

Number of faces of cube
↓
$6 \times \underbrace{9 \text{ ft}^2}_{\text{Area of each face}} = 54 \text{ ft}^2$

You need enough paint to cover 54 ft².

Some boxes have tops and others are open the top. Note that the box in Example 1 has a top. In the next example, we consider a box without a top and a can with a bottom but not a top.

Example 2 Find the outside surface area

Find the outside surface area.

a.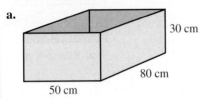

b.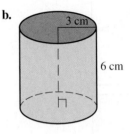

Solution

a. We find the sum of the areas of all the faces:

Front:	$30 \times 50 = 1,500$	
Back:	$1,500$	*Same as front*
Side:	$80 \times 30 = 2,400$	
Side:	$2,400$	*Sides are the same size.*
Bottom:	$80 \times 50 = 4,000$	
Total:	$11,800 \text{ cm}^2$	

b. To find the surface area, find the area of a circle (the bottom) and think of the sides of the can as being "rolled out." The length of the resulting rectangle is the circumference of the can and the width of the rectangle is the height of the can.

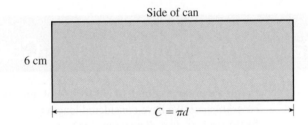
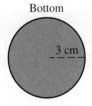

Side:	$A = \ell w = (\pi d)w = \pi(6)(6) = 36\pi$
Bottom:	$A = \pi r^2 = \pi(3)^2 = 9\pi$
Surface area:	$36\pi + 9\pi = 45\pi \approx 141.37167$

The surface area is about 141 cm².

Example 3 Amount of paint needed

Pólya's Method

You want to paint 200 ft of a three-rail fence, and you need to know how much paint to purchase.

Solution   We use Pólya's problem-solving guidelines for this example.

Understand the Problem. There is some additional information you need to gather before you can answer this question. You want to paint 200 ft of a three-rail fence that is made up of three boards, each 6 inches wide.

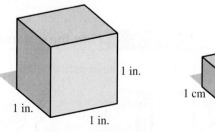

You want to know:
a. the number of square feet on one side of the fence.
b. the number of square feet to be painted if the posts and edges of the boards comprise 100 ft².
c. the number of gallons of paint to purchase if each gallon covers 325 ft².

Devise a Plan. The paint coverage is found on the paint can (325 ft² for this problem). The next thing we need to know is the number of square feet to be painted. We estimate the number of square feet to be painted on the posts and the edges to be 100 ft². Now, calculate the number of square feet to be painted and divide by 325 (the number of square feet per gallon).

Carry Out the Plan.
a. We first calculate A, the number of square feet to be painted on one side of the fence.

$$A = (200 \text{ ft}) \times (6 \text{ in.}) \times 3$$

$$= (200 \text{ ft}) \times \left(\frac{1}{2}\text{ft}\right) \times 3$$

$$= 300 \text{ ft}^2$$

b.   AMOUNT TO BE PAINTED $= 2(\text{AMOUNT ON ONE SIDE}) + \text{EDGES AND POSTS}$
$$= 2(300 \text{ ft}^2) + 100 \text{ ft}^2$$
$$= 700 \text{ ft}^2$$

c. $\begin{pmatrix} \text{NUMBER OF SQUARE} \\ \text{FEET PAINTED} \end{pmatrix} = \begin{pmatrix} \text{NUMBER OF SQUARE} \\ \text{FEET PER GALLON} \end{pmatrix} \begin{pmatrix} \text{NUMBER OF} \\ \text{GALLONS} \end{pmatrix}$

$$700 = 325 \begin{pmatrix} \text{NUMBER OF} \\ \text{GALLONS} \end{pmatrix}$$

$$2.15 \approx \begin{pmatrix} \text{NUMBER OF} \\ \text{GALLONS} \end{pmatrix} \quad \text{Divide both sides by 325.}$$

Look Back. If paint must be purchased by the gallon (as implied by the question), the amount to purchase is 3 gallons.

Volume

To measure area, we covered a region with square units and then found the area by using a mathematical formula. A similar procedure is used to find the amount of space inside a solid object, which is called its **volume.** We can imagine filling the space with **cubes.** A **cubic inch** and a **cubic centimeter** are shown in Figure 9.12.

If the solid is not a cube but is a box (called a **rectangular parallelepiped**) with edges of different lengths, the volume can be found similarly.

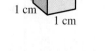

a. 1 cubic inch
(1 cu in. or 1 in.³)

b. 1 cubic centimeter
(1 cu cm, cc, or 1 cm³)

FIGURE 9.12 Common units of measuring volume

Volume of a Cube

The volume, V, of a cube with edge s is $V = s^3$.

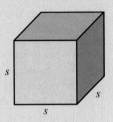

Example **4** **Find the volume of a box**

Find the volume of a box that measures 4 ft by 6 ft by 4 ft.

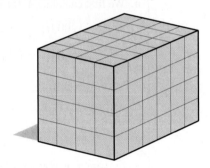

Solution   There are 24 cubic feet on the bottom layer of cubes. Do you see how many layers of cubes will fill the solid? Since there are four layers with 24 cubes in each, the total is

$$4 \times 24 = 96$$

The volume is 96 ft^3.

This example leads us to the following formula for the volume of a box.

Volume of a Box

The volume V of a box (parallelepiped) with edges ℓ, w, and h is

$$V = \ell w h$$

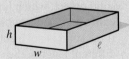

Example **5** **Find the volume of a solid**

Find the volume of each solid.

a.

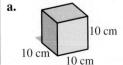

10 cm
10 cm
10 cm

b.

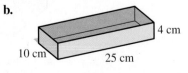

4 cm
10 cm
25 cm

c.

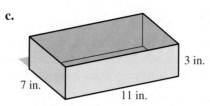

7 in.

11 in.

3 in.

Solution

a. $V = s^3$

$= (10 \text{ cm})^3$

$= (10 \times 10 \times 10) \text{ cm}^3$

$= 1,000 \text{ cm}^3$

b. $V = \ell wh$

$= (25 \text{ cm})(10 \text{ cm})(4 \text{ cm})$

$= (25 \times 10 \times 4) \text{ cm}^3$

$= 1,000 \text{ cm}^3$

c. $V = \ell wh$

$= (11 \text{ in.})(7 \text{ in.})(3 \text{ in.})$

$= (11 \times 7 \times 3) \text{ in.}^3$

Sometimes the dimensions for the volume we are finding are not all given in the same units. In such cases, you must convert all units to a common unit. The common conversions are as follows:

1 ft = 12 in.	To convert feet to inches, multiply by 12.
	To convert inches to feet, divide by 12.
1 yd = 3 ft	To convert yards to feet, multiply by 3.
	To convert feet to yards, divide by 3.
1 yd = 36 in.	To convert yards to inches, multiply by 36.
	To convert inches to yards, divide by 36.

| Example   **6**   **Find the amount of concrete to order** |

Suppose you are pouring a rectangular driveway with dimensions 24 ft by 65 ft. The depth of the driveway is 3 in. and concrete is ordered by the yard. By a "yard" of concrete, we mean a cubic yard. You cannot order part of a yard of concrete. How much concrete should you order?

Solution   There are three different measurements in this problem: inches, feet, and yards. Since we want the answer in cubic yards, we will convert all of these measurements to yards:

$65 \text{ ft} = (65 \div 3) \text{ yd} = \frac{65}{3} \text{ yd}$ *This is the length, ℓ.*

$24 \text{ ft} = (24 \div 3) \text{ yd} = 8 \text{ yd}$ *This is the width, w.*

$3 \text{ in.} = (3 \div 36) \text{ yd} = \frac{3}{36} \text{ yd} = \frac{1}{12} \text{ yd}$ *This is the height (depth), h.*

$V = \ell wh$

$= \dfrac{65 \times 8 \times 1}{3 \times 12}$

$= \dfrac{130}{9}$

$= 14\dfrac{4}{9}$

You must order 15 cubic yards of concrete.

Do you know where the expression "the whole 9 yards" comes from? It's related to concrete. A standard-size "cement mixer" has a capacity of 9 cubic yards of concrete. Thus, a job requiring the mixer's full capacity demands "the whole 9 yards."

Capacity

One of the most common applications of volume involves measuring the amount of liquid a container holds, which we refer to as **capacity.** For example, if a container is 2 ft by 2 ft by 12 ft, it is fairly easy to calculate the volume:

$$2 \times 2 \times 12 = 48 \text{ ft}^3$$

But this still doesn't tell us how much water the container holds. The capacities of a can of cola, a bottle of milk, an aquarium tank, the gas tank in your car, and a swimming pool can all be measured by the amount of fluid they can hold.

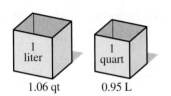

Standard Capacity Units

U.S. System	Metric System
gallon (gal.)	**liter (L)**
ounce $\left(\text{oz}, \frac{1}{128} \text{ gal.} \right)$	kiloliter (kl, 1,000 L)
cup (c, 8 oz)	milliliter $\left(\text{ml}, \frac{1}{1,000} \text{L} \right)$
quart $\left(\text{qt}, \frac{1}{4} \text{ gal}, 32 \text{ oz} \right)$	

1.06 qt   0.95 L

FIGURE 9.13 Standard capacities

Most containers of liquid that you buy have capacities stated in both milliliters and ounces, or quarts and liters (see Figure 9.13).

Some capacity statements from purchased products are listed in Table 9.1. The U.S. Bureau of Alcohol, Tobacco, and Firearms has made metric bottle sizes for liquor mandatory, so the half-pint, fifth, and quart have been replaced by 200-mL, 750-mL, and 1-L sizes. A typical dose of cough medicine is 5 mL, and 1 kL is 1,000 L, or about the amount of water one person would use for all purposes in two or three days.

⚠ CAUTION

You should remember some of these references for purposes of estimation. For example, remember that a can of Coke is 355 ml, and the size of a liter of milk is about the same as a quart of milk. A cup of coffee is about 300 ml and a spoonful of medicine is about 5 ml.

TABLE 9.1

Capacities of Common Grocery Items, as Shown on Labels

Item	U. S. Capacity	Metric Capacity
Milk	$\frac{1}{2}$ gal	1.89L
Milk	1.06 qt	1 L
Budweiser	12 oz	355 mL
Coke	67.6 oz	2 L
Hawaiian Punch	1 qt	0.95 L
Del Monte pickles	1 pt 6 oz	651 mL

Since it is common practice to label capacities in both U.S. and metric measuring units, it will generally not be necessary for you to make conversions from one system to another. But if you do, it is easy to remember that a liter is just a little larger than a quart, just as a meter is a little larger than a yard. To measure capacity, you use a measuring cup.

Example 7 Measure the quantity of a liquid

Measure the amount of liquid in the measuring cup in Figure 9.14, both in the U.S. system and in the metric system.

Solution

Metric: 240 mL
U.S.:   About 1 c or 8 oz

5 marks = 100 ml, so each mark is 20 ml.

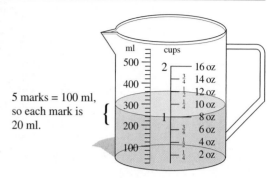

FIGURE 9.14 Standard measuring cup with both metric and U.S. measurements

Some common relationships among volume and capacity measurements in the U.S. system are shown in Figure 9.15.

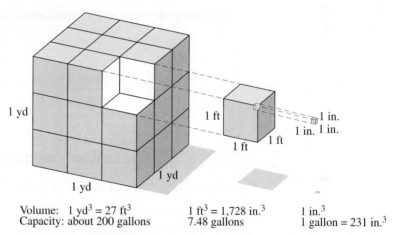

Volume: 1 yd³ = 27 ft³   1 ft³ = 1,728 in.³   1 in.³
Capacity: about 200 gallons   7.48 gallons   1 gallon = 231 in.³

FIGURE 9.15 U.S. measurement relationship between volume and capacity

In the U.S. system of measurement, the relationship between volume and capacity is not particularly convenient. One gallon of capacity occupies 231 in.³. This means that, since the box in part **c** of Example 5 has a volume of 231 in.³, we know that it will hold exactly 1 gallon of water.

To find the capacity of the 2-ft by 2-ft by 12-ft box mentioned earlier, we must change 48 ft³ to cubic inches:

$$48 \text{ ft}^3 = 48 \times (1 \text{ ft}) \times (1 \text{ ft}) \times (1 \text{ ft})$$
$$= 48 \times 12 \text{ in.} \times 12 \text{ in.} \times 12 \text{ in.}$$
$$= 82{,}944 \text{ in.}^3 \qquad \text{A calculator would help here.}$$

Since 1 gallon is 231 in.³, the final step is to divide 82,944 by 231 to obtain approximately 359 gallons.

The relationship between volume and capacity in the metric system is easier to remember. One cubic centimeter is one-thousandth of a liter. Notice that this is the same as a milliliter. For this reason, you will sometimes see cc used to mean cm³ or mL. These relationships are shown in Figure 9.16.

ESTIMATE: Since 1 ft³ ≈ 7.5 gal, 48 ft³ ≈ 50 ft³ ≈ (50 × 7.5) gal ≈ 375 gal

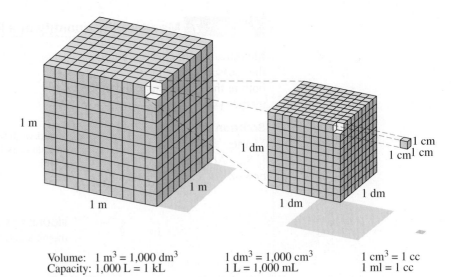

Volume: $1 \text{ m}^3 = 1{,}000 \text{ dm}^3$ $1 \text{ dm}^3 = 1{,}000 \text{ cm}^3$ $1 \text{ cm}^3 = 1 \text{ cc}$
Capacity: $1{,}000 \text{ L} = 1 \text{ kL}$ $1 \text{ L} = 1{,}000 \text{ mL}$ $1 \text{ ml} = 1 \text{ cc}$

FIGURE 9.16 Metric measurement relationship between volume and capacity

---

Relationship Between Volume and Capacity

$1 \text{ liter} = 1{,}000 \text{ cm}^3$

$1 \text{ gallon} = 231 \text{ in}^3$

$1 \text{ ft.} \approx 7.48 \text{ gal}$

Example   **8**   **Capacity of a container**

How much water would each of the following containers hold?

a.
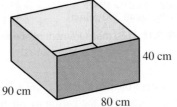
40 cm
90 cm
80 cm

b.
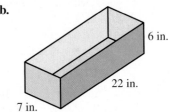
6 in.
22 in.
7 in.

Solution

a. $V = 90 \text{ cm} \times 80 \text{ cm} \times 40 \text{ cm}$

$= 288{,}000 \text{ cm}^3$

Since each $1{,}000 \text{ cm}^3$ is 1 liter,

$$\frac{288{,}000}{1{,}000} = 288$$

This container would hold 288 liters.

b. $V = 7 \text{ in.} \times 22 \text{ in.} \times 6 \text{ in.}$

$= 924 \text{ in.}^3$

Since each 231 in.^3 is 1 gallon,

$$\frac{924}{231} = 4$$

This container would hold 4 gallons.

Note that there is no need to estimate part **a** since the arithmetic is fairly easy. However, for part **b,** the estimate might be $\frac{1}{2} \text{ ft} \times 2 \text{ ft} \times \frac{1}{2} \text{ ft} = \frac{1}{2} \text{ ft}^3 \approx \left(\frac{1}{2} \times 7.5 \text{ gal}\right) = 3.75 \text{ gal}$.

| Example | 9 | **Swimming pool capacity** |

An ecology swimming pool is advertised as being 20 ft × 25 ft × 5 ft. How many gallons will it hold?

Solution $V = 20 \text{ ft} \times 25 \text{ ft} \times 5 \text{ ft} = 2,500 \text{ ft}^3$

Since $1 \text{ ft}^3 \approx 7.48 \text{ gal}$, the swimming pool contains

$$2,500 \times 7.48 = 18,700 \text{ gallons}$$

Problem Set 9.3

Level 1

1. **IN YOUR OWN WORDS**   Contrast length, area, and volume.
2. **IN YOUR OWN WORDS**   What do we mean by surface area?
3. **IN YOUR OWN WORDS**   Contrast volume and capacity.
4. Compare the sizes of a cubic inch and a cubic centimeter.
5. Compare the sizes of a quart and a liter.
6. Compare a meter and a yard.

In Problems 7–8, find the volume of each solid by counting the number of cubic centimeters in each box.

7. 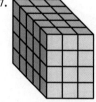   **8.**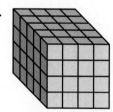

Find the volume of each solid in Problems 9–16.

9.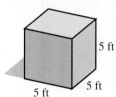
5 ft
5 ft
5 ft

10.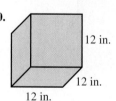
12 in.
12 in.
12 in.

11.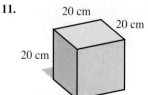
20 cm
20 cm
20 cm

12.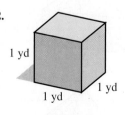
1 yd
1 yd
1 yd

13.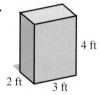
4 ft
2 ft
3 ft

14.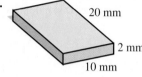
20 mm
2 mm
10 mm

15.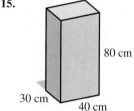
80 cm
30 cm
40 cm

16.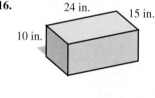
24 in.
15 in.
10 in.

Measure each amount given in Problems 17–21.

Container A   Container B

17. a. Container A in cups
 b. Container A in ounces

18. a. Container B in ounces
   b. Container B in milliliters

Container C   Container D

19. a. Container C in ounces
   b. Container C in milliliters

20. a. Container D in cups
   b. Container D in ounces
   c. Container D in milliliters

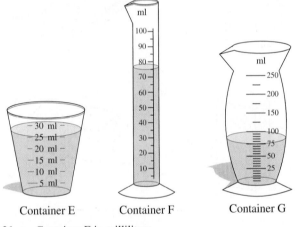

Container E   Container F   Container G

21. a. Container E in milliliters
   b. Container F in milliliters
   c. Container G in milliliters

Level 2

The ability to estimate capacities is an important skill to develop. Without measuring, pick the best answer in Problems 22–32.

22. An average cup of coffee is about
   A. 250 mL   B. 750 mL   C. 1 L

23. If you want to paint a small bookshelf, how much paint would you probably need?
   A. 5 mL   B. 500 mL   C. 5 L

24. A six-pack of beer would contain about
   A. 2 mL   B. 200 mL   C. 2 L

25. The dose of a strong cough medicine might be
   A. 5 mL   B. 500 mL   C. 5 L

26. A glass of water served at a restaurant is about
   A. 200 mL   B. 2 mL   C. 2 L

27. Enough gas to fill your car's empty tank would be about
   A. 15 L   B. 200 mL   C. 70 L

28. You order some champagne for yourself and one companion. You would most likely order
   A. 2 mL   B. 700 mL   C. 20 L

29. 50 kL of water would be about enough for
   A. taking a bath   B. taking a swim
   C. supplying the drinking water for a large city

30. The prefix *centi-* means
   A. one thousand   B. one-thousandth   C. one-hundredth

31. The prefix *milli-* means
   A. one thousand   B. one-thousandth   C. one-hundredth

32. The prefix *kilo-* means
   A. one thousand   B. one-thousandth   C. one-hundredth

33. How much water will a 7-m by 8-m by 2-m swimming pool contain (in kiloliters)?

34. How much water will a 21-ft by 24-ft by 4-ft swimming pool contain (rounded to the nearest gallon)?

Find the outside surface area in Problems 35–46. Note that some boxes have tops and others do not.

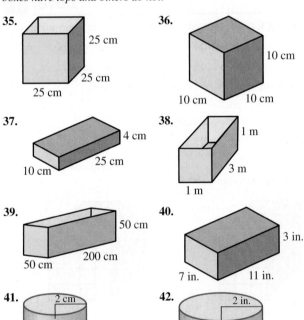

35. 25 cm / 25 cm / 25 cm

36. 10 cm / 10 cm / 10 cm

37. 4 cm / 25 cm / 10 cm

38. 1 m / 3 m / 1 m

39. 50 cm / 200 cm / 50 cm

40. 3 in. / 7 in. / 11 in.

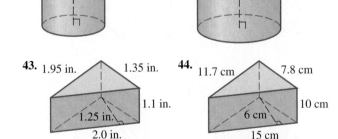

41. 2 cm / 8 cm

42. 2 in. / 4 in.

43. 1.95 in. / 1.35 in. / 1.1 in. / 1.25 in. / 2.0 in.

44. 11.7 cm / 7.8 cm / 10 cm / 6 cm / 15 cm

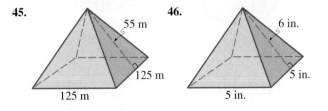

45. 55 m / 125 m / 125 m

46. 6 in. / 5 in. / 5 in.

What is the capacity for each of the containers in Problems 47–54?
(Give answers in the U.S. system to the nearest tenth of a gallon or
in metric to the nearest tenth of a liter.)

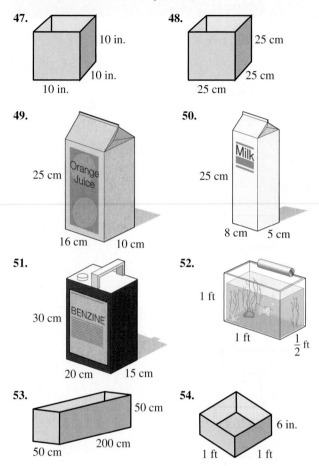

47.
10 in.
10 in.
10 in.

48.
25 cm
25 cm
25 cm

49.
Orange Juice
25 cm
16 cm   10 cm

50.
Milk
25 cm
8 cm   5 cm

51.
BENZINE
30 cm
20 cm   15 cm

52.
1 ft
1 ft $\frac{1}{2}$ ft

53.
50 cm
200 cm
50 cm

54.
6 in.
1 ft   1 ft

Level 3

55. The exterior dimensions of a refrigerator/freezer are shown in Figure 9.17.

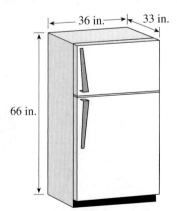

36 in.   33 in.

66 in.

FIGURE 9.17 Refrigerator/freezer

a. How many cubic feet are contained within the refrigerator/freezer?
b. If it is advertised as a 19-cu-ft refrigerator, how much space is taken up by the motor, insulation, and so on?

56. The exterior dimensions of a freezer are 48 inches by 36 inches by 24 inches, and it is advertised as being 27.0 cu ft. Is the advertised volume correctly stated?

57. Suppose that you must order concrete for a sidewalk 50 ft by 4 ft to a depth of 4 in. How much concrete is required? (Answer to the nearest $\frac{1}{2}$ cubic yard.)

58. Use the plot plan shown in Figure 9.18 and give your answers to the nearest *cubic yard*.

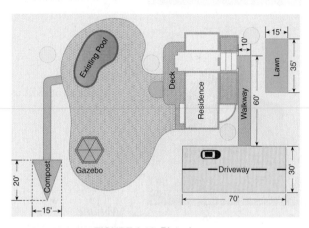

FIGURE 9.18 Plot plan

a. How many cubic yards of sawdust are needed for preparation of the lawn area if it is to be spread to a depth of six inches?*
b. How many cubic yards of gravel are necessary if it is to be placed to a depth of three inches?
c. How much compost is available if the pit is six inches deep?
d. Suppose that you wish to pave the driveway. How much concrete is needed if it is to be poured to a depth of four inches?

Problem Solving 3

59. The total human population of the earth is about 6.2×10^9.
a. If each person has the room of a prison cell (50 sq ft), and if there are about 2.8×10^7 sq ft in a square mile, how many people could fit into a square mile?
b. How many square miles would be required to accommodate the entire human population of the earth?
c. If the total land area of the earth is about 5.2×10^7 sq mi, and if all the land area were divided equally, how many acres of land would each person be allocated ($1\,\text{sq mi} = 640$ acres)?

60. a. Guess what percentage of the world's population could be packed into a cubical box measuring $\frac{1}{2}$ mi on each side. [*Hint:* The volume of a typical person is about 2 cu ft.]
b. Now calculate the answer to part **a**, using the earth's population as given in Problem 59.

*In practice, you would not round to the nearest cubic yard, but rather would round up to ensure that you had enough material. However, for consistency in this book, we will round according to the rules developed in the first chapter.

9.4 | Miscellaneous Measurements

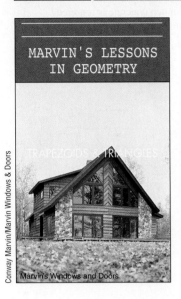

In this chapter we have been discussing measurement.

Length
If you are measuring length, you will use linear measures:

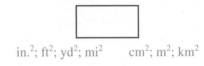

in.; ft; yd; mi   cm; m; km

Area
For area, you will use square measures:

in.²; ft²; yd²; mi²   cm²; m²; km²

Volume
For volumes, you will use cubic measures:

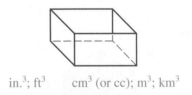

in.³; ft³   cm³ (or cc); m³; km³

Volume of Solids

In the previous section we found the volume and the capacity of boxes; now we can extend this to other solids. In Figure 9.19, we show some of the more common solids, along with the volume formulas. We will use B in each case to signify the area of the base and h for the height.

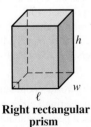

Right rectangular prism

$V = Bh$

$S = 2\ell w + 2wh + 2\ell h$

Right circular cylinder

$V = Bh$

$S = 2\pi r^2 + 2\pi rh$

Pyramid

$V = \frac{1}{3}Bh$

$S = s^2 + s\sqrt{s^2 + 4h^2}$

Right circular cone

$V = \frac{1}{3}Bh$

$S = \pi r\sqrt{r^2 + h^2} + \pi r^2$

Sphere

$V = \frac{4}{3}\pi r^3$

$S = 4\pi r^2$

FIGURE 9.19 Common solids, with accompanying volume and surface area formulas

Example **1** **Find the volume of a cylinder**

Find the volume of the solid shown in Figure 9.20 to the nearest cubic unit.

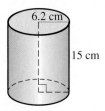

FIGURE 9.20 Volume of cylinder

Solution   This is a right circular cylinder. Hence,

$V = Bh$ For this example, $B = \pi r^2$, where r is the radius of the circular base.

$\quad = \pi(6.2)^2(15)$ Note that $r = 6.2$, $h = 15$.

$\quad = 1{,}811.4423$ Use a calculator.

The volume is $1{,}811$ cm^3.

Example **2** **Find the volume of a prism**

Find the volume of the solid shown in Figure 9.21 to the nearest cubic unit.

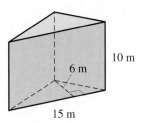

FIGURE 9.21 Volume of a triangular prism

Solution   Figure 9.19 shows a right rectangular prism, and Figure 9.21 shows a right triangular prism. The procedure shown here works for all prisms.

$V = Bh$ For this example, $B = \dfrac{1}{2}ba$.

$\quad = \dfrac{1}{2}(15)(6)(10)$ Note $a = 6$, $b = 15$, and $h = 10$.

$\quad = 450$

The volume is 450 m^3.

Example **3** **Find the volume of a pyramid**

Find the volume of the solid shown in Figure 9.22 to the nearest cubic unit.

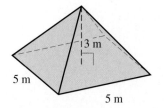

FIGURE 9.22 Volume of a pyramid

Solution   This is a pyramid. Hence,

$$V = \frac{1}{3}Bh \qquad \text{For this example, } B = s^2.$$

$$= \frac{1}{3}(5)^2(3) \qquad \text{Note } s = 5, \text{ and } h = 3.$$

$$= 25$$

The volume is 25 m^3.

Example **4** **Find the surface area and volume of a sphere**

Find the surface area and volume of the sphere shown in Figure 9.23. Give your answers to the nearest whole unit, paying attention to the unit.

FIGURE 9.23 Surface area and volume of a sphere

Solution   For the suface area,

$$S = 4\pi r^2 \qquad \text{Note that } r = 8.2.$$

$$= 4\pi(8.2)^2$$

$$\approx 844.96276 \qquad \text{By calculator}$$

For the volume,

$$V = \frac{4}{3}\pi r^3$$

$$= \frac{4}{3}\pi(8.2)^3$$

$$\approx 2,309.564878$$

The surface area is 845 cm^2 and the volume is 2,310 cm^3.

Example **5** **Volume of a cone**

Find the surface area (to the nearest square unit), and also find the volume (to the nearest cubic unit) of the solid shown in Figure 9.24.

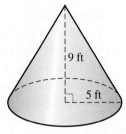

FIGURE 9.24 Surface area and volume of a right circular cone

Solution   We calculate the requested values.

$$S = \pi r \sqrt{r^2 + h^2} + \pi r^2$$
$$= \pi(5)\sqrt{5^2 + 9^2} + 5^2\pi \qquad \text{Notice that } r = 5 \text{ and } h = 9.$$
$$= 5\pi\sqrt{106} + 25\pi$$
$$\approx 240.26$$

$$V = \frac{1}{3} Bh \qquad \text{For this example, } B = \pi r^2, \text{ where } r = 5.$$

$$= \frac{1}{3}\pi(5)^2 9$$

$$= 75\pi$$

$$\approx 235.62$$

The surface area is 240 ft^2 and the volume is 236 ft^3.

Comparisons

The following example compares the areas and volumes of similarly shaped figures.

Example **6** **Compare areas**

a. Compare the area of a square whose side is tripled with the area of the original square.
b. Compare the area of a circle after its radius is doubled.
c. Compare the volume of a sphere after its radius is doubled.
d. Compare the volume of a cube when the length of its side is multiplied by 5.

Solution
a. Let s be the length of the side of the square; then the area is $A = s^2$. If the side is tripled to $3s$, then the area is $A = (3s)^2 = 9s^2$. We see that the new area is *nine times as large* as the original area.
b. Let r be the radius of the original circle; then the area is $A = \pi r^2$. If the radius is doubled to $2r$, then the area is $A = \pi(2r)^2 = 4\pi r^2$. We see that the new area is *increased fourfold.*
c. Let r be the radius of the original sphere; then the area is $A = \frac{4}{3}\pi r^3$. If the radius is doubled to $2r$, then the area is $A = \frac{4}{3}\pi(2r)^3 = \frac{4}{3}\pi(8r^3) = \frac{32}{3}\pi r^3$. We see that the volume is *eight times as large.*
d. Use patterns based on parts **a–c** to *guess* the result without calculation. Since the formula cubes the variable that is multiplied by 5, we guess the volume is $5^3 = 125$ times as large as the original volume.

The measurements of length, area, volume, and capacity were discussed in the previous section. Two additional measurements we need to consider are mass and temperature.

Mass

The **mass** of an item is the amount of matter it comprises. The **weight** of an item is the heaviness of the matter.* The U.S. and metric units of measurement for mass or weight are given in the following box. Notice that, in the U.S. measurement system, an ounce is used as a weight measurement; this is not the same use of an ounce as a capacity measurement that we used earlier. The basic unit of measurement for mass in the metric system is the **gram,** which is defined as the mass of 1 cm^3 of water, as shown in Figure 9.25.

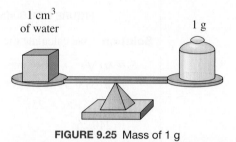

1 cm^3
of water

1 g

FIGURE 9.25 Mass of 1 g

Standard Weight Units

U.S. System	Metric System
ounce (oz)	**gram (g)**
pound (lb, 16 oz)	kilogram (kg, 1,000 g)
ton (2,000 lb, 32,000 oz)	milligram $\left(\text{mg.} \dfrac{1}{1,000}\,\text{g} \right)$

Standard units of weight are shown in Figure 9.26.

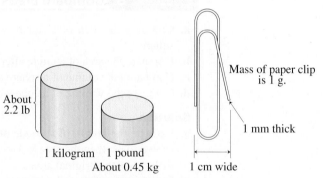

About
2.2 lb

Mass of paper clip
is 1 g.

1 mm thick

1 kilogram   1 pound
About 0.45 kg   1 cm wide

FIGURE 9.26 Standard weight units

A paper clip weighs about 1 g, a cube of sugar about 3 g, and a nickel about 5 g. This book weighs about 1 kg, and an average-size person weighs from 50 to 100 kg. The weights of some common grocery items are shown in Table 9.2.

*Technically, weight is the force of gravity acting on a body. The mass of an object is the same on the moon as on Earth, whereas the weight of that same object would be different on the moon and on Earth, since they have different forces of gravity. For our purposes, you can use either word, *mass* or *weight*, because we are weighing things only on Earth.

TABLE 9.2		
Weights of Common Grocery Items, as Shown on Labels		
Item	**U.S. Weight**	**Metric Weight**
Kraft cheese spread	5 oz	142 g
Del Monte tomato sauce	8 oz	227 g
Campbell's cream of chicken soup	$10\frac{3}{4}$ oz	305 g
Kraft marshmallow cream	11 oz	312 g
Bag of sugar	**5 lb**	2.3 kg
Bag of sugar	22 lb	**10 kg**

We use a scale to measure weight. To weigh items, or ourselves, in metric units, we need only replace our U.S. weight scales with metric weight scales. As with other measures, we need to begin to think in terms of metric units, and to estimate the weight of various items. The multiple-choice questions in the problem set are designed to help you do this.

Temperature

The final quantity of measure that we'll consider in this chapter is **temperature,** which is the degree of hotness or coldness.

Standard Temperature	
U.S. System	*Metric System*
Fahrenheit (°F)	**Celsius (°C)**

To work with temperatures, it is necessary to have some reference points.

Temperature Properties		
	U.S. System	*Metric System*
Water freezes	32°F	0°C
Water boils	212°F	100°C

Cengage Learning

We are usually interested in measuring temperature in three areas: atmospheric temperature (usually given in weather reports), body temperature (used to determine illness), and oven temperature (used in cooking). The same scales are used, of course, for measuring all of these temperatures. But notice the difference in the ranges of temperatures we're considering. The comparisons for Fahrenheit and Celsius (formerly called *centigrade*) are shown in Figure 9.27.

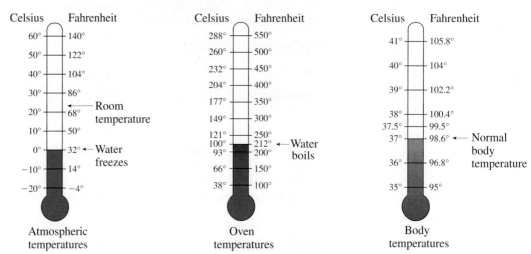

FIGURE 9.27 Temperature comparisons between Celsius and Fahrenheit

Converting Units

As we have mentioned, one of the advantages (if not the chief advantage) of the metric system is the ease with which you can remember and convert units of measurement. The three basic metric units are related as shown in Figure 9.28.

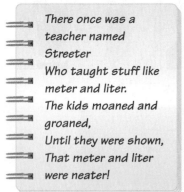

There once was a teacher named Streeter
Who taught stuff like meter and liter.
The kids moaned and groaned,
Until they were shown,
That meter and liter were neater!

Go Metric—
    Be a liter bug!

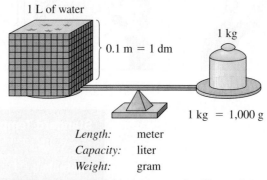

FIGURE 9.28 Relationship among meter, liter, and gram

These units are combined with certain prefixes:

$$milli\text{- means } \frac{1}{1,000} \qquad centi\text{- means } \frac{1}{100} \qquad kilo\text{- means } 1,000$$

Other metric units are used less frequently, but they should be mentioned:

$$deci\text{- means } \frac{1}{10} \qquad deka\text{- means } 10 \qquad hecto\text{- means } 100$$

A listing of metric units (in order of size) is given in Table 9.3.

TABLE 9.3

			Metric Measurements	
Length	**Capacity**	**Weight**	**Meaning**	**Memory Aid**
kilometer (km)	**kilo**liter (kL)	**kilo**gram (kg)	1,000 units	**K**arl
hectometer (hm)	**hecto**liter (hL)	**hecto**gram (hg)	100 units	**H**as
dekameter (dkm)	**deka**liter (dkL)	**deka**gram (dkg)	10 units	**D**eveloped
meter (m)	**liter** (L)	**gram** (g)	1 unit	**M**y
decimeter (dm)	**deci**liter (dL)	**deci**gram (dg)	0.1 unit	**D**ecimal
centimeter (cm)	**centi**liter (cL)	**centi**gram (cg)	0.01 unit	**C**raving for
millimeter (mm)	**milli**liter (mL)	**milli**gram (mg)	0.001 unit	**M**etrics

Basic Unit: (indicates the meter/liter/gram row)

To convert from one metric unit to another, you **simply move the decimal point.** For example, if the height of a book is 0.235 m, then it is also

Larger prefixes
$$\begin{cases} 0.000235 \text{ km} & \leftarrow \text{ Three decimal places} \\ 0.00235 \text{ hm} & \leftarrow \text{ Two decimal places} \\ 0.0235 \text{ dkm} & \leftarrow \text{ One decimal place} \end{cases}$$

Smaller prefixes
$$\begin{cases} 2.35 \text{ dm} & \leftarrow \text{ One decimal places} \\ 23.5 \text{ cm} & \leftarrow \text{ Two decimal places} \\ 235 \text{ mm} & \leftarrow \text{ Three decimal place} \end{cases}$$

Example 7   Converting metric units

Write each metric measurement using each of the other prefixes.
a. 43 km   **b.** 60 L   **c.** 14.1 cg

Solution

a. All prefixes are smaller, so the numbers of units become larger:

43 km	**Given**
430 hm	One place
4,300 dkm	Two places
43,000 m	Three places
430,000 dm	Four places
4,300,000 cm	Five places
43,000,000 mm	Six places

b. Larger prefixes, smaller numbers:

60 L	**Given**
6 dkL	One place
0.6 hL	Two places
0.06 kL	Three places

Smaller prefixes, larger numbers:

60 L	**Given**
600 dL	One place
6,000 cL	Two places
60,000 mL	Three places

   c. Larger prefixes:

14.1 cg	**Given**
1.41 dg	*One place*
0.141 g	*Two places*
0.0141 dkg	*Three places*
0.00141 hg	*Four places*
0.000141 kg	*Five places*

   Smaller prefix:

14.1 cg	**Given**
141 mg	*One place*

To make a single conversion, use the pattern illustrated by Example 7 and simply count the number of decimal places.

Example **8** **Find metric conversions**

Make the indicated conversions.
a. 287 cm to km   **b.** 1.5 kL to L   **c.** 4.8 kg to g

Solution

a. From cm to km is five places; a larger prefix implies a smaller number, so move the decimal point to the left:

   287 cm = 0.00287 km

b. From kL to L is three places; a smaller prefix implies a larger number, so move the decimal point to the right:

   1.5 kL = 1,500 L

c. From kg to g is three places; a smaller prefix implies a larger number, so move the decimal point to the right:

   4.8 kg = 4,800 g

Problem Set 9.4

Level 1

1. **IN YOUR OWN WORDS**   Discuss the merits of the metric system as opposed to the U.S. measurement system. Why do you think the metric system has not yet been adopted in the United States?

2. **IN YOUR OWN WORDS**   Discuss the relationships among measuring length, capacity, and weight.

3. **IN YOUR OWN WORDS**   Explain how we change units within the metric system.

Name the metric unit you would use to measure each of the quantities in Problems 4–9.

4. **a.** The distance from New York to Chicago
   b. The distance around your waist

5. **a.** Your height
   b. The height of a building

6. **a.** The capacity of a wine bottle
   b. The amount of gin in a martini

7. **a.** The capacity of a car's gas tank
   b. The amount of water in a swimming pool

8. **a.** The weight of a pencil
   b. The weight of an automobile

9. **a.** The outside temperature
   b. The temperature needed to bake a cake

Without measuring, pick the best choice in Problems 10–18 by estimating.

10. A hamburger patty would weigh about
    A. 170 g   B. 240 mg   C. 2 kg

11. A can of carrots at the grocery store most likely weighs about
   A. 40 kg   B. 4 kg   C. 0.4 kg

12. A newborn baby would weigh about
   A. 490 mg   B. 4 kg   C. 140 kg

13. You have invited 15 people for Thanksgiving dinner. You should buy a turkey that weighs about
   A. 795 mg   B. 4 kg   C. 12 kg

14. Water boils at
   A. 0°C   B. 100°C   C. 212°C

15. If it is 32°C outside, you would most likely find people
   A. ice skating   B. water skiing

16. If the doctor says that your child's temperature is 37°C, your child's temperature is
   A. low   B. normal   C. high

17. You would most likely broil steaks at
   A. 120°C   B. 500°C   C. 290°C

18. John tells you he weighs 150 kg. If John is an adult, he is
   A. underweight   B. about average   C. overweight

Without measuring, pick the best choice in Problems 19–29.

19. A kilogram is ___?___ a pound.
   A. more than   B. about the same as   C. less than

20. The prefix used to mean 1,000 is
   A. centi-   B. milli-   C. kilo-

21. The prefix used to mean $\frac{1}{1,000}$ is
   A. centi-   B. milli-   C. kilo-

22. The prefix used to mean $\frac{1}{100}$ is
   A. centi-   B. milli-   C. kilo-

23. 15 kg is a measure of
   A. length   B. capacity
   C. weight   D. temperature

24. 28.5 m is a measure of
   A. length   B. capacity
   C. weight   D. temperature

25. 6 L is a measure of
   A. length   B. capacity
   C. weight   D. temperature

26. 38°C is a measure of
   A. length   B. capacity
   C. weight   D. temperature

27. 7 mL is a measure of
   A. length   B. capacity
   C. weight   D. temperature

28. 68 km is a measure of
   A. length   B. capacity
   C. weight   D. temperature

29. 14.3 cm is a measure of
   A. length   B. capacity
   C. weight   D. temperature

Write each measurement given in Problems 30–37 using all of the metric prefixes.

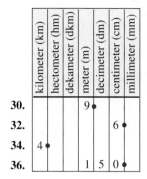

	kilometer (km)	hectometer (hm)	dekameter (dkm)	meter (m)	decimeter (dm)	centimeter (cm)	millimeter (mm)
30.				9•			
32.						6•	
34.	4•						
36.					1	5	0•

	kiloliter (kl)	hectoliter (hl)	dekaliter (dkl)	liter (L)	deciliter (dl)	centiliter (cl)	milliliter (ml)
31.						6	3•
33.			3	•5			
35.				8•			
37.		3	•1				

Level 2

In Problems 38–42 make the indicated conversions.

38. a. 1 cm = ___ mm   **b.** 1 cm = ___ m
   c. 1 cm = _____ km   **d.** 1 mm = ___ cm

39. a. 1 mm = _____ m   **b.** 1 mm = _____ km
   c. 1 m = _____ mm   **d.** 1 m = ___ cm

40. a. 1 m = _____ km   **b.** 1 mL = ____ dL
   c. 1 mL = _____ L   **d.** 1 mL = _____ kL

41. a. 1 L = _____ mL   **b.** 1 L = __ dL
   c. 1 L = _____ kL   **d.** 1 kL = _____ mL

42. a. 1 kL = _____ dL   **b.** 1 kL = ___ L
   c. 1 g = _____ mg   **d.** 1 g = ___ cg

Find the surface areas of the solids in Problems 43–46 correct to the nearest square unit. Note that some of the containers do not have tops.

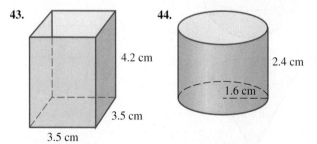

43.

4.2 cm

3.5 cm

3.5 cm

44.

2.4 cm

1.6 cm

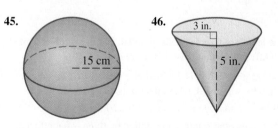

45.

15 cm

46.

3 in.

5 in.

Find the volumes of the solids in Problems 47–50 correct to the nearest unit.

47.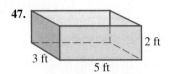
2 ft
3 ft
5 ft

48.
3 in. 13 in.
5 in. 12 in.

49.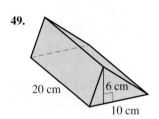
20 cm 6 cm
10 cm

50.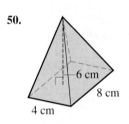
6 cm
8 cm
4 cm

Find the surface areas and volumes of the solids in Problems 51–52 correct to the nearest unit.

51.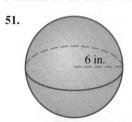
6 in.

52.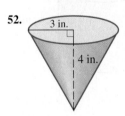
3 in.
4 in.

53. If the length of a rectangle is doubled, and the width is tripled, what effect does that have on the area?

54. If the length of a box is doubled, the width is tripled, and the height is doubled, what effect does that have on the volume?

55. If the radius of a sphere is multiplied by three, what effect does that have on the surface area?

56. If the height of a square pyramid is doubled, what effect does that have on the volume?

Level 3

57. If the diameter of a circle is doubled, what effect does that have on the area?

58. Suppose it takes 10 hours to fill a rectangular swimming pool of uniform depth. How long will it take to fill a similarly shaped swimming pool if all of the dimensions are doubled?

Problem Solving 3

59. Show how a spreadsheet could be used to print a table of temperature conversions between the Fahrenheit and Celsius scales. Your table should consist of a column showing Celsius temperatures from 0 to 40 in steps of 5 degrees, and a column to the right, with corresponding Fahrenheit temperatures.

60. A polyhedron is a simple closed surface in space whose boundary is composed of polygonal regions (see Figure 9.29). A rather surprising relationship exists among the number of vertices, edges, and faces of polyhedra. See if you can discover it by looking for patterns in the figures and filling in the blanks.

| Triangular pyramid | Quadrilateral pyramid | Pentagonal pyramid | Regular tetrahedron |

| Regular hexahedron (cube) | Regular octahedron | Regular dodecahedron | Regular icosahedron |

FIGURE 9.29 Some common polyhedra

	Figure	*# of faces*	*# of vertices*	*# of edges*
a.	Triangular pyramid	4	4	6
b.	Quadrilateral pyramid	5	_____	8
c.	Pentagonal pyramid	_____	6	10
d.	Regular tetrahedron	4	4	_____
e.	Cube	_____	_____	12
f.	Regular octahedron	_____	6	_____
g.	Regular dodecahedron	_____	_____	30
h.	Regular icosahedron	_____	_____	30

9.5 | U.S.–Metric Conversions

This section may be for reference. The emphasis in this chapter is on everyday use of the metric system. However, certain specialized applications require more precise conversions than we've considered. On the other hand, you don't want to become bogged down with arithmetic to the point where you say "nuts to the metric system."

The most difficult obstacle involving the change from the U.S. system to the metric system is not mathematical but psychological. However, if you understood the presentation in this chapter, you now realize that *working within the metric system is much easier than working within the U.S. system.*

With these ideas firmly in mind, and realizing that your everyday work with the metric system is discussed in Sections 9.1–9.4, we present a list of conversion factors between these measurement systems. Many calculators will perform these conversions for you; check your owner's manual.

Length Conversions

U.S. to Metric			Metric to U.S.		
When you know...	**multiply by...**	**to find:**	**When you know...**	**multiply by...**	**to find:**
in.	2.54	cm	cm	0.39370	in.
ft	30.48	cm	m	39.37	in.
ft	0.3048	m	m	3.28084	ft
yd	0.9144	m	m	1.09361	yd
mi	1.60934	km	km	0.62137	mi

Capacity Conversions

U.S. to Metric			Metric to U.S.		
When you know...	**multiply by...**	**to find:**	**When you know...**	**multiply by...**	**to find:**
tsp	4.9289	mL	mL	0.20288	tsp
tbsp	14.7868	mL	mL	0.06763	tbsp
oz	29.5735	mL	mL	0.03381	oz
c	236.5882	mL	mL	0.00423	c
pt	473.1765	mL	mL	0.00211	pt
qt	946.353	mL	mL	0.00106	qt
qt	0.9464	L	L	1.05672	qt
gal	3.7854	L	L	0.26418	gal

Weight Conversions

U.S. to Metric			Metric to U.S.		
When you know...	**multiply by...**	**to find:**	**When you know...**	**multiply by...**	**to find:**
oz	28.3495	g	g	0.0352739	oz
lb	453.59237	g	g	0.0022046	lb
lb	0.453592	kg	kg	2.2046226	lb
T	907.18474	kg			

Temperature Conversions

U.S. to Metric			Metric to U.S.		
When you know...	**multiply by...**	**to find:**	**When you know...**	**multiply by...**	**to find:**
°F	$\frac{5}{9}$, after subtracting 32°	°C	°C	$\frac{9}{5}$, then add 32°	°F

9.6 | CHAPTER SUMMARY

*Many arts there are
which beautify the mind
of people; of all other
none do more garnish
and beautify it than those
arts which are called
mathematical.*

H. BILLINGSLEY

Important Ideas

Metric system; length (meter), capacity (liter), and mass (gram) [9.2–9.4]
Accuracy of measurements [9.1]
Perimeter and circumference [9.1]
Formulas for rectangles, squares, parallelograms, triangles, trapezoids, and circles [9.2]
Acre [9.2]
Volume formulas for boxes (parallelepipeds), right rectangular prisms, right circular cylinders, pyramids, right circular cones, and spheres [9.3]
Capacity measurements [9.3]
Relationship between volume and capacity [9.3]
Fahrenheit and Celsius temperatures for both water freezing and water boiling [9.4]

Take some time getting ready to work the review problems in this section. First review these important ideas. Look back at the definition and property boxes. If you look online, you will find a list of important terms introduced in this chapter, as well as the types of problems that were introduced. You will maximize your understanding of this chapter by working the problems in this section only after you have studied the material.

You will find some review help online at **www.mathnature.com.** There are links giving general test help in studying for a mathematics examination, as well as specific help for reviewing this chapter.

Chapter | 9 | Review Questions

1. a. Without any measuring device, draw a segment approximately 10 cm long.
 b. Measure this segment to the nearest tenth cm.
 _____

2. a. What are the units of measurement in both the U.S. and metric measurement systems that you would use to measure the height of a building?
 b. Estimate the temperature on a hot summer day in both the U.S. and metric measurement systems.
 c. What is the width of your index finger in both the metric and U.S. systems?

3. a. What does the prefix *milli-* mean?
 b. What does the measurement 8.6 mL measure?
 c. How many centimeters are equivalent to 10 km?

Use the region shown in Figure 9.30 for Problems 4–7.

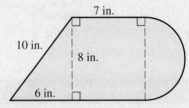

FIGURE 9.30 Region with triangular and semicircular ends

4. What is the distance around (to the nearest inch)?

5. What is the area (to the nearest square inch)?

6. Use estimation: if the figure is the top of a prism of height 1 ft, will the volume (in square feet) be greater than or less than 1 ft³?

7. If this is the top of a prism of height 1 ft, what is the volume of the prism?

8. For a sphere with radius 1 ft,
   a. What is the exact surface area?
   b. What is the surface area (to the nearest tenth)?

9. For a sphere with radius 1 ft,
   a. What is the exact volume?
   b. What is the volume (to the nearest tenth)?

Use the figures shown in Figure 9.31 for Problems 10–13.

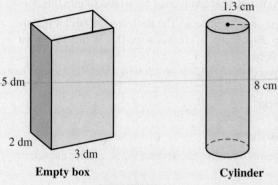

Empty box   **Cylinder**

FIGURE 9.31

10. a. What is the volume?
    b. What is the capacity of the box?

11. What is the outside surface area of the box?

12. a. What are the dimensions of this box in centimeters?
    b. If the measurements of this box were 50 in. by 20 in. by 30 in., what is the capacity of the box?

13. What is the outside surface area (to the nearest unit) of the cylinder in Figure 9.31?

14. Consider the following advertisement from Round Table Pizza.

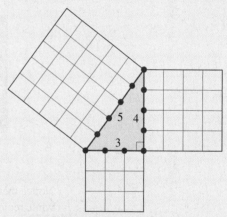

Pizzas

Made with fine natural cheeses, the choicest meats and the freshest vegetables, plus our own spicy sauce.

Original Style

Lavish toppings on a thin, crisp crust.

Mini	Small	Medium	Large
Serves	Serves	Serves	Serves
1	1–2	2–3	3–4
3.71	7.95	11.18	13.88

King Arthur's Supreme

Cheeses, pepperoni, sausage, salami, beef, linguica, mushrooms, green peppers, onions and black olives. Round Table's finest.

Shrimp and anchovies available upon request.

a. If you order a pizza, what size should you order if you want the best price per square inch? (That is, compare the price per square inch for the various pizza sizes; assume the diameters for the four sizes are 6 in., 10 in., 12 in., and 14 in.)

b. Suppose you were the owner of a pizza restaurant and were going to offer a 16-in.-diameter pizza. What would you charge for the pizza if you wanted the price to be comparable with the prices of the other sizes?

A rectangular pool is 80 ft by 30 ft with a depth of 3 ft. Use this information to answer the questions in Problems 15–17.

15. What is the surface area of the top of the pool?

16. What is the volume of the pool?

17. What is the capacity of the pool (to the nearest gallon)?

18. How many square yards of carpet are necessary to carpet an 11-ft by 16-ft room? (Assume that you cannot purchase part of a square yard.)

19. The Pythagorean theorem (Chapter 7) states that, for a right triangle with sides a and b and hypotenuse c,

$$a^2 + b^2 = c^2$$

In this chapter, we saw that a^2 is the area of a square of side length a. We can, therefore, illustrate the Pythagorean theorem for a triangle with sides 3, 4, and 5, as shown in Figure 9.32.

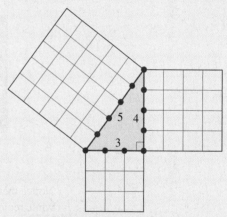

FIGURE 9.32 Geometric interpretation of the Pythagorean theorem

a. Illustrate the Pythagorean theorem for a right triangle with sides 5, 12, and 13.

b. On the other hand, if you draw a triangle with sides measuring 2, 3, and 4 units, you would find that $2^2 + 3^2 = 4 + 9 = 13$ and $4^2 = 16$ so $2^2 + 3^2 \neq 4^2$. Also notice that such triangles are not right triangles. Is it possible to form a right triangle with sides of 25, 312, and 313?

20. The distributive law states that

$$ab + ac = a(b + c)$$

In this chapter, we say ab and ac represent the areas of two rectangles, one with sides a and b, and the other with sides a and c. Give a geometric justification for the distributive property.

Group RESEARCH PROJECTS

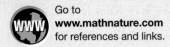

Working in small groups is typical of most work environments, and learning to work with others to communicate specific ideas is an important skill. Work with three or four other students to submit a single report based on each of the following questions.

G31. Suppose a house has an 8-ft ceiling in all rooms except the living room, which has a 10-ft cathedral ceiling. Approximately how many marbles would fit into this house?

G32. Suppose you wish to build a spa on a wood deck. The deck is to be built 4 ft above level ground. It is to be 50 ft by 30 ft and is to contain a spa that is circular with a 14-ft diameter. The spa is 4 ft deep.
  a. How much water will the spa contain, and how much will it weigh? Assume that the spa itself weighs 550 lb.
  b. Draw plans for the wood deck.
  c. Draw up a materials list.
  d. Estimate the cost for this installation.

G33. This investigation is an extension of Problem 60, Section 9.2. Consider the square shown in Figure 9.33a.

 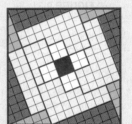

     a. 16 regions   **b. 17 regions**

FIGURE 9.33 Pieces arranged to find an extra square inch

Notice there are four green regions, four pink regions, four yellow regions and four white regions. Rearrange these same pieces as shown in Figure 9.33b, and notice that now there is an **extra square inch** in the center. Now arrange these pieces back to their original position. . . . what happened to the extra square inch?[*]

G34. HISTORICAL QUEST Here is a simple formula for finding Pythagorean triples (numbers *a, b,* and *c* that satisfy the Pythagorean theorem). It was given to me in an elevator by my friend Bert Liberi (who is also a great mathematician). If *m* is any natural number greater than 1, then

$$\frac{1}{m} + \frac{1}{m+2} = \frac{a}{b}$$

The reduced fraction $\frac{a}{b}$ will have the property that the set $\{a, b, c\}$ is a Pythagorean triple. For example, if $m = 2$, then

$$\frac{1}{2} + \frac{1}{2+2} = \frac{1}{2} + \frac{1}{4} = \frac{3}{4}$$

Thus, the first two numbers of the triple are 3 and 4. For the third number in the triple, we find

$$c = \sqrt{3^2 + 4^2} = \sqrt{9 + 16} = \sqrt{25} = 5$$

so the set is $\{3, 4, 5\}$. Find ten sets of Pythagorean triples.

*Problem invented by Martin Gardner, printed in *The College Mathematics Journal*, Vol. 40, No. 3, May 2009, p. 158. Martin Gardner was the author of the Mathematical Games Department in the *Scientific American* for over 25 years.

Individual RESEARCH PROJECTS

www.mathnature.com

Learning to use sources outside your classroom and textbook is an important skill, and here are some ideas for extending some of the ideas in this chapter. You can find references to these projects in a library or at **www.mathnature.com.**

PROJECT 9.1 Historical Quest   In Chapter 1, we introduced Pascal's triangle. The reproduction in Figure 9.34 is from a 14th-century Chinese manuscript, and in this form is sometimes called *Yang Hui's triangle.*

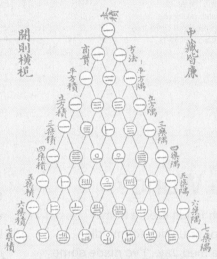

FIGURE 9.34 Yang Hui's triangle

Even though we have not discussed these ancient Chinese numerals, see if you can reconstruct the basics of their numeration system. Use your imagination.

PROJECT 9.2 Historical Quest   The Ionic Greek numeration system (approximately 3000 B.C.) counts as follows:

$$\alpha, \beta, \gamma, \delta, \epsilon, F, \zeta, \eta, \theta, \iota, \iota\alpha, \iota\beta, \iota\gamma, \iota\delta, \iota\epsilon, \iota F, \iota\zeta, \iota\eta, \iota\theta, \kappa, \kappa\alpha, \ldots$$

Other numbers are λ (for 30), μ (for 40), ν (for 50), ξ (for 60), ρ (for 70), π (for 80), ψ (for 90), ρ (for 100), σ (for 200), τ (for 300), ν (for 400), ϕ (for 500), χ (for 600), ψ (for 700), ω (for 800), and II (for 900). Try to reconstruct the basics of the Ionic Greek numeration system. Use your imagination.

PROJECT 9.3 Write a paper on Pythagorean triples (numbers *a, b,* and *c* that satisfy the Pythagorean theorem).

PROJECT 9.4 Construct models for the regular polyhedra.

PROJECT 9.5 What solids occur in nature? Find examples of each of the five regular solids.

BOOK REPORTS

Write a 500-word report on one of these books:

Flatland, a Romance of Many Dimensions, by a Square, Edwin A. Abbott (New York: Dover Publications, original edition printed in 1880). It can also be found on the Web at: **http://www.alcyone.com/max/lit/flatland/**

Flatterland: Like Flatland, Only More So, Ian Stewart (Cambridge, MA: Perseus Publishing, 2002).

10 THE NATURE OF GROWTH

What in the World?

"Phoenix's smart growth funds create home ownership and rental opportunities for middle-income families," proclaimed Jay. "I've made some real bucks, not only for myself but for those whom I've helped with their portfolios."

"I don't know, I really like having my money in an insured account," said Cecily. "Slow and steady, that's my motto."

"The funds I'm talking about bring together institutional investors and developers to meet the needs of critical workforce employees, including firefighters, police officers, teachers, nurses, and office workers facing few housing choices and long commutes," said Jay. "The growth has been exponential for the investors so far!"

"I know what you are talking about!" added Cecily. "We've just finished studying that topic in an evening class I'm taking. The long-term differences between linear and exponential growth are unbelievable. Our teacher said understanding this difference could be the key to our financial success."

Overview

One of the most important ideas in understanding the nature of the world around us is to understand applications of growth and decay. Ross Honsberger, a contemporary mathematics professor, said in his book *Mathematical Morsels* (Washington, D.C.: Mathematical Association of America, p. vii), "Mathematics abounds in bright ideas. No matter how long and hard one pursues her, mathematics never seems to run out of exciting surprises." Nothing illustrates that concept more than the study of growth and decay, which involves the understanding of both exponential and logarithmic equations. In this chapter we investigate both of these important mathematical ideas.

10.1 Exponential Equations

Little can be understood of even the simplest phenomena of nature without some knowledge of mathematics, and the attempt to penetrate deeper into the mysteries of nature compels simultaneous development of the mathematical processes.

J. W. A. YOUNG

CHAPTER **CHALLENGE**

See if you can fill in the question mark.

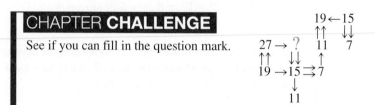

The measurement of growth and decay often involves the study of relatively large or relatively small quantities. Difficulty with scaling measurements is often one of our primary concerns when describing and measuring figures and data. Recall from Section 1.3 that large and small numbers are often represented in exponential form and scientific notation. In this section, we investigate solving equations known as *exponential equations.*

> ### Exponential Equation
>
> An equation of the form $b^x = N$ in which an unknown value is included as part of the exponent is called an **exponential equation.**

An **exponential** is an expression of the form b^x; we begin by using a calculator to evaluate some exponentials. The number b is called the *base.* In Chapter 1, we defined b^x for integer values of x. In more advanced courses, b^x is defined for all real numbers. We will use a calculator to approximate these values. We will also frequently approximate two irrational numbers

$$\pi \approx 3.1416 \quad \text{and} \quad e \approx 2.7183$$

Example 1   Evaluate exponentials

Evaluate the given exponentials (correct to two decimal places). Show the calculator steps.

a. 2^8 **b.** $3^{2.5}$ **c.** $4^{\sqrt{2}}$ **d.** π^3 **e.** e^2 **f.** e^{π} **g.** π^e

Solution

Given	Keys Pressed	Evaluation	Approximation
a. 2^8	2 ⌃ 8 =	256 (exact)	256.00
b. $3^{2.5}$	3 ⌃ 2.5 =	15.58845727	15.59
c. $4^{\sqrt{2}}$	4 ⌃ √ 2 =	7.102993301	7.10
d. π^3	π ⌃ 3 =	31.00627668	31.01
e. e^2	e ⌃ 2 =	7.389056099	7.39

Don't skip this example! Press each of these on your calculator because these evaluations set the groundwork for the rest of this chapter. In other words, actually press keys on your own calculator to verify that you obtain what is shown here.

Given	Keys Pressed	Evaluation	Approximation

Note: If you want to evaluate e, find e^1:

	$\boxed{e}\;\boxed{\wedge}\;\boxed{1}\;\boxed{=}$	2.718281828	2.72
f. e^π	$\boxed{e}\;\boxed{\wedge}\;\boxed{\pi}\;\boxed{=}$	23.14069263	23.14
g. π^e	$\boxed{\pi}\;\boxed{\wedge}\;\boxed{e}\;\boxed{=}$	22.45915772	22.46

Let's solve the exponential equation $2^x = 14$. To solve an equation means to find the replacement(s) for the variable that make the equation true. You might try certain values:

$$x = 1: \quad 2^x = 2^1 = 2 \qquad \textit{Too small}$$
$$x = 2: \quad 2^x = 2^2 = 4 \qquad \textit{Too small}$$
$$x = 3: \quad 2^x = 2^3 = 8 \qquad \textit{Still too small}$$
$$x = 4: \quad 2^x = 2^4 = 16 \qquad \textit{Too big}$$

It seems as if the number you are looking for is between 3 and 4. Our task in this section is to find both an approximate as well as an exact value for x. To answer this problem, we need some preliminary information.

Definition of Logarithm

STOP Understanding this development is essential to understanding the meaning of logarithm.

The solution of the equation $2^x = 14$ seeks an x-value. What is this x-value? We express the idea in words:

x is the exponent on a base 2 that gives the answer 14

This can be abbreviated as

x = exp on base 2 to give 14

We further shorten this notation to

$x = \exp_2 14$

This statement is read, "x is the exponent on a base 2 that gives the answer 14." It appears that the equation is now solved for x, but this is simply a notational change. The expression "exponent of 14 to the base 2" is called, for historical reasons, "the log of 14 to the base 2." That is,

$x = \exp_2 14$ and $x = \log_2 14$

mean exactly the same thing. This leads us to the following definition of logarithm.

Spend some time with this definition.

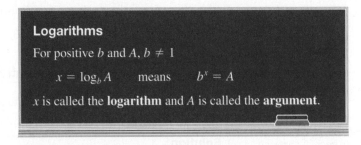

Logarithms

For positive b and A, $b \neq 1$

$$x = \log_b A \qquad \text{means} \qquad b^x = A$$

x is called the **logarithm** and A is called the **argument**.

The statement $x = \log_b A$ should be read as "x is the log (exponent) on a base b that gives the value A." *Do not forget that a logarithm is an exponent.*

Example 2   Write exponentials in logarithmic form

Write in logarithmic form:
a. $5^2 = 25$ **b.** $\frac{1}{8} = 2^{-3}$ **c.** $\sqrt{64} = 8$

Solution

a. In $5^2 = 25$, 5 is the base and 2 is the exponent, so we write

$$2 = \log_5 25$$

Remember, the logarithmic expression "solves" for the exponent.

b. With $\frac{1}{8} = 2^{-3}$, the base is 2 and the exponent is -3:

$$-3 = \log_2 \frac{1}{8}$$

c. With $\sqrt{64} = 8$, the base is 64 and the exponent is $\frac{1}{2}$ (since $\sqrt{64} = 64^{1/2}$):

$$\frac{1}{2} = \log_{64} 8$$

Example 3   Write logarithmic forms in exponential form

Tell what each expression means, and then rewrite in exponential form.
a. $\log_{10} 100$ **b.** $\log_{10} \frac{1}{1,000}$ **c.** $\log_3 1$

Solution

a. $\log_{10} 100$ is the exponent on a base 10 that gives 100. We see that the exponent is 2, so we write $\log_{10} 100 = 2$; the base is 10 and the exponent is 2, so $10^2 = 100$.

b. $\log_{10} \frac{1}{1,000}$ is the exponent on a base 10 that gives $\frac{1}{1,000}$; this exponent is -3. The base is 10 and the exponent is -3, so we write

$$10^{-3} = \frac{1}{1,000}$$

c. $\log_3 1$ is the exponent on a base 3 that gives 1; this exponent is 0. The base is 3 and the exponent is 0, so we write

$$3^0 = 1$$

Historical NOTE

In his history book, F. Cajori wrote, "The miraculous powers of modern calculation are due to three inventions: the Arabic notation, decimal fractions, and logarithms." Today, we would no doubt add a fourth invention to this list—namely, the handheld calculator. Nevertheless, the idea of a logarithm was a revolutionary idea that forever changed the face of mathematics. Simply stated, *a logarithm is an exponent*, and even though logarithms are no longer important as an aid in calculation, the ideas of logarithm and exponent are extremely important in advanced mathematics, particularly in growth and decay applications.

Example 4   Solve simple exponential equations

Solve for x:
a. $3^x = 5$ **b.** $10^x = 2$ **c.** $e^x = 0.56$

Solution

a. $x = \log_3 5$ **b.** $x = \log_{10} 2$ **c.** $x = \log_e 0.56$

In elementary work, the most commonly used base is 10, so we call a logarithm to the base 10 a **common logarithm,** and we agree to write it without using a subscript 10. That is, $\log x$ is a *common logarithm.* A logarithm to the base e is called a **natural logarithm** and is denoted by $\ln x$. The expression $\ln x$ is often pronounced "ell en x" or "lon x."

Logarithmic Notations

Common logarithm: $\log x$ means $\log_{10} x$
Natural logarithm: $\ln x$ means $\log_e x$

The solution for the equation $10^x = 2$ is $x = \log 2$, and the solution for the equation $e^x = 0.56$ is $x = \ln 0.56$.

Evaluating Logarithms

To **evaluate** a logarithm means to find a numerical value for the given logarithm. Calculators have, to a large extent, eliminated the need for logarithm tables. You should find two logarithm keys on your calculator. One is labeled LOG for common logarithms, and the other is labeled LN for natural logarithms.

Example 5	**Use a calculator to evaluate a logarithm**

Use a calculator to evaluate:
a. log 5.03   **b.** ln 3.49   **c.** log 0.00728

Use your own calculator to verify these answers because the number of digits shown may vary.

Solution   Calculator answers are more accurate than were the old table answers, but it is important to realize that any answer (whether from a table or a calculator) is only as accurate as the input numbers. However, in this book we will not be concerned with significant digits, but instead will use all the accuracy our calculator gives us, rounding only once (if requested) at the end of the problem.
a. $\log 5.03 \approx 0.7015679851$
b. $\ln 3.49 \approx 1.249901736$
c. $\log 0.00728 \approx -2.137868621$

Example 5 shows fairly straightforward evaluations, since the problems involve common or natural logarithms and because your calculator has both LOG and LN keys. However, suppose we wish to evaluate a logarithm to some base *other than* base 10 or base *e*. The first method uses the definition of logarithm (as in Example 3), and the second method uses what is called the **change of base theorem.** Before we state this theorem, we consider its plausibility with the following example.

Example 6	**Evaluate logarithmic expressions**

Evaluate the given expressions.

a. $\log_2 8, \dfrac{\log 8}{\log 2},$ and $\dfrac{\ln 8}{\ln 2}$ **b.** $\log_3 9, \dfrac{\log 9}{\log 3},$ and $\dfrac{\ln 9}{\ln 3}$

Solution
a. From the definition of logarithm, $\log_2 8 = x$ means $2^x = 8$ or $x = 3$. Thus, $\log_2 8 = 3$. By calculator,

$$\frac{\log 8}{\log 2} \approx \frac{0.903089987}{0.3010299957} \approx 3$$

Also,

$$\frac{\ln 8}{\ln 2} \approx \frac{2.079441542}{0.6931471806} \approx 3$$

b. $\log_3 9 = x$ means $3^x = 3^2$, so that $x = \log_3 9 = 2$. By calculator,

$$\frac{\log 9}{\log 3} \approx \frac{0.9542425094}{0.4771212547} \approx 2$$

and

$$\frac{\ln 9}{\ln 3} \approx \frac{2.197224577}{1.098612289} \approx 2$$

You no doubt noticed that the answers to each part of Example 6 are the same. This result is summarized with the following theorem, which is proved on page 487 (in Section 10.2).

STOP

Remember this formula so you can evaluate logarithms to bases other than 10 or e.

Change of Base

$$\log_a x = \frac{\log_b x}{\log_b a}$$

Example 7 | Evaluate logarithms with a change of base

Evaluate (round to the nearest hundredth):
a. $\log_7 3$ **b.** $\log_3 3.84$

Solution

a. $\log_7 3 = \dfrac{\log 3}{\log 7} \approx \dfrac{0.4771212547}{0.84509804} \approx 0.5645750341 \approx 0.56$

This is all done by calculator and not on paper.

b. $\log_3 3.84 = \dfrac{\log 3.84}{\log 3} \approx \dfrac{0.5843312244}{0.4771212547} \approx 1.224701726 \approx 1.22$

Calculator work

We now return to the problem of solving

$$2^x = 14 \qquad \text{Given equation}$$
$$x = \log_2 14 \qquad \text{Solution}$$

We call $\log_2 14$ the **exact solution** for the equation, and Example 8 finds an approximate solution.

Example 8 | Solve an exponential equation

Solve $2^x = 14$ (correct to the nearest hundredth).

Solution   We use the definition of logarithm and the change of base theorem to write

$$x = \log_2 14 = \frac{\log 14}{\log 2} \approx 3.807354922 \approx 3.81$$

Calculator work

Exponential Equations

We now turn to solving *exponential equations.* Exponential equations will fall into one of three types:

Common log	*Natural log*	*Arbitrary*
base 10	base e	base b
Example: $10^x = 5$	Example: $e^{-0.06x} = 3.456$	Example: $8^x = 156.8$

The following example illustrates the procedure for solving each type of exponential equation.

Example 9 | Solve exponential equations with common and natural logs

Solve the following exponential equations:
a. $10^x = 5$ **b.** $e^{-0.06x} = 3.456$ **c.** $8^x = 156.8$

Solution   Regardless of the base, we use the definition of logarithm to solve an exponential equation.

a. $10^x = 5$ Given equation

   $x = \log 5$ Definition of logarithm; this is the exact answer.

   ≈ 0.6989700043 Approximate calculator answer

b. $e^{-0.06x} = 3.456$ Given equation

   $-0.06x = \ln 3.456$ Definition of logarithm

   $x = \dfrac{\ln 3.456}{-0.06}$ Exact answer; this can be simplified to $x = -\dfrac{50}{3}\ln 3.456$.

   ≈ -20.66853085 Approximate calculator answer

c. $8^x = 156.8$ Given equation

   $x = \log_8 156.8$ Definition of logarithm; this is the exact answer.

   ≈ 2.43092725 Approximate calculator answer. Use the change of base theorem:

   $$\log_8 156.8 = \frac{\log 156.8}{\log 8}.$$

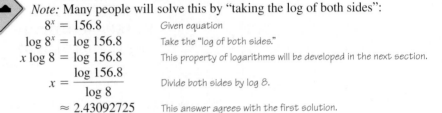

Note: Many people will solve this by "taking the log of both sides":

   $8^x = 156.8$ Given equation

   $\log 8^x = \log 156.8$ Take the "log of both sides."

   $x \log 8 = \log 156.8$ This property of logarithms will be developed in the next section.

   $x = \dfrac{\log 156.8}{\log 8}$ Divide both sides by log 8.

   ≈ 2.43092725 This answer agrees with the first solution.

Did you notice that the results of these two calculations in part **c** are the same? It simply involves several extra steps and some additional properties of logarithms. It is rather like solving quadratic equations by completing the square each time instead of using the quadratic formula. You can see that, before calculators, there were good reasons to avoid representations such as $\log_8 156.8$. Whenever you *see* an expression such as $\log_8 156.8$, you *know* how to calculate it: log 156.8/log 8.

We now consider some more general exponential equations. We will follow the general procedure illustrated in Example 9. Some algebraic steps will be required to put the equations into the correct form. That is, before we use the definition of logarithm, we first solve for the exponential form.

Example 10 Solve logarithmic equations

Solve:

a. $\dfrac{10^{5x+3}}{5} = 39$ **b.** $1 = 2e^{-0.000425x}$ **c.** $8 \cdot 6^{3x+2} = 1{,}600$

Solution   Note that, in each case, we use the definition of logarithm.

a. $\dfrac{10^{5x+3}}{5} = 39$ Given equation

   $10^{5x+3} = 195$ Multiply both sides by 5. (Solve for exponential.)

   $5x + 3 = \log 195$ Definition of logarithm

   $x = \dfrac{\log 195 - 3}{5}$ Solve linear equation for x. This is the exact answer.

   ≈ -0.1419930777 Approximate calculator answer

b. $1 = 2e^{-0.000425x}$ Given equation

   $\dfrac{1}{2} = e^{-0.000425x}$ Divide both sides by 5. (Solve for exponential.)

   $-0.000425x = \ln \dfrac{1}{2}$ Definition of logarithm

   $x = \dfrac{\ln 0.5}{-0.000425}$ Solve linear equation for x. This is the exact answer.

   $\approx 1{,}630.934542$ Approximate calculator answer

Historical NOTE

**Jhone Neper
(1550–1617)**

The history of logarithms is interesting reading. Jhone Neper (1550–1617) is more commonly known using the modern spelling of John Napier in the context of Napier's bones discussed in Chapter 4. He is usually credited with the discovery of logarithms because he was the first to publish a work on logarithms, called *Descriptio*, in 1614. However, similar ideas were developed independently by Jobst Burgi around 1588. Some of the sources on the history of logarithms are "Logarithms" by J. W. L. Glaisher in the *Encyclopaedia Britannica*, 11th ed. Vol. 16, pp. 868–877, and Florian Cajori, "History of the Exponential and Logarithmic Concepts," *American Mathematical Monthly*, Vol. 20 (1913). In Glaisher's article he says, "The invention of logarithms and the calculation of the earlier tables form a very striking episode in the history of exact science, and, with the exception of the *Principia* of Newton, there is no mathematical work published in the country which has produced such important consequences, or to which so much interest attaches as to Napier's *Descriptio*."

c. $8 \cdot 6^{3x+2} = 1{,}600$ *Given equation*

$6^{3x+2} = 200$ *Divide both sides by 8. (Solve for exponential.)*

$3x + 2 = \log_6 200$ *Definition of logarithm*

$$x = \frac{\log_6 200 - 2}{3}$$ *Solve linear equation for x. This is the exact answer.*

≈ 0.3190157417 *Approximate calculator answer. You will need the change of base theorem.*

Example 10b illustrates exponential decay. Growth and decay problems are common examples of exponential equations. We will consider growth and decay applications in Section 10.3.

In Chapter 1 we considered large numbers. The following example involves large numbers, volumes, and exponential equations.

Example 11 Large number puzzle

Pólya's Method

The earth will fit into a cube that is 12,740 km on a side. How many generations of an organism that doubles each generation and takes up one μm^3 of space would it take to fill up this earth-sized cube? Note that the symbol μm stands for the length of a **micrometer,** which is defined to be one-millionth of a meter.

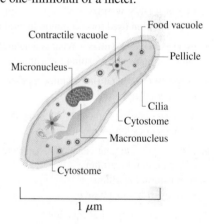

Food vacuole
Contractile vacuole
Pellicle
Micronucleus
Cilia
Cytostome
Macronucleus
Cytostome

$1\ \mu\text{m}$

Solution   We use Pólya's problem-solving guidelines for this example.

Understand the Problem. Do you know what a μm is? It is one-millionth of a meter, so $(1\ \mu\text{m})^3$ equals the size of a cube whose dimensions are:

$1\ \mu\text{m} \times 1\ \mu\text{m} \times 1\ \mu\text{m}$

Devise a Plan. The plan we will use is first to find the number of organisms of size $1\ \mu\text{m}^3$ that are contained in this earth-sized cube, and then to find the number of generations by solving an exponential equation.

Carry Out the Plan. First, solve $V = s^3$, where s is the length of a side of a cube.

$V = s^3$

$= (12{,}740 \text{ km})^3$ *This is the earth-sized cube.*

$= (2.07 \times 10^{12}) \text{ km}^3$

$= (2.07 \times 10^{12})(1{,}000 \text{ m})^3$

$= 2.07 \times 10^{12} \times (10^3)^3 \text{ m}^3$

$= 2.07 \times 10^{12} \times 10^9 \text{ m}^3$

$= 2.07 \times 10^{21} \text{ m}^3$

$= (2.07 \times 10^{21})(1{,}000{,}000\ \mu\text{m})^3$

$= 2.07 \times 10^{21} \times (10^6)^3\ \mu\text{m}^3$

$= 2.07 \times 10^{21} \times 10^{18}\ \mu\text{m}^3$

$= 2.07 \times 10^{39}\ \mu\text{m}^3$

Since the organism doubles each generation, we want to find n so that

$$2^n = 2.07 \times 10^{39}$$

Using the definition of logarithm, we find

$$n = \log_2(2.07 \times 10^{39}) \approx 130.6$$

It would take about 131 generations.

Look Back. Since n is the number of doublings, we find

$$2^{131} \approx 2.72 \times 10^{39}$$

which is the appropriate magnitude.

Problem Set | **10.1**

Level **1**

1. IN YOUR OWN WORDS   What is the definition of logarithm?

2. IN YOUR OWN WORDS   What is a common logarithm? What is the notation used for a common logarithm?

3. IN YOUR OWN WORDS   What is a natural logarithm? What is the notation used for natural logarithm?

4. IN YOUR OWN WORDS   Outline a procedure for solving exponential equations.

5. a. What does log N mean?   **b.** What does ln N mean?
c. What does $\log_b N$ mean?

6. How do you use your calculator to evaluate the following?
a. a common logarithm
b. a natural logarithm
c. a logarithm to a base b

7. What is an exponential equation?

8. What are the three types of exponential equations?

Write the equations in Problems 9–12 in logarithmic form.

9. a. $64 = 2^6$ **b.** $100 = 10^2$ **c.** $m = n^p$

10. a. $1{,}000 = 10^3$ **b.** $81 = 9^2$ **c.** $\dfrac{1}{e} = e^{-1}$

11. a. $\dfrac{1}{10} = 10^{-1}$ **b.** $36 = 6^2$ **c.** $s = t^n$

12. a. $125 = 5^3$ **b.** $9 = \left(\frac{1}{3}\right)^{-2}$ **c.** $a = b^c$

In Problems 13–18 use the definition of logarithm to simplify each expression.

13. a. $\log_{10} 10$ **b.** $\log_{10} 1{,}000$ **c.** $\log_{10} 10^{-5}$

14. a. $\log 100$ **b.** $\log_5 5$ **c.** $\log_5 25$

15. a. $\log_5 5^{-3}$ **b.** $\log 0.1$ **c.** $\log_e e$

16. a. $\log_e e^2$ **b.** $\log_e e^{-4}$ **c.** $\log 10^{-3}$

17. a. $\log_b b$ **b.** $\log_b b^3$ **c.** $\log_b b^{-6}$

18. a. $\log_2 128$ **b.** $\log 10^n$ **c.** $\ln e^n$

Solve each exponential equation in Problems 19–22. Give the exact value for x.

19. a. $4^x = \frac{1}{16}$ **b.** $5^x = 8$
c. $6^x = 4.5$ **d.** $4^x = 5$

20. a. $125^x = 25$ **b.** $(\sqrt{2})^x = 5$
c. $e^x = 9$ **d.** $\pi^x = 10$

21. a. $216^x = 36$ **b.** $10^x = 2.5$
c. $10^x = 45$ **d.** $10^x = 15$

22. a. $27^x = \frac{1}{81}$ **b.** $e^x = 1.8$
c. $e^x = 34.2$ **d.** $e^x = 6$

Evaluate the given expressions in Problems 23–28 (to two decimal places).

23. a. $\log 4.27$ **b.** $\log_b b^2$
c. $\log_t t^3$ **d.** $\ln 10$

24. a. $\log 1.08$ **b.** $\log_e e^4$
c. $\log_\pi \sqrt{\pi}$ **d.** $\ln 100$

25. a. $\log 71{,}600$ **b.** $\log_3 9$
c. $\log_{19} 1$ **d.** $\ln 1{,}000$

26. a. $\log 18.9$ **b.** $\log_2 32$
c. $\log_7 1$ **d.** $\log 0.042$

27. a. $\ln 2.27$ **b.** $\ln 16.77$
c. $\ln 7.3$ **d.** $\ln 0.321$

28. a. $\log e$ **b.** $\log e^2$
c. $\log e^3$ **d.** $\log 0.0532$

Write each expression in Problems 29–32 in terms of common logarithms, and then give a calculator approximation (correct to four decimal places).

29. a. $\log_3 45$　　　　　　　**b.** $\log_5 91$

30. a. $\log_6 10$　　　　　　　**b.** $\log_5 304$

31. $\log_7 13$　　　　　　　　**b.** $\log_2 556$

32. $\log_4 3.05$　　　　　　　**b.** $\log_2 1{,}513$

Write each expression in Problems 33–36 in terms of natural logarithms, and then give a calculator approximation (correct to four decimal places).

33. a. $\log_2 0.0056$　　　　　**b.** $\log_{8.3} 105$

34. a. $\log_8 10$　　　　　　　**b.** $\log_\pi e^2$

35. a. $\log_7 56$　　　　　　　**b.** $\log_2 10.85$

36. a. $\log_8 e$　　　　　　　　**b.** $\log_{1.08} 5{,}450$

Level 2

Solve the exponential equations in Problems 37–44. Show the approximation you obtain with your calculator without rounding.

37. a. $8^x = 3$　　　　　　　**b.** $64^x = 5$

38. a. $e^x = 4$　　　　　　　**b.** $e^x = 25$

39. a. $10^x = 42$　　　　　　**b.** $10^x = 0.0234$

40. a. $10^{5x} = 5$　　　　　　**b.** $10^{3x} = 0.45$

41. a. $e^{-x} = 8$　　　　　　**b.** $10^{-x} = 125$

42. a. $10^{-2x} = 50$　　　　　**b.** $e^{-2x} = 450$

43. a. $10^{5-3x} = 0.041$　　　**b.** $10^{2x-1} = 515$

44. a. $e^{1-2x} = 3$　　　　　　**b.** $e^{1-5x} = 25$

Scientific research has shown that the risk of having an automobile accident increases exponentially as the concentration of alcohol in the blood increases. The blood alcohol concentration is modeled by the formula

$$a = \ln x - 0.14$$

where x is the percent of risk. In Problems 45–48, find the alcohol concentration (correct to the nearest hundredth) for the given percent of risk.

45. The risk of an accident is 25%.

46. The risk of an accident is 15%.

47. The risk of an accident is 10%.

48. The risk of an accident is 2%.

49. In 2009, the minimum wage in Arizona was $7.25. If we assume the growth of this minimum wage is exponential and is growing at the rate of 2%, then the length of time until the minimum wage reaches *d* dollars is modeled by the formula

$$t = 50 \ln\left(\frac{d}{7.25}\right)$$

In what year would you expect the minimum wage to be $8.25?

50. In 2009, the minimum wage in California was $8.00. If we assume the growth of this minimum wage is exponential and is growing at the rate of 2%, then the length of time until the minimum wage reaches *d* dollars is modeled by the formula

$$t = 50 \ln\left(\frac{d}{8}\right)$$

In what year would you expect the minimum wage to be $10.00?

51. In 2010, the minimum wage in Washington state was $8.55. If we assume the growth of this minimum wage is exponential and is growing at the rate of 2%, then the length of time until the minimum wage reaches *d* dollars is modeled by the formula

$$t = 50 \ln\left(\frac{d}{8.55}\right)$$

In what year would you expect the minimum wage to be $10.00?

52. In 2009, the minimum wage in Florida was $7.21. If we assume the growth of this minimum wage is exponential and is growing at the rate of 2%, then the length of time until the minimum wage reaches *d* dollars is modeled by the formula

$$t = 50 \ln\left(\frac{d}{7.21}\right)$$

In what year would you expect the minimum wage to be $9.00?

Level 3

Solve the exponential equations in Problems 53–58. Show the approximation you obtain with your calculator without rounding.

53. $2 \cdot 3^x + 7 = 61$

54. $3 \cdot 5^x + 30 = 105$

55. $\left(1 + \frac{0.08}{360}\right)^{360x} = 2$

56. $\left(1 + \frac{0.055}{12}\right)^{12x} = 2$

57. Solve $P = P_0 e^{rt}$ for t

58. Solve $I = I_0 e^{-rt}$ for r

59. If the earth's radius is approximately 3,963 mi, and if a hypothetical perpetual water-making machine pours out 1 gallon in the first minute and then doubles its output each minute, in which minute would this hypothetical machine pour out a single quantity of water that would be enough to fill the earth with water?

60. If the weight of the earth is 5.9×10^{21} metric tons, and if its weight were cut in half each minute, how long would it take for the earth to weigh less than one g? *Note:* A metric ton is 1 million grams.

10.2 | Logarithmic Equations

In 2005, an earthquake measuring 7.6 on the Richter scale struck Pakistan, India, and the Kashmir region. The intensity of this quake was similar to the one that devastated San Francisco in 1906, but the death toll for this earthquake was measured in the tens of thousands, and the number of homeless was measured in the millions. What is the amount of energy released (in ergs) by this earthquake? How would you answer this question? Where would you begin?

Fundamental Properties

With a little earthquake research, you could find information on the *Richter scale*, which was developed by Gütenberg and Richter. The formula relating the energy E (in ergs) to the magnitude of the earthquake M is given by

$$M = \frac{\log E - 11.8}{1.5}$$

This equation is called a *logarithmic equation*, and the topic of this section is to solve such equations. To answer the question, we first solve for $\log E$, and then we use the definition of logarithm to write this as an exponential equation.

$$M = \frac{\log E - 11.8}{1.5} \qquad \text{Given equation}$$

$$1.5M = \log E - 11.8 \qquad \text{Multiply both sides by 1.5.}$$

$$1.5M + 11.8 = \log E \qquad \text{Add 11.8 to both sides.}$$

$$10^{1.5M + 11.8} = E \qquad \text{Definition of logarithm}$$

We can now answer the question. Since $M = 7.6$,

$$E = 10^{1.5(7.6)+11.8} \approx 1.58 \times 10^{23}$$

However, to solve certain logarithmic equations, we must first develop some properties of logarithms.

We begin with *two fundamental properties of logarithms*. If you understand the definition of logarithm, you can see that these two properties are self-evident, so we call these the **Grant's tomb properties** of logarithms.

 STOP Spend some time making sure you understand these properties. Do you see why we called them Grant's tomb properties?

Fundamental Properties of Logarithms

1. $\log_b b^x = x$
> In words, x is the exponent on a base b that gives b^x.
> That is, $b^x = b^x$.

2. $b^{\log_b x} = x \qquad x > 0$
> In words, $\log_b x$ is the exponent on a base b that gives x,
> which is the definition of logarithm.

Example 1 | Evaluate a logarithmic expression

Evaluate the given expressions.
a. $\log 10^3$
b. $\ln e^2$
c. $10^{\log 8.3}$
d. $e^{\ln 4.5}$
e. $\log_8 8^{4.2}$

Solution

a. $\log 10^3 = 3$ $\log 10^3$ is the exponent on 10 that gives 10^3; obviously it is 3.
b. $\ln e^2 = 2$ $\ln e^2$ is the exponent on e that gives e^2; obviously it is 2.
c. $10^{\log 8.3} = 8.3$ Consider $\log 8.3$; what is this? It is the exponent on a base 10 that gives the answer 8.3.
d. $e^{\ln 4.5} = 4.5$ Consider $\ln 4.5$; what is this? It is the exponent on a base e that gives the answer 4.5.
e. $\log_8 8^{4.2} = 4.2$ What is the exponent on 8 that gives $8^{4.2}$? Obviously, it is 4.2.

Logarithmic Equations

A **logarithmic equation** is an equation for which there is a logarithm on one or both sides. The key to solving logarithmic equations is the following theorem, which we will call the **log of both sides theorem**.

The proof of this theorem is not difficult, and it depends on the two fundamental properties of logarithms given in the previous subsection.

Log of Both Sides Theorem

If A, B, and b are positive real numbers with $b \neq 1$, then

$$\log_b A = \log_b B \text{ is equivalent to } A = B$$

Basically, all logarithmic equations in this book fall into one of four types:

		Example:
Type I:	The unknown is the logarithm.	$\log_2 \sqrt{3} = x$
Type II:	The unknown is the base.	$\log_x 6 = 2$
Type III:	The logarithm of an unknown is equal to a number.	$\ln x = 5$
Type IV:	The logarithm of an unknown is equal to the log of a number.	$\log_5 x = \log_5 72$

The following example illustrates the procedure for solving each type of logarithmic equation.

Example 2 | Solve four types of logarithmic equations

Solve the following logarithmic equations:
a. Type I: $\log_2 \sqrt{3} = x$
b. Type II: $\log_x 6 = 2$
c. Type III: $\ln x = 5$
d. Type IV: $\log_5 x = \log_5 72$

Solution

a. Type I: $\log_2 \sqrt{3} = x$. If the logarithmic expression does not contain a variable, you can use your calculator to evaluate. Remember, this type was evaluated in Section 10.1. If it is a common logarithm (base 10), use the $\boxed{\text{LOG}}$ key; if it is a natural logarithm (base e), use the $\boxed{\text{LN}}$ key; if it has another base, use the change of base theorem:

$$\log_a N = \frac{\log N}{\log a} \quad \text{or} \quad \log_a N = \frac{\ln N}{\ln a}$$

For this example, we see

$$x = \log_2 \sqrt{3} = \frac{\log \sqrt{3}}{\log 2} \approx 0.7924812504$$

b. Type II: $\log_x 6 = 2$. If the unknown is the base, then use the definition of logarithm to write an equation that is not a logarithmic equation.

$$\log_x 6 = 2 \qquad \textit{Given}$$
$$x^2 = 6 \qquad \textit{Definition of logarithm}$$
$$x = \pm\sqrt{6} \qquad \textit{Solve quadratic equation.}$$

When solving logarithmic equations, make sure your answers are in the domain of the variable. Remember that, by definition, the base must be positive. For this example, $x = -\sqrt{6}$ is not in the domain, so the solution is $x = \sqrt{6}$.

c. Type III: The third and fourth types of logarithmic equations are the most common, and both involve the logarithm of an unknown quantity on one side of an equation. For the third type, use the definition of logarithm (and a calculator for an approximate solution, if necessary).

$$\ln x = 5 \qquad \textit{Given}$$
$$e^5 = x \qquad \textit{5 is the exponent on e that gives x.}$$

This is the exact solution. An approximate solution is $x \approx 148.4131591$.

d. Type IV: When a logarithm occurs on both sides, use the log of both sides theorem. *Make sure the log on both sides has the same base:*

$$\log_5 x = \log_5 72$$
$$x = 72$$

Historical **NOTE**

For years, logarithms were used as a computational aid, but today they are more important in solving problems. Perhaps we have come full circle: The Babylonians first used logarithms to solve problems and not to do calculations.

The modern basis of logarithms was developed by Tycho Brahe (1546–1601) to disprove the Copernican theory of planetary motion. The name he used for the method was *prostaphaeresis.* In 1590, a storm brought together Brahe and John Craig, who in turn told Napier about Brahe's method. It was John Napier (1550–1617) who was the first to use the word *logarithm.* Napier was the Isaac Asimov of his day, having envisioned the tank, the machine gun, and the submarine. He also predicted that the end of the world would occur between 1688 and 1700. He is best known today as the inventor of logarithms, which, until the advent of the calculator, were used extensively with complicated calculations. Today, the use of logarithms goes far beyond numerical calculations.

Example 2 illustrates the procedures for solving logarithmic equations, but most logarithmic equations are not as easy as those in Example 2. Usually, you must do some algebraic simplification to put the problem into the form of one of the four types of logarithmic equations. You might also have realized that Type IV is a special case of Type III. For example, to solve

$$\log_5 x = \log_3 72$$

which looks like Example 2d, we see that we cannot use the log of both sides theorem because the bases are not the same. We can, however, treat this as a Type III equation by using the definition of logarithm to write

$$x = 5^{\log_3 72}$$

This can be evaluated using a calculator. However, you may find it easier to visualize if we write

$$\log_5 x = \log_3 72 \approx 3.892789261$$

so that 3.892789261 is the exponent on a base 5 that gives x. In other words,

$$x \approx 5^{3.892789261} \approx 525.9481435$$

Laws of Logarithms

To simplify logarithmic expressions, we remember that a logarithm is an exponent and the laws of exponents correspond to the **laws of logarithms.**

Laws of Logarithms

If A, B, and b are positive numbers, p is any real number, and $b \neq 1$:

First Law (Additive)

$\log_b (AB) = \log_b A + \log_b B$ The log of the product of two numbers is the sum of the logs of those numbers.

Second Law (Subtractive)

$\log_b \left(\frac{A}{B}\right) = \log_b A - \log_b B$ The log of the quotient of two numbers is the log of the numerator minus the log of the denominator.

Third Law (Multiplicative)

$\log_b A^p = p \log_b A$ The log of the pth power of a number is p times the log of that number.

The proofs of these laws of logarithms are easy. The additive law of logarithms comes from the additive law of exponents:

$$b^x b^y = b^{x+y}$$

Let $A = b^x$ and $B = b^y$, so that $AB = b^{x+y}$. Then, from the definition of logarithm, these three equations are equivalent to

$$x = \log_b A, \quad y = \log_b B, \quad \text{and} \quad x + y = \log_b(AB)$$

Therefore, by putting these pieces together, we have

$$\log_b(AB) = x + y = \log_b A + \log_b B$$

Similarly, for the subtractive law of logarithms,

$$\frac{A}{B} = \frac{b^x}{b^y}$$

$$= b^{x-y} \qquad \text{Subtractive law of exponents}$$

$$x - y = \log_b\left(\frac{A}{B}\right) \qquad \text{Definition of logarithm}$$

$$\log_b A - \log_b B = \log_b\left(\frac{A}{B}\right) \qquad \text{Since } x = \log_b A \text{ and } y = \log_b B$$

The proof of the multiplicative law of logarithms follows from the multiplicative law of exponents and you are asked to do this in the problem set. We can also prove this multiplicative law by using the additive law of logarithms for p a positive integer:

$$\log_b A^p = \log_b(\underbrace{A \cdot A \cdot A \cdot \cdots \cdot A}_{p \text{ factors}}) \qquad \text{Definition of } A^p$$

$$= \underbrace{\log_b A + \log_b A + \log_b A + \cdots + \log_b A}_{p \text{ terms}} \qquad \text{Additive law of logarithms}$$

When logarithms were used for calculations, the laws of logarithms were used to expand an expression such as $\log\left(\frac{6.45 \cdot 62^2}{84.2}\right)$. Calculators have made such problems obsolete. Today, logarithms are important in solving equations, and the procedure for solving logarithmic equations requires that we take an algebraic expression involving logarithms and write it as a single logarithm. We might call this *contracting* a logarithmic expression.

Be sure the bases are the same before you use the laws of logarithms.

Example 3 Use logarithmic properties to contract

Write each statement as a single logarithm.
a. $\log x + 5 \log y - \log z$ **b.** $\log_2 3x - 2 \log_2 x + \log_2(x + 3)$

Solution

a.
$$\log x + 5 \log y - \log z = \log x + \log y^5 - \log z \qquad \text{Third law}$$
$$= \log xy^5 - \log z \qquad \text{First law}$$
$$= \log \frac{xy^5}{z} \qquad \text{Second law}$$

b.
$$\log_2 3x - 2 \log_2 x + \log_2(x + 3) = \log_2 3x - \log_2 x^2 + \log_2(x + 3)$$
$$= \log_2 \frac{3x(x + 3)}{x^2}$$
$$= \log_2 \frac{3(x + 3)}{x}$$

Example 4 Solve a logarithmic equation

Solve: $\log_8 3 + \frac{1}{2} \log_8 25 = \log_8 x$

Solution The goal here is to make this look like a Type IV logarithmic equation so that there is a single log expression on both sides.

$$\log_8 3 + \frac{1}{2} \log_8 25 = \log_8 x \qquad \text{Given equation}$$

$$\log_8 3 + \log_8 25^{1/2} = \log_8 x \qquad \text{Third law of logarithms}$$

$$\log_8 3 + \log_8 5 = \log_8 x \qquad 25^{1/2} = (5^2)^{1/2} = 5$$

$$\log_8(3 \cdot 5) = \log_8 x \qquad \text{First law of logarithms}$$

$$15 = x \qquad \text{Log of both sides theorem}$$

The solution is 15. (Check to be sure 15 is in the domain of the variable.)

When solving logarithmic equations, you must look for extraneous solutions because the logarithm requires that the arguments be positive, but when solving an equation, we may not know the signs of the arguments. For example, if you solve an equation involving $\log x$ and obtain two answers (for example, $x = 3$ and $x = -4$), then the value $x = -4$ must be extraneous because $\log(-4)$ is not defined.

Example 5 Solve logarithmic equations

Solve:
a. $\log 15 + 2 = \log(x + 250)$ **b.** $\log 5 + \log(2x^2) = \log x + \log 15$

Solution Use the laws of logarithms to combine the log statements. We have chosen two examples that are quite similar but whose solutions require slightly different procedures. For part **a,** rewrite the expression so that all parts involving logarithms are on one side. In part **b,** rewrite the expression so that all logarithms involving the variable are on one side.

a. This is a Type III logarithmic equation. Use the definition of logarithm.

$$\log 15 + 2 = \log(x + 250) \qquad \text{Given equation}$$

$$2 = \log(x + 250) - \log 15 \qquad \text{Subtract log 15 from both sides.}$$

$$2 = \log \frac{x + 250}{15} \qquad \text{Second law of logarithms}$$

When solving logarithmic equations, you must be mindful of extraneous solutions.

$$10^2 = \frac{x + 250}{15} \qquad \text{Definition of logarithm}$$

$$1{,}500 = x + 250 \qquad \text{Multiply both sides by 15.}$$

$$1{,}250 = x \qquad \text{Subtract 250 from both sides.}$$

The solution is $x = 1{,}250$.

b. This is a Type IV logarithmic equation. Use the log of both sides theorem.

$$\log 5 + \log(2x^2) = \log x + \log 15 \qquad \text{Given equation}$$

$$\log(2x^2) - \log x = \log 15 - \log 5 \qquad \text{Subtract log } x \text{ and log 5 from both sides.}$$

$$\log \frac{2x^2}{x} = \log \frac{15}{5} \qquad \text{Second law of logarithms}$$

$$\log(2x) = \log 3 \qquad \text{Log of both sides theorem}$$

$$2x = 3$$

$$x = \frac{3}{2}$$

The solution is $x = \frac{3}{2}$.

Example 6 Prove change of base theorem

Prove: $\log_a x = \dfrac{\log_b x}{\log_b a}$

Solution   Let $y = \log_a x$.

$$a^y = x \qquad \text{Definition of logarithm}$$

$$\log_b a^y = \log_b x \qquad \text{Log of both sides theorem}$$

$$y \log_b a = \log_b x \qquad \text{Third law of logarithms}$$

$$y = \frac{\log_b x}{\log_b a} \qquad \text{Divide both sides by } \log_b a \ (\log_b a \neq 0).$$

Thus, by substitution, $\log_a x = \dfrac{\log_b x}{\log_b a}$.

In Chapter 1, we asked you to name the largest number that can be written with three digits. The next example considers the size of this huge number.

Example 7 Largest number puzzle

Pólya's Method

The largest number that you can write with three digits is 9^{9^9}. Write this number as a power of 10.

Solution   We use Pólya's problem-solving guidelines for this example.

Understand the Problem. Do you know what 9^{9^9} means? Try using your calculator to find this number.

First try: $9\wedge9\wedge9 \approx 1.966270505 \times 10^{77}$

What the calculator is giving is $(9^9)^9$. Is this the largest possible number with three digits?

Second try: 9^(9^9) produces an *overflow*, so we know this number is larger than (9^9)^9.

Devise a Plan. Find the number 9^(9^9) as a power of 10, which we write as 10^n, and the plan is to use logarithms to find n.

Carry Out the Plan.

$$9\text{^}(9\text{^}9) = 10^n \qquad \text{For some number } n$$

$$9^{387,420,489} = 10^n \qquad \text{Use a calculator for } 9^9.$$

$$n = \log 9^{387,420,489} \qquad \text{Definition of logarithm}$$

$$= 387,420,489 \log 9 \qquad \text{Property of logarithms}$$

$$\approx 369,693,099.6 \qquad \text{Calculator approximation}$$

Look Back. This means $9^{9^9} \approx 10^{369,693,099.6}$.

Problem Set 10.2

Level 1

1. **IN YOUR OWN WORDS**   What is a logarithmic equation?

2. **IN YOUR OWN WORDS**   What are the four types of logarithmic equations?

3. **IN YOUR OWN WORDS**   Outline a procedure for solving logarithmic equations.

Classify each of the statements in Problems 4–20 as true or false. If a statement is false, explain why you think it is false.

4. log 500 is the exponent on 10 that gives 500.

5. A common logarithm is a logarithm in which the base is 2.

6. A natural logarithm is a logarithm in which the base is 10.

7. In $\log_b N$, the exponent is N.

8. To evaluate $\log_5 N$, divide log 5 by log N.

9. If $2\log_3 81 = 8$, then $\log_3 6{,}561 = 8$.

10. If $2\log_3 81 = 8$, then $\log_3 81 = 4$.

11. If $\log_{1.5} 8 = x$, then $x^{1.5} = 8$.

12. $\ln\dfrac{x}{2} = \dfrac{\ln x}{2}$

13. $\log_b(A + B) = \log_b A + \log_b B$

14. $\log_b AB = (\log_b A)(\log_b B)$

15. $\dfrac{\log_b A}{\log_b B} = \log_b \dfrac{A}{B}$ 16. $\dfrac{\log A}{\log B} = \dfrac{\ln A}{\ln B}$

17. $\dfrac{\log_b A}{\log_b B} = \log_b(A - B)$

18. $\dfrac{\log_b A}{\log_b B} = \log_b A - \log_b B$

19. $\log_b N$ is negative when N is negative.

20. log N is negative when $N > 1$.

Find a simplified value for x in Problems 21–26 by inspection. Do not use a calculator.

21. **a.** $e^{\ln 23}$ **b.** $10^{\log 3.4}$
    c. $4^{\log_4 x}$ **d.** $\log_b b^x$

22. **a.** $\log 10^{4.2}$ **b.** $\ln e^3$
    c. $\log_6 6^x$ **d.** $b^{\log_b x}$

23. **a.** $\log_5 25 = x$ **b.** $\log_2 128 = x$
    c. $\log_3 81 = x$ **d.** $\log_4 64 = x$

24. **a.** $\log\frac{1}{10} = x$ **b.** $\log 10{,}000 = x$
    c. $\log 1{,}000 = x$ **d.** $\log \frac{1}{1{,}000} = x$

25. **a.** $\log x = 5$ **b.** $\log_x e = 1$
    c. $\ln x = 2$ **d.** $\ln x = 3$

26. **a.** $\ln x = 4$ **b.** $\ln x = \ln 14$
    c. $\ln 9.3 = \ln x$ **d.** $\ln 109 = \ln x$

Contract the expressions given in Problems 27–30. That is, use the properties of logarithms to write each expression as a single logarithm with a coefficient of 1.

27. **a.** $\log 2 + \log 3 + \log 4$
    b. $\log 40 - \log 10 - \log 2$
    c. $2\ln x + 3\ln y - 4\ln z$

28. **a.** $3\ln 4 - 5\ln 2 + \ln 3$
    b. $3\ln 4 - 2\ln(2 + 2)$
    c. $3\ln 4 - 5(\ln 2 + \ln 3)$

29. **a.** $\ln 3 - 2\ln 4 + \ln 8$
    b. $\ln 3 - 2\ln(4 + 8)$
    c. $\ln 3 - 2(\ln 4 + \ln 8)$

30. **a.** $\log(x^2 - 9) - \log(x + 3)$
    b. $\log(x^2 - x - 6) - \log(x + 2)$
    c. $\ln(x^2 - 4) - \ln(x + 2)$

The pH of a substance measures its acidity or alkalinity. It is found by the formula

$$pH = -\log[H^+]$$

where $[H^+]$ is the concentration of hydrogen ions in an aqueous solution given in moles per liter. Find the pH (to the nearest tenth) for the solutions in Problems 31–34.

31. lemon with $[H^+] = 2.86 \times 10^{-4}$

32. vinegar with $[H^+] = 3.98 \times 10^{-3}$

33. rainwater with $[H^+] = 6.31 \times 10^{-7}$

34. toothpaste with $[H^+] = 1.26 \times 10^{-10}$

35. An advertising agency conducted a survey and found that the number of units sold, N, is related to the amount a spent on advertising (in dollars) by the following formula:

$$N = 1,500 + 300 \ln a \ (a \geq 1)$$

How many units are sold after spending $1,000?

36. An advertising agency conducted a survey and found that the number of units sold, N, is related to the amount a spent on advertising (in dollars) by the following formula:

$$N = 1,500 + 300 \ln a \ (a \geq 1)$$

How many units are sold after spending $50,000?

Level 2

Solve the equations in Problems 37–55 by finding the exact solution.

37. a. $\frac{1}{2}x - 2 = 2$
   b. $\frac{1}{2}\log x - \log 100 = 2$

38. a. $3 + 2x = 11$
   b. $\ln e^3 + 2\log x = 11$

39. a. $\frac{1}{2}x = 3 - x$
   b. $\frac{1}{2}\log_b x = 3\log_b 5 - \log_b x$

40. a. $x - 2 = 2$
   b. $\log 10^x - 2 = \log 100$

41. a. $1 = x - 1$
   b. $\ln e = \ln \frac{\sqrt{2}}{x} - \ln e$

42. a. $1 = \frac{3}{2} - x$
   b. $\log 10 = \log \sqrt{1,000} - \log x$

43. a. $3 - x = 1$
   b. $\ln e^3 - \ln x = 1$

44. a. $0 + x = 2$
   b. $\ln 1 + \ln e^x = 2$

45. $\log(\log x) = 1$

46. $\ln[\log(\ln x)] = 0$

47. $x^2 5^x = 5^x$

48. $x^2 3^x = 9(3^x)$

49. $\log x = 1.8 + \log 4.8$

50. $\ln x = 1.8 - \ln 4.8$

51. $\ln x - \ln 8 = 12$

52. $\log x + \log 8 = 12$

53. $\log 2 = \frac{1}{4}\log 16 - x$

54. $\log_8 5 + \frac{1}{2}\log_8 9 = \log_8 x$

55. $\log x + \log(x - 3) = 2$

Level 3

56. An advertising agency conducted a survey and found that the number of units sold, N, is related to the amount a spent on advertising (in dollars) by the following formula:

$$N = 1,500 + 300 \ln a(a \geq 1)$$

How much needs to be spent (to the nearest hundred dollars) to sell 4,000 units?

57. The "forgetting curve" for memorizing nonsense syllables is given by

$$R = 80 - 27 \ln t(t \geq 1)$$

where R is the percentage who remember the syllables after t seconds.
   a. In how many seconds would only 10% $(R = 10)$ of the students remember?
   b. Solve for t.

58. The "learning curve" describes the rate at which a person learns certain tasks. If a person sets a goal of typing N words per minute (wpm), the length of time t (in days) to achieve this goal is given by

$$t = -62.5 \ln \left(1 - \frac{N}{80}\right)$$

   a. According to this formula, what is the maximum number of words per minute?
   b. Solve for N.

Problem Solving 3

59. HISTORICAL QUEST   In 1935 Charles Richter and Beno Gütenberg of Cal Tech used a new scale for measuring earthquakes in California. This scale is now universally used to measure the strength of earthquakes. It relates the energy E (in ergs) to the magnitude of the earthquake M by the formula

$$M = \frac{\log E - 11.8}{1.5}$$

   a. A small earthquake is one that releases 15^{15} ergs of energy. What is the magnitude (to the nearest hundredth) of such an earthquake on the Richter scale?
   b. A large earthquake is one that releases 10^{25} ergs of energy. What is the magnitude (to the nearest hundredth) of such an earthquake on the Richter scale?
   c. How much energy is released in an 8.0 earthquake?
   d. Solve for E.

60. Prove the multiplicative law of logarithms using the multiplicative law of exponents. That is, prove

$$\log_b A^p = p \log_b A$$

10.3 | Applications of Growth and Decay

In populations, bank accounts, and radioactive decay, where the *rate of change* is held constant, the average growth rate is a constant percent of the current value. In calculus, a **growth formula** is derived, and this formula is presented in the following box.

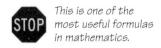

This is one of the most useful formulas in mathematics.

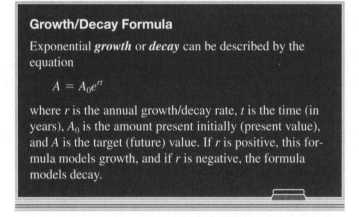

Growth/Decay Formula

Exponential *growth* or *decay* can be described by the equation

$$A = A_0 e^{rt}$$

where r is the annual growth/decay rate, t is the time (in years), A_0 is the amount present initially (present value), and A is the target (future) value. If r is positive, this formula models growth, and if r is negative, the formula models decay.

Note: You can use this formula as long as the units of time are the same. That is, if the time is measured in years, then the growth/decay rate is in years, but if the time is measured in days, then the growth/decay rate is a daily growth/decay rate.

Population Growth

Example 1 Find date of 7 billion people on earth

In 2009, the world population reached 7 billion. If we assume a growth rate of 1.5%, when will the population reach 8 billion?

Solution We are given the growth rate $r = 0.015$, the initial population $A_0 = 7$ (in billions), and the target population $A_0 = 8$ (in billions).

$$A = A_0 e^{rt} \qquad \text{\textit{Growth formula}}$$

$$8 = 7e^{0.015t} \qquad \text{\textit{Substitute known values.}}$$

$$\frac{8}{7} = e^{0.015t} \qquad \text{\textit{Solve for the exponential.}}$$

$$0.015t = \ln \frac{8}{7} \qquad \text{\textit{Definition of logarithm}}$$

$$t = \frac{\ln 8/7}{0.015} \qquad \text{\textit{Divide both sides by 0.015.}}$$

$$\approx 8.9020928 \qquad \text{\textit{Approximate answer by calculator}}$$

This means that we should pass the 8 billion mark in 2018.

Suppose we do not know the growth rate but have some population data. Consider the following example, assuming an exponential growth model.

Example 2 | Compare actual with predicted population

Phoenix, Arizona, had a population of 983,403 in 1990 and 1,321,045 in 2000. Predict the population in Phoenix in 2003. Compare this calculated number with the actual 2003 population of 1,388,416.

Solution   We use the growth/decay formula, where $A_0 = 983,403$ and $A = 1,321,045$ for $t = 10$ (years).

$$A = A_0 e^{rt} \qquad \text{\textit{Growth formula}}$$

$$1,321,045 = 983,403 e^{10r} \qquad \text{\textit{Substitute known values.}}$$

$$\frac{1,321,045}{983,403} = e^{10r} \qquad \text{\textit{Solve for the exponential.}}$$

$$10r = \ln\left(\frac{1,321,045}{983,403}\right) \qquad \text{\textit{Definition of logarithm}}$$

$$r = \frac{1}{10}\ln\left(\frac{1,321,045}{983,403}\right) \qquad \text{\textit{Solve for the unknown, r.}}$$

$$\approx 0.029515936343 \qquad \text{\textit{Calculator approximation}}$$

For 2003, we use the population in 2000 and calculate A using the above value for r:

$$A = A_0 e^{rt}$$
$$= 1,321,045 e^{r(3)}$$
$$\approx 1,443,356 \qquad \text{\textit{Round to the nearest unit.}}$$

Thus, the predicted 2003 population is 1,443,356. Since the actual population, 1,388,416, is less than the predicted number, we conclude that the growth rate between 2000 and 2003 has decreased a bit from the growth rate in 1990–2000.

Example 3 | Estimate AIDS fatalities

According to the Centers for Disease Control in Atlanta, Georgia, at the end of January 1992, the total number of AIDS-related U.S. deaths for all ages was 209,693. At that time, it was predicted that by January 1996 there would be from 400,000 to 450,000 cumulative deaths from the disease. Assuming these numbers are correct, estimate the cumulative number of AIDS-related deaths at the end of January 2002.

Solution   We assume that the growth rate will remain constant over the years of our study and also that the growth takes place continuously. From the end of January 1992 to January 1996 is 4 years, so $t = 4$. Furthermore, we will work with the more conservative estimate of cumulative deaths; that is, we let $A_0 = 209,693$ and $A = 400,000$.

$$A = A_0 e^{rt} \qquad \text{\textit{Growth formula}}$$

$$400,000 = 209,693 e^{r(4)} \qquad \text{\textit{Substitute known values.}}$$

$$\frac{400,000}{209,693} = e^{4r} \qquad \text{\textit{Solve for the exponential.}}$$

$$4r = \ln\left(\frac{400,000}{209,693}\right) \qquad \text{\textit{Definition of logarithm}}$$

$$r = \frac{1}{4}\ln\left(\frac{400,000}{209,693}\right) \qquad \text{\textit{Solve for the unknown.}}$$

$$\approx 0.1614549977 \qquad \text{\textit{Approximate answer by calculator}}$$

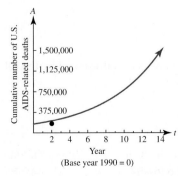

FIGURE 10.1 Cumulative deaths from AIDS, 1992–2002

Thus, at the end of January 2002, we have (for $t = 10$), $A = 209,693 e^{10r} \approx 1,053,839$. A graph is shown in Figure 10.1.

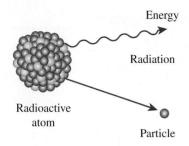

Energy

Radiation

Radioactive atom

Particle

FIGURE 10.2 Radioactive decay

Radioactive Decay

We now consider another application involving decay. Radioactive materials decay over time (see Figure 10.2). Each substance has a different decay rate. In the following example, when we say the decay rate of neptunium-239 is 31%, we imply that the rate is negative—that is, growth implies positive rate and decay implies negative rate; or, saying it another way, positive rate is growth and negative rate is decay.

Example 4 | Find a decay rate

If 100.0 mg of neptunium-239 (^{239}Np) decays to 73.36 mg after 24 hours, find the value of r in the growth/decay formula for t expressed in days.

Solution Since $A = 73.36$, $A_0 = 100.0$, and $t = 1$ (day), we have

$A = A_0 e^{rt}$	*Growth/decay formula*
$73.36 = 100 e^{r(1)}$	*Substitute known values.*
$0.7336 = e^r$	*Solve for the exponential.*
$r = \ln 0.7336$	*Definition of logarithm*
≈ -0.309791358	*Approximate answer by calculator*

Thus, the daily decay rate is approximately 31%.

We often specify radioactive decay in terms of what is called the **half-life.** This is the time that it takes for a particular radioactive substance to decay to half of its original amount.

Example 5 | Find the decay rate for carbon-14

Carbon-14, used for archeological dating, has a half-life of 5,730 years. Find the decay rate for carbon-14.

Solution If 100 mg of carbon-14 ($A_0 = 100$) is present, we are given that in 5,730 years ($t = 5,730$) there will be 50 mg present ($A = 50$). We use the growth/decay formula:

$A = A_0 e^{rt}$	*Growth/decay formula*
$50 = 100 e^{r(5,730)}$	*Substitute known values.*
$0.5 = e^{5,730r}$	*Solve for the exponential.*
$5,730r = \ln 0.5$	*Definition of logarithm*
$r = \dfrac{\ln 0.5}{5,730}$	*Solve for the unknown.*
$\approx -1.209680943\text{E} - 4$	*Approximate answer by calculator*

Use this value of r for carbon-14 calculations.

CAUTION

We will use the decay rate we find in this example in other examples and in the problem set.

We can use the decay rate for carbon-14 to date an artifact, as illustrated by the following example.

Example 6 | Estimate the age of a fossil

An archaeologist has found a fossil in which the ratio of ^{14}C to ^{12}C is 20% of the ratio found in the atmosphere. Approximately how old is the fossil?

Solution As the radioactive isotope carbon-14 (denoted by ^{14}C) decays, it changes to a stable form of carbon, called carbon-12. Once again, we use the growth/decay formula with the ratio of ^{14}C to ^{12}C to be 20% as follows:

$$A = A_0 e^{rt} \qquad \text{\textit{Growth/decay formula}}$$

$$\frac{A}{A_0} = e^{rt} \qquad \text{\textit{Divide both sides by } A_0.}$$

$$0.20 = e^{rt} \qquad \text{\textit{The given ratio is 20\%.}}$$

$$rt = \ln 0.20 \qquad \text{\textit{Definition of logarithm}}$$

$$t = \frac{\ln 0.20}{r} \qquad \text{\textit{Solve for the unknown; use the value of r from Example 5.}}$$

$$\approx 13{,}304.64798 \qquad \text{\textit{Approximate answer by calculator}}$$

The fossil is approximately 13,000 yr old.

Logarithmic Scales

Growth and decay examples are exponential models, but as we have seen, a logarithm is an exponent and is therefore directly related to growth and decay. A **logarithmic scale** is a scale in which logarithms are used to make data more manageable by expanding small variations and compressing large ones.

For example, prior to 1935, *seismographs* were used to record the amount of earth movement generated by an earthquake's seismic wave; this movement was recorded on a *seismogram*. The *amplitude* of a seismogram is the vertical distance between the peak or valley of the recording of the seismic wave and a horizontal line formed if there is no earth movement. This amplitude is measured using a very small unit, *micrometer* (denoted by μm), which is one-millionth of a meter. This small movement on a seismograph is used to measure a very large amount of energy released by an earthquake, and it must be done in such a way that the location of the seismograph relative to the earthquake's location (called the *epicenter*) is not relevant. You have, no doubt, heard of the well-known *Richter scale* used today as a means of measuring the magnitude, M, of an earthquake.

In 1935, Charles F. Richter, a seismologist at the California Institute of Technology, declared that the magnitude M of an earthquake with amplitude A on a seismograph was

$$M = \log \frac{A}{A_0}$$

where A_0 is the amplitude of a "standard earthquake." This number M is called the **Richter number** or **Richter scale** to denote the size of an earthquake. Richter measured a large number of extremely small southern California earthquakes, and *defined* $\log A_0$ to be -1.7 for a seismograph located 20 km from the epicenter. Using the properties of logarithms, we know

$$M = \log A - \log A_0$$
$$= \log A - (-1.7)$$
$$= \log A + 1.7$$

If the seismograph is located 300 km from the epicenter, then $\log A_0$ is *defined* to be -4.0, so

$$M = \log A - \log A_0$$
$$= \log A + 4.0$$

Since the calculation of the magnitude depends on the distance of the seismograph from the epicenter, and since for a particular earthquake the distance of the seismograph from the epicenter is not known, an actual earthquake is usually measured at three different

TABLE 10.1		
Magnitudes of Major World Earthquakes		
Date	Location	Magnitude
1906	San Francisco	8.3
1906	Valparaiso	8.6
1915	Avezzano	7.5
1920	Gansu, China	8.6
1933	Japan	8.9
1946	Honshu	8.4
1960	Chile	9.5
1971	San Fernando	6.6
1985	Mexico	8.1
1989	Loma Prieta	7.1
1994	Northridge	6.6
1998	Balleny Islands	8.3
1999	Turkey	7.4
2003	Colima, Mexico	7.8
2005	Indonesia	8.6
2007	Chile	7.7
2007	Indonesia	8.6
2010	Chile	8.8

Note: It is customary to give Richter scale readings rounded to the nearest tenth.

locations in order to determine the epicenter. The Richter scale ratings for some well-known quakes are shown in Table 10.1, but it should be noted that if you search the Web you will find that earthquakes are common. In 1998 there were more than 10 quakes recorded with a magnitude of over 7.0.

Example **7** **Find a Richter scale reading**

A seismograph 300 km from the epicenter of an earthquake recorded a maximum amplitude of $4.9 \times 10^3 \mu m$. Find the earthquake's magnitude. What would be the Richter scale reading for a seismograph 20 km from the epicenter of the same earthquake?

Solution

$$
\begin{aligned}
M &= \log A + 4.0 \\
&= \log(4.9 \times 10^3) + 4.0 \\
&\approx 7.7
\end{aligned}
$$

For a seismograph 20 km from the epicenter,

$$M = \log(4.9 \times 10^3) + 1.7 \approx 5.4$$

Why is the Richter scale called logarithmic? The reason is that if you increase the magnitude by 1, then the quake is 10 times stronger; if you increase the magnitude by 2, then the quake is $10^2 = 100$ times stronger; if you increase the magnitude by 3, then the quake is 10^3 times stronger. We illustrate an increase of magnitude by 4 with the following example.

Example **8** **Compare the strength of two earthquakes**

Compare the strengths of earthquakes with magnitudes 4 and 8.

Solution Let M_1 and M_2 be the two given magnitudes. Then

$$
\begin{aligned}
M_1 - M_2 &= (\log A_1 - \log A_0) - (\log A_2 - \log A_0) \\
&= \log A_1 - \log A_2 \\
&= \log \frac{A_1}{A_2}
\end{aligned}
$$

For this problem, in particular, we have

$$
\begin{aligned}
8 - 4 &= \log \frac{A_1}{A_2} \\
4 &= \log \frac{A_1}{A_2} \\
10^4 &= \frac{A_1}{A_2} \\
A_1 &= 10{,}000 A_2
\end{aligned}
$$

A doubling of the magnitude from 4 to 8 means that the stronger earthquake's amplitude is $10^4 = 10{,}000$ times greater.

The amount of earth movement from an earthquake is measured by the amount of energy released by the earthquake. Consider the following example and compare with Example 8.

Example 9 Find the energy released from an earthquake

Compare the amount of earth movement (the energy released) by earthquakes of magnitudes of 4 and 8.

Solution Use the formula

$$M = \frac{\log E - 11.8}{1.5}$$

from Section 10.2. We solve for E:

$$1.5M = \log E - 11.8 \qquad \text{Multiply both sides by 1.5.}$$
$$1.5M + 11.8 = \log E \qquad \text{Add 11.8 to both sides.}$$
$$E = 10^{1.5M+11.8} \qquad \text{Definition of logarithm}$$

Let $M_1 = 4$ and $M_2 = 8$ with corresponding energies E_1 and E_2, respectively. We now compare these energies:

$$\frac{E_1}{E_2} = \frac{10^{1.5(4)+11.8}}{10^{1.5(8)+11.8}} \qquad \text{Same bases, subtract exponents.}$$
$$E_1 = 10^{-6}E_2 \qquad \text{Multiply both sides by } E_2.$$
$$10^6 E_1 = E_2 \qquad \text{Multiply both sides by } 10^6.$$

The earthquake with magnitude 8 releases a million times more energy than an earthquake with magnitude 4.

A second example of a logarithmic scale is the decibel rating used for measuring the intensity of sounds. To measure the intensity of sound, we need to understand that a sound is a vibration received by the ear and processed by the brain. We can place a listening device in the path of the sound and measure the amount of energy on that device per unit of area per second. This listening device acts like an eardrum, but the problem is that experiments have shown that humans perceive loudness on the basis of the ratio of intensities of two different sounds. For this reason, the unit of measurement for measuring sounds, called the **decibel,** in honor of Alexander Graham Bell, the inventor of the telephone, is defined as a ratio of the intensity of one sound, I, and another sound, $I_0 \approx 10^{-16}$ watt/cm^2, which is the intensity of a barely audible sound for a person with normal hearing.

The issue is further complicated by the fact that a human ear can hear an incredible range of sounds. A painful sound is 10^4 (100 trillion) times more intense than a barely audible sound. This leads us to define the number of decibels, D, by

$$D = 10 \log \frac{I}{I_0}$$

for a sound of intensity I. This is called the **decibel formula.** The decibel rating, abbreviated dB, for various sounds is shown in Table 10.2.

TABLE 10.2

Decibel Ratings

Sound	dB Rating
threshold of hearing	0
recording studio	20
whisper	25
quiet room	30
conversation	60
traffic	70
train	100
orchestra	110
rock music	115
pain threshold	120
rocket	125

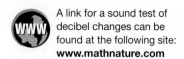

A link for a sound test of decibel changes can be found at the following site: **www.mathnature.com**

Example 10 Find a decibel rating

The background sound of a study room was measured to be 10^{-10} watt/cm^2. Find the decibel rating for the room.

Solution We are given $I = 10^{-10}$ and $I_0 = 10^{-16}$, so

$$D = 10 \log \frac{I}{I_0} \qquad \text{\textit{Decibel formula}}$$

$$= 10 \log \frac{10^{-10}}{10^{-16}} \qquad \text{\textit{Substitute known values.}}$$

$$= 10 \log 10^6 \qquad \text{\textit{For division, subtract exponents.}}$$

$$= 60 \qquad \text{\textit{Grant's tomb property: } log 10^6 = 6}$$

A scale for loudness of sounds begins at 0 dB (threshold of hearing) and extends to the threshold of pain, 120 dB. Each increase of 10 decibels is perceived as a doubling of loudness. A sound of 70 dB is twice as loud as a 60 dB sound.

Example 11 Compare sound intensities

The noise in a classroom varies from 50 dB to 62 dB. Find the corresponding variation in intensities.

Solution We are given $D_1 = 50$ and $D_2 = 62$, and we wish to compare I_1 and I_2:

$$D_2 - D_1 = 10 \log \frac{I_2}{I_0} - 10 \log \frac{I_1}{I_0}$$

$$= 10 \log I_2 - 10 \log I_0 - 10 \log I_1 + 10 \log I_0$$

$$= 10(\log I_2 - \log I_1)$$

$$= 10 \log \frac{I_2}{I_1}$$

$$62 - 50 = 10 \log \frac{I_2}{I_1} \qquad \text{\textit{Substitute given values.}}$$

$$1.2 = \log \frac{I_2}{I_1}$$

$$10^{1.2} = \frac{I_2}{I_1} \qquad \text{\textit{Definition of logarithm}}$$

$$15.85 \approx \frac{I_2}{I_1} \qquad \text{\textit{Calculate approximation of } 10^{1.2}}$$

$$I_2 \approx 16 I_1$$

The louder room is about 16 times noisier than the quiet room.

Problem Set 10.3

Level 1

1. What is the growth/decay formula?

2. **IN YOUR OWN WORDS**   What do we mean by half-life?

3. **IN YOUR OWN WORDS**   What is a logarithmic scale?

4. **IN YOUR OWN WORDS**   What is a micrometer?

Find the number of decibels for the power of the sound given in Problems 5–10. Round to the nearest decibel.

5. A whisper, 10^{-13} watts/cm^2

6. A traffic jam, 10^{-8} watts/cm^2

7. A rock concert, 5.23×10^{-6} watts/cm^2

8. A conversation, 3.16×10^{-10} watts/cm^2

9. A rocket engine, 2.53×10^{-5} watts/cm^2

10. According to the *Guinness Book of World Records*, the world's loudest shout, by Skipper Kenny Leader, was 10^{-5} watts/cm^2.

Level 2

11. Example 3 calculates the growth rate of AIDS-related U.S. deaths and uses this information to predict the cumulative number of AIDS-related deaths at the end of January 2002. Repeat this example for the number of deaths at the turn of the century, January 2001.

12. Example 3 calculates the growth rate of AIDS-related U.S. deaths and uses this information to predict the cumulative number of AIDS-related deaths at the end of January 2002 by using the conservative estimate of 400,000 in 1996. Repeat this example using the less conservative estimate of 450,000.

13. If the half-life of cesium-137 is 30 years, find the decay constant, r.

14. If the half-life of plutonium-238 is 86 years, find the decay constant, r.

15. Find the half-life (to the nearest year) of strontium-90 if $r = -0.0246$.

16. Find the half-life (to the nearest year) of krypton if $r = -0.0641$.

17. A seismograph 300 km from the epicenter of an earthquake recorded a maximum amplitude of $5.1 \times 10^2 \, \mu$m. Find this earthquake's magnitude.

18. A seismograph 20 km from the epicenter of an earthquake recorded a maximum amplitude of $8.2 \times 10^5 \, \mu$m. Find this earthquake's magnitude.

19. Compare the strengths of earthquakes with magnitudes 4 and 6.

20. Compare the strengths of earthquakes with magnitudes 7 and 8.

21. The energy released by an earthquake is approximated by

$$\log E = 11.8 + 1.5M$$

What is the energy released by the 1906 San Francisco quake, which measured 8.3 on the Richter scale? This energy, it is estimated, would be sufficient to provide the entire world's food requirements for a day.

22. The Dead Sea Scrolls were written on parchment in about 100 BC. What percentage of ^{14}C originally contained in the parchment remained when the scrolls were discovered in 1947?

23. Tests of an artifact discovered at the Debert site in Nova Scotia show that 28% of the original ^{14}C is still present. What is the probable age of the artifact?

24. The half-life of ^{234}U, uranium-234, is 2.52×10^5 yr. If 97.3% of the uranium in the original sample is present, what length of time (to the nearest thousand years) has elapsed?

25. The half-life of ^{22}Na, sodium-22, is 2.6 yr. If 15.5 g of an original 100-g specimen remains, how much time has elapsed (to the nearest year)?

26. How much more energy was released by the Loma Prieta quake (Richter scale 7.1) than by the Northridge quake (Richter scale 6.6)?

27. How much more energy was released by the 1960 Chile quake (Richter scale 9.5) than the 1906 San Francisco quake (Richter scale 8.3)?

FIGURE 10.3 Archeological dig

28. An artifact was found and tested for its carbon-14 content. If 12% of the original carbon-14 was still present, what is its probable age (to the nearest 100 years)?

29. An artifact was found and tested for its carbon-14 content. If 85% of the original carbon-14 was still present, what is its probable age (to the nearest 100 years)?

30. The radioactive substance neptunium-139 decays to 73.36% of its original amount after 24 hours. How long (to the nearest hour) would it take for 43% of the original neptunium to be present? What is the half-life of neptunium-139?

31. The 1989 World Series San Francisco quake was initially reported to have magnitude 7.0, but later this was revised to 7.1. How much energy released corresponds to this increase in magnitude?

32. Compare the amount of earth movement (energy released) by earthquakes of magnitudes of 3 and 6.

33. Compare the amount of earth movement (energy released) by earthquakes of magnitudes 7 and 8.

If an object at temperature B is surrounded by air at temperature A, it will gradually cool so that the temperature T, t minutes later, is given by Newton's law of cooling:

$$T = A + (B - A)10^{-kt}$$

The constant k depends on the particular object and can be found by solving for k

$$k = \frac{1}{t} \log \frac{B - A}{T - A}$$

Use this information for Problems 34–37.

34. You draw a tub of hot water $(k = 0.01)$ for a bath. The water is 100°F when drawn and the room is 72°F. If you are called away to the phone, what is the temperature of the water 20 minutes later when you get in?

35. You take a batch of chocolate-chip cookies from the oven $(375°F)$ when the room temperature is 74°F. If the cookies cool for 10 minutes and $k = 0.075$, what is the temperature of the cookies?

36. It is known that the temperature of a newly baked brick falls from 120°C to 70°C in an hour when placed in 20°C air. What is the temperature of the brick after 30 minutes?

37. The temperature of a piece of cooked chicken is initially 100°C. In air of 22°C it cools to 45° in 30 minutes. What is the temperature in 40 minutes?

38. The atmospheric pressure P in pounds per square inch (psi) is given by

$$P = 14.7e^{-0.21a}$$

where a is the altitude above sea level (in miles). If a city has an atmospheric pressure of 13.23 psi, what is its altitude? (Recall that 1 mi = 5,280 ft.)

39. The atmospheric pressure P in pounds per square inch (psi) is approximated by

$$P = 14.7e^{-0.21a}$$

where a is the altitude above sea level in miles. If the atmospheric pressure of Denver is 11.9 psi, estimate Denver's altitude. (Recall that 1 mi = 5,280 ft.)

40. The atmospheric pressure P in pounds per square inch (psi) is approximated by

$$P = 14.7e^{-0.21a}$$

where a is the altitude above sea level in miles. If the pressure gauge in a small plane shows 10.2 psi, estimate the plane's altitude in feet. (Recall that 1 mi = 5,280 ft.)

41. A satellite has an initial radioisotope power supply of 50 watts (W). The power output in watts is given by

$$P = 50e^{-t/250}$$

where t is the time in days. Solve for t to find the time when the power supply is 30 W.

42. A satellite has an initial radioisotope power supply of 50 watts. The power output in watts is given by the equation

$$P = 50e^{-t/250}$$

where t is the time in days. If the satellite will operate if there is at least 10 watts of power, how long would we expect the satellite to operate?

Level 3

Find the following information for each of the cities described in Problems 43–50.

a. *What is the growth rate from 1990 to 2000?*
b. *Predict the population for 1994, using the growth rate for the period 1990–2000.*
c. *Compare the prediction with the 1994 actual, and comment on possible reasons for discrepancies, if any.*
d. *Predict the population for the year 2006.*

City	1990	2000	1994
43. Jacksonville	636,070	735,616	665,070
44. Sacramento	369,365	1,223,449	373,964
45. Nashville	488,374	545,524	504,505
46. El Paso	515,342	679,622	579,307
47. Honolulu	365,272	876,593	385,881
48. Boston	574,283	589,141	547,725
49. St. Paul	198,518	287,151	262,071
50. Shreveport	272,235	200,145	196,982

51. Between 1980 and 1990, the population of Los Angeles grew from 2,968,528 to 3,485,557, but by 1994 had dropped to 3,448,613. Predict the population of Los Angeles in 2004 using the given assumptions.
   a. Using the 1980 to 1990 growth rate
   b. Using the 1990 to 1994 growth rate
   c. Using the 1980 to 1994 growth rate

52. A bacteria culture had a population of 10 million at 10:00 A.M., and by 2:00 P.M. had grown to 18 million.
   a. Predict the population at 6:00 P.M. that same day.
   b. When will the population double in size?

53. **IN YOUR OWN WORDS** In July 1990, the world population was about 5.3 billion.
   a. If we assume a growth rate of 0.98%, when did the population reach 6 billion? Answer to the nearest month.
   b. Compare this with the actual time 6 billion was reached (October 1999).
   c. Formulate a better growth rate and a prediction about when the world will reach 7 billion. Answer to the nearest month.

54. **IN YOUR OWN WORDS** According to the World POPClock, the world population on 7/1/99 was 5,996,215,340 and on 7/1/06 it was 6,525,486,603. Use these numbers to find the annual growth rate for this period. Use this growth rate to project the world population on the day you work this problem. Check the World POPClock for the day's population and compare with your answer. Comment on the difference. Here is the Web address:

 http://www.census.gov/cgi-bin/ipc/popclockw

As usual, you can access this Web address through the address for this text: **www.mathnature.com**

55. A certain artifact is tested by carbon dating and found to contain 73% of its original carbon-14. As a cross-check, it is also dated using radium, and was found to contain 32% of the original amount. Assuming the dating procedures were accurate, what is the half-life of radium?

56. The radioactive isotope gallium-67 (symbol ^{67}Ga) used in the diagnosis of malignant tumors has a half-life of 46.5 hours. If we start with 100 mg of ^{67}Ga, what percentage is lost between the 30th and 35th hours? Is this the same as the percentage lost in any other 5-hour period?

Problem Solving 3

57. HISTORICAL QUEST   The Shroud of Turin (see Figure 10.4) is a rectangular linen cloth kept in the Chapel of the Holy Shroud in the cathedral of St. John the Baptist in Turin, Italy. It shows the image of a man whose wounds correspond with the biblical accounts of the crucifixion.

In 1389, Pierre d'Arcis, the Bishop of Troyes, wrote a memo to the Pope, accusing a colleague of passing off "a certain cloth, cunningly painted" as the burial shroud of Jesus Christ. Despite this early testimony of forgery, the so-called Shroud of Turin has survived as a famous relic. In 1988, a small sample of the Shroud of Turin was taken, and scientists from Oxford University, the University of Arizona, and the Swiss Federal Institute of Technology were permitted to test it. It was determined that 92.3% of the Shroud's original ^{14}C still remained. According to this information, how old was the Shroud in 1988?

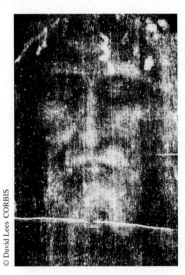

© David Lees CORBIS

FIGURE 10.4 Shroud of Turin

58. IN YOUR OWN WORDS   Construct a logarithmic scale. On a sheet of paper, draw a line 10 in. long and mark it in tenths and hundredths. Place the edge of a second sheet along this line and use the first as a ruler to mark off logs on the second. At 0, mark 1, since log 1 = 0. Mark 2 at 0.301 because log 2 ≈ 0.301. Mark 3 at 0.477, and so on until you have something similar to Figure 10.5.

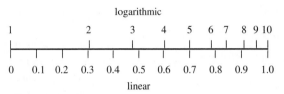

FIGURE 10.5 Logarithmic scale

Fill in the scale between 1 and 2 for every tenth, and the other units at least at the halves. Using two of these log scales, you can make a simple slide rule that you can use to multiply and divide. Explain how to do this and illustrate it to at least one other person.

59. In 1986, it was determined that the *Challenger* disaster was caused by failure of the primary O-rings. Linda Tappin gives a formula in "Analyzing Data Relating to the *Challenger* Disaster" (*The Mathematics Teacher*, Vol. 87, No. 6, Sept. 1994, pp. 423–426) that relates the temperature x (in degrees Fahrenheit) around the O-rings and the expected number y of eroded or leaky primary O-rings:

$$y = \frac{6e^{5.085 - 0.1156x}}{1 + e^{5.085 - 0.1156x}}$$

a. What is the predicted number of eroded or leaky O-rings at a temperature of 75°F?

b. What is the predicted number of eroded or leaky O-rings at a temperature of 32°F?

60. The Arrhenius equation is used to relate the viscosity η of a fluid (the fluid's internal friction, which is what makes it resist a tendency to flow) to its absolute temperature T:

$$\frac{1}{\eta} = Ae^{-E/(RT)}$$

where A is a constant specific to that fluid and R is the ideal gas constant. Solve this equation for T. The resulting formula is one you could use to investigate the viscosity of different grades of motor oil at different temperatures.

10.4 | CHAPTER SUMMARY

Mathematics abounds in bright ideas. No matter how long and hard one pursues her, mathematics never seems to run out of exciting surprises.

ROSS ONSBERGER

Important Ideas

A logarithm is an exponent. That is, $\log_b A$ is the exponent on a base b that gives the result A. [10.1]

Change of base theorem [10.1]

Fundamental properties of logarithms [10.2]

Log of both sides theorem [10.2]

Laws of logarithms—additive, subtractive, and multiplicative laws [10.2]

Take some time getting ready to work the review problems in this section. First review these important ideas. Look back at the definition and property boxes. If you look online, you will find a list of important terms introduced in this chapter, as well as the types of problems that were introduced. You will maximize your understanding of this chapter by working the problems in this section only after you have studied the material.

You will find some review help online at **www.mathnature.com**. There are links giving general test help in studying for a mathematics examination, as well as specific help for reviewing this chapter.

Chapter 10 Review Questions

Simplify each expression in Problems 1–5 without using a calculator or computer.

1. $\log 100 + \log \sqrt{10}$

2. $\ln e + \ln 1 + \ln e^{542}$

3. $\log_8 4 + \log_8 16 + \log_8 8^{2.3}$

4. $10^{\log 0.5}$

5. $\ln e^{\log 1,000}$

Evaluate each expression in Problems 6–8 rounded to two decimal places.

6. **a.** $\log 8.43$ **b.** $\log 9,760$

7. **a.** $\ln 2$ **b.** $\ln 0.125$

8. **a.** $\log_2 10$ **b.** $\log_{\pi} \frac{1}{\pi}$

Solve each equation in Problems 9–12 rounded to calculator accuracy.

9. $10^x = 85$ 10. $e^x = 500$

11. $435^x = 890$ 12. $e^{3x+1} = 45$

Give the exact solution for each equation in Problems 13–18.

13. $\log_6 x = 4$

14. $2^{3x-1} = 6$

15. $10^{2x} = 5$

16. $\log(x + 1) = 2 + \log(x - 1)$

17. $3 \ln \frac{e}{\sqrt[3]{5}} = 3 - \ln x$

18. Solve $A = P(1 + i)^x$ for x.

19. A healing law for skin wounds states that $A = A_0 e^{-0.1t}$, where A is the number of square centimeters of unhealed skin after t days when the original area of the wound was A_0. How many days does it take for half the wound to heal?

20. In 1992, it was reported that the number of teenagers with AIDS doubles every 14 months. Find an equation to model the number of teenagers that may be infected over the next 10 years.

BOOK REPORTS

Write a 500-word report on one of these books:

The Mathematical Experience, Philip J. Davis and Reuben Hersh (Boston: Houghton Mifflin, 1981).

Ethnomathematics: A Multicultural View of Mathematical Ideas, Marcia Ascher (Pacific Grove, CA: Brooks/Cole, 1991).

Group RESEARCH PROJECTS

Go to
www.mathnature.com
for references and links.

Karl J. Smith library

Working in small groups is typical of most work environments, and learning to work with others to communicate specific ideas is an important skill. Work with three or four other students to submit a single report based on each of the following questions.

G35. Before Hurricane Katrina in 2005, the entrance of the Aquarium of Americas in New Orleans had a gigantic building-size curve called a *logarithmic spiral*. Find out how to construct a logarithmic spiral, and write a paper about what you learned. Why do you suppose it would appear on the front of an aquarium?

G36. If we assume that the world population grows exponentially, then it is also reasonable to assume that the use of some nonrenewable resource (such as petroleum) will also grow exponentially. In calculus it is shown that for some constant k, under these assumptions, the formula for the amount of the resource, A, consumed from time $t = 0$ to $t = T$ is given by the formula

$$A = \frac{A_0}{k}(e^{rT} - 1)$$

where r is the relative growth rate of annual consumption.

a. Solve this equation for T to find a formula for the life expectancy of a particular resource.

b. According to the Energy Information Administration, the annual world production (in billions of barrels per day) of petroleum is shown in the following table:

Year	1975	1980	1985	1990	1995	2000	2003
Quantity	52.42	62.39	52.97	60.90	61.85	66.03	67.00

Find an exponential equation for these data.

c. If in 1998, the world petroleum reserves were 2.8 trillion barrels, estimate the life expectancy for petroleum.

Individual RESEARCH PROJECTS

www.mathnature.com

Learning to use sources outside your classroom and textbook is an important skill, and here are some ideas for extending some of the ideas in this chapter. You can find references to these projects in a library or at **www.mathnature.com.**

PROJECT 10.1 HISTORICAL QUEST   Write an essay on John Napier. Include what he is famous for today and what he considered to be his crowning achievement. Also include a discussion of "Napier's bones."

PROJECT 10.2 Write an essay on earthquakes. In particular, discuss the Richter scale for measuring earthquakes. What is its relationship to logarithms?

PROJECT 10.3 Write a paper using population analysis. See the Web pages mentioned in Section 10.3 for this individual project.

PROJECT 10.4 From your local chamber of commerce, obtain the population figures for your city for the years 1980, 1990, and 2000. Find the rate of growth for each period. Forecast the population of your city for the year 2010. Include charts and graphs. List some factors, such as new zoning laws, that could change the growth rate of your city.

PROJECT 10.5 Write an essay on carbon-14 dating. What is its relationship to logarithms?

11

THE NATURE OF FINANCIAL MANAGEMENT

Outline

What in the World?

"I think we should pay off our home loan," proclaimed Lorraine. "We have had this loan for almost 20 years and we only owe $20,000. What do you think?"

"I agree. I'd really like to save the $195 per month payment that we are making. Even though it is smaller than most house payments, it would be wonderful to be debt free," said Ron.

"I checked with our bank, and they told me that to pay off the loan, we must pay $20,000 plus a 'prepayment penalty' of 3%!"

"Oh, don't worry about that," chimed Ron. "Most lenders will waive this penalty. I'll call and request them to do this. But I still don't know if it is a wise financial move. Let's think about it . . . " added Lorraine.

Austin MacRae

Overview

The stated goal of this book is to strengthen your ability to solve problems—not the classroom type of problems, but those problems that you may encounter as an employee, a manager, or in everyday living. You can apply your problem-solving ability to your financial life. A goal of this chapter might well be to put some money into your bank account that you would not have had if you had not read this chapter. As a preview to this chapter, consider the question asked in Example 2 of Section 11.5 (page 551):

> Suppose you are 21 years old and will make monthly deposits to a bank account paying 10% annual interest compounded monthly. Which is the better option?
>
> *Option I:* Pay yourself $200 per month for 5 years and then leave the balance in the bank until age 65. (Total amount of deposits is $200 × 5 × 12 = $12,000.)
>
> *Option II:* Wait until you are 40 years old (the age most of us start thinking seriously about retirement) and then deposit $200 per month until age 65. (Total amount of deposits is $200 × 25 × 12 = $60,000)
>
> Warning!! The wrong answer to this question could cost you $4,000/mo for the rest of your life!

11.1   Interest

Finance is the art of passing money from hand to hand until it finally disappears.

ROBERT W. SARNOFF

It seems that everyone has money problems . . . . Either we have too much (you've read stories of people who don't know what to do with all their money) or we have too little. Very few people believe that they have "just the right amount of money."

Money . . . too much or too little?

Certain arithmetic skills enable us to make intelligent decisions about how we spend the money we earn.

Amount of Simple Interest

One of the most fundamental mathematical concepts that consumers, as well as business-people, must understand is *interest.* Simply stated, **interest** is money paid for the use of money. We receive interest when we let others use our money (when we deposit money in a savings account, for example), and we pay interest when we use the money of others (for example, when we borrow from a bank).

The amount of the deposit or loan is called the **principal** or **present value,** and the interest is stated as a percent of the principal, called the **interest rate.** The **time** is the length of time for which the money is borrowed or lent. The interest rate is usually an *annual interest rate,* and the time is stated in years unless otherwise given. These variables are related in what is known as the **simple interest formula.**

STOP   Remember this fundamental formula.

Simple Interest Formula

INTEREST = PRESENT VALUE × RATE × TIME

$$I = Prt$$

I = AMOUNT OF INTEREST
P = PRESENT VALUE (or PRINCIPAL)
r = ANNUAL INTEREST RATE
t = TIME (in years)

Suppose you save 20¢ per day, but only for a year. At the end of a year you will have saved $73. If you then put the money into a savings account paying 3.5% interest, how much interest will the bank pay you after one year? The present value (P) is $73, the rate ($r$) is 3.5% = 0.035, and the time (t, in years) is 1. Therefore,

$$I = Prt$$
$$= 73(0.035)(1)$$
$$= 2.555 \qquad \text{You can do this computation on a calculator: } \boxed{73}\boxed{\times}\boxed{.035}\boxed{=}$$

Round money answers to the nearest cent: After one year, the interest is $2.56.

Example 1 Find the amount of interest

How much interest will you earn in three years with an initial deposit of $73?

Solution $I = Prt = 73(0.035)(3) = 7.665$. After three years, the interest is $7.67.

Future Value

There is a difference between asking for the amount of interest, as illustrated in Example 1, and asking for the **future value.** The future value is the amount you will have after the interest is added to the principal, or present value. Let A = FUTURE VALUE. Then

$$A = P + I$$

Example 2 Find an amount of interest

Suppose you see a car with a price of $12,436 that is advertised at $290 per month for 5 years. What is the amount of interest paid?

Solution The present value is $12,436. The future value is the total amount of all the payments:

Monthly payment   Number of years

$$\underbrace{\$290}_{} \qquad \underbrace{\times\ 12}_{\text{Number of payments per year}} \qquad \times\ 5 \qquad = \$17,400$$

Therefore, the amount of interest is

$$I = A - P$$
$$= 17,400 - 12,436$$
$$= 4,964$$

The amount of interest is $4,964.

Interest for Part of a Year

The numbers in Example 2 were constructed to give a "nice" answer, but the length of time for an investment is not always a whole number of years. There are two ways to convert a number of days into a year:

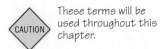
CAUTION

These terms will be used throughout this chapter.

Exact interest: 365 days per year

Ordinary interest: 360 days per year

Most applications and businesses use ordinary interest. So in this book, unless it is otherwise stated, assume ordinary interest; that is, use 360 for the number of days in a year:

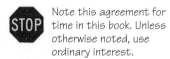

Note this agreement for time in this book. Unless otherwise noted, use ordinary interest.

Time (Ordinary Interest)

$$t = \frac{\text{ACTUAL NUMBER OF DAYS}}{360}$$

Calculating the time using ordinary interest often requires a calculator, as illustrated by Example 3.

Example 3 Time necessary to achieve a financial goal

Suppose you want to save $3,650, and you put $3,000 in the bank at 8% simple interest. How long must you wait?

Solution   Begin by identifying the variables: $P = 3,000$; $r = 0.08$; $I = 3,650 - 3,000 = 650$.

$$I = Prt$$
$$650 = 3,000(0.08)t$$
$$650 = 240t$$
$$\frac{650}{240} = t$$

This is 2 years plus some part of a year. To change the fractional part to days, do not clear your calculator, but subtract 2 (the number of years); then multiply the fractional part by 360:

$$\boxed{650} \; \boxed{\div} \; \boxed{240} \; \boxed{-} \; \boxed{2} \; \boxed{=} \; \boxed{\times} \; \boxed{360} \; \boxed{=}$$ Display: 255

The time you must wait is 2 years, 255 days.

Example 4 Find the amount to repay a loan

Suppose that you borrow $1,200 on March 25 at 21% simple interest. How much interest accrues by September 15 (174 days later)? What is the total amount that must be repaid?

Solution   We are given $P = 1,200$, $r = 0.21$, and $t = \dfrac{174 \leftarrow \text{Actual number of days}}{360 \leftarrow \text{Assume ordinary interest.}}$

$$I = Prt \qquad \text{Simple interest formula}$$
$$= 1,200(0.21)\left(\frac{174}{360}\right) \qquad \text{Substitute known values.}$$
$$= 121.8 \qquad \text{Use a calculator to do the arithmetic.}$$

The amount of interest is $121.80. To find the amount that must be repaid, find the future value:

$$A = P + I = 1,200 + 121.80 = 1,321.80$$

The amount that must be repaid is $1,321.80.

It is worthwhile to derive a formula for future value because sometimes we will not calculate the interest separately as we did in Example 4.

$$\text{FUTURE VALUE} = \text{PRESENT VALUE} + \text{INTEREST}$$
$$A = P + I$$
$$= P + Prt \qquad \text{Substitute } I = Prt.$$
$$= P(1 + rt) \qquad \text{Distributive property}$$

This is also known as the present value formula for simple interest when solved for the variable P.

Future Value Formula (Simple Interest)

$$A = P(1 + rt)$$

Example 5 Find a future value

If \$10,000 is deposited in an account earning $5\frac{3}{4}\%$ simple interest, what is the future value in 5 years?

Solution We identify $P = 10,000$, $r = 0.0575$, and $t = 5$.

$$A = P(1 + rt)$$
$$= 10,000(1 + 0.0575 \times 5)$$
$$= 10,000(1 + 0.2875)$$
$$= 10,000(1.2875)$$
$$= 12,875$$

The future value in 5 years is \$12,875.

Example 6 Find the amount necessary to retire

Suppose you have decided that you will need \$4,000 per month on which to live in retirement. If the rate of interest is 8%, how much must you have in the bank when you retire so that you can live on interest only?

Solution We are given $I = 4,000$, $r = 0.08$, and $t = \frac{1}{12}$ (one month $= \frac{1}{12}$ year):

$$I = Prt \qquad \text{Simple interest formula}$$
$$4,000 = P(0.08)\left(\frac{1}{12}\right) \qquad \text{Substitute known values.}$$
$$48,000 = 0.08P \qquad \text{Multiply both sides by 12.}$$
$$600,000 = P \qquad \text{Divide both sides by 0.08.}$$

You must have \$600,000 on deposit to earn \$4,000 per month at 8%.

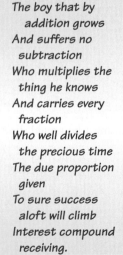

The boy that by
*   addition grows*
And suffers no
*   subtraction*
Who multiplies the
*   thing he knows*
And carries every
*   fraction*
Who well divides
*   the precious time*
The due proportion
*   given*
To sure success
*   aloft will climb*
Interest compound
*   receiving.*

Compounding Interest

Most banks do not pay interest according to the simple interest formula; instead, after some period of time, they add the interest to the principal and then pay interest on this new, larger amount. When this is done, it is called **compound interest.**

Example 7 Compare simple and compound interest

Compare simple and compound interest for a $1,000 deposit at 8% interest for 3 years.

Solution   Identify the known values: $P = 1,000$, $r = 0.08$, and $t = 3$. Next, calculate the future value using simple interest:

$$A = P(1 + rt)$$
$$= 1,000(1 + 0.08 \times 3)$$
$$= 1,000(1.24)$$
$$= 1,240$$

With simple interest, the future value in 3 years is $1,240.

Next, assume that the interest is **compounded annually.** This means that the interest is added to the principal after 1 year has passed. This new amount then becomes the principal for the following year. Since the time period for each calculation is 1 year, we let $t = 1$ for each calculation.

First year $(t = 1)$: $A = P(1 + r)$
$$= 1,000(1 + 0.08)$$
$$= 1,080$$
$$\downarrow$$

Second year $(t = 1)$: $A = P(1 + r)$ *One year's principal is previous year's total.*
$$= 1,080(1 + 0.08)$$
$$= 1,166.40$$
$$\downarrow$$

Third year $(t = 1)$: $A = P(1 + r)$
$$= 1,166.40(1 + 0.08)$$
$$= 1,259.71$$

With interest compounded annually, the future value in 3 years is $1,259.71. The earnings from compounding are $19.71 more than from simple interest.

The problem with compound interest relates to the difficulty of calculating it. Notice that, to simplify the calculations in Example 7, the variable representing time t was given the value 1, and the process was repeated three times. Also notice that, after the future value was found, it was used as the principal in the next step. What if we wanted to compound annually for 20 years instead of for 3 years? Look at Example 7 to discover the following pattern:

Simple interest (20 years): $A = P(1 + rt)$
$$= 1,000(1 + 0.08 \times 20)$$
$$= 1,000(1 + 1.6)$$
$$= 1,000(2.6)$$
$$= 2,600$$

Annual compounding (20 years): $A = \underbrace{P(1 + r \cdot 1)}$ *First year*
$$\downarrow$$
$$= P(1 + r)(1 + r) \qquad \text{\small Second year; for } P \text{ use end value from first year.}$$
$$= \underbrace{P(1 + r)^2} \qquad \text{\small Second year simplified.}$$
$$\downarrow$$
$$= P(1 + r)^2(1 + r) \qquad \text{\small Third year}$$
$$= P(1 + r)^3$$
$$\vdots$$
$$= P(1 + r)^{20} \qquad \text{\small Twentieth year}$$

For a period of 20 years, starting with $1,000 at 8% compounded annually, we have

$$A = 1,000(1.08)^{20}$$

To complete this chapter successfully, you will need to know how to do this on your calculator. Use this as a test problem.

The difficulty lies in calculating this number. For years we relied on extensive tables for obtaining numbers such as this, but the availability of calculators has made such calculations accessible to all. You will need an exponent key. These are labeled in different ways, depending on the brand of calculator. It might be y^x or x^y or $\wedge$. In this book we will show exponents by using $\wedge$, but you should press the appropriate key on your own brand of calculator.

$\boxed{1000}\ \boxed{\times}\ \boxed{1.08}\ \boxed{\wedge}\ \boxed{20}\ \boxed{=}$ Display: 4660.957144

Money makes money, and the money that money makes makes more money.

Benjamin Franklin

Round money answers to the nearest cent: $4,660.96 is the future value of $1,000 compounded annually at 8% for 20 years. This compounding yields $2,060.96 *more* than simple interest.

Most banks compound interest more frequently than once a year. For instance, a bank may pay interest as follows:

Semiannually: twice a year or every 180 days

Quarterly: 4 times a year or every 90 days

Monthly: 12 times a year or every 30 days

Daily: 360 times a year

If we repeat the same steps for more frequent intervals than annual compounding, we again begin with the simple interest formula $A = P(1 + rt)$.

Semiannually, then $t = \dfrac{1}{2}$: $A = P\left(1 + r \cdot \frac{1}{2}\right)$

Quarterly, then $t = \dfrac{1}{4}$: $A = P\left(1 + r \cdot \frac{1}{4}\right)$

Monthly, then $t = \dfrac{1}{12}$: $A = P\left(1 + r \cdot \frac{1}{12}\right)$

Daily, then $t = \dfrac{1}{360}$: $A = P\left(1 + r \cdot \frac{1}{360}\right)$

We now compound for t years and introduce a new variable, n, as follows:

Annual compounding, $n = 1$: $A = P(1 + r)^t$

Semiannual compounding, $n = 2$: $A = P\left(1 + r \cdot \frac{1}{2}\right)^{2t}$

Quarterly compounding, $n = 4$: $A = P\left(1 + r \cdot \frac{1}{4}\right)^{4t}$

Monthly compounding, $n = 12$: $A = P\left(1 + r \cdot \frac{1}{12}\right)^{12t}$

Daily compounding, $n = 360$: $A = P\left(1 + r \cdot \frac{1}{360}\right)^{360t}$

We are now ready to state the future value formula for compound interest, which is sometimes called the **compound interest formula.** For these calculations you will need access to a calculator with an exponent key.

Future Value Formula (Compound Interest)

$$A = P\left(1 + \frac{r}{n}\right)^{nt}$$

The variables we use in this formula are presented in a separate box because these variables will be used throughout this chapter. We contrast simple and compound interest in the same summary box.

Contrast Simple and Compound Interest

If interest is withdrawn from the amount in the account, use the *simple interest formula.* If the interest is deposited into the account to accrue future interest, then use the *compound interest formula.*

You will need to know the variables used with the interest formulas in this chapter, so we summarize those in the following definition box.

Interest Variables

A = FUTURE VALUE   This is the principal plus interest.
P = PRESENT VALUE   This is the same as the principal.
r = INTEREST RATE   This is the *annual* interest rate.
t = TIME   This is the time in *years.*
n = NUMBER OF COMPOUNDING PERIODS EACH YEAR
m = PERIODIC PAYMENT   This is usually monthly.

Spend some time here; you will need to remember what these variables represent. These are the variables used in this chapter.

Example 8 Compare compounding methods

Find the future value of $1,000 invested for 10 years at 8% interest
a. compounded annually.
b. compounded semiannually.
c. compounded quarterly.
d. compounded daily.

Solution   Identify the variables: $P = 1{,}000$, $r = 0.08$, $t = 10$.

a. $n = 1$: $A = \$1{,}000(1 + 0.08)^{10} = \$2{,}158.92$

b. $n = 2$: $A = \$1{,}000\left(1 + \frac{0.08}{2}\right)^{2 \cdot 10} = \$2{,}191.12$

c. $n = 4$: $A = \$1{,}000\left(1 + \frac{0.08}{4}\right)^{4 \cdot 10} = \$2{,}208.04$

d. $n = 360$: $A = \$1{,}000\left(1 + \frac{0.08}{360}\right)^{360 \cdot 10} = \$2{,}225.34$

Continuous Compounding

A reasonable extension of the current discussion is to ask the effect of more frequent compounding. To model this situation, consider the following contrived example. Suppose $1 is invested at 100% interest for 1 year compounded at different intervals. The compound interest formula for this example is

$$A = \left(1 + \frac{1}{n}\right)^n$$

where n is the number of times of compounding in 1 year. The calculations of this formula for different values of n are shown in the following table.

Number of Periods	Formula	Amount
Annual, $n = 1$	$\left(1 + \frac{1}{1}\right)^1$	$2.00
Semiannual, $n = 2$	$\left(1 + \frac{1}{2}\right)^2$	$2.25
Quarterly, $n = 4$	$\left(1 + \frac{1}{4}\right)^4$	$2.44
Monthly, $n = 12$	$\left(1 + \frac{1}{12}\right)^{12}$	$2.61
Daily, $n = 360$	$\left(1 + \frac{1}{360}\right)^{360}$	$2.71

Looking only at this table, you might (incorrectly) conclude that as the number of times the investment is compounded increases, the amount of the investment increases without bound. Let us continue these calculations for even larger n:

Number of Periods	Amount
$n = 8{,}640$ (compounding every hour)	2.718124536
$n = 518{,}400$ (every minute)	2.718279142
$n = 1{,}000{,}000$	2.718280469
$n = 10{,}000{,}000$	2.718281693
$n = 100{,}000{,}000$	2.718281815

The spreadsheet we are using for these calculations can no longer distinguish the values of $(1 + 1/n)^n$ for larger n. These values are approaching a particular number. This number, it turns out, is an irrational number, and it does not have a convenient decimal representation. (That is, its decimal representation does not terminate and does not repeat.) Mathematicians, therefore, have agreed to denote this number by using the symbol e. This number is called the **natural base** or **Euler's number.**

> ### The Number *e*
> As n increases without bound, the **number e** is the irrational number that is the limiting value of the formula
> $$\left(1 + \frac{1}{n}\right)^n$$

In Section 5.5 we noted that the number e is irrational and consequently does not have a terminating or repeating decimal representation. This same irrational number was used extensively in Chapter 10 as the base number in growth/decay applications as well as in evaluating natural logarithms. Although you must wait until you study the concept of limit in calculus for a formal definition of e, the preceding discussion should be enough to convince you that

Remember this approximate value.

$$e \approx 2.7183$$

Even though you will not find a bank that compounds interest every minute, you will find banks that use this limiting value to compound **continuously.** When using this model, we assume the year has 365 days.

Future Value Formula (Continuous Compounding)

The future value, A, of an investment of P, *compounded continuously* at a rate of r for t years, is found by

$$A = Pe^{rt}$$

It is easy to see how this formula follows from the compound interest formula if we let $\frac{1}{m} = \frac{r}{n}$, so that $n = mr$.

$$A = P\left(1 + \frac{r}{n}\right)^{nt}$$

$$= P\left(1 + \frac{1}{m}\right)^{mrt}$$

$$= P\left[\left(1 + \frac{1}{m}\right)^{m}\right]^{rt}$$

As m gets large, $\left(1 + \frac{1}{m}\right)^{m}$ approaches e, so we have

$$A = Pe^{rt}$$

Example 9 Find future value using continuous compounding

Find the future value of $890 invested at 21.3% for 3 years, 240 days, compounded continuously.

Solution   We use the formula $A = Pe^{rt}$ where $P = 890$, $r = 0.213$. For continuous compounding, use a 365-day year, so

$$3 \text{ years, } 240 \text{ days} = 3 + \frac{240}{365} = 3.657534247 \text{ years}$$

Remember that t is in years, and also remember to use this calculator value and not a rounded value.

$$A = 890e^{0.213t} \approx 1,939.676057$$

The future value is $1,939.68.

Example 10 Compare daily and continuous compounding

Let P dollars be invested at an annual rate of r for t years. Then the future value A depends on the number of times the money is compounded each year. How long will it take for $1,250 to grow to $2,000 if it is invested at
a. 8% compounded daily? Use exact interest.
b. 8% compounded continuously? Give your answer to the nearest day (assume that one year is 365 days).

Solution
a. Use the formula $A = P\left(1 + \frac{r}{365}\right)^{365t}$, and let $n = 365$ (exact interest); $P = \$1,250$; $A = \$2,000$; and $r = 0.08$; t is the unknown.

$$A = P\left(1 + \frac{r}{365}\right)^{365t} \qquad \text{\small Given formula}$$

$$2,000 = 1,250\left(1 + \frac{0.08}{365}\right)^{365t} \qquad \text{\small Substitute known values.}$$

$$1.6 = \left(1 + \frac{0.08}{365}\right)^{365t} \qquad \text{\small Divide both sides by 1,250.}$$

$$365t = \log_{(1+0.08/365)}1.6$$

Definition of logarithm

$$t = \frac{\log_{(1+0.08/365)}1.6}{365}$$

Divide both sides by 365. Evaluate as
$$\frac{\log 1.6}{\log(1 + 0.08/365)} \div 365.$$

$$\approx 5.875689182$$

Approximate solution

We see that it is almost 6 years, but we need the answer to the nearest day. The time is 5 years + 0.875689182 year. Multiply 0.875689182 by 365 to find 319.6265516. This means that on the 319th day of the 6th year, we are still a bit short of $2,000, so the time necessary is 5 years, 320 days.

b. For continuous compounding, use the formula $A = Pe^{rt}$.

$$A = Pe^{rt}$$

Given formula

$$2,000 = 1,250e^{0.08t}$$

Substitute known values.

$$1.6 = e^{0.08t}$$

Divide both sides by 1,250.

$$0.08t = \ln 1.6$$

Definition of logarithm

$$t = \frac{\ln 1.6}{0.08}$$

Divide both sides by 0.08. This is the exact solution.

$$\approx 5.875045366$$

Approximate solution

To find the number of days, we once again subtract 5 and multiply by 365 to find that the time necessary is 5 years, 320 days. (*Note:* 319.39 means that on the 319th day, you do not quite have the $2,000. The necessary time is 5 years, 320 days.) Notice that for this example, it does not matter whether we compound daily or continuously.

Inflation

Don Mason/CORBIS

Any discussion of compound interest is incomplete without a discussion of **inflation.** The same procedure we used to calculate compound interest can be used to calculate the effects of inflation. The government releases reports of monthly and annual inflation rates. In 1980 the inflation rate was over 14%, but in 2010 it was less than 1.25%. Keep in mind that inflation rates can vary tremendously and that the best we can do in this section is to assume different constant inflation rates. For our purposes in this book, we will assume continuous compounding when working inflation problems.

Example **11** Find future value due to inflation

If your salary today is $55,000 per year, what would you expect your salary to be in 20 years (rounded to the nearest thousand dollars) if you assume that inflation will continue at a constant rate of 6% over that time period?

Solution   Inflation is an example of continuous compounding. The problem with estimating inflation is that you must "guess" the value of future inflation, which in reality does not remain constant. However, if you look back 20 years and use an average inflation rate for the past 20 years—say, 6%—you may use this as a reasonable estimate for the next 20 years. Thus, you may use $P = 55,000$, $r = 0.06$, and $t = 20$ to find

$$A = Pe^{rt} = 55,000e^{0.06(20)}$$

Note: Be sure to use parentheses for the exponent:

$$\boxed{55000}\ \boxed{\times}\ \boxed{e^{\char`\^}}\ \boxed{(}\ \boxed{.06}\ \boxed{\times}\ \boxed{20}\ \boxed{)}\ \boxed{=}\quad \text{Display:}\quad 182606.430751$$

The answer means that, if inflation continues at a constant 6% rate, an annual salary of $183,000 (rounded to the nearest thousand dollars) will have about the same purchasing power in 20 years as a salary of $55,000 today.

Present Value

Sometimes we know the future value of an investment and we wish to know its present value. Such a problem is called a *present value problem*. The formula follows directly from the future value formula (by division).

Monday, March 1, 1982
6-mo. adjustable rate **17.5%**

Tuesday, March 1, 1983
6-mo. adjustable rate **12.5%**

Friday, March 1, 1985
3-mo. adjustable rate **10.5%**

Monday, March 1, 1993
3 mo. adjustable rate **4.5%**

Tuesday, March 1, 1994
3-mo. adjustable rate **3.0%**

Friday, March 1, 1996
3-mo. adjustable rate **4.9%**

Wednesday, March 1, 2000
3-mo. adjustable rate **5.2%**

Saturday, March 1, 2003
6-mo. adjustable rate **1.6%**

Monday, March 1, 2010
3-mo. adjustable rate **0.9%**

Present Value Formula

$$P = A \div \left(1 + \frac{r}{n}\right)^{nt} = A\left(1 + \frac{r}{n}\right)^{-nt}$$

Example 12 Find a present value for a Tahiti trip

Suppose that you want to take a trip to Tahiti in 5 years and you decide that you will need $5,000. To have that much money set aside in 5 years, how much money should you deposit now into a bank account paying 6% compounded quarterly?

Solution   In this problem, P is unknown and A is given: $A = 5,000$. We also have $r = 0.06$, $t = 5$, and $n = 4$.
Calculate:

$$P = A\left(1 + \frac{r}{n}\right)^{-nt}$$
$$= 5,000\left(1 + \frac{0.06}{4}\right)^{-20}$$
$$= \$3,712.35$$

Example 13 Find the value of an insurance policy

An insurance agent wishes to sell you a policy that will pay you $100,000 in 30 years. What is the value of this policy in today's dollars if we assume a 9% inflation rate, compounded annually?

Solution   This is a present value problem for which $A = 100,000$, $r = 0.09$, $n = 1$, and $t = 30$.
To find the present value, calculate:

$$P = 100,000(1 + 0.09)^{-30}$$
$$= \$7,537.11$$

This means that the agent is offering you an amount comparable to $7,537.11 in terms of today's dollars.

Example 14 Make your child a millionaire   Pólya's Method

Your first child has just been born. You want to give her 1 million dollars when she retires at age 65. If you invest your money on the day of her birth, how much do you need to invest so that she will have $1,000,000 on her 65th birthday?

Solution   We use Pólya's problem-solving guidelines for this example.

Understand the Problem. You want to make a single deposit and let it be compounded for 65 years, so that at the end of that time there will be 1 million dollars. Neither the rate of return nor the compounding period is specified. We assume daily compounding, a constant rate of return over the time period, and we need to determine whether we can find an investment to meet our goals.

Devise a Plan. With daily compounding, $n = 360$. We will use the present value formula, and experiment with different interest rates:

$$P = 1,000,000\left(1 + \frac{r}{360}\right)^{-(360 \cdot 65)}$$

Carry Out the Plan. We determine the present values based on the different interest rates using a calculator and display the results in tabular form.

Interest Rate	Formula	Value of P
2%	$1,000,000\left(1 + \frac{0.02}{360}\right)^{-(360 \cdot 65)}$	272,541.63
5%	$1,000,000\left(1 + \frac{0.05}{360}\right)^{-(360 \cdot 65)}$	38,782.96
8%	$1,000,000\left(1 + \frac{0.08}{360}\right)^{-(360 \cdot 65)}$	5,519.75
12%	$1,000,000\left(1 + \frac{0.12}{360}\right)^{-(360 \cdot 65)}$	410.27
20%	$1,000,000\left(1 + \frac{0.20}{360}\right)^{-(360 \cdot 65)}$	2.27

Look Back. Look down the list of interest rates and compare the rates with the amount of deposit necessary. Generally, the greater the risk of an investment, the higher the rate. Insured savings accounts may pay lower rates, bonds may pay higher rates for long-term investments, and other investments in stamps, coins, or real estate may pay the highest rates. The amount of the investment necessary to build an estate of 1 million dollars is dramatic!

Problem Set 11.1

Level 1

1. **IN YOUR OWN WORDS** What is interest?

2. **IN YOUR OWN WORDS** Contrast amount of interest and interest rate.

3. **IN YOUR OWN WORDS** Compare and contrast simple and compound interest.

4. **IN YOUR OWN WORDS** Compare and contrast present value and future value.

5. **IN YOUR OWN WORDS** What is the subject of Example 14? Discuss the real life application of this example in your own life.

Use estimation to select the best response in Problems 6–15. Do not calculate.

6. If you deposit $100 in a bank account for a year, then the amount of interest is likely to be
   A. $2   B. $5
   C. $102   D. impossible to estimate

7. If you deposit $100 in a bank account for a year, then the future value is likely to be
   A. $2   B. $5
   C. $102   D. impossible to estimate

8. If you purchase a new automobile and finance it for four years, the amount of interest you might pay is
   A. $400   B. $100
   C. $4,000   D. impossible to estimate

9. What is a reasonable monthly income when you retire?
   A. $300   B. $10,000
   C. $500,000   D. impossible to estimate

10. In order to retire and live on the interest only, what is a reasonable amount to have in the bank?
    A. $300   B. $10,000
    C. $500,000   D. impossible to estimate

11. If $I = Prt$ and $P = \$49,236.45$, $r = 10.5\%$, and $t = 2$ years, estimate I.
    A. $10,000   B. $600
    C. $50,000   D. $120,000

12. If $I = Prt$ and $I = \$398.90$, $r = 9.85\%$, and $t = 1$ year, estimate P.
A. \$400   B. \$40
C. \$40,000   D. \$4,000

13. If $t = 3.52895$, then the time is about 3 years and how many days?
A. 30   B. 300
C. 52   D. 200

14. If a loan is held for 450 days, then t is about
A. 450   B. 3
C. $1\frac{1}{4}$ D. 5

15. If a loan is held for 180 days, then t is about
A. 180   B. $\frac{1}{2}$
C. $\frac{1}{4}$ D. 3

In Problems 16–19, calculate the amount of simple interest earned.

16. \$1,000 at 8% for 5 years

17. \$5,000 at 10% for 3 years

18. \$2,000 at 12% for 5 years

19. \$1,000 at 14% for 30 years

In Problems 20–25, find the future value, using the future value formula and a calculator.

20. \$350 at $4\frac{3}{4}\%$ simple interest for 2 years

21. \$835 at 3.5% compounded semiannually for 6 years

22. \$575 at 5.5% compounded quarterly for 5 years

23. \$9,730.50 at 7.6% compounded monthly for 7 years

24. \$45.67 at 3.5% compounded daily for 3 years

25. \$119,400 at 7.5% compounded continuously for 30 years

In Problems 26–30, find the present value, using the present value formula and a calculator.

26. Achieve \$5,000 in three years at 3.5% simple interest.

27. Achieve \$2,500 in five years at 8.2% interest compounded monthly.

28. Achieve \$420,000 in 30 years at 6% interest compounded monthly.

29. Achieve a million dollars in 30 years at 6% interest compounded continuously.

30. Achieve \$225,500 at 8.65% compounded continuously for 8 years, 135 days.

31. If \$12,000 is invested at 4.5% for 20 years, find the future value if the interest is compounded:
a. annually
b. semiannually
c. quarterly
d. monthly
e. daily
f. every minute ($N = 525,600$)
g. continuously
h. simple (not compounded)

32. If \$34,500 is invested at 6.9% for 30 years, find the future value if the interest is compounded:
a. annually
b. semiannually
c. quarterly
d. monthly
e. daily
f. every minute ($N = 525,600$)
g. continuously
h. simple (not compounded)

Level 2

Find the total amount that must be repaid on the notes described in Problems 33–34.

33. \$1,500 borrowed at 21% simple interest. What is the total amount to be repaid 55 days later?

34. \$8,553 borrowed at 16.5% simple interest. What is the total amount to be repaid 3 years, 125 days later?

35. Find the cost of each item in 5 years, assuming an inflation rate of 9%.
a. cup of coffee, \$1.75
b. Sunday paper, \$1.25
c. Big Mac, \$1.95
d. gallon of gas, \$2.95
e. HDTV set, \$1,600
f. small car, \$19,000
g. car, \$28,000
h. tuition, \$16,000

36. Find the cost of each item in 10 years, assuming an inflation rate of 5%.
a. movie admission, \$7.00
b. CD, \$14.95
c. textbook, \$90.00
d. electric bill, \$105
e. phone bill, \$45
f. pair of shoes, \$85
g. new suit, \$570
h. monthly rent, \$800

37. How much would you have in 5 years if you purchased a \$1,000 5-year savings certificate that paid 4% compounded quarterly?

38. What is the interest on \$2,400 for 5 years at 12% compounded continuously?

39. What is the future value after 15 years if you deposit \$1,000 for your child's education and the interest is guaranteed at 16% compounded continuously?

40. Suppose you see a car with an advertised price of \$18,490 at \$480 per month for 5 years. What is the amount of interest paid?

41. Suppose you see a car with an advertised price of \$14,500 at \$410.83 per month for 4 years. What is the amount of interest paid?

42. Suppose you buy a home and finance $285,000 at $2,293.17 per month for 30 years. What is the amount of interest paid?

43. Suppose you buy a home and finance $170,000 at $1,247.40 per month for 30 years. What is the amount of interest paid?

44. Find the cost of a home in 30 years, assuming an annual inflation rate of 10%, if the present value of the house is $125,000.

45. Find the cost of the monthly rent for a two-bedroom apartment in 30 years, assuming an annual inflation rate of 10%, if the current rent is $850.

46. Suppose that an insurance agent offers you a policy that will provide you with a yearly income of $50,000 in 30 years. What is the comparable salary today, assuming an inflation rate of 6% compounded annually?

47. If a friend tells you she earned $5,075 interest for the year on a 5-year certificate of deposit paying 5% simple interest, what is the amount of the deposit?

48. If Rita receives $45.33 interest for a deposit earning 3% simple interest for 240 days, what is the amount of her deposit?

49. If John wants to retire with $10,000 per month, how much principal is necessary to generate this amount of monthly income if the interest rate is 15%?[*]

50. If Melissa wants to retire with $50,000 per month, how much principal is necessary to generate this amount of monthly income if the interest rate is 12%?[*]

51. If Jack wants to retire with $1,000 per month, how much principal is necessary to generate this amount of monthly income if the interest rate is 6%?

Level 3

52. In 2009, the U.S. national soared to 11.0 trillion dollars.
   a. If this debt is shared equally by the 300 million U.S. citizens, how much would it cost each of us (rounded to the nearest thousand dollars)?
   b. If the interest rate is 6%, what is the interest on the national debt *each second?* Assume a 365-day year.

 You can check on the current national debt at **www.brillig.com/debt_clock/** This link, as usual, can be accessed through **www.mathnature.com**

In Problems 53–56, calculate the time necessary to achieve an investment goal. Give your answer to the nearest day. Use a 365-day year.[†]

53. $1,000 at 8% simple interest; deposit $750

54. $3,500 at 6% simple interest; deposit $3,000

55. $5,000 at 5% daily interest; deposit $3,500

---

56. $5,000 at 4.5% compounded continuously; deposit $3,500

57. Suppose that $1,000 is invested at 7% interest compounded monthly. Use the formula

$$A = P\left(1 + \frac{r}{n}\right)^{nt}$$

   a. How long (to the nearest month) before the value is $1,250?
   b. How long (to the nearest month) before the money doubles?
   c. What is the interest rate (compounded monthly and rounded to the nearest percent) if the money doubles in 5 years?

58. Suppose that $1,000 is invested at 5% interest compounded continuously. Use the formula

$$A = Pe^{rt}$$

   a. How long (to the nearest day) before the value is $1,250?
   b. How long (to the nearest day) before the money doubles?
   c. What is the interest rate (compounded continuously and rounded to the nearest tenth of a percent) if the money doubles in 5 years?

Problem Solving 3

59. The News Clip below is typical of what you will see in a newspaper.

> **NEW, HIGHEST INTEREST RATE EVER ON INSURED SAVINGS**
>
> **8.33%** annual yield on
> **8%** interest compounded daily
> Annual yield based on daily compounding when funds and interest remain on deposit a year. Note: Federal regulations require a substantial interest penalty for early withdrawal of principal from Certificate Accounts.

The Clip gives two rates, the *annual yield* or *effective rate* (8.33%) and a *nominal rate* (8%). Since banks pay interest compounded for different periods (quarterly, monthly, daily, for example), they calculate a rate for which annual compounding would yield the same amount at the end of 1 year. That is, for an 8% rate:

Nominal Rate	Effective Rate
8%, annual compounding	8%
8%, semiannual compounding	8.16%
8%, quarterly compounding	8.24%
8%, monthly compounding	8.30%
8%, daily compounding	8.33%

To find a formula for effective rate, we recall that the compound interest formula is

$$A = P\left(1 + \frac{r}{n}\right)^{nt}$$

and the future value formula for simple interest is

$$A = P(1 + Yt)$$

The effective rate, Y, is a rate such that, at the end of one year $(t = 1)$, the future value for the simple interest is equal to the future value for the compound interest rate r with n compounding periods. That is,

$$P\left(1 + \frac{r}{n}\right)^n = P(1 + Y)$$

Find a formula for effective (annual) rate, Y, for which the future value for compound interest is equal to the future value for simple interest at the end of 1 year.

60. Find the effective yield for the following investments (see Problem 59). Round to the nearest hundredth of a percent.
   a. 6%, compounded quarterly
   b. 6%, compounded monthly
   c. 4%, compounded semiannually
   d. 4%, compounded daily

11.2 | Installment Buying

Two types of consumer credit allow you to make installment purchases. The first, called **closed-end,** is the traditional installment loan. An **installment loan** is an agreement to pay off a loan or a purchase by making equal payments at regular intervals for some specific period of time. In this book, it is assumed that all installment payments are made monthly.

There are two common ways of calculating installment interest. The first uses simple interest and is called *add-on interest*, and the second uses compound interest and is called *amortization*. We discuss the simple interest application in this section, and the compound interest application in Section 11.6.

In addition to closed-end credit, it is common to obtain a type of consumer credit called **open-end, revolving credit,** or, more commonly, a **credit card** loan. MasterCard, VISA, and Discover cards, as well as those from department stores and oil companies, are examples of open-end loans. This type of loan allows for purchases or cash advances up to a specified maximum **line of credit** and has a flexible repayment schedule.

Add-On Interest

The most common method for calculating interest on installment loans is by a method known as **add-on interest.** It is nothing more than an application of the simple interest formula. It is called *add-on interest* because the interest is *added to* the amount borrowed so that both interest and the amount borrowed are paid for over the length of the loan. You should be familiar with the following variables:

$P =$ AMOUNT TO BE FINANCED (present value)

$r =$ ADD-ON INTEREST RATE

$t =$ TIME (in years) TO REPAY THE LOAN

$I =$ AMOUNT OF INTEREST

$A =$ AMOUNT TO BE REPAID (future value)

$m =$ AMOUNT OF THE MONTHLY PAYMENT

$N =$ NUMBER OF PAYMENTS

Installment Loan Formulas	
AMOUNT OF INTEREST:	$I = Prt$
AMOUNT REPAID:	$A = P + I$ or $A - P(1 + rt)$
NUMBER OF PAYMENTS:	$N = 12t$
AMOUNT OF EACH PAYMENT:	$m = \frac{A}{N}$

STOP *Do you see why this is called add-on interest? Do these formulas make sense to you?*

Example 1 | Find a monthly payment

You want to purchase a computer that has a price of $1,399, and you decide to pay for it with installments over 3 years. The store tells you that the interest rate is 15%. What is the amount of each monthly payment?

Solution   You ask the clerk how the interest is calculated, and you are told that the store uses add-on interest. Thus, $P = 1,399$, $r = 0.15$, $t = 3$, and $N = 36$.

Two-step solution	*One-step solution*
$I = Prt$	$A = P(1 + rt)$
$\quad = 1,399(0.15)(3)$	$\quad = 1,399(1 + 0.15 \cdot 3)$
$\quad = 629.55$	$\quad = 2,028.55$
$A = P + I$	
$\quad = 1,399 + 629.55$	
$\quad = 2,028.55$	

$$m = \frac{2,028.55}{36}$$
$$\approx 56.35$$

The amount of each monthly payment is $56.35.

The most common applications of installment loans are for the purchase of a car or a home. Interest for purchasing a car is determined by the add-on method, but interest for purchasing a home is not. We will, therefore, delay our discussion of home loans until after we have discussed periodic payments with compound interest. The next example shows a calculation for a car loan.

Example 2 | Find a monthly payment   Pólya's Method

Visual Mining / Alamy

Suppose that you have decided to purchase a Toyota Prius and want to determine the monthly payment if you pay for the car in 4 years. The value of your trade-in is $4,100.

Solution   We use Póylya's problem-solving guidelines for this example.

Understand the Problem.   Not enough information is given, so you need to ask some questions of the car dealer:

> Sticker price of the car (as posted on the window): $22,720
>
> Dealer's preparation charges (as posted on the window): $350.00
>
> Total asking (sticker) price: $23,070
>
> Tax rate (determined by the state):   8%
>
> Add-on interest rate: 6%
>
> You need to make an offer.

Devise a Plan.   The plan is to offer the dealer 2% over dealer's cost. Assuming that the dealer accepts that offer, then we will calculate the monthly payment.

Carry Out the Plan.   If you are serious about getting the best price, find out the **dealer's cost**—the price the dealer paid for the car you want to buy. In this book, we will tell you the dealer's cost, but in the real world you will need to do some homework to find it (consult an April issue of *Consumer Reports*). Assume that the dealer's cost for this car is $19,993.45. You decide to offer the dealer 2% *over* this cost. We will call this a 2% **offer:**

$$\$19,993.45(1 + 0.02) = \$20,393.32$$

You will notice that we ignored the sticker price and the dealer's preparation charges. Our offer is based only on the *dealer's cost.* Most car dealers will accept an offer that is between 5% and 10% over what they actually paid for the car. For this example, we will assume that the dealer accepted a price of $20,400. We also assume that we have a trade-in with a value of $4,100. Here is a list of calculations shown on the sales contract:

Sale price of Prius:	$20,394.00
Destination charges:	200.00
Subtotal:	20,594.00
Tax (8% rate)	1,647.52
Less trade-in	4,100.00
Amount to be financed:	18,141.52

We now calculate several key amounts:

Interest: $I = Prt = 18{,}141.52(0.06)(4) = 4{,}353.96$

Amount to be repaid: $A = P + I = 18{,}141.52 + 4{,}353.96 = 22{,}495.48$

Monthly payment: $m = \dfrac{22{,}495.48}{48} \approx 468.66$

Look Back. The monthly payment for the car is $468.66.

Annual Percentage Rate (APR)

An important aspect of add-on interest is that the actual rate you pay exceeds the quoted add-on interest rate. The reason for this is that you do not keep the entire amount borrowed for the entire time. For the car payments calculated in Example 2, the principal used was $14,090, but you do not *owe* this entire amount for 4 years. After the first payment, you will owe *less* than this amount. In fact, after you make 47 payments, you owe only $293.32; but the calculation shown in Example 2 assumes that the principal remains constant for 4 years. To see this a little more clearly, consider a simpler example.

Suppose you borrow $2,000 for 2 years with 10% add on-interest. The amount of interest is

$$\$2{,}000 \times 0.10 \times 2 = \$400$$

Now if you pay back $2,000 + $400 at the end of two years, the annual interest rate is 10%. However, if you make a partial payment of $1,200 at the end of the first year and $1,200 at the end of the second year, your total paid back is still the same ($2,400), but you have now paid a higher annual interest rate. Why? Take a look at Figure 11.1.

On the left we see that the interest on $2,000 is $400. But if you make a partial payment (figure on the right), we see that $200 for the first year is the correct interest, but the remaining $200 interest piled on the remaining balance of $1,000 is 20% interest (not the stated 10%). Note that since you did not owe $2,000 for 2 years, the interest rate, r, necessary to give $400 interest can be calculated using $I = Prt$:

$$(2{,}000)r(1) + (1{,}000)r(1) = 400$$
$$3{,}000r = 400$$
$$r = \frac{400}{3{,}000}$$
$$\approx 0.13333 \quad \text{or} \quad 13.3\%$$

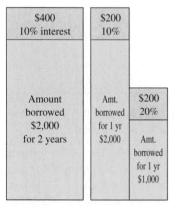

FIGURE 11.1 Interest on a $2,000 two-year loan

This number, 13.3%, is called the *annual percentage rate.* This number is too difficult to calculate, as we have just done here, if the number of months is very large. We will, instead, give an approximation formula.

APR Formula

The **annual percentage rate,** or **APR,** is the rate paid on a loan when that rate is based on the actual amount owed for the length of time that it is owed. It can be approximated for an add-on interest rate, r, with N payments by using the formula

$$APR \approx \frac{2Nr}{N + 1}$$

We can verify this formula for the example illustrated in Figure 11.1:

$$APR \approx \frac{2(2)(0.10)}{2 + 1} \approx 0.1333$$

In 1968, a Truth-in-Lending Act was passed by Congress; it requires all lenders to state the true annual interest rate, which is called the *annual percentage rate* (APR) and is based on the actual amount owed. Regardless of the rate quoted, when you ask a salesperson what the APR is, the law requires that you be told this rate. This regulation enables you to compare interest rates *before* you sign a contract, which must state the APR even if you haven't asked for it.

Example 3 Find an APR for a installment purchase

In Example 1, we considered the purchase of a computer with a price of $1,399, paid for in installments over 3 years at an add-on rate of 15%. Use the given APR formula (rounded to the nearest tenth of a percent) to approximate the APR.

Solution Knowing the amount of the purchase is not necessary when finding the APR. We need to know only N and r. Since N is the number of payments, we have $N = 12(3) = 36$, and r is given as 0.15:

$$APR \approx \frac{2(36)(0.15)}{36 + 1} \approx 0.292$$

The APR is approximately 29.2%.

Example 4 Find an APR for a car

Consider a Blazer with a price of $18,436 that is advertised at a monthly payment of $384.00 for 60 months. What is the APR (to the nearest tenth of a percent)?

Solution We are given $P = 18{,}436$, $m = 384$, and $N = 60$. The APR formula requires that we know the rate r. The future value is the total amount to be repaid $(A = P + I)$ and the amount of interest is $I = Prt$. Now, $A = 384(60) = 23{,}040$, so $I = A - P = 23{,}040 - 18{,}436 = 4{,}604$. Since $N = 12t$, we see that $t = 5$ when $N = 60$.

$$I = Prt \qquad \text{Simple interest formula}$$
$$4{,}604 = 18{,}436(r)(5) \qquad \text{Substitute known values.}$$
$$920.8 = 18{,}436r \qquad \text{Divide both sides by 5.}$$
$$0.0499457583 \approx r \qquad \text{Divide both sides by 18,436.}$$

Finally, from the APR formula, $APR \approx \frac{2Nr}{N + 1}$,

$$APR \approx \frac{2(60)(0.0499457583)}{61} \approx 0.098$$

Don't round until the last step. You will really need a calculator for a problem like this one, so we will show you the appropriate steps:

Find A.

| 384 | × | 60 | − | 18436 | = | ÷ | 5 | ÷ | 18436 | = | × | 2 | × | 60 | ÷ | 61 | = |

Find I.   Find r.   Find APR.

Display: .0982539508   The APR is approximately 9.8%.

Many automobiles are being offered at 0% interest! Good deal? It seems so, but buyer beware! Consider the following example.

Example 5 Compare APR and add-on rate

A local car dealer offered a 2002 Dodge 3/4 ton 4 × 4 pick-up truck with a manufacturer's suggested retail price (MSRP) of $33,29, less factory and dealer rebates of $5,489, for an "out-the-door no-haggle price of $27,801." The advertisement also offered "0% APR for 60 months" and in smaller print "In lieu of rebate." Suppose you can get a 2.5% add-on rate from the credit union. Should you choose the 0% APR or the 2.5% add-on rate? What is the credit union's APR rate?

Solution   The 0% APR would finance $33,290 for 60 months; the monthly payment would be

$$\frac{\$33{,}290}{60} = \$554.83/\text{mo}$$

The credit union rate is 2.5% for 60 months, so

$$I = Prt = \$27{,}801(0.025)(5) = \$3{,}475.13$$
$$A = P + I = \$31{,}276.13$$

The monthly payment would be

$$\frac{\$31{,}276.13}{60} = \$521.27$$

If we convert the 2.5% credit union add-on rate to an APR, we find

$$APR \approx \frac{2(0.025)(60)}{61} \approx 4.9\%$$

Note that the 0% APR financing is more costly than the credit union's 4.9% APR.

Open-End Credit

The most common type of open-end credit used today involves credit cards issued by VISA, MasterCard, Discover, American Express, department stores, and oil companies. Because you don't have to apply for credit each time you want to charge an item, this type of credit is very convenient.

When comparing the interest rates on loans, you should use the APR. Earlier, we introduced a formula for add-on interest; but for credit cards, the stated interest rate *is* the APR. However, the APR on credit cards is often stated as a daily or a monthly rate. For credit cards, we use a 365-day year rather than a 360-day year.

CAUTION   Don't forget: credit cards use a 365-day year (exact interest).

WARNING

Credit Card Trap
Suppose you "max out" your credit charges at $1,500 and decide that you will not use it again until it is paid off. How long do you think it will take you to pay it off if you make the minimum required payment (which would be $30 for the beginning balance of $1,500 and would decrease as the remaining balance decreases)?
A.   5 months
B.   2 years
C.   8 years
D.   12 years
E.   20 years
Even though we cannot mathematically derive this result, it does seem appropriate to consider this real-life question. Go ahead, make a guess!
   If you keep the payment at a constant $30, the length of time for repayment is 8 years. However, if you make only the required minimum payment, this "maxed" credit card will take 20 years to pay off! The correct answer is E.

Example 6 Find APR from a given rate

Convert the given credit card rate to APR (rounded to the nearest tenth of a percent).
a. $1\frac{1}{2}\%$ per month
b. Daily rate of 0.05753%

Solution

a. Since there are 12 months per year, multiply a monthly rate by 12 to get the APR:

$$1\frac{1}{2}\% \times 12 = 18\% \text{ APR}$$

b. Multiply the daily rate by 365 to obtain the APR:

$$0.05753\% \times 365 = 20.99845\%$$

Rounded to the nearest tenth, this is equivalent to 21.0% APR.

Many credit cards charge an annual fee; some charge $1 every billing period the card is used, whereas others are free. These charges affect the APR differently, depending on how much the credit card is used during the year and on the monthly balance. If you always pay your credit card bill in full as soon as you receive it, the card with no yearly fee would obviously be the best for you. On the other hand, if you use your credit card to stretch out your payments, the APR is more important than the flat fee. For our purposes, we won't use the yearly fee in our calculations of APR on credit cards. Like annual fees, the interest rates or APRs for credit cards vary greatly. Because VISA and MasterCard are issued by many different banks, the terms can vary greatly even in one locality.

Credit Card Interest

An interest charge added to a consumer account is often called a **finance charge.** The finance charges can vary greatly even on credit cards that show the *same* APR, depending on the way the interest is calculated. There are three generally accepted methods for calculating these charges: *previous balance, adjusted balance,* and *average daily balance.*

Consumer Reports magazine states that, according to one accounting study, the interest costs for average daily balance and previous balance methods are about 16% higher than those for the adjusted balance method.

Procedures for Calculating Credit Card Interest

For credit card interest, use the simple interest formula, $I = Prt$.

Previous balance method   Interest is calculated on the previous month's balance. With this method, P = previous balance, r = annual rate, and $t = \frac{1}{12}$.

Adjusted balance method   Interest is calculated on the previous month's balance *less* credits and payments. With this method, P = adjusted balance, r = annual rate, and $t = \frac{1}{12}$.

Average daily balance method   Add the outstanding balances for *each day* in the billing period, and then divide by the number of days in the billing period to find what is called the *average daily balance*. With this method, P = average daily balance, r = annual rate, and t = number of days in the billing period divided by 365.

In Example 7 we compare the finance charges on a $1,000 credit card purchase, using these three different methods.

Example 7 Contrast methods for calculating credit card interest

Calculate the interest on a $1,000 credit card bill that shows an 18% APR, assuming that $50 is sent on January 3 and is recorded on January 10. Contrast the three methods for calculating the interest.

Solution The three methods are the *previous balance method*, *adjusted balance method*, and *average daily balance method*. All three methods use the formula $I = Prt$.

	Method		
	Previous Balance	**Adjusted Balance**	**Average Daily Balance**
P:	$1,000	$1,000 − $50 = $950	Balance is $1,000 for 10 days of the 31-day month; balance is $950 for 21 days of the 31-day month: $$\frac{10 \times \$1,000 + 21 \times \$950}{31}$$ $= \$966.13$
r:	0.18	0.18	0.18
t:	$\dfrac{1}{12}$	$\dfrac{1}{12}$	$\dfrac{31}{365}$
$I = Prt$	$\$1,000(0.18)\left(\dfrac{1}{12}\right)$	$\$950(0.18)\left(\dfrac{1}{12}\right)$	$\$966.13(0.18)\left(\dfrac{31}{365}\right)$
	$= \$15.00$	$= \$14.25$	$= \$14.77$

You can sometimes make good use of credit cards by taking advantage of the period during which no finance charges are levied. Many credit cards charge no interest if you pay in full within a certain period of time (usually 20 or 30 days). This is called the **grace period.** On the other hand, if you borrow cash on your credit card, you should know that many credit cards have an additional charge for cash advances—and these can be as high as 4%. This 4% is *in addition* to the normal finance charges.

Problem Set 11.2

Level 1

1. **IN YOUR OWN WORDS** What is add-on interest?

2. **IN YOUR OWN WORDS** What is APR?

3. **IN YOUR OWN WORDS** Compare and contrast open-end and closed-end credit.

4. **IN YOUR OWN WORDS** Discuss the methods of calculating credit card interest.

5. **IN YOUR OWN WORDS** If you have a credit card, describe the method of calculating interest on your card. Name the bank issuing the credit card. If you do not have a credit card, contact a bank and obtain an application to answer this question. Name the bank.

6. **IN YOUR OWN WORDS** Describe a good procedure for saving money with the purchase of an automobile.

Use estimation to select the best response in Problems 7–24. Do not calculate.

7. If you purchase a $2,400 item and pay for it with monthly installments for 2 years, the monthly payment is
   A. $100 per month
   B. more than $100 per month
   C. less than $100 per month

8. If you purchase a $595.95 item and pay for it with monthly installments for 1 year, the monthly payment is
   A. about $50
   B. more than $50
   C. less than $50

9. If I do not pay off my credit card each month, the most important cost factor is
   A. the annual fee   B. the APR   C. the grace period

10. If I pay off my credit card balance each month, the most important cost factor is
A. the annual fee
B. the APR
C. the grace period

11. The method of calculation most advantageous to the consumer is the
A. previous balance method
B. adjusted balance method
C. average daily balance method

12. If you purchase an item for $1,295 at an interest rate of 9.8%, and you finance it for 1 year, then the amount of add-on interest is about
A. $13.00   B. $500   C. $130

13. If you purchase an item for $1,295 at an interest rate of 9.8%, and you finance it for 4 years, then the amount of add-on interest is about
A. $13.00   B. $500   C. $130

14. If you purchase a new car for $10,000 and finance it for 4 years, the amount of interest you would expect to pay is about
A. $4,000   B. $400   C. $24,000

15. A reasonable APR to pay for a 3-year installment loan is
A. 1%   B. 12%   C. 32%

16. A reasonable APR to pay for a 3-year automobile loan is
A. 6%   B. 40%   C. $2,000

17. If you wish to purchase a car with a sticker price of $10,000, a reasonable offer to make to the dealer is:
A. $10,000   B. $9,000   C. $11,000

18. A reasonable APR for a credit card is
A. 1%   B. 30%   C. 12%

19. In an application of the average daily balance method for the month of August, t is
A. $\frac{1}{12}$ B. $\frac{30}{365}$ C. $\frac{31}{365}$

20. When using the average daily balance method for the month of September, t is
A. $\frac{1}{12}$ B. $\frac{30}{365}$ C. $\frac{31}{365}$

21. If your credit card balance is $650 and the interest rate is 12% APR, then the credit card interest charge is
A. $6.50   B. $65   C. $8.25

22. If your credit card balance is $952, you make a $50 payment, the APR is 12%, and the interest is calculated according to the previous balance method, then the finance charge is
A. $9.52   B. $9.02   C. $9.06

23. If your credit card balance is $952, you make a $50 payment, the APR is 12%, and the interest is calculated according to the adjusted balance method, then the finance charge is
A. $9.52   B. $9.02   C. $9.06

24. If your credit card balance is $952, you make a $50 payment, the APR is 12%, and the interest is calculated according to the average daily balance method, then the finance charge is
A. $9.52   B. $9.02   C. $9.06

Convert each credit card rate in Problems 25–30 to the APR*.

25. Oregon, $1\frac{1}{4}$% per month

26. Arizona, $1\frac{1}{3}$% per month

27. New York, $1\frac{1}{2}$% per month

28. Tennessee, 0.02740% daily rate

29. Ohio, 0.02192% daily rate

30. Nebraska, 0.03014% daily rate

Calculate the monthly finance charge for each credit card transaction in Problems 31–34. Assume that it takes 10 days for a payment to be received and recorded, and that the month is 30 days long.

31. $300 balance, 18%, $50 payment
a. previous balance method
b. adjusted balance method
c. average daily balance method

32. $300 balance, 18%, $250 payment
a. previous balance method
b. adjusted balance method
c. average daily balance method

33. $3,000 balance, 15%, $50 payment
a. previous balance method
b. adjusted balance method
c. average daily balance method

34. $3,000 balance, 15%, $2,500 payment
a. previous balance method
b. adjusted balance method
c. average daily balance method

Round your answers in Problems 35–38 to the nearest dollar.

35. Make a 6% offer on a Chevrolet Corsica that has a sticker price of $14,385 and a dealer cost of $13,378.

36. Make a 5% offer on a Ford Escort that has a sticker price of $13,205 and a dealer cost of $12,412.70.

37. Make a 10% offer on a Saturn that has a sticker price of $19,895 and a dealer cost of $17,250.

38. Make a 10% offer on a Nissan Pathfinder that has a sticker price of $32,129 and a dealer cost of $28,916.

Level 2

Find the APR (rounded to the nearest tenth of a percent) for each of the loans described in Problems 39–42.

39. Purchase a living room set for 3,600 at 12% add-on interest for 3 years.

40. Purchase a stereo for $2,500 at 13% add-on interest for 2 years.

41. Purchase an oven for $650 at 11% add-on interest for 2 years.

42. Purchase a refrigerator for $2,100 at 14% add-on interest for 3 years.

*These rates were the listed finance charges on purchases of less than $500 on a Citibank VISA statement.

Assume the cars in Problems 43–46 can be purchased for 0% down for 60 months (in lieu of rebate).

a. *Find the monthly payment if financed for 60 months at 0% APR.*
b. *Find the monthly payment if financed at 2.5% add-on interest for 60 months.*
c. *Use the APR approximation formula for part* **b.**
d. *State whether the 0% APR or the 2.5% add-on rate should be preferred.*

43. A Dodge Ram that has a sticker price of $20,650 with factory and dealer rebates of $2,000

44. A BMW that has a sticker price of $62,490 with factory and dealer rebates of $6,000

45. A car with a sticker price of $42,700 with factory and dealer rebates of $5,100

46. A car with a sticker price of $36,500 with factory and dealer rebates of $4,200

Level 3

For each of the car loans described in Problems 47–52, give the following information.

a. *Amount to be paid*
b. *Amount of interest*
c. *Interest rate*
d. *APR (rounded to the nearest tenth percent)*

47. A newspaper advertisement offers a $9,000 car for nothing down and 36 easy monthly payments of $317.50.

48. A newspaper advertisement offers a $4,000 used car for nothing down and 36 easy monthly payments of $141.62.

49. A newspaper advertisement offers a $14,350 car for nothing down and 48 easy monthly payments of $488.40.

50. A car dealer will sell you the $16,450 car of your dreams for $3,290 down and payments of $339.97 per month for 48 months.

51. A car dealer will sell you a used car for $6,798 with $798 down and payments of $168.51 per month for 48 months.

52. A car dealer will sell you the $30,450 car of your dreams for $6,000 down and payments of $662.06 per month for 60 months.

53. A car dealer carries out the following calculations:

List price	$5,368.00
Options	$1,625.00
Destination charges	$ 200.00
Subtotal	$7,193.00
Tax	$ 431.58
Less trade-in	$2,932.00
Amount to be financed	$4,692.58
8% interest for 48 months	$1,501.63
Total	$6,194.21
MONTHLY PAYMENT	$ 129.05

What is the annual percentage rate?

54. A car dealer carries out the following calculations:

List price	$15,428.00
Options	$ 3,625.00
Destination charges	$ 350.00
Subtotal	$19,403.00
Tax	$ 1,164.18
Less trade-in	$ 7,950.00
Amount to be financed	$12,617.18
5% interest for 48 months	$ 2,523.44
Total	$15,140.62
MONTHLY PAYMENT	$ 315.43

What is the annual percentage rate?

55. A car dealer carries out the following calculations:

List price	$ 9,450.00
Options	$ 1,125.00
Destination charges	$ 300.00
Subtotal	$10,875.00
Tax	$ 652.50
Less trade-in	$ .00
Amount to be financed	$11,527.50
11% interest for 48 months	$ 5,072.10
Total	$16,599.60
MONTHLY PAYMENT	$ 345.83

What is the annual percentage rate?

Problem Solving 3

56. The finance charge statement on a Sears Revolving Charge Card statement is shown here. Why do you suppose that the limitation on the 50¢ finance charge is for amounts less than $28.50?

> **SEARS ROEBUCK AND CO.**
> **SEARSCHARGE SECURITY AGREEMENT**
>
> 4. FINANCE CHARGE. If I do not pay the entire New Balance within 30 days (28 days for February statements) of the monthly billing date, a FINANCE CHARGE will be added to the account for the current monthly billing period. **THE FINANCE CHARGE** will be either a minimum of $0.50 if the Average Daily Balance is $28.50 or less, or a periodic rate of 1.75% per month (**ANNUAL PERCENTAGE RATE of 21%**) on the Average Daily Balance.

57. Marsha needs to have a surgical procedure done and does not have the $3,000 cash necessary for the operation. Upon talking to an administrator at the hospital, she finds that it will accept MasterCard, VISA, and Discover credit cards. All of these credit cards have an APR of 18%, so she figures that it does not matter which card she uses, even though she plans to take a year to pay off the loan. Assume that Marsha makes a payment of $300 and then receives a bill. Show the interest from credit cards of 18% APR according to the previous balance, adjusted balance, and average daily balance methods. Assume that the month has 31 days and that it takes 14 days for Marsha's payment to be mailed and recorded.

58. Karen and Wayne need to buy a refrigerator because theirs just broke. Unfortunately, their savings account is depleted, and they will need to borrow money in order to buy a new one. The bank offers them a personal loan at 21% (APR), and Sears offers them an installment loan at 15% (add-on rate). Suppose that the refrigerator at Sears costs $1,598 plus 5% sales tax, and Karen and Wayne plan to pay for the refrigerator for 3 years. Should they finance it with the bank or with Sears?

59. Karen and Wayne need to buy a refrigerator because theirs just broke. Unfortunately, their savings account is depleted, and they will need to borrow money in order to buy a new one. Sears offers them an installment loan at 15% (add-on rate). If the refrigerator at Sears costs $1,598 plus 5% sales tax, and Karen and Wayne plan to pay for the refrigerator for 3 years, what is the monthly payment?

60. Rule of 78 With a typical installment loan, you are asked to sign a contract stating the terms of repayment. If you pay off the loan early, you are entitled to an interest rebate. For example, if you finance $500 and are charged $90 interest (APR 8.46%), the total to be repaid is $590 with

24 monthly payments of $24.59. After 1 year, you decide to pay off the loan, so you figure that the rebate should be $45 (half of the interest for 2 years), but instead you are told the interest rebate is only $23.40. What happened? Look at the fine print on the contract. It says interest will be refunded according to the Rule of 78. The formula for the rebate is as follows:

$$\text{INTEREST REBATE} = \frac{k(k+1)}{n(n+1)} \times \text{FINANCE CHARGE}$$

where k is the number of payments remaining and n is the total number of payments. Determine the interest rebate on the following:

a. $1,026 interest on an 18-month loan; pay off loan after 12 months.

b. $350 interest on a 2-year loan with 10 payments remaining.

c. $10,200 borrowed at 11% on a 4-year loan with 36 months remaining.

d. $51,000 borrowed at 10% on a 5-year loan with 18 payments remaining.

11.3 | Sequences

In the previous sections we considered financial applications dealing with a lump sum, or with applications based on simple or compound interest. However, many financial problems in real life deal with periodic payments in which the interest is calculated on a month-by-month basis. In order to consider such financial applications, it is necessary to understand two important mathematical concepts. The first, sequences, is considered in this section, and the second, series, is considered in the next.

Sequences or Progressions

Patterns are sometimes used as part of an IQ test. It used to be thought that an IQ test measured "innate intelligence" and that a person's IQ score was fairly constant. Today, it is known that this is not the case. IQ test scores can be significantly changed by studying the types of questions asked. Even if you have never taken an IQ test, you have taken (or will take) tests that ask pattern-type questions.

An example of an IQ test is shown below. It is a so-called "quickie" test, but it illustrates a few mathematical patterns (see Problems 1–6 on this IQ test). The purpose of this section is to look at some simple patterns to become more proficient in recognizing them, and then to apply them to develop some important financial formulas.

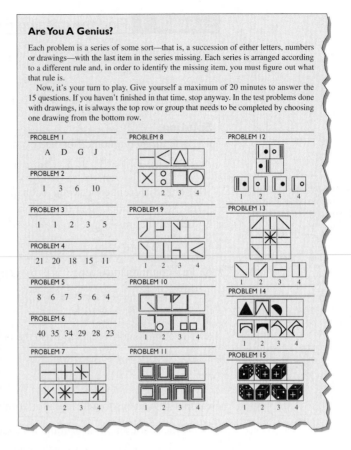

Are You A Genius?

Each problem is a series of some sort—that is, a succession of either letters, numbers or drawings—with the last item in the series missing. Each series is arranged according to a different rule and, in order to identify the missing item, you must figure out what that rule is.

Now, it's your turn to play. Give yourself a maximum of 20 minutes to answer the 15 questions. If you haven't finished in that time, stop anyway. In the test problems done with drawings, it is always the top row or group that needs to be completed by choosing one drawing from the bottom row.

PROBLEM 1

A   D   G   J

PROBLEM 2

1   3   6   10

PROBLEM 3

1   1   2   3   5

PROBLEM 4

21   20   18   15   11

PROBLEM 5

8   6   7   5   6   4

PROBLEM 6

40   35   34   29   28   23

PROBLEM 7

PROBLEM 8

PROBLEM 9

PROBLEM 10

PROBLEM 11

PROBLEM 12

PROBLEM 13

PROBLEM 14

PROBLEM 15

This test was distributed by MENSA, the high-IQ society. Here is the scoring for this test:

Give yourself 1 point for each correct answer. If you completed the test in 15 minutes or less, give yourself an additional 4 points.

1. M 2. 15 3. 8 4. 6 5. 5 6. 2
7. 2 8. 3 9. 2 10. 4 11. 4
12. 4 13. 2 14. 3 15. 3

Arithmetic Sequences

Perhaps the simplest pattern is the counting numbers themselves: 1, 2, 3, 4, 5, 6, ... A list of numbers having a first number, a second number, a third number, and so on is called a **sequence.**

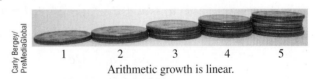

Carly Bergey/
PreMediaGlobal

1   2   3   4   5
Arithmetic growth is linear.

The numbers in a sequence are called the **terms** of the sequence. The sequence of counting numbers is formed by adding 1 to each term to obtain the next term. Your math assignments may well have been identified by some sequence: "Do the multiples of 3 from 3 to 30," that is, 3, 6, 9, . . . , 27, 30. This sequence is formed by adding 3 to each term. Sequences obtained by adding the same number to each term to obtain the next term are called *arithmetic sequences* or *arithmetic progressions*.

Arithmetic Sequence

An **arithmetic sequence** is a sequence whose consecutive terms differ by the same real number, called the **common difference.**

Example 1 Find the missing term of an arithmetic sequence

Show that each sequence is arithmetic, and find the missing term.
a. $1, 4, 7, 10, 13, \underline{\hspace{1cm}}, \ldots$
b. $20, 14, 8, 2, -4, -10, \underline{\hspace{1cm}}, \ldots$
c. $a_1, a_1 + d, a_1 + 2d, a_1 + 3d, a_1 + 4d, \underline{\hspace{1cm}}, \ldots$

Solution
a. Look for a common difference by subtracting each term from the succeeding term:

$$4 - 1 = 3, \quad 7 - 4 = 3, \quad 10 - 7 = 3, \quad 13 - 10 = 3$$

If the difference between each pair of consecutive terms of the sequence is the same number, then that number is the common difference; in this case it is 3. To find the missing term, simply add the common difference. The next term is

$$13 + 3 = 16$$

b. The common difference is -6. The next term is found by adding the common difference:

$$-10 + (-6) = -16$$

c. The common difference is d, so the next term is

$$(a_1 + 4d) + d = a_1 + 5d$$

Do not check only the first difference. All the differences must be the same.

Example 1c leads us to a formula for arithmetic sequences:

a_1 is the first term of an arithmetic sequence.

a_2 is the second term of an arithmetic sequence, and

$$a_2 = a_1 + d$$

a_3 is the third term of an arithmetic sequence, and

$$a_3 = a_2 + d = (a_1 + d) + d = a_1 + 2d$$

a_4 is the fourth term of an arithmetic sequence, and

$$a_4 = a_3 + d = (a_1 + 2d) + d = a_1 + 3d$$

a_{43} is the 43rd term of an arithmetic sequence, and

$$a_{43} = a_1 + \underbrace{42}d$$

One less than the term number

$$\vdots$$

This pattern leads to the following formula.

General Term of an Arithmetic Sequence

The *general term* of an arithmetic sequence $a_1, a_2, a_3, \ldots, a_n$, with common difference d is

$$a_n = a_1 + (n - 1)d$$

Example 2 Find a sequence given a general term

If $a_n = 26 - 6n$, list the sequence.

Solution

$$a_1 = 26 - 6(1) = 20 \qquad a_1 \text{ means evaluate } 26 - 6n \text{ for } n = 1.$$

$$a_2 = 26 - 6(2) = 14$$

$$a_3 = 26 - 6(3) = 8$$
$$a_4 = 26 - 6(4) = 2$$

The sequence is 20, 14, 8, 2, ...

Geometric Sequences

A second type of sequence is the *geometric sequence* or *geometric progression*. If each term is *multiplied* by the same number (instead of *added* to the same number) to obtain successive terms, a geometric sequence is formed.

1　　2　　4　　8　　16

Geometric growth is exponential.

> ### Geometric Sequence
>
> A **geometric sequence** is a sequence whose consecutive terms have the same quotient, called the **common ratio.**

If the sequence is geometric, the number obtained by dividing any term into the following term of that sequence will be the same nonzero number. No term of a geometric sequence may be zero.

Example 3　Find the common ratio of a geometric sequence

Show that each sequence is geometric, and find the common ratio.
a. 2, 4, 8, 16, 32, ———, . . .　　**b.** $10, 5, \frac{5}{2}, \frac{5}{4}, \frac{5}{8}$, ———, . . .
c. $g_1, g_1 r, g_1 r^2, g_1 r^3, g_1 r^4$, ———, . . .

Solution
a. First, verify that there is a common ratio:

$$\frac{4}{2} = 2, \quad \frac{8}{4} = 2, \quad \frac{16}{8} = 2, \quad \frac{32}{16} = 2$$

The common ratio is 2, so to find the next term, multiply the common ratio by the preceding term. The next term is

$$32(2) = 64$$

b. The common ratio is $\frac{1}{2}$ (be sure to check *each* ratio). The next term is found by multiplication:

$$\frac{5}{8}\left(\frac{1}{2}\right) = \frac{5}{16}$$

c. The common ratio is r, and the next term is

$$g_1 r^4(r) = g_1 r^5$$

As with arithmetic sequences, we denote the terms of a geometric sequence by using a special notation. Let $g_1, g_2, g_3, \ldots, g_n, \ldots$ be the terms of a *geometric* sequence. Example 3c leads us to a formula for geometric sequences.

$$g_2 = rg_1$$
$$g_3 = rg_2 = r(rg_1) = r^2g_1$$
$$g_4 = rg_3 = r(rg_2) = r(r^2g_1) = r^3g_1$$
$$g_5 = rg_4 = r(r^3g_1) = r^4g_1$$
$$\vdots$$

one less than the term number
$$g_{92} = r^{\overbrace{91}}g_1$$

Look for patterns to find the following formula.

Historical NOTE

**Leonardo Fibonacci
(ca. 1175–1250)**

Fibonacci, also known as Leonardo da Pisa, visited a number of Eastern and Arabic cities, where he became interested in the Hindu-Arabic numeration system we use today. He wrote *Liber Abaci,* in which he strongly advocated the use of the Hindu-Arabic numeration system. Howard Eves, in his book *In Mathematical Circles,* states: "Fibonacci sometimes signed his work with the name *Leonardo Bigollo.* Now *bigollo* has more than one meaning; it means both 'traveler' and 'blockhead.' In signing his work as he did, Fibonacci may have meant that he was a great traveler, for so he was. But a story has circulated that he took pleasure in using this signature because many of his contemporaries considered him a blockhead (for his interest in the new numbers), and it pleased him to show these critics what a blockhead could accomplish."

General Term Geometric Sequence

For a geometric sequence $g_1, g_2, g_3, \ldots, g_n, \ldots$, with common ratio r, the *general term* is

$$g_n = g_1 r^{n-1}$$

Example 4 Find the terms of a geometric sequence

List the sequence generated by $g_n = 50(2)^{n-1}$.

Solution

$$g_1 = 50(2)^{1-1} = 50$$
$$g_2 = 50(2)^{2-1} = 100$$
$$g_3 = 50(2)^{3-1} = 200$$
$$g_4 = 50(2)^{4-1} = 400$$

The sequence is 50, 100, 200, 400, ...

Fibonacci-Type Sequences

Even though our attention is focused on arithmetic and geometric sequences, it is important to realize that there can be other types of sequences.

The next type of sequence came about, oddly enough, by looking at the birth patterns of rabbits. In the 13th century, Leonardo Fibonacci wrote a book, *Liber Abaci,* in which he discussed the advantages of the Hindu-Arabic numerals over Roman numerals. In this book, one problem was to find the number of rabbits alive after a given number of generations. Let us consider what he did with this problem.

Example 5 Find the number of rabbits

Pólya's
Method

Suppose a pair of rabbits will produce a new pair of rabbits in their second month, and thereafter will produce a new pair every month. The new rabbits will do exactly the same. Start with one pair. How many pairs will there be in 10 months?

Solution   We use Pólya's problem-solving guidelines for this example.

Understand the Problem.   We can begin to understand the problem by looking at the following chart:

Number of Months	Number of Pairs	Pairs of Rabbits (the pairs shown in color are ready to reproduce in the next month)
Start	1	
1	1	
2	2	
3	3	
4	5	
5	8	
⋮	⋮	Same pair (rabbits never die)

Devise a Plan.   We look for a pattern with the sequence 1, 1, 2, 3, 5, 8, . . . ; it is not arithmetic and it is not geometric. It looks as if (after the first two months) each new number can be found by adding the two previous terms.

Carry Out the Plan. $1 + 1 = 2$

$$1 + 2 = 3$$
$$2 + 3 = 5$$
$$3 + 5 = 8$$
$$5 + 8$$

Do you see the pattern? The sequence is 1, 1, 2, 3, 5, 8, 13, 21, 34, 55, 89, . . .

Look Back. Using this pattern, Fibonacci was able to compute the number of pairs of rabbits alive after 10 months (it is the tenth term after the first 1): 89. He could also compute the number of pairs of rabbits after the first year or any other interval. Without a pattern, the problem would indeed be a difficult one.

Fibonacci Sequence

A **Fibonacci-type sequence** is a sequence in which any two numbers form the first and second terms, and subsequent terms are found by adding the previous two terms. The **Fibonacci sequence** is that sequence for which the first two terms are both 1.

If we state this definition symbolically with first term s_1, second term s_2, and general term s_n, we have a formula for the general term.

General Term of a Fibonacci Sequence

For a Fibonacci sequence $s_1, s_2, s_3, \ldots, s_n, \ldots$, the general term is given by

$$s_n = s_{n-1} + s_{n-2}$$

for $n \geq 3$.

Example 6 Find the terms of a Fibonacci-type sequence

a. If $s_1 = 5$ and $s_2 = 2$, list the first five terms of this Fibonacci-type sequence.
b. If s_1 and s_2 represent any first numbers, list the first eight terms of this Fibonacci-type sequence.

Solution

a. $5, 2, \underset{5+2}{7,} \overset{2+7}{\underset{7+9}{9,}} 16$

b. s_1 and s_2 are given.

$$s_3 = s_2 + s_1$$
$$s_4 = s_3 + s_2 = (s_1 + s_2) + s_2 = s_1 + 2s_2$$
$$s_5 = s_4 + s_3 = (s_1 + 2s_2) + (s_1 + s_2) = 2s_1 + 3s_2$$
$$s_6 = s_5 + s_4 = (2s_1 + 3s_2) + (s_1 + 2s_2) = 3s_1 + 5s_2$$
$$s_7 = s_6 + s_5 = (3s_1 + 5s_2) + (2s_1 + 3s_2) = 5s_1 + 8s_2$$
$$s_8 = s_7 + s_6 = (5s_1 + 8s_2) + (3s_1 + 5s_2) = 8s_1 + 13s_2$$

Look at the coefficients in the algebraic simplification and notice that the Fibonacci sequence 1, 1, 2, 3, 5, 8, . . . is part of the construction of any Fibonacci-type sequence regardless of the terms s_1 and s_2.

Historically, there has been much interest in the Fibonacci sequence. It is used in botany, zoology, business, economics, statistics, operations research, archeology, architecture, education, and sociology. There is even an official Fibonacci Association.

An example of Fibonacci numbers occurring in nature is illustrated by a sunflower. The seeds are arranged in spiral curves as shown in Figure 11.2.

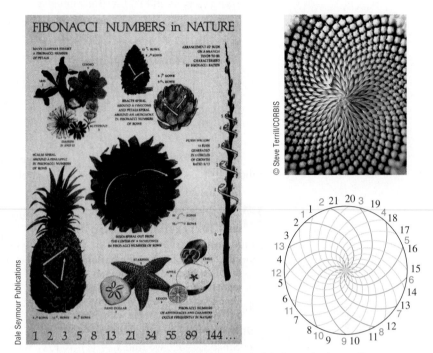

FIGURE 11.2 The arrangement of the pods (phyllotaxy) of a sunflower illustrates a Fibonacci sequence.

If we count the number of counterclockwise spirals (13 and 21 in this example), they are successive terms in the Fibonacci sequence. This is true of all sunflowers and, indeed, of the seed head of any composite flower such as the daisy or aster.

Example 7   Classify sequences

Classify the given sequences as arithmetic, geometric, Fibonacci-type, or none of the above. Find the next term for each sequence and give its general term if it is arithmetic, geometric, or Fibonacci-type.

a. 15, 30, 60, 120, . . .   **b.** 15, 30, 45, 60, . . .   **c.** 15, 30, 45, 75, . . .
d. 15, 20, 26, 33, . . .   **e.** 3, 3, 3, 3, . . .   **f.** 15, 30, 90, 360, . . .

Solution

a. 15, 30, 60, 120, . . . does not have a common difference, but a common ratio of 2, so this is a geometric sequence. The next term is $120(2) = 240$. The general term is $g_n = 15(2)^{n-1}$.

b. 15, 30, 45, 60, . . . has a common difference of 15, so this is an arithmetic sequence. The next term of the sequence is $60 + 15 = 75$. The general term is $a_n = 15 + (n - 1)15 = 15 + 15n - 15 = 15n$.

c. 15, 30, 45, 75, . . . does not have a common difference or a common ratio. Next, we check for a Fibonacci-type sequence by adding successive terms: $15 + 30 = 45$; $30 + 45 = 75$, so this is a Fibonacci-type sequence. The next term is $45 + 75 = 120$. The general term is $s_n = s_{n-1} + s_{n-2}$ where $s_1 = 15$ and $s_2 = 30$.

d. 15, 20, 26, 33, . . . does not have a common difference or a common ratio. Since $15 + 20 \neq 26$, we see it is not a Fibonacci-type sequence. We do see a pattern, however, when looking at the differences:

$$20 - 15 = 5; \quad 26 - 20 = 6; \quad 33 - 26 = 7$$

The next difference is 8. Thus, the next term is $33 + 8 = 41$.

e. 3, 3, 3, 3, . . . has a common difference of 0 and the common ratio is 1, so it is both arithmetic and geometric. The next term is 3. The general term is $a_n = 3 + (n - 1)0 = 3$ or $g_n = 3(1)^{n-1} = 3$.

f. 15, 30, 90, 360, . . . does not have a common difference or a common ratio. We see a pattern when looking at the ratios:

$$\frac{30}{15} = 2; \quad \frac{90}{30} = 3; \quad \frac{360}{90} = 4$$

The next ratio is 5. Thus, the next term is $360(5) = 1{,}800$.

General terms can be given for sequences that are not arithmetic, geometric, or Fibonacci-type. Consider the following example.

Example **8** **Find the terms of a given sequence**

Find the first four terms for the sequences with the given general terms. If the general term defines an arithmetic, geometric, or Fibonacci-type sequence, so state.

a. $s_n = n^2$ **b.** $s_n = (-1)^n - 5n$ **c.** $s_n = s_{n-1} + s_{n-2}$, where $s_1 = -4$ and $s_2 = 6$
d. $s_n = 2n$ **e.** $s_n = 2n + (n - 1)(n - 2)(n - 3)(n - 4)$

Solution
a. Since $s_n = n^2$, we find $s_1 = (1)^2 = 1$; $s_2 = (2)^2 = 4$, $s_3 = (3)^2, \ldots$.
The sequence is 1, 4, 9, 16, ...
b. Since $s_n = (-1)^n - 5n$, we find

$$s_1 = (-1)^1 - 5(1) = -6$$
$$s_2 = (-1)^2 - 5(2) = -9$$
$$s_3 = (-1)^3 - 5(3) = -16$$
$$s_4 = (-1)^4 - 5(4) = -19$$

The sequence is $-6, -9, -16, -19, \ldots$.
c. $s_n = s_{n-1} + s_{n-2}$ is the form of a Fibonacci-type sequence. Using the two given terms, we find:

$$s_1 = -4 \qquad \textit{Given}$$
$$s_2 = 6 \qquad \textit{Given}$$
$$s_3 = s_2 + s_1 = 6 + (-4) = 2$$
$$s_4 = s_3 + s_2 = 2 + 6 = 8$$

The sequence is $-4, 6, 2, 8, \ldots$.
d. $s_n = 2n$; $s_1 = 2(1) = 2$; $s_2 = 2(2) = 4$; $s_3 = 2(3) = 6$; . . . .
The sequence is 2, 4, 6, 8, . . . ; this is an arithmetic sequence.
e. $s_n = 2n + (n - 1)(n - 2)(n - 3)(n - 4)$

$$s_1 = 2(1) + (1 - 1)(1 - 2)(1 - 3)(1 - 4) = 2$$
$$s_2 = 2(2) + (2 - 1)(2 - 2)(2 - 3)(2 - 4) = 4$$
$$s_3 = 2(3) + (3 - 1)(3 - 2)(3 - 3)(3 - 4) = 6$$
$$s_4 = 2(4) + (4 - 1)(4 - 2)(4 - 3)(4 - 4) = 8$$

The sequence is 2, 4, 6, 8, ...

Examples 8d and 8e show that if only a finite number of successive terms is known and no general term is given, then a *unique* general term cannot be given. That is, if we are given the sequence

2, 4, 6, 8, . . .

the next term is probably 10 (if we are thinking of the general term of Example 8d), but it *may* be something different. In Example 8e,

$$s_5 = 2(5) + (5 - 1)(5 - 2)(5 - 3)(5 - 4) = 34$$

This gives the unlikely sequence 2, 4, 6, 8, 34, ... In general, you are looking for the simplest general term; nevertheless, you must remember that answers are not unique *unless the general term is given.*

COMPUTATIONAL WINDOW

If you have access to a spreadsheet program, it is easy to generate the terms of a sequence. The spreadsheet below shows the first 19 terms for the sequences given in Example 8.

Spreadsheet Application

	A	B	C	D	E	F	G
1	Term	a	b	c	d	e	
2	1	1	-6	-4	2	2	
3	2	4	-9	6	4	4	
4	3	9	-16	2	6	6	
5	4	16	-19	8	8	8	
6	5	25	-26	10	10	34	
7	6	36	-29	18	12	132	
8	7	49	-36	28	14	374	
9	8	64	-39	46	16	856	
10	9	81	-46	74	18	1698	
11	10	100	-49	120	20	3044	
12	11	121	-56	194	22	5062	
13	12	144	-59	314	24	7944	
14	13	169	-66	508	26	11906	
15	14	196	-69	822	28	17188	
16	15	225	-76	1330	30	24054	
17	16	256	-79	2152	32	32792	
18	17	289	-86	3482	34	43714	
19	18	324	-89	5634	36	57156	
20	19	361	-96	9116	38	73478	

The formulas for the cells correspond to the formulas in Example 8.
Term: 1; +A2+1; replicate

    a: +A2^2; +A3^2; replicate

    b: +(−1)^A2 − 5*A2; +(−1)^A3 − 5*A3; replicate

    c: −4; 6; +D2+D3; replicate

    d: +2*A2; +2*A3; replicate

    e: +2*A2+(A2 − 1)*(A2 − 2)*(A2 − 3)*(A2 − 4); replicate

We conclude this section by summarizing the procedure for classifying sequences.

Procedure for Classifying Sequences

To classify the sequence $s_1, s_2, s_3, \ldots, s_n$:

Step 1   Check to see if it is *arithmetic*. Find the successive differences:

$$s_2 - s_1 = d_1$$
$$s_3 - s_2 = d_2$$
$$s_4 - s_3 = d_3$$
$$\vdots$$

If all these differences

$$d_1 = d_2 = d_3 = \ldots$$

are the same, then it is an arithmetic sequence.

Step 2   Check to see if it is *geometric*. Find successive ratios

$$\frac{s_2}{s_1} = r_1; \quad \frac{s_3}{s_2} = r_2; \quad \frac{s_4}{s_3} = r_3, \ldots$$

If all these ratios

$$r_1 = r_2 = r_3 = \ldots$$

are the same, then it is a geometric sequence.

Step 3   Check to see if it is *Fibonacci-type*. Check to see if

$$s_3 = s_2 + s_1$$
$$s_4 = s_3 + s_2$$
$$\vdots$$

If these sums check, then it is a Fibonacci-type sequence.

Step 4   If the sequence is not arithmetic, geometric, Fibonacci, or Fibonacci-type, then we classify it by saying it is none of the above.

Problem Set 11.3

Level 1

1. **IN YOUR OWN WORDS**   What is a sequence?

2. **IN YOUR OWN WORDS**   What do we mean by a general term?

3. **IN YOUR OWN WORDS**   What is an arithmetic sequence?

4. **IN YOUR OWN WORDS**   What is a geometric sequence?

5. **IN YOUR OWN WORDS**   What is a Fibonacci sequence?

In Problems 6–31,
a. *Classify the sequences as arithmetic, geometric, Fibonacci, or none of these.*
b. *If arithmetic, give d; if geometric, give r; if Fibonacci, give the first two terms; and if none of these, state a pattern using your own words.*
c. *Supply the next term.*

6. $2, 4, 6, 8,$ _____, . . .

7. $2, 4, 8, 16,$ _____, . . .

8. $2, 4, 6, 10,$ _____, . . .

9. $5, 15, 25,$ _____, . . .

10. $5, 15, 45,$ _____, . . .

11. $5, 15, 20,$ _____, . . .

12. $1, 5, 25,$ _____, . . .

13. $25, 5, 1,$ _____, . . .

14. $9, 3, 1,$ _____, . . .

15. $1, 3, 9,$ _____, . . .

16. $21, 20, 18, 15, 11,$ _____, . . .

17. $8, 6, 7, 5, 6, 4,$ _____, . . .

18. $2, 5, 8, 11, 14,$ _____, . . .

19. $3, 6, 12, 24, 48,$ _____, . . .

20. $5, -15, 45, -135, 405,$ _____, . . .

21. $10, 10, 10,$ _____, . . .

22. $2, 5, 7, 12,$ _____, . . .

23. 3, 6, 9, 15, _____, ...

24. 1, 8, 27, 64, 125, _____, ...

25. 8, 12, 18, 27, _____, ...

26. 3^2, 3^5, 3^8, 3^{11}, _____, ...

27. 4^5, 4^4, 4^3, 4^2, _____, ...

28. $\frac{1}{2}$, $\frac{1}{3}$, $\frac{2}{3}$, $\frac{1}{4}$, $\frac{3}{4}$, $\frac{1}{5}$, $\frac{2}{5}$, $\frac{3}{5}$, $\frac{4}{5}$, $\frac{1}{6}$, _____, ...

29. $\frac{1}{10}$, $\frac{1}{5}$, $\frac{3}{10}$, $\frac{2}{5}$, $\frac{1}{2}$, _____, ...

30. $\frac{4}{3}$, 2, 3, $4\frac{1}{2}$, _____, ...

31. $\frac{7}{12}$, $\frac{2}{3}$, $\frac{3}{4}$, $\frac{5}{6}$, _____, ...

Level 2

In Problems 32–47,

a. *Find the first three terms of the sequences whose nth terms are given.*

b. *Classify the sequence as arithmetic (give d), geometric (give r), both, or neither.*

32. $s_n = 4n - 3$　　**33.** $s_n = -3 + 3n$

34. $s_n = 10n$　　**35.** $s_n = 2 - n$

36. $s_n = 7 - 3n$　　**37.** $s_n = 10 - 10n$

38. $s_n = \frac{2}{n}$

39. $s_n = 1 - \frac{1}{n}$

40. $s_n = \frac{n-1}{n+1}$

41. $s_n = \frac{1}{2}n(n+1)$

42. $s_n = \frac{1}{4}n^2(n+1)^2$

43. $s_n = (-1)^n$

44. $s_n = -5$

45. $s_n = \frac{2}{3}$

46. $s_n = (-1)^{n+1}$

47. $s_n = (-1)^n(n+1)$

Find the requested terms in Problems 48–55.

48. Find the 15th term of the sequence $s_n = 4n - 3$

49. Find the 69th term of the sequence $s_n = 7 - 3n$

50. Find the 20th term of the sequence $s_n = (-1)^n(n+1)$

51. Find the 3rd term of the sequence $s_n = (-1)^{n+1}5^{n+1}$

52. Find the first five terms of the sequence where
$s_1 = 2$ and $s_n = 3s_{n-1}$, $n \geq 2$

53. Find the first five terms of the sequence where
$s_1 = 3$ and $s_n = \frac{1}{3}s_{n-1}$, $n \geq 2$

54. Find the first five terms of the sequence where
$s_1 = 1$, $s_2 = 1$, and $s_n = s_{n-1} + s_{n-2}$, $n \geq 3$

55. Find the first five terms of the sequence where
$s_1 = 1$, $s_2 = 2$, and $s_n = s_{n-1} + s_{n-2}$, $n \geq 3$

Problem Solving 3

56. Is the following sequence a Fibonacci sequence?
a_n is one more than the nth term of the Fibonacci sequence.

57. Is the following sequence a Fibonacci sequence?
1, 1, 2, 3, 5, 8, . . . , a_n, where a_n is the integer nearest to
$$\frac{1}{\sqrt{5}}\left[\frac{1 + \sqrt{5}}{2}\right]^n$$

58. Apartment blocks of n floors are to be painted blue and yellow, with the rule that no two adjacent floors can be blue. (They can, however, be yellow.) Let a_n be the number of ways to paint a block with n floors.*

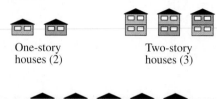

One-story　　　　Two-story
houses (2)　　　　houses (3)

Three-story
houses (5)

Four-story houses (8)

Does the sequence $a_1, a_2, a_3, \ldots$ form a Fibonacci sequence?

59. Consider the following magic trick. The magician asks the audience for any two numbers (you can limit it to the counting numbers between 1 and 10 to keep the arithmetic manageable). Add these two numbers to obtain a third number. Add the second and third numbers to obtain a fourth. Continue until ten numbers are obtained. Then ask the audience to add the ten numbers, while the magician instantly gives the sum.

Consider the example shown here:

First number:	5
Second number:	9
(3) Add:	14
(4)	23
(5)	37
(6)	60
(7)	97
(8)	157
(9)	254
(10)	411
Add column:	1,067

*From "Fibonacci Forgeries" by Ian Stewart, *Scientific American*, May 1995, p. 104. Illustration by Johnny Johnson. © 1995 by Scientific American, Inc. All rights reserved.

The trick depends on the magician's ability to multiply quickly and mentally by 11. Consider the following pattern of multiplication by 11:

$11 \times 10 = 110$
$11 \times 11 = 121$
$11 \times 12 = 132$

Sum of the two digits
↓
$11 \times 13 = 14\underset{\uparrow\ \uparrow}{3}$
Original two digits

$11 \times 52 = 572$

Sum of the two digits may be a two-digit number.
↓
$11 \times 74 = 7\boxed{11}4$
$\quad\quad\quad = 814$

Do the usual carry.

Consider 11 times the 7th number in the example in the pattern:

$$11 \times 97 = 9\boxed{16}7 = 1{,}067$$

Note that this is the sum of the ten numbers. Explain why this magician's trick works, and why it is called *Fibonacci's magic trick*.

60. Fill in the blanks so that

$$\_\_\_,\ 8,\ \_\_\_,\ \_\_\_,\ 27,\ \_\_\_,\ \ldots$$

is

a. an arithmetic sequence.
b. a geometric sequence.
c. a sequence that is neither arithmetic nor geometric, for which you are able to write a general term.

11.4 Series

In the last section we looked at a list of numbers having a first number, a second number, a third number, and so on. We called it a sequence, and in this section we look at the sum of terms of a sequence. If the terms of a sequence are added, the expression is called a **series.** We first consider a finite sequence along with its associated sum. Note that a capital letter is used to indicate the sum.

CAUTION Write a lowercase s on your paper; now write a capital S. Do they look different? You need to distinguish between the lowercase s and capital S in your own written work.

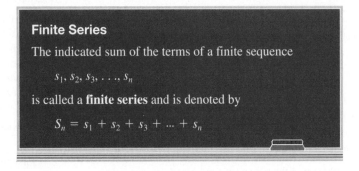

> **Finite Series**
>
> The indicated sum of the terms of a finite sequence
>
> $$s_1, s_2, s_3, \ldots, s_n$$
>
> is called a **finite series** and is denoted by
>
> $$S_n = s_1 + s_2 + s_3 + \ldots + s_n$$

Example 1 Find the sum of terms of a sequence

a. Find S_4, where $s_n = 26 - 6n$.
b. Find S_3, where $s_n = (-1)^n n^2$.

Solution

a. $S_4 = \underset{s_1}{s_1} + \underset{s_2}{s_2} + \underset{s_3}{s_3} + \underset{s_4}{s_4}$

$\quad = \underset{s_1}{\left[26 - 6(1)\right]} + \underset{s_2}{\left[26 - 6(2)\right]} + \underset{s_3}{\left[26 - 6(3)\right]} + \underset{s_4}{\left[26 - 6(4)\right]}$

$\quad = 20 + 14 + 8 + 2$

$\quad = 44$

b. $S_3 = s_1 + s_2 + s_3$

$$= \overbrace{[(-1)^1(1)^2]}^{s_1} + \overbrace{[(-1)^2(2)^2]}^{s_2} + \overbrace{[(-1)^3(3)^2]}^{s_3}$$

$$= -1 + 4 + (-9)$$

$$= -6$$

The terms of the sequence in Example 1b alternate in sign: $-1, 4, -9, 16, \ldots.$ A factor of $(-1)^n$ or $(-1)^{n+1}$ in the general term will cause the sign of the terms to alternate, creating a series called an **alternating series.**

Summation Notation

Historical NOTE

Once when walking past a lounge in the University of Chicago that was filled with a loud crowd watching TV, [Antoni Zygmund] asked one of his students what was going on. The student told him that the crowd was watching the World Series and explained to him some of the features of this baseball phenomenon. Zygmund thought about it all for a few minutes and commented, "I think it should be called the World Sequence."

Ronald Coifman and Peter Duren,
*A Century of Mathematics in
America, Part III*
Providence, RI: American
Mathematical Society, 1980, p. 348.

Before we continue to discuss finding the sum of the terms of a sequence, we need a handy notation, called **summation notation.** In Example 1a we wrote

$$S_4 = s_1 + s_2 + s_3 + s_4$$

Using summation notation (or, as it is sometimes called, **sigma notation**), we could write this sum using the Greek letter Σ:

$$S_4 = \sum_{k=1}^{4} s_k = s_1 + s_2 + s_3 + s_4$$

The sigma notation evaluates the expression (s_k) immediately following the sigma (Σ) sign, first for $k = 1$, then for $k = 2$, then for $k = 3$, and finally for $k = 4$, and then adds these numbers. That is, the expression is evaluated for *consecutive counting numbers* starting with the value of k listed at the bottom of the sigma $(k = 1)$ and ending with the value of k listed at the top of the sigma $(k = 4)$. For example, consider $s_k = 2k$ with $k = 1, 2, 3, 4, 5, 6, 7, 8, 9, 10$. Then

This is the last natural number in the domain. It is called the upper limit.
$$\downarrow$$
$$\sum_{k=1}^{10} 2k \} \quad \leftarrow \text{This is the function being evaluated. It is called the general term.}$$
$$\uparrow$$
This is the first natural number in the domain. It is called the lower limit.

Thus, $\displaystyle\sum_{k=1}^{10} 2k = 2(1) + 2(2) + 2(3) + 2(4) + 2(5) + 2(6) + 2(7) + 2(8) + 2(9) + 2(10)$

$$= 110.$$

The words **evaluate** and **expand** are both used to mean "write out an expression in summation notation, and then sum the resulting terms, if possible."

Example **2**   **Evaluate a summation**

a. Evaluate: $\displaystyle\sum_{k=3}^{6} (2k + 1)$

b. Expand: $\displaystyle\sum_{k=3}^{n} \frac{1}{2^k}$

Solution

a. $\displaystyle\sum_{k=3}^{6} (2k + 1) = \underbrace{(2 \cdot 3 + 1)}_{} + \underbrace{(2 \cdot 4 + 1)}_{k\,=\,4} + \underbrace{(2 \cdot 5 + 1)}_{k\,=\,5} + \underbrace{(2 \cdot 6 + 1)}_{k\,=\,6}$

Evaluate the expression $2k + 1$ for $k = 3$

$= 7 + 9 + 11 + 13$

$= 40$

b. $\displaystyle\sum_{k=3}^{n} \frac{1}{2^k} = \frac{1}{2^3} + \frac{1}{2^4} + \frac{1}{2^5} + \frac{1}{2^6} + \cdots + \frac{1}{2^{n-1}} + \frac{1}{2^n}$

Arithmetic Series

An **arithmetic series** is the sum of the terms of an arithmetic sequence. Let us consider a rather simple-minded example. How many blocks are shown in the stack in Figure 11.3? We can answer this question by simply counting the blocks: There are 34 blocks. Somehow it does not seem like this is what we have in mind with this question. Suppose we ask a better question: How many blocks are in a similar building with n rows?

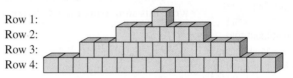

Row 1:
Row 2:
Row 3:
Row 4:

FIGURE 11.3 How many blocks?

We notice that the number of blocks (counting from the top) in each row forms an arithmetic series:

$$1 + 6 + 11 + 16 + \cdots$$

Look for a pattern:

Denote one row by $A_1 = 1$ block.

Two rows: $A_2 = 1 + 6 = 7$

Three rows: $A_3 = 1 + 6 + 11 = 18$

Four rows: $A_4 = 1 + 6 + 11 + 16 = 34$ (shown in Figure 11.3)

$$\vdots$$

What about 10 rows?

$$A_{10} = 1 + 6 + 11 + 16 + 21 + 26 + 31 + 36 + 41 + 46$$

Instead of adding all these numbers directly, let us try an easier way. Write down A_{10} twice, once counting from the top and once counting from the bottom:

$$A_{10} = 1 + 6 + 11 + 16 + 21 + 26 + 31 + 36 + 41 + 46$$
$$\updownarrow \ \updownarrow \ \updownarrow \ \updownarrow \ \updownarrow \ \updownarrow \ \updownarrow \ \updownarrow \ \updownarrow \ \updownarrow$$
$$A_{10} = 46 + 41 + 36 + 31 + 26 + 21 + 16 + 11 + 6 + 1$$

Add these equations:

$$2A_{10} = 47 + 47 + 47 + 47 + 47 + 47 + 47 + 47 + 47 + 47$$

$$\overset{\text{Number of terms}}{2A_{10} = \overbrace{10}\,(47)} = 470$$

$$A_{10} = 10\underbrace{\left(\tfrac{47}{2}\right)}_{\text{Average of 1st and last terms}} = 235 \qquad \text{Divide both sides by 2.}$$

Repeated term

This pattern leads us to a formula for n terms. We note that the number of blocks is an arithmetic sequence with $a_1 = 1$, $d = 5$; thus since $a_n = a_1 + (n-1)d$, we have for the blocks in Figure 11.3 a stack starting with 1 and ending (in the nth row) with

$$a_n = 1 + (n-1)5 = 1 + 5n - 5 = 5n - 4$$

Thus,   Number of blocks in n rows   number of rows
$$A_n = n \left[\frac{1 + (5n - 4)}{2} \right]$$
Average of 1st and last terms

$$= n \left[\frac{5n - 3}{2} \right]$$

$$= \frac{1}{2}(5n^2 - 3n)$$

This formula can be used for the number of blocks for any number of rows. Looking back, we see

$n = 1$: $A_1 = \dfrac{1}{2}[5(1)^2 - 3(1)] = 1$

$n = 4$: $A_4 = \dfrac{1}{2}[5(4)^2 - 3(4)] = 34$ (Figure 11.3)

$n = 10$: $A_{10} = \dfrac{1}{2}[5(10)^2 - 3(10)] = 235$

If we carry out these same steps for A_n where $a_n = a_1 + (n-1)d$, we derive the following formula for the sum of the terms of an arithmetic sequence.

YIELD

These formulas are not as difficult as they look. The first formula tells us that the sum of n terms of an arithmetic sequence is n times the average of the first and last terms.

Arithmetic Series Formula

The sum of the terms of an arithmetic sequence $a_1, a_2, a_3, \ldots, a_n$ with common difference d is

$$A_n = \sum_{k=1}^{n} a_k = n \left(\frac{a_1 + a_n}{2} \right) \text{ or } A_n = \frac{n}{2}[2a_1 + (n-1)d]$$

The last part of the formula for A_n is used when the last term is not explicitly stated or known. To derive this formula, we know $a_n = a_1 + (n-1)d$ so

$$A_n = n \left(\frac{a_1 + a_n}{2} \right)$$

$$= n \left(\frac{a_1 + [a_1 + (n-1)d]}{2} \right)$$

$$= \frac{n}{2}[a_1 + a_1 + (n-1)d]$$

$$= \frac{n}{2}[2a_1 + (n-1)d]$$

Example 3 Find the sum of the numbers of a sequence

In a classroom of 35 students, each student "counts off" by threes (i.e., 3, 6, 9, 12, . . .). What is the sum of the students' numbers?

Solution   We recognize the sequence 3, 6, 9, 12, . . . as an arithmetic sequence with the first term $a_1 = 3$ and the common difference $d = 3$. The sum of these numbers is denoted by A_{35} since there are 35 students "counting off":

$$A_{35} = \frac{35}{2}[2(3) + (35 - 1)3] = 1,890$$

Geometric Series

A **geometric series** is the sum of the terms of a geometric sequence. To motivate a formula for a geometric series, we once again consider an example. Suppose Charlie Brown receives a chain letter, and he is to copy this letter and send it to six of his friends.

You may have heard that chain letters "do not work." Why not? Consider the number of people who could become involved with this chain letter if we assume that everyone carries out their task and does not break the chain. The first mailing would consist of six letters. The second mailing involves 42 letters since the second mailing of 36 letters is added to the total: $6 + 36 = 42$.

The number of letters in each successive mailing is a number in the geometric sequence

$$6, 36, 216, 1296, \ldots \quad \text{or} \quad 6, 6^2, 6^3, 6^4, \ldots$$

How many people receive letters with 11 mailings, assuming that no person receives a letter more than once? To answer this question, consider the series associated with a geometric sequence. We begin with a pattern:

Denote one mailing by	$G_1 = 6$
Two mailings:	$G_2 = 6 + 6^2$
Three mailings:	$G_3 = 6 + 6^2 + 6^3$
$\vdots$	
Eleven mailings:	$G_{11} = 6 + 6^2 + \cdots + 6^{11}$

We could probably use a calculator to find this sum, but we are looking for a formula, so we try something different. In fact, this time we will work out the general formula. Let

$$G_n = g_1 + g_2 + g_3 + \cdots + g^n$$
$$G_n = g_1 + g_1 r + g_1 r^2 + \cdots + g_1 r^{n-1}$$

Multiply both sides of the latter equation by r:

$$rG_n = g_1 r + g_1 r^2 + g_1 r^3 + \cdots + g_1 r^n$$

Notice that, except for the first and last terms, all the terms in the expansions for G_n and rG_n are the same, so that if we subtract one equation from the other, we have

$$G_n - rG_n = g_1 - g_1 r^n$$
$$(1 - r)G_n = g_1(1 - r^n) \qquad \text{We solve for } G_n.$$
$$G_n = \frac{g_1(1 - r^n)}{1 - r} \qquad \text{if } r \neq 1$$

For Charlie Brown's chain letter problem, $g_1 = 6$, $n = 11$, and $r = 6$, so we find

$$G_{11} = \frac{6(1 - 6^{11})}{1 - 6} = \frac{6}{5}(6^{11} - 1) = 435{,}356{,}466$$

This is more than the number of people in the United States! The number of letters in only two more mailings would exceed the number of men, women, and children in the whole world.

Geometric Series Formula

The sum of the terms of a geometric sequence $g_1, g_2, g_3, \ldots, g_n$ with common ratio r **(where $r \neq 1$) is**

$$G_n = \frac{g_1(1 - r^n)}{1 - r}$$

Example 4 **Make a financial decision**

Suppose some eccentric millionaire offered to hire you for a month (say, 31 days) and offered you the following salary choice. She will pay you $500,000 per day or else will pay you 1¢ for the first day, 2¢ for the second day, 4¢ for the third day, and so on for the 31 days. Which salary should you accept?

Solution   If you are paid $500,000 per day, your salary for the 31 days is

$500,000(31) = \$15,500,000$

Now, if you are paid using the doubling scheme, your salary is (in cents)

$$1 + 2 + 4 + 8 + \cdots + \overset{\text{31st day}}{\overline{\text{last day}}} \quad \text{or} \quad 2^0 + 2^1 + 2^2 + 2^3 + \cdots + 2^{30}$$

We see that this is the sum of the geometric sequence where $g_1 = 1$ and $r = 2$. We are looking for G_{31}:

$$G_{31} = \frac{1(1 - 2^{31})}{1 - 2} = -(1 - 2^{31}) = 2^{31} - 1$$

Using a calculator, we find this to be 2,147,483,647 cents or $21,474,836.47. You should certainly accept the doubling scheme (starting with 1¢, but do not ask for any days off).

Example 5 **Find the number of games**

Pólya's Method

The NCAA men's basketball tournament has 64 teams. How many games are necessary to determine a champion?

Solution   We use Pólya's problem-solving guidelines for this example.

Understand the Problem. Most tournaments are formed by drawing an elimination schedule similar to the one shown in Figure 11.4. This is sometimes called a *two-team elimination tournament.*

Devise a Plan. We could obtain the answer to the question by direct counting, but instead we will find a general solution, working backward. We know there will be 1 championship game and 2 semifinal games; continuing to work backward, there are 4 quarter-final games, ... :

$$1 + 2 + 2^2 + 2^3 + \cdots$$

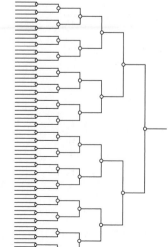

FIGURE 11.4 NCAA playoffs

We recognize this as a geometric series. We note that since there are 64 teams, the first round has $64/2 = 32$ games. Also, we know that $2^5 = 32$, so we must find

$$1 + 2 + 2^2 + 2^3 + 2^4 + 2^5$$

Carry Out the Plan. We note that $g_1 = 1$, $r = 2$, and $n = 6$:

$$G_6 = \frac{1(1 - 2^6)}{1 - 2} = 2^6 - 1 = 63$$

Thus, the NCAA playoffs will require 63 games.

Look Back. We see not only that the NCAA tournament has 63 games for a playoff tournament, but, in general, that if there are 2^n teams in a tournament, there will be $2^n - 1$ games. However, do not forget to check by estimation or by using common sense. Each game eliminates one team, and so 63 teams must be eliminated to crown a champion.

Infinite Geometric Series

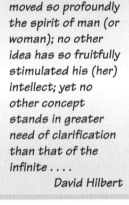

> The infinite! No other question has ever moved so profoundly the spirit of man (or woman); no other idea has so fruitfully stimulated his (her) intellect; yet no other concept stands in greater need of clarification than that of the infinite . . . .
>
> David Hilbert

We have just found a formula for the sum of the first n terms of a geometric sequence. Sometimes it is also possible to find the sum of an entire infinite geometric series. Suppose an infinite geometric series

$$g_1 + g_2 + g_3 + g_4 + \cdots$$

is denoted by G. This is an **infinite series.**

The **partial sums** are defined by

$$G_1 = g_1; \quad G_2 = g_1 + g_2; \quad G_3 = g_1 + g_2 + g_3; \ldots$$

Consider the partial sums for an infinite geometric series with $g_1 = \frac{1}{2}$ and $r = \frac{1}{2}$. The geometric sequence is $\frac{1}{2}, \frac{1}{4}, \frac{1}{8}, \frac{1}{16}, \ldots$. The first few partial sums can be found as follows:

$$G_1 = \frac{1}{2}; \quad G_2 = \frac{1}{2} + \frac{1}{4} = \frac{3}{4}; \quad G_3 = \frac{1}{2} + \frac{1}{4} + \frac{1}{8} = \frac{7}{8}$$

Does this series have a sum if you add *all* its terms? It does seem that as you take more terms of the series, the sum is closer and closer to 1. Analytically, we can check the partial sums on a calculator or spreadsheet (see below). Geometrically, you can see (Figure 11.5) that if the terms are laid end-to-end as lengths on a number line, each term is half the remaining distance to 1.

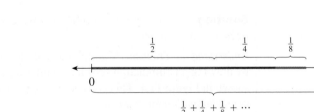

FIGURE 11.5 Series $\frac{1}{2} + \frac{1}{4} + \frac{1}{8} + \cdots$

It appears that the partial sums are getting closer to 1 as n becomes larger. We *can* find the sum of an infinite geometric sequence.

COMPUTATIONAL WINDOW

It is easy to use a spreadsheet (review Section 6.3) to look at the partial sums of a sequence. For the sequence $\frac{1}{2}, \frac{1}{4}, \frac{1}{8}, \ldots$, define the cells as shown:

Replicate subsequent rows. The output is:

n	Term	Partial sum
1	.5	.5
2	.25	.75
3	.125	.875
4	.0625	.9375
5	.03125	.96875
6	.015625	.984375
7	.0078125	.9921875
8	.00390625	.99609375
9	.00195313	.99804688
10	.00097656	.99902344
⋮	⋮	⋮
20	.00000095	.99999905

Spreadsheet Application ⬚ _ □ ✕

	A	B	C
1	n	term	partial sum
2	1	0.5	+B2
3	+A2+1	+B2*.5	+C2+B3
4	+A3+1	+B3*.5	+C3+B4

Consider

$$G_n = \frac{g_1(1 - r^n)}{1 - r} = \frac{g_1 - g_1 r^n}{1 - r} = \frac{g_1}{1 - r} - \frac{g_1}{1 - r} r^n$$

Now, g_1, r, and $1 - r$ are fixed numbers. If $|r| < 1$, then r^n approaches 0 as n grows, and thus G_n approaches $\frac{g_1}{1 - r}$.

Infinite Geometric Series Formula

If $g_1, g_2, g_3, \ldots, g_n, \ldots$ is an infinite geometric sequence with a common ratio r such that $|r| < 1$, then its sum is denoted by G and is found by

$$G = \frac{g_1}{1 - r}$$

If $|r| \geq 1$, the infinite geometric series has no sum.

Example 6   Find the distance a pendulum travels

The path of each swing of a pendulum is 0.85 as long as the path of the previous swing (after the first). If the path of the tip of the first swing is 36 in. long, how far does the tip of the pendulum travel before it eventually comes to rest?

Solution

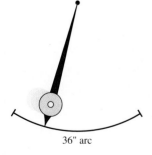

36" arc

$$
\begin{aligned}
\text{TOTAL DISTANCE} \quad &= 36 + 36(0.85) + 36(0.85)^2 + \cdots \\
&= \frac{36}{1 - 0.85} \\
&= 240
\end{aligned}
$$

Infinite geometric series; $g_1 = 36$; $r = 0.85$

The tip of the pendulum travels 240 in.

Summary of Sequence and Series Formulas

We conclude this section by repeating the important formulas related to sequences and series. Given a sequence of numbers, a *series* arises by considering the *sum* of the terms of the sequence. The formulas for the *general term of a sequence* and for the *sum* (of terms of the sequence; that is, a *series*) are given in Table 11.1.

Distinguishing a Series from a Sequence

Type	Definition	Notation	Formula
Sequences	A list of numbers having a first term, a second term, . . .	s_n	
Arithmetic	A sequence with a common difference, d	a_n	$a_n = a_1 + (n-1)d$
Geometric	A sequence with a common ratio, r	g_n	$g_n = g_1 r^{n-1}$
Fibonacci	A sequence with first two terms given, and subsequent terms the sum of the two previous terms		$s_n = s_{n-1} + s_{n-2}, n \geq 3$
Series	The indicated sum of terms of a sequence	S_n	
Arithmetic	Sum of the terms of an arithmetic sequence: $$A_n = \sum_{k=1}^{n} a_k = a_1 + a_2 + a_3 + \cdots + a_n$$	A_n	$A_n = n\left(\dfrac{a_1 + a_n}{2}\right)$ or $A_n = \dfrac{n}{2}[2a_1 + (n-1)d]$
Geometric	Sum of the terms of a geometric sequence: $$G_n = \sum_{k=1}^{n} g_k = g_1 + g_2 + g_3 + \ldots + g_n$$ Sum of the terms of an infinite geometric sequence: $$G = g_1 + g_2 + g_3 + \cdots$$	G_n G	$G_n = \dfrac{g_1(1 - r^n)}{1 - r}, r \neq 1$ $G = \dfrac{g_1}{1 - r}, |r| < 1$

Problem Set 11.4

Level 1

1. **IN YOUR OWN WORDS**   Distinguish a sequence from a series.

2. **IN YOUR OWN WORDS**   Explain summation notation.

3. **IN YOUR OWN WORDS**   What is a partial sum?

4. **IN YOUR OWN WORDS**   Distinguish a geometric series and an infinite geometric series.

Find the requested values in Problems 5–10.

5. S_5 when $s_n = 15 - 3n$

6. S_8 when $s_n = 5n$

7. S_4 when $s_n = 5 \cdot 2^n$

8. S_6 when $s_n = (-1)^n$

9. S_7 when $s_n = (-1)^n$

10. S_3 when $s_n = 8 \cdot 5^n$

Evaluate the expressions in Problems 11–18.

11. $\displaystyle\sum_{k=3}^{5} k$

12. $\displaystyle\sum_{k=1}^{4} k^2$

13. $\displaystyle\sum_{k=2}^{6} k^2$

14. $\displaystyle\sum_{k=2}^{5} (100 - 5k)$

15. $\sum_{k=1}^{10}[1^k + (-1)^k]$ **16.** $\sum_{k=1}^{5}(-2)^{k-1}$

17. $\sum_{k=0}^{4}3(-2)^k$ **18.** $\sum_{k=1}^{3}(-1)^k(k^2 + 1)$

Level 2

If possible, find the sum of the infinite geometric series in Problems 19–24.

19. $1 + \frac{1}{2} + \frac{1}{4} + \cdots$

20. $1 + \frac{3}{2} + \frac{9}{4} + \cdots$

21. $1 + \frac{1}{3} + \frac{1}{9} + \cdots$

22. $100 + 50 + 25 + \cdots$

23. $-20 + 10 - 5 + \cdots$

24. $-100 + 50 - 25 + \cdots$

25. Find the sum of the first 5 odd positive integers.

26. Find the sum of the first 5 even positive integers.

27. Find the sum of the first 5 positive integers.

28. Find the sum of the first 10 odd positive integers.

29. Find the sum of the first 10 even positive integers.

30. Find the sum of the first 10 positive integers.

31. Find the sum of the first 100 odd positive integers.

32. Find the sum of the first 100 even positive integers.

33. Find the sum of the first 100 positive integers.

34. Find the sum of the first *n* odd positive integers.

35. Find the sum of the first *n* even positive integers.

36. Find the sum of the first *n* positive integers.

37. Find the sum of the first 20 terms of the arithmetic sequence whose first term is 100 and whose common difference is 50.

38. Find the sum of the first 50 terms of the arithmetic sequence whose first term is -15 and whose common difference is 5.

39. Find the sum of the even integers between 41 and 99.

40. Find the sum of the odd integers between 48 and 136.

The game of pool uses 15 balls numbered from 1 to 15 (see Figure 11.6). In the game of rotation, a player attempts to "sink" a ball in a pocket of the table and receives the number of points on the ball. Answer the questions in Problems 41–44.

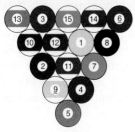

FIGURE 11.6 Pool balls

41. How many points would a player who "runs the table" receive? (To "run the table" means to sink all the balls.)

42. Suppose Missy sinks balls 1 through 8, and Shannon sinks balls 9 through 15. What are their respective scores?

43. Suppose Missy sinks the even-numbered balls and Shannon sinks the odd-numbered balls. What are their respective scores?

44. Suppose we consider a game of "super pool," which has 30 consecutively numbered balls on the table. How many points would a player receive to "run the table"?

45. The *Peanuts* cartoon (p. 549) expresses a common feeling regarding chain letters. Consider the total number of letters sent after a particular mailing:

1st mailing:	6
2nd mailing:	$6 + 36 = 42$
3rd mailing:	$6 + 36 + 216 = 258$

Determine the total number of letters sent in five mailings of the chain letter.

46. How many blocks would be needed to build a stack like the one shown in Figure 11.7 if the bottom row has 28 blocks?

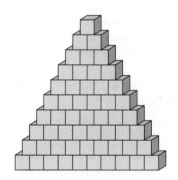

FIGURE 11.7 How many blocks?

47. Repeat Problem 46 if the bottom row has 87 blocks.

48. Repeat Problem 46 if the bottom row has 100 blocks.

49. A pendulum is swung 20 cm and allowed to swing freely until it eventually comes to rest. Each subsequent swing of the bob of the pendulum is 90% as far as the preceding swing. How far will the bob travel before coming to rest?

50. The initial swing of the tip of a pendulum is 25 cm. If each swing of the tip is 75% of the preceding swing, how far does the tip travel before eventually coming to rest?

51. A flywheel is brought to a speed of 375 revolutions per minute (rpm) and allowed to slow and eventually come to rest. If, in slowing, it rotates three-fourths as fast each subsequent minute, how many revolutions will the wheel make before returning to rest?

52. A rotating flywheel is allowed to slow to a stop from a speed of 500 rpm. While slowing, each minute it rotates two-thirds as many times as in the preceding minute. How many revolutions will the wheel make before coming to rest?

53. Advertisements say that a new type of superball will rebound to 9/10 of its original height. If it is dropped from a height of 10 ft, how far, based on the advertisements, will the ball travel before coming to rest?

54. A tennis ball is dropped from a height of 10 ft. If the ball rebounds 2/3 of its height on each bounce, how far will the ball travel before coming to rest?

Level 3

55. A culture of bacteria increases by 100% every 24 hours. If the original culture contains 1 million bacteria ($a_0 = 1$ million), find the number of bacteria present after 10 days.

56. Use Problem 55 to find a formula for the number of bacteria present after d days.

57. How many games are necessary for a two-team elimination tournament with 32 teams?

58. Games like "*Wheel of Fortune*" and "*Jeopardy*" have one winner and two losers. A three-team game tournament is illustrated by Figure 11.8. If "Jeopardy" has a Tournament of Champions consisting of 27 players, what is the necessary number of games?

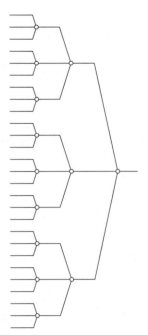

FIGURE 11.8 A three-team tournament

59. How many games are necessary for a three-team elimination tournament with 729 teams?

Problem Solving 3

60. a. How many blocks are there in the solid figure shown in Figure 11.9?

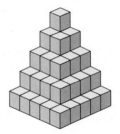

FIGURE 11.9 How many blocks?

b. How many blocks are there in a similar figure with 50 layers?

11.5 Annuities

Sir Isaac Newton, one of the greatest mathematicians of all time, worked at the mint, but apparently did not like to apply mathematics to money. In a book by Rev. J. Spence published 1858, Spence writes, "Sir Isaac Newton, though so deep in algebra and fluxions, could not readily make up a common account: and, when he was Master of the Mint, used to get somebody else to make up his accounts for him." There is an important lesson here.

If you do not like to work with money, hire someone to help you. When Albert Einstein was asked what was the most amazing formula he knew, you might think he would have given his famous formula $E = mc^2$, but he did not. He thought the most amazing formula was the compound interest formula.

Ordinary Annuities

One of the most fundamental mathematical concepts for businesspeople and consumers is the idea of interest. We have considered present- and future-value problems involving a fixed amount, so we refer to such problems as **lump-sum problems.** On the other hand, it is far more common to encounter financial problems based on monthly or other periodic payments, so we refer to such problems as **periodic-payment problems.** Consider the situation in which the *monthly payment* is known and the future value is to be determined. A sequence of payments into or out of an interest-bearing account is called an **annuity.** If the payments are made into an interest-bearing account at the end of each time period, and if the frequency of payments is the same as the frequency of compounding, the annuity is called an *ordinary annuity.* The amount of an annuity is the sum of all payments made plus all accumulated interest. In this book we will assume that all annuities are ordinary annuities.

The best way to understand what we mean by an annuity is to consider an example. Suppose you decide to give up smoking and save the $2 per day you spend on cigarettes. How much will you save in 5 years? (Assume that each year has 360 days.) If you save the money without earning any interest, you will have

$$\$2 \times 360 \times 5 = \$3,600$$

However, let us assume that you save $2 per day, and at the end of each month you deposit the $60 (assume that all months are 30 days; that is, assume ordinary interest) into an account earning 12% interest compounded monthly. Now, how much will you have in 5 years? This is an example of an annuity. We could solve the problem using a spreadsheet to simulate our bank statement. Part of such a spreadsheet is shown in the margin.

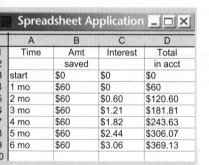

	A	B	C	D
	Time	Amt	Interest	Total
		saved		in acct
3	start	$0	$0	$0
4	1 mo	$60	$0	$60
5	2 mo	$60	$0.60	$120.60
6	3 mo	$60	$1.21	$181.81
7	4 mo	$60	$1.82	$243.63
8	5 mo	$60	$2.44	$306.07
9	6 mo	$60	$3.06	$369.13

Notice that even though the interest on $60 for one month is $0.60, the total monthly increase in interest is not linear, because the interest is compounded.

To derive a formula for annuities, let us consider this problem for a period of 6 months, and calculate the amounts plus interest for *each* deposit separately. We need a new variable to represent the amount of periodic deposit. Since this is usually a monthly payment, we let m = periodic payment. The **monthly payment** is a periodic payment that is made monthly.

We are given $m = 60$, $r = 0.12$, and $n = 12$ ("monthly deposit" means "monthly compounding" for an ordinary annuity). Let $i = r/n$. For this example, $i = 0.12/12 = 0.01$. The time, t, varies for each deposit. Remember that the deposit comes at the *end* of the month.

First deposit will earn 5 months' interest:	$60(1 + 0.01)^5 = 63.06$
Second deposit will earn 4 months' interest:	$60(1 + 0.01)^4 = 62.44$
Third deposit will earn 3 months' interest:	$60(1 + 0.01)^3 = 61.82$
Fourth deposit will earn 2 months' interest:	$60(1 + 0.01)^2 = 61.21$
Fifth deposit will earn 1 month's interest:	$60(1 + 0.01)^1 = 60.60$
Sixth deposit will earn no interest:	60.00
TOTAL IN THE ACCOUNT	369.13

Historical NOTE

**Isaac Newton
(1642–1727)**

Newton was one of the greatest mathematicians of all time. He was a genius of the highest order but was often absent-minded. One story about Newton is that, when he was a boy, he was sent to cut a hole in the bottom of the barn door for the cats to go in and out. He cut two holes—a large one for the cat and a small one for the kittens. Newton's influence on mathematics was so great that the field is sometimes divided into pre-Newtonian mathematics and post-Newtonian mathematics. Post-Newtonian mathematics is characterized by the changing and the infinite rather than by the static and the finite. As an example of his tremendous ability, when Newton was 74 years old the mathematician Leibniz (see the Historical Note on p. 85) posed a challenge problem to all the mathematicians in Europe. Newton received the problem after a day's work at the mint (remember he was 74!) and solved the problem that evening. His intellect was monumental.

This leads us to the following pattern (using variables). The total after 6 months is:

$$A = m + m(1 + i)^1 + m(1 + i)^2 + m(1 + i)^3 + m(1 + i)^4 + m(1 + i)^5$$

$$= \sum_{k=1}^{6} m(1 + i)^{k-1}$$

We are using the following formula for the sum of the terms of a geometric sequence: That is, it is the sum of the terms of a geometric sequence with $g_1 = m$ and common ratio $(1 + i)$. The sum, G_n, is the future value A. Thus (from the formula for the sum of a geometric sequence), we have

$$A = \frac{m[1 - (1 + i)^{nt}]}{1 - (1 + i)}$$

For the total after nt periods, we recognize this as a geometric series with first term m and common ratio 1 + i.

$$= \frac{m[1 - (1 + i)^{nt}]}{-i}$$

$$= \frac{m[(1 + i)^{nt} - 1]}{-i}$$

 This is one of the most useful formulas for you to use in your personal financial planning.

Ordinary Annuity Formula

The future value, A, of an annuity is found with the formula

$$A = m\left[\frac{\left(1 + \frac{r}{n}\right)^{nt} - 1}{\frac{r}{n}}\right]$$

where r is the annual rate, m the periodic payment, t the time (in years), and n is the number of payments per year.

Example 1   Find the value of an ordinary annuity

How much do you save in 5 years if you deposit $60 at the end of each month into an account paying 12% compounded monthly?

Solution   Begin by identifying the problem type as well as the variables: *annuity; $t = 5$, $m = \$60$, $r = 0.12$, and $n = 12$*. Next, evaluate the formula:

$$A = m\left[\frac{\left(1 + \frac{r}{n}\right)^{nt} - 1}{\frac{r}{n}}\right]$$

$$= 60\left[\frac{(1 + 0.01)^{60} - 1}{0.01}\right]$$

$$\approx 4,900.180191$$

The future value is $4,900.18. A graph is shown in Figure 11.10.

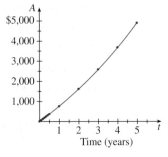

FIGURE 11.10 Growth of an account with a $60 monthly deposit. The graph shows both deposits and interest compounded monthly at 12% annual rate.

The following example is called the Financial Independence Example. It answers the question, "Why should I take math?"

Financial Independence

| Example 2 | **Retirement calculation** |

Suppose you are 21 years old and will make monthly deposits to a bank account paying 10% annual interest compounded monthly.

Option I: Pay yourself $200 per month for 5 years and then leave the balance in the bank until age 65. (Total amount of deposits is $200 × 5 × 12 = $12,000.)

Option II: Wait until you are 40 years old (the age most of us start thinking seriously about retirement) and then deposit $200 per month until age 65. (Total amount of deposits is $200 × 25 × 12 = $60,000.)

Compare the amounts you would have from each of these options.

Solution   We use Pólya's problem-solving guidelines for this example.

Understand the Problem. When most of us are 21 years old, we do not think about retirement. However, if we do, the results can be dramatic. With this example, we investigate the differences if we save early (for 5 years) or later (for 25 years).

Devise a Plan. We calculate the value of the annuity for the 5 years of the first option, and then calculate the effect of leaving the value at the end of 5 years (the annuity) in a savings account until retirement. This part of the problem is a future value problem because it becomes a lump sum problem when deposits are no longer made (after 5 years). For the second option, we calculate the value of the annuity for 25 years.

Carry Out the Plan. Option I: $200 per month for 5 years at 10% annual interest is an annuity with $m = 200$, $r = 0.10$, $t = 5$, and $n = 12$; this is an *annuity.*

$$A = 200\left[\frac{\left(1 + \frac{0.1}{12}\right)^{12(5)} - 1}{\frac{0.1}{12}}\right] \approx 15{,}487.41443$$

At the end of 5 years the amount in the account is $15,487.41. This money is left in the account, so it is now a future value problem for 39 years (ages 26 to 65); this part of the problem is a **future value problem** with $P = 15{,}487.41$, $r = 0.1$, $t = 39$, and $n = 12$, so that

$$A = P\left(1 + \frac{r}{n}\right)^{nt} = 15{,}487.41\left(1 + \frac{0.1}{12}\right)^{12(39)} \approx 752{,}850.86$$

(It is 752,851.08 if you use the calculator value for A from the first part of the problem.) The amount you would have at age 65 from 5 years of payments to yourself starting at age 21 is $752,850.86.

   Option II: If you wait until you are 40 years old, this is an *annuity* problem with $m = 200$, $r = 0.10$, $t = 25$, and $n = 12$:

$$A = 200\left[\frac{\left(1 + \frac{0.1}{12}\right)^{12(25)} - 1}{\frac{0.1}{12}}\right]$$

$$\approx 265{,}366.6806$$

The amount you would have at age 65 from 25 years of payments to yourself starting at age 40 is $265,366.68.

Look Back. Retirement Options I and II are compared graphically in Figure 11.11.

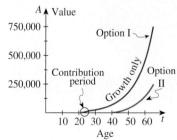

FIGURE 11.11 Comparison of Options I and II

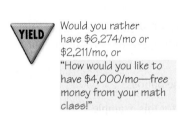

Example 3 | Why study math? For real!

The previous example gave you a choice:

Option I: Pay yourself $200 per month for 5 years beginning when you are 21 years old (this is comparable to the car payments you might make).

Option II: Wait until age 40, then make payments of $200 per month to yourself for the next 25 years. If you were to select one of these options, what effect would it have on your retirement monthly income, assuming you decide to live on the interest only and that the interest rate is 10%?

Solution   Option I gives you a retirement fund of $752,850.86. This will provide a monthly income as determined by the simple interest formula (because the interest is not left to accumulate):

$$I = Prt \qquad \text{Simple interest formula}$$

$$= 752{,}850.86(0.10)\left(\frac{1}{12}\right) \qquad \text{Substitute values.}$$

$$= 6{,}273.76 \qquad \text{Simplify.}$$

Option II gives you a retirement fund of $265,366.68. This will provide a monthly income found as follows:

$$I = Prt \qquad \text{Repeat the steps from Option I.}$$

$$= 265{,}366.68(0.10)\left(\frac{1}{12}\right) \qquad \text{Substitute values.}$$

$$= 2{,}211.39 \qquad \text{Simplify.}$$

The better choice (Option I) would mean that you would have about $4,000 *more* each and every month that you live beyond age 65. Did you ever ask the question "Why should I take math?"

Sinking Funds

We now consider the situation in which we need (or want) to have a lump sum of money (a future value) in a certain period of time. The present value formula will tell us how much we need to have today, but we frequently do not have that amount available. Suppose your goal is $10,000 in 5 years. You can obtain 8% compounded monthly, so the present value formula yields

$$P = A\left(1 + \frac{r}{n}\right)^{-nt} = \$10{,}000\left(1 + \frac{0.08}{12}\right)^{-5(12)} \approx \$6{,}712.10$$

However, this is more than you can afford to put into the bank now. The next choice is to make a series of small equal investments to accumulate at 8% compounded monthly, so that the end result is the same—namely, $10,000 in 5 years. The account you set up to receive those investments is called a **sinking fund.**

To find a formula for a sinking fund, we begin with the formula for an ordinary annuity, and solve for *m* (which is the unknown for a sinking fund). Once again, we let $i = r/n$:

$$A = m\left[\frac{(1 + i)^{nt} - 1}{i}\right] \qquad \text{Annuity formula}$$

$$Ai = m[(1 + i)^{nt} - 1] \qquad \text{Multiply both sides by } i.$$

$$\frac{Ai}{(1 + i)^{nt} - 1} = m \qquad \text{Divide both sides by } [(1 + i)^{nt} - 1].$$

Sinking Fund Formula

If the future value (A) is known, and you wish to find the amount of the periodic payment (m), use the *sinking fund formula*

$$m = \frac{A\left(\frac{r}{n}\right)}{\left(1 + \frac{r}{n}\right)^{nt} - 1}$$

where r is the annual rate, t is the time (in years), and n is the number of times per year the payments are made.

CAUTION Spend a few minutes with this idea; in your own words, can you explain when you would use this formula?

"Sure it was your idea, Bob, but I don't like naming it 'The Sinking Fund.'"

Harley Schwardron © Tribune Media Services. Used with permission.

Example 4 | Find the monthly payment to save a given amount

Suppose you want to have $10,000 in 5 years and decide to make monthly payments into an account paying 8% compounded monthly. What is the amount of each monthly payment?

Solution   You might begin by estimation. If you are paid *no* interest, then since you are making 60 equal deposits, the amount of each deposit would be

$$\frac{10,000}{60} \approx \$166.67$$

Therefore, since you are paid interest, you would expect the amount of each deposit to be somewhat less than this estimation. This is a **sinking fund problem.** We identify the known values: $A = \$10,000$ (future value), $r = 0.08$, $t = 5$, and $n = 12$. Finally, use the appropriate formula:

$$m = \frac{A\left(\frac{r}{n}\right)}{\left(1 + \frac{r}{n}\right)^{nt} - 1} = \frac{10,000\left(\frac{0.08}{12}\right)}{\left(1 + \frac{0.08}{12}\right)^{12(5)} - 1} \approx 136.0972762$$

The necessary monthly payment to the account is $136.10.

COMPUTATIONAL WINDOW

Most graphing calculators have the ability to accept programs. Modern calculators require no special ability to write a program. We input the values of the variables into the memory locations as shown in the margin:

When using a calculator, generally the only difficulty is the proper use of parentheses:

Variable	Input to Storage
P	STO P
A	STO A
m	STO M
r	STO R
n	STO N

Type	Formula	Calculator Notation
Future value	$A = P\left(1 + \dfrac{r}{n}\right)^{nt}$	P*(1+R/N)^(N*T)
Present value	$P = A\left(1 + \dfrac{r}{n}\right)^{-nt}$	A*(1+R/N)^(−N*T)
Annuity	$A = m\left[\dfrac{\left(1 + \frac{r}{n}\right)^{nt} - 1}{\frac{r}{n}}\right]$	M*(((1+R/N)^(N*T)−1)/(R/N))
Sinking fund	$m = \dfrac{A\left(\frac{r}{n}\right)}{\left(1 + \frac{r}{n}\right)^{nt} - 1}$	(A*R/N)/((1+R/N)^(N*T)−1)

To write a program, press the PRGM key and then input the formula. Most calculator programs also require some statement that forces the output of an answer. Check with your owner's manual, but this is usually a one-step process. To use your calculator for one of these formulas, you simply need to identify the type of financial problem, input the values for the appropriate variables, and then run the program containing the formula you wish to evaluate.

Problem Set 11.5

Level 1

1. **IN YOUR OWN WORDS** What do we mean by a lump-sum problem?

2. **IN YOUR OWN WORDS** Why should we call an annuity a periodic payment problem?

3. **IN YOUR OWN WORDS** What is an annuity?

4. **IN YOUR OWN WORDS** What is a sinking fund?

5. **IN YOUR OWN WORDS** Distinguish an annuity from a sinking fund.

6. **IN YOUR OWN WORDS** Describe Example 3 and comment on its possible relevance.

*Use a calculator to evaluate an **ordinary annuity formula***

$$A = m\left[\frac{\left(1 + \frac{r}{n}\right)^{nt} - 1}{\frac{r}{n}}\right]$$

for m, r, and t (respectively) given in Problems 7–22. Assume monthly payments.

7. $50; 5%; 3 yr		8. $50; 6%; 3 yr	
9. $50; 8%; 3 yr		10. $50; 12%; 3 yr	
11. $50; 5%; 30 yr		12. $50; 6%; 30 yr	
13. $50; 8%; 30 yr		14. $50; 12%; 30 yr	

15. $100; 5%; 10 yr

16. $100; 6%; 10 yr

17. $100; 8%; 10 yr

18. $100; 12%; 10 yr

19. $150; 5%; 35 yr

20. $150; 6%; 35 yr

21. $650; 5%; 30 yr

22. $650; 6%; 30 yr

In Problems 23–34, find the value of each annuity at the end of the indicated number of years. Assume that the interest is compounded with the same frequency as the deposits.

	Amount of Deposit m	Frequency n	Rate r	Time t
23.	$500	annually	8%	30 yr
24.	$500	annually	6%	30 yr
25.	$250	semiannually	8%	30 yr
26.	$600	semiannually	2%	10 yr
27.	$300	quarterly	6%	30 yr
28.	$100	monthly	4%	5 yr
29.	$200	quarterly	8%	20 yr
30.	$400	quarterly	11%	20 yr
31.	$30	monthly	8%	5 yr
32.	$5,000	annually	4%	10 yr
33.	$2,500	semiannually	8.5%	20 yr
34.	$1,250	quarterly	3%	20 yr

Find the amount of periodic payment necessary for each deposit to a sinking fund in Problems 35–46.

	Amount Needed A	Frequency n	Rate r	Time t
35.	$7,000	annually	8%	5 yr
36.	$25,000	annually	11%	5 yr
37.	$25,000	semiannually	12%	5 yr
38.	$50,000	semiannually	14%	10 yr
39.	$165,000	semiannually	2%	10 yr
40.	$3,000,000	semiannually	3%	20 yr
41.	$500,000	quarterly	8%	10 yr
42.	$55,000	quarterly	10%	5 yr
43.	$100,000	quarterly	8%	8 yr
44.	$35,000	quarterly	8%	12 yr
45.	$45,000	monthly	7%	30 yr
46.	$120,000	quarterly	7%	30 yr

Level 2

47. Self-employed persons can make contributions for their retirement into a special tax-deferred account called a *Keogh account.* Suppose you are able to contribute $20,000 into this account at the end of each year. How much will you have at the end of 20 years if the account pays 8% annual interest?

48. The owner of Sebastopol Tree Farm deposits $650 at the end of each quarter into an account paying 8% compounded quarterly. What is the value of the account at the end of 5 years?

49. The owner of Oak Hill Squirrel Farm deposits $1,000 at the end of each quarter into an account paying 8% compounded quarterly. What is the value at the end of 5 years, 6 months?

50. Clearlake Optical has a $50,000 note that comes due in 4 years. The owners wish to create a sinking fund to pay this note. If the fund earns 8% compounded semiannually, how much must each semiannual deposit be?

51. A business must raise $70,000 in 5 years. What should be the size of the owners' quarterly payment to a sinking fund paying 8% compounded quarterly?

52. A lottery offers a $1,000,000 prize to be paid in 20 equal installments of $50,000 at the end of each year. What is the future value of this annuity if the current annual rate is 5%?

© AP/Wide World Photos

Level **3**

53. A lottery offers a $1,000,000 prize to be paid in 29 equal annual installments of $20,000 with a 30th final payment of $420,000. What is the total value of this annuity if the current annual rate is 5%?

54. John and Rosamond want to retire in 5 years and can save $150 every three months. They plan to deposit the money at the end of each quarter into an account paying 6.72% compounded quarterly. How much will they have at the end of 5 years?

55. In 2005 the maximum Social Security deposit by an individual was $6,885. Suppose you are 25 and make a deposit of this amount into an account at the end of each year. How much would you have (to the nearest dollar) when you retire if the account pays 5% compounded annually and you retire at age 65?

56. You want to retire at age 65. You decide to make a deposit to yourself at the end of each year into an account paying 13%, compounded annually. Assuming you are now 25 and can spare $1,200 per year, how much will you have when you retire at age 65?

57. Repeat Problem 56 using your own age.

58. Repeat Problem 55 using your own age.

Problem Solving **3**

59. Clearlake Optical has developed a new lens. The owners plan to issue a $4,000,000 30-year bond with a contract rate of 5.5% paid annually to raise capital to market this new lens. This means that Clearlake will be required to pay 5.5% interest each year for 30 years. To pay off the debt, Clearlake will also set up a sinking fund paying 8% interest compounded annually. What size annual payment is necessary for interest and sinking fund combined?

60. The owners of Bardoza Greeting Cards wish to introduce a new line of cards but need to raise $200,000 to do it. They decide to issue 10-year bonds with a contract rate of 6% paid semiannually. This means Bardoza must make interest payments to the bondholders each 6 months for 10 years. They also set up a sinking fund paying 8% interest compounded semiannually. How much money will they need to make the semiannual interest payments as well as payments to the sinking fund?

11.6 | Amortization

In the previous sections we considered the lump-sum formulas for present value and for future value. We also considered the periodic payment formula, which provided the future value of periodic payments into an interest-bearing account. With the annuity formula, we assumed that we knew the amount of the periodic payment.

Present Value of an Annuity

In the real world, if you need or want to purchase an item and you do not have the entire amount necessary, you can generally save for the item, or purchase the item and pay for it with monthly payments. Let us consider an example. Suppose you can afford $275 per month for an automobile. One possibility is to put the $275 per month into a bank account until you have enough money to buy a car. To find the future value of this monthly payment, you would need to know the interest rate (say, 5.2%) and the length of time you will make these deposits (say, 48 months or monthly payments for 4 years). The future value in the account is found using the annuity formula:

$$A = m\left[\frac{\left(1 + \frac{r}{n}\right)^{nt} - 1}{\frac{r}{n}}\right] = 275\left[\frac{\left(1 + \frac{0.052}{12}\right)^{48} - 1}{\frac{0.052}{12}}\right] \approx 14{,}638.04004$$

This means that if you make these payments to yourself, in 4 years you will have saved $275 \times 48 = \$13,200$, but because of the interest paid on this account, the total amount in the account is \$14,638.04.

However, it is more common to purchase a car and agree to make monthly payments to pay for it. Given the annuity calculation above, can we conclude that we can look for a car that costs \$14,638? No, because if we borrow to purchase the car we need to *pay* interest charges rather than *receive* the interest as we did with the annuity. Well, then, how much can we borrow when we assume that we know the monthly payment, the interest rate, and the length of time we will pay? This type of financial problem is called the **present value of an annuity.**

The financial problem called the present value of an annuity is not the same as a present value financial problem, and it is also not the same as an annuity.

Continuing with this car payment example, we need to ask, "What is the present value of 14,638.04004?" This is found using the formula $P = A\left(1 + \frac{r}{n}\right)^{-nt}$:

$$P = 14{,}638.04004\left(1 + \frac{0.052}{12}\right)^{-48} \approx 11{,}894.46293$$

This means we could finance a loan in the amount of \$11,894.46 and pay it off in 48 months with payments of \$275, provided the interest rate is 5.2% compounded monthly.

We will now derive a formula for the present value of an annuity. For convenience, we let $i = r/n$. Since $A = P(1 + i)^{nt}$ and $A = m\left[\frac{(1 + i)^{nt} - 1}{i}\right]$, we see

$$P(1 + i)^{nt} = m\left[\frac{(1 + i)^{nt} - 1}{i}\right] \qquad \text{Both the left and right sides are equal to A.}$$

$$P = \frac{m}{i}\left[\frac{(1 + i)^{nt} - 1}{(1 + i)^{nt}}\right] \qquad \text{Divide both sides by } (1 + i)^{nt}.$$

$$= \frac{m}{i}\left[1 - \frac{1}{(1 + i)^{nt}}\right]$$

$$= \frac{m}{i}\left[1 - (1 + i)^{-nt}\right]$$

COMPUTATIONAL WINDOW

A formula for the present value of an annuity would be input into a calculator program as follows:
M*((1 − (1 + R/N)^(−N*T))/(R/N))

Present Value of an Annuity

If the periodic payment is known (m) and you wish to find the *present value of those periodic payments*, use the *present value of an annuity* formula:

$$P = m\left[\frac{1 - \left(1 + \frac{r}{n}\right)^{-nt}}{\frac{r}{n}}\right]$$

where P is the present value of the annuity, r is the annual interest rate, and n is the number of payments per year.

Example **1**   **Find the amount you can afford for a car**

You look at your budget and decide that you can afford \$250 per month for a car. What is the *maximum loan* you can afford if the interest rate is 13% and you want to repay the loan in 4 years?

Solution   This requires the *present value of an annuity* formula. First, determine the given variables: $m = 250$, $r = 0.13$, $t = 4$, $n = 12$; thus,

$$P = m\left[\frac{1 - \left(1 + \frac{r}{n}\right)^{-nt}}{\frac{r}{n}}\right] = 250\left[\frac{1 - \left(1 + \frac{0.13}{12}\right)^{-12(4)}}{\frac{0.13}{12}}\right] \approx 9{,}318.797438$$

This means that you can afford to finance about \$9,319. To keep the payments at \$250 (or less), a down payment should be made, if necessary, to lower the price of the car to the finance amount.

| Example 2 | Find the home you can afford |

Pólya's Method

Suppose your budget (or your banker) tells you that you can afford $1,575 per month for a house payment. If the current interest rate for a home loan is 10.5% and you will finance the home for 30 years, what is the expected price of the home you can afford? Assume that you have saved the money to make the required 20% down payment.

Solution   We use Pólya's problem-solving guidelines for this example.

Understand the Problem. Most textbook problems give you a price for a home and ask for the monthly payment. This problem takes the real-life situation in which you know the monthly payment and wish to know the price of the house you can afford.

Devise a Plan. We first find the amount of the loan, and then find the down payment by subtracting the amount of the loan from the price of the home we wish to purchase.

Carry Out the Plan. We first find the amount of the loan; this is the **present value of an annuity.** We know $m = 1,575$, $r = 0.105$, $t = 30$, $n = 12$; thus,

$$P = m\left[\frac{1 - \left(1 + \frac{r}{n}\right)^{-nt}}{i}\right] = 1,575\left[\frac{1 - \left(1 + \frac{0.105}{12}\right)^{-12(30)}}{\frac{0.105}{12}}\right] \approx 172,180.2058$$

The loan that you can afford is about $172,180.

What about the down payment? If there is a 20% down payment, then the amount of the loan is 80% of the price of the home. We know the amount of the loan, so the problem is now to determine the home price such that 80% of the price is equal to the amount of the loan. In symbols,

$$0.80x = 172,180.2058$$

which means that you simply divide the calculator output for the amount of the loan by 0.8 to find 215,225.2573.

Look Back. Given the conditions of the problem, we can look back to check our work (rounded to the nearest dollar for the check):

Purchase price of home:	$215,225
Less 20% down payment:	$ 43,045
Amount of loan:	$172,180

| Example 3 | |

Theresa's Social Security benefit is $450 per month if she retires at age 62 instead of age 65. What is the present value of an annuity that would pay $450 per month for 3 years if the current interest rate is 10% compounded monthly?

Solution   Here, $m = 450$, $r = 0.10$, $t = 3$, and $n = 12$; then,

$$P = m\left[\frac{1 - (1 + i)^{-N}}{i}\right] = 450\left[\frac{1 - \left(1 + \frac{0.10}{12}\right)^{-36}}{\frac{0.10}{12}}\right] \approx 13,946.05601$$

The present value of this decision is $13,946.06. This represents the value of the additional 3 years of Social Security payments.

Monthly Payments

The process of paying off a debt by systematically making partial payments until the debt (principal) and the interest are repaid is called **amortization.** If the loan is paid off in regular equal installments, then we use the formula for the present value of an annuity to find the monthly payments by algebraically solving for m, where we let $i = r/n$:

$$P = m\left[\frac{1 - (1 + i)^{-nt}}{i}\right]$$ *Present value of an annuity formula*

$$Pi = m\left[1 - (1 + i)^{-nt}\right]$$ *Multiply both sides by i.*

$$\frac{Pi}{1 - (1 + i)^{-nt}} = m$$ *Divide both sides by $1 - (1 + i)^{-nt}$.*

COMPUTATIONAL WINDOW

To program the amortization formula into a calculator input it as

$(P*R/N)/(1 - (1+R/N)^{\wedge}(-N*T))$

Amortization Formula

If the amount of the loan is known (P), and you wish to find the amount of the periodic payment (m), use the formula

$$m = \frac{P\left(\frac{r}{n}\right)}{1 - \left(1 + \frac{r}{n}\right)^{-nt}}$$

where r is the annual rate, t is the time (in years), and n is the number of payments per year.

Example 4   Find the monthly payment for a house

In 2010, the average price of a new home was $291,000 and the interest rate was 6%. If this amount is financed for 30 years at 6% interest, what is the monthly payment?

Solution   This is an *amortization* problem with $P = 291{,}000$, $r = 0.06$, $t = 30$, and $n = 12$; then

$$m = \frac{P\left(\frac{r}{n}\right)}{1 - \left(1 + \frac{r}{n}\right)^{-nt}} = \frac{291{,}000\left(\frac{0.06}{12}\right)}{1 - \left(1 + \frac{0.06}{12}\right)^{-12(30)}} \approx 1{,}744.69202819$$

The monthly payment is $1,744.69. This is the payment for interest and principal to pay off a 30-year 6% loan of $291,000.

It is noteworthy to see how much interest is paid for the home loan of Example 4. There are 360 payments of $1,744.69, so the total amount repaid is

$$360(1{,}744.69) = 628{,}088.40$$

Since the loan was for $291,000, the interest is

$$\$628{,}088.40 - \$291{,}000 = \$337{,}088.40$$

The interest is more than the amount of the loan!

The amount of interest paid can be reduced by making a larger down payment. For Example 4, if a 20% down payment (which is fairly standard) was made, then $291,000(0.80) = \$232,800$ is the amount to be financed. If this amount is amortized over 20 years (instead of 30), the monthly payments are $1,667.85. The total amount paid on *this* loan is

Amount of each payment

$$240\overbrace{(1{,}667.85)} + \underbrace{58{,}200} = 458{,}484$$

Number of payments   Down payment

The savings yielded over the term of the loan by making a 20% down payment for 20 years rather than financing the entire amount for 30 years is $628,088.40 − $458,484 = $169,604.40.

It is interesting to see how much a difference in rates will affect the costs in Example 4. Suppose the interest rate rises by $\frac{1}{2}$%. If $P = \$291,000$, $r = 0.065$, $t = 30$, and $n = 12$, then (using the amortization formula)

$$m = \frac{P\left(\frac{r}{n}\right)}{1 - \left(1 + \frac{r}{n}\right)^{-nt}} = \frac{291,000\left(\frac{0.065}{12}\right)}{1 - \left(1 + \frac{0.065}{12}\right)^{-12(30)}} \approx 1,839.32$$

The interest on the 30-year 6% loan in Example 4 was $337,088.40; now at 6.5% interest for 30 years the total interest is

$$360(\$1,839.32) - \$291,000 = \$371,155.20$$

an additional cost of almost $34,100 ($371,155.20 − $337,088.40 = $34,066.80).

Most of the time when you obtain a home loan, you will need to make a choice about which loan offer is best for your particular situation, and a difference of $\frac{1}{2}$% between different choices is common.

Example 5 Make the better home loan choice

Suppose you are considering a $400,000 30-year home loan and are considering two possibilities:

 Loan A: 7.25% + 0 pts   Loan B: 6.875% + 2.375 pts

Which is the better loan if the home is sold in 10 years?

Solution You might think that the lower rate is always the better choice, but you need to take into consideration the loan fees. One point on a loan is 1% of the amount of the loan. This means that 2.375 points represents a loan charge of

$$0.02375(\$400,000) = \$9,500$$

First, use the amortization formula for each loan:

$$\text{Loan A:} \quad m = \frac{\$400,000\left(\frac{0.0725}{12}\right)}{1 - \left(1 + \frac{0.0725}{12}\right)^{-12(30)}} \approx \$2,728.71$$

$$\text{Loan B:} \quad m = \frac{\$400,000\left(\frac{0.06875}{12}\right)}{1 - \left(1 + \frac{0.06875}{12}\right)^{-12(30)}} \approx \$2,627.72$$

The difference in payments is $100.99/mo. Now, find the present value of an annuity with $m = \$100.99$ using the 6.875% rate for 10 years:

$$A = \$100.99\left[\frac{1 - \left(1 + \frac{0.06875}{12}\right)^{-12(10)}}{\frac{0.06875}{12}}\right]$$

$$\approx \$8,746.35$$

Since the fees ($9,500) are greater than the present value of this annuity, the better choice is Loan A, even though it has the higher rate.

We should note that even though both amortization and add-on interest calculate the monthly (or periodic) payment, you should pay attention to which procedure or formula to use.

Distinguish Add-On Simple Interest and Amortization

To calculate the monthly (or periodic) payment, consider one of the following methods:

Simple interest Use this with installment loans or loans that use the words "add-on interest." To calculate the payment:

Step 1 Calculate the interest using $I = Prt$.

Step 2 Add the interest to the amount of the loan using $A = P + I$.

Step 3 Divide this result by the number of payments (N):

$$m = \frac{A}{N}$$

The interest rate, r, used for the add-on rate is not the same as the annual interest rate (APR).

Compound interest Use this with long-term loans, such as when purchasing real estate. Those loans are *amortized*, which means the interest rate, r, is the same as the annual interest rate, and the payments pay off both the principal and the interest.

Step 1 Identify the variables P, r, n, and t.

Step 2 Use the *amortization formula*:

$$m = \frac{P\left(\frac{r}{n}\right)}{1 - \left(1 + \frac{r}{n}\right)^{-nt}}$$

Problem Set | 11.6

Level 1

1. **IN YOUR OWN WORDS** What does amortization mean?

2. **IN YOUR OWN WORDS** Describe when you would use the present value of an annuity formula.

3. The variables $m, n, r, t, A,$ and P are used in the various financial formulas. Tell what each of these variables represents.

*Use a calculator to evaluate the **present value of an annuity** formula*

$$P = m\left[\frac{1 - \left(1 + \frac{r}{n}\right)^{-nt}}{\frac{r}{n}}\right]$$

for the values of the variables m, r, and t (respectively) given in Problems 4–11. Assume n = 12.

4. $50; 5%; 5 yr

5. $50; 6%; 5 yr

6. $50; 8%; 5 yr

7. $150; 5%; 30 yr

8. $150; 6%; 30 yr

9. $150; 8%; 30 yr

10. $1,050; 5%; 30 yr

11. $1,050; 6%; 30 yr

*Use a calculator to evaluate the **amortization** formula*

$$m = \frac{P\left(\frac{r}{n}\right)}{1 - \left(1 + \frac{r}{n}\right)^{-nt}}$$

for the values of the variables P, r, and t (respectively) given in Problems 12–19. Assume n = 12.

12. $14,000; 5%; 5 yr

13. $14,000; 10%; 5 yr

14. $14,000; 19%; 5 yr

15. $150,000; 8%; 30 yr

16. $150,000; 9%; 30 yr

17. $150,000; 10%; 30 yr

18. $260,000; 12%; 30 yr

19. $260,000; 9%; 30 yr

Find the present value of the ordinary annuities in Problems 20–31.

	Amount of Deposit m	Frequency n	Rate r	Time t
20.	$500	annually	8%	30 yr
21.	$500	annually	6%	30 yr
22.	$250	semiannually	8%	30 yr

	Amount of Deposit *m*	Frequency *n*	Rate *r*	Time *t*
23.	$600	semiannually	2%	10 yr
24.	$300	quarterly	6%	30 yr
25.	$100	monthly	4%	5 yr
26.	$200	quarterly	8%	20 yr
27.	$400	quarterly	11%	20 yr
28.	$ 30	monthly	8%	5 yr
29.	$ 75	monthly	4%	10 yr
30.	$ 50	monthly	8.5%	20 yr
31.	$100	quarterly	3%	20 yr

Level 2

Find the monthly payment for the loans in Problems 32–45.

32. $500 loan for 12 months at 12%

33. $100 loan for 18 months at 18%

34. $4,560 loan for 20 months at 21%

35. $3,520 loan for 30 months at 19%

36. Used-car financing of $2,300 for 24 months at 15%

37. New-car financing of 2.9% on a 30-month $12,450 loan

38. Furniture financed at $3,456 for 36 months at 23%

39. A refrigerator financed for $985 at 17% for 15 months

40. A $112,000 home bought with a 20% down payment and the balance financed for 30 years at 11.5%

41. A $108,000 condominium bought with a 30% down payment and the balance financed for 30 years at 12.05%

42. Finance $450,000 for a warehouse with a 12.5% 30-year loan

43. Finance $859,000 for an apartment complex with a 13.2% 20-year loan

44. How much interest (to the nearest dollar) would be saved in Problem 40 if the home were financed for 15 rather than 30 years?

45. How much interest (to the nearest dollar) would be saved in Problem 41 if the condominium were financed for 15 rather than 30 years?

46. Melissa agrees to contribute $500 to the alumni fund at the end of each year for the next 5 years. Shannon wants to match Melissa's gift, but he wants to make a lump-sum contribution. If the current interest rate is 12.5% compounded annually, how much should Shannon contribute to equal Melissa's gift?

47. A $1,000,000 lottery prize pays $50,000 per year for the next 20 years. If the current rate of return is 12.25%, what is the present value of this prize?

48. You look at your budget and decide that you can afford $250 per month for a car. What is the maximum loan you can afford if the interest rate is 13% and you want to repay the loan in 4 years?

Level 3

49. Suppose you have an annuity from an insurance policy and you have the option of being paid $250 per month for 20 years or having a lump-sum payment of $25,000. Which has more value if the current rate of return is 10%, compounded monthly?

50. An insurance policy offers you the option of being paid $750 per month for 20 years or a lump sum of $50,000. Which has more value if the current rate of return is 9%, compounded monthly, and you expect to live for 20 years?

51. I recently found a real-life advertisement in the newspaper:

(Only the phone number has been changed.) Suppose that you have won a $20,000,000 lottery, paid in 20 annual installments. How much would be a fair price to be paid today for the assignment of this prize? Assume the interest rate is 5%.

52. The bottom notice (not circled) states that $6,000 is needed for 90 days, and that the advertiser is willing to pay 20% interest. How much would you expect to receive in 90 days if you lent this party the $6,000?

53. Suppose your gross monthly income is $5,500 and your current monthly payments are $625. If the bank will allow you to pay up to 36% of your gross monthly income (less current monthly payments) for a monthly house payment, what is the maximum loan you can obtain if the rate for a 30-year mortgage is 9.65%?

Problem Solving 3

54. As the interest rate increases, determine whether each of the given amounts increases or decreases. Assume that all other variables remain constant.
   a. future value
   b. present value
   c. annuity

55. As the interest rate increases, determine whether each of the given amounts increases or decreases. Assume that all other variables remain constant.
   a. monthly value for a sinking fund
   b. monthly value for amortization
   c. present value of an annuity

56. Suppose your gross monthly income is $4,550 and your spouse's gross monthly salary is $3,980. Your monthly bills are $1,235. The home you wish to purchase costs $355,000 and the loan is an 11.85% 30-year loan. How much down payment (rounded to the nearest hundred dollars) is necessary to be able to afford this home? This down payment is what percent of the cost of the home? Assume the bank will allow you to pay up to 36% of your gross monthly income (less current monthly payments) for house payments.

57. Suppose you want to purchase a home for $225,000 with a 30-year mortgage at 10.24% interest. Suppose also that you can put down 25%. What are the monthly payments? What is the total amount paid for principal and interest? What is the amount saved if this home is financed for 15 years instead of for 30 years?

58. The McBertys have $30,000 in savings to use as a down payment on a new home. They also have determined that they can afford between $1,500 and $1,800 per month for mortgage payments. If the mortgage rates are 11% per year compounded monthly, what is the price range for houses they should consider for a 30-year loan?

59. Rework Problem 58 for a 20-year loan instead of a 30-year loan.

60. Rework Problem 58 if interest rates go down to 10.2%.

11.7 Summary of Financial Formulas

Definitions of variables:

P = present value (principal)
A = future value
I = amount of interest
r = annual interest rate
t = number of years
n = periods; number of times compounded each year

For many students, the most difficult part of working with finances (other than the problem of lack of funds) is in determining *which* formula to use. If we wish to free ourselves from the problem of lack of funds, we must use our knowledge in a variety of situations that occur outside the classroom.

We will now outline a procedure for using the financial formulas we have introduced in this chapter that you can use to unlock your financial security. First, review the meaning of the variables, as shown in the margin. Next, study the following procedure for working financial problems.

Classification of Financial Problems

When classifying financial formulas:

Step 1 Is the interest compounded?

If the interest is *removed,* use the **simple interest formula.**
If the interest is *added* to the account, use the **compound interest formula.**

Step 2 Is it a lump sum problem? What is the unknown?

If the future value is unknown, then it is a **future value problem.**
If the present value is unknown, then it is a **present value problem.**

Step 3 Is it a periodic payment problem with the periodic payment known?

If the future value is desired, then it is an **ordinary annuity.**
If the present value is desired, then it is a **present value of an annuity.**

Step 4 Is it a periodic payment problem with the periodic payment unknown?

If the future value is known, then it is a **sinking fund.**
If the present value is known, then it is an **amortization** problem.

The rest of this section discusses this procedure box with a bit more detail.

Is the interest compounded?

If the interest is *removed* from the account, then use the *simple interest formula.*

$$I = Prt$$

Use this for amount of interest; any one of the variables I, P, r, or t, may be unknown.

Examples: amount of interest; add-on interest; car payments; using the r from the simple interest formula when calculating the APR; finding the APR when given the purchase price and the monthly payment; credit card interest

If the interest is *added* to the account, then use a *compound interest formula*. In such a case there are two remaining questions to ask when classifying a compound interest problem.

Is it a lump-sum problem? If it is, then what is the unknown?

If FUTURE VALUE is the unknown, then it is a *future value* problem.

$$A = P\left(1 + \frac{r}{n}\right)^{nt} \quad \text{or} \quad A = Pe^{rt}$$

Examples: future value; inflation; retirement planning

If PRESENT VALUE is the unknown, then it is a *present value* problem.

$$P = A\left(1 + \frac{r}{n}\right)^{-nt}$$

Examples: present value; retirement planning

Is it a periodic payment problem? If it is, then is the periodic payment known?

PERIODIC PAYMENT KNOWN and you want to find the future value, then it is an *ordinary annuity* problem.

$$A = m\left[\frac{\left(1 + \frac{r}{n}\right)^{nt} - 1}{\frac{r}{n}}\right]$$

Examples: future value of periodic payments made to an interest-bearing account; retirement planning

PERIODIC PAYMENT KNOWN and you want to find the present value, then it is a *present value of an annuity* problem.

$$P = m\left[\frac{1 - \left(1 + \frac{r}{n}\right)^{-nt}}{\frac{r}{n}}\right]$$

Examples: if the monthly payment, time, and rate are known, how much loan is possible? the current value of a known annuity (as in selling off the proceeds from an insurance policy or winning the lottery)

PERIODIC PAYMENT UNKNOWN and you know the future value, then it is a *sinking fund* problem.

$$m = \frac{A\left(\frac{r}{n}\right)}{\left(1 + \frac{r}{n}\right)^{nt} - 1}$$

Example: the periodic payment necessary to make to an interest-bearing account in order to reach some known financial goal

PERIODIC PAYMENT UNKNOWN and you know the present value, then it is an *amortization* problem.

$$m = \frac{P\left(\frac{r}{n}\right)}{1 - \left(1 + \frac{r}{n}\right)^{-nt}}$$

Example 1 Monthly payment for an amortized loan

Figure 11.12 reproduces part of an advertisement to refinance a loan.

Here is an example of how some of my clients are using this new program:

Bills	Current Debt	Current Pmt.	New Loan
Mortgage	$378,000	$2,389	**$434,200**
Home Equity Loan	$36,000	$345	
Visa	$8,000	$190	
Master Card	$5,000	$150	**New Payment**
Sears	$3,500	$105	
Department Store	$3,700	$145	**$1,397**
Total	**$434,200**	**$3,324**	

FIGURE 11.12 Super duper brand new mortgage program

Assume a 6.5% annual mortgage rate. Find the monthly payment for both the old mortgage and for the new loan.

Solution The present value is known, and we see the monthly payment, so this is an amortization problem. Here are the variable values:

$$m = \frac{P\left(\frac{r}{n}\right)}{1 - \left(1 + \frac{r}{n}\right)^{-nt}}$$

Mortgage	P	r	n	t	
Old	$378,000	0.065	12	30	$m = \$2{,}389.21$
New	$434,200	0.065	12	30	$m = \$2{,}744.43$

Wait a minute! The advertisement shows a new monthly payment of $1,397. The new loan is at a new "teaser" rate. These loans were particularly responsible for the financial melt-down of 2009. The person who did not read the fine print and do some math calculations became trapped when the teaser rate increased to an unaffordable rate.

Example 2 Classifying financial problems

Classify each formula as a financial problem type, and give one possible example which would use the given formula.

a. $y = 500(1.07)^{20}$ **b.** $y = 280{,}000\left(1 + \frac{0.04}{4}\right)^{-60}$ **c.** $y = 80(0.05)(20)$

d. $y = 480\left[\dfrac{\left(1 + \frac{0.04}{12}\right)^{120} - 1}{\frac{0.04}{12}}\right]$

Solution

a. Future value problem; specifically the future value of $500 invested at 7% interest for 20 years.
b. Present value problem; how much would need to be invested today to yield $280,000 in 15 years if the money is invested at 4%?
c. Simple interest formula; what is the future value of $80 invested at 5% for 20 years?
d. Ordinary annuity; what is the future value of $480 monthly payment invested at 4% for 10 years?

Problem Set 11.7

Level 1

1. **IN YOUR OWN WORDS** What are a reasonable down payment and monthly payment for a home in your area?

2. **IN YOUR OWN WORDS** What are a reasonable down payment and monthly payment for a new automobile?

3. **IN YOUR OWN WORDS** Outline a procedure for identifying the type of financial formula for a given applied problem.

4. What is the formula for the present value of an annuity? What is the unknown?

5. What is the amortization formula? What is the unknown?

Classify the type of financial formula for the information given in Problems 6–11.

Lump-sum Problems

	P	A
6.	known	unknown
7.	unknown	known

Periodic Payment Problems

	P	A	m
8.	unknown		known
9.		unknown	known
10.		known	unknown
11.	known		unknown

In Problems 12–15, match each formula in Column A with the type of financial problem in Column B.

Column A   **Column B**

12. $A = m\left[\dfrac{\left(1 + \frac{r}{n}\right)^{nt} - 1}{\frac{r}{n}}\right]$ Annuity

13. $P = m\left[\dfrac{1 - \left(1 + \frac{r}{n}\right)^{-nt}}{\frac{r}{n}}\right]$ Amortization

14. $m = \dfrac{A\left(\frac{r}{n}\right)}{\left(1 + \frac{r}{n}\right)^{nt} - 1}$ Present value of an annuity

15. $m = \dfrac{P\left(\frac{r}{n}\right)}{1 - \left(1 + \frac{r}{n}\right)^{-nt}}$ Sinking fund

Classify the financial problems in Problems 16–19, and then answer each question by assuming a 12% interest rate compounded annually.

16. Find the value of a $1,000 certificate in 3 years.

17. Deposit $300 at the end of each year. What is the total in the account in 10 years?

18. An insurance policy pays $10,000 in 5 years. What lump-sum deposit today will yield $10,000 in 5 years?

19. What annual deposit is necessary to give $10,000 in 5 years?

Level 2

In Problems 20–52: **a.** *State the type; and* **b.** *Answer the question.*

20. Find the value of a $1,000 certificate in $2\frac{1}{2}$ years if the interest rate is 12% compounded monthly.

21. You deposit $300 at the end of each year into an account paying 12% compounded annually. How much is in the account in 10 years?

22. A 5-year term insurance policy has an annual premium of $300, and at the end of 5 years, all payments and interest are refunded. What lump-sum deposit is necessary to equal this amount if you assume an interest rate of 10% compounded annually?

23. What annual deposit is necessary to have $10,000 in 5 years if all the money is deposited at 9% interest compounded annually?

24. A $5,000,000 apartment complex loan is to be paid off in 10 years by making 10 equal annual payments. How much is each payment if the interest rate is 14% compounded annually?

25. The price of automobiles has increased at 6.25% per year. How much would you expect a $20,000 automobile to cost in 5 years if you assume annual compounding?

26. The amount to be financed on a new car is $9,500. The terms are 7% for 4 years. What is the monthly payment?

27. What deposit today is equal to 33 annual deposits of $500 into an account paying 8% compounded annually?

28. If you can afford $875 for your house payments, what is the loan you can afford if the interest rate is 6.5% and the monthly payments are made for 30 years?

29. What is the monthly payment for a home costing $125,000 with a 20% down payment and the balance financed for 30 years at 12%?

30. Ricon Bowling Alley will need $80,000 in 4 years to resurface the lanes. What lump sum would be necessary today if the owner of the business can deposit it in an account that pays 9% compounded semiannually?

31. Rita wants to save for a trip to Tahiti, so she puts $2.00 per day into a jar. After 1 year she has saved $730 and puts the money into a bank account paying 10% compounded annually. She continues to save in this manner and makes her annual $730 deposit for 15 years. How much does she have at the end of that time period?

32. Karen receives a $12,500 inheritance that she wants to save until she retires in 20 years. If she deposits the money in a fixed 11% account, compounded daily ($n = 365$), how much will she have when she retires?

33. You want to give your child a million dollars when he retires at age 65. How much money do you need to deposit into an account paying 9% compounded monthly if your child is now 10 years old?

34. An accounting firm agrees to purchase a computer for $150,000 (cash on delivery) and the delivery date is in 270 days. How much do the owners need to deposit in an account paying 18% compounded quarterly so that they will have $150,000 in 270 days?

35. For 5 years, Thompson Cleaners deposits $900 at the end of each quarter into an account paying 8% compounded quarterly. What is the value of the account at the end of 5 years?

36. What is the necessary amount of monthly payments to an account paying 18% compounded monthly in order to have $100,000 in $8\frac{1}{3}$ years if the deposits are made at the end of each month?

37. Mark's Grocery Store is going to be remodeled in 5 years, and the remodeling will cost $300,000. How much should be deposited now in order to pay for this remodeling if the account pays 12% compounded monthly?

38. If an apartment complex will need painting in $3\frac{1}{2}$ years and the job will cost $45,000, what amount needs to be deposited into an account now in order to have the necessary funds? The account pays 12% interest compounded semiannually.

39. Teal and Associates needs to borrow $45,000. The best loan they can find is one at 12% that must be repaid in monthly installments over the next $3\frac{1}{2}$ years. How much are the monthly payments?

40. Certain Concrete Company deposits $4,000 at the end of each quarter into an account paying 10% interest compounded quarterly. What is the value of the account at the end of $7\frac{1}{2}$ years?

41. Major Magic Corporation deposits $1,000 at the end of each month into an account paying 18% interest compounded monthly. What is the value of the account at the end of $8\frac{1}{3}$ years?

42. What is the future value of $112,000 invested for 5 years at 14% compounded monthly?

43. What is the future value of $800 invested for 1 year at 10% compounded daily?

44. What is the future value of $9,000 invested for 4 years at 20% compounded monthly?

45. If $5,000 is compounded annually at 5.5% for 12 years, what is the future value?

46. If $10,000 is compounded annually at 8% for 18 years, what is the future value?

47. You owe $5,000 due in 3 years, but you would like to pay the debt today. If the present interest rate is compounded annually at 11%, how much should you pay today so that the present value is equivalent to the $5,000 payment in 3 years?

48. Sebastopol Movie Theater will need $20,000 in 5 years to replace the seats. What deposit should be made today in an account that pays 9% compounded semiannually?

49. The Fair View Market must be remodeled in 3 years. It is estimated that remodeling will cost $200,000. How much should be deposited now (to the nearest dollar) to pay for this remodeling if the account pays 10% compounded monthly?

50. A laundromat will need seven new washing machines in $2\frac{1}{2}$ years for a total cost of $2,900. How much money (to the nearest dollar) should be deposited now to pay for these machines? The interest rate is 11% compounded semiannually.

51. A computerized checkout system is planned for Able's Grocery Store. The system will be delivered in 18 months at a cost of $560,000. How much should be deposited today (to the nearest dollar) into an account paying 7.5% compounded daily?

52. A lottery offers you a choice of $1,000,000 per year for 5 years or a lump-sum payment. What lump-sum payment (rounded to the nearest dollar) would equal the annual payments if the current interest rate is 14% compounded annually?

Level 3

Problems 53–55 are based on a 30-year fixed-rate home loan of $185,500 with an interest rate of 7.75%.

53. What is the monthly payment?

54. What is the total amount of interest paid?

55. Suppose you reduce the term to 20 years. What is the total amount of interest paid, and what is the savings over the 30-year loan?

Problems 56–58 are based on a 30-year fixed-rate home loan of $418,500 with an interest rate of 8.375%.

56. What is the monthly payment?

57. What is the total amount of interest paid?

58. Suppose you reduce the term to 22 years. What is the total amount of interest paid, and what is the savings over the 30-year loan?

Problem Solving 3

59. A contest offers the winner $50,000 now or $10,000 now and $45,000 in one year. Which is the better choice if the current interest rate is 10% compounded monthly and the winner does not intend to use any of the money for one year?

60. In 1982 the inflation rate hit 16%. Suppose that the average cost of a textbook in 1982 was $15. What is the expected cost in the year 2012 if we project this rate of inflation on the cost? (Assume continuous compounding.) If the average cost of a textbook in 2010 was $120, what is the actual inflation rate (rounded to the nearest tenth percent)?*

*This problem requires Chapter 10 (solving exponential equations).

11.8   CHAPTER SUMMARY

To get money is difficult, to keep it is more difficult, but to spend it wisely is most difficult of all.

ANONYMOUS

Important Ideas

Simple interest formula and the future value formula for compound interest [11.1]

Ordinary and exact interest [11.1]

Interest formula variables [11.1]

Credit card interest [11.2]

Arithmetic, geometric, and Fibonacci sequences [11.3]

Distinguish between sequences and series [11.3, 11.4]

Retirement decisions [11.5]

Add-on interest vs. amortization [11.2; 11.6]

Distinguish financial problems [11.7]

Take some time getting ready to work the review problems in this section. First review these important ideas. Look back at the definition and property boxes. If you look online, you will find a list of important terms introduced in this chapter, as well as the types of problems that were introduced. You will maximize your understanding of this chapter by working the problems in this section only after you have studied the material.

You will find some review help online at **www.mathnature.com**. There are links giving general test help in studying for a mathematics examination, as well as specific help for reviewing this chapter.

Chapter 11   Review Questions

1. **IN YOUR OWN WORDS**   Explain the difference between a sequence and a series. Include in your discussion how to distinguish arithmetic, geometric, and Fibonacci-type sequences, and give the formulas for their general terms. What are the relevant formulas for arithmetic and geometric series?

2. **IN YOUR OWN WORDS**   Outline a procedure for identifying financial formulas. Include in your discussion future value, present value, ordinary annuity, present value of an annuity, amortization, and sinking fund classifications.

3. Classify each sequence as arithmetic, geometric, Fibonacci-type, or none of these. Find an expression for the general term if it is one of these types; otherwise, give the next two terms.
   a. 5, 10, 15, 20, . . .
   b. 5, 10, 20, 40, . . .
   c. 5, 10, 15, 25, . . .
   d. 5, 10, 20, 35, . . .
   e. 5, 50, 500, 5,000, . . .
   f. 5, 50, 5, 50, . . .

4. a. Evaluate: $\displaystyle\sum_{k=1}^{3}\left(k^2 - 2k + 1\right)$

   b. Expand: $\displaystyle\sum_{k=1}^{4}\frac{k-1}{k+1}$

5. Suppose someone tells you she has traced her family tree back 10 generations. What is the minimum number of people on her family tree if there were no intermarriages?

6. A certain bacterium divides into two bacteria every 20 minutes. If there are 1,024 bacteria in the culture now, how many will there be in 24 hours, assuming that no bacteria die? Leave your answer in exponential form.

7. Suppose that the car you wish to purchase has a sticker price of $22,730 with a dealer cost of $18,579. Make a 5% offer for this car (rounded to the nearest hundred dollars).

8. Suppose that the amount to be financed for a car purchase is $13,500 at an add-on interest rate of 2.9% for 2 years.
   a. What are the monthly installment and the amount of interest that you will pay?
   b. Use the APR formula for this loan.

9. If a car with a cash price of $11,450 is offered for nothing down with 48 monthly payments of $353.04, what is the APR?

10. From the consumer's point of view, which method of calculating interest on a credit card is most advantageous? Illustrate the three types of calculating interest for a purchase of $525 with 9% APR for a 31-day month in which it takes 7 days for your $100 payment to be received and recorded.

11. Suppose that you want to have $1,000,000 in 50 years. To achieve this goal, how much do you need to deposit today if you can earn 9% interest compounded monthly?

12. **a.** What is the monthly payment for a home loan of $154,000 if the rate is 8% and the time is 20 years?

 b. If you paid a 20% down payment, what is the price of the home?

 c. What would you expect this home to be worth in 30 years if you assume an inflation rate of 4%?

13. Suppose that you expect to receive a $100,000 inheritance when you reach 21 in three years and four months. What is the present value of your inheritance if the current interest rate is 6.4% compounded monthly?

Use the advertisement shown in Figure 11.13 as a basis for answering Problems 14–20. Assume the current interest rate is 5%.

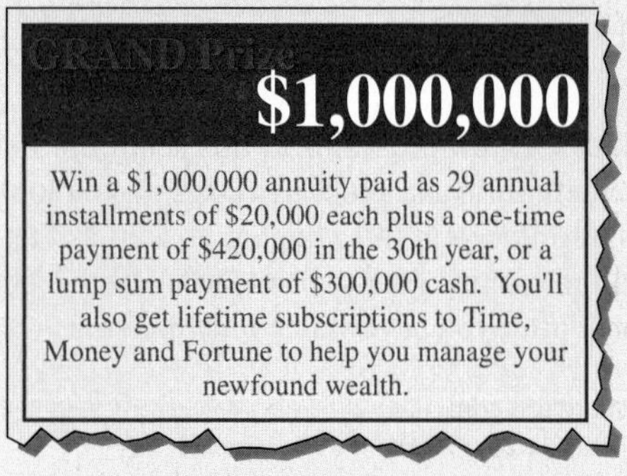

GRAND Prize

$1,000,000

Win a $1,000,000 annuity paid as 29 annual installments of $20,000 each plus a one-time payment of $420,000 in the 30th year, or a lump sum payment of $300,000 cash. You'll also get lifetime subscriptions to Time, Money and Fortune to help you manage your newfound wealth.

FIGURE 11.13 Contest prize notice

14. What is the present value of the $420,000 payment to be received in 30 years?

15. What is the actual value of the annuity portion of the prize?

16. What is the present value of the annuity?

17. Suppose you take the $300,000 cash payment and still want to have $420,000 in 30 years. How much of the $300,000 do you need to set aside now to have $420,000 in 30 years?

18. Suppose you take the $300,000 cash payment and set aside $100,000 for savings. You plan on using the interest on the remaining $200,000. What are your annual earnings from interest?

19. Which option should you take and why?

20. If the current interest rate is 10%, which option should you take and why?

BOOK REPORTS

Write a 500-word report on this book:

What Are Numbers? (Glenview, IL: Scott, Foresman and Company, 1969) by Louis Auslander. Here is a problem from this book (p. 112): "Suppose a ball has the property that whenever it is dropped, it bounces up 1/2 the distance that it fell. For instance, if we drop it from 6 feet, it bounces up 3 feet. Let us hold the ball 6 feet off the ground and drop it. How far will the ball go before coming to rest?" Include the solution of this problem as part of your report.

Group | RESEARCH PROJECTS

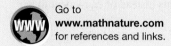

Go to
www.mathnature.com
for references and links.

Working in small groups is typical of most work environments, and learning to work with others to communicate specific ideas is an important skill. Work with three or four other students to submit a single report based on each of the following questions.

G37. Suppose you have just inherited $30,000 and need to decide what to do with the money. Write a paper discussing your options and the financial implications of those options. The paper you turn in should offer several alternatives and then the members of your group should reach a consensus of the best course of action.

G38. Suppose you were hired for a job paying $21,000 per year and were given the following options:

Option A: Annual salary increase of $1,200

Option B: Semiannual salary increase of $300

Option C: Quarterly salary increase of $75

Option D: Monthly salary increase of $10

Each person should write the arithmetic series for the total amount of money earned in 10 years under a different option. Your group should reach a consensus as to which is the best option. Give reasons and show your calculations in the paper that your group submits.

G39. It is not uncommon for the owner of a home to receive a letter similar to the one shown below. Write a paper based on this letter. Different members of your group can work on different parts of the question, but you should submit one paper from your group.

```
Dear Customer:

Previously we provided you an enrollment package for our BiWeekly
Advantage Plan.  We have had an excellent response from our mortgagors,
yet we have not received your signed enrollment form and program fee to
start your program.

How effective is the BiWeekly Advantage Plan?  Reviewing your loan as of
the week of this letter we estimate:

            MORTGAGE INTEREST SAVINGS OF $99,287.56
            LOAN TERM REDUCED BY 7 YEARS.

Many financial advisors agree it is a sound investment practice to pay off
a mortgage early, particularly in light of growing economic uncertainty.

Act today!  Don't let procrastination or indecision deny you and your
family the opportunity to obtain mortgage-free homeownership sooner than
you ever expected.

Please call us toll-free at 1-800-555-6060 if you would like more
information about our BiWeekly Advantage Plan and a free personalized
mortgage analysis.  Our telephone representatives are available to
serve you from 9 am to 5 pm (EST).

                              Sincerely,
```

Interest Rate:	8.3750
Current Balance:	$ 416,640.54
Monthly Payment:	$ 3,180.91

Homeownership: Remaining Terms		
	Years	Months
Current Payment Plan	29	5
BiWeekly Plan	22	5
REDUCTION	7	0

Total Remaining **Principal & Interest Payments:**	
Current Payment Plan	$ 1,122,849.24
BiWeekly Plan	$ 923,561.68
INTEREST SAVINGS	$ 199,287.56

a. What is the letter about?

b. A computer printout (shown above) was included with the letter. Assuming that these calculations are correct, discuss the advantages or disadvantages of accepting this offer.

   c. The plan as described in the letter costs $375 to sign up. I called the company and asked what their plan would do that I could not do myself by simply making 13 payments a year to my mortgage holder. The answer I received was that the plan would do nothing more, but the reason people do sign up is because they do not have the self-discipline to make the bimonthly payments by themselves. Why is a biweekly payment equivalent to 13 annual payments instead of equivalent to a monthly payment?

   d. The representative of the company told me that more than 250,000 people have signed up. How much income has the company received from this offer?

   e. You calculated the income the company has received from this offer in part **d**, but that is not all it receives. It acts as a bonded and secure "holding company" for your funds (because the mortgage company does not accept "two-week" payments). This means that the company receives the use (interest value) of your money for two weeks out of every month. This is equivalent to half the year. Let's assume that the average monthly payment is $1,000 and that the company has 250,000 payments that it holds for half the year. If the interest rate is 5% (a secure guaranteed rate), how much potential interest can be received by this company?

Individual RESEARCH PROJECTS

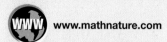
www.mathnature.com

Learning to use sources outside your classroom and textbook is an important skill, and here are ideas for extending some of the ideas in this chapter. You can find references to these projects in a library or at **www.mathnature.com.**

PROJECT 11.1 Conduct a survey of banks, savings and loan companies, and credit unions in your area. Prepare a report on the different types of savings accounts available and the interest rates they pay. Include methods of payment as well as interest rates.

PROJECT 11.2 Do you expect to live long enough to be a millionaire? Suppose that your annual salary today is $52,000. If inflation continues at 6%, how long will it be before $52,000 increases to an annual salary of a million dollars?

PROJECT 11.3 Consult an almanac or some government source, and then write a report on the current inflation rate. Project some of these results to the year of your own expected retirement.

PROJECT 11.4 Historical Quest   Write a short paper about Fibonacci numbers. You might check *The Fibonacci Quarterly,* particularly "A Primer on the Fibonacci Sequence," Parts I and II, in the February and April 1963 issues. The articles, written by Verner Hogatt and S. L. Basin, are considered classic articles on the subject. One member of your group should investigate the relationship of the Fibonacci numbers to nature, another the algebraic properties of the sequence, and another the history of the sequence.

PROJECT 11.5 Karen says that she has heard something about APR rates but doesn't really know what the term means. Wayne says he thinks it has something to do with the prime rate, but he isn't sure what. Write a short paper explaining APR to Karen and Wayne.

PROJECT 11.6 Some savings and loan companies advertise that they pay interest *continuously.* Do some research to explain what this means.

PROJECT 11.7 Select a car of your choice, find the list price, and calculate 5% and 10% price offers. Check out available money sources in your community, and prepare a report showing the different costs for the same car. Back up your figures with data.

PROJECT 11.8 Outline a program for your own retirement. In the process of writing this paper, answer the following questions. You will need to state your assumptions about interest and inflation rates.

   a. What monthly amount of money today would provide you a comfortable living?
   b. Using the answer to part **a**, project that amount to your own retirement, calculating the effects of inflation. Use your own age and assume that you will retire at age 65.
   c. How much money would you need to have accumulated to provide the amount you found in part **b**, if you decide to live on the interest only?
   d. If you set up a sinking fund to provide the amount you found in part **c**, how much would you need to deposit each month?
   e. Offer some alternatives to the sinking fund you considered in part **d**.
   f. Draw some conclusions about your retirement.

12 THE NATURE OF COUNTING

What in the World?

"Hi, guys. I'm glad you came to my party!" said Bobby. "My dad is starting a contest at his hardware store tomorrow. It is a 'Guess the jar's contents' puzzle. Whoever gets the number closest to the number of gum balls in the jar will win $50."

"Fifty dollars!!!" shouted Bess. "That's more than I make in a year! Let me get a closer look. . . . I'm going to guess 185."

Jeannie, chimed in, "I'm sitting closer to this jar and it looks more like 2,450 gum balls. Yep, that is my guess, 2,450."

Bobby, the boy with the notebook, took a more academic approach. "I need to help dad get this right, so I measured the jar. It is 10 inches high and 5 inches in diameter. The jar's volume is $V = Bh = \pi(2.5 \text{ in.})^2(10 \text{ in.}) \approx 196.35 \text{ in}^3$. Now all I need to do is find out how many gum balls are in a cubic inch. . . ." (See Problem 50, Section 12.3, p. 597.)

Overview

Counting is one of the most fundamental ideas in mathematics and, along with language, is one of the most essential skills each child must master. But don't let the fact that it is one of the first concepts learned by a child fool you. Counting can be as simple or as complicated as you like. In this chapter, we will learn some of the fundamental ideas of counting, and in the next chapter we will apply those ideas to the theory of probability and chance.

The risks and rewards of insurance, arguments involving the size of the universe, and the complexity of intelligent design all require that you have an advanced notion of counting. If you want to build a safer highway system or analyze the risks of hormone therapy, you will use the ideas presented in this chapter.

Courtesy of Orca School Santa Barbara, CA; photo by © Austin MacRae

12.1 | Permutations

Music is the pleasure the human soul experiences from counting without being aware that it is counting.

GOTTFRIED LEIBNIZ, MATHEMATICAL MAXIMS AND MINIMS, BY N. ROSE (RALEIGH: ROME PRESS, INC., 1988).

> ## CHAPTER **CHALLENGE**
> See if you can fill in the question mark.
> *J   T   I   R   H   P   X   G   N   U   F   L*   ?

"I must, each day say o'er the very same,
Counting no old thing old, thou mine, I thine,
Even as when first I hallow'd thy fair name."
  Sonnet CVIII, William Shakespeare

We now turn to an investigation of the nature of counting. It may seem like a simple topic for a college math book, since everyone knows how to count in the usual "one, two, three, . . ." method. However, there are many times we need to know "How many?" but can't find out by direct counting.

Election Problem

Consider a club with five members:

$A = \{$Alfie, Bogie, Calvin, Doug, Ernie$\}$

In how many ways could they elect a president and secretary? We call this the *election problem*. There are several ways to solve this problem. The first, and perhaps the easiest, is to make a picture, known as a **tree diagram,** listing all possibilities (see Figure 12.1).

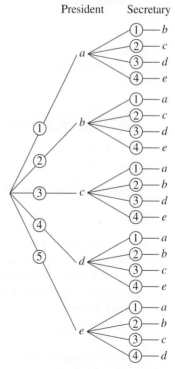

FIGURE 12.1 Tree diagram for five choices

We see there are 20 possibilities for choosing a president and a secretary from five club members. This method is effective for "small" tree diagrams, but the technique quickly gets out of hand. For example, if we wished to see how many ways the club could elect a president, secretary, and treasurer, this technique would be very lengthy.

A second method of solution is to use boxes or "pigeonholes" representing each choice separately. Here we determine the number of ways of choosing a president, and then the number of ways of choosing a secretary.

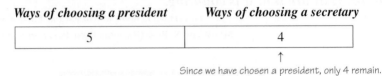

Ways of choosing a president	Ways of choosing a secretary
5	4

↑
Since we have chosen a president, only 4 remain.

We multiply the numbers in the pigeonholes,

$$5 \cdot 4 = 20$$

and we see that the result is the same as from the tree diagram.

A third method is to use the fundamental counting principle (from Section 2.4). We repeat this principle here for review.

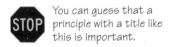

You can guess that a principle with a title like this is important.

Fundamental Counting Principle

The fundamental counting principle gives the number of ways of performing two or more tasks. If task A can be performed in m ways, and if after task A is performed, a second task B can be performed in n ways, then task A followed by B can performed in $m \times n$ ways.

Example 1 Determine the number of ways of choosing three

In how many ways could the given club choose a president, secretary, and treasurer?

Solution

5	4	3
↑	↑	↑
President	Secretary	Treasurer

The answer is $5 \times 4 \times 3 = 60$.

Example 2 Determine the number of ways of choosing four

In how many ways could this club choose a president, vice president, secretary, and treasurer?

Solution $5 \times 4 \times 3 \times 2 = 120$

When set symbols { } are used, the order in which the elements are listed is not important. Suppose now that you wish to select elements from $A = \{a, b, c, d, e\}$ by picking them in a certain order. The selected elements are enclosed in parentheses to signify order and are called an **arrangement** of elements of A. For example, if the elements a and b are selected from A, then there are two pairs, or arrangements:

   (a, b) and (b, a)

These arrangements are called **ordered pairs.** Remember, when parentheses are used, the order in which the elements are listed *is* important. This example shows ordered pairs, but you could also select an **ordered triple** such as (*d, c, a*). These arrangements are said to be selected *without repetition,* since a symbol cannot be used twice in the same arrangement.

If we are given some set of elements and we are to choose *without repetition* a certain number of elements from the set, we can choose them so that the order in which the choices are made is important, or so that it is not. If the *order is important*, we call our result a **permutation,** and if the order is not important, it is called a **combination.** We now consider permutations, and we will consider combinations in the next section.

Suppose we consider the election problem for a larger set than *A;* we wish to elect a president, secretary, and treasurer from among

{Frank, George, Hans, Iris, Jane, Karl}

We could use one of the previously mentioned three methods, but we wish to generalize, so we ask, "Is it possible to count the number of arrangements without actually counting them?"

In answering this question, we note that we are selecting 3 persons in order from a group of 6 people. If an arbitrary finite set *S* has *n* elements and *r* elements are selected without repetition from *S* (where $r \leq n$), then an arrangement of the *r* selected elements is called a *permutation.*

> ### Permutation
>
> A **permutation** of *r* elements selected from a set *S* with *n* elements is an ordered arrangement of those *r* elements selected without repetition.

Remember that since we are considering permutations, the order is important; that is, (a, b) is NOT the same as (b, a).

Example 3   Number of permutations of two chosen from six

How many permutations of two elements can be selected from a set of six elements?

Solution   Let $B = \{a, b, c, d, e, f\}$ and select two elements:

$(a, b), (a, c), (a, d), (a, e), (a, f), (b, a), (b, c), (b, d), (b, e), (b, f),$

$(c, a), (c, b), (c, d), (c, e), (c, f), (d, a), (d, b), (d, c), (d, e), (d, f),$

$(e, a), (e, b), (e, c), (e, d), (e, f), (f, a), (f, b), (f, c), (f, d), (f, e)$

There are 30 permutations of two elements selected from a set of six elements.

Example 3 brings up two difficulties. The first is the lack of notation for the phrase, "the number of permutations of two elements selected from a set of six elements," and the second is the inadequacy of relying on direct counting, especially if the sets are very large.

> ### Notation for Permutations
>
> The symbol $_nP_r$ is used to denote the **number of permutations** of *r* elements selected from a set of *n* elements.

Example 3 can now be shortened by writing

$$_6P_2 = 30 \qquad \text{Actual number of permutations is 30.}$$

$$\uparrow \uparrow$$

Total number in set · Number we are selecting from the set

From the fundamental counting principle, we can also write $_6P_2 = \underbrace{6 \cdot 5}_{\text{Two factors}} = 30.$

The next example leads us to a method for calculating permutations.

Example 4 Evaluate permutations

Evaluate: **a.** $_{52}P_2$ **b.** $_7P_3$ **c.** $_{10}P_4$ **d.** $_{10}P_1$ **e.** $_nP_r$

Solution We use the fundamental counting principle:

a. $_{52}P_2 = \underbrace{52 \cdot 51}_{\text{Two factors}} = 2{,}652$ **b.** $_7P_3 = \underbrace{7 \cdot 6 \cdot 5}_{\text{Three factors}} = 210$

c. $_{10}P_4 = \underbrace{10 \cdot 9 \cdot 8 \cdot 7}_{\text{Four factors}} = 5{,}040$ **d.** $_{10}P_1 = \underbrace{10}_{\text{One factor}}$

e. $_nP_r = \underbrace{n(n-1)(n-2) \cdots \cdots (n-r+1)}_{r \text{ factors}}$

Example 5 Find the nature of permutation variables

a. Can r be larger than n? **b.** Can r be 0? **c.** Can $r = n$?

Solution
a. No, since n is the total number of objects.
b. You must be careful. This example does not exactly fit our definition, so we must interpret it separately. Now, $_6P_0$ means the number of ways you can permute six objects by taking none of them. There is only one way to take no objects, and that is by not taking any. Thus, we define $_nP_0 = 1$ for any counting number n.
c. Yes; consider

$$_5P_5 = 5 \cdot 4 \cdot 3 \cdot 2 \cdot 1 = 120$$
$$_6P_6 = 6 \cdot 5 \cdot 4 \cdot 3 \cdot 2 \cdot 1 = 720$$

Factorial

In our work with permutations, we will frequently encounter products such as (see Example 5c)

$$6 \cdot 5 \cdot 4 \cdot 3 \cdot 2 \cdot 1 \quad \text{or} \quad 10 \cdot 9 \cdot 8 \cdot 7 \cdot 6 \cdot 5 \cdot 4 \cdot 3 \cdot 2 \cdot 1 \quad \text{or}$$
$$52 \cdot 51 \cdot 50 \cdot 49 \cdots \cdots 4 \cdot 3 \cdot 2 \cdot 1$$

Since these are rather lengthy, we use *factorial notation*.

Factorial notation was first used by Christian Kramp in 1808.

> **Factorial**
>
> For any counting number n, the **factorial of n** is defined by
>
> $$n! = n(n-1)(n-2) \cdots \cdots 3 \cdot 2 \cdot 1$$
>
> Also, $0! = 1$.

Example 6 Evaluate factorials

Evaluate factorials for the numbers $0, 1, \cdots, 10$

Solution

$0! = 1$ $1! = 1$

$2! = 2 \cdot 1 = 2$ $3! = 3 \cdot 2 \cdot 1 = 6$

$4! = 4 \cdot 3 \cdot 2 \cdot 1 = 24$ $5! = 5 \cdot 4 \cdot 3 \cdot 2 \cdot 1 = 120$

$6! = 6 \cdot 5 \cdot 4 \cdot 3 \cdot 2 \cdot 1 = 720$ $7! = 7 \cdot 6 \cdot 5 \cdot 4 \cdot 3 \cdot 2 \cdot 1 = 5{,}040$

$8! = 8 \cdot 7 \cdot 6 \cdot 5 \cdot 4 \cdot 3 \cdot 2 \cdot 1 = 40{,}320$

$9! = 9 \cdot 8 \cdot 7 \cdot 6 \cdot 5 \cdot 4 \cdot 3 \cdot 2 \cdot 1 = 362{,}880$

$10! = 10 \cdot 9 \cdot 8 \cdot 7 \cdot 6 \cdot 5 \cdot 4 \cdot 3 \cdot 2 \cdot 1 = 3{,}628{,}800$

Notice (from Example 6) that these numbers get big pretty fast. We need not wonder why the notation "!" was chosen! For example,

$$52! \approx 8.065817517 \times 10^{67}$$

If you actually carry out the calculations in Example 6, you will discover a useful property of factorials: $3! = 3 \cdot 2!$, $4! = 4 \cdot 3!$, $5! = 5 \cdot 4!$; that is, to calculate 11! you would not need to "start over" but simply multiply 11 by the answer found for 10!. This property is called the **multiplication property of factorials** or the **count-down property.**

Count-down Property

$$n! = n \cdot (n - 1)!$$

Example 7 Evaluate factorial expressions

Evaluate: **a.** $6! - 4!$ **b.** $(6 - 4)!$ **c.** $\frac{8!}{7!}$ **d.** $\frac{12!}{9!}$ **e.** $\frac{100!}{98!}$

Solution

a. $6! - 4! = 720 - 24 = 696$

b. $(6 - 4)! = 2! = 2$ *Compare order of operations in parts a and b.*

c. $\dfrac{8!}{7!} = \dfrac{8 \cdot 7!}{7!} = 8$ *Count-down property*

d. $\dfrac{12!}{9!} = \dfrac{12 \cdot 11 \cdot 10 \cdot 9!}{9!} = 12 \cdot 11 \cdot 10 = 1{,}320$ *Repeated use of the count-down property*

e. $\dfrac{100!}{98!} = \dfrac{100 \cdot 99 \cdot 98!}{98!} = 9{,}900$ *This example shows that you need the count-down property even if you are using a calculator.*

Using factorials, we can find a formula for $_nP_r$.

$$_nP_r = n(n - 1)(n - 2) \cdot \cdots \cdot (n - r + 1)$$

$$= n(n - 1)(n - 2) \cdot \cdots \cdot (n - r + 1) \cdot \frac{(n - r)!}{(n - r)!}$$

$$= \frac{n(n - 1)(n - 2) \cdot \cdots \cdot (n - r + 1)(n - r)!}{(n - r)!}$$

$$= \frac{n!}{(n - r)!}$$

This is the general formula for $_nP_r$.

Permutation Formula

$$_nP_r = \frac{n!}{(n-r)!}$$

Example **8** **Evaluate permutations by formula**

Evaluate: **a.** $_{10}P_2$ **b.** $_nP_0$ **c.** $_8P_8$

Solution

a. $_{10}P_2 = \dfrac{10!}{(10-2)!} = \dfrac{10!}{8!} = \dfrac{10 \cdot 9 \cdot 8!}{8!} = 90$

b. $_nP_0 = \dfrac{n!}{(n-0)} = \dfrac{n!}{n!} = 1$

c. $_8P_8 = \dfrac{8!}{(8-8)!} = \dfrac{8!}{0!} = 8! = 40,320$

Example **9** **Find the number of drivers**

A dispatcher is assigning cars to taxi drivers. There are eight possible cars and five drivers. In how many ways can the assignments be done?

Solution   Five of the eight cars will be assigned to a driver, and since the car one receives could make a difference, the order of selection is important, so this is a permutation problem:

$$_8P_5 = \frac{8!}{(8-5)!} = \frac{8!}{3!} = \frac{8 \cdot 7 \cdot 6 \cdot 5 \cdot 4 \cdot 3!}{3!} = 6,720$$

Example **10** **Naming the Königsberg bridges**

In Section 8.1 we identified the seven bridges of Königsberg by numbering these bridges. Historical documents tell us these bridges are named the Blacksmith's Bridge, Connecting Bridge, High Bridge, Green Bridge, Honey Bridge, Merchant's Bridge, and Wooden Bridge. In how many ways could the seven bridges be named?

Solution   We see seven bridges and we have seven names and the order in which we name these bridges is important. We see that this is a permutation problem:

$$_7P_7 = 7! = 5,040$$

Example **11** **Determine the number of ways of choosing six**

We return to the situation at the beginning of this section. In how many ways can we select a president, secretary, and treasurer from a club of six members?

Solution   The election is a permutation problem:

$$_6P_3 = \frac{6!}{(6-3)!} = 120$$

Distinguishable Permutations

We now consider a generalization of permutations in which one or more of the selected items is *indistinguishable* from the others.

Example 12 Find the number of distinguishable permutations

Find the number of arrangements of letters in the given word:
a. MATH   **b.** HATH

Solution

a. This is a permutation of four objects taken four at a time:

$$_4P_4 = 4! = 24$$

b. This is different from part **a** because not all the letters in the word HATH are distinguishable; that is, the first and last letters are both H's, so they are *indistinguishable*. If we make the letters distinguishable by labeling them as HATH, we see that now the list is the same as in part **a**. Let us list the possibilities as shown in the margin.

    We see that there are 24 possibilities, but notice how we have arranged this listing. If we consider H = H (that is, the H's are indistinguishable), we see that the first and the second columns are the same. Since there are *two* indistinguishable letters, we divide the total by 2:

$$\frac{4!}{2} = \frac{24}{2} = 12$$

Possibilities if H's are indistinguishable
(First and second columns are the same.)

HATH	HATH
HAHT	HAHT
HTAH	HTAH
HTHA	HTHA
HHTA	HHTA
HHAT	HHAT
AHHT	AHHT
AHTH	AHTH
ATHH	ATHH
THHA	THHA
THAH	THAH
TAHH	TAHH

Example 13 Number of letters in a word with a repeated letter

How many permutations are there of the letters in the word ASSIST?

Solution   There are six letters, and if you consider the letters as *distinguishable,* as in

$$AS_1S_2IS_3T$$

there are $_6P_6 = 6! = 720$ possibilities. However,

$$AS_1S_2IS_3T, \quad AS_1S_3IS_2T, \quad AS_2S_1IS_3T, \quad AS_2S_3IS_1T, \quad AS_3S_1IS_2T, \quad AS_3S_2IS_1T$$

are all indistinguishable, so you must divide the total by $3! = 6$:

$$\frac{6!}{3!} = \frac{6 \cdot 5 \cdot 4 \cdot 3!}{3!} = 120$$

There are 120 permutations of the letters in the word ASSIST.

These examples can be generalized to include several categories.

Example 14 Number of letters in a word with two repetitions

Find the number of permutations of the letters in the word ATTRACT.

Solution   The total number of arrangements of the letters in the word is 7!; this is divided by factorials of the numbers of repeated letters in subcategories:

$$\frac{7!}{3! \; 2! \; 1! 1!}$$

Letter *T* occurs three times. ↑ ↑ Letters *R* and *C* occur once each.

Letter *A* occurs twice.

These examples suggest a general result.

Formula for Distinguishable Permutations

The number of distinguishable permutations of n objects which n_1 are of one kind, n_2 are of another kind, . . . and n_k are of a further kind, so that $n = n_1 + n_2 + \cdots + n_k$ is denoted by $\binom{n}{n_1,\, n_2,\, \ldots,\, n_k}$ and is defined by the formula

$$\binom{n}{n_1,\, n_2,\, \ldots,\, n_k} = \frac{n!}{n_1!\, n_2!\, \cdots\, n_k!}$$

Example 15 Find the number of distinguishable permutations

What is the number of distinguishable permutations of the letters in the words NATURE OF MATHEMATICS?

Solution   There are 19 letters: A and T (3 times each), E and M (2 times each), and 9 letters each appearing once. We use the formula for distinguishable permutations.

$$\frac{19!}{3!3!2!2!1!1!1!1!1!1!1!1!1!} = 8.447576417 \times 10^{14}$$

By calculator; note the sum of the numbers on the bottom is 19.

If we also consider the spaces and where they occur (which would be necessary if you were programming this on a computer), then the number of possibilities is found by

$$\frac{21!}{3!3!2!2!2!1!1!1!1!1!1!1!1!1!} = 1.773991048 \times 10^{17}$$

Problem Set 12.1

Level 1

1. IN YOUR OWN WORDS   What is a permutation? What is the formula for permutations?

2. IN YOUR OWN WORDS   What is a distinguishable permutation? What is the formula for distinguishable permutations?

Evaluate each expression in Problems 3–30.

3. $_9P_1$

4. $_{100}P_1$

5. $_9P_2$

6. $_9P_3$

7. $_9P_4$

8. $_5P_4$

9. $_{52}P_3$

10. $_4P_4$

11. $_{12}P_5$

12. $_5P_3$

13. $_8P_4$

14. $_8P_0$

15. $_{92}P_0$

16. $_{52}P_1$

17. $_7P_5$

18. $_{16}P_3$

19. $_7P_3$

20. $_5P_5$

21. $_{50}P_4$

22. $_mP_3$

23. $_8P_3$

24. $_{10}P_2$

25. $_{11}P_4$

26. $_nP_4$

27. $_gP_h$

28. $_nP_5$

29. $_5P_r$

30. $_xP_y$

How many permutations are there of the words given in Problems 31–40?

31. HOLIDAY

32. ANNEX

33. ESCHEW

34. OBFUSCATION

35. MISSISSIPPI

36. CONCENTRATION

37. BOOKKEEPING

38. GRAMMATICAL

39. APOSIOPESIS

40. COMBINATORICS

Level 2

41. In how many ways can a president, vice president, secretary, and treasurer be elected from a group of five people?

42. In how many ways can a president and a vice president be elected from a group of 15 people?

43. In how many ways can a president, a vice president, and a secretary be elected from a group of 10 people?

44. In how many ways can a chairperson and a vice chairperson be selected from a committee of 25 senators?

45. How many outfits consisting of a suit and a tie can a man select if he has two suits and eight ties?

46. In how many different ways can eight books be arranged on a shelf?

47. In how many ways can you select and read three books from a shelf of eight books?

48. In how many ways can a row of three contestants for a TV game show be selected from an audience of 362 people?

49. How many seven-digit telephone numbers are possible if the first two digits cannot be ones or zeros?

50. Foley's Village Inn offers the following menu in its restaurant:

Main course	Dessert	Beverage
Prime rib	Ice cream	Coffee
Steak	Sherbet	Tea
Chicken	Cheesecake	Milk
Ham		Decaf
Shrimp		

In how many different ways can someone order a meal consisting of one choice from each category?

51. Most ATMs require that you enter a four-digit code, using the digits 0 to 9. How many four-digit codes are available?

52. Some automobiles have five-button locks. How many different five-digit codes are available for these locks?

53. A museum wishes to display eight paintings next to one another on a wall. In how many ways may this be done?

54. If there are nine baseball players who need to be arranged in a batting order, in how many ways might this be done?

55. The "Pick 3" at horse racetracks requires that a person select the winning horse for three consecutive races. If the first race has nine entries, the second race eight entries, and the third race ten entries, how many different possible tickets might be purchased?

56. Tarot cards are used for telling fortunes, and in a reading, the arrangement of the cards is as important as the cards themselves. How many different readings are possible if three cards are selected from a set of seven Tarot cards?

Level 3

57. A DNA molecule is shown here.

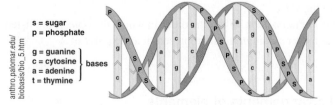

A nucleotide is represented by one of the letters $\{a, t, c, g\}$. A *codon* is defined to be a triplet of nucleotides. The order of the nucleotides in a codon is significant, and the nucleotides can be repeated. These codons are important in determining how DNA is transformed into protein. How many different codons are possible?

58. My favorite Moroccan restaurant offers the following menu:

> *Cold appetizer* (no choice): spicy carrots, khobiza, and zalouk
>
> *Appetizer* (select one): spicy chicken wings, charmoula chicken drumsticks, maaquda, shrimp maaquda, or lamb borek
>
> *Entree* (select two): free range chicken, baked kefta, or shekshouka
>
> *Dessert* (no choice): almond semolina flour cookies

How many different meals are possible if the cold appetizer and dessert are served with all meals and the entries are served in the order of selection?

59. A world-famous prix fixe menu includes the following choices:

> *Beverage* (choose one): wine, soft drink, coffee, tea, milk, juice, or cappuccino
>
> *First course* (no choice): baguette, bruschetta, whole roasted garlic, and butter
>
> *Second course* (choose one): quesadilla, onion soup, salad, Caesar salad, lettuce wedge, spinach salad, or romaine salad
>
> *Third course* (choose one): rib eye steak, swordfish, rack of lamb, whitefish, grilled chicken, southern fried chicken, lobster, or vegetarian plate.
>
> If you choose steak, then choose one of the following: fries, mashed potatoes, creamed corn, spinach, broccoli, or grilled onions.
>
> *Fourth course* (choose one): chocolate cake, cinnamon tart, apple tart, pecan tart, crème brulee, mango sorbet, or chocolate tart.

How many different meals are possible?

Problem Solving 3

60. Suppose you flip a coin and keep a record of the results. In how many ways could you obtain at least one head if you flip the coin six times?

12.2 Combinations

In the last section, we introduced the situation in which we are given some set of elements and we are to choose *without repetition* a certain number of elements from the set. We can choose them so that the order in which the choices are made is important, or so that it is not. If the *order is important,* the selection was called a permutation, and if the order is not important it was called a combination. We consider combinations in this section.

Committee Problem

Reconsider the club example given at the beginning of the previous section:

$$A = \{\text{Alfie, Bogie, Calvin, Doug, Ernie }\}$$

In how many ways could they elect a committee of two persons? We call this the *committee problem.*

One method of solution is easy (but tedious) and involves the enumeration of all possibilities:

$$\{a, b\} \quad \{b, c\} \quad \{c, d\} \quad \{d, e\}$$
$$\{a, c\} \quad \{b, d\} \quad \{c, e\}$$
$$\{a, d\} \quad \{b, e\}$$
$$\{a, e\}$$

We see there are ten possible two-member committees. Do you see why we cannot use the fundamental counting principle for this committee problem in the same way we did for the election problem?

We have presented two different types of counting problems, the *election problem* and the *committee problem.* For the election problem, we found an easy numerical method of counting, using the fundamental counting principle. For the committee problem, we did not; further investigation is necessary.

Let us compare the committee and election problems. We saw in the preceding section, while considering the election problem, that the following arrangements are *different*:

President	Secretary	Treasurer
Alfie	Bogie	Calvin
Bogie	Alfie	Calvin

That is, the order was important. However, in the committee problem, the three-member committees listed above are the *same.*

In this case, the order *is not important.* When the order in which the objects are listed is not important, we call the list a *combination* and represent it as a *subset* of A.

Example 1 List all arrangements of elements

Select two elements from the set $A = \{a, b, c, d, e\}$, and list all possible arrangements as well as all possible subsets.

Solution

Permutations—arrangements **Order important**				Combinations—subsets **Order not important**			
(a, b),	(a, c),	(a, d),	(a, e),	$\{a, b\}$,	$\{a, c\}$,	$\{a, d\}$,	$\{a, e\}$,
(b, a),	(b, c),	(b, d),	(b, e),		$\{b, c\}$,	$\{b, d\}$,	$\{b, e\}$,
(c, a),	(c, b),	(c, d),	(c, e),	$\uparrow$		$\{c, d\}$,	$\{c, e\}$,
(d, a),	(d, b),	(d, c),	(d, e),				$\{d, e\}$
(e, a),	(e, b),	(e, c),	(e, d)				

Do not list $\{b, a\}$ since
$\{a, b\} = \{b, a\}$.

Note that ordered pair notation is used.
There are 20 permutations.

Note that set notation is used.
There are 10 combinations.

We now state a definition of a combination.

Combination

A **combination** of r elements of a finite set S of n elements is a subset of S that contains r distinct elements. The notations $\binom{n}{r}$ and $_nC_r$ are both used to denote the number of combinations of r elements selected from a set of n elements $(r \le n)$.

The notation $_nC_r$ is similar to the notation used for permutations, but since $\binom{n}{r}$ is more common in your later work in mathematics, we will usually use this notation for combinations. The notation $\binom{n}{r}$ is read as "n choose r." The formula for the number of permutations leads directly to a formula for the number of combinations since each subset of r elements has $r!$ permutations of its members. Thus, by the fundamental counting principle,

$$\binom{n}{r} \cdot r! = {}_nP_r \quad \text{so} \quad \binom{n}{r} = \frac{{}_nP_r}{r!} = \frac{n!}{r!(n-r)!}$$

Combination Formula

$$\binom{n}{r} = \frac{n!}{r!(n-r)!}$$

Example 2 Evaluate combinations

Evaluate: **a.** $\binom{10}{3}$ **b.** $\binom{n}{0}$ **c.** $\binom{m-1}{2}$

Solution

a. $\binom{10}{3} = \frac{10!}{3!7!} = \frac{10 \cdot 9 \cdot 8 \cdot 7!}{3 \cdot 2 \cdot 7!} = 120$

b. $\binom{n}{0} = \frac{n!}{0!(n-0)!} = 1$

c. $\displaystyle\binom{m-1}{2} = \frac{(m-1)!}{2!(m-1-2)!}$

$\displaystyle\phantom{\binom{m-1}{2}} = \frac{(m-1)(m-2)(m-3)!}{2\cdot 1\cdot(m-3)!}$

$\displaystyle\phantom{\binom{m-1}{2}} = \frac{(m-1)(m-2)}{2}$

Example 3 Find the number of possible committees

In how many ways can a club of five members select a three-person committee?

Solution   In an election, the order of selection is important, but in choosing a committee, the order of selection is not important, so this is a combination. Three members are selected from five, so we have "5 choose 3":

$$\binom{5}{3} = \frac{5!}{3!2!} = \frac{5\cdot 4}{2!} = 10$$

Many applications deal with an ordinary **deck of cards,** as shown in Figure 12.2.

Hearts (red cards)   Spades (black cards)

Diamonds (red cards)   Clubs (black cards)

FIGURE 12.2 A deck of cards

Example 4 Find the number of five-hand cards that can be drawn

Find the number of five-card hands that can be drawn from an ordinary deck of cards.

Solution

$$\binom{52}{5} = \frac{52!}{5!47!} = \frac{52\cdot 51\cdot 50\cdot 49\cdot 48\cdot 47}{5\cdot 4\cdot 3\cdot 2\cdot 1\cdot 47!} = 2,598,960$$

Example 5 Find the number of ways for a heart flush

In how many ways can a heart flush be drawn in poker? (A heart flush is any hand of five hearts.)

Solution   We need to determine the number of combinations of 13 objects (hearts) taken 5 at a time. Thus,

$$\binom{13}{5} = \frac{13!}{5!8!} = \frac{13 \cdot 12 \cdot 11 \cdot 10 \cdot 9 \cdot 8!}{5 \cdot 4 \cdot 3 \cdot 2 \cdot 1 \cdot 8!} = 1{,}287$$

Example   6   Find the number of possible flush hands

In how many ways can a flush be drawn in poker? (A flush is five cards of one suit.)

Solution   There are four possible suits, so we use the fundamental counting principle, along with the result from Example 5.

Number of suits
$$\overbrace{4} \cdot \underbrace{1{,}287} = 5{,}148$$
Number of ways for a particular suit

Example   7   Find the number of possible ways for a full house

In how many ways can a full house of three tens and two queens be dealt?

Solution   Begin with the fundamental counting principle:

$$\boxed{\text{Ways of obtaining two queens}} \cdot \boxed{\text{Ways of obtaining three tens}}$$

Each of the numbers in each pigeonhole is a combination (since the order in which the cards are dealt is not important):

Queens Tens
$$\overbrace{\binom{4}{2}} \cdot \overbrace{\binom{4}{3}} = \frac{4!}{2!2!} \cdot \frac{4!}{3!1!} = \frac{4 \cdot 3 \cdot 4}{2} = 24$$

Pascal's Triangle

For many applications, if n and r are relatively small (which will be the case for most of the problems for this course), you should notice the following relationship:

$$\binom{0}{0} = 1$$

$$\binom{1}{0} = 1 \qquad \binom{1}{1} = 1$$

$$\binom{2}{0} = 1 \qquad \binom{2}{1} = 2 \qquad \binom{2}{2} = 1$$

$$\binom{3}{0} = 1 \qquad \binom{3}{1} = 3 \qquad \binom{3}{2} = 3 \qquad \binom{3}{3} = 1$$

$$\binom{4}{0} = 1 \qquad \binom{4}{1} = 4 \qquad \binom{4}{2} = 6 \qquad \binom{4}{3} = 4 \qquad \binom{4}{4} = 1$$

$$\vdots$$

That is, $\binom{n}{r}$ is the number in the nth row, rth diagonal of Pascal's triangle. We repeat the triangle (from Section 1.1) here for easy reference.

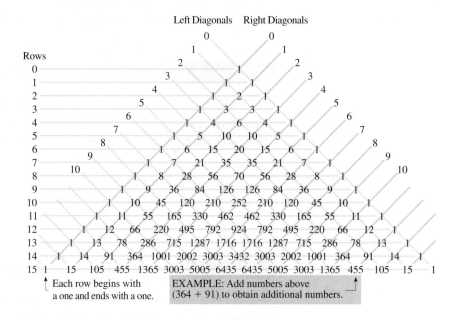

Left Diagonals   **Right Diagonals**

Rows

Each row begins with a one and ends with a one.

EXAMPLE: Add numbers above $(364 + 91)$ to obtain additional numbers.

Example **8**   **Evaluate combinations using Pascal's triangle**

Find

a. $\binom{8}{3}$ **b.** $\binom{6}{5}$ **c.** $\binom{14}{6}$ **d.** $_{12}C_{10}$

Solution

a. $\binom{8}{3} = 56$ row 8, diagonal 3   **b.** $\binom{6}{5} = 6$ row 6, diagonal 5

c. $\binom{14}{6} = 3{,}003$ row 14, diagonal 6   **d.** $_{12}C_{10} = 66$ row 12, diagonal 10

Counting with the Binomial Theorem

In Section 6.1 we introduced the binomial theorem, and we saw that Pascal's triangle gives the coefficients of the expansion of a power of a binomial. If you look back at the notation we used in Chapter 6, you will see that we chose to use combination notation (even though we did not call it such) back then. You see now that this was no coincidence. We can now restate the binomial theorem using combination notation.

Binomial Theorem

For any positive integer n,

$$(a + b)^n = \binom{n}{0}a^n + \binom{n}{1}a^{n-1}b + \binom{n}{2}a^{n-2}b^2 + \cdots + \binom{n}{n-1}ab^{n-1} + \binom{n}{n}b^n$$

where $\binom{n}{r}$ is the number in the nth row, rth diagonal of Pascal's triangle.

| Example | 9 | **Number of birth orders** | *Pólya's Method* |

If a family has five children, how many different birth orders could the parents have if there are three boys and two girls?

Solution   We use Pólya's problem-solving guidelines for this example.

Understand the Problem. Part of understanding this problem might involve estimation. For example, if a family has one child, there are two ways (boy and girl). If a family has two children, there are four ways (BB, BG, GB, GG); for three children, eight ways; for four children, 16 ways; and for five children, a total of 32 ways. This means that an answer of 140 ways, for example, is an unreasonable answer, since for 5 children the number of *all* possibilities is 32.

Devise a Plan. The first plan might try direct enumeration: BBBGG, BBGBG, BBGGB, This seems to be too tedious. Look for a pattern:

One child:	B ← One way
	G ← One way
Two children:	BB ← One way
	BG ⎫
	GB ⎭ ← Two ways
	GG ← One way
Three children:	BBB ← One way
	BBG ⎫
	BGB ⎬ ← Three ways
	GBB ⎭
	BGG ⎫
	GBG ⎬ ← Three ways
	GGB ⎭
	GGG ← One way

If we write this as follows, it should look familiar:

One child:			1		1	
Two children:		1		2		1	
Three children:	1		3		3		1

If we write $(B + G)^3 = B^3 + 3B^2G + 3BG^2 + G^3$, we see that the exponents correspond to the genders of the children and the coefficients give the related number of ways these gender combinations can occur. Thus, our plan is to use the binomial theorem as a counting technique.

Carry Out the Plan. We find

$$(B + G)^5 = B^5 + 5B^4G + 10B^3G^2 + 10B^2G^3 + 5BG^4 + G^5$$

This says that the family could have three boys and two girls (look at the coefficient of B^3G^2) a total of 10 ways.

Look Back. We see that our plan led us to an answer that is consistent with the pattern we used in the first step (understanding the problem).

| Example | 10 | **Find the number of subsets** |

How many subsets of $\{1, 2, 3, 4, 5\}$ are there?

Solution   There are $\binom{5}{5}$ subsets of 5 elements, $\binom{5}{4}$ subsets of 4 elements, and so forth. Thus, the *total* number of subsets is

$$\binom{5}{0} + \binom{5}{1} + \binom{5}{2} + \binom{5}{3} + \binom{5}{4} + \binom{5}{5}$$

Instead of carrying out all of the arithmetic in evaluating these combinations, we note that

$$(a + b)^5 = \binom{5}{0}a^5 + \binom{5}{1}a^4b + \binom{5}{2}a^3b^2 + \binom{5}{3}a^2b^3 + \binom{5}{4}ab^4 + \binom{5}{5}b^5$$

so that if we let $a = 1$ and $b = 1$, we have

$$(1 + 1)^5 = \binom{5}{0}1^5 + \binom{5}{1}1^41 + \binom{5}{2}1^31^2 + \binom{5}{3}1^21^3 + \binom{5}{4}1 \cdot 1^4 + \binom{5}{5}1^5$$

$$2^5 = \binom{5}{0} + \binom{5}{1} + \binom{5}{2} + \binom{5}{3} + \binom{5}{4} + \binom{5}{5}$$

We see that there are $2^5 = 32$ subsets of a set containing five elements.

Number of Subsets

A set of n distinct elements has 2^n subsets.

Example 11 Find the number of subsets of a set of 10 elements

How many subsets are there for a set containing 10 elements?

Solution There are $2^{10} = 1,024$ subsets.

Problem Set 12.2

Level 1

1. **IN YOUR OWN WORDS** What is a combination? What is the formula for combinations?

2. **IN YOUR OWN WORDS** What is the relationship between combinations and Pascal's triangle?

Evaluate each expression in Problems 3–28.

3. $\binom{9}{1}$

4. $\binom{9}{2}$

5. $\binom{9}{3}$

6. $\binom{9}{4}$

7. $\binom{9}{0}$

8. $\binom{5}{4}$

9. $\binom{52}{3}$

10. $\binom{7}{2}$

11. $\binom{4}{4}$

12. $\binom{100}{1}$

13. $\binom{7}{3}$

14. $\binom{5}{3}$

15. $\binom{50}{48}$

16. $\binom{25}{1}$

17. $\binom{g}{h}$

18. $\binom{n}{4}$

19. $_kP_4$

20. $_nC_5$

21. $_mC_n$

22. $_mP_n$

23. $\binom{10}{4, 3, 3}$

24. $\binom{7}{2, 2, 3}$

25. $\binom{9}{2, 3, 4}$

26. $\binom{9}{1}\binom{8}{4}\binom{4}{4}$

27. $\binom{7}{1}\binom{6}{4}\binom{2}{2}$

28. $\binom{10}{6}\binom{4}{2}\binom{2}{1}$

Level 2

29. A bag contains 12 pieces of candy. In how many ways can five pieces be selected?

30. In how many ways can three aces be drawn from a deck of cards?

31. In how many ways can two kings be drawn from a deck of cards?

32. In how many ways can four aces be drawn from a deck of cards?

33. In how many ways can a club flush be obtained?

34. In how many ways can a full house of three aces and two kings be obtained?

35. In how many ways can a full house of three jacks and a pair of twos be obtained?

36. In how many ways can a committee of five be formed from a group of 15 people?

37. If a family has five children, in how many ways could the parents have two boys and three girls?

38. If a family has six children, in how many ways could the parents have three boys and three girls?

39. How many different subsets can be chosen from a set of seven elements?

40. How many different subsets can be chosen from the U.S. Senate?

Level

In Problems 41–55, decide whether you would use a permutation, a combination, or neither. Next, write the solution using permutation notation or combination notation, if possible, and, finally, answer the question.

41. A club with 30 members is to select a committee of five persons. In how many ways can this be done?

42. A club with 30 members is to select five officers (president, vice president, secretary, treasurer, and historian). In how many ways can this be done?

43. Tom is leading a workshop for 20 people. He wants each person to meet every other person (shake hands once). How many handshakes occur?

44. Rita is asked to answer three of five essay questions on an exam. How many choices does Rita have?

45. An ice cream parlor has 31 different flavors. A super sundae allows a person to select 3 different flavors. How many different selections are possible?

Sandra O'Claire/istockphoto.com

46. An ice cream parlor has 31 different flavors. A super sundae allows a person to select 3 different flavors and 3 (out of 8) different toppings. How many different selections are possible?

47. How many different arrangements are there of the letters in the word CORRECT?

48. There are three boys and three girls at a party. In how many ways can they be seated in a row if they must sit alternating boys and girls?

49. If there are 10 employees in a club, in how many ways can they choose a dishwasher and a bouncer?

50. In how many ways can you be dealt two cards from an ordinary deck of 52 cards?

51. In how many ways can five taxi drivers be assigned to six cars?

52. How many arrangements are there of the letters in the word GAMBLE?

53. In how many ways can a group of seven choose a committee of four?

54. In how many ways can seven books be arranged on a bookshelf?

55. In how many ways can you choose two books to read from a bookshelf containing seven books?

Problem Solving

56. **a.** Draw three points on a circle. How many triangles can you draw using these points as vertices?
   b. Draw four points on a circle. How many triangles can you draw by using any three of these points as vertices?
   c. Draw six points on a circle. How many triangles can you draw by using any three of these points as vertices?
   d. Draw n points on a circle ($n \geq 3$). How many triangles can you draw by using any three of these points as vertices?

57. Draw n points on a circle ($n \geq 5$). How many pentagons can you draw by connecting any five of these points?

58. A club consists of 16 men and 19 women. In how many ways can they choose a president, vice president, treasurer, and secretary, along with an advisory committee of six people?

59. In Problem 58, how many ways can the selection be made if exactly two of the officers must be women?

60. Prove: $\binom{n}{r} = \binom{n}{n-r}$

12.3 | Counting without Counting

"Five trillion, four hundred eighty billion, five hundred twenty-three million, two hundred ninety-seven thousand, one-hundred and sixty-two . . ."

STOP *Remember: Permutation: Order is important. Combination: Order is not important.*

One of the important topics of this chapter is to understand that there are more efficient ways of counting than the old "one, two, three, . . ." technique. There are many ways of counting. We saw that some counting problems can be solved by using the fundamental counting principle, some by using permutations, and some by using combinations. Other problems required that we combine some of these ideas. However, it is important to keep in mind that not all counting problems fall into one of these neat categories.

Keep in mind, therefore, as we go through this section, that, although some problems may be permutation problems and some may be combination problems, there are many counting problems that are neither.

In practice, we are usually required to decide whether a given counting problem is a permutation or a combination before we can find a solution. For the sake of review, recall the difference between the election and committee problems of the previous sections. The election problem is a permutation problem, and the committee problem is a combination problem. We summarize the definitions of permutation and combination.

Permutations and Combinations

A *permutation* of a set of objects is an arrangement of certain of these objects in a *specific order*. A *combination* of a set of objects is an arrangement of certain of these objects *without regard to their order.*

Example 1 Classifying permutations and combinations

Classify the following as permutations, combinations, or neither.
a. The number of three-letter "words" that can be formed using the letters $\{m, a, t, h\}$
b. The number of ways you can change a $1 bill with 5 nickels, 12 dimes, and 6 quarters
c. The number of ways a five-card hand can be drawn from a deck of cards
d. The number of different five-numeral combinations on a combination lock
e. The number of license plates possible in Florida (routine plates have three numerals followed by three letters)

Solution
a. Permutation, since *mat* is different from *tam*.
b. Combination, since "2 quarters and 5 dimes" is the same as "5 dimes and 2 quarters."
c. Combination, since the order in which you receive the cards is unimportant.
d. Permutation, since "5 to the L, 6 to the R, 3 to the L, . . ." is different from "6 to the R, 5 to the L, 3 to the L, ..." We should not be misled by everyday usage of the word *combination*. We made a strict distinction between combination and permutation—one that is not made in everyday terminology. (The correct terminology would require that we call these "permutation locks.")
e. Neither; even though the *order* in which the elements are arranged is important, this does not actually fit the definition of a permutation because the objects are separated into two categories (pigeonholes). The arrangement of letters is a permutation and the arrangement of numerals is a permutation, but to count the actual number of arrangements for this problem would require permutations *and* the fundamental counting principle.

License Plate Problem

States issue vehicle license plates, and as the population increases, new plates are designed with more numerals and letters. For example, the state of California has some plates consisting of three letters followed by three digits. When they ran out of these possibilities, they began making license plates with three digits followed by three letters. Most recently they have issued plates with one digit followed by three letters in turn followed by three more digits.

Example 2 | Find the number of possible license plates

How many possibilities are there for each of the following license plate schemes?
a. Three letters followed by three digits
b. Three digits followed by three letters
c. One digit followed by three letters followed by three digits

Solution
a. $26 \times 26 \times 26 \times 10 \times 10 \times 10 = 17{,}576{,}000$ *Fundamental counting principle*
b. This is the same as part **a.** However, if both schemes are in operation at the same time, then the effective number of possible plates is

$$2 \times 17{,}576{,}000 = 35{,}152{,}000$$

c. The addition of one digit yields ten times the number of possibilities in part **a**:

$$10 \times 17{,}576{,}000 = 175{,}760{,}000$$

Note: A state, such as California, that decides to leave the old plates in use as they move through parts **a, b,** and **c** could have the following number of license plates:

$$17{,}576{,}000 + 17{,}576{,}000 + 175{,}760{,}000 = 210{,}912{,}000$$

Notice that the fundamental counting principle is not used for all counting problems.

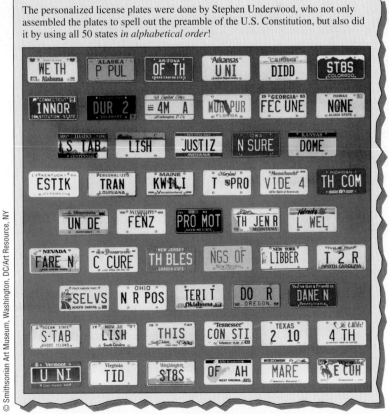

The personalized license plates were done by Stephen Underwood, who not only assembled the plates to spell out the preamble of the U.S. Constitution, but also did it by using all 50 states *in alphabetical order*!

Sandra Lord

Example   **3**   **Find the number of license plates without repetitions - I**

How many license plates can be formed if repetition of letters or digits is not allowed and the state uses the scheme of three numerals followed by three letters?

Solution   To make sure you understand the question, which of the following plates would be considered a success (no repetition)?

123ABC   *Success*

122ABC   *Failure; repeated digit*

456AAB   *Failure; repeated letter*

111AAA   *Failure; repeated letter and repeated numeral*

890XYZ   *Success*

We can use the fundamental counting principle to count the number of license plates that *do not* have a repetition:

Number of digits   *Letters in the alphabet*
$$10 \times 9 \times 8 \times 26 \times 25 \times 24 = 11{,}232{,}000$$
Digits left   *Letters left after the first one (no repetitions)*

Which Method?

We have now looked at several counting schemes: tree diagrams, pigeonholes, the fundamental counting principle, permutations, distinguishable permutations, and combinations. In practice, you will generally not be told what type of counting problem you are dealing with—you will need to decide. Table 12.1 should help with that decision. Remember, tree diagrams and pigeonholes are applications of the fundamental counting principle, so they are not listed separately in the table.

TABLE 12.1

			Counting Methods
Fundamental Counting Principle	**Permutations**	**Combinations**	**Distinguishable Permutations**
Counts the total number of separate tasks.	Number of ways of selecting r items out of n items.		n elements divided into k categories
Repetitions are allowed.	Repetitions are not allowed.		Repetitions are not allowed.
If tasks 1, 2, 3, $\cdot$, k, can be performed in n_1, n_2, n_3, ..., n_k ways, respectively, then the total number of ways the tasks can be done is $n_1 \cdot n_2, \cdot n_3, \ldots n_k$	Order important, called *arrangements*	Order not important, called *subsets*	Order of categories is important
	Formula: $$_nP_r = \frac{n!}{(n-r)!}$$	Formula: $$\binom{n}{r} = \frac{n!}{r!(n-r)!}$$	$$\binom{n}{n_1, n_2, \ldots, n_k} = \frac{n!}{n_1! n_2! \ldots n_k!}$$ where $n = n_1 + n_2 + \ldots + n_k$

Example   **4**   **Find the number of license plates without repetitions - II**

What is the number of license plates possible if each license plate consists of three letters followed by three digits, and we add the condition that repetition of letters or digits is not permitted?

Solution   This is a permutation problem since the *order* in which the elements are arranged is important, and the choice is without repetition. The number is found by using permutations along with the fundamental counting principle:

$$_{26}P_3 \cdot {}_{10}P_3 = 26 \cdot 25 \cdot 24 \cdot 10 \cdot 9 \cdot 8 = 11{,}232{,}000$$

By comparing the solutions to Examples 3 and 4, you can see that alternate approaches still provide the same result.

Example 5   Find the number of different quartets

A quartet is to be selected from a choir. There is to be one soprano selected from a group of six sopranos, two tenors selected from a group of five tenors, and a bass selected from three basses.
a. In how many ways can the quartet be formed?
b. In how many ways can the quartet be formed if one of the tenors is designated lead tenor?

Solution
a. Begin with the fundamental counting principle:

$$\overbrace{\binom{6}{1}}^{\text{Soprano}} \cdot \overbrace{\binom{5}{2}}^{\text{Two tenors}} \cdot \overbrace{\binom{3}{1}}^{\text{Bass}} = 6 \cdot 10 \cdot 3 = 180$$

b. Since the order of selecting the tenors is important, the middle factor from part **a** is replaced by

$$_{5}P_2 = \frac{5!}{3!} = 20$$

The number of ways of selecting the quartet is $6 \cdot 20 \cdot 3 = 360$.

Example 6   Find the number of sales representatives

Suppose that the Sharp Investment Company has 15 sales representatives who are to be reassigned to three geographical areas as follows: four in the North, five in the South, and six in the West. In how many ways could the sales representatives be assigned to the geographical areas?

Solution   Begin with the fundamental counting principle to fill three pigeonholes with combinations (order not important):

$$\overbrace{\binom{15}{4}}^{\text{North}} \cdot \overbrace{\binom{11}{5}}^{\text{South}} \cdot \overbrace{\binom{6}{6}}^{\text{West}} = \frac{15!}{4!11!} \cdot \frac{11!}{5!6!} \cdot \frac{6!}{6!0!} = \frac{15!}{4!5!6!} = 630{,}630$$

Note the connection between combinations and distinguishable permutations. We could also model this problem by noting that the geographical areas divide up all the sales representatives ($4 + 5 + 6 = 15$):

$$\binom{15}{4,\,5,\,6} = \frac{15!}{4!5!6!} = 630{,}630$$

Example 7 | Find the number of ways

A club with 42 members wants to elect a president, a vice president, and a treasurer. From the remaining members, an advisory committee of five people is to be selected. In how many ways can this be done?

Solution   This is both a permutation and a combination problem, with the final result calculated by using the fundamental counting principle.

$$\underset{\text{Officers}}{\underbrace{_{42}P_3}} \cdot \underset{\text{Committee}}{\underbrace{\binom{39}{5}}} = \frac{42!}{39!} \cdot \frac{39!}{5!34!} = 39{,}658{,}142{,}160$$

Sometimes it is easier to count what we are not counting than to enumerate all those items with which we are concerned. For example, suppose we wish to know the number of four-member committees that can be appointed from the club:

{Alfie, Bogie, Calvin, Doug, Ernie}

1. We could count directly:
   a. Alfie, Bogie, Calvin, Doug
   b. Alfie, Bogie, Calvin, Ernie
   c. Alfie, Bogie, Doug, Ernie
   $\vdots$

2. We could use the formula

$$\binom{5}{4} = \frac{_5P_4}{4!} = \cdots$$

3. We could count those we are not counting: For each four-member committee, there is one person left out. We can leave out Alfie, Bogie, Calvin, Doug, or Ernie. Therefore, there are five four-member committees.

Example 8 | Number of ways of obtaining at least one head

Suppose you flip a coin three times and keep a record of the result. In how many ways can you obtain at least one head?

Solution   Consider all possibilities:

H H H

H H T

H T H

H T T

T H H

T H T

T T H

T T T

You could count directly to obtain the answer, but you could also count what you are not counting by noticing that in only one out of eight possibilities do you obtain no heads. Thus, there are

$$8 - 1 = 7$$

possibilities in which you obtain at least one head.

The principle of counting without counting is particularly useful when the results become more complicated, as illustrated by the following example.

Example 9   Find the number of ways of obtaining a diamond

Find the number of ways of obtaining at least one diamond when drawing five cards from an ordinary deck of cards.

Solution   This is very difficult if we proceed directly, but we can compute the number of ways of not drawing a diamond:

$$\binom{39}{5} = \frac{39 \cdot 38 \cdot 37 \cdot 36 \cdot 35}{5 \cdot 4 \cdot 3 \cdot 2 \cdot 1} = 575{,}757$$

From Example 4, Section 12.2, there is a total of 2,598,960 possibilities, so the number of ways of drawing at least one diamond is

$$2{,}598{,}960 - 575{,}757 = 2{,}023{,}203$$

Problem Set | 12.3

Level 1

1. **IN YOUR OWN WORDS**   State the fundamental counting principle, and explain it in your own words.

2. **IN YOUR OWN WORDS**   Explain the difference between a permutation and a combination.

3. How many skirt-blouse outfits can a woman wear if she has three skirts and five blouses?

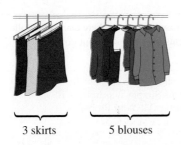

3 skirts   5 blouses

4. If a state issued license plates using the scheme of one letter followed by five digits, how many plates could it issue?

5. Repeat Problem 4 if repetitions are not allowed.

6. In how many ways can a group of 15 people elect a president, vice president, and secretary?

7. How many two-member committees can be formed from a group of seven people?

8. New York license plates consist of three letters followed by three numerals, and 245 letter arrangements are not allowed. How many plates can New York issue?

9. Boats often relay messages by using flags. How many messages can be made using five flags out of a package of 40 different flags?

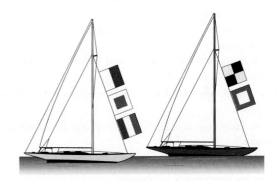

10. If a family has seven children, in how many ways could the parents have four boys and three girls?

11. A certain lock has four tumblers, and each tumbler can assume six positions. How many different possibilities are there?

12. Suppose you flip a coin and keep a record of the results. In how many ways could you obtain at least one head if you flip the coin four times?

13. You flip a coin five times and keep a record of the results. In how many ways could you obtain at least one tail?

14. You flip a coin six times and keep a record of the results. In how many ways could you obtain at least one tail?

15. You flip a coin seven times and keep a record of the results. In how many ways could you obtain at least one tail?

16. You flip a coin *n* times and keep a record of the results. In how many ways could you obtain at least one tail?

17. In how many ways could a club of 15 members choose a president, vice president, and treasurer?

18. In how many ways could a club of 30 members choose a president, vice president, and secretary?

19.

Many states offer personalized license plates. The state of California, for example, allows personalized plates with seven spaces for numerals or letters, or one of the following four symbols:

What is the total number of license plates possible using this counting scheme? (Assume that each available space is occupied by a numeral, letter, symbol, or space.)

20. How many different ways (boy-girl patterns) can a family have five children?

21. Suppose a restaurant offers the following prix fixe menu:
    Main course: prime rib, steak, chicken, filet of sole, shrimp
    Side dish: soup, salad, crab cakes
    Dessert: cheesecake, chocolate delight, ice cream
    Beverage: coffee, tea, milk
    In how many ways can someone order a meal consisting of one choice from each category?

22. In how many ways could a club of 15 appoint a committee of 4 people?

23. In how many ways could a club of 30 appoint a committee of 4 people?

In Problems 24–39, classify each as a permutation, a combination, or neither, and then answer the question.

24. At Mr. Furry's Dance Studio, every man must dance the last dance. If there are five men and eight women, in how many ways can dance couples be formed for the last dance?

25. Martin's Ice Cream Store sells sundaes with chocolate, strawberry, butterscotch, or marshmallow toppings, nuts, and whipped cream. If you can choose exactly three of these, how many possible sundaes are there?

26. Five people are to dine together at a rectangular table, but the hostess cannot decide on a seating arrangement. In how many ways can the guests be seated?

27. In how many ways can three hearts be drawn from a deck of cards?

28. In how many ways can four diamonds be drawn from a deck of cards?

29. A shipment of 100 TV sets is received. Six sets are to be chosen at random and tested for defects. In how many ways can six sets be chosen?

30. A night watchman visits 125 offices every night. To prevent others from knowing when he will be at a particular office, he varies the order of his visits. In how many ways can this be done?

31. A certain manufacturing process calls for the mixing of six chemicals. One liquid is to be poured into the vat, and then the others are to be added in turn. All possibilities must be tested to see which gives the best results. How many tests are required?

32. There are three boys and three girls at a party. In how many ways can they be seated if they can sit down four at a time?

33. What is the number of distinguishable arrangements of the letters in the word KARL?

34. What is the number of distinguishable arrangements in the letters in the word SMITH?

35. The receiving line at a wedding consists of the bride, the groom, parents of the bride, parents of the groom, the best man, and the maid of honor (all in that order). In addition, there are five groomsmen (in any order) followed by five bridesmaids (in any order). In how many ways can the receiving line be arranged?

36. A space shuttle mission consists of a commander, a pilot, 3 engineers, a doctor, a scientist, and a civilian. If there are 7 people from whom the commander and pilot must be chosen, 18 possible engineers, 4 possible doctors, 7 scientists and 3 candidates for the civilian crew member, in how many ways can a crew be formed?

37. How many three-digit numbers can be formed using the numerals {3, 4, 6, 7} without repetition?

38. How many three-digit numbers can be formed using the numbers {1, 2, 3, 4, 5, 6, 7} without repetition?

39. A student is asked to answer 10 out of 12 questions on an exam. In how many ways can she select the questions to be answered?

Level 2

40. The advertisement shown here claims the following 12 mix-and-match items give 122 different outfits.

The wardrobe consists of 4 blouses, 2 slacks, 2 skirts, 1 sweater, 2 jackets, and 1 scarf. Assume that the model must choose one top and one bottom item of clothing. She may or may not choose a sweater or jacket or a scarf. How many different outfits are possible?

41. Consider selecting two elements, say, *a*, and *b*, from the set $A = \{a, b, c, d, e\}$. List all possible subsets of those two elements, as well as all possible arrangements.

42. Consider selecting three elements, say, *c*, *d*, and *e*, from the set $A = \{a, b, c, d, e\}$. List all possible subsets of those three elements. How many arrangements are there?

43. Consider selecting four elements, say, *a*, *b*, *c*, and *d*, from the set $A = \{a, b, c, d, e\}$. List all possible subsets of those four elements. How many arrangements are there?

44. Consider selecting five elements, say, *a*, *b*, *c*, *d*, and *e*, from the set $A = \{a, b, c, d, e\}$. List all possible subsets of those five elements. How many arrangements are there?

45. A typical Social Security number is 555-47-5593. How many Social Security numbers are possible if the first two digits cannot be 0?

46. A club consists of four men and three women. In how many ways can this club elect a president, vice president, and secretary, in that order, if the president must be a woman and the other two officers must be men?

47. A history teacher gives a 20-question true-false exam. In how many ways can the test be answered if the possible answers are T or F or possibly to leave the answer blank?

48. A JVC advertisement appeared in several national periodicals. It claimed that the SEA graphic equalizer system can create 371,293 different sounds from the thirteen possible zone levels on each of five different controls. Can this claim be substantiated from the advertisement and the knowledge of this chapter? If so, explain how; if not, tell why this number is impossible.

49. In how many ways can 26 books be arranged on a bookshelf?

Level 3

50.

"I did it without counting. There are – – – beans in the jar."

Outline a procedure that would allow the fellow in the cartoon to proclaim that he "did it without counting." Estimate the number assuming that the container is a cylinder 15 in. tall and 10 in. across, and that it is filled with kidney beans.

51. The old lady jumping rope in the cartoon at the beginning of this section (see p. 590) is counting one at a time. Assume that she jumps the rope 50 times per minute and that she jumps 8 hours a day, 5 days a week, 50 weeks a year. Estimate the length of time necessary for her to jump the rope the number of times indicated in the cartoon.

52. A marginal note in this section describes the longest paper-link chain (see p. 594). If there are 12 paper links per foot, and it takes 1 minute to construct each link, estimate the length of time necessary to build this paper-link chain.

53. What numerical property is exhibited by the arrangement of billiard balls shown in Figure 12.3?

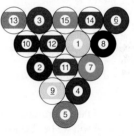

FIGURE 12.3 Billiard balls

Problem Solving 3

54. A die can be held so that one, two, or three of its faces can be seen at any one time.

One face showing a 5

Two faces showing a total of 9

Three faces showing a total of 10

Is it possible to hold a die in different ways so that at different times the visible numbers add up to every number from 1 through 15?

55. The News Clip below presents an argument proving that at least two New Yorkers must have *exactly* the same number of hairs on their heads! For this problem, just *suppose* you knew the number

of hairs on your head, as well as the *exact* number of hairs on anyone else's head. Now multiply the number of hairs on your head by the number of hairs on your neighbor's head. Take this result and multiply by the number of hairs on the heads of each person in your town or city. Continue this process until you have done this for everyone in the entire world! Make a guess (you can use scientific notation, if you like) about the size of this answer. (*Hint:* The author has worked this out and claims to know the exact answer.)

An example about reaching a conclusion about the number of objects in a set is given by M. Cohen and E. Nagel in *An Introduction to Logic.* They conclude that there are at least two people in New York City who have the same number of hairs on their heads. This conclusion was reached not through counting the hairs on the heads of 8 million inhabitants of the city, but through studies revealing that: (1) the maximum number of hairs on the human scalp could never be as many as 5,000 per square centimeter, and (2) the maximum area of the human scalp could never reach 1,000 square centimeters. We can now conclude that no human head could contain even

$$5,000 \times 1,000 = 5,000,000$$

hairs. Since this number is less than the population of New York City, it follows that at least two New Yorkers must have the same number of hairs on their heads!

The following transcribed television advertisement is the basis for Problems 56–59.

56. If we accept the facts of the advertisement and order one pizza per day, how long will we be able to order different pizzas before we are forced to order a pizza previously ordered?

LITTLE CAESAR'S PIZZA PIZZA!!!

Customer: So what's this new deal?
Pizza chef: Two pizzas
Customer [Toward four-year-old boy]:
    Two pizzas. Write that down.
Pizza chef: And on the two pizzas choose any toppings—up to five [*from a list of* 11 *toppings*].
Older boy: Do you . . .
Pizza chef: . . . have to pick the same toppings on each pizza? NO!
Four-year-old-math whiz: Then the possibilities are endless.
Customer: What do you mean? Five plus five are ten.
Math whiz: Actually, there are 1,048,576 possibilities.
Customer: Ten was just a ballpark figure.
Old man: You got that right.

First aired on national television on November 8, 1993.

57. How many different pizza orders are possible?

58. Jean Sherrod of Little Caesar's Enterprises, Inc., explains that an order where the first pizza is pepperoni and the second pizza is ham would be considered different from an order in which the first pizza is ham and the second pizza is pepperoni. Using this scheme, how many pizza orders are possible?

59. Double toppings are not included in the calculation in the advertisement. However, Little Caesar's does allow double toppings. If double toppings are possible, how many different pizza orders are possible?

60. Show that the sum of the entries of the nth row of Pascal's triangle is 2^n.

12.4 Rubik's Cube and Instant Insanity

This optional section discusses two famous puzzles, the Instant Insanity and Rubik's Cube puzzles.* If you are not familiar with these fascinating puzzles, check them out on the web. You can reach them via **www.mathnature.com.**

Rubik's Cube

Rubik's Cube:
www.schubart.net/rc/

In the summer of 1974, a Hungarian architect invented a three-dimensional object that could rotate about *all three axes* (sounds impossible, doesn't it?). He wrote up the details of the cube and obtained a patent in 1975. The cube is now known worldwide as **Rubik's cube.** In case you have not seen one of these cubes, it is shown in Figure 12.4.

*The term *Rubik's Cube* is a trademark of Ideal Toy Corporation, Hollis, New York. The term *Rubik's cube* as used in this book means the cube puzzle sold under any trademark. "Instant Insanity" is a trademark of Parker Brothers, Inc.

Bettmann/CORBIS

Historical NOTE

The puzzle known as Rubik's Cube was designed by Ernö Rubik, an architect and teacher in Budapest, Hungary. It was also invented independently by Terutoshi Ishige, an engineer in Japan. Both applied for patents in the mid-1970s. The cubes were first manufactured in Hungary. In 1978, a Hungarian mathematics professor brought several with him to the International Congress of Mathematics in Helsinki, Finland. This formally introduced the cube to Europe and the rest of the world. They have been widely available in the United States since 1980. They are often used today in mathematics classes to illustrate mathematical ideas, particularly in group theory.

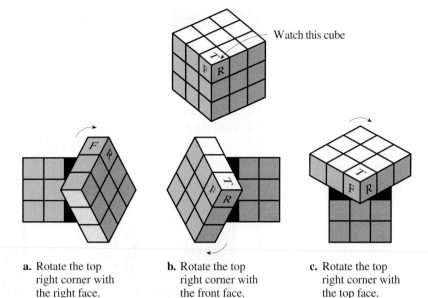

Watch this cube

a. Rotate the top right corner with the right face.

b. Rotate the top right corner with the front face.

c. Rotate the top right corner with the top face.

FIGURE 12.4 Rubik's cube

When you purchase the cube, it is arranged so that each face is showing a different color, but after a few turns it seems next to impossible to return to the start. In fact, the manufacturer claims there are 8.86×10^{22} possible arrangements, of which there is only one correct solution. This claim is incorrect; there are 2,048 possible solutions among the 8.86×10^{22} claimed arrangements (actually, $8.85801027 \times 10^{22}$). This means there is one solution for each 4.3×10^{19} (actually 43,252,003,274,489,856,000) arrangements.

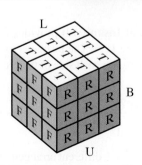

Example 1 How long to solve a Rubik's cube?

If it took you one-half second for each arrangement, how long would it take you to move the cube into all arrangements?

Solution

4.3×10^{19}	Arrangements
2.15×10^{19}	Divide by 2 for the number of seconds.
3.58×10^{17}	Divide by 60 for the number of minutes.
5.97×10^{15}	Divide by 60 for the number of hours.
2.49×10^{14}	Divide by 24 for the number of days.
6.81×10^{11}	Divide by 365.25 for the number of years.
6.81×10^{9}	Divide by 100 for the number of centuries.
6.81×10^{8}	Divide by 10 for the number of millennia.

Our problem ends here because it would require over 681,000,000 *millennia*! If a high-speed computer listed 1,000,000 arrangements per minute, it would take over 80,000 millennia to print out the possibilities. Remember, it has been only 2 millennia since Christ!

FIGURE 12.5 Standard-position Rubik's Cube

Let's consider what we call the *standard-position cube*, as shown in Figure 12.5. Label the faces Front (F), Right (R), Left (L), Back (B), Top (T), and Under (U), as shown. Hold the cube in your left hand with T up and F toward you so that L is against your left palm. Now describe the results of the moves in Example 2.

Example 2 Rubik's cube moves

a. Rotate the right face 90° clockwise; denote this move by R. Return the cube to standard position; we denote this move by R^{-1}.

b. Rotate the right face 180° clockwise; denote this by R^2. Return the cube to standard position by doing another R^2. Notice that $R^2 R^2 = R^4$, which returns the cube to standard position.

c. Rotate the top 90° clockwise; call this T. Return the cube by doing T^{-1}.

d. *TR* means rotate the top face 90° clockwise, *then* rotate the right face 90° clockwise. Describe the steps necessary to return the cube to standard position.

Solution

a.

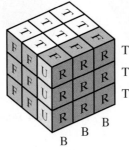

b.

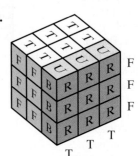

c.

d.

To return this cube to standard position, you need $R^{-1}T^{-1}$.

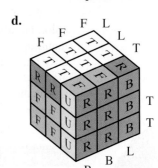

Now, a rearrangement of the 54 colored faces of a small cube is a *permutation*. We discussed permutations of a square in Problem 60, Section 5.6, at which time we called them *symmetries of a square*.

Instant Insanity

An older puzzle, simpler than Rubik's cube, is called **Instant Insanity.** It provides four cubes colored red, white, blue, and green, as indicated in Figure 12.6. The puzzle is to assemble them into a $1 \times 1 \times 4$ block so that all four colors appear on each side of the block.

Example 3 How long to solve Instant Insanity?

In how many ways can this Instant Insanity puzzle be arranged?

Solution A common mistake is to assume that a cube can have 6 different arrangements (because of our experience with dice). In the case of Instant Insanity, we are interested not only in the top, but also in the other sides. There are 6 possible faces for the top and *then* 4 possible faces for the front. Thus, by the fundamental counting principle, there are

$6 \cdot 4 = 24$ arrangements for one cube

FIGURE 12.6 Instant Insanity

Instant Insanity:
http://mysite.verizon
.net/m5mammek/
markspuz.html

Now, for four cubes, again use the fundamental counting principle to find

$$24 \cdot 24 \cdot 24 \cdot 24 = 331{,}776$$

This is *not* the number of *different* possibilities. You are asked to find that number in the problem set.

Problem Set 12.4

Level 1

1. Example 3 shows that Instant Insanity has 331,776 possibilities (not all different). If you could make one different move per second, how long do you think it would take you to go through all these arrangements? Estimate your answer first, then calculate the correct answer.

2. Suppose a computer could print out 1,000 arrangements in Problem 1 every minute. How long would it take for this computer to print out all possibilities?

Show the result of the moves on Rubik's cube indicated in Problems 3–29. Remember that R, F, L, B, T, and U mean rotate 90° clockwise the right, front, left, back, top, and under faces, respectively. Use the standard Rubik's cube shown in Figure 12.5 as your starting point. Consider each clockwise rotation by first turning the cube so that the side you are rotating is facing you.

3. F	**4.** B	**5.** U	**6.** L	**7.** B^{-1}
8. F^{-1}	**9.** L^{-1}	**10.** T^{-1}	**11.** F^2	**12.** T^2
13. L^2	**14.** B^2	**15.** F^3	**16.** R^3	**17.** T^3
18. U^3	**19.** RL	**20.** TU	**21.** FB^{-1}	**22.** $F^{-1}B$
23. RT	**24.** FT	**25.** FT^{-1}	**26.** FR^{-1}	**27.** F^5
28. U^5	**29.** $F^{-1}T^{-1}$			

Level 2

Determine whether each of the figures in Problems 30–37 will be a solution to an Instant Insanity puzzle.

30.

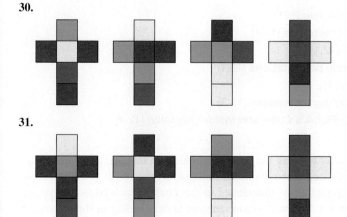

31.

32.

33.

34.

35.

36.

37.

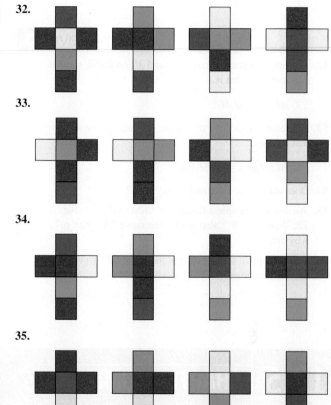

Name the move or moves that will return the Rubik's cube to standard position after making the move shown in Problems 38–51.

38. F

39. U

40. L^{-1}

41. B^{-1}

42. T^2

43. U^2

44. F^3

45. B^3

46. TU

47. BU

48. FB^{-1}

49. FT^{-1}

50. $U^{-1}T^{-1}$

51. $F^{-1}B^{-1}$

Level 3

52. Are two consecutive moves M_1 and M_2 on Rubik's cube commutative; that is,

$$M_1M_2 = M_2M_1?$$

53. Are three consecutive moves M_1, M_2, and M_3, associative; that is, does

$$M_1(M_2M_3) = (M_1M_2)M_3?$$

54. Does each move have an inverse move?

55. We saw in Example 2 that $R^{-1}T^{-1}$ reversed the move TR. Would $T^{-1}R^{-1}$ also reverse the move TR? Why or why not?

56. Does the move $L^{-1}R^{-1}$ reverse the move LR? Would it also reverse the move RL? Why or why not?

Problem Solving 3

57. Find the number of *different* arrangements for the Instant Insanity blocks.

58. **IN YOUR OWN WORDS** The most difficult part of understanding a solution to Rubik's cube is understanding the notation used by the author in stating the solution. In one solution, the Ledbetter Nering algorithm (see the references in the Group Research Projects), the authors describe a sequence of moves that they call THE MAD DOG. This series of moves can be described using our notation by

$$(RTF)^5$$

Start with the standard-position cube and show the result after carrying out THE MAD DOG.

59. **IN YOUR OWN WORDS** David Singmaster, in his book *Notes on Rubik's "Magic Cube"* (see the references in the Group Research Projects), describes a sequence of moves to put the upper corners in place (after certain other moves). Using our notation, this set of moves is

$$FU^2F^{-1}U^2R^{-1}UR$$

Start with the standard-position cube and show the result after carrying out this sequence of moves.

60. In *The Simple Solution to Rubik's Cube* (see the references in the Group Research Projects), James Nourse describes a process to orient the bottom corner cubes. Using our notation, this set of moves is

$$R^{-1}U^{-1}RU^{-1}R^{-1}U^2RU^2$$

Start with the standard-position cube and show the result after carrying out this sequence of moves.

12.5 CHAPTER SUMMARY

There is no problem in all mathematics that cannot be solved by direct counting. But with the present implements of mathematics many operations can be performed in a few minutes which without mathematical methods would take a lifetime.

ERNST MACH

STOP

WWW

Important Ideas

Fundamental counting principle [12.1]
Multiplication property of factorials [12.1]
Formula for permutation [12.1]
Formula for distinguishable permutations [12.1]
Formula for combinations [12.2]
Classify and identify counting techniques [12.3]
Applications of counting (Rubik's Cube and Instant Insanity) [12.4]

Take some time getting ready to work the review problems in this section. First review these important ideas. Look back at the definition and property boxes. If you look online, you will find a list of important terms introduced in this chapter, as well as the types of problems that were introduced. You will maximize your understanding of this chapter by working the problems in this section only after you have studied the material.

You will find some review help online at **www.mathnature.com.** There are links giving general test help in studying for a mathematics examination, as well as specific help for reviewing this chapter.

Chapter **12**   Review Questions

Simplify each expression in Problems 1–5.

1. $8! - 3!$ **2.** $8 - 3!$ **3.** $(8 - 3)!$ **4.** $\left(\dfrac{8}{2}\right)!$ **5.** $\left(\dbinom{8}{2}\right)!$

Find the numerical value of each expression in Problems 6–10.

6. $\dbinom{5}{3}$ **7.** $_8P_3$ **8.** $_{12}P_0$ **9.** $_{14}C_4$ **10.** $_{100}P_3$

11. In how many ways can a three-member committee be chosen from a group of 12 people?

12. In how many ways can five people line up at a bank teller's window?

13. How many distinguishable permutations are there of the letters of the words HAPPY and COLLEGE?

14. A jar contains four red and six white balls. Three balls are drawn at random. In how many ways can at least one red ball be drawn?

15. If the Senate is to form a new committee of five members, in how many different ways can the committee be chosen if all 100 senators are available to serve on this committee?

16. Advertisements for Wendy's Old Fashioned Hamburgers claim that you can have your hamburgers 256 ways. If Wendy's offers catsup, onion, mustard, pickles, lettuce, tomato, mayonnaise, and relish, is the advertisement's claim correct? Explain why or why not.

Bloomberg/Getty Images

17. a. A certain mathematics test consists of 10 questions. In how many ways can the test be answered if the possible answers are "true" and "false"?

b. Answer the question if the possible answers are "true," "false," and "maybe."

c. Answer the question if the possible answers are (a), (b), (c), (d), and (e); that is, the test is multiple choice.

18. a. One variation of Instant Insanity is a puzzle with five blocks instead of four. How many arrangements are possible?

b. If you carry out one arrangement every second, how much time is required to show the number of arrangements in part **a**?

19. The March 2005 cover of a magazine from the editors of *Gardening* declares "Plant Combinations . . . 65 Design Plans." Is it possible that the number of combinations of any number of items (more than one item) equals 65?

20. The Wednesday Luncheon Club, consisting of ten members, decided to celebrate its first anniversary by having lunch at a fancy restaurant. When the members arrived and were ready to take their seats, they could not decide where to sit.

Just when they were ready to leave because of their embarrassment at being unable to decide how to sit around the rectangular table, the manager come to the rescue and told them to sit down just where they were standing. Then the secretary of the club was to write down where they were sitting, and they were to return the following day and sit in a different order. If they continued this until they had tried all arrangements, the manager would give them anything on the menu, free of charge.

It sounded like a very good deal, so they agreed. However, the day of the free lunch never came. Can you explain why?

BOOK REPORTS

Group   RESEARCH PROJECTS

Go to
www.mathnature.com
for references and links.

FIGURE 12.7 A hexahexaflexagon

Working in small groups is typical of most work environments, and learning to work with others to communicate specific ideas is an important skill. Work with three or four other students to submit a single report based on each of the following questions.

G40. HISTORICAL QUEST   An interesting pastime began in the fall of 1939 when Princeton University graduate student Arthur Stone from England trimmed an inch from his American notebook sheets to fit his English binder. After folding the trimmed-off strips, he came up with the first *hexaflexagon*. This led to a more interesting variant known as a *hexahexaflexagon*, as shown in Figure 12.7.

As you "flex" your device, you can cause each of the numerals 1 through 6 to show on one side (Figure 12.7 shows the numeral 4). For this problem you are asked first to construct one, and then you are asked questions about your completed device. Prepare a strip of paper as shown in Figure 12.8a. Make sure each triangle is equilateral. Turn it over and mark the other side as shown in Figure 12.8a.

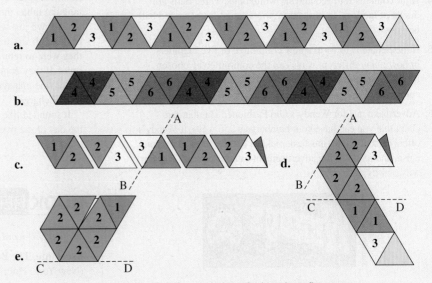

FIGURE 12.8 Construction of a hexahexaflexagon

Starting from the left of Figure 12.8b, fold the 4 onto the 4, the 5 onto the 5, 6 onto 6, 4 onto 4, and so on. Continue by folding 1 onto the 1 from the front, by folding the 1 onto the 1 from the back, and finally by bringing the 1 up from the bottom so that it rests on top of the 1 on the top. Your paper should look like the one shown in Figure 12.8c and Figure 12.8d. Paste the blank onto the blank, and the result is called a **hexahexaflexagon.**

With a little practice you'll be able to "flex" your hexahexaflexagon (see Figure 12.9) so that you can obtain a side with all 1s, another with all 2s, . . . , and another with all 6s.

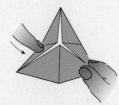
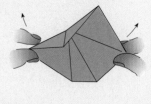

FIGURE 12.9 To "flex" your hexahexaflexagon, pinch together two of the triangles (left two figures). The inner edge may then be opened with the other hand (rightmost picture). If the hexahexaflexagon cannot be opened, an adjacent pair of triangles is pinched. If it opens, turn it inside out, finding a side that was not visible before. Be careful not to tear the hexahexaflexagon by forcing the flex.

After you have become fairly proficient at "flexing," count the number of flexes required to obtain all six "sides." What do you think is the fewest number of flexes necessary to obtain all six sides?

G41. A puzzle sold under the name The Avenger is pictured in Figure 12.10.

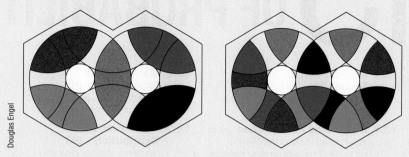

Douglas Engel

FIGURE 12.10 The Avenger Puzzle

There are four problems posed in the article shown in the reference. Write a report on this article.

REFERENCE: "Group Theory, Rubik's Cube and The Avenger," *Games*, June/July 1987, pp. 44–45.

G42. Consult one of the references listed below and learn to solve Rubik's cube. Demonstrate your skill to the class. Nourse names the following categories:

    20 minutes: WHIZ
    10 minutes: SPEED DEMON
     5 minutes: EXPERT
     3 minutes: MASTER OF THE CUBE

a. Stage a contest in front of the class to see which members of your group can complete one face of a Rubik's cube.

b. Stage a contest to see which member of your group can solve the Rubik's cube puzzle the fastest. Report the results to the class.

REFERENCES: Ledbetter and Nering, *The Solution to Rubik's Cube* (Rohnert Park, CA, Noah's Ark Enterprises, 1980).

James G. Nourse, *The Simple Solution to Rubik's Cube* (New York: Bantam Books, 1981).

David Singmaster, *Notes on Rubik's "Magic Cube,"* 5th ed. (Hillside, NJ: Enslow Publishers, 1980).

Individual RESEARCH PROJECTS

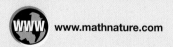

www.mathnature.com

*Learning to use sources outside your classroom and textbook is an important skill, and here are some ideas for extending some of the ideas in this chapter. You can find references to these projects in a library or at **www.mathnature.com.***

PROJECT 12.1 Historical Quest Write a paper on the famous *Tower of Hanoi* problem.

PROJECT 12.2 How can all the constructions of Euclidean geometry be done by folding paper?

    What assumptions are made when paper is folded to construct geometric figures?

    What is a hexaflexagon? How many flexes are possible?

    What is origami?

PROJECT 12.3 Find a solution for the Instant Insanity puzzle.

13 THE NATURE OF PROBABILITY

What in the World?

"Probability is the mathematics of uncertainty," proclaimed Kevin.

"Wait a minute," interrupted Sammy. "I thought that mathematics was absolute and that there was no uncertainty about it!"

"Well, Sammy, suppose you are playing a game of Monopoly™. Do you know which property you'll land on for your next turn?" asked Kevin.

"No, but neither do you!"

"That's right, but mathematics can tell us where you are *most likely* to land. There is nothing uncertain about probability; rather, probability is a way of describing uncertainty. For example, what is the probability of tossing a coin and obtaining heads?"

"That's easy; it's one half," answers Sammy.

"Right, but *why* do you say it is one-half?" (See Problem 48 in Problem Set 13.1.)

Overview

We see examples of probability every day. Weather forecasts, stock market analyses, contests, children's games, political polls, game shows, and gambling all involve ideas of probability. Probability is the mathematics of uncertainty.

If you have ever played the lottery (hundreds of billions are wagered legally each year) or bought a life insurance policy, you have indirectly used probability. Animals and plants are bred to enhance certain traits to be passed through generations, and these breeding techniques are directed by the probability of genetics. Research in human inherited diseases such as cystic fibrosis, sickle-cell anemia, and Huntington's disease involves probability.

In this chapter, we'll investigate the definition of probability and some of the procedures for dealing with probability.

Austin MacRae

13.1 | Introduction to Probability

It is a truth very certain that, when it is not in our power to
determine what is true, we ought to follow what is most probable.

RENÉ DESCARTES

In this chapter, we introduce the mathematics that describes the likelihood of some event whose occurrence may not be certain. We begin by introducing some necessary terminology.

Terminology

An **experiment** is an observation of any physical occurrence. The **sample space** of an experiment is the set of all its possible outcomes. An **event** is a subset of the sample space. If an event is the empty set, it is called the **impossible event;** and if it has only one element, it is called a **simple event.**

Example 1 Find a sample space

Suppose a researcher wishes to study how the color of a child's eyes is related to the parents' eye color.
a. List the sample space.
b. Give an example of a simple event.
c. Give an example of an impossible event.

Solution
a. The sample space is a listing of all possible eye colors: {green, blue, brown, hazel}.
b. A simple event is an event with only one element, say {blue}.
c. An impossible event is one that is empty, say, {purple}.

The next example involves a **die,** which is a cube showing six faces marked 1, 2, 3, 4, 5, and 6. If an outcome of an experiment has the same chance of occurring as any other outcome, then they are said to be **equally likely outcomes.**

Example 2 Find a sample space and list events

a. What is the sample space for the experiment of tossing a coin and then rolling a die?
b. List the following events for the sample space in part **a:**

$E = \{$rolling an even number on the die$\}$

$H = \{$tossing a head$\}$

$X = \{$rolling a six and tossing a tail$\}$

Which of these (if any) are simple events?

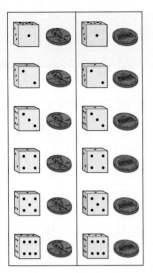

FIGURE 13.1 Sample space for tossing a coin and rolling a die

Solution A coin is considered *fair* if the outcomes of head and tail are *equally likely*. We also note that a fair die is one for which the outcomes from rolling it are *equally likely*. A die for which one outcome is more likely than the others is called a *loaded die*. In this book, we will assume fair dice and fair coins unless otherwise noted.

a. You can visualize the sample space as shown in Figure 13.1. If the sample space is small, it is sometimes worthwhile to make a direct listing, as shown in Figure 13.1. Sometimes, it is helpful to build sample spaces by using a *tree diagram:*

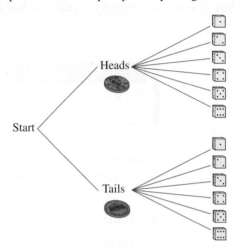

We see that S (the sample space) is

$$S = \{H1, H2, H3, H4, H5, H6, T1, T2, T3, T4, T5, T6\}$$

b. An event must be a *subset* of the sample space. Notice that E, H, and X are all subsets of S. Therefore,

$$E = \{H2, H4, H6, T2, T4, T6\}$$
$$H = \{H1, H2, H3, H4, H5, H6\}$$
$$X = \{T6\}$$

X is a simple event because it has only one element.

Two events E and F are said to be **mutually exclusive** if $E \cap F = \varnothing$, that is, they cannot both occur simultaneously.

Example 3 Sample space for rolling a die

Suppose that you perform an experiment of rolling a die. Find the sample space, and then let $E = \{1, 3, 5\}$, $F = \{2, 4, 6\}$, $G = \{1, 3, 6\}$, and $H = \{2, 4\}$. Which of these are mutually exclusive?

Solution The sample space for a single die is shown in Figure 13.2.

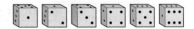

FIGURE 13.2 Sample space for a single die

The sample space is found by looking at the spots on top:

$$S = \{1, 2, 3, 4, 5, 6\}$$

Evidently,

E and F are mutually exclusive, since $E \cap F = \varnothing$.

G and H are mutually exclusive, since $G \cap H = \varnothing$.

E and H are mutually exclusive, since $E \cap H = \varnothing$.

But

F and H are *not* mutually exclusive.

G and F are *not* mutually exclusive.

E and G are *not* mutually exclusive.

Probability

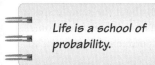

A **probabilistic model** deals with situations that are random in character and attempts to predict the outcomes of events with a certain stated or known degree of accuracy. For example, if we toss a coin, it is impossible to predict in advance whether the outcome will be a head or a tail. Our intuition tells us that it is equally likely to be a head or a tail, and somehow we sense that if we repeat the experiment of tossing a coin a large number of times, a head will occur "about half the time." To check this out, I recently flipped a coin 1,000 times and obtained 460 heads and 540 tails. The percentage of heads is $\frac{460}{1,000} = 0.46 = 46\%$, which is called the *relative frequency*. If an experiment is repeated n times and an event occurs m times, then

$$\frac{m}{n} \text{ is called the } \textbf{relative frequency} \text{ of the event}$$

Our task is to create a model that will assign a number p, called the *probability of an event*, which will predict the relative frequency. This means that for a *sufficiently large number of repetitions* of an experiment,

$$p \approx \frac{m}{n}$$

Probabilities can be obtained in one of three ways:

1. **Empirical probabilities** (also called *a posteriori* models) are obtained from experimental data. For example, an assembly line producing brake assemblies for General Motors produces 1,500 items per day. The probability of a defective brake can be obtained by experimentation. Suppose the 1,500 brakes are tested and 3 are found to be defective. Then the empirical probability is the relative frequency of occurrence, namely,

$$\frac{3}{1,500} = 0.002 \text{ or } 0.2\%$$

2. **Theoretical probabilities** (also called *a priori* models) are obtained by logical reasoning according to stated definitions. For example, the probability of rolling a die and obtaining a 3 is $\frac{1}{6}$, because there are six possible outcomes, each with an equal chance of occurring, so a 3 should appear $\frac{1}{6} \approx 17\%$ of the time.

3. **Subjective probabilities** are obtained by experience and indicate a measure of "certainty" on the part of the speaker. These probabilities are not necessarily arrived at through experimentation or theory. For example, a dad says to his daughter, "After that outburst the probability that you will go to the dance is almost 0%."

Our focus in this chapter is on theoretical probabilities, but we should keep in mind that our theoretical model should be predictive of the results obtained by experimentation (empirical probabilities); if they are not consistent, and we have been careful about our record keeping in arriving at an empirical probability, we would conclude that our theoretical model is faulty. The reason for this conclusion is called the **law of large numbers.**

Law of Large Numbers

As an experiment is repeated more and more times, the empirical probability (that is, the proportion of outcomes favorable to any particular event) will tend to come closer and closer to the theoretical probability of that event.

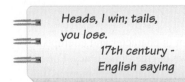

www.mathnature.com
has a link to an interactive coin-flipping site.

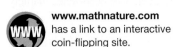

This is the basic definition for this chapter.

This law of large numbers keeps us "honest" in maintaining our records and in setting up models for finding the theoretical probability.

The simplest probability model makes certain assumptions about the sample space. If the sample space can be divided into *mutually exclusive* and *equally likely* outcomes, we can define the probability of an event. Let's consider the experiment of tossing a single coin. A suitable sample space is

$$S = \{\text{heads, tails}\}$$

Suppose we wish to consider the event of obtaining heads; we'll call this event A. Then

$$A = \{\text{heads}\}$$

and this is a simple event.

We wish to define the probability of event A, which we denote by $P(A)$. Notice that the outcomes in the sample space are mutually exclusive; that is, if one occurs, the other cannot occur. If we flip a coin, there are two possible outcomes, and *one and only one* outcome can occur on a toss. If each outcome in the sample space is equally likely, we define the probability of A as

$$P(A) = \frac{\text{NUMBER OF SUCCESSFUL RESULTS}}{\text{NUMBER OF POSSIBLE RESULTS}}$$

A "successful" result is a result that corresponds to the event whose probability we are seeking—in this case, {heads}. Since we can obtain a head (success) in only one way, and the total number of possible outcomes is two, the probability of heads is given by this definition as

$$P(\text{heads}) = P(A) = \frac{1}{2}$$

This must correspond to the empirical results you would obtain if you repeated the experiment a large number of times. In Problem 48, you are asked to repeat this experiment 100 times.

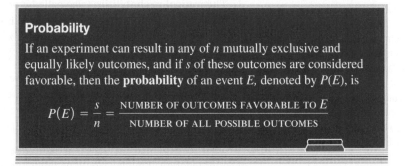

Probability

If an experiment can result in any of n mutually exclusive and equally likely outcomes, and if s of these outcomes are considered favorable, then the **probability** of an event E, denoted by $P(E)$, is

$$P(E) = \frac{s}{n} = \frac{\text{NUMBER OF OUTCOMES FAVORABLE TO } E}{\text{NUMBER OF ALL POSSIBLE OUTCOMES}}$$

Example 4   **Find the probability of a spinner event**

Use the definition of probability to find first the probability of white and second the probability of black, using the spinner shown and assuming that the arrow will never lie on a border line.

Solution   Looking at the spinner, we note that it is divided into three areas of the same size. We assume that the spinner is equally likely to land in any of these three areas.

$$P(\text{white}) = \frac{2}{3} \quad \leftarrow \text{Two sections are white.}$$
$$\phantom{P(\text{white}) = \frac{2}{3}} \leftarrow \text{Three sections all together}$$

$$P(\text{black}) = \frac{1}{3} \quad \leftarrow \text{One section is black.}$$
$$\phantom{P(\text{black}) = \frac{1}{3}} \leftarrow \text{Three sections all together}$$

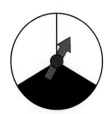

These are theoretical probabilities.

FIGURE 13.3 Marbles in jar

Example **5** **Find the probability of drawing from an urn**

Consider a jar that contains marbles as shown in Figure 13.3. Suppose that each marble has an equal chance of being picked from the jar. Find:
a. $P(\text{blue})$ **b.** $P(\text{green})$ **c.** $P(\text{yellow})$

Solution

a. $P(\text{blue}) = \dfrac{4}{12}$ ← 4 blue marbles in jar   **b.** $P(\text{green}) = \dfrac{7}{12}$ **c.** $P(\text{yellow}) = \dfrac{1}{12}$
    ← 12 marbles in jar

  $= \dfrac{1}{3}$ Reduce fractions.

This is a theoretical probability; also note that the probabilities of all the simple events sum to

$$\frac{1}{3} + \frac{7}{12} + \frac{1}{12} = 1$$

Reduced fractions are used to state probabilities when the fractions are fairly simple. If, however, the fractions are not simple and you have a calculator, it is acceptable to state the probabilities as decimals, as shown in Example 6.

Example **6** **Find an empirical probability**

Suppose that, in a certain study, 46 out of 155 people showed a certain kind of behavior. Assign a probability to this behavior.

Solution $P(\text{behavior shown}) = \frac{46}{155} \approx 0.30$

This is an empirical probability because it was arrived at through experimentation.

Example **7** **Select the better spinner**

Consider two spinners as shown in Figure 13.4

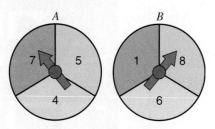

FIGURE 13.4 Two spinners

You and an opponent are to spin a spinner simultaneously, and the one with the higher number wins. Which spinner should you choose, and why?

Solution   We begin by listing the sample space:

$A\backslash B$	1	6	8	
4	(4, 1)	(4, 6)	(4, 8)	The times that A wins are highlighted.
5	(5, 1)	(5, 6)	(5, 8)	
7	(7, 1)	(7, 6)	(7, 8)	

$$P(A \text{ wins}) = \frac{4}{9}; \quad P(B \text{ wins}) = \frac{5}{9}$$

We would choose spinner B because it has a greater probability of winning.

Example **8** **Probability of drawing a card**

Suppose that a single card is selected from an ordinary deck of 52 cards. Find:
a. $P(\text{ace})$ **b.** $P(\text{heart})$ **c.** $P(\text{face card})$

Solution   The sample space for a *deck of cards* is shown in Figure 12.2, page 584.

a. An ace is a card with one spot. $P(\text{ace}) = \dfrac{4}{52} = \dfrac{1}{13}$

b. $P(\text{heart}) = \dfrac{13}{52} = \dfrac{1}{4}$

c. $P(\text{face card}) = \dfrac{12}{52}$ ← A face card is a jack, queen, or king.
   ← Number of cards in the sample space

   $= \dfrac{3}{13}$

The following example uses two **dice** (plural form of **die,** which is a common randomizing device).

Example **9** **Find the probability of landing on a RR in Monopoly**

Suppose you are just beginning a game of Monopoly. (See "What in the World?" at the beginning of this chapter.) You roll a pair of dice. What is the probability that you land on a railroad on the first roll of the dice?

Solution   You need to know the locations of the railroads on a Monopoly board. There are four railroads, which are positioned so that only one can be reached on one roll of a pair of dice. The required number to roll is a 5. We begin by listing the sample space.

You might try $\{2, 3, 4, 5, 6, 7, 8, 9, 10\ 11, 12\}$, but these possible outcomes are not equally likely, which you can see by considering a tree diagram. The roll of the first die has 6 possibilities, and then *each* of these in turn can combine with any of 6 possibilities on the second die, for a total of 36 possibilities. The sample space is summarized in Figure 13.5.

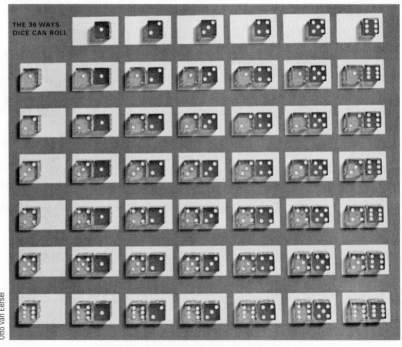

FIGURE 13.5 Sample space for a pair of dice

Thus, $n = 36$ for the definition of probability. We need to look at Figure 13.5 to see how many possibilities there are for obtaining a 5. We find (1, 4) (2, 3) (3, 2), and (4, 1), so $s = 4$. Then

$$P(\text{five}) = \frac{4}{36} \quad \leftarrow 4 \text{ ways to obtain a 5}$$
$$\phantom{P(\text{five})} \quad \leftarrow 36 \text{ ways to roll a pair of dice}$$
$$= \frac{1}{9}$$

You should use Figure 13.5 when working probability problems dealing with rolling a pair of dice. You will be asked to perform an experiment (Problem 49) in which you roll a pair of dice 100 times and then compare the results you obtain with the probabilities you calculate by using Figure 13.5.

When using the definition of probability, you must be careful not to overlook the restrictions on s and n. The requirement that the elements listed in the sample space are *mutually* exclusive simply means that each outcome of the experiment will be counted in only one category (that is easy). The importance of the second part of the restriction, that the n possible outcomes are *equally likely*, is evident in Example 9. If we roll a pair of dice and list the outcomes as 2, 3, 4, 5, 6, 7, 8, 9, 10, 11, and 12, we note these are *not equally likely* possibilities and do not lead to the correct theoretical probabilities. If we use the sample space in Figure 13.5 for a pair of dice, we note these *are equally likely possibilities*, so these 36 possibilities are used for n when rolling a pair of dice. The following example illustrates the importance of looking at the correct sample space.

Example 10   Probability of a winning by switching doors

A contest game show has a large prize hidden behind one of three doors. There is a small joke prize behind each of the other doors. You are asked to pick one of the three doors, and then the emcee opens a different door to reveal a small joke prize. You are then asked if you want to stick with your initial choice or switch to the remaining door. You win whatever is behind the door of your choice. If you switch, what is your probability of winning the big prize?

There is a similar but different problem, known as the "Monty Hall Dilemma." An interactive link for this variation can be found at **www.mathnature.com.**

Solution   It is an *error* to reason as follows: the probability of winning the large prize is 50% because there are two prizes left, one large and one small, so the sample space has two possibilities, one of which is success. If you reason this way, you are not considering equally likely possibilities in the sample space. (Problem 54 gives an empirical experiment to illustrate this fact.) Here is a correct arrangement of the sample space.

Player Choice	Emcee Choice
Switch and win	You picked a door hiding small prize #1. The emcee *must* select the other small prize.
Switch and win	You picked a door hiding small prize #2. The emcee *must* select the other small prize.
Switch and lose	You picked a door hiding the big prize. The emcee can select either of the other doors.

Thus, we see if you switch, $P(\text{big prize}) = \frac{2}{3}$ and if you don't switch, $P(\text{big prize}) = \frac{1}{3}$. Thus, the correct choice is to switch.

We know that a probability of $\frac{2}{3}$ is greater than a probability $\frac{1}{3}$. But we might seek to find the smallest and largest probabilities. The probability of the empty set is 0, which means that the event cannot occur. In the problem set, you will be asked to show that the probability of an event that *must* occur is 1. These are the two extremes. All other probabilities fall somewhere in between. The closer a probability is to 1, the more likely the event is to occur; the closer a probability is to 0, the less likely the event is to occur.

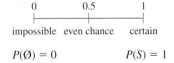

$P(\varnothing) = 0$ $P(S) = 1$

If the probability you are finding has a sample space that is easy to count, and if all simple events in the sample space are equally likely, then there is a simple procedure that we can use to find probabilities.

Finding Probabilities by Counting

If the experiment you are finding has a sample space that is easy to count, and if all simple events in the sample space are equally likely, then use the following steps to find the probability of an event of interest:

Step 1   Describe and identify the sample space, *S*. The number of elements in *S* is *n*.

Step 2   Count the number of occurrences that satisfy *E*, the event of interest (we call this success); denote it by *s*.

Step 3   Compute the probability of the event using the formula

$$P(E) = \frac{s}{n}$$

Pay attention to this procedure; it will suffice for much of what you will do in this chapter.

Keep in mind that this procedure does not apply to every situation. If it doesn't, you need a more complicated model, or else you must proceed experimentally.

Probabilities of Unions and Intersections

The word *or* is translated as $\cup$ (union), and the word *and* is translated as $\cap$ (intersection). We will find the probabilities of compound events involving the words *or* and *and* by finding unions and intersections of events.

Example 11 Comparing *or* and *and*

Suppose that a single card is selected from an ordinary deck of cards.
a. What is the probability that it is a two or a king? (See Figure 13.6a.)
b. What is the probability that it is a two or a heart? (See Figure 13.6b.)

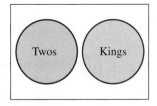 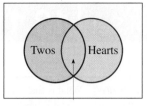

a. Mutually exclusive events   b. Not mutually exclusive

FIGURE 13.6 Finding probabilities

c. What is the probability that it is a two and a heart?

d. What is the probability that it is a two and a king?

Solution

a. P(two or a king) $= P$(two $\cup$ king). Look at Figure 12.2, p. 584. if you are not familiar with a deck of cards.

> two $= \{$two of hearts, two of spades, two of diamonds, two of clubs$\}$
>
> king $= \{$king of hearts, king of spades, king of diamonds, king of clubs$\}$
>
> two $\cup$ king $= \{$two of hearts, two of spades, two of diamonds, two of clubs, king of hearts, king of spades, king of diamonds, king of clubs$\}$

There are 8 possibilities for success. It is usually not necessary to list all of these possibilities to know that there are 8 possibilities (four twos and four kings):

$$P(\text{two} \cup \text{king}) = \frac{8}{52} = \frac{2}{13}$$

b. This seems to be very similar to part **a**, but there is one important difference. Look at the sample space and notice that although there are 4 twos and 13 hearts, the total number of successes is *not* $4 + 13 = 17$, *but rather* 16.

> two $= \{$**two of hearts,** two of spades, two of diamonds, two of clubs$\}$
>
> heart $= \{$ace of hearts, **two of hearts,** three of hearts, . . . , king of hearts$\}$
>
> two $\cup$ heart $= \{$**two of hearts,** two of spades, two of diamonds, two of clubs, ace of hearts, three of hearts, four of hearts, five of hearts, six of hearts, seven of hearts, eight of hearts, nine of hearts, ten of hearts, jack of hearts, queen of hearts, king of hearts$\}$.

It is not necessary to list these possibilities. The purpose of doing so in this case was to reinforce the fact that there are *actually* 16 (not 17) possibilities. In Chapter 2, we found the cardinality of a union, which we now use for |two $\cup$ hearts|:

$$|T \cup H| = |T| + |H| - |T \cap H|$$
$$= 4 + 13 - 1$$
$$= 16$$

Thus, we see

$$P(\text{two} \cup \text{heart}) = \frac{16}{52} = \frac{4}{13}$$

c. For this part, we seek two $\cap$ heart $= \{$two of hearts$\}$ so there is one element in the intersection:

$$P(\text{two} \cap \text{heart}) = \frac{1}{52}$$

d. Finally, two $\cap$ king $= \varnothing$, so there are no elements in the intersection:

$$P(\text{two} \cap \text{king}) = \frac{0}{52} = 0$$

Problem Set | 13.1

Level **1**

1. **IN YOUR OWN WORDS** What is the difference between empirical and theoretical probabilities?
2. **IN YOUR OWN WORDS** Define probability.
3. **IN YOUR OWN WORDS** Write a simple argument to convince someone why all probabilities must be between 0 and 1 (including 0 and 1).

For the spinners in Problems 4–7, assume that the pointer can never lie on a border line.

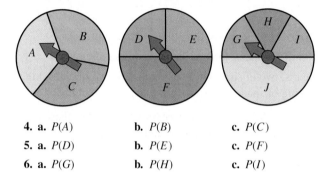

4. **a.** $P(A)$ **b.** $P(B)$ **c.** $P(C)$
5. **a.** $P(D)$ **b.** $P(E)$ **c.** $P(F)$
6. **a.** $P(G)$ **b.** $P(H)$ **c.** $P(I)$
7. **a.** $P(J)$ **b.** $P(G \text{ or } H)$ **c.** $P(I \text{ or } J)$

Give the probabilities in Problems 8–11 in decimal form (correct to two decimal places). A calculator may be helpful with these problems.

8. Last year, 1,485 calculators were returned to the manufacturer. If 85,000 were produced, assign a number to specify the probability that a particular calculator would be returned.
9. Last semester, a certain professor gave 13 A's out of 285 grades. If one of the 285 students is to be selected at random, what is the probability that his or her grade is an A?
10. Last year in Ferndale, California, it rained on 75 days. What is the probability of rain on a day selected at random?
11. The campus vets club is having a raffle and is selling 1,500 tickets. If the people on your floor of the dorm bought 285 of those tickets, what is the probability that someone on your floor will hold the winning ticket?
12. Consider the jar containing marbles shown in Figure 13.7. Suppose each marble has an equal chance of being picked from the jar.

FIGURE 13.7 A jar of marbles

Find: **a.** $P(\text{yellow})$ **b.** $P(\text{blue})$ **c.** $P(\text{green})$

Poker is a common game in which players are dealt five cards from a deck of cards. It can be shown that there are 2,598,960 different possible poker hands. The winning hands (from highest to lowest) are shown in Table 13.1. Find the requested probabilities in Problems 13–22. Use a calculator, and show your answers to whatever accuracy possible on your calculator.

TABLE 13.1

Poker Hands*

Royal flush 4 hands	
Other straight flush 36 hands	
Four of a kind 624 hands	
Full house 3,744 hands	
Flush 5,108 hands	
Straight 10,200 hands	
Three of a kind 54,912 hands	
Two pair 123,552 hands	
One pair 1,098,240 hands	
Other hands 1,302,540 hands	

13. $P(\text{royal flush})$
14. $P(\text{other straight flush})$
15. $P(\text{four of a kind})$
16. $P(\text{full house})$
17. $P(\text{flush})$
18. $P(\text{straight})$

*All of these probabilities are mutually exclusive. That is, the 36 straight flushes do not include the 4 royal flushes, and the 5,108 flush hands do not include the better hands of straight flush or royal flush.

19. P(three of a kind)   **20.** P(two pair)

21. P(one pair)   **22.** P(no pair or better)

40. A plays B **41.** A plays C **42.** B plays C

43. D plays E **44.** E plays F **45.** D plays F

46. B plays D **47.** F plays C

Level 2

A single card is selected from an ordinary deck of cards. The same space is shown in Figure 12.2. Find the probabilities in Problems 23–26.

23. a. P(five of clubs)   **b.** P(five)
  c. P(club)

24. a. P(jack)   **b.** P(spade)
  c. P(jack of spades)

25. a. P(five and a jack)   **b.** P(five or a jack)

26. a. P(heart and a jack)   **b.** P(heart or a jack)

Suppose that you toss a coin and roll a die in Problems 27–30. The sample space is shown in Figure 13.1.

27. What is the probability of obtaining:
  a. Tails *and* a five?   **b.** Tails *or* a five?
  c. Heads *and* a two?

28. What is the probability of obtaining:
  a. Tails?   **b.** Heads *or* a two?
  c. One, two, three, *or* four?

29. What is the probability of obtaining:
  a. Heads *and* an odd number?
  b. Heads *or* an odd number?

30. What is the probability of obtaining:
  a. Heads *and* a five?   **b.** Heads *or* a five?

Use the sample space shown in Figure 13.5 to find the probabilities in Problems 31–39 for the experiment of rolling a pair of dice.

31. P(five)   **32.** P(six)   **33.** P(seven)

34. P(eight)   **35.** P(nine)   **36.** P(two)

37. P(four *or* five)   **38.** P(even)   **39.** P(eight *or* ten)

Suppose you and an opponent each pick one of the spinners shown here. A "win" means spinning a higher number. Construct a sample space to answer each question, and tell which of the two spinners given in Problems 40–47 you would choose in each case.

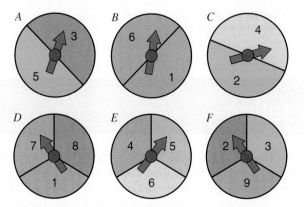

Level 3

Perform the experiments in Problems 48–51, tally your results, and calculate the probabilities (to the nearest hundredth).

48. Toss a coin 100 times.

Make sure that, each time the coin is flipped, it rotates several times in the air and lands on a table or on the floor. Keep a record of the results of this experiment. Based on your experiment, what is P(heads)?

49. Roll a pair of dice 100 times. Keep a record of the results. Based on your experiment, find:
  a. P(two)   **b.** P(three)   **c.** P(four)
  d. P(five)   **e.** P(six)   **f.** P(seven)
  g. P(eight)   **h.** P(nine)   **i.** P(ten)
  j. P(eleven)   **k.** P(twelve)

50. Flip three coins simultaneously 100 times, and note the results. The possible outcomes are:
  a. three heads
  b. two heads and one tail
  c. two tails and one head
  d. three tails
  Based on your experiment, find the probabilities of each of these events. Do these appear to be equally likely?

51. Simultaneously toss a coin and roll a die 100 times, and note the outcome of each trial. The possible outcomes are H1, H2, H3, H4, H5, H6, T1, T2, T3, T4, T5, and T6. Do these appear to be equally likely outcomes?

52. Suppose it is certain that an earthquake will occur someday. What is the probability (to the nearest percent) that it will occur while you are at work? Assume you are at work 8 hours per day, 240 days per year.

53. Suppose it is certain that an earthquake will occur someday. What is the probability (to the nearest percent) that it will occur while you are at school? Assume you are at school 5 hours per day, 174 days per year.

54. Prepare three cards that are identical except for the color. One card is black on both sides, one is white on both sides, and one is black on one side and white on the other.

Black on   Black on   White on
both sides   one side,   both sides
                  white on
                  the other

One side of one card is selected at random and placed flat on the table. You will see either a black or a white card; record the color of the face (the top). This is *not* the event with which we are concerned; rather, we are interested in finding the probability of the *other* side (the side you cannot see) being black or white. Record the underside, as shown in the following table. Repeat the experiment 50 times and find the probability of occurrence with respect to the known color. Do these appear to be equally likely outcomes?

Color of Face	Frequency	Outcome (Underside Color)	Probability
White		White	
		Black	
Black		White	
		Black	

55. Dice is a popular game in gambling casinos. Two dice are tossed, and various amounts are paid according to the outcome.

B.C. By permission of Johnny Hart and Creators Syndicate, Inc.

If a seven or eleven occurs on the first roll, the player wins.
a. What is the probability of winning on the first roll?
b. The player loses if the outcome of the first roll is a two, three, or twelve. What is the probability of losing on the first roll?

56. In dice, a pair of ones is called *snake eyes*. What is the probability of losing a dice game by rolling snake eyes?

57. Consider a die with only four sides, marked one, two, three, and four.
    a. Write out a sample space similar to the one in Figure 13.5 for rolling a pair of these dice.

Assuming equally likely outcomes, find the probability that the sum of the dice is the given number.
    b. P(two)   **c.** P(three)   **d.** P(four)
    e. P(five)   **f.** P(six)   **g.** P(seven)

58. The game of Dungeons and Dragons uses nonstandard dice. Consider a die with eight sides marked one, two, three, four, five, six, seven, and eight. Write a sample space similar to the one in Figure 13.5 for rolling a pair of these dice.

Problem Solving 3

59. A mad scientist has captured you and is showing you around his foul-smelling laboratory. He motions to an opaque, formalin-filled jar. "This jar contains one organ, either a kidney or a brain," he cheerily informs you. His voice seems an octave too high as he gives you a twisted leer. You watch as the madman grabs a brain lying on his worktable and drops it into the jar as well. He then shakes the jar and quickly withdraws a single organ. It proves to be a brain. He turns to you and says, "What is now the chance of removing another brain?" Fearing that the scientist might remove *your* brain in his next ghoulish experiment, you want to give him the right answer. What is your response?[*]

60. IN YOUR OWN WORDS This question is to test your intuition. *Read* the following exercises (do *not* carry out the experiment).
    A. Open a phone book to any page in the white pages. Select 100 consecutive phone numbers and tally the number of times each of the digits 0, 1, 2, 3, 4, 5, 6, 7, 8, 9 occurs as a first digit.
    B. Repeat A for the third digit.
    C. Repeat A for the last digit.
    a. Do you think the occurrence of each digit will be about the same for each of these experiments?
    b. Do you think the results of all three experiments will be about the same?
    c. *Now,* after answering parts **a** and **b**, carry out the experiments described in A, B, and C.

[*]Problem by Clifford A. Pickover from *Discover*®, published by Walt Disney Magazine Publishing Group Inc., March 1997, p. 94. Reprinted with permission.

13.2 | Mathematical Expectation

One of the best ways to understand the idea of probability is to associate it with a possible "payoff" if a certain event occurs. You and a friend flip a coin to see who pays for the coffee you buy each morning, or you purchase a lottery ticket hoping to wind the "big spin"; these are obvious examples of payoffs associated with certain events. However, purchasing an insurance policy and deciding which realtor to use to sell your property are also examples of this same idea.

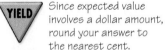

Expected Value

Choke-up toothpaste is giving away $10,000. All you must do to have a chance to win is send a postcard with your name on it (the fine print says you do not need to buy a tube of toothpaste). Is it worthwhile to enter?

Suppose the contest receives 1 million postcards (a conservative estimate). We wish to compute the **expected value** (or your **expectation**) of winning this contest. We find the expectation for this contest by multiplying the amount to win by the probability of winning:

$$\text{EXPECTATION} = (\text{AMOUNT TO WIN}) \times (\text{PROBABILITY OF WINNING})$$

$$= \$10,000 \times \frac{1}{1,000,000}$$

$$= \$0.01$$

What does this expected value mean? It means that if you were to play this "game" a large number of times, you would expect your *average winnings per game* to be $0.01.

> **Fair Game**
>
> A game is said to be **fair** if the expected value is 0. If the expected value is positive, then the game is in your favor; and if the expected value is negative, then the game is not in your favor.

Is the "Choke-up toothpaste giveaway" game fair? If the toothpaste company charges you 1¢ to play the game, then it is fair. But how much does the postcard cost? If you include this cost, then there is a negative expectation. We see that this is not a fair game.

Example 1 Find the expected value for drawing a card

Suppose that you draw a card from a deck of cards and are paid $10 if it is an ace. What is the expected value?

Solution $\text{EXPECTATION} = \$10 \times \frac{4}{52} \approx \$0.77$

YIELD ⚠ Since expected value involves a dollar amount, round your answer to the nearest cent.

Sometimes there is more than one payoff, and we define the expected value (or expectation) as the sum of the expected values from each separate payoff.

Example 2 Find the expectation for a contest

A recent contest offered one grand prize worth $10,000, two second prizes worth $5,000 each, and ten third prizes worth $1,000 each. What is the expected value if you assume that there are 1 million entries and you can win only one prize?

Solution

$$P(\text{1st prize}) = \tfrac{1}{1,000,000}; \quad P(\text{2nd prize}) = \tfrac{2}{1,000,000}; \quad P(\text{3rd prize}) = \tfrac{10}{1,000,000}$$

$$\text{EXPECTATION} = \underbrace{\$10,000}_{\text{amount of 1st prize}} \times \underbrace{\frac{1}{1,000,000}}_{P(\text{1st prize})} + \$5,000 \times \overbrace{\frac{2}{1,000,000}}^{\text{2nd prize}} + \$1,000 \times \underbrace{\frac{10}{1,000,000}}_{\text{3rd prize}}$$

$$= \$0.01 + \$0.01 + \$0.01$$

$$= \$0.03$$

> *Let the king prohibit gambling and betting in his kingdom, for these are vices that destroy the kingdoms of princes.*
> *The Code of Manu, ca.*
> *A.D. 100.*

You can use mathematical expectation to help you make a decision, as suggested by the following example.

Example 3 Select the best game to play

You are offered two games:

Game A: Two dice are rolled. You will be paid $3.60 if you roll two ones, and will not receive anything for any other outcome.

Game B: Two dice are rolled. You will be paid $36.00 if you roll any pair, but you must pay $3.60 for any other outcome.

Which game should you play?

Solution   You might say, "I'll play the first game because, if I play that game, I cannot lose anything." This strategy involves *minimizing your losses.* On the other hand, you can use a strategy that *maximizes your winnings.* In this book, we will base our decisions on maximizing the winnings. That is, we wish to select the game that provides the larger expectation.

Game A: EXPECTATION = $3.60 $\times \frac{1}{36}$ = $0.10

Game B: When calculating the expected value with a charge (a loss), write that charge as a negative number (a negative payoff is a loss).

$$\text{EXPECTATION} = \$36.00 \times \frac{6}{36} + (-\$3.60) \times \frac{30}{36}$$
$$= \$6.00 + (-\$3.00)$$
$$= \$3.00$$

This means that, if you were to play each game 100 times, you would expect your winnings for Game A to be 100 ($0.10) or about $10 and those from playing Game B to be 100($3.00) or about $300. You should choose to play Game B.

> **CAUTION** Study this formula to be sure you understand what it is saying.

Mathematical Expectation

If an event e has several possible outcomes with probabilities $p_1, p_2, p_3, \ldots, p_n$, and if for each of these outcomes the amount that can be won is $a_1, a_2, a_3, \ldots a_n$, respectively, then the **mathematical expectation** (or expected value) of E is

$$\text{EXPECTATION} = a_1 p_1 + a_2 p_2 + a_3 p_3 + \cdots + a_n p_n$$

Since we know that $p_1 + p_2 + \cdots + p_n = 1$, we note that in many examples some of the probabilities may be 0. For Example 1, we might have said that there are two different payoffs: $10 if you draw an ace $\left(\text{probability } \frac{4}{52}\right)$ and $0 otherwise $\left(\text{probability } 1 - \frac{4}{52} = \frac{48}{52}\right)$, so that

$$\text{EXPECTATION} = \$10\left(\frac{4}{52}\right) + \$0\left(\frac{48}{52}\right) \approx \$0.77$$

Example **4** **Find the expected value for a contest**

A contest offered the prizes shown here. What is the expected value for this contest?

Solution

We note the following values:

$a_1 = \$15,000; \quad p_1 = 0.000008$

$a_2 = \$1,000; \quad p_2 = 0.000016$

$a_3 = \$625; \quad p_3 = 0.000016$

$a_4 = \$525; \quad p_4 = 0.000016$

$a_5 = \$390; \quad p_5 = 0.000032$

$a_6 = \$250; \quad p_6 = 0.000032$

$ WIN • WIN • WIN $

PRIZE	VALUE	PROBABILITY
Grand Prize Trip	$15,000	0.000008
Samsonite Luggage	$1,000	0.000016
Magic Chef Range	$625.00	0.000016
Murray Bicycle	$525.00	0.000016
Lawn Boy Mower	$390.00	0.000032
Weber Kettle	$250.00	0.000032

$$\text{EXPECTATION} = \$15,000(0.000008) + \$1,000(0.000016) + \$625(0.000016)$$
$$+ \$525(0.000016) + \$390(0.000032) + \$250(0.000032)$$
$$\approx \$0.17$$

It is not necessary that the expectation or expected value be reported in terms of dollars and cents. Consider the following example.

Example **5** **Expected number of girls**

Suppose a family has three children. What is the expected number of girls?

Solution The sample space for this problem can be found using a tree diagram: {GGG, GGB, GBG, GBB, BGG, BGB, BBG, BBB}. All possible outcomes and their probabilities are listed in the following table:

Number of Girls	Probability	Product
0	$\frac{1}{8}$	0
1	$\frac{3}{8}$	$\frac{3}{8}$
2	$\frac{3}{8}$	$\frac{6}{8}$
3	$\frac{1}{8}$	$\frac{3}{8}$

$$\text{EXPECTED VALUE} = 0 + \tfrac{3}{8} + \tfrac{6}{8} + \tfrac{3}{8} = \tfrac{12}{8} = 1.5$$

Notice from Example 5 that the expected value may be a number that can never occur; it is obvious that 1.5 girls will never occur. An expected value of 1.5 simply means that if we record the numbers of girls in a large number of different three-child families, the *average* number of girls for all these families will be 1.5. We will discuss averages in the next chapter.

Expectation with a Cost of Playing

Many games charge you a fee to play. If you must pay to play, this cost of playing should be taken into consideration when you calculate the expected value. Remember, if the expected value is 0, it is a fair game; if the expected value is positive, you should play, but if it is negative, you should not.

Expectation with a Cost of Playing

If there is a cost of playing a game, the cost of playing must be subtracted.

EXPECTATION = (AMT. TO WIN)(PROB OF WINNING) − COST OF PLAYING

Example **6**   **Decide to play or not to play**

Consider a game that consists of drawing a card from a deck of cards. If it is a face card, you win $20. Should you play the game if it costs $5 to play?

Solution

$$\text{EXPECTATION} = \overbrace{\$20}^{\text{Winnings}} \overbrace{\left(\frac{12}{52}\right)}^{\text{Prob. of winning}} - \overbrace{\$5}^{\text{Amt. to play}}$$

$$\approx -\$0.38$$

You should not play this game, because it has a negative expectation.

Example **7**   **Decide to take a listing**   **Pólya's Method**

Eva, who is a realtor, knows that if she takes a listing to sell a house, it will cost her $1,000. However, if she sells the house, she will receive 6% of the selling price. If another realtor sells the house, Eva will receive 3% of the selling price. If the house remains unsold after 3 months, Eva will lose the listing and receive nothing. Suppose that the probabilities for selling a particular $200,000 house are as follows: The probability that Eva will sell the house is 0.4; the probability that another agent will sell the house is 0.2; and the probability that the house will remain unsold is 0.4. What is Eva's expectation if she takes this listing?

Solution   We use Pólya's problem-solving guidelines for this example.

Understand the Problem. What are the "payoffs"? If the house is sold by the listing agent, the payoff is 6% of the selling price. If the house is sold by another agent, the payoff is 3% of the selling price. Finally, if the house is not sold, there is no payoff.

$$6\% \text{ of } \$200,000 = 0.06(\$200,000) = \$12,000$$
$$3\% \text{ of } \$200,000 = 0.03(\$200,000) = \$6,000$$

Devise a Plan. First, we must decide whether this is an "entrance fee" problem. Is Eva required to pay the $1,000 before the "game" of selling the house is played? The answer is yes, so the $1,000 must be subtracted from the expectation.

Carry Out the Plan. We calculate the expectation.

$$\text{EXPECTATION} = \overbrace{(\$12,000)(0.4)}^{\text{Eva sells the house.}} + \overbrace{(\$6,000)(0.2)}^{\text{Another agent sells.}} - \overbrace{(\$1,000)}^{\text{Cost of playing}}$$

$$= \$5,000$$

Look Back. The expectation for the realtor, Eva, is $5,000.

Sometimes it is easier to subtract the cost of playing as you go. For example, you could calculate the expectation of Example 7 as follows:

$$\text{EXPECTATION} = \overbrace{(\$12{,}000 - \$1{,}000)(0.4)}^{\text{Eva sells the house.}} + \overbrace{(\$6{,}000 - \$1{,}000)(0.2)}^{\text{Another agent sells.}} + \overbrace{(-\$1{,}000)(0.04)}^{\text{House doesn't sell.}}$$

$$= \$11{,}000(0.4) + \$5{,}000(0.2) - \$1{,}000(0.4)$$

$$= \$4{,}400 + \$1{,}000 - \$400$$

$$= \$5{,}000$$

You must understand the nature of the game to know whether to subtract the cost of playing, as we did in Example 7. If you surrender your money to play, then it must be subtracted, but if you "leave it on the table," then you do not subtract it. An example of the latter is a U.S. roulette game, in which your bet is placed on the table but is not collected until after the play of the game and it is determined that you lost. A U.S. roulette wheel has 38 numbered slots (1–36, 0, and 00), as shown in Figure 13.8.

Here is how bets are placed on the roulette table

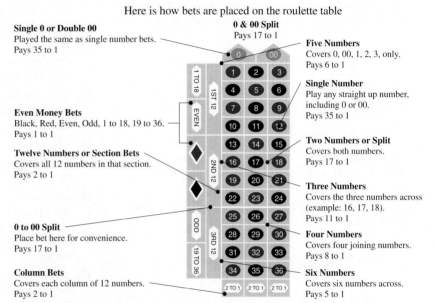

Single 0 or Double 00
Played the same as single number bets.
Pays 35 to 1

0 & 00 Split
Pays 17 to 1

Five Numbers
Covers 0, 00, 1, 2, 3, only.
Pays 6 to 1

Single Number
Play any straight up number, including 0 or 00.
Pays 35 to 1

Even Money Bets
Black, Red, Even, Odd, 1 to 18, 19 to 36.
Pays 1 to 1

Twelve Numbers or Section Bets
Covers all 12 numbers in that section.
Pays 2 to 1

Two Numbers or Split
Covers both numbers.
Pays 17 to 1

Three Numbers
Covers the three numbers across (example: 16, 17, 18).
Pays 11 to 1

0 to 00 Split
Place bet here for convenience.
Pays 17 to 1

Four Numbers
Covers four joining numbers.
Pays 8 to 1

Column Bets
Covers each column of 12 numbers.
Pays 2 to 1

Six Numbers
Covers six numbers across.
Pays 5 to 1

FIGURE 13.8 U.S. roulette wheel and board

Some of the more common roulette bets and payoffs are shown. If the payoff is listed as 6 to 1, you would receive $6 for each $1 bet. In addition, you would keep the $1 you originally wagered. One play consists of having the croupier spin the wheel and a little ball in opposite directions. As the ball slows to a stop, it lands in one of the 38 numbered slots, which are colored black, red, or green. A single number bet has a payoff of 35 to 1.

Example 8 Find a roulette expectation

What is the expectation for playing roulette if you bet $1 on number 5?

Solution   The $1 you bet is collected only if you lose. Now, you can calculate the expected value:

$$\text{EXPECTATION} = \overbrace{\$35}^{\text{Payoff for a win}} \underbrace{\left(\frac{1}{38}\right)}_{\text{Probability of winning}} + \overbrace{(-1)}^{\text{Payoff for a loss}} \underbrace{\left(\frac{37}{38}\right)}_{\text{Probability of losing}}$$

$$\approx -\$0.05$$

The expected loss is about 5¢ per play.

Problem Set | 13.2

Level 1

1. **IN YOUR OWN WORDS** True or false? In roulette, if you bet on black, the probability of winning is $\frac{1}{2}$ because there are equal numbers of black and red spots. Explain.

2. **IN YOUR OWN WORDS** True or false? An expected value of $5 means that you should expect to win $5 each time you play the game. Explain.

3. **IN YOUR OWN WORDS** True or false? If the expected value of a game is positive, then it is a game you should play. Explain.

4. **IN YOUR OWN WORDS** If you were asked to choose between a sure $10,000, or an 80% chance of winning $15,000 and a 20% chance of winning nothing, which would you take?

   Game A:

   $$E = \$10,000(1) = \$10,000$$

   This is a sure thing.

   Game B:

   $$E = \$15,000(0.8) + \$0(0.2) = \$12,000$$

   True or false? The better choice, according to expected value, is to take the 80% chance of winning $15,000. Explain.

5. **IN YOUR OWN WORDS** Suppose that you buy a lottery ticket for $1. The payoff is $50,000, with a probability of winning 1/1,000,000. Therefore, the expected value is

   $$E = \$50,000\left(\frac{1}{1,000,000}\right) \approx \$0.05$$

   Is this statement true or false? Explain.

Use estimation to select the best response in Problems 6–11. Do not calculate.

6. The expectation from playing a game in which you win $950 by correctly calling heads or tails when you flip a coin is about
   A. $500   B. $50   C. $950

7. The expectation from playing a game in which you win $950 by correctly calling heads or tails on each of five flips of a coin is about
   A. $500   B. $50   C. $950

8. If the expected value of playing a $1 game of blackjack is $0.04, then after playing the game 100 times you should have netted about
   A. $104   B. −$4   C. $4

9. If your expected value when playing a $1 game of roulette is −$0.05, then after playing the game 100 times you should have netted about
   A. −$105   B. −$5   C. $5

10. The probability of correctly guessing a telephone number is about
    A. 1 out of 100
    B. 1 out of 1,000
    C. 1 out of 1,000,000

11. Winning over $10 million in a super lottery is about as probable as
    A. having a car accident
    B. having an item fall out of the sky into your yard
    C. being a contestant on *Jeopardy*

12. Suppose that you roll two dice. You will be paid $5 if you roll a double. You will not receive anything for any other outcome. How much should you be willing to pay for the privilege of rolling the dice?

13. A magazine subscription service is having a contest in which the prize is $80,000. If the company receives 1 million entries, what is the expectation of the contest?

14. A box contains one each of $1, $5, $10, $20, and $100 bills. You reach in and withdraw one bill. What is the expected value?

15. A box contains one each of $1, $5, $10, $20, and $100 bills. It costs $20 to reach in and withdraw one bill. What is the expected value?

16. Suppose that you have 5 quarters, 5 dimes, 10 nickels, and 5 pennies in your pocket. You reach in and choose a coin at random so that you can tip your barber. What is the barber's expectation? What tip is the barber most likely to receive?

17. A game involves tossing two coins and receiving 50¢ if they are both heads. What is a fair price to pay for the privilege of playing?

18. Krinkles potato chips is having a "Lucky Seven Sweepstakes." The one grand prize is $70,000; 7 second prizes each pay $7,000; 77 third prizes each pay $700; and 777 fourth prizes each pay $70. What is the expectation of this contest, if there are 10 million entries?

19. A punch-out card contains 100 spaces. One space pays $100, five spaces pay $10, and the others pay nothing. How much should you pay to punch out one space?

What is the expectation for the $1 bets in Problems 20–29 on a U.S. roulette wheel? See Figure 13.8 on page 623.

20. Black   21. Odd
22. Single-number bet   23. Double-number bet
24. Three-number bet   25. Four-number bet
26. Five-number bet   27. Six-number bet
28. Twelve-number bet   29. Column bet

Level 2

Consider the spinners in Problems 30–33. Determine which represent fair games. Assume that the cost to spin the wheel once is $5.00 and that you will receive the amount shown on the spinner after it stops.

30.
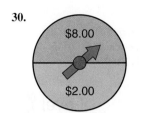

31.
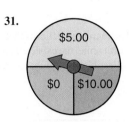

32.
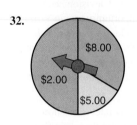

33.
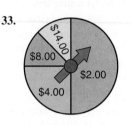

34. Assume that a dart is randomly thrown at the dartboard shown here and strikes the board every time. The payoffs are listed on the board. How much should you be willing to pay for the opportunity to play this game?

```
┌─────────────────┬───────────────────┐
│                 │      $6.00        │
│                 ├──────────┬────────┤
│     $1.00       │  $8.00   │        │
│                 ├──────────┤ $10.00 │
│                 │  $4.00   │        │
└─────────────────┴──────────┴────────┘
```

35. Assume that a dart is randomly thrown at the dartboard shown here and strikes the board every time. The payoffs are listed on the board. How much should you be willing to pay for the opportunity to play this game?

```
┌──────────────────┬───────────────┐
│      $16         │               │
│                  │     -$5       │
│   ┌────┬────┐    │               │
│$8 │ $2 │ $1 │    │               │
│   ├────┴────┤    │               │
│   │   $4    │    │               │
└───┴─────────┴────┴───────────────┘
```

36. In old gangster movies on TV, you often hear of "number runners" or the "numbers racket." This numbers game, which is still played today, involves betting $1 on the last three digits of the number of stocks sold on a particular day in the future as reported in *The Wall Street Journal*. If the payoff is $500, what is the expectation for this numbers game?

37. In a TV game show, four prizes are hidden on a game board that contains 20 spaces. One prize is worth $10,000, two prizes are worth $5,000, and the other prize is worth $1,000. The remaining spaces contain no prizes. The game show host offers a sure prize of $1,000 not to play this game. Should the contestant choose the sure prize or play the game?

38. Suppose it costs $2,400 to advertise and list a $350,000 house for sale. The listing agent will earn 5% of the selling price if the listing agent sells the property, but only 2.5% if the house is sold by another agent. If the house is unsold after 4 months, the listing (and the cost of advertising) will be lost. Suppose that the probabilities for selling the house are as follows:

Event	Probability
Sells the house alone	0.40
Sells through another agent	0.30
Does not sell in 4 months	0.30

What is the expected profit from listing on this house?

39. A realtor who takes the listing on a house to be sold knows that she will spend $800 trying to sell the house. If she sells it herself, she will earn 6% of the selling price. If another realtor sells a house from her list, the first realtor will earn only 3% of the price. If the house remains unsold after 6 months, she will lose the listing. Suppose that probabilities are as follows:

Event	Probability
Sells the house alone	0.50
Sells through another realtor	0.30
Does not sell in 6 months	0.20

What is the expected profit from listing a $185,000 house?

40. An oil-drilling company knows that it costs $25,000 to sink a test well. If oil is hit, the income for the drilling company will be $425,000. If only natural gas is hit, the income will be $125,000. If nothing is hit, there will be no income. If the probability of hitting oil is 1/40 and if the probability of hitting gas is 1/20, what is the expectation for the drilling company? Should the company sink the test well?

41. In Problem 40, suppose that the income for hitting oil is changed to $825,000 and the income for gas to $225,000. Now what is the expectation for the drilling company? Should the company sink the test well?

42. Consider the following game in which a player rolls a single die. If a prime (2, 3, or 5) is rolled, the player wins $2. If a square (1 or 4) is rolled, the player wins $1. However, if the player rolls a perfect number (6), it costs the player $11. Is this a good deal for the player or not?

43. A game involves drawing a single card from an ordinary deck. If an ace is drawn, you receive 50¢; if a face card is drawn, you receive 25¢; if the two of spades is drawn, you receive $1. If the cost of playing is 10¢, should you play?

44. A company held a contest, and the following information was included in the fine print:

Prize	Number of Prizes	Probability of Winning
$10,000	13	0.000005
$1,000	52	0.00002
$100	520	0.0002
$10	28,900	0.010886
TOTAL	29,485	0.011111

Read this information carefully, and calculate the expectation (to the nearest cent) for this contest.

45. A company held a bingo contest for which the following chances of winning were given:

Playing One Card, Your Chances of Winning Are at Least:

	1 *Time*	7 *Times*	13 *Times*
$25	1 in 21,252	1 in 3,036	1 in 1,635
$3	1 in 2,125	1 in 304	1 in 163
$1	1 in 886	1 in 127	1 in 68
Any prize	1 in 609	1 in 87	1 in 47

What is the expectation (to the nearest cent) from playing one card 13 times?

46. Heights (in inches) obtained by a group of people in a random survey are reported in the following table:

Heights	Probability
55	0.001
60	0.022
65	0.136
70	0.341
75	0.341
80	0.136
85	0.022
90	0.001

What is the expected height (in inches)?

47. In a certain school, the probabilities of the number of students who are reported tardy are shown in the following table:

Number tardy:	0	1	2	3	4
Probability:	0.15	0.25	0.31	0.21	0.08

What is the expected number of tardies (rounded to two decimal places)?

48. Calculate the expectation (to the nearest cent) for the *Reader's Digest* sweepstakes described. Assume there are 197,000,000 entries.

49. A merchant is considering a purchase of autographed sports memorabilia for resale to collectors. The probabilities of

various possible incomes from reselling these collectibles are estimated in the following table:

Income	Probability
$1,000	0.12
$800	0.38
$500	0.45
$200	0.05

The merchant believes that, in order to make this purchase worthwhile, the expected profit must be at least $200. Should the collectibles be purchased if the cost is $500?

Level 3

50. Consider a state lottery that has a weekly television show. On this show, a contestant receives the opportunity to win $1 million. The contestant picks from four hidden windows. Behind each is one of the following: $150,000, $200,000, $1 million, or a "stopper." Before beginning, the contestant is offered $100,000 to stop. Mathematically speaking, should the contestant take the $100,000?

51. Consider a state lottery that has a weekly television show. On this show, a contestant receives the opportunity to win $1 million. The contestant picks from four hidden windows. Behind each is one of the following: $150,000, $200,000, $1 million, or a "stopper." If the contestant picks the window containing $150,000 or $200,000, the contestant is asked whether he or she wishes to quit or continue. Suppose a certain contestant has picked both the $150,000 and $200,000 windows. Mathematically speaking, should the contestant pick "one more time"?

52. **IN YOUR OWN WORDS** Read Problem 51. Discuss what you would do. Consider this situation: Suppose you owned a $350,000 home free and clear; would you gamble it on a 50/50 chance to win $1,000,000? According to the mathematical expectation, what should you do? Discuss.

53. **IN YOUR OWN WORDS** Suppose you are in class and your instructor makes you the following legitimate offer. Each student receives a ballot with the following two choices, without communicating with each other:

☐ I share the wealth and will receive $1,000 if *everyone* in the class checks this box.

☐ I will not share the wealth and want a certain $100.

If *everyone* checks the first box, then all will receive $1,000. If *anyone* checks the second box, then only those who check the second box will receive $100. Which box would you check, and why?

54. **IN YOUR OWN WORDS** Repeat Problem 53, except change the stakes to $110 and $100, respectively.

55. **IN YOUR OWN WORDS** Repeat Problem 53, except change the stakes to $10,000 and $10, respectively.

56. **IN YOUR OWN WORDS** Repeat Problem 53, except change the stakes to be an A grade in the class if *everyone* checks the first box, but if anyone checks the second box, the following will occur: Those who check it will have a 50% chance of an A and a 50% chance of an F but those who do not check the second box will be given an F in this course.

Problem Solving 3

HISTORICAL QUEST *The Swiss mathematican Daniel Bernoulli (1700–1782) was part of the famous Bernoulli family (see Historical Note in Section 9.3). In 1738 he published the book* St. Petersburg Academy Proceedings *in which the problem known as the* **St. Petersburg Paradox** *was first published. To understand this paradox you must first work Problems 57–60 together. The paradox is seen after you answer the question in Problem 60.*

57. Suppose you toss a coin and will win $1 if it comes up heads. If it comes up tails, you toss again. This time you will receive $2 if it comes up heads. If it comes up tails, toss again. This time you will receive $4 if it is heads and nothing if it comes up tails. What is the mathematical expectation for this game?

58. Suppose you toss a coin and will win $1 if it comes up heads. If it comes up tails, you toss again. This time you will receive $2 if it comes up heads. If it comes up tails, toss again. This time you will receive $4 if it is heads. Continue in this fashion for a total of 10 flips of the coin, after which you receive nothing if it comes up tails. What is the mathematical expectation for this game?

59. Suppose you toss a coin and will win $1 if it comes up heads. If it comes up tails, you toss again. This time you will receive $2 if it comes up heads. If it comes up tails, toss again. This time you will receive $4 if it is heads. Continue in this fashion for a total of 1,000 flips of the coin, after which you receive nothing if it comes up tails. What is the mathematical expectation for this game?

60. IN YOUR OWN WORDS Suppose you toss a coin and will win $1 if it comes up heads. If it comes up tails, you toss again. This time you will receive $2 if it comes up heads. If it comes up tails, toss again. This time you will receive $4 if it is heads. You continue in this fashion until you finally toss a head. Would you pay $100 for the privilege of playing this game? What is the mathematical expectation for this game?

13.3 | Probability Models

Historical NOTE

Bettmann/CORBIS

Engraving by Darcis: *Le Trente-et-un.*

The engraving depicts gambling in 18th-century France. The mathematical theory of probability arose in France in the 17th century when a gambler, Chevalier de Méré, became interested in adjusting the stakes so that he could win more often than he lost. In 1654 he wrote to Blaise Pascal, who in turn sent his questions to Pierre de Fermat. Together they developed the first theory of probability.

Complementary Probabilities

In Section 13.1 we looked at the probability of an event E that consists of n mutually exclusive and equally likely outcomes where s of these outcomes are considered favorable. Now we wish to expand our discussion. Let

$s = $ NUMBER OF GOOD OUTCOMES (successes)

$f = $ NUMBER OF BAD OUTCOMES (failures)

$n = $ TOTAL NUMBER OF POSSIBLE OUTCOMES $(s + f = n)$

Then the probability that event E occurs is

$$P(E) = \frac{s}{n}$$

The probability that event E does not occur is

$$P(\overline{E}) = \frac{f}{n}$$

An important property is found by adding these probabilities:

$$P(E) + P(\overline{E}) = \frac{s}{n} + \frac{f}{n} = \frac{s+f}{n} = \frac{n}{n} = 1$$

 Don't underestimate the importance of this simple property, for it has far-reaching applications.

Property of Complements

$$P(E) = 1 - P(\overline{E}) \quad \text{or} \quad P(\overline{E}) = 1 - P(E)$$

The probabilities $P(E)$ and $P(\overline{E})$ are called **complementary probabilities.** In other words, the property of complements says that probabilities whose sum is 1 are complementary.

Example 1 Find the probability of a pair

Use Table 13.1 (page 616) to find the probability of not obtaining one pair with a poker hand.

Solution   From Table 13.1, we see that $P(\text{pair}) = \dfrac{1{,}098{,}240}{2{,}598{,}960} \approx 0.42$, so that

$$
\begin{aligned}
P(\text{no pair}) &= P(\overline{\text{pair}}) \\
&= 1 - P(\text{pair}) \\
&\approx 1 - 0.42 \\
&= 0.58
\end{aligned}
$$

Example 2 Find the probability of at least one head

What is the probability of obtaining at least one head in three flips of a coin?

Solution   Let $F = \{\text{obtain at least one head in three flips of a coin}\}$.

METHOD I: Work directly; use a tree diagram to find the possibilities.

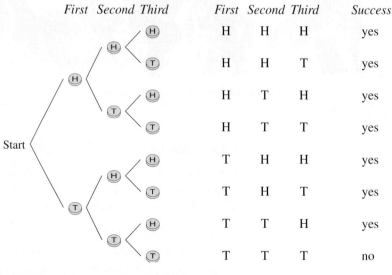

First	Second	Third	Success
H	H	H	yes
H	H	T	yes
H	T	H	yes
H	T	T	yes
T	H	H	yes
T	H	T	yes
T	T	H	yes
T	T	T	no

$$P(F) = \tfrac{7}{8}$$

Girolamo Cardano (1501–1576)

Cardano was a physician, mathematician, and astrologer. He investigated many interesting problems in probability and wrote a gambler's manual, probably because he was a compulsive gambler. He was the illegitimate son of a jurist and was at one time imprisoned for heresy—for publishing a controversial horoscope of Christ's life. He had a firm belief in astrology and predicted the date of his death. However, when the day came, he was alive and well—so he drank poison toward the end of the day to make his prediction come true!

METHOD II: For one coin, there are 2 outcomes (heads and tails); for three coins we have

$$2 \times 2 \times 2 = 8 \text{ possibilities}$$ *Fundamental counting principle*

We answer the question by finding the complement; $\overline{F}$ is the event of receiving no heads (that is, of obtaining all tails). *Without* drawing the tree diagram, we note that there is only one way of obtaining all tails (TTT). Thus,

$$P(F) = 1 - P(\overline{F}) = 1 - \tfrac{1}{8} = \tfrac{7}{8}$$

Odds

Related to probability is the notion of odds. Instead of forming ratios

$$P(E) = \frac{s}{n} \quad \text{and} \quad P(\overline{E}) = \frac{f}{n}$$

we form the following ratios:

Odds in favor of an event E: $\dfrac{s}{f}$ (ratio of success to failure)

Odds against an event E: $\dfrac{f}{s}$ (ratio of failure to success)

Recall: $s = $ NUMBER OF SUCCESSES
$f = $ NUMBER OF FAILURES
$n = $ NUMBER OF POSSIBILITIES

Example 3 Find odds against

If a jar has 2 quarters, 200 dimes, and 800 pennies, and a coin is to be chosen at random, what are the odds against picking a quarter?

Solution   We are interested in picking a quarter, so let $s = 2$; a failure is not obtaining a quarter, so $f = 1{,}000$. (Find f by adding 200 and 800.)

Odds in favor of obtaining a quarter: $\dfrac{s}{f} = \dfrac{2}{1{,}000} = \dfrac{1}{500}$

Odds against obtaining a quarter: $\dfrac{f}{s} = \dfrac{1{,}000}{2} = \dfrac{500}{1}$ *Do not write $\frac{500}{1}$ as 500.*

The odds against picking a quarter when a coin is selected at random are 500 to 1.

Example 4 Visualize odds

Pólya's Method

The odds against winning a lottery are 50,000,000 to 1. Make up an example to help visualize these odds.

Solution   We use Pólya's problem-solving guidelines for this example.

Understand the Problem. It is difficult for the human mind to comprehend numbers like 50 million to 1, so the point of this problem is to help us "visualize" the magnitude of that number.

Devise a Plan. We will fill a house with 50 million ping-pong balls, and then paint one of them red.

Carry Out the Plan. Imagine one red ping-pong ball and 50,000,000 white ping-pong balls. Winning a lottery is equivalent to reaching into a container containing these 50,000,001 ping-pong balls and obtaining the red one. To visualize this, consider the size container you would need. Assume that a ping-pong ball takes up a volume of 1 in.^{3}. A home of 1,200 ft^2 with an 8-ft ceiling has a volume of

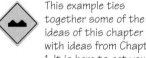

The odds are in on the Academy Awards

LAS VEGAS

The Odds Are Against You...

Your chances of becoming are only 1 out of the book

This example ties together some of the ideas of this chapter with ideas from Chapter 1. It is here to get you to think twice before purchasing a lottery ticket.

$$1,200 \text{ ft}^2 \times 8 \text{ ft} = 9,600 \text{ ft}^3$$
$$= 9,600 \times (12 \text{ in.} \times 12 \text{ in.} \times 12 \text{ in.})$$
$$= 16,588,800 \text{ in.}^3$$

Look Back. Even though the number 16,588,800 is still not directly comprehensible, you should be able to imagine a house filled with ping-pong balls. However, we can now say we would need *to fill* about 3 homes with ping-pong balls and imagine that one ball is painted red. We do not even know which house contains the red ball, but we can imagine picking one of the three houses at random, and then reaching into that one house and choosing one ball. If the red one is chosen, we win the lottery!

Sometimes you know the probability and want to find the odds, or you may know the odds and want to find the probability. These relationships are easy if you remember:

$$s + f = n$$

Do not confuse odds and probability.

Odds, Given Probability

Suppose that you *know* $P(E)$.

Odds in favor of an event E: $\dfrac{P(E)}{P(\overline{E})}$

Odds against an event E: $\dfrac{P(\overline{E})}{P(E)}$

You can show these formulas to be true:

$$\frac{P(E)}{P(\overline{E})} = \frac{\frac{s}{n}}{\frac{f}{n}} = \frac{s}{n} \cdot \frac{n}{f} = \frac{s}{f} = \text{odds in favor}$$

The verification of the second formula is left for the problem set.

Probability, Given Odds

Suppose that you know the odds in favor of an event E: s to f or the odds against an event E: f to s. Then

$$P(E) = \frac{s}{s + f} \quad \text{and} \quad P(\overline{E}) = \frac{f}{s + f}$$

Example   5   Find the odds in favor

If the probability of an event is 0.45, what are the odds in favor of the event?

Solution $P(E) = 0.45 = \frac{45}{100} = \frac{9}{20}$ This is given.
$$P(\overline{E}) = 1 - \frac{9}{20} = \frac{11}{20}$$

Then the odds in favor of E are

$$\frac{P(E)}{P(\overline{E})} = \frac{\frac{9}{20}}{\frac{11}{20}} = \frac{9}{20} \cdot \frac{20}{11} = \frac{9}{11}$$

The odds in favor are 9 to 11.

Example 6 **Find a probability given the odds against**

If the odds against you are 20 to 1, what is the probability of the event?

Solution   Odds against an event are f to s, so $f = 20$ and $s = 1$; thus,

$$P(E) = \frac{1}{20 + 1} = \frac{1}{21} \approx 0.048$$

Example 7 **Find a probability, given the odds in favor**

If the odds in favor of some event are 2 to 5, what is the probability of the event?

Solution   Odds in favor are s to f, so $s = 2$ and $f = 5$; thus,

$$P(E) = \frac{2}{2 + 5} = \frac{2}{7} \approx 0.286$$

Conditional Probability

Frequently, we wish to compute the probability of an event but we have additional information that will alter the sample space. For example, suppose that a family has two children. What is the probability that the family has two boys?

$$P(2 \text{ boys}) = \frac{1}{4} \qquad \text{Sample space: BB, BG, GB, GG; 1 success out of 4 possibilities}$$

Now, let's complicate the problem a little. Suppose that we know that the older child is a boy. We have altered the sample space as follows:

> *Original sample space:* BB, BG, GB, GG; but we need to cross out the last two possibilities because we *know* that the older child is a boy:

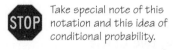

> *Altered sample space:*   BB, BG,   ~~GB~~, ~~GG~~;   therefore,

$$P(2 \text{ boys given the older is a boy}) = \frac{1}{2}$$

This is a problem involving a **conditional probability**—namely, a *probability of an event given that another event F has occurred*. We denote this by

Take special note of this notation and this idea of conditional probability.

$P(E|F)$ *Read this as: "probability of E given F."*

If the sample space consists of mutually exclusive and equally likely possibilities, then $P(E|F)$ represents the number of outcomes corresponding to $E \cap F$ over the number of outcomes corresponding to F. That is, in "E given F" we are taking F as "given" by putting only the outcomes in F into the denominator in the definition of probability. In the numerator, we have the outcomes of E that are also in F. This means that

$$P(E|F) = \frac{|E \cap F|}{|F|}$$

We call this the *conditional probability formula*. While this formula is useful in probability theory, we will generally find conditional probability by looking at altered sample spaces, as illustrated by the following example.

Example **8**   **Find a probability using an altered sample space**

Suppose that you toss two coins (or a single coin twice). What is the probability that two heads are obtained if you know that at least one head is obtained?

Solution   Consider an altered sample space: HH, HT, TH, ~~TT~~. The probability is $\frac{1}{3}$.

Example **9**   **Find conditional probabilities**   **Pólya's Method**

In a life science experiment, it is necessary to examine fruit flies and to determine their sex and whether they have mutated after exposure to a certain dose of radiation. Suppose a single fruit fly is selected at random from the radiated fruit flies. Find the following probabilities (correct to the nearest hundredth).
a. It is male.   **b.** It is a normal male.   **c.** It is normal, given that it is a male.
d. It is a male, given that it is normal.   **e.** It is mutated, given that it is a male.

Solution   We use Pólya's problem-solving guidelines for this example.

Understand the Problem. We obviously need more information to find the desired probabilities. We need to know the number of fruit flies that are part of the experiment. Assume that the total number is 1,000. We also need some data. Upon examination, we find that for 1,000 fruit flies examined, there were 643 females and 357 males. Also, 403 of the females were normal and 240 were mutated; of the males, 190 were normal and 167 were mutated.

Devise a Plan. We will display the data in table form to make it easy to calculate the desired probabilities.

Carry Out the Plan.

	Mutated	Normal	Total
Male	167	190	357
Female	240	403	643
Total	407	593	1,000

a. $P(\text{male}) = \frac{357}{1,000} \approx 0.36$ Round answers to the nearest hundredth.
b. $P(\text{normal male}) = \frac{190}{1,000} = 0.19$
c. $P(\text{normal} \mid \text{male}) = \frac{190}{357} \approx 0.53$ Note the altered sample space.
d. $P(\text{male} \mid \text{normal}) = \frac{190}{593} \approx 0.32$ Yet another altered sample space
e. $P(\text{mutated} \mid \text{male}) = \frac{167}{357} \approx 0.47$

Look Back. All probabilities seem reasonable, and all are between 0 and 1.

Example **10**   **Find the probability of drawing two cards**

CAUTION This is a crucial example. Spend some time working through all the parts.

Two cards are drawn from a deck of cards. Find the probability of the given events.
a. The first card drawn is a heart.
b. The first card drawn is not a heart.
c. The second card drawn is a heart if the first card drawn was a heart.
d. The second card drawn is not a heart if the first card drawn was a heart.
e. The second card drawn is a heart if the first card drawn was not a heart.
f. The second card drawn is not a heart if the first card drawn was not a heart.
g. The second card drawn is a heart.
h. Use a tree diagram to represent the indicated probabilities.

Solution   Let $H_1 = \{$first card drawn is a heart$\}$, $H_2 = \{$second card drawn is a heart$\}$.

a. $P(H_1) = \frac{13}{52} = \frac{1}{4}$ *There are 13 hearts in a deck of 52 cards.*

b. $P(\overline{H}_1) = \frac{39}{52} = \frac{3}{4}$ *There are 39 nonhearts in a deck of 52 cards.*

c. $P(H_2 | H_1) = \frac{12}{51} = \frac{4}{17}$ *If a heart is obtained on the first draw, then for the second draw there are 12 hearts in the deck of 51 remaining cards.*

d. $P(\overline{H}_2 | H_1) = \frac{39}{51} = \frac{13}{17}$ *If a heart is obtained on the first draw, then for the second draw there are 39 nonhearts in a deck of 51 remaining cards.*

e. $P(H_2 | \overline{H}_1) = \frac{13}{51}$ *If a heart is not obtained on the first draw, then for the second draw there are 13 hearts in a deck of 51 remaining cards.*

f. $P(\overline{H}_2 | \overline{H}_1) = \frac{38}{51}$ *If a nonheart is obtained on the first draw, then for the second draw there are 38 nonhearts in a deck of 51 cards.*

g. $P(H_2) = \frac{1}{4}$ *If we do not know what happened on the first draw, then this is not a conditional probability, so we consider this probability to be the same as $P(H_1)$.*

h. We use a tree diagram to illustrate this situation.

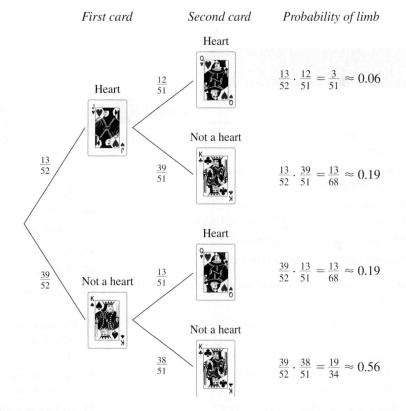

Note that we use the word *limb* to mean "a sequence of branches that starts at the beginning." Also note that we have filled in the probabilities on each branch. To find the probabilities, we multiply as we move horizontally across a limb, and we add as we move vertically from limb to limb. Verify that the probabilities for the first card add to 1:

$$P(H_1) + P(\overline{H}_1) = \frac{13}{52} + \frac{39}{52} = \frac{52}{52} = 1$$

Also, for the second card, the probabilities for each branch add to 1. For $P(H_2)$, we can confirm our reasoning in part **g** by considering the first and third limbs of the tree diagram:

$$P(H_2) = \frac{13}{52} \cdot \frac{12}{51} + \frac{39}{52} \cdot \frac{13}{51} = \frac{663}{2,652} = \frac{1}{4}$$

Notice that conditional probabilities are found in the tree diagram by beginning at their condition.

Verify that all of the information in parts **a–g** can be found directly from the tree diagram. For this reason, we often use tree diagrams to assist us in finding probabilities and conditional probabilities.

Procedure for Using Tree Diagrams

You will find using tree diagrams a very helpful procedure for finding probabilities.

CAUTION

To use a tree diagram to find an unconditional probability: Start at the beginning of the tree.

Step 1   Multiply when moving horizontally across a limb.

Step 2   Add when moving vertically from limb to limb.

To use a tree diagram to find a conditional probability.

Step 3   Start at the condition.

Problem Set | 13.3

Level 1

1. **IN YOUR OWN WORDS**   What is the fundamental counting principle?

2. **IN YOUR OWN WORDS**   Explain what we mean by "conditional probability."

3. How many different sequences are there for the first five moves in a game of tick-tack-toe? Give reasons for your answer.

4. What is the property of complements?

5. Compare and contrast odds and probability.

Use estimation to select the best response in Problems 6–11. Do not calculate.

6. Of these three means of travel, the safest is
   A. car   B. train   C. plane

7. Which of the following is a conditional probability?
   A. $P(\text{normal female})$ B. $P(\overline{A})$
   C. $P(\text{normal}|\text{female})$

8. Which of the following is more probable?
   A. Flipping a coin 3 times and obtaining at least 2 heads
   B. Flipping a coin 4 times and obtaining at least 2 heads

9. Which of the following is more probable?
   A. Rolling a die 3 times and obtaining a six 2 times
   B. Rolling a die 3 times and obtaining a six at least 2 times

10. Which of the following is more probable?
    A. Correctly guessing all the answers on a 20-question true-false examination
    B. Flipping a coin 20 times and obtaining all heads

11. Which of the following is more probable?
    A. Correctly guessing all the answers on a 10-question 5-part multiple-choice test

B. Your living room is filled with white ping pong balls. There is also one red ping-pong ball in the room. You reach in and select a ping-pong ball at random and select the red ping-pong ball.

Find the requested probabilities in Problems 12–15.

12. $P(\overline{A})$ if $P(A) = 0.6$

13. $P(\overline{B})$ if $P(B) = \frac{4}{5}$

14. $P(C)$ if $P(\overline{C}) = \frac{9}{13}$

15. $P(D)$ if $P(\overline{D}) = 0.005$

16. A card is selected from an ordinary deck. What is the probability that it is not a face card?

17. Choose a natural number between 1 and 100, inclusive. What is the probability that the number chosen is not a multiple of 5?

18. Three fair coins are tossed. What is the probability that at least one is a head?

19. Find the probability of obtaining at least one head in four flips of a coin.

20. What are the odds in favor of drawing an ace from an ordinary deck of cards?

21. What are a four-child family's odds against having four boys?

22. The probability of drawing a heart from a deck of cards is $\frac{1}{4}$; what are the odds in favor of drawing a heart?

23. Suppose the probability of an event is 0.81. What are the odds in favor of this event?

24. The odds against an event are ten to one. What is the probability of this event?

25. Racetracks quote the approximate odds for each race on a large display board called a *tote board*.

Here's what it might say for a particular race:

Horse Number	Odds
1	18 to 1
2	3 to 2
3	2 to 1
4	7 to 5
5	1 to 1

What would be the probability of winning for each of these horses? *Note:* The odds stated are for the horse's *losing*. Thus,

$$P(\text{horse 1 losing}) = \frac{18}{18+1} = \frac{18}{19}$$

so

$$P(\text{horse 1 winning}) = 1 - \frac{18}{19} = \frac{1}{19}.$$

26. Suppose the odds in favor are 9 to 1 that a man will be bald by the time he is 60. State this as a probability.

27. Suppose the odds are 33 to 1 that someone will lie to you at least once in the next seven days. State this as a probability.

28. Suppose that a family wants to have four children.
   a. What is the sample space?
   b. What is the probability of 4 girls? 4 boys?
   c. What is the probability of 1 girl and 3 boys? 1 boy and 3 girls?
   d. What is the probability of 2 boys and 2 girls?
   e. What is the sum of your answers in parts **b** through **d**?

29. What is the probability of obtaining exactly three heads in four flips of a coin, given that at least two are heads?

Level 2

30. The Emory Harrison family of Tennessee had 13 boys.

   a. What is the probability of a 13-child family having 13 boys?
   b. What is the probability that the next child of the Harrison family will be a boy (giving them 14 boys in a row)?

A single card is drawn from a standard deck of cards. Find the probabilities if the given information is known about the chosen card in Problems 31–36. A face card is a jack, queen, or king.

31. $P(\text{face card} \mid \text{jack})$ **32.** $P(\text{jack} \mid \text{face card})$

33. $P(\text{heart} \mid \text{not a spade})$ **34.** $P(\text{two} \mid \text{not a face card})$

35. $P(\text{black} \mid \text{jack})$ **36.** $P(\text{jack} \mid \text{black})$

Two cards are drawn from a standard deck of cards, and one of the two cards is noted and removed. Find the probabilities of the second card, given the information about the removed card provided in Problems 37–42.

37. $P(\text{ace} \mid \text{two})$ **38.** $P(\text{king} \mid \text{king})$

39. $P(\text{heart} \mid \text{heart})$ **40.** $P(\text{heart} \mid \text{spade})$

41. $P(\text{black} \mid \text{red})$ **42.** $P(\text{black} \mid \text{black})$

43. What is the probability of getting a license plate that has a repeated letter or digit if you live in a state in which license plates have two letters followed by four numerals?

44. What is the probability of getting a license plate that has a repeated letter or digit if you live in a state in which license plates have one numeral followed by three letters followed by three numerals?

45. What is the probability of getting a license plate that has a repeated letter or digit if you live in a state where the scheme is three numerals followed by three letters?

46. Consider this letter to Dear Abby.[*]

> **DEAR ABBY:** My husband and I just had our eighth child. Another girl, and I am really one disappointed woman. I suppose I should thank God she was healthy, but, Abby, this one was supposed to have been a boy. Even the doctor told me the law of averages was in our favor 100 to one. What is the probability of our next child being a boy?

What is the family's probability of having eight girls in a row? What are the odds against?

47. Consider the following table showing the results of a survey of TV network executives.

| Network | Opinion of Current Programming | | Total |
	Satisfied, S	Not Satisfied, $\bar{S}$	
NBC, N	18	7	25
CBS, C	21	9	30
ABC, A	15	10	25
Total	54	26	80

Suppose one network executive is selected at random. Find the indicated probabilities.

a. What is the probability that it is an NBC executive?

b. What is the probability that the selected person is satisfied?

c. What is the probability the selected person is from CBS if we know the person is satisfied with current programming?

d. What is the probability the selected person is satisfied if we know the person is from CBS?

48. Suppose a single die is rolled. Find the probabilities.

a. 6, given that an odd number was rolled

b. 5, given that an odd number was rolled

49. Suppose a single die is rolled. Find the probabilities.

a. odd, given that the rolled number was a 6

b. odd, given that the rolled number was a 5

Level 3

50. Suppose a pair of dice are rolled. Consider the sum of the numbers on the top of the dice and find the probabilities.

a. 7, given that the sum is odd

b. odd, given that a 7 was rolled

c. 7, given that at least one die came up 2

51. Suppose a pair of dice are rolled. Consider the sum of the numbers on the top of the dice and find the probabilities.

a. 5, given that exactly one die came up 2

b. 3, given that exactly one die came up 2

c. 2, given that exactly one die came up 2

52. Suppose a pair of dice are rolled. Consider the sum of the numbers on the top of the dice and find the probabilities.

a. 8, given that a double was rolled

b. a double, given that an 8 was rolled

53. Show that the odds against an event E can be found by computing $P(\overline{E})/P(E)$.

54. Two cards are drawn from a deck of cards. Find the requested probabilities.

a. The first card drawn is a club.

b. The first card drawn is not a club.

c. The second card drawn is a club if the first card drawn was a club.

d. The second card drawn is not a club if the first card drawn was a club.

e. The second card drawn is a club if the first card drawn was not a club.

f. The second card drawn is not a club if the first card drawn was not a club.

g. The second card drawn is a club.

h. Use a tree diagram to represent the indicated probabilities.

55. A sorority has 35 members, 25 of whom are full members and 10 are pledges. Two persons are selected at random from the membership list of the sorority. Find the requested probabilities.

a. The first person selected is a pledge.

b. The first person selected is not a pledge.

c. The second person selected is a pledge if the first person selected was also a pledge.

d. The second person selected was a full member if the first person selected was a pledge.

e. The second person selected is a pledge if the first person selected was a full member.

f. The second person selected is a full member if the first person selected was also a full member.

g. The second person selected was a pledge.

h. Use a tree diagram to represent the indicated probabilities.

56. **IN YOUR OWN WORDS**   The odds against winning a certain lottery are a million to one. Make up an example to help visualize these odds.

57. **IN YOUR OWN WORDS**   The odds against winning a certain lottery are ten million to one. Make up an example to help visualize these odds.

58. **IN YOUR OWN WORDS**   The odds against winning a certain "Power Ball" lotto are 150 million to 1. Make up an example to help visualize these odds.

Problem Solving 3

59. **IN YOUR OWN WORDS**   One betting system for roulette is to bet $1 on black. If black comes up on the first spin of the wheel, you win $1 and the game is over. If black does not come up, double your bet ($2). If you win, the game is over and your net winnings for two spins is still $1 (show this). If black does not come up, double your bet again ($4). If you win, the game is over and your net winnings for three spins is still $1 (show this). Continue in this doubling procedure until you eventually win; in every case your net winnings amount to $1 (show this). What is the fallacy with this betting system?

60. **IN YOUR OWN WORDS**   "Last week I won a free Big Mac at McDonald's. I sure was lucky!" exclaimed Charlie. "Do you go there often?" asked Pat. "Only twenty or thirty times a month. And the odds of winning a Big Mac were only 20 to 1." "Don't you mean 1 to 20?" queried Pat. "Don't confuse me with details. I don't even understand odds at the racetrack, and I go all the time." Is Charlie or Pat correct about the odds?

13.4   Calculated Probabilities

In this section, we formulate some probability problems using tree diagrams, combinations, permutations, and the fundamental counting principle. We will also consider some additional models that will allow us to break up more complicated probability problems into some simpler types.

Keno Games

A very popular lotto game in several states, as well as in most casinos, is a game called **Keno.** The game consists of a player's trying to guess in advance which numbers will be selected from a pot containing 80 numbers. The person's choices are marked on a *Keno card*, and those choices are registered before the game starts. Twenty numbers are then selected at random. The player may choose from 1 to 15 spots and gets paid according to Table 13.2 on below.

YOU CAN WIN $50,000

KENO

1	2	3	4	5	6	●	●	●	●
11	●	●	●	15	16	17	18	19	20
21	22	23	24	25	26	27	28	29	30
31	32	33	34	35	36	37	38	39	40
41	42	43	44	●	●	●	48	49	50
51	52	53	54	55	56	57	58	59	60
61	62	63	64	65	66	67	68	69	70
71	72	●	●	●	●	77	78	79	80

TABLE 13.2

Keno Payoffs

MARK 1 SPOT

Catch	Play $1.00	Play $3.00	Play $5.00
Win	3.00	9.00	15.00

MARK 2 SPOTS

Catch	Play $1.00	Play $3.00	Play $5.00
2 Win	12.00	36.00	60.00

MARK 3 SPOTS

Catch	Play $1.00	Play $3.00	Play $5.00
2 Win	1.00	3.00	5.00
3 Win	40.00	120.00	200.00

MARK 4 SPOTS

Catch	Play $1.00	Play $3.00	Play $5.00
2 Win	1.00	3.00	5.00
3 Win	3.00	9.00	15.00
4 Win	113.00	339.00	565.00

MARK 5 SPOTS

Catch	Play $1.00	Play $3.00	Play $5.00
3 Win	1.00	3.00	5.00
4 Win	10.00	30.00	50.00
5 Win	750.00	2,250.00	3,750.00

MARK 6 SPOTS

Catch	Play $1.00	Play $3.00	Play $5.00
3 Win	1.00	3.00	5.00
4 Win	3.00	9.00	15.00
5 Win	90.00	270.00	450.00
6 Win	1,480.00	4,400.00	7,400.00

MARK 7 SPOTS

Catch	Play $1.00	Play $3.00	Play $5.00
4 Win	1.00	3.00	5.00
5 Win	18.00	54.00	90.00
6 Win	400.00	1,200.00	2,000.00
7 Win	8,000.00	24,000.00	40,000.00

MARK 8 SPOTS

Catch	Play $1.00	Play $3.00	Play $5.00
5 Win	8.00	24.00	40.00
6 Win	100.00	300.00	500.00
7 Win	1,480.00	4,440.00	7,400.00
8 Win	17,000.00	50,000.00	50,000.00

MARK 9 SPOTS

Catch	Play $1.00	Play $3.00	Play $5.00
5 Win	3.00	9.00	15.00
6 Win	42.00	126.00	210.00
7 Win	350.00	1,050.00	1,750.00
8 Win	4,200.00	12,600.00	21,000.00
9 Win	18,000.00	50,000.00	50,000.00

MARK 10 SPOTS

Catch	Play $1.00	Play $3.00	Play $5.00
5 Win	2.00	6.00	10.00
6 Win	20.00	60.00	100.00
7 Win	130.00	390.00	650.00
8 Win	900.00	2,700.00	4,500.00
9 Win	4,500.00	13,500.00	22,500.00
10 Win	20,000.00	50,000.00	50,000.00

MARK 11 SPOTS

Catch	Play $1.00	Play $3.00	Play $5.00
6 Win	9.00	27.00	45.00
7 Win	75.00	225.00	375.00
8 Win	380.00	1,140.00	1,900.00
9 Win	2,000.00	6,000.00	10,000.00
10 Win	12,500.00	37,500.00	50,000.00
11 Win	21,000.00	50,000.00	50,000.00

MARK 12 SPOTS

Catch	Play $1.00	Play $3.00	Play $5.00
6 Win	5.00	15.00	25.00
7 Win	28.00	84.00	140.00
8 Win	200.00	600.00	1,000.00
9 Win	850.00	2,550.00	4,250.00
10 Win	2,400.00	7,200.00	12,000.00
11 Win	13,000.00	39,000.00	50,000.00
12 Win	25,000.00	50,000.00	50,000.00

MARK 13 SPOTS
($2.00 minimum)

Catch	Play $2.00	Play $3.00	Play $5.00
6 Win	6.00	9.00	15.00
7 Win	24.00	36.00	60.00
8 Win	150.00	225.00	375.00
9 Win	1,400.00	2,100.00	3,500.00
10 Win	4,000.00	6,000.00	10,000.00
11 Win	18,000.00	27,000.00	45,000.00
12 Win	28,000.00	42,000.00	50,000.00
13 Win	50,000.00	50,000.00	50,000.00

MARK 14 SPOTS
($2.00 minimum)

Catch	Play $2.00	Play $3.00	Play $5.00
6 Win	4.00	6.00	10.00
7 Win	16.00	24.00	40.00
8 Win	64.00	96.00	160.00
9 Win	600.00	900.00	1,500.00
10 Win	1,600.00	2,400.00	4,000.00
11 Win	5,000.00	7,500.00	12,500.00
12 Win	24,000.00	36,000.00	50,000.00
13 Win	36,000.00	50,000.00	50,000.00
14 Win	50,000.00	50,000.00	50,000.00

MARK 15 SPOTS
($2.00 minimum)

Catch	Play $2.00	Play $3.00	Play $5.00
6 Win	2.00	3.00	5.00
7 Win	14.00	21.00	35.00
8 Win	42.00	63.00	105.00
9 Win	200.00	300.00	500.00
10 Win	800.00	1,200.00	2,000.00
11 Win	4,000.00	6,000.00	10,000.00
12 Win	16,000.00	24,000.00	40,000.00
13 Win	24,000.00	36,000.00	50,000.00
14 Win	50,000.00	50,000.00	50,000.00
15 Win	50,000.00	50,000.00	50,000.00

Example   1   Decide if a game is fair

Suppose a person picks one number and is paid $3.00 if the number picked is among the 20 chosen for the game. The cost for playing this game (the cost is collected in advance) is $1.00. Is this a fair game?

Solution   In this problem we are picking 1 out of 80 and we will win if the one number we pick is among the 20 numbers chosen from the pot. Since the order in which the 20 numbers are picked from the pot does not matter, we see this is a combination.

It is common to work this example incorrectly. Note that this is WRONG:

$(\$3.00)\left(\frac{1}{4}\right) + (-\$1.00)\left(\frac{3}{4}\right) = 0$

or $\$3.00\left(\frac{1}{4}\right) = \$0.75.$

On the other hand,

$(\$2.00)\left(\frac{1}{4}\right) - \$1.00\left(\frac{3}{4}\right) = -\$0.25$

is acceptable.

$$P(\text{picking one number}) = \frac{\text{NUMBER OF WAYS OF CHOOSING 1 NO. FROM 20}}{\text{NUMBER OF WAYS OF CHOOSING 1 NO. FROM 80}}$$

$$= \frac{\binom{20}{1}}{\binom{80}{1}}$$

$$= \frac{20}{80}$$

$$= \frac{1}{4}$$

We now calculate the mathematical expectation:

$$\text{EXPECTATION} = \$3.00\left(\frac{1}{4}\right) - \$1.00 = -\$0.25$$

Subtract because you pay first.

No, it is not a fair game, since the expectation is negative.

The game of Keno becomes more complicated when picking more numbers.

Example 2   Find the mathematical expectation of a three-spot Keno

What is the mathematical expectation for playing a three-spot $1.00 Keno ticket?

Solution   There are four possibilities for a three-spot ticket: pick 0, 1, 2, or 3. We take these one at a time. "Pick 2" means picking *exactly* two numbers.

$$P(\text{pick 3}) = \frac{s}{n}$$

Where s is the number of ways of picking 3 numbers from the 20 winning numbers, and n is the number of ways of picking 3 numbers from the 80 possible numbers

$$= \frac{\binom{20}{3}}{\binom{80}{3}}$$

$$= \frac{20 \cdot 19 \cdot 18}{80 \cdot 79 \cdot 78}$$ *By calculator*

$$\approx 0.0138753651$$

$$P(\text{pick 2}) = \frac{s}{n}$$

Where s is the number of ways of picking 2 numbers from the 20 winning numbers, and n is the number of ways of picking 3 numbers from the 80 possible numbers

$$= \frac{\binom{20}{2} \cdot \binom{60}{1}}{\binom{80}{3}}$$

$$= \frac{20 \cdot 19 \cdot 60 \cdot 3}{80 \cdot 79 \cdot 78}$$ *By calculator*

$$\approx 0.1387536514$$

Similarly,

$$P(\text{pick 1}) = \frac{s}{n} \qquad\qquad P(\text{pick 0}) = \frac{s}{n}$$

$$= \frac{\binom{20}{1} \cdot \binom{60}{2}}{\binom{80}{3}} \qquad\qquad = \frac{\binom{60}{3}}{\binom{80}{3}}$$

$$\approx 0.4308666018 \qquad\qquad \approx 0.4165043817$$

We now find the expected value of one $1.00 play using the payoffs from Table 13.2 for a three-spot ticket:

EXPECTED VALUE = AMT. TO WIN · PROB. OF WINNING − AMT. TO PLAY

$$= \$40(\text{probability of picking 3}) + \$1(\text{probability of picking 2}) - \$1$$
$$= \$40(0.0138753651) + \$1(0.1387536514) - \$1$$
$$= -\$0.3062$$

An alternate method is to subtract the amount you pay from each payoff as you go.

As you can see, either method gives the same answer. The loss is about $0.31 for each play.

Independent Events

Consider the following problem dealing with four cards. Suppose four cards are taken from an ordinary deck of cards, and we form two stacks—one with an ace (one) and a deuce (two) and the other with an ace and a jack. If a card is drawn at random from each pile, what is the probability that a blackjack will occur (an ace and a jack)?

The tree diagram illustrates the possibilities. Notice that the probability of obtaining an ace from the first stack is 1/2 and the probability of obtaining a jack from the second stack is also 1/2.

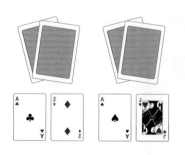

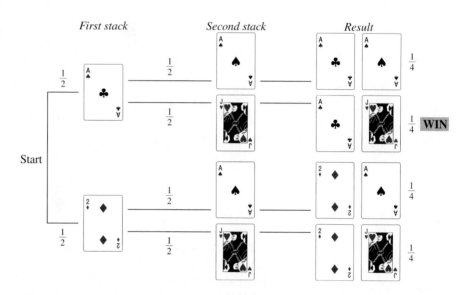

The probability of a blackjack is $\frac{1}{4}$. Notice *for this example* that

PROBABILITY OF ACE FROM FIRST STACK $= \dfrac{1}{2}$

PROBABILITY OF JACK FROM SECOND STACK $= \dfrac{1}{2}$

PROBABILITY OF BLACKJACK $= \dfrac{1}{2} \times \dfrac{1}{2} = \dfrac{1}{4}$

If one event (draw from the first stack) has no effect on the outcome of the second event (draw from the second stack), then we say that the events are **independent.** Thus, if events E and F are independent, $P(E|F) = P(E)$, since the occurrence of event F has no effect on the occurrence of E.

To understand how independence is determined, consider the following alternative problem. Suppose that all four cards are put together into one pile. This is an entirely different situation. There are four possibilities for the first draw, and three possibilities for the second draw, as illustrated here.

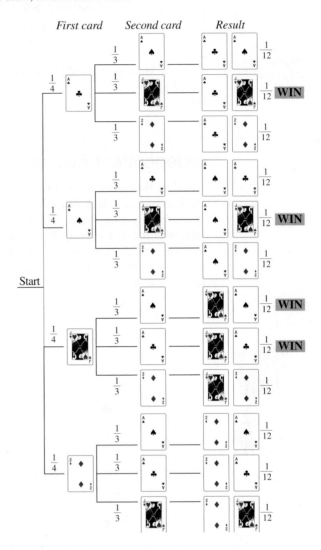

The probability of a blackjack is now $\frac{4}{12} = \frac{1}{3}$. We find this by looking at the sample space of equally likely possibilities. We *cannot* find this probability by multiplying the component parts for the first and second draws, as we did when considering independent stacks.

For this second situation we see that the first and second draws are **dependent** because the probabilities of selecting a particular second card are influenced by the draw on the first card.

Probability of an Intersection

If two events are independent, then we can find the probability of an intersection by multiplication. This property is called the **multiplication property of probability** or the *probability of an intersection.* If there are *n* mutually exclusive and equally likely possibilities, then by definition

$$P(E \cap F) = \frac{|E \cap F|}{n}$$

$$= \frac{P(E|F)|F|}{n} \qquad \text{By the conditional probability formula}$$

$$= P(E|F) \cdot \frac{|F|}{n}$$

$$= P(E|F) \cdot P(F) \qquad \text{By definition of probability}$$

$$= P(E) \cdot P(F) \qquad P(E|F) = P(E) \text{ since } E \text{ and } F \text{ are independent.}$$

This leads us to the following property.

Multiplication Property of Probability

If E and F are independent events, then

$$P(E \text{ and } F) = P(E \cap F)$$
$$= P(E|F) \cdot P(F)$$
$$= P(E) \cdot P(F)$$

STOP Note the conditions for this property: E and F must be independent.

Example 3 Find a roulette probability

What is the probability of black occurring on two successive plays on a U.S. roulette wheel?

Solution   Recall (Figure 13.8, page 623) that a U.S. roulette wheel has 38 compartments, 18 of which are black. Since the results of the spin of the wheel for one game are independent of the results of another spin, the plays are independent. Let $B_1 = \{\text{black on first spin}\}$ and $B_2 = \{\text{black on second spin}\}$. Then

$$P(B_1 \cap B_2) = P(B_1) \cdot P(B_2) = \frac{18}{38} \cdot \frac{18}{38} \approx 0.2243767313$$

Gambling casinos like to take advantage of this fallacy in the thinking of their clients. There used to be a pair of dice displayed on a red velvet pillow behind glass in the lobby of the Desert Inn in Las Vegas. This pair of dice made 28 straight passes (winning tosses) in 1950. The man rolling them won only $750; if he had bet $1 and let his winnings ride (double for each pass), he would have won $134,217,727.

The idea of independence is sometimes a difficult idea to communicate outside of the classroom. On July 8, 1995, in a preliminary hearing in the O. J. Simpson murder trial, an objection was raised that an adequate foundation had not been laid to establish the independence (and consequently the use of the multiplication property of probability) of blood-type factors. It was a perfect place for the mathematical principle of independence to be introduced, but instead the response focused on how many previous times the multiplication principle had been allowed by the court.

Another example involving a misuse of the multiplication principle involves gamblers. If you have ever been at a casino and watched players bet on black or red on a roulette game, you have seen that if one color comes up several times in a row, players start betting on the other color because they think it is "due." The spins of a roulette wheel are independent and it is a fallacy to think that previous spins have any effect on any future spins. For example, let's continue with Example 3 and ask, What is the probability of black on the *next* or third spin?

$$P(B_3) = \frac{18}{38} \approx 0.4736842105 \qquad \text{Where } B_3 = \{\text{black on third spin}\}$$

But if we ask, what is the probability of obtaining black on the *next* three consecutive spins, it is

$$P(B_1 \cap B_2 \cap B_3) = P(B_1) \cdot P(B_2) \cdot P(B_3) = \left(\frac{18}{38}\right)^3 \approx 0.1062837148$$

You might also note that the multiplication property can be used for two *or more* independent events.

Independence and the multiplication property are often used with the idea of complementary events. We now consider the mathematics associated with an experiment called the **birthday problem.**

See **www.mathnature** **.com** for links to an interactive site exploring this problem.

| Example 4 | **Birthday problem** | **Pólya's Method** |

What is the probability that at least 2 of 4 unrelated persons share the same birthday? This example is considered in more detail as a Group Research Project at the end of this chapter.

Solution We use Pólya's problem-solving guidelines for this example.

Understand the Problem. We assume that the birthdays of the individuals are independent, and we ignore leap years. We also are not considering the birthday year. We want to know, for example, whether any 2 (or more) of the 4 persons have the same birthday—say, August 3.

Devise a Plan. The first person's birthday can be any day (365 possibilities out of 365 days). We will find the probability that the second person's birthday is *different* from the first person's birthday. This is

$$1 - \frac{1}{365} = \frac{364}{365}$$

The probability that the third person's birthday is different from the first two birthdays is $\frac{363}{365}$, and the probability that the fourth person's birthday is different is $\frac{362}{365}$. Since these are independent events, we use the multiplication property of probability.

Carry Out the Plan.

$$P(\text{match}) = 1 - P(\text{no match})$$
$$= 1 - \frac{365}{365} \times \frac{364}{365} \times \frac{363}{365} \times \frac{362}{365}$$
$$\approx 0.0163559125$$

Look Back. There are about 2 chances out of 100 that there will be a birthday match with 4 persons in the group. It is worth noting that if we consider this for 23 persons, we have

$$1 - \frac{_{365}P_{23}}{365^{23}} \approx 0.5072972343$$

This means that if you look at a group of more than 23 persons, it is more likely than not that there will be a birthday match.

Probability of a Union

The multiplication property of probability is used to find the probability of an intersection (E and F); now let's turn to the *probability of a union* (E or F).

$$P(E \text{ or } F) = P(E \cup F) \qquad \text{Translate the word "or" as a union.}$$
$$= \frac{|E \cup F|}{n} \qquad \text{Definition of probability}$$
$$= \frac{|E| + |F| - |E \cap F|}{n} \qquad \text{Cardinality of a union}$$
$$= P(E) + P(F) - P(E \cap F) \qquad \text{Definition of probability}$$

We have derived a property called the **addition property of probability.**

Addition Property of Probability

For any events E and F,

$$P(E \text{ or } F) = P(E \cup F)$$
$$= P(E) + P(F) - P(E \cap F)$$

Example **5** **Find the probability of a union**

Suppose a coin is tossed and a die is simultaneously rolled. What is the probability of tossing a tail or rolling a 4? Is this event more or less likely than flipping a coin and obtaining a head?

Solution Let $T = \{\text{tail is tossed}\}$ and $F = \{4 \text{ is rolled}\}$.

$$
\begin{aligned}
P(T \cup F) &= P(T) + P(F) - P(T \cap F) &&\text{Addition property} \\
&= P(T) + P(F) - P(T) \cdot P(F) &&\text{Multiplication property, events are independent.} \\
&= \frac{1}{2} + \frac{1}{6} - \frac{1}{2} \cdot \frac{1}{6} &&\text{Calculate the probabilities.} \\
&= \frac{7}{12} &&\text{Simplify.}
\end{aligned}
$$

Since $\frac{7}{12} \approx 0.58$, we see that this event is more likely than flipping a coin and obtaining a head.

Example **6** **Find a calculated probability**

If A, B, and C are independent events so that $P(A) = 0.2$, $P(B) = 0.6$, and $P(C) = 0.3$, find $P[\overline{(A \cup B) \cap C}]$.

Solution We need to do some algebra before we evaluate.

$$
\begin{aligned}
P[\overline{(A \cup B) \cap C}] &= 1 - P[(A \cup B) \cap C] &&\text{Complementary probabilities} \\
&= 1 - P(A \cup B) \cdot P(C) &&\text{Multiplication property of probability} \\
&= 1 - [P(A) + P(B) - P(A \cap B)] \cdot P(C) &&\text{Addition property} \\
&= 1 - [P(A) + P(B) - P(A) \cdot P(B)] \cdot P(C) &&\text{Multiplication property} \\
&= 1 - P(A) \cdot P(C) - P(B) \cdot P(C) + P(A) \cdot P(B) \cdot P(C) \\
&&&\text{Distributive property (watch signs)}
\end{aligned}
$$

We now evaluate:

$$
\begin{aligned}
P[\overline{(A \cup B) \cap C}] &= 1 - (0.2)(0.3) - (0.6)(0.3) + (0.2)(0.6)(0.3) \\
&= 0.796
\end{aligned}
$$

Drawing With and Without Replacement

Take some time with this example; it illustrates the main ideas of this section.

To highlight the difference between a combination model and a permutation model, we focus on the idea of *replacement*. Drawing **with replacement** means choosing the first item, noting the result, and then replacing the item in the sample space before selecting the second item. Drawing **without replacement** means selecting the first item, noting the result, and then selecting a second item *without* replacing the first item.

Example **7** Find probabilities with and without replacement

Find the probability of each event when drawing two cards from an ordinary deck of cards.
a. Drawing a spade on the first draw and a heart on the second draw with replacement
b. Drawing a spade on the first draw or a heart on the second draw with replacement
c. Drawing two hearts with replacement
d. Drawing a spade on the first draw and a heart on the second draw without replacement
e. Drawing a spade on the first draw or a heart on the second draw without replacement
f. Drawing two hearts without replacement

Rank the events from the most probable to the least probable.

Solution We state the decimal approximations for each probability for ease of comparison. Let $S_1 = \{$draw a spade on the first draw$\}$ and $H_2 = \{$draw a heart on the second draw$\}$.

With replacement
a. With replacement, the events are independent. Thus

$$P(S_1 \cap H_2) = P(S_1) \cdot P(H_2) = \frac{1}{4} \cdot \frac{1}{4} = \frac{1}{16} = 0.0625$$

b. $P(S_1 \cup H_2) = P(S_1) + P(H_2) - P(S_1 \cap H_2)$

$$= \frac{1}{4} + \frac{1}{4} - \frac{1}{16} \qquad \text{From part } \mathbf{a}$$

$$= 0.4375$$

c. With replacement, the events are independent. Then

$$P(\text{two hearts}) = \frac{1}{4} \cdot \frac{1}{4} = \frac{1}{16} = 0.0625$$

Without replacement
d. We use the permutation model since the order is important. The number of possibilities is $_{52}P_2 = 52 \cdot 51$; the number of successes (fundamental counting principle) is $13 \cdot 13$. Thus,

$$P(S_1 \cap H_2) = \frac{13 \cdot 13}{_{52}P_2} = \frac{13 \cdot 13}{52 \cdot 51} = \frac{13}{204} \approx 0.0637254902$$

e. $P(S_1 \cup H_2) = P(S_1) + P(H_2) - P(S_1 \cap H_2)$

$$= \frac{1}{4} + \frac{1}{4} - \frac{13}{204} \qquad \text{From part } \mathbf{a}$$

$$\approx 0.4362745098$$

f. Without replacement, draw both cards at once. We use combinations since the order is not important.

$$P(\text{two hearts}) = \frac{\binom{13}{2}}{\binom{52}{2}} = \frac{\frac{13 \cdot 12}{2}}{\frac{52 \cdot 51}{2}} = \frac{1}{17} \approx 0.0588235294$$

The ranking is: (1) part **b**, (2) part **e** (probabilities of parts **b** and **e** are about the same), (3) part **d**, (4) parts **a** and **c** (tie), (5) part **f**. The probabilities of parts **a, c, d,** and **f** are about the same.

Tree Diagrams

CAUTION

Note the first three words of this subsection.

A powerful tool in handling probability problems is a device we have frequently used—a tree diagram. If the events are independent, we find the probabilities using the multiplication property of probability.

Example **8** **Deciding to play. Or not?**

Consider a game consisting of at most three cuts with a deck of cards. You win and the game is over if a face card turns up on any of the cuts, but you lose if a face card does not turn up. (A face card is a jack, queen, or king.) If you stand to win or lose $1 on this game, should you play?

Solution

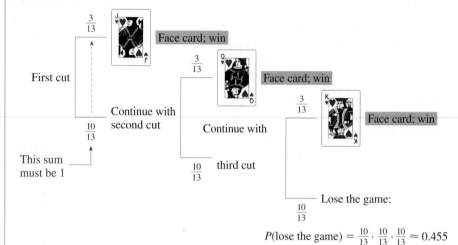

$P(\text{lose the game}) = \frac{10}{13} \cdot \frac{10}{13} \cdot \frac{10}{13} \approx 0.455$

$P(\text{win}) = 1 - P(\text{lose}) \approx 1 - 0.455 = 0.545$

EXPECTATION = AMOUNT TO WIN $\cdot$ $P(\text{win})$ + AMOUNT TO LOSE $\cdot$ $P(\text{lose})$

$\approx \$1 \cdot (0.545) + (-\$1) \cdot (0.455)$

$= \$0.09$

Since the mathematical expectation is positive, the recommendation is to play the game. If you played this game 1,000 times, you could expect to be ahead about 1,000($0.09) = $90.00

Guest Essay: Extrasensory Perception

Another presumed kind of extrasensory perception is the predictive dream. Everyone has an Aunt Matilda who had a vivid dream of a fiery car crash the night before Uncle Mortimer wrapped his Ford around a utility pole. I'm my own Aunt Matilda: When I was a kid I once dreamed of hitting a grand-slam home run and two days later I hit a bases-loaded triple. (Even believers in precognitive experiences don't expect an exact correspondence.) When one has such a dream and the predicted event happens, it's hard not to believe in precognition. But as the following derivation shows, such experiences are more rationally accounted for by coincidence.

Assume the probability to be one out of 10,000 that a particular dream matches a few vivid details of some sequence of events in real life. This is a pretty unlikely occurrence, and means that the chances of a nonpredictive dream are an overwhelming 9,999 out of 10,000. Also assume that whether or not a dream matches experience one day is independent of whether or not some other dream matches experience some other day. Thus, the probability of having two successive nonmatching dreams is, by the multiplication principle for probability, the product of 9,999/10,000 and 9,999/10,000. Likewise, the probability of having N straight nights of nonmatching dreams is $(9,999/10,000)^N$; for a year's worth of nonmatching or nonpredictive dreams, the probability is $(9,999/10,000)^{365}$.

Since $(9,999/10,000)^{365}$ is about .964, we can conclude that about 96.4 percent of the people who dream every night will have only nonmatching dreams during a one-year span. But that means that about 3.6 percent of the people who dream every night will have a predictive dream. 3.6 percent is not such a small fraction; it translates into millions of apparently precognitive dreams every year. Even if we change the probability to one in a million for such a predictive dream, we'll still get huge numbers of them by chance alone in a country the size of the United States. There's no need to invoke any special parapsychological abilities; the ordinariness of apparently predictive dreams does not need any explaining. What would need explaining would be the nonoccurrence of such dreams.

John Paulos, *Innumeracy*, pp. 54–55.

Problem Set 13.4

Level 1

1. **IN YOUR OWN WORDS** What do we mean by independent events?
2. **IN YOUR OWN WORDS** What is the birthday problem?
3. What is the formula for the probability of an intersection?
4. What is the formula for the probability of a union?

Suppose events A, B, and C are independent and

$$P(A) = \frac{1}{2} \qquad P(B) = \frac{1}{3} \qquad P(C) = \frac{1}{6}$$

Find the probabilities in Problems 5–22.

5. $P(\overline{A})$
6. $P(\overline{B})$
7. $P(\overline{C})$
8. $P(A \cap B)$
9. $P(A \cap C)$
10. $P(B \cap C)$
11. $P(A \cup B)$
12. $P(A \cup C)$
13. $P(B \cup C)$
14. $P(\overline{A \cap B})$
15. $P(\overline{A \cap C})$
16. $P(\overline{B \cap C})$
17. $P(\overline{A \cup B})$
18. $P(\overline{A \cup C})$
19. $P(\overline{B \cup C})$
20. $P(A \cap B \cap C)$
21. $P(\overline{A \cap B \cap C})$
22. $P[(A \cup B) \cap C]$

In Problems 23–34, suppose a die is rolled twice and let

$A = \{\text{first toss is a prime}\}$ $\quad B = \{\text{first toss is a 3}\}$
$C = \{\text{second toss is a 2}\}$ $\quad D = \{\text{second toss is a 3}\}$

Answer the questions or find the requested probabilities.

23. Are *A* and *B* independent?
24. Are *A* and *C* independent?
25. Are *A* and *D* independent?
26. Are *B* and *C* independent?
27. Are *B* and *D* independent?
28. Are *C* and *D* independent?
29. $P(A \cap B)$
30. $P(A \cap C)$
31. $P(A \cup B)$
32. $P(A \cup C)$
33. $P(B \cup D)$
34. $P(C \cup D)$

Level 2

35. A certain slot machine has three identical independent wheels, each with 13 symbols as follows: 1 bar, 2 lemons, 2 bells, 3 plums, 2 cherries, and 3 oranges. Suppose you spin the wheels and one of the 13 symbols on each wheel is selected. Find the probabilities and leave your answers in decimal form.
   a. P(3 bars)
   b. P(3 oranges)
   c. P(3 plums)
   d. P(cherry on the first wheel)
   e. P(cherries on the first 2 wheels)
36. Suppose a slot machine has three independent wheels as shown in Figure 13.9. Find the probabilities.

First Wheel	*Second Wheel*	*Third Wheel*
Orange	Cherry	Cherry
Cherry	Bar	Plum
Lemon	Plum	Cherry
Bell	Bar	Plum
Plum	Orange	Cherry
Orange	Bell	Cherry
Bar	Cherry	Plum
Plum	Bell	Bar
Orange	Cherry	Orange
Lemon	Plum	Bell
Orange	Orange	Cherry
Bar	Bell	Plum
Lemon	Cherry	Bar
Orange	Bell	Orange
Lemon	Bar	Plum
Plum	Bell	Cherry
Lemon	Bar	Orange
Orange	Bell	Cherry
Cherry	Cherry	Plum
Lemon	Lemon	Cherry

FIGURE 13.9 Slot machine wheels

   a. P(3 bars)
   b. P(3 bells)
   c. P(3 cherries)
   d. P(cherries on the first wheel)
   e. P(cherries on the first 2 wheels)
37. The payoffs for the slot machine shown in Figure 13.9 are as follows:

First one cherry	2 coins
First two cherries	5 coins
First two wheels are cherries and the third wheel a bar	10 coins
Three cherries	10 coins
Three oranges	14 coins
Three plums	14 coins
Three bells	20 coins
Three bars (jackpot)	50 coins

   What is the mathematical expectation for playing the game with the coin being a quarter dollar? Assume the three wheels are independent.
38. A "high rollers" Keno is offered in which you pay \$749 to play a one-spot ticket. If you catch your number, you are paid \$2,247. What is your expectation for this game?

39. A special "catch all" Keno ticket allows you to play a six-spot ticket that pays only if you pick all 6 numbers. It costs $5 to play and pays $27,777 if you win. What is your expectation for this game?

In Problems 40–42, use Table 13.2 for the payoffs.

40. What is the expectation for playing Keno by picking two numbers and paying $1.00 to play?

41. What is the expectation for playing a three-spot Keno and paying $3.00 to play?

42. What is the expectation for playing a four-spot Keno and paying $5.00 to play?

43. What is the probability that at least two of five unrelated persons share the same birthday?

44. What is the probability that at least two people in your class (assume a class of 30 students) have the same birthday?

45. What is the probability of obtaining five tails when a coin is flipped five times?

46. What is the probability of obtaining at least one tail when a coin is flipped five times?

47. Assume a jar has five red marbles and three black marbles. Draw out two marbles with and without replacement. Find the requested probabilities.
   a. P(two red marbles)
   b. P(two black marbles)
   c. P(one red and one black marble)
   d. P(red on the first draw and black on the second draw)

48. Suppose that in an assortment of 20 calculators there are 5 with defective switches. Draw with and without replacement.
   a. If one machine is selected at random, what is the probability it has a defective switch?
   b. If two machines are selected at random, what is the probability that both have defective switches?
   c. If three machines are selected at random, what is the probability that all three have defective switches?

49. a. A game consists of at most three cuts with a deck of 52 cards. You win $1 and the game is over if a heart turns up, but lose $1 otherwise. Should you play?
   b. Repeat this game, but remove the card that turns up on the cut.

50. a. A game consists of removing a card from a deck of 52 cards. If the card is a face card, you win $1 and the game is over. If it is not a face card, remove another card. If that one is a face card, you win $1. Repeat the process again, and then again (for a total of 4 times). If you have not removed any face card, then you lose the game, and must pay $1. Should you play this game?
   b. Repeat this game, but replace the card before doing the following draw.

Level 3

51. IN YOUR OWN WORDS Comment on extrasensory perception (ESP) as described in the guest essay by John Paulos on page 54.

52. IN YOUR OWN WORDS Compare and contrast the following two betting schemes for roulette, which are sometimes used by gamblers.

Double on each loss Bet $1 on black. If black comes up, you win $1, and the game is over. If black does not come up, double your bet ($2). If black now comes up, your net winnings are $1 (lose $1, then win $2), and the game is over. If black does not come up, double your bet ($4). If black now comes up, your net winnings are $1 (lose $1, lose $2, then win $4). Continue with the doubling procedure until you eventually win; in every case, your net winnings are $1.

Double on each win Bet $1 on black. If black does not come up, you lose $1, and the game is over. If black comes up, double your bet ($2), and play again. If black does not come up on the next spin, you lose $1 (win $1, then lose $2), and the game is over. If black comes up, double your bet ($4) and play again. If black does not come up on the next spin, you lose $1 ($+1 + 2 - 4$), and the game is over. If black comes up, double your bet ($8), and play again. If black does not come up on the next spin, you lose $1 ($+1 + 2 + 4 - 8$), and the game is over. If black does come up, double your bet ($16) one last time. If black does not come up on the next spin, you lose $1 ($+1 + 2 + 4 + 8 - 16$), and the game is over. If black does come up, the game is over and you win $31 ($+1 + 2 + 4 + 8 + 16$). Using this scheme, you are risking a maximum loss of $1 for a chance at winning $31.

Many states conduct lotteries; a typical lottery (Keno) payoff ticket is shown here. Assume there are 20 numbers chosen from a set of 80 possible numbers. Use this information for Problems 53–57.

KENO PAYOUTS

NUMBER OF SPOTS MATCHED	PLAY $1.00...WIN:

6 SPOT GAME

MATCH	PRIZE
6	$1,000
5	$25
4	$4
3	$1

Overall odds of winning a prize in this game — 1:6.2

5 SPOT GAME

MATCH	PRIZE
5	$250
4	$10
3	$2

Overall odds of winning a prize in this game — 1:10.3

4 SPOT GAME

MATCH	PRIZE
4	$50
3	$4
2	$1

Overall odds of winning a prize in this game — 1:3.9

3 SPOT GAME

MATCH	PRIZE
3	$20
2	$2

Overall odds of winning a prize in this game — 1:6.6

2 SPOT GAME

MATCH	PRIZE
2	$8

Overall odds of winning a prize in this game — 1:16.5

1 SPOT GAME

MATCH	PRIZE
1	$2

Overall odds of winning a prize in this game — 1:4.0

Source: The California Lottery

53. What is the expectation for a three-spot game?

54. What is the expectation for a four-spot game?

55. What is the expectation for a five-spot game?

56. What is the expectation for a six-spot game?

57. Which game illustrated on this Keno payoff ticket has the best expectation?

Problem Solving 3

58. We know that in a game of U.S. roulette, the probability that the ball drops into any one slot is 1 out of 38. Suppose that there are two balls spinning at once, and that it is physically possible for two balls to fit into the same slot. What is the probability that *both* balls would drop into the same slot?

59. Chevalier de Méré used to bet that he could get at least one 6 in four rolls of a die. He also bet that, in 24 tosses of a pair of dice, he would get at least one 12. He found that he won more often than he lost with the first bet, but not with the second. He did not know why, so he wrote to Pascal seeking the probabilities of these events. What are the probabilities for winning these two games?

60. In a book by John Fisher, *Never Give a Sucker an Even Break* (Pantheon Books, 1976), we find the following problem:

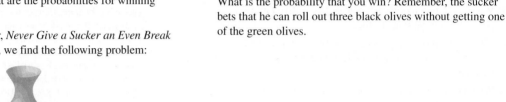

Take a small opaque bottle and seven olives, two of which are green, five black. The green ones are considered the "unlucky" ones. Place all seven olives in the bottle, the neck of which should be of such a size that it will allow only one olive to pass through at a time. Ask the sucker to shake them and then wager that he will not be able to roll out three olives without getting an unlucky green one amongst them. If a green olive shows, he loses.

What is the probability that you win? Remember, the sucker bets that he can roll out three black olives without getting one of the green olives.

13.5 | The Binomial Distribution

Many situations model events that have two possible outcomes. For example, when a coin is flipped, the outcome is either a head or a tail; when a child is born, it is a boy or it is a girl; the child is either born in December or it is not; when a magician guesses a card selected at random from a deck of cards, the magician is either correct or incorrect. In this section, we consider a topic that models this idea.

Random Variables

Managers at Mercedes Benz want to introduce a "zero defects" advertising campaign on a new line of automobiles, so they decide to do extensive testing of parts coming off the assembly line. Suppose 150 parts come off the assembly line every day, and inspectors keep track of the number of items with any type of defect. If X is the number of defective items, then X can obviously assume any of the values 0, 1, 2, 3, ..., 149, 150. This variable X is an example of a *random variable*. A **random variable** X associated with the sample space S of an experiment is an assignment of a real number to each simple event in S. If S has a finite number of outcomes, then X is called *a finite* (or *discrete*) *random variable;* if X can assume any real value on an interval, then it is called a *continuous random variable.*

Consider now a common type of experiment—one with only two outcomes, A and $\overline{A}$. Suppose that $P(A) = p$ (call this *success*) and $P(\overline{A}) = q = 1 - p$ (call this *failure*). We are interested in n independent repetitions of the experiment. If $P(A)$ remains the same for each repetition and we let X represent the number of times that event A has occurred, then we call X a **binomial random variable.**

Toss a coin four times and keep track of the results. The sample space is shown below.

Number of Heads	Outcomes				
4	HHHH				
3	HHHT	HHTH	HTHH	THHH	
2	HHTT	HTHT	HTTH	THTH	THHT	TTHH
1	TTTH	TTHT	THTT	HTTT	
0	TTTT				

Let the random variable X represent the number of heads that have occurred. That is, $X = 0$ if we obtain no heads; $X = 1$ means one head is obtained; $X = 2$ means two heads are obtained; $X = 3, X = 4$ mean three and four heads are obtained, respectively. Since there are 16 equally likely possibilities,

$$P(x = 4) = \frac{1}{16}$$

$$P(x = 3) = \frac{4}{16} = \frac{1}{4}$$

$$P(x = 2) = \frac{6}{16} = \frac{3}{8}$$

$$P(x = 1) = \frac{4}{16} = \frac{1}{4}$$

$$P(x = 0) = \frac{1}{16}$$

Note that the sum of these probabilities is one:

$$\frac{1}{16} + \frac{4}{16} + \frac{6}{16} + \frac{4}{16} + \frac{1}{16} = 1$$

This introduction illustrates a common probability model called the **binomial distribution,** which is a list of outcomes and probabilities for what we call a *binomial experiment.*

Binomial Experiments

Before you apply the results of this section, you need to determine whether the application is a binomial experiment. The four conditions to check are summarized in the following box.

> **Binomial Experiment**
>
> A **binomial experiment** is an experiment that satisfies four conditions:
> 1. There must be a fixed number of trials. Denote this number by n.
> 2. There must be exactly two possible mutually exclusive outcomes for each trial. Call them *success* and *failure.*
> 3. The trials must be independent. That is, the outcome of a particular trial is not affected by the outcome of any other trial.
> 4. The probabilities of success and failure must remain constant for each trial.

Consider the manufacture of computer chips. Suppose three items are chosen at random from a day's production and are classified as defective (F) or nondefective (S). We are interested in the number of successes obtained. Suppose that an item has a probability of 0.1 of being defective and therefore a probability of 0.9 of being nondefective. We will assume that these probabilities remain the same throughout the experiment and that the classification of any particular item is independent of the classification of any other item. The sample space and probabilities are shown in the margin.

If we let X = the number of successes obtained, then we can calculate the probabilities using the information from the margin:

$$P(X = 0) = (0.1)^3 \qquad \text{Line 1 of the sample space}$$

$$P(X = 1) = 3(0.9)(0.1)^2 \qquad \text{This is found from lines 2, 3, and 4.}$$
$$\text{FFS} \quad (0.9)(0.1)^2$$
$$\text{FSF} \quad (0.9)(0.1)^2$$
$$\text{SFF} \quad (0.9)(0.1)^2$$

$$P(X = 2) = 3(0.9)^2(0.1) \qquad \text{This is found from lines 5, 6, and 7.}$$
$$\text{FSS} \quad (0.9)^2(0.1)$$
$$\text{SFS} \quad (0.9)^2(0.1)$$
$$\text{SSF} \quad (0.9)^2(0.1)$$

$$P(X = 3) = (0.9)^3 \qquad \text{Line 8 of the sample space}$$

Sample space	Associated probabilities
1. FFF	$(0.1)(0.1)(0.1) = (0.1)^3$
2. FFS	$(0.1)(0.1)(0.9) = (0.9)(0.1)^2$
3. FSF	$(0.1)(0.9)(0.1) = (0.9)(0.1)^2$
4. SFF	$(0.9)(0.1)(0.1) = (0.9)(0.1)^2$
5. FSS	$(0.1)(0.9)(0.9) = (0.9)^2(0.1)$
6. SFS	$(0.9)(0.1)(0.9) = (0.9)^2(0.1)$
7. SSF	$(0.9)(0.9)(0.1) = (0.9)^2(0.1)$
8. SSS	$(0.9)(0.9)(0.9) = (0.9)^3$

Note that this same result can be achieved by simply considering the following binomial expansion:*

$$(0.1 + 0.9)^3 = (0.1)^3 + 3(0.1)^2(0.9) + 3(0.1)(0.9)^2 + (0.9)^3$$

This leads us to the **binomial distribution theorem.**

Binomial Distribution Theorem

Let X be a random variable for the number of successes in n independent and identical repetitions of an experiment with two possible outcomes, success and failure. If p is the probability of success, then

$$P(X = k) = \binom{n}{k} p^k (1 - p)^{n-k}$$

where $k = 0, 1, \ldots, n$.

Example 1 Distinguishing binomial probabilities

In Problem 30 of Section 13.3 we asked for the probability that a family will have 13 boys. The Grady family also has 13 children, 10 girls and 3 boys.
a. What is the probability of a 10-child family having 10 girls?
b. What is the probability of a 3-child family having 3 boys?
c. What is the probability (rounded to the nearest thousandth) of a 13-child family having 10 girls and 3 boys?

Solution
a. $P(10 \text{ girls}) = \left(\frac{1}{2}\right)^{10} = \frac{1}{1,024}$ **b.** $P(3 \text{ boys}) = \left(\frac{1}{2}\right)^3 = \frac{1}{8}$
c. This is a binomial distribution because there are a fixed number of trials ($n = 13$).
There are two mutually exclusive outcomes (girl/boy).
The trials are independent.
The probabilities of a girl/boy remain the same for each trial.
Thus, let X represent the number of girls in a 13-child family.

---

*You might wish to review the binomial theorem in Section 6.1.

$$P(X = 10) = \binom{13}{10}\left(\frac{1}{2}\right)^{10}\left(\frac{1}{2}\right)^{3}$$

$$= \frac{13!}{10!3!} \cdot \frac{1}{2^{13}}$$

$$\approx 0.035$$

Example 2   Finding binomial probabilities

Suppose a sociology teacher always gives true-false tests with 10 questions.
a. What is the probability of getting exactly 70% correct by guessing?
b. What is the probability of getting 70% or more correct by guessing?
c. If a student can be sure of getting five questions correct, but must guess at the others, what is the probability of getting 70% or more correct?

Solution   For this problem $P(\text{success}) = \frac{1}{2}$ and $P(\text{failure}) = \frac{1}{2}$.

a. $P(X = 7) = \binom{10}{7}\left(\frac{1}{2}\right)^{7}\left(\frac{1}{2}\right)^{3}$

$$= \frac{10!}{7!3!} \cdot \frac{1}{2^{10}}$$

$$= \frac{120}{1,024} \approx 0.117$$

b. $P(X = 8) = \binom{10}{8}\left(\frac{1}{2}\right)^{8}\left(\frac{1}{2}\right)^{2}$ $P(X = 9) = \binom{10}{9}\left(\frac{1}{2}\right)^{9}\left(\frac{1}{2}\right)$

$$= \frac{10!}{8!2!} \cdot \frac{1}{2^{10}} \qquad\qquad\qquad = \frac{10!}{9!1!} \cdot \frac{1}{2^{10}}$$

$$= \frac{45}{1,024} \approx 0.044 \qquad\qquad = \frac{10}{1,024} \approx 0.098$$

$P(X = 10) = \binom{10}{10}\left(\frac{1}{2}\right)^{10}$ Thus, $P(X \geq 7) = \frac{120}{1,024} + \frac{45}{1,024} + \frac{10}{1,024} + \frac{1}{1,024}$

$$= 1 \cdot \frac{1}{2^{10}} \qquad\qquad\qquad\qquad\qquad = \frac{176}{1,024}$$

$$= \frac{1}{1,024} \approx 0.001 \qquad\qquad\qquad = \frac{11}{64} \approx 0.172$$

c. $P(2 \leq X \leq 5) = \binom{5}{2}\left(\frac{1}{2}\right)^{5} + \binom{5}{3}\left(\frac{1}{2}\right)^{5} + \binom{5}{4}\left(\frac{1}{2}\right)^{5} + \binom{5}{5}\left(\frac{1}{2}\right)^{5}$

$$= 10 \cdot \frac{1}{32} + 10 \cdot \frac{1}{32} + 5 \cdot \frac{1}{32} + 1 \cdot \frac{1}{32}$$

$$= \frac{13}{16} = 0.8125$$

Binomial experiments are also sometimes called **Bernoulli trials,** after Jacob Bernoulli (1654–1705).

Historical NOTE

Karl Smith library

Jacob Bernoulli (1654–1705)

Jacob Bernoulli is part of the famous Bernoulli family (see Historical Note in Section 9.3). He was a mathematics professor at the University of Basel from 1683 until his death, at which time his brother Johann was chosen to fill his chair. He and his brother were bitter rivals and he stated (in print) that Johann only repeated what his brother (Jacob) had taught him. His biographer wrote about Jacob: "He was self-willed, obstinate, aggressive, vindictive, beset by feelings of inferiority, and yet convinced of his own abilities. With these characteristics, he necessarily had to collide with his similarly disposed brother." His achievements included his work in probability (after which Bernoulli trials are named), logic, algebra, geometry, and calculus (after which the Bernoulli equation is named).

Example 3 Find the probability of Bernoulli trials

If the probabilities for the two outcomes of n Bernoulli trials are p and $q = 1 - p$, and S denotes success and F denotes failure, find a formula for the probability of k successes.

Solution   Sometimes Bernoulli trials are cast in terms of randomly drawing balls from an urn filled with r red balls and b black balls. The proportion of

$$\text{red balls is } p = \frac{r}{r + b} \quad \text{and} \quad \text{black balls is } q = \frac{b}{r + b}$$

If each ball is replaced after drawing (and the urn is stirred up to make sure the next ball is drawn at random), then p and q will remain constant for each drawing. If the balls are not replaced, then p and q will change after each drawing. For example, if balls are drawn 8 times with replacement and the sequence is

　　RBRBBBBB

the probability of this sequence is

$$pqpqqqqq = p^2 q^6$$

On the other hand, if the order is not important and we want the probability of obtaining 2 red balls and 6 black balls, then each sequence such as

　　RBRBBBBB　　or　　RBBBBBBR or . . .

occurs with probability $p^2 q^6$. Further, the number of sequences in which 2 red balls and 6 black balls are drawn is given as the coefficient of $p^2 q^6$ in the binomial expansion of $p + q$. This number is

$$\binom{8}{2} \quad \text{so that the probability is} \quad \binom{8}{2} p^2 q^6$$

The probability of k successes in n trials of a Bernoulli experiment is

$$P(k \text{ successes}) = \binom{n}{k} p^k q^{n-k}$$

for $k = 0, 1, 2, \ldots, n$.

Example 4 Find the probability of hitting a target

A missile has a probability of 0.1 of penetrating enemy defenses and reaching its target. If five missiles are aimed at the same target, what is the probability that exactly one will hit its target? What is the probability that at least one will hit its target?

Solution $P(X = 1) = \binom{5}{1}\left(\frac{1}{10}\right)^1 \left(\frac{9}{10}\right)^4$

$$= 0.32805 \qquad \text{\small By calculator}$$

$$P(X \geq 1) = 1 - P(X = 0) \qquad \text{\small Complementary probabilities}$$

$$= 1 - \binom{5}{0}\left(\frac{1}{10}\right)^0 \left(\frac{9}{10}\right)^5$$

$$= 0.40951$$

The probability that one missile will hit its target is 0.33 (or about 1/3 of time), and the probability that at least one will hit its target is 0.41 (or about 40% of the time).

Example 5 **Find the probability of errors**

Suppose the probability that a page of a term paper is properly prepared is 0.25. What is the probability that there will be exactly one page with at least one error in a term paper of five pages?

Solution   The probability of at least one error on a page is $1 - 0.25 = 0.75$. The solution is given by

$$P(X = 1) = \binom{5}{1}(0.75)^1(0.25)^4$$

$$= 0.0146484375 \qquad \text{By calculator}$$

The probability that there is exactly one page with at least one error is about 15%.

Problem Set 13.5

Level 1

1. IN YOUR OWN WORDS   What is a binomial random variable?

2. IN YOUR OWN WORDS   What is a Bernoulli trial?

Find the binomial probabilities in Problems 3–10.

3. $n = 5, X = 3, p = 0.30$ **4.** $n = 4, X = 3, p = 0.25$

5. $n = 12, X = 6, p = 0.65$ **6.** $n = 10, X = 4, p = 0.80$

7. $n = 6, X = 6, p = 0.50$ **8.** $n = 8, X = 8, p = 0.75$

9. $n = 7, X = 5, p = 0.10$ **10.** $n = 15, X = 13, p = 0.40$

11. Find the probability of obtaining exactly three heads on five tosses of a fair coin.

12. Find the probability of obtaining exactly four heads on five tosses of a fair coin.

13. Find the probability that a family of six children has 4 boys and 2 girls.

14. Find the probability that a family of eight children has 5 boys and 3 girls.

15. Find the probability of obtaining exactly two threes on five rolls of a fair die.

16. Find the probability of obtaining exactly three twos on five rolls of a fair die.

17. A couple plans to have four children. Find the probability that they will have more than two girls.

18. A couple plans to have five children. Find the probability that they will have more than three girls.

Suppose you are taking a true-false test with ten questions. If you guess at the answers on this test, find the probabilities in Problems 19–22.

19. Exactly eight correct answers

20. Exactly nine correct answers

21. Fewer than two correct answers

22. More than eight correct answers

Suppose a jar contains 3 pens: 1 red, 1 blue, and 1 green. Three people sign a document, one at a time. Each person selects a pen, signs the document, and then replaces the pen before the next person selects a pen at random. Find the probabilities in Problems 23–26.

23. The red pen is never selected.

24. Exactly one person uses the red pen.

25. Exactly two people use the red pen.

26. All three people select the red pen.

All the cows in a certain herd are white-faced. The probability that a white-faced calf will be born by mating with a certain bull is 0.9. Suppose four cows are bred to the same bull. Find the probabilities in Problems 27–32.

27. Four white-faced calves

28. Three white-faced calves

29. No white-faced calves

30. One white-faced calf

31. At least three white-faced calves

32. No more than two white-faced calves

A researcher chooses three leaves from a target environment and classifies each sample as fungus free or contaminated. Suppose that a leaf has a probability of 0.2 of being infected. In Problems 33–38, find the requested probabilities.

33. No infected leaves are found.

34. All three samples are infected.

35. Two are found to be infected.

36. One is infected.

37. One or more is infected.

38. Numerically, how are your answers for Problems 33–36 related?

Level 2

39. The probability that children in a specific family will have blond hair is $\frac{3}{4}$. If there are four children in the family, what is the probability that two of them are blond?

40. The probability that the children in a specific family will have brown eyes is $\frac{3}{8}$. If there are four children in the family, what is the probability that at least two of them have brown eyes?

41. What is the probability that if a pair of dice is rolled six times, exactly one sum of seven is rolled?

42. What is the probability that if a pair of dice is rolled twelve times, exactly two sums of seven are rolled?

43. If Marcy can make 70% of free throws in a basketball game, what is the probability that she makes five of the next six free throws?

44. If Bart can make only 40% of free throws in a basketball game, what is the probability that he will make exactly one of the next three free throws?

45. Lymnozyme cures most infections in Koi fish caused by bacteria; in fact, it has been shown to be 96% effective if used according to the directions. If five Koi with a bacterial infection are treated, what is the probability that four of the fish are cured?

46. Lymnozyme cures most infections in Koi fish caused by bacteria; in fact, it has been shown to be 96% effective if used according to the directions. If 15 Koi with a bacterial infection are treated, what is the probability that 14 of the fish are cured?

47. It is known that 85% of the graduates of Foley's School of Motel Management are placed in a job within 6 months of graduation. If a class has 20 graduates, what is the probability that 15 will be placed within 6 months?

48. If a graduating class from Problem 47 has 10 graduates, what is the probability that at least 9 will be placed within 6 months?

49. Suppose the National League team has a probability of $\frac{3}{5}$ of winning the World Series and the American League team has a probability of $\frac{2}{5}$ The series is over as soon as one team wins four games. Find the probability that it takes seven games to decide the winner.

50. Suppose the National League team has a probability of $\frac{3}{5}$ of winning a World Series game and the American League team has a probability of $\frac{2}{5}$. The series is over as soon as one team wins four games. Find the probability that the National League team wins the series in seven games.

51. The batting average of the star baseball player Mike Thompson is 0.300 (that is, $P(\text{hit}) = 0.3$). What is the probability that Thompson will get at least three hits in four times at bat?

52. Eighty percent of the widgets of the Ampex Widget Company meet the specifications of its customers. If a sample of six widgets is tested, what is the probability that three or more of them would fail to meet the specifications?

Level 3

53. Suppose that research has shown that the probability that a missile penetrates enemy defenses and reaches its target is 0.1. Find the smallest number of identical missiles that are necessary in order to be 80% certain of hitting the target at least once.

54. Suppose a person claims to have extrasensory perception (ESP) and says she can read mental images 75% of the time. We set up an experiment where she must determine whether we are looking at a picture of a circle or a square. We agree that if she can correctly identify the mental image at least five out of six times, we will grant her claim.

  a. What is the probability of granting her claim if she is in fact only guessing?
  b. What is the probability of denying her claim if she really does have the ability she claims?

55. Repeat Problem 54, except that the person claims to read mental images 90% of the time.

56. Fire frequency has a great impact on national parks. If the fire frequency is too severe, generally species that occur in later successional habitats will be excluded. Likewise, the absence of fire will lead to a paucity of species that require early successional habitats. If, on average, fire occurs in a site forest once every 10 years, then the probability is 10% that this site will burn in a given year. If there are 100 such site forests, what is the probability that fire will occur in at least one site over the next year? Assume that all sites are independent of each other; that is, fire does not spread into neighboring patches.

Problem Solving 3

57. In a certain office, three men determine who will pay for coffee by each flipping a coin. If one of them has an outcome that is different from the other two, he must pay for the coffee. What is the probability that in any play of the game there will be an "odd man out"?

58. In the game described in Problem 57, what is the probability that there will be an odd man out in a particular play if there are four players?

59. In the game of Problem 57, what is the probability that there will be an odd man out in a particular play if there are five players?

60. Generalize Problem 57 for n people playing the game of "odd man out."

13.6   CHAPTER SUMMARY

The mathematics curriculum should include the continued study of probability so that all students can use experimental or theoretical probability, as appropriate, to represent and solve problems involving uncertainty.

NCTM STANDARDS

Important Ideas

Probability by counting [13.1]
Expectation with and without a cost for playing [13.2]
Finding odds (given probability) [13.3]
Finding probability (given odds) [13.3]
Procedure for using tree diagrams [13.3]
Probabilities of unions and intersections [13.4]
Drawing with and without replacement [13.4]
Tree diagrams to find probabilities [13.4]
Probabilities of binomial experiments [13.5]

Take some time getting ready to work the review problems in this section. First review these important ideas. Look back at the definition and property boxes. If you look online, you will find a list of important terms introduced in this chapter, as well as the types of problems that were introduced. You will maximize your understanding of this chapter by working the problems in this section only after you have studied the material.

You will find some review help online at **www.mathnature.com.** There are links giving general test help in studying for a mathematics examination, as well as specific help for reviewing this chapter.

Chapter 13 Review Questions

1. Which of the following is the most probable?
   A. Winning the grand prize in a state lottery
   B. Being struck by lightning
   C. Appearing on the *Tonight Show*

JIM RUYMEN/UPI/Landov

2. Which of the following events has odds of about a million to one?
   A. Selecting a particular Social Security number by picking the digits at random
   B. Selecting a particular phone number at random if the first digit is known
   C. Winning first prize in a state lottery

3. If a sample from an assembly line reveals four defective items out of 1,000 sampled items, what is the probability that any one item is defective?

4. What is the probability of obtaining a prime in a single roll of a die?

5. A pair of dice is rolled. What is the probability that the resulting sum is eight?

6. A card is selected from an ordinary deck of cards. What is the probability that it is a jack or better? (A jack or better is a jack, queen, king, or ace.)

7. A single card is drawn from a standard deck of cards. What is the probability that it is an ace, if you know that one ace has already been removed from the deck?

8. For a roll of a pair of dice, find the following.
   a. P(5 on at least one of the dice)
   b. P(5 on one die or 4 on the other)
   c. P(5 on one die and 4 on the other)

9. **a.** If the probability of dropping an egg is 0.01, what is the probability of not dropping the egg?
   b. If $P(E) = 0.9$, what are the odds in favor of E?
   c. If the odds against an event are 1,000 to 1, what is the probability of the event?

10. A game consists of rolling a die and receiving $12 if a one is rolled and nothing otherwise. What is the mathematical expectation?

11. A box contains three orange balls and two purple balls, and two are selected at random. What is the probability of the second ball being orange if we know that the first draw was orange and we draw the second ball:
    a. after replacing the first ball?
    b. without replacing the first ball?

12. If A and B are independent events with $P(A) = \frac{2}{3}$ and $P(B) = \frac{3}{5}$, what is $P(\overline{A \cup B})$?

13. A manufacturer of water test kits knows that about 0.1% of the test strips in the kits are defective. Write the binomial probability formula that would be used to determine the probability that exactly x test strips in a kit of n strips are defective.

The Weather Channel makes accurate predictions about 85% of the time. Use this information to answer the questions in Problems 14–15.

14. What is the probability that Weather Channel meterologists are accurate on exactly 3 of the next 5 days?

15. What is the probability that they are accurate exactly 4 of the next 4 days?

16. A friend says she has two new baby cats to show you, but she doesn't know whether they're both male, both female, or a pair. You tell her that you want only a male, and she telephones the vet and asks, "Is at least one a male?" The vet answers, "Yes!" What is the probability that if you select one kitten, then it is a male? Assume that a baby cat is equally likely to be female or male.

17. The probability that an individual who is selected at random from a population has blue eyes has been estimated to be 0.35. Find the probability that at least one of four people selected at random has blue eyes.

18. Consider the set of octahedral dice shown in Figure 13.10. Suppose your opponent chooses die D. Which die would you choose, and why, if the game consists of each player rolling his or her chosen die once and the higher number on top is the winner?

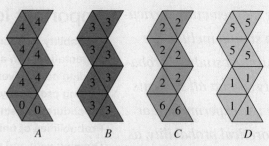

FIGURE 13.10 Octahedral dice

19. The following problem appeared in an "Ask Marilyn" column, by Marilyn vos Savant:

    There are five cars on display as prizes, and their five ignition keys are in a box. You get to pick one key out of the box and try it in the ignition of one car. If it fits, you win the car. What are your chances of winning a car?

    Answer this question.

20. Answer the question in Problem 19 assuming that you are in the audience and want the probability that the contestant will win the car you have selected.

BOOK REPORTS

Write a 500-word report on one of these books:

How to Take a Chance. Darrell Huff and Irving Geis (New York: Norton, 1959).

Beat the Dealer. Edward O. Thorp (New York: Vintage Books, 1966).

The Power of Logical Thinking, Marilyn vos Savant (New York: St. Martin's Press, 1997). This book deals with people's perception of probability and statistics.

Group | RESEARCH PROJECTS

Go to
www.mathnature.com
for references and links.

Working in small groups is typical of most work environments, and learning to work with others to communicate specific ideas is an important skill. Work with three or four other students to submit a single report based on each of the following questions.

G43. a. In how many possible ways can you land on jail (just visiting) on your first turn when playing a Monopoly game?

b. Is it possible to make it from GO to Park Place on your first roll of the dice in a Monopoly game? If so, what is the probability that not only would that happen, but also that you would obtain a 2 on your next roll to complete a set (a monopoly)?

Austin MacRae

See the links for this problem at
www.mathnature.com

G44. Birthday problem:

a. Experiment: Consider the birthdates of some famous mathematicians:

Abel	August 5, 1802
Cardano	September 24, 1501
Descartes	March 31, 1596
Euler	April 15, 1707
Fermat	August 17, 1601
Galois	October 25, 1811
Gauss	April 30, 1777
Newton	December 25, 1642
Pascal	June 19, 1623
Riemann	September 17, 1826

Add to this list the birthdates of the members of your class. *But before you compile this list, guess the probability that at least two people in this group will have exactly the same birthday (not counting the year).* Be sure to make your guess *before* finding out the birthdates of your classmates. The answer, of course, depends on the number of people on the list. Ten mathematicians are listed and you may have 20 people in your class, giving 30 names on the list.

b. Find the probability of at least one birthday match among 3 randomly selected people. (See Example 4, Section 13.4.)

c. Find the probability of at least one birthday match among 23 randomly selected people. Have each person in your group pick 23 names at random from a biographical dictionary or a *Who's Who,* and verify empirically the probability you calculated.

    d. Draw a graph showing the probability of a birthday match given a group of n people. How many people are necessary for the probability actually to reach 1?

    e. In the previous parts of this problem we interpreted two people having the same birthday as meaning at least 2 have the same birthday (see Example 4, Section 13.4). We now refine this idea. Find the following probabilities for a group of 5 randomly selected people:

        Exactly 2 of the 5 have the same birthday.

        Exactly 3 have the same birthday.

        Exactly 4 have the same birthday.

        All 5 have the same birthday.

        There are exactly two pairs sharing
          (a different) birthday.

        There is a full house of birthdays
          (that is, three share one birthday,
          and two share another).

    Show that the questions of this problem account for all the possibilities; that is, show that the sum of the probabilities for all of these possibilities is the same as for the original birthday problem involving 5 persons: What is the probability of a birthday match among 5 randomly selected people?

G45. Consider the following classroom activity. Suppose the floor consists of square tiles 9 in. on each side. The players will toss a circular disk onto the floor. If the disk comes to rest on the edge of any tile, the player loses $1. Otherwise, the player wins $1. What is the probability of winning if the disk is:

a. a dime **b.** a quarter **c.** a disk with a diameter of 4 in. **d.** Now, the real question: What size should the disk be so that the probability that the player wins is 0.45?

Individual RESEARCH PROJECTS

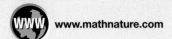
www.mathnature.com

Learning to use sources outside your classroom and textbook is an important skill, and here are ideas for extending some of the ideas in this chapter. You can find references to these projects in a library or at **www.mathnature.com.**

PROJECT 13.1 Devise a fair scheme for eliminating coins in this country.

PROJECT 13.2 Find the probability that the 13th day of a randomly chosen month will be a Friday.

PROJECT 13.3 You have five alternatives from which to choose. List your preferences for the alternatives from best to worst.

1. Sure win of $5 and no chance of loss

2. 6.92% chance to win $20 and 93.08% chance to win $3.98

3. 27.52% chance to win $20 and 72.48% chance to lose 69 cents

4. 61.85% chance to win $20 and 38.15% chance to lose $19.31

5. 90.46% chance to win $20 and 9.54% chance to lose $137.20

a. Answer the question based on your own feelings.
b. Answer the question using mathematical expectation as a basis for selecting your answer.
c. Conduct a survey of at least 10 people and summarize your results.
d. What are the conclusions of the study?

PROJECT 13.4

a. Choose between A and B:
   A. A sure gain of $240
   B. 25% chance to gain $1,000 and 75% chance to gain $0

b. Choose between C and D:
   C. A sure loss of $700
   D. 75% chance to lose $1,000 and 25% chance to lose nothing

c. Choose between E and F:
   E. Imagine that you have decided to see a concert and have paid the admission price of $10. As you enter the concert hall, you discover that you have lost your ticket. Would you pay $10 for another ticket?
   F. Imagine that you have decided to see a concert where the admission is $10. As you begin to enter the concert ticket line, you discover that you have lost one of your $10 bills. Would you still pay $10 for a ticket to the concert?

d. Answer each of the questions[*] based on your own feelings.
e. Answer questions **a** and **b** using mathematical expectation as a basis for selecting your answers.
f. Conduct a survey of at least 10 people and summarize your results.

PROJECT 13.5 Do some research on Keno probabilities. Write a paper on playing Keno.

*These questions are from "A Bird in the Hand," by Carolyn Richbart and Lynn Richbart, *The Mathematics Teacher*, November 1996, pp. 674–676.

14

THE NATURE OF STATISTICS

What in the World?

"I signed up for Hunter in History 17," said Ben.

"Why did you sign up for her?" asked Ted. "Don't you know she has given only three A's in the last 14 years?!"

"That's just a rumor. Hunter grades on a curve," Ben replied.

"Don't give me that 'on a curve' stuff," continued Ted, "I'll bet you don't even know what that means. Anyway, my sister-in-law had her last year and said it was so bad that . . . ."

Overview

Undoubtedly, you have some idea about what is meant by the term *statistics.* For example, we hear about:

- Statistics on population growth
- Information about the depletion of the rain forests or the ozone layer
- The latest statistics on the cost of living
- The Gallup Poll's use of statistics to predict election outcomes
- The Nielsen ratings, which indicate that one show has 30% more viewers than another
- Baseball or sports statistics

We could go on and on, but you can see from these examples that there are two main uses for the word **statistics.** First, we use the term to mean a mass of data, including charts and tables. This is the everyday, nontechnical use of the word. Second, the word refers to a methodology for collecting, analyzing, and interpreting data. In this chapter, we'll examine some of these statistical methods. We do not intend to present a statistics course, but rather to prepare you to use statistics in your everyday life, as well as possibly to prepare you to take a college-level statistics course.

Digital Vision/Getty Images

14.1 | Frequency Distributions and Graphs

There are three kinds of lies: lies, damned lies, and statistics.

G. B. HALSTED

TABLE 14.1		
Frequency Distribution for 50 Rolls of a Pair of Dice		
Outcome	**Tally**	**Frequency**
2	|	1
3	|||	3
4	||||	5
5	||	2
6	|||| ||||	9
7	|||| |||	8
8	|||| |	6
9	|||| |	6
10	||||	4
11	||||	5
12	|	1

CHAPTER CHALLENGE

See if you can fill in the question mark.

We are sometimes confronted with large amounts of data, which can be difficult to understand. Computers, spreadsheets, and simulation programs have done a lot to help us deal with hundreds or thousands of pieces of information at the same time.

Frequency Distribution

We can deal with large batches of data by organizing them into groups, or **classes.** The difference between the lower limit of one class and the lower limit of the next class is called the **interval** of the class. After determining the number of values within a class, termed the **frequency,** you can use this information to summarize the data. The end result of this classification and tabulation is called a **frequency distribution.** For example, suppose that you roll a pair of dice 50 times and obtain these outcomes:

3, 2, 6, 5, 3, 8, 8, 7, 10, 9, 7, 5, 12, 9, 6, 11, 8, 11, 11, 8, 7, 7, 7, 10, 11, 6, 4, 8,

8, 7, 6, 4, 10, 7, 9, 7, 9, 6, 6, 9, 4, 4, 6, 3, 4, 10, 6, 9, 6, 11

We can organize these data in a convenient way by using a frequency distribution, as shown in Table 14.1.

TABLE 14.2						
High Temperatures on Lake Erie for the Month of October						
74	79	79	85	69	69	65
75	81	82	74	70	71	61
75	80	86	70	62	71	69
77	78	82	82	71	75	64
77	82	86				

Example 1   Make a frequency diagram

Make a frequency distribution for the high temperatures (in °F) for October on Lake Erie as summarized in Table 14.2.

Solution   First, notice that the included temperatures range from a low of 61 to a high of 86. If we were to list each temperature and make tally marks for each, we would have a long table with few entries for each temperature. Instead, we arbitrarily divide the temperatures into eight categories. Make the tallies and total the number in each group. The result is called a **grouped frequency distribution.**

Temperature	Tally	Number
below 60		
60–64	|||	3
65–69	||||	4
70–74	|||| ||	7
75–79	|||| |||	8
80–84	|||| |	6
85–89	|||	3
90 and above		

A graphical method useful for organizing large sets of data is a **stem-and-leaf plot.** Consider the data shown in Table 14.3.[*] For these data we form stems representing decades of the ages of the actors. For example, Jamie Foxx, who won the 2004 best actor award in 2005 for his role in *Ray*, was 38 years of age when he won the award. We would say that the stem for this age is 3 (for 30), and the leaf is 8.

TABLE 14.3

Best Actors, 1928–2009

Year	Actor	Age	Movie	Year	Actor	Age	Movie
1928	Emil Jannings	44	*The Way of All Flesh*	1969	John Wayne	62	*True Grit*
1929	Warner Baxter	38	*In Old Arizona*	1970	George C. Scott	43	*Patton*
1930	George Arliss	46	*Disraeli*	1971	Gene Hackman	40	*The French Connection*
1931	Lionel Barrymore	53	*A Free Soul*	1972	Marlon Brando	48	*The Godfather*
1932	Fredric March ⎫ tie	35	*Dr. Jekyll and Mr. Hyde*	1973	Jack Lemmon	48	*Save the Tiger*
	Wallace Beery ⎭	47	*The Champ*	1974	Art Carney	56	*Harry and Tonto*
1933	Charles Laughton	34	*The Private Life of Henry VIII*	1975	Jack Nicholson	38	*One Flew over the Cuckoo's Nest*
1934	Clark Gable	33	*It Happened One Night*	1976	Peter Finch	60	*Network*
1935	Victor McLaglen	49	*The Informer*	1977	Richard Dreyfuss	32	*The Goodbye Girl*
1936	Paul Muni	41	*The Story of Louis Pasteur*	1978	Jon Voight	40	*Coming Home*
1937	Spencer Tracy	37	*Captains Courageous*	1979	Dustin Hoffman	42	*Kramer vs Kramer*
1938	Spencer Tracy	38	*Boys' Town*	1980	Robert De Niro	37	*Raging Bull*
1939	Robert Donat	34	*Goodbye Mr. Chips*	1981	Henry Fonda	76	*On Golden Pond*
1940	James Stewart	32	*The Philadelphia Story*	1982	Ben Kingsley	39	*Gandhi*
1941	Gary Cooper	40	*Sergeant York*	1983	Robert Duvall	55	*Tender Mercies*
1942	James Cagney	48	*Yankee Doodle Dandy*	1984	F. Murray Abraham	45	*Amadeus*
1943	Paul Lukas	48	*Watch on the Rhine*	1985	William Hurt	35	*Kiss of the Spider Woman*
1944	Bing Crosby	43	*Going My Way*	1986	Paul Newman	61	*The Color of Money*
1945	Ray Milland	40	*The Lost Weekend*	1987	Michael Douglas	33	*Wall Street*
1946	Fredric March	49	*The Best Years of Our Lives*	1988	Dustin Hoffman	51	*Rainman*
1947	Ronald Colman	56	*A Double Life*	1989	Daniel Day-Lewis	31	*My Left Foot*
1948	Laurence Olivier	41	*Hamlet*	1990	Jeremy Irons	42	*Reversal of Fortune*
1949	Broderick Crawford	38	*All the King's Men*	1991	Anthony Hopkins	55	*Silence of the Lambs*
1950	Jose Ferrer	38	*Cyrano de Bergerac*	1992	Al Pacino	52	*Scent of a Woman*
1951	Humphrey Bogart	52	*The African Queen*	1993	Tom Hanks	37	*Philadelphia*
1952	Gary Cooper	51	*High Noon*	1994	Tom Hanks	38	*Forrest Gump*
1953	William Holden	35	*Stalag 17*	1995	Nicholas Cage	31	*Leaving Las Vegas*
1954	Marlon Brando	30	*On the Waterfront*	1996	Geoffrey Rush	45	*Shine*
1955	Ernest Borgnine	38	*Marty*	1997	Jack Nicholson	59	*As Good As It Gets*
1956	Yul Brynner	41	*The King and I*	1998	Roberto Benigni	46	*Life Is Beautiful*
1957	Alec Guinness	43	*The Bridge on the River Kwai*	1999	Kevin Spacey	41	*American Beauty*
1958	David Niven	49	*Separate Tables*	2000	Russell Crowe	36	*Gladiator*
1959	Charlton Heston	35	*Ben Hur*	2001	Denzel Washington	47	*Training Day*
1960	Burt Lancaster	47	*Elmer Gantry*	2002	Adrien Brody	29	*The Pianist*
1961	Maximilian Schell	31	*Judgment at Nuremburg*	2003	Sean Penn	43	*Mystic River*
1962	Gregory Peck	46	*To Kill a Mockingbird*	2004	Jamie Foxx	38	*Ray*
1963	Sidney Poitier	39	*Lilies of the Field*	2005	Philip Seymour Hoffman	38	*Capote*
1964	Rex Harrison	56	*My Fair Lady*	2006	Forest Witaker	46	*The Last Kind of Scotland*
1965	Lee Marvin	41	*Cat Ballou*	2007	Daniel Day-Lewis	50	*There Will Be Blood*
1966	Paul Scofield	44	*A Man for All Seasons*	2008	Sean Penn	48	*Milk*
1967	Rod Steiger	42	*In the Heat of the Night*	2009	Jeff Bridges	60	*Crazy Heart*
1968	Cliff Robertson	43	*Charly*				

Example 2   Make a stem-and-leaf plot

Construct a stem-and-leaf plot for the best actor ages in Table 14.3.

Solution   Stem-and-leaf plot of ages of best actor, 1928–2009.

```
2 | 9
3 | 0 1 1 1 2 2 3 3 4 4 5 5 5 5 6 7 7 7 8 8 8 8 8 8 8 8 8 9 9
4 | 0 0 0 0 1 1 1 1 1 2 2 2 3 3 3 3 3 4 4 5 5 6 6 6 6 7 7 7 8 8 8 8 9 9 9
5 | 0 1 1 2 2 3 5 5 6 6 6 9
6 | 0 1 2
7 | 6
```

This plot is useful because it is easy to see that most best actor winners received the award in their forties.

---

*The idea for this example came from "Ages of Oscar-Winning Best Actors and Actresses" by Richard Brown and Gretchen Davis, *The Mathematics Teacher*, February 1990, pp. 96–102.

The data sets {32, 56, 47, 30, 41} and {3.2, 5.6, 4.7, 3.0, 4.1} have the same stem-and-leaf plots, with the first set having a leaf unit of 1 and the second set having a leaf unit of 0.1.

To help us understand the relationship between and among variables, we use a diagram called a **graph.** In this section, we consider *bar graphs, line graphs, circle graphs, and pictographs.*

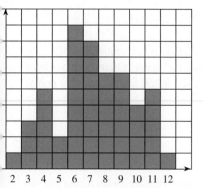

E 14.1 Outcomes of experiment of a pair of dice

Bar Graphs

A **bar graph** compares several related pieces of data using horizontal or vertical bars of uniform width. There must be some sort of scale or measurement on both the horizontal and vertical axes. An example of a bar graph is shown in Figure 14.1, which shows the data from Table 14.1.

Example **3** **Make a bar graph**

Construct a bar graph for the Lake Erie temperatures given in Example 1. Use the grouped categories.

Solution To construct a bar graph, draw and label the horizontal and vertical axes, as shown in part **a** of Figure 14.2. It is helpful (although not necessary) to use graph paper. Next, draw marks indicating the frequency, as shown in part **b** of Figure 14.2. Finally, complete the bars and shade them as shown in part **c.**

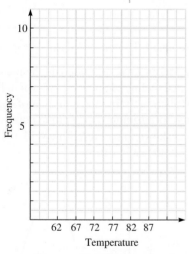

a. Draw and label axes and scales.

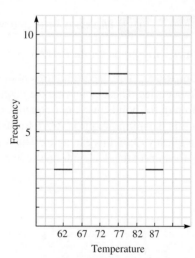

b. Mark the frequency levels.

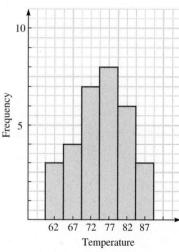

c. Complete and shade the bars.

FIGURE 14.2 Bar graph for Lake Erie temperatures

You will frequently need to look at and interpret bar graphs in which bars of different lengths are used for comparison purposes.

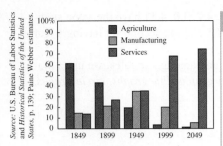

Source: U.S. Bureau of Labor Statistics and *Historical Statistics of the United States*, p. 139; Paine Webber estimates.

FIGURE 14.3 Share of U.S. employment in agriculture, manufacturing, and services from 1849 to 2049

Example 4 Read a bar graph

Refer to Figure 14.3 to answer the following questions.
a. What was the share of U.S. employment in manufacturing in 1999?
b. In which year(s) was U.S. employment in services approximately equal to employment in manufacturing?
c. From the graph, form a conclusion about U.S. employment in agriculture for the period 1849–2049.

Solution
a. We see the bar representing manufacturing in 1999 is approximately 20%. (Use a straightedge for help.)
b. Employments in services and manufacturing were approximately equal in 1949.
c. Conclusions, of course, might vary, but one obvious conclusion is that there has been a dramatic decline of employment in agriculture in the United States over this period of time.

Line Graphs

A graph that uses a broken line to illustrate how one quantity changes with respect to another is called a **line graph.** A line graph is one of the most widely used kinds of graph.

Example 5 Make a line graph

Draw a line graph for the data given in Example 1. Use the previously grouped categories.

Solution The line graph uses points instead of bars to designate the locations of the frequencies. These points are then connected by line segments, as shown in Figure 14.4. To plot the points, use the frequency distribution to find the midpoint of each category (62 is the midpoint of the 60–64 category, for example); then plot a point showing the frequency (3, in this example). This step is shown in part **a** of Figure 14.4. The last step is to connect the dots with line segments, as shown in part **b.**

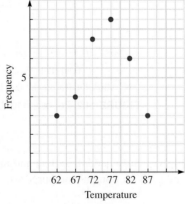 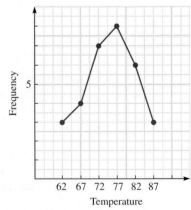

a. Plot the points to represent frequency levels.   **b.** Connect the dots with line segments.

FIGURE 14.4 Constructing a line graph for Lake Erie temperatures

Just as with bar graphs, you need to be able to read and interpret line graphs.

Example 6 Read a line graph

Refer to Figure 14.5 to answer the following questions.

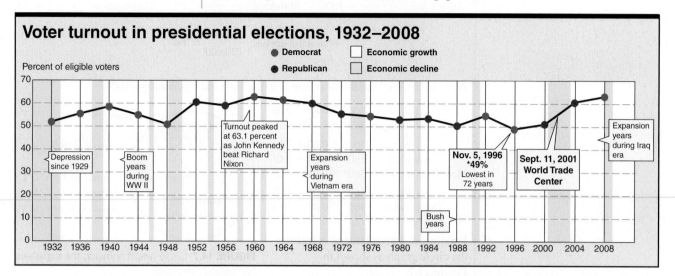

FIGURE 14.5 Voter turnout in presidential elections, 1932–2008

a. In which year was the voter turnout in a presidential election the greatest?

b. During the period 1932–2008, were there more periods of economic growth or economic decline?

c. What was the voter turnout in 1996? Did a Republican or Democrat win that election? In which year was the voter turnout the closest to that in 1996? Did a Republican or a Democrat win in that election?

Solution

a. Voter turnout was the greatest (63.1%) in 1960.

b. Growth is shown as a white region and decline as a shaded region; there are 13 periods of each, but the growth periods are of greater duration.

c. In 1996 the voter turnout was 49% and a Democrat won the election. Voter turnout had not been that low since 1988 (about 51%), when the winner was a Republican, or since 1948 (about 52%), when the winner was a Democrat.

Circle Graphs

Another type of commonly used graph is the **circle graph,** also known as a **pie chart.** This graph is particularly useful in illustrating how a whole quantity is divided into parts—for example, income or expenses in a budget.

To create a circle graph, first express the number in each category as a percentage of the total. Then convert this percentage to an angle in a circle. Remember that a circle is divided into 360°, so we multiply the percent by 360 to find the number of degrees for each category. You can use a protractor to construct a circle graph, as shown in the following example.

Example 7 Make a circle graph

The 2010 expenses for Karlin Enterprises are shown in Figure 14.6. Construct a circle graph showing the expenses for Karlin Enterprises.

Solution The first step in constructing a circle graph is to write the ratio of each entry to the total of the entries, as a percent. This is done by finding the total ($120,000) and then dividing each entry by that total:

KARLIN ENTERPRISES	EXPENSE REPORT FY 2007
es	$ 72,000
taxes, insurance	$ 24,000
es	$ 6,000
ising	$ 12,000
age	$ 1,200
als and supplies	$ 1,200
iation	$ 3,600
L	$120,000

E 14.6 Karlin Enterprises expenses

Salaries: $\dfrac{72,000}{120,000} = 60\%$ Rents: $\dfrac{24,000}{120,000} = 20\%$ Utilities: $\dfrac{6,000}{120,000} = 5\%$

Advertising: $\dfrac{12,000}{120,000} = 10\%$ Shrinkage: $\dfrac{1,2000}{120,000} = 1\%$ Depreciation: $\dfrac{3,600}{120,000} = 3\%$

Materials/supplies: $\dfrac{1,200}{120,000} = 1\%$

A circle has $360°$, so the next step is to multiply each percent by $360°$:

Salaries:	$360° \times 0.60 = 216°$
Advertising:	$360° \times 0.10 = 36°$
Rents:	$360° \times 0.20 = 72°$
Shrinkage:	$360° \times 0.01 = 3.6°$
Materials/supplies:	$360° \times 0.01 = 3.6°$
Utilities:	$360° \times 0.05 = 18°$
Depreciation:	$360° \times 0.03 = 10.8°$

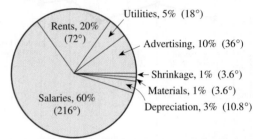

Finally, use a protractor to construct the circle graph as shown in Figure 14.7.

FIGURE 14.7 Circle graph showing the expenses for Karlin Enterprises

Pictographs

A **pictograph** is a representation of data that uses pictures to show quantity. Consider the raw data shown in Table 14.4.

TABLE 14.4				
Marital Status* (for persons age 65 and older)				
	Married	**Widowed**	**Divorced**	**Never Married**
Women	5.4	7.2	0.4	0.8
Men	7.5	1.4	0.3	0.6

*In millions, rounded to the nearest 100,000.

A pictograph uses a picture to illustrate data; it is normally used only in popular publications, rather than for scientific applications. For the data in Table 14.4, suppose that we draw pictures of a woman and a man so that each picture represents 1 million persons, as shown in Figure 14.8.

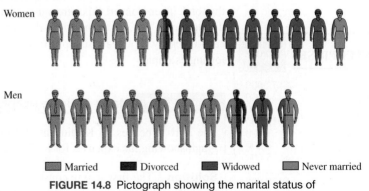

FIGURE 14.8 Pictograph showing the marital status of persons age 65 and older

Misuses of Graphs

The most misused type of graph is the pictograph. Consider the data from Table 14.4. Such data can be used to determine the height of a three-dimensional object, as in part **a** of Figure 14.9. When an object (such as a person) is viewed as three-dimensional, differences seem much larger than they actually are. Look at part **b** of Figure 14.9 and notice that, as the height and width are doubled, the volume is actually increased eightfold.

a. Pictograph representing values as heights   **b.** Doubling height and width increases volume 8-fold

FIGURE 14.9 Examples of misuses in pictographs

This pictograph fallacy carries over to graphs of all kinds, especially since software programs easily change two-dimensional scales to three-dimensional scales. For example, percentages are represented as heights on the scale shown in Figure 14.10a, but the software has incorrectly drawn the graph as three-dimensional bars. Another fallacy is to choose the scales to exaggerate or diminish real differences. Even worse, graphs are sometimes presented with no scale whatsoever, as illustrated in a graphical "comparison" between Anacin and "regular strength aspirin" shown in Figure 14.10b.

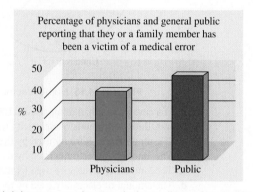

a. Heights are used to graph three-dimensional objects.

Source: Blendon, RJ. *New England Journal of Medicine*, 2002: 347–1940.

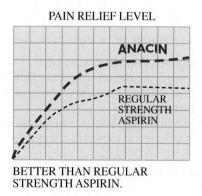

b. No scale is shown in this Anacin advertisment.

FIGURE 14.10 Misuse of graphs

Problem Set 14.1

Level 1

1. IN YOUR OWN WORDS   Name and describe several different types of graphs.

2. IN YOUR OWN WORDS   Distinguish between a frequency distribution and a stem-and-leaf plot.

3. IN YOUR OWN WORDS   Write down what you think the following advertisement means:

Nine out of ten dentists recommend Trident for their patients who chew gum.

4. IN YOUR OWN WORDS   Write down what you think the following advertisement means:

Eight out of ten owners said their cats prefer Whiskas.

5. Consider the line graph shown in Figure 14.11.

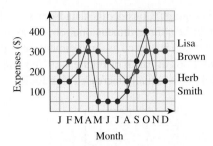

FIGURE 14.11 Expenses for two sales people of the Leadwell Pencil Company

a. During which month did Herb incur the most expenses?
b. During which month did Lisa incur the least expenses?

6. Consider the graph shown in Figure 14.12.

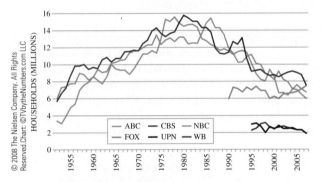

FIGURE 14.12 Major TV network ratings

a. How many networks are tracked?
Can you think of any other networks that might have been included? If so, why do you think they have not been included?
b. In which year(s) were the various networks tracked?
c. Which network showed a trend of increased numbers of viewers in 2008?

7. Consider the graph shown in Figure 14.13.

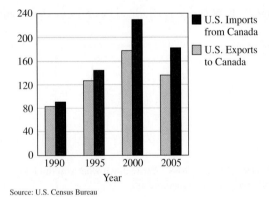

Source: U.S. Census Bureau

FIGURE 14.13 U.S. trade with Canada

a. What is being illustrated in this graph?
b. Did the United States have more imports or exports for the years 1990–2005?

8. Consider the graph shown in Figure 14.14.

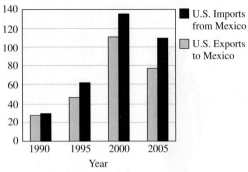

Source: U.S. Census Bureau

FIGURE 14.14 U.S. trade with Mexico

a. What is being illustrated in this graph?
b. Did the United States have more imports or exports for the years 1990–2005?

9. The amount of electricity used in a typical all-electric home is shown in the circle graph in Figure 14.15.

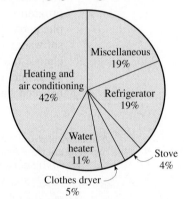

FIGURE 14.15 Home electricity usage

If, in a certain month, a home used 1,100 kWh (kilowatt-hours), find the amounts of electricity used from the graph.
a. The amount of electricity used by the water heater
b. The amount of electricity used by the stove
c. The amount of electricity used by the refrigerator
d. The amount of electricity used by the clothes dryer

10. At a well-known major university, the career choices of 1,500 graduates from an MS program are shown in Figure 14.16.

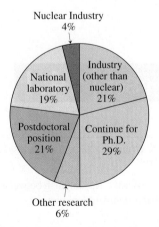

FIGURE 14.16 Career choices of graduates

a. How many plan to get a job in a national laboratory?

b. How many are planning to continue for a Ph.D.?

c. How many are planning to enter the nuclear or other industry?

d. How many are going to continue for further research (including Ph.D. and postdoctoral)?

11. In Section 14.4 we will discuss scatter diagrams, which are used to study the relationship between two variables, such as the students' final exam score and final grade in Smith's Math 10 class (see Figure 14.17).

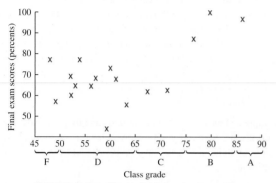

FIGURE 14.17 Class grade vs exam grade

Approximately what is the lowest final exam score received by a student awarded a final grade of C or better?

A. 50%   B. 55%   C. 60%   D. 65%

12. In Section 14.4 we will discuss scatter diagrams, which are used to study the relationship between two variables, such as incomes from crops and livestock for selected states in the United States, as shown in Figure 14.18.

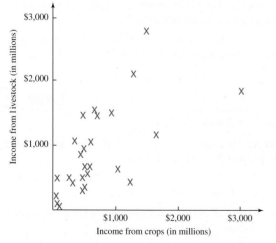

FIGURE 14.18 State income crops vs livestock

For states with income from crops greater than a billion dollars, the number of states with income from livestock greater than a billion dollars is about:

A. 16   B. 5   C. 10   D. 22

Decide whether something is wrong with each of the graphs shown in Problems 13–17. Explain your reasoning.

13. Consider the graph shown in Figure 14.19. Clearly, Anacin is better.

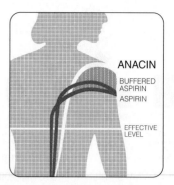

FIGURE 14.19 Advertisement for a pain reliever

14. Consider the graph shown in Figure 14.20. Chevy trucks are clearly better.

FIGURE 14.20 Advertisement for Chevy trucks

15. Consider the graph shown in Figure 14.21. The potential commercial forest growth, as compared with the current commercial forest growth, is almost double (44.9 to 74.2).

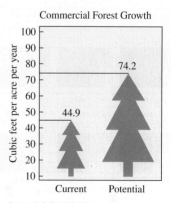

FIGURE 14.21 From the U.S. Forest Service

16. Consider the graph shown in Figure 14.22.

How much is "in the bank"?
Conflicting estimates of undiscovered oil and gas reserves—which can be recovered and produced.

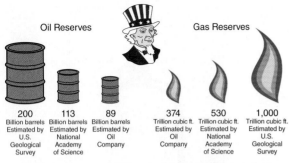

Oil Reserves			Gas Reserves		
200	113	89	374	530	1,000
Billion barrels Estimated by U.S. Geological Survey	Billion barrels Estimated by National Academy of Science	Billion barrels Estimated by Oil Company	Trillion cubic ft. Estimated by Oil Company	Trillion cubic ft. Estimated by National Academy of Science	Trillion cubic ft. Estimated by U.S. Geological Survey

FIGURE 14.22 Pictograph showing conflicting estimates of oil and gas reserves

17. The graph shown in Figure 14.23 was circulated throughout the mathematics department of a leading college.

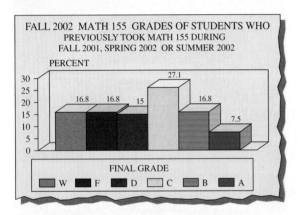

FALL 2002 MATH 155 GRADES OF STUDENTS WHO PREVIOUSLY TOOK MATH 155 DURING FALL 2001, SPRING 2002 OR SUMMER 2002

PERCENT: 16.8, 16.8, 15, 27.1, 16.8, 7.5

FINAL GRADE: W F D C B A

FIGURE 14.23 Math 155 final grades

When a person in California renews the registration for an automobile, the bar graph shown in Figure 14.24 is included with the bill.

BAC Zones:	90 to 109 lbs.	110 to 129 lbs.	130 to 149 lbs.	150 to 169 lbs.
TIME FROM 1st DRINK	TOTAL DRINKS 1 2 3 4 5 6 7 8	TOTAL DRINKS 1 2 3 4 5 6 7 8	TOTAL DRINKS 1 2 3 4 5 6 7 8	TOTAL DRINKS 1 2 3 4 5 6 7 8
1 hr				
2 hrs				
3 hrs				
4 hrs				

BAC Zones:	170 to 189 lbs.	190 to 209 lbs.	210 to 229 lbs.	230 lbs. & up
TIME FROM 1st DRINK	TOTAL DRINKS 1 2 3 4 5 6 7 8	TOTAL DRINKS 1 2 3 4 5 6 7 8	TOTAL DRINKS 1 2 3 4 5 6 7 8	TOTAL DRINKS 1 2 3 4 5 6 7 8
1 hr				
2 hrs				
3 hrs				
4 hrs				

SHADINGS IN THE CHARTS ABOVE MEAN:

☐ (.01% – .04%) Seldom illegal   ▨ (.05% – .09%) May be illegal   ■ (.10% Up) Definitely illegal

▨ (.05% – .09%) Illegal if under 18 yrs. old

Prepared by the Department of Motor Vehicles in cooperation with the Highway Patrol.

FIGURE 14.24 Blood Alcohol Concentration (BAC) charts

Use this bar graph to answer the questions in Problems 18–23. The following statement is included with the graph:

> *There is no safe way to drive after drinking. These charts show that a few drinks can make you an unsafe driver. They show that drinking affects your BLOOD ALCOHOL CONCENTRATION (BAC). The BAC zones for various numbers of drinks and time periods are printed in white, gray, and red. HOW TO USE THESE CHARTS: First, find the chart that includes your weight. For example, if you weigh 160 lbs., use the "150 to 169" chart. Then look under "Total Drinks" at the "2" on this "150 to 169" chart. Now look below the "2" drinks, in the row for 1 hour. You'll see your BAC is in the gray shaded zone. This means that if you drive after 2 drinks in 1 hour, you could be arrested. In the gray zone, your chances of having an accident are 5 times higher than if you had no drinks. But if you had 4 drinks in 1 hour, your BAC would be in the red shaded area . . . and your chances of having an accident 25 times higher.*

18. Suppose that you weigh 115 pounds and that you have two drinks in 2 hours. If you then drive, how much more likely are you to have an accident than if you had refrained from drinking?

19. Suppose that you weigh 115 pounds and that you have four drinks in 3 hours. If you then drive, how much more likely are you to have an accident than if you had refrained from drinking?

20. Suppose that you weigh 195 pounds and have two drinks in 2 hours. According to Figure 14.24, are you seldom illegal, maybe illegal, or definitely illegal?

21. Suppose that you weigh 195 pounds and that you have four drinks in 3 hours. According to Figure 14.24, are you seldom illegal, maybe illegal, or definitely illegal?

22. If you weigh 135 pounds, how many drinks in 3 hours would you need to be definitely illegal?

23. If you weigh 185 pounds and drink a six-pack of beer during a 3-hour baseball game, can you legally drive home?

Level 2

Consider the following data sets:

Data set A: The annual wages of employees at a small accounting firm are given in thousands of dollars.

25	25	25	30	30	35	50	60
16	14	18	18	20			

Data set B: The numbers of cars registered to homes in a certain neighborhood are given.

1	2	1	1	0	3	4	1
2	2	2	3	2	2	2	2

Data set C: The heights (rounded to the nearest inch) of 30 students are given.

66	68	65	70	67	67	68	64	64	66
64	70	72	71	69	64	63	70	71	63
68	67	67	65	69	65	67	66	69	69

Data set D: The temperatures (°F) at a particular location for 30 days are given.

57	50	58	45	49	50	53	52	43	55
39	53	50	49	57	45	49	40	47	55
52	58	50	44	59	54	51	49	43	54

Prepare a frequency distribution for each data set in Problems 24–27.

24. Data set A   **25.** Data set B

26. Data set C   **27.** Data set D

Draw a stem-and-leaf plot for each data set in Problems 28–31.

28. Data set A   **29.** Data set B

30. Data set C   **31.** Data set D

Draw a bar graph for each data set in Problems 32–35.

32. Data set A   **33.** Data set B

34. Data set C   **35.** Data set D

Draw a line graph for each data set in Problems 36–39.

36. Data set A   **37.** Data set B

38. Data set C   **39.** Data set D

40. The amounts of time it takes for three leading pain relievers to reach a person's bloodstream are as follows: Brand A, 480 seconds; Brand B, 500 seconds; Brand C, 490 seconds.

a. Draw a bar graph using the shown scale.

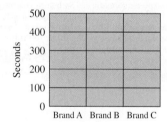

b. Draw a bar graph using the shown scale.

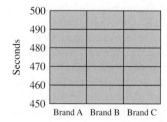

c. If you were an advertiser working on a promotion campaign for Brand A, which graph would seem to give your product a more distinct advantage?

41. The numbers of trucks from three automobile manufacturers that are still on the road 10 years after they are originally sold are Brand A, 98%; Brand B, 97.5%; Brand C, 96.4%.

a. Draw a bar graph using the shown scale.

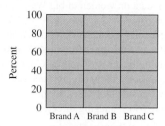

b. Draw a bar graph using the shown scale.

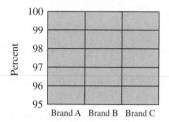

c. If you were an advertiser working on a promotion campaign for Brand A, which graph would seem to give your product a more distinct advantage?

42. The financial report from a small college shows the following income:

Tuition	$1,388,000
Development	90,000
Fund-raising	119,000
Athletic Dept.	44,000
Interest	75,000
Tax subsidy	29,000
TOTAL	$1,745,000

Draw a circle graph showing this income.

43. The financial report from a small college shows the following expenses:

Salaries	$1,400,000
Academic	$68,000
Plant Operations	$196,000
Student activities	$31,000
Athletic programs	$50,000
TOTAL	$1,745,000

Draw a circle graph showing this income.

44. The national debt is of growing concern in the United States. It reached $1 billion in 1916 during World War I and climbed to $278 billion by the end of World War II. It reached its first trillion on October 1, 1981, and rose to $2 trillion on April 3, 1986. The third trillion milestone was reached on April 4, 1990. However, the magnitude of the national debt became headline news when it reached $4 trillion and became a presidential campaign slogan for Ross Perot in 1992. It passed $5 trillion in 1996, $6 trillion in 2002, and on January 15, 2004, it reached $7 trillion. We hardly noticed the other milestones:

$8 trillion on October 18, 2005;
$9 trillion on March 6, 2006;
$10 trillion on October 7, 2008; and
$11 trillion on March 22, 2009.

Represent this information on a line graph.

45. **Top Women on Wall Street** Draw a pictograph to represent the following data on women in big Wall Street firms. In 1996, women comprised approximately 11% of the top-tier executives. Regina Dolan, the chief financial officer of Paine Webber, is one of 46 women among 465 managing directors. Robin Neustein, the chief of staff at Goldman Sachs, is one of 9 women among 173 managing directors. Theresa Lang, the company treasurer for Merrill Lynch, is one of 76 women among 694 managing directors.

46. Using Figure 14.13 (Problem 7), what was the approximate balance of trade with Canada in 2005?

47. Using Figure 14.14 (Problem 8), what was the approximate balance of trade with Mexico in 2005?

48. Using Figure 14.12 (Problem 6), which network had the best market share in 2008?

49. Using Figure 14.12 (Problem 6), what is the maximum rating share by any of the tracked networks?

50. Using Figure 14.12 (Problem 6), what is the minimum rating share by any of the tracked networks?

51. Using Figure 14.12 (Problem 6), form a conclusion.

Level 3

A newspaper article discussing whether Social Security could be cut offered the information shown in Figure 14.25. Use the information in these graphs to answer the questions in Problems 52–58.

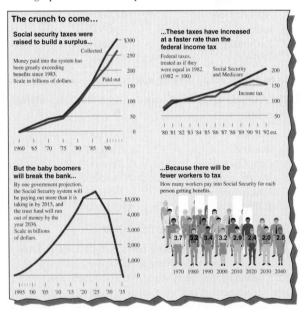

FIGURE 14.25 Social Security reform

52. What is the difference between the amount of money collected from Social Security taxes and the amount paid out in 1992?

53. How many workers pay into Social Security for each person getting benefits in 2020?

54. What are "baby boomers"?

55. When will the baby boomers break the bank?

56. How much money is projected for the Social Security system for the year 2012?

57. There seems to be an error in one of the three line graphs. What is it?

58. There seems to be an error in the pictograph. What is it?

59. Find the error in the Saab advertisement.

Problem Solving 3

60. The following advertisements (adapted from "Caveat Emptor" by Robert Leighton, *Games,* January, 1988, p. 38) show some vastly overpriced items. Read between the lines and identify the item each ad is describing.

   a. Miniature, sparkling white crystals harvested in exotic Hawaii make perfect specimens for studying the wonders of science! Each educationally decorated packet contains hundreds of refined crystals. Completely safe and nontoxic. Only **$3.95.**

   b. It's like having U.S. history in your pocket! Miniature portraits of America's best-loved presidents, Washington sites, and more. Richly detailed copper and silver engraving. Set of four, just **$4.50.**

   c. Today multimillion-dollar computers need special programs to generate random numbers. But this GEOMETRICALLY DESIGNED device ingeniously bypasses computer technology to provide random numbers quickly and easily! Exciting? You bet! Batteries not required. Only **$4.00.**

d. They said we couldn't do it, but we are! For a limited time, our LEADING HOLLYWOOD PROP HOUSE is dividing up and selling—yes, SELLING—parcels of this famous movie setting. You've seen it featured

in *Lawrence of Arabia, Beach Blanket Bingo,* and *Ishtar*! Our warehouse must be cleared! Own a vial of movie history—a great conversation piece! Only **$4.50.**

e. As you command water—or any liquid—to flow upward! Could you be tampering with the secrets of the Earth? Your friends will be amazed! Just **$31.99.**

14.2 | Descriptive Statistics

Descriptive statistics is concerned with the accumulation of data, measures of central tendency, and dispersion.

Measures of Central Tendency

In Section 14.1, we organized data into a frequency distribution and then discussed their presentation in graphical form. However, some properties of data can help us interpret masses of information. We will use the accompanying *Peanuts* cartoon to introduce the notion of *average.*

Do you suppose that Violet's dad bowled better on Monday nights (185 avg) than on Thursday nights (170 avg)? Don't be too hasty to say "yes" before you look at the scores that make up these averages:

	Monday Night	*Thursday Night*
Game 1	175	180
Game 2	150	130
Game 3	160	161
Game 4	180	185
Game 5	160	163
Game 6	183	185
Game 7	287	186
Totals	1,295	1,190

To find the averages used by Violet in the cartoon, we divide these totals by the number of games:

Monday Night   *Thursday Night*

$$\frac{1,295}{7} = 185 \qquad \frac{1,190}{7} = 170$$

If we consider the averages, Violet's dad did better on Mondays; but if we consider the games separately, we see that Violet's dad typically did better on Thursday (five out of seven games). Would any other properties of the bowling scores tell us this fact?

Since we must often add up a list of numbers in statistics, as we did above, we use the symbol Σx to mean *the sum of all the values that x can assume.* Similarly, Σx^2 means to square each value that x can assume, and then add the results; $(\Sigma x)^2$ means to first add the values and then square the result. The symbol Σ is the Greek capital letter sigma (which is chosen because S reminds us of "sum").

The average used by Violet is only one kind of statistical measure that can be used. It is the measure that most of us think of when we hear someone use the word *average.* It is called the *mean*. Other statistical measures, called **averages** or **measures of central tendency,** are defined in the following box.

These ideas are essential for understanding many concepts.

The *mean* is the most sensitive average. It reflects the entire distribution and is the most common average.

The *median* gives the middle value. It is useful when there are a few extraordinary values to distort the mean.

The *mode* is the average that measures "popularity." It is possible to have no mode or more than one mode.

Measures of Central Tendency

1. Mean   The number found by adding the data and then dividing by the number of data values. The mean is usually denoted by $\bar{x}$:

$$\bar{x} = \frac{\Sigma x}{n}$$

2. Median   The middle number when the numbers in the data values are arranged in order of size. If there are two middle numbers (in the case of an even number of data), the median is the mean of these two middle numbers.

3. Mode   The value that occurs most frequently. If no number occurs more than once, there is no mode. It is possible to have more than one mode.

Violet used the *mean* and called it the average. Let us consider other measures of central tendency for Violet's dad's bowling scores.

MEDIAN   Rearrange the data values from smallest to largest when finding the median:

Monday Night		Thursday Night
150		130
160		161
160		163
175	← Middle number is the median. →	**180**
180		185
183		185
287		186

MODE   Look for the number that occurs most frequently:

Monday Night		Thursday Night
150		130
160 ⎫		161
160 ⎭ ← Most frequent is the mode.		163
175		180
180	Most frequent is the mode. →	⎰ **185**
183		⎱ **185**
287		186

If we compare the three measures of central tendency for the bowling scores, we find the following:

	Monday Night	Thursday Night
Mean	185	170
Median	175	180
Mode	160	185

We are no longer convinced that Violet's dad did better on Monday nights than on Thursday nights. (See highlighted parts for the winning night, according to each measure of central tendency—Thursday wins two out of three.)

Example 1 | Find the mean, median, and mode

Find the mean, median, and mode for the following sets of numbers.
a. 3, 5, 5, 8, 9   **b.** 4, 10, 9, 8, 9, 4, 5   **c.** 6, 5, 4, 7, 1, 9

Solution

Many calculators have built-in statistical functions. The method of inputting data varies from brand to brand, and you should check with your owner's manual. Once the data are entered your calculator will find the mean, median, and sometimes the mode when you press a single statistical button.

a. *Mean*: $\dfrac{\text{SUM OF TERMS}}{\text{NUMBER OF TERMS}} = \dfrac{3 + 5 + 5 + 8 + 9}{5} = \dfrac{30}{5} = 6$

   Median: Arrange in order: 3, 5, **5**, 8, 9. The middle term is the median: 5.

   Mode: The most frequently occurring term is the mode: 5.

b. *Mean*: $\dfrac{4 + 10 + 9 + 8 + 9 + 4 + 5}{7} = \dfrac{49}{7} = 7$

   Median: 4, 4, 5, **8**, 9, 9, 10; the median is 8.

   Mode: The data have two modes: 4 and 9. If data have two modes, we say they are **bimodal.**

c. *Mean*: $\dfrac{6 + 5 + 4 + 7 + 1 + 9}{6} = \dfrac{32}{6} \approx 5.33$

   Median: 1, 4, <u>5, 6</u>, 7, 9; the median is $\frac{11}{2} = 5.5$.

   $\frac{5 + 6}{2} = \frac{11}{2}$

   Mode: There is **no mode** because no term appears more than once.

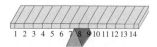

FIGURE 14.26 Fulcrum and plank model for mean

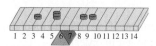

FIGURE 14.27 Balance point for the data set 3, 5, 5, 8, 9 is 6

A rather nice physical model illustrates the idea of the mean. Consider a seesaw that consists of a plank and a movable support (called *a fulcrum*). We assume that the plank has no weight and is marked off into units as shown in Figure 14.26.

Now let's place some 1-lb weights in the position of the numbers in some given distribution. The balance point for the plank is the mean. For example, consider the data from part **a** of Example 1: 3, 5, 5, 8, 9. If weights are placed on these locations on the plank, the balance point is 6, as shown in Figure 14.27.

COMPUTATIONAL WINDOW

It is relatively easy to illustrate the basic operations using a calculator. However, as the processes become more complex, calculators not only become more essential but at the same time become more dependent upon the particular model of calculator you might own. In this book, we have made the decision not to be dependent on a particular model or brand of calculator. For this reason, at this point we recommend you consult the owner's manual or Web site for your model of calculator. Many calculators have a [STAT] menu that finds the mean, median, and mode for a given data set.

Example 2 | Interpret a bar graph

The graph shown in Figure 14.28 represents the distribution of grades in Clark's math class.

Which of the statements is true?
A. The mean and the mode are the same.
B. The mean is greater than the mode.
C. The median is less than the mode.
D. The mean is less than the mode.

Notice that you are given data values but are not given the corresponding frequencies or relative frequencies. Therefore, we are not able to compute the values of the mean and median.

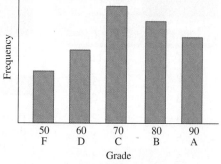

Figure 14.28 Grade distribution for Clark's math class

Solution

The only measure of central tendency we can determine exactly is the mode: *The mode is the value under the tallest bar.* We see the mode is 70. We now estimate the mean. In finding the mean, the number 70 would be added most frequently. If the value of 60 occurred the same number of times as 80, the data would have a mean of 70, but we see there are more 80s than there are 60s, so the mean must be greater than 70. Similarly, we see that adding the 50s and 90s would give us a mean of more than 70. Thus, the mean is greater than 70, and the correct answer is B.

Example 3 | Mean, median, and mode for table values

Consider Table 14.5, which shows the number of days one must wait for a marriage license in the various states in the United States What are the mean, the median, and the mode for these data?

Solution

Mean: To find the mean, we could, of course, add all 50 individual numbers, but instead, notice that

0 occurs 25 times, so write	0×25
1 occurs 1 time, so write	1×1
2 occurs 1 time, so write	2×1
3 occurs 19 times, so write	3×19
4 occurs 1 time, so write	4×1
5 occurs 3 times, so write	5×3

Thus, the mean is

$$\bar{x} = \frac{0 \times 25 + 1 \times 1 + 2 \times 1 + 3 \times 19 + 4 \times 1 + 5 \times 3}{50} = \frac{79}{50} = 1.58$$

Median: Since the median is the middle number and there are 50 values, the median is the mean of the 25th and 26th numbers (when they are arranged in order):

$$\begin{matrix} \text{25th term is } 0 \\ \text{26th term is } 1 \end{matrix} \Big\} \quad \frac{0+1}{2} = \frac{1}{2}$$

Mode: The mode is the value that occurs most frequently, which is 0.

When finding the mean from a frequency distribution, you are finding what is called a *weighted mean*.

TABLE 14.5

Wait Time for a U.S. Marriage License

Day's Wait	Frequency
0	25
1	1
2	1
3	19
4	1
5	3
Total	50

Weighted Mean

If a list of scores $x_1, x_2, x_3, \ldots, x_n$ occurs $w_1, w_2, \ldots, w_n$ times, respectively, then the **weighted mean** is

$$\bar{x} = \frac{\Sigma(w \cdot x)}{\Sigma w}$$

Example 4 | Find a weighted mean

A sociology class is studying family structures and the professor asks each student to state the number of children in his or her family. The results are summarized in Table 14.6.

What is the average number of children in the families of students in this sociology class?

TABLE 14.6

Family Data

Number of Children	Number of Students
1	11
2	7
3	3
4	2
5	1
6	1

Solution   We need to find the weighted mean, where x represents the number of students and w the population (number of families).

$$\bar{x} = \frac{\Sigma(w \cdot x)}{\Sigma w}$$

$$= \frac{1 \cdot 11 + 2 \cdot 7 + 3 \cdot 3 + 4 \cdot 2 + 5 \cdot 1 + 6 \cdot 1}{11 + 7 + 3 + 2 + 1 + 1} = \frac{53}{25}$$

$$= 2.12$$

There is an average of two children per family.

Measures of Position

The median divides the data into two equal parts, with half the values above the median and half below the median, so the median is called a **measure of position.** Sometimes we use benchmark positions that divide the data into more than two parts. **Quartiles,** denoted by Q_1 (first quartile), Q_2 (second quartile), and Q_3 (third quartile), divide the data into four equal parts. **Deciles** are nine values that divide the data into ten equal parts, and **percentiles** are 99 values that divide the data into 100 equal parts. For example, when you take the Scholastic Assessment Test (SAT), your score is recorded as a percentile score. If you scored in the 92nd percentile, it means that you scored better than approximately 92% of those who took the test.

Example 5   Divide exam scores into quartiles

The test results for Professor Hunter's midterm exam are summarized in Table 14.7.

TABLE 14.7

		Grade Distribution
Scores	Number (frequency)	Actual Scores
91–100	4	95, 95, 92, 91
81–90	7	90, 89, 88, 85, 85, 84, 81
65–80	16	80, 78, 78, 75, 74, 74, 74, 72, 70, 70, 70, 69, 68, 65, 65, 65
0–64	3	62, 59, 48

Divide these scores into quartiles.

Solution   The quartiles are the three scores that divide the data into four parts. The first quartile is the data value that separates the lowest 25% of the scores from the remaining scores; the 2nd quartile is the value that separates the lower 50% of the scores from the remainder. Note that the 2nd quartile is the same as the median since the median divides the scores so that 50% are above and 50% are below. The 3rd quartile is the value that separates the lower 75% of the scores from the upper 25%.

Begin by noting the number of scores: $4 + 7 + 16 + 3 = 30$.

First quartile: $0.25(30) = 7.5$, so Q_1 (the first quartile) is the 8th lowest score. From Table 14.7, we see that this score is 69.

Second quartile: Q_2 the second quartile score, is the median, which is the mean of the 15th and 16th scores from the bottom. $\frac{74 + 75}{2} = 74.5$.

Third quartile: $0.75(30) = 22.5$, so Q_3 (the third quartile score) is 23 scores from the bottom (or the 8th from the top). From Table 14.7, we see this score is 85.

Measures of Dispersion

The measures we've been discussing can help us interpret information, but they do not give the entire story. For example, consider these sets of data:

Set A: $\{8, 9, 9, 9, 10\}$ *Mean:* $\dfrac{8 + 9 + 9 + 9 + 10}{5} = 9$

Median: 9
Mode: 9

Set B: $\{2, 9, 9, 12, 13\}$ *Mean:* $\dfrac{2 + 9 + 9 + 12 + 13}{5} = 9$

Median: 9
Mode: 9

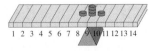

a. $A = \{8, 9, 9, 9, 10\}$

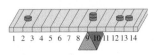

b. $B = \{2, 9, 9, 12, 13\}$

FIGURE 14.29 Visualization of dispersion of sets of data

Notice that, for sets A and B, the measures of central tendency do not distinguish the data. However, if you look at the data placed on planks, as shown in Figure 14.29, you will see that the data in Set B are relatively widely dispersed along the plank, whereas the data in Set A are clumped around the mean.

We'll consider three **measures of dispersion**: the *range,* the *standard deviation,* and the *variance.*

Range

The **range** of a set of data is the difference between the largest and the smallest numbers in the set.

Example 6 Find the range

Find the ranges for the data sets in Figure 14.29:
a. Set $A = \{8, 9, 9, 9, 10\}$ **b.** Set $B = \{2, 9, 9, 12, 13\}$

Solution   Notice from Figure 14.29 that the mean for each of these sets of data is the same. The range is found by comparing the difference between the largest and smallest values in the set.
a. $10 - 8 = 2$ **b.** $13 - 2 = 11$

Notice that the range is determined by only the largest and the smallest numbers in the set; it does not give us any information about the other numbers.

The range is used, along with quartiles, to construct a statistical tool called a *box plot.* For a given set of data, a **box plot** consists of a rectangular box positioned above a numerical scale, drawn from Q_1 (the first quartile) to Q_3 (the third quartile). The median (Q_2, or second quartile) is shown as a dashed line, and a segment is extended to the left to show the distance to the minimum value; another segment is extended to the right for the maximum value. Figure 14.30 shows a box plot for the data in Example 5.

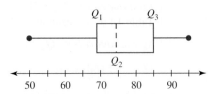

FIGURE 14.30 Box plot for grade distribution

Example **7** **Draw a box plot**

Draw a box plot for the ages of the actors who received the Best Actor award at the Academy Awards, as reported in Table 14.3, page 662.

Solution   We find the quartiles by first finding Q_2, the median. Since there are 83 entries, the median is the middle item. For this example, the middle (42nd) item is 42.

Next, we find Q_1 (first quartile), which is the median of all items below Q_2. Since there are 42 elements in this lower portion of the data set, the median of this group is the mean of the 21st and 22nd values, or 38.

Finally, find Q_3 (third quartile), which is the median of all items above Q_2. Since there are 42 elements in this upper portion of the data set, the median of this group is 48. The minimum age is 29 (Adrien Brody) and the maximum age is 76 (Henry Fonda), so we have a box plot as shown in Figure 14.31.

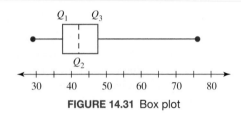

FIGURE 14.31 Box plot

Sometimes a box plot is called a *box-and-whisker plot*. Its usefulness should be clear when you look at Figure 14.31. A box plot shows:

1. the median (a measure of central tendency);
2. the location of the middle half of the data (represented by the extent of the box);
3. the range (a measure of dispersion);
4. the skewness (the nonsymmetry of both the box and the whiskers).

The *variance* and *standard deviation* are measures that use all the numbers in the data set to give information about the dispersion. When finding the variance, we must make a distinction between the **variance of the entire population** and the **variance of a random sample** from the population. When the variance is based on a set of sample scores, it is denoted by s^2; and when it is based on all scores in a population, it is denoted by σ^2 (σ is the lowercase Greek letter sigma). We will discuss populations and samples in Section 14.5, but for this section, we will assume the given data represents a sample rather than the entire population. The variance for a random sample is found by

$$s^2 = \frac{\Sigma(x - \bar{x})^2}{n - 1}$$

To understand this formula for the sample variance, we will consider an example before summarizing a procedure. Again, let's use the data sets we worked with in Example 6.

Set $A = \{8, 9, 9, 9, 10\}$ Set $B = \{2, 9, 9, 12, 13\}$
Mean is 9.   Mean is 9.

Find the deviations by subtracting the mean from each term:

$8 - 9 = -1$	$2 - 9 = -7$
$9 - 9 = 0$	$9 - 9 = 0$
$9 - 9 = 0$	$9 - 9 = 0$
$9 - 9 = 0$	$12 - 9 = 3$
$10 - 9 = 1$	$13 - 9 = 4$
↑	↑
Mean	Mean

If we sum these deviations (to obtain a measure of the total deviation), in each case we obtain 0, because the positive and negative differences "cancel each other out." Next we calculate the *square of each of these deviations:*

Set $A = \{8, 9, 9, 9, 10\}$	Set $B = \{2, 9, 9, 12, 13\}$
$(8 - 9)^2 = (-1)^2 = 1$	$(2 - 9)^2 = (-7)^2 = 49$
$(9 - 9)^2 = 0^2 = 0$	$(9 - 9)^2 = 0^2 = 0$
$(9 - 9)^2 = 0^2 = 0$	$(9 - 9)^2 = 0^2 = 0$
$(9 - 9)^2 = 0^2 = 0$	$(12 - 9)^2 = 3^2 = 9$
$(10 - 9)^2 = 1^2 = 1$	$(13 - 9)^2 = 4^2 = 16$

Finally, we find the sum of these squares and divide by one less than the number of items to obtain the variance:

Set *A*:

Set *B*:

$$s^2 = \frac{1 + 0 + 0 + 0 + 1}{5 - 1} = \frac{2}{4} = 0.5 \qquad s^2 = \frac{49 + 0 + 0 + 9 + 16}{5 - 1} = \frac{74}{4} = 18.5$$

The larger the variance, the more dispersion there is in the original data. However, we will continue to develop a true picture of the dispersion. Since we squared each difference (to eliminate the canceling effect of positive and negative differences), it seems reasonable that we should find the square root of the variance as a more meaningful measure of dispersion. This number, called the **standard deviation,** is denoted by the lowercase Greek letter sigma (σ) when it is based on a population and by *s* when it is based on a sample. You will need a calculator to find square roots. For the data sets of Example 6:

Set *A*: $s = \sqrt{0.5}$ Set *B*: $s = \sqrt{18.5}$

≈ 0.707 ≈ 4.301

We summarize these steps in the following box.

Standard Deviation

The standard deviation of a sample, denoted by *s*, is the square root of the variance. To find it, carry out these steps:

Step 1 Determine the mean of the set of numbers.

Step 2 Subtract the mean from each number in the set.

Step 3 Square each of these differences.

Step 4 Find the sum of the squares of the differences.

Step 5 Divide this sum by one less than the number of pieces of data. This gives the *variance* of the sample.

Step 6 Take the square root of the variance. This is the *standard deviation* of the sample.

STOP Spend some time with this box; make sure you understand not only the procedure, but also the concept.

Example **8** **Find the standard deviation for a math test**

Suppose that Hannah received the following test scores in a math class: 92, 85, 65, 89, 96, and 71. Find *s*, the standard deviation, for her test scores.

Solution

Step 1 $\bar{x} = \dfrac{92 + 85 + 65 + 89 + 96 + 71}{6} = 83$ This is the mean.

Steps 2–4 We summarize these steps in table format:

Score	Square of the Deviation from the Mean
92	$(92 - 83)^2 = 9^2 = 81$
85	$(85 - 83)^2 = 2^2 = 4$
65	$(65 - 83)^2 = (-18)^2 = 324$
89	$(89 - 83)^2 = 6^2 = 36$
96	$(96 - 83)^2 = 13^2 = 169$
71	$(71 - 83)^2 = (-12)^2 = 144$

Step 5 Divide the sum by 5 (one less than the number of scores):

$$\frac{81 + 4 + 324 + 36 + 169 + 144}{6 - 1} = \frac{758}{5} = 151.6$$

We note that this number, 151.6, is called the variance. If you do not have access to a calculator, you can use the variance as a measure of dispersion. However, we assume you have a calculator and can find the standard deviation.

Step 6 $s = \sqrt{\dfrac{758}{5}} \approx 12.31$

How can we use the standard deviation? We will begin the discussion in Section 14.3 with this question. For now, however, we will give one example. Suppose that Hannah obtained 65 on an examination for which the mean was 50 and the standard deviation was 15, whereas Søren in another class scored 74 on an examination for which the mean was 80 and the standard deviation was 3. Did Hannah or Søren do better in her respective class? We see that Hannah scored one standard deviation *above* the mean $(50 + 15 = 65)$, whereas Søren scored two standard deviations *below* the mean $(80 - 2 \times 3 = 74)$; therefore, Hannah did better compared to her classmates than did Søren.

> Baseball, as everyone knows, is played on a field with a bat and a ball. But baseball is also played on paper and computers with numbers and decimals. The name of this game-within-a-game is statistics, and to some fans, it is more engrossing and more real than the action on the field. Most sports keep statistics, but baseball statistics are in a league by themselves.

Problem Set 14.2

Level 1

1. IN YOUR OWN WORDS   What do we mean by *average*?

2. IN YOUR OWN WORDS   What is a measure of central tendency?

3. IN YOUR OWN WORDS   What is a measure of dispersion?

4. IN YOUR OWN WORDS   Compare and contrast mean, median, and mode.

5. IN YOUR OWN WORDS   In 1989, Andy Van Slyke's batting average was better than Dave Justice's; and in 1990 Andy once again beat Dave. Does it follows that Andy's combined 1989–1990 batting average is better than Dave's?

	Andy				Dave		
	Hits	AB	Avg.		Hits	AB	Avg.
1989	113	476	0.237		12	51	0.235
1990	140	493	0.284		124	439	0.282

6. IN YOUR OWN WORDS   A professor gives five exams. Two students' scores have the same mean, although one student did better on all the tests except one. Give an example of such scores.

7. IN YOUR OWN WORDS   A professor gives six exams. Two students' scores have the same mean, although one student's scores have a small standard deviation and the other student's scores have a large standard deviation. Give an example of such scores.

8. IN YOUR OWN WORDS   Comment on the following quotation, which was printed in the May 28, 1989, issue of *The Community, Technical, and Junior College TIMES*: "Half the American students are below average."

9. IN YOUR OWN WORDS   Comment on the following quotation: "I just received my SAT scores and I scored in the 78th percentile in English, and 50th percentile in math."

10. Is the standard deviation always smaller than the variance?

11. Suppose that a variance is zero. What can you say about the data?

In Problems 12–21, find the three measures of central tendency (the mean, median, and mode).

12. 1, 2, 3, 4, 5

13. 17, 18, 19, 20, 21

14. 103, 104, 105, 106, 107

15. 765, 766, 767, 768, 769

16. 3, 5, 8, 13, 21

17. 1, 4, 9, 16, 25

18. 79, 90, 95, 95, 96

19. 70, 81, 95, 79, 85

20. 1, 2, 3, 3, 3, 4, 5

21. 0, 1, 1, 2, 3, 4, 16, 21

In Problems 22–31, find the range and the standard deviation (correct to two decimal places). If you do not have a calculator, find the range and the variance.

22. 1, 2, 3, 4, 5

23. 17, 18, 19, 20, 21

24. 103, 104, 105, 106, 107

25. 765, 766, 767, 768, 769

26. 3, 5, 8, 13, 21

27. 1, 4, 9, 16, 25

28. 79, 90, 95, 95, 96

29. 70, 81, 95, 79, 85

30. 1, 2, 3, 3, 3, 4, 5

31. 0, 1, 1, 2, 3, 4, 16, 21

32. By looking at Problems 12–15 and 22–25, and discovering a pattern, answer the following questions about the numbers 217,849, 217,850, 217,851, 217,852, and 217,853.
   a. What is the mean?
   b. What is the standard deviation?

33. The 2001 monthly cost for health care for an employee and two dependents is given in the following table.

Provider	Cost	Provider	Cost
Maxicare	$415.24	Kaiser	$433.50
Cigna	$424.77	Aetna	$436.11
Health Net	$427.48	Blue Shield	$442.28
Pacific Care	$428.05	Omni Healthcare	$457.86
Redwoods Plan	$431.52	Lifeguard	$457.94

Draw a box plot for the monthly costs of health care (ranked from lowest to highest).

34. Find the mean, the median, and the mode of the following salaries of employees of the Moe D. Lawn Landscaping Company:

Salary	Frequency
$25,000	4
$28,000	3
$30,000	2
$45,000	1

PictureNet/Corbis

35. G. Thumb, the leading salesperson for the Moe D. Lawn Landscaping Company, turned in the following summary of sales for the week of October 23–28:

Date	No. of Clients
Oct. 23	12
Oct. 24	9
Oct. 25	10
Oct. 26	16
Oct. 27	10
Oct. 28	21

Find the mean, median, and mode.

36. Find the mean, the median, and the mode of the following scores:

Test Score	Frequency
90	1
80	3
70	10
60	5
50	2

37. A class obtained the following scores on a test:

Test Score	Frequency
90	1
80	6
70	10
60	4
50	3
40	1

Find the mean, the median, the mode, and the range for the class.

38. A class obtained the following test scores:

Test Score	Frequency
90	2
80	4
70	9
60	5
50	3
40	1
30	2
0	4

Find the mean, median, mode, and range for the class.

39. The county fair reported the following total attendance (in thousands).

Year	2005	2004	2003	2002	2001	2000
Attendance	366	391	358	373	346	364

Find the mean, median, and mode for the attendance figures (rounded to the nearest thousand).

40. The following salaries for the executives of a certain company are known:

Position	Salary
President	$170,000
1st VP	$140,000
2nd VP	$120,000
Supervising manager	$54,000
Accounting manager	$40,000
Personnel manager	$40,000
Dept. manager	$30,000
Dept. manager	$30,000

Find the mean, the median, and the mode. Which measure seems to best describe the average executive salary for the company?

41. The graph in Figure 14.32 shows the distribution of scores on an examination.

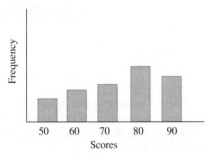

FIGURE 14.32 Scores on an examination

Which of the following statements is true about the distribution?
A. The mean is less than the mode.
B. The median is the same as the mode.
C. The median is greater than the mode.
D. The mean is the same as the mode.

42. The graph in Figure 14.33 shows the salaries of the employees for a certain company.

FIGURE 14.33 Company salaries

Which of the following statements is true about the distribution?
A. The mean is less than the mode.
B. The median is the same as the mode.
C. The median is greater than the mode.
D. The mean is the same as the mode.

43. Shoe size for a team of football players ranges from size 10 to 14. The majority of players wear size 12, and the number of players wearing size 11 is the same as the number wearing size 13. It is also true that the number of players wearing size 10 is the same as the number wearing size 14. Which statement is true about the distribution?
A. The median is less than the mode.
B. The mean is less than the mode.
C. The mean is the same as the median.
D. The median is greater than the mean.

44. In a survey, members of a math class were asked how many minutes per week they exercised. There were three possible responses: 10 minutes or less (70% checked this), between 10 and 60 minutes (20% checked this one), and 60 minutes or more (10%). Which statement is true about the distribution?
A. The mode is less than the mean.
B. The median is less than the mode.
C. The mean is less than the median.
D. The mean is the same as the mode.

Use the nutritional information about candy bars given in Table 14.8 to answer the questions in Problems 45 and 46.

TABLE 14.8				
	Nutritional Information about Popular Candy Bars			
Company	**Serving Size (g)**	**Total Size (g)**	**Calories from Fat**	**Total Fat (g)**
Almond Joy	36	180	90	10
Baby Ruth	60	280	110	12
Butterfinger	42	200	70	8
Clark Bar	50	240	90	10
5th Avenue	57	280	110	12
Heath Bar	40	210	110	13
Hershey	43	230	120	13
Hershey w/alm.	41	230	120	14
100 Grand	43	200	70	8
Kit Kat	42	220	110	12
Milky Way	61	280	100	11
Mr. Good Bar	49	280	160	18
Nestle milk choc.	41	220	110	13
Nut Rageous	45	250	140	15
Peppermint Patty	42	170	35	4
Reese's Cup	45	240	130	14
Rolo	54	230	110	12
Snickers	59	280	120	14
3 Musketeers	60	260	70	8
Twix	57	280	130	14

45. a. What is the mean calories from fat?
  b. What is the median (in grams) of total fat?
  c. What is the mode (in grams) of the serving size?

46. a. Find the three quartiles for serving size.
  b. Draw a box plot for the calories from fat.
  c. What are the mean and standard deviation for total size in Table 14.8?

Find the standard deviation (rounded to the nearest unit) for the data indicated in Problems 47–52.

47. Problem 35

48. Problem 34

49. Problem 37

50. Problem 36

51. Problem 39

52. Problem 38

53. The number of miles driven on each of five tires was 17,000, 19,000, 19,000, 20,000, and 21,000. Find the mean, the range, and the standard deviation (rounded to the nearest unit) for these mileages.

54. IN YOUR OWN WORDS Roll a single die until all six numbers occur at least once. Repeat the experiment 20 times. Find the mean, the median, the mode, and the range of the number of tosses.

55. IN YOUR OWN WORDS Roll a pair of dice until all 11 possible sums occur at least once. Repeat the experiment 20 times. Find the mean, the median, the mode, and the range of the number of tosses.

56. When you take the Scholastic Assessment Test (SAT), your score is recorded as a percentile score. If you scored in the 92nd percentile, it means that you scored better than approximately 92% of those who took the test.
   a. If Lisa's score was 85 and that score was the 23rd score from the top in a class of 240 scores, what is Lisa's percentile rank?
   b. Lee has received a percentile rank of 85% in a class of 50 students. What is Lee's rank in the class?

Level 3

57. The mean defined in the text is sometimes called the *arithmetic mean* to distinguish it from other possible means. For example, a different mean, called the *harmonic mean* (H.M.), is used to find average speeds. This mean is defined to be the sum of the reciprocals of all scores divided into the number of scores. For example, the harmonic mean of the numbers 4, 5, 6, 6, 7, 8 is

$$\text{H.M.} = \frac{n}{\Sigma\frac{1}{x}}$$

$$= \frac{6}{\frac{1}{4} + \frac{1}{5} + \frac{1}{6} + \frac{1}{6} + \frac{1}{7} + \frac{1}{8}}$$

$$\approx 5.7$$

   a. Find the arithmetic and harmonic mean of the numbers 2, 2, 5, 5, 7, 8, 8, 9, 9, 10.
   b. A trip from San Francisco to Disneyland is approximately 460 miles. If the southbound trip averaged 52 mph and the return trip averaged 61 mph, what is the average speed for the round trip? Compare the arithmetic and harmonic means.

58. The mean defined in the text is sometimes called the *arithmetic mean* to distinguish it from other possible means. For example, a different mean, called the *geometric mean* (G.M.), is used in business and economics for finding average

rates of change, average rates of growth, or average ratios. This mean is defined to be the nth root of the product of the numbers. For example, the geometric mean of 4, 5, and 6 is

$$\text{G.M.} = (4 \cdot 5 \cdot 6)^{1/3}$$

$$= \sqrt[3]{120}$$

$$\approx 4.9$$

   a. Find the arithmetic and geometric means for the numbers 2, 3, 5, 7, 7, 8, and 10.
   b. The growth rates for three cities are 1.5%, 2.0%, and 0.9%. Compare the arithmetic and geometric means.

Problem Solving 3

59. Journal Problem IN YOUR OWN WORDS The example about averages of bowling scores at the beginning of this section is an instance of what is known as *Simpson's paradox.* The following example illustrates this paradox.[*]

	Player A		
	At Bat	*Hits*	*Average*
Against right-handed pitchers	202	45	0.223
Against left-handed pitchers	250	71	0.284
Overall	452	116	0.257

	Player B		
	At Bat	*Hits*	*Average*
Against right-handed pitchers	250	58	0.232
Against left-handed pitchers	108	32	0.296
Overall	358	90	0.251

Notice that Player A has a better overall batting average than Player B, but yet is worse against both right-handed pitchers and left-handed pitchers. The following is an algebraic statement of Simpson's Paradox.

> *Consider two populations for which the overall rate r of occurrence of some phenomenon in population A is greater than the corresponding rate R in population B. Suppose that each of the two populations is composed of the same two categories C_1 and C_2, and the rates of occurrence of the phenomenon for the two categories in population A are r_1 and r_2, and in population B are R_1 and R_2. If $r_1 < R_1$ and $r_2 < R_2$, despite the fact that $r > R$, then Simpson's paradox is said to have occurred.*

   a. Relate the variables in this statement to the numbers in the example.
   b. An example of Simpson's paradox from real life is the fact that the overall federal income tax rate increased from 1974 to 1978, but decreased for each bracket. Make up a fictitious example to show how this might be possible.

60. Let Q_1, Q_2, and Q_3 be the quartiles for a large population of scores. Is the following true or false (with reasons):

$$Q_2 - Q_1 = Q_3 - Q_2$$

[*]From "Instances of Simpson's Paradox," by Thomas R. Knapp, *College Mathematics Journal*, July 1985, pp. 209–211.

14.3 | The Normal Curve

Sometimes we represent frequencies in a cumulative way, especially when we want to find the position of one case relative to the performance of the group.

Cumulative Distributions

A **cumulative frequency** is the sum of all preceding frequencies in which some order has been established.

Example 1 | Find the mean, median, and mode

TABLE 14.9

Number of Previous Appearances

Number	Percent	Cumulative Percent
0	21%	21%
1	25%	46%
2	24%	70%
3	13%	83%
4	8%	91%
5	5%	96%
6 or more	4%	100%

A judge ordered a survey to determine how many of the offenders appearing in her court during the past year had three or more previous appearances. The accumulated data are shown in Table 14.9.

Note the last column shows the cumulative percent. This is usually called the *cumulative relative frequency*. Use the cumulative relative frequency to find the percent who had three or more previous appearances, and then find the mean, median, and mode.

Solution   From the cumulative relative frequency we see that 70% had 2 or fewer court appearances; we see that since the total is 100%, 30% must have had 3 or more previous appearances.

The mean is found (using the idea of a weighted mean). Note that the sum is 100%, or 1:

$$\bar{x} = \frac{0(0.21) + 1(0.25) + 2(0.24) + 3(0.13) + 4(0.08) + 5(.05) + 6(0.04)}{1}$$

$$= 1.93$$

The median is the number of court appearances for which the cumulative percent first exceeds 50%; we see that this is 2 court appearances.

The mode is the number of court appearances that occurs most frequently; we see that this is 1 court appearance.

Bell-Shaped Curves

Cartoon by David Pascal.

The cartoon in the margin suggests that most people do not like to think of themselves or their children as having "normal intelligence." But what do we mean by *normal* or *normal intelligence*?

Suppose we survey the results of 20 children's scores on an IQ test. The scores (rounded to the nearest 5 points) are 115, 90, 100, 95, 105, 95, 105, 105, 95, 125, 120, 110, 100, 100, 90, 110, 100, 115, 105, and 80. We can find $\bar{x} = 103$ and $s \approx 10.93$. A frequency graph of these data is shown in part **a** of Figure 14.34. If we consider 10,000 scores instead of only 20, we might obtain the frequency distribution shown in part **b** of Figure 14.34.

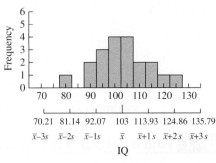

a. IQs of 20 children

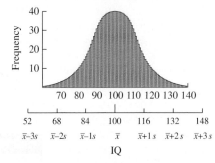

b. IQs of 10,000 children

FIGURE 14.34 Frequency distributions for IQ scores

The data illustrated in Figure 14.34 approximate a commonly used curve called a *normal frequency curve*, or simply a **normal curve.** (See Figure 14.35.)

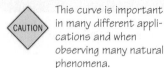

This curve is important in many different applications and when observing many natural phenomena.

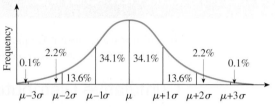

FIGURE 14.35 A normal curve

If we obtain the frequency distribution of a large number of measurements (as with IQ), the corresponding graph tends to look normal, or **bell-shaped.** The normal curve has some interesting properties. In it, the mean, the median, and the mode all have the same value, and all occur exactly at the center of the distribution; we denote this value by the Greek letter mu (μ). The standard deviation for this distribution is σ (sigma). Roughly 68% of all values lie within the region from 1 standard deviation below to 1 standard deviation above the mean. About 95% lie within 2 standard deviations on either side of the mean, and virtually all (99.8%) values lie within 3 standard deviations on either side. These percentages are the same regardless of the particular mean or standard deviation.

Hallgrímskirkja, a modern Lutheran church in Reykjavik, Iceland, models a normal curve.

The normal distribution is a **continuous** (rather than a discrete) **distribution,** and it extends indefinitely in both directions, never touching the *x*-axis. It is symmetric about a vertical line drawn through the mean, μ. Graphs of this curve for several choices of σ are shown in Figure 14.36.

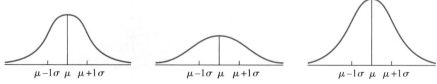

FIGURE 14.36 Variations of normal curves

Example **2** **Find a normal distribution**

Predict the distribution of IQ scores of 1,000 people if we assume that IQ scores are normally distributed, with a mean of 100 and a standard deviation of 15.

Solution   First, find the breaking points around the mean. For $\mu = 100$ and $\sigma = 15$:

$$\mu + \sigma = 100 + 15 = 115 \qquad \mu - \sigma = 100 - 15 = 85$$
$$\mu + 2\sigma = 100 + 2(15) = 130 \qquad \mu - 2\sigma = 100 - 2(15) = 70$$
$$\mu + 3\sigma = 100 + 3(15) = 145 \qquad \mu - 3\sigma = 100 - 3(15) = 55$$

We use Figure 14.35 to find that 34.1% of the scores will be between 100 and 115 (i.e., between μ and $\mu + 1\sigma$):

$$0.341 \times 1,000 = 341$$

About 13.6% will be between 115 and 130 (between $\mu + 1\sigma$ and $\mu + 2\sigma$):

$$0.136 \times 1,000 = 136$$

About 2.2% will be between 130 and 145 (between $\mu + 2\sigma$ and $\mu + 3\sigma$):

$$0.022 \times 1,000 = 22$$

About 0.1% will be above 145 (more than $\mu + 3\sigma$):

$$0.001 \times 1,000 = 1$$

The distribution for intervals below the mean is identical, since the normal curve is the same to the left and to the right of the mean. The distribution is shown in the margin.

Scores	Percent	Expected Number
Above 55 to 70	2.2%	22
Above 70 to 85	13.6%	136
Above 85 to 100	34.1%	341
Above 100 to 115	34.1%	341
Above 115 to 130	13.6%	136
Above 130 to 145	2.2%	22
Above 145	0.1%	1
Totals	100.0%	1,000

Example 3   Find a grading distribution

Suppose that an instructor "grades on a curve." Show the grading distribution on an examination of 45 students, if the scores are normally distributed with a mean of 73 and a standard deviation of 9.

Solution   Grading on a curve means determining students' grades according to the percentages shown in Figure 14.37.

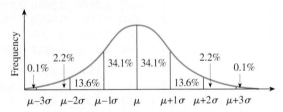

FIGURE 14.37 Grade distribution for a class "graded on a curve"

We calculate these numbers as shown:

Calculation	Scores	Grade	Calculation	Number
Two or more std. dev. above μ	91–100	A	$0.023 \times 45 = 1.04$	1
$\mu + 2\sigma = 73 + 2 \cdot 9 = 91$ $\mu + 1\sigma = 73 + 1 \cdot 9 = 82$	82–90	B	$0.136 \times 45 = 6.12$	6
Mean: $\mu = 73$	64–81	C	$0.682 \times 45 = 30.69$	31
$\mu - 1\sigma = 73 - 1 \cdot 9 = 64$ $\mu - 2\sigma = 73 - 2 \cdot 9 = 55$	55–63	D	$0.136 \times 45 = 6.12$	6
Two or more std. dev. below μ	0–54	F	$0.023 \times 45 = 1.04$	1

Grading on a curve means that the person with the top score in this class of 45 receives an A; the next 6 ranked persons (from the top) receive B grades; the bottom score in the class receives an F; the next 6 ranked persons (from the bottom) receive D grades; and, finally, the remaining 31 persons receive C grades. Notice that the majority of the class (34.1% + 34.1% = 68.2%) will receive an "average" C grade.

z-Scores

Sometimes we want to know the percent of occurrence for scores that do not happen to be 1, 2, or 3 standard deviations from the mean. For Example 3, we can see from Figure 14.37 that 34.1% of the scores are between the mean and 1 standard deviation above the mean. Suppose we wish to find the percent of scores that are between the mean and 1.2 standard deviations above the mean. To find this percent, we use Table 14.10.

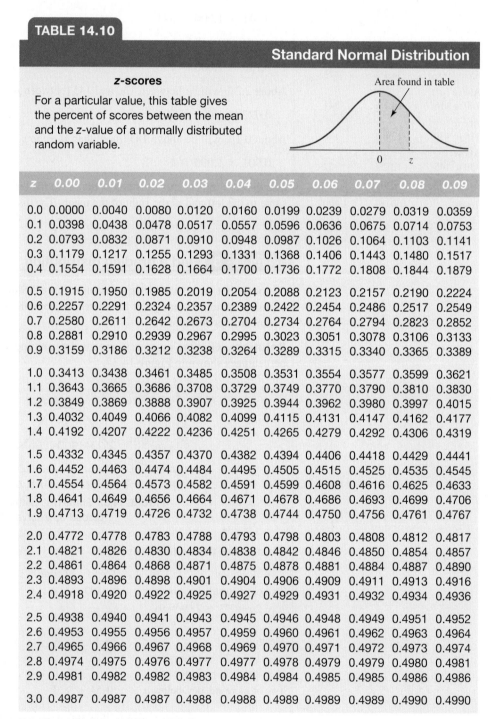

TABLE 14.10

Standard Normal Distribution

z-scores

For a particular value, this table gives the percent of scores between the mean and the z-value of a normally distributed random variable.

z	0.00	0.01	0.02	0.03	0.04	0.05	0.06	0.07	0.08	0.09
0.0	0.0000	0.0040	0.0080	0.0120	0.0160	0.0199	0.0239	0.0279	0.0319	0.0359
0.1	0.0398	0.0438	0.0478	0.0517	0.0557	0.0596	0.0636	0.0675	0.0714	0.0753
0.2	0.0793	0.0832	0.0871	0.0910	0.0948	0.0987	0.1026	0.1064	0.1103	0.1141
0.3	0.1179	0.1217	0.1255	0.1293	0.1331	0.1368	0.1406	0.1443	0.1480	0.1517
0.4	0.1554	0.1591	0.1628	0.1664	0.1700	0.1736	0.1772	0.1808	0.1844	0.1879
0.5	0.1915	0.1950	0.1985	0.2019	0.2054	0.2088	0.2123	0.2157	0.2190	0.2224
0.6	0.2257	0.2291	0.2324	0.2357	0.2389	0.2422	0.2454	0.2486	0.2517	0.2549
0.7	0.2580	0.2611	0.2642	0.2673	0.2704	0.2734	0.2764	0.2794	0.2823	0.2852
0.8	0.2881	0.2910	0.2939	0.2967	0.2995	0.3023	0.3051	0.3078	0.3106	0.3133
0.9	0.3159	0.3186	0.3212	0.3238	0.3264	0.3289	0.3315	0.3340	0.3365	0.3389
1.0	0.3413	0.3438	0.3461	0.3485	0.3508	0.3531	0.3554	0.3577	0.3599	0.3621
1.1	0.3643	0.3665	0.3686	0.3708	0.3729	0.3749	0.3770	0.3790	0.3810	0.3830
1.2	0.3849	0.3869	0.3888	0.3907	0.3925	0.3944	0.3962	0.3980	0.3997	0.4015
1.3	0.4032	0.4049	0.4066	0.4082	0.4099	0.4115	0.4131	0.4147	0.4162	0.4177
1.4	0.4192	0.4207	0.4222	0.4236	0.4251	0.4265	0.4279	0.4292	0.4306	0.4319
1.5	0.4332	0.4345	0.4357	0.4370	0.4382	0.4394	0.4406	0.4418	0.4429	0.4441
1.6	0.4452	0.4463	0.4474	0.4484	0.4495	0.4505	0.4515	0.4525	0.4535	0.4545
1.7	0.4554	0.4564	0.4573	0.4582	0.4591	0.4599	0.4608	0.4616	0.4625	0.4633
1.8	0.4641	0.4649	0.4656	0.4664	0.4671	0.4678	0.4686	0.4693	0.4699	0.4706
1.9	0.4713	0.4719	0.4726	0.4732	0.4738	0.4744	0.4750	0.4756	0.4761	0.4767
2.0	0.4772	0.4778	0.4783	0.4788	0.4793	0.4798	0.4803	0.4808	0.4812	0.4817
2.1	0.4821	0.4826	0.4830	0.4834	0.4838	0.4842	0.4846	0.4850	0.4854	0.4857
2.2	0.4861	0.4864	0.4868	0.4871	0.4875	0.4878	0.4881	0.4884	0.4887	0.4890
2.3	0.4893	0.4896	0.4898	0.4901	0.4904	0.4906	0.4909	0.4911	0.4913	0.4916
2.4	0.4918	0.4920	0.4922	0.4925	0.4927	0.4929	0.4931	0.4932	0.4934	0.4936
2.5	0.4938	0.4940	0.4941	0.4943	0.4945	0.4946	0.4948	0.4949	0.4951	0.4952
2.6	0.4953	0.4955	0.4956	0.4957	0.4959	0.4960	0.4961	0.4962	0.4963	0.4964
2.7	0.4965	0.4966	0.4967	0.4968	0.4969	0.4970	0.4971	0.4972	0.4973	0.4974
2.8	0.4974	0.4975	0.4976	0.4977	0.4977	0.4978	0.4979	0.4979	0.4980	0.4981
2.9	0.4981	0.4982	0.4982	0.4983	0.4984	0.4984	0.4985	0.4985	0.4986	0.4986
3.0	0.4987	0.4987	0.4987	0.4988	0.4988	0.4989	0.4989	0.4989	0.4990	0.4990

Note: For values of z above 3.09, use 0.4999.

First, we introduce some terminology. We use *z-scores* (sometimes called *standard scores*) to determine how far, in terms of standard deviations, a given score is from the mean of the distribution.

| Example | 4 | **Find a probability** |

Find the probability that for a given normal distribution, a score is between the mean and the indicated number of standard deviations above that mean.
a. 1 standard deviation
b. 1.2 standard deviations
c. 1.68 standard deviations

Solution
a. Since this is a normal curve, we know from Figure 14.35 that the percentage is 34.1%. You can also find this on Table 14.10. Look in the row labeled (at the left) 1.0 and in the column headed 0.00: The entry is 0.3413, which is 34.13%. This is denoted by saying that the percentage for $z = 1$ is 34.13%.
b. We are given $z = 1.2$. Look at the entry in the row marked 1.2 in Table 14.10 and the 0.00 column: It is 0.3849.
c. $z = 1.68$; look in Table 14.10 at the row labeled 1.6 and the column headed 0.08 to find the entry 0.4535. This means that 45.35% of the values in a normal distribution are between the mean and 1.68 standard deviations above the mean.

We use the z-score to translate any normal curve into a standard normal curve (the particular normal curve with a mean of 0 and a standard deviation of 1) by using the definition.

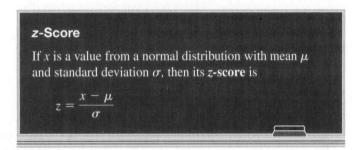

z-Score

If x is a value from a normal distribution with mean μ and standard deviation σ, then its **z-score** is

$$z = \frac{x - \mu}{\sigma}$$

| Example | 5 | **Find probabilities in a normal distribution** |

The Eureka Lightbulb Company tested a new line of lightbulbs and found their lifetimes to be normally distributed, with a mean life of 98 hours and a standard deviation of 13 hours.
a. What percentage of bulbs will last less than 72 hours?
b. What percentage of bulbs will last less than 100 hours?
c. What is the probability that a bulb selected at random will last longer than 111 hours?
d. What is the probability that a bulb will last between 106 and 120 hours?

Solution   Draw a normal curve with mean 98 and standard deviation 13, as shown in Figure 14.38.

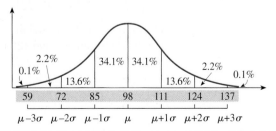

FIGURE 14.38 Lightbulb lifetimes are normally distributed

For this example, we are given $\mu = 98$ and $\sigma = 13$.

a. For $x = 72$, $z = \frac{72 - 98}{13} = -2$.

This is 2 standard deviations below the mean. The percentage we seek is shown in blue in Figure 14.39a.

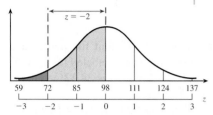
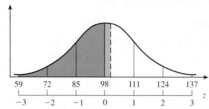
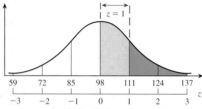

a. 2 standard deviations below the mean   **b.** 0.15 standard deviations above the mean   **c.** 1 standard deviation above the mean

FIGURE 14.39 Using z-scores

About 2.3% (2.2% + 0.1% = 2.3%) will last less than 72 hours. We can also use the z-score to find the percentage. From Table 14.10, the area to the left of $z = -2$ is $0.5 - 0.4772 = 0.0228$, so the probability is about 2.3%.

b. For $x = 100$, $z = \frac{100 - 98}{13} \approx 0.15$.

This is 0.15 standard deviation above the mean. From Table 14.10, we find 0.0596 (this is shown in green in Figure 14.39b). Since we want the percent of values less than 100, we must add 50% for the numbers below the mean (shown in blue). The percentage we seek is

$0.5000 + 0.0596 = 0.5596$ or about 56.0%.

c. For $x = 111$, $z = \frac{111 - 98}{13} = 1$.

This is one standard deviation above the mean, which is the same as the z-score. The percentage we seek is shown in blue (see Figure 14.39c). We know that about 15.9% (13.6% + 2.2% + 0.1% = 15.9%) of the bulbs will last longer than 111 hours, so

$P(\text{bulb life} > 111 \text{ hours}) \approx 0.159$

We can also use Table 14.10. For $z = 1.00$, the table entry is 0.3413, and we are looking for the area to the right, so we compute

$0.5000 - 0.3413 = 0.1587$

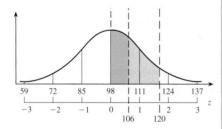

d. We first find the z-scores using $\mu = 98$ and $\sigma = 13$. For $x = 106$, $z = \frac{106 - 98}{13} \approx 0.62$. From Table 14.10, the area between the z-score and the mean is 0.2324 (shown in green in the margin). For $x = 120$, $z = \frac{120 - 98}{13} \approx 1.69$. From Table 14.10, the area between this z-score and the mean is 0.4545. The desired answer (shown in yellow) is approximately

$0.4545 - 0.2324 = 0.2221$

Since percent and probability are the same, we see the probability that the life of the bulb is between 106 and 120 hours is about 22.2%.

Example 5 leads us to the observation that there are three equivalent ideas associated with the normal curve. These are summarized in the following box.

Normal Curve Interpretations

In a standard normal curve, the following three quantities are equivalent:

Probability of a randomly chosen item lying in an interval;

Percentage of total items that lie in an interval;

Area under a normal curve along an interval.

We will come back to the idea of the area under this curve in Section 18.4.

Sometimes data do not fall into a normal distribution, but are **skewed,** which means their distribution has more tail on one side or the other. For example, Figure 14.40a shows that the 1941 scores on the SAT exam (when the test was first used) were normally distributed. However, by 1990, the scale had become skewed to the left, as shown in Figure 14.40b.

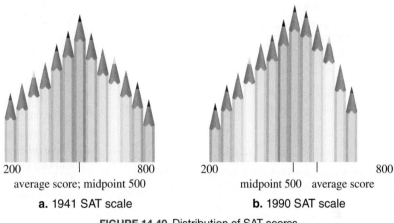

200 800	200 800
average score; midpoint 500	midpoint 500 average score
a. 1941 SAT scale	**b.** 1990 SAT scale

FIGURE 14.40 Distribution of SAT scores

In a normal distribution, the mean, median, and mode all have the same value, but if the distribution is skewed, the relative positions of the mean, median, and mode would be as shown in Figure 14.41.

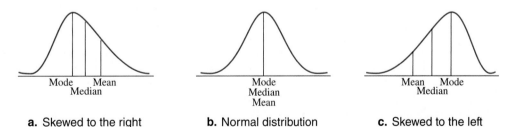

a. Skewed to the right (positive skew)	**b.** Normal distribution	**c.** Skewed to the left (negative skew)

FIGURE 14.41 Comparison of three distributions

Problem Set 14.3

Level 1

1. **IN YOUR OWN WORDS** What do we mean by a cumulative frequency?

2. **IN YOUR OWN WORDS** What does it mean to "grade on a curve"?

3. **IN YOUR OWN WORDS** What is a normal curve?

4. **IN YOUR OWN WORDS** What does the z-score, or standard score, represent?

5. What is a distribution that is skewed to the right?

6. When will a z-score be negative?

Find the cumulative distribution, the mean, median, and mode for each of the tables in Problems 7–12.

7. The number of bedrooms of homes in a certain community is shown on the table below.

Number of Bedrooms	Percent
0	5%
1	11%
2	29%
3	34%
4	15%
5	5%
6 or more	1%

8. The cumulative grade point average of graduating seniors at a small liberal arts college is shown in the following table.

GPA	Interval Mean	Percent
3.25–4.00	3.625	23%
2.50–3.24	2.87	41%
1.75–2.49	2.12	19%
1.00–1.74	1.37	6%
0.00–0.99	0.495	11%

9. The table gives the distribution of the number of exemptions claimed by employees of a large university.

Number of Exemptions	Percent
0	1%
1	11%
2	35%
3	21%
4	20%
5	6%
6	3%
7	2%
8	1%

10. The table shows the distribution of the number of motor vehicles per household.

No. of Motor Vehicles	Proportion
0	0.18
1	0.24
2	0.32
3	0.16
4	0.09
5 or more	0.01

11. A store orders tubes of glue in the proportions shown in the following table.

Size (oz)	Proportion
2	0.28
6	0.12
8	0.35
12	0.06
16	0.19

12. The distribution of number of years of education of employees of a large company is shown in the following table.

Number of Years of School	Proportion of Employees
11 or fewer	0.01
12	0.27
13	0.13
14	0.07
15	0.05
16	0.32
17	0.02
18	0.11
19 or more	0.02

What percent of the total population is found between the mean and the z-score given in Problems 13–24?

13. $z = 1.4$ 14. $z = 0.3$ 15. $z = 2.43$

16. $z = 1.86$ 17. $z = 3.25$ 18. $z = -0.6$

19. $z = -2.33$ 20. $z = -0.50$ 21. $z = -0.46$

22. $z = -1.19$ 23. $z = -2.22$ 24. $z = -3.41$

In Problems 25–29, suppose that people's heights (in centimeters) are normally distributed, with a mean of 170 and a standard deviation of 5. We find the heights of 50 people.

25. **a.** How many would you expect to be between 165 and 175 cm tall?
    b. How many would you expect to be taller than 168 cm?

26. **a.** How many would you expect to be between 170 and 180 cm?
    b. How many would you expect to be taller than 176 cm?

27. What is the probability that a person selected at random is taller than 163 cm?

28. What is the variance in heights for this experiment?

29. What is the cumulative distribution?

In Problems 30–34 suppose that, for a certain exam, a teacher grades on a curve. It is known that the mean is 50 and the standard deviation is 5. There are 45 students in the class.

30. **a.** How many students should receive a C?
    b. How many students should receive an A?

31. What score would be necessary to obtain an A?

32. If an exam paper is selected at random, what is the probability that it will be a failing paper?

33. What is the variance in scores for this exam?

34. What is the cumulative distribution?

Level **2**

In Problems 35–39, suppose that, for a certain mathematics class, the scores are normally distributed with a mean of 75 and a standard deviation of 8. The teacher wishes to give A's to the top 6% of the students and F's to the bottom 6%. The next 16% in either direction will be given B's and D's, with the other students receiving C's. Find the bottom cutoff for the grades in Problems 35–38.

35. A   36. B

37. C   38. D

39. What is the cumulative distribution?

40. In a distribution that is skewed to the right, which has the greatest value—the mean, median, or mode? Explain why this is the case.

41. In a distribution that is skewed to the left, which has the greatest value—the mean, median, or mode? Explain why this is the case.

A normal distribution has a mean of 85.7 and a standard deviation of 4.85. Find data values corresponding to the values of z given in Problems 42–45.

42. $z = 0.85$

43. $z = 2.55$

44. $z = -1.25$

45. $z = -3.46$

46. Suppose that the breaking strength of a rope (in pounds) is normally distributed, with a mean of 100 pounds and a standard deviation of 16. What is the probability that a certain rope will break before being subjected to 130 pounds?

47. The diameter of an electric cable is normally distributed, with a mean of 0.9 inch and a standard deviation of 0.01 inch. What is the probability that the diameter will exceed 0.91 inch?

48. Suppose that the annual rainfall in Ferndale, California, is known to be normally distributed, with a mean of 35.5 inches and a standard deviation of 2.5 inches. About 2.3% of the years, the annual rainfall will exceed how many inches?

The Victorian Village of Ferndale, CA

49. About what percent of the years will it rain more than 36 inches in Ferndale (see Problem 48)?

50. In Problem 48, what is the probability that the rainfall in a given year will exceed 30.5 inches in Ferndale?

51. The breaking strength (in pounds) of a certain new synthetic is normally distributed, with a mean of 165 and a variance of 9. The material is considered defective if the breaking strength is less than 159 pounds. What is the probability that a single, randomly selected piece of material will be defective?

52. The diameter of a pipe is normally distributed, with a mean of 0.4 inch and a variance of 0.0004. What is the probability that the diameter of a randomly selected pipe will exceed 0.44 inch?

53. The diameter of a pipe is normally distributed, with a mean of 0.4 inch and a variance of 0.0004. What is the probability that the diameter of a randomly selected pipe will exceed 0.41 inch?

54. Suppose the neck size of men is normally distributed, with a mean of 15.5 inches and a standard deviation of 0.5 inch. A shirt manufacturer is going to introduce a new line of shirts. Assume that if your neck size falls between two shirt sizes, you purchase the next larger shirt size. How many of each of the following sizes should be included in a batch of 1,000 shirts?
a. 14   **b.** 14.5   **c.** 15
d. 15.5   **e.** 16   **f.** 16.5
g. 17

55. A package of Toys Galore Cereal is marked "Net Wt. 12 oz." The actual weight is normally distributed, with a mean of 12 oz and a variance of 0.04.
a. What percent of the packages will weigh less than 12 oz?
b. What weight will be exceeded by 2.3% of the packages?

56. Instant Dinner comes in packages with weights that are normally distributed, with a standard deviation of 0.3 oz. If 2.3% of the dinners weigh more than 13.5 oz, what is the mean weight?

Level 3

57. Journal Problem   The graph shown in Figure 14.42 is from the February 1991 issue of *Scientific American*. If the curve in the middle is a standard normal curve, describe the variance of the

upper curve (labeled **a**) and then describe the variance of the lower curve (labeled **b**).

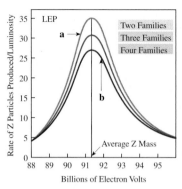

FIGURE 14.42 Normal curves

Problem Solving 3

58. The equation for the *standard normal* curve is an exponential equation:

$$y = \frac{e^{-x^2/2}}{\sqrt{2\pi}}$$

Evaluate this formula for

$$x = -4, -3, \ldots, 3, 4.$$

59. IN YOUR OWN WORDS   Calculate values for the equation $y = 2^{-x^2}$ and compare with the standard normal curve in Problem 58. List some similarities and some differences.

60. The equation for a general normal curve with mean μ and standard deviation σ is

$$y = \frac{e^{-(x-\mu)^2/(2\sigma^2)}}{\sigma\sqrt{2\pi}}$$

Calculate values for $x = 20, 30, \ldots, 70, 80$ where $\mu = 50$ and $\sigma = 10$. Note that setting $\mu = 0$ and $\sigma = 1$ gives the equation for the standard normal curve (Problem 58).

14.4 Correlation and Regression

In mathematical modeling, it is often necessary to deal with numerical data and to make assumptions regarding the relationship between two variables. For example, you may want to examine the relationship between

IQ and salary
Study time and grades
Age and heart disease
Runner's speed and runner's brand of shoe
Math grades in the 8th grade and amount of TV viewing
Teachers' salaries and beer consumption

All are attempts to relate two variables in some way or another.

Regression Analysis

If it is established that there is a **correlation** between two variables, then the next step in the modeling process is to identify the nature of the relationship. This is called **regression analysis.** In this section we consider only linear relationships. It is assumed that you are familiar with the slope-intercept form of the equation of a line (namely, $y = mx + b$). If you need a review of graphing lines, you should look at Section 15.1.

We are interested in finding a *best-fitting line* by a technique called the **least squares method.** The derivations of the results and the formulas given in this section are, for the most part, based on calculus and are therefore beyond the scope of this book. We will attempt to focus instead on how to use and interpret the formulas.

The first consideration is one of correlation. We want to know whether two variables are related. Let us call one variable x and the other y. These variables can be represented as ordered pairs (x, y) in a graph called a **scatter diagram.**

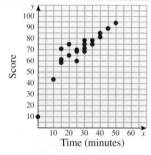

FIGURE 14.43 Correlation between study time and grade

Example 1 Draw a scatter diagram

A survey of 20 students compared the grade received on an examination with the length of time (to the nearest minutes) the student studied. Draw a scatter diagram to represent the data in the table.

Student number	1	2	3	4	5	6	7	8	9	10	11	12	13	14	15	16	17	18	19	20
Study time length	30	40	30	35	45	15	15	50	30	0	20	10	25	25	25	30	40	35	20	15
Grade (100 poss.)	72	85	75	78	89	58	71	94	78	10	75	43	68	60	70	68	82	75	65	62

Solution   Let x be the study time (in minutes) and let y be the grade (in points). The graph is shown in Figure 14.43.

Correlation is a measure to determine whether there is a statistically significant relationship between two variables. Intuitively, it should assign a measure consistent with the scatter diagrams as shown in Figure 14.44.

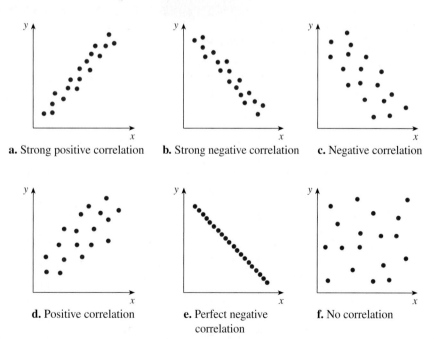

a. Strong positive correlation   **b.** Strong negative correlation   **c.** Negative correlation

d. Positive correlation   **e.** Perfect negative correlation   **f.** No correlation

FIGURE 14.44 Examples of correlation

Such a measure, called the *linear correlation coefficient, r*, is defined so that it has the following properties:

1. r measures the correlation between x and y.
2. r is between -1 and 1.
3. If r is close to 0, it means there is little correlation.
4. If r is close to 1, it means there is a strong positive correlation.
5. If r is close to -1, it means there is a strong negative correlation.

To write a formula for r, let n denote the number of pairs of data present; and as before:

$\sum x$	denotes the sum of the x-values.
$\sum x^2$	means square the x-values and then sum.
$(\sum x)^2$	means sum the x-values and then square.
$\sum xy$	means multiply each x-value by the corresponding y-value and then sum.
$n\sum xy$	means multiply n times $\sum xy$.
$(\sum x)(\sum y)$	means multiply $\sum x$ times $\sum y$.

Formula for Linear Correlation Coefficient

The **linear correlation coefficient, r,** is

$$r = \frac{n\sum xy - (\sum x)(\sum y)}{\sqrt{n(\sum x^2) - (\sum x)^2}\sqrt{n(\sum y^2) - (\sum y)^2}}$$

Example 2 Find a correlation coefficient

Find r for the following data (from Example 1).

Student number	1	2	3	4	5	6	7	8	9	10	11	12	13	14	15	16	17	18	19	20
Study time length	30	40	30	35	45	15	15	50	30	0	20	10	25	25	25	30	40	35	20	15
Grade (100 poss.)	72	85	75	78	89	58	71	94	78	10	75	43	68	60	70	68	82	75	65	62

Solution

Study Time, x	Score, y	xy	x^2	y^2
30	72	2,160	900	5,184
40	85	3,400	1,600	7,225
30	75	2,250	900	5,625
35	78	2,730	1,225	6,084
45	89	4,005	2,025	7,921
15	58	870	225	3,364
15	71	1,065	225	5,041
50	94	4,700	2,500	8,836
30	78	2,340	900	6,084
0	10	0	0	100
20	75	1,500	400	5,625
10	43	430	100	1,849
25	68	1,700	625	4,624
25	60	1,500	625	3,600
25	70	1,750	625	4,900
30	68	2,040	900	4,624
40	82	3,280	1,600	6,724
35	75	2,625	1,225	5,625
20	65	1,300	400	4,225
15	62	930	225	3,844

Total: $\sum x = 535$ $\sum y = 1,378$ $\sum xy = 40,575$ $\sum x^2 = 17,225$ $\sum y^2 = 101,104$

$$r = \frac{n\sum xy - (\sum x)(\sum y)}{\sqrt{n(\sum x^2) - (\sum x)^2}\sqrt{n(\sum y^2) - (\sum y)^2}}$$

$$= \frac{20(40{,}575) - (535)(1{,}378)}{\sqrt{20(17{,}225) - (535)^2}\sqrt{20(101{,}104) - (1{,}378)^2}}$$

$$= \frac{74{,}270}{\sqrt{58{,}275}\sqrt{123{,}196}}$$

$$\approx 0.877$$

TABLE 14.11

Correlation coefficient, r

n	$\alpha = 0.05$	$\alpha = 0.01$
4	0.950	0.999
5	0.878	0.959
6	0.811	0.917
7	0.754	0.875
8	0.707	0.834
9	0.666	0.798
10	0.632	0.765
11	0.602	0.735
12	0.576	0.708
13	0.553	0.684
14	0.532	0.661
15	0.514	0.641
16	0.497	0.623
17	0.482	0.606
18	0.468	0.590
19	0.456	0.575
20	0.444	0.561
25	0.396	0.505
30	0.361	0.463
35	0.335	0.430
40	0.312	0.402
45	0.294	0.378
50	0.279	0.361
60	0.254	0.330
70	0.236	0.305
80	0.220	0.286
90	0.207	0.269
100	0.196	0.256

The derivation of this table is beyond the scope of this course. It shows the critical values of the Pearson correlation coefficient.

COMPUTATIONAL WINDOW

Most scientific computers have standard operating procedures for finding not only the correlation coefficient, but also the best fitting-line (discussed later in this section). Look at your owner's manual to find the procedure for properly entering the data and then for accessing the STAT functions available on your calculator. The problem you will encounter is learning the correct *syntax* for asking your calculator to do what you want it to do. For example, many calculators use LinReg to find linear regression and LnReg to find logarithmic regression. If you don't know which buttons to press, you will not be able to obtain the values you seek.

Example 2 shows a very strong positive correlation. But if r for this example had been 0.46, would we still have been able to conclude that there is a strong correlation? This question is a topic of major concern in statistics. The term **significance level** is used to denote the cutoff between results attributed to chance and results attributed to significant differences. Table 14.11 gives *critical values* for determining whether two variables are correlated. If $|r|$ is greater than the given table value, then we may assume a correlation exists between the variables. If we use the column labeled $\alpha = 0.05$, then we say the significance level is 5%. This means that the probability is 0.05 that we will say the variables are correlated when, in fact, the results should be attributed to chance, and similarly for a significance level of 1% ($\alpha = 0.01$). For Example 2, since $n = 20$, we see in Table 14.11 that $r = 0.877$ shows a significant linear correlation at both the 1% and 5% levels. On the other hand, if $r = 0.46$ and $n = 20$, then there is a significant linear correlation at a 5% level, but not at a 1% level.

Example 3   Find a critical value

Find the critical value of the linear correlation coefficient for 13 pairs of data and a significance level of 0.05.

Solution   From Table 14.11, the critical value is 0.553. For $n = 13$, any value greater than $r = 0.553$ or less than -0.553 is evidence of linear correlation.

Example 4   Determine if variables are correlated

If $r = -0.85$ and $n = 13$, are the variables correlated at a significance level of 1%?

Solution   For $n = 13$ and $\alpha = 0.01$, the Table 14.11 table entry is 0.684. Since r is negative and $|r| > 0.684$, we see that there is a negative linear correlation.

Once again, we remind you to use a calculator to carry out calculations such as the ones shown in Example 5. You should check the work shown in this example on your own calculator. Keep in mind that many calculators have built-in function keys for finding the correlation.

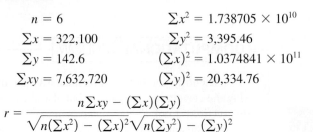

| Example | 5 | Find a correlation coefficient |

The following table shows a sample of some past annual mean salaries for college professors, along with the annual per capita beer consumption (in gallons) for Americans. Find the correlation coefficient.

Year	1995	1993	1991	1990	1989	1988
Salary	$58,400	$57,400	$55,800	$53,200	$50,100	$47,200
Consumption	22.6	22.8	23.1	24.0	25.4	24.7

Solution

$$n = 6 \qquad\qquad \Sigma x^2 = 1.738705 \times 10^{10}$$
$$\Sigma x = 322{,}100 \qquad\qquad \Sigma y^2 = 3{,}395.46$$
$$\Sigma y = 142.6 \qquad\qquad (\Sigma x)^2 = 1.0374841 \times 10^{11}$$
$$\Sigma xy = 7{,}632{,}720 \qquad\qquad (\Sigma y)^2 = 20{,}334.76$$

$$r = \frac{n\Sigma xy - (\Sigma x)(\Sigma y)}{\sqrt{n(\Sigma x^2) - (\Sigma x)^2}\sqrt{n(\Sigma y^2) - (\Sigma y)^2}}$$

$$= \frac{6(7{,}632{,}720) - (322{,}100)(142.6)}{\sqrt{6(1.738705 \times 10^{10}) - 1.0374841 \times 10^{11}}\sqrt{6(3{,}395.46) - 20{,}334.76}}$$

$$\approx -0.9151$$

The number $r \approx -0.9151$ is statistically significant at the 5% level, but not at the 1% level. This means that there is a negative correlation between these two phenomena.

The significance of the negative correlation in Example 5 implies that professors' salaries and beer drinking are related, but be careful! In any event, the techniques in this chapter can be used only to establish a *statistical* linear relationship. *We cannot establish the existence or absence of any inherent cause-and-effect relationship on the basis of a correlation analysis.*

Best-Fitting Line

The final step in our discussion of correlation is to find the best-fitting line.[*] That is, we want to find a line $y' = mx + b$ so that the sum of the squared distances of the data points from this line will be as small as possible. (We use y' instead of y to distinguish between the actual second component, y, and the predicted y-value, y'.) Since some of these distances may be positive and some negative, and since we do not want opposites to "cancel each other out," we minimize the sum of the *squares* of the distances. Therefore, the regression line is sometimes called the **least squares line.**

Least Squares Line

The **least squares** (or *regression*) **line** is $y' = mx + b$, where

$$m = \frac{n(\Sigma xy) - (\Sigma x)(\Sigma y)}{n(\Sigma x^2) - (\Sigma x)^2} \qquad b = \frac{\Sigma y - m(\Sigma x)}{n}$$

This is the line of *best fit.*

*If you need a review of graphing lines, see Section 15.1.

Example **6**   **Find a best-fitting line**

Find the best-fitting line for the data in Example 1.

Student number	1	2	3	4	5	6	7	8	9	10	11	12	13	14	15	16	17	18	19	20
Study time length	30	40	30	35	45	15	15	50	30	0	20	10	25	25	25	30	40	35	20	15
Grade (100 poss.)	72	85	75	78	89	58	71	94	78	10	75	43	68	60	70	68	82	75	65	62

Solution   From Example 2, $n = 20$, $\Sigma x = 535$, $\Sigma y = 1{,}378$, $\Sigma xy = 40{,}575$, $\Sigma x^2 = 17{,}225$, and $\Sigma y^2 = 101{,}104$. Thus,

$$m = \frac{n(\Sigma xy) - (\Sigma x)(\Sigma y)}{n(\Sigma x^2) - (\Sigma x)^2} \qquad \text{and} \qquad b = \frac{\Sigma y - m(\Sigma x)}{n}$$

$$= \frac{20(40{,}575) - (535)(1{,}378)}{20(17{,}225) - (535)^2} \qquad\qquad = \frac{1{,}378 - 1.27447(535)}{20}$$

$$\approx 1.27447 \qquad\qquad\qquad\qquad \approx 34.8078$$

Then we approximate the best-fitting line as the line with equation $y' = 1.3x + 35$. This line (along with the data points) is shown in Figure 14.45.

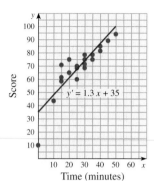

y' = 1.3x + 35

FIGURE 14.45 Regression line

Example **7**   **Use regression to make a prediction**

Use the regression line in Example 6 to predict the score of a person who studied $\frac{1}{2}$ hour.

Solution $x = 30$ minutes, so $y' = 1.3x + 35 = 1.3(30) + 35 = 74$.

A Final Word of Caution.   You should use the regression line only if r indicates that there is a significant linear correlation, as shown in Table 14.11.

Problem Set   **14.4**

Level **1**

1. IN YOUR OWN WORDS   What do we mean by *correlation*?

2. IN YOUR OWN WORDS   What is a least squares line?

3. How do you find a linear correlation coefficient?

4. How do you determine whether there is a linear correlation between two variables x and y?

5. Discuss the correlation shown by the following chart.

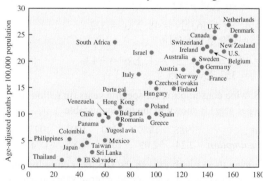

6. Discuss the correlation shown by the following chart.

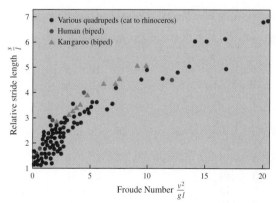

In Problems 7–18, a sample of paired data gives a linear correlation coefficient r. In each case, use Table 14.11 to determine whether there is a significant linear correlation.

7. $n = 10$, $r = 0.7$; 1% level

8. $n = 30$, $r = 0.4$; 1% level

9. $n = 30$, $r = 0.4$; 5% level

10. $n = 15$, $r = -0.732$; 5% level

11. $n = 35$, $r = -0.413$; 1% level

12. $n = 50$, $r = -0.3416$; 1% level

13. $n = 25$, $r = 0.521$, 1% level

14. $n = 100$, $r = -0.4109$; 1% level

15. $n = 40$, $r = 0.416$; 5% level

16. $n = 20$, $r = -0.214$; 5% level

17. $n = 10$, $r = -0.56$; 1% level

18. $n = 10$, $r = 0.7$; 5% level

Draw a scatter diagram and find r for the data shown in each table in Problems 19–24.

19.

x	y
4	0
5	−10
10	−10
10	−20

20.

x	y
1	1
2	5
3	8
4	13

21.

x	y
1	30
3	22
3	19
5	15
8	10

22.

x	y
0	25
1	19
2	16
3	12
4	10

23.

x	y
85	80
90	40
100	30
102	28
105	25

24.

x	y
10	20
20	48
30	60
30	58
50	70
60	75

Find the regression line for the data points in Problems 25–30.

25.

x	y
4	0
5	−10
10	−10
10	−20

26.

x	y
1	1
2	5
3	8
4	13

27.

x	y
1	30
3	22
3	19
5	15
8	10

28.

x	y
0	25
1	19
2	16
3	12
4	10

29.

x	y
85	80
90	40
100	30
102	28
105	25

30.

x	y
10	20
20	48
30	60
30	58
50	70
60	75

Match the equation and correlation in Problems 31–36 with a graph.

31. $y = 0.6x + 2$; $r = 0.9$

32. $y = 0.5x + 2$; $r = 0.7$

33. $y = 0.4x + 2$; $r = 0.5$

34. $y = -0.4x + 2$; $r = -0.4$

35. $y = -0.5x + 2$; $r = -0.6$

36. $y = -0.7x + 2$; $r = -0.8$

A.

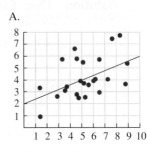

B.

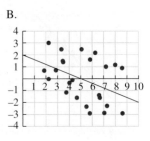

C.

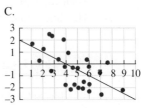

D.

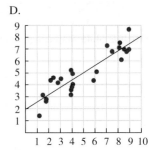

E.

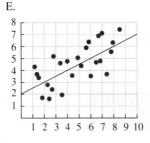

F.

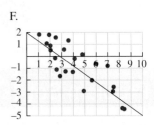

Level 2

Given the information in Problems 37–40, find the equation for the least squares line as well as the correlation coefficient.

37. $n = 8$, $\Sigma x = 40$, $\Sigma y = 48$, $\Sigma xy = 240$, $\Sigma x^2 = 210$, $\Sigma y^2 = 288$

38. $n = 10$, $\Sigma x = 30$, $\Sigma y = 20$, $\Sigma xy = 60$, $\Sigma x^2 = 90$, $\Sigma y^2 = 40$

39. $n = 6$, $\Sigma x = 36$, $\Sigma y = 48$, $\Sigma xy = 322$, $\Sigma x^2 = 250$, $\Sigma y^2 = 418$

40. $n = 10$, $\Sigma x = 57$, $\Sigma y = 1$, $\Sigma xy = -13$, $\Sigma x^2 = 353$, $\Sigma y^2 = 45$

41. U.S. wine consumption for specific years (in gallons per person per year) is shown in the following table.

Year	1988	1989	1990	1991	1993	1995
Consumption	2.24	2.09	2.05	1.85	1.74	1.80

Compare these numbers with the U.S. beer consumption (found in Example 5, page 698) for those years in which both numbers are reported. Are beer drinking and wine drinking correlated? What is the correlation coefficient?

42. A group of ten people were selected and given a standard IQ test. These scores were then compared with their high school grades.

IQ	117	105	111	96	135	81	103	99	107	109
GPA	3.1	2.8	2.5	2.8	3.4	1.9	2.1	3.2	2.9	2.3

Find r and determine whether it is statistically significant at the 1% level.

43. A new computer circuit was tested and the times (in nanoseconds) required to carry out different subroutines were recorded.

Difficulty	1	2	2	3	4	5	5	5
Time	10	11	13	8	15	18	21	19

Find r and determine whether it is statistically significant at the 1% level.

44. Find the regression line for the data in Problem 42. Assume x is the IQ and y is the GPA.

45. Find the regression line for the data in Problem 43. Assume x is the difficulty level and y is the time.

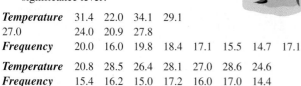

"What do you mean, I can't predict the weather!"

46. The following data are measurements of temperature ($x =$ °F) and chirping frequency ($y =$ chirps per second) for the striped ground cricket.

Is there a correlation between temperature and chirping, and if so, is it significant at the 5% or the 1% significance level?

Temperature	31.4	22.0	34.1	29.1
	27.0	24.0	20.9	27.8
Frequency	20.0	16.0	19.8	18.4	17.1	15.5	14.7	17.1

Temperature	20.8	28.5	26.4	28.1	27.0	28.6	24.6
Frequency	15.4	16.2	15.0	17.2	16.0	17.0	14.4

47. The following data are the number of years of full-time education (x) and the annual salary in thousands of dollars (y) for 15 persons. Is there a correlation between education and salary, and if so, is it significant at the 5% or the 1% significance level?

Education	20	27	28	18	13	18	9	16
Salary	35.2	24.6	23.7	33.3	24.4	33.4	11.2	32.3

Education	16	12	12	19	16	14	13
Salary	25.1	22.1	18.9	37.8	25.9	28.4	29.6

48. A researcher chooses and interviews a group of 15 male workers in an automobile plant. The researcher then gives a score (x) ranging from 1 to 20 based on a scale of patriotism—the higher the score, the more patriotic the person appears to be. Each person is then given a written test and is scored (y) on their patriotism. Is there a correlation between the researcher score and the test score, and if so, is it significant at the 5% or the 1% significance level?

Researcher	10	14	15	17	17	18	18	19
Test	15	12	19	8	9	16	17	6

Researcher	16	18	20	12	14	9	17
Test	11	14	12	11	10	12	6

49. A bank records the number of mortgage applications and its own prevailing interest rate (at the first of the month) for each of 16 consecutive months. Is there a correlation between the interest rate (x) and the number of applicants (y), and if so, is it significant at the 5% or the 1% significance level?

Interest	9.5	9.9	10.0	10.5	11.0	11.5	11.0	12.0
Number	27	29	25	25	19	20	17	13

Interest	12.0	12.5	13.0	13.5	13.0	12.5	11.5	11.5
Number	15	10	10	6	5	5	11	14

50. Find the best-fitting line for the data in Problem 46.

51. Find the best-fitting line for the data in Problem 47.

52. Find the best-fitting line for the data in Problem 48.

53. Find the best-fitting line for the data in Problem 49.

Level 3

Problems 54–58 are based on a 100-mile footrace, called the Hardrock 100, held in Colorado each year. For each of these problems, the distances are in miles and the times are given in hours and minutes, but when answering these questions, convert them to decimal hours. For example,*

$$2{:}57 \text{ is 2 hours 57 minutes} = 2 \text{ hr} + \frac{57}{60} \text{ hr}$$

$$= 2.95 \text{ hr}$$

54. The winner of the 1994 race was Scott Hirst. Here are his arrival times at the first 6 aid stations:

Time	2:57	5:03	7:22	9:26	10:10	11:16
Distance	12.2	19.0	28.0	33.4	36.6	43.8

Predict his arrival time at the next aid station at 51.7 miles.

55. a. Predict when Scott Hirst (see Problem 54) will reach the end of the race (100.1 miles).
b. Scott's actual time of arrival was 32:00. What is the difference between the predicted and actual arrival times? Comment on the results.

56. The last-place runner who completed the 1994 race was John DeWalt. Here are his arrival times at the first 6 aid stations:

Time	3:53	6:56	10:36	14:12	15:26	17:50
Distance	12.2	19.0	28.0	33.4	36.6	43.8

Predict his arrival time at the next aid station at 51.7 miles.

57. a. Predict when John DeWalt (see Problem 56) will reach the end of the race (100.1 miles).
b. John's actual time of arrival was 47:50. What is the difference between the predicted and actual arrival times? Comment on the results.

58. The ages of the runners in the 1994 race and their finishing positions are given in the following ordered pairs. We write (x, y), where x is the finish position and y is the person's age.†

(1, 33), (2, 46), (3, 32), (4, 31), (5, 42), (6, 38), (7, 43), (8, 39), (9, 43), (10, 44), (11, 35), (12, 38), (13, 42), (14, 27), (15, 49), (16, 49), (17, 40), (18, 40), (19, 43), (20, 47), (21, 41), (22, 36), (23, 59), (24, 48), (25, 51), (26, 36), (27, 31), (28, 33), (29, 51), (30, 43), (31, 34), (32, 56), (33, 54), (34, 46), (36, 35), (37, 58)

Is there a significant correlation between age and running times?

*This information is from "Data Analysis and the Hardrock 100," by Marny Frantz and Sylvia Lazarnick in *The Mathematics Teacher*, April 1997, pp. 274–276.
†The 35th-place finisher had no age given, so we omitted the data for this runner for this problem.

59. The ages of the runners in the 1994 race and their finishing times are given in the following table.*

(33, 32:00), (46, 32:20), (32, 32:29), (31, 33:57),
(42, 34:36), (38, 35:29), (43, 35:52), (39, 36:54),
(43, 38:04), (44, 38:04), (35, 38:20), (38, 38:43),
(42, 39:18), (27, 39:21), (49, 41:50), (49, 42:06),
(40, 42:06), (40, 42:06), (43, 42:59), (47, 43:41),
(41, 43:41), (36, 43:41), (59, 44:46), (48, 44:46),
(51, 44:46), (36, 45:21), (31, 46:19), (33, 46:26),
(51, 46:49), (43, 47:08), (34, 47:21), (56, 47:27),
(54, 47:43), (46, 47:46), (35, 47:50), (58, 47:50)

Is there a significant correlation between age and running times?

Problem Solving 3

60. IN YOUR OWN WORDS   A study was made to see whether a correlation existed between students' grades in a course and students' rankings of the professor's teaching.† The results of this survey are shown in Table 14.12. Use these results to formulate a conclusion. Remove the last data point from the table (class no. 47). This was a class who rated the professor's teaching low, and the students' grades were also very low. Does this change your conclusion? Elaborate on your answer.

*Technically, times should be changed to hours. However, for this problem it is acceptable to enter these as decimal hours. For example, 32:20 is 32 hours 20 minutes = 32 hours + 20/60 hours = 32.33, but for purposes of convenience, you can use 32.20.
†"The Correlation Coefficient and Influential Data Points," by Donald J. Dessart, *The Mathematics Teacher*, March 1997, pp. 242–246.

TABLE 14.12

Mean Student Academic Grades and Mean Teacher Ratings for Forty-seven Classes

Student Class Rank	Grade	Teacher Rating	Student Class Rank	Teacher Grade	Rating
1	3.19	1.00	25	2.68	2.22
2	3.58	1.06	26	2.36	2.25
3	3.89	1.17	27	3.35	1.00
4	2.50	1.30	28	3.27	1.00
5	3.11	1.38	29	3.11	1.13
6	2.81	1.39	30	3.18	1.27
7	3.73	1.40	31	3.16	1.31
8	3.06	1.43	32	2.83	1.33
9	3.55	1.44	33	3.05	1.50
10	3.63	1.50	34	3.61	1.50
11	3.71	1.55	35	2.92	1.53
12	2.00	1.56	36	3.17	1.58
13	3.19	1.60	37	2.18	1.58
14	3.23	1.70	38	3.14	1.73
15	3.38	1.70	39	3.55	1.80
16	3.21	1.79	40	2.69	1.82
17	3.25	1.82	41	2.55	1.90
18	2.63	1.83	42	3.08	1.91
19	3.27	1.85	43	3.00	1.92
20	2.58	2.00	44	3.67	2.20
21	3.13	2.00	45	3.67	2.27
22	3.19	2.00	46	2.95	2.60
23	3.46	2.07	47	1.68	2.80
24	3.00	2.10			

14.5 | Sampling

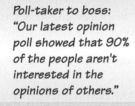

Poll-taker to boss:
"Our latest opinion poll showed that 90% of the people aren't interested in the opinions of others."

The first three sections of this chapter dealt with what is called *descriptive statistics,* which is concerned with the accumulation of data, measures of central tendency, and dispersion. A second branch of statistics is **inferential statistics,** which is concerned with making generalizations or predictions about a population based on a sample from that population.

Sampling Methods

A **sample** is a group of items chosen to represent a larger group. The larger group is called a **population.** Thus, the sample is a proper subset of the population. The population to be considered for a statistical application is called the **target population.** The sample is analyzed; then, based on this analysis, some conclusion about the entire population is made. Sampling necessarily involves some error, because the sample and the population are not identical. A great deal of effort in statistics is devoted to the methodology of sampling. Here are some common sampling procedures:

Simple random sampling: This sample is obtained in a way that allows every member of the target population to have the same chance of being chosen. This is the type of sampling we assumed so far.

Systematic sampling: The sample is obtained by drawing every *k*th item on a list or production line. The first item is determined by using a random number.

Cluster sampling: This procedure is applied on a geographical basis. The result is sometimes known as an *area sample*.

Stratified sampling: The entire population is divided into parts, called *strata*, according to some factor (such as gender, age, or income). When a population has varied characteristics, it is desirable to separate the population into homogeneous strata and then take a random sample from each stratum.

To choose the most appropriate procedure for selecting an unbiased sample from a target population, we must take into account two considerations:

1. *Is the procedure random?* For example, asking people to voluntarily mail or phone in their preferences is not random, since the respondents select themselves (and there is nothing to prevent someone from sending in more than one response). Likewise, selecting names from an alphabetical listing may introduce an ethnic bias and is not random.

2. *Does the procedure take into account the intended target population?* For example, if you wish to know how many students at a certain college are interested in expanding the hours of operation at the college library, then it would not make sense to send the survey forms to the entire population of the city. Any procedure that would be appropriate in this case would ensure that only college students were surveyed.

Example 1   Discuss methods for obtaining an unbiased sample

A long-distance phone provider wants to sell a new "unlimited minutes" plan, and wishes to conduct a survey to find the degree of interest its customers have in such a plan. Which of the following methods would be most appropriate for obtaining an unbiased sample?

A. Survey the current customers whose names are chosen from an alphabetical listing of all customers on the west coast.
B. Survey a random selection of customers whose names are chosen from an alphabetical listing of all customers.
C. Survey the first 1,000 customers whose names are chosen from an alphabetical listing of all the customers.
D. Survey the first 1,000 customers whose names are chosen from a list purchased from a competing phone company.
E. Survey 1,000 persons whose names are chosen from the local telephone directory.

Solution   The solution depends on identifying the target population; it is the current customers of the long-distance phone company.

Choices D and E are not appropriate, because they do not reach the target population. Choice C hits the target population, but is not sufficiently random. Choice A hits only part of the target population. Therefore, choice B is the most appropriate since it addresses the target population in a random manner.

Bettmann/Corbis

The inference drawn from a poll can, of course, be wrong, so statistics is also concerned with estimating the error involved in predictions based on samples. In 1936, the *Literary Digest* predicted that in the race for president, Alfred Landon would defeat Franklin D. Roosevelt—who was subsequently reelected president by a landslide. (The magazine ceased publication the following year.) In 1948, the *Chicago Daily Tribune* drew an incorrect conclusion from its polls and declared in a headline that Thomas Dewey had just been elected president over Harry S. Truman. And in 1976, the *Milwaukee Sentinel* printed the erroneous headline shown here about the Wisconsin Democratic primary, again based on the result of its polls.

In an attempt to minimize error in their predictions, statisticians follow very careful procedures.

Inference Procedure

Procedure for making statistical inferences:

Step 1   Propose some hypothesis about a population.

Step 2   Gather a sample from the population.

Step 3   Analyze the data.

Step 4   Accept or reject the hypothesis.

Hypothesis Testing

Suppose that you want to decide whether a certain coin is a "fair" coin. You decide to test the hypothesis, "This is a fair coin," by flipping the coin 100 times. This provides a *sample*. Suppose the result is

Heads: 55   Tails: 45

Do you accept or reject the hypothesis that "This coin is fair"? The expected number of heads is 50, but certainly a fair coin might well produce the results obtained.

As you can readily see, two types of errors are possible in any hypothesis-testing situation.

Types of Error

Type I: Rejection of the hypothesis when it is true

Type II: Acceptance of the hypothesis when it is false

How can we minimize the possibility of making either error? Let's carry this example further, and repeat the experiment of flipping the coin 100 times:

Trial number	1	2	3	4	5	6	...
Number of heads	55	52	54	57	59	55	...

If the coin is fair and we repeat the experiment a large number of times, it can be shown mathematically that the distribution of the number of heads should be normal, with a mean of 50 and a standard deviation of 5, as shown in Figure 14.46. The question is whether to accept the *unknown* test coin as fair.

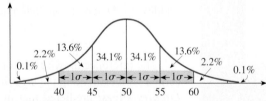

FIGURE 14.46 Given normal distribution for the number of heads upon flipping a fair coin 100 times

Suppose that you are willing to accept the coin as fair only if the number of heads falls between 45 and 55 (that is, within 1 standard deviation of the expected number). If you adopt this standard, you know you will be correct 68% of the time if the coin is fair. How do you know this? Look at Figure 14.46, and note that 34.1% of the results are within $+1\sigma$ and 34.1% are within -1σ of the mean; the total is $34.1\% + 34.1\% = 68.2\%$.

But a friend says, "Yes, you will be correct 68% of the time, but you will also be rejecting a lot of fair coins!" You respond, "But suppose that a coin really is a bad coin (it really favors heads), with a mean number of heads of 60 and a standard deviation of 5. If I adopted the same standard $(\pm 1\sigma)$, I'd be accepting all the coins in the shaded region of Figure 14.47."

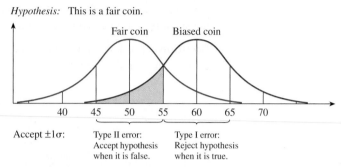

FIGURE 14.47 Comparison of Type I and Type II errors

Example 2 Classify errors

Consider the hypothesis that the mean grade for a class is 75. We conduct a survey of five people from the class and find that their grades are 85, 65, 70, 52, and 96. There are four possibilities:

We decide to reject the conclusion that the mean grade is 75 and:
    A. The actual mean is 75.
    B. The actual mean is not 75.

We decide to accept the conclusion that the mean grade is 75 and:
    C. The actual mean is 75.
    D. The actual mean is not 75.

For each of the four possibilities, state whether the conclusion is true or which type of error has been made.

Solution
A. reject a true conclusion; Type I error
B. correct decision
C. correct decision
D. accept a false conclusion; Type II error

As you can see, decreasing the Type I error probability increases the Type II error probability, and vice versa. Deciding which type of error to minimize depends on the stakes involved and on some statistical calculations that go beyond the scope of this course.

Consider a company that produces two types of valves. The first type is used in jet aircraft, and the failure of this valve might cause many deaths. A sample of the valves is taken and tested, and the company must accept or reject the entire shipment on the basis of these test results. Under these circumstances, the company would rather reject many good valves than accept a bad one. On the other hand, the second valve is used in toy airplanes; the failure of this valve would merely cause the crash of the model. In this case, the company wouldn't want to reject too many good valves, so it would minimize the probability of rejecting the good valves.

Many times we read of a poll in the newspaper or hear of it on the evening news, and a percentage is given. For example, "The candidate was favored by 48% of those sampled." What is not often stated (or is stated only in small print) is that there is a margin of error and a confidence level. For example, "The margin of error is 4 percent at a confidence level of 95%." This means that we are 95% confident that the actual percent of people who favor the candidate is between 44% and 52%.

Example **3** **Find error in faulty reasoning**

My mother, who is 78, has been a smoker since she was 18 years old. Even though she has emphysema, she claims that she will live well into her eighties, which is proof that the smoking warning claims do not mean a thing. What is wrong with this reasoning?

Solution It is known that the average nonsmoker lives about $5\frac{1}{2}$ years longer than the average smoker. Consider the normal curves shown in Figure 14.48.

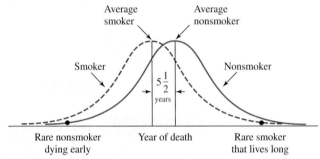

FIGURE 14.48 Life expectancies for smokers and nonsmokers

As you can see, sometimes the nonsmoker will die early (analogous to a Type II error) and sometimes the smoker will live long (analogous to a Type I error).

Many times you may hear of people reasoning or forming a conclusion by looking at one particular case, which may be an exception. This occurs so often, in fact, that we give this type of reasoning a name—we call it the **fallacy of exceptions.**

Problem Set 14.5

Level 1

1. **IN YOUR OWN WORDS** When you hear the word *statistics*, what comes to mind? Look up the word in a dictionary, and see if you wish to supplement or change any part of your answer.

2. **IN YOUR OWN WORDS** Explain the difference between descriptive and inferential statistics.

3. **IN YOUR OWN WORDS** Explain the fallacy of the exception.

4. An ice cream store wants to use a survey to determine which newspapers its customers regularly read. Which of the following procedures would be most appropriate for obtaining a statistically unbiased sample?
   A. Survey a randomly selected sample of people living in the city's limits.
   B. Survey the first 100 customers on an alphabetical listing of all the ice cream store's customers.
   C. Survey 100 customers whose names are randomly chosen from an alphabetical listing of all customers of the ice cream store.
   D. Survey 100 people whose names are randomly chosen from the local telephone directory.

5. In a certain geographical market, Sprint wishes to increase its market share by taking a survey of customers of its competitor, Pacific Bell, in an effort to find out which parts of its service are least satisfactory. Which of the following procedures would be most appropriate for obtaining a statistically unbiased sample?
   A. Survey a random selection of people chosen from a list of Sprint's customers.
   B. Survey a random selection of people chosen from a list of Pacific Bell's customers.

C. Place a newspaper advertisement asking for survey participants, and then randomly choose 100 people from those who respond.

D. Call local phone numbers and ask people who answer if they are satisfied with their phone service.

6. A movie theater would like to use a survey to determine which factors are most important to those who come to their theater. Which of the following procedures would be most appropriate for obtaining a statistically unbiased sample?

A. Survey a random selection of people passing by the theater in the mall in front of the theater.

B. Survey a random selection of shoppers in the mall.

C. Survey a random selection of people who are leaving the theater after watching a movie.

D. Induce people to answer survey questions by offering discount coupons for various city events, including the movie theater.

7. A local congressional candidate would like to use a survey to determine the issues of importance to the voters in her congressional district. Which of the following procedures would be most appropriate for obtaining a statistically unbiased sample?

A. Survey a selection of people whose names are randomly chosen from a local list of registered voters who are party members.

B. Survey a selection of people whose names are randomly chosen from a local list of all registered voters.

C. Survey a selection of people whose names are randomly chosen from the telephone directory.

D. Survey a selection of people whose names are randomly chosen from a national list of party members.

8. A school board wishes to determine opinions of parents regarding the assigning of homework in mathematics classes. Which of the following procedures would be most appropriate for obtaining a statistically unbiased sample?

A. Survey a selection of parents from the official school roster of parents.

B. Survey a selection of people whose names are randomly chosen from the telephone directory.

C. Survey people chosen from the first 1000 names on an alphabetical listing of parents of mathematics students.

D. Survey a selection of people whose names are randomly chosen from a listing of parents of mathematics students.

9. An environmental group wants to use a survey to determine the extent to which its members would be willing to protest a recent decision of the governor of a particular state. Which of the following procedures would be most appropriate for obtaining a statistically unbiased sample?

A. Survey a selection of people whose names are randomly chosen from the telephone directory.

B. Have signups of which workers gather signatures at local malls throughout the state.

C. Survey members attending the annual convention of the environmental group.

D. Survey a selection of people whose names are randomly chosen from a list of members of the environmental group.

10. Cal Trans wants to use a survey to determine the extent of public approval for a project to construct a new freeway through a particular neighborhood. Which of the following procedures

would be most appropriate for obtaining a statistically unbiased sample?

A. Go door-to-door in the affected neighborhood, soliciting opinions.

B. Hold a public forum in the affected neighborhood, and distribute questionnaires there.

C. Place advertisements in newspapers in the affected neighborhood, asking readers to mail in their opinions.

D. Survey a selection of people whose names are randomly chosen from the city telephone directory.

11. Betty's Koi Emporium has decided to conduct a survey to determine the number of its customers who order supplies on the Internet. Which of the following procedures would be most appropriate for obtaining a statistically unbiased sample?

A. Survey a random selection of customers chosen from a list of all people who have purchased an item from Betty's Koi Emporium.

B. Distribute questionnaires to people who make a purchase in the next 30 days.

C. Post notices soliciting the necessary information on bulletin boards in the Koi Emporium.

D. Survey people coming out of a nearby shopping mall.

12. IN-N-OUT BURGER® is considering building a franchise in a particular city. It will use a survey to determine the extent of interest among residents of the city. Which of the following procedures would be most appropriate for obtaining a statistically unbiased sample?

AP Photo/Adam Lau

A. Survey a random selection of people whose names are obtained from national sales records of IN-N-OUT BURGER customers.

B. Survey a selection of people who are listed on the residential roll sheets of the city.

C. Survey a selection of people whose names are randomly chosen from the telephone directory of the target city.

D. Survey a random selection of customers coming out of other donut shops in the city.

13. An espresso coffee franchise is considering expanding its services to a new location in the city, but prior to doing so it will use a survey to determine the extent to which its customers are interested in a second location. Which of the following procedures would be most appropriate for obtaining a statistically unbiased sample?

A. Survey a selection of people whose names are randomly chosen from the telephone directory for the city.

B. Ask customers who drive through to voluntarily phone in their preferences.

C. Ask customers who drive through to voluntarily mail in their preferences.

D. Survey a selection of people whose names are randomly chosen from a list of all customers.

Level 2

IN YOUR OWN WORDS *In Problems 14–23, decide on a reasonable means for conducting the survey to obtain the desired information.*

14. A newsstand wishes to use a survey to determine which out-of-town newspapers it should regularly stock.

15. Sprint would like to take customers from AT&T, one of its competitors in a certain geographical area. Sprint would like to survey AT&T's customers in order to find out which elements of its service are least satisfactory.

16. A pet store would like to use a survey to determine which factors are most important to cat owners in determining the brand of cat food that they purchase.

17. A local political party would like to use a survey to determine the level of support among party members for the party's candidates in the upcoming primary election.

18. A health and fitness club would like to survey its members in order to determine which new equipment they would prefer.

19. A union wants to use a survey to determine the extent members approve of a newly negotiated collective bargaining agreement.

20. The city council wants to use a survey to determine the extent of public approval for a project to construct a new playground in a certain residential neighborhood.

21. The college student government will use a survey to determine the level of support among students for a proposal to lower the student fees.

22. A retailer is considering offering extended warranty policies on video recorders. It will use a survey to determine the extent of interest among owners of the video recorders.

23. A supermarket is considering expanding its services by adding a florist, but prior to doing so, it will use a survey to determine the extent to which its customers are interested in such a service.

Level 3

24. Suppose you were interested in knowing the mean score for a certain test in a mathematics class with 20 students. You sample four members of the class and find their test scores to be 49, 56, 72, and 96. The instructor reports that the mean was 72, but you are not sure that is truthful. List the four possibilities and categorize the conclusions including the types of possible errors.

25. Suppose you were interested in knowing the mean score for a certain test in a mathematics class with 20 students. You sample four members of the class and find their test scores to be 52, 56, 96, and 99. The instructor reports that the mean was 72, but you are not sure that is truthful. List the four possibilities and categorize the conclusions including the types of possible errors.

26. Suppose that of 80 workers randomly selected and interviewed, 55 were opposed to an increase in Social Security taxes. Would you accept or reject the hypothesis that the majority of workers are opposed to the increase in taxes? List the four possibilities and categorize the conclusion including the types of possible errors.

Problem Solving 3

27. Conduct a survey asking the following questions.
   a. Do you doodle?
   b. If you doodle, which of the following do you doodle?

   > (1) circles, curves, or spirals
   > (2) squares, rectangles, or other straight-line designs
   > (3) people or human features
   > (4) symbols, such as stars, arrows, or other objects
   > (5) words, letters, or numerals
   > (6) other

   c. When do you most often doodle?

   > (1) while at a meeting or class
   > (2) while on the phone at home
   > (3) when solving a problem
   > (4) when you have nothing else to do
   > (5) when writing a letter
   > (6) at meals

28. Conduct a survey to determine the major worry of college students.

29. Conduct a survey to determine whether there is a significant correlation between math scores in the 8th grade and amount of TV viewing.

30. Suppose that John hands you a coin to flip and wants to bet on the outcome. Now, John has tried this sort of thing before, and you suspect that the coin is "rigged." You decide to test this hypothesis by taking a sample. You flip the coin twice, and it is heads both times. You say, "Aha, I knew it was rigged!" John replies, "Don't be silly. Any coin can come up heads twice in a row."

    The following scheme was devised by mathematician John von Neumann to allow fair results even if the coin is somewhat biased. The coin is flipped twice. If it comes up heads both times or tails both times, it is flipped twice again. If it comes up heads-tails, this will decide the outcome in favor of the first party; and if it comes up tails-heads, this will decide the outcome in favor of the second party. Show that this will result in a fair toss even if the coins are biased.

14.6   CHAPTER SUMMARY

Important Ideas

Bar graphs, line graphs, circle graphs, and pictographs [14.1]

Misuses of graphs [14.1]

Measures of central tendency [14.2]

Mean, weighted mean, median, and mode [14.2]

Standard deviation [14.2]

Normal curve [14.3]

***z*-scores** [14.3]

Linear correlation coefficient [14.4]

Slope and *y*-intercept of the least squares (or regression) line [14.4]

Descriptive and inferential statistics [14.5]

Sample vs. population [14.5]

Type I and Type II sampling error [14.5]

Take some time getting ready to work the review problems in this section. First review these important ideas. Look back at the definition and property boxes. If you look online, you will find a list of important terms introduced in this chapter, as well as the types of problems that were introduced. You will maximize your understanding of this chapter by working the problems in this section only after you have studied the material.

You will find some review help online at **www.mathnature.com.** There are links giving general test help in studying for a mathematics examination, as well as specific help for reviewing this chapter.

Chapter 14 Review Questions

1. Make a frequency table for the following results of tossing a coin 40 times:

   HTTTT HHTHH TTHHT THTHT HHTTH THHTH
   HHTHT TTTTT

2. Draw a bar graph for the number of heads and tails given in Problem 1.

3. Three students, Andrew, Barbara, and Carla, were assessed for their reading proficiency, as shown in Table 14.13.

TABLE 14.13
Grade Level from a Standardized Reading Test

	Word Recognition	Spelling	Vocabulary	Comprehension
A (Andrew)	10.0	8.0	7.0	8.0
B (Barbara)	3.0	3.0	8.0	8.0
C (Carla)	5.0	5.0	6.0	8.0

Draw a line graph showing this information.

Use the data set {5, 21, 21, 25, 30, 40} in Problems 4–8 to find the requested value.

4. mean
5. median
6. mode
7. range
8. standard deviation

9. A small grocery store stocked several sizes of Copycat cola last year. The sales figures are shown in Table 14.14. Find the mean, the median, and the mode. If the store manager decides to cut back the variety and stock only one size, which measure of central tendency will be most useful in making this decision?

TABLE 14.14
Sales of Copycat Cola

Size	Number of Case Sold
6 oz	5 cases
10 oz	10 cases
12 oz	35 cases

10. A student's scores in a certain math class are 72, 73, 74, 85, and 91. Find the mean, the median, and the mode. Which measure of central tendency is most representative of the student's scores?

11. The 2003–2004 school superintendents' salaries in Sonoma County, California, are shown in Table 14.15.

TABLE 14.15
Sonoma Country Superintendents' Salaries

District	Salary
Bellevue	$102,686
Bennett Valley Elementary	$96,402
Cotati-Rohnert Park Unified	$148,209
Forestville Elementary	$96,565
Geyserville Unified	$90,094
Gravenstein Elementary	$92,978
Healdsburg Unified	$114,517
Mark West Elementary	$119,000
Oak Grove Elementary	$91,870
Old Adobe Elementary	$110,750
Petaluma City Schools	$129,064
Piner-Olivet Elementary	$115,145
Ricon Valley Elementary	$117,899
Roseland Elementary	$128,092
Santa Rosa City School	$143,760
Sebastopol Elementary	$110,000
Sonoma Valley Unified	$124,879
Twin Hills Elementary	$93,500
West Sonoma County High	$106,466
Windsor Unified	$145,000
Wright Elementary	$105,000

Source: California Teachers Association, Research and Finance Department

Find the mean, the median, and the mode (to the nearest dollar). Which measure of central tendency is most representative of the superintendents' salaries?

12. The blockbuster movie *Titanic* brought about a renewed interest in that disaster. Table 14.16 shows actual passenger survival for that fateful voyage.

TABLE 14.16
Passengers and Survivors on the *Titanic*

Passenger Category	No. of Passengers	No. of Survivors
Children, 1st class	6	6
Children, 2nd class	24	24
Children, 3rd class	79	27
Women, 1st class	144	140
Women, 2nd class	93	80
Women, 3rd class	165	76
Men, 1st class	175	57
Men, 2nd class	168	14
Men, 3rd class	462	75

a. Was a difference in the survival rates related to the class of passenger? Draw a bar graph of survival rates of first-, second-, and third-class passengers.

b. Was a difference in the survival rates related to the gender or age of the passenger? Draw a bar graph of survival rates of men, women, and children.

13. Consider the graph shown in Figure 14.49.

a. Which mode of transportation cost the most in October?

b. Which mode of transportation in Figure 14.49 generally has the least monthly cost for May–December?

c. In which month were the transportation costs for Ocean and TL about the same?

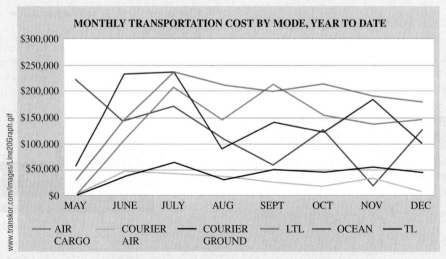

FIGURE 14.49 Monthly transportation cost by mode

The fastest growing cities in California are shown with their populations in Table 14.17.

TABLE 14.17
Fastest-Growing California Cities

City	Population
Coalinga	15,200
Brisbane	4,060
Brentwood	23,100
Dublin	32,500
Rio Vista	4,850
Cupertino	52,900
La Quinta	24,250
Rocklin	35,250
Lincoln	9,675
Temecula	53,800

Use this information for Problems 14–16.

14. Draw a bar graph showing the population of each of these cities.

15. What is the mean, median, and mode for these populations?

16. What is the standard deviation?

17. Table 14.18 shows the U.S. government's expenditures for defense.

TABLE 14.18
U.S. Defense Expenditures

Year	Billions of Dollars
1993	$279
1994	$269
1995	$260
1996	$253

a. Draw a bar graph to represent these data.

b. Suppose that you had the following viewpoint: *Defense spending down only $0.3 trillion from 1993 to 1996.* Draw a graph that shows very little decrease in the expenditures.

c. Suppose you had the following viewpoint: *Defense expenditures drastically cut $26,000,000,000 from 1993 to 1996.* Draw a graph that shows a tremendous decrease in the expenditures.

d. In view of parts **b** and **c,** discuss the possibilities of using statistics to mislead or support different views.

18. Table 14.19 compares age with blood pressure.

TABLE 14.19

Age and Blood Pressure								
Age (in years)	20	25	30	40	50	35	68	55
Blood Pressure	85	91	84	93	100	86	94	92

Draw a scatter diagram and find the regression line for this set of data.

19. Find the linear correlation coefficient (rounded to the nearest thousandth) for the data in Table 14.19 and determine whether the variables are significantly correlated at either the 1% or 5% level.

20. Read the *Dear Abby* column. Abby's answer was consoling and gracious, but not very statistical. If pregnancy durations have a normal distribution with a mean of 266 and a standard deviation of 16 days, what is the probability of having a 314-day pregnancy?

Dear Abby:
You wrote in your column that a woman is pregnant for 266 days. Who said so? I carried my baby for ten months and five days, and there is no doubt about it because I know the exact date my baby was conceived. My husband is in the Navy and it could not have possibly been conceived any other time because I saw him only once for an hour, and I didn't see him again until the day before the baby was born. I don't drink or run around, and there is no way this baby isn't his, so please print a retraction about that 266-day carrying time because otherwise I am in a lot of trouble.
San Diego Reader

Dear Reader:
The average gestation period is 266 days. Some babies come early. Others come late. Yours was late.

BOOK REPORTS

Write a 500-word report on this book:

How to Lie with Statistics, Darrell Huff (New York: Norton, 1954).

Group RESEARCH PROJECTS

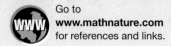

Go to
www.mathnature.com
for references and links.

Working in small groups is typical of most work environments, and learning to work with others to communicate specific ideas is an important skill. Work with three or four other students to submit a single report based on each of the following questions.

G46. Toss a toothpick onto a hardwood floor 1,000 times as described in Figure 14.50, or toss 1,000 toothpicks, one at a time, onto the floor. Let l be the length of the toothpick and d be the distance between the parallel lines determined by the floorboards.

Equipment needed: A box of toothpicks (of uniform length) and a large sheet of paper with equidistant parallel lines. A hardwood floor works very well instead of using a sheet of paper. The length of a toothpick should be less than the perpendicular distance between the parallel lines.

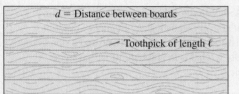

FIGURE 14.50 Buffon's needle problem

a. Guess the probability p that a toothpick will cross a line. *Do this before you begin the experiment.* The members of your group should reach a consensus before continuing.

b. Perform the experiment and find p empirically. That is, to find p, divide the number of toothpicks crossing a line by the number of toothpicks tossed (1,000 in this case).

c. By direct measurement, find l and d.

d. Calculate $2l$ and pd, and $\frac{2l}{pd}$.

e. Formulate a conclusion. This is an experiment known as Buffon's needle problem.

G47. You are interested in knowing the number and ages of children (0–18 years) in a part (or all) of your community. You will need to sample 50 families, finding the number of children in each family and the age of each child. It is important that you select the 50 families at random. How to do this is the subject of a course in statistics. For this problem, however, follow these steps:

Step 1   Determine the geographic boundaries of the area with which you are concerned.

Step 2   Consider various methods for selecting the families at random. For example, could you:
  (i) select the first 50 homes at which someone is at home when you call?
  (ii) select 50 numbers from a phone book that covers the same geographic boundaries as those described in step 1?
  (iii) Using (i) or (ii) could result in a biased sample. Can you guess why this might be true? In a statistics course, you might explore other ways of selecting the homes. For this problem, use one of these methods.

Step 3   Consider different ways of asking the question. Can the way the family is approached affect the response?

Step 4   Gather your data.

Step 5   Organize your data. Construct a frequency distribution for the children, with integral values from 0 to 18.

Step 6   Find out the number of families who actually live in the area you've selected. If you can't do this, assume that the area has 1,000 families.
  a. What is the average number of children per family?
  b. What percent of the children are in the first grade (age 6)?
  c. If all the children aged 12–15 are in junior high, how many are in junior high for the geographic area you are considering?
  d. See if you can actually find out the answers to parts **b** and **c**, and compare these answers with your projections.
  e. What other inferences can you make from your data?

Individual | RESEARCH PROJECTS

www.mathnature.com

Learning to use sources outside your classroom and textbook is an important skill, and here are some ideas for extending some of the ideas discussed in this chapter. You can find references to these projects in a library or at **www.mathnature.com.**

PROJECT 14.1 Collect examples of good statistical graphs and examples of misleading graphs. Use some of the leading newspapers, national magazines, and web sites.

PROJECT 14.2 Carry out the following experiment: *A cat has two bowls of food; one bowl contains Whiskas and the other contains some other brand. The cat eats Whiskas and leaves the other untouched.* Make a list of possible reasons why the cat ignored the second bowl. Describe the circumstances under which you think the advertiser could claim: *Eight out of ten owners said their cat preferred Whiskas.*

PROJECT 14.3 Roll a pair of dice 36 times, and draw a bar graph of the outcomes.
  a. Find the mean, the variance, and the standard deviation for this model.
  b. Repeat the experiment.
  c. Compare the results of parts **a** and **b**.

PROJECT 14.4 Prepare a report or exhibit showing how statistics are used in baseball.

PROJECT 14.5 Prepare a report or exhibit showing how statistics are used in educational testing.

PROJECT 14.6 Prepare a report or exhibit showing how statistics are used in psychology.

PROJECT 14.7 Prepare a report or exhibit showing how statistics are used in business. Use a daily report of transactions on the New York Stock Exchange. What inferences can you make from the information reported?

PROJECT 14.8 Investigate the work of Adolph Quetelet, Francis Galton, Karl Pearson, R. A. Fisher, and Florence Nightingale. Prepare a report or an exhibit of their work in statistics.

PROJECT 14.9 Historical Quest "We need privacy and a consistent wind," said Wilbur. "Did you write to the Weather Bureau to find a suitable location?" "Well," replied Orville, "I received this list of possible locations and Kitty Hawk, North Carolina, looks like just what we want. Look at this. . .."

However, Orville and Wilbur spent many days waiting in frustration after they arrived in Kitty Hawk, because the winds weren't suitable. The Weather Bureau's information gave the averages, but the Wright brothers didn't realize that an acceptable average can be produced by unacceptable extremes. Write a paper explaining how it is possible to have an acceptable average produced by unacceptable extremes.

PROJECT 14.10 Select something that you think might be normally distributed (for example, the ring size of students at your college). Next, select 100 people and make the appropriate measurements (in this example, ring size). Calculate the mean and standard deviation. Illustrate your findings using a bar graph. Do your data appear to be normally distributed?

18 THE NATURE OF CALCULUS

Outline

What in the World?

"Wow! That is much bigger than I imagined. I think I'll let you guys ride and I'll watch," said Charlie.

"Come on . . . it will be fun. We have made this our goal, man! This is *Colossus*. We studied it in our math class. It is 120 ft tall and is the world's tallest wooden coaster. It's been open almost 30 years and thousands have ridden safely," said Miguel with a reassuring tap on the arm. "Remember, we calculated its top velocity . . . "

Overview

In this chapter, we move from elementary mathematics to presenting an overview of the underpinnings of college mathematics. Some of the basic building blocks for advanced mathematics are functions, functional notation, and limits. These ideas are, in turn, the building blocks for a discussion of the nature of calculus.

It is not the intent of this chapter to teach calculus, but to discuss the *nature* of calculus in a context that will enable you to see why its invention was not only necessary, but inevitable.

© 2005 Joel Rogers www.CoasterGallery.com

18.1 | What Is Calculus?

The language of calculus has spread to all scientific fields; the insight it conveys about the nature of change is something that no educated person can afford to be without.

EVERYBODY COUNTS, NATIONAL RESEARCH COUNCIL

CHAPTER **CHALLENGE**

See if you can fill in the question mark.

Historical NOTE

Isaac Newton (1642–1727)

Gottfreid Leibniz (1646–1716)

The invention of calculus is credited jointly to Isaac Newton and Gottfried Leibniz. At the time of their respective break-through publications using calculus (around 1685), there was a bitter controversy throughout Europe as to whose work had been done first, along with accusations that each stole the idea from the other. Part of the explanation for this is the fact that each had done his actual work earlier, and another part can be attributed to the rivalry between mathematicians in England who championed Newton and those in Europe who supported Leibniz. The fact is, however, that the intellectual climate for the invention of calculus was ripe and actually inevitable.

If there is an event that marked the coming of age of mathematics in Western culture, it must be the essentially simultaneous development of the calculus by Newton and Leibniz in the 17th century. Before this remarkable synthesis, mathematics had often been viewed as merely a strange but harmless pursuit, indulged in by those with an excess of leisure time. After the calculus, mathematics became virtually the only acceptable language for describing the physical universe. This view of mathematics and its association with the scientific method has come to dominate the Western view of how the world ought to be explained.

What distinguishes calculus from algebra, geometry, and trigonometry is the transition from static or discrete applications (see Figure 18.1) to those that are dynamic or continuous (see Figure 18.2). For example, in elementary mathematics we consider the slope of a line, but in calculus we define the (nonconstant) slope of a nonlinear curve. In elementary mathematics we find average values of position and velocity, but in calculus we can find instantaneous values of changes of velocity and acceleration. In elementary mathematics we find the average of a finite collection of numbers, but in calculus we can find the average value of a function with infinitely many values over an interval.

1. Slope of a line

2. Tangent line to a circle

3. Area of a region bounded by line segments

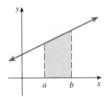

4. Average position and velocity

5. Average of a finite collection of numbers

FIGURE 18.1 Topics from elementary mathematics

Calculus is the mathematics of motion and change, which is why calculus is a prerequisite for many courses. Whenever we move from the static to the dynamic, we would consider using calculus.

The development of calculus in the 17th century by Newton and Leibniz was the result of their attempt to answer some fundamental questions about the world and the way things work. These investigations led to two fundamental concepts of calculus, namely, the idea of a *derivative* and that of an *integral*. The breakthrough in the development of these concepts was the formulation of a mathematical tool called a *limit*.

1. **Limit**: The limit is a mathematical tool for studying the *tendency* of a function as its variable *approaches* some value. Calculus is based on the concept of limit. We shall introduce the limit of a function informally in the next section.

2. **Derivative:** The derivative is defined as a certain type of limit, and it is used initially to compute rates of change and slopes of tangent lines to curves. The study of derivatives is called *differential calculus*. Derivatives can be used in sketching graphs and in finding the extreme (largest and smallest) values of functions. We shall discuss derivatives in Section 18.3.

3. **Integral:** The integral is found by taking a special limit of a sum of terms, and the study of this process is called *integral calculus*. Area, volume, arc length, work, and hydrostatic force are a few of the many quantities that can be expressed as integrals. We shall discuss integrals in Section 18.4.

In this section, let us take an intuitive look at each of these three essential ideas of calculus.

The Limit: Zeno's Paradox

In Chapter 6 we first introduced Zeno's paradox. Zeno (ca. 500 BC) was a Greek philosopher known primarily for his famous paradoxes. One of those concerns a race between Achilles, a legendary Greek hero, and a tortoise. When the race begins, the (slower) tortoise is given a head start, as shown in Figure 18.3.

FIGURE 18.3 Achilles and the tortoise

Is it possible for Achilles to overtake the tortoise? Zeno pointed out that by the time Achilles reaches the tortoise's starting point, $a_1 = t_0$, the tortoise will have moved ahead to a new point t_1. When Achilles gets to this next point, a_2, the tortoise will be at a new point, t_2. The tortoise, even though much slower than Achilles, keeps moving forward. Although the distance between Achilles and the tortoise is getting smaller and smaller, the tortoise will apparently always be ahead.

Of course, common sense tells us that Achilles will overtake the slow tortoise, but where is the error in reasoning in the previous paragraph that always gives the tortoise the lead? The error is in the assumption that an infinite amount of time is required to cover a distance divided into an infinite number of segments. This discussion is getting at an essential idea in calculus—namely, the notion of a limit.

Consider the successive positions for both Achilles and the tortoise:

↓ Starting position

Achilles: $a_0, a_1, a_2, a_3, \ldots$

Tortoise: $t_0, t_1, t_2, t_3, \ldots$

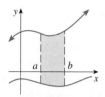
After the start, the positions for Achilles, as well as those for the tortoise, form sets of positions that are ordered with positive integers. Such ordered listings are called *sequences* (see Section 11.3).

For Achilles and the tortoise we have two sequences $\{a_1, a_2, a_3, \ldots, a_n, \ldots\}$ and $\{t_1, t_2, t_3, \ldots, t_n, \ldots\}$, where $a_n < t_n$ for all values of n. Both the sequence for Achilles' position and the sequence for the tortoise's position have limits, and it is precisely at that limit point that Achilles overtakes the tortoise. The idea of limit will be discussed in the next section, and it is this limit idea that allows us to define the other two basic concepts of calculus: the derivative and the integral. Even if the solution to Zeno's paradox using limits seems unnatural at first, do not be discouraged. It took over 2000 years to refine the ideas of Zeno and provide conclusive answers to those questions about limits. The following example will provide an intuitive preview of a limit, which we consider in more detail in Section 18.2.

Example 1 Intuitively find a limit

The sequence $\frac{1}{2}, \frac{2}{3}, \frac{3}{4}, \frac{4}{5}, \cdots$ can be described by writing a *general term*: $\frac{n}{n+1}$, where $n = 1, 2, 3, 4, \ldots$.* Can you guess the limit, L, of this sequence? We will say that L is the number that the sequence with general term $\frac{n}{n+1}$ tends toward as n becomes large without bound. We will define a notation to summarize this idea:

$$L = \lim_{n \to \infty} \frac{n}{n+1}$$

Solution   As you consider larger and larger values for n, you find a sequence of fractions:

$$\frac{1}{2}, \frac{2}{3}, \frac{3}{4}, \ldots, \frac{1{,}000}{1{,}001}, \frac{1{,}001}{1{,}002}, \ldots, \frac{9{,}999{,}999}{10{,}000{,}000}, \cdots$$

It is reasonable to guess that the sequence of fractions is approaching the number 1. This number is called the **limit.**

The Derivative: The Tangent Problem

A tangent line (or, if the context is clear, simply say "tangent") to a circle at a given point P is a line that intersects the circle at P and only at P (see Figure 18.4a). This characterization does not apply for curves in general, as you can see by looking at Figure 18.4b. If we wish to have one tangent line, which line shown should be called the tangent to the curve?

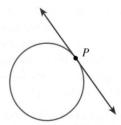

a. At each point P on a circle, there is one line that intersects the circle exactly once.

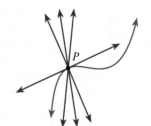

b. At a point P on a curve, there may be several lines that intersect that curve only once.

FIGURE 18.4 Tangent line

---

*You might wish to review sequences and general terms in Section 11.3.

To find a tangent line, begin by considering a line that passes through two points P and Q on the curve, as shown in Figure 18.5a. This line is called a **secant line.**

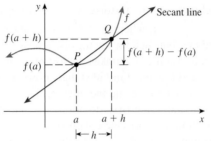 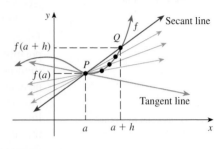

a. Locate points P and Q and draw secant line   **b.** Let point Q "slide" toward P and draw lines

FIGURE 18.5 Tangent line is the "limiting line" as Q approaches P

The coordinates of the two points P and Q are $P(a, f(a))$ and $Q(a + h, f(a + h))$. The slope of the secant line is

$$m = \frac{\text{RISE}}{\text{RUN}} = \frac{f(a + h) - f(a)}{h}$$

Now imagine that Q moves along the curve toward P, as shown in Figure 18.5b. You can see that the secant line approaches a limiting position as h approaches zero. We define this limiting position to be the **tangent line.** The slope of the tangent line is defined as a limit of the sequence of slopes of a set of secant lines. Once again, we can use limit notation to summarize this idea: We say that the slopes of the secant lines, as h becomes small, tend toward a number that we call the slope of the tangent line. We will define the following notation to summarize this idea:

$$\lim_{h \to 0} \frac{f(a + h) - f(a)}{h}$$

Why would the slope of a tangent line be important? That is what we investigate in the first half of a calculus course, and it forms the definition of derivative, which is the foundation for what is called **differential calculus** (see Section 18.3).

The Integral: The Area Problem

You probably know the formula for the area of a circle with radius r:

$$A = \pi r^2$$

The Egyptians were the first to use this formula over 5,000 years ago, but the Greek Archimedes (ca. 300 B.C.) showed how to derive the formula for the area of a circle by using a limiting process. Consider the areas of inscribed polygons, as shown in Figure 18.6.

Since we know the areas of these polygons, we can use them to estimate the area of a circle. Even though Archimedes did not use the following notation, here is the essence of what he did, using a method called "exhaustion":

Let A_3 be the area of the inscribed equilateral triangle;

A_4 be the area of the inscribed square;

A_5 be the area of the inscribed regular pentagon.

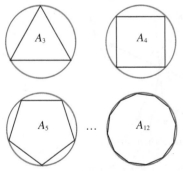

FIGURE 18.6 Approximating the area of a circle

How can we find the area of this circle? As you can see from Figure 18.6, if we consider the area of A_3, then A_4, then A_5, . . . , we should have a sequence of areas such that each successive area more closely approximates that of the circle. We write this idea as a limit statement:

$$A = \lim_{n \to \infty} A_n$$

In this course we will use limits in yet a different way to find the areas of regions enclosed by curves. For example, consider the area shown in color in Figure 18.7.

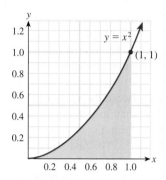

FIGURE 18.7 Area under a curve

We can approximate the area by using rectangles. If A_n is the area of the nth rectangle, then the total area can be approximated by finding the sum

$$A_1 + A_2 + A_3 + \cdots + A_{n-1} + A_n$$

The process is shown in Figure 18.8.

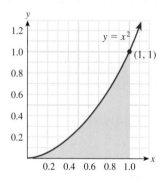 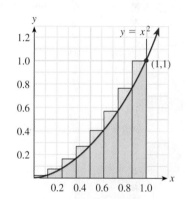

a. 8 approximating rectangles   **b.** 16 approximating rectangles

FIGURE 18.8 Approximating the area using rectangles

Notice that in the Figure 18.8**a**, the area of the rectangles approximates the area under the curve, but there is some error. In Figure 18.8**b**, there is a little less error than in the first approximation. As more rectangles are included, the error decreases.

The area problem leads to a process called *integration*, and the study of integration forms what is called **integral calculus** (see Section 18.4). Similar reasoning allows us to calculate such quantities as volume, the length of a curve, the average value, or the amount of work required for a particular task.

Mathematical Modeling

A real-life situation is usually far too complicated to be precisely and mathematically defined. When confronted with a problem in the real world, therefore, it is usually necessary to develop a mathematical framework based on certain assumptions about the real world. This framework can then be used to find a solution to the real-world problem. The process of developing this body of mathematics is referred to as **mathematical modeling.** Most mathematical models are dynamic (not static) and are continually being revised (modified) as additional relevant information becomes known.

Some mathematical models are quite accurate, particularly those used in the physical sciences. For example, one of the first models we will consider in calculus is a model for the path of a projectile. Other rather precise models predict such things as the time of

How Global Climate Is Modeled

We find a good example of mathematical modeling by looking at the work being done with weather prediction. In theory, if the correct assumptions could be programmed into a computer, along with appropriate mathematical statements of the ways global climate conditions operate, we would have a model to predict the weather throughout the world. In the global climate model, a system of equations calculates time-dependent changes in wind as well as temperature and moisture changes in the atmosphere and on the land. The model may also predict alterations in the temperature of the ocean's surface. At the National Center for Atmospheric Research, a CRAY supercomputer is used to do this modeling.

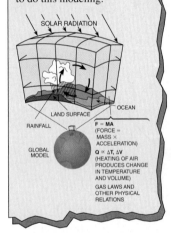

sunrise and sunset, or the speed at which an object falls in a vacuum. Some mathematical models, however, are less accurate, especially those that involve examples from the life sciences and social sciences. Only recently has modeling in these disciplines become precise enough to be expressed in terms of calculus.

What, precisely, is a mathematical model? Sometimes, mathematical modeling can mean nothing more than a textbook word problem. But mathematical modeling can also mean choosing appropriate mathematics to solve a problem that has previously been unsolved. In this book, we use the term *mathematical modeling* to mean something between these two extremes. That is, it is a process we will apply to some real-life problem that does not have an obvious solution. It usually cannot be solved by applying a single formula.

The first step of what we call mathematical modeling involves *abstraction.*

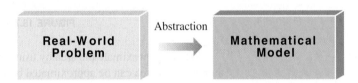

With the method of abstraction, certain assumptions about the real world are made, variables are defined, and appropriate mathematics is developed. The next step is to simplify the mathematics or derive related mathematical facts from the mathematical model.

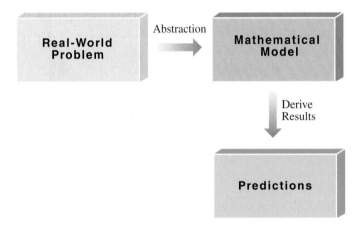

The results derived from the mathematical model should lead us to some predictions about the real world. The next step is to gather data from the situation being modeled, and then to compare those data with the predictions. If the two do not agree, then the gathered data are used to modify the assumptions used in the model.

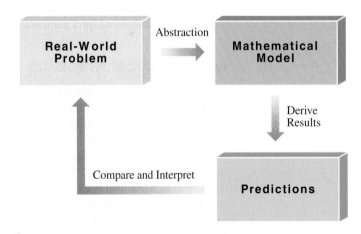

Mathematical modeling is an ongoing process. As long as the predictions match the real world, the assumptions made about the real world are regarded as correct, as are the defined variables. On the other hand, as discrepancies are noticed, it is necessary to construct a closer and a more dependable mathematical model. You might wish to read the article in *Scientific American* presented in the News Clip.

Example 2 Formulate a budget to illustrate modeling

Develop a monthly budget for your personal expenses. You can assume any family situation (for example, number of people and amount of income and expenses), but this budget requires that you are self-sufficient and that income and expenses must always be equal. *This example is not about budgets, but about the process of modeling.*

Solution The first step in modeling is to understand the problem. Let us assume that we are budgeting for one person living at home with his or her parents. We begin by setting up a mathematical model to describe the income and expenses as shown in the margin.

This budget satisfies the condition that income equals expenses. However, this model does not reflect the real world. We need to compare and interpret the results of this budget with reality. What about taxes and FICA? Let us make a second attempt at abstraction.

MONTHLY BUDGET

INCOME		$275.00
EXPENSES		
Car payments	$125.00	
Savings	$ 50.00	
Movies	$ 40.00	
Gasoline	$ 60.00	
		$275.00

MONTHLY BUDGET

INCOME

Monthly Income before taxes	$640.00	
From parents	$530.00	
TOTAL MONTHLY INCOME	$1,170.00	

EXPENSES

Fixed		Variable	
Rent/mortgage		Food	
Car payments	$125.00	Gasoline	$70.00
Income tax withholding	$170.00		
FICA	$85.00	Utilities	
Retirement	$110.00	Electricity	
Contributions	$65.00	Gas	
		Water/Sewer	
Installments		Telephone	$35.00
Wells Fargo	$35.00	Cable TV	
Sears	$25.00	Other	
VISA	$25.00		
		Entertainment	$30.00
Savings			
Wells Fargo	$400.00	Other	
Other			
Total Fixed Expenses	$1,040.00	Total Variable Expenses	$135.00
Total Variable Expenses	$135.00		
TOTAL MONTHLY EXPENSE	$1,175.00		

We can see that the budget is a rough estimate of actual real-world spending, but it is not quite balanced (which it needs to be) and does not include "hidden" income and expenses, such as value for the room and board at home. "But my parents pay for that!" is the response.

A model must reflect the true income and expenses for the individual, so we must go back and create a more realistic model. Even though "cash" is not exchanged for room and board, "value" is given and must be taken into account. What about income or expenses that do not occur monthly? How about clothes, car repairs, or a vacation? Once again, the modeling process compares and interprets, and then refines with another abstraction.

MONTHLY BUDGET

INCOME			
Monthly Income before taxes	$640.00	Other nonmonthly income	$500.00
From parents	$530.00		$215.00
Credit cards	$125.00		
Total other/12	$59.58	Total other	$715.00
TOTAL MONTHLY INCOME	$1,354.58		

EXPENSES	Fixed		Variable			Nonmonthly
Rent/mortgage	$350.00	Food	$280.00	Car repairs		$290.00
Car payments	$125.00	Gasoline	$70.00	Home repairs		
Income tax withholding	$170.00			Car license		$85.00
FICA	$85.00	Utilities		Medical		
Retirement	$110.00	Electricity	$25.00	Dental		
Contributions	$65.00	Gas	$15.00	Clothing		$350.00
Insurance	$105.00	Water/Sewer		Gifts		$400.00
Installments		Telephone	$35.00	Contributions		
Wells Fargo	$35.00	Cable TV	$20.00	Vacation		$800.00
Sears	$25.00	Other		Professional		
VISA	$25.00					
		Entertainment	$30.00	Insurance:		
Savings		Other				
Wells Fargo	$65.00			Other:		
Other						
Total Fixed Expenses	$1,160.00	Total Variable Expense	$475.00	Total Nonmonthly	$1,925.00	
Total Variable Expenses	$475.00					
1/12 Nonmonthly	$160.42					
TOTAL MONTHLY EXPENSE	$1,795.42					

Now this iteration of the budget is much more realistic, but it does not "balance." Each successive iteration should more closely model the real world, and to complicate the process, the real-world situation of income and expenses is always changing, which may require further revisions of the model shown below. As we can see, the process of modeling is ongoing and may never be "complete."

MONTHLY BUDGET

INCOME			
Monthly Income before taxes	$640.00	Other nonmonthly income	$500.00
From parents	$580.00		$215.00
Credit cards	$125.00		
Second job	$240.00		
Total other/12	$59.58	Total other	$715.00
TOTAL MONTHLY INCOME	$1,644.58		

EXPENSES	Fixed		Variable			Nonmonthly
Rent/mortgage	$350.00	Food	$250.00	Car repairs		$290.00
Car payments	$125.00	Gasoline	$70.00	Home repairs		
Income tax withholding	$170.00			Car license		$85.00
FICA	$85.00	Utilities		Medical		
Retirement	$110.00	Electricity	$25.00	Dental		
Contributions	$65.00	Gas	$15.00	Clothing		$350.00
Insurance	$105.00	Water/Sewer		Gifts		$300.00
Installments		Telephone	$35.00	Contributions		
Wells Fargo	$35.00	Cable TV	$20.00	Vacation		
Sears	$25.00	Other		Professional		
VISA	$25.00					
		Entertainment	$30.00	Insurance:		
Savings						
Wells Fargo	$19.16					
Other		Other		Other		
Total Fixed Expenses	$1,114.16					
Total Variable Expenses	$445.00	Total Variable Expenses	$445.00	Total Nonmonthly	$1,025.00	
1/12 Nonmonthly	$85.42					
TOTAL MONTHLY EXPENSE	$1,644.58					

Problem Set 18.1

Level **1**

1. **IN YOUR OWN WORDS** What are the three main topics of calculus?

2. **IN YOUR OWN WORDS** What is a mathematical model? Why are mathematical models necessary or useful?

3. **IN YOUR OWN WORDS** An analogy to Zeno's tortoise paradox can be made as follows.

> A woman standing in a room cannot walk to a wall. To do so, she would first have to go half the distance, then half the remaining distance, and then again half of what still remains. This process can always be continued and can never be ended.

Draw an appropriate figure for this problem and then present an argument using sequences to show that the woman will, indeed, reach the wall.

4. **IN YOUR OWN WORDS** Zeno's paradoxes remind us of an argument that might lead to an absurd conclusion:

> Suppose I am playing baseball and decide to steal second base. To run from first to second base, I must first go half the distance, then half the remaining distance, and then again half of what remains. This process is continued so that I never reach second base. Therefore it is pointless to steal base.

Draw an appropriate figure for this problem and then present a mathematical argument using sequences to show that the conclusion is absurd.

See **www.mathnature.com** for a link to a site on Zeno's paradox.

5. Consider the sequence 0.3, 0.33, 0.333, 0.3333, ... What do you think is the appropriate limit of this sequence?

6. Consider the sequence 6, 6.6, 6.66, 6.666, ... What do you think is the appropriate limit of this sequence?

7. Consider the sequence 0.9, 0.99, 0.999, 0.9999, ... What do you think is the appropriate limit of this sequence?

8. Consider the sequence 0.2, 0.27, 0.272, 0.2727, ... What do you think is the appropriate limit of this sequence?

9. Consider the sequence 3, 3.1, 3.14, 3.141, 3.1415, 3.14159, 3.141592, ... What do you think is the appropriate limit of this sequence?

Copy the figures in Problems 10–15 on your paper. Draw what you think is an appropriate tangent line for each curve at the point P by using the secant method.

10.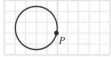

11.

12.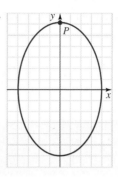

13.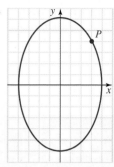

14.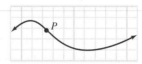

15.

Level **2**

*In Problems 16–21, **guess** the requested limits.*

16. $\lim_{n\to\infty} \dfrac{2n}{n+4}$

17. $\lim_{n\to\infty} \dfrac{2n}{3n+1}$

18. $\lim_{n\to\infty} \dfrac{n+1}{n+2}$

19. $\lim_{n\to\infty} \dfrac{n+1}{2n}$

20. $\lim_{n\to\infty} \dfrac{3n}{n^2+2}$

21. $\lim_{n\to\infty} \dfrac{3n^2+1}{2n^2-1}$

Estimate the area in each figure shown in Problems 22–27.

22.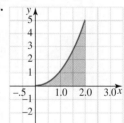

23.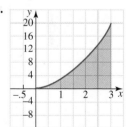

24.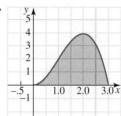

25.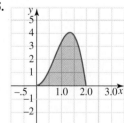

26.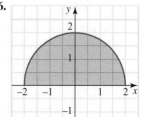

27.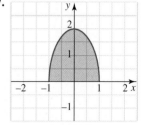

Level 3

28. IN YOUR OWN WORDS   Prepare a personal budget.

29. IN YOUR OWN WORDS   Prepare a budget for a family of four with an annual income of $100,000.

Problem Solving 3

30. a. Calculate the sum of the areas of the rectangles shown in Figure 18.8a.

  b. Calculate the sum of the areas of the rectangles shown in Figure 18.8b.

  c. Make a guess about the shaded area under the curve.

18.2 Limits

The making of a motion picture is a complex process, and editing all the film into a movie requires that all the frames of the action be labeled in chronological order. For example, R21–435 might signify the 435th frame of the 21st reel. If you have used a movie-making program, you may look at the frames in a film of a bouncing ball, as shown in Figure 18.9.

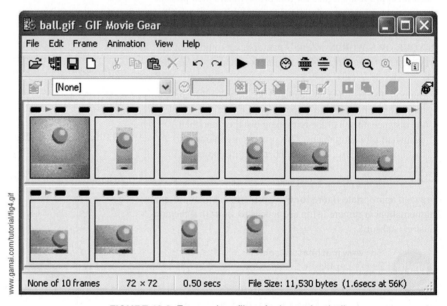

www.gamai.com/tutorial/fig4.gif

FIGURE 18.9 Frames in a film of a bouncing ball

A mathematician might refer to the movie editor's labeling procedure by saying the frames are arranged in a *sequence*. We introduced sequences in Chapter 11, but let's briefly review that idea. A **sequence** is a succession of numbers that are listed according to a given prescription or rule. Specifically, if n is a positive integer, the sequence whose nth term is the number a_n can be written as

$$a_1, a_2, \ldots, a_n, \ldots$$

or, more simply,

$$\{a_n\}$$

The number a_n is called the **general term** of the sequence. We will deal only with infinite sequences, so each term a_n has a **successor** a_{n+1} and for $n > 1$, a **predecessor** a_{n-1}. For example, by associating each positive integer n with its reciprocal $\frac{1}{n}$, we obtain the sequence denoted by

$\left\{\dfrac{1}{n}\right\}$, which represents the succession of numbers $1, \dfrac{1}{2}, \dfrac{1}{3}, \ldots, \dfrac{1}{n}, \ldots$

The general term is denoted by $a_n = \frac{1}{n}$. The following examples illustrate the notation and terminology used in connection with sequences.

Example 1   Find terms of a sequence

Find the 1st, 2nd, and 15th terms of the sequence $\{a_n\}$ where the general term is

$$a_n = \left(\frac{1}{2}\right)^{n-1}$$

Solution   If $n = 1$, then $a_1 = \left(\frac{1}{2}\right)^{1-1} = 1$. Similarly,

$$a_2 = \left(\frac{1}{2}\right)^{2-1} = \frac{1}{2}$$

$$a_{15} = \left(\frac{1}{2}\right)^{15-1} = \left(\frac{1}{2}\right)^{14} = 2^{-14}$$

The Limit of a Sequence

CAUTION
Even though we write a_n, this is a function, $a(n) = \frac{n}{n+1}$, for which the domain is the set of nonnegative integers.

It is often desirable to examine the behavior of a given sequence $\{a_n\}$ as n gets arbitrarily large. For example, consider the sequence

$$a_n = \frac{n}{n+1}$$

Because $a_1 = \frac{1}{2}, a_2 = \frac{2}{3}, a_3 = \frac{3}{4}, \ldots$, we can plot the terms of this sequence on a number line as shown in Figure 18.10a, or in two dimensions as shown in Figure 18.10b.

Even though we know these tracks to be parallel (they do not actually approach each other), they appear to meet at the horizon. This concept of convergence at a point is used in calculus to give specific values for otherwise immeasurable quantities.

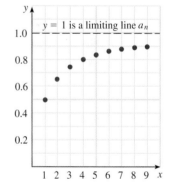

a. Graphing a sequence in one dimension   **b.** Graphing a sequence in two dimensions

FIGURE 18.10 Graphing the sequence $a_n = \frac{n}{n+1}$

Limits to Infinity

By looking at either graph in Figure 18.10, we can see that the terms of the sequence are approaching 1. We say that the *limit of this sequence* is $L = 1$.

In general, we say that the sequence **converges** to the limit L and write

$$\lim_{n \to \infty} a_n = L$$

if L is a *unique* number that the value of the expression a_n is moving ever and ever closer to as $n \to \infty$. The number L is called the **limit of the sequence**. The value L does not need to exist. If a limit does not exist, we say the sequence **diverges.**

Example 2 Find a limit to infinity

Find the limit, if it exists.

a. $\displaystyle\lim_{n\to\infty} \frac{n}{n+1}$ **b.** $\displaystyle\lim_{n\to\infty} n$ **c.** $\displaystyle\lim_{n\to\infty} \frac{10{,}000}{n}$ **d.** $\displaystyle\lim_{n\to\infty} \frac{2n-1}{n}$

Solution

a. $\displaystyle\lim_{n\to\infty} \frac{n}{n+1}$ This sequence is shown in Figure 18.10. We see

$$\lim_{n\to\infty} \frac{n}{n+1} = 1$$

b. $\displaystyle\lim_{n\to\infty} n$ This sequence is 1, 2, 3, 4, . . . , and we see that the values are not approaching any limit, so we say this limit does not exist.

c. $\displaystyle\lim_{n\to\infty} \frac{10{,}000}{n}$ The values of this expression are: 10,000 ($n = 1$); 5,000 ($n = 2$); 3,333.33 ($n = 3$); 2,500 ($n = 4$); 2,000 ($n = 5$); . . . . For large n we see that the value of the sequence is close to 0:

$$n = 10{,}000: \quad \frac{10{,}000}{n} = \frac{10{,}000}{10{,}000} = 1$$

$$n = 10{,}000{,}000: \quad \frac{10{,}000}{n} = \frac{10{,}000}{10{,}000{,}000} = 0.001$$

$$\vdots$$

It appears that $\displaystyle\lim_{n\to\infty} \frac{10{,}000}{n} = 0$.

d. $\displaystyle\lim_{n\to\infty} \frac{2n-1}{n}$ This sequence can be shown with the following table for selected values for n:

n	1	10	100	1,000	10,000	100,000
$\dfrac{2n-1}{n}$	1	1.9	1.99	1.999	1.9999	1.99999

In this case it appears that $\displaystyle\lim_{n\to\infty} \frac{2n-1}{n} = 2$.

Example 2 leads to some conclusions.

Limit of $\frac{1}{n}$ to Infinity

$$\lim_{n\to\infty} \frac{1}{n} = 0$$

Furthermore, for any A, and k a positive integer,

$$\lim_{n\to\infty} \frac{A}{n^k} = 0$$

Example 3 Find the limit of a convergent sequence

Find the limit of each of these convergent sequences:

a. $\left\{\dfrac{100}{n}\right\}$ **b.** $\left\{\dfrac{2n^2 + 5n - 7}{n^3}\right\}$ **c.** $\left\{\dfrac{3n^4 + n - 1}{5n^4 + 2n^2 + 1}\right\}$

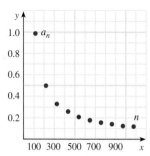

FIGURE 18.11 Graphical representation of $a_n = 100/n$

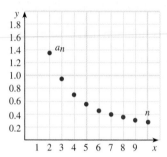

FIGURE 18.12 Graph of
$$a_n = \frac{2n^2 + 5n - 7}{n^3}$$

Solution

a. As n grows arbitrarily large, $\frac{100}{n}$ gets smaller and smaller. Thus,

$$\lim_{n \to \infty} \frac{100}{n} = 0$$

A graphical representation is shown in Figure 18.11.

b. We begin by algebraically rewriting the expression:

$$\frac{2n^2 + 5n - 7}{n^3} = \frac{2}{n} + \frac{5}{n^2} + \frac{-7}{n^3}$$

We now use the limit to infinity property:

$$\lim_{n \to \infty} \frac{2n^2 + 5n - 7}{n^3} = \lim_{n \to \infty} \left[\frac{2n^2}{n^3} + \frac{5n}{n^3} + \frac{-7}{n^3} \right]$$

If we assume that the limit of a sum is the sum of the limits, then we have

$$= 0 + 0 + 0$$
$$= 0$$

A graph is shown in Figure 18.12.

c. Divide the numerator and denominator by n^4 to obtain

$$\lim_{n \to \infty} \frac{3n^4 + n - 1}{5n^4 + 2n^2 + 1} = \lim_{n \to \infty} \frac{\frac{3n^4}{n^4} + \frac{n}{n^4} - \frac{1}{n^4}}{\frac{5n^4}{n^4} + \frac{2n^2}{n^4} + \frac{1}{n^4}} \qquad \textit{Divide each term by } n^4.$$

$$= \lim_{n \to \infty} \frac{3 + \frac{1}{n^3} - \frac{1}{n^4}}{5 + \frac{2}{n^2} + \frac{1}{n^4}}$$

$$= \frac{3}{5}$$

Example 4   Illustrate a divergent sequence

Find the limit (if it exists) of the sequence with general term $\{(-1)^n\}$.

Solution   The sequence defined by $\{(-1)^n\}$ is $-1, 1, -1, 1, \ldots$, and this sequence diverges by oscillation because the nth term is always either 1 or -1. Thus a_n cannot approach one specific number L as n grows large. The graph is shown in Figure 18.13. We call this a *divergent sequence*.

n	a_n
1	-1
2	1
3	-1
4	1
5	-1
6	1
7	-1
8	1
9	-1
10	1

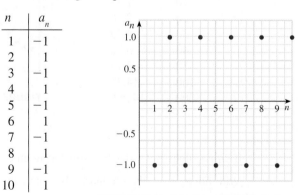

FIGURE 18.13 A limit to infinity that diverges

Infinite Limits

A limit to infinity is the limit of a sequence when $n \to \infty$. We now consider the possibility that the limit itself increases without limit.

Example **5** | **Find a limit of a rational expression**

Find each limit (if it exists).

a. $\lim\limits_{n \to 2} \dfrac{1}{n - 2}$ **b.** $\lim\limits_{n \to \infty} \dfrac{n}{n + 1}$ **c.** $\lim\limits_{n \to \infty} \dfrac{n^5 + n^3 + 2}{7n^4 + n^2 + 3}$

Solution

a. For $\lim\limits_{n \to 2} \dfrac{1}{n-2}$, we see that as n approaches 2 the denominator $n - 2$ approaches 0. As

this denominator approaches 0, the quotient $\dfrac{1}{n-2}$ gets infinitely large, so we say

that $\lim\limits_{n \to 2} \dfrac{1}{n-2} \to \infty$. In words, we say the sequence diverges.

b. The given limit can be written as

$$\lim_{n \to \infty} \frac{n}{n + 1} = \lim_{n \to \infty} \frac{1}{1 + \frac{1}{n}} = 1$$

c. $\lim\limits_{n \to \infty} \dfrac{n^5 + n^3 + 2}{7n^4 + n^2 + 3} = \lim\limits_{n \to \infty} \dfrac{1 + \frac{1}{n^2} + \frac{2}{n^5}}{\frac{7}{n} + \frac{1}{n^3} + \frac{3}{n^5}}$

The numerator tends toward 1 as $n \to \infty$, and the denominator approaches 0. Hence the quotient gets ever larger and larger, passing by all numbers, so it cannot approach a specific number L; thus the sequence must diverge. The graph is shown in Figure 18.14.

n	a_n
1	0.36
2	0.35
3	0.47
4	0.60
5	0.74
6	0.88
7	1.02
8	1.16
10	1.44
20	2.86
30	4.29
40	5.72
50	7.15
100	14.29
1,000	142.86

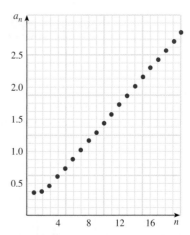

FIGURE 18.14 An infinite limit

If $\lim\limits_{n \to \infty} a_n$ does not exist because the numbers a_n become arbitrarily large as $n \to \infty$, write $\lim\limits_{n \to \infty} a_n = \infty$. We summarize this more precisely in the following box.

Infinite Limits

The **infinite limit** $\lim\limits_{n\to\infty} a_n = \infty$ means that for any real number A, we have $a_n > A$ for all sufficiently large n. The **infinite limit** $\lim\limits_{n\to\infty} b_n = -\infty$ means that for any real number B, we have $b_n < B$ for all sufficiently large n.

The answer to Example 5c can be rewritten in this notation as

$$\lim_{n\to\infty} \frac{n^5 + n^3 + 2}{7n^4 + n^2 + 3} = \infty$$

Also notice that $\lim\limits_{n\to\infty}(-5n) = -\infty$ (that is, exceeds all bounds), whereas $\lim\limits_{n\to\infty}(-1)^n$ does not exist. Thus, the answer to Example 4 is *not* ∞ or $-\infty$.

Problem Set 18.2

Level 1

1. **IN YOUR OWN WORDS**   What do we mean by the limit of a sequence?

2. **IN YOUR OWN WORDS**   Outline a procedure for finding the limit of a sequence.

Write out the first five terms (beginning with $n = 1$) of the sequences given in Problems 3–6.

3. $\{1 + (-1)^n\}$

4. $\left\{\left(\dfrac{-1}{2}\right)^{n+2}\right\}$

5. $\left\{\dfrac{3n + 1}{n + 2}\right\}$

6. $\left\{\dfrac{n^2 - n}{n^2 + n}\right\}$

Find each limit in Problems 7–10, if it exists.

7. $\lim\limits_{n\to\infty} \dfrac{8,000n}{n + 1}$

8. $\lim\limits_{n\to\infty} \dfrac{8,000}{n - 1}$

9. $\lim\limits_{n\to\infty} \dfrac{2n + 1}{3n - 4}$

10. $\lim\limits_{n\to\infty} \dfrac{4 - 7n}{8 + n}$

Compute the limit of the convergent sequences in Problems 11–16.

11. $\left\{\dfrac{5n + 8}{n}\right\}$

12. $\left\{\dfrac{5n}{n + 7}\right\}$

13. $\left\{\dfrac{8n^2 + 800n + 5,000}{2n^2 - 1,000n + 2}\right\}$

14. $\left\{\dfrac{100n + 7,000}{n^2 - n - 1}\right\}$

15. $\left\{\dfrac{8n^2 + 6n + 4,000}{n^3 + 1}\right\}$

16. $\left\{\dfrac{n^3 - 6n^2 + 85}{2n^3 - 5n + 170}\right\}$

Level 2

Find the limit (if it exists) as $n \to \infty$ for each of the sequences in Problems 17–28.

17. $\left\{\dfrac{1}{3n}\right\}$

18. $\left\{\dfrac{1}{2^n}\right\}$

19. $0.69, 0.699, 0.6999, 0.69999, \ldots$

20. $5, 5\frac{1}{2}, 5\frac{2}{3}, 5\frac{3}{4}, 5\frac{4}{5}, \ldots$

21. $\lim\limits_{n\to\infty} \dfrac{3n^2 - 7n + 2}{5n^4 + 9n^2}$

22. $\lim\limits_{n\to\infty} \dfrac{4n^4 + 10n - 1}{9n^3 - 2n^2 - 7n + 3}$

23. $\lim\limits_{n\to\infty} \dfrac{2n^4 + 5n^2 - 6}{3n + 8}$

24. $\lim\limits_{n\to\infty} \dfrac{10n^3 + 13}{7n^2 - n + 2}$

25. $\lim\limits_{n\to\infty} \dfrac{15 + 9n - 6n^2}{2n^2 - 4n + 1}$

26. $\lim\limits_{n\to\infty} \dfrac{n^4 + 5n^3 + 8n^2 - 4n + 12}{2n^4 - 7n^2 + 3n}$

27. $\lim\limits_{n\to\infty} \dfrac{-21n^3 + 52}{-7n^3 + n^2 + 20n - 9}$

28. $\lim\limits_{n\to\infty} \dfrac{12n^5 + 7n^4 - 3n^2 + 2n}{n^5 + 8n^3 + 14}$

Level 3

29. Twenty-four milligrams of a drug is administered into the body. At the end of each hour, the amount of drug present is half what it was at the end of the previous hour. What amount of the drug is present at the end of 5 hours? At the end of n hours?

Problem Solving 3

30. HISTORICAL QUEST **The Fibonacci Rabbit Problem**

The general term in a sequence can be defined in a number of ways. In this problem, we consider a sequence whose nth term is defined by a *recursion formula*—that is, a formula in which the nth term is given in terms of previous terms in the sequence. The problem was originally examined by Leonardo Pisano (also called Fibonacci) in the 13th century, and was first introduced in Chapter 11. Recall that we suppose rabbits breed in such a way that each pair of adult rabbits produces a pair of baby rabbits each month.

Number of Months	Number of Pairs	Pairs of Rabbits (the pairs shown in color are ready to reproduce in the next month)
Start	1	
1	1	
2	2	
3	3	
4	5	
5	8	
⋮	⋮	Same pair (rabbits never die)

Further assume that young rabbits become adults after two months and produce another pair of offspring at that time. A rabbit breeder begins with one adult pair. Let a_n denote the number of adult pairs of rabbits in this "colony" at the end of n months.

a. Explain why $a_1 = 1$, $a_2 = 1$, $a_3 = 2$, $a_4 = 3$, and, in general, $a_{n+1} = a_{n-1} + a_n$ for $n = 2, 3, 4, \ldots$

b. The *growth rate* of the colony during the $(n + 1)$st month is

$$r_n = \frac{a_{n+1}}{a_n}$$

Compute r_n for $n = 1, 2, 3, \ldots, 10$.

c. Assume that the growth rate sequence $\{r_n\}$ defined in part **b** converges, and let $L = \lim_{n \to \infty} r_n$. Use the recursion formula in part **a** to show that

$$\frac{a_{n+1}}{a_n} = 1 + \frac{a_{n-1}}{a_n}$$

and conclude that L must satisfy the equation

$$L = 1 + \frac{1}{L}$$

Use this information to compute L.

18.3 Derivatives

Elementary mathematics focuses on formulas and relationships among variables but does not include the analysis of quantities that are in a state of constant change. For example, the following problem might be found in elementary mathematics. If you drive at 55 miles per hour (mph) for a total of 3 hours, how far did you travel? This example uses a formula, $d = rt$ (distance = rate · time), and the answer is $d = 55(3) = 165$. However, this model does not adequately describe the situation in the real world. You could drive for 3 hours at an *average rate* of 55 mph, but you probably could not drive for 3 hours at a *constant rate* of 55 mph—in reality, the rate would be in a state of constant change. Other applications in which rates may not remain constant quickly come to mind:

> Profits changing with sales
>
> Population changing with the growth rate
>
> Property taxes changing with the tax rate
>
> Tumor sizes changing with chemotherapy
>
> The speed of falling objects changing over time
>
> The slope of a nonlinear curve

The rate at which one quantity changes relative to another is mathematically described by using a concept called a *derivative*. We will see that the rate of change of one quantity relative to another is mathematically determined by finding the slope of a line drawn tangent to a curve.

Average Rate of Change

We begin with a simple example. Consider the speed of a moving object, say, a car. By *speed* we mean the rate at which the distance traveled varies with time. It has magnitude, but no direction. We can measure speed in two ways: *average speed* and *instantaneous speed*. We begin with the average speed of a commuter driving from home to the office.

To find the average speed, we use the formula $d = rt$ or $r = \frac{d}{t}$; that is, we divide the distance traveled by the elapsed time:

$$\text{AVERAGE SPEED} = \frac{\text{DISTANCE TRAVELED}}{\text{ELAPSED TIME}}$$

TABLE 18.1	
Distance and Time for a Commuter Car	
Time	Distance from Home
6:09 A.M.	0 miles
6:25	16
6:30	21
6:34	25
6:36	26.7
6:37	27.7
6:39	28.5
7:01	34
7:03	35
7:09	38
7:15	43
7:28	50

Example 1 Find the average speed

A commuter left home one morning and set the odometer to zero. The times and distances were noted as shown in Table 18.1.
Find the average speed for the requested time intervals.
a. 6:09 A.M. to 6:39 A.M. (0–30 min)
b. 6:39 A.M. to 7:09 A.M. (30–60 min)
c. 7:01 A.M. to 7:28 A.M. (52–79 min)

Solution

a. 0–30 min: AVERAGE SPEED $= \dfrac{28.5 - 0}{\frac{1}{2} - 0} = 57$ mph

b. 30–60 min: AVERAGE SPEED $= \dfrac{38 - 28.5}{1 - \frac{1}{2}} = 19$ mph

c. 52–79 min: AVERAGE SPEED $= \dfrac{50 - 34}{\frac{79}{60} - \frac{52}{60}} \approx 35.6$ mph

We can generalize the work done in Example 1 by writing the following formula:

$$\text{AVERAGE SPEED} = \frac{d_2 - d_1}{t_2 - t_1} \text{ mph}$$

where the car travels d_1 miles in t_1 hours and d_2 miles in t_2 hours. The relationship is shown graphically in Figure 18.15, where the average speed is the slope of the line joining (t_1, d_1) and (t_2, d_2).

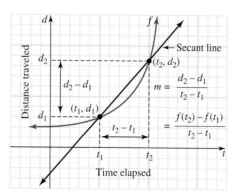

FIGURE 18.15 Geometrical interpretation of average speed

The line connecting (d_1, t_1) and (d_2, t_2) in Figure 18.15 is sometimes called the **secant line.** We see that the average rate of change is

$$\frac{f(t_2) - f(t_1)}{t_2 - t_1}$$

It is worthwhile to state this formula in terms of the usual variable x. Let $x = t_1$ and let $h = t_2 - t_1$, which is the length of the time interval over which we are finding an average. From these substitutions it follows that

$$h = t_2 - x$$
$$h + x = t_2$$

This leads to a general statement of the average rate of change.

Average Rate of Change

The average rate of change of a function f with respect to x over an interval from x to $x + h$ is given by the formula

$$\frac{f(x + h) - f(x)}{h}$$

This is called the difference quotient.

Example 2 | Find an average rate of change

Find the average rate of change of f with respect to x between $x = 3$ and $x = 5$ if $f(x) = x^2 - 4x + 7$.

Solution   Given $x = 3$ and $h = 5 - 3 = 2$:

$$f(x + h) = f(3 + 2) = f(5) = 5^2 - 4(5) + 7 = 12$$
$$f(x) = f(3) = 3^2 - 4(3) + 7 = 4$$

$$\text{AVERAGE RATE OF CHANGE} = \frac{f(x + h) - f(x)}{h}$$

$$= \frac{12 - 4}{2}$$

$$= 4$$

The slope of the secant line is 4, which is confirmed geometrically by looking at Figure 18.16.

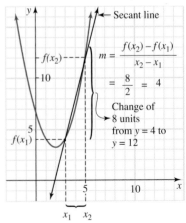

FIGURE 18.16 Average rate of change of f, where $f(x) = x^2 - 4x + 7$, between $x = 3$ and $x = 5$

We consider an example from the social sciences.

Example 3 **Find an average divorce rate**

Table 18.2 shows the number of divorces (in millions) for selected years from 1987 to 1996. Find the average divorce rate for each of the time periods given.

a. 1987–1992　　**b.** 1987–1996　　**c.** 1992–1996　　**d.** 1995–1996

Solution　The average rate of change of f is defined by

$$\frac{f(x+h) - f(x)}{h}$$

TABLE 18.2	
Number of U.S. Divorces	
Year	**Number**
1987	1.157
1992	1.215
1995	1.169
1996	1.150

a. $x = 1987$ and $h = 5$:　$\dfrac{f(1992) - f(1987)}{5} = \dfrac{1.215 - 1.157}{5}$

$$= 0.0116$$

The average divorce rate (rate of change in number of divorces) between 1987 and 1992 is an average increase of 11,600 per year.　*Note: 0.0116 million is 11,600.*

b. $x = 1987$ and $h = 9$:　$\dfrac{f(1996) - f(1987)}{9} = \dfrac{1.150 - 1.157}{9}$

$$= -0.000\overline{7}$$

A negative rate of change indicates a decrease. The average divorce rate between 1987 and 1996 is a decrease of 778 divorces per year.

c. $x = 1992$ and $h = 4$:　$\dfrac{f(1996) - f(1992)}{4} = \dfrac{1.150 - 1.215}{4}$

$$= -0.01625$$

The average divorce rate between 1992 and 1996 is an average decrease of 16,250 divorces per year.

d. $x = 1995$ and $h = 1$:　$\dfrac{f(1996) - f(1995)}{1} = 1.150 - 1.169$

$$= -0.019$$

Between 1995 and 1996, the number of divorces decreased by 19,000 or about 2%.

Instantaneous Rate of Change

Suppose in Example 3 we are interested in the divorce rate in 1996. To estimate this rate, we can use the slopes of the secant lines (labeled **b**, **c**, and **d**), as shown in Figure 18.17. Which one would you use as a more accurate estimate?

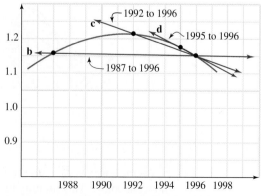

FIGURE 18.17 U.S. divorces and secant lines at 1996

To find the rate *at a particular time*, imagine that a curve called f is defined by a roller-coaster track. Consider a fixed location on this track; call it $(x, f(x))$. Now imagine the front car of the roller coaster at some point on the track a horizontal distance of h units from x. The slope of the secant line between these points is the average rate of change of the function f between the two points.

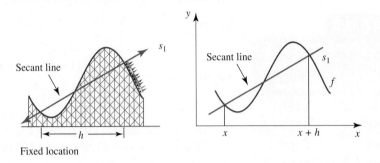

Also imagine the roller coaster car moving along the track toward the fixed location. That is, let $h \rightarrow 0$. As the car moves along the track, we obtain a sequence of secant lanes. The roller coaster is shown at the left and the curve with the secant lines is shown at the right.

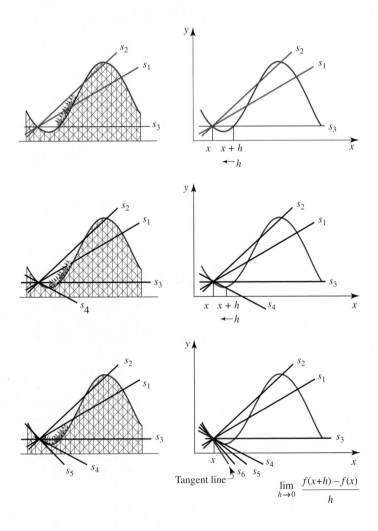

As the car gets close to the fixed location, the secant lines approach a limiting line (shown in color). This limiting line for which $h \rightarrow 0$ is called the *tangent line*.

Instantaneous Rate of Change

Given a function f and the graph of $y = f(x)$, the tangent line at the point $(x, f(x))$ is the line that passes through this point with slope

$$\lim_{h \to 0} \frac{f(x + h) - f(x)}{h}$$

if this limit exists. The slope of the tangent line is also referred to as the **instantaneous rate of change** of the function f with respect to x.

If you look at Figure 18.17, you can measure the slope of the tangent line to approximate the instantaneous rate of change in 1996 to be -0.02 or about -2%.

Example 4 Find an instantaneous rate of change

Find the instantaneous rate of change of the function $f(x) = 5x^2$ with respect to x.

Solution Find $\lim\limits_{h \to 0} \frac{f(x + h) - f(x)}{h}$. We evaluate this expression in small steps.

1. $f(x) = 5x^2$ is given.
2. $f(x + h) = 5(x + h)^2 = 5x^2 + 10xh + 5h^2$
3. $f(x + h) - f(x) = (5x^2 + 10xh + 5h^2) - 5x^2 = 10xh + 5h^2$
4. $\dfrac{f(x + h) - f(x)}{h} = \dfrac{10xh + 5h^2}{h} = 10x + 5h \quad (h \neq 0)$
5. $\lim\limits_{h \to 0} \dfrac{f(x + h) - f(x)}{h} = \lim\limits_{h \to 0} (10x + 5h) = 10x$

We will refer to the small steps (shown in boldface) as the *five-step process* for finding the instantaneous rate of change.

Derivative

The concept of the derivative is a very powerful mathematical idea, and the variety of applications is almost unlimited. The instantaneous rate of change of a function f per unit of change in x is only one of many applications using that specific limiting idea.

STOP This is one of the most important definitions in all of mathematics.

Derivative

For a given function f, we define the **derivative of f at x**, denoted by $f'(x)$, to be

$$f'(x) = \lim_{h \to 0} \frac{f(x + h) - f(x)}{h}$$

provided this limit exists. If the limit exists, we say f is a **differentiable function** of x.

If the limit does not exist, we say that f is *not differentiable* at x. We have now developed all of the techniques necessary to apply the definition of derivative to a variety of functions. Also note that as $h \to 0$, $h \neq 0$, so we assume $h \neq 0$ without stating so when using the five-step process for finding the derivative.

Example 5 | Use the definition to find a derivative

Use the definition to find the derivative of $y = 2x^2$.

Solution   We carry out the five-step process:

1. $f(x) = 2x^2$ is given.

2. $f(x + h) = 2(x + h)^2$ Evaluate f at x + h.
   $= 2x^2 + 4xh + 2h^2$

3. $f(x + h) - f(x) = (2x^2 + 4xh + 2h^2) - 2x^2$ Subtract f(x) from f(x + h).
   $= 4xh + 2h^2$

4. $\dfrac{f(x + h) - f(x)}{h} = \dfrac{4xh + 2h^2}{h}$ Now, divide by h.
   $= \dfrac{h(4x + 2h)}{h}$
   $= 4x + 2h$

5. $\displaystyle\lim_{h \to 0} \dfrac{f(x + h) - f(x)}{h} = \lim_{h \to 0}(4x + 2h)$ Finally, take the limit.
   $= 4x$

The derivative of $y = 2x^2$ is $4x$. Sometimes we write $y' = 4x$ or $f'(x) = 4x$.

In more advanced courses, it is shown that the exponential function has a very important property involving derivatives. *The exponential function is its own derivative.*

Definition of e^x

If $f(x) = e^x$, then $f'(x) = e^x$, and if $f(x) = e^{ax}$, then $f'(x) = ae^{ax}$.

Example 6 | Find the derivative on an exponential

Find the derivative of the given functions: **a.** $y = e^{3x}$ **b.** $y = e^{-5x}$.

Solution   a. If $y = e^{3x}$, then $y' = 3e^{3x}$.   **b.** If $y = e^{-5x}$, then $y' = -5e^{-5x}$

Tangent Line

Given a function $y = f(x)$, we know the slope of the tangent line is $y = f'(x)$. Recall, from Problem 58, Section 15.1, that a line with slope m passing through the point (h, k) is

$$y - k = m(x - h)$$

This gives us the following **tangent-line formula.**

Tangent Line

The equation of the line tangent to the curve defined by $y = f(x)$ at the point (h, k) is

$$y - k = f'(x)(x - h)$$

Example 7 | Find the equation of a tangent line

Find the standard-form equation of the line tangent to the graph of $y = 2x^2$ at the point where $x = -1$.

Solution First, find the coordinates of the point where $x = -1$:

$$y = 2x^2$$
$$= 2(-1)^2$$
$$= 2$$

The point is $(-1, 2)$. The slope at this point is $y' = f'(x)$ and from Example 5, the derivative of $y(x) = 2x^2$ is $y'(x) = 4x$. Thus, at the point $x = -1$, we find the derivative

$$y' = 4x$$
$$= 4(-1)$$
$$= -4$$

The equation of the tangent line is

$$y - k = f'(x)(x - h)$$
$$y - 2 = -4(x - (-1))$$
$$y - 2 = -4(x + 1)$$
$$y - 2 = -4x - 4$$
$$y = -4x - 2$$
$$4x + y + 2 = 0$$

It is worth noting that if the derivative of a function is positive, then the curve is rising at that point (since the slope of the tangent line is positive); and if the derivative is negative at a point, then the curve is falling at that point. If the derivative is 0, then the tangent line is horizontal. In Example 7, we found the equation of the tangent line of the function $y = 2x^2$ at $(-1, 2)$. This tangent line is shown in Figure 18.18.

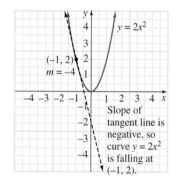

FIGURE 18.18 Graph of $y = 2x^2$ with tangent line at $(-1, 2)$

Velocity

A final application of derivative we will consider is the **velocity**, which is defined as an instantaneous rate of change. Consider the following example.

Example 8 | Find an instantaneous rate

Consider a "free-fall" ride at an amusement park that involves falling 100 feet in 2.5 seconds. As you are falling, you pass the 16-ft mark at one second and the 64-ft mark at two seconds. What is the velocity at the *instant* the ride passes the 100-ft mark (as measured from the top)?

Solution Let $s(t)$ be the distance in feet from the top t seconds after release. Also assume that $s(t) = 16t^2$.*

$$\frac{s(1) - s(0)}{1} = \frac{16 - 0}{1}$$
$$= 16\text{ft/s}$$
$$\frac{s(2) - s(1)}{1} = \frac{16(4) - 16}{1}$$
$$= 48\text{ ft/s}$$

*This is the formula for free fall in a vacuum; it is sufficiently accurate for our purposes in this problem.

Paramount's Great America

$t = 0$ — 0 ft

$t = 1$ — 16 ft

$t = 2$ — 64 ft

$t = 2.5$ — 100 ft

$t = 3$ — 144 ft

$$\frac{s(3) - s(2)}{1} = \frac{16(9) - 16(4)}{1}$$
$$= 80 \text{ ft/s}$$

In general, the average velocity from time $t = t_1$ to $t = t_1 + h$ is given by the following formula (where $h \neq 0$).

$$\text{AVERAGE VELOCITY} = \frac{\text{CHANGE IN POSITION}}{\text{CHANGE IN TIME}}$$

$$= \frac{s(t_1 + h) - s(t_1)}{h}$$

$$= \frac{16(t_1 + h)^2 - 16t_1^2}{h}$$

$$= \frac{16t_1^2 + 32t_1h + 16h^2 - 16t_1^2}{h}$$

$$= \frac{32t_1h + 16h^2}{h}$$

$$= 32t_1 + 16h$$

Now, to find the velocity at a particular instant in time, we simply need to consider the limit as $h \to 0$:

$$\text{INSTANTANEOUS VELOCITY} = \lim_{h \to 0} \frac{s(t_1 + h) - s(t_1)}{h}$$

$$= \lim_{h \to 0} (32t_1 + 16h)$$

$$= 32t_1$$

We can now use these formulas to find various average and instantaneous velocities. (We do this to show you how to apply the formula we have just derived as practice before answering the question asked in this example.)

Time t		Average Velocity $32t + 16h$	Instantaneous-Velocity $32t$
$t = 0$	$h = 1$	$32(0) + 16(1) = 16$	$32(0) = 0$
	From $t = 0$ to $t = 1$		
$t = 1$	$h = 2$	$32(1) + 16(2) = 64$	$32(1) = 32$
	From $t = 0$ to $t = 2$		
	$h = 1$	$32(1) + 16(1) = 48$	
	From $t = 1$ to $t = 2$		
$t = 2$	$h = 3$	$32(2) + 16(3) = 112$	$32(2) = 64$
	From $t = 0$ to $t = 3$		
	$h = 2$	$32(2) + 16(2) = 96$	
	From $t = 1$ to $t = 3$		
	$h = 1$	$32(2) + 16(1) = 80$	
	From $t = 2$ to $t = 3$	$\vdots$	

We are given that the ride passes 100 ft at 2.5 seconds, so the instantaneous velocity at that instant is

$$32(2.5) = 80 \text{ ft/s}$$

(By the way, 80 ft/s is approximately 55 mph.)

Problem Set **18.3**

Level **1**

1. The graph in Figure 18.19 shows the height h of a projectile after t seconds. Find the average rate of change of height (in feet) with respect to the requested changes in time t (in seconds).

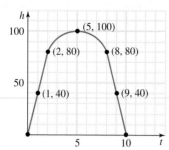

FIGURE 18.19 Height of a projectile in feet t seconds after fired

   a. 1 to 8
   b. 1 to 5
   c. 1 to 2
   d. What do you think the rate of change is at $t = 1$?

2. The graph in Figure 18.20 shows company output as a function of the number of workers. Find the average rate of change of output for the given change in the number of workers.

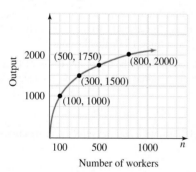

FIGURE 18.20 Output in production relative to the number of employees at Kampbell Construction

   a. 100 to 800
   b. 100 to 500
   c. 100 to 300
   d. What do you think the rate of change is at $n = 100$?

3. Table 18.1 gives some distances and commute times for a typical daily commute. Find the average speed for the requested time intervals.
   a. 6:09 to 6:36
   b. 6:36 to 7:03
   c. 7:03 to 7:28
   d. 6:09 to 7:28

4. The SAT scores of entering first-year college students are shown in Figure 18.21. Find the average yearly rate of change of the scores for the requested periods.

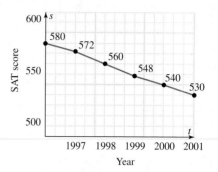

FIGURE 18.21 SAT scores at Riveria College

   a. 1997 to 2001
   b. 1998 to 2001
   c. 1999 to 2001
   d. 2000 to 2001

5. Table 18.3 shows the Gross National Product (GNP) in trillions of dollars for the years 1960–1996. Find the average yearly rate of change of the GNP for the requested years.

TABLE 18.3	
Gross National Product	
Year	**Dollars**
1960	0.5153
1970	1.0155
1980	2.7320
1990	5.5461
1995	7.2654
1996	7.6360

   a. 1960 to 1996
   b. 1970 to 1996
   c. 1980 to 1996
   d. 1990 to 1996
   e. 1995 to 1996
   f. How fast do you think the GNP is changing in 1996?

Trace the curves in Problems 6–11 onto your own paper and draw the secant line passing through P and Q. Next, imagine h → 0 and draw the tangent line at P assuming that Q moves along the curve to the point P. Finally, estimate the slope of the curve at P using the slope of the tangent line you have drawn.

6.
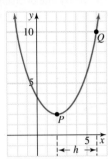

7.
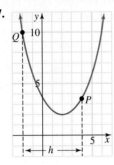

8.
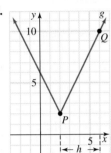

9.
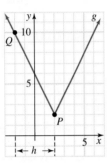

10.
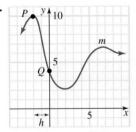

11.
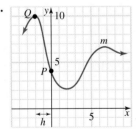

Level **2**

Find the average and instantaneous rates of change of the functions in Problems 12–15.

12. $f(x) = 4 - 3x$
   a. for $x = -3$ to $x = 2$
   b. at $x = -3$

13. $f(x) = 5$
   a. for $x = -3$ to $x = 3$
   b. at $x = -3$

14. $f(x) = 3x^2$
   a. for $x = 1$ to $x = 3$
   b. at $x = 1$

15. $f(x) = -2x^2 + x + 4$
   a. for $x = 4$ to $x = 9$
   b. at $x = 4$

Find the derivative, $f'(x)$, of each of the functions in Problems 16–21 by using the derivative definition or the derivative of the exponential function.

16. $f(x) = \frac{1}{2}x^2$

17. $f(x) = \frac{1}{3}x^3$

18. $f(x) = e^{1.2x}$

19. $y = e^{-6x}$

20. $y = 25 - 250x$

21. $f(x) = 3 + 2x - 3x^2$

Find an equation of the line tangent to the curves in Problems 22–25 at the given point.

22. $y = 5x^2$ at $x = -3$

23. $y = 2x^2$ at $x = 4$

24. $y = 4 - 5x$ at $x = -2$

25. $y = 3x^2 + 4x$ at $x = 0$

26. A freely falling body experiencing no air resistance falls $s(t) = 16t^2$ feet in t seconds. Express the body's velocity at time $t = 2$ as a limit. Evaluate this limit.

27. If you toss a ball from the top of the Tower of Pisa directly upward with an initial speed of 96 ft/s, the height h at time t is given by

$$h(t) = -16t^2 + 96t + 176$$

Figure 18.22 shows h and the velocity, v, at various times, t.

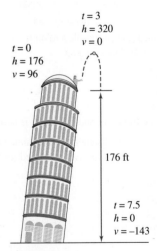

FIGURE 18.22 Tower of Pisa

   a. What is the height of the tower?
   b. Find the velocity as a function of time.

28. Suppose the profit, P, measured in thousands of dollars, for a manufacturer is a function of the number of units produced, x, and behaves according to the model

$$P(x) = 50x - x^2$$

Also suppose that the present production is 20 units.
   a. What is the per-unit increase in profit if production is increased from 20 to 30 units?
   b. Repeat for 20 to 25 units.
   c. Repeat for 20 to 21 units.
   d. What is the rate of change at $x = 20$?

29. The cost, C, in dollars, for producing x items is given by

$$C(x) = 30x^2 - 100x$$

   a. Find the average rate of change of cost as x increases from 100 to 200 items.
   b. Repeat for 100 to 110 items.
   c. Repeat for 100 to 101 items.
   d. Repeat for 100 to $(100 + h)$ items.

30. Suppose the number (in millions) of bacteria present in a culture at time t is given by the formula

$$N(t) = 2t^2 - 200t + 1{,}000$$

   a. Derive a formula for the instantaneous rate of change of the number of bacteria with respect to time.
   b. Find the instantaneous rate of change of the number of bacteria with respect to time at time $t = 3$.
   c. Find the instantaneous rate of change of the number of bacteria with respect to time at the beginning of this experiment.

18.4 | Integrals

We considered several simple area formulas in Chapter 9. We found those formulas by filling regions with square units and then counting them in some convenient manner. Sir Isaac Newton (see page 871) described area by using calculus. We will use his ideas, along with modern notation, to give you a glimpse of what we mean by *integral calculus*.

Area Function

We begin by defining an *area function*.

STOP Study this definition until you understand this function.

> **Area Function**
>
> The **area function,** $A(t)$, is the area bounded below by the x-axis, above by a function $y = f(x)$, on the left by the y-axis, and on the right by the vertical line $x = t$.

The area function is shown in Figure 18.23.

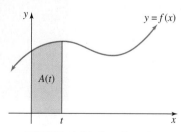

FIGURE 18.23 Area function

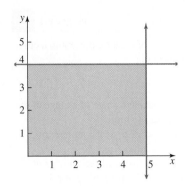

FIGURE 18.24 Area function

Example 1 Evaluate the area function

Let $y = 4$. Find: **a.** $A(5)$ **b.** $A(6)$ **c.** $A(50)$ **d.** $A(x)$

Solution
a. We draw the region as shown in Figure 18.24.

The number $A(5)$ is the area below the line $y = 4$, above the x-axis, to the right of the y-axis, and to the left of the line $x = 5$. This is a rectangle, so we know

$$A(5) = 4(5) = 20$$

b. $A(6) = 4(6) = 24$
c. $A(50) = 4(50) = 200$
d. $A(x) = 4x$

The task before us is to find $A(x)$ for any function $y = f(x)$.

Example 2 Find the area function for a given line

Let $y = 4x$. Find: **a.** $A(5)$ **b.** $A(10)$ **c.** $A(50)$ **d.** $A(x)$

Solution   Begin by drawing the region as shown in Figure 18.25.
a. Figure 18.25 shows the area function for $A(5)$. We see the area we seek is a triangle:

$$A = \frac{1}{2}bh \qquad \text{Area of a triangle}$$

$$= \frac{1}{2}(5)(20) \qquad \text{The base is } b = 5. \text{ For the height, use } y = 4x \text{ for } x = 5; h = 4(5) = 20.$$

$$= 50$$

b. For $A(10)$, $x = 10$, so $b = 10$ and $h = 4(10) = 40$ and

$$A = \frac{1}{2}bh = \frac{1}{2}(10)(40) = 200$$

c. $A(50) = \frac{1}{2}(50)(200) = 5{,}000$

d. $A(x) = \frac{1}{2}(x)(4x) = 2x^2$

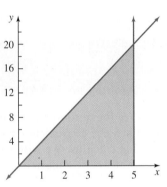

FIGURE 18.25 Area of a triangle

The answer to Example 2d, namely, $A(x) = 2x^2$, is called the *antiderivative* of the original function $f(x) = 4x$ because if we take the derivative of $A(x) = 2x^2$ we obtain $4x$ (see Example 5 of Section 18.3). An antiderivative of the function defined by $f(x)$ is written as $\int f(x)\, dx$.

Antiderivative

If the derivative of a function defined by $y = F(x)$ is $f(x)$, then the **antiderivative** of $f(x)$ is

$$\int f(x)\, dx = F(x)$$

This means that $F'(x) = f(x)$.

We have not yet discussed a procedure for *finding* an antiderivative, but we can check to see whether a particular function *is* an antiderivative, as shown by the following example.

Example 3 Find an antiderivative

Show:

a. $\int x\,dx = \dfrac{x^2}{2}$ **b.** $\int dx = x$

Solution   We use the five-step process:

a.   Let $f(x) = \dfrac{x^2}{2}$.

$$f(x+h) = \frac{(x+h)^2}{2}$$

$$= \frac{x^2 + 2xh + h^2}{2}$$

$$f(x+h) - f(x) = \frac{2xh + h^2}{2}$$

$$\frac{f(x+h) - f(x)}{h} = x + \frac{h}{2}$$

$$\lim_{h\to 0}\frac{f(x+h) - f(x)}{h} = \lim_{h\to 0}\left(x + \frac{h}{2}\right)$$

$$= x$$

Thus, $\int x\,dx = \dfrac{x^2}{2}$.

b.   Let $f(x) = x$.

$$f(x+h) = x + h$$

$$f(x+h) - f(x) = h$$

$$\frac{f(x+h) - f(x)}{h} = 1$$

$$\lim_{h\to 0}\frac{f(x+h) - f(x)}{h} = \lim_{h\to 0} 1$$

$$= 1$$

Thus, $\int dx = x$.

The next example shows that a given function may have more than one antiderivative.

Example 4 Show that a function may have more than one antiderivative

Show that $x^3 + 8$ and $x^3 - 2$ are antiderivatives of $3x^2$.

Solution   We carry out the five-step process for both functions.

1. $f(x)$:

$x^3 + 8$

$x^3 - 2$

2. $f(x + h)$

$(x+h)^3 + 8$

$(x+h)^3 - 2$

3. $f(x + h) - f(x)$:

$(x+h)^3 + 8 - (x^3 + 8)$
$= x^3 + 3x^2h + 3xh^2 + h^3 + 8 - x^3 - 8$

$(x+h)^3 - 2 - (x^3 - 2)$
$= x^3 + 3x^2h + 3xh^2 + h^3 - 2 - x^3 + 2$

4. $\dfrac{f(x+h) - f(x)}{h}$:

$\dfrac{3x^2h + 3xh^2 + h^3}{h} = 3x^2 + 3xh + h^2$

$\dfrac{3x^2h + 3xh^2 + h^3}{h} = 3x^2 + 3xh + h^2$

5. $\displaystyle\lim_{h\to 0}\dfrac{f(x+h) - f(x)}{h}$

$\displaystyle\lim_{h\to 0}(3x^2 + 3xh + h^2) = 3x^2$

$\displaystyle\lim_{h\to 0}(3x^2 + 3xh + h^2) = 3x^2$

We see that both of the given functions are antiderivatives of $3x^2$. We write this as

$$\int 3x^2\,dx = x^3 + 8 \quad \text{and} \quad \int 3x^2\,dx = x^3 - 2$$

STOP Pay attention to the statement shown in italics.

Notice from Example 4 that antiderivatives are not unique. If we study the steps of this example, we see that *antiderivatives are the same except for a constant that is added or subtracted.*

If we represent this constant as C, we write

$$\int 3x^2\, dx = x^3 + C$$

where C is any constant. This says if you select any constant value for C and find the derivative of $f(x) = x^3 + C$, the result will be $3x^2$.

The exponential function $f(x) = e^x$ is the function that is equal to its own derivative, so it follows that it is a function that is its own antiderivative. We state this result in formula form.

Antiderivative of e^x

If $f(x) = e^x$, then

$$\int e^x\, dx = e^x + C \qquad \text{and} \qquad \int e^{ax}\, dx = \frac{e^{ax}}{a} + C$$

More advanced courses derive some properties of antiderivatives, which we summarize in the following box.

Properties of Antiderivatives

Constant multiple: $\displaystyle \int a f(x)\, dx = a \int f(x)\, dx$

Antiderivative (integral) of a sum:

$$\int [f(x) + g(x)]\, dx = \int f(x)\, dx + \int g(x)\, dx$$

Newton found that the derivative of the area function $A(x)$ is $f(x)$. In other words, the area function is an antiderivative of f; we write

$$\int f(x)\, dx = A(x)$$

Antiderivatives as Areas

We can find an antiderivative by using areas, as illustrated by the following example.

FIGURE 18.26 Trapezoid with $h = x,\ b = 3,\ B = 2x + 3$

Example 5 Find the antiderivative of a linear function

Find $\int (2x + 3)\, dx$.

Solution   Let $f(x) = 2x + 3$; consider the region bounded by the x-axis, the y-axis, the line $y = 2x + 3$, and the vertical line x units to the right of the origin, as shown in Figure 18.26. We recognize this as a trapezoid, so we can write

$$\int (2x + 3)\, dx = \frac{1}{2}h(b + B)$$

$$= \frac{1}{2}x[3 + (2x + 3)]$$

$$= x^2 + 3x$$

Since we know that antiderivatives are not unique, but differ only by a constant, we write

$$\int (2x + 3)\,dx = x^2 + 3x + C$$

We can also use properties of an antiderivative, as shown here:

$$\int (2x + 3)\,dx = \int 2x\,dx + \int 3\,dx \qquad \text{Antiderivative of a sum}$$

$$= 2\int x\,dx + 3\int dx \qquad \text{Constant multiple}$$

$$= 2\frac{x^2}{2} + 3x \qquad \text{Antiderivative of } x \text{ is } \tfrac{1}{2}x^2 \text{ (Example 3a) and antiderivative of 1 is } x \text{ (Example 3b).}$$

$$= x^2 + 3x$$

The Definite Integral

We note that the antiderivative of a function is a function. We now consider a concept called the *definite integral*, which although it can be related to an antiderivative, is an important concept in its own right. We introduce the definite integral with an example.

Suppose that between 2000 and 2005, the rate of oil consumption (in billions of barrels) is given by the formula

$$R(t) = 76e^{-0.0185t}$$

where t is measured in years after 2000. If we use this consumption rate, we can predict the amount of oil that will be consumed from 2005 to 2010. Since the derivative is the rate of change of a function, we can say that the total consumption since 2000 is given by a function we will call T with the property that $T'(t) = R(t)$. Then

$$T(t) = \int T'(t)\,dt = \int R(t)\,dt$$

$$= \int 76e^{-0.0185t}\,dt$$

$$= 76\int e^{-0.0185t}\,dt \qquad \text{Constant multiple property}$$

$$= 76\frac{e^{-0.0185t}}{-0.0185} + C \qquad \text{From the formula for antiderivative of } e^x.$$

$$\approx -4{,}108e^{-0.0185t} + C$$

We can find C because if $t = 0$, then oil consumption is also 0, so

$$T(0) = -4{,}108e^{-0.0185t} + C = 0$$

$$C = 4{,}108$$

Thus, $T(t) = -4{,}108e^{-0.0185t} + 4{,}108$ is the total consumption for t years after 2000. This means that $T(5)$ is the total consumption from 2005 to 2010, and $T(10)$ is the total consumption from 2000 to 2010.

$$T(10) = -4{,}108e^{-0.0185(10)} + 4{,}108 \approx 693.824$$

$$T(5) = -4{,}108e^{-0.0185(5)} + 4{,}108 \approx 362.945$$

We can find the total consumption from 2000 to 2005 by subtraction:

$$T(10) - T(5) \approx 694 - 363 = 331$$

We would predict approximately 331 billion barrels to be consumed, assuming the rate of consumption does not change.

Let us take a closer look at what we have done. Since $T(t)$ is an antiderivative of $R(t)$ over the interval from 2000 to 2010, we see that $T(10) - T(5)$ is the *net change* of the function T over the interval. In general, if $F(x)$ is any function and a and b are real numbers with $a < b$, then the *net change of $F(x)$ over the interval* $[a, b]$ is the number

$$F(b) - F(a)$$

The quantity $F(b) - F(a)$ is often abbreviated by $F(x)|_a^b$. This discussion leads to the following definition.

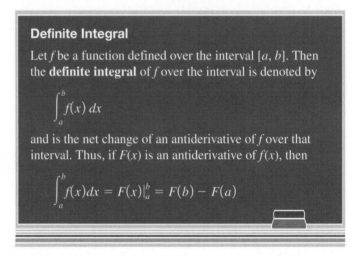

Definite Integral

Let f be a function defined over the interval $[a, b]$. Then the **definite integral** of f over the interval is denoted by

$$\int_a^b f(x)\,dx$$

and is the net change of an antiderivative of f over that interval. Thus, if $F(x)$ is an antiderivative of $f(x)$, then

$$\int_a^b f(x)\,dx = F(x)|_a^b = F(b) - F(a)$$

CAUTION *The definite integral is a real number. The antiderivative is a function.*

The function f is called the **integrand** and the constants a and b are called the **limits of integration.** The letter x is called a **dummy variable** because the definite integral is a fixed number and not a function of x. To contrast a definite integral with its corresponding antiderivative, an antiderivative is sometimes called an **indefinite integral.**

Areas as Antiderivatives

In Example 5 we found an antiderivative by using a known area formula, but the real power of calculus is to use an antiderivative to find an area.

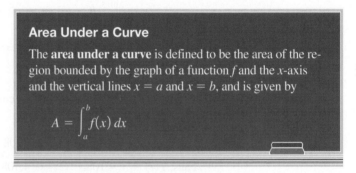

Area Under a Curve

The **area under a curve** is defined to be the area of the region bounded by the graph of a function f and the x-axis and the vertical lines $x = a$ and $x = b$, and is given by

$$A = \int_a^b f(x)\,dx$$

This area function is illustrated in Figure 18.27.

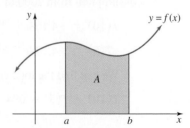

FIGURE 18.27 Area under a curve

Example 6 Find the area under a parabola in more than one way

Find the area under the curve $y = x^2$ on $[1, 5]$.

Solution   We now have two ways for finding this area; the first is to use an antiderivative, and the second is to approximate the actual area using known area formulas.

(1) *Using an Antiderivative*

Suppose you know that the derivative of $\frac{1}{3}x^3$ is x^2 (for example, from Problem 17, Problem Set 18.3). Then,

$$\int_1^5 x^2 dx = \left.\frac{x^3}{3}\right|_1^5 = \frac{125}{3} - \frac{1}{3} = \frac{124}{3} \approx 41.3$$

(2) *Using Areas*

We do not have a known area formula for the region shown in Figure 18.28, but we can approximate this area using rectangles.

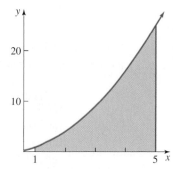

FIGURE 18.28 Area bounded by $y = x^2$, the *x*-axis, and the lines $x = 1$ and $x = 5$

If we draw one rectangle determined by the left endpoint of the region (namely, $x = 1$), we obtain an approximation we call A_1:

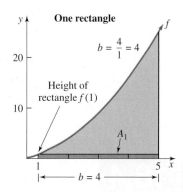

This rectangle has area $A = bh$, where $b = 4$ and $h = f(1) = 1$:

$$\int_1^5 x^2 dx \approx A_1 = 4 \cdot f(1) = 4 \cdot 1^2 = 4$$

If we draw two rectangles determined by the left endpoints we obtain:

$$\int_1^5 x^2 dx \approx A_1 + A_2 = b \cdot f(1) + b \cdot f(3) = 2 \cdot 1^2 + 2 \cdot 3^2 = 20$$

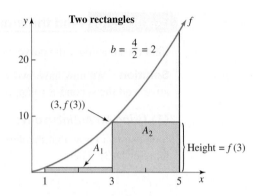

Two rectangles

We continue the process; for example, draw four rectangles:

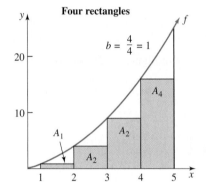

Four rectangles

$$\int_1^5 x^2 dx \approx A_1 + A_2 + A_3 + A_4$$

$$= b \cdot f(1) + b \cdot f(2) + b \cdot f(3) + b \cdot f(4)$$

$$= 1 \cdot 1^2 + 1 \cdot 2^2 + 1 \cdot 3^2 + 1 \cdot 4^2$$

$$= 30$$

If we continue for 8 rectangles, we would approximate the area as 35.5. In fact, using a computer, we could generate the following table of values for a given number (#) of subintervals:

Type	#	Estimate Over [1, 5]
Left endpt	4	30.0000000000
Left endpt	8	35.5000000000
Left endpt	16	38.3750000000
Left endpt	32	39.8437500000
Left endpt	64	40.5859375000
Left endpt	128	40.9589843750
Left endpt	256	41.1459960938
Left endpt	512	41.2396240234
Left endpt	1024	41.2864685059
Left endpt	2048	41.3098983765
Left endpt	4096	41.3216152191
Left endpt	10000	41.3285334400

If we take the limit as the number of rectangles increases, we would find this number approaches $41.\overline{3} = \frac{124}{3}$.

Problem Set 18.4

Level 1

Evaluate the area function for the functions given in Problems 1–6.

1. Let $y = 5$; find $A(8)$.

2. Let $y = 8.3$; find $A(x)$.

3. Let $y = 3x$; find $A(4)$.

4. Let $y = 2x$; find $A(x)$.

5. Let $y = 3x + 2$; find $A(3)$.

6. Let $y = x + 5$; find $A(x)$.

Find the antiderivative by using areas in Problems 7–12.

7. $\int 6 \, dx$

8. $\int 3 \, dx$

9. $\int (x + 5)\, dx$

10. $\int (x + 6)\, dx$

11. $\int (3x + 4)\, dx$

12. $\int (2x + 5)\, dx$

27. $\int_1^5 (1 + 6x)\, dx$

28. $\int_0^2 (e^x + e^2)\, dx$

Level 2

13. IN YOUR OWN WORDS   What is a derivative?

14. IN YOUR OWN WORDS   What is an integral?

15. IN YOUR OWN WORDS   What is the area function?

16. IN YOUR OWN WORDS   Contrast a definite integral and an indefinite integral.

17. Show that x^2 is an antiderivative of $2x$.

18. Show that $\frac{1}{2}x^2 - 5$ is an antiderivative of x.

19. Show that $2x^3$ is an antiderivative of $6x^2$.

20. Show that $2x^3 - 3$ is an antiderivative of $6x^2$.

Find the area under the curves in Problems 21–24 on the given intervals.

21. $y = x^2$ on $[1, 9]$

22. $y = x^2$ on $[2, 5]$

23. $y = 3x$ on $[3, 8]$

24. $y = 2x + 1$ on $[1, 4]$

Evaluate the integrals given in Problems 25–28.

25. $\int_0^2 e^{0.5x}\, dx$

26. $\int_{-1}^3 (9 - x^2)\, dx$

Level 3

29. A culture is growing at an hourly rate of

$$R'(t) = 200e^{0.5t}$$

for $0 \le t \le 10$. Find the area between the graph of this equation and the t-axis. What do you think this area represents?

Problem Solving 3

30. Suppose the accumulated cost of a piece of equipment is $C(t)$ and the accumulated revenue is $R(t)$, where both of these are measured in thousands of dollars and t is the number of years since the piece of equipment was installed. If it is known that

$$C'(t) = 18 \quad \text{and} \quad R'(t) = 21e^{-0.01t}$$

find the area (to the nearest unit) between the graphs of C' and R'. Do not forget that $t \ge 0$. What do you think this area represents?

18.5   CHAPTER SUMMARY

The development of the calculus represents one of the great intellectual accomplishments in human history.

NATIONAL COUNCIL OF TEACHERS
OF MATHEMATICS STANDARDS

Important Ideas

Nature of calculus [18.1]
Mathematical modeling [18.1]
Limit of a sequence [18.2]
Limits to infinity [18.2]
Derivative [18.3]
Derivative of e^x [18.3]
Area function [18.4]
Antiderivative [18.4]
Antiderivative of e^x [18.4]

Take some time getting ready to work the review problems in this section. First review the important ideas listed at the beginning of this chapter (see page 872). Look back at the definition and property boxes. If you look online, you will find a list of important terms introduced in this chapter, as well as the types of problems that were introduced. You will maximize your understanding of this chapter by working the problems in this section only after you have studied the material.

You will find some review help online at **www.mathnature.com**. There are links giving general test help in studying for a mathematics examination, as well as specific help for reviewing this chapter.

Chapter **18**   Review Questions

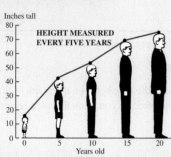

A rough estimate of a man's growth from childhood to maturity is obtained by measuring his height every five years. The straight lines between heads show the rate of growth.

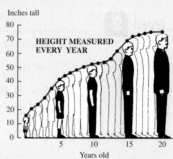

A more accurate approximation of growth is obtained by taking yearly measurements. The straight lines connecting the heads now begin to blend into a continuous curve.

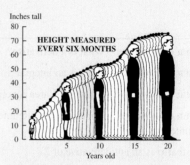

Measurements taken every six months give a still more accurate rate-of-growth line. With increasingly smaller intervals, the lines merge into an ever-smoother curve.

FIGURE 18.29 Predicting the growth rate of an adult by looking at a child's growth

1. What are the main ideas of calculus? Briefly describe each of these main ideas.

2. There are three pictures in Figure 18.29 that can be used to find the growth rate of the boy. Explain what is being illustrated with this sequence of drawings. Why do you think we might title this sequence of illustrations "instantaneous growth rate"?

Evaluate the limits in Problems 3–7.

3. $\lim\limits_{n\to\infty} \frac{1}{n}$

4. $\lim\limits_{n\to\infty} (-1)^n$

5. $\lim\limits_{n\to\infty} \frac{3n^4 + 20}{7n^4}$

6. $\lim\limits_{n\to\infty} (2n + 3)$

7. $\lim\limits_{n\to\infty} \left(1 + \frac{1}{n}\right)^n$

Use the definition of derivative to find the derivatives in Problems 8–12.

8. $f(x) = 9$

9. $f(x) = 10 - 2x$

10. $f(x) = x - 10$

11. $f(x) = 6 - 4x^2$

12. $f(x) = 44e^{0.5x}$

Find the antiderivative in Problems 13–14.

13. $\int 16x \, dx$

14. $\int \pi \, dx$

Evaluate the integrals in Problems 15–17.

15. $\int_2^3 (6x^2 - 2x) \, dx$

16. $\int_0^2 (x - 3x^2) \, dx$

17. $\int_1^2 e^x \, dx$ (correct to two decimal places)

18. Find the equation of the line tangent to $y = x^2$ at $x = 1$.

19. An object moving in a straight line travels d miles in t hours according to the formula

$$d(t) = \frac{1}{5}t^2 + 5t$$

What is the object's velocity when $t = 5$?

20. The rate of consumption (in billions of barrels per year) for oil conforms to the formula

$$R(t) = 32.4e^{6t/125}$$

for t years after 2000. If the total oil still left in the earth is estimated to be 670 billion barrels, estimate the length of time before all available oil is consumed if the rate does not change and no new oil reserves are discovered.

BOOK REPORTS

Write a 500-word report on this book:

To Infinity and Beyond: A Cultural History of the Infinite. Eli Maor (Boston: Birkhauser, 1987).

MATHEMATICS IN THE NATURAL SCIENCES, SOCIAL SCIENCES, AND HUMANITIES

We began this book with a prologue that asked the question, "Why study math?" so it seems appropriate that we end the book with an epilogue asking, "Why not study math?" Mathematics is the foundation and lifeblood of nearly all human endeavors. The German philosopher John Frederick Herbart (1776–1841) summarizes this idea:

All quantitative determinations are in the hands of mathematics, and it at once follows from this that all speculation which is heedless of mathematics, which does not enter into partnership with it, which does not seek its aid in distinguishing between the manifold modifications that must of necessity arise by a change of quantitative determinations, is either an empty play of thoughts, or at most a fruitless effort.

Historically, mathematics has always been at the core of a liberal arts education. This is a book of mathematics for the liberal arts. Karl Gauss, one of the greatest mathematicians of all time, called mathematics the "Queen of the Sciences," but mathematics goes beyond the sciences. Bertrand Russell claimed, "Mathematics, rightly viewed, possesses not only truth, but supreme beauty. . . . " And finally, Maxine Bôcher concludes, "I like to look at mathematics almost more as an art than as a science. . . . " If fact, the mathematics degree at UCLA is classified as an art, not a science. Mathematics seems to be part of the structure of our minds, more akin to memory than to a learnable discipline. Enjoyment and use of mathematics are not dependent on "book learning," and even a casual perusal of the topics in this book will clearly illustrate that mathematics is many things to many people.

Like music, mathematics resists definition. Bertrand Russell had this to say about mathematics: "Mathematics may be defined as the subject in which we never know what we are talking about, nor whether what we are saying is true." Einstein, with his customary mildness, tells us, "So far as the theorems of mathematics are about reality, they are not certain; so far as they are certain, they are not about reality." Aristotle, who was as sure of everything as anyone can be of anything, thought mathematics to be the study of quantity, whereas Russell, in a less playful mood, thinks of it as the "class of all propositions of the type 'p implies q,'" which seems to have little to do with quantity." Willard Gibbs thought of mathematics as a language; Hilbert thought of it as a game. Hardy stressed its uselessness, Hogben its practicality. Mill thought it an empirical science, whereas to Sullivan it was an art, and to the wonderful J. J. Sylvester, it was "the music of reason."

> **math·e·mat·ics**\\'math-ə-'mat-iks \\ *n pl but usu sing in constr* **1:** the science of numbers and their operations, interrelations, combinations, generalizations, and abstractions and of space configurations and their structure, measurement, transformations, and generalizations **2:** a branch of, operation in, or use of mathematics
>
> This is what it is.

This ambiguity should be consoling. It suggests that mathematics has so many mansions that there is room for everyone. In this epilogue we will discuss mathematics in the natural sciences, in the social sciences, and in the humanities.

Mathematics in the Natural Sciences

When most of us think about applications of mathematics, we think of those in the natural sciences, or those sciences that deal with matter, energy, and the interrelations and transformations. Some of the major categories include (alphabetically) astronomy, biology, chemistry (and biochemistry), computer science, ecology, geology, medicine, meteorology, physics (and biophysics), statistics, and zoology.

We began this text by looking at reasoning, both inductive and deductive reasoning. Inductive reasoning in the natural sciences is called the *scientific method*. With this type of reasoning, we are trying to recognize and formulate hypotheses, and then collect data through observation and experimentation in an effort to test our hypotheses and to make further conclusions or conjectures based on the obtained data.

Using mathematics to make real-life predictions involves a process known as **mathematical modeling,** which we discussed in Chapter 18. Since most models are dynamic, a feedback and reevaluation process is part of the formulation of a good model. It involves *abstraction, deriving results, interpretation*, and then *verification.* An illustration of what we mean is shown in Figure 1.

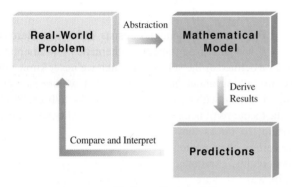

FIGURE 1 Mathematical modeling

A good example of mathematics modeling in the natural sciences, but with implications to history and the social sciences (not to mention golf!), is modeling the path of a cannonball. The *real-world* problem is to hit a target.

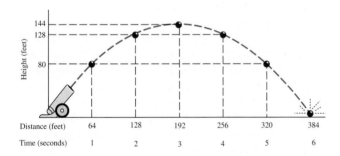

Fire a cannonball at 63 m/s at a 57° angle. We might begin by using trial and error. One possible trial is shown in Figure 2.

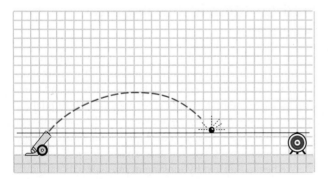

FIGURE 2 Trying to hit a target by trial and error

You can try this experiment (it s fun!) at **www**
.mathnature. com. *Choose "Epilogue links.*

The first use of a cannon was primarily to demoralize the enemy by its overwhelming power with little hope of actually hitting a specific target. In World War I, the trajectory of a cannon was trial and error, and we all remember old World War II movies where a "spotter" would phone directions to a gunner to adjust the angle of the cannon. We now need to make some assumptions; this is called *abstraction.* We assume that we are on earth (the earth's gravitational acceleration is 32 ft/s^2 or 9.8 m/s^2; we also assume that the density of the cannonball is constant, and that wind and friction are negligible). When we look at the path of a cannonball, we recognize it as having a parabolic shape, so our first attempt at modeling the path is to guess that it has an equation like $y = ax^2$ for an appropriate a. We also know that it opens downward, so a is negative. We now need to *derive some results* for the modeling process. We begin with the equation of a parabola that opens down (from Section 15.4): $x^2 = -4cy$. This equation assumes that the vertex of the parabola is at $(0, 0)$. It can be shown that if the vertex is at a point (h, k), then the equation has the form

$$(x - h)^2 = -4c(y - h)$$

The vertex of the parabola formed by the path of the cannonball has a vertex at $(192, 144)$, so the equation we seek has the form

$$(x - 192)^2 = -4c(y - 144)$$

We also see that the curve passes through $(0, 0)$, so

$$(0 - 192)^2 = -4c(0 - 144)$$
$$36,864 = 576c$$
$$c = 64$$

We guess that the equation for the path of the cannonball can be described by the equation

$$(x - 192)^2 = -256(y - 144)$$

The graph is shown in Figure 3.

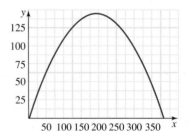

FIGURE 3 Path of a cannonball

Modeling is an iterative process, so we now must *interpret and compare.* Even though the path of the cannonball seems correct (it passes through all of the appropriate points), it does not correctly model the given information. The path should somehow be dependent on the angle of elevation as well as the speed of the cannonball.

In order to model the path as a function of the angle of trajectory and initial velocity, we need some concepts from calculus, specifically Newton's laws of motion. Even though the derivation of these equations is beyond the scope of this course, we can understand these equations once they are stated as follows:

$$x = (v_0 \cos \theta)t \qquad y = h_0 + (v_0 \sin \theta)t - 16t^2$$

where h_0 is the initial height off the ground, and v_0 is the initial velocity of the projectile in the direction of θ with the horizontal ($0° \leq \theta \leq 180°$). You might wish to review the meaning of $\cos \theta$ and $\sin \theta$ discussed in Section 7.5. The variables x and y represent the horizontal and vertical distances, respectively, measured in feet, and the variable t represents the time, in seconds, after firing the projectile.

Example 1 Graph of a projectile

Graph the equations

$$x = (v_0 \cos \theta)t \qquad y = h_0 + (v_0 \sin \theta)t - 4.9t^2$$

where $\theta = 57°$, $v_0 = 63$ ft/s, and $h_0 = 0$.

Solution   The desired equations are

$$x = (63 \cos 57°)t \qquad y = (63 \sin 57°)t - 4.9t^2$$

We can set up a table of values (as shown in the margin) or use a graphing calculator to sketch the curve as shown in Figure 4.

t	x	y
0	0	0
1	34	48
2	69	86
3	103	114
4	137	133
5	172	142
6	206	141
7	240	130
8	274	109
9	309	79
10	343	38

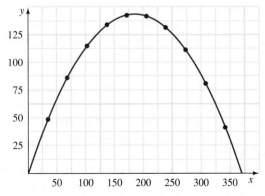

FIGURE 4 Graph of the path of a cannonball

We compare Figure 4 with the real-world problem and it seems to model the situation correctly. In modeling, it may be necessary to go back and modify the assumptions several times before obtaining a result that appropriately models what we seek to find.

Astronomy is the study of objects and matter outside the earth's atmosphere, their motions and paths, as well as their physical and chemical properties. Science often deals with the very small (such as cells, atoms, and electrons) or the very large (as with planets and galaxies). In the text, we defined scientific notation and developed the laws of exponents to help us deal with large and small numbers.

Example 2 Number of miles in a light year

One light-year is approximately

$$5,870,000,000,000 = 5.87 \times 10^{12} \text{ miles.}$$

The closest star is Alpha Centauri, a distance of 4.35 light-years from the sun. How far is this in miles?

Solution   We use the properties of exponents:

$$
\begin{aligned}
4.35 \text{ light-years} &= 4.35 \times (1 \text{ light-year}) \\
&= 4.35 \times (5.87 \times 10^{12} \text{ miles}) \\
&= 25.5345 \times 10^{12} \text{ miles} \\
&= 2.55 \times 10^{13} \text{ miles}
\end{aligned}
$$

Note: Because we are working with approximate numbers, all of these equal signs would generally be "$\approx$," meaning "approximately equal to," but common usage is to write equal signs.

The subsection on scientific notation and estimation (Section 1.3) is particularly important in astronomy. These large numbers are necessary in Example 8, p. 739, Section 15.4, which finds the equation for the earth's orbit around the sun. This example is revisited in the following example.

Example 3 Investigate the orbit of a planet

Describe the meaning of the equation

$$\frac{x^2}{8.649 \times 10^{15}} + \frac{y^2}{8.647 \times 10^{15}} = 1$$

which models the path of the earth's orbit around the sun. The given distances are in miles. Find the eccentricity. The closer the eccentricity is to zero, the more circular the orbit.

Solution   We compare this with the equation of a *standard-form ellipse* centered at the origin.

$$\frac{x^2}{a^2} + \frac{y^2}{b^2} = 1$$

We note that $a = \sqrt{8.649 \times 10^{15}} \approx 93{,}000{,}000$ and $b = \sqrt{8.647 \times 10^{15}} \approx 92{,}990{,}000$. Since $a \approx b$ we note that the elliptic orbit is almost circular. The measure of the circularity of the orbit is the *eccentricity*, which is defined as

$$\epsilon = \frac{c}{a} = \sqrt{1 - \frac{b^2}{a^2}}$$

$$= \sqrt{1 - \frac{8.647 \times 10^{15}}{8.649 \times 10^{15}}} \approx 0.015$$

Note: In reality, ϵ for the earth is about 0.016678, which we could obtain with better accuracy on the measurement of the lengths of the sides. The *aphelion* (the greatest distance from the sun) is $a \approx 93{,}000{,}000$ miles and the *perihelion* (the closest distance from the sun) is $b \approx 92{,}990{,}000$.

Biology is defined as that branch of knowledge that deals with living organisms and vital processes and includes the study of the plant and animal life of a region or environment. In Section 6.3, we considered a significant example from biology, specifically an application from genetics.

Example   4
Find the proportion of genotypes and phenotypes

Suppose a certain population has two eye color genes: *B* (brown eyes, dominant) and *b* (blue eyes, recessive). Suppose we have an isolated population in which 60% of the genes in the gene pool are dominant *B*, and the other 40% are recessive *b*. What fraction of the population has each genotype? What percent of the population has each phenotype?

Solution   Let $p = 0.6$ and $q = 0.4$. Since p and q give us 100% of all the genes in the gene pool, we see that $p + q = 1$. Since

$$(p + q)^2 = p^2 + 2pq + q^2$$

we can find the percents:

Genotype *BB*: $p^2 = (0.6)^2 = 0.36,$
    so 36% have *BB* genotype.

Genotype *bB* or *Bb*: $2pq = 2(0.6)(0.4) = 0.48,$ so 48% have this genotype.

Genotype *bb*: $q^2 = (0.4)^2 = 0.16,$
    so 16% have *bb* genotype.

Check genotypes: $0.36 + 0.48 + 0.16 = 1.00$

As for the phenotypes, we look only at outward appearances, and since brown is dominant, *BB, bB,* and *Bb* all have brown eyes; this accounts for 84%, leaving 16% with blue eyes.

Another area of biology that relates closely with mathematics is the issue of *scale*. For example, we say that to predict a child's height as an adult, use a *scaling factor* of 2, which means if a two-year-old is 31 in. tall, then we predict that this child will be 62 in. tall as an adult. If we scale up a linear measure by a factor of two, then a two-dimensional measure will increase by a factor of $2^2 = 4$, and a three-dimensional measure will increase by a factor of $2^3 = 8$. We considered this idea relative to scaling on graphs in Section 14.1.

Example   5
Discuss the effect of area due to scaling factor changes

Consider a tissue sample as shown in Figure 5.

FIGURE 5 Small tissue sample

If this sample is scaled up by a factor of 2, how does the area of the new sample compare with the area of the original? Repeat for scaling up by a factor of 3.

Solution   Since the area function is a square $(A = s^2)$, we know that the area of the sample is scaled up by a factor of $2^2 = 4$ when doubled or $3^2 = 9$ when tripled.

Doubling the width and length results in increasing the area by 4.

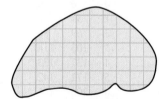

Tripling the width and length results in increasing the area by 9.

Chemistry is the science that deals with the composition, structure, and properties of substances and with the transformations that they undergo. A popular and exciting natural science called **biochemistry** deals with the chemical compounds and processes occurring in organisms.

Computer science is the study of computers, their design, and programming and includes artificial intelligence, networking, computer graphics, and computer languages. Computers, calculators, and spreadsheets are discussed throughout the text.

Geology is the science that deals with the history of the earth and its life, especially as recorded in rocks. **Ecology** is the science that is concerned with the interrelationship of organisms and their environments. **Medicine** is the science and art dealing with the maintenance of health and the prevention, alleviation, or cure of disease. **Meteorology** is the science that deals with the atmosphere and its phenomena, and especially with weather and weather forecasting.

Physics is the science that deals with matter and energy and their interactions. **Biophysics** deals with the application of physical principles and methods to biological systems and problems.

Statistics is a branch of mathematics dealing with the collection, analysis, interpretation, and presentation of masses of numerical data. Chapter 14 of this text introduces some topics in statistics—namely, frequency distributions and graphs; measures of central tendency, position, and dispersion; normal curves; and correlation and regression analysis.

And last, but not least, on our list of major categories of the natural sciences is **zoology,** which is concerned with classification and properties of animals, including their structure, function, growth, origin, evolution, and distribution. Zoology is often considered to be a branch of biology.

An excellent discussion of mathematics in modeling a projectile's path can be found in the following article in the classic book, *World of Mathematics* (New York, Simon and Schuster, 1956): "Mathematics of Motion," by Galileo Galilei. You will find other references to mathematics in the natural sciences on the World Wide Web at

 **www.mathnature.com**

Mathematics in the Social Sciences

The social sciences are the study of human society and of individual relationships in and to society. Some of the major categories include (arranged alphabetically) anthropology, archaeology, civics, history, languages, political science, psychology, and sociology.

Anthropology refers to the study of human beings in relation to distribution, origin, classification, and relationship of races, physical character, environmental and social relations, and culture. In its broadest sense, it also includes theology.

Example 6 | Discussion

Discuss the invention of the zero symbol.

Solution The term zero, written 0, is defined to be the *additive identity*, which is that number with the property that $x + 0 = 0 + x$ for any number x. The Babylonians used written symbols for thousands of years before they invented a symbol for 0, which was initially introduced as a position

marker to differentiate between numbers such as 123 and 1230. The first documented use of the *number* 0 is found around the first century A.D. in the Mayan numeration system. The Hindus customarily wrote numbers in columns and used a zero symbol to represent a blank column, which is necessary for our present place-value numeration system.

Another application of mathematics to anthropology is carbon dating of artifacts using the decay formula. Problem 57 of Problem Set 10.3, p. 499 asked you to use the gathered data to set a probable date for the Shroud of Turin. If you worked this problem, you found that it is not dated from the time of Christ.

Example 7 | Age of Shroud of Turin

In 1988, a small sample of the Shroud of Turin was taken and scientists from Oxford University, the University of Arizona, and the Swiss Federal Institute of Technology were permitted to test it. Suppose the cloth contained 90.7% of the original amount of carbon. According to this information, how old is the Shroud?

Solution We use the decay formula $A = A_0 e^{rt}$, where $A/A_0 = 0.907$ and $r = -1.209680943\mathrm{E} - 4$ (from Example 5, p. 492, Section 10.3):

$$A = A_0 e^{rt} \qquad \text{Decay formula}$$

$$\frac{A}{A_0} = e^{rt} \qquad \text{Divide both sides by } A_0.$$

$$0.907 = e^{rt} \qquad \text{Remember, } t \text{ is the unknown; } r \text{ is known.}$$

$$rt = \ln 0.907 \qquad \text{Definition of logarithm}$$

$$t = \frac{\ln 0.907}{r} \qquad r = -1.209680943\mathrm{E} - 4$$

$$\approx 806.9 \qquad \text{This is the most probable age (in years).}$$

The probable date for the Shroud is about A.D. 1200. Since the first recorded evidence of the Shroud is about 1389, we see that A.D. 1200 is not only possible, but plausible.

"Human sacrifices and human compassion. Greek artists and American cannibals. The birth of language and the death of civilizations. Mongol hordes and ancient toys. Missing aviators and sunken ships. Egyptian politicians and modern saints. The first human and the latest war. The past hides the key to the future. Find it in. . . ." This advertisement for the periodical *Discovering Archaeology* does a good job of describing **archaeology.**

Civics is the subject that deals with the rights and duties of citizens. **History** refers to a chronological record of significant events, often including an explanation of their causes. (The prologue of this book focused on mathematical history.) **Linguistics** is the study of human speech, including the units, nature, structure, and modification of language, whereas **language** refers to the actual knowledge of the words and vocabulary used to communicate ideas and feelings.

Political science is a branch of social science that is the study of the description and analysis of political and especially governmental institutions and processes. Related to this idea is the problem of representation in government, namely, the issue of voting and apportionment. In the text we examined different forms of representative government, including dictatorship, majority, and plurality rules. What effect does a runoff election or a third party have on the results of an election?

Apportionment was an important issue for the framers of the U.S. Constitution, and historically there were four plans that are important to government planning today. These plans were advanced by Thomas Jefferson, John Quincy Adams, Daniel Webster, and Alexander Hamilton. We discussed apportionment at length in the text.

Psychology, broadly defined, refers to the science of mind and behavior. Mathematical tools used in psychology include sets, statistics, and the analysis of data, which we discuss at length in Chapter 14. The organization of data and surveys using Venn diagrams is introduced in Chapter 2.

Sociology is the study of society, social institutions, and social relationships, specifically the systematic study of the development, structure, interaction, and collective behavior of organized groups of human beings. Throughout the book, we encountered a secret sect of intellectuals called the Pythagoreans. Every evening each member of the Pythagorean Society had to reflect on three questions:

PYTHAGOREAN CREED

1. What good have I done today?
2. What have I failed at today?
3. What have I not done today that I should have done?

Example 8 | Show that 28 is perfect

The Pythagoreans studied *perfect numbers*. A perfect number is a counting number that is equal to the sum of all its divisors that are less than the number itself. Show that 28 is a perfect number.

Solution   We first find all divisors of 28 that are less than 28: 1, 2, 4, 7, and 14. Their sum is

$$1 + 2 + 4 + 7 + 14 = 28$$

So we see that 28 is perfect!

Understanding population growth is an essential part of understanding the underlying principles used in the social sciences. Population growth, along with a working knowledge of the number e, is used to predict population sizes for different growth rates. Recall the growth formula:

$$A = A_0 e^{rt}$$

which gives the future population A after t years for an initial population A_0 and a growth rate of r; this relationship was considered extensively in Chapter 10.

The principal use of mathematics in the social sciences is in its heavy use of statistics. In the text, we studied surveys and sampling, which are important to social scientists. To choose an unbiased sample from a target population, we must ask two questions: *Is the procedure random?* and *Does the procedure take into account the target population?* Finally, matrices, an important topic in mathematics, are used to organize and manipulate data in the social sciences. This topic is introduced and discussed extensively in the text.

Social Choice

The study of the decision-making process by which individual preferences are translated into a single group choice is known as *social choice theory*. It is an important topic in mathematics (see Chapter 17) and culminates with the surprising mathematical result that it is impossible to find a suitable technique for voting.

Related to voting is the process by which a representative government is chosen, called *apportionment*, which is also discussed in Chapter 17.

Stable Marriages

Suppose there are n graduates of a well-known medical school, and those graduates will be matched to n medical centers to serve their internship programs. Each graduate must select his or her preferences for medical school, and in turn, each medical school must make a list of preferences for graduates. We will call this pairing a **marriage.** A pairing is called **unstable** if there are a graduate and a school who have not picked each other and who would prefer to be married to each other rather than to their current selection. Otherwise, the pairing is said to be **stable.**

Example 9 | Showing that a marriage possibility is stable

Suppose there are three graduates, *a, b, c,* and three schools, *A, B, C,* with choices as follows:

   $a(ABC)$ means that a ranks the schools as A, first choice; B, second choice; and C, third choice.

   $b(ACB)$ means that b ranks the schools A, C, and B, respectively.

   $c(ABC)$ means that c ranks the schools A, B, and then C.

On the other hand, the schools' rankings of candidates are:

   $A(cab)$, $B(cba)$,   and $C(bca)$

How many possible marriages are there, and how many of these are stable?

Solution   We begin by representing the choices for this example in matrix form (we considered matrices in Section 16.3):

$$\begin{array}{c} \\ a \\ b \\ c \end{array} \begin{array}{ccc} A & B & C \\ \left[\begin{array}{ccc} (1,2) & (2,3) & (3,3) \\ (1,3) & (3,2) & (2,1) \\ (1,1) & (2,1) & (3,2) \end{array}\right] \end{array}$$

The entry (1, 2) means that a picks A as its 1st choice, and A picks a as its 2nd choice.

To answer the first question, we use the fundamental counting principle. Person a can be paired with any of three schools, and then (since the matching is one school to one student) there are two schools left for person b, and finally one for person c:

Number of possibilities: $3 \cdot 2 \cdot 1 = 6$

We show these choices using a tree diagram in Figure 6.

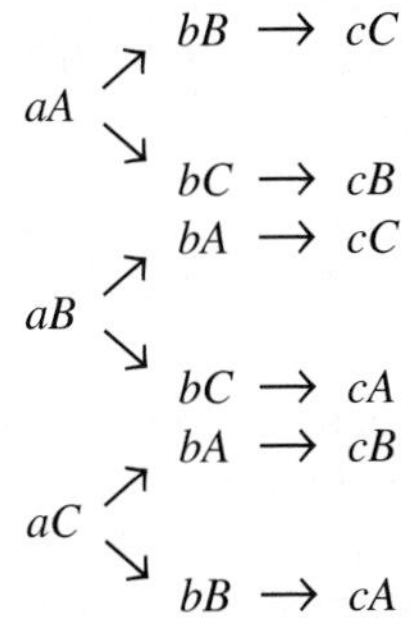

FIGURE 6 Possible pairings

We need to consider each of these six possibilities, one at a time. The first one listed is aA, bB, and cC. We repeat the above matrix, this time with these highlighted pairings:

$$\begin{array}{c} \\ a \\ b \\ c \end{array} \begin{array}{ccc} A & B & C \\ \left[\begin{array}{ccc} (1,2) & (2,3) & (3,3) \\ (1,3) & (3,2) & (2,1) \\ (1,1) & (2,1) & (3,2) \end{array}\right] \end{array}$$

We look for a pairing in which both partners would rather be paired with someone else. This will not occur for a row in which the pairing shows a first choice, so we do not look at row a to find dissatisfaction; instead we look at row b. The pairing $(3, 2)$ tells us that b is associated with his or her third choice. We are looking for another pair *in the same row* where the second component is smaller than its corresponding choice in *its column*. Note that in row b the entry $(2, 1)$ has a second component of 1, which means that C would rather be paired with b than with its current pairing of c. We also see that person b would also prefer to be married to c (second choice) than with its current pairing, so this is *unstable*.

Let's move to the second pairing shown in Figure 6. It is aA, bC, and cB. We show this pairing:

$$\begin{array}{c} \\ a \\ b \\ c \end{array} \begin{array}{ccc} A & B & C \\ \left[\begin{array}{ccc} (1,2) & (2,3) & (3,3) \\ (1,3) & (3,2) & (2,1) \\ (1,1) & (2,1) & (3,2) \end{array}\right] \end{array}$$

Row a is OK because a is paired with his or her first choice.

Row b is OK because b is paired with his or her second choice, but C is paired with a first choice.

Row c shows that c is paired with his or her second choice, but would rather be with A. A would also rather be with c, so this marriage is unstable.

The other possibilities are listed here:

$$\begin{array}{c} \\ a \\ b \\ c \end{array} \begin{array}{ccc} A & B & C \\ \left[\begin{array}{ccc} (1,2) & (2,3) & (3,3) \\ (1,3) & (3,2) & (2,1) \\ (1,1) & (2,1) & (3,2) \end{array}\right] \end{array}$$

Not stable; a prefers A, and A prefers a to b.

$$\begin{array}{c} \\ a \\ b \\ c \end{array} \begin{array}{ccc} A & B & C \\ \left[\begin{array}{ccc} (1,2) & (2,3) & (3,3) \\ (1,3) & (3,2) & (2,1) \\ (1,1) & (2,1) & (3,2) \end{array}\right] \end{array}$$

a prefers A, but A is paired with first choice.
b prefers A, but A is paired with first choice.
Stable

$$\begin{array}{c} \\ a \\ b \\ c \end{array} \begin{array}{ccc} A & B & C \\ \left[\begin{array}{ccc} (1,2) & (2,3) & (3,3) \\ (1,3) & (3,2) & (2,1) \\ (1,1) & (2,1) & (3,2) \end{array}\right] \end{array}$$

c prefers A, and A prefers c, so this is unstable.

$$\begin{array}{c} \\ a \\ b \\ c \end{array} \begin{array}{ccc} A & B & C \\ \left[\begin{array}{ccc} (1,2) & (2,3) & (3,3) \\ (1,3) & (3,2) & (2,1) \\ (1,1) & (2,1) & (3,2) \end{array}\right] \end{array}$$

b prefers C, and C prefers b, so this is unstable.

Example 9 has five unstable pairings and one stable one. It can be proved that every marriage has at least one stable pairing. In the problem set you are asked to consider the situation where the number of candidates is greater than the number of schools.

An excellent discussion of mathematics and the social sciences can be found in the following list of articles in the classic book, *World of Mathematics* (New York, Simon and Schuster, 1956):

"Gustav Theodor Fechner," a commentary on the founder of psychophysics, by Edwin G. Boring. Fechner's contribution was to introduce measurement as a tool.

"Mathematics of Population and Food," by Thomas Robert Malthus, explores the thesis that all animated life tends to increase beyond the nourishment prepared for it.

"A Mathematical Approach to Ethics," by George Birkhoff

 You will find other references to mathematics in the social sciences on the World wide at **www.mathnature.com**

Mathematics in the Humanities

By humanities we mean art, music, and literature. One of the major themes of this book has been the analysis of shapes and patterns, and shapes and patterns are the basis for a great deal of art and music.

There is an interdependence of mathematics and art (see Chapter 7). The shape that seems to be the most appealing to human intellect is the so-called golden rectangle, which was used in the design of the Parthenon in Athens and in *La Parade* by the French impressionist Georges Seurat.

Example 10 Verify a golden ratio

An unfinished canvas by Leonardo da Vinci entitled *St. Jerome* was painted about 1481. Find a golden rectangle that fits neatly around a prominent part of this painting. Measure the length and width of the rectangle and then find the ratio of the width to length.

Solution A portion of the work of art is shown below (with a rectangle superimposed). The width and length of the rectangle will vary with the size of the reproduction, but the ratio of width to length is about 0.61.

St. Jerome by Leonardo da Vinci, 1481

A golden rectangle

In order to draw realistic-looking drawings, mathematical ideas from projective geometry are necessary. Early attempts at three-dimensional art failed miserably because they did not use projective geometry. We cite as examples Duccio's *Last Supper* (Figure 7.63, p. 376) or the Egyptian art shown in Figure 7.

FIGURE 7 *Judgment of the Dead-Pai on papyrusca.* 14th century B.C.

Projective geometry was an early mathematical attempt to represent three-dimensional objects on a canvas (see Dürer's *Recumbent Woman* on page 377). Finally, a plan for perspective was developed, as shown in Figure 8.

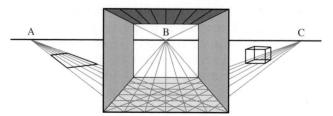

FIGURE 8 Perspective in Art

The rule of optical perspective begins with the horizon (line $\overleftrightarrow{AC}$). One point B on line $\overleftrightarrow{AC}$ is selected to be the vanishing point, and all other lines recede directly from the viewer and converge at the vanishing point. All other lines (except verticals and parallels to the horizon) have their own vanishing points, governed by their particular angle to the plane of the picture. The vanishing point for the rectangle in Figure 8 is point A, and for the cube at the right, it is point C.

Music is very mathematical, and the Greeks included music when studying arithmetic, geometry, and astronomy. Today, with a great deal of music created on a synthesizer, it is not uncommon for a computer programmer to be a musician and for a musician to be a computer programmer. The modern musical scale is divided into 13 tones, as shown in Figure 9.

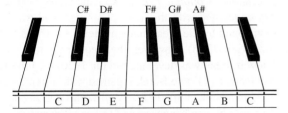

FIGURE 9 Piano keyboard showing one octave

The Greeks plucked strings and found what types of string vibrations made pleasing sounds. They found that the sound was pleasing if it was plucked at a location that was in the ratio of 1 to 2 (which we call an *octave*), 2 to 3 (*a fifth*), 3 to 4 (*a fourth*), and so on. You might notice that the piano keyboard is divided into 8 white keys (diatonic scale) and 5 black keys (pentatonic scale). These numbers 5 and 8 are two consecutive Fibonacci numbers. The middle C is tuned so the string vibrates 264 times per second, whereas the note A above middle C vibrates 440 times per second. Note that the ratio of 440 to 264 reduces to 5 to 3, two other Fibonacci numbers.

A mathematical concept known as the sine function (see Section 7.5) can be used to record the motion of the vibrations over time. These vibrations (known as sound) are shown in the following example.

Example 11 Graph tuning fork sound wave

A tuning fork vibrates at 264 Hz [frequency $f = 264$ and Hz is an abbreviation for Hertz meaning "cycles per unit of time," after the physicist Heinrich Hertz (1857–1894)]. This sound produces middle C on the musical scale and can be described by an equation of the form

$$y = 0.0050 \sin(528 \cdot 360x)$$

Use a graphing calculator to look at this curve.

Solution The graph is shown in Figure 10, along with the necessary input values. Technology does not do a good job of graphing an equation such as this, and you may

obtain many different looking graphs, depending on the window values.

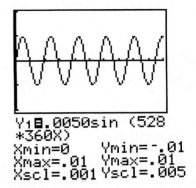

FIGURE 10 Graph of a musical note described by $y = 0.0050 \sin(528 \cdot 360x)$

A composition by Mozart is interesting not only for its music, but also because it illustrates the mathematical idea of an inversion transformation, introduced in Section 7.1. This unique piece for violin, reproduced in Figure 11, can be played simultaneously by two musicians facing each other and reading the music, laid flat between them, in opposite directions. The two parts, though different, will be in perfect harmony, without violating a single rule of classical composition. Another piece, Bach's *Art of the Fugue,* made use of mirror reflections in passages between the upper treble and lower bass figures.

FIGURE 11 Mozart composition

An excellent discussion of mathematics and the humanities can be found in the following list of articles from the classic book, *World of Mathematics* (New York, Simon and Schuster, 1956):

"Mathematics of Aesthetics," by George David Birkhoff

"Mathematics as an Art," by John William Navin Sullivan

"Mathematics and Music," by James Jeans

"Geometry in the South Pacific," by Sylvia Townsend Warner. The story is an example of mathematics in literature.

 You will find other references to mathematics in the humanities on the World Wide Web at **www.mathnature.com**

Mathematics in Business and Economics

The focus of Chapter 11 in the text forms the foundations for business and economics. A fundamental, and essential, idea in business and economics is the concept of **interest.** Interest is an amount paid for the use of someone else's money. It can be compounded once, or it can be compounded at regular intervals. Related to interest are annuities (future value of interest and deposits), sinking funds (monthly payment when the future value is known), present value of an annuity (the amount you can borrow, given the monthly payment), and amortization (the monthly payment when the present value is known). All of these concepts are essential ideas for understanding the processes of business, borrowing, and economics, and they are all important topics in mathematics.

An excellent discussion of mathematics in business and economics can be found in the following list of articles from the classic book, *World of Mathematics* (New York, Simon and Schuster, 1956):

"Mathematics of Value and Demand," by Augustin Cournot

"Theory of Political Economy," by William Stanley Jevons. This essay begins with the declaration, "It is clear that economics, if it is to be a science at all, must be a mathematical science. . . . Wherever the things treated are capable of being *greater or less*, there the laws and relations must be mathematical in nature."

"The Theory of Economic Behavior," by Leonid Hurwicz, introduces the important application of *theory of games and economic behavior.* Game theory is primarily concerned with the logic of strategy and was first envisioned by the great mathematician Gottfried Leibniz (1646–1716). However, the theory of games as we know it today was developed in the 1920s by John von Neumann and Emile Borel. It gained wide acceptance in 1944 in a book entitled *Theory of Games and Economic Behavior* by von Neumann and Oskar Morgenstern.

"Theory of Games," by S. Vajda provides further study in game theory.

Epilogue Problem Set

1. **IN YOUR OWN WORDS**   Describe some similarities between the scientific method and mathematical modeling.

2. **IN YOUR OWN WORDS**   Pick one of the areas that is of most interest to you: natural sciences, social sciences, humanities, or business and economics. Write a paper discussing at least one application of mathematics to this area that was not discussed in this epilogue.

3. Classify each of the disciplines as humanities, a natural science, or a social science.
   a. psychology   **b.** physics
   c. meteorology   **d.** music
   e. history

4. Classify each of the disciplines as humanities, a natural science, or a social science.
   a. chemistry   **b.** art
   c. French   **d.** geology
   e. computer programming

5. Classify each of the disciplines as humanities, a natural science, or a social science.
   a. civics   **b.** medicine
   c. anthropology   **d.** political science
   e. biology   **f.** ecology

6. Classify each of the disciplines as humanities, a natural science, or a social science.
   a. dance   **b.** theology
   c. women's studies   **d.** social work
   e. linguistics   **f.** conflict studies

7. Classify each of the disciplines as humanities, a natural science, or a social science.
   a. plant pathology   **b.** geography
   c. horticulture   **d.** food science
   e. entomology   **f.** forestry

8. Classify each of the disciplines as humanities, a natural science, or a social science.
   a. literature   **b.** Jewish studies
   c. human nutrition   **d.** agriculture
   e. gender studies   **f.** kinesiology

9. If a star is located at a distance of 858,000,000,000,000,000,000 miles, what is this distance in light-years?

10. If a star is located at a distance of 5.61 light-years, what is this distance in miles?

11. The eccentricities of the planets are given in the following table:

    Mercury, 0.21
    Venus, 0.01
    Earth, 0.02
    Mars, 0.09
    Jupiter, 0.04
    Saturn, 0.06
    Uranus, 0.05
    Neptune, 0.01
    Pluto, 0.24

    Which planet has the most circular orbit?

12. Each person has 23 pairs of chromosomes. If we inherit one of each pair from each parent, what is the number of possibilities?

13. An average chicken egg is 2 in. from top to bottom, and an ostrich egg is 6 in. from top to bottom. What is the scaling factor? If a chicken egg weighs 2 oz, what is the expected weight of an ostrich egg?

14. Eggs are graded for both quantity and size. Jumbo eggs must have a minimum weight of 56 lb per 30-doz case, and small eggs must have a minimum weight of 34 lb per 30-doz case. What is the scaling factor for the weight of one small egg compared to one jumbo egg?

15. Suppose a pea has two skin characteristics: S (smooth, dominant) and w (wrinkled, recessive). Suppose we have an isolated population in which 55% of the genes in the gene pool are dominant S and the other 45% are recessive w. What fraction of the population has each genotype?

16. Suppose a pea has two skin characteristics: S (smooth, dominant) and w (wrinkled, recessive). Suppose we have an isolated population in which 48% of the genes in the gene pool are dominant S and the other 52% are recessive w. What fraction of the population has each genotype?

17. The area of a cell magnified 2,000 times is about 20 cm². What is the area of the unmagnified cell?

18. **a.** If you change the angle in Example 1 to 45°, would you expect the cannonball to go farther or shorter than shown in the example?
    b. Use a calculator to sketch the path of a cannonball with initial velocity 63 ft/s at a 45° angle. Was your conjecture in part **a** correct?

19. What is the path of a cannonball fired at 64 ft/s at an angle of 60°?

20. Describe the meaning of the equation

$$\frac{x^2}{4.5837 \times 10^{15}} + \frac{y^2}{4.5835 \times 10^{15}} = 1$$

which models the path of Venus' orbit around the sun. The given distances are in miles.
   a. What is the eccentricity?
   b. Graph this equation.

21. Describe the meaning of the equation

$$\frac{x^2}{2.015 \times 10^{16}} + \frac{y^2}{1.995 \times 10^{16}} = 1$$

which models the path of Mars' orbit around the sun. The given distances are in miles.
   a. What is the eccentricity?
   b. Graph this equation.

22. Four professors, A, B, C, and D, are each going to hire a student. Six students have applied. How many possible marriages are there?

23. Each of four women is going to marry one of four men. How many possible marriages are there?

24. Four professors, A, B, C, and D, are each going to hire a student. Six students have applied. Here are the preferences:

$$
\begin{array}{ll}
A(a, f, b, c, d, e) & a(C, B, D, A) \\
B(a, b, f, e, d, c) & b(C, D, B, A) \\
C(b, a, e, c, d, f) & c(B, A, C, D) \\
D(a, b, e, f, c, d) & d(A, B, D, C) \\
& e(A, B, C, D) \\
& f(B, D, A, C)
\end{array}
$$

   a. Show the matrix of their choices.
   b. List one pairing and state whether it is stable or unstable.

25. Four women are going to marry four men. Here are their preferences:

$$
\begin{array}{ll}
a(A, D, B, C) & A(d, a, b, c) \\
b(D, A, C, B) & B(a, d, c, b) \\
c(D, A, B, C) & C(a, b, d, c) \\
d(D, B, A, C) & D(b, a, c, d)
\end{array}
$$

   a. Show the matrix of their choices.
   b. List one pairing and state whether it is stable or unstable.

26. An equation for middle C with frequency $f = 261.626$ and amplitude 8 is

$$y = 8 \sin(360fx)$$

Use a calculator to graph this note.

27. Graph the sound wave with equation

$$y = 8 \sin(360fx) + 4 \sin(720fx)$$

for $f = 261.626$.

28. Suppose there are three graduates, a, b, c, and three schools, A, B, C, with choices as follows:

$$
\begin{array}{lll}
a(CAB) & b(CAB) & c(CBA) \\
A(abc) & B(bac) & C(acb)
\end{array}
$$

   a. How many possible marriages are there?
   b. How many of these are stable?

29. Suppose there are three graduates, a, b, c, and three schools A, B, C, with choices as follows:

$$
\begin{array}{lll}
a(BCA) & b(CAB) & c(ABC) \\
A(abc) & B(bca) & C(cab)
\end{array}
$$

   a. How many possible marriages are there?
   b. How many of these are stable?

30. The moon's orbit is elliptical with the earth at one focus. The point at which the moon is farthest from the earth is called the *apogee,* and the point at which it is closest is called the *perigee.* If the moon is 199,000 miles from the earth at apogee and the length of the major axis of its orbit is 378,000 miles, what is the eccentricity of the moon's orbit?

Abelian group   A group that is also commutative.

Abscissa   The horizontal coordinate in a two-dimensional system of rectangular coordinates, usually denoted by x.

Absolute value   The absolute value of a number is the distance of that number from the origin. Symbolically,

$$|n| = \begin{cases} n & \text{if } n \geq 0 \\ -n & \text{if } n < 0 \end{cases}$$

Accuracy   One speaks of an *accurate statement* in the sense that it is true and correct, or of an *accurate computation* in the sense that it contains no numerical error. *Accurate to a certain decimal place* means that all digits preceding and including the given one are correct.

Acre   A unit commonly used in the United States system for measuring land. It contains 43,560 ft^2.

Acute angle   An angle whose measure is smaller than a right angle.

Acute triangle   A triangle with three acute angles.

Adams' apportionment plan   An apportionment plan in which the representation of a geographical area is determined by finding the quotient of the number of people in that area divided by the total number of people and then rounding the result as follows: Any quotient with a decimal portion must be rounded up to the next whole number.

Addition   One of the fundamental undefined operations applied to the set of counting numbers.

Addition law of exponents   To multiply two numbers with like bases, add the exponents; that is,

$$b^m \cdot b^n = b^{m+n}$$

Addition law of logarithms   The log of the product of two numbers is the sum of the logs of those numbers. In symbols,

$$\log_b (AB) = \log_b A + \log_b B$$

Addition method   The method of solution of a system of equations in which the coefficients of one of the variables are opposites so that when the equations are added, one of the variables is eliminated.

Addition of integers   If the integers to be added have the same sign, the answer will also have that same sign and will have a magnitude equal to the sum of the absolute values of the given integers. If the integers to be added have opposite signs, the answer will have the sign of the integer with the larger absolute value, and will have a magnitude equal to the difference of the absolute values. Finally, if one or both of the given integers is 0, use the property that $n + 0 = n$ for any integer n.

Addition of matrices $[M] + [N] = [S]$ if and only if $[M]$ and $[N]$ are the same order and the entries of $[S]$ are found by adding the corresponding entries of $[M]$ and $[N]$.

Addition of rational numbers

$$\frac{a}{b} + \frac{c}{d} = \frac{ad}{bd} + \frac{bc}{bd} = \frac{ad + bc}{bd}$$

Addition principle   A method of representing numbers by repeating a symbol in a numeration system and repeatedly adding the symbol's value. For example, $\cap\cap\cap||||$ means $10 + 10 + 10 + 1 + 1 + 1 + 1$.

Addition property (of equations)   The solution of an equation is unchanged by adding the same number to both sides of the equation.

Addition property of inequality   The solution of an inequality is unchanged if you add the same number to both sides of the inequality.

Addition property of probabilities   For any events E and F, the probability of their union can be found by

$$P(E \cup F) = P(E \text{ or } F)$$
$$= P(E) + P(F) - P(E \cap F)$$

Additive identity   The number 0, which has the property that $a + 0 = a$ for any number a.

Additive inverse   See *Opposites*. The additive inverse of a matrix $[M]$ is denoted by $[-M]$ and is defined by $(-1)[M]$.

Add-on interest   A method of calculating interest and installments on a loan. The amount of interest is calculated according to the formula $I = Prt$ and is then added to the amount of the loan. This sum, divided by the number of payments, is the amount of monthly payment.

Address   Designation of the location of data within internal memory or on a magnetic disk or tape.

Adjacent angles   Two angles are adjacent if they share a common side.

Adjacent side   In a right triangle, an acute angle is made up of two sides; one of those sides is the hypotenuse, and the other side is called the adjacent side.

Adjusted balance method   A method of calculating credit card interest using the formula $I = Prt$ in which P is the balance owed after the current payment is subtracted.

Alabama paradox   An increase in the total numbers of items to be apportioned resulting in a loss for a group is called the Alabama paradox.

Algebra   A generalization of arithmetic. Letters called variables are used to denote numbers, which are related by laws that hold (or are assumed) for any of the numbers in the set. The four main processes of algebra are (1) simplify, (2) evaluate, (3) factor, and (4) solve.

Algebraic expression   Any meaningful combination of numbers, variables, and signs of operation.

Alternate exterior angles   Two *alternate angles* are angles on opposite sides of a transversal cutting two parallel lines, each having one of the lines for one of its sides. They are *alternate exterior angles* if neither lies between the two lines cut by the transversal.

Alternate interior angles   Two *alternate angles* are angles on opposite sides of a transversal cutting two parallel lines, each having one of the lines for one of its sides. They are *alternate interior angles* if both lie between the two lines cut by the transversal.

Alternating series   A series that alternates in sign.

Amortization   The process of paying off a debt by systematically making partial payments until the debt (principal) and interest are repaid.

Amortization schedule   A table showing the schedule of payments of a loan detailing the amount of each payment that goes to repay the principal and how much goes to pay interest.

Amortized loan   A loan that is fully paid off with the last periodic payment.

Analytic geometry   The geometry in which position is represented analytically (or by coordinates) and algebraic methods of reasoning are used for the most part.

And   See *Conjunction*. In everyday usage, it is used to join together elements that are connected in two sets simultaneously.

AND-gate   An electrical circuit that simulates conjunction; that is, the circuit is on when two switches are on.

Angle   Two rays or segments with a common endpoint.

Angle of depression   The angle between the line of sight to an object below measured from a horizontal.

Angle of elevation   The angle between the line of sight to an object above measured from a horizontal.

Annual compounding   In the compound interest formula, it is when $n = 1$.

Annual percentage rate   The percentage rate charged on a loan based on the actual amount owed and the actual time it is owed. The approximation formula for annual percentage rate (APR) is

$$APR = \frac{2Nr}{N + 1}$$

Annuity   A sequence of payments into or out of an interest-bearing account. If the payments are made into an interest-bearing account at the end of each time period, and if the frequency of payments is the same as the frequency of compounding, the annuity is called an *ordinary annuity*.

Antecedent   See *Conditional*.

Antiderivative   If the derivative of a function defined by $y = F(x)$ is $f(x)$, then the *antiderivative* of $f(x)$ is

$$\int f(x)\,dx = F(x)$$

This means that $F'(x) = f(x)$.

Apportionment   process of dividing the representation in a legislative body according to some plan.

Approval voting   The approval voting method allows each voter to cast one vote for each candidate who meets with his or her approval. The candidate with the most votes is declared the winner.

APR   Abbreviation for annual percentage rate. See *Annual percentage rate*.

Arc   (1) Part of the circumference of a circle. (2) In networks, it is a connection between two vertices.

Area   A number describing the two-dimensional content of a set. Specifically, it is the number of square units enclosed in a plane figure.

Area formulas   Square, s^2; rectangle, ℓw; parallelogram, bh; triangle, $\frac{1}{2}bh$; circle, πr^2; trapezoid, $\frac{1}{2}h(b_1 + b_2)$.

Area function   The function that is the area bounded below by the x-axis, above by a function $y = f(x)$, on the left by the y-axis, and on the right by the vertical line $x = t$.

Area under a curve   The area of the region bounded by the graph of a function f, the x-axis, and the vertical lines $x = a$ and $x = b$ is given by

$$A = \int_a^b f(x)\,dx$$

Argument   (1) The statements and conclusion as a form of logical reasoning. (2) In a logarithmic expression $\log_b N$, it is the number N.

Arithmetic mean   The arithmetic mean of the numbers a and b is $\dfrac{a + b}{2}$.

Arithmetic sequence   A sequence, each term of which is equal to the sum of the preceding term and a constant, written $a_1, a_2 = a_1 + d, a_3 = a_1 + 2d, \ldots$; the nth term of an arithmetic sequence is

$$a_1 + (n - 1)d$$

where a_1 is the first term and d is the *common difference*. Also called an *arithmetic progression*. See also *Sequence*.

Arithmetic series   The indicated sum of the terms of an arithmetic sequence. The sum of n terms is denoted by A_n and

$$A_n = \tfrac{n}{2}(a_1 + a_n) \text{ or } A_n = \tfrac{n}{2}[2a_1 + (n - 1)d]$$

Arrangement   Same as *Permutation*.

Array   An arrangement of items into rows and columns. See *Matrix*.

Arrow's impossibility theorem   No social choice rule satisfies all six of the following conditions.

1. **Unrestricted domain**   Any set of rankings is possible; if there are n candidates, then there are $n!$ possible rankings.
2. **Decisiveness**   Given any set of individual rankings, the method produces a winner.
3. **Symmetry and transitivity**   The voting system should be symmetric and transitive over the set of all outcomes.
4. **Independence of irrelevant alternatives**   If a voter prefers A to B with C as a possible choice, then the voter still prefers A to B when C is not a possible choice.
5. **Pareto principle**   If each voter prefers A over B, then the group chooses A over B.
6. There should be **no dictator.**

Artificial intelligence   A field of study devoted to computer simulation of human intelligence.

ASCII code   A standard computer code used to facilitate the interchange of information among various types of computer equipment.

Assignment   A computer term for setting the value of one variable to match the value of another.

Associative property   A property of grouping that applies to certain operations (addition and multiplication, for example, but not to subtraction or division): If a, b, and c are real numbers, then

$$(a + b) + c = a + (b + c)$$

and

$$(ab)c = a(bc)$$

Assuming the antecedent   Same as *direct reasoning*.

Assuming the consequent   A logical fallacy; same as the *fallacy of the converse*.

Augmented matrix   A matrix that results after affixing an additional column to an existing matrix.

Average   A single number that is used to typify or represent a set of numbers. In this book, it refers to the *mean, median*, or *mode*.

Average daily balance method   A method of calculating credit card interest using the formula $I = Prt$ in which P is the average daily balance owed for a current month, and t is the number of days in the month divided by 365.

Average rate of change   The average rate of change of a function f from x to $x + h$ is

$$\frac{f(x + h) - f(x)}{h}$$

Axes   The intersecting lines of a Cartesian coordinate system. The horizontal axis is called the x-axis, and the vertical axis is called the y-axis. The axes divide the plane into four parts called *quadrants*.

Axiom   A statement that is accepted without proof.

Axis of a parabola   The line through the focus of a parabola drawn perpendicular to the directrix.

Axis of symmetry   A curve is symmetric with respect to a line, called the axis of symmetry, if for any point P on the curve, there is a point Q also on the curve such that the axis of symmetry is the perpendicular bisector of the line segment $\overline{PQ}$.

Balinski and Young's impossibility theorem   Any apportionment plan that does not violate the quota rule must produce paradoxes. And any apportionment method that does not produce paradoxes must violate the quota rule.

Balloon payment   A single larger payment made at the end of the time period of an installment loan that is not amortized.

Bar graph   See *Graph*.

Base   (1) See *Exponent*. (2) In a percent problem, it is the whole quantity.

Base angles   In an isosceles triangle, the angles formed by the base with each of the equal sides.

Base b numeration system   A numeration system with b elements.

Base of an exponential   In $y = b^x$, the *base* is $b(b \neq 1)$.

Base of a triangle   In an isosceles triangle, the side that is not the same length as the sides whose lengths are the same.

Because   A logical operator for p because q, which is defined to mean

$$(p \wedge q) \wedge (q \rightarrow p).$$

Bell-shaped curve   See *Normal curve*.

Belong to a set   To be an element of a set.

Bernoulli trial   A Bernoulli trial is an experiment with two possible, mutually exclusive outcomes (usually called success and failure).

Biconditional   A logical operator for simple statements p and q which is true when p and q have the same truth values, and is false when p and q have different truth values.

Billion   A name for $10^9 = 1,000,000,000$.

Bimodal   A distribution with two modes.

Binary numeration system   A numeration system with two numerals, 0 and 1.

Binary voting   Selecting a winner by pairing two competitors and then pairing up the winners to select the final victor.

Binomial   A polynomial with exactly two terms.

Binomial distribution theorem   Let X be a random variable for the number of successes in n independent and identical repetitions of an experiment with two possible outcomes, success and failure. If p is the probability of success, then

$$P(X = k) = \binom{n}{k}p^k(1 - p)^{n-k}$$

where $k = 0, 1, \ldots, n$.

Binomial experiment   A *binomial experiment* is an experiment that meets four conditions:
1. There must be a fixed number of trials. Denote this number by n.
2. There must be two possible mutually exclusive outcomes for each trial. Call them *success* and *failure*.
3. Each trial must be independent. That is, the outcome of a particular trial is not affected by the outcome of any other trial.
4. The probability of success and failure must remain constant for each trial.

Binomial random variable   A random variable that counts the number of successes in a binomial experiment.

Binomial theorem   For any positive integer n,

$$(a + b)^n = \sum_{k=0}^{n}\binom{n}{k}a^{n-k}b^k$$

Birthday problem   The probability that two unrelated people will have their birthdays on the same day (not counting the year of birth).

Bisect   To divide into two equal or congruent parts.

Bit   Binary digit, the smallest unit of data storage with a value of either 0 or 1, thereby representing whether a circuit is open or closed.

Bond   An interest-bearing certificate issued by a government or business, promising to pay the holder a specified amount (usually $1,000) on a certain date.

Boot   To start a computer. To do this the computer must have access to an operating system on a CD or hard disk.

Borda count   A common way of determining a winner when there is no majority is to assign a point value to each voter's ranking. If there are n candidates, then n points are assigned to the first choice for each voter, with $n - 1$ points for the next choice, and so on. The points for each candidate are added and if one has more votes, that candidate is declared the winner.

Boundary   See *Half-plane*.

Box plot   A rectangular box positioned above a numerical scale associated with a given set of data. It shows the maximum value, the minimum value, and the quartiles for the data.

Braces   See *Grouping symbols*.

Brackets   See *Grouping symbols*.

Bug   An error in the design or makeup of a computer program (software bug) or a hardware component of the system (hardware bug).

Bulletin boards   An online means of communicating with others on a particular topic.

Byte   The fundamental block of data that can be processed by a computer. In most microcomputers, a byte is a group of eight adjacent bits and the rough equivalent of one alphanumeric symbol.

Calculus   The field of mathematics that deals with differentiation and integration of functions, and related concepts and applications.

Canceling   The process of reducing a fraction by dividing the same number into both the numerator and the denominator.

Canonical form   When a given number is written as a product of prime factors in ascending order, it is said to be in canonical form.

Capacity   A measurement for the amount of liquid a container holds.

Cardinal number   A number that designates the manyness of a set; the number of units, but not the order in which they are arranged.

Cardinality   The number of elements in a set.

Cards   A deck of 52 matching objects that are identical on one side and on the other side are divided into four suits (hearts, diamonds, spades, and clubs). The objects, called cards, are labeled A, 2, . . . , 9, 10, J, Q, and K in each suit.

Cartesian coordinate system   Two intersecting lines, called *axes*, used to locate points in a plane called a *Cartesian plane*. If the intersecting lines are perpendicular, the system is called a *rectangular coordinate system*.

Cartesian plane   See *Cartesian coordinate system*.

CD-ROM   A form of mass storage. It is a cheap read-only device, which means that you can only use the data stored on it when it was created, but it can store a massive amount of material, such as an entire encyclopedia.

Cell   A specific location on a spreadsheet. It is designated using a letter (column heading) followed by a numeral (row heading). A cell can contain a letter, word, sentence, number, or formula.

Celsius   A metric measurement for temperature for which the freezing point of water is 0° and the boiling point of water is 100°.

Center of a circle   See *Circle*.

Center of an ellipse   The midpoint of the line segment connecting the foci of the ellipse.

Center of a hyperbola   The midpoint of the line segment connecting the foci of the hyperbola.

Centi-   A prefix that means 1/100.

Centigram   One hundredth of a gram.

Centiliter   One hundredth of a liter.

Centimeter   One hundredth of a meter.

Change of base theorem

$$\log_a x = \frac{\log_b x}{\log_b a}$$

Ciphertext   A secret or coded message.

Circle   The set of points in a plane that are a given distance from a given point. The given point is called the *center,* and the given distance is called the *radius*. The diameter is twice the radius. The *unit circle* is the circle with center at (0, 0) and $r = 1$.

Circle graph   See *Graph*.

Circuit   A complete path of electrical current including a power source and a switch. It may include an indicator light to show when the circuit is complete.

Circular definition   A definition that relies on the use of the word being defined, or other words that rely on the word being defined.

Circumference   The distance around a circle. The formula for finding the circumference is $C = \pi D$ or $C = 2\pi r$.

Classes   One of the groupings when organizing data. The difference between the lower limit of one class and the lower limit of the next class is called the *interval* of the class. The number of values within a class is called the *frequency*.

Closed   See *Closure property*.

Closed curve   A curve that has no endpoints.

Closed-ended loan   An installment loan.

Closed half-plane   See *Half-plane*.

Closed network   A network that connects each point.

Closed set   A set that satisfies the closure property for some operation.

Closing   The process of settlement on a real estate loan.

Closing costs   Costs paid at the closing of a real estate loan.

Closure property   A set S is *closed* for an operation ∘ if $a \circ b$ is an element of S for all elements a and b in S. This property is called the *closure property*.

Coefficient   Any factor of a term is said to be the coefficient of the remaining factors. The *numerical coefficient* is the numerical part of the term, usually written before the variable part. In $3x$, it is the number 3, in $9x^2y^3$ it is the number 9. Generally, the word *coefficient* is taken to be the numerical coefficient of the variable factors.

Column   A vertical arrangement of numbers or entries of a matrix. It is denoted by letters $A, B, C, \ldots$ on a spreadsheet.

Combination   A selection of objects from a given set without regard to the order in which they are selected. Sometimes it refers to the number of ways this selection can be done and is denoted by $_nC_r$ or $\binom{n}{r}$ and is pronounced "n choose r." The formula for finding it is

$$\binom{n}{r} = \frac{n!}{r!(n-r)!}$$

Common denominator   For two or more fractions, a common multiple of the denominators.

Common difference   The difference between successive terms of an arithmetic sequence.

Common factor   A factor that two or more terms of a polynomial have in common.

Common fraction   Fractions written in the form of one integer divided by a whole number are common fractions. For example, 1/10 is common fraction representation and 0.1 is the decimal representation of the same number.

Common logarithm   A logarithm to the base 10; written log N.

Common ratio   The ratio between successive terms of a geometric sequence.

Communication matrix A square matrix in which the entries symbolize the occurrence of some facet or event with a 1 and the nonoccurrence with a 0.

Communications package A program that allows one computer to communicate with another computer.

Commutative group A group that also satisfies the property that

$$a \circ b = b \circ a$$

for some operation $\circ$ and elements a and b in the set.

Commutative property A property of order that applies to certain operations (addition and multiplication, for example, but not to subtraction and division). If a and b are real numbers, then

$$a + b = b + a$$

and

$$ab = ba$$

Comparison property For any two numbers x and y, exactly one of the following is true: (1) $x = y$; x is equal to y (the same as) (2) $x > y$; x is greater than y (bigger than) (3) $x < y$; x is less than y (smaller than). This is sometimes known as the *trichotomy property*.

Comparison rate for home loans A formula for comparing terms of a home loan. The formula is

$$\text{APR} + 0.125\left(\text{POINTS} + \frac{\text{ORIGINATION FEE}}{\text{AMOUNT OF LOAN}}\right)$$

Compass An instrument for scribing circles or for measuring distances between two points.

Compiler An operating system program that converts an entire program written in a higher-level language into machine language before the program is executed.

Complement (1) Two numbers less than 1 are called complements if their sum is 1. (2) The complement of a set is everything not in the set relative to a given universe.

Complementary angles Two angles are complementary if the sum of their measures is $90°$.

Complementary probabilities

$$P(E) = 1 - P(\overline{E})$$

that is, two probabilities are *complementary* if

$$P(E) + P(\overline{E}) = 1$$

Completely factored An expression is completely factored if it is a product and there are no common factors and no difference of squares—that is, if no further factoring is possible.

Complex decimal A form that mixes decimal and fractional form, such as $0.12\frac{1}{2}$.

Complex fraction A rational expression a/b where a or b (or both) have fractional form.

Components See *Ordered pair*.

Composite number Sometimes simply referred to as a *composite*; it is a positive integer that has more than two divisors.

Compound interest A method of calculating interest by adding the interest to the principal at the end of the compounding period so that this sum is used in the interest calculation for the next period.

Compound interest formula

$$A = P(1 + i)^N$$

where A = future value; P = present value (or principal); r = annual interest rate (APR); t = number of years; n = number of times compounded per year; $i = \frac{r}{n}$; and $N = nt$.

Compound statement A statement formed by combining simple statements with one or more operators.

Compounding The process of adding interest to the principal so that in the next time period the interest is calculated on this sum.

Computer A device which, under the direction of a program, can process data, alter its own program instructions, and perform computations and logical operations without human intervention.

Computer abuse A misuse of a computer.

Computer program A set of step-by-step directions that instruct a computer how to carry out a certain task.

Conclusion The statement that follows (or is to be proved to follow) as a consequence of the hypothesis of the theorem. Also called a *logical conclusion*.

Conditional The statement "if p, then q," symbolized by $p \rightarrow q$. The statement p is called the *antecedent* and q is called the *consequent*.

Conditional equation See *Equation*.

Conditional inequality See *Inequality*.

Conditional probability A probability that is found on the condition that a certain event has occurred. The notation $P(E \mid F)$ is the probability of event E *on the condition* that event F has occurred.

Condorcet candidate A candidate who wins all the one-to-one matchups.

Condorcet criterion If a candidate is favored when compared one on one with every other candidate, then that candidate should be declared the winner.

Condorcet's paradox There are three citizens, A, B, and C, and each citizen ranks three different policies as shown:

$$A: x > y > z$$
$$B: y > z > x$$
$$C: z > x > y$$

Then two citizens prefer x to y, two prefer y to z, and two prefer z to x, which is not transitive. Each voter is consistent, but the social choice is inconsistent. This is known as Condorcet's paradox.

Cone A solid with a circle for its base and a curved surface tapering evenly to an apex so that any point on this surface is in a straight line between the circumference of the base and the apex.

Congruent Of the same size and shape; if one is placed on top of the other, the two figures will coincide exactly in all their parts.

Congruent angles Two angles that have the same measure.

Congruent modulo m Two real numbers a and b are congruent modulo m, written $a \equiv b \pmod m$, if a and b differ by a multiple of m.

Congruent triangles Two triangles that have the same size and shape.

Conic sections The set of points that results from the intersection of a cone and a plane is a two-dimensional curve known as a conic section. The conic sections include the parabola, ellipse (special case, circle), and hyperbola.

Conjecture   A guess or prediction based on incomplete or uncertain evidence.

Conjugate axis   The line passing through the center perpendicular to the transverse axis of a hyperbola.

Conjunction   The conjunction of two simple statements p and q is true whenever both p and q are true, and is false otherwise. The common translation of conjunction is "and."

Connected network   A network that connects each point.

Connective   A rule that operates on one or two simple statements. Some examples of connectives are *and, or, not,* and *unless.*

Consecutive integers   Integers that differ by 1.

Consequent   See *Conditional.*

Consistent system   If a system of equations has at least one solution, it is consistent; otherwise it is said to be *inconsistent.*

Constant   Symbol with exactly one possible value.

Constant function   A function of the form $f(x) = c$.

Constant multiple   The following property of integrals:

$$\int af(x)dx = a\int f(x)dx$$

Constraint   A limitation placed on an objective function. See *Linear programming.*

Construct   The process of drawing a figure that will satisfy certain given conditions.

Contained in a set   An element is contained in a set if it is a member of the set.

Continuous compounding   If we let the number of compounding periods in a year increase without limit, the result is called continuous compounding. The formula is

$$A = Pe^{rt}$$

for a present value of P, future value A at a rate of r for t years.

Continuous distribution   A probability distribution that includes all x-values (as opposed to a discrete distribution, which allows a finite number of x-values).

Contradiction   An open equation for which the solution set is empty.

Contrapositive   For the implication $p \to q$, the contrapositive is $\sim q \to \sim p$.

Converge   To draw near to. A series is said to converge when the sum of the first n terms approaches a limit as n increases without bound. We say that the sequence converges to a limit L if the values of the successive terms of the sequence get closer and closer to the number L as $n \to \infty$.

Converse   For the implication $p \to q$, the converse is $q \to p$.

Convex set   A set that contains the line segment joining any two of its points.

Coordinate plane   See *Cartesian coordinate system.*

Coordinates   A numerical description for a point. Also see *Ordered pair.*

Correlation   The interdependence between two sets of numbers. It is a relationship between two quantities, such that when one changes the other does (simultaneous increasing or decreasing is called *positive correlation;* and one increasing, the other decreasing, *negative correlation*).

Corresponding angles   Angles in different triangles that are similarly related to the rest of the triangle.

Corresponding parts   Points, angles, lines, etc., in different figures, similarly related to the rest of the figures.

Corresponding sides   Sides of different triangles that are similarly related to the rest of the triangle.

Cosine   In a right triangle ABC with right angle C,

$$\cos A = \frac{\text{LENGTH OF ADJACENT SIDE OF } A}{\text{LENGTH OF HYPOTENUSE}}$$

Countable set   A set with cardinality $\aleph_0$. That is, a set that can be placed into a one-to-one correspondence with the set of counting numbers.

Count-down property $n! = n(n - 1)!$

Counterclockwise   In the direction of rotation opposite to that in which the hands move around the dial of a clock.

Counterexample   An example that is used to disprove a proposition.

Counting numbers   See *Natural numbers.*

CPU   Central Processing Unit, the primary section of the computer that contains the memory, logic, and arithmetic procedures necessary to process data and perform computations. The CPU also controls the functions performed by the input, output, and memory devices.

Credit card   A card signifying that the person or business issued the card has been approved for open-ended credit. It can be used at certain restaurants, airlines, and stores accepting that card.

Cryptography   The writing or deciphering of messages in code.

Cube   (1) A solid with six equal square sides. (2) In an expression such as x^3, which is pronounced "x cubed," it means xxx.

Cube root   See *Root of a number.*

Cubed   See *Cube.*

Cubic unit   A three-dimensional unit. It is the result of cubing a unit of measurement.

Cumulative frequency   A frequency that is the total number of cases having any given score or a score that is lower.

Cup   A unit of measurement in the United States measurement system that is equivalent to 8 fluid ounces.

Cursor   Indicator (often flashing) on a computer or calculator display to designate where the next character input will be placed.

Cylinder   Suppose we are given two parallel planes and two simple closed curves C_1 and C_2 in these planes for which lines joining corresponding points of C_1 and C_2 are parallel to a given line L. A cylinder is a closed surface consisting of two bases that are plane regions bounded by such curves C_1 and C_2 and a lateral surface that is the union of all line segments joining corresponding points of C_1 and C_2.

Daily compounding   In the compound interest formula, it is when $n = 365$ (exact interest) or when $n = 360$ (ordinary interest). In this book, use ordinary interest unless otherwise indicated.

Data processing   The recording and handling of information by means of mechanical or electronic equipment.

Database   A collection of information. A data-base manager is a program that is in charge of the information stored in a database.

Database manager   A computer program that allows a user to interface with a database.

Dealer's cost   The actual amount that a dealer pays for the goods sold.

Debug   The organized process of testing for, locating, and correcting errors within a program.

Decagon   A polygon having ten sides.

Decay formula   Refers to exponential decay. It is described by the equation

$$A = A_0 e^{rt}$$

where r is the annual decay rate (and consequently is negative), t is the time (in years), A_0 is the amount present initially (present value), and A is the future value. If r is positive, this formula models growth, and if r is negative, the formula models decay.

Deci-   A prefix that means 1/10.

Decibel   A unit of measurement for measuring sounds. It is defined as a ratio of the intensity of one sound, I, and another sound $I_0 \approx 10^{-16}$ watt/cm^2, the intensity of a barely audible sound for a person with normal hearing.

Deciles   Nine values that divide a data set into ten equal parts.

Decimal   Any number written in decimal notation. The digits represent powers of ten with whole numbers and fractions being separated by a period, called a *decimal point*. Sometimes called a Hindu-Arabic numeral.

Decimal fraction   A number in decimal notation that has fractional parts, such as 23.25. If a common fraction p/q is written as a decimal fraction, the result will be either a *terminating decimal* as with $\frac{1}{4} = 0.25$ or a *repeating decimal* as with $\frac{2}{3} = 0.6666\ldots$.

Decimal notation   The representation of a number using the decimal number system. See *Decimal*.

Decimal numeration system   A numeration system with base 10.

Decimal point   See *Decimal*.

Decisiveness   Given any set of individual rankings, the method produces a winner when voting.

Decoding key   A key that allows one to unscramble a coded message.

Deductive reasoning   A formal structure based on a set of axioms and a set of undefined terms. New terms are defined in terms of the given undefined terms and new statements, or *theorems*, are derived from the axioms by proof.

Definite integral   Let f be a function defined over the interval $[a, b]$. Then the definite integral of f over the interval is denoted by

$$\int_a^b f(x)\,dx$$

and is the net change of an antiderivative of f over that interval. Thus, if $F(x)$ is an antiderivative of $f(x)$, then

$$\int_a^b f(x)\,dx = F(x)\big|_a^b = F(b) - F(a)$$

Degree   (1) The degree of a term in one variable is the exponent of the variable, or it is the sum of the exponents of the variables if there are more than one. The degree of a polynomial is the degree of its highest-degree term. (2) A unit of measurement of an angle that is equal to 1/360 of a revolution.

Deka-   A prefix that means 10.

Deleted point   A single point that is excluded from the domain.

De Morgan's laws   For sets X and Y,

$$\overline{X \cup Y} = \overline{X} \cap \overline{Y}$$

and

$$\overline{X \cap Y} = \overline{X} \cup \overline{Y}$$

Demand   The number of items that can be sold at a given price.

Denominator   See *Rational number*.

Dense set   A set of numbers with the property that between any two points of the set, there exists another point in the set that is between the two given points.

Denying the antecedent   A logical fallacy; same as the *fallacy of the inverse*.

Denying the consequent   Same as *indirect reasoning*.

Dependent events   Two events are dependent if the occurrence of one influences the probability of the occurrence of the other.

Dependent system   If *every* ordered pair satisfying one equation in a system of equations also satisfies every other equation of the given system, then we describe the system as dependent.

Dependent variable   The variable associated with the second component of an ordered pair.

Derivative   One of the fundamental operations of calculus; it is the instantaneous rate of change of a function with respect to the variable. Formally, for a given function f, we define the derivative of f at x, denoted by $f'(x)$, to be

$$f'(x) = \lim_{h \to 0} \frac{f(x + h) - f(x)}{h}$$

provided this limit exists. If the limit exists, we say f is a differentiable function of x.

Description method   A method of defining a set by describing the set (as opposed to listing its elements).

Descriptive statistics   Statistics that is concerned with the accumulation of data, measures of central tendency, and dispersion.

Diagonal form   A matrix with the terms arranged on a diagonal, from upper left to lower right, and zeros elsewhere.

Diameter   See *Circle*.

Dice   Plural for the word *die*, which is a small, marked cube used in games of chance.

Dictatorship   A selection process where one person alone makes a decision.

Die   See *Dice*.

Difference   The result of a subtraction.

Difference quotient   If f is a function, then the *difference quotient* is defined to be the function

$$\frac{f(x + h) - f(x)}{h}$$

Difference of squares   A mathematical expression in the form $a^2 - b^2$.

Differentiable function   See *Derivative*.

Differential calculus   That branch of calculus that deals with the derivative and applications of the derivative.

Dimension   (1) A configuration having length only is said to be of one dimension; area and not volume, two dimensions; volume, three dimensions. (2) In reference to matrices, dimension is the numbers of rows and columns.

Direct reasoning   One of the principal forms of logical reasoning. It is an argument of the form

$$[(p \to q) \wedge p] \to q$$

Directrix   See *Parabola*.

Discount   A reduction from a usual or list price.

Discrete mathematics   That part of mathematics that deals with sets of objects that can be counted or processes that consist of a sequence of individual steps.

Disjoint sets   Sets that have no elements in common.

Disjunction   The disjunction of two simple statements p and q is false whenever both p and q are false, and is true otherwise. The common translation of disjunction is "or."

Disk drive   A mechanical device that uses the rotating surface of a magnetic disk for the high-speed transfer and storage of data.

Distinguishable permutation   The number of distinguishable permutations of n objects in which n_1 are of one kind, n_2 are of another kind, . . . , and n_k are of a further kind, so that $n = n_1 + n_2 + \cdots + n_k$ is denoted by $\begin{pmatrix} n \\ n_1, n_2, \ldots, n_k \end{pmatrix}$ and is defined by

$$\begin{pmatrix} n \\ n_1, n_2, \ldots, n_k \end{pmatrix} = \frac{n!}{n_1!\, n_2! \cdots \cdots n_k!}$$

Distributive law of exponents

$$(1)\ (ab)^m = a^m b^m; \qquad (2)\ \left(\frac{a}{b}\right)^m = \frac{a^m}{b^m}$$

Distributive property (for multiplication over addition)   If a, b, and c are real numbers, then $a(b + c) = ab + ac$ and $(a + b)c = ac + bc$ for the basic operations. That is, the number outside the parentheses indicating a sum or difference is distributed to each of the numbers inside the parentheses.

Diverge   A sequence that does not converge is said to diverge.

Dividend   The number or quantity to be divided. In a/b, the dividend is a.

Divides   See *Divisibility*.

Divine proportion   If two lengths h and w satisfy the proportion

$$\frac{h}{w} = \frac{w}{h + w}$$

then the lengths are said to be in a *divine proportion*.

Divisibility   If m and d are counting numbers, and if there is a counting number k so that $m = d \cdot k$, we say that d is a divisor of m, d is a factor of m, d divides m, and m is a multiple of d.

Division $\frac{a}{b} = x$ is $a \div b = x$ and means $a = bx$.

Division by zero   In the definition of division, $b \neq 0$, because if $b = 0$, then $bx = 0$, regardless of the value of x. If $a \neq 0$, then there is no such number. On the other hand, if $a = 0$, then $0/0 = 1$ checks from the definition, and so also does $0/0 = 2$, which means that $1 = 2$, another contradiction. Thus, division by 0 is excluded.

Division of integers   The quotient of two integers is the quotient of the absolute values, and is positive if the given integers have the same sign, and negative if the given numbers have opposite signs. Furthermore, division by zero is not possible and division into 0 gives the answer 0.

Division of rational numbers

$$\frac{a}{b} \div \frac{c}{d} = \frac{ad}{bc} \quad (c \neq 0)$$

Division property (of equations)   The solution of an equation is unchanged by dividing both sides of the equation by the same nonzero number.

Division property of inequality   See *Multiplication property of inequality*.

Divisor   The quantity by which the dividend is to be divided. In a/b, b is the divisor.

Dodecagon   A polygon with 12 sides.

Domain   The *domain* of a variable is the set of replacements for the variable. The *domain* of a graph of an equation with two variables x and y is the set of permissible real-number replacements *for x*.

Double negative $-(-a) = a$

Double subscripts   Two subscripts on a variable as in a_{12}. (Do not read this as "twelve.")

Down payment   An amount paid at the time a product is financed. The purchase price minus the down payment is equal to the amount financed.

Download   The process of copying a program form the network to your computer.

Dummy variable   A variable in a mathematical expression whose only function is as a placeholder.

e   Euler's number, defined by

$$e = \lim_{n \to \infty} \left(1 + \frac{1}{n}\right)^n$$

Eccentricity   For a conic section, it is defined as the ratio

$$\epsilon = \frac{c}{a}$$

For the ellipse, $0 \leq \epsilon < 1$, where ϵ measures the amount of roundness. If $\epsilon = 0$, then the conic is a circle. For the parabola, $\epsilon = 1$; and for the hyperbola, $\epsilon > 1$.

Edge   A line or a line segment that is the intersection of two plane faces of a geometric figure, or that is in the boundary of a plane figure.

Either . . . or   A logical operator for "either p or q," which is defined to mean

$$(p \lor q) \land \sim(q \land p)$$

Element   One of the individual objects that belong to a set.

Elementary operations   Refers to the operations of addition, subtraction, multiplication, and division.

Elementary row operations   There are four elementary row operations for producing equivalent matrices: (1) *RowSwap*: Interchange any two rows. (2) *Row+*: Row addition—add a row to any other row. (3) **Row*: Scalar multiplication—multiply (or divide) all the elements of a row by the same nonzero real number. (4) **Row+*: Multiply all the entries of a row (*pivot row*) by a nonzero real number and add each resulting product to the corresponding entry of another specified row (*target row*).

Ellipse   The set of all points in a plane such that, for each point on the ellipse, the sum of its distances from two fixed points (called the **foci**) is a constant.

Elliptic geometry   A non-Euclidean geometry in which a Saccheri quadrilateral is constructed with summit angles obtuse.

e-mail   Electronic mail sent from one computer to another.

Empirical probability   A probability obtained empirically by experimentation.

Empty set   See *Set*.

Encoding key   A key that allows one to scramble, or encode a message.

Encrypt   To scramble a message so that it cannot be read by an unwanted person.

Equal angles   Two angles that have the same measure.

Equal matrices Two matrices are equal if they are the same order (dimension) and also the corresponding elements are the same (equal).

Equal sets Sets that contain the same elements.

Equal to Two numbers are equal if they represent the same quantity or are identical. In mathematics, a relationship that satisfies the axioms of equality.

Equality, axioms of For $a, b, c \in \mathbb{R}$,
Reflexive: $a = a$
Symmetric: If $a = b$, then $b = a$.
Transitive: If $a = b$ and $b = c$, then $a = c$.
Substitution: If $a = b$, then a may be replaced throughout by b (or b by a) in any statement without changing the truth or falsity of the statement.

Equally likely outcomes Outcomes whose probabilities of occurring are the same.

Equation A statement of equality. If always true, an equation is called an *identity*; if always false, it is called a *contradiction*. If it is sometimes true and sometimes false, it is called a *conditional equation*. Values that make an equation true are said to *satisfy* the equation and are called *solutions* or *roots* of the equation. Equations with the same solutions are called *equivalent equations*.

Equation of a graph Every point on the graph has coordinates that satisfy the equation, and every ordered pair that satisfies the equation has coordinates that lie on the graph.

Equation properties There are four equation properties:
(1) *Addition property:* Adding the same number to both sides of an equation results in an equivalent equation.
(2) *Subtraction property:* Subtracting the same number from both sides of an equation results in an equivalent equation.
(3) *Multiplication property:* Multiplying both sides of a given equation by the same nonzero number results in an equivalent equation.
(4) *Division property:* Dividing both sides of a given equation by the same nonzero number results in an equivalent equation.

Equilateral triangle A triangle whose three sides all have the same length.

Equilibrium point A point for which the supply and demand are equal.

Equivalent equations See *Equation*.

Equivalent matrices Matrices that represent equivalent systems.

Equivalent sets Sets that have the same cardinality.

Equivalent systems Systems that have the same solution set.

Estimate An approximation (usually mental) of size or value used to form an opinion.

Euclidean geometry The study of geometry based on the assumptions of Euclid. These basic assumptions are called Euclid's postulates.

Euclid's postulates 1. A straight line can be drawn from any point to any other point. 2. A straight line extends infinitely in either direction. 3. A circle can be described with any point as center and with a radius equal to any finite straight line drawn from the center. 4. All right angles are equal to each other. 5. Given a straight line and any point not on this line, there is one and only one line through that point that is parallel to the given line.

Euler circles The representation of sets using interlocking circles.

Euler circuit Begin at some vertex of a graph, travel on each edge exactly once, and return to the starting vertex. The path that is a trace of the tip is called an *Euler circuit*.

Euler's circuit theorem Every vertex on a graph which is an Euler circuit has an even degree, and conversely, if in a connected graph every vertex has an even degree, then the graph is an Euler circuit.

Euler's number It is the number e.

Evaluate To *evaluate* an expression means to replace the variables by given numerical values and then simplify the resulting numerical expression. To *evaluate* a trigonometric ratio means to find its approximate numerical value. To *evaluate* a summation means to find its value.

Even vertex In a network, a vertex with even degree—that is, with an even number of arcs or line segments connected at that vertex.

Event A subset of a sample space.

Exact interest The calculation of interest assuming that there are 365 days in a year.

Exact solution The simplified value of an expression before approximation by calculator.

Exclusive or A translation of *p or q* which includes *p* or *q*, but not both. In this book we translate the *exclusive or* as "either *p or q*."

Expand To simplify by carrying out the given operations.

Expand a summation To write out a summation notation showing the individual terms without a sigma.

Expanded notation A way of writing a number that lists the meaning of each grouping symbol and the number of items in that group. For example, 382.5 written in expanded notation is

$$3 \times 10^2 + 8 \times 10^1 + 2 \times 10^0 + 5 \times 10^{-1}$$

Expectation See *Mathematical expectation*.

Expected value See *Mathematical expectation*.

Experiment An observation of any physical occurrence.

Exponent Where b is any nonzero real number and n is any natural number, *exponent* is defined as follows:

$$b^n = \underbrace{b \cdot b \cdot \cdots \cdot b}_{n \text{ factors}} \qquad b^0 = 1 \qquad b^{-n} = \frac{1}{b^n}$$

b is called the *base*, n is called the *exponent*, and b^n is called a *power* or *exponential*.

Exponential See *Exponent*.

Exponential curve The graph of an exponential equation. It indicates an increasingly steep rise, and passes through the point $(0, 1)$.

Exponential equation An equation of the form $y = b^x$, where b is positive and not equal to 1.

Exponential function A function that can be written as $f(x) = b^x$, where $b > 0, b \neq 1$.

Exponential notation A notation involving exponents.

Exponentiation The process of raising a number to some power. See *Exponent*.

Expression Numbers, variables, functions, and their arguments that can be evaluated to obtain a single result.

Extended order of operations 1. Perform any operations enclosed in parentheses. 2. Perform any operations that involve raising to a power. 3. Perform multiplications and

divisions as they occur by working from left to right.
4. Finally, perform additions and subtractions as they occur by working from left to right.

Exterior angle   An exterior angle of a triangle is the angle on the other side of an extension on one side of the triangle.

Extraneous root   A number obtained in the process of solving an equation that is not a root of the equation to be solved.

Extremes   See *Proportion*.

Factor   (*noun*) Each of the numbers multiplied to form a product is called a factor of the product. (*verb*) To write a given number as a product.

Factor tree   The representation of a composite number showing the steps of successive factoring by writing each new pair of factors under the composite.

Factorial   For a natural number n, the product of all the positive integers less than or equal to n. It is denoted by $n!$ and is defined by

$$n! = n(n-1)(n-2) \cdot \cdots \cdot 4 \cdot 3 \cdot 2 \cdot 1$$

Also, $0! = 1$.

Factoring   The process of determining the factors of a product.

Factorization   The result of factoring a number or an expression.

Fahrenheit   A unit of measurement in the United States system for measuring temperature based on a system where the freezing point of water is $32°$ and the boiling point of water is $212°$.

Fair coin   A coin for which heads and tails are equally likely.

Fair game   A game for which the mathematical expectation is zero.

Fair voting principles   See *Fairness criteria*.

Fairness criteria   Properties that would seem to be desirable in any voting system.

Majority criterion: If a candidate receives a majority of the first-place votes, then that candidate should be declared the winner.

Condorcet criterion: If a candidate is favored when compared one-on-one with every other candidate, then that candidate should be declared the winner.

Monotonicity criterion: A candidate who wins a first election and then gains additional support, without losing any of the original support, should also win a second election.

Irrelevant alternatives criterion: If a candidate is declared the winner of an election, and in a second election one or more of the other candidates is removed, then the previous winner should still be declared the winner.

Fallacy   An invalid form of reasoning.

Fallacy of exceptions   Reasoning or forming a conclusion by looking at one particular case, which may be an exception.

Fallacy of the converse   An invalid form of reasoning that has the form $[(p \rightarrow q) \wedge q]$ and reaches the incorrect conclusion p.

Fallacy of the inverse   An invalid form of reasoning that has the form

$$[(p \rightarrow q) \wedge (\sim p)]$$

and reaches the incorrect conclusion $\sim q$.

False chain pattern   An invalid form of reasoning that has the form

$$[(p \rightarrow q) \wedge (p \rightarrow r)]$$

and reaches the incorrect conclusion $q \rightarrow r$.

Feasible solution   A set of values that satisfies the set of constraints in a linear programming problem.

Fibonacci sequence   The sequence 1, 1, 2, 3, 5, 8, 13, 21, ... The general term is $s_n = s_{n-1} + s_{n-2}$, for any given s_1 and s_2.

Fibonacci-type sequence   A sequence with general term $s_n = s_{n-1} + s_{n-2}$, for any given s_1 and s_2. *The* Fibonacci sequence has first terms 1, 1, . . . , but *a* Fibonacci-type sequence can have any two first terms.

Field   A set with two operations satisfying the closure, commutative, associative, identity, and inverse properties for both operations. A field also satisfies a distributive property combining both operations.

Finance charge   A charge made for the use of someone else's money.

Finite series   A series with n terms, where n is a counting number.

Finite set   See *Set*.

First component   See *Ordered pair*.

First-degree equation   With one variable, an equation of the form $ax + b = 0$; with two variables, an equation of the form $y = mx + b$.

Five-percent offer   An offer made that is 105% of the price paid by the dealer. That is, it is an offer that is 5% over the cost.

Fixed-point form   The usual decimal representation of a number. It is usually used in the context of writing numbers in scientific notation or in floating-point form. See *Floating-point form*.

Floating-point form   A calculator or computer variation of scientific notation in which a number is written as a number between one and ten times a power of ten where the power of ten is understood. For example, 2.678×10^{11} is scientific notation and 2.678E 11 or 2.678 + 11 are floating-point representations. The fixed-point representation is the usual decimal representation of 267,800,000,000.

Floor-plan problem   Given a floor plan of some building you wish to find a path from room to room that will proceed through all of the rooms exactly once.

Floppy disk   Storage medium that is a flexible platter ($3\frac{1}{2}$ or $5\frac{1}{4}$ inches in diameter) of mylar plastic coated with a magnetic material. Data are represented on the disk by electrical impulses.

Foci   Plural *for focus*.

Focus   See *Parabola, Ellipse,* and *Hyperbola*.

FOIL   (1) A method for multiplying binomials that requires First terms, Outer terms + Inner terms, Last terms:

$$(a + b)(c + d) = ac + (ad + bc) + bd$$

(2) A method for factoring a trinomial into the product of two binomials.

Foot   A unit of linear measure in the United States system that is equal to 12 inches.

Foreclose   If the scheduled payments are not made, the lender takes the right to redeem the mortgage and keeps the collateral property.

Formula   A general answer, rule, or principle stated in mathematical notation.

Fractal   A family of shapes involving chance whose irregularities are statistical in nature. They are shapes used, for example, to model coastlines, growth, and boundaries of clouds. Fractals model curves as well as surfaces. The term *fractal set* is also used in place of the word *fractal*.

Fractal geometry   The branch of geometry that studies the properties of fractals.

Fraction   See *Rational number*.

Frequency   See *Classes*.

Frequency distribution   For a collection of data, the tabulation of the number of elements in each class.

Function   A rule that assigns to each element in the domain a single (unique) element.

Function machine   A device used to help us understand the nature of functions. It is the representation of a function as a machine into which some number is input and "processed" through the machine; the machine then outputs a single value.

Functional notation   The representation of a function f using the notation $f(x)$.

Fundamental counting principle   If one task can be performed in m ways and a second task can be performed in n ways, then the number of ways that the tasks can be performed one after the other is mn.

Fundamental operators   In symbolic logic, the fundamental operators are the connectives *and, or,* and *not.*

Fundamental property of equations   If P and Q are algebraic expressions, and k is a real number, then each of the following is equivalent to $P = Q$:

Addition	$P + k = Q + k$
Subtraction	$P - k = Q - k$
Nonzero multiplication	$kP = kQ, k \neq 0$
Nonzero division	$\dfrac{P}{k} = \dfrac{Q}{k}, k \neq 0$

Fundamental property of fractions   If both the numerator and denominator are multiplied by the same nonzero number, the resulting fraction will be the same. That is,

$$\frac{PK}{QK} = \frac{P}{Q} \quad (Q, K \neq 0)$$

Fundamental property of inequalities   If P and Q are algebraic expressions, and k is a real number, then each of the following is equivalent to $P < Q$:

Addition	$P + k < Q + k$
Subtraction	$P - k < Q - k$
Positive multiplication	$kP < kQ, k > 0$
Positive division	$\frac{P}{k} < \frac{Q}{k}, k > 0$
Negative multiplication	$kP > kQ, k < 0$
Negative division	$\frac{P}{k} > \frac{Q}{k}, k < 0$

This property also applies for $\leq$, $>$, and $\geq$.

Fundamental theorem of arithmetic   Every counting number greater than 1 is either a prime or a product of primes, and the prime factorization is unique (except for the order in which the factors appear).

Future value   See *Compound interest formula*.

Future value formula   For simple interest: $A = P(1 + rt)$; for compound interest:

$$A = P(1 + i)^N$$

Fuzzy logic   A relatively new branch of logic used in computer programming that does not use the law of the excluded middle.

Gallon   A measure of capacity in the United States system that is equal to 4 quarts.

Gates   In circuit logic, it is a symbolic representation of a particular circuit.

Gauss-Jordan elimination   A method for solving a system of equations that uses the following steps. *Step 1*: Select as the first pivot the element in the first row, first column, and pivot. *Step 2*: The next pivot is the element in the second row, second column; pivot. *Step 3*: Repeat the process until you arrive at the last row, or until the pivot element is a zero. If it is a zero and you can interchange that row with a row below it, so that the pivot element is no longer a zero, do so and continue. If it is zero and you cannot interchange rows so that it is not a zero, continue with the next row. The final matrix is called the **row-reduced form**.

g.c.f.   An abbreviation for *greatest common factor*.

General form   In relation to second-degree equations (or conic sections), it refers to the form

$$Ax^2 + Bxy + Cy^2 + Dx + Ey + F = 0$$

where $A, B, C, D, E,$ and F are real numbers and (x, y) is any point on the curve.

General term   The nth term of a sequence or series.

Genus   The number of cuts that can be made without cutting a figure into two pieces. The genus is equivalent to the number of holes in the object.

Geometric mean   The geometric mean of the numbers a and b is $\sqrt{ab}$.

Geometric sequence   A sequence for which the ratio of each term to the preceding term is a constant, written $g_1, g_2, g_3, \ldots$. The nth term of a geometric sequence is $g_n = g_1 r^{n-1}$, where g_1 is the first term and r is the *common ratio*. It is also called a *geometric progression*.

Geometric series   The indicated sum of the terms of a geometric sequence. The sum of n terms is denoted by G_n and

$$G_n = \frac{g_1(1 - r^n)}{1 - r}, r \neq 1$$

If $|r| < 1$, then $G = \frac{g_1}{1 - r}$, where G is the sum of the infinite geometric series. If $|r| \geq 1$, the infinite geometric series has no sum.

Geometry   The branch of mathematics that treats the shape and size of things. Technically, it is the study of invariant properties of given elements under specified groups of transformations.

GIGO   <u>G</u>arbage <u>I</u>n, <u>G</u>arbage <u>O</u>ut, an old axiom regarding the use of computers.

Golden ratio   The division of a line segment $\overline{AB}$ by an interior point P so that

$$\frac{|\overline{AB}|}{|\overline{AP}|} = \frac{|\overline{AP}|}{|\overline{PB}|}$$

It follows that this ratio is a root of the equation $x^2 - x - 1 = 0$, or $x = \frac{1}{2}(1 + \sqrt{5})$. This ratio is called the golden ratio and is considered pleasing to the eye.

Golden rectangle   A rectangle R with the property that it can be divided into a square and a rectangle similar to R; a rectangle whose sides form a golden ratio.

Googol   The number with 1 followed by 100 zeros—that is, 10,000,000,000,000,000,000,000,000,000,000,000,000, 000,000,000,000,000,000,000,000,000,000,000,000,000, 000,000,000,000,000,000.

Grace period   A period of time between when an item is purchased and when it is paid during which no interest is charged.

Gram   A unit of weight in the metric system. It is equal to the weight of one cubic centimeter of water at 4°C.

Grant's tomb properties   Two fundamental properties of logarithms:

1. $\log_b b^x = x$
2. $b^{\log_b x} = x, x > 0$

Graph   (1) In statistics, it is a drawing that shows the relation between certain sets of numbers. Common forms are bar graphs, line graphs, pictographs, and pie charts (circle graphs). (2) A drawing that shows the relation between certain sets of numbers. It may be one-dimensional ($\mathbb{R}$), two-dimensional ($\mathbb{R}^2$), or three-dimensional ($\mathbb{R}^3$). (3) A set of vertices connected by arcs or line segments.

Graph of an equation   See *Equation of a graph.*

Graphing method   A method of solving a system of equations that finds the solution by looking at the intersection of the individual graphs. It is an approximate method of solving a system of equations, and depends on the accuracy of the graph that is drawn.

Great circle   A circle on a sphere that has its diameter equal to that of the sphere.

Greater than   If a lies to the right of b on a number line, then a is greater than b, $a > b$. Formally, $a > b$ if and only if $a - b$ is positive.

Greater than or equal to   Written $a \geq b$ means $a > b$ or $a = b$.

Greatest common factor   The largest divisor common to a given set of numbers.

Group   A set with one defined operation that satisfies the closure, associative, identity, and inverse properties.

Grouped frequency distribution   If the data are grouped before they are tallied, then the resulting distribution is called a *grouped frequency distribution.*

Grouping symbols   Parentheses ( ), brackets [ ], and braces { } indicate the order of operations and are also sometimes used to indicate multiplication, as in $(2)(3) = 6$. Also called *symbols of inclusion.*

Growth formula   Refers to exponential growth. It is described by the equation

$$A = A_0 e^{rt}$$

where r is the annual growth rate (and consequently is positive), t is the time (in years), A_0 is the amount present initially (present value), and A is the future value. If r is positive, this formula models growth, and if r is negative, the formula models decay.

Half-life   The time that it takes for a particular radioactive substance to decay to half of its original amount.

Half-line   A ray, with or without its endpoint. The half-line is said to be closed if it includes the endpoint, and open if it does not include the endpoint.

Half-plane   The part of a plane that lies on one side of a line in the plane. It is a *closed* half-plane if the line is included. It is an *open* half-plane if the line is not included. The line is the *boundary* of the half-plane in either case.

Hamiltonian cycle   A path that begins at some vertex and then visits each vertex exactly once, ending up at the original vertex.

Hamilton's apportionment plan   An apportionment plan in which the representation of a geographical area is determined by finding the quotient of the number of people in that area divided by the total number of people and then the result is rounded as follows: Allocate the remainder, one at a time, to districts on the basis of the decreasing order of the decimal portion of the quotients. This method is sometimes called the *method of the largest fractions.*

Hare method   Each voter votes for one candidate. If a candidate receives a majority of the votes, that candidate is declared to be a *first-round winner.* If no candidate receives a majority of the votes, then with the *Hare method,* also known as the *plurality with an elimination runoff method,* the candidate(s) with the fewest number of first-place votes is (are) eliminated. Each voter votes for one candidate in the second-round. If a candidate receives a majority, that candidate is declared to be a *second-round winner.* If no candidate receives a majority of the second-round votes, then eliminate the candidate(s) with the fewest number of votes is (are) eliminated. Repeat this process until a candidate receives a majority.

Hecto-   A prefix meaning 100.

Heptagon   A polygon having seven sides.

Hexagon   A polygon having six sides.

HH method   See *Huntington-Hill's plan.*

Higher-level language   A computer programming language (e.g., BASIC, PASCAL, LOGO) that approaches the syntax of English and is easier both to use and to learn than machine language. It is also not system-dependent.

Hindu-Arabic numerals   Same as the usual decimal numeration system that is in everyday use.

Horizontal ellipse   An ellipse whose major axis is horizontal.

Horizontal hyperbola   A hyperbola whose transverse axis is horizontal.

Horizontal line   A line with zero slope. Its equation has the form $y = $ constant.

Hundred   Ten 10s.

Huntington-Hill's plan   An apportionment plan currently in use by the U.S. legislature. It is based on the geometric mean of two numbers, a and b, where a is the value of the exact ratio rounded down and b is the value of the exact ratio rounded up. It rounds down if the exact quota is less than the geometric mean and rounds up if it is greater than the geometric mean.

Hyperbola   The set of all points in a plane such that, for each point on the hyperbola, the difference of its distances from two fixed points (called the *foci*) is a constant.

Hyperbolic geometry   A non-Euclidean geometry in which a Saccheri quadrilateral is constructed with summit angles acute.

Hypotenuse   The longest side in a right triangle.

Hypothesis   An assumed proposition used as a premise in proving something else.

Identity   (1) A statement of equality that is true for all values of the variable. It also refers to a number I so that for some operation ∘, $I \circ a = a \circ I = a$ for every number a in a given set. (2) An open equation that is true for all replacements of the variable.

Identity matrix   A matrix satisfying the identity property. It is a square matrix consisting of ones along the main diagonal and zeros elsewhere.

If-then   In symbolic logic, it is a connective also called *implication.*

Implication   A statement that follows from other statements. It is also a proposition formed from two given propositions by connecting them with an "if . . . , then . . . " form. It is symbolized by $p \rightarrow q$.

Impossible event   An event for which the probability is zero—that is, an event that cannot happen.

Improper fraction   A fraction for which the numerator is greater than the denominator.

Improper subset   See *Subset*.

Inch   A linear measurement in the United States system equal in length to the following segment: _____

Inclusive or   The same as *Disjunction*. The compound statement "*p* or *q*" is called the *inclusive or*.

Inconsistent system   A system for which no replacements of the variable make the equations true simultaneously.

Indefinite integral   An antiderivative.

Independence of irrelevant alternatives   If a voter prefers A to B with C as a possible choice, we assume that the voter still prefers A to B when C is not a possible choice.

Independent events   Events *E* and *F* are *independent* if the occurrence of one in no way affects the occurrence of the other.

Independent system   A system of equations such that no one of them is necessarily satisfied by a set of values of the variables that satisfy all the others.

Independent variable   The variable associated with the first component of an ordered pair.

Indirect reasoning   One of the principal forms of logical reasoning. It is an argument of the form

$$[(p \rightarrow q) \wedge \sim q] \rightarrow \sim p$$

Inductive reasoning   A type of reasoning accomplished by first observing patterns and then predicting answers for more complicated similar problems.

Inequality   A statement of order. If always true, an inequality is called an *absolute inequality*; if always false, an inequality is called a *contradiction*. If sometimes true and sometimes false, it is called a *conditional inequality*. Values that make the statement true are said to satisfy the inequality. A *string of inequalities* may be used to show the order of three or more quantities.

Inequality symbols   The symbols $>$, $\geq$, $<$, and $\leq$. Also called *order symbols*.

Inferential statistics   Statistics that is concerned with making generalizations or predictions about a population based on a sample from that population.

Infinite series   The indicated sum of an infinite sequence.

Infinite set   See *Set*.

Infinity symbol ∞

Inflation   An increase in the amount of money in circulation, resulting in a fall in its value and a rise in prices. In this book, we assume annual compounding with the future value formula; that is, use $A = P(1 + r)^n$, where r is the projected annual inflation rate, n is the number of years, and P is the present value.

Information retrieval   The locating and displaying of specific material from a description of its content.

Input   A method of putting information into a computer. *Input* includes downloading a program, typing on a keyboard, pressing on a pressure-sensitive screen.

Input device   Component of a system that allows the entry of data or a program into a computer's memory.

Installment loan   A financial problem in which an item is paid for over a period of time. It is calculated using add-on interest or compound interest.

Installments   Part of a debt paid at regular intervals over a period of time.

Instant Insanity   A puzzle game consisting of four blocks with different colors on the faces. The object of the game is to arrange the four blocks in a row so that no color is repeated on one side as the four blocks are rotated through 360°.

Instantaneous rate of change   The instantaneous rate of change of a function *f* from *x* to $x + h$ is

$$\lim_{h \to 0} \frac{f(x + h) - f(x)}{h}$$

Integers $\mathbb{Z} = \{\ldots, -3, -2, -1, 0, 1, 2, 3, \ldots\}$, composed of the natural numbers, their opposites, and 0.

Integral   A fundamental concept of calculus, which involves finding the area bounded by a curve, the *x*-axis, and two vertical lines. Let *f* be a function defined over the interval $[a, b]$. Then the definite integral of *f* over the interval is denoted by

$$\int_a^b f(x) \, dx$$

and is the net change of an antiderivative of *f* over that interval. The numbers *a* and *b* are called the *limits of integration*.

Integral calculus   That branch of calculus that involves applications of the integral.

Integral of a sum   The following property of integrals:

$$\int [f(x) + g(x)] \, dx = \int f(x) \, dx + \int g(x) \, dx$$

Integrand   In an integral, it is the function to be integrated.

Integrated circuit   The plastic or ceramic body that contains a chip and the leads connecting it to other components.

Interactive   Software that allows continuous two-way communication between the user and the program.

Intercept form of the equation of a line   The form

$$\frac{x}{a} + \frac{y}{b} = 1$$

of a linear equation where the *x*-intercept is *a* and the *y*-intercept is *b*.

Intercepts   The point or points where a line or a curve crosses a coordinate axis. The *x*-intercepts are sometimes called the *zeros* of the equation.

Interest   An amount of money paid for the use of another's money. See *Compound interest*.

Interest-only loan   A loan in which periodic payments are for interest only so that the principal amount of the loan remains the same.

Interest rate   The percentage rate paid on financial problems. In this book it is denoted by *r* and is assumed to be an annual rate unless otherwise stated.

Interface   The electronics necessary for a computer to communicate with a peripheral.

Internet   A network of computers from all over the world that are connected together.

Intersection   The *intersection* of sets A and B, denoted by $A \cap B$ is the set consisting of elements in *both* A and B.

Interval   See *Classes*.

Invalid argument   An argument that is not valid.

Inverse   (1) In symbolic logic, for an implication $p \rightarrow q$, the inverse is the statement $\sim p \rightarrow \sim q$. (2) For addition, see *Opposites*.

For multiplication, see *Reciprocal*. (3) For matrices, if [A] is a square matrix, and if there exists a matrix $[A]^{-1}$ such that

$$[A]^{-1}[A] = [A][A]^{-1} = [I]$$

where [I] is the identity matrix for multiplication, then $[A]^{-1}$ is called the inverse of [A] for multiplication.

Inverse cosine   See *Inverse trigonometric ratios*.

Inverse property   For each $a \in \mathbb{R}$, there is a unique number $(-a) \in \mathbb{R}$, called the *opposite* (or *additive inverse*) of a, so that

$$a + (-a) = -a + a = 0$$

Inverse sine   See *Inverse trigonometric ratios*.

Inverse tangent   See *Inverse trigonometric ratios*.

Inverse trigonometric ratios   The inverse sine, inverse cosine, and inverse tangent are the *inverse trigonometric ratios*. For θ an acute angle in a right triangle,

$$\sin^{-1}\left(\frac{\text{OPP}}{\text{HYP}}\right) = \theta; \quad \cos^{-1}\left(\frac{\text{ADJ}}{\text{HYP}}\right) = \theta,$$

$$\tan^{-1}\left(\frac{\text{OPP}}{\text{ADJ}}\right) = \theta$$

Invert   In relation to the fraction a/b, it means to interchange the numerator and the denominator to obtain the fraction b/a.

Irrational number   A number that can be expressed as a nonrepeating, nonterminating decimal; the set of irrational numbers is denoted by $\mathbb{Q}'$.

Irrelevant alternatives criterion   If a candidate is declared the winner of an election, and in a second election one or more of the other candidates is removed, then the previous winner should still be declared the winner.

Isosceles triangle   A triangle with two sides the same length.

Isosceles triangle property   If two sides of a triangle have the same length, then angles opposite them are equal.

Jefferson's apportionment plan   An apportionment plan in which the representation of a geographical area is determined by finding the quotient of the number of people in that area divided by the total number of people and then rounding the result as follows: Any quotient with a decimal portion must be rounded down to the previous whole number.

Jordan curve   Also called a *simple closed curve*. For example, a curve such as a circle or an ellipse or a rectangle that is closed and does not intersect itself.

Juxtaposition   When two variables, a number and a variable, or a symbol and a parenthesis, are written next to each other with no operation symbol, as in *xy, 2x,* or $3(x + y)$. Juxtaposition is used to indicate multiplication.

K   The symbol represents 1,024 (or 2^{10}). For example, 48K bytes of memory is the same as $48 \times 1,024$ or 49,152 bytes. It is sometimes used as an approximation for 1,000.

Keno   A lottery game that consists of a player trying to guess in advance which numbers will be selected from a pot containing n numbers. A certain number, say m, where $m < n$, of selections is randomly made from the pot of n numbers. The player then gets paid according to how many of the m numbers were selected.

Keyboard   Typewriter-like device that allows the user to input data and commands into a computer.

Kilo-   A prefix that means 1,000.

Kilogram   1,000 grams.

Kiloliter   1,000 liters.

Kilometer   1,000 meters.

Kruskal's algorithm   To construct a minimum spanning tree from a weighted graph: 1. Select any edge with minimum weight. 2. Select the next edge with minimum weight among those not yet selected. 3. Continue to choose edges of minimum weight from those not yet selected, but make sure not to select any edge that forms a circuit. 4. Repeat this process until the tree connect all of the vertices of the original graph.

Laptop   A small portable computer.

Law of contraposition   A conditional may always be replaced by its contrapositive without having its truth value affected.

Law of detachment   Same as *direct reasoning*.

Law of double negation $\sim(\sim p) \Leftrightarrow p$

Law of the excluded middle   Every simple statement is either true or false.

Laws of exponents   There are five laws of exponents.

Addition law	$b^m \cdot b^n = b^{m+n}$
Multiplication law	$(b^n)^m = b^{mn}$
Subtraction law	$\dfrac{b^m}{b^n} = b^{m-n}$
Distributive laws	$(ab)^m = a^m b^m$
	$\left(\dfrac{a}{b}\right)^m = \dfrac{a^m}{b^m}$

Laws of logarithms   If A, B, and b are positive numbers, p any real number, and $b \neq 1$, then

Addition law	$\log_b (AB) = \log_b A + \log_b B$
Subtraction law	$\log_b \frac{A}{B} = \log_b A - \log_b B$
Multiplication law	$\log_b A^p = p \log_b A$

Laws of square roots   There are 4 laws of square roots.

(1) $\sqrt{0} = 0$ (2) $\sqrt{a^2} = a$

(3) $\sqrt{ab} = \sqrt{a}\sqrt{b}$ (4) $\sqrt{\frac{a}{b}} = \frac{\sqrt{a}}{\sqrt{b}}$

l.c.d.   An abbreviation for least common denominator.

l.c.m.   An abbreviation for least common multiple.

Least common denominator (l.c.d.)   The smallest number that is exactly divisible by each of the given numbers.

Least common multiple (l.c.m.)   The smallest number that each of a given set of numbers divides into.

Least squares line   A line $y = mx + b$ so that the sum of the squares of the vertical distances of the data points from this line will be as small as possible.

Least squares method   A method based on the principle that the best prediction of a quantity that can be deduced from a set of measurements or observations is that for which the sum of the squares of the deviations of the observed values (from predictions) is a minimum.

Leg of a triangle   One of the two sides of a right triangle that are not the hypotenuse.

Length   A measurement of an object from end to end.

Less than   If a is to the left of b on a number line, then a is less than b, $a < b$. Formally, $a < b$ if and only if $b > a$.

Less than or equal to   Written $a \leq b$, means $a < b$ or $a = b$.

Like terms   Terms that differ only in their numerical coefficients. Also called *similar terms*.

Limit   The formal definition of a limit is beyond the scope of this course. Intuitively, it is the tendency of a function to approach some value as its variable approaches a given value.

Limit of a sequence   The formal definition of a limit of a sequence is beyond the scope of this course. Intuitively, it is an accumulation point such that there are an infinite number of terms of the sequence arbitrarily close to the accumulation point.

Limits of integration   See *Integral*.

Line   In mathematics, it is an undefined term. It is a curve that is straight, so it is sometimes referred to as a *straight line*. It extends in both directions and is considered one-dimensional, so it has no thickness.

Line graph   See *Graph*.

Line of credit   A preapproved credit limit on a credit account. The maximum amount of credit to be extended to a borrower. That is, it is a promise by a lender to extend credit up to some predetermined amount.

Line of symmetry   A line with the property that for a given curve, any point P on the curve has a corresponding point Q (called the reflection point of P) so that the perpendicular bisector of $\overline{PQ}$ is on the line of symmetry.

Line segment   A part of a line between two points on the line.

Linear   (1) A first-degree polynomial. (2) Pertaining to a line. In two variables, a set of points satisfying the equation $Ax + By + C = 0$.

Linear combination method   See *Addition method*.

Linear correlation coefficient   A measure to determine whether there is a statistically significant linear relationship between two variables.

Linear equation   An equation of the form

$ax + b = 0$ (one variable) or
$Ax + By + C = 0$ (two variables)

A first-degree equation with one or two variables. For example, $x + 5 = 0$ and $x + y + 5 = 0$ are linear. An equation is linear in a certain variable if it is first-degree in that variable. For example, $x + y^2 = 0$ is linear in x, but not y.

Linear function   A function whose equation can be written in the form $f(x) = mx + b$.

Linear inequality   A first-degree inequality with one or two variables.

Linear polynomial   A first-degree polynomial.

Linear programming   A type of problem that seeks to maximize or minimize a function called the *objective function* subject to a set of *restrictions* (linear inequalities) called *constraints*.

Linear programming theorem   A linear expression in two variables, $c_1 x + c_2 y$, defined over a convex set S whose sides are line segments, takes on its maximum value at a corner point of S and its minimum value at a corner point of S. If S is unbounded, there may or may not be an optimum value, but if there is, then it must occur at a corner point.

Linear system   A system of equations, each of which is first degree.

Liter   The basic unit of capacity in the metric system. It is the capacity of 1 cubic decimeter.

Literal equation   An equation with more than one variable.

Logarithm   For $A > 0, b > 0, b \neq 1$ $x = \log_b A$ means $b^x = A$ x is the called the logarithm and A is called the argument.

Logarithmic equation   An equation for which there is a logarithm on one or both sides.

Logarithmic function

$$f(x) = \log_b x, \quad b > 0, \quad x > 0$$

Logarithmic scale   A scale in which logarithms are used to make data more manageable by expanding small variations and compressing large ones.

Logic   The science of correct reasoning.

Logical conclusion   The statement that follows logically as a consequence of the hypotheses of a theorem.

Logical equivalence   Two statements are *logically equivalent* if they have the same truth values.

Logical fallacy   An invalid form of reasoning.

Log of both sides theorem   If A, B, and b are positive real numbers with $b \neq 1$, then $\log_b A = \log_b B$ is equivalent to $A = B$

Lower quota   In apportionment, the result of a quota found by rounding down.

Lowest common denominator   For two or more fractions, the smallest common multiple of the denominators. It is the same as the *lowest common multiple*.

Lump-sum problem   A financial problem that deals with a single sum of money, called a *lump sum*. Contrast with *a periodic payment* problem.

Main diagonal   The entries $a_{11}, a_{22}, a_{33}, \ldots$ in a matrix.

Major axis   In an ellipse, the line passing though the foci.

Majority   Voting to find an alternative that receives more than 50% of the vote.

Majority criterion   If a candidate receives a majority of the first-place votes, then that candidate should be declared the winner.

Marriage   A pairing of one couple where each partner is taken from a separate group. A marriage is *stable* if each partner is satisfied with the pairing and *unstable* if one or the other (or both) would prefer to be paired with another.

Mass   In this course, it is the amount of matter an object comprises. Formally, it is a measure of the tendency of a body to oppose changes in its velocity.

Mathematical expectation   A calculation defined as the product of an amount to be won and the probability that it is won. If there is more than one amount to be won, it is the sum of the expectations of all the prizes. It is also called the *expected value* or *expectation*.

Mathematical modeling   An iterative procedure that makes assumptions about real-world problems to formulate the problem in mathematical terms. After the mathematical problem is solved, it is tested for accuracy in the real world, and revised for the next step in the iterative process.

Mathematical system   A set with at least one defined operation and some developed properties.

Matrix   A rectangular array of terms called *elements*.

Matrix equation   An equation whose elements are matrices.

Maximum loan   In this book, it refers to the maximum amount of loan that can be obtained for a home with a given amount of income and a given amount of debt. To find this amount, use the present value of an annuity formula.

Mean   The number found by adding the data and dividing by the number of values in the data set. The sample mean is usually denoted by $\bar{x}$.

Means   See *Proportion*.

Measure   Comparison to some unit recognized as standard.

Measures of central tendency   Refers to the averages of mean, median, and mode.

Measures of dispersion   Refers to the measures of range, standard deviation, and variance.

Measures of position   Measures that divide a data set by position, which include median, quartiles, deciles, and percentiles.

Median   The middle number when the numbers in the data are arranged in order of size. If there are two middle numbers (in the case of an even number of data values), the median is the mean of these two middle numbers.

Member   See *Set*.

Meter   The basic unit for measuring length in the metric system.

Metric system   A decimal system of weights and measures in which the gram, the meter, and the liter are the basic units of mass, length, and capacity, respectively. One gram is the mass of one cm^3 of water and one liter is the same as 1,000 cm^3. In this book, the metric system refers to SI metric system as revised in 1960.

Micrometer   One millionth of a meter, denoted by μm.

Mile   A unit of linear measurement in the United States system that is equal to 5,280 ft.

Milli-   A prefix that means 1/1,000.

Milligram   1/1,000 of a gram.

Milliliter   1/1,000 of a liter.

Millimeter   1/1,000 of a meter.

Million   A name for $10^6 = 1,000,000$.

Minicomputer   An everyday name for a personal computer.

Minimum spanning tree   A spanning tree for which the sum of the numbers with the edges is a minimum.

Minor axis   In an ellipse, the axis perpendicular to the major axis passing though the center of the ellipse.

Minus   Refers to the operation of subtraction. The symbol "$-$" means minus only when it appears between two numbers, two variables, or between numbers and variables.

Mixed number   A number that has both a counting number part and a proper fraction part; for example, $3\frac{1}{2}$.

Mode   The value in a data set that occurs most frequently. If no number occurs more than once, there is no mode. It is possible to have more than one mode.

Modem   A device connected to a computer that allows the computer to communicate with other computers using electric cables, phone lines, or wireless.

Modified quotient   Adjust the standard divisor so that the desired number of seats are used. This adjusted number is known as the modified quotient.

Modular codes   A code based on modular arithmetic.

Modulo 5   A mathematical system consisting of five elements having the property that every number is equivalent to one of these five elements if they have the same remainder when divided by 5.

Modulo n   A mathematical system consisting of n elements having the property that every number is equivalent to one of these n elements if they have the same remainder when divided by n.

Modus ponens   Same as *direct reasoning*.

Modus tollens   Same as *indirect reasoning*.

Monitor   An output device for communicating with a computer. It is similar to a television screen.

Monomial   A polynomial with one and only one term.

Monotonicity criterion   A candidate who wins a first election and then gains additional support, without losing any of the original support, should also win a second election.

Monthly compounding   In the compound interest formula, it is when $n = 12$.

Monthly payment   In an installment application, it is a periodic payment that is made once every month.

Mortgage   An agreement, or loan contract, in which a borrower pledges a home or other real estate as security.

Mouse   A small plastic "box" usually with two buttons on top and a ball on the bottom so that it can be rolled around on a pad. It is attached to the computer by a long cord and is used to take over some of the keyboard functions.

Multiple   See *Divisibility*.

Multiplication   For $b \neq 0$, $a \times b$ means

$$\underbrace{b + b + b + \cdots + b}_{a \text{ addends}}$$

If $a = 0$, then $0 \times b = 0$.

Multiplication law of equality   If $a = b$, then $ac = bc$. Also called the *multiplication property of equality* or a *fundamental property of equations*.

Multiplication law of exponents   To raise a power to a power, multiply the exponents. That is, $(b^n)^m = b^{mn}$.

Multiplication law of inequality

1. If $a > b$ and $c > 0$, then $ac > bc$.
2. If $a > b$ and $c < 0$, then $ac < bc$.

Multiplication law of logarithms   The log of the pth power of a number is p times the log of that number. In symbols,

$$\log_b A^p = p \log_b A$$

Multiplication of integers   If the integers to be multiplied both have the same sign, the result is positive and the magnitude of the answer is the product of the absolute values of the integers. If the integers to be multiplied have opposite signs, the product is negative and has magnitude equal to the product of the absolute values of the given integers. Finally, if one or both of the given integers is 0, the product is 0.

Multiplication of matrices   Let [M] be an $m \times r$ matrix and [N] an $r \times n$ matrix. The product matrix [M][N] = [P] is an $m \times n$ matrix. The entry in the ith row and jth column of [M][N] is the sum of the products formed by multiplying each entry of the ith row of [M] by the corresponding element in the jth column of [N].

Multiplication of rational numbers

$$\frac{a}{b} \times \frac{c}{d} = \frac{ac}{bd}$$

Multiplication principle   In a numeration system, multiplication of the value of a symbol by some number. Also, see *Fundamental counting principle*.

Multiplication property (of equations)   Both sides of an equation may be multiplied or divided by any nonzero number to obtain an equivalent equation.

Multiplication property of factorials

$$n! = n(n - 1)!$$

Multiplication property of inequality   Both sides of an inequality may be multiplied or divided by a positive number, and the order of the inequality will remain unchanged. The order is reversed if both sides are multiplied or divided by a negative number. That is, if $a < b$ then $ac < bc$ if $c > 0$ and $ac > bc$ if $c < 0$. This also applies to $\leq$, $>$, and $\geq$.

Multiplication property of probability   If events E and F are independent events, then we can find the probability of an intersection as follows:

$$P(E \cap F) = P(E \text{ and } F) = P(E) \cdot P(F)$$

Multiplicative identity   The number 1, with the property that $1 \cdot a = a$ for any real number a

Multiplicative inverse   (1) See *Reciprocal.* (2) If [A] is a square matrix and if there exists a matrix $[A]^{-1}$ such that

$$[A]^{-1}[A] = [A][A]^{-1} = [I]$$

where [I] is the identity matrix for multiplication, then $[A]^{-1}$ is called the inverse of [A] for multiplication.

Multiplicity   If a root for an equation appears more than once, it is called a *root of multiplicity.* For example,

$$(x - 1)(x - 1)(x - 1)(x - 2)(x - 2)(x - 3) = 0$$

has roots 1, 2, and 3. The root 1 has multiplicity three and root 2 has multiplicity two.

Mutually exclusive   Events are *mutually exclusive* if their intersection is empty.

Natural base   The natural base is e; it refers to an exponential with a base e.

Natural logarithm   A logarithm to the base e, written ln N.

Natural numbers $\mathbb{N} = \{1, 2, 3, 4, 5, \ldots\}$, the positive integers, also called the *counting numbers.*

Negation   A logical connective that changes the truth value of a given statement. The negation of p is symbolized by $\sim p$.

Negative of a conditional   The negative of a conditional, $p \rightarrow q$ is found by

$$\sim(p \rightarrow q) \Leftrightarrow p \wedge \sim q$$

Negative number   A number less than zero.

Negative sign   The symbol "—" when used in front of a number, as in -5. Do not confuse with the same symbol used for subtraction.

Neither . . . nor   A logical operator for "neither p nor q," which is defined to mean $\sim(p \vee q)$

Network   (1) A linking together of computers. (2) A set of points connected by arcs or by line segments.

New states paradox   When a reapportionment of an increased number of seats causes a shift in the apportionment of the existing states, it is known as the *new states paradox.*

n-gon   A polygon with n sides.

No p is q   A logical operator for "no p is q," which is defined to mean $p \rightarrow \sim q$.

Nonagon   A polygon with 9 sides.

Nonconformable matrices   Matrices that cannot be added or multiplied because their dimensions are not compatible.

Non-Euclidean geometries   A geometry that results when Euclid's fifth postulate is not accepted.

Nonrepeating decimal   A decimal representation of a number that does not repeat.

Nonsingular matrix   A matrix that has an inverse.

Nonterminating decimal   A decimal representation of a number that does not terminate.

Normal curve   A graphical representation of a normal distribution. Its high point occurs at the mean, it is symmetric with respect to this mean, and each side of the mean has an area that includes 34.1% of the population within on standard deviation, 13.6% from one to two standard deviations, and

about 2.3% of the population more than two standard deviations from the mean.

Not   A common translation for the connective of negation.

NOT-gate   A logical gate that changes the truth value of a given statement.

Null set   See *Set.*

Number   A *number* represents a given quantity, as opposed to a *numeral,* which is the symbol for the number. In mathematics, it generally refers to a specific set of numbers—for example, counting numbers, whole numbers, integers, rationals, or real numbers. If the set is not specified, the assumed usage is to the set of real numbers.

Number line   A line used to display a set of numbers graphically (the axis for a one-dimensional graph).

Numeral   Symbol used to denote a number.

Numeration system   A system of symbols with rules of combination for representing all numbers.

Numerator   See *Rational number.*

Numerical coefficient   See *Coefficient.*

Objective function   The function to be maximized or minimized in a linear programming problem.

Obtuse angle   An angle that is greater than a right angle and smaller than a straight angle.

Obtuse triangle   A triangle with one obtuse angle.

Octagon   A polygon with eight sides.

Octal numeration system   A numeration system with eight symbols.

Odd vertex   In a network, a vertex of odd degree—that is, with an odd number of arcs or line segments connected at that vertex.

Odds   If $s + f = n$, where s is the number of outcomes considered favorable to an event E and n is the total number of possibilities, then the *odds in favor* of E is s/f and the *odds against E* is f/s.

One   The first counting number; it is also called the *identity element for multiplication;* that is, it satisfies the property that

$$x \cdot 1 = 1 \cdot x = x$$

for all numbers x.

One-dimensional coordinate system   A real number line.

One-to-one correspondence   Between two sets A and B, this means each element of A can be matched with exactly one element of B and also each element of B can be matched with exactly one element of A.

Online   To be connected to a computer network or to the Internet.

Open-ended loan   A preapproved line of credit that the borrower can access as long as timely payments are made and the credit line is not exceeded. It is usually known as a credit card loan.

Open equation   An equation that has at least one variable.

Open half-plane   See *Half-plane.*

Operator   A rule, such as negation, that modifies the value of a simple statement, or a rule that combines two simple statements, such as conjunction or disjunction.

Opposite side   In a right triangle, an acute angle is made up of two sides. The opposite side of the angle refers to the third side that is not used to make up the sides of the angle.

Opposites   Opposites x and $-x$ are the same distance from 0 on the number line but in opposite directions; $-x$ is also called the *additive inverse* of x. Do not confuse the negative symbol "$-$"

meaning opposite with the same symbol as used to mean subtraction or negative.

Optimum solution The maximum or minimum value in a linear programming problem.

Or A common translation for the connective of disjunction.

OR-gate An electrical circuit that simulates disjunction. That is, the circuit is on when either of two switches is on.

Order Refers to the direction that an inequality symbol points. In reference to matrices, it refers to the number of rows and columns in a matrix. When used in relation to a matrix, it is the same as the *dimension* of the matrix.

Order of an inequality Refers to $>, \geq, <,$ or $\leq$ relationship.

Order of operations If no grouping symbols are used in a numerical expression, first perform all multiplications and divisions from left to right, and then perform all additions and subtractions from left to right.

Order symbols Refers to $>, \geq, <, \leq$ in an inequality. Also called inequality symbols.

Ordered pair A pair of numbers, written (x, y), in which the order of naming is important. The numbers x and y are sometimes called the *first* and *second components* of the pair and are called the *coordinates* of the point designated by (x, y).

Ordered triple Three numbers, written (x, y, z), in which the order of the components is important.

Ordinary annuity See *Annuity*.

Ordinary interest The calculation of interest assuming a year has 360 days. In this book, we assume ordinary interest unless otherwise stated.

Ordinate The vertical coordinate in a two-dimensional system of rectangular coordinates, usually denoted by y.

Origin The point designating 0 on a number line. In two dimensions, the point of intersection of the coordinate axes; the coordinates are $(0, 0)$.

Origination fee A fee paid to obtain a real estate loan.

Ounce (1) A unit of capacity in the United States system that is equal to 1/128 of a gallon. (2) A unit of mass in the United States system that is equal to 1/16 of a pound.

Output A method of getting information out of a computer.

Output device Component of a system that allows the output of data. The most common output device is a printer.

Overlapping sets Sets whose intersection is not empty.

Pairwise comparison method In the *pairwise comparison method* of voting, the voters rank the candidates. The method consists of a series of comparisons in which each candidate is compared to each of the other candidates. If choice A is preferred to choice B, then A receives 1 point. If B is preferred to A, then B receives 1 point. If the candidates tie, each receives $\frac{1}{2}$ point. The candidate with the most points is the winner.

Parabola A set of points in the plane equidistant from a given point (called the *focus*) and a given line (called the *directrix*). It is the path of a projectile. The *axis of symmetry* is the axis of the parabola. The point where the axis cuts the parabola is the *vertex*.

Parallel circuit Two switches connected together so that if either of the two switches is turned on, the circuit is on.

Parallel lines Two nonintersecting straight lines in the same plane.

Parallelepiped A polyhedron, all of whose faces are parallelograms.

Parallelogram A quadrilateral with its opposite sides parallel.

Parentheses See *Grouping symbols*.

Pareto principle If each voter prefers A over B, then the group chooses A over B.

Partial sum If $s_1, s_2, s_3, \ldots$ is a sequence, then the partial sums are

$$S_1 = s_1, S_2 = s_1 + s_2, S_3 = s_1 + s_2 + s_3, \ldots.$$

Pascal's triangle A triangular array of numbers that is bordered by ones and the sum of two adjacent numbers in one row is equal to the number in the next row between the two numbers.

$$
\begin{array}{ccccccccccc}
 & & & & & 1 & & & & & \\
 & & & & 1 & & 1 & & & & \\
 & & & 1 & & 2 & & 1 & & & \\
 & & 1 & & 3 & & 3 & & 1 & & \\
 & 1 & & 4 & & 6 & & 4 & & 1 & \\
1 & & 5 & & 10 & & 10 & & 5 & & 1 \\
\end{array}
$$
$$\vdots$$

Password A word or set of symbols that allows access to a computer account.

Pattern recognition A computer function that entails the automatic identification and classification of shapes, forms, or relationships.

Pearson correlation coefficient A number between -1 and $+1$ that indicates the degree of linear relationship between two sets of numbers. In the text, we call this the *linear correlation coefficient*.

Pentagon A polygon with five sides.

Percent The ratio of a given number to 100; hundredths; denoted by %; that is, 5% means 5/100.

Percent markdown The percent of an original price used to find the amount of discount.

Percent problem A is $P\%$ of W is formulated as a proportion

$$\frac{P}{100} = \frac{A}{W}$$

Percentage The given amount in a percent problem.

Percentile Ninety-nine values that divide a data set into one hundred equal parts.

Perfect number An integer that is equal to the sum of all of its factors except the number itself. For example, 28 is a perfect number since

$$28 = 1 + 2 + 4 + 7 + 14$$

Perfect square $1^2 = 1, 2^2 = 4, 3^2 = 9, \ldots$, so the perfect squares are $1, 4, 9, 16, 25, 36, 49, \ldots$.

Perimeter The distance around a polygon.

Periodic payment problem A financial problem that involves monthly or other periodic payments.

Peripheral A device, such as a printer, that is connected to and operated by a computer.

Permutation A selection of objects from a given set with regard to the order in which they are selected. Sometimes it refers to the number of ways this selection can be done and is denoted by $_nP_r$. The formula for finding it is

$$_nP_r = \frac{n!}{(n-r)!}$$

Perpendicular lines Two lines are perpendicular if they meet at right angles.

Personal computer   A computer kept for and used by an individual.

Pi (π)   A number that is defined as the ratio of the circumference to the diameter of a circle. It cannot be represented exactly as a decimal, but it is a number between 3.1415 and 3.1416.

Pictograph   See *Graph*.

Pie chart   See *Graph*.

Pirating   Stealing software by copying it illegally for the use of someone other than the person who paid for it.

Pivot   A process that uses elementary row operations to carry out the following steps: 1. Divide all entries in the row in which the pivot appears (called the *pivot row*) by the nonzero pivot element so that the pivot entry becomes a 1. This uses elementary row operation 3. 2. Obtain zeros above and below the pivot element by using elementary row operation 4.

Pivot row   In an elementary row operation, it is the row that is multiplied by a constant. See *Elementary row operations*.

Pivoting   See *Pivot*.

Pixel   Any of the thousands (or millions) of tiny dots that make up a computer or calculator image.

Place-value names   Trillions, hundred billions, ten billions, billions, hundred millions, ten millions, millions, hundred thousands, ten thousands, thousands, hundreds, tens, units, tenths, hundredths, thousandths, ten-thousandths, hundred-thousandths, and millionths (from large to small).

Planar curve   A curve completely contained in a plane.

Plane   In mathematics, it is an undefined term. It is flat and level and extends infinitely in horizontal and vertical directions. It is considered two-dimensional.

Plot a point   To mark the position of a point.

Plurality rule   The winner of an election is the candidate with the highest number of votes.

Point   (1) In the decimal representation of a number, it is a mark that divides the whole number part of a number from its fractional part. (2) In relation to a home loan, it represents 1% of the value of a loan, so that 3 points would be a fee paid to a lender equal to 3% of the amount of the loan. (3) In geometry, it is an undefinedword that signifies a position, but that has no dimension or size.

Point-slope form   An algebraic form of an equation of a line that is given in terms of a point (x_1, y_1) and slope m of a given line:

$$y - y_1 = m(x - x_1)$$

Police patrol problem   Suppose a police car needs to patrol a gated subdivision and would like to enter the gate, cruise all the streets exactly once, and then leave by the same gate.

Polygon   A geometric figure that has three or more straight sides that all lie in a plane so that the starting point and the ending point are the same.

Polynomial   An algebraic expression that may be written as a sum (or difference) of terms. Each *term* of a polynomial contains multiplication only.

Population   The total set of items (actual or potential) defined by some characteristic of the items.

Population growth   The population, P, at some future time can be predicted if you know the population P_0 at some time, and the annual growth rate, r. The predicted population t years after the given time is $P = P_0 e^{rt}$.

Population paradox   When there is a fixed number of seats, a reapportionment may cause a state to lose a seat to another state, even though the percent increase in the population of the state that loses the seat is larger than the percent increase of the state that wins the seat. When this occurs, it is known as the *population paradox*.

Positional system   A numeration system in which the position of a symbol in the representation of a number determines the meaning of that symbol.

Positive number   A number greater than 0.

Positive sign   The symbol "+" when used in front of a number or an expression.

Positive square root   The symbol $\sqrt{x}$ is the positive number that, when multiplied by itself, gives the number x. The symbol "$\sqrt{}$" is always positive.

Postulate   A statement that is accepted without proof.

Pound   A unit of measurement for mass in the United States system. It is equal to 16 oz.

Power   See *Exponent*.

Precision   The accuracy of the measurement; for example, a measurement is taken to the nearest inch, nearest foot, or nearest mile. It is not to be confused with accuracy that applies to the calculation.

Predecessor   In a sequence, the predecessor of an element a_n is the preceding element, a_{n-1}.

Premise   A previous statement or assertion that serves as the basis for an argument.

Present value   See *Compound interest formula*.

Present value formula   The present value, P, of a known future value A invested at an annual interest rate of r for t years compounded n times per year is found by the formula

$$P = A\left(1 + \frac{r}{n}\right)^{-nt}$$

Present value of an annuity   A financial formula that seeks the present value from periodic payments over a period of time. the formula is

$$P = m\left[\frac{1 - \left(1 + \frac{r}{n}\right)^{-nt}}{\frac{r}{n}}\right]$$

Previous balance method   A method of calculating credit card interest using the formula $I = Prt$ in which P is the balance owed before the current payment is subtracted.

Prime factorization   The factorization of a number so that all of the factors are primes and so that their product is equal to the given number.

Prime number $P = \{2, 3, 5, 7, 11, 13, 17, 19, 23, \ldots\}$: a number with exactly two factors: 1 and the number itself.

Principal   See *Compound interest formula*.

Printer   An output device for a computer.

Prism   In this book, it refers to a right prism, which is also called a parallelepiped or more commonly a box.

Probabilistic model   A model that deals with situations that are random in character and attempts to predict the outcomes of events with a certain stated or known degree of accuracy.

Probability   If an experiment can result in any of $n (n \geq 1)$ mutually exclusive and equally likely outcomes, and if s of these are considered favorable to event E, then $P(E) = s/n$.

Probability function   A function P that satisfies the following properties: $0 \leq P(E) \leq 1$, $P(S) = 1$, and if E and F are mutually exclusive events, then $P(E \cup F) = P(E) + P(F)$.

Problem-solving procedure   1. *Read the problem.* Note what it is all about. Focus on processes rather than numbers. You can't work a problem you don't understand. 2. *Restate the problem.* Write a verbal description of the problem using operation signs and an equal sign. Look for equality. If you can't find equal quantities, you will never formulate an equation. 3. *Choose a variable.* If there is a single unknown, choose a variable. 4. *Substitute.* Replace the verbal phrases by known numbers and by the variable. 5. *Solve the equation.* This is the easy step. Be sure your answer makes sense by checking it with the original question in the problem. Use estimation to eliminate unreasonable answers. 6. *State the answer.* There were no variables defined when you started, so $x = 3$ is not an answer. Pay attention to units of measure and other details of the problem. Remember to answer the question that was asked.

Product The result of a multiplication.

Profit formula $P = S - C$, where P represents the profit, S represents the selling price (or revenue), and C the cost (or overhead).

Program A set of step-by-step instructions that instruct a computer what to do in a specified situation.

Progression See *Sequence*

Projective geometry The study of those properties of geometric configurations that are invariant under projection. It was developed to satisfy the need for depth in works of art.

Prompt In a computer program, a prompt is a direction that causes the program to print some message to help the user understand what is happening at a particular time.

Proper divisor A divisor of a number that is less than the number itself.

Proper fraction A fraction for which the numerator is less than the denominator.

Proper subset See *Subset*.

Proof A logical argument that establishes the truth of a statement.

Property of complements See *Complementary probabilities*.

Property of proportions If the product of the means equals the product of the extremes, then the ratios form a proportion. Also, if the ratios form a proportion, then the product of the means equals the product of the extremes.

Property of rational expressions Let $P, Q, R, S,$ and K be any polynomials such that all values of the variable that cause division by zero are excluded from the domain.

Equality $\dfrac{P}{Q} = \dfrac{R}{S}$ if and only if $PS = QR$.

Fundamental property $\dfrac{PK}{QK} = \dfrac{P}{Q}$

Addition $\dfrac{P}{Q} + \dfrac{R}{S} = \dfrac{PS + QR}{QS}$

Subtraction $\dfrac{P}{Q} - \dfrac{R}{S} = \dfrac{PS - QR}{QS}$

Multiplication $\dfrac{P}{Q} \cdot \dfrac{R}{S} = \dfrac{PR}{QS}$

Division $\dfrac{P}{Q} \div \dfrac{R}{S} = \dfrac{PS}{QR}$

Property of zero $AB = 0$ if and only if $A = 0$ or $B = 0$ (or both). Also called the *zero-product rule*.

Proportion A statement of equality between two ratios. For example,

$$\frac{a}{b} = \frac{c}{d}$$

For this proportion, a and d are called the *extremes; b* and *c* are called the *means.*

Protractor A device used to measure angles.

Pseudosphere The surface of revolution of a tractrix about its asymptote. It is sometimes called a "four-dimensional sphere."

Pyramid A solid figure having a polygon as a base, the sides of which form the bases of triangular surfaces meeting at a common vertex.

Pythagorean theorem If a triangle with legs a and b and hypotenuse c is a right triangle, then $a^2 + b^2 = c^2$.

Quadrant See *Axes*.

Quadratic A second-degree polynomial

Quadratic equation An equation of the form $ax^2 + bx + c = 0, a \neq 0$.

Quadratic formula If $ax^2 + bx + c = 0$ and $a \neq 0$, then

$$x = \frac{-b \pm \sqrt{b^2 - 4ac}}{2a}$$

The radicand $b^2 - 4ac$ is called the *discriminant* of the quadratic.

Quadratic function

$$f(x) = ax^2 + bx + c, a \neq 0$$

Quadrilateral A polygon having four sides.

Quart A measure of capacity in the United States system equal to 1/4 of a gallon.

Quarterly compounding In the compound interest formula, it is when $n = 4$.

Quartile Three values that divide a data set into four equal parts.

Quota rule The number assigned to each represented unit must be either the standard quota rounded down to the nearest integer, or the standard quota rounded up to the nearest integer.

Quotient The result of a division.

Radical form The $\sqrt{\ }$ symbol in an expression such as $\sqrt{2}$. The number 2 is called the *radicand* and an expression involving a radical is called a *radical expression.*

Radicand See *Radical form.*

Radius The distance of a point on a circle from the center of the same circle.

RAM Random-Access Memory, or memory where each location is uniformly accessible, often used for the storage of a program and the data being processed.

Random variable A *random variable X* associated with the sample space S of an experiment is a function that assigns a real number to each simple event in S.

Range (1) In statistics, it is the difference between the largest and the smallest numbers in the data set. (2) The *range* of a graph of an equation with two variables x and y is the set of permissible real-number replacements for y.

Rate (1) In percent problems, it is the percent. (2) In tax problems, it is the level of taxation, written as a percent. (3) In financial problems, it refers to the APR.

Ratio The quotient of two numbers or expressions.

Rational equation An equation that has at least one variable in the denominator.

Rational number A number belonging to the set $\mathbb{Q}$ defined by

$$\mathbb{Q} = \left\{ \frac{a}{b} \middle| a \text{ is an integer, } b \text{ is a nonzero integer} \right\}$$

a is called the *numerator* and b is called the *denominator.* A rational number is also called *a fraction.*

Ray   If P is a point on a line, then a ray from the point P is all points on the line on one side of P.

Real number line   A line on which points are associated with real numbers in a one-to-one fashion.

Real numbers   The set of all rational and irrational numbers, denoted by $\mathbb{R}$.

Reciprocal   The reciprocal of n is $\frac{1}{n}$, also called the *multiplicative inverse of n*.

Rectangle   A quadrilateral whose angles are all right angles.

Rectangular coordinate system   See *Cartesian coordinate system*.

Rectangular coordinates   See *Ordered pair*.

Rectangular parallelepiped   In this book, it refers to a box all of whose angles are right angles.

Reduced fraction   A fraction so that the numerator and denominator have no common divisors (other than 1).

Reducing fractions   The process by which we make sure that there are no common factors (other than 1) for the numerator and denominator of a fraction.

Reflection   Given a line L and a point P, we call the point P' the *reflection* about the line L if $\overline{PP'}$ is perpendicular to L and is also bisected by L.

Region   In a network, a separate part of the plane.

Regression analysis   The analysis used to determine the relationship between two variables.

Regular polygon   A polygon with all sides the same length.

Relation   A set of ordered pairs.

Relative frequency   If an experiment is repeated n times and an event occurs m times, then the relative frequency is the ratio m/n.

Relatively prime   Two integers are relatively prime if they have no common factors other than ± 1; two polynomials are relatively prime if they have no common factors except constants.

Remainder   When an integer m is divided by a positive integer n, and a quotient q is obtained for which $m = nq + r$ with $0 \le r < n$, then r is the remainder.

Repeating decimal   See *Decimal fraction*.

Repetitive system   numeration system for which a single symbol is repeated to represent a given number. For example, $\cap\cap\cap$ in the Babylonian system means
$10 + 10 + 10 = 30$

Replication   On a spreadsheet, the operation of copying a formula from one place to another.

Resolution   The number of dots (or pixels) determines the clarity, or resolution, of the image on the monitor.

Revolving credit   It is the same as open-ended or credit card credit.

Rhombus   A parallelogram with adjacent sides equal.

Richter number   A number used to denote the magnitude or size of an earthquake.

Richter scale   Same as *Richter number*.

Right angle   An angle of 90°.

Right circular cone   A cone with a circular base for which the base is perpendicular to its axis.

Right circular cylinder   A cylinder with a circular base for which the base is perpendicular to its axis.

Right prism   A prism whose base is perpendicular to the lateral edges.

Right triangle   A triangle with one right angle.

Rise   See *Slope*.

ROM   Read-Only Memory, or memory that cannot be altered either by the user or a loss of power. In microcomputers, the ROM usually contains the operating system and system programs.

Root of a number   An nth root (n is a natural number) of a number b is a only if $a^n = b$. If $n = 2$, then the root is called a *square root*; if $n = 3$, it is called a *cube root*.

Root of an equation   See *Solution*.

Roster method   A method of defining a set by listing its members.

Rounding a number   Dropping decimals after a certain significant place. The procedure for rounding is: 1. Locate the rounding place digit. 2. Determine the rounding place digit: It stays the same if the first digit to its right is a 0, 1, 2, 3, or 4; it increases by 1 if the digit to the right is a 5, 6, 7, 8, or 9. 3. Change digits: all digits to the left of the rounding digit remain the same (unless there is a carry) and all digits to the right of the rounding digit are changed to zeros. 4. Drop zeros: If the rounding place digit is to the left of the decimal point, drop all trailing zeros; if the rounding place digit is to the right of the decimal point, drop all trailing zeros to the right of the rounding place digit.

Row   A horizontal arrangement of numbers or entries of a matrix. It is denoted by numerals 1, 2, 3, . . . on a spreadsheet.

Row+   An elementary row transformation that causes one row of a matrix (called the *pivot row*) to be added to another row (called the *target row*). The answer to this addition replaces the entries in the target row, entry by entry.

Row-reduced form   The final matrix after the process of Gauss-Jordan elimination.

RowSwap   An elementary row operation that causes two rows of a matrix to be switched, entry-by-entry.

Rubik's cube   A three-dimensional cube that can rotate about all three axes. It is a puzzle that has the object of returning the faces to a single-color position.

Rules of divisibility   A number N is divisible by:

1	
2	if the last digit is divisible by 2.
3	if the sum of the digits is divisible by 3.
4	if the number formed by the last two digits is divisible by 4.
5	if the last digit is 0 or 5.
6	if the number is divisible by 2 and by 3.
8	if the number formed by the last three digits is divisible by 8.
9	if the sum of the digits is divisible by 9.
10	if the last digit is 0.
12	if the number is divisible by 3 and by 4.

Run   See *Slope*.

Runoff election   An attempt to obtain a majority vote by eliminating one or more alternatives and voting again on the remaining choices.

Saccheri quadrilateral   A rectangle with base angles A and B right angles, and with sides $\overline{AC}$ and $\overline{BD}$ the same length.

Sales price   A reduced price usually offered to stimulate sales. It can be found by subtracting the discount from the original price, or by multiplying the original price by the complement of the markdown.

Sales tax   A tax levied by government bodies that is based on the sale price of an item.

Sample   A finite portion of a population.

Sample space   The set of possible outcomes for an experiment.

Satisfy   See *Equation* or *Inequality*.

Scalar   A real number.

Scalar multiplication   The multiplication of a real number and a matrix.

Scalene triangle   A triangle with no two sides having the same length.

Scatter diagram   A diagram showing the frequencies with which joint values of variables are observed. One variable is indicated along the x-axis and the other along the y-axis.

Scientific notation   Writing a number as the product of a number between 1 and 10 and a power of 10: For any real number n, $n = m \cdot 10^c$, $1 \leq m < 10$, and c is an integer. Calculators often switch to scientific notation to represent large or small numbers. The usual notation is 8.234 05, where the space separates the number from the power; thus 8.234 05 means 8.234×10^5.

Secant line   A line passing through two points of a given curve.

Second component   See *Ordered pair*.

Second-degree equation   With one variable, an equation of the form $ax^2 + bx + c = 0$; with two variables, an equation of the form

$$Ax^2 + Bxy + Cy^2 + Dx + Ey + F = 0$$

Semiannual compounding   In the compound interest formula, it is when $n = 2$.

Semicircle   Half a circle.

Sequence   An *infinite sequence* is a function whose domain is the set of counting numbers. It is sometimes called *a progression*. A *finite sequence* with n terms is a function whose domain is the set of numbers $\{1, 2, 3, \ldots, n\}$.

Sequential voting   A runoff election procedure that has one vote followed by another.

Series   The indicated sum of a finite or an infinite sequence of terms.

Series circuit   Two switches connected together so that the circuit is on only if both switches are on.

Set   A collection of particular things, called the *members* or *elements* of the set. A set with no elements is called the *null set* or *empty set* and is denoted by the symbol $\emptyset$. All elements of a *finite set* may be listed, whereas the elements of an *infinite set* continue without end.

Set-builder notation   A technical notation for defining a set. For example,

$$\{a \mid a \in \mathbb{Z}, 5 < a < 100\}$$

means "the set of all elements a such that a is an integer between 5 and 100."

Set theory   The branch of mathematics that studies sets.

SI system   See *Metric system*.

Sieve of Eratosthenes   A method for determining a set of primes less than some counting number n. Write out the consecutive numbers from 1 to n. Cross out 1, since it is not classified as a prime number. Draw a circle around 2, the smallest prime number. Then cross out every following multiple of 2, since each is divisible by 2 and thus is not prime. Draw a circle around 3, the next prime number. Then cross out each succeeding multiple of 3. Some of these numbers, such as 6 and 12, will already have been crossed out because they are also multiples of 2. Circle the next open prime, 5, and cross

out all subsequent multiples of 5. The next prime number is 7; circle 7 and cross out multiples of 7. Continue this process until you have crossed out the primes up to $\sqrt{n}$. All of the remaining numbers on the list are prime.

Sigma notation   Sigma, the Greek letter corresponding to S, is written Σ. It is used to indicate the process of summing the first to the nth terms of a set of numbers $s_1, s_2, s_3, \ldots, s_n$, which is written as

$$\sum_{k=1}^{n} s_k$$

This notation is also called *summation notation*.

Signed number   An integer.

Significance level   Deviations between hypothesis and observations that are so improbable under the hypothesis as not to be due merely to sampling error or random fluctuations are said to be *statistically significant*. The significance level is set at an acceptable level for a deviation to be statistically significant.

Similar figures   Two geometric figures are similar if they have the same shape, but not necessarily the same size.

Similar terms   Terms that differ only in their numerical coefficients.

Similar triangle theorem   Two triangles are similar if two angles of one triangle are equal to two angles of the other triangle. If the triangles are similar, then their corresponding sides are proportional.

Similar triangles   Triangles that have the same shape.

Similarity   Two geometric figures are *similar* if they have the same shape.

Simple curve   A curve that does not intersect itself.

Simple event   An event that contains only one element of the sample space.

Simple grouping system   A numeration system is a grouping system if the position of the symbols is not important, and each symbol larger than 1 represents a group of other symbols.

Simple interest formula $I = Prt$

Simple statement   A statement that does not contain a connective.

Simplify   (1) A *polynomial*: combine similar terms and write terms in order of descending degree. (2) A fraction (a rational expression): Simplify numerator and denominator, factor if possible, and eliminate all common factors. (3) A square root: The *radicand* (the number under the radical sign) has no factor with an exponent larger than 1 when it is written in factored form; the radicand is not written as a fraction or by using negative exponents; there are no square root symbols used in the denominators of fractions.

Simulation   Use of a computer program to simulate some real-world situation.

Simultaneous solution   The solution of a simultaneous system of equations.

Sine   In a right triangle ABC with right angle C,

$$\sin A = \frac{\text{LENGTH OF OPPOSITE SIDE OF } A}{\text{LENGTH OF HYPOTENUSE}}$$

Singular matrix   A matrix that does not have an inverse.

Sinking fund   A financial problem in which the monthly payment must be found to obtain a known future value. The formula is

$$m = \frac{A\left(\frac{r}{n}\right)}{\left(1 + \frac{r}{n}\right)^{nt} - 1}$$

Skewed distribution   A statistical distribution that is not symmetric, but favors the occurrence on one side of the mean or the other.

Slant asymptotes   In graphing a hyperbola, the diagonal lines passing through the corners of the central rectangle.

Slope   The slope of a line passing through (x_1, y_1) and (x_2, y_2) is denoted by m, and is found by

$$m = \frac{y_2 - y_1}{x_2 - x_1} = \frac{\text{VERTICAL CHANGE}}{\text{HORIZONTAL CHANGE}} = \frac{\text{RISE}}{\text{RUN}}$$

Slope-intercept form $y = mx + b$

Slope point   A point that is found after counting out the rise and the run from the y-intercept.

Software   The routines, programs, and associated documentation in a computer system.

Software package   A commercially available computer program that is written to carry out a specific purpose, for example, a database program or a word-processing program.

Solution   The values or ordered pairs of values for which an equation, a system of equations, inequality, or system of inequalities is true. Also called *roots*.

Solution set   The set of all solutions to an equation.

Solve a proportion   To find the missing term of a proportion. Procedure: First, find the product of the means or the product of the extremes, whichever does not contain the unknown term; next, divide this product by the number that is opposite the unknown term.

Solve an equation   To find the values of the variable that make the equation true.

Solve an inequality   To find the values of the variable that make the inequality true.

Some   A word used to mean "at least one."

Spanning tree   A tree that is created from another graph by removing edges while keeping a path to each vertex.

Sphere   The set of all points in space that are a given distance from a given point.

Spreadsheet   A rectangular grid used to collect and perform calculations on data. *Rows* are horizontal and are labeled with numbers and *columns* are vertical and are labeled with letters to designate *cells* such as A4, P604. Each cell can contain text, numbers, or formulas.

Square   (1) A quadrilateral with all sides the same length and all angles right angles. (2) In an expression such as x^2, which is pronounced "x-squared," it means xx.

Square matrix   A matrix with the same number of rows and columns.

Square number   Numbers that are squares of the counting numbers: 1, 4, 9, 16, 25, 36, 49, 64, 81, 100, 121, 144, 169, ...

Square root   See *Root of a number.*

Square unit   A two-dimensional unit. It is the result of squaring a unit of measurement.

Stable marriage   A pairing in which both partners are satisfied.

Standard deviation   It is a measure of the variation of a data set. In particular, it is the square root of the mean of the squares of the deviations from the mean.

Standard divisor

$$\text{STANDARD DIVISOR} = \frac{\text{TOTAL POPULATION}}{\text{NUMBER OF SHARES}}$$

Standard form   The standard form of the equation of a line is

$$Ax + By + C = 0.$$

Standard quota

$$\text{STANDARD QUOTA} = \frac{\text{TOTAL POPULATION}}{\text{STANDARD DIVISOR}}$$

Statement   A declarative sentence that is either true or false, but not both true and false.

Statistics   Methods of obtaining and analyzing quantitative data.

Stem-and-leaf plot   A procedure for organizing data that can be divided into two categories. The first category is listed at the left, and the second category at the right.

Sticker price   In this book, it refers to the manufacturer's total price of a new automobile as listed on the window of the car.

Straight angle   An angle whose rays point in opposite directions; an angle whose measure is 180°.

Straightedge   A device used as an aid in drawing a straight line segment.

Straw vote   A nonbinding vote taken before all the discussion has taken place. It precedes the actual vote.

Street problem   A problem that asks the number of possible routes from one location to another along some city's streets. The assumptions are that we always move in the correct direction and that we do not cut through the middle of a block, but rather stay on the streets or alleys.

Subjective probability   A probability obtained by experience and used to indicate a measure of "certainty" on the part of the speaker. These probabilities are not necessarily arrived at through experimentation or theory.

Subscript   A small number or letter written below and to the right or left of a letter as a mark of distinction.

Subset   A set contained within a set. There are 2^n subsets of a set with n distinct elements. A subset is *improper* if it is equivalent to the given set; otherwise it is *proper*.

Substitution method   The method of solution of a system of equations in which one of the equations is solved for one of the variables and substituted into another equation.

Substitution property   The process of replacing one quantity or unknown by another quantity. That is, if $a = b$, then a may be substituted for b in any mathematical statement without affecting the truth or falsity of the given mathematical statement.

Subtraction   The operation of subtraction is defined by:

$$a - b = x \text{ means } a = b + x$$

Subtraction law of exponents   To divide two numbers with the same base, subtract the exponents. That is,

$$\frac{b^m}{b^n} = b^{m-n}$$

Subtraction law of logarithms   The log of the quotient of two numbers is the log of the numerator minus the log of the denominator. In symbols,

$$\log_b\left(\frac{A}{B}\right) = \log_b A - \log_b B$$

Subtraction of integers

$$a - b = a + (-b)$$

Subtraction of matrices $[M] - [N] = [S]$ if and only if $[M]$ and $[N]$ are the same order and the entries of $[S]$ are found by subtracting the corresponding entries of $[M]$ and $[N]$.

Subtraction of rational numbers

$$\frac{a}{b} - \frac{c}{d} = \frac{ad}{bd} - \frac{bc}{bd} = \frac{ad - bc}{bd}$$

Subtraction principle In reference to numeration systems, it is subtracting the value of some symbol from the value of the other symbols. For example, in the Roman numeration system IX uses the subtraction principle because the position of the I in front of the X indicates that the value of I (which is 1) is to be subtracted from the value of X (which is 10). IX = 9.

Subtraction property of equations The solution of an equation is unchanged by subtracting the same number from both sides of the equation.

Subtraction property of inequality *See Addition property of inequality.*

Successor In a sequence, the successor of an element a_n is the following element, a_{n+1}.

Sum The result of an addition.

Summation notation See *Sigma notation.*

Supercomputer A large, very fast mainframe computer used especially for scientific computations.

Superfluous constraint In a linear programming problem, a constraint that does not change the outcome if it is deleted.

Supermarket problem Set up the shelves in a market or convenience store so that it is possible to enter the store at one door and travel each aisle once (and only once) and leave by the same door.

Supplementary angles Two angles whose sum is $180°$.

Supply The number of items that can be supplied at a given price.

Surface In mathematics, it is an undefined term. It is the outer face or exterior of an object; it has an extent or magnitude having length and breadth, but no thickness.

Surface area The area of the outside faces of a solid.

Syllogism A logical argument that involves three propositions, usually two premises and a conclusion, the conclusion necessarily being true if the premises are true.

Symbols of inclusion See *Grouping symbols.*

Symmetric property of equality If $a = b$, then $b = a$.

Symmetry (1) In geometry, a graph or picture is *symmetric with respect to a line* if the graph is a mirror reflection along the line. (2) In voting, it means that if one voter prefers A to B and another B to A, then the votes should cancel each other out.

Syntax error The breaking of a rule governing the structure of the programming language being used.

System of equations A set of equations that are to be solved *simultaneously.* A brace symbol is used to show the equations belonging to the system.

System of inequalities A set of inequalities that are to be solved simultaneously. The solution is the set of all ordered pairs (x, y) that satisfy all the given inequalities. It is found by finding the intersection of the half-planes defined by each inequality.

Tangent In a right triangle ABC with right angle C,

$$\tan A = \frac{\text{LENGTH OF OPPOSITE SIDE OF } A}{\text{LENGTH OF ADJACENT SIDE OF } A}$$

Tangent line A tangent line to a circle is a line that contains exactly one point of the circle. The tangent line to a curve at a point P is the limiting position, if this exists, of the secant line through a fixed point P on the curve and a variable point P' on the curve so that P' approaches P along the curve.

Target population The population to be considered for a statistical application.

Target row In an elementary row operation, it is the row that is changed. See *Elementary row operations.*

Tautology A compound statement is a tautology if all values on its truth table are true.

Temperature The degree of hotness or coldness.

Ten A representation for "ΔΔΔΔΔΔΔΔΔΔ" objects

Term (1) A number, a variable, or a product of numbers and variables. See *Polynomial.* (2) A *term of a sequence* is one of the elements of that sequence.

Terminating decimal See *Decimal fraction.*

Tessellation A mosaic, repetitive pattern.

Test point A point that is chosen to find the appropriate half-plane when graphing a linear inequality in two variables.

Theorem A statement that has been proved. See *Deductive reasoning.*

Theoretical probability A probability obtained by logical reasoning according to stated definitions.

Time In a financial problem, the length of time (in years) from the present value to the future value.

*Row An elementary row operation that multiplies each entry of a row of a matrix (called the *target row*) by some number, called a *scalar.* The elements of the row are replaced term-by-term by the products. It is denoted by *Row.

*Row+ An elementary row operation that multiplies each entry of a row of a matrix (called the *pivot row*) by some number (called a *scalar*), and then adds that product, term by term, to the numbers in another row (called the *target row*). The results replace the entries in the target row, term by term. It is denoted by *Row+.

Ton A measurement of mass in the United States system; it is equal to 2,000 lb.

Topologically equivalent Two geometric figures are said to be *topologically equivalent* if one figure can be elastically twisted, stretched, bent, shrunk, or straightened into the same shape as the other. One can cut the figure, provided at some point the cut edges are "glued" back together again to be exactly the same as before.

Topology That branch of geometry that deals with the *topological properties* of figures. If one figure can be transformed into another by stretching or contracting, then the figures are said to be *topologically equivalent.*

Tournament method A method of selecting a winner by pairing candidates head-to-head with the winner of one facing a new opponent for the next election.

Trailing zeros Sometimes zeros are placed after the decimal point or after the last digit to the right of the decimal point, and if these zeros do not change the value of the number, they are called *trailing zeros.*

Transformation A passage from one figure or expression to another, such as a reflection, translation, rotation, contraction, or dilation.

Transformational geometry The geometry that studies transformations.

Transitive law If A beats B, and B beats C, then A should beat C. In symbols, equality: If $a = b$ and $b = c$, then $a = c$; *inequality:* If $a > b$ and $b > c$, then $a > c$. Also holds for $\geq, <,$ and $\leq$.

Transitive reasoning   If $a = b$ and $b = c$, then $a = c$.

Translating symbols   The process of writing an English sentence in mathematical symbols.

Transversal   A line that intersects two parallel lines.

Transverse axis   The line passing through the foci is called the *transverse axis*.

Trapezoid   A quadrilateral that has two parallel sides.

Traveling salesperson problem (TSP)   A salesperson starts at home and wants to visit several cities without going through any city more than once and then returning to the starting city.

Traversable network   A network is said to be *traversable* if it can be traced in one sweep without lifting the pencil from the paper and without tracing the same edge more than once. Vertices may be passed through more than once.

Tree   A graph that is connected and has no circuits.

Tree diagram   A device used to list all the possibilities for an experiment.

Triangle   A polygon with three sides.

Trichotomy   Exactly one of the following is true, for any real numbers a and b: $a < b$, $a > b$, or $a = b$.

Trigonometric functions   The same as the *trigonometric ratios*.

Trigonometric ratios   The sine, cosine, and tangent ratios are known as the *trigonometric ratios*.

Trillion   A name for $10^{12} = 1{,}000{,}000{,}000{,}000$.

Trinomial   A polynomial with exactly three terms.

Truth set   The set of values that makes a given statement true.

Truth table   A table that shows the truth values of all possibilities for compound statements.

Truth value   The truth value of a simple statement is true or false. The truth value of a compound statement is true or false and depends only on the truth values of its simple component parts. It is determined by using the rules for connecting those parts with well-defined operators.

Two-point form   The equation of a line passing through (x_1, y_1) and (x_2, y_2) is

$$y - y_1 = \left(\frac{y_2 - y_1}{x_2 - x_1}\right)(x - x_1)$$

Type I error   Rejection of a hypothesis based on sampling when, in fact, the hypothesis is true.

Type II error   Acceptance of a hypothesis based on sampling when, in fact, it is false.

Undefined terms   To avoid circular definitions, it is necessary to include certain terms without specific mathematical definition.

Union   The union of sets A and B, denoted by $A \cup B$, is the set consisting of elements in A or in B or in both A and B.

Unit circle   A circle with radius 1 centered at the origin.

Unit distance   The distance between 0 and 1 on a number line.

Unit scale   The distance between the points marked 0 and 1 on a number line.

United States system   The measurement system used in the United States.

Universal set   The set that contains all of the elements under consideration for a problem or a set of problems.

Unless   A logical operator for "p unless q" that is defined to mean $\sim q \rightarrow p$.

Unrestricted domain   Any set of rankings is possible; if there are n candidates, then there are $n!$ possible rankings.

Unstable marriage   A pairing in which one (or both) of the partners would prefer to be paired with another partner.

Upload   The process of copying a program from your computer to the network.

Upper quota   In apportionment, the result of a quota found by rounding up.

User-friendly   A term used to describe software that is easy to use. It includes built-in safeguards to keep the user from changing important parts of a program.

Valid argument   In logic, refers to a correctly inferred logical argument.

Variable   A symbol that represents unspecified elements of a given set. On a calculator, it refers to the name given to a location in the memory that can be assigned a value.

Variable expression   An expression that contains at least one variable.

Variance   The square of the standard deviation. When the variance is based on a set of sample scores, it is called the *variance of a random sample* and is denoted by s^2, and when it is based on the entire population, it is called the *variance of the population* and is denoted by σ^2. The formulas for variance are

$$s^2 = \frac{\Sigma(x - \bar{x})^2}{n - 1} \qquad \sigma^2 = \frac{\Sigma(x - \mu)^2}{n}$$

Velocity   An instantaneous rate of change; a directed speed.

Venn diagram   A diagram used to illustrate relationships among sets.

Vertex   (1) A *vertex* of a polygon is a corner point, or a point of intersection of two sides. (2) A *vertex* of a parabola is the lowest point for a parabola that opens upward; the highest point for one that opens downward; the leftmost point for one that opens to the right; and the rightmost point for one that opens to the left.

Vertex angle   The angle included between the legs of an isosceles triangle.

Vertical angles   Two angles such that each side of one is a prolongation through the vertex of a side of the other.

Vertical ellipse   An ellipse whose major axis is vertical.

Vertical hyperbola   A hyperbola whose transverse axis is vertical.

Vertical line   A line with undefined slope. Its equation has the form $x = $ constant.

Vertical line test   Every vertical line passes through the graph of a function in at most one point. This means that if you sweep a vertical line across a graph and it simultaneous intersects the curve at more than one point, then the curve is not the graph of a function.

Volume   A number describing three-dimensional content of a set. Specifically, it is the number of cubic units enclosed in a solid figure.

Vote   A decision by a group on a proposal, resolution, law, or a choice between candidates for office.

Water-pipe problem   Consider a network of water pipes to be inspected. Is it possible to pass a hand over each pipe exactly once without lifting it from a pipe, and without going over the samepipe more than once?

Webster's apportionment plan   An apportionment plan in which the representation of a geographical area is determined by finding the quotient of the number of people in that area divided by the total number of people and then rounding the result as follows: Any quotient with a decimal portion must be rounded to the nearest whole number.

Weight (1) In everyday usage, the heaviness of an object. In scientific usage, the gravitational pull on a body. (2) In a network or graph, a cost associated with an edge.

Weighted graph A graph for which all its edges have weight.

Weighted mean If the scores $x_1, x_2, x_3, \ldots, x_n$ occur $w_1, w_2, \ldots, w_n$ times, respectively, then the *weighted mean* is

$$\bar{x} = \frac{\Sigma(w \cdot x)}{\Sigma w}$$

Well-defined set A set for which there is no doubt about whether a particular element is included in the given set.

Whole numbers The positive integers and zero; $\mathbb{W} = \{0, 1, 2, 3, \ldots\}$.

Windows A graphical environment for IBM format computers.

With replacement If there is more than one step for an experiment, to perform the experiment with replacement means that the object chosen on the first step is replaced before the next steps are completed.

Without replacement If there is more than one step for an experiment, to perform the experiment without replacement means that the object chosen on the first step is not replaced before the next steps are completed.

Word processing The process of creating, modifying, deleting, and formatting textual materials.

World Wide Web A network that connects together computers from all over the world. It is abbreviated by www.

WYSIWYG: An acronym for What You See Is What You Get.

x-axis The horizontal axis in a Cartesian coordinate system.

x-intercept The place where a graph passes through the x-axis.

y-axis The vertical axis in a Cartesian coordinate system.

y-intercept The place where a graph passes through the y-axis. For a line

$$y = mx + b$$

it is the point $(0, b)$.

Yard A linear measure in the United States system; it has the same length as 3 ft.

Zero The number that separates the positive and negative numbers; it is also called the *identity element* for addition; that is, it satisfies the property that

$$x + 0 = 0 + x = x$$

for all numbers x.

Zero matrix A matrix with all entries equal to 0.

Zero multiplication If a is any real number, then $a \cdot 0 = 0 \cdot a = 0$.

Zero-one matrix A square matrix in which the entries symbolize the occurrence of some facet or event with a 1 and the nonoccurrence with a 0.

Zero-product rule If $a \cdot b = 0$, then either $a = 0$ or $b = 0$.

z-score A measure to determine the distance (in terms of standard deviations) that a given score is from the mean of a distribution.

PROLOGUE

Prologue Problem Set, page P11

9. 181 **11.** 1,099 **13.** February and March **15.** 30 **17.** March 2011
19. 119 **21.** 945 **23.** $\frac{3}{8}$ **25.** 2 **27. a.** tetrahedron **b.** cube
(hexahedron) **c.** octahedron **d.** dodecahedron **e.** icosahedron
29. 1,499 **31.** More than a million trees must be cut down to print a trillion one-dollar bills. **33.** The area is approximately 13.7 in.^{2}. **35.** $\frac{1}{3}$
37. She had $22 at the start of the first day. **39.** The cost is $390 if 100
make the trip. **41.** Let x be the number of weeks enrolled, and y be the
total cost. Then $y = 45(10 - x)$. The graph is shown.

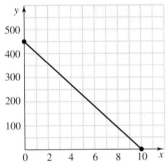

43. The rate at which percentage of alcohol is changing with respect to
time is $\frac{dC}{dt} = -0.15e^{-t/2}$. **45.** There are 33 shortest paths. **47.** It will
take until March 2125. **49.** $b = 2^{2/3}$ **51.** Answers vary; at least three
cubes can be seen. **53.** $\frac{1}{120}$ **55.** This pattern of numbers has the property
that the sum of the numbers in any row, column, or diagonal is the same
(namely, 15). **57.** The population in 2000 was 153,000. The graph is shown.

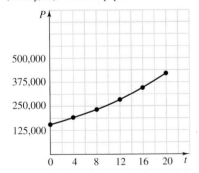

59. $2^{30} - 1$ dollars (more than a billion dollars)

CHAPTER 1

1.1 Problem Solving, page 14

5. first diagonal **7.** Yes; yes; answer is the same because the triangle is
symmetric. **9. a.** 35 **b.** 56 **11.** 2 up, 3 over; 10 paths **13.** 4 up,
6 over; 210 paths **15.** 4 down, 3 over; 35 paths **17.** 2 up, 4 over; 15
paths **19.** Answers vary. **21.** There is a total of 27 boxes. **23.** The
blind person says, "I need a pair of scissors." **25. a.** 2 **b.** 4 **c.** 8 **d.** 16
27. 49 paths **29.** 284 paths **31.** 12 **33.** 7 **35.** 11 **37.** 12:30

39. 54 **41.** 4 mm **43.** Answers vary. Put one lump in the first cup, four
lumps in the second cup, and 5 lumps in the third cup. Now, put the first
cup (containing the lump) inside the second cup. **45.** Answers vary.
47. Answers vary. **49.** The fly flies 20 miles **51.** 8 miles or 12 miles
53. puff **55.** 1.62 **57.** The number of routes from the start to the finish
minus the number of routes passing through the barricade. **a.** 26 **b.** 23
c. 27 **59.** Answers vary.

1.2 Inductive and Deductive Reasoning, page 24

7. a. 17 **b.** 13 **9. a.** 45 **b.** 45 **11. a.** 14 **b.** 6 **13. a.** 50 **b.** 2
15. a. 38 **b.** 54 **17. a.** 13 **b.** 12 **19.** inductive reasoning; answers
vary **21.** 3, 6, 9, . . . **23.** 5, 1, 6, 2, 7, 3, 8, 4, 9, . . . **25. a.** 625
b. 62,500
27.

2	7	6
9	5	1
4	3	8

29. valid **31.** valid **33.** valid **35.** not
valid **37.** not valid **39.** not valid **41.** east;
deductive reasoning **43.** probably Las Vegas;
inductive reasoning **45.** eight persons;
deductive reasoning **47.** probably Alabama; inductive reasoning
49. There is no door seen, so this means that the bus is heading
west (or to the left); this is an example of deductive reasoning.
51. a. $9 \times 54,321 - 1$; 488,888 **b.** 8,888,888,888 **c.** 98,888,888,888
53. $34^2 = 1156$; $334^2 = 111556$; $3334^2 = 11115556$; $\cdots$;
111111111555555556, so the sum is $9(1) + 8(5) + 6 = 55$; inductive
reasoning **55.** 91 squares **57.** Balance 3 with 3 to determine which
group is heavier. Repeat with the heavier group. **59. a.** One diagonal is
38,307; other sums are 21,609. **b.** One diagonal is 13,546,875; other
sums are 20,966,014.

1.3 Scientific Notation and Estimation, page 40

7. a. 3.2×10^3; 3.2 03 **b.** 4×10^{-4}; 4 −04 **c.** 6.4×10^{10}; 6.4 10
9. a. 5.629×10^3; 5.629 03 **b.** 6.3×10^5; 6.3 05
c. 3.4×10^{-8}; 3.4 −08 **11. a.** 49 **b.** 72,000,000,000 **c.** 4,560
13. a. 216 **b.** 0.00000 041 **c.** 0.00000 048 **15.** 9.3×10^7; 3.65×10^2
17. 2.2×10^8; 2.5×10^{-6} **19.** 3,600,000 **21.** 0.00000 003
Estimates in Problems 23–30 *may vary.* **23.** 35 mi **25.** Estimate
$2,000 \times 500 = 1,000,000$; $1,850 \times 487 = 900,950$ **27.** Estimate
$20 \times 15 = 300$; $23 \times 15 = 345$ **29.** Estimate $600,000 \div 60$
$= 10,000$ hours; $10,000 \div 25 = 400$ days; estimate one year.
$525,600 \div 60 \div 24 = 365$ **31. a.** 1.2×10^9 **b.** 3×10^2
33. a. 1.2×10^{-3} **b.** 2×10^7 **35. a.** 4×10^{-8} **b.** 1.2 **37.** Estimate
10/cm^2, so about 120 oranges. **39.** Estimate 270 chairs **41.** Assume
one balloon is about 1 ft^3 and also assume a typical classroom is
30 ft $\times$ 50 ft $\times$ 10 ft $= 15,000$ ft^3, so $10^6 \div (1.5 \times 10^4) \approx 67$; we
estimate that it would take 60–70 classrooms. **43.** Assume there are 233
$1 bills per inch. A million dollars would form a stack about 119 yd high.
45. Assume there are 2,260,000 grains in a pound of sugar. 1.356×10^{17}
grains **47.** Assume that a box of chalk contains 12 four inch pieces, so one
box is about 4 ft of chalk. Also assume that a box of chalk costs about
$1. 400,000 ft $\approx$ 75 miles. **49. a.** Answers vary. **b.** 9^{9^9} **c.** $10^{369,692,902}$
to $10^{387,000,000}$ **51.** $3.36271772 \times 10^{23}$ **53.** Assume a brick is 8 in. by $3\frac{1}{2}$
in. by 2 in. Answers vary; the wall will contain between 27,000 and 31,000
bricks. **55.** 48 minutes, 13 seconds **57.** 192 zeros **59. a.** D
b. 701.8 ft^2/person

Chapter 1 Review Questions, page 44

1. Understand the problem, devise a plan, carry out the plan, and then look back. **2.** 126 **3.** 330 paths **4.** 12,345,678,987,654,321 **5.** Order of operations: (1) First, perform any operations enclosed in parentheses. (2) Next, perform multiplications and divisions as they occur by working from left to right. (3) Finally, perform additions and subtractions as they occur by working from left to right. **6.** The scientific notation for a number is that number written as a power of 10 times another number x, such that $1 \le x < 10$. **7.** Inductive reasoning; the answer was found by looking at the pattern of questions. **8.** Answers vary. **a.** $\boxed{2}\;\boxed{y^x}\;\boxed{63}\;\boxed{=}$ or $\boxed{2}\;\boxed{\wedge}$ $\boxed{63}\;\boxed{\text{ENTER}}$ 9.223372037E18 **b.** $\boxed{9.22}\;\boxed{\text{EE}}\;\boxed{18}\;\boxed{\div}\;\boxed{6.34}\;\boxed{\text{EE}}\;\boxed{6}\;\boxed{=}$ 1.454258675E12 **9.** C **10.** $\frac{3}{4}$ hr $\times$ 365 = 273.75 hr. This is approximately $2\frac{3}{4}$ hr/book. Could not possibly be a complete transcription of each book. **11.** Answers vary. It is larger than the capacity of all of lakes behind all the world's dams, even the world's largest one to be completed in 2009 on the Yangzi River in China. **12.** Each person's share is about $44,516. **13.** Arrange the cards as $\begin{bmatrix} 2 & 7 & 6 \\ 9 & 5 & 1 \\ 4 & 3 & 8 \end{bmatrix}$. **14.** Fill the classroom with $100 bills; it will take 857 classrooms. **15.** not valid **16.** not valid **17.** valid **18.** valid **19.** Answers vary. **20.** 7^{1000} ends in 1.

CHAPTER 2

2.1 Sets, Subsets, and Venn Diagrams, page 57

9. a. well defined **b.** not well defined **11. a.** not well defined **b.** not well defined **13. a.** $\{m, a, t, h, e, i, c, s\}$ **b.** {Barack Obama} **15. a.** $\{7, 8, 9, 10, \ldots\}$ **b.** $\{1, 2, 3, 4, 5\}$ **17. a.** $\{p, i, e\}$ **b.** $\{151, 152, 153, \ldots\}$ *Answers to Problems 19–24 may vary.* **19.** counting numbers less than 10 **21.** multiples of 10 between 0 and 101 **23.** odd numbers between 100 and 170 **25.** the set of all numbers, x, such that x is an odd counting number; $\{1, 3, 5, 7, \ldots\}$ **27.** the set of all natural numbers, x, except 8; $\{1, 2, 3, 4, 5, 6, 7, 9, 10, \ldots\}$ **29.** The set of all whole numbers, x, such that x is less than 8; $\{0, 1, 2, 3, 4, 5, 6, 7\}$ **31. a.** $\varnothing$ **b.** $\varnothing, \{1\}$ **c.** $\varnothing, \{1\}, \{2\}, \{1, 2\}$ **d.** $\varnothing, \{1\}, \{2\}, \{3\}, \{1, 2\}, \{1, 3\}, \{2, 3\}, \{1, 2, 3\}$ **e.** $\varnothing, \{1\}, \{2\}, \{3\}, \{4\}\{1, 2\}, \{1, 3\}, \{1, 4\}, \{2, 3\}, \{2, 4\}, \{3, 4\}, \{1, 2, 3\}, \{1, 2, 4\}, \{1, 3, 4\}, \{2, 3, 4\}, \{1, 2, 3, 4\}$
33. 32 subsets; yes

35.
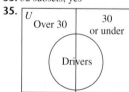

37.
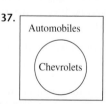

39. a. $|A| = 3, |B| = 1, |C| = 3, |D| = 1, |E| = 1, |F| = 1$ **b.** $A \leftrightarrow C; B \leftrightarrow D \leftrightarrow E \leftrightarrow F$ **c.** $A = C; D = E = F$
41. a. true **b.** false **43. a.** true **b.** false **45.** true **47.** true
49. false **51.** false **53.** true **55.** Answers vary; the set of real numbers, or the set of people alive in China today. **57.** There will be five different ways, since each group of four will have one missing person. **59.** The set of Chevrolets is a subset of the set of automobiles.

2.2 Operations with Sets, page 62

3. a. or **b.** and **c.** not **5.** $\{2, 6, 8, 10\}$ **7.** $\{2, 3, 5, 6, 8, 9\}$
9. $\{3, 4, 5\}$ **11.** $\{1, 3, 4, 5, 6, 7, 10\}$ **13.** $\{2, 4, 6, 8, 10\}$
15. $\{2, 3, 4, 6, 8, 9, 10\}$ **17.** $\{1, 3, 5, 7, 9\}$ **19.** $\{x \mid x \text{ is a nonzero integer}\}$ **21.** $\mathbb{W}$ **23.** $\varnothing$ **25.** $\varnothing$ **27.** U **29.** $\{1, 2, 3, 4, 5, 6\}$
31. $\{1, 2, 3, 4, 5, 7\}$ **33.** $\{5\}$ **35.** $\{5, 6, 7\}$ **37.** $\{1, 2, 4, 6\}$
39. $\{1, 2, 3, 4\}$ **41.** $\{1, 2, 3, 4, 6, 7\}$ **43.** A **45.** $A \cap B$ **47.** $\overline{A \cap B}$

49.

51.
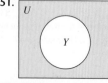

53. yes **55.** 3 **57.** 51,000 **59.** 2^n regions

2.3 Applications of Sets, page 70

1. $\overline{X \cup Y} = \overline{X} \cap \overline{Y}$ and $\overline{X \cap Y} = \overline{X} \cup \overline{Y}$ **3.** $\{3, 5\}$ **5.** $\{7\}$
7. $\{5, 6, 7\}$ **9.** $\{1, 2, 4, 6, 7\}$ **11.** $\overline{X} \cup \overline{Y}$ **13.** $\overline{X \cap Y}$
15. $X \cup Y$ **17.** $\overline{X} \cap \overline{Y}$

19.
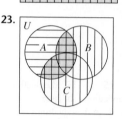

21.
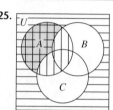

23.
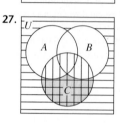

25.
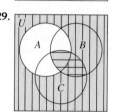

27.
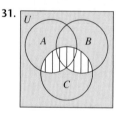

29.
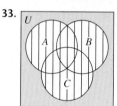

31.
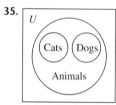

33.
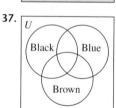

35.
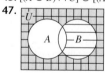

37.

39.
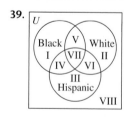

I: 24,985,000
II: 203,825,000
III: 15,780,000
IV: 5,260,000
V: 1,315,000
VI: 2,630,000
VII: 0
VIII: 9,205,000

41. Steffi Graf, II; Michael Stich, V; Martina Navratilova, III; Stefan Edberg, V; Chris Evert Lloyd, VI; Boris Becker, V; Evonne Goolagong, II; Pat Cash, V; Virginia Wade, II; John McEnroe, IV **43.** $\overline{A} \cap B$
45. $[(\overline{A \cup B}) \cap C] \cup [(A \cap B) \cap C]$

47.

It is true.

49.

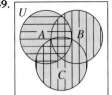 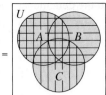

It is true.

51.

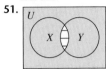

It is true.

53. No, there were 102 persons polled, not 100.

55. a.

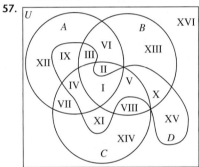

$E = \{$favor Prop. 8$\}$
$T = \{$favor Prop. 13$\}$
$F = \{$favor Prop. 5$\}$
b. 12   **c.** 7

57.

59. a. region 32   **b.** region 2

2.4 Finite and Infinite Sets, page 77

5. $A \times B = \{(c, w), (c, x), (d, w), (d, x), (f, w), (f, x)\}$ **7.** $\{(1, a), (1, b),$
$(1, c), (2, a), (2, b), (2, c), (3, a), (3, b), (3, c), (4, a), (4, b), (4, c), (5, a),$
$(5, b), (5, c)\}$ **9.** $\{(2, a), (2, e), (2, i), (2, o), (2, u), (3, a), (3, e), (3, i),$
$(3, o), (3, u), (4, a), (4, e), (4, i), (4, o), (4, u)\}$ **11.** 143   **13.** 85
15. 21   **17.** 0   **19.** $\aleph_0$ **21.** 460   **23.** $\aleph_0$ **25.** 260   **27.** 5,000
29. $\{m, a, t\}$; $\{m, a, t\}$; there are others.

$\updownarrow \updownarrow \updownarrow \quad \updownarrow \updownarrow \updownarrow$

$\{1, 2, 3\}$; $\{3, 2, 1\}$

31. $\{1, \quad 2, \ldots, \quad n, \quad n+1, \ldots, \quad 353, 354, 355, 356 \ldots, 586,$
$587\}$

$\updownarrow, \updownarrow \qquad \updownarrow \qquad \updownarrow \qquad \qquad \updownarrow \quad \updownarrow$

$\{550, 551, \ldots, n + 549, n + 550, \ldots, 902, 903\}$

Thus, the sets do not have the same cardinality.

33. a. finite   **b.** infinite

35. $\{1, \quad 2, \quad 3, \ldots, \quad n, \ldots\}$

$\updownarrow \quad \updownarrow \quad \updownarrow \qquad \updownarrow$

$\{-1, -2, -3, \ldots, -n, \ldots\}$

Since these sets can be put into a one-to-one correspondence, they
have the same cardinality—namely, $\aleph_0$.

37. $\{1, 2, 3, \ldots, n, \ldots\}$

$\updownarrow \updownarrow \updownarrow \qquad \updownarrow$

$\{\frac{1}{1}, \frac{1}{2}, \frac{1}{3}, \ldots, \frac{1}{n}, \ldots\}$

Since these sets can be put into a one-to-one correspondence, they
have the same cardinality—namely, $\aleph_0$.

39. $\mathbb{W} = \{0, 1, 2, 3, \ldots, \quad n, \ldots\}$

$\updownarrow \updownarrow \updownarrow \updownarrow \qquad \updownarrow$

$\{1, 2, 3, 4, \ldots, n + 1, \ldots\}$

Since these sets can be put into a one-to-one correspondence, they
have the same cardinality—namely, $\aleph_0$.

41. $\{1, 3, 5, \ldots, 2n + 1, \ldots\}$

$\updownarrow \updownarrow \updownarrow \qquad \updownarrow$

$\{1, 2, 3, \ldots, \quad n, \ldots\}$

Since these sets can be put into a one-to-one correspondence, they
have the same cardinality—namely, $\aleph_0$.

43. $\{1, 2, 3, \ldots, \quad n, \ldots\}$

$\updownarrow \updownarrow \updownarrow \qquad \updownarrow$

$\{1, 3, 9, \ldots, \quad 3^{n-1}, \ldots\}$

Since these sets can be put into a one-to-one correspondence, they
have the same cardinality—namely, $\aleph_0$.

45. $\mathbb{W} = \{0, 1, 2, 3, \ldots, \quad n, \ldots\}$

$\updownarrow \updownarrow \updownarrow \updownarrow \qquad \updownarrow$

$\{1, 2, 3, 4, \ldots, n + 1, \ldots\}$

Since $\mathbb{W}$ can be put into a one-to-one correspondence with a proper
subset of itself, it is infinite.

47. $\{12, 14, 16, \ldots, \quad n, \ldots\}$

$\updownarrow \updownarrow \updownarrow \qquad \updownarrow$

$\{14, 16, 18, \ldots, n + 2, \ldots\}$

Thus, this set is infinite.

49. $\mathbb{W} = \{0, 1, 2, 3, \ldots, \quad n, \ldots\}$

$\updownarrow \updownarrow \updownarrow \updownarrow \qquad \updownarrow$

$\{1, 2, 3, 4, \ldots, n + 1, \ldots\}$

Thus, $\mathbb{W}$ is countably infinite.

51. $\mathbb{Q}$ is countably infinite because it can be put into a one-to-one correspondence with the counting numbers (see Example 3).

53. $\{2, 4, 8, \ldots, 2^n, \ldots\}$

$\updownarrow \updownarrow \updownarrow \qquad \updownarrow$

$\{1, 2, 3, \ldots, n, \ldots\}$

Thus, the set $\{2, 4, 8, 16, 32, \ldots\}$ is countably infinite.

55. If a set is uncountable, then it is infinite, by definition; the statement is
true.   **57.** Some infinite sets have cardinality $\aleph_0$. Whereas other infinite
sets may have larger cardinality; the statement is false.   **59.** Answers
vary; odd numbers + even numbers = counting numbers.

Chapter 2 Review Questions, page 80

1. a. $\{1, 2, 3, 4, 5, 6, 7, 9, 10\}$ **b.** $\{9\}$ **2. a.** $\{1, 3, 5, 7, 8\}$ **b.** 0
3. a. 10   **b.** 5   **4. a.** 25   **b.** 50   **5. a.** $\{1, 2, 3, 4, 5, 6, 7, 8, 10\}$
b. $\{1, 2, 3, 4, 5, 6, 7, 8, 10\}$ **c.** yes, from De Morgan's law
6. $\{2, 4, 6, 8, 10\}$ **7.** $\{2, 4, 6, 10\}$ **8. a.** $\subset$ or $\subseteq$ **b.** $\in$
c. $=$ or $\subseteq$ **d.** $=$ or $\subseteq$ **e.** $\subset$ or $\subseteq$ **9. a.** 25   **b.** 41
c. 20   **d.** 46

10. a. 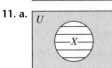   **b.**

11. a. 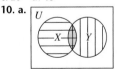   **b.**

12. a. 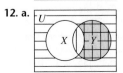   **b.**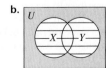

13. a. 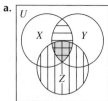 **b.**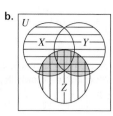

14. 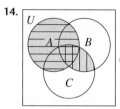 $\neq$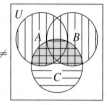

The statement has been disproved.

15. 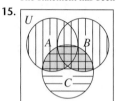 $=$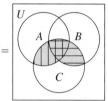

The statement has been proved.

16. a. The set of rational numbers is the set of all numbers of the form $\frac{a}{b}$ such that a is an integer and b is a counting number.
b. $\frac{2}{3}$; answers vary.

17. $\{5,\ 10,\ 15,\ \ldots,\ 5n,\ \ldots\}$
$\qquad \updownarrow \quad \updownarrow \quad \updownarrow \qquad \updownarrow$
$\{10, 20, 30, \ldots, 10n, \ldots\}$

Since the second set is a proper subset of the first set, we see that the set F is infinite.

18. a. $\{n \mid (-n) \in \mathbb{N}\}$; answers vary.
b. $\{\ 1,\quad 2,\quad 3, \ldots,\quad n, \ldots\}$
$\qquad \updownarrow \quad \updownarrow \quad \updownarrow \qquad \updownarrow$
$\{-1, -2, -3, \ldots, -n, \ldots\}$

Since the first set is the set of counting numbers, it has cardinality $\aleph_0$; so the given set also has cardinality $\aleph_0$ since it can be put into a one-to-one correspondence with the set of counting numbers.

19. Ten students had none of the items.

```
U
   C          TV
  5    20     5
     15
   2     10
     3
        B
              10
```

20. a. $A^-, AB^-, B^-, A^+, AB^+, B^+, O^+, O^-$
b. $A^-, AB^-, B^-, A^+, AB^+, B^+, O^-$ **c.** A, B, O^+

CHAPTER 3

3.1 Deductive Reasoning, page 89

5. a, b, and d are statements. **7.** a, b, and d are statements. **9.** Claire
11. Charles, Joe **13.** T **15.** F **17.** T **19.** F **21.** Some dogs do not have fleas. **23.** Some triangles are squares. **25.** Some counting numbers are not divisible by 1. **27.** All integers are odd. **29. a.** T **b.** F **c.** T
d. F **31. a.** Prices or taxes will rise. **b.** Prices will not rise and taxes will rise. **c.** Prices will rise or taxes will not rise. **d.** Prices will not rise or taxes will not rise. **33. a.** T **b.** F **c.** T **d.** F **35. a.** T **b.** F
37. a. F **b.** T **39. a.** T **b.** T **41.** $e \wedge d \wedge t$ **43.** $\sim t \wedge \sim m$
45. $(j \vee i) \wedge \sim p$ **47.** $d \vee e$ **49. a.** $(s \wedge t) \vee c$ **b.** $s \wedge (t \vee c)$ **51.** T
53. T **55.** T **57.** T **59.** Melissa did change her mind.

3.2 Truth Tables and the Conditional, page 98

5.

p	q	$\sim p$	$\sim p \vee q$
T	T	F	T
T	F	F	F
F	T	T	T
F	F	T	T

7.

p	q	$p \wedge q$	$\sim(p \wedge q)$
T	T	T	F
T	F	F	T
F	T	F	T
F	F	F	T

9.

r	$\sim r$	$\sim(\sim r)$
T	F	T
F	T	F

11.

p	q	$\sim q$	$p \wedge \sim q$
T	T	F	F
T	F	T	T
F	T	F	F
F	F	T	F

13.

p	q	$\sim p$	$\sim p \wedge q$	$\sim q$	$(\sim p \wedge q) \vee \sim q$
T	T	F	F	F	F
T	F	F	F	T	T
F	T	T	T	F	T
F	F	T	F	T	T

15.

p	q	$p \rightarrow q$	$p \vee (p \rightarrow q)$
T	T	T	T
T	F	F	T
F	T	T	T
F	F	T	T

17.

p	q	$p \vee q$	$p \wedge (p \vee q)$	$[p \wedge (p \vee q)] \rightarrow p$
T	T	T	T	T
T	F	T	T	T
F	T	T	F	T
F	F	F	F	T

19.

p	q	r	$p \vee q$	$(p \vee q) \vee r$
T	T	T	T	T
T	T	F	T	T
T	F	T	T	T
T	F	F	T	T
F	T	T	T	T
F	T	F	T	T
F	F	T	F	T
F	F	F	F	F

21.

p	q	r	$p \vee q$	$\sim r$	$(p \vee q) \wedge \sim r$	$[(p \vee q) \wedge \sim r] \wedge r$
T	T	T	T	F	F	F
T	T	F	T	T	T	F
T	F	T	T	F	F	F
T	F	F	T	T	T	F
F	T	T	T	F	F	F
F	T	F	T	T	T	F
F	F	T	F	F	F	F
F	F	F	F	T	F	F

23. If the slash through "P" means no parking, then a slash through "No Parking" must mean park here (law of double negation).
25. Statement: $\sim p \rightarrow \sim q$; Converse: $\sim q \rightarrow \sim p$
Inverse: $p \rightarrow q$; Contrapositive: $q \rightarrow p$
27. Statement: $\sim t \rightarrow \sim s$; Converse: $\sim s \rightarrow \sim t$
Inverse: $t \rightarrow s$; Contrapositive: $s \rightarrow t$
29. Statement: If I get paid, then I will go Saturday.
Converse: If I go Saturday, then I got paid.
Inverse: If I do not get paid, then I will not go Saturday.
Contrapositive: If I do not go Saturday, then I do not get paid.
31. If it is a triangle, then it is a polygon. **33.** If you are a good person, then you will go to heaven. **35.** If we make a proper use of these means

which the God of Nature has placed in our power, then we are not weak.   **37.** If it is work, then it is noble.   **39.** T   **41.** T   **43. a.** T   **b.** F   **c.** F   **45. a.** F   **b.** T   **c.** T   **47.** $(q \wedge \sim d) \to n$, where q: The qualifying person is a child; d: This child is your dependent; n: You enter your child's name.   **49.** $(b \vee c) \to n$, where b: The amount on line 32 is \$86,025; c: The amount on line 32 is less than \$86,025; n: You multiply the number of exemptions by \$2,500.   **51.** $(m \vee d) \to s$, where m: You are a student; d: You are a person with disabilities; s: you see line 6 of instructions.   **53. a.** $a \to (e \vee f)$ **b.** $(a \wedge e) \to q$ **c.** $(a \wedge f) \to q$ **55.** $[(m \vee t \vee w \vee h) \wedge s] \to q$ **57.** $t \to (m \wedge s \wedge p)$ **59.** This is not a statement, since it gives rise to a paradox and is neither true nor false.

3.3 Operators and Laws of Logic, page 105

7. a. 9 hours   **b.** if you look at the middle sign, 10 hours; if you look at the lower sign, 15 minutes   **9.** no   **11.** no   **13.** yes   **15.** no

17.

p	q	$p \vee q$	$\sim(p \vee q)$
T	T	T	F
T	F	T	F
F	T	T	F
F	F	F	T

19.

p	q	$\sim q$	$p \to \sim q$
T	T	F	F
T	F	T	T
F	T	F	T
F	F	T	T

21. Let h: I will buy a new house; p: All provisions of the sale are clearly understood. Not h unless p: $\sim p \to \sim h$.   **23.** Let r: I am obligated to pay the rent; s: I signed the contract. r because s: $(r \wedge s) \wedge (s \to r)$ **25.** Let m: It is a man; i: It is an island. No m is i: $m \to \sim i$ **27.** Let n: You are nice to people on your way up; m: You will meet people on your way down. n because m: $(n \wedge m) \wedge (m \to n)$.   **29.** Let f: The majority, by mere force of numbers, deprives a minority of a clearly written constitutional right; r: Revolution is justified. If f then r: $f \to r$. *Answers to Problems 30–35 may vary.*   **31.** The cherries have not turned red or they are ready to be picked.   **33.** If Hannah watches Jon Stewart, then she watches the NBC late-night orchestra.   **35.** If the money is not available, then I will not take my vacation.

37.

p	$\sim p$	$\sim(\sim p)$	$p \leftrightarrow \sim(\sim p)$
T	F	T	T
F	T	F	T

Thus, $p \Leftrightarrow \sim(\sim p)$.

39.

p	q	$p \vee q$	$\sim(p \vee q)$	$\sim p$	$\sim q$	$\sim p \wedge \sim q$	$\sim(p \vee q) \leftrightarrow (\sim p \wedge \sim q)$
T	T	T	F	F	F	F	T
T	F	T	F	F	T	F	T
F	T	T	F	T	F	F	T
F	F	F	T	T	T	T	T

Thus, $\sim(p \vee q) \Leftrightarrow \sim p \wedge \sim q$.

41.

p	q	$p \to q$	$\sim p$	$\sim p \vee q$	$(p \to q) \leftrightarrow (\sim p \vee q)$
T	T	T	F	T	T
T	F	F	F	F	T
F	T	T	T	T	T
F	F	T	T	T	T

Thus, $(p \to q) \Leftrightarrow (\sim p \vee q)$.

43. $\sim(p \to \sim q) \Leftrightarrow \sim(\sim p \vee \sim q) \Leftrightarrow \sim(\sim p) \wedge \sim(\sim q) \Leftrightarrow p \wedge q$ **45.** $\sim(\sim p \to \sim q) \Leftrightarrow \sim[\sim(\sim p) \vee \sim q] \Leftrightarrow \sim p \wedge \sim(\sim q) \Leftrightarrow \sim p \wedge q$ **47.** Jane did not go to the basketball game and she did not go to the soccer game.   **49.** Missy is on time or she did not miss the boat.   **51.** You are out of Schlitz and you have beer.   **53.** $x - 5 = 4$ and $x \neq 1$.   **55.** $x = 1$ and $y = 2$, and $2x + 3y \neq 8$.   **57.** Let d: You purchase your ticket between January 5 and February 15; f: You fly round trip between February 20 and May 3; m: You depart on Monday; t: You depart on Tuesday; w: You depart on Wednesday; h: You return on Tuesday; i: You return on Wednesday; j: You return on Thursday; s: You stay over a Saturday night; e: You obtain 40% off regular fare. Symbolic statement: $[d \wedge f \wedge (m \vee t \vee w) \wedge (h \vee i \vee j) \wedge s] \to e$ **59.** Let ℓ: The tenant lets the premises; m: The tenant lets a portion of the premises; s: The tenant sublets the premises; t: The tenant

sublets a portion of the premises; p: Permission is obtained. Symbolic statement: $\sim p \to [\sim(\ell \vee m) \wedge \sim(s \vee t)]$

3.4 The Nature of Proof, page 113

1. reasoning that can be written in the form $[(p \to q) \wedge p] \to q$ **3.** reasoning that can be written in the form $[(p \to q) \wedge (q \to r)] \to (p \to r)]$ **7. a.** valid; indirect   **b.** invalid; fallacy of the converse   **9. a.** invalid; fallacy of the inverse   **b.** valid; direct   **11.** valid; direct   **13.** valid; transitive   **15.** valid; direct   **17.** valid; direct   **19.** invalid; false chain   **21.** valid; indirect   **23.** valid; indirect   **25.** invalid; fallacy of the converse   **27.** valid; transitive   **29.** valid; transitive   **31.** valid; indirect   **33.** valid; contrapositive and transitive   **35.** false chain pattern   **37.** If you learn mathematics, then you understand human nature. (*transitive*)   **39.** We do not go to the concert. (*indirect*)   **41.** $b = 0$ (*excluded middle*)   **43.** We do not interfere with the publication of false information. (*indirect*)   **45.** I will not eat that piece of pie. (*indirect*)   **47.** We will go to Europe. (*transitive; direct*)   **49.** Babies cannot manage crocodiles. (*transitive*)   **51.** None of my poultry are officers. (*transitive, contrapositive* twice, *indirect*)   **53.** Airsecond Aircraft Company suffers financial setbacks. (*direct; indirect*)   **55.** The janitor could not have taken the elevator because the building fuses were blown.   **57.** Guinea pigs do not appreciate Beethoven.   **59.** This is not an easy problem.

3.5 Problem Solving Using Logic, page 121

1. a. The playing board is an undefined term. The directions simply present the board, but do not define it.   **b.** This is a stated rule of the game, so we would call this an axiom.   **c.** Passing "GO" is an undefined term, because the rules of the game simply refer to GO.   **d.** The "chance" cards are part of the definitions, so this is a defined term.   **e.** This playing piece is simply given as part of the game, so it qualifies as an undefined term.   **3. a.** The playing pieces in *Sorry* are not specifically defined, so this would be an undefined term.   **b.** This is a stated rule of the game, so we would call this an axiom.   **c.** "Home" is a defined term in *Sorry*.   **d.** This is a stated rule of the game, so we would call this an axiom.   **e.** This term is not defined, so it is an undefined term.   **5. a.** This is a stated rule of the game, so we would call this an axiom.   **b.** This is a stated rule of the game, so we would call this an axiom.   **c.** It is assumed that you know what is meant by a "die," so we take this as an undefined term.   **d.** A door's location is precisely located on the playing board, but the notion of a door itself is not defined, so this is an undefined term.   **e.** This is a stated rule of the game, so we would call this an axiom.   **7. a.** While the number, size, and specifications of permitted players are all part of the rules, the word "player" is not defined, so this is an undefined term.   **b.** This is a stated rule of the game, so we would call this an axiom.   **c.** This is a defined term.   **d.** This is a defined term.   **e.** This is a consequence of the rules of the game, so this would be considered a theorem.

9.

Statements	Reasons
MIII	Given
MIIIIII	Doubling (Rule 2)
MUU	Substitution (Rule 3)
M	Deletion (Rule 4)

11.

Statements	Reasons
MI	Given
MII	Doubling (Rule 2)
MIIII	Doubling (Rule 2)
MIIIIIIII	Doubling (Rule 2)
MIUIU	Substitution (Rule 3)

13.

Statements	Reasons
MIIIUUIIIII	Given
MIIIIIIII	Deletion (Rule 4)
MIIUU	Substitution (Rule 3)
MII	Deletion (Rule 4)
MIIU	Addition (Rule 1)

15.

Statements	Reasons
EA	Given
AE	Rule 3
AEE	Rule 1
A	Rule 5

17.

Statements	Reasons
AI	Given
AIII	Rule 1
AE	Rule 4
AEE	Rule 1
A	Rule 5

19.

Statements	Reasons
AE	Given
AEE	Rule 1
A	Rule 5

21. The first is a knight and the second is also a knight.   **23.** Bob is a knave and Cary is a knight.   **25.** white   **27.** need to choose three socks   **29.** There must be 49 red and 1 yellow.   **31.** Alice has an ace on her head.   **33.** Ben and Cole have kings and Alice has a lower card.   **35.** Ben says, "I lose" and Cole says, "I win." Cole has a queen and the others have lower cards. *You should take Problems 37–54 together.*   **37.** 6   **39.** 3   **41.** 6   **43.** 7   **45.** 3   **47.** 3   **49.** 4   **51.** 2   **53.** 2   **55.** Moe knows that his hat must be black.   **57.** Gary was 17 and had 15 marbles; Harry was 10 and had 12 marbles; Iggy was 3 and had 12 marbles; Jack was 18 and had 9 marbles. They shot in the order of Gary, Jack, Harry, and then Iggy.   **59.** three weighings

3.6 Logic Circuits, page 128

5. light   **7.** AND-gate   **9.** NOT-gate   *Answers to Problems 10–15 may vary.*   **11.** $\sim p$ **13.** p or q; $p \vee q$ **15.** $p \wedge \sim q$

17.

p	q	$\sim p$	$\sim p \wedge q$
T	T	F	F
T	F	F	F
F	T	T	T
F	F	T	F

19.

p	q	$\sim q$	$p \vee \sim q$	$\sim(p \vee \sim q)$
T	T	F	T	F
T	F	T	T	F
F	T	F	F	T
F	F	T	T	F

21. $p \vee q$

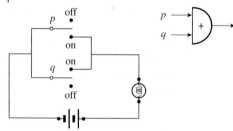

23. $p \wedge \sim q$

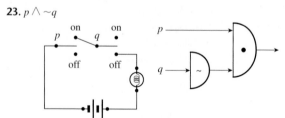

25. $\sim(p \wedge q)$

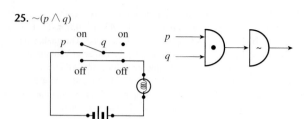

27. $q \to p$

p	q	$q \to p$
T	T	T
T	F	T
F	T	F
F	F	T

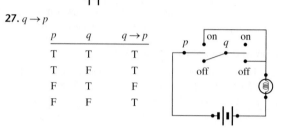

29. $q \to \sim p$ (note that this is the contrapositive of Problem 28, so the truth table and circuits are the same.)

p	q	$\sim p$	$q \to \sim p$
T	T	F	F
T	F	F	T
F	T	T	T
F	F	T	T

31. $\sim(q \to p)$

p	q	$q \to p$	$\sim(q \to p)$
T	T	T	F
T	F	T	F
F	T	F	T
F	F	T	F

33. $(p \wedge q) \vee r$ **35.** $\sim(\sim p \vee q)$

37. $(\sim p \wedge q) \vee (p \vee \sim q)$

39.

p	q	r	$p \vee q$	$(p \vee q) \vee r$
T	T	T	T	T
T	T	F	T	T
T	F	T	T	T
T	F	F	T	T
F	T	T	T	T
F	T	F	T	T
F	F	T	F	T
F	F	F	F	F

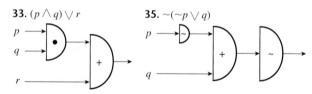

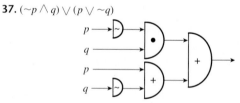

41.

p	q	r	q ∨ r	p ∨ (q ∨ r)
T	T	T	T	T
T	T	F	F	T
T	F	T	F	T
T	F	F	F	T
F	T	T	T	T
F	T	F	F	F
F	F	T	F	F
F	F	F	F	F

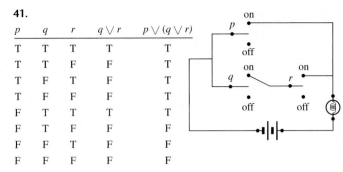

43.

p	q	r	p ∧ q	(p ∧ q) ∨ r
T	T	T	T	T
T	T	F	T	T
T	F	T	F	T
T	F	F	F	F
F	T	T	F	T
F	T	F	F	F
F	F	T	F	T
F	F	F	F	F

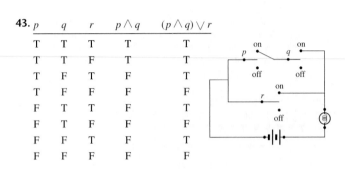

45.

p	q	r	q ∧ r	p ∧ (q ∨ r)
T	T	T	T	T
T	T	F	F	F
T	F	T	F	F
T	F	F	F	F
F	T	T	F	F
F	T	F	F	F
F	F	T	F	F
F	F	F	F	F

Note the circuit for this problem is the same as the one for Problem 38. This fact is sometimes called the *associative property for conjunction*. That is,

$$(p \wedge q) \wedge r = p \wedge (q \wedge r)$$

47. Let b_1: the first rest room is occupied; b_2: the second rest room is occupied; b_3: the third rest room is occupied.
a. $(b_1 \vee b_2) \vee b_3$; note that this is equivalent to $b_1 \vee (b_2 \vee b_3)$.
b.

b_1	b_2	b_3	$b_1 \vee b_2$	$(b_1 \vee b_2) \vee b_3$
T	T	T	T	T
T	T	F	T	T
T	F	T	T	T
T	F	F	T	T
F	T	T	T	T
F	T	F	T	T
F	F	T	F	T
F	F	F	F	F

c.

49. Let a: the thermostat is set on automatic mode; d: all the doors are closed; m: the thermostat is set on manual mode; c: the proper authorization code has been entered. A logical statement might be: $(a \wedge d) \vee (m \wedge c)$.

51.

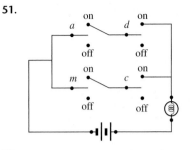

53.

p	q	p → q	~p	~p ∨ q	(p → q) ↔ (~p ∨ q)
T	T	T	F	T	T
T	F	F	F	F	T
F	T	T	T	T	T
F	F	T	T	T	T

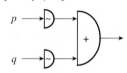

$(p \rightarrow q) \Leftrightarrow (\sim p \vee q)$

55. Notice that $p \rightarrow \sim q \Leftrightarrow \sim p \vee \sim q$.

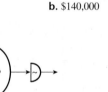

57. a.   **b.** $140,000

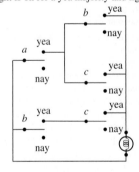

59. Let the committee members be a, b, and c, respectively. Light *on* represents a majority. Light is on for a yea majority and light is off otherwise.

Chapter 3 Review Questions, page 130

1. a. A *logical statement* is a declarative sentence that is either true or false.   **b.** A *tautology* is a logical statement in which the conclusion is equivalent to its premise.   **c.** $(p \rightarrow q) \leftrightarrow (\sim q \rightarrow \sim p)$ or, in words: A conditional may always be replaced by its contrapositive without having its truth value affected.

2.

p	q	~p	p ∧ q	p ∨ q	p → q	p ↔ q
T	T	F	T	T	T	T
T	F	F	F	T	F	F
F	T	T	F	T	T	F
F	F	T	F	F	T	T

3.

p	q	$p \wedge q$	$\sim(p \wedge q)$
T	T	T	F
T	F	F	T
F	T	F	T
F	F	F	T

4.

p	q	$\sim q$	$p \vee \sim q$	$\sim p$	$(p \vee \sim q) \wedge \sim p$	$[(p \vee \sim q) \wedge \sim p] \to \sim q$
T	T	F	T	F	F	T
T	F	T	T	F	F	T
F	T	F	F	T	F	T
F	F	T	T	T	T	T

5.

p	q	r	$p \wedge q$	$(p \wedge q) \wedge r$	$[(p \wedge q) \wedge r] \to p$
T	T	T	T	T	T
T	T	F	T	F	T
T	F	T	F	F	T
T	F	F	F	F	T
F	T	T	F	F	T
F	T	F	F	F	T
F	F	T	F	F	T
F	F	F	F	F	T

6.

p	q	$\sim p$	$\sim q$	$p \wedge q$	$\sim(p \wedge q)$	$\sim p \vee \sim q$	$\sim(p \wedge q) \leftrightarrow (\sim p \vee \sim q)$
T	T	F	F	T	F	F	T
T	F	F	T	F	T	T	T
F	T	T	F	F	T	T	T
F	F	T	T	F	T	T	T

7. $p \to q$

p

∴ q

p	q	$p \to q$	$(p \to q) \wedge p$	$[(p \to q) \wedge p] \to q$
T	T	T	T	T
T	F	F	F	T
F	T	T	F	T
F	F	T	F	T

8. Answers vary.

> If you study hard, then you will get an *A*.
> You do not get an *A*.
> Therefore, you did not study hard.

9. Answers vary; fallacy of the converse, fallacy of the inverse, or false chain pattern; for example:

> If you study hard, then you will get an *A*.
> You do not study hard, so you do not get an *A*. *(fallacy of the inverse.)*

10. Yes; it is indirect reasoning.   **11. a.** Some birds do not have feathers.
b. No apples are rotten.   **c.** Some cars have two wheels. **d.** All smart people attend college.   **e.** You go on Tuesday and you can win the lottery.
12. a. T   **b.** T   **c.** T   **d.** T   **e.** T   **13. a.** If P is a prime number, then $P + 2$ is a prime number.   **b.** Either P or $P + 2$ is a prime number.
14. a. Let p: This machine is a computer; q: This machine is capable of self-direction. $p \to \sim q$ **b.** Contrapositive: $\sim(\sim q) \to \sim p$; $q \to \sim p$. If this machine is capable of self-direction, then it is not a computer.   **15.** Let p: There are a finite number of primes; q: There is some natural number, greater than 1, that is not divisible by any prime. Then the argument in symbolic form is:

$$p \to q$$
$$\sim q$$
$$\therefore \sim p$$

Conclusion: There are infinitely many primes (assuming that "not a finite number" is the same as "infinitely many"). This is indirect reasoning.
16. a.

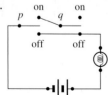

b.

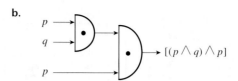

17. I do not attend to my duties.   **18.** No prune is an artificial sweetener.
19. All squares are polygons.   **20.** North: Maureen (baker); East: Terry (hiker); South: Josie (gardener); West: Warren (surfer)

CHAPTER 4

4.1 Early Numeration Systems, page 143

7. a. Roman, Egyptian   **b.** Roman, Babylonian   **c.** Egyptian, Roman, Babylonian   **d.** Egyptian, Roman, Babylonian   **e.** Roman, Babylonian
f. Roman   **9. a.** $\frac{2}{3}$ **b.** 100,010   **11. a.** $24\frac{1}{2}$ **b.** $\frac{1}{100}$ **13. a.** 1,100
b. 1,997   **15. a.** 1,999   **b.** 2,001   **17. a.** 7,600   **b.** 9,712   **19. a.** 71
b. 671   **21. a.** 47   **b.** 25   **23.** 133   **25.** 1
27. ∩ ∩ ∩ ∩ ∩ ∩ ∩ |||||

29. 99999∩∩| **31.** 𝕷 𝕷 ||||||| **33.** LXXV
35. DXXI   **37.** MMVII   **39.** ▼ ◁ ▼▼▼▼▼
41. ▼▼▼▼▼▼▼ ◁◁◁◁ ▼
43. ◁◁◁▼▼▼◁◁▼▼▼▼▼▼▼
45. 𝕷999999999∩∩∩∩∩∩∩∩||||| ||
47. ◁◁◁◁◁   **49.** ◁◁▼▼▼▼▼   **51.** VIII
53. LXXXIX   **55.** MMMCMXCIX   **57. a.** MDCLXVI
b. MCDXLIV   **59.** ▼▼▼▼▼ by ◁▼▼

4.2 Hindu–Arabic Numeration System, page 147

9. 5 units   **11.** 5 hundred-thousandths   **13.** 5 thousandths
15. a. 100,000   **b.** 1,000   **17. a.** 0.0001   **b.** 0.001   **19. a.** 5,000
b. 500   **21. a.** 0.06   **b.** 0.00009   **23.** 10,234   **25.** 521,658
27. 7,000,000.03   **29.** 500,457.34   **31.** 20,600.40769
33. a. $9 \times 10^{-2} + 6 \times 10^{-3} + 4 \times 10^{-4} + 2 \times 10^{-5} + 1 \times 10^{-6}$
b. $2 \times 10^1 + 7 \times 10^0 + 5 \times 10^{-1} + 7 \times 10^{-2} + 2 \times 10^{-3}$
35. a. $6 \times 10^3 + 2 \times 10^2 + 4 \times 10^1 + 5 \times 10^0$
b. $2 \times 10^6 + 3 \times 10^5 + 5 \times 10^3 + 6 \times 10^2 + 8 \times 10^1 + 1 \times 10^0$
37. $5 \times 10^{-6} + 2 \times 10^{-7} + 7 \times 10^{-8}$
39. $8 \times 10^2 + 9 \times 10^1 + 3 \times 10^0 + 1 \times 10^{-4}$
41. $6 \times 10^5 + 7 \times 10^4 + 8 \times 10^3 + 1 \times 10^{-2}$
43. $5 \times 10^4 + 7 \times 10^3$
$+ 2 \times 10^2 + 8 \times 10^1 + 5 \times 10^0 + 9 \times 10^{-1}$
$+ 3 \times 10^{-2} + 6 \times 10^{-3} + 1 \times 10^{-4}$ **45.** 3,201   **47.** 5,001,005
49. 8,009,026
51.

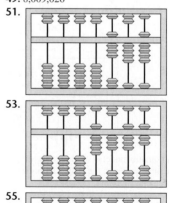

53.

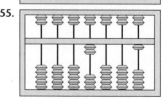

55.

57.

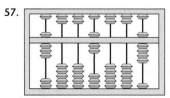

59. 22, 23, 24, 30, 31, 32, 33, 34, 40, . . .

4.3 Different Numeration Systems, page 152

5. a. 9   **b.** 14_{five} **c.** 100_{three} **d.** 11_{eight} **e.** 1001_{two} **f.** 10_{nine}
7. $6 \times 8^2 + 4 \times 8^1 + 3 \times 8^0$ **9.** $1 \times 2^5 + 1 \times 2^4 + 1 \times 2^2 + 1 \times 2^1$
$+ 1 \times 2^0 + 1 \times 2^{-1} + 1 \times 2^{-4}$ **11.** $6 \times 8^7 + 4 \times 8^6 + 2 \times 8^5$
$+ 5 \times 8^1 + 1 \times 8^0$ **13.** $3 \times 4^5 + 2 \times 4^4 + 3 \times 4^3 + 2 \times 4^{-1}$
15. $3 \times 5^0 + 4 \times 5^{-1} + 2 \times 5^{-3} + 3 \times 5^{-4} + 1 \times 5^{-5}$ **17.** 343
19. 4,307   **21.** 116   **23.** 11.625   **25.** 807   **27.** 66   **29.** 351.125
31. a. 10344_{five} **b.** 21310_{four} **33. a.** 3122_{five} **b.** 1147_{eight}
35. a. 1000000000_{two} **b.** 1002110_{three} **37. a.** 1132_{eight} **b.** 1030_{four}
39. 7 weeks, 3 days   **41.** 4 ft, 7 in.   **43.** 3 gross, 5 doz, 8 units
45. $84 = 314_{five}$, so eight coins are needed   **47.** $954_{twelve} = 1,360$ pencils
49. $44 = 62_{seven}$; 6 weeks, 2 days   **51.** $29 = 15_{twenty-four}$; 1 day, 5 hours
53. 6 years, 3 months   **55.** 16 ft, 1 in.   **57.** 4 gross, 6 dozen, 3 units
59. 4 years, 11 months, 6 days

4.4 Binary Numeration System, page 157

3. 39   **5.** 167   **7.** 13   **9.** 11   **11.** 29   **13.** 27   **15.** 99   **17.** 184
19. 1101_{two} **21.** 100011_{two} **23.** 110011_{two} **25.** 1000000_{two}
27. 10000000_{two} **29.** 1100011011_{two} **31.** 68 79   **33.** 69 78 68
35. HAVE   **37.** STUDY   **39.** 101_{two} **41.** 1101_{two} **43.** 10_{two}
45. 100011_{two} **47.** The computer is correct because there must be either
war or peace.   **49. a.** 101_{two} **b.** 110_{two} **51. a.** $001\ 110\ 111_{two}$
b. $110\ 010\ 100_{two}$ **53. a.** 5_{eight} **b.** 7_{eight} **55.** 007750_{eight}
57. 773521_{eight}

4.5 History of Calculating Devices, page 168

11. a. 27   **b.** 63   **c.** 54   **13. a.** 243   **b.** 432   **c.** 504
15. ENIAC, UNIVAC, Cray, Altair, Apple
Answers to Problems 16–33 may vary.
17. Aristophanes developed a finger-counting system about 500 B.C.
19. Babbage built a calculating machine in the 19th century.   **21.** Berry
helped Atanasoff build the first computer.   **23.** Cray developed the first
supercomputer.   **25.** Engelbart invented the computer mouse.
27. Jobs was the cofounder of Apple Computer and codesigned the
Apple II computer.   **29.** Leibniz built a calculating device in 1695.
31. Napier invented a calculating device to do multiplication (in 1617).
33. Wozniak codesigned the Apple II computer.   **35.** yes; speed,
complicated computations   **37.** yes; speed, repetition   **39.** yes;
repetition **41.** yes; ability to make corrections easily   **43.** yes (but not
completely); it can help with some of the technical aspects   **45.** yes;
speed, complicated computations, repetition   **47.** E   **49.** D

Chapter 4 Review Questions, page 171

1. Answers vary. The position in which the individual digits are listed is
relevant; examples will vary.   **2.** Answers vary. Addition is easier in a
simple grouping system; examples will vary.   **3.** Answers vary. It uses ten
symbols; it is positional; it has a placeholder symbol (0); and it uses 10 as
its basic unit for grouping.   **4.** Answers vary. Should include finger
calculating, Napier's rods, Pascal's calculator, Leibniz' reckoning machine,
Babbage's difference and analytic engines, ENIAC, UNIVAC, Atanasoff's
and Eckert and Mauchly's computers, and the dispute they had in proving
their position in the history of computers. Should also include the role and
impact of the Apple and Macintosh computers, as well as the supercomputers (such as the Cray).   **5.** Answers vary. Should include illegal (breaking
into another's computer, adding, modifying, or destroying information;
copying programs without authorization or permission) and ignorance
(assuming that output information is correct, or not using software for

purposes for which it was intended).   **6.** Answers vary.   **a.** the physical
components (mechanical, magnetic, electronic) of a computer system
b. the routine programs, and associated documentation in a computer system   **c.** the process of creating, modifying, deleting, and formatting text
and materials   **d.** a system for communication between computers using
electronic cables, phone lines, or wireless devices   **e.** electronic mail—
that is, messages sent along computer modems   **f.** R̲andom-A̲ccess
Memory or memory where each location is uniformly accessible, often
used for the storage of a program and data being processed   **g.** an
electronic place to exchange information with others, usually on a particular topic   **h.** a computer component that reads and stores data   **7.** 10^9
8. $4 \times 10^2 + 3 \times 10^1 + 6 \times 10^0 + 2 \times 10^{-1} + 1 \times 10^{-5}$
9. $5 \times 8^2 + 2 \times 8^1 + 3 \times 8^0$ **10.** $1 \times 2^6 + 1 \times 2^3 + 1 \times 2^2 + 1 \times 2^1$
11. 4,020,005.62   **12.** 29   **13.** 123   **14.** 17   **15.** 1,177   **16.** 1100_{two}
17. 110100_{two} **18.** 11111010111_{two} **19.** $11110100001001000000_{two}$
20. a. $92E_{twelve}$ **b.** 400_{five}

CHAPTER 5

5.1 Natural Numbers, page 180

11. a. $4 + 4 + 4$ **b.** $3 + 3 + 3 + 3$ **13. a.** $184 + 184$
b. $2 + 2 + \cdots + 2$ (a total of 184 terms)   **15. a.** $y + y + y + \cdots + y$
(a total of x terms)   **b.** $x + x + x + \cdots + x$ (a total of y terms)
17. commutative   **19.** associative   **21.** commutative   **23.** commutative
25. commutative   **27.** commutative   **29.** none are associative
31. Answers vary; dictionary definitions are often circular.
33. a. $-i$ **b.** -1 **c.** 1   **d.** $-i$ **e.** i **f.** -1 **35.** Answers vary; yes
37. Answers vary;   **a.** yes   **b.** yes   **39.** yes; answers vary
41. a. $\square$ **b.** $\circ$ **c.** yes; no $[\circ \bullet \triangle \neq \triangle \bullet \circ]$ **d.** yes
43. Check: $a \downarrow (b \rightarrow c) = (a \downarrow b) \rightarrow (a \downarrow c)$. Try several examples; it holds.
45. a. $5 \times (100 - 1) = 500 - 5 = 495$ **b.** $4 \times (90 - 2) = 360 - 8 = 352$
c. $8 \times (50 + 2) = 400 + 16 = 416$ **47.** yes   **49.** yes; even + even = even
51. yes; odd $\times$ odd = odd   **53.** yes; let a and b be any elements in S. Then,
$a \odot b = 0 \cdot a + 1 \cdot b = b$ and we know that b is an element of S.   **55.** It is
commutative but not associative: $2 \oslash 3 = 10$ and $3 \oslash 2 = 10$; try other
examples. $(2 \oslash 3) \oslash 4 \neq 2 \oslash (3 \oslash 4)$ **57.** yes   **59.** Answers vary;
$0 + 1 + 2 - 3 - 4 + 5 - 6 + 7 + 8 - 9 = 1$

5.2 Prime Numbers, page 194

7. a. prime; use sieve   **b.** not prime; $3 \cdot 19$ **c.** not prime; only 1
divisor   **d.** prime; check primes under 45   **9. a.** prime; use sieve
b. prime; use sieve   **c.** not prime; $3^2 \cdot 19$ **d.** not prime; $3^2 \cdot 223$
11. a. T   **b.** F   **c.** T   **d.** F   **13. a.** F   **b.** T   **c.** T   **d.** T   **15.** Use the
sieve of Eratosthenes; you need only to cross out the multiples of primes
up to 17. The list of primes is: 2, 3, 5, 7, 11, 13, 17, 19, 23, 29, 31, 37, 41,
43, 47, 53, 59, 61, 67, 71, 73, 79, 83, 89, 97, 101, 103, 107, 109, 113, 127,
131, 137, 139, 149, 151, 157, 163, 167, 173, 179, 181, 191, 193, 197, 199,
211, 223, 227, 229, 233, 239, 241, 251, 257, 263, 269, 271, 277, 281, 283,
and 293.   **17. a.** $2^3 \cdot 3$ **b.** $2 \cdot 3 \cdot 5$ **c.** $2^2 \cdot 3 \cdot 5^2$ **d.** $2^4 \cdot 3^2$
19. a. $2^3 \cdot 3 \cdot 5$ **b.** $2 \cdot 3^2 \cdot 5$ **c.** $3 \cdot 5^2$ **d.** $3 \cdot 5^2 \cdot 13$ **21.** prime
23. prime   **25.** $13 \cdot 29$ **27.** $3 \cdot 5 \cdot 7$ **29.** prime   **31.** $3^2 \cdot 5 \cdot 7$
33. $3^4 \cdot 7$ **35.** $19 \cdot 151$ **37.** 12; 360   **39.** 3; 2,052   **41.** 1; 252
43. 15; 1,800   **45.** The least common multiple of 6 and 8 is 24, so the
next night off together is in 3 weeks, 3 days.   **47.** Answers vary.
49. 1, 3, 7, 9, 13, 15, 21, 25, 31, 33, 37, 43, 45, 49, 51, 63, 67, 69, 73, 75,
79, 87, 93, and 99   **51.** Answers vary.   **53. a.** yes   **b.** yes   **c.** yes
55. $2 \cdot 3 \cdot 5 \cdot 7 \cdot 11 \cdot 13 + 1 = 30{,}031$, which is not prime,
since $30{,}031 = 59 \cdot 509$ **57.** Answers vary; some other possibilities are:
2, 17, 37, 101, 197, 257, 401, 577, 677, 1297, 1601, 2917, 3137, 4357,
5477, 7057, 8101, 8837. A computer could also be used to answer this
question.   **59.** If $N = 5$, $2^{5-1}(2^5 - 1) = 496$.

5.3 Integers, page 203

7. a. 30   **b.** 30   **c.** -30 **d.** 0   **9. a.** 8   **b.** -2 **11. a.** 10   **b.** -4
13. a. -7 **b.** 12   **15. a.** 3   **b.** -3 **17. a.** -18 **b.** -20
19. a. -70 **b.** 70   **21. a.** -3 **b.** 7   **23. a.** 132   **b.** -1 **25. a.** -56

b. −75   **27. a.** −4   **b.** 4   **29. a.** −12   **b.** 15   **31. a.** 150   **b.** −19
33. a. 5   **b.** 10   **35. a.** 0   **b.** −8   **37. a.** 0   **b.** −23   **39. a.** −7
b. 0   **41. a.** 8   **b.** −2   **43. a.** 6   **b.** 7   **45. a.** 1   **b.** 22   **47. a.** 4
b. 16   **49. a.** 44   **b.** 16   **51. a.** 1   **b.** 1   **c.** 1   **d.** 1   **e.** 1   **f.** 1
53. a. 1   **b.** −1   **c.** −1   **d.** 1   **e.** −1   **f.** −1   **55. a.** An operation ○ is
commutative for a set S if $a \circ b = b \circ a$ for all elements a and b in S.   **b.**
yes   **c.** no   **d.** yes   **e.** no   **59.** 12345668765433

5.4 Rational Numbers, page 210

7. a. $\frac{1}{5}$ **b.** $\frac{1}{4}$ **9. a.** 2   **b.** 2   **11. a.** 3   **b.** $\frac{2}{3}$ **13. a.** $\frac{1}{8}$ **b.** $\frac{1}{3}$
15. a. $\frac{6}{35}$ **b.** $\frac{3}{20}$ **17. a.** $\frac{17}{21}$ **b.** $\frac{7}{8}$ **19. a.** $\frac{-92}{105}$ **b.** $\frac{5}{6}$ **21. a.** $\frac{1}{9}$
b. $\frac{71}{63}$ **23. a.** $\frac{20}{7}$ **b.** $\frac{-7}{27}$ **25. a.** $\frac{8}{15}$ **b.** $\frac{7}{10}$ **27. a.** $\frac{10}{21}$ **b.** $\frac{1}{20}$ **29. a.** 1
b. −1   **31. a.** 36   **b.** −25   **33. a.** $\frac{18}{35}$ **b.** $\frac{4}{45}$ **35. a.** $\frac{80}{27}$ **b.** $\frac{-15}{56}$
37. a. 20   **b.** 6   **39. a.** $\frac{11}{15}$ **b.** $\frac{39}{5}$ **41. a.** $\frac{-3}{4}$ **b.** $\frac{-3}{4}$ **43.** $\frac{2{,}137}{10{,}800}$
45. $\frac{971}{3{,}060}$ **47.** $\frac{10{,}573}{13{,}020}$ **55.** $\frac{3}{4} = \frac{1}{2} + \frac{1}{4}$ **57.** $\frac{67}{120} = \frac{1}{3} + \frac{1}{8} + \frac{1}{10}$ or $\frac{1}{2} + \frac{1}{20} + \frac{1}{120}$
59. It checks.

5.5 Irrational Numbers, page 218

7. true   **9. a.** 30   **b.** 36   **c.** 807   **d.** 169   **11. a.** a **b.** xy **c.** $4b$
d. $40w$ **13. a.** irrational; 3.162   **b.** irrational; 5.477   **c.** irrational; 9.870
d. irrational; 7.389   **15. a.** rational; 13   **b.** rational; 20
c. irrational; 23.141   **d.** irrational; 22.459   **17. a.** rational; 32
b. rational; 44   **c.** irrational; 1.772   **d.** irrational; 1.253   **19. a.** $10\sqrt{10}$
b. $20\sqrt{7}$ **c.** $8\sqrt{35}$ **d.** $21\sqrt{10}$ **21. a.** $\frac{1}{2}\sqrt{2}$ **b.** $\frac{1}{3}\sqrt{3}$
c. $\frac{1}{5}\sqrt{15}$ **d.** $\frac{1}{7}\sqrt{21}$ **23. a.** $\frac{1}{2}\sqrt{2}$ **b.** $-\frac{1}{3}\sqrt{3}$ **c.** $\frac{2}{5}\sqrt{5}$ **d.** $\frac{1}{2}\sqrt{10}$
25. a. simplified   **b.** $x + 2$ **27. a.** $10\sqrt{2}$ **b.** 0   **29. a.** $3 + \sqrt{5}$
b. $2 - \sqrt{3}$ **31. a.** $1 - 3\sqrt{x}$ **b.** $-3 - \sqrt{x}$ **33. a.** $\frac{2x}{5y}\sqrt{y}$
b. $\frac{1}{4x}\sqrt{5xy}$ **35.** $\frac{1 - \sqrt{19}}{6}$ **37.** $6 + \sqrt{37}$ **39.** 24 ft   **41.** 10 ft
43. $\sqrt{8}$ in. or $2\sqrt{2}$ in.   **45.** 200 ft   **47.** 15 ft   **49.** Answers vary;
1.2323323332...   **51.** Answers vary; 0.0919919991...   **53.** Solution
comes from Pythagorean theorem.   **a.** 2-in. square   **b.** same   **c.** 7-in.
square   **d.** 9-in. square   **e.** 15-in. square

55.

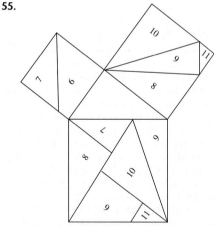

57. Answers vary; any number can be written as the sum of three triangular
numbers. He must have meant 3 or fewer triangular numbers because to
represent 2, for example, we write $1 + 1$.   **59.** Consider the second
diagonal: 1, 3, 6, 10, 15, 21, ... If we take the sum of any 2 adjacent terms,
we obtain square numbers: 4, 9, 16, 25, ... (*Note:* Don't forget that we start
counting with the 0th diagonal.)

5.6 Groups, Fields, and Real Numbers, page 228

5. Let $\mathbb{S}$ be any set, let ○ be any operation, and let a, b, and c be any
elements of $\mathbb{S}$. We say that $\mathbb{S}$ is a **group** for the operation of ○ if the
following properties are satisfied: 1. The set $\mathbb{S}$ is *closed* for ○: $(a \circ b) \in \mathbb{S}$.
2. The set $\mathbb{S}$ is *associative* for ○: $(a \circ b) \circ c = a \circ (b \circ c)$. 3. The set $\mathbb{S}$ satisfies the *identity* for ○: There exists a number $I \in \mathbb{S}$ so that
$x \circ I = I \circ x = x$ for *every* $x \in \mathbb{S}$. 4. The set $\mathbb{S}$ satisfies the *inverse* for ○:
For *each* $x \in \mathbb{S}$, there exists a corresponding $x^{-1} \in \mathbb{S}$ so that
$x \circ x^{-1} = x^{-1} \circ x = I$, where I is the identity element in $\mathbb{S}$.
7. a. $\mathbb{N}, \mathbb{Z}, \mathbb{Q}, \mathbb{R}$ **b.** $\mathbb{Q}, \mathbb{R}$ **c.** $\mathbb{Q}', \mathbb{R}$ **d.** $\mathbb{Q}', \mathbb{R}$ **e.** $\mathbb{Q}, \mathbb{R}$ **f.** $\mathbb{Q}, \mathbb{R}$
g. $\mathbb{Q}, \mathbb{R}$ **h.** $\mathbb{N}, \mathbb{Z}, \mathbb{Q}, \mathbb{R}$ **i.** $\mathbb{Z}, \mathbb{Q}, \mathbb{R}$ **9. a.** 1.5   **b.** 0.7   **c.** 0.6   **d.** 1.8
11. a. $0.\overline{6}$ **b.** $2.\overline{153846}$ **c.** 5   **d.** 1.09   **13. a.** $\frac{1}{2}$ **b.** $\frac{4}{3}$ **15. a.** $\frac{9}{20}$
b. $\frac{117}{500}$ **17. a.** $\frac{987}{10}$ **b.** $\frac{63}{100}$ **19. a.** $\frac{153}{10}$ **b.** $\frac{139}{10}$ **21. a.** 3.179   **b.** −5.504
c. 1.901   **d.** −6.31   **23. a.** −0.13112   **b.** −65.415   **c.** 12.4   **d.** 0.85
25. a. 7.46   **b.** −5.15   **c.** 72.6   **d.** 45.96   **27. a.** commutative
b. distributive   **29. a.** commutative   **b.** associative   **31.** identity
33. x **35.** no   **37.** $\mathbb{N}$ is not a group for − since it is not closed.
39. $\mathbb{W}$ is not a group for × since the inverse property is not satisfied.
41. $\mathbb{Z}$ is not a group for × since the inverse property is not satisfied.
43. $\mathbb{Q}$ is not a group for × since there is no multiplicative inverse for the
rational number 0.

45. a.

×	1	2	3	4
1	1	2	3	4
2	2	4	6	8
3	3	6	9	12
4	4	8	12	16

b.

*	1	2	3	4
1	2	2	2	2
2	4	4	4	4
3	6	6	6	6
4	8	8	8	8

c. Operation ×: not closed, associative, identity is 1, no inverse property,
commutative; Operation *: not closed, not associative, no identity, no inverse property, not commutative

Answers to Problems 46–50 may vary.
47. 2.5; $\frac{2\pi}{3}$ **49.** $4.\overline{5}$; 4.545545554...
51. $\left\{\frac{1}{2}, \frac{2}{3}, \frac{1}{3}, \frac{3}{4}, \frac{1}{4}, \frac{3}{5}, \frac{2}{5}, \frac{4}{5}, \frac{1}{6}, \frac{5}{6}, \frac{1}{7}, \frac{2}{7}, \ldots\right\}$
53. $a \circ b = ab$ **55.** $a \circ b = a + b + 1$ **57.** $a \circ b = 2(a + b)$
59. $a \circ b = a^2 + 1$

5.7 Discrete Mathematics, page 238

5. a. 3   **b.** 10   **c.** 3   **d.** 2   **7. a.** 12   **b.** 8   **c.** 8   **d.** 3
9. a. 8   **b.** 1   **11. a.** T   **b.** F   **13. a.** T   **b.** T   **15. a.** T   **b.** T
17. a. 0, (mod 5)   **b.** 8, (mod 12)   **c.** 2, (mod 5)   **d.** 3, (mod 5)
19. a. 2, (mod 5)   **b.** 2, (mod 8)   **21. a.** 3, (mod 5)   **b.** 10, (mod 11)
23. a. 4, (mod 7)   **b.** 4, (mod 5)   **25. a.** 5, (mod 6)   **b.** 3, (mod 7)
27. a. 1, 3, (mod 4)   **b.** 2, (mod 9)   **29. a.** 0, (mod 4)   **b.** 2, (mod 4)
31. all x's (mod 2)   **33. a.** Sunday   **b.** Thursday   **c.** Wednesday
d. Monday   **35.** The possible distances are 4, 14, 24, 34, ... or
9, 19, 29, 39, ...   **37.** She lives 14 miles from your house.
39. It is a group for addition modulo 7.
41.

+	0	1	2	3	4	5	6	7	8	9	10
0	0	1	2	3	4	5	6	7	8	9	10
1	1	2	3	4	5	6	7	8	9	10	0
2	2	3	4	5	6	7	8	9	10	0	1
3	3	4	5	6	7	8	9	10	0	1	2
4	4	5	6	7	8	9	10	0	1	2	3
5	5	6	7	8	9	10	0	1	2	3	4
6	6	7	8	9	10	0	1	2	3	4	5
7	7	8	9	10	0	1	2	3	4	5	6
8	8	9	10	0	1	2	3	4	5	6	7
9	9	10	0	1	2	3	4	5	6	7	8
10	10	0	1	2	3	4	5	6	7	8	9

×	0	1	2	3	4	5	6	7	8	9	10
0	0	0	0	0	0	0	0	0	0	0	0
1	0	1	2	3	4	5	6	7	8	9	10
2	0	2	4	6	8	10	1	3	5	7	9
3	0	3	6	9	1	4	7	10	2	5	8
4	0	4	8	1	5	9	2	6	10	3	7
5	0	5	10	4	9	3	8	2	7	1	6
6	0	6	1	7	2	8	3	9	4	10	5
7	0	7	3	10	6	2	9	5	1	8	4
8	0	8	5	2	10	7	4	1	9	6	3
9	0	9	7	5	3	1	10	8	6	4	2
10	0	10	9	8	7	6	5	4	3	2	1

43. no **45.** no

47.

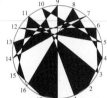

49.

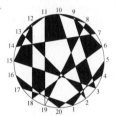

51.

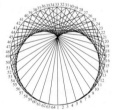

53. a. 8 **b.** 5 **55.** The check digit is 10. **57.** Assign eleven teams the numbers 1 to 11. On the first day, all teams whose sum is 1, (mod 11) play; day 2, matches those whose sum is 2, (mod 11); . . . . The last team plays the left over numbered team so each team plays every day. The schedule follows:

Day 1:	1-11;	2-10;	3-9;	4-8;	5-7;	6-12
Day 2:	2-11;	3-10;	4-9;	5-8;	6-7;	1-12
Day 3:	3-11;	4-10;	5-9;	6-8;	1-2;	7-12
Day 4:	4-11;	5-10;	6-9;	7-8;	1-3;	2-12
Day 5:	5-11;	6-10;	7-9;	1-4;	2-3;	8-12
Day 6:	6-11;	7-10;	8-9;	1-5;	2-4;	3-12
Day 7:	7-11;	8-10;	1-6;	2-5;	3-4;	9-12
Day 8:	9-10;	1-7;	2-6;	3-5;	8-11;	4-12
Day 9:	9-11;	1-8;	2-7;	3-6;	4-5;	10-12
Day 10:	10-11;	1-9;	2-8;	3-7;	4-6;	5-12
Day 11:	1-10;	2-9;	3-8;	4-7;	5-6;	11-12;

Note: The answer is not unique. **59.** The smallest result is 785.

5.8 Cryptography, page 244

1. 14-5-22-5-18-29-19-1-25-29-14-5-22-5-18-28 **3.** 25-15-21-29-2-5-20-29-25-15-21-18-29-12-9-6-5-28 **5.** ARE WE HAVING FUN YET
7. FAILURE TEACHES SUCCESS **9.** Divide by 8. **11.** Subtract 2 and then divide by 20. **13.** Divide by 2, then subtract 2, then divide by 4.
15. MOHOY .CQ MOHOYZ **17.** ZFREIOPEZFRLE WQOC
19. HUMPTY DUMPTY IS A FALL GUY. **21.** MIDAS HAD A GILT COMPLEX.

23. D UPEOTEHEQAUXUJGEWAUOEHEQHGHEXAUNTDDCRZEDN UULEHRQECEPCLLED UPEVUJEHERCRNUOXJGTAB
25. ANYONE WHO SLAPS CATSUP, MUSTARD, AND RELISH ON HIS HOT DOG IS TRULY A MAN FOR ALL SEASONINGS.
27. CRYSTAL-CLEAR AIR OF ROCKY MOUNTAINS PROVIDES IDEAL ENVIRONMENT FOR WEATHER STATION.
29. $3{,}915 + 15 + 4{,}826 = 8{,}756$

Chapter 5 Review Questions, page 246

1. -19 **2.** $\frac{71}{63}$ **3.** $\frac{1{,}396}{3{,}465}$ **4.** -10 **5.** 2 **6.** $\frac{7}{72}$ **7.** $-\frac{7}{9}$ **8.** $\frac{-3 + \sqrt{33}}{4}$
9. $\frac{8}{3}$ (*Note:* $2\frac{2}{3}$ is mixed-number form, but $\frac{8}{3}$ is reduced.) **10.** $\frac{8}{9}$
11. $\frac{4}{33}$ **12.** $\frac{8}{9}$ **13.** $\frac{3}{7}$ **14.** 0.375, rational **15.** $2.\overline{3}$, rational
16. $0.\overline{428571}$, rational **17.** 6.25, rational **18.** $0.\overline{230769}$, rational
19. prime **20.** prime **21.** prime **22.** $7 \cdot 11 \cdot 13$
23. $3 \cdot 5^2 \cdot 7 \cdot 13$ **24.** $x \equiv 2$, (mod 8) **25.** $x \equiv 5$, (mod 7)
26. $x \equiv 1$, (mod 2) **27.** 7 **28.** 343,343 **29.** 119 **30.** 2 **31.** For $b \neq 0$, multiplication is defined as: $a \times b$ means $\underbrace{b + b + b + \cdots + b}_{a \text{ addends}}$.

If $a = 0$, then $0 \times b = 0$. **32.** $a - b = a + (-b)$ **33.** $\frac{a}{b} = m$ means $a = bm$ where $b \neq 0$. **34.** If we use the definition of division, $\frac{x}{0} = m$, then $0 \cdot m = x$. If $x \neq 0$, then there is no solution because $0 \cdot m = 0$ for all numbers m. Also, if $\frac{0}{0} = m$, then $0 = 0 \cdot m$ which is true for *every* number m. Thus, $\frac{0}{0} = 0$ checks, and $\frac{0}{0} = 1$ checks, and since two numbers equal to the same number must also be equal, we obtain the statement $0 = 1$; thus, we say $\frac{0}{0}$ is indeterminate. **35.** 34.1011011101111011 . . .
36. See p. 228. **37.** It is not a group. **38.** BMI $= \frac{703w}{h^2}$ **a.** 27.5
b. 25.1 **c.** $w = 155$ lb **39.** There are 31 doors left open. **40.** 12 steps are necessary; length of segments is 20 ft; length of diagonal is $4\sqrt{13}$ ft.

CHAPTER 6

6.1 Polynomials, page 257

9. $-x - 8$; 1st-degree binomial **11.** $3x - 6y - 4z$; 1st-degree trinomial
13. $5x^2 - 7x - 7$; 2nd-degree trinomial **15.** $-x^2 - 5x + 3$; 2nd-degree trinomial **17.** $x - 31$; 1st-degree binomial **19.** $16x^2 + 9x - 3$; 2nd-degree trinomial **21.** $-3x^2 + 15x - 17$; 2nd-degree trinomial
23. a. $x^2 + 5x + 6$ **b.** $y^2 + 6y + 5$ **c.** $z^2 + 4z - 12$ **d.** $s^2 + s - 20$
25. a. $c^2 - 6c - 7$ **b.** $z^2 + 2z - 15$ **c.** $2x^2 - x - 1$ **d.** $2x^2 - 5x + 3$
27. a. $x^2 + 2xy + y^2$ **b.** $x^2 - 2xy + y^2$ **c.** $x^2 - y^2$ **d.** $a^2 - b^2$
29. a. $x^2 + 8x + 16$ **b.** $y^2 - 6y + 9$ **c.** $s^2 + 2st + t^2$
d. $u^2 - 2uv + v^2$ **31.** $6x^3 + x^2 - 12x + 5$ **33.** $5x^4 - 9x^3 + 13x^2 + 3x$
35. $-4x^3 + 25x^2 - 19x + 22$ **37.** $7x^2$
39.

	$x + 4$				
	x^2	x	x	x	x
$x + 2$	x	1	1	1	1
	x	1	1	1	1

$(x + 2)(x + 4) = x^2 + 6x + 8$

41.

	$x + 4$				
	x^2	x	x	x	x
$x + 3$	x	1	1	1	1
	x	1	1	1	1
	x	1	1	1	1

$(x + 3)(x + 4) = x^2 + 7x + 12$

43.

	$3x + 2$				
	x^2	x^2	x^2	x	x
$2x + 3$	x^2	x^2	x^2	x	x
	x	x	x	1	1
	x	x	x	1	1
	x	x	x	1	1

$(2x + 3)(3x + 2) = 6x^2 + 13x + 6$

45. $(x - 1)^3 = x^3 - 3x^2 + 3x - 1$
47. $(x + y)^6 = x^6 + 6x^5y + 15x^4y^2 + 20x^3y^3 + 15x^2y^4 + 6xy^5 + y^6$
49. $(x - y)^8 = x^8 - 8x^7y + 28x^6y^2 - 56x^5y^3 + 70x^4y^4 - 56x^3y^5 + 28x^2y^6 - 8xy^7 + y^8$ **51.** $(2x - 3y)^4 = 16x^4 - 96x^3y + 216x^2y^2 - 216xy^3 + 81y^4$

53. $91x^2y^{12} + 14xy^{13} + y^{14}$ **55.** $(6x + 2)(51x - 7) = 306x^2 + 60x - 14$
57. $(6b + 15)(10 - 2b) = 150 + 30b - 12b^2$

6.2 Factoring, page 263

3. $2x(5y - 3)$ **5.** $2x(4y - 3)$ **7.** $(x - 3)(x - 1)$ **9.** $(x - 3)(x - 2)$
11. $(x - 4)(x - 3)$ **13.** $(x - 6)(x + 5)$ **15.** $(x - 7)(x + 5)$
17. $(3x + 10)(x - 1)$ **19.** $(2x - 1)(x - 3)$ **21.** $(3x + 1)(x - 2)$
23. $(2x + 1)(x + 4)$ **25.** $(3x - 2)(x + 1)$ **27.** $x(5x - 3)(x + 2)$
29. $x^2(7x + 3)(x - 2)$ **31.** $(x - 8)(x + 8)$ **33.** $25(x^2 + 2)$
35. $(x - 1)(x + 1)(x^2 + 1)$ **37.** $x^2 + 5x + 6 = (x + 2)(x + 3)$
39. $x^2 + 4x + 3 = (x + 1)(x + 3)$ **41.** $x^2 + 6x + 8 = (x + 2)(x + 4)$
43. $x^2 - 1 = (x - 1)(x + 1)$ **45.** $x^2 + x - 2 = (x - 1)(x + 2)$
47. $x^2 - x - 2 = (x + 1)(x - 2)$ **49.** The dimensions are $x - 13$ feet by
$x + 11$ feet. **51.** The time is $2x - 1$ hours. **53.** $(3x + 2)(2x + 1)$
55. $x^2(x - 3)(x + 3)(x - 2)(x + 2)$ **57.** $x^2(4y + 5z)(5y - 2z)$
59. Let $x - 1$, x, and $x + 1$ be the three integers. Then
$(x - 1)(x + 1) = x^2 - 1$ so the square of the middle integer is one more
than the product of the first and third.

6.3 Evaluation, Applications, and Spreadsheets, page 271

5. a. +(2/3)*A1^2 **b.** +5*A1^2 − 6*A2^2 **7. a.** +12*(A1^2+4)
b. +(15*A1+7)/2 **9. a.** +(5 − A1)*(A1+3)^2
b. +6*(A1+3)*(2*A1 − 7)^2 **11. a.** +(1/4)*A1^2−(1/2)*A1+12
b. +(2/3)*A1^2+(1/3)*A1 − 17 **13. a.** $4x + 3$ **b.** $5x^2 - 3x + 4$
15. a. $\frac{5}{4}x + 14^2$ **b.** $(\frac{5}{4}x + 14)^2$ **17. a.** $\frac{x}{y}(z)$ **b.** $\frac{x}{yz}$ **19.** $A = 13$
21. $C = 8$ **23.** $E = 21$ **25.** $G = 19$ **27.** $I = 2$ **29.** $K = 4$
31. $M = 9$ **33.** $P = 20$ **35.** $R = 11$ **37.** $T = 17$ **39.** $V = 48$
41. $X = 100$ **45. a.** 8 **b.** 5 **c.** 9 **d.** 3.5

47. a.

	A	B	C	D	E
1	1	3	4	5	6
2	1	3	3	3	3
3					
4					
5					

b.

49. a.

	A	B	C
1	7	0	5
2	2	4	6
3	3	8	1

b.

	A	B	C
1	107	100	105
2	102	104	106
3	103	108	101

51.

	A	B	C	D	E
1	NAME	SALES	COST	PROFIT	COMMISSION
2				+B2-C2	+.08*D2
3	Replicate Row	2 for Rows 3 to 21.			
4					
5					

53. Genotype: black, 42.25%; black (recessive brown), 45.5%; brown,
12.25% Phenotype: black, 87.75%; brown, 12.25% **55.** Genotypes and
phenotypes are the same: red, 4%; pink, 32%; white, 64% **57.** $B = 50\%$
and $b = 50\%$ **59.** Answers vary. For the general population, FF is 49%,
Ff is 42%, and ff is 9%; free hanging is 91% and attached is 9%. In other
words, F is 70% and f is 30%.

6.4 Equations, page 281

3. a. 15 **b.** 8 **c.** −4 **d.** −8 **5. a.** 0 **b.** 0 **c.** 8 **d.** 2 **7. a.** $\frac{1,111}{10,000}$
b. $\frac{47}{90}$ **9. a.** $\frac{13}{33}$ **b.** $\frac{28}{45}$ **11. a.** $X = \frac{1}{5}$ **b.** $C = 10$ **13. a.** $F = 1$
b. $G = 6$ **15. a.** $J = 15$ **b.** $K = 9$ **17. a.** $N = -6$ **b.** $P = -8$
19. $S = 8$ **21.** $U = 3$ **23.** $W = -15$ **25.** $Z = 14$ **27.** 0, 10

29. $11, -6$ **31.** $0, 2, -2$ **33.** $-\frac{3}{2}, \frac{3}{2}$ **35.** $\frac{-7 \pm \sqrt{41}}{2}$ **37.** $\frac{5 \pm \sqrt{37}}{2}$
39. $3 \pm \sqrt{2}$ **41.** $\frac{-5 \pm \sqrt{73}}{6}$ **43.** no real values **45.** $4, -\frac{1}{3}$ **47.** $0, \frac{5}{6}$
49. 1.41, 2.83 **51.** 0.91, −2.33 **53.** −41.54, −0.01 **55. a.** 45.5 mg
b. 20 mg **57.** On the last step, dividing by $a + b - c$ is not allowed
because $a + b - c$ so $a + b - c = 0$; cannot divide by 0.
59. a. $-13, 3$ **b.** 3

6.5 Inequalities, page 287

3.
5.
7.
9.
11.
13.
15.
17. $x \geq -4$ **19.** $x \geq -2$ **21.** $y > -6$ **23.** $s > -2$ **25.** $m < 5$
27. $x > -1$ **29.** $x \leq 1$ **31.** $s < -6$ **33.** $a \geq 2$ **35.** $s > 2$ **37.** $u \leq 2$
39. $w < -1$ **41.** $A > 0$ **43.** $C > \frac{19}{2}$ **45.** $E < -1$ **47.** $G > 7$
49. $I > \frac{5}{4}$ **51.** any number greater than −5 **53.** any number less
than −4 **55.** 2 **57.** less than 2 **59.** height and width must be 18 in. or
less and the length is 36 in. or less

6.6 Algebra in Problem Solving, page 297

3. a. $3 + 2 \times 4 = 11$ **b.** $3(2 + 4) = 18$ **5. a.** $8 \times 9 + 10 = 82$
b. $8(9 + 10) = 152$ **7. a.** $3^2 + 2^3 = 17$ **b.** $3^3 - 2^2 = 23$
9. a. $4^2 + 9^2 = 97$ **b.** $(4 + 9)^2 = 169$ **11. a.** $3(n + 4) = 16; n = \frac{4}{3}$;
open, conditional **b.** $5(n + 1) = 5n + 5$; open, identity
13. a. $3^2 + 4^2 = 25$; true **b.** $1^2 + 2^2 + 3^2 = n; n = 14$; open, conditional
15. a. $6n + 12 = 6(n + 2)$; open, identity **b.** $n(7 + n) = 0; n = 0, -7$;
open, conditional **17.** $A = bh$ **19.** $A = \frac{1}{2}pq$ **21.** $V = s^3$
23. $V = \frac{1}{3}\pi r^2 h$ **25.** $V = \frac{4}{3}\pi r^3$ **27.** 2(NUMBER) − 12 = 6; The number
is 9. **29.** 3(NUMBER) − 6 = 2(NUMBER); the number is 6.
31. EVEN INTEGER + NEXT EVEN INTEGER = 94; The integers are 46 and 48.
33. INTEGER + NEXT INTEGER + THIRD CONSECUTIVE INTEGER + FOURTH
CONSECUTIVE INTEGER = 74; the integers are 17, 18, 19, and 20.
35. PRICE OF FIRST CABINET + PRICE OF SECOND CABINET = 4,150; the prices
of the cabinets are $830 and $3,320. **37.** 4(AMOUNT OF MONEY TO START)
− 72 = 48; The gambler started with $30. **39.** $8^2 + 14^2 = $ (LENGTH OF
BRACE)^2. The exact length of the brace is $2\sqrt{65}$; this is approximately 16 ft.
41. DIST FROM N TO M + DIST FROM M TO C + DIST FROM C TO D = TOTAL
DISTANCE; it is 250 mi from Cincinnati to Detroit. **43.** DIST OF SLOWER
RUNNER + HEAD START = DIST OF FASTER RUNNER; the faster runner will
catch the slower one in 12.5 seconds. **45.** DIST OF CAR + HEAD START
= DIST OF POLICE CAR; it will take the police car 6 minutes to catch the
speeder. **47.** DIST OF 1ST JOGGER + DIST OF 2ND JOGGER = TOTAL DISTANCE;
the jogger's rates are 6 mph and 8 mph. **49.** The length of the shorter side
is 5 cm. **51.** The base is 3.6 ft and the height is 1.6 ft. **53.** Need to obtain
about 41.4% interest. **55.** The center would be approximately 10.5 ft
above the ground. **57.** There are 120 lilies. **59.** The second monkey
jumped 50 cubits into the air.

6.7 Ratios, Proportions, and Problem Solving, page 306

5. 20 to 1 **7.** 53 to 50 **9.** 18 to 1 **11. a.** yes **b.** yes **c.** yes
13. a. yes **b.** no **c.** no **15. a.** > **b.** < **c.** > **17. a.** < **b.** >
c. < **d.** < **19.** $A = 30$ **21.** $C = 10$ **23.** $E = 8$ **25.** $G = 21$
27. $I = 16$ **29.** $K = \frac{15}{2}$ or 7.5 **31.** $M = 4$ **33.** $P = 3$ **35.** $R = 9$
37. $T = \frac{1}{4}$ or 0.25 **39.** $V = \frac{5}{2}$ or 2.5 **41.** $Y = 22$ **43.** $0.91
45. 2 gallons **47.** 27 minutes **49.** 2,475 calories **51.** 159.25
53. $\frac{2}{3}$ **55.** $3\frac{1}{3}$ ft or 3 ft, 4 in. **57.** $780 **59.** Since she started with

32 gallons and ended with 1 pt (of pure soft drink), the amount of soft drink served was 31 gal, 3 qt, and 1 pt.

6.8 Percents, page 316

1. $\frac{3}{4}$; 75% **3.** $\frac{2}{5}$; 0.4 **5.** $0.\overline{3}$; $33\frac{1}{3}$% **7.** $\frac{17}{20}$; 85% **9.** 0.375; 37.5%
11. $\frac{6}{5}$; 1.2 **13.** $\frac{1}{20}$; 5% **15.** $0.1\overline{6}$; $16\frac{2}{3}$% **17.** $\frac{2}{9}$; $0.\overline{2}$ **19.** $\frac{7}{40}$; 17.5%
21. $\frac{1}{400}$; 0.25% *Estimates in Problems 22–28 may vary.* **23. a.** 9,500
b. 8.56 **25. a.** 200 **b.** 200 **27. a.** 40 **b.** 8,100 **29.** $\frac{15}{100} = \frac{A}{64}$; 9.6
31. $\frac{14}{100} = \frac{21}{W}$; 150 **33.** $\frac{P}{100} = \frac{10}{5}$; 200% **35.** $\frac{P}{100} = \frac{4}{5}$; 80%
37. $\frac{P}{100} = \frac{9}{12}$; 75% **39.** $\frac{35}{100} = \frac{49}{W}$; 140 **41.** $\frac{120}{100} = \frac{16}{W}$; $13\frac{1}{3}$
43. $\frac{33\frac{1}{3}}{100} = \frac{12}{W}$; 36 **45.** $\frac{6}{100} = \frac{A}{8,150}$; $489 **47.** 23.1 million **49.** $10.86
51. 5% **53.** 40% **55.** The tax withheld is $2,624. **57.** 90%
59. The old wage was $1,250 and the new wage is $1,350.

6.9 Modeling Uncategorized Problems, page 323

3. A **5.** A **7.** A **9.** B **11.** D **13.** C **15.** A **17.** A **19.** A
21. C **23.** C **25.** C **27.** B **29.** A **33.** 15 and 17 **35.** 1 and 2
37. a. 55% **b.** 122% **39.** Standard Oil bldg., 1,136 ft; Sears, 1,454 ft.
41. 10,100 **43.** 70 **45.** 180 hours **47.** $x = 2\ell - \frac{1}{30}c\ell - 40$ **49.** Should invest $64,000 in savings and $36,000 in annuities. **51.** Cut off 2 in.; the box is $8 \times 9 \times 12$ in. **53.** The wind speed is 21 mph. **55.** $2,350 with 30 additional persons **57.** The reading would not change.
59. 24,000.

Chapter 6 Review Questions, page 327

1. a. Algebra refers to a structure as a set of axioms that forms the basis for what is accepted and what is not. **b.** The four main processes are *simplify* (carry out all operations according to the order-of-operations agreement and write the result in a prescribed form), *evaluate* (replace the variable(s) with specified numbers and then simplify), *factor* (write the expression as a product), and *solve* (find the replacement(s) for the variable(s) that make the equation true). **2. a.** $x^3 + x^2 + 6x - 8$ **b.** $x^4 - x^3y - x^2y + xy$
3. a. -19 **b.** -19 **4. a.** $3(x-3)(x+3)$ **b.** $(x-6)(x+1)$
5. a. $\frac{3}{2}$ **b.** $-2 \pm \sqrt{10}$
6.

	x^2	x	x	x
x	1	1	1
x	1	1	1

with $x + 3$ across the top and $x + 2$ on the left.

$x^2 + 5x + 6 = (x+3)(x+2)$

7. a. 4 **b.** $\frac{1}{4}$ **8. a.** 9 **b.** $-\frac{7}{3}$ **9. a.** $P = 15$ **b.** $W = 3.75$ **10.** 5, -1
11. $\frac{3 \pm \sqrt{5}}{4}$ **12.** $x < -3$ **13.** $x \leq -2$ **14.** $x \leq 2$ **15.** $x < 3$
16. a. < **b.** = **c.** < **d.** < **e.** > **17.** $12\frac{1}{2}$ cups of flour
18. The integers are 22, 24, 26, and 28. **19.** Genotypes: 27% tall, 50% tall (recessive short), 23% short; Phenotypes: 77% tall, 23% short
20. The distance from New York to Chicago is 720 mi, from Chicago to San Francisco is 1,860 mi, and from San Francisco to Honolulu is 2,400 mi.

CHAPTER 7

7.1 Geometry, page 337

1. both **3.** Answers vary; the point that is being made is that although we may use a figure in geometry to help us understand a problem, we cannot use what we see in a figure as a basis for our reasoning.
9. reflection

11. $P \quad Q$ **13.** $P \quad Q$ **15.** $P \quad Q$
17. $P \quad Q$ **19.** $P \quad\quad\quad S$

21. symmetric **23.** symmetric **25.** symmetric **27.** not symmetric
41. not symmetric **43.** symmetric (except for the words)
45. symmetric **47.** symmetric

49.   **51.** **53.**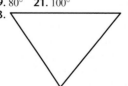

55. C **57.** C **59.** (1) letters with no symmetry; (2) letters with horizontal line symmetry; (3) letters with vertical line symmetry; (4) letters with symmetry around both vertical and horizontal lines

7.2 Polygons and Angles, page 346

9. a. quadrilateral **b.** pentagon **11. a.** triangle **b.** hexagon
13. a. quadrilateral **b.** heptagon **15. a.** T **b.** T **17. a.** T **b.** F
19. a. F **b.** F **21. a.** yes **b.** yes **c.** yes **d.** yes **e.** yes
23. a. no **b.** no **c.** no **d.** no **e.** no
25. **27.**

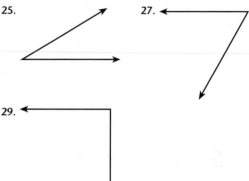

29.

31. a. acute **b.** right **33. a.** right **b.** obtuse **35. a.** acute **b.** acute
37. adjacent and supplementary **39.** adjacent and supplementary
41. vertical **43.** alternate interior angles **45.** F **47.** T **49.** F
51. $m\angle 1 = m\angle 3 = m\angle 5 = m\angle 7 = 115°$;
$m\angle 2 = m\angle 4 = m\angle 6 = m\angle 8 = 65°$
53. $m\angle 1 = m\angle 3 = m\angle 5 = m\angle 7 = 153°$;
$m\angle 2 = m\angle 4 = m\angle 6 = m\angle 8 = 27°$ **55.** $m\angle 1 = m\angle 3 = m\angle 5 = m\angle 7 = 163°$; $m\angle 2 = m\angle 4 = m\angle 6 = m\angle 8 = 17°$ **57. a.** obtuse
b. obtuse **c.** acute **d.** acute **e.** straight **f.** right **g.** acute
h. straight **i.** acute **j.** $\angle AOC$
59.

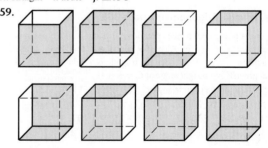

7.3 Triangles, page 354

5. $\overline{AB} \cong \overline{ED}$; $\overline{AC} \cong \overline{EF}$; $\overline{CB} \cong \overline{FD}$; $\angle A \cong \angle E$; $\angle B \cong \angle D$; $\angle C \cong \angle F$
7. $\overline{RS} \cong \overline{TU}$; $\overline{RT} \cong \overline{TR}$; $\overline{ST} \cong \overline{UR}$; $\angle SRT \cong \angle UTR$; $\angle S \cong \angle U$; $\angle STR \cong \angle URT$ **9.** $\overline{JL} \cong \overline{PN}$; $\overline{LK} \cong \overline{NM}$; $\overline{JK} \cong \overline{PM}$; $\angle J \cong \angle P$; $\angle L \cong \angle N$; $\angle K \cong \angle M$ **11.** 88° **13.** 145° **15.** 56° **17.** 75°
19. 80° **21.** 100°
23. **25.**

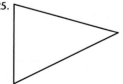

27.

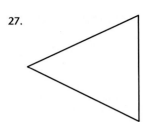

29. 20°   **31.** 21°   **33.** 6°   **35.** 60°; 60°; 60°   **37.** 50°; 60°; 70°
39. 13°; 53°; 114°   **41.** 50°   **43.** 120°   *Answers to Problems 45–52 may vary.*

53.   **55.**

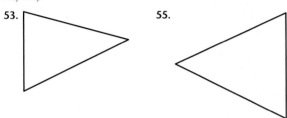

57. a. $x = 135°$, $y = 45°$ **b.** $x = 132.5°$; $y = 47.5°$ **c.** $x = 136°$; $y = 44°$ **59.** 540°

7.4 Similar Triangles, page 359

3. similar   **5.** not similar   **7.** similar
9. $m\angle A = m\angle A' = 25°$; $m\angle B = m\angle B' = 75°$; $m\angle C = m\angle C' = 80°$
11. $m\angle A = m\angle D = 38°$; $m\angle B = m\angle E = 68°$; $m\angle C = m\angle F = 74°$
13. $m\angle A = m\angle A' = 54°$; $m\angle B = m\angle B' = 36°$; $m\angle C = m\angle C' = 90°$
15. $|\overline{AC}| = |\overline{AB}| = 11$; $|\overline{A'C'}| = |\overline{A'B'}| = 22$; $|\overline{BC}| = 5$; $|\overline{B'C'}| = 10$
17. $|\overline{GH}| = 14$; $|\overline{HI}| = 12$; $|\overline{GI}| = 16$; $|\overline{DF}| = 20$; $|\overline{DE}| = 17.5$; $|\overline{EF}| = 15$ **19.** $|\overline{AB}| = 10$; $|\overline{AC}| = 6$; $|\overline{BC}| = 8$; $|\overline{B'C'}| = 5$;
$|\overline{A'B'}| = 6.25$; $|\overline{A'C'}| = 3.75$ **21.** $\sqrt{80}$ or $4\sqrt{5}$ **23.** 4   **25.** $\frac{16}{3}$
27. $\frac{8}{3}$ **29.** 22   **31.** 7.8   **33.** 13.7   **39.** The lake is 125 ft long.
41. The height of the building is 24 ft.   **43.** The building is 45 ft tall.
45. The bell tower is 70 ft tall.   **47.** The tree is 17 ft tall.   **49.** The tree is 6 ft 8 in. tall.   **51.** Denver to New Orleans is 1,080 mi. and Chicago to Denver is 1,020 mi.   **53. a.** ΔHSP and ΔBTP **b.** yes; answers vary.
c. The footbridge is 21 ft above the base.

7.5 Right-Triangle Trigonometry, page 368

1. For any right triangle with sides with lengths a and b and hypotenuse with length c, $a^2 + b^2 = c^2$. Also, if a, b, and c are lengths of sides of a triangle so that $a^2 + b^2 = c^2$, then the triangle is a right triangle.   **3.** In a right triangle ABC with right angle at C, $\cos A = \frac{\text{length of adjacent side of } A}{\text{length of hypotenuse}}$.
5. Rope A would form a right triangle.   **7.** b **9.** a **11.** $\frac{a}{c}$ **13.** $\frac{b}{c}$
15. $\frac{a}{b}$ **17.** 0.8290   **19.** 0.8746   **21.** 0.5878   **23.** 0   **25.** 0.4452
27. 3.7321   **29.** 30°   **31.** 45°   **33.** 60°   **35.** 56°   **37.** 37°
39. Yes; $\sin A = \frac{12}{13}$; $\cos A = \frac{5}{13}$; $\tan A = \frac{12}{5}$ **41.** No   **43.** Yes;
$\sin A = \frac{1}{6}$; $\cos A = \frac{\sqrt{35}}{6}$; $\tan A = \frac{1}{\sqrt{35}}$ **45.** Yes; $\sin A$; $\cos A = \frac{\sqrt{3}}{2}$;
$\tan A = \frac{1}{\sqrt{3}}$ **47.** No   **49.** The height is 109 ft.   **51.** The distance is 199 m.
53. The top of the ladder is 12 ft 7 in.   **55.** The chimney stack is 1,251 ft tall.   **57.** $\cos 30° = \frac{\sqrt{3}}{2}$, $\sin 30° = \frac{1}{2}$, and $\tan 30° = \frac{\sqrt{3}}{3}$ **59.** 12 in.

7.6 Mathematics, Art, and Non-Euclidean Geometries, page 380

5. The divine proportion is the relationship $\frac{h}{w} = \frac{w}{h + w}$.   **7.** It is the number $\frac{1 + \sqrt{5}}{2}$.   **9.** Answers vary; for example, start with 4 and 4.
a. The sequence is 4, 4, 8, 12, 20, 32, ...   **b.** The ratios are
$\frac{4}{4} = 1, \frac{8}{4} = 2, \frac{12}{8} = 1.5, \frac{20}{12} = 1.\overline{6}, \frac{32}{20} = 1.6$. It seems that these ratios oscillate around $\tau \approx 1.62$.   **11.** $1.61\overline{6}$; it is close to τ.   **13.** $\frac{b}{2}$ to h is 1.62, about the same; b to s is 1.7, close   **15.** 2.17, not too close
17. Euclidean   **19.** Euclidean   **21.** elliptic   **23.** hyperbolic   **25.** It is a Saccheri quadrilateral.   **27.** It is not a Saccheri quadrilateral.

29. a. 5 in. by 3 in.   **b.** 1.67; about the same   **31. a.** 10.25 in. by 8 in.
b. 1.28; not close   **33.** Answers vary.   **35. a.** 2   **b.** 8   **c.** 144
d. 1, 2, 1.50, 1.67, 1.60, 1.63, 1.62, . . . ; close to τ.   **37.** close
39. 3.1 ft or 8.1 ft   **41.** 15 cm   **43.** B   **45.** C   **47.** Their sum is greater than 180°.   **49. a.** A great circle is a circle on a sphere with a diameter equal to the diameter of the sphere.   **b.** ℓ is a line (a great circle), but m is not.   **c.** yes   **51.** Lines are great circles, and the circle labeled m is not a great circle.   **53.** True; answers vary.
55. The length-to-width ratio will remain unchanged if $\frac{L}{W} = \sqrt{2}$.
57. a. $x = \frac{1 \pm \sqrt{5}}{2}$ **b.** They are negative reciprocals.   **59.** Just down from the North Pole there is a parallel that has a circumference of exactly one mile. If you begin anywhere on the circle that is 300 feet north of this parallel, the conditions are satisfied.

Chapter 7 Review Questions, page 384

1. a. 0   **b.** 8   **c.** 24   **d.** 24   **e.** 8   **2.** two men (rotate the picture 180°)
3. rotation   **4.** the photograph   **5.** the photograph   **6. a.** Flag is symmetric; picture is not (if you include the flagpole).   **b.** no   **c.** no   **d.** yes
7. a. a rectangle satisfying the divine proportion   **b.** yes
8. 173 ft   **9.** $C = \frac{\ell w}{15}$ **10. a.** obtuse angle   **b.** adjacent angles or supplementary angles   **c.** vertical angles   **d.** acute angle   **e.** obtuse angle   **f.** vertical angles   **11. a.** 41°   **b.** 131°   **c.** 41°   **12.** 36°
13. The width is about 6 in.   **14.** The other leg is 5 in.   **15.** $5\sqrt{2}$ in.
16. a. 0.8572   **b.** 0   **c.** 0.9511   **d.** 7.1154   **17.** The distance is about 753 ft.
18.

19. A Saccheri quadrilateral is a quadrilateral $ABCD$ with base angles A and B right angles and with sides AC and BD with equal lengths.
20. If the summit angles C and D of a quadrilateral $ABCD$ are right angles, then the result is Euclidean geometry. If they are acute, then the result is hyperbolic geometry. If they are obtuse, the result is elliptic geometry.

CHAPTER 8

8.1 Euler Circuits and Hamiltonian Cycles, page 397

7. not an Euler circuit; traversable   **9.** not an Euler circuit; not traversable
11. not an Euler circuit; not traversable   **13.** Hamiltonian cycle;
$A \rightarrow B \rightarrow C \rightarrow D \rightarrow A$ **15.** Hamiltonian cycle; $A \rightarrow B \rightarrow C \rightarrow G \rightarrow H \rightarrow F \rightarrow E \rightarrow D \rightarrow A$ **17.** Hamiltonian cycle; $A \rightarrow B \rightarrow C \rightarrow D \rightarrow A$
Possible paths in Problems 18–23 may vary.   **19.** not traversable
21. not traversable   **23.** traversable   **25.** not an Euler circuit; not traversable   **27.** not an Euler circuit; not traversable   **29.** no Hamiltonian cycle   **31.** no Hamiltonian cycle   **33.** Answers may vary. $1 \rightarrow 2 \rightarrow 3 \rightarrow 10 \rightarrow 11 \rightarrow 12 \rightarrow 4 \rightarrow 5 \rightarrow 14 \rightarrow 13 \rightarrow 19 \rightarrow 18 \rightarrow 17 \rightarrow 9 \rightarrow 8 \rightarrow 7 \rightarrow 16 \rightarrow 20 \rightarrow 15 \rightarrow 6 \rightarrow 1$ **35.** no   **37.** It is traversable; begin at either Queens or Manhattan.   **39.** yes; 2 odd vertices   **41.** No; there are 8 odd vertices.   **43. a.** NYC to Boston to Washington, D.C. to NYC; 892 mi; forms a loop without including Cleveland   **b.** NYC to Boston to Washington, D.C. to Cleveland to NYC; 1,513 mi   **45.** 4,627 mi
47. 2,699 mi   **49.** 720°   **51. a.** Room 1: 1 path; room 2: 2 paths; room 3: 3 paths; room 4: 5 paths   **b.** 89   **c.** 377   **53. a.** Start at the top (or at the right) at one of the two odd vertices.   **b.** The chance that you will choose one of the two odd vertices at random is small.   **55.** The result is a twisted band twice as long and half as wide as the band with which you began.   **57.** The result is two interlocking pieces, one twice as long as the other.   **59.** One edge; one side; after cutting, you will have a single loop with a knot.

8.2 Trees and Minimum Spanning Trees, page 409

5. no   **7.** no   **9.** no   **11.** yes
13.

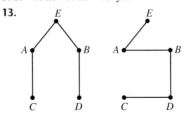

15.

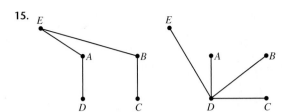

17.

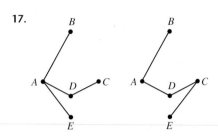

19.

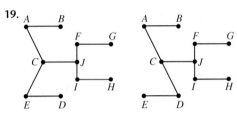

21.

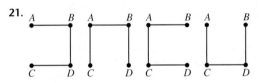

23.

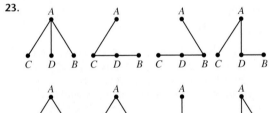

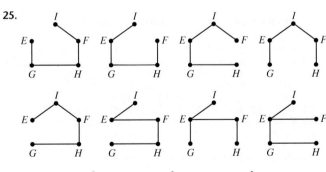

25.

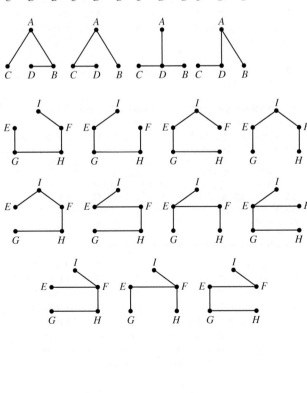

27. The propane molecule is a tree.

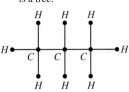

29. The isobutane molecule is a tree.

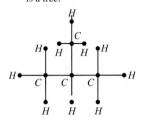

31. Not a tree; because there is a circuit.

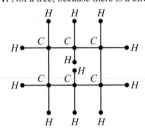

33. 60   **35.** 45   **37.** 180   **39.** 65   **41.** 30   **43.** 64   **45.** no

47.

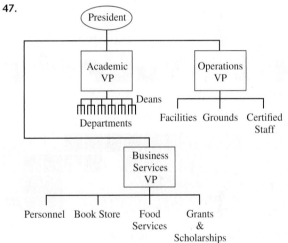

49. 49   **51.** $38,500   **53.** 375 miles

55. a.   **b.**   **c.** $4,000,000

57. a.

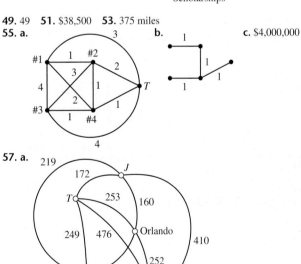

b.

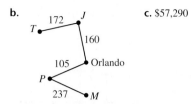

c. $57,290

59. a. $E = v - 1$; a tree with n vertices has $n - 1$ edges **b.** 0
c. one edge **d.** two edges

8.3 Topology and Fractals, page 418

3. A and F; B and G; C, D and H **5.** A and F are simple closed curves.
7. C, E, F, G, H, I, J, K, L, M, N, S, T, U, V, W, X, Y, and Z are the same
class; A, D, O, P, Q, and R are another class; B is a third class. **9.** A, C, E,
and F are the same class; B and D are the same class; G is different.
11. A and C are inside; B is outside. **13.** A and B are inside; C is outside.
15. It will need only two colors.

19.

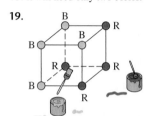

21.

25. Answers vary.

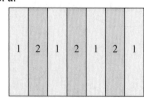

29. a. No, it will now fit the left hand. **b.** yes

Chapter 8 Review Questions, page 422

1. no **2.** yes **3.** yes **4.** yes **5. a.** no **b.** yes
6. a. **b.**

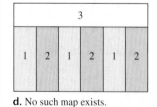

c. **d.** No such map exists.

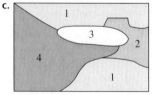

7. 20,160 ways **8.** 1,596 miles **9.** 635

10. Weighted graph:

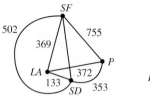

The minimum spanning tree is:

The minimum distance is 855 miles.
11. yes **12.** no **13.** 20 **14.** 47 **15.** 4.3589×10^{10} **16.** 39
17. no **18.** yes
19. Answers vary; here is one:

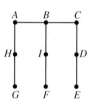

20. There are two edges and two sides; the result after cutting is two
interlocking loops.

CHAPTER 9

9.1 Perimeter, page 433

3. $P = 4s$; $P = 2(\ell + w)$; $P = 3s$; $P = 5s$ **5.** The number π is the
ratio of the circumference to the diameter of a circle.
7. a. ———————— **b.**
c. ——— **9.** A **11.** B **13.** A **15.** C **17.** B **19.** B **21.** C
23. C **25.** C **27.** A **29.** C **31.** A **33.** A **35. a.** 3 cm
b. 3.4 cm **c.** 1 in. **d.** $1\frac{3}{8}$ in. **37.** 18 in. **39.** 9 dm **41.** 15.08 m
43. 397.08 ft **45.** 46 ft **47.** 15.2 cm **49.** 35.7 in. **51.** 11.14 cm
53. 300 m **55.** The sides are 26 in., 39 in., and 52 in. **57.** probably
smaller **59.** The approximate measurements of the ark are 158 m long,
26 m wide, and 16 m high.

9.2 Area, page 442

3. a. 3 cm² **b.** 3 cm² **5.** 8 cm² **7.** 11 cm² **9.** C **11.** A **13.** C
15. A **17.** C **19.** 15 in.² **21.** 1,196 m² **23.** 100 mm² **25.** 7,560 ft²
27. 136.5 dm² **29.** 5,550 cm² **31.** 28 in.² **33.** 314.2 in.²
35. 78.5 in.² **37.** 307.9 in.² **39.** 7.6 cm² **41.** A; $2.86/ft² for Lot A
and $3.19/ft² for Lot B. **43.** 22.2 acres **45.** 216 in.² **47.** $93\frac{1}{2}$ in.²
49. 20 pounds are necessary; $117 **51.** $s = 11.5$; $A \approx 20$ ft²
53. $s = 230$; area of triangle $\approx 9{,}591.66$; area of figure $\approx 52{,}792$ ft²
55. 172 in.² **57.** 483 in.² **59.** 79 in.²

9.3 Surface Area, Volume, and Capacity, page 453

5. A liter is a bit larger than a quart. **7.** 60 cm³ **9.** 125 ft³
11. 8,000 cm³ **13.** 24 ft³ **15.** 96,000 cm³ **17. a.** 2 c **b.** 16 oz
19. a. 13 oz **b.** 380 mL **21. a.** 25 mL **b.** 75 mL **c.** 75 mL **23.** B
25. A **27.** C **29.** B **31.** B **33.** 112 kL **35.** 3,125 cm²
37. 780 cm² **39.** 35,000 cm² **41.** 36π cm² $\approx$ 113.1 cm²
43. 7.8 in.² **45.** 29,375 m² **47.** 4.3 gal **49.** 4 L **51.** 9 L
53. 500 L **55. a.** 45.375 ft³ **b.** 26.375 ft³ **57.** 2.5 yd³
59. a. 560,000 people/mi² **b.** 11,071 mi²; this is about the size of the
state of Maryland. **c.** about 5.4 acres per person

9.4 Miscellaneous Measurements, page 464

5. a. centimeter **b.** meter **7. a.** liter **b.** kiloliter **9. a.** Celsius
b. Celsius **11.** C **13.** C **15.** B **17.** C **19.** A **21.** B **23.** C
25. B **27.** B **29.** A **31.** 0.000063 kiloliter; 0.00063 hectoliter;
0.0063 dekaliter; 0.063 liter; 0.63 deciliter; 6.3 centiliter; **63 milliliter**
33. 0.0035 kiloliter; 0.035 hectoliter; 0.35 dekaliter; **3.5 liter;**
35 deciliter; 350 centiliter; 3,500 milliliter **35.** 0.08 kiloliter;
0.8 hectoliter; **8 dekaliter;** 80 liter; 800 deciliter; 8,000 centiliter;
80,000 milliliter **37.** 0.31 kiloliter; **3.1 hectoliter;** 31 dekaliter;

310 liter; 3,100 deciliter; 31,000 centiliter; 310,000 milliliter   **39. a.** 0.001
b. 0.000001   **c.** 1,000   **d.** 100   **41. a.** 1,000   **b.** 10   **c.** 0.001
d. 1,000,000   **43.** 71 cm²   **45.** 2,827 cm²   **47.** 30 ft³   **49.** 600 cm³
51. $S = 452$ in.²; $V = 905$ in.³   **53.** The area is increased six-fold.
55. The area is increased nine-fold.   **57.** The area is increased four-fold.
59. Answers vary; use the formula $F = \frac{9}{5}C + 32$ in the column headed
Fahrenheit.

Chapter 9 Review Questions, page 468

1. a. ——————————————
b. 1.9 cm   **2. a.** feet and meters   **b.** Answers vary; 40°C; 100°F
c. Answers vary; $\frac{1}{2}$ in.; 1.5 cm   **3. a.** one-thousandth   **b.** capacity
c. 1,000,000 cm   **4.** s 43 in.   **5.** 105 in.²   **6.** less than   **7.** 1,262 in.³
8. a. 4π ft²   **b.** 12.6 ft²   **9. a.** $\frac{4}{3}\pi$ ft³   **b.** 4.2 ft³   **10. a.** 30 dm³
b. 30 L   **11.** 56 dm²   **12. a.** 30 cm × 20 cm × 50 cm   **b.** 130 gal
13. 76 cm²   **14. a.** The best value is the large size.   **b.** Answers vary;
$17.95.   **15.** 2,400 ft²   **16.** 7,200 ft³   **17.** 53,856 gal   **18.** You need to
purchase 20 square yards.
19. a.

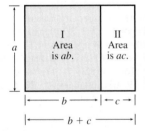

b. yes

20. The diagram gives a geometric justification of the distributive law.

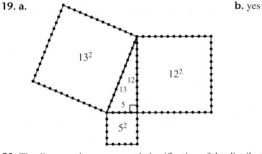

Area of rectangle I: ab
Area of rectangle II: ac
Area of I + II is $ab + ac$
Area of large rectangle is
$a(b+c)$

CHAPTER 10

10.1 Exponential Equations, page 480

5. a. $\log N$ is the exponent on a base 10 that gives N.   **b.** $\ln N$ is the
exponent on a base e that gives N.   **c.** $\log_b N$ is the exponent on a base b
that gives N.   **7.** An exponential equation is an equation where the variable
appears as an exponent.   **9. a.** $6 = \log_2 64$ **b.** $2 = \log 100$
c. $p = \log_n m$ **11. a.** $-1 = \log \frac{1}{10}$ **b.** $2 = \log_6 36$ **c.** $n = \log_t s$
13. a. 1   **b.** 3   **c.** −5   **15. a.** −3   **b.** −1   **c.** 1   **17. a.** 1   **b.** 3
c. −6   **19. a.** −2   **b.** $\log_5 8$ **c.** $\log_6 4.5$ **d.** $\log_4 5$ **21. a.** $\frac{2}{3}$
b. $\log 2.5$ **c.** $\log 45$ **d.** $\log 15$ **23. a.** 0.63   **b.** 2.00   **c.** 3.00   **d.** 2.30
25. a. 4.85   **b.** 2.00   **c.** 0.00   **d.** 6.91   **27. a.** 0.82   **b.** 2.82
c. 1.99   **d.** −1.14   **29.** 3.4650   **b.** 2.8028   **31. a.** 1.3181   **b.** 9.1189
33. a. −7.4804   **b.** 2.1991   **35. a.** 2.0686   **b.** 3.4396
37. a. 0.528320834   **b.** 0.386988016   **39. a.** 1.623249290
b. −1.630784143   **41. a.** −2.07944154168   **b.** −2.09691001301
43. a. 2.129072048   **b.** 1.855903615   **45.** 3.07   **47.** 2.16   **49.** 2015
51. 2016   **53.** 3   **55.** 8.665302427   **57.** $t = \frac{1}{r} \ln \frac{P}{P_0}$ **59.** 1 hour, 18 min

10.2 Logarithmic Equations, page 488

5. F   **7.** F   **9.** T   **11.** F   **13.** F   **15.** F   **17.** F   **19.** F   **21. a.** 23
b. 3.4   **c.** x **d.** x **23. a.** 2   **b.** 7   **c.** 4   **d.** 3   **25. a.** 10^5 **b.** e **c.** e^2
d. e^3 **27. a.** $\log 24$ **b.** $\log 2$ **c.** $\ln \frac{x^2 y^3}{z^4}$ **29. a.** $\ln \frac{3}{2}$ **b.** $\ln \frac{1}{48}$
c. $\ln \frac{3}{1,024}$ **31.** 3.5   **33.** 6.2   **35.** 3,572   **37. a.** 8   **b.** 10^8 **39. a.** 2

b. 25   **41. a.** 2   **b.** $\frac{\sqrt{2}}{e^2}$ **43. a.** 2   **b.** e^2 **45.** 10^{10} **47.** 1, −1
49. $4.8 \cdot 10^{18}$ **51.** $8e^{12}$ **53.** 0   **55.** $\frac{3 \pm \sqrt{409}}{2}$ **57. a.** 13 sec
b. $t = e^{(80 - R)/27}$ **59. a.** 3.89   **b.** 8.8   **c.** $10^{23.8}$ **d.** $10^{1.5M + 11.8}$

10.3 Applications of Growth and Decay, page 497

5. 30 dB   **7.** 107 dB   **9.** 114 dB   **11.** 896,716   **13.** −0.023104906
15. The half-life is about 28 years.   **17.** 6.7   **19.** An earthquake with
magnitude 6 is 100 times stronger than an earthquake of magnitude 4.
21. $10^{24.25}$ ergs   **23.** The artifact is about 10,500 years old.   **25.** The
elapsed time is about 7 years.   **27.** 63 times more energy   **29.** The
artifact is about 1,300 years old.   **31.** 1.41 times more energy
33. 31.6 times more energy   **35.** The cookies are about 128° F.
37. 37°C   **39.** 5,313 ft   **41.** The satellite will operate for about
128 days.   **43. a.** 1.5%   **b.** 674,161   **d.** 802,673   **45. a.** 1.1%
b. 510,478   **d.** 582,976   **47. a.** 8.8%   **b.** 518,429   **d.** 1,482,201
49. a. 0.5%   **b.** 278,106   **d.** 296.490   **51. a.** Growth rate of about 1.6%;
predict the 2004 population to be 4,049,226.   **b.** Growth rate of
−0.27%; predict the 2004 population to be 3,357,957.   **c.** Growth rate of
about 1.07%; predict the 2004 population to be about 3,838,310.
53. a. February 2003   **b.** The actual growth rate is greater than the
assumed growth rate.   **c.** April 2011   **55.** 1,583 yr   **57.** In 1988, the
Shroud was about 662 years old.   **59.** When $x = 75$, $y \approx 0.16$ or no
O-ring failures; when $x = 32$, $y \approx 4.80$ or 5 O-ring failures.

Chapter 10 Review Questions, page 500

1. $2\frac{1}{2}$ **2.** 543   **3.** 4.3   **4.** 0.5   **5.** 3   **6. a.** 0.93   **b.** 3.99   **7. a.** 0.69
b. −2.08   **8. a.** 3.32   **b.** −1   **9.** 1.929418926   **10.** 6.214608098
11. 1.1178328654   **12.** 0.935554163   **13.** 6^4 **14.** $\frac{\log_2 6 + 1}{3}$ **15.** $\frac{\log 5}{2}$
16. $\frac{101}{99}$ **17.** 5   **18.** $\log_{(1+i)} \frac{A}{P}$ **19.** about 7 days   **20.** Equation is
$A = A_0 e^{0.59t}$, where A_0 is the number of teens infected and t is the
number of years after 1992.

CHAPTER 11

11.1 Interest, page 514

7. C; interest rate is not stated, but you should still recognize a
reasonable answer.   **9.** B is the most reasonable.   **11.** A   **13.** D
15. B   **17.** $1,500   **19.** $4,200   **21.** $1,028.25   **23.** $16,536.79
25. $1,132,835.66   **27.** $1,661.44   **29.** $165,298.89   **31. a.** $28,940.57
b. $29,222.27   **c.** $29,367.30   **d.** $29,465.60   **e.** $29,513.58
f. $29,515.24   **g.** $29,515.24   **h.** $22,800   **33.** $1,548.13   **35. a.** $2.74
b. $1.96   **c.** $3.06   **d.** $4.63   **e.** $2,509.30   **f.** $29,797.93
g. $43,912.74   **h.** $25,092.99   **37.** $1,220.19   **39.** $11,023.18
41. $5,219.84   **43.** $279,064   **45.** $17,072.71   **47.** $P = 101,500$
49. $800,000   **51.** $200,000   **53.** 4 years, 61 days   **55.** 7 years, 49 days
57. a. 3 years, 3 months   **b.** 10 years   **c.** The interest rate is about 14%.
59. $Y = \left(1 + \frac{r}{n}\right)^n - 1$

11.2 Installment Buying, page 523

7. B   **9.** B   **11.** B   **13.** B   **15.** B   **17.** B   **19.** C   **21.** A   **23.** B
(it should be the least expensive)   **25.** 15%   **27.** 18%   **29.** 8%
31. a. $4.50   **b.** $3.75   **c.** $3.95   **33. a.** $37.50   **b.** $36.88   **c.** $36.58
35. $14,181   **37.** $18,975   **39.** 23.4%   **41.** 21.1%   **43. a.** $344.17
b. $349.69   **c.** 4.9%   **d.** 0% is better   **45. a.** $711.67   **b.** $705.00
c. 4.9%   **d.** 2.5% is better   **47. a.** $11,430   **b.** $2,430   **c.** 0.09
d. 17.5%   **49. a.** $23,443.20   **b.** $9,093.20   **c.** 0.1584181185
d. 31.0%   **51. a.** $8,886.48   **b.** $2,088.48   **c.** 0.08702   **d.** 17.0%
53. 8% add-on rate; APR is about 15.7%   **55.** 11% add-on rate; APR is
about 21.6%   **57.** previous balance method, $45; adjusted balance
method, $40.50; average daily balance method, $43.35   **59.** $67.58

11.3 Sequences, page 536

7. a. geometric   **b.** $r = 2$ **c.** 32   **9. a.** arithmetic   **b.** $d = 10$ **c.** 35
11. a. Fibonacci-type   **b.** $s_1 = 5, s_2 = 15$ **c.** 35   **13. a.** geometric
b. $r = \frac{1}{5}$ **c.** $\frac{1}{5}$ **15. a.** geometric   **b.** $r = 3$ **c.** 27   **17. a.** none of the
classified types   **b.** differences are $-2, 1, -2, 1, -2, \ldots$ **c.** 5

19. a. geometric   **b.** $r = 2$ **c.** 96   **21. a.** both arithmetic and geometric
b. $d = 0$ or $r = 1$ **c.** 10   **23. a.** Fibonacci-type   **b.** $s_1 = 3, s_2 = 6$
c. 24   **25. a.** geometric   **b.** $r = \frac{3}{2}$ **c.** $\frac{81}{2}$ **27. a.** geometric
b. $r = 4^{-1}$ or $\frac{1}{4}$ **c.** 4   **29. a.** arithmetic   **b.** $d = \frac{1}{10}$ **c.** $\frac{3}{5}$
31. a. arithmetic   **b.** $d = \frac{1}{12}$ **c.** $\frac{11}{12}$ **33. a.** 0, 3, 6   **b.** arithmetic; $d = 3$
35. a. 1, 0, -1 **b.** arithmetic; $d = -1$ **37. a.** 0, $-10, -20$ **b.** arithmetic; $d = -10$ **39. a.** 0, $\frac{1}{2}, \frac{2}{3}$ **b.** neither   **41. a.** 1, 3, 6
b. neither   **43. a.** $-1, 1, -1$ **b.** geometric, $r = -1$ **45. a.** $\frac{2}{3}, \frac{2}{3}, \frac{2}{3}$
b. both; $d = 0$ or $r = 1$ **47. a.** $-2, 3, -4$ **b.** neither   **49.** -200
51. 625   **53.** 3, 1, $\frac{1}{3}, \frac{1}{9}, \frac{1}{27}$ **55.** 1, 2, 3, 5, 8   **57.** It is Fibonacci.
59.

1st number:	x
2nd number:	y
3rd number:	$x + y$
4th number:	$x + 2y$
5th number:	$2x + 3y$
6th number:	$3x + 5y$
7th number:	$5x + 8y$
8th number:	$8x + 13y$
9th number:	$13x + 21y$
10th number:	$21x + 34y$
SUM:	$55x + 88y = 11(5x + 8y)$

11.4 Series, page 546

5. 30   **7.** 150   **9.** -1 **11.** 12   **13.** 90   **15.** 10   **17.** 33   **19.** 2
21. $\frac{3}{2}$ **23.** $-\frac{40}{3}$ **25.** 25   **27.** 15   **29.** 110   **31.** 10,000   **33.** 5,050
35. $n(n + 1)$ **37.** 11,500   **39.** 2,030   **41.** 120   **43.** 56; 64
45. 9,330   **47.** 3,828 blocks   **49.** 200 cm   **51.** 1,500   **53.** 190 ft
55. 1,024,000,000 present after 10 days   **57.** 31 games   **59.** 364 games

11.5 Annuities, page 554

7. $1,937.67   **9.** $2,026.78   **11.** $41,612.93   **13.** $74,517.97
15. $15,528.23   **17.** $18,294.60   **19.** $170,413.86   **21.** $540,968.11
23. $56,641.61   **25.** $59,497.67   **27.** $99,386.46   **29.** $38,754.39
31. $2,204.31   **33.** $252,057.07   **35.** $1,193.20   **37.** $1,896.70
39. $7,493.53   **41.** $8,277.87   **43.** $2,261.06   **45.** $36.89
47. $915,239.29   **49.** $27,298.98   **51.** $2,880.97   **53.** $1,666,454.24
55. $831,706   **57.** Answer depends on your age.   **59.** $255,309.73

11.6 Amortization, page 560

3. m is the amount of a periodic payment (usually a monthly payment);
n is the number of payments made each year; t is the number of years;
r is the annual interest rate; A is the future value; and P is the present value
5. $2,586.28   **7.** $27,942.24   **9.** $20,442.52   **11.** $175,131.20
13. $297.46   **15.** $1,100.65   **17.** $1,316.36   **19.** $2,092.02
21. $6,882.42   **23.** $10,827.33   **25.** $5,429.91   **27.** $12,885.18
29. $7,407.76   **31.** $5,999.44   **33.** $6.38   **35.** $148.31   **37.** $430.73
39. $73.35   **41.** $780.54   **43.** $10,186.47   **45.** $117,238
47. $367,695.71   **49.** The annuity is the better choice.
51. $12,462,210.34   **53.** $206,029.43   **55. a.** decreases   **b.** increases
c. decreases   **57.** $1,510.92; $543,931.20; $213,046.20   **59.** price range
of $175,322.31 to $204,386.77

11.7 Summary of Financial Formulas, page 565

5. $m = \dfrac{P\left(\frac{r}{n}\right)}{1 - \left(1 + \frac{r}{n}\right)^{-nt}}$; the unknown is m **7.** present value   **9.** annuity
11. amortization   **13.** present value of an annuity   **15.** amortization
17. annuity; $5,264.62   **19.** sinking fund; $1,574.10   **21. a.** annuity
b. $5,264.62   **23. a.** sinking fund   **b.** $1,670.92   **25. a.** future value
b. $27,081.62   **27. a.** present value of an annuity   **b.** $5,756.94
29. a. amortization   **b.** $1,028.61   **31. a.** annuity   **b.** $23,193.91
33. a. present value   **b.** $7,215.46   **35. a.** annuity   **b.** $21,867.63
37. a. present value   **b.** $165,134.88   **39. a.** amortization   **b.** $1,317.40
41. a. annuity   **b.** $228,803.04   **43. a.** future value   **b.** $884.12
45. a. future value   **b.** $9,506.04   **47. a.** present value   **b.** $3,655.96
49. a. present value   **b.** $148,348   **51. a.** present value   **b.** $500,420

53. $1,328.94   **55.** $179,986.40, total; $112,932 savings   **57.** $726,624
59. Take $10,000 now and $45,000 in one year.

Chapter 11 Review Questions, page 567

1. A sequence is a list of numbers having a first term, a second term,
and so on; a series is the indicated sum of the terms of a sequence.
An arithmetic sequence is one that has a common difference,
$a_n = a_1 + (n - 1)d$; a geometric sequence is one that has a common
ratio, $g_n = g_1 r^{n-1}$; and a Fibonacci-type sequence, $s_n = s_{n-1} + s_{n-2}$, is
one that, given the first two terms, the next is found by adding the
previous two terms. The sum of an arithmetic sequence is $A_n = n\left(\frac{a_1 + a_n}{2}\right)$ or
$A_n = \frac{n}{2}[2a_1 + (n - 1)d]$, and the sum of a geometric sequence is
$G_n = \frac{g_1(1 - r^n)}{1 - r}$.   **2.** A good procedure is to ask a series of questions.
Is it a lump-sum problem? If it is, then what is the unknown? If FUTURE
VALUE is the unknown, then it is a *future value* problem. If PRESENT VALUE
is the unknown, then it is a *present value* problem. **Is it a periodic
payment problem?** If it is, then is the periodic payment known? If the
PERIODIC PAYMENT IS KNOWN and you want to find the future value, then it
is an *ordinary annuity* problem. If the PERIODIC PAYMENT IS KNOWN and
you want to find the present value, then it is a *present value of an annuity*
problem. If the PERIODIC PAYMENT IS UNKNOWN and you know the future
value, then it is a *sinking fund* problem. If the PERIODIC PAYMENT IS
UNKNOWN and you know the present value, then it is an *amortization* problem.   **3. a.** arithmetic; $a_n = 5n$ **b.** geometric; $g_n = 5 \cdot 2^{n-1}$
c. Fibonacci-type; $s_1 = 5, s_2 = 10, s_n = s_{n-1} + s_{n-2}, n \geq 3$ **d.** none of
these (add 5, 10, 15, 20, . . .); 55, 80   **e.** geometric; $g_n = 5 \cdot 10^{n-1}$
f. none of these (alternate terms); 5, 50   **4. a.** 5   **b.** $\frac{43}{30}$ or $1.4\overline{3}$ **5.** There
are a minimum of 2,047 people.   **6.** 2^{82} **7.** You should offer $19,500 for
the car.   **8. a.** The total interest is $783, and the monthly payment
is $595.13.   **b.** The APR is 5.568%.   **9.** The APR is about 23.5%.
10. The adjusted balance method is most advantageous to the consumer.
The previous balance method's finance charge is $3.94. The adjusted bal-
ance method's finance charge is $3.19. The average daily balance
method's finance charge is $3.42.   **11.** Deposit $11,297.10 to have
a million dollars in 50 years.   **12. a.** The monthly payments are $1,288.12.
b. The price of the home is $192,500.   **c.** $639,122.51   **13.** $80,834.49
14. $97,178.53   **15.** $1,246,454.24   **16.** $302,821.47   **17.** You need to
set aside $97,178.53.   **18.** $10,000   **19.** Take installments.   **20.** Take
the one-time payment.

CHAPTER 12

12.1 Permutations, page 580

3. 9   **5.** 72   **7.** 3,024   **9.** 132,600   **11.** 95,040   **13.** 1,680   **15.** 1
17. 2,520   **19.** 210   **21.** 5,527,200   **23.** 336   **25.** 7,920
27. $\frac{g!}{(g - h)!}$ **29.** $\frac{5!}{(5 - r)!}$ **31.** 5,040   **33.** 360   **35.** 34,650
37. 4,989,600   **39.** 831,600   **41.** 120   **43.** 720   **45.** 16   **47.** 336
49. 6,400,000   **51.** 10,000   **53.** 40,320   **55.** 720   **57.** 64   **59.** 4,459

12.2 Combinations, page 588

3. 9   **5.** 84   **7.** 1   **9.** 22,100   **11.** 1   **13.** 35   **15.** 1,225
17. $\frac{g!}{h!(g - h)!}$ **19.** $\frac{k!}{(k - 4)!}$ **21.** $\frac{m!}{n!(m - n)!}$ **23.** 4,200
25. 1,260   **27.** 105   **29.** 792   **31.** 6   **33.** 1,287   **35.** 24   **37.** 10
39. 128   **41.** combination; $\binom{30}{5}$; 142,506   **43.** combination; $\binom{20}{2}$; 190
45. combination; $\binom{31}{3}$; 4,495   **47.** neither; distinguishable permutation;
1,260   **49.** permutation; $_{10}P_2$; 90   **51.** permutation; $_6P_5$; 720
53. combination; $\binom{7}{4}$; 35   **55.** permutation; $_7P_2$; 42   **57.** $\binom{n}{5}$
59. $_{16}P_2 \cdot {}_{19}P_2 \cdot \binom{31}{6}$; approximately $6.043394448 \times 10^{10}$

12.3 Counting without Counting, page 595

3. 15   **5.** 786,240   **7.** 21   **9.** 78,960,960   **11.** 1,296   **13.** 31
15. 127   **17.** 2,730   **19.** 1.95×10^{11} **21.** 135   **23.** 27,405

25. permutation; 120 **27.** combination; 286 **29.** combination; 1,191,052,400 **31.** permutation; 720 **33.** distinguishable permutation; 24 **35.** FCP, permutation, and combination; 57,600 **37.** permutation; 24 **39.** combination; 66 **41.** Subsets: $\{a, b\}$; Arrangements: $(a, b), (b, a)$ **43.** Subsets: $\{a, b, c, d\}$; 24 arrangements **45.** 810,000,000 **47.** 3,486,784,401 **49.** 26! **51.** 9,134.2 centuries **53.** Subtract each pair of billiard balls to find the value of the ball located below the pair. **55.** 0 **57.** 524,800 **59.** 6,161,805

12.4 Rubik's Cube and Instant Insanity, page 601

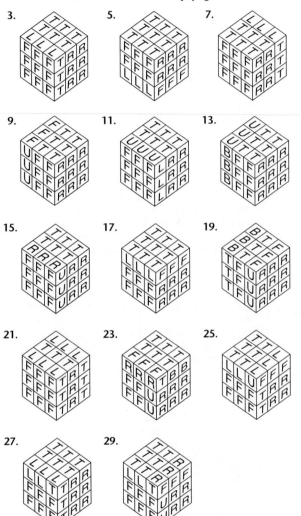

31. yes **33.** no **35.** no **37.** yes **39.** U^{-1} **41.** B **43.** U^2 or $(U^{-1})^2$ **45.** B or $(B^{-1})^3$ **47.** $U^{-1}B^{-1}$ **49.** TF^{-1} **51.** BF **53.** no **55.** No, answers vary. **57.** 41,472

Chapter 12 Review Questions, page 603

1. 40,314 **2.** 2 **3.** 120 **4.** 24 **5.** 28! **6.** 10 **7.** 336 **8.** 1 **9.** 1,001 **10.** 970,200 **11.** They can form 220 different committees. **12.** They can form 120 different lineups at the teller's window. **13.** happy: 60; college: 1,260 **14.** At least one red ball can be drawn in 100 ways. **15.** The 100 senators can be formed into 75,287,520 different five-member committees. **16.** The claim is correct. **17. a.** 1,024 **b.** 59,049 **c.** 9,765,625 **18. a.** 7,962,624 **b.** The result is more than 92 days (nonstop). **19.** It is not possible. **20.** The number of arrangements is $_{10}P_{10} = 10! = 3{,}628{,}800$. This is almost 10,000 years, so that day will never come for the members of this club.

CHAPTER 13

13.1 Introduction to Probability, page 616

5. a. $\frac{1}{4}$ **b.** $\frac{1}{4}$ **c.** $\frac{1}{2}$ **7. a.** $\frac{1}{2}$ **b.** $\frac{1}{3}$ **c.** $\frac{2}{3}$ **9.** about 0.05 **11.** 0.19 **13.** $P(\text{royal flush}) \approx 0.000001539077169$ **15.** $P(\text{four of a kind}) \approx 0.0002400960384$ **17.** $P(\text{flush}) \approx 0.00196540155$ **19.** $P(\text{three of a kind}) \approx 0.02112845138$ **21.** $P(\text{one pair}) \approx 0.42256902761$ **23. a.** $P(\text{five of clubs}) = \frac{1}{52}$ **b.** $P(\text{five}) = \frac{1}{13}$ **c.** $P(\text{club}) = \frac{1}{4}$ **25. a.** $P(\text{five and a jack}) = 0$ **b.** $P(\text{five or a jack}) = \frac{2}{13}$ **27. a.** $\frac{1}{12}$ **b.** $\frac{7}{12}$ **c.** $\frac{1}{12}$ **29. a.** $\frac{1}{4}$ **b.** $\frac{3}{4}$ **31.** $P(\text{five}) = \frac{1}{9}$ **33.** $P(\text{seven}) = \frac{1}{6}$ **35.** $P(\text{nine}) = \frac{1}{9}$ **37.** $P(\text{four } or \text{ five}) = \frac{7}{36}$ **39.** $P(\text{eight } or \text{ ten}) = \frac{2}{9}$ **41.** Pick A. **43.** Pick D. **45.** Pick F. **47.** Pick F. **49.** Answers vary. **51.** Answers vary; yes **53.** 10% **55. a.** $\frac{2}{9}$ **b.** $\frac{1}{9}$
57. a.

	1	2	3	4
1	(1, 1)	(1, 2)	(1, 3)	(1, 4)
2	(2, 1)	(2, 2)	(2, 3)	(2, 4)
3	(3, 1)	(3, 2)	(3, 3)	(3, 4)
4	(4, 1)	(4, 2)	(4, 3)	(4, 4)

b. $\frac{1}{16}$ **c.** $\frac{1}{8}$ **d.** $\frac{3}{16}$ **e.** $\frac{1}{4}$ **f.** $\frac{3}{16}$ **g.** $\frac{1}{8}$ **59.** $\frac{2}{3}$

13.2 Mathematical Expectation, page 624

7. B **9.** B **11.** B **13.** $0.08 **15.** $7.20 **17.** $0.25 for two plays of the game **19.** $1.50 **21.** $-$0.05 **23.** $-$0.05 **25.** $-$0.05 **27.** $-$0.05 **29.** $-$0.05 **31.** fair game **33.** not a fair game **35.** $2.84 **37.** play **39.** $6,415 **41.** $6,875; yes **43.** $0.02; yes **45.** $0.05 **47.** 1.82 **49.** no **51.** yes **57.** $1.50 **59.** $500.00

13.3 Probability Models, page 634

3. 15,120 ways **5.** The probability of event E is the ratio of s (success) to n (total number), whereas the odds in favor of that same event is the ratio of s (success) to f (failure) where $s + f = n$. **7.** C **9.** B **11.** B **13.** $\frac{1}{5}$ **15.** 0.995 **17.** $\frac{4}{5}$ **19.** $\frac{15}{16}$ **21.** 15 to 1 **23.** 9 to 2 **25.** $P(\#1) = \frac{1}{19}$; $P(\#2) = \frac{2}{5}$; $P(\#3) = \frac{1}{3}$; $P(\#4) = \frac{5}{12}$; $P(\#5) = \frac{1}{2}$ **27.** $\frac{33}{34}$ **29.** $\frac{4}{11}$ **31.** 1 **33.** $\frac{1}{3}$ **35.** $\frac{1}{2}$ **37.** $\frac{4}{51}$ **39.** $\frac{4}{17}$ **41.** $\frac{26}{51}$ **43.** 51.5% **45.** 36.1% **47. a.** 0.3125 **b.** 0.675 **c.** 0.389 **d.** 0.70 **49. a.** 0 **b.** 1 **51. a.** $\frac{1}{5}$ **b.** $\frac{1}{5}$ **c.** 0
53. $\dfrac{P(\overline{E})}{P(E)} = \dfrac{\frac{f}{n}}{\frac{s}{n}} = \dfrac{f}{n} \cdot \dfrac{n}{s} = \dfrac{f}{s}$ = odds against **55. a.** $\frac{2}{7}$ **b.** $\frac{5}{7}$ **c.** $\frac{9}{34}$
d. $\frac{25}{34}$ **e.** $\frac{5}{17}$ **f.** $\frac{12}{17}$ **g.** $\frac{2}{7}$

h. *First selection*   *Second selection*

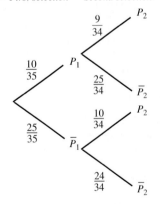

13.4 Calculated Probabilities, page 646

3. $P(E \cap F) = P(E) \cdot P(F)$, provided E and F are independent. **5.** $\frac{1}{2}$ **7.** $\frac{5}{6}$ **9.** $\frac{1}{12}$ **11.** $\frac{2}{3}$ **13.** $\frac{4}{9}$ **15.** $\frac{11}{12}$ **17.** $\frac{1}{3}$ **19.** $\frac{5}{9}$ **21.** $\frac{35}{36}$ **23.** no **25.** yes **27.** yes **29.** $\frac{1}{6}$ **31.** $\frac{1}{2}$ **33.** $\frac{11}{36}$ **35. a.** 0.000455 **b.** 0.01229 **c.** 0.01229 **d.** 0.1538 **e.** 0.0237 **37.** $-$0.08 **39.** $-$1.42 **41.** $-$0.92 **43.** 2.7% **45.** $\frac{1}{32}$ **47. a.** $\frac{25}{64}$; $\frac{5}{14}$

b. $\frac{9}{64}; \frac{3}{38}$ **c.** $\frac{15}{32}; \frac{15}{28}$ **d.** $\frac{15}{64}; \frac{15}{56}$ **49. a.** play **b.** play **53.** $-\$0.44$
55. $-\$0.55$ **57.** one spot **59.** First bet: 0.5177; second bet: 0.4914

13.5 The Binomial Distribution, page 653

3. 0.132 **5.** 0.128 **7.** 0.016 **9.** 1.701×10^{-4} **11.** 0.3125 **13.** .234375
15. 0.161 **17.** 0.3125 **19.** 0.044 **21.** 0.0107 **23.** 0.296 **25.** 0.222
27. 0.656 **29.** 0.0001 **31.** 0.9477 **33.** 0.512 **35.** 0.096 **37.** 0.488
39. 0.2109 **41.** 0.4019 **43.** 0.3025 **45.** 0.1699 **47.** 0.1028
49. 0.2765 **51.** 0.0837 **53.** 16 **55. a.** 0.109375 **b.** 0.114265
57. 0.75 **59.** 0.3125

Chapter 13 Review Questions, page 655

1. C **2.** B **3.** 0.004 **4.** $\frac{1}{2}$ **5.** $\frac{5}{36}$ **6.** $\frac{4}{13}$ **7.** $\frac{1}{17}$ **8. a.** $\frac{11}{36}$ **b.** $\frac{5}{9}$
c. $\frac{1}{18}$ **9. a.** 0.99 **b.** 9 to 1 **c.** $\frac{1}{1,001}$ **10.** \$2 **11. a.** $\frac{3}{5}$ **b.** $\frac{1}{2}$
12. $\frac{2}{15}$ **13.** $P(X = x) = \binom{n}{x}(0.001)^x(0.999)^{n-x}$ **14.** 0.138 **15.** 0.522
16. $\frac{2}{3}$ **17.** 0.821 **18.** Choose C; $P(C \text{ wins}) = \frac{5}{8}$ **19.** $\frac{1}{5}$ **20.** $\frac{1}{25}$

CHAPTER 14

14.1 Frequency Distributions and Graphs, page 667

5. a. October **b.** August **7. a.** U.S. trade with Canada for 1990–2005
b. more imports **9. a.** 121 kwh **b.** 44 kwh **c.** 209 kwh **d.** 55 kwh
11. C **13.** Graph is meaningless without a scale. **15.** Graphs are based
on height, but the impression is that of area. **17.** Three-dimensional bars
are used to represent linear data, which is not appropriate. **19.** 25 times
21. Maybe DUI (illegal if under 18 yr old) **23.** no

25.

Number of Cars	Tally	Frequency
0	|	1
1	||||	4
2	|||| |||	8
3	||	2
4	|	1

27.

Temperature	Tally	Frequency
39	|	1
40	|	1
41		
42		
43	||	2
44	|	1
45	||	2
46		
47	|	1
48		
49	||||	4
50	||||	4
51	|	1
52	||	2
53	||	2
54	||	2
55	||	2
56		
57	||	2
58	||	2
59	|	1

29.

0	0	1	1	1	1	2	2	2	2	2	2	2	2	3	3	4

31.

3	9																
4	0	3	3	4	5	5	7	9	9	9	9						
5	0	0	0	0	1	2	2	3	3	4	4	5	5	7	7	8	8	9

33.

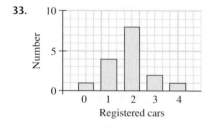

35.

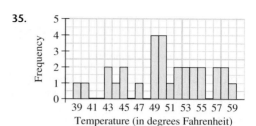

37.

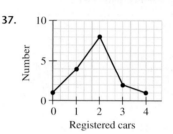

39.

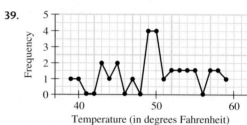

41. a.

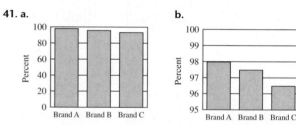

b.

c. The graph in part **b**.

43.

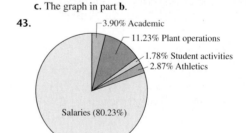

45. Each figure represents 30 managing directors.

Paine Webber: 46 out of 465, or about 10%.

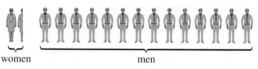

women   men

Goldman, Sachs: 9 out of 173, or about 5%.

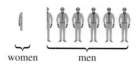

women   men

Merrill Lynch: 76 out of 694, or about 11%

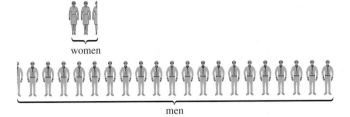

women

men

47. −$32 billion   **49.** CBS   **51.** The ratings for all three
networks (except for FOX) have generally declined from 1986 to 2008.
53. 2.4 persons   **55.** 2036   **57.** The line graph at the right begins at
50 rather than 0.   **59.** The advertisement says that the car is 57 in. wide on
the outside, but a full 5 ft across on the inside.

14.2 Descriptive Statistics, page 681

11. The data values are all the same.   **13.** mean = 19; median = 19;
no mode   **15.** mean = 767; median = 767; no mode   **17.** mean = 11;
median = 9; no mode   **19.** mean = 82; median = 81; no mode
21. mean = 6; median = 2.5; mode = 1   **23.** range = 4; var. = 2.5;
$s \approx 1.58$ **25.** range = 4; var. = 2.5; $s \approx 1.58$ **27.** range = 24;
var. = 93.5; $s \approx 9.67$ **29.** range = 25; var. = 83; $s \approx 9.11$
31. range = 21; var. = 62.86; $s \approx 7.93$ **33.** median = 432.51;
$Q_1 = 427.48$; $Q_3 = 442.28$

Q_1 Q_2 Q_3

420   450

35. mean = 13; median = 11; mode = 10   **37.** mean = 68;
median = 70; mode = 70; range = 50   **39.** mean $\approx$ 366; median = 365; no
mode   **41.** A   **43.** C   **45. a.** mean = 105.25   **b.** median = 12
c. mode = 42   **47.** 5   **49.** 12   **51.** 15   **53.** mean = 19,200;
range = 4,000; $s \approx 1,483$ **57. a.** $\bar{x} = 6.5$; H.M. = 4.7
b. $\bar{x} = 56.5$ mph; H.M. = 56.1 mph

14.3 The Normal Curve, page 692

5. The mean is to the right of the median.   **7.** 5%, 16%, 45%, 79%, 94%,
99%, 100%; $\bar{x} = 2.62$, median = 3, mode = 3   **9.** 1%, 12%, 47%, 68%,
88%, 94%, 97%, 99%, 100%; $\bar{x} = 2.94$, median = 3, mode = 2
11. 28%, 40%, 75%, 81%, 100%; $\bar{x} = 7.84$; median = 8; mode = 8
13. 41.92%   **15.** 49.25%   **17.** 49.99%   **19.** 49.01%   **21.** 17.72%
23. 48.68%   **25. a.** 34 people   **b.** 34 people   **27.** 91.92%
29.

Height	Number	Cumulative
155		0.1%
160	1	2.3%
165	7	15.9%
170	17	50.0%
175	17	84.1%
180	7	97.7%
185	1	99.9%
190		100.0%

31. 60 or above   **33.** 25   **35.** 87

37. 69   **39.**

Grade	Score	Cumulative
A	87 and above	6%
B	80–86	22%
C	70–79	78%
D	65–69	94%
F	64 and below	100.0%

41. mode

43. $x = 98.0675$ **45.** $x = 68.919$ **47.** 0.1587   **49.** 42.07%
51. 0.0228 or 2.28%   **53.** 0.3085   **55. a.** 50%   **b.** 12.4 oz
57. Graph **a** has less variance and graph **b** has more variance.

59.

−4	0.00001
−3	0.00195
−2	0.06250
−1	0.50000
0	1.00000
1	0.50000
2	0.06250
3	0.00195
4	0.00001

14.4 Correlation and Regression, page 699

3. $r = \dfrac{n\Sigma xy - (\Sigma x)(\Sigma y)}{\sqrt{n(\Sigma x^2) - (\Sigma x)^2}\ \sqrt{n(\Sigma y)^2 - (\Sigma y)^2}}$ **5.** Strong positive
correlation   **7.** no   **9.** yes   **11.** no   **13.** yes   **15.** yes   **17.** no
19. $r = -0.765$ **21.** $r = -0.954$ **23.** $r = -0.890$
25. $y' = -1.95x + 4.146$ **27.** $y' = -2.7143x + 30.0571$
29. $y' = -2.380x + 270.00$ **31.** D   **33.** A   **35.** C   **37.** $y' = 6$; $r = 0$
39. $y' = x + 2$, $r = 1$ **41.** $r = 0.8830$, significant at 5%
43. $r = 0.8421$, significant at 1%
45. $y' = 2.4545x + 6.0909$

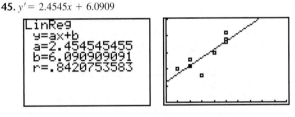

47. $r = 0.358$, no significant
correlation

49. $r = -0.936$, significant at 1%

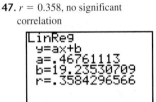

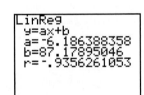

51. $y' = 0.468x + 19.235$ **53.** $y' = -6.186x + 87.179$

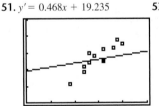

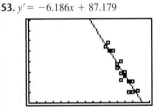

55. a. Time 27:31 or 27 hr 31 min; the difference is 4 hr 29 min.
57. a. Time is 44:21 or 44 hr 21 min.   **59.** $r = 0.380$, significant at the
5% level

14.5 Sampling, page 706

5. B   **7.** B   **9.** D   **11.** A   **13.** D   **25.** (1) You accept that 72 is the
mean, and it is the mean. (2) You accept that 72 is the mean, and it is not
the mean. This is Type II error (accept a false hypotheses). (3) You do not
accept that 72 is the mean, and it is the mean. This is Type I error (reject a
true hypothesis). (4) You do not accept that 72 is the mean, and it is not the
mean.

Chapter 14 Review Questions, page 709

1. Heads: |||| |||| |||| ||| ||| (18); Tails |||| |||| |||| |||| || (22)

2.

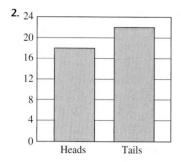

3.

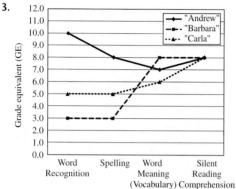

4. $23\frac{2}{3}$ **5.** 23 **6.** 21 **7.** 35 **8.** 11.59 **9.** mean = 11; median = 12; mode = 12; mode is the most appropriate measure. **10.** mean = 79; median = 74; no mode; mean is the most appropriate measure.
11. mean = \$113,423; median = \$110,750; no mode; the median is the most appropriate measure.

12. a. yes

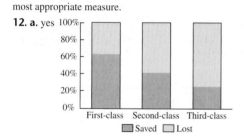

b. yes

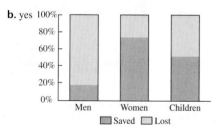

13. a. LTL **b.** Courier Air **c.** October
14.

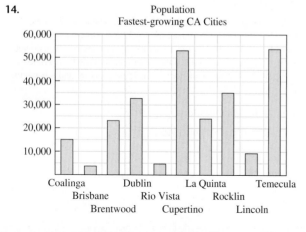

15. mean = $\dfrac{255{,}585}{10}$ = 25,558.5; median = 23,675; no mode
16. 18,091
17. a.

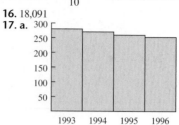

b.

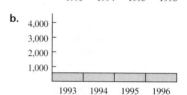

c.

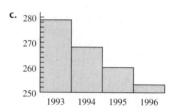

d. Impressions can be greatly influenced by using faulty or inappropriate scale (or even worse, no scale at all).
18.

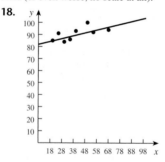

The best-fitting line is $y' = 0.22x + 81.8$
19. $r = 0.658$; not significant at 1% or 5% levels **20.** Approximately 1 in 1,000 pregnancies will have a 314-day duration.

CHAPTER 15

15.1 Cartesian Coordinates and Graphing Lines, page 722

1. a. Aphrodite **b.** Maxwell Montes **c.** Atalanta Planitia **d.** Rhea Mons **e.** Lavina Planitia *Ordered pairs in Problems 7–18 may vary.*
7. (0, 5), (1, 6), (2, 7) **9.** (0, 5), (1, 7), (−1, 3) **11.** (0, −1), (1, 0), (2, 1) **13.** (0, 1), (1, −1), (−1, 3) **15.** (0, 1), (1, 3), (2, 5) **17.** (0, 2), $(2, \frac{1}{2})$, (4, −1)
19. 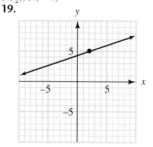 **21.**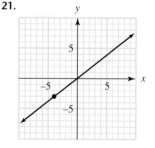

23.

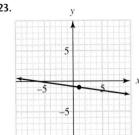

25.

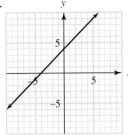

27.

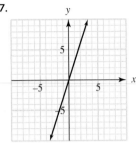

29.

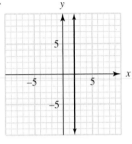

31.

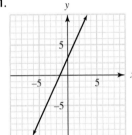

33.

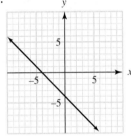

35.

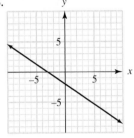

37.

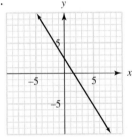

39.

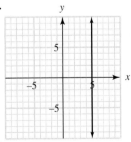

41.

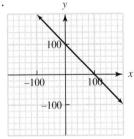

43.

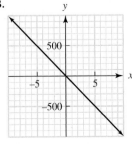

45. C   **47.** A   **49.** E

51. a. nearly   **b.** $y = -0.14x + 5.8$ **c.** 0.9
53. a.   **b.** 800   **c.** 10,000

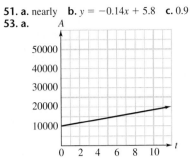

55. a.

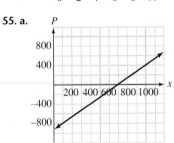

b. loss of $850   **c.** 1.25; it is the profit increase corresponding to each unit increase in number of items sold   **57.** 22.3 million

15.2 Graphing Half-Planes, page 725

3. F   **5.** F   **7.** T   **9.** T   **11.** T

13.

15.

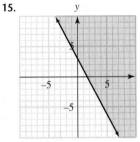

17.

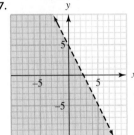

19.

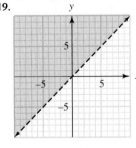

21.

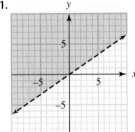

23.

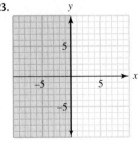

25.

27.

19.

21.

29.

23.

25.

15.3 Graphing Curves, page 730

3.

5.

27.

29.

7.

9.

31.

33.

11.

13.

35.

37.

15.

17.

39.

41.

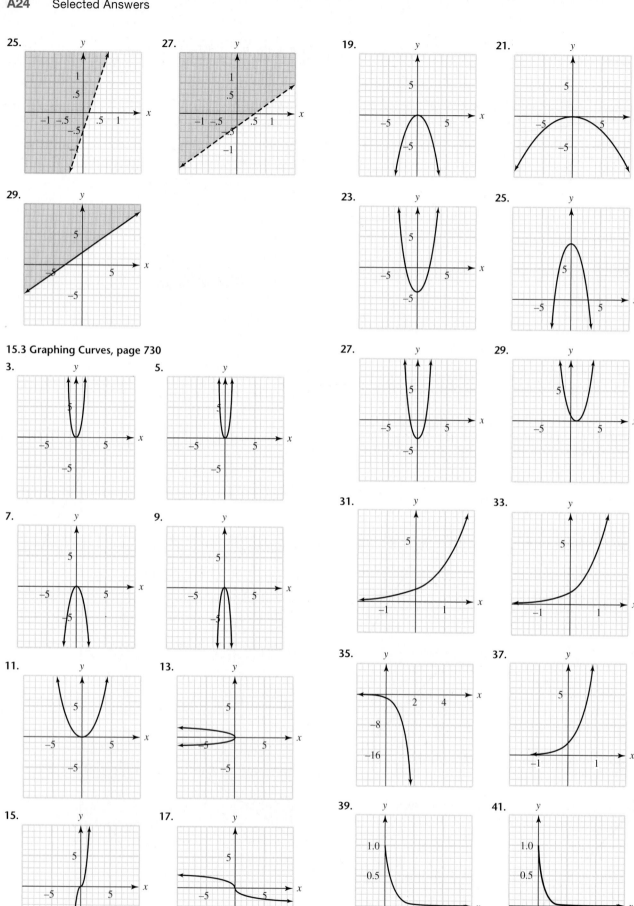

43. a.

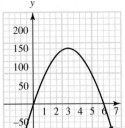

b. downward

15.4 Conic Sections, page 743

9. *A* and *C* have opposite signs.   **11.** $A = 0$ and $C \neq 0$, or $A \neq 0$ and $C = 0$ **13.** $A = C = 0$

15.

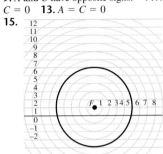

c.

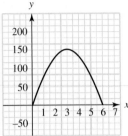

45.

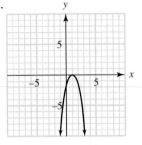

17.

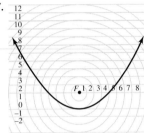

47.

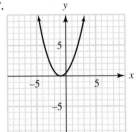

49.

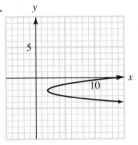

19.

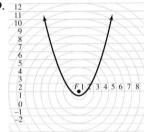

51.

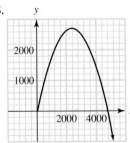

53.

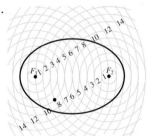

If $x = 4{,}000$, $y = 1{,}560$, positive; $x = 4{,}700$, $y = 188$, positive; $x = 5{,}000$, $y = -550$, negative. The crossover point $(y = 0)$ is between 4,700 and 5,000, so he will make it across.

21.

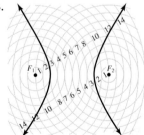

55.

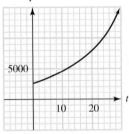

23.

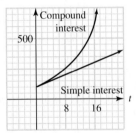

57.

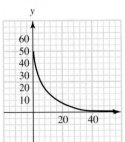

59.

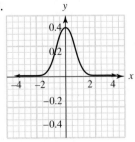

25. a. line   **b.** ellipse   **c.** parabola   **27. a.** parabola   **b.** hyperbola   **c.** ellipse

29.

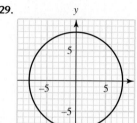

31.

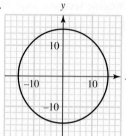

33.

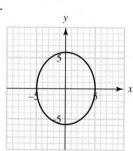

35.

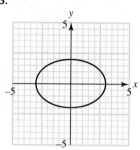

37.

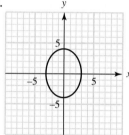

39.

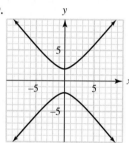

41.

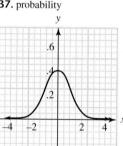

43.

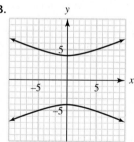

45.

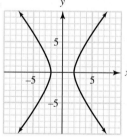

47.

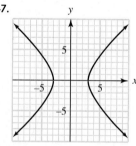

49.

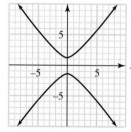

51.

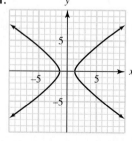

53. aphelion is 1.5×10^8 mi; perihelion is 1.3×10^8 mi. **55.** $\epsilon \approx 0.053$
57. 2.25 m **59.** 8 in.

15.5 Functions, page 750

3. function **5.** function **7.** not a function **9.** function **11.** function
13. not a function **15. a.** 18 **b.** 22 **c.** -6 **d.** 11 **e.** $2(t+5)$
17. a. 17 **b.** 37 **c.** 65 **d.** $1\frac{1}{4}$ **e.** t^2+1 **19. a.** 7 **b.** 11 **c.** -17
d. 0 **e.** $2t-1$ **21. a.** 8 **b.** -16 **c.** $p-7$ **23. a.** -1 **b.** -31
c. $3a-1$ **25. a.** 5 **b.** -2 **c.** $\frac{3}{2}$ **27.** not a function; domain:
$-1 \le x \le 1$; range: $-3 \le y \le 3$ **29.** not a function; domain: $x \ge -3$;
range: $\mathbb{R}$ **31.** function; domain: $-2 \le x \le 3$; range: $-8 \le y \le 4$

33. quadratic

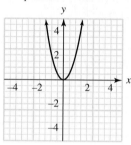

35. logarithmic

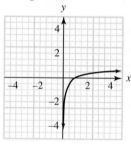

37. probability

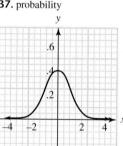

39. 3 **41.** $3x^2 + 3xh + h^2$

43. $\dfrac{-1}{x(x+h)}$ **45. a.** 64 **b.** 96 **c.** 128 **d.** 256 **e.** 512
47. a. 1,430 **b.** 1,050 **c.** 670 **d.** 290 **e.** 100 **49. a.** 4 **b.** 0
c. 25 **51.** $y = 3x - 5$; domain: $\mathbb{R}$ **53.** $y = \sqrt{5-x}$; domain: $x \le 5$
55. Toss a rock into the well and measure the time it takes to hit the
bottom. It will take 18 seconds. **57.** $f(x) = 10x + 25$
59. $A = \left(\frac{P}{4}\right)^2$

Chapter 15 Review Questions, page 753

1.

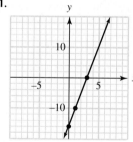

2.

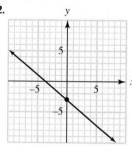

3.

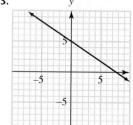

4.

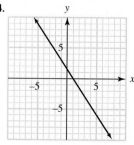

5.

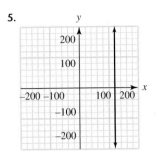

6.

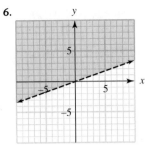

20.

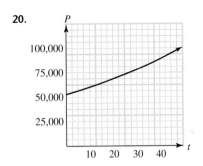

7.

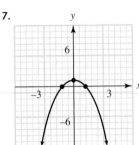

8.

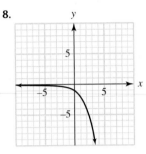

CHAPTER 16

16.1 Systems of Linear Equations, page 761

7. $(4, -1)$ **9.** inconsistent   **11.** $(-1, 1)$ **13.** $(1, 1)$ **15.** $(-5, -3)$
17. $(-1, -2)$ **19.** $(2, 10)$ **21.** $(9, -4)$ **23.** $(4, -1)$ **25.** $(6, -2)$
27. $(525, 35)$ **29.** $(5, -2)$ **31.** dependent   **33.** $(s_1, s_2) = (3, 4)$
35. $(u, v) = (5, -2)$ **37.** $(\frac{3}{5}, \frac{1}{2})$ **39.** $(5, -3)$ **41.** $(3, 1)$ **43.** $(3, 1)$
45. $(4, -1)$ **47.** $(\frac{1}{3}, -\frac{2}{3})$ **49.** $(1, 1)$ **51.** inconsistent   **53.** $(8, 2)$
55. $(-\frac{8}{5}, -\frac{21}{5})$
57. 2,601 days after the cat's birth—namely, June 2, 1999.
59. $\left(\dfrac{a + \sqrt{a^2 - 4}}{2}, \dfrac{a - \sqrt{a^2 - 4}}{2}, \left(\dfrac{a - \sqrt{a^2 - 4}}{2}, \dfrac{a + \sqrt{a^2 - 4}}{2}\right)\right.$

9.

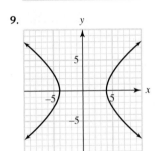

10.

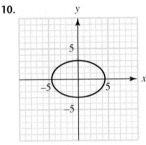

16.2 Problem Solving with Systems, page 770

1. The box has 31 nickels and 56 dimes.   **3.** There are 18 nickels.
5. The plane's speed in still air is 390 mph.   **7.** The equilibrium point for
the system is (50, 250 000).   **9.** The optimum price for the items is $3.
11. Goldie Hawn was born in 1945.   **13.** Mixture **a** has 7 lb of micoden.
15. Mixture **c** has 0.5p L of bixon.   **17.** Mixture **a** contains $46\frac{2}{3}\%$
micoden.   **19.** There are 40 oz of the base metal.   **21.** Mix 45 gal of
milk with 135 gal of cream.   **23.** The equilibrium point is (58, 9200).
25. The equilibrium point is (2, 10,000).   **27.** There are 14 dimes
and 28 quarters.   **29.** Clint Eastwood was born in 1930.
31. Matt Damon was born in 1970.   **33.** Add $\frac{1}{3}$ gal water.   **35.** TX is
262,134 sq mi and FL is 54,090 sq mi.   **37.** There are 37 quarters.
39. There are 30 dimes.   **41.** The robbery included 14 $5 bills,
70 $10 bills, and 31 $20 bills.   **43.** The plane's speed is 304.5 mph.
45. Mix together 28.5 gal of milk with 1.5 gal of cream.   **47.** The Bank
of America building is 779 ft and the Transamerica Tower is 853 ft.
49. The length of the Verrazano Narrows bridge is 4,260 ft.   **51.** The
equilibrium point is $3.50 for 4,000 items.   **53.** Mix 20 g of pure silver
and 80 g of sterling silver.   **55.** Add in 25 gallons of cream.
57. Replace $\frac{5}{8}$ qt of the punch with 7-UP.   **59. a.** 125 items would be
supplied; 75 would be demanded   **b.** No items would be supplied at
$200.   **c.** No items would be demanded at $400.   **d.** The equilibrium
price is $233.33.   **e.** The number of items produced at the equilibrium
price is 83.

11.

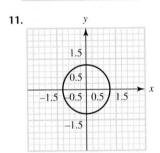

12.

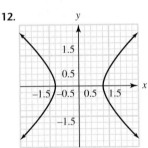

16.3 Matrix Solution of a System of Equations, page 781

5. false   **7.** true   **9.** true   **11.** false   **13.** false

15. a. $\begin{cases} 6x + 7y + 8z = 3 \\ x + 2y + 3z = 4 \\ y + 3z = 4 \end{cases}$ **b.** $\begin{cases} x_1 = 3 \\ x_2 + 2x_3 = 4 \end{cases}$ **c.** $\begin{cases} x_1 = 32 \\ x_2 = 27 \\ x_3 = -5 \\ 0 = 3 \end{cases}$

13. Yes   **14.** 20   **15.** -3 **16.** 75   **17.** 5

18.

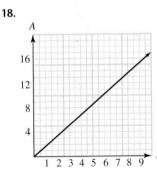

19.

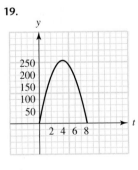

17. RowSwap([B], 1, 2); $\begin{bmatrix} 1 & 0 & 2 & | & -8 \\ -2 & 3 & 5 & | & 9 \\ 0 & 1 & 0 & | & 5 \end{bmatrix}$ **19.** *Row(1/5, [D], 1);

$$\begin{bmatrix} 1 & 4 & 3 & | & \frac{6}{5} \\ 7 & -5 & 3 & | & 2 \\ 12 & 0 & 1 & | & 4 \end{bmatrix}$$

21. *Row+(3, [B], 1, 2); $\begin{bmatrix} 1 & 3 & -5 & | & 6 \\ 0 & 13 & -14 & | & 20 \\ 0 & 5 & 1 & | & 3 \end{bmatrix}$

23. *Row+(−2, [D], 1, 2), Row+(−3, [Ans], 1, 3);

$$\begin{bmatrix} 1 & 5 & 3 & | & 2 \\ 0 & -7 & -7 & | & 0 \\ 0 & -13 & -8 & | & -6 \end{bmatrix}$$

25. *Row(1/3, [B], 2); $\begin{bmatrix} 1 & 5 & -3 & | & 5 \\ 0 & 1 & 3 & | & -5 \\ 0 & 2 & 1 & | & 5 \end{bmatrix}$

27. *Row+(−1, [D], 3, 2); $\begin{bmatrix} 1 & 3 & -2 & | & 0 \\ 0 & 1 & -4 & | & 8 \\ 0 & 3 & 6 & | & 1 \end{bmatrix}$

29. *Row+(−3, [B], 2, 1), *Row+(2, [Ans], 2, 3); $\begin{bmatrix} 1 & 0 & 12 & | & 27 \\ 0 & 1 & -2 & | & -5 \\ 0 & 0 & -2 & | & -4 \end{bmatrix}$

31. *Row+(−5, [D], 2, 1), *Row+(−1, [Ans], 2, 3),

*Row+(−2, [Ans], 2, 4); $\begin{bmatrix} 1 & 0 & -26 & -8 & | & 8 \\ 0 & 1 & 5 & 2 & | & 0 \\ 0 & 0 & -1 & -2 & | & 5 \\ 0 & 0 & -13 & -3 & | & 7 \end{bmatrix}$

33. *Row(1/8, [B], 3); $\begin{bmatrix} 1 & 0 & 4 & | & -5 \\ 0 & 1 & 3 & | & 6 \\ 0 & 0 & 1 & | & 1.5 \end{bmatrix}$

35. *Row(1/2, [D], 3); $\begin{bmatrix} 1 & 0 & -8 & 2 & | & 8 \\ 0 & 1 & -1 & 3 & | & 2 \\ 0 & 0 & 1 & 0 & | & 5 \\ 0 & 0 & -2 & 1 & | & 6 \end{bmatrix}$

37. *Row+(−4, [B], 3, 2), *Row+(3, [Ans], 3, 1); $\begin{bmatrix} 1 & 0 & 0 & | & 7 \\ 0 & 1 & 0 & | & -7 \\ 0 & 0 & 1 & | & 3 \end{bmatrix}$

39. *Row+(8, [D], 3, 1), *Row+(−4, [Ans], 3, 2),
*Row+(1, [Ans], 3, 4); $\begin{bmatrix} 1 & 0 & 0 & 2 & | & 24 \\ 0 & 1 & 0 & 2 & | & -8 \\ 0 & 0 & 1 & 0 & | & 2 \\ 0 & 0 & 0 & 1 & | & 9 \end{bmatrix}$ **41.** (5, −3)   **43.** (3, 1)   **45.** (4, −1)

47. (4, −1)   **49.** (2, 0, 1)   **51.** (3, 2, 5)   **53.** (2, −3, −1)   **55.** (2, −3, 2)
57. $\left(1, 2, -\frac{1}{2}\right)$ **59.** Mix 3 containers of Spray I with 4 containers of Spray II.

16.4 Inverse Matrices, page 794

9. a. $\begin{bmatrix} 2 & 4 & 2 \\ 6 & -2 & 4 \\ 2 & 2 & 5 \end{bmatrix}$ **b.** $\begin{bmatrix} -6 & -1 & -2 \\ 3 & -7 & -3 \\ 4 & -7 & -2 \end{bmatrix}$ **11. a.** $\begin{bmatrix} 20 & 21 & 34 \\ 29 & 16 & 15 \\ 7 & 48 & 5 \end{bmatrix}$

b. $\begin{bmatrix} -1 & 37 & 32 \\ 4 & 14 & 45 \\ 27 & 9 & 28 \end{bmatrix}$ **13. a.** $\begin{bmatrix} 14 & 14 \\ -7 & 7 \end{bmatrix}$ **b.** $\begin{bmatrix} 1 & 0 & 0 & 0 \\ 0 & 1 & 0 & 0 \\ 0 & 0 & 1 & 0 \\ 0 & 0 & 0 & 1 \end{bmatrix}$

15. a. not conformable   **b.** not conformable   **17.** $\begin{cases} x + 2y + 4z = 13 \\ -3x + 2y + z = 11 \\ 2x + z = 0 \end{cases}$

19. [A][B] = [B][A] = $\begin{bmatrix} 1 & 0 \\ 0 & 1 \end{bmatrix}$ **21.** $\begin{bmatrix} 2 & 7 \\ 1 & 4 \end{bmatrix}$ **23.** $\begin{bmatrix} 9 & -4 & -2 \\ -18 & 9 & 4 \\ -4 & 2 & 1 \end{bmatrix}$

25. $\begin{bmatrix} 1 & 0 & -1 & 0 \\ 0 & \frac{1}{2} & 0 & 0 \\ -2 & 0 & 2 & 1 \\ 0 & 0 & 1 & 0 \end{bmatrix}$ **27.** (3, 2)   **29.** (5, −4)   **31.** (38, 21)

33. (3, −2)   **35.** (3, −5)   **37.** (−4, 1)   **39.** (3, 1)   **41.** (−2, 2)

43. (1, −8)   **45.** (5, 6, 1)   **47.** (−26, 52, 15)   **49.** (88, −176, −38)

51. (1, 1, 1)   **53.** (2, −4, −1)   **55. a.** $P = \begin{bmatrix} 1 & 0 & 0 & 0 & 0 \\ -1 & 1 & 0 & 0 & 0 \\ 1 & -2 & 1 & 0 & 0 \\ -1 & 3 & -3 & 1 & 0 \\ 1 & -4 & 6 & -4 & 1 \end{bmatrix}$

b. It is the same as the Pascal's matrix except that the signs of the terms

alternate.   **57.** $[A]^3 = \begin{bmatrix} 2 & 4 & 1 & 3 \\ 4 & 2 & 3 & 4 \\ 1 & 3 & 0 & 1 \\ 3 & 4 & 1 & 2 \end{bmatrix}$ **59. a.** Riesling costs 24;

Charbono, 25; and Rosé, 51   **b.** Outside bottling, 230; produced and bottled at winery, 520; estate bottled, 280   **c.** It is the total cost of production of all three wines.

16.5 Modeling with Linear Programming, page 804

5. a. 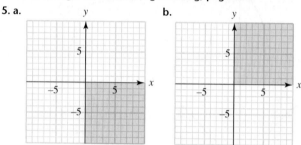 **b.**

7. 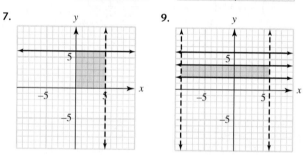 **9.**

11. 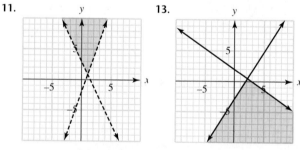 **13.**

15. 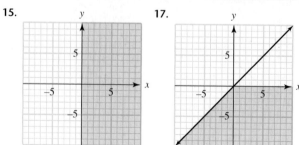 **17.**

19. a. no   **b.** yes   **c.** no   **d.** no   **e.** no   **f.** no

21.

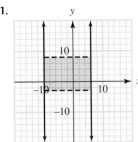

23.

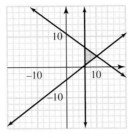

25.

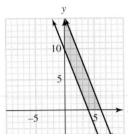

27.

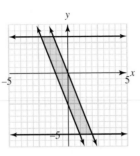

29.

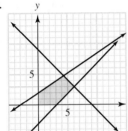

31.

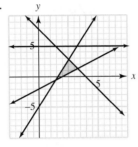

33.

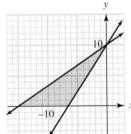

35. $(0, 0)$, $\left(0, \frac{9}{2}\right)$, $(5, 2)$, $(6, 0)$

37. $(0, 0)$, $(0, 4)$, $(2, 3)$, $(4, 0)$ **39.** $(0, 0)$, $(0, 4)$, $(4, 4)$, $(6, 2)$, $(6, 0)$

41. $(0, 0)$, $\left(0, \frac{8}{5}\right)$, $\left(\frac{24}{13}, \frac{16}{13}\right)$, $\left(\frac{8}{3}, 0\right)$ **43.** $(50, 0)$, $\left(\frac{200}{7}, \frac{60}{7}\right)$, $(8, 24)$,

$(0, 40)$ **45.** $(3, 2)$, $(5, 5)$, $(7, 5)$, $\left(\frac{10}{3}, \frac{4}{3}\right)$ **47.** maximum $W = 190$ at

$(5, 2)$ **49.** maximum $P = 500$ at $(2, 3)$ **51.** minimum $A = -12$ at $(0, 4)$

53. Let x = number of grams of food A, y = number of grams of food B;

Minimize $C = 0.29x + 0.15y$, Subject to $\begin{cases} x \geq 0, y \geq 0 \\ 10x + 5y \geq 200 \\ 2x + 5y \geq 100 \\ 3x + 4y \geq 20 \end{cases}$

55. Let x = amount invested in stock (in millions of dollars), y = amount invested in bonds (in millions of dollars); Maximize $T = 0.12x + 0.08y$,

Subject to $\begin{cases} x \geq 0, \\ x \leq 8, y \geq 2 \\ x + y \leq 10 \\ x \leq 3y \end{cases}$ **57.** The maximum profit $P = \$14{,}300$ is

achieved with all 100 acres planted in corn.   **59.** $70,000 in stocks and $30,000 in bonds for a yield of $17,000

Chapter 16 Review Questions, page 807

1. $\begin{bmatrix} 4 & -5 & 0 \\ 7 & -1 & 2 \\ -1 & -2 & 4 \end{bmatrix}$ **2.** $\begin{bmatrix} 3 & -2 \\ 5 & -5 \end{bmatrix}$ **3.** not conformable

4. $\begin{bmatrix} 4 & -2 & 0 \\ 8 & -5 & -2 \\ 1 & -1 & -1 \end{bmatrix}$ **5.** $\begin{bmatrix} 1 & 0 \\ 2 & -1 \end{bmatrix}$ **6.** $\begin{bmatrix} -1 & -2 \\ 3 & 4 \end{bmatrix}$

7. $\begin{bmatrix} 7 & -3 & -3 \\ -1 & 1 & 0 \\ -1 & 0 & 1 \end{bmatrix}$ **8.** $(3, 4)$ **9.** $(0, 1)$ **10.** $(-1, 3)$ **11.** $(-1, 2, 1)$

12. $(-33, 79)$ **13.** $(-3, 1)$ **14.** $(5, 0)$ **15.** $(4, 2, -1)$ **16.** $(11, 9)$

17.

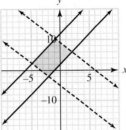

18. Use 7 bars of Product I and 3 bars of Product II.   **19.** The number of product I to be manufactured is 100; product II, 150; and product III, 250. **20.** Plant 100 acres of corn and 200 acres of wheat (200 acres are left unplanted).

CHAPTER 17

17.1 Voting, page 822

13. 12   **15. a.** "(ACB)" means that the voter ranks three candidates in the order of A first, C next, and candidate B last.   **b.** It means that in the election there were 4 voters who ranked the candidates in the order ACB. **17. a.** "(BCA)" means that the voter ranks three candidates in the order of B first, C next, and candidate A last.   **b.** It means that in the election there were no voters who ranked the candidates in the order BCA. **19.** 10   **21. a.** "(ADBC)" means the voter picks A as the first choice, followed by D, then B, with C in last place.   **b.** It means that in the election there were 8 voters who ranked the candidates in the order ADBC. **23. a.** 24   **b.** (ABCD), (ABDC), (ACBD), (ACDB), (ADCB), (BACD), (BCAD), (BCDA), (BDAC), (CABD), (CADB), (CBAD), (CBDA), (CDAB), (CDBA), (DABC), (DBAC), (DBCA), (DCBA)   **25. a.** 6 **b.** 24   **c.** 120   **27.** $n!$ **29.** A wins   **31.** A wins   **33.** A wins **35.** snacks   **37.** Howard Dean   **39.** no winner   **41.** A wins **43.** A wins   **45.** no winner   **47.** C wins   **49.** C wins   **51. a.** B **b.** plurality vote   **53. a.** C   **b.** plurality vote   **55.** 120   **57.** C wins **59.** D wins

17.2 Voting Dilemmas, page 836

11. a. California Teachers Association   **b.** no   **13. a.** There is no majority winner; A wins plurality.   **b.** A   **c.** no   **15. a.** A wins **b.** C wins   **c.** yes   **17. a.** A   **b.** There is no majority winner; B wins plurality. This violates the Condorcet criterion.   **19. a.** A wins **b.** B wins   **c.** yes   **21.** There is no majority winner; C wins plurality. This violates the Condorcet criterion.   **23.** C wins; yes   **25.** A is the Condorcet candidate.   **27.** A wins; no   **29.** A wins; no   **31.** E wins **33.** E wins; no   **35.** M wins; yes   **37.** Lillehammer wins; no **39. a.** Beijing wins   **b.** no   **41. a.** Chirac and Le Pen   **b.** Answers vary.   **c.** Answers vary.   **43.** Betty wins; no   **45.** Betty wins; none **47.** D (Dave) wins; no   **49.** D; no   **51. a.** no winner   **b.** A, C, or D could win depending on the way they are paired.   **c.** Yes, both of these methods violate the condition of decisiveness.   **53. a.** B wins   **b.** yes **55. a.** no majority; C wins plurality   **b.** There is no winner because B and C tie.   **c.** B wins   **d.** C wins; yes

17.3 Apportionment, page 856

7. a. 3; 4 **b.** 3.5 **c.** 3.46 **d.** 4; 4 **9. a.** 1; 2 **b.** 1.5 **c.** 1.41
d. 1; 2 **11. a.** 2; 3 **b.** 2.5 **c.** 2.45 **d.** 2; 3 **13. a.** 1,695; 1,696
b. 1,695.5 **c.** 1,695.50 **d.** 1,695; 1,695 **15.** 6,500 **17.** 126
19. 120,833.33 **21.** 184,000

Year	d	Manhattan	Bronx	Brooklyn	Queens	Staten Island
23. 1800	10,125	6.02	0.20	0.59	0.69	0.49
25. 1900	429,750	4.30	0.47	2.72	0.36	0.16
27. 1990	915,500	1.63	1.32	2.51	2.13	0.41

29. a. 11,600 **b.** 3.02, 1.81, 1.03, 4.14 **c.** 9 **d.** 10,000
31. a. 80,000 **b.** 1.69, 2.89, 1.48, 3.95 **c.** 7 **d.** 65,000
33. a. 11,600 **b.** 3.02, 1.81, 1.03, 4.14 **c.** 13 **d.** 13,000
35. a. 80,000 **b.** 1.69, 2.89, 1.48, 3.95 **c.** 11 **d.** 110,000

CT DE GA KY ME MD MA NH NJ NY NC PA RI SC VT VA
37. 6.41 1.59 2.23 1.99 2.61 8.62 10.21 3.83 4.97 9.17 10.65 11.69 1.86 6.72 2.30 20.16

39. 96 **41.** N, 3; S, 2; E, 3; W, 2 **43.** N, 4; S, 2; E, 3; W, 1 **45.** N, 4; S,
2; E, 3; W, 1 **47.** N, 8; S, 5; E, 7; W, 6 **49.** N, 8; S, 5; E, 7; W, 6 **51.**
N, 5; S, 3; E, 4; W, 4 **53.** N, 5; S, 3; E, 5; W, 3 **55.** N, 5; S, 3; E, 5; W, 3
57. N, 52; NE, 84; E, 50; SE, 57; S, 34; SW, 71; W, 95; NW, 32 **59.** N,
52; NE, 84; E, 50; SE, 57; S, 34; SW, 71; W, 94; NW, 33

17.4 Apportionment Paradoxes, page 863

5. State A violates the quota rule. **7.** State A violates the quota rule.
9. State D violates the quota rule. **11.** State A illustrates the Alabama
paradox. **13.** State B illustrates the Alabama paradox. **15.** State C
illustrates the population paradox. **17.** State C illustrates the population
paradox. **19.** State A illustrates the new states paradox. **21.** State B
illustrates the new states paradox. **23. a.** 62.78 **b.** A: 199.43;
B: 72.55; C: 12.93; D: 15.08 **c.** A: 199, 200; B: 72, 73; C: 12, 13;
D: 15, 16 **d.** modified quota is 62.4; A: 200, B: 72, C: 13, D: 15 **e.** no
25. a. 203.3 **b.** Uptown (U): 83.52; Downtown (D): 16.48 **c.** U: 84;
D: 16 **d.** U: 83; D: 17; New: 12 **e.** yes **27.** yes **29.** Adams' or
Webster's plan gives the correct apportionment of the horses.

Chapter 17 Review Questions, page 866

1. no majority; plurality vote goes to C **2.** 3 **3.** A **4.** B **5.** Hare
method violates the Condorcet criterion. **6.** A **7.** yes, B **8.** Carr (280),
Crouch (770), Dorsey (638), Freeney (42), Grossman (708),
Harrington (364), McKinnie (116), Peppers (41), El (267), Williams (146);
Crouch wins **9.** A **10.** A **11.** C; yes **12.** 7.9 **13.** EA: 11.39; MC:
27.22; M: 33.92; P: 16.84; SR: 10.63. The lower and upper quotas are EA:
11, 12; MC: 27, 28; M: 33, 34; P: 16, 17 SR: 10, 11.
14. EA: 12; MC: 27; M: 33; P: 17; SR: 11 **15.** EA: 11; MC: 28;
M: 34; P: 17; SR: 10 **16.** EA: 11; MC: 27; M: 34; P: 17; SR: 11 **17.**
EA: 11; MC: 27; M: 34; P: 17; SR: 11 **18.** EA: 11; MC: 27;
M: 34; P: 17; SR: 11 **19.** no **20.** Downtown: 70; Fairground: 29;
Columbus Square: 28; Downtown West: 27; Peabody: 26

CHAPTER 18

18.1 What Is Calculus?, page 879

5. $\frac{1}{3}$ **7.** 1 **9.** π **11.**

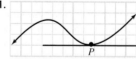

13.

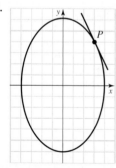

15.

No single tangent line exists.

17. (continued)

17. $\frac{2}{3}$ **19.** $\frac{1}{2}$ **21.** $\frac{3}{2}$ **23.** 20 **25.** 4 or 5 **27.** 3

18.2 Limits, page 855

3. 0, 2, 0, 2, 0 **5.** $\frac{4}{3}, \frac{7}{4}, 2, \frac{13}{6}, \frac{16}{7}$ **7.** 8,000 **9.** $\frac{2}{3}$ **11.** 5 **13.** 4
15. 0 **17.** 0 **19.** 0.7 **21.** 0 **23.** ∞ **25.** -3 **27.** 3 **29.** 0.75 mg;
$24\left(\frac{1}{2}\right)^n$ mg

18.3 Derivatives, page 895

1. a. 5.71 ft/s **b.** 15 ft/s **c.** 40 ft/s **d.** 40 ft/s **3. a.** 59 mph
b. 18 mph **c.** 36 mph **d.** 38 mph **5.** Let \$ be trillions of dollars.
a. 0.198 \$/yr **b.** 0.255 \$/yr **c.** 0.307 \$/yr **d.** 0.348 \$/yr **e.** 0.371 \$/yr
f. It is changing at the rate of \$371 billion per year (or greater).

7.

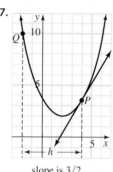

slope is 3/2

9.

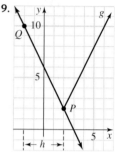

slope is -2

11.

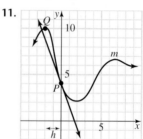

slope is -2

13. a. 0 **b.** 0

15. a. -25 **b.** -15 **17.** x^2 **19.** $-6e^{-6x}$ **21.** $2 - 6x$
23. $16x - y - 32 = 0$ **25.** $4x - y = 0$ **27. a.** The height of the
tower is 176 ft. **b.** The velocity is $-32t + 96$. **29. a.** \$8,900/item
b. \$6,200/item **c.** \$5,930/item **d.** $5,900 + 30h$

18.4 Integrals, page 904

1. 40 **3.** 24 **5.** 19.5 **7.** $6x + C$ **9.** $\frac{1}{2}x^2 + 5x + C$
11. $\frac{3}{2}x^2 + 4x + C$ **21.** $242\frac{2}{3}$ **23.** 82.5 **25.** $2e - 2$
27. 76 **29.** 58,965.3; this represents the increase in the number of
bacteria after 10 hours.

Chapter 18 Review Questions, page 906

1. limits, derivatives, and integrals **2.** The closer the measurements are
together, the better the approximation at a particular point. **3.** 0
4. does not exist **5.** $\frac{3}{7}$ **6.** ∞ **7.** e **8.** 0 **9.** -2 **10.** 1 **11.** $-8x$
12. $22e^{0.5x}$ **13.** $8x^2 + C$ **14.** $\pi x + C$ **15.** 33 **16.** -6 **17.** 4.67
18. $2x - y - 1 = 0$ **19.** 7 mph **20.** Oil reserves will be depleted
in 2014.

EPILOGUE

Epilogue Problem Set, page E10

3. a. social science **b.** natural science **c.** natural science
d. humanities **e.** social science **5. a.** social science **b.** natural science
c. social science **d.** social science **e.** natural science
f. natural science **7.** all natural science **9.** 1.46×10^8
11. Venus and Neptune **13.** 54 oz.
15. genotype $SS = 0.3025$; genotype wS or $Sw = 0.4950$;
genotype $ww = 0.2025$

17. 0.45 cm² **19.**

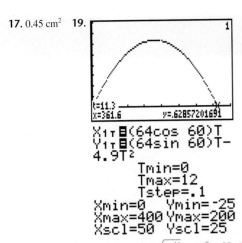

$X_{1T} \boxminus (64\cos 60)T$
$Y_{1T} \boxminus (64\sin 60)T-$
$4.9T^2$
Tmin=0
Tmax=12
Tstep=.1
Xmin=0   Ymin=-25
Xmax=400 Ymax=200
Xscl=50 Yscl=25

21. a. 0.0996   **b.** Let each unit be $\sqrt{10^{16}} = 10^8$.   **23.** 24

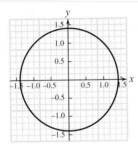

25. a.

	A	B	C	D
a	(1, 2)	(3, 1)	(4, 1)	(2, 2)
b	(2, 3)	(4, 4)	(3, 2)	(1, 1)
c	(2, 4)	(3, 3)	(4, 4)	(1, 3)
d	(3, 1)	(2, 2)	(4, 3)	(1, 4)

b. Answers vary.

27.

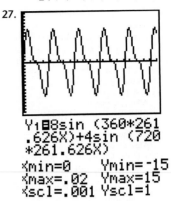

$Y_1 \boxminus 8\sin (360*261$
$.626X)+4\sin (720$
$*261.626X)$
Xmin=0   Ymin=-15
Xmax=.02 Ymax=15
Xscl=.001 Yscl=1

29. A. 6   **b.** 3

Index